# Future perspectives in broiler rearing

## Incorporating the broiler point of view

It is widely recognised that agriculture is a significant contributor to global warming and climate change. Agriculture needs to reduce its environmental impact and adapt to current climate change whilst still feeding a growing population, i.e. become more 'climate-smart'. Burleigh Dodds Science Publishing is playing its part in achieving this by bringing together key research on making the production of the world's most important crops and livestock products more sustainable.

Based on extensive research, our publications specifically target the challenge of climate-smart agriculture. In this way we are using 'smart publishing' to help achieve climate-smart agriculture.

Burleigh Dodds Science Publishing is an independent and innovative publisher delivering high quality customer-focused agricultural science content in both print and online formats for the academic and research communities. Our aim is to build a foundation of knowledge on which researchers can build to meet the challenge of climate-smart agriculture.

For more information about Burleigh Dodds Science Publishing simply call us on +44 (0) 1223 839365, email info@bdspublishing.com or alternatively please visit our website at www.bdspublishing.com.

**Related titles:**

Achieving sustainable production of poultry meat Volume 1
Print (ISBN 978-1-78676-064-7); Online (ISBN 978-1-78676-066-1, 978-1-78676-067-8)

Achieving sustainable production of poultry meat Volume 2
Print (ISBN 978-1-78676-068-5); Online (ISBN 978-1-78676-070-8, 978-1-78676-071-5)

Achieving sustainable production of poultry meat Volume 3
Print (ISBN 978-1-78676-072-2); Online (ISBN 978-1-78676-074-6, 978-1-78676-075-3)

Advances in precision livestock farming
Print (ISBN 978-1-78676-471-3); Online (ISBN 978-1-78676-473-7, 978-1-78676-474-4)

Embryo development and hatchery practice in poultry production
Print (ISBN 978-1-80146-252-5); Online (ISBN 978-1-80146-253-2, 978-1-80146-254-9)

Understanding the behaviour and improving the welfare of chickens
Print (ISBN 978-1-78676-422-5); Online (ISBN 978-1-78676-424-9, 978-1-78676-425-6)

Chapters are available individually from our online bookshop:
https://www.bdspublishing.com/webshop/

climate•SMART•publishing
IN AGRICULTURAL SCIENCE

BURLEIGH DODDS SERIES IN AGRICULTURAL SCIENCE
NUMBER 179

# Future perspectives in broiler rearing

## Incorporating the broiler point of view

Edited by Dr Henry van den Brand

burleigh dodds
SCIENCE PUBLISHING

Published by Burleigh Dodds Science Publishing Limited
82 High Street, Sawston, Cambridge CB22 3HJ, UK
www.bdspublishing.com

Burleigh Dodds Science Publishing, 1518 Walnut Street, Suite 900, Philadelphia, PA 19102-3406, USA

First published 2026 by Burleigh Dodds Science Publishing Limited


Library of Congress Control Number: 2026935175

British Library Cataloguing in Publication Data
A catalogue record for this book is available from the British Library

ISBN 978-1-83545-149-6 (print)
ISBN 978-1-83545-150-2 (ePub)
ISBN 978-1-83545-151-9 (PDF)
ISSN 2059-6936 (print)
ISSN 2059-6944 (online)

DOI: 10.19103/AS.2025.0165

Typeset by Deanta Global Publishing Services, Dublin, Ireland

# Contents


Series list xiv
Preface xxvi
Acknowledgements xxviii
Introduction xxix

**Part 1 Setting the scene in broiler production**

1 Limitations and challenges in broiler production: ethical considerations beyond welfare issues 3
*Mona F. Giersberg, Utrecht University, The Netherlands*

1 Introduction 3
2 Contemporary animal welfare concepts 5
3 Major welfare challenges in broiler production 8
4 The link between welfare and ethics 11
5 Ethical perspectives beyond welfare-centred approaches 13
6 Case study: the broiler breeder paradox 16
7 Conclusion and future trends in research 18
8 Where to look for further information 19
9 References 20

2 Managing co-products in broiler production: repurposing poultry manure and litter 25
*C. W. Ritz, University of Georgia, USA*

1 Introduction 25
2 Nutrient management planning: decades of best management practices 29
3 Nutrient management plan challenges 39
4 Air quality challenges 41
5 Sustainability/circularity 43
6 Improvement opportunities 44
7 Conclusion 45
8 References 46

3 The importance of data sharing between different links in the broiler production chain 49
*Hasitha Priyashantha, University of Ruhuna, Sri Lanka; Maleeka Nambapana, Uva Wellassa University, Sri Lanka; Ashen Epa Arachchi, University of Peradeniya, Sri Lanka; Shishanthi Jayarathna, University of Ruhuna, Sri Lanka; and Janak K. Vidanarachchi, University of Peradeniya, Sri Lanka*

1 Introduction 49
2 The broiler chicken supply chain: an overview 55
3 Levels of integration in the broiler chicken supply chain 62
4 Data sharing and integration: case studies 70
5 From data sharing to better products for consumers 73
6 Future trends and recommendations 74
7 Conclusion 76
8 Conflict of interest 76
9 Acknowledgements 76
10 Funding 76
11 Data availability statement 76
12 References 77

**Part 2 Creating the best day-old broiler chick**

4 Breeding for improved health and welfare in broiler chickens 85
*Pascal Duenk and Mario Calus, Wageningen University & Research, The Netherlands*

1 Introduction 85
2 History of broiler breeding 86
3 Global developments and consequences for broiler welfare 88
4 Challenges, opportunities and future directions for broiler breeding 90
5 Conclusion and summary 103
6 Where to look for further information 103
7 Acknowledgments 103
8 References 103

5 Chick quality as affected by the broiler breeder and breeder house management 111
*Michael J. Wineland, North Carolina State University, USA*

1 Introduction 111
2 Methods of measuring chick quality typically used 112
3 Nutrients available to the developing embryo 113
4 Egg component characteristics 116
5 Influence of eggshell conductance on chick quality 117
6 Influence of egg size and breeder age on chick quality 119

7 Influence of egg storage on chick quality 123
8 Eggshell quality and its influence on chick quality 124
9 Egg contamination and egg sweating influence on chick quality 126
10 Influence of eggs set upside down for incubation and the resulting chick quality 127
11 Conclusion 128
12 References 128

6 Day-old chick quality 137
*Itallo Conrado Sousa Araújo, Federal University of Minas Gerais, Brazil; and Roos Molenaar, Aart Lammers and Henry van den Brand, Wageningen University & Research, The Netherlands*

1 Introduction 137
2 Physical day-old chick quality 138
3 Immunological day-old chick quality: immunoglobulins 146
4 Microbiological chick quality 150
5 Conclusion 155
6 Where to look for further information 157
7 References 157

7 Early feeding from a broiler perspective 163
*H. J. Wijnen and C. W. van der Pol, HatchTech, The Netherlands; and R. Molenaar and H. van den Brand, Wageningen University & Research, The Netherlands*

1 Introduction 163
2 Delayed feeding 165
3 Residual yolk 168
4 Growth rate 172
5 Health 174
6 Welfare 177
7 Early feeding case studies 180
8 Future trends in early feeding 183
9 Where to look for further information 184
10 References 185

8 Optimising in ovo delivery of nutrients and bioactive compounds in broiler production 193
*M. M. Y. Meijer, Wageningen University & Research, The Netherlands and The University of Queensland, Australia; E. Roura, The University of Queensland, Australia; and H. van den Brand, Wageningen University & Research, The Netherlands*

1 Introduction 193
2 Methodological challenges and variability 196
3 Case study: in ovo delivery of essential oils 224

4 Conclusion and future trends 227
5 Acknowledgements 229
6 Where to look for further information 229
7 References 229

**Part 3 Managing the needs of broilers**

9 Understanding the behavioral needs of broilers 237
*Leonie Jacobs, Virginia Tech, USA*

1 Introduction 237
2 Behavior and behavioral needs 238
3 Normal broiler chicken behavior 239
4 Abnormal behaviors 250
5 The impact of genetics on behavior 252
6 Conclusion 254
7 Where to look for further information 255
8 References 256

10 Understanding the nutritional needs of broilers 265
*E. Delezie and M. Lourenço, Flanders Research Institute for Agriculture, Fisheries and Food (ILVO), Belgium*

1 Introduction 265
2 Feed formulation and use of raw materials 267
3 Nutritional requirements of broilers 274
4 What will broiler diets look like in 10 years? 282
5 Conclusion 284
6 Future trends in research 285
7 Where to look for further information 285
8 References 285

11 Understanding the environmental needs of broilers 293
*Edgar Orlando Oviedo-Rondón, North Carolina State University, USA*

1 Introduction 293
2 Temperature 295
3 Ventilation 303
4 Air quality 307
5 Variability in environmental parameters 310
6 Flooring systems 313
7 Lighting 314
8 Space management 316
9 Modern technologies to achieve environmental needs 319
10 Conclusion 319
11 References 320

## Part 4 Challenges in broiler production

12 Challenges in achieving antibiotic-free broiler production 333
*Elijah G. Kiarie, University of Guelph, Canada and North-West University, South Africa; Emily Kim, New-Life Mills – Division of Parrish & Heimbecker Ltd., Canada; Aizwarya Thanabalan, Jones Feed Mill, Canada; Munene Kithama, Pwani University, Kenya; and Lisa Bean-Hodgins, BSC Animal Nutrition Inc., Canada*

1 Introduction 333
2 Antibiotic use in broiler production: a brief history 334
3 Genesis of antibiotic-free broiler chicken production 335
4 State of antibiotic-free broiler production 336
5 Challenges for achieving antibiotic-free broiler production 337
6 Opportunities for achieving antibiotic-free broiler production 339
7 Conclusion and future perspectives 349
8 Author contributions 349
9 Conflict of interest statement 349
10 Acknowledgments 349
11 Where to look for further information 350
12 References 351

13 Concepts of circular broiler production 361
*Ollie van Hal, Ben van Selm, Asaah Ndambi and Simon Oosting, Wageningen University & Research, The Netherlands*

1 Introduction 361
2 Circularity indicators and circularity of broiler production 363
3 Strategies for more circular broiler production 369
4 The potential of broilers in a more circular food system 379
5 Conclusion 382
6 Where to look for further information 382
7 References 383

14 Challenges in achieving low-input, extensive broiler production 391
*Claire Bonnefous and Veronika Maurer, Research Institute of Organic Agriculture (FiBL), Switzerland; and Elisabeth Duval and Anne Collin, Biologie des Oiseaux et Aviculture – INRAE and Université de Tours, France*

1 Introduction 391
2 Outdoor ranges: characteristics and use by chickens 392
3 Providing outdoor range access: risks and potential solutions 395
4 Providing outdoor range access: welfare issues 398
5 Impacts of outdoor rearing systems on the environment 399
6 Conclusion 400
7 References 401

15 Heat stress in broiler production 407
*Victoria Anthony Uyanga, Lincoln University of Missouri, USA; Semiu Folaniyi Bello, Agriculture and Agri-Food Canada, Canada; and Folasade Olukemi Oke, Okanlawon Mohammed Onagbesan and Oyegunle Emmanuel Oke, Federal University of Agriculture, Nigeria*

1 Introduction 407
2 Understanding heat stress in broiler production 409
3 Metabolic and physiological changes during heat stress 414
4 Genetic basis of heat resistance in broilers 419
5 Impacts of heat stress on broiler's growth and productivity 423
6 Impacts of heat stress on broiler's behavior and welfare 429
7 Conclusion and future trends 434
8 Where to look for further information 435
9 References 435

**Part 5 Precision farming technologies in broiler production**

16 Sensor technologies for monitoring broiler welfare indicators 449
*Malou van der Sluis, Wageningen University & Research, The Netherlands*

1 Introduction 449
2 Types of sensor technologies 451
3 Implementing sensors for recording welfare indicators 460
4 Case studies 464
5 Conclusion and future trends 470
6 Where to look for further information 471
7 References 471

17 Use of vocalisation technology for early detection of challenges in broiler production 477
*Gerardo José Ginovart-Panisello, CEALVET S.L.U. and La Salle Ramon Llull University, Spain; Silvia Riva and Tesa Panisello Monjo, CEALVET S.L.U., Spain; and Rosa Ma Alsina-Pagès, CEALVET S.L.U. and La Salle Ramon Llull University, Spain*

1 Introduction 477
2 Acoustic features of broilers 478
3 Animal disease detector 484
4 Tracking of animal behaviour 488
5 Case study 491
6 Conclusion and future trends 494
7 Where to look for further information 497
8 References 497

18 The potential of using volatile organic compounds (VOCs) to monitor health issues in broiler production 503
*Nelle Lauwers, Maarten Hertog, Nadia Everaert and Matthias Corion, Katholieke University of Leuven, Belgium*

1 Introduction 503
2 Volatile organic compounds: primary metabolic pathways 506
3 Volatile organic compounds: secondary metabolic pathways 513
4 Sampling volatile organic compounds in chickens 515
5 Pathogen influence on volatile organic compounds 520
6 General challenges in identifying reliable pathogen biomarkers 529
7 Internal factors influencing volatile organic compound fingerprints 530
8 External factors influencing volatile organic compound fingerprints 533
9 Volatile organic compound measurement and analysis 535
10 Conclusion and future trends 540
11 Where to look for further information 541
12 References 542

**Part 6 Maintaining the health of broilers**

19 Vaccination strategies in broilers 555
*Alessandra Piccirillo, University of Padua, Italy; Giacomo Berto, Veterinary Services Local Health Unit (ULSS) 8 "Berica" – Vicenza, Italy; Davide Giovanardi, Tre Valli Laboratory, Italy; Luca Bano, Experimental Zooprophylactic Institute of the Venezie, Italy; and Andrea Laconi, University of Padua, Italy*

1 Introduction 555
2 The immune system in broilers 558
3 Vaccination strategies in broiler production 567
4 Types of vaccines used in broilers 579
5 Vaccine delivery methods 585
6 Factors influencing vaccination in broilers 591
7 New developments and future perspectives 603
8 Conclusion 604
9 Where to look for further information 604
10 References 605

20 Managing health issues in broilers: first-week mortality 627
*Serdar Özlü, Ankara University, Turkey*

1 Introduction 627
2 Mortality patterns 628
3 Causes of first-week mortality: animal-related factors 629
4 Causes of first-week mortality: egg and incubation-related factors 633
5 Causes of first-week mortality: external factors 639

6 Conclusion 647
7 Where to look for future information 648
8 References 649

21 Managing health issues in broilers: leg health and leg problems 659
*Wendy I. Muir and Peter J. Groves, The University of Sydney, Australia; and Rodney L. Reece, Veterinary Consultant, Australia*

1 Introduction 659
2 Broiler leg problems 661
3 Assessment of leg problems 681
4 Improving leg health and reducing leg problems 681
5 Future trends in research 700
6 Conclusion 701
7 Where to look for further information 702
8 References 703

22 Managing health issues in broilers: respiratory health 715
*Elena Catelli, Giulia Graziosi and Caterina Lupini, University of Bologna, Italy*

1 Introduction 715
2 Main respiratory conditions of broilers 716
3 Diagnosis and lab testing 729
4 Key tools for respiratory disease prevention and control 733
5 Case study: Molecular approach to respiratory pathogens epidemiology and control 738
6 Future trends in research 739
7 Conclusion 741
8 Where we look for further information 742
9 References 742

**Part 7 Managing the end of life stage in broiler production**

23 Humane end-of-life management of broilers: catching, crating and loading 763
*Helen Louton, Jenny Wessel and Julia Mönch, Ludwig Maximilian University of Munich, Germany*

1 Introduction 763
2 Legislation and recommendations for catching and loading 765
3 Possible methods and perceptions of farmers 766
4 Manual catching: different methods 768
5 Mechanical loading systems 772
6 Loading-related stress reactions 777
7 Loading-related injuries 779
8 Factors impacting loading quality or loading-related injuries 785

| | | |
|---|---|---|
| 9 | Loading staff: issues to consider | 789 |
| 10 | Crating | 790 |
| 11 | Conclusion | 791 |
| 12 | Summary | 791 |
| 13 | Where to look for further information | 792 |
| 14 | References | 793 |

24 Managing breast myopathies in broiler production 799
*Martina Bordini, Massimiliano Petracci and Francesca Soglia, University of Bologna, Italy*

| | | |
|---|---|---|
| 1 | Introduction | 799 |
| 2 | Suggested mechanisms and predisposing factors underlying the onset of breast myopathies | 806 |
| 3 | Potential strategies to mitigate breast myopathies | 812 |
| 4 | Implication on broiler welfare | 825 |
| 5 | Conclusion and future trends in research | 825 |
| 6 | Where to look for further information | 826 |
| 7 | References | 827 |

Index 839

# Series list

| Title | Series number |
|---|---|
| **Achieving sustainable cultivation of maize - Vol 1**<br>From improved varieties to local applications<br>Edited by: Dr Dave Watson, CGIAR Maize Research Program Manager, CIMMYT, Mexico | 001 |
| **Achieving sustainable cultivation of maize - Vol 2**<br>Cultivation techniques, pest and disease control<br>Edited by: Dr Dave Watson, CGIAR Maize Research Program Manager, CIMMYT, Mexico | 002 |
| **Achieving sustainable cultivation of rice - Vol 1**<br>Breeding for higher yield and quality<br>Edited by: Prof. Takuji Sasaki, Tokyo University of Agriculture, Japan | 003 |
| **Achieving sustainable cultivation of rice - Vol 2**<br>Cultivation, pest and disease management<br>Edited by: Prof. Takuji Sasaki, Tokyo University of Agriculture, Japan | 004 |
| **Achieving sustainable cultivation of wheat - Vol 1**<br>Breeding, quality traits, pests and diseases<br>Edited by: Prof. Peter Langridge, University of Adelaide, Australia | 005 |
| **Achieving sustainable cultivation of wheat - Vol 2**<br>Cultivation techniques<br>Edited by: Prof. Peter Langridge, University of Adelaide, Australia | 006 |
| **Achieving sustainable cultivation of tomatoes**<br>Edited by: Dr Autar Mattoo, USDA-ARS, USA and Prof. Avtar Handa, Purdue University, USA | 007 |
| **Achieving sustainable production of milk - Vol 1**<br>Milk composition, genetics and breeding<br>Edited by: Dr Nico van Belzen, International Dairy Federation (IDF), Belgium | 008 |
| **Achieving sustainable production of milk - Vol 2**<br>Safety, quality and sustainability<br>Edited by: Dr Nico van Belzen, International Dairy Federation (IDF), Belgium | 009 |
| **Achieving sustainable production of milk - Vol 3**<br>Dairy herd management and welfare<br>Edited by: Emeritus Prof. John Webster, University of Bristol, UK | 010 |
| **Ensuring safety and quality in the production of beef - Vol 1**<br>Safety<br>Edited by: Prof. Gary Acuff, Texas A&M University, USA and Prof. James Dickson, Iowa State University, USA | 011 |
| **Ensuring safety and quality in the production of beef - Vol 2**<br>Quality<br>Edited by: Emeritus Prof. Michael Dikeman, Kansas State University, USA | 012 |
| **Achieving sustainable production of poultry meat - Vol 1**<br>Safety, quality and sustainability<br>Edited by: Prof. Steven C. Ricke, University of Arkansas, USA | 013 |
| **Achieving sustainable production of poultry meat - Vol 2**<br>Breeding and nutrition<br>Edited by: Prof. Todd Applegate, University of Georgia, USA | 014 |

**Achieving sustainable production of poultry meat - Vol 3** 015
Health and welfare
Edited by: Prof. Todd Applegate, University of Georgia, USA

**Achieving sustainable production of eggs - Vol 1** 016
Safety and quality
Edited by: Prof. Julie Roberts, University of New England, Australia

**Achieving sustainable production of eggs - Vol 2** 017
Animal welfare and sustainability
Edited by: Prof. Julie Roberts, University of New England, Australia

**Achieving sustainable cultivation of apples** 018
Edited by: Dr Kate Evans, Washington State University, USA

**Integrated disease management of wheat and barley** 019
Edited by: Prof. Richard Oliver, Curtin University, Australia

**Achieving sustainable cultivation of cassava - Vol 1** 020
Cultivation techniques
Edited by: Dr Clair Hershey, formerly International Center for Tropical Agriculture (CIAT), Colombia

**Achieving sustainable cultivation of cassava - Vol 2** 021
Genetics, breeding, pests and diseases
Edited by: Dr Clair Hershey, formerly International Center for Tropical Agriculture (CIAT), Colombia

**Achieving sustainable production of sheep** 022
Edited by: Prof. Johan Greyling, University of the Free State, South Africa

**Achieving sustainable production of pig meat - Vol 1** 023
Safety, quality and sustainability
Edited by: Prof. Alan Mathew, Purdue University, USA

**Achieving sustainable production of pig meat - Vol 2** 024
Animal breeding and nutrition
Edited by: Prof. Julian Wiseman, University of Nottingham, UK

**Achieving sustainable production of pig meat - Vol 3** 025
Animal health and welfare
Edited by: Prof. Julian Wiseman, University of Nottingham, UK

**Achieving sustainable cultivation of potatoes - Vol 1** 026
Breeding improved varieties
Edited by: Prof. Gefu Wang-Pruski, Dalhousie University, Canada

**Achieving sustainable cultivation of oil palm - Vol 1** 027
Introduction, breeding and cultivation techniques
Edited by: Prof. Alain Rival, Center for International Cooperation in Agricultural Research for Development (CIRAD), France

**Achieving sustainable cultivation of oil palm - Vol 2** 028
Diseases, pests, quality and sustainability
Edited by: Prof. Alain Rival, Center for International Cooperation in Agricultural Research for Development (CIRAD), France

**Achieving sustainable cultivation of soybeans - Vol 1** 029
Breeding and cultivation techniques
Edited by: Prof. Henry T. Nguyen, University of Missouri, USA

**Achieving sustainable cultivation of soybeans - Vol 2** 030
Diseases, pests, food and other uses
Edited by: Prof. Henry T. Nguyen, University of Missouri, USA

**Achieving sustainable cultivation of sorghum - Vol 1** 031
Genetics, breeding and production techniques
Edited by: Prof. William Rooney, Texas A&M University, USA

**Achieving sustainable cultivation of sorghum - Vol 2** 032
Sorghum utilization around the world
Edited by: Prof. William Rooney, Texas A&M University, USA

**Achieving sustainable cultivation of potatoes - Vol 2** 033
Production, storage and crop protection
Edited by: Dr Stuart Wale, Potato Dynamics Ltd, UK

**Achieving sustainable cultivation of mangoes** 034
Edited by: Prof. Víctor Galán Saúco, Instituto Canario de Investigaciones Agrarias (ICIA), Spain and Dr Ping Lu, Research Institute for the Environment and Livelihoods, Charles Darwin University, Australia

**Achieving sustainable cultivation of grain legumes - Vol 1** 035
Advances in breeding and cultivation techniques
Edited by: Dr Shoba Sivasankar et al., formerly International Crops Research Institute for the Semi-Arid Tropics (ICRISAT), India

**Achieving sustainable cultivation of grain legumes - Vol 2** 036
Improving cultivation of particular grain legumes
Edited by: Dr Shoba Sivasankar et al., formerly International Crops Research Institute for the Semi-Arid Tropics (ICRISAT), India

**Achieving sustainable cultivation of sugarcane - Vol 1** 037
Cultivation techniques, quality and sustainability
Edited by: Prof. Philippe Rott, University of Florida, USA

**Achieving sustainable cultivation of sugarcane - Vol 2** 038
Breeding, pests and diseases
Edited by: Prof. Philippe Rott, University of Florida, USA

**Achieving sustainable cultivation of coffee** 039
Breeding and quality traits
Edited by: Dr Philippe Lashermes, Institut de Recherche pour le Développement (IRD), France

**Achieving sustainable cultivation of bananas - Vol 1** 040
Cultivation techniques
Edited by: Prof. Gert H. J. Kema, Wageningen University and Research, The Netherlands and Prof. André Drenth, University of Queensland, Australia

**Global tea science** 041
Current status and future needs
Edited by: Dr V. S. Sharma, formerly UPASI Tea Research Institute, India and Dr M. T. Kumudini Gunasekare, formerly Tea Research Institute, Sri Lanka

**Integrated weed management for sustainable agriculture** 042
Edited by: Emeritus Prof. Robert L. Zimdahl, Colorado State University, USA

**Achieving sustainable cultivation of cocoa** 043
Edited by: Prof. Pathmanathan Umaharan, Cocoa Research Centre – The University of the West Indies, Trinidad and Tobago

**Robotics and automation for improving agriculture** 044
Edited by: Prof. John Billingsley, University of Southern Queensland, Australia

**Water management for sustainable agriculture** 045
Edited by: Prof. Theib Oweis, International Centre for Agricultural Research in the Dry Areas (ICARDA), Jordan

**Improving organic animal farming** 046
Edited by: Dr Mette Vaarst, Aarhus University, Denmark and Dr Stephen Roderick, Duchy College, UK

**Improving organic crop cultivation** 047
Edited by: Prof. Ulrich Köpke, University of Bonn, Germany

**Managing soil health for sustainable agriculture - Vol 1** 048
Fundamentals
Edited by: Dr Don Reicosky, Soil Scientist Emeritus USDA-ARS and University of Minnesota, USA

**Managing soil health for sustainable agriculture - Vol 2** 049
Monitoring and management
Edited by: Dr Don Reicosky, Soil Scientist Emeritus USDA-ARS and University of Minnesota, USA

**Rice insect pests and their management** 050
Prof. Elvis A. Heinrichs, Dr Francis E. Nwilene, Prof. Michael Stout, Dr Buyung A. R. Hadi and Dr Thais Freitas

**Improving grassland and pasture management in temperate agriculture** 051
Edited by: Prof. Athole Marshall and Dr Rosemary Collins, IBERS, Aberystwyth University, UK

**Precision agriculture for sustainability** 052
Edited by: Dr John Stafford, Silsoe Solutions, UK

**Achieving sustainable cultivation of temperate zone tree fruit and berries - Vol 1** 053
Physiology, genetics and cultivation
Edited by: Prof. Gregory A. Lang, Michigan State University, USA

**Achieving sustainable cultivation of temperate zone tree fruit and berries - Vol 2** 054
Case studies
Edited by: Prof. Gregory A. Lang, Michigan State University, USA

**Agroforestry for sustainable agriculture** 055
Edited by: Prof. María Rosa Mosquera-Losada, Universidade de Santiago de Compostela, Spain and Dr Ravi Prabhu, World Agroforestry Centre (ICRAF), Kenya

**Achieving sustainable cultivation of tree nuts** 056
Edited by: Prof. Ümit Serdar, Ondokuz Mayis University, Turkey and Emeritus Prof. Dennis Fulbright, Michigan State University, USA

**Assessing the environmental impact of agriculture** 057
Edited by: Prof. Bo P. Weidema, Aalborg University, Denmark

**Critical issues in plant health: 50 years of research in African agriculture** 058
Edited by: Dr Peter Neuenschwander and Dr Manuele Tamò, International Institute of Tropical Agriculture (IITA), Benin

**Achieving sustainable cultivation of vegetables** 059
Edited by: Prof. George Hochmuth, University of Florida, USA

**Advances in breeding techniques for cereal crops** 060
Edited by: Prof. Frank Ordon, Julius Kühn Institute (JKI), Germany and Prof. Wolfgang Friedt, Justus-Liebig University of Giessen, Germany

**Advances in Conservation Agriculture - Vol 1** 061
Systems and Science
Edited by: Prof. Amir Kassam, University of Reading, UK and Moderator, Global Conservation Agriculture Community of Practice (CA-CoP), FAO, Rome, Italy

**Advances in Conservation Agriculture - Vol 2** 062
Practice and Benefits
Edited by: Prof. Amir Kassam, University of Reading, UK and Moderator, Global Conservation Agriculture Community of Practice (CA-CoP), FAO, Rome, Italy

**Achieving sustainable greenhouse cultivation** 063

Edited by: Prof. Leo Marcelis and Dr Ep Heuvelink, Wageningen University, The Netherlands

**Achieving carbon-negative bioenergy systems from plant materials** 064
Edited by: Dr Chris Saffron, Michigan State University, USA

**Achieving sustainable cultivation of tropical fruits** 065
Edited by: Prof. Elhadi M. Yahia, Universidad Autónoma de Querétaro, Mexico

**Advances in postharvest management of horticultural produce** 066
Edited by: Prof. Chris Watkins, Cornell University, USA

**Pesticides and agriculture** 067
Profit, politics and policy
Dave Watson

**Integrated management of diseases and insect pests of tree fruit** 068
Edited by: Prof. Xiangming Xu and Dr Michelle Fountain, NIAB-EMR, UK

**Integrated management of insect pests** 069
Current and future developments
Edited by: Emeritus Prof. Marcos Kogan, Oregon State University, USA and Emeritus Prof. E. A. Heinrichs, University of Nebraska-Lincoln, USA

**Preventing food losses and waste to achieve food security and sustainability** 070
Edited by: Prof. Elhadi M. Yahia, Universidad Autónoma de Querétaro, Mexico

**Achieving sustainable management of boreal and temperate forests** 071
Edited by: Dr John A. Stanturf, Estonian University of Life Sciences, Estonia

**Advances in breeding of dairy cattle** 072
Edited by: Prof. Julius van der Werf, University of New England, Australia and Prof. Jennie Pryce, Agriculture Victoria and La Trobe University, Australia

**Improving gut health in poultry** 073
Edited by: Prof. Steven C. Ricke, University of Arkansas, USA

**Achieving sustainable cultivation of barley** 074
Edited by: Prof. Glen Fox, University of California-Davis, USA and The University of Queensland, Australia and Prof. Chengdao Li, Murdoch University, Australia

**Advances in crop modelling for a sustainable agriculture** 075
Edited by: Emeritus Prof. Kenneth Boote, University of Florida, USA

**Achieving sustainable crop nutrition** 076
Edited by: Prof. Zed Rengel, University of Western Australia, Australia

**Achieving sustainable urban agriculture** 077
Edited by: Prof. Johannes S. C. Wiskerke, Wageningen University, The Netherlands

**Climate change and agriculture** 078
Edited by Dr Delphine Deryng, NewClimate Institute/Integrative Research Institute on Transformations of Human-Environment Systems (IRI THESys), Humboldt-Universität zu Berlin, Germany

**Advances in poultry genetics and genomics** 079
Edited by: Prof. Samuel E. Aggrey, University of Georgia, USA, Prof. Huaijun Zhou, University of California-Davis, USA, Dr Michèle Tixier-Boichard, INRAE, France and Prof. Douglas D. Rhoads, University of Arkansas, USA

**Achieving sustainable management of tropical forests** 080
Edited by: Prof. Jürgen Blaser, Bern University of Applied Sciences, Switzerland and Patrick D. Hardcastle, Forestry Development Specialist, UK

**Improving the nutritional and nutraceutical properties of wheat and other cereals** 081
Edited by: Prof. Trust Beta, University of Manitoba, Canada

**Achieving sustainable cultivation of ornamental plants** 082
Edited by: Emeritus Prof. Michael Reid, University of California-Davis, USA

**Improving rumen function** 083
Edited by: Dr C. S. McSweeney, CSIRO, Australia and Prof. R. I. Mackie, University of Illinois, USA

**Biostimulants for sustainable crop production** 084
Edited by: Youssef Rouphael, Patrick du Jardin, Patrick Brown, Stefania De Pascale and Giuseppe Colla

**Improving data management and decision support systems in agriculture** 085
Edited by: Dr Leisa Armstrong, Edith Cowan University, Australia

**Achieving sustainable cultivation of bananas - Vol 2** 086
Germplasm and genetic improvement
Edited by: Prof. Gert H. J. Kema, Wageningen University, The Netherlands and Prof. Andrè Drenth, The University of Queensland, Australia

**Reconciling agricultural production with biodiversity conservation** 087
Edited by: Prof. Paolo Bàrberi and Dr Anna-Camilla Moonen, Institute of Life Sciences - Scuola Superiore Sant'Anna, Italy

**Advances in postharvest management of cereals and grains** 088
Edited by: Prof. Dirk E. Maier, Iowa State University, USA

**Biopesticides for sustainable agriculture** 089
Edited by: Prof. Nick Birch, formerly The James Hutton Institute, UK and Prof. Travis Glare, Lincoln University, New Zealand

**Understanding and improving crop root function** 090
Edited by: Emeritus Prof. Peter J. Gregory, University of Reading, UK

**Understanding the behaviour and improving the welfare of chickens** 091
Edited by: Prof. Christine Nicol, Royal Veterinary College - University of London, UK

**Advances in measuring soil health** 092
Edited by: Prof. Wilfred Otten, Cranfield University, UK

**The sustainable intensification of smallholder farming systems** 093
Edited by: Dr Dominik Klauser and Dr Michael Robinson, Syngenta Foundation for Sustainable Agriculture, Switzerland

**Advances in horticultural soilless culture** 094
Edited by: Prof. Nazim S. Gruda, University of Bonn, Germany

**Reducing greenhouse gas emissions from livestock production** 095
Edited by: Dr Richard Baines, Royal Agricultural University, UK

**Understanding the behaviour and improving the welfare of pigs** 096
Edited by: Emerita Prof. Sandra Edwards, Newcastle University, UK

**Genome editing for precision crop breeding** 097
Edited by: Dr Matthew R. Willmann, Cornell University, USA

**Understanding the behaviour and improving the welfare of dairy cattle** 098
Edited by: Prof. Marcia Endres, University of Minnesota, USA

**Transforming food systems** 099
The quest for sustainability
Dr Dave Watson

**Plant genetic resources** 100
A review of current research and future needs
Edited by: Dr M. Ehsan Dulloo, Bioversity International, Italy

**Developing animal feed products** 101
Edited by: Dr Navaratnam Partheeban, formerly Royal Agricultural University, UK

**Improving dairy herd health** 102
Edited by: Prof. Emeritus Émile Bouchard, University of Montreal, Canada

**Understanding gut microbiomes as targets for improving pig gut health** 103
Edited by: Prof. Mick Bailey and Emeritus Prof. Chris Stokes, University of Bristol, UK

**Advances in Conservation Agriculture – Vol 3** 104
Adoption and Spread
Edited by: Professor Amir Kassam, University of Reading, UK and Moderator, Global Conservation Agriculture Community of Practice (CA-CoP), FAO, Rome, Italy

**Advances in precision livestock farming** 105
Edited by: Prof. Daniel Berckmans, Katholieke University of Leuven, Belgium

**Achieving durable disease resistance in cereals** 106
Edited by: Prof. Richard Oliver, formerly Curtin University, Australia

**Seaweed and microalgae as alternative sources of protein** 107
Edited by: Prof. Xin Gen Lei, Cornell University, USA

**Microbial bioprotectants for plant disease management** 108
Edited by: Dr Jürgen Köhl, Wageningen University & Research, The Netherlands and Dr Willem J. Ravensberg, Koppert, The Netherlands

**Improving soil health** 109
Edited by: Prof. William R. Horwath, University of California-Davis, USA

**Improving integrated pest management in horticulture** 110
Edited by: Prof. Rosemary Collier, Warwick University, UK

**Climate-smart production of coffee** 111
Improving social and environmental sustainability
Edited by: Prof. Reinhold Muschler, CATIE, Costa Rica

**Developing smart agri-food supply chains** 112
Using technology to improve safety and quality
Edited by: Prof. Louise Manning, Royal Agricultural University, UK

**Advances in integrated weed management** 113
Edited by: Prof. Per Kudsk, Aarhus University, Denmark

**Understanding and improving the functional and nutritional properties of milk** 114
Edited by: Prof. Thom Huppertz, Wageningen University & Research, The Netherlands and Prof. Todor Vasiljevic, Victoria University, Australia

**Energy-smart farming** 115
Efficiency, renewable energy and sustainability
Edited by: Emeritus Prof. Ralph E. H. Sims, Massey University, New Zealand

**Understanding and optimising the nutraceutical properties of fruit and vegetables** 116
Edited by: Prof. Victor R. Preedy, King's College London, UK and Dr Vinood B. Patel, University of Westminster, UK

**Advances in plant phenotyping for more sustainable crop production** 117
Edited by: Prof. Achim Walter, ETH Zurich, Switzerland

**Optimising pig herd health and production** 118
Edited by: Prof. Dominiek Maes, Ghent University, Belgium and Prof. Joaquim Segalés, Universitat Autònoma de Barcelona and IRTA-CReSA, Spain

**Optimising poultry flock health** 119
Edited by: Prof. Sjaak de Wit, Royal GD and University of Utrecht, The Netherlands

**Advances in seed science and technology for more sustainable crop production** 120
Edited by: Dr Julia Buitink, INRAE, France and Prof. Olivier Leprince, L'Institut Agro Rennes Angers, France

**Understanding and fostering soil carbon sequestration** 121
Edited by: Dr Cornelia Rumpel, CNRS, Sorbonne University, Institute of Ecology and Environmental Sciences Paris, France

**Advances in sensor technology for sustainable crop production** 122
Edited by: Dr Craig Lobsey, University of Southern Queensland, Australia and Prof. Asim Biswas, University of Guelph, Canada

**Achieving sustainable cultivation of bananas - Vol 3** 123
Diseases and pests
Edited by: Prof. André Drenth, University of Queensland, Australia and Prof. Gert H. J. Kema, Wageningen University, The Netherlands

**Developing drought-resistant cereals** 124
Edited by: Prof. Roberto Tuberosa, University of Bologna, Italy

**Achieving sustainable turfgrass management** 125
Edited by: Prof. Michael Fidanza, Pennsylvania State University, USA

**Promoting pollination and pollinators in farming** 126
Edited by: Emeritus Prof. Peter Kevan and Dr D. Susan Willis Chan, University of Guelph, Canada

**Improving poultry meat quality** 127
Edited by: Prof. Massimiliano Petracci, Alma Mater Studiorum - Università di Bologna, Italy and Dr Mario Estévez, Universidad de Extremadura, Spain

**Advances in monitoring of native and invasive insect pests of crops** 128
Edited by: Dr Michelle Fountain, NIAB-EMR, UK and Dr Tom Pope, Harper Adams University, UK

**Advances in understanding insect pests affecting wheat and other cereals** 129
Edited by: Prof. Sanford Eigenbrode, University of Idaho and Dr Arash Rashed, Virginia Tech, USA

**Understanding and improving crop photosynthesis** 130
Edited by: Dr Robert Sharwood, Western Sydney University, Australia

**Modelling climate change impacts on agricultural systems** 131
Edited by: Prof. Claas Nendel, Leibniz Centre for Agricultural Landscape Research (ZALF), Germany

**Understanding and minimising fungicide resistance** 132
Edited by: Dr Francisco J. Lopez-Ruiz, Curtin University, Australia

**Advances in sustainable dairy cattle nutrition** 133
Edited by: Prof. Alexander N. Hristov, The Pennsylvania State University, USA

**Embryo development and hatchery practice in poultry production** 134
Edited by: Dr Nick French

**Developing circular agricultural production systems** 135
Edited by: Prof. (UZ) Dr Barbara Amon, University of Zielona Góra, Poland and Leibniz Institute for Agricultural Engineering and Bioeconomy (ATB), Germany

**Advances in ensuring the microbiological safety of fresh produce** 136
Edited by: Prof. Karl R. Matthews, Rutgers University, USA

**Frontiers in agri-food supply chains** 137

Frameworks and case studies
Edited by: Prof. Sander de Leeuw, Dr Renzo Akkerman and Dr Rodrigo Romero-Silva, Wageningen University, The Netherlands

**Improving water management in agriculture** 138
Irrigation and food production
Edited by: Prof. Jerry W. Knox, Cranfield University, UK

**Advances in agri-food robotics** 139
Edited by: Prof. Eldert van Henten, Wageningen University, The Netherlands and Prof. Yael Edan, Ben-Gurion University of the Negev, Israel

**Key issues in agricultural ethics** 140
Edited by: Prof. Emeritus Robert L. Zimdahl, Colorado State University, USA

**Advances in plant factories** 141
New technologies in indoor vertical farming
Edited by: Toyoki Kozai and Eri Hayashi

**Improving the quality of apples** 142
Edited by: Prof. Fabrizio Costa, University of Trento, Italy

**Protecting natural capital and biodiversity in the agri-food sector** 143
Edited by: Prof. Jill Atkins, Cardiff University, UK

**Consumers and food** 144
Understanding and shaping consumer behaviour
Edited by: Professor Marian Garcia Martinez, The University of Kent, UK

**Advances in cultured meat technology** 145
Edited by: Prof. Mark Post, Maastricht University, The Netherlands, Prof. Che Connon, Newcastle University, UK and Dr Chris Bryant, University of Bath and Bryant Research, UK

**Understanding and preventing soil erosion** 146
Edited by: Dr Manuel Seeger, University of Trier, Germany

**Smart farms** 147
Improving data-driven decision making in agriculture
Edited by: Prof. Claus Grøn Sørensen, Aarhus University, Denmark

**Improving standards and certification in agri-food supply chains** 148
Ensuring safety, sustainability and social responsibility
Edited by: Prof. Louise Manning, University of Lincoln, UK

**Managing biodiversity in agricultural landscapes** 149
Conservation, restoration and rewilding
Edited by: Dr Nick Reid and Dr Rhiannon Smith, University of New England, Australia

**Improving nitrogen use efficiency in crop production** 150
Edited by: Prof. Jagdish Kumar Ladha, University of California-Davis, USA

**Understanding and utilising soil microbiomes for a more sustainable agriculture** 151
Edited by: Prof. Kari Dunfield, University of Guelph, Canada

**Advances in pig breeding and reproduction** 152
Edited by: Prof. Jason Ross, Iowa State University, USA

**Advances in organic dairy cattle farming** 153
Edited by: Dr Mette Vaarst, Aarhus University, Denmark, Dr Stephen Roderick, Duchy College, UK and Dr Lindsay Whistance, Organic Research Centre, UK

**Insects as alternative sources of protein for food and feed** 154
Edited by: Adriana Casillas, Tebrio, Spain

**Advances in pig nutrition** 155
Edited by: Emeritus Prof. Julian Wiseman, University of Nottingham, UK

**Advances in temperate agroforestry** 156
Edited by: Prof. María Rosa Mosquera-Losada, Universidade de Santiago de Compostela, Spain, Dr Ladislau Martin, Embrapa, Brazil, Prof. Anastasia Pantera, Agricultural University of Athens, Greece and Dr Allison Chatrchyan, Cornell University, USA

**Sustainable production and postharvest handling of avocado** 157
Emeritus Prof. Elhadi M. Yahia, Autonomous University of Querétaro, Mexico

**Advances in bioprotection of plants against diseases** 158
Edited by: Prof. Shashi Sharma, Murdoch University, Australia and Dr Minshad Ansari, Bionema, UK

**Advances in poultry nutrition** 159
Edited by: Prof. Todd J. Applegate, University of Georgia, USA

**Weed management in Conservation Agriculture systems** 160
Edited by: Prof. Gottlieb Basch, Prof. Emilio Gonzalez-Sanchez, Mr John Geraghty, Dr Seyed Vahid Eslami, Prof. Sjoerd Willem Duiker, Dr Saidi Mkomwa and Dr Marie Bartz

**Improving poultry meat safety and sustainability** 161
Edited by: Prof. Steven C. Ricke, University of Wisconsin-Madison, USA

**Advances in sheep production** 162
Edited by: Lesley Stubbings, Consultant, UK and Kate Phillips, Consultant, UK

**Women and smallholder farming** 163
Addressing global inequities in agriculture
Edited by: Prof. Emerita Carolyn Sachs, Pennsylvania State University, USA and Dr Paige Castellanos, Oxfam America (OUS), USA

**Advances in biocontrol of crop insect pests** 164
Edited by: Prof. Travis Glare, Lincoln Agritech Ltd and Lincoln University, New Zealand and Prof. Johannes A. Jehle, Institute of Biological Control - Federal Research Centre for Cultivated Plants (JKI), Germany

**Advances in improving the safety and quality of eggs** 165
Edited by: Prof. Cristina Alamprese, University of Milan, Italy, Dr Yves Nys, formerly INRAE, France, Prof. Yoshinori Mine, University of Guelph, Canada, and Dr Vincent Guyonnet, University of Montreal, Canada

**Contaminants in agricultural soils** 166
Challenges and solutions
Edited by Distinguished Laureate Prof. Ravi Naidu, University of Newcastle, Australia and CRC for Contamination Assessment and Remediation of the Environment (crcCARE), Australia

**Precision agriculture for sustainability: Second edition** 167
Edited by: Dr John Stafford, Silsoe Solutions, UK

**The science and practice of agroecology** 168
Pathway to sustainable food systems
Edited by: Prof. Laurie Drinkwater, Cornell University, USA

**Advances in temperate grassland science and management** 169
Edited by: Dr Agnes van den Pol-van Dasselaar

**Current crop models** 170
State-of-the-art and future developments
Edited by: Prof. Gerrit Hoogenboom, University of Florida, USA

**Achieving net-zero dairy farming** 171
Edited by: Emeritus Prof. John Webster, University of Bristol, UK

**Chemical contaminants in foods** 172
Understanding and managing risks
Edited by: Dr Rob Theelen

**Measuring and assessing the biological health of soils** 173
Edited by: Prof. Jeanette Norton, Utah State University, USA, Prof. Josh Schimel, University of California – Santa Barbara, USA and Prof. Zoë Lindo, University of Western Ontario, Canada

**The European spruce bark beetle** 174
Connecting ecology, human aspects and practical forest management
Edited by: Dr Sigrid Netherer, University of Natural Resources and Life Sciences Vienna (BOKU), Austria and Dr Maartje Klapwijk, Swedish University of Agricultural Sciences, Sweden

**Achieving sustainable cultivation of hemp** 175
Edited by: Prof. Emerit Jay Stratton Noller, Oregon State University, USA

**Cover crops for sustainable soil management** 176
Edited by: Prof. Sjoerd W. Duiker, The Pennsylvania State University, USA and Dr Andy Clark, formerly University of Maryland and USDA Sustainable Agriculture Research and Education (SARE) USA

**Global trends in pesticide resistance** 177
Edited by: Prof. Richard Oliver, University of Nottingham, UK

**Advances in controlled environment agriculture** 178
Edited by: Dr Oliver Körner, Leibniz Institute of Vegetable and Ornamental Crops, Germany and Prof. Gundula Proksch, University of Washington, USA

**Future perspectives in broiler rearing** 179
Incorporating the broiler point of view
Edited by: Dr Henry van den Brand

**Optimising bio-based fertilisation for a more sustainable agriculture** 180
Edited by: Prof. Ruben Sakrabani, Cranfield University, UK

**Advances in precision dairy and beef farming technologies** 181
Edited by: Prof. Daniel Berckmans and Prof. Tomás Norton, Katholieke University of Leuven, Belgium

**Developing alternative animal feed products** 182
Edited by: Dr Nico van Belzen, ScienceConsult, The Netherlands

**Agriculture, land use and the path to net zero** 183
Edited by: Prof. Robert M. Rees, Scotland's Rural College (SRUC), UK and Prof. Jørgen Eivind Olesen, Aarhus University, Denmark

**Advances in precision pig farming technologies** 184
Edited by: Prof. Lisa Collins, University of Surrey, UK

**Advances in improving animal feed safety** 185
Edited by: Dr Regiane R. Santos, Schothorst Feed Research, The Netherlands

**Optimising breeding of dairy cows: Leveraging genetics for more sustainable production** 186
Edited by: Prof. Donagh Berry, Teagasc, Ireland

**Improving the processing quality of wheat** 187
Edited by: Prof. Kurt A. Rosentrater, Iowa State University, USA

**Advances in integrated wheat disease management** 188
Edited by: Prof. Stephen Wegulo, University of Nebraska-Lincoln, USA

**Farming and land use change to achieve net zero: Achieving multi-functional landscapes** 189
Edited by: Emeritus Prof. Guy Robinson, Adelaide University, Australia

**Advances in precision poultry farming technologies** 190
Edited by: Prof. María Cambra-López, Polytechnic University of Valencia, Spain and Dr Yang Zhao, University of Tennessee, USA

# Preface

Global broiler production exceeds 70 billion chickens annually with poultry remaining the world's second most consumed meat. This growth reflects the efficiency, affordability and relatively low environmental footprint of poultry compared with other terrestrial livestock species. Yet this success has also intensified scrutiny. Modern broiler systems are increasingly evaluated not only for productivity and profitability, but also increasingly against criteria such as environmental sustainability, contribution to antimicrobial resistance and capacity to safeguard animal welfare.

The broiler of today is the result of decades of genetic selection for rapid growth and feed efficiency as well as advances in precision nutrition, housing and management systems. These achievements have delivered remarkable improvements in performance. However, they have also introduced new biological vulnerabilities, system dependencies and societal tensions. Fast growth challenges skeletal integrity and metabolic balance; globalised feed sourcing could create nutrient imbalances and environmental externalities; and intensive systems amplify disease transmission risks. Public expectations have evolved from minimising suffering to enabling positive welfare states and environmental stewardship.

This volume approaches these challenges from a central organising principle: incorporating the broiler's biological and behavioural perspective into production design. Rather than viewing the chicken solely as a production unit within an economic system, this book considers the broiler as a developing organism whose genetic, maternal, environmental, nutritional and social contexts interact dynamically across its lifecycle. Sustainable broiler production, therefore, requires alignment between biological resilience, management practices, technological innovation, welfare and ethical priorities.

This book is structured to follow the life trajectory of the broiler while embedding that trajectory within broader system-level considerations. Part 1 reframes broiler production within ethical, environmental and data system contexts, establishing the societal and ecological boundaries within future systems must operate. Part 2 focuses on the biological foundations of broiler life. It examines breeding strategies, transgenerational influences, incubation biology as well as in ovo and early feeding. Chapters emphasise developmental programming and early-life resilience.

Part 3 shifts the focus to the growing broiler, analysing behavioural, nutritional and environmental requirements; it highlights how management decisions influence welfare, performance and system sustainability. Part 4 addresses structural and emerging challenges, including: antibiotic-free production, developing more circular production systems, optimising extensive and small-scale production, and the particular challenge of climate-induced heat stress. Part 5 explores precision livestock technologies that enable continuous monitoring and real-time decision-making, positioning data-driven management as a cornerstone of future broiler systems.

Part 6 examines preventative health management, focusing on key areas such as vaccination strategies, early mortality, skeletal and respiratory health. Part 7 closes the broiler production lifecycle with end-of-life management and its impact on meat quality.

Overall, the book provides an integrated, forward-looking synthesis of biological science and management practice. By aligning productivity with resilience, welfare and sustainability, this volume aims to support the transition towards broiler production systems that remain economically viable while meeting the evolving expectations of society. This volume cannot, of course, address every aspect of modern broiler production. Readers might also already be familiar with some aspects. However, we hope all readers will discover new topics and perspectives that will deepen their understanding and promote improved practices to take the sector forward. For those readers who would like more detailed information on what the book contains, the following sections provide concise summaries of what each chapter covers.

# Acknowledgements

We wish to acknowledge Dr Nick French for his help in reviewing particular chapters, his contribution to the book is greatly appreciated.

# Introduction

For those readers who would like more detailed information on what the book contains than is contained in the Preface, the following sections in this Introduction provide concise summaries of what each chapter covers.

## Part 1 Setting the scene in broiler production

Chapter 1 examines contemporary animal welfare science and its ethical foundations. It introduces major welfare concepts, such as the shift from avoiding harm to achieving positive affective states. The chapter also explores the inseparability of ethics and welfare assessment. While a welfare-centred perspective is related to the experiences of the animal, there are also ethical challenges in broiler production that are more closely related to how humans experience certain production practices or the use of animals in food production systems in general. The chapter also analyses ways of understanding and resolving key ethical issues, such as the broiler breeder paradox (which describes the challenge of achieving an acceptable reproductive performance and health status in broiler breeders without restrictive feeding, a practice which then raises welfare issues). In doing so, it frames the moral landscape needed to guide future production systems.

In addition to welfare, sustainability is an issue of growing importance in the face of climate change. As Chapter 2 indicates, poultry production produces large volumes of 'waste' material that can either pose a serious environmental risk or provide valuable co-products for re-use. As an example, broiler litter is a mixture of broiler manure, feathers, waste feed and bedding material that contains nitrogen, phosphate, potash and other nutrients essential for plant growth. Broiler litter is thus a great natural fertiliser but, if not properly managed, can cause contamination of surface and ground waters as well as air pollution. The chapter explores the use of nutrient management plans (NMPs) to optimise use of poultry litter as a fertiliser, including best management practices for storage, testing of litter and soil, calibration and use of spreading equipment as well as site practices such as buffer strips to prevent runoff.

Building on the previous chapters, Chapter 3 explores the role of data integration across the broiler production chain (including breeding companies, hatcheries, growers, processors and retailers) as a foundation for improving productivity, animal welfare and sustainability outcomes. It evaluates the transformative potential of IoT-enabled sensing systems and AI-driven analytics for real-time monitoring, predictive modelling and adaptive decision making. At the same time, it addresses structural challenges such as data ownership, interoperability, cybersecurity risks, and unequal access to technology.

The chapter frames data sharing as not just a technical issue, stressing that transparency, trust and equitable participation are also essential to success. It argues that future production systems should treat data as a shared resource that supports collective optimisation rather than isolated competitive advantage.

## Part 2 Creating the best day-old broiler chick

In Chapter 4 the evolution of broiler breeding goals is traced from a narrow focus on rapid growth and feed efficiency toward more balanced strategies incorporating health, robustness and welfare traits. It examines advances in high-throughput phenotyping, genomic selection and breeding for traits such as disease resistance, positioning genetics as a central lever for system resilience. The chapter also addresses the limits to intensive selection for growth characteristics, including consequences such as skeletal disorders and metabolic stress. It argues for a redefinition of "genetic progress" with breeding objectives that align productivity with long-term health, adaptability and reduced reliance on therapeutic interventions.

Chapter 5 examines how breeder age and nutrition, feed restriction strategies, egg composition, eggshell quality and egg storage conditions influence embryonic development and subsequent chick resilience. The chapter highlights how management decisions at the breeder and egg stages can create either resilience or vulnerability in the next generation. It argues for the need to prioritise developmental robustness and lifetime welfare outcomes rather than solely hatchability or short-term performance metrics.

Following on from the preceding chapter, Chapter 6 reviews key issues affecting day-old chicks which can then have profound effects on subsequent development. It reviews ways of assessing chick quality, including physical characteristics (such as chick size), immunological and microbiological measurements. Immunological quality is determined by maternal antibody transfer which is critical during the early post-hatch period. Microbiological measurements, such as tracking early colonisation of the gut microbiota, are of growing importance in assessing chick quality and later development. As the chapter shows, chick quality is a multi-factorial concept that goes beyond simple physical traits.

Chapter 7 compares immediate post-hatch feeding systems with traditional delayed feeding whilst hatchlings are transported to poultry farms. It examines the available research on potential benefits of early feeding on growth, mortality rates, thermoregulation, gut development and welfare. It explores how early nutritional access influences metabolic programming and resilience to environmental challenges. The chapter also considers logistical and economic constraints in commercial systems, identifying practical strategies to optimise nutrition for the developing chick. It argues that early-life care should favour minimising deprivation and supporting physiological stability during the transition from embryonic to independent life.

Complementing Part 2, Chapter 8 reviews the emerging field of in-ovo delivery of nutrients, vaccines and bioactive compounds (such as essential oils), focusing on timing, dosage, delivery technologies and broader integration with existing hatchery practices. It evaluates the potential to enhance immune development, gut health and early performance, while acknowledging variability in outcomes and scalability challenges. The chapter emphasises the need for further research and robust validation before widespread adoption. More broadly, it advocates for holistic approaches that support natural developmental processes rather than just targeting particular health problems.

## Part 3 Managing the needs of broilers

Chapter 9 examines broiler behaviour as an interaction between genetic predisposition and environmental conditions, with particular focus on highly-motivated natural behaviours associated with positive affective states. These behaviours include: foraging, dustbathing, preening, exploration, play and social interaction. It analyses how rapid growth rates, increased body weight, and lameness can constrain behavioural expression, potentially leading to poorer welfare as well as health. The chapter argues for management and environmental design that enable meaningful behavioural opportunities, framing behavioural expression not as a luxury but as a core component of ethical and resilient system design.

Chapter 10 explores diet formulation strategies, including ingredient selection, feed formulation and amino acid optimisation in the context of modern broiler genetics and production systems. It also evaluates the potential of novel protein sources and feed additives as well as factors such as feed particle size and texture to improve feed use efficiency and gut health. The chapter frames nutrition as a balancing act between factors such as biological needs, resource use efficiency and broader food system sustainability.

Chapter 11 examines environmental management within broiler housing, focusing on factors such as temperature, ventilation and air quality and the ways they can be optimised in poultry barns. The chapter also reviews research on optimising flooring systems, lighting and space management, stocking density, litter quality and climate control. The chapter highlights the importance of maintaining stable and appropriate microclimates to prevent stress, reduce pathogen proliferation and support welfare. It argues that environmental control systems should be designed not only for productivity but also to provide conditions that respect the biological and welfare needs of the growing broiler.

## Part 4 Challenges in broiler production

Building on themes in the preceding Parts of the book, Chapter 12 examines the transition toward antibiotic-free broiler production, focusing on strategies to maintain gut health, enhance immune competence and prevent enteric diseases such as necrotic enteritis and colibacillosis. It evaluates nutritional interventions, including probiotics and prebiotics, as well as preventative vaccination approaches. The chapter acknowledges the complexity and risk associated with removing antibiotics, emphasising that success depends on system-wide improvements rather than single interventions. It frames antibiotic reduction as a public health priority to preserve antimicrobial efficacy for future generations.

Chapter 13 explores the integration of broiler production within wider circular and sustainable food systems. It addresses challenges such as nutrient imbalances, limited utilisation of recycled or alternative feed ingredients, and competition with use of crops for food. The chapter discusses issues in redesigning systems that prioritise resource efficiency, waste valorisation, use of low opportunity-cost biomass as an alternative feed source and, where possible, re-integration of crop and livestock production systems. It shows that sustainability needs to be evaluated across entire supply chains to create production systems that contribute positively to food security while meeting environmental and social goals.

In contrast to the intensive systems that have characterised much of the sector, Chapter 14 examines the trade-offs between productivity, animal welfare and environmental impacts in extensive production systems. It compares these outdoor systems with conventional intensive models, highlighting differences in land use, resource efficiency, animal welfare and health outcomes. The chapter explores ways of optimising extensive systems, such as focussing on local breeds suited to particular conditions. However, it challenges simplistic assumptions about "better" systems, emphasising the need for evidence-based assessment frameworks that acknowledge inherent trade-offs and support informed decision making.

Closing Part 4 on key challenges facing the sector, Chapter 15 analyses the impacts of heat stress on broilers in the context of climate change. It reviews the wide range of factors affecting vulnerability to heat stress, metabolic and physiological changes caused by heat stress and impacts on broiler growth, behaviour, welfare and meat quality. The chapter also evaluates mitigation strategies such as genetic selection for heat tolerance. The chapter positions heat stress as a growing systemic risk, emphasising the urgency of developing resilient production systems capable of maintaining performance and welfare under increasingly variable environmental conditions.

## Part 5 Precision farming technologies in broiler production

Precision livestock farming technologies promise to revolutionise livestock production. Chapter 16 reviews wearable and remote sensor technologies to allow continuous monitoring of broiler behaviour and differing health and welfare indicators. It compares body-worn devices with non-invasive remote systems, assessing their accuracy, scalability and practical application in both research and commercial environments. Case studies illustrate implementation challenges, including cost, data integration and user adoption. The chapter frames precision monitoring as a tool for proactive management, enabling earlier intervention and more individualised care within large populations of chickens.

Following on from the previous chapter, Chapter 17 explores bioacoustic monitoring as a method for assessing broiler health and welfare through analysis of vocalisations. It demonstrates how changes in vocal patterns can reflect stress, disease and emotional state. The chapter highlights the potential of this non-invasive approach for early detection of welfare and health issues, reducing reliance on visual inspection alone. It also addresses challenges related to signal interpretation, background noise and system calibration in ensuring accuracy and reliability in measurement.

Chapter 18 evaluates electronic nose technologies for non-invasive detection of health and welfare status in broiler production systems. It examines how volatile organic compounds emitted by chickens or their environment can serve as biomarkers for disease or stress. The chapter discusses technological limitations, data interpretation challenges and integration with other monitoring systems. It positions these tools as part of a broader shift toward PLF, where continuous data streams inform more rapid, responsive and informed management decisions.

## Part 6 Maintaining the health of broilers

Opening Part 6, Chapter 19 reviews the role of modern vaccination strategies within broiler production. It examines how vaccination programs interact with biosecurity, nutrition and management of the poultry environment to prevent major infectious diseases. The chapter emphasises the importance of timing, coverage and methods of vaccine delivery, as well as the need for integrated approaches that consider the entire production system. It frames disease prevention (rather than treatment after infection has occurred) as a cornerstone of both animal welfare and production efficiency.

Building on themes discussed in Part 2, Chapter 20 examines first-week mortality as an indicator of early-life stress. It analyses contributing factors such as strain, sex and breeder flock age as well as egg condition, incubation environment, transportation and holding arrangements, broiler flock size, stocking density, ventilation, nutrition, drinker type and litter quality. The chapter highlights ways to improve each of these variables to improve survivability, highlighting early mortality not just as a separate metric but as a reflection of overall system performance.

Reflecting the discussion in Chapter 4 as well as the chapters in Part 3, Chapter 21 analyses skeletal development and the prevalence of leg disorders in fast-growing broilers, exploring genetic, nutritional and environmental risk factors. It evaluates preventive strategies aimed at maintaining mobility, reducing pain and improving overall welfare. The chapter highlights the particular importance of early preventative measures such as improving selection to minimise leg problems, managing breeder age and diet, in-ovo administration of nutrients and incubation conditions which can all significantly affect subsequent leg weakness.

Chapter 22 addresses respiratory diseases affecting broilers, examining their complex aetiology, impacts on health and productivity, as well as challenges in diagnosis and prevention. It reviews advances in diagnostic technologies and integrated prevention strategies, including biosecurity, environmental management, good animal husbandry, the role of vaccination, and both conventional and alternative therapeutic strategies. The chapter emphasises the importance of early detection and intervention to avoid or minimise disease impact and thus reduce reliance on therapeutic treatments.

## Part 7 Managing the end of life stage in broiler production

Opening the final Part of the book, Chapter 23 evaluates manual and mechanical loading systems used during the end of life stage for broilers, examining their effects on stress and injury risk. It reviews regulatory frameworks and industry guidelines designed to ensure humane handling. The chapter emphasises the importance of training, equipment design and process optimisation to minimise suffering during this critical phase. It frames end of life management as a key indicator that reflects the values and practices embedded within the production system as a whole.

Drawing on themes from earlier chapters, Chapter 24 concludes the book by examining growth-related muscle disorders, including their underlying risk factors and implications for meat quality. It links these conditions to factors such as genetic selection, growth rates, nutrition and management practices, highlighting the interconnected nature of production decisions in determining meat quality. The chapter explores mitigation strategies ranging from genetics through to incubation, diet and management interventions such as lighting and stocking densities.

# Part 1

## Setting the scene in broiler production

# Chapter 1

## Limitations and challenges in broiler production: ethical considerations beyond welfare issues

*Mona F. Giersberg, Utrecht University, The Netherlands*

1 Introduction
2 Contemporary animal welfare concepts
3 Major welfare challenges in broiler production
4 The link between welfare and ethics
5 Ethical perspectives beyond welfare-centred approaches
6 Case study: the broiler breeder paradox
7 Conclusion and future trends in research
8 Where to look for further information
9 References

### 1 Introduction

These days, the agricultural sector is confronted with major global challenges, such as climate change, depleted natural resources and biodiversity loss (FAO, 2017). In response to these issues, several transformative, system-changing developments are in progress, including a shift towards more circular production systems (De Boer and Van Ittersum, 2018). The future role of food-producing animals in these systems is uncertain. This applies particularly to high-yielding monogastric animals like fast-growing broiler chickens which rely on grains that are often directly edible for humans or that are grown on land suitable for human food production (Mottet and Tempio, 2017). In addition, societal acceptance of using animals for food production no longer depends on environmental, economic and public health criteria only (Kanis et al., 2003). An important factor in this context is animal welfare. Research has shown that citizens and consumers are generally concerned about the welfare of animals in food production systems (Carnovale et al., 2021; Clark et al., 2016; Estévez-Moreno et al., 2022). Based on the concepts of humane treatment and

http://dx.doi.org/10.19103/AS.2025.0165.02

naturalness, people seem to consider the farming conditions of laying hens and broiler chickens to be worse in terms of animal welfare than those of dairy cows, with pigs in between (Clark et al., 2016).

Welfare concerns in broiler chickens are not just a public perception. Several welfare challenges can be substantiated by scientific evidence. The recently published Scientific Opinion of the European Food Safety Authority's Panel on Animal Health and Welfare (EFSA AHAW Panel et al., 2023) combines a comprehensive review of the literature with expert assessments regarding the welfare of broilers on farm. It states 19 highly relevant welfare consequences for broiler chickens kept in current production systems (EFSA AHAW Panel et al., 2023). These include, for instance, the inability to perform comfort and foraging behaviours, locomotory disorders, resting problems and integument damage. These examples show that the EFSA AHAW Panel focusses on animal experiences that affect their welfare negatively by causing discomfort, stress, fear, frustration or pain.

However, the present understanding of animal welfare is not limited to the prevention of harm (Giersberg, 2024). It also includes enabling animals to experience positive affective states (Mellor, 2015) and to lead an overall 'good life' (Edgar et al., 2013). Whether positive or negative, the identification of welfare consequences is always based on an underlying theoretical concept of animal welfare. Furthermore, our treatment of and relationship with animals may lead to concerns which cannot be understood through concepts of animal welfare. The aim of this chapter is to present and reconcile seemingly disparate academic perspectives on animal welfare and broiler welfare in particular. Understanding discipline-specific assumptions and traditions is the first step that enables researchers and professionals to join forces and genuinely collaborate for the benefit of broiler chickens and other animals. The following part of this chapter is dedicated to an introduction to contemporary animal welfare concepts to offer an insight into the complexity of such concepts. This is followed by a discussion of a number of major welfare challenges in broiler production. The fourth part of this chapter examines the link between welfare and ethics and addresses some limitations of a welfare-centred perspective. While a welfare-centred perspective is related to the experiences of the animal, there are also challenges in broiler production that are more related to how humans experience certain production practices or the use of animals in food production systems in general. The fifth section of this chapter, therefore, explores several ethical challenges of broiler production that go beyond animal welfare concerns. This chapter includes a case study on the broiler breeder paradox. It closes with a summary of key points and future trends in research in the field of animal welfare and ethics.

## 2 Contemporary animal welfare concepts

It is not accidental that we consider, for instance, resting problems and integument damage as welfare consequences in broiler chickens and measure broiler disturbance and footpad dermatitis as animal-based welfare indicators. It is rather related to the underlying theoretical concepts of animal welfare. These concepts are dynamic in nature; they emerge, evolve or may even become obsolete over time (Harfeld, 2024). Animal welfare concepts are the product of researchers who study fundamental ideas and of those who shape them through practical work, without intending to participate in the concepts' theoretical development (Harfeld, 2024). From the large number of animal welfare concepts that have been developed and have been published mainly in the Western world in recent decades, this section introduces three frameworks or concepts in detail: the popular models of Mellor and Fraser and the recently published dynamic animal welfare concept by Arndt et al.

The Five Domains Model was first introduced by Mellor and Reid (1994) and uses the well-known Five Freedoms as a basis for assessing welfare compromise. It can more accurately be described as a welfare assessment framework rather than a theoretical concept of animal welfare. The idea behind the model was to include the ethical dimension in the generation of knowledge in animal-based sciences and to advance the interests of the animals in these contexts. In its initial form, the model contained the following domains: thirst/hunger/malnutrition (domain 1), environmental challenge (domain 2), disease/injury/functional impairment (domain 3), behavioural/interactive restriction (domain 4) and domain 5, anxiety/fear/pain/distress, which represents a cumulation of negative affective states arising from compromise in domains 1–4 (Mellor and Reid, 1994). The integrated assessment results of each domain describe the overall welfare status of the animal.

As animal welfare science progressed, the Five Domains Model has been continuously adapted over the years. A major update extended the model to cover not only welfare compromise but also positive experiences of animals in the respective domains that potentially enhance welfare (Mellor and Beausoleil, 2015). The physical or functional domains (1–4) were therefore renamed to terms that can capture both negative and positive aspects: nutrition (domain 1), environment (domain 2), health (domain 3) and behaviour (domain 4). The affective experience domain (5) was labelled 'mental state'. It summarises the negative and positive affects resulting from restrictions or opportunities in the other domains. The overall affective experiences in the fifth domain represent the welfare status of the animal (Mellor and Beausoleil, 2015). It is important to note that the Five Domains Model does not focus on the presence of welfare-enhancing resources, such as environmental enrichment, but on the actual behavioural utilisation of these resources by the animal. A resource (e.g. a strawbale for perching) that is available but unused could indicate a welfare compromise (e.g. pain caused by skeletal disorders). In this way, the model allows for describing the interaction of positive and negative states (Mellor and Beausoleil, 2015).

In the latest published update of the Five Domains Model, domain 4 was renamed to 'behavioural interactions' (Mellor et al., 2020). It now focusses explicitly on goal-directed behavioural interactions of animals with the environment, other animals and humans. With this addition, it is possible to assess the negative and positive implications of a wide range of human-animal interactions for animal welfare. However, the model also has several constraints (Mellor, 2017). First, like any animal welfare assessment framework, it is limited to affects that can be identified by humans. Second, if this criterion is met, the extent to which the model represents reality depends on the confidence with which a human can assess a particular affect in an animal. For broiler chickens, there are, for instance, validated tests to determine fear (Giersberg et al., 2021), whereas this is not the case for assessing contentment or pleasure. Finally, when weighing the negative and positive affects to obtain the overall welfare status of an animal, it is not possible to determine the relative impacts of the individual affects (Mellor, 2017).

Around the same period, Fraser et al. (1997) introduced a concept of animal welfare based on and emphasising ethical concerns about the quality of life of animals. They argued that in contrast to other disciplines, the emergence of animal welfare science was not motivated by scientific curiosity but by ethical concerns of society about the quality of life of animals. If animal welfare science aimed to address these concerns, the concepts of animal welfare adopted by researchers would need to reflect the full spectrum of these concerns (Fraser et al., 1997). According to the authors, the three major ethical concerns regarding the quality of life of animals are: (1) natural living – animals should be allowed to have an ontogenic development considered 'normal' for the species and to live in a way that matches with their adaptations, (2) subjective experiences – animals should be free from negative affective states and experience 'normal pleasures' and (3) biological functioning – animals should function well in terms of health, physiology and behaviour.

As models based on only one of these concerns would have major limitations, Fraser et al. (1997) proposed an integrated concept of animal welfare. This concept distinguished three categories of issues that may arise when the adaptations of an animal do not completely match the challenges present in its environment. The first category includes adaptations that no longer serve an important function in the animal's current environment. These superfluous adaptations can lead to frustrations, but do not necessarily affect biological functioning. For instance, a broiler chicken may experience a desire to forage, even though it receives adequate nutrition through concentrated, pelleted feed offered in a feeding trough. Issues in this category may lead to concerns about subjective experiences and natural living. The second category contains environmental challenges to which the animal is not able to adapt and which may lead to functional problems. For instance, when kept on wet litter, broiler chickens may develop footpad dermatitis, i.e. the natural protective and barrier function of the skin is destroyed in the affected areas. Problems like this may lead to concerns about biological functioning and natural living. The third category describes challenges for which the animal is equipped with corresponding adaptations. Issues may arise when the degree of the challenge exceeds the animal's ability to adapt. For instance, in hot environments broiler chickens pant to lose heat. However, if the ambient temperature exceeds the chickens' thermoregulatory capabilities, they collapse. From this category, concerns about subjective experiences and biological functioning may arise.

In his later work, Fraser also stressed the hybrid nature of the conception of animal welfare: it can be investigated scientifically (science-based concept), but at the same time it depends on value-based ideas about what is important for animals to lead a good life (value-based concept) (Fraser, 2008). Thus to be widely accepted, welfare-improving initiatives would need to be both scientifically sound and in line with major value positions. These value positions stem from the three major ethical concerns outlined above (Fraser et al., 1997) and were rephrased to (1) natural living, (2) affective states and (3) basic health and functioning (Fraser, 2008).

The recently published dynamic animal welfare concept (DAWCon) by Arndt et al. (2022) focusses on the capability, ability and opportunity of an animal to adequately adapt to pleasant or adverse internal or environmental stimuli to reach a state that it experiences as positive. With its consideration of affective experiences, the concept is in line with the Five Domains Model (Mellor and Beausoleil, 2015). It is also obvious that the concept proposed by Arndt et al. (2022) reflects the three major value positions described by Fraser (2008): an animal should live under circumstances that match its adaptations (natural living), it should experience its state as positive (affective states), and it should be in good health to be able to adequately react to environmental circumstances (basic health and functioning).

DAWCon adds to earlier animal welfare concepts by emphasising that both the adaptations of an animal and the environmental conditions are dynamic and may change over time. The adaptations of broiler chickens subjected individually to an unfamiliar environment (novel environment test) illustrate this dynamic. Young chickens will show high frequencies of vocalisations in this situation as seeking social reinstatement is a priority at a young age (Giersberg et al., 2020; Suarez and Gallup, 1983). At an older age, however, broiler chickens are more likely to be silent and show freezing behaviour as avoiding a predatory threat becomes predominant (Giersberg et al., 2020; Suarez and Gallup, 1983). This means that certain behavioural adaptations may be adequate at one specific age but not at another. It is therefore important to take an animal's previous experience into account when evaluating the appropriateness of behavioural responses and to measure certain indicators repeatedly over time to draw valid conclusions about the welfare status of an animal (Arndt et al., 2022).

## 3 Major welfare challenges in broiler production

With the notions of what animal welfare entails in mind, this section briefly discusses major challenges in current broiler production. The aim is not to provide an exhaustive and detailed account of all welfare problems that may occur in large-scale broiler production, but to introduce concrete examples that help to understand the practical relevance of the theoretical concepts outlined above and the broader ethical considerations of the following sections. The examples were chosen based on the list of 19 highly relevant welfare consequences which were identified by the EFSA Task Force on animal welfare for different categories of birds (day-old chickens, broiler chickens on farm for meat production and broiler breeders) and husbandry systems (EFSA AHAW Panel et al., 2023). From this list, the welfare challenges that are of particular relevance for broiler chickens on farm, in intensive production systems, were selected and are discussed in detail below.

Most welfare problems currently observed in broiler chickens can be attributed to the intense genetic selection for fast growth rates and feed efficiency, and to the enormous intensification of the poultry sector over the past decades (EFSA AHAW Panel et al., 2023; Hartcher and Lum, 2020). At 35 days of age, a fast-growing broiler can reach a bodyweight of 2.30 kg with an average daily weight gain of 64 g and a feed conversion ratio of 1.38 (Aviagen, 2022). The enormous intensification of the sector is reflected by the global numbers of slaughtered animals: in 2023, about 76 billion chickens were slaughtered worldwide (FAO, 2023). This situation increases the chance that the animals will be confronted with internal and external challenges for which they possess no adaptations or which exceed their ability to adapt. As shown above, this can lead to lasting negative affective states and impaired biological functioning.

In broilers reared for meat, major welfare problems include thermoregulatory stress, gastrointestinal disorders and infectious diseases, locomotory disorders, integument damage and behavioural restrictions, such as reduced activity and limited performance of resting, foraging and comfort behaviours (EFSA AHAW Panel et al., 2023). Thermoregulatory stress occurs when an animal is exposed to effective temperatures below (cold stress) or above (heat stress) its thermal comfort zone. Particularly, young chickens are at risk of experiencing cold stress. As their thermoregulatory abilities are not fully developed in the first days of life, they depend on the ambient temperature to maintain their body temperature (Nichelmann and Tzschentke, 2002). Undercooling can lead to death by itself or through an increased susceptibility to infectious diseases. However, in young chickens also huddling, which is an adequate adaptative behaviour to cold temperatures, can be lethal if it exceeds certain levels. Chickens that are clumping tightly together in an attempt to reduce heat loss are less active and less likely to find feed and water in the barn, which can lead to starvation and dehydration, and thus increased first-week mortality (Heier et al., 2002). At an older age, broiler chickens are at a higher risk of experiencing heat stress. Especially fast-growing hybrids have a high metabolic rate which leads to increased heat production and sensitivity to high ambient temperatures (Deeb and Cahaner, 2002). Broiler chickens adapt to high ambient temperatures by reducing feed intake and activity, and increasing water consumption. As broilers cannot sweat, they also pant and hold their wings away from the body to increase respiratory and convective heat dissipation, respectively (EFSA AHAW Panel et al., 2023). High stocking densities, inadequate ventilation and cooling systems of indoor barns and heat waves can contribute to a situation in which broiler chickens reach the limits of their thermoregulatory capacities.

Disfunctions of the gastrointestinal tract are common in commercial broiler flocks. They are often caused by unspecific imbalances of the gut microbiome (Wickramasuriya et al., 2022). However, they can also be the result of bacterial (mainly necrotic enteritis and *E. coli* infections (Fancher et al., 2021; Moore, 2016), viral or parasitic infections (mainly coccidiosis (Blake and Tomley, 2014)). Depending on the cause, these diseases can lead to reduced performance, lethargy, diarrhoea and increased mortality, particularly during the first week. In such cases, the animal's ability to adapt to, for instance, altered feed components or to react with an adequate immune response has been exceeded.

Locomotory disorders or lameness occur particularly in broiler hybrids with a high growth rate (Dixon, 2020). Reduced mobility in broiler chickens can be caused by conditions such as footpad dermatitis or severe bone deformations (EFSA AHAW Panel et al., 2023). Besides causing acute or lasting pain, lameness can prevent broiler chickens from performing various behaviours, such as exploration and foraging. In severe cases, chickens may also be reluctant to move to feeders and drinkers. Environmental factors such as inadequate feed, high stocking density, reduced litter quality and barren environments add to the risk of broilers developing lameness (Liu et al., 2023).

Integument damage affects broiler chickens in current production systems mainly in the form of footpad dermatitis and hock burn (Hartcher and Lum, 2020). These types of contact dermatitis can be described as painful, inflammatory and necrotic lesions. Fast-growing hybrids have a predisposition to develop footpad dermatitis due to lower levels of activity and more time spent sitting on litter (Bessei, 2006; Dixon, 2020; Hartcher and Lum, 2020). However, contact dermatitis is caused by an interplay of multiple factors, which include type and quality of litter material, stocking density, climate conditions, barn design, nutrition and animal health factors (Shepherd and Fairchild, 2010). Most of the factors can be summarised into the 'wet litter' complex (Dunlop et al., 2016). It is important to note that the focus needs to be on preventive measures as once footpad dermatitis and hock burn are present in a flock, it is difficult to apply any correcting or mitigating measures during the lifespan of the animals (EFSA AHAW Panel et al., 2023).

With regard to the frequently observed behavioural restrictions in commercial broiler production, again both internal animal-related and external environment-related factors can play a causal role. Reduced overall activity can, for instance, be attributed to an inability to walk comfortably due to lameness or to an inability to move freely due to a high stocking density (EFSA AHAW Panel et al., 2023). The same applies to restrictions of foraging and comfort behaviour, such as dustbathing. In addition, the possibility to forage and dustbathe depends on the availability of adequate substrate (Baxter et al., 2018; Riber et al., 2018). As mentioned above, foraging is an adaptation in broiler chickens that does not serve an important function under commercial production conditions. However, as broiler chickens are still motivated to forage, a lack of the ability or opportunity to do so can lead to negative affective states, such as frustration. Similarly, broiler chickens are still motivated to rest on elevated structures (Güz et al., 2021; Riber et al., 2018), which are often not provided under intensive production conditions. When offered, both fast- and slow-growing hybrids use elevated structures for resting during the day and during night-time (Güz et al., 2021; Malchow et al., 2019). In addition, high stocking densities and the absence of separate functional zones in commercial barns restrict resting behaviour as resting broiler chickens are disturbed by active conspecifics that walk over or close to them (Forslind et al., 2021). Disrupted resting patterns are likely to cause negative affective states. However, other potential effects of sleep deprivation on biological functioning have not been investigated yet in broiler chickens (Forslind et al., 2021).

The examples above show that both the welfare consequences and their causations cannot be regarded in isolation. Often multiple genetic and environmental factors (e.g. high growth rate, inadequate litter type, high stocking density) cause a welfare consequence (e.g. footpad dermatitis), which in turn can lead to other welfare consequences (e.g. lameness and reduced activity). However, broiler chickens that are lame due to other reasons (e.g. bone deformations) also move less, have prolonged contact with the litter, and thus develop footpad dermatitis. These relationships illustrate that it is presently not possible to capture all influencing factors, their interactions and their relative effects on welfare with practical animal welfare assessment tools. These tools remain – to a varying degree – a simplification of a complex reality.

## 4 The link between welfare and ethics

Animal welfare science and ethics cannot be separated. There are two main reasons for this. First, as we have seen above, animal welfare science developed out of ethical concerns about the quality of life of animals (Fraser et al., 1997). Concerns about the quality of life of certain animals originate in the notion that these animals are morally significant (McBride and Baugh, 2022; Tannenbaum, 1991). If they were not, why would we study, assess, maintain or improve their welfare? Second, any claims about animal welfare, whether they refer to natural living, subjective experiences, biological functioning or other concerns, are value-laden (Fraser et al., 1997). Approaches to animal welfare imply assumptions about which condition is better or worse for an animal. The fact that we can substantiate these assumptions with empirical findings for many conditions is irrelevant regarding the value-load.

In the case of footpad dermatitis in broiler chickens, we assume that 'large ulcers or scrabs, signs of haemorrhages or deep dermatitis' are worse for the animal's welfare compared to 'small superficial lesions, slight discolouration on a limited area, mild hyperkeratosis' (Welfare Quality®, 2009). This assumption can be substantiated by empirical findings: visual scoring of footpad dermatitis reflects histological observations of inflammatory and necrotic lesions of the soft tissue (Piller et al., 2020). It is commonly agreed that such lesions are painful and become more painful with increasing size (EFSA AHAW Panel et al., 2023). However, there is more controversy in other cases, such as whether access to an outdoor area contributes positively or negatively to the overall welfare status of a broiler chicken (Riber et al., 2018). Is a broiler chicken experiencing predator stress in an outdoor system worse off than a chicken in an indoor system that lacks the opportunity to forage in soil?

In such cases, animal scientists tend to reduce complex, value-laden concepts of animal welfare to parameters which are empirically accessible (Stafleu et al., 1996). In the example of the outdoor area, they may compare fear-related behaviours, corticosterone levels or other parameters of broilers kept in an outdoor system with those of broilers kept indoors. By this, animal welfare seems to become fully empirically accessible, when in fact only the concept is narrowed down to a version that omits the relevant ethical aspects. Stafleu et al. (1996) call this the 'erosion of a moral concept'.

A systematic erosion of the concept of animal welfare in practice has led to the commonly held positivist view in animal sciences that even subjective animal experiences can be measured through behavioural (e.g. preferences, vocalisations) or physiological (e.g. hormone levels) indicators. At the same time, animal ethicists have approached the aim of tackling concerns about the quality of life of animals by using such different concepts, assumptions and language that we may speak of 'two cultures' (Fraser, 1999). On the one side, there are animal scientists who, following a positivist tradition, seek to measure all aspects of animal welfare, while on the other side, there are animal ethicists who often engage in philosophical discourses with little practical relevance or ignore empirical knowledge about animal biology. More than two decades ago, Fraser (1999) noted an increasing convergence of scientific and philosophical approaches to animal welfare. However, the separation mentioned above is still very much present in many academic institutions, associations or joint research projects.

Even if we acknowledge that neither empirical knowledge nor ethical considerations alone can tackle quality-of-life concerns, a perspective purely centred on animal welfare has several limitations. Current large-scale broiler production is not compatible with the consequent application of contemporary animal welfare concepts, such as those proposed by Mellor and Reid (1994), Fraser et al. (1997) or Arndt et al. (2022). These concepts are centred on the individual animal, whereas in commercial broiler production, thousands of animals are kept together. One may argue that for instance for the domain nutrition, all chickens of a group have the same needs, as they have been selected for uniformity and are of the same age. In an ideal situation, it may even be possible to allow each individual adequate access to the feed.

However, the situation is more complex regarding other aspects of animal welfare. When we assess footpad dermatitis in a subset of individuals from the same flock, we often find a variation in severity, ranging from no lesions in some chickens to severe lesions in others (e.g. Giersberg et al., 2021). As outlined above, in this case it is quite straightforward to weigh which condition is 'good' (no lesion) and which is 'bad' (severe lesion) for the welfare of an individual broiler chicken. For groups of chickens, however, this weighing is impossible: is it better if all broiler chickens of a flock have mild footpad dermatitis and mild pain or if the majority of a flock has no lesions and no pain, but there are several individuals with severe lesions and severe pain? An additional problem arises from the low economic value and the short lifespan of an individual broiler chicken. Even if we were able to assess and weigh footpad dermatitis for all individuals in a flock, there would be no realistic option to treat injured individual animals in a way that they return to a state that they experience as positive. All adjustments of housing and management conditions that may reduce the risk of footpad dermatitis would only benefit future flocks of broiler chickens.

This does not mean that there is no possibility to assess and improve at least some aspects of animal welfare for at least some of the individuals in current large-scale broiler production systems. Practical welfare assessment tools have been developed (e.g. BenSassi et al., 2019; De Jong et al., 2016) that cover indicators for important welfare challenges in broiler chickens, such as locomotory disorders. The application of such tools and the implementation of recommendations to improve animal welfare, such as those proposed by the EFSA (EFSA AHAW Panel et al., 2023), can benefit the quality of life of broiler chickens. However, it is important to be aware that these represent a reduction to empirically accessible parameters that do not reflect the full spectrum of aspects outlined in contemporary animal welfare concepts.

An additional limitation of an animal welfare-centred perspective is that it only refers to the experiences (e.g. affective states, adaptations to overcome certain states) of the animal. However, there are also challenges in our relationship with animals that are more related to how we as humans experience certain practices. These challenges beyond welfare issues are explored in the following section.

## 5 Ethical perspectives beyond welfare-centred approaches

To illustrate a situation that raises questions beyond welfare issues, we revisit the example of footpad dermatitis. Keeping broiler chickens with a genetic predisposition in an environment that makes them prone to develop footpad dermatitis may be no issue if the chickens are preventively and curatively provided with an analgetic. The analgetic is provided via drinkers in different locations of the barn to ensure that all animals can access it. The chickens can also choose between drinkers offering only water and drinkers offering water plus analgetic. As a result, the chickens do not experience any pain due to footpad dermatitis and can engage in highly motivated behaviours, such as locomotion and foraging. The analgetic represents an adequate adaptation to overcome a negative affective state. As a consequence, the chickens reach a state that they experience as positive.

The fact that we do not consider the routine administration of an analgetic to be appropriate has other reasons than purely economic and practical ones. Our intuitive rejection of the fictional practice described in the previous paragraph is more related to concerns regarding the treatment of animals that go beyond quality-of-life concerns. In the previous section, we explored the link between and the inseparability of animal welfare concepts and ethics. This section goes a step further by introducing ethical perspectives regarding our treatment of animals that are not covered by welfare-centred approaches. It is important to emphasise that this overview serves as an introduction which includes the necessary simplifications of ethical perspectives. These are much more complex and debated in philosophical discourses.

We may object to systematically keeping broiler chickens with footpad dermatitis, although their quality of life is not impaired as a result of the administration of an analgetic, because their integrity is violated. The concept of animal integrity has been applied as an argument against genetic modifications of animals (Bovenkerk, 2020), but also in assessments of animal ethics committees in animal experimentation (Röcklinsberg et al., 2014), or more broadly to evaluate current animal farming practices (Millar and Morton, 2017). Similar to welfare, integrity can be regarded as a hybrid concept, including both empirically accessible and morally relevant components. In our example of footpad dermatitis, the chickens are not biologically intact, their skin is damaged, which we can measure empirically. As the same kind of damage may occur due to spontaneous injuries, it is not the damage per se that violates the integrity of the chickens. It is rather the practice of deliberately and systematically keeping damaged animals that forms the integrity issue in the example. In this sense, integrity does not simply refer to biological intactness but also to a perceived intactness of an animal, i.e. how we as humans experience an animal should be (Bovenkerk et al., 2002).

If we take it a step further, we can use the concept of animal integrity not only to reject the practice in the fictional example but also to question current commercial broiler production in general. A fast-growing broiler hybrid that is not able to reach maturity due to its high growth rate and can no longer reproduce and sustain its species may be viewed as inherently compromised in its integrity. This perspective links to Fraser's notion of natural living which entails that animals should be allowed to have a 'normal' ontogenic development (Fraser, 2008; Fraser et al., 1997). In a similar way to natural living, the concept of integrity remains vague (Bovenkerk et al., 2002; Röcklinsberg et al., 2014). There are many conflicting ways when interpreting the concept, for instance, whether a violation of integrity is always absolute, and if not, which criteria should be used to evaluate and compare different kinds of violations (Bovenkerk et al., 2002; Millar and Morton, 2017). Nevertheless, the concept of animal integrity has been found useful in practice (Bovenkerk, 2020; Röcklinsberg et al., 2014). That is because interpreting integrity as a 'basic zone of untouchableness' (Röcklinsberg et al., 2014) adds a deontological dimension to human perceptions of relationships with animals, which is not captured by animal welfare concepts.

As outlined above, the concept of animal integrity is related to how humans experience how an animal should be. A set of other ethical perspectives is more related to how humans experience how an animal should be viewed or treated. These perspectives are termed instrumentalisation, objectification and de-animalisation, which are closely related, but originate from different theoretical backgrounds. The notion of instrumentalisation has been applied to animals in reference to Kant's ethics (Bos et al., 2018). It has been defined as treating someone (including animals) as a means only instead of an end in itself (Bos et al., 2018). Objectification derives from feminist theories, for instance from Nussbaum's work, and means – when applied to an animal – treating it as an object despite not being an object (Bos et al., 2018). Treating animals as objects includes, for instance, denying their autonomy and subjectivity.

In the fictional scenario at the beginning of this section, broiler chickens with footpad dermatitis are – due to the routinely administered analgetic – not able to feel pain. This practice can be viewed as treating them as inert objects, as mere means to produce meat for human consumption. If we would see them as ends in themselves, we would provide them with an environment that would not systematically lead to skin damage in the first place. Similar to integrity, we do not need to stay with this fictional example. We can also use the concepts of instrumentalisation or objectification to voice concerns about large-scale commercial broiler production as a whole. This is possible even if we do not go as far as to regard any form of using animals for food production as objectification. The current system of broiler production does, for instance, not allow for individual veterinary care, due to the large groups of chickens kept together and the low economic value of the individual. This practice and these considerations deny the subjectivity of the sick or injured chicken and can be described as treating it as a mere means for meat production. In contrast, in dairy production, a veterinarian would be called if a cow was sick or injured. We would do the same for a dog we keep as a companion.

Harfeld et al. (2016) elaborated on the concept of de-animalisation, which expresses concerns about turning animals into objects in contrast to treating them as objects. The authors argue that animals in current food production systems are placed out of their (semi-) natural context, which would allow humans to see them as individuals with subjective experiences and desires (Harfeld et al., 2016). Instead they become parts, i.e. objects, of a system that reduces them to their functions, particularly to functions in which humans are interested (Harfeld et al., 2016). Large-scale broiler production can be viewed as a typical example of this process. The chickens are part of vertically integrated production chains that operate on strict time schedules to allow for optimum use of hatcheries, farms, transport capacities and slaughterhouses (EFSA AHAW Panel et al., 2023). In this highly efficient system, there is no room for veterinary or other additional care for individuals that cannot adapt to the system. Harfeld et al. (2016) considered this reduction of animals to production units problematic, because it prevents us from understanding the 'animality and singularity' of an animal and thus to give it proper moral consideration.

The perspectives introduced in this section may seem – similar to more holistic concepts of animal welfare – vague to people who are more familiar with concepts that are empirically accessible. However, although arguments such as integrity, instrumentalisation, objectification or de-animalisation are open to debate and disagreement, they cover important values and meanings people attach to our relationship with animals (Bovenkerk, 2020). These values and meanings are not covered by many animal welfare concepts, or at least not by their eroded versions that are often used in practice.

## 6 Case study: the broiler breeder paradox

This section addresses an inherent, often hidden challenge in the production of fast-growing broiler chickens which is known as the broiler breeder paradox (Decuypere et al., 2006, 2010). Broiler breeders are the parent generation of broiler chickens reared for meat. Although this book focusses on the latter, the broiler breeder paradox provides an excellent case to reflect on the animal welfare concepts and the broader ethical perspectives that have been introduced in this chapter.

In poultry, there is a genetically determined antagonism between bodyweight gain and reproductive performance. Conventional broiler breeders have been selected to produce offspring with high growth rates. As a consequence, their own reproductive performance has been reduced (Decuypere et al., 2010). The term 'broiler breeder paradox' describes the impossibility of achieving an acceptable reproductive performance and health status in broiler breeders without feeding them restrictively, a practice which in turn causes welfare issues (Decuypere et al., 2006, 2010). This situation is a direct consequence of the genetic selection for producing offspring with fast growth rates and high daily weight gain. Feed restriction in breeder hens, particularly during rearing, can lead to chronic hunger, frustration and stereotypic behaviours (De Jong and Guémené, 2011). However, if broiler breeders were fed ad libitum, reproductive performance would decrease, while body weight, locomotory disorders and mortality would increase.

If we apply the updated Five Domains Model (Mellor and Beausoleil, 2015) to this context, welfare compromise can mainly be identified in domain 1 (nutrition) and domain 5 (mental state). Due to a restriction in feed intake (1), broiler breeders may feel chronic hunger (5) and frustration (5). The possible occurrence of stereotypic behaviours can in this case be interpreted as a consequence of frustration rather than a lack of behavioural opportunities (domain 4). It is likely that the negative states induced by hunger and frustration contribute to an overall compromised welfare status. However, the extent of the compromise is difficult to assess and depends on restrictions or opportunities in the other domains (e.g. are the breeders in good health? Can they use nest boxes or resting areas?). Paradoxically, the restriction in the nutrition domain can have positive effects in the health domain, such as 'appropriate body condition' and 'good fitness level' (Mellor and Beausoleil, 2015). On the other hand, chronic hunger may lead to over-drinking (De Jong and Guémené, 2011), which may lead to more liquid faeces and wet litter, which in turn may cause footpad dermatitis and a compromise in domain 3 (health).

In terms of the dynamic animal welfare concept (DAWCon, Arndt et al., 2022), broiler breeders face an adverse internal stimulus (hunger) to which they cannot adapt adequately through their behaviour (eating) in their present environment (restrictive feed supply). It is therefore likely that their welfare is compromised. In practice, different strategies are applied to alleviate the negative consequences of feed restriction for the breeders. These include the provision of diluted diets which contain components with lower nutritional value, such as fibres (Riber et al., 2017). If such diets are offered ad libitum or less restrictively, this may be an opportunity for the broiler breeders to overcome the negative internal stimulus (hunger) through adaptive behaviour (eating) without suffering the long-term consequences of excessive body weight (De Jong et al., 2005).

The measure of feeding diluted diets seems also adequate to tackle the concerns regarding subjective experiences/affective states and basic health and functioning outlined in Fraser's concept of animal welfare (Fraser, 2008; Fraser et al., 1997). However, this does not hold for the third concern, i.e. natural living. First, this concern entails that animals should be allowed to live according to their genetically encoded 'natures' (Fraser et al., 1997). If broiler breeders were granted this, they would eat detrimental amounts of feed to adapt to their feeling of hunger (Decuypere et al., 2006). This feed would in contrast to for instance ruminants not 'naturally' contain high levels of fibre. Second, it can be questioned whether fast-growing broiler breeder hybrids can have an ontogenic development that is considered 'normal' for the species at all, or whether they are still 'normal' representatives of their species.

These questions lead on to the ethical perspectives discussed above, which focus more on how humans experience how an animal should be or should be treated. Fast-growing broiler breeder hybrids are a typical example of an animal integrity issue. As these hybrids cannot sustain themselves in an environment suitable to their species (which would certainly not include high-fibre diets or other measures to alleviate hunger and over-eating), their perceived intactness can be considered to be violated (Bovenkerk, 2020). A question that follows is why these hybrids then have been so intensively selected that they can no longer live under conditions considered as 'natural' or 'normal'. In response to this question it may be argued that this selection was necessary to provide a growing human population with affordable, high-quality animal protein or that the offspring of those breeders is characterised by a high feed efficiency, which contributes to environmental sustainability. It is beyond the scope of this chapter to judge the validity of those arguments. However, it is clear that they do not focus on the animal as a subject or as an end in itself. Broiler breeders are means or instruments to achieve goals such as affordable human nutrition or efficient use of natural resources. One may also argue that they have become objects in a system in which their 'animalness' has become more of a hurdle than something that has value in itself.

There are many recommendations to overcome the problem of the broiler breeder paradox (EFSA AHAW Panel et al., 2023). As shown above, some of them, such as feeding diluted diets, aim to alleviate welfare issues in terms of negative experiences of the animals. Others, such as using slower-growing broiler breeders that require feed restriction to a lesser extend or not at all, may also cover broader concerns about how animals should be, how we should view them, and how we should treat them to a certain extent.

## 7 Conclusion and future trends in research

This chapter places practices in commercial, large-scale broiler production into a broader academic and societal perspective. The current understanding of animal welfare goes beyond identifying harmful conditions in animals, such as footpad dermatitis in broiler chickens. As animal welfare science builds on societal concerns about an animal's quality of life, it cannot be separated from ethics. This means that holistic accounts of animal welfare are hybrid in nature: they integrate scientific and moral considerations. Therefore, certain aspects of welfare are empirically accessible, such as the extent of skin damage, while others, such as the question of what constitutes a good life for a broiler chicken, are not. It is important to realise that attempts to make animal welfare entirely measurable are rather an erosion or a simplification of a complex concept. That does not mean that we should not assess certain indicators of animal welfare in practice and take preventive or corrective measures to improve the lives of broiler chickens in the current system. As we have seen, there are still many basic welfare challenges in large-scale broiler production that need to be addressed, such as thermoregulatory stress, gastrointestinal disorders, lameness or skin damage.

In this chapter, several ethical perspectives are outlined that are not about how a broiler chicken experiences its life, but about how humans experience a broiler chicken's life. This chapter is not a manual that defines how a broiler chicken should be, how we should view it in the right way and how we should treat it accordingly. However, knowing these different perspectives enables us to understand people's views and to take a stance in the public debate about using animals in food production systems. The main focus here is on the value we attach to an animal and its life, not on other, potentially conflicting values regarding, for instance, food safety and security, economic viability or environmental sustainability.

A current trend in animal welfare science is the study of so-called 'positive animal welfare', which focusses on enabling and assessing positive states and experiences in animals instead of focussing on the prevention of harm (e.g.

Laurijs et al., 2021; Lawrence et al., 2024; Rault et al., 2025). However, as detailed above, this idea is not new and dates back at least to the early 1990s (e.g. Sandøe and Simonsen, 1992). This trend is associated with the increasing use of modern AI- and sensor-based technologies to assess animal behaviour more accurately and more continuously (e.g. Van der Eijk et al., 2022, 2023; Wurtz et al., 2019). It is interesting that most of this research follows a positivist tradition with the ultimate aim of making animal welfare entirely measurable.

At the same time, animal ethicists integrate, for instance, theories of care ethics with strategies from the field of arts to foster decision-making processes regarding the use of animals in food production systems (Rooijakkers et al., 2024). Other researchers draw on the above-mentioned feminist approaches (Thompson, 2023) or develop Kantian ethics further for animal contexts (Müller, 2022). In the light of these research trends, Fraser's (1999) observation of an increasing conversion of scientific and philosophical approaches to animal welfare seems to become a distant prospect. However, to tackle the societal challenges regarding animal welfare or our relationship with animals in a broader sense in broiler production and in other contexts we keep animals, we need to join forces and move beyond discipline-specific assumptions and traditions. Together with previous work (Giersberg et al., 2022; Giersberg, 2024), this chapter forms a step in this direction: to reconcile seemingly disparate perspectives to enable researchers and professionals to genuinely collaborate for the benefit of broiler chickens and other animals.

## 8 Where to look for further information

The following resources provide a good overview of the subject:

A comprehensive overview of welfare issues in commercial broiler production and recommendations to tackle them can be found in the Scientific Opinion of the European Food Safety Authority (EFSA): EFSA AHAW Panel, 2023. Welfare of broilers on farm. EFS2 21. https://doi.org/10.2903/j.efsa.2023.7788.

A contemporary outline of human-animal relationships, their transformation and implications for ideas and practices of animal welfare is the book *Human/Animal Relationships in Transformation: Scientific, Moral and Legal Perspectives*. A. Vitale, & S. Pollo (Eds.). Palgrave Macmillan, 2022. https://doi.org/10.1007/978-3-030-85277-1.

Key research in this area can be found at the following organisations:

The Centre for Sustainable Animal Stewardship (CenSAS) aims to bring together science and society in a constructive dialogue on animal welfare-related topics (www.censas.org).

The European Society for Agricultural and Food Ethics (EurSafe) aims to encourage academic education and research, and international debate on the ethical issues involved in agriculture and food supply. EurSafe organises an international conference with several sessions on animal ethics every two years (www.eursafe.org).

The Universities Federation for Animal Welfare (UFAW) is an independent charity that works with the animal welfare science community worldwide to develop and promote improvements in the welfare of animals through scientific and educational activity (www.ufaw.org.uk).

## 9 References

Arndt, S.S., et al., 2022. A dynamic concept of animal welfare: the role of appetitive and adverse internal and external factors and the animal's ability to adapt to them. *Frontiers in Animal Science* 3, 908513. https://doi.org/10.3389/fanim.2022.908513.

Aviagen, 2022. Ross 308 Performance Objectives. Accessed 25 March 2024. https://aviagen.com/assets/Tech_Center/Ross_Broiler/RossxRoss308-BroilerPerformanceObjectives2022-EN.pdf.

Baxter, M., et al., 2018. An evaluation of potential dustbathing substrates for commercial broiler chickens. *Animal* 12, 1933–1941. https://doi.org/10.1017/S1751731117003408.

BenSassi, N., et al., 2019. On-farm broiler chicken welfare assessment using transect sampling reflects environmental inputs and production outcomes. *PLoS One* 14, e0214070. https://doi.org/10.1371/journal.pone.0214070.

Bessei, W., 2006. Welfare of broilers: a review. *World's Poultry Science Journal* 62, 455–466. https://doi.org/10.1079/WPS2005108.

Blake, D.P. and Tomley, F.M., 2014. Securing poultry production from the ever-present Eimeria challenge. *Trends in Parasitology* 30, 12–19. https://doi.org/10.1016/j.pt.2013.10.003.

Bos, J.M., et al., 2018. The quantified animal: precision livestock farming and the ethical implications of objectification. *Food Ethics* 2, 77–92. https://doi.org/10.1007/s41055-018-00029-x.

Bovenkerk, B., 2020. Ethical perspectives on modifying animals: beyond welfare arguments. *Animal Frontiers* 10, 45–50. https://doi.org/10.1093/af/vfz055.

Bovenkerk, B., et al., 2002. Brave new birds: the use of "animal integrity" in animal ethics. *The Hastings Center Report* 32, 16. https://doi.org/10.2307/3528292.

Carnovale, F., et al., 2021. Chinese public attitudes towards, and knowledge of, animal welfare. *Animals* 11, 855. https://doi.org/10.3390/ani11030855.

Clark, B., et al., 2016. A systematic review of public attitudes, perceptions and behaviours towards production diseases associated with farm animal welfare. *Journal of Agricultural and Environmental Ethics* 29, 455–478. https://doi.org/10.1007/s10806-016-9615-x.

De Boer, I.J.M. and Van Ittersum, M.K., 2018. *Circularity in agricultural production.* Wageningen University & Research.

De Jong, I.C., et al., 2005. Do low-density diets improve broiler breeder welfare during rearing and laying? *Poultry Science* 84, 194–203. https://doi.org/10.1093/ps/84.2.194.

De Jong, I.C. and Guémené, D., 2011. Major welfare issues in broiler breeders. *World's Poultry Science Journal* 67, 73–82. https://doi.org/10.1017/S0043933911000067.

De Jong, I.C., et al., 2016. Simplifying the Welfare Quality® assessment protocol for broiler chicken welfare. *Animal* 10, 117–127. https://doi.org/10.1017/S1751731115001706.

Decuypere, E., et al., 2006. Broiler breeder paradox: a project report. *World's Poultry Science Journal* 62, 443–453. https://doi.org/10.1079/WPS2005107.

Decuypere, E., et al., 2010. The broiler breeder paradox: ethical, genetic and physiological perspectives, and suggestions for solutions. *British Poultry Science* 51, 569–579. https://doi.org/10.1080/00071668.2010.519121.

Deeb, N. and Cahaner, A., 2002. Genotype-by-environment interaction with broiler genotypes differing in growth rate. 3. Growth rate and water consumption of broiler progeny from weight-selected versus nonselected parents under normal and high ambient temperatures. *Poultry Science* 81, 293–301. https://doi.org/10.1093/ps/81.3.293.

Dixon, L.M., 2020. Slow and steady wins the race: the behaviour and welfare of commercial faster growing broiler breeds compared to a commercial slower growing breed. *PLoS One* 15, e0231006. https://doi.org/10.1371/journal.pone.0231006.

Dunlop, M.W., et al., 2016. The multidimensional causal factors of 'wet litter' in chicken-meat production. *Science of The Total Environment* 562, 766–776. https://doi.org/10.1016/j.scitotenv.2016.03.147.

Edgar, J., et al., 2013. Towards a 'good life' for farm animals: development of a resource tier framework to achieve positive welfare for laying hens. *Animals* 3, 584–605. https://doi.org/10.3390/ani3030584.

EFSA AHAW Panel (EFSA Panel on Animal Health and Welfare), et al., 2023. Welfare of broilers on farm. *EFSA Journal* 21, e07788. https://doi.org/10.2903/j.efsa.2023.7788.

Estévez-Moreno, L.X., et al., 2022. Consumer attitudes towards farm animal welfare in Argentina, Chile, Colombia, Ecuador, Peru and Bolivia: a segmentation-based study. *Meat Science* 187, 108747. https://doi.org/10.1016/j.meatsci.2022.108747.

Fancher, C.A., et al., 2021. Prevalence and molecular characteristics of avian pathogenic escherichia coli in "no antibiotics ever" broiler farms. *Microbiology Spectrum* 9, e00834-21. https://doi.org/10.1128/Spectrum.00834-21.

FAO, 2017. *The future of food and agriculture: trends and challenges*. Food and Agriculture Organization of the United Nations.

FAO (Food and Agriculture Organization of the United Nations), 2023. Accessed 25 March 2025. http://www.fao.org.

Forslind, S., et al., 2021. Disturbance of resting behaviour of broilers under different environmental conditions. *Applied Animal Behaviour Science* 242, 105425. https://doi.org/10.1016/j.applanim.2021.105425.

Fraser, D., 1999. Animal ethics and animal welfare science: bridging the two cultures. *Applied Animal Behaviour Science* 65, 171–189. https://doi.org/10.1016/S0168-1591(99)00090-8.

Fraser, D., 2008. Understanding animal welfare. *Acta Veterinaria Scandinavica* 50, S1. https://doi.org/10.1186/1751-0147-50-S1-S1.

Fraser, D., et al., 1997. A scientific conception of animal welfare that reflects ethical concerns. *Animal Welfare* 6, 187–205. https://doi.org/10.1017/S0962728600019795.

Giersberg, M.F., 2024. Animal-centred AI: on the need of value-sensitive innovation in the context of animal production, in: Giersberg, M., et al. (Eds.), *EurSafe2024*

*Proceedings*. Brill | Wageningen Academic, pp. 254–260. https://doi.org/10.1163/9789004715509_042.

Giersberg, M.F., et al., 2020. Comparative assessment of general behaviour and fear-related responses in hatchery-hatched and on-farm hatched broiler chickens. *Applied Animal Behaviour Science* 232, 105100. https://doi.org/10.1016/j.applanim.2020.105100.

Giersberg, M.F., et al., 2021. Effects of hatching system on the welfare of broiler chickens in early and later life. *Poultry Science* 100, 100946. https://doi.org/10.1016/j.psj.2020.12.043.

Giersberg, M.F., et al., 2022. 80. How smart should resilience be? On the need of a transdisciplinary approach to transform pig production systems, in: D.Bruce, and Bruce, A. (Eds.), *Transforming food systems: ethics, innovation and responsibility*. Wageningen Academic Publishers, pp. 513–518.

Güz, B.C., et al., 2021. Effects of pen enrichment on leg health of fast and slower-growing broiler chickens. *PLoS One* 16, e0254462. https://doi.org/10.1371/journal.pone.0254462.

Harfeld, J.L., 2024. The history of animal welfare and its connection to ethics, in: *EurSafe2024 proceedings: back to the future - sustainable innovations for ethical food production and consumption*. Brill | Wageningen Academic. https://doi.org/10.1163/9789004715509.

Harfeld, J.L., et al., 2016. Seeing the animal: on the ethical implications of de-animalization in intensive animal production systems. *Journal of Agricultural and Environmental Ethics* 29, 407–423. https://doi.org/10.1007/s10806-016-9611-1.

Hartcher, K.M. and Lum, H.K., 2020. Genetic selection of broilers and welfare consequences: a review. *World's Poultry Science Journal* 76, 154–167. https://doi.org/10.1080/00439339.2019.1680025.

Heier, B.T., et al., 2002. Factors associated with mortality in Norwegian broiler flocks. *Preventive Veterinary Medicine*, 53(1–2), 147–158.

Kanis, E., et al., 2003. Societal concerns about pork and pork production and their relationships to the production system. *Journal of Agricultural and Environmental Ethics* 16, 137–162. https://doi.org/10.1023/A:1022985913847.

Laurijs, K.A., et al., 2021. Vocalisations in farm animals: a step towards positive welfare assessment. *Applied Animal Behaviour Science* 236, 105264. https://doi.org/10.1016/j.applanim.2021.105264.

Lawrence, A.B., et al., 2024. Positive welfare: what does it add to the debate over pig welfare?, in: Lawrence, A.B., et al. (Eds.), *Advances in pig welfare*. Elsevier, pp. 83–112. https://doi.org/10.1016/B978-0-323-85676-8.00009-2.

Liu, K., et al., 2023. Leg disorders in broiler chickens: a review of current knowledge. *Animal Biotechnology* 34, 5124–5138. https://doi.org/10.1080/10495398.2023.2270000.

Malchow, J., et al., 2019. Perches or grids? What do rearing chickens differing in growth performance prefer for roosting? *Poultry Science* 98, 29–38. https://doi.org/10.3382/ps/pey320.

McBride, E.A. and Baugh, S., 2022. Animal welfare in context: historical, scientific, ethical, moral and one welfare perspectives, in: Vitale, A. and Pollo, S. (Eds.), *Human/Animal relationships in transformation*, The Palgrave Macmillan Animal Ethics Series. Springer International Publishing, pp. 119–147. https://doi.org/10.1007/978-3-030-85277-1_7.

Mellor, D., 2015. Positive animal welfare states and reference standards for welfare assessment. *New Zealand Veterinary Journal* 63, 17–23. https://doi.org/10.1080/00480169.2014.926802.

Mellor, D., 2017. Operational details of the five domains model and its key applications to the assessment and management of animal welfare. *Animals* 7, 60. https://doi.org/10.3390/ani7080060.

Mellor, D. and Beausoleil, N., 2015. Extending the 'five domains' model for animal welfare assessment to incorporate positive welfare states. *Animal Welfare* 24, 241–253. https://doi.org/10.7120/09627286.24.3.241.

Mellor, D.J. and Reid, C.S.W., 1994. *Concepts of animal well-being and predicting the impact of procedures on experimental animals*. Department of Physiology and Anatomy, Massey University Palmerston North.

Mellor, D.J., et al., 2020. The 2020 five domains model: including human–animal interactions in assessments of animal welfare. *Animals* 10, 1870. https://doi.org/10.3390/ani10101870.

Millar, K. and Morton, D., 2017. Animal integrity in modern farming, in: Gunning, J., et al. (Eds.), *Ethics, law and society*. Routledge, pp. 19–30. https://doi.org/10.4324/9781315094328.

Moore, R.J., 2016. Necrotic enteritis predisposing factors in broiler chickens. *Avian Pathology* 45, 275–281. https://doi.org/10.1080/03079457.2016.1150587.

Mottet, A. and Tempio, G., 2017. Global poultry production: current state and future outlook and challenges. *World's Poultry Science Journal* 73, 245–256. https://doi.org/10.1017/S0043933917000071.

Müller, N.D., 2022. *Kantianism for animals: a radical kantian animal ethic*, The Palgrave Macmillan Animal Ethics Series. Springer International Publishing. https://doi.org/10.1007/978-3-031-01930-2.

Nichelmann, M. and Tzschentke, B., 2002. Ontogeny of thermoregulation in precocial birds. *Comparative Biochemistry and Physiology Part A: Molecular & Integrative Physiology* 131, 751–763. https://doi.org/10.1016/S1095-6433(02)00013-2.

Piller, A., et al., 2020. Validation of histological and visual scoring systems for foot-pad dermatitis in broiler chickens. *Animal Welfare* 29, 185–196. https://doi.org/10.7120/09627286.29.2.185.

Rault, J.-L., et al., 2025. A consensus on the definition of positive animal welfare. *Biology Letters* 21, 20240382. https://doi.org/10.1098/rsbl.2024.0382.

Riber, A.B., et al., 2017. Environmental enrichment for broiler breeders: an undeveloped field. *Frontiers in Veterinary Science* 4, 86. https://doi.org/10.3389/fvets.2017.00086.

Riber, A.B., et al., 2018. Review of environmental enrichment for broiler chickens. *Poultry Science* 97, 378–396. https://doi.org/10.3382/ps/pex344.

Röcklinsberg, H., et al., 2014. A case for integrity: gains from including more than animal welfare in animal ethics committee deliberations. *Laboratory Animals* 48, 61–71. https://doi.org/10.1177/0023677213514220.

Rooijakkers, M., et al., 2024. Implementing animal care ethics through the arts of visual thinking strategies, in: Giersberg, M., et al. (Eds.), *EurSafe2024 proceedings: back to the future - sustainable innovations for ethical food production and consumption*. Brill | Wageningen Academic. https://doi.org/10.1163/9789004715509.

Sandøe, P. and Simonsen, H.B., 1992. Assessing animal welfare: where does science end and philosophy begin?. Animal welfare, 1(4), 257–267.

Shepherd, E.M. and Fairchild, B.D., 2010. Footpad dermatitis in poultry. *Poultry Science* 89, 2043–2051. https://doi.org/10.3382/ps.2010-00770.

Stafleu, F.R., et al., 1996. Animal welfare: evolution and erosion of a moral concept. *Animal Welfare* 5, 225–234. https://doi.org/10.1017/S0962728600018819.

Suarez, S.D. and Gallup, G.G., 1983. Social reinstatement and open-field testing in chickens. *Animal Learning & Behavior* 11, 119–126. https://doi.org/10.3758/BF03212318.

Tannenbaum, J., 1991. Ethics and animal welfare: the inextricable connection. *Journal of the American Veterinary Medical Association* 198, 1360–1376.

Thompson, M.S., 2023. Alternative visions of "ethical" dairying: changing entanglements with calves, cows and care. *Agriculture and Human Values* 40, 693–707. https://doi.org/10.1007/s10460-022-10384-5.

Van der Eijk, J.A.J., et al., 2022. Seeing is caring – automated assessment of resource use of broilers with computer vision techniques. *Frontiers in Animal Science* 3, 945534. https://doi.org/10.3389/fanim.2022.945534.

Van der Eijk, J.A.J., et al., 2023. Individuality of a group: detailed walking ability analysis of broiler flocks using optical flow approach. *Smart Agricultural Technology* 5, 100298. https://doi.org/10.1016/j.atech.2023.100298.

Welfare Quality®, 2009. *Welfare Quality® assessment protocol for poultry (broilers, laying hens).* Welfare Quality® Consortium.

Wickramasuriya, S.S., et al., 2022. Role of physiology, immunity, microbiota, and infectious diseases in the gut health of poultry. Vaccines 10, 172. https://doi.org/10.3390/vaccines10020172.

Wurtz, K., et al., 2019. Recording behaviour of indoor-housed farm animals automatically using machine vision technology: a systematic review. *PLoS One* 14, e0226669. https://doi.org/10.1371/journal.pone.0226669.

# Chapter 2

# Managing co-products in broiler production: repurposing poultry manure and litter

*C. W. Ritz, University of Georgia, USA*


1 Introduction

2 Nutrient management planning: decades of best management practices

3 Nutrient management plan challenges

4 Air quality challenges

5 Sustainability/circularity

6 Improvement opportunities

7 Conclusion

8 References


## 1 Introduction

Broiler production, the process of raising chickens for meat, is a significant component of the global agricultural meat industry. With the increasing demand for poultry products, broiler production has expanded rapidly, but with this growth has come various environmental challenges associated with that sustained growth. Addressing these issues is crucial for ensuring the sustainability of the industry and minimizing its impact on the environment. In the United States, poultry is a major source of agricultural income. In 2023, according to the National Agricultural Statistics Service, the combined value of production from broilers, eggs, and turkeys contributed nearly $80 billion to the U.S. economy (Table 1).

The consumption of broiler products per capita in the United States reached a high of 99.5 lbs in 2023. With the export and domestic use of poultry, egg, and turkey products, the United States is a world leader in poultry production.

Such impressive growth and economic impact are accompanied by an additional yearly legacy. Every increase in poultry production increases the production of manure, mortality, and wastewater from broiler, hatchery, egg

http://dx.doi.org/10.19103/AS.2025.0165.03

**Table 1** Broilers, eggs, turkeys, chickens (excluding broilers), and total value in the United States between 2013 and 2023 (USDA-NASS, 2024)

| Year | Broilers | Eggs | Turkeys | Chickens (excluding broilers) | Total (selected commodities) |
|---|---|---|---|---|---|
| **2023** | 42.6 | 17.9 | 6.57 | 0.062 | 67.132 |
| **2022** | 50.45 | 19.4 | 7.12 | 0.0747 | 77.0447 |
| **2021** | 31.52 | 8.71 | 5.91 | 0.0149 | 46.1549 |
| **2020** | 21.27 | 8.35 | 5.19 | 0.0188 | 34.8288 |
| **2019** | 28.31 | 10.45 | 4.31 | 0.0376 | 43.1076 |
| **2018** | 31.75 | 9.77 | 4.9 | 0.0343 | 46.4543 |
| **2017** | 30.23 | 9.94 | 4.79 | 0.0396 | 45.0096 |
| **2016** | 25.94 | 8.44 | 5.2 | 0.0401 | 39.6201 |
| **2015** | 28.72 | 7.74 | 5.09 | 0.0336 | 41.5836 |
| **2014** | 32.73 | 8.04 | 5.4 | 0.0315 | 46.2015 |
| **2013** | 30.76 | 8.08 | 4.84 | 0.0305 | 43.7105 |

Value of U.S. poultry production by commodity, 2013–2023 (in billions of U.S. dollars).

layer, and live-chicken processing operations. These co-products (material generated during the production process, such as manure and dead chickens) must be safely utilized or disposed of to ensure that they do not lead to environmental pollution. The challenge for the broiler industry is where and how to use these broiler co-products to both benefit and protect the environment. The rapid growth of the broiler industry internationally only augments the challenge, as does the clustering of broiler operations near food processing plants or large urban markets.

The traditional use for broiler co-products is land application as a fertilizer and soil amendment but functioning as a land resource is not always sufficient. Expanded or new uses for broiler co-products must continue to be sought after: for example, broiler co-products can and have been used as an ingredient in organic fertilizers, as a horticultural growing medium, and for energy generation. Indeed, a continuing search for additional uses is part of the challenge of modern broiler production.

The broiler industry is committed to protecting water and air quality, the environment, and natural resources and shares responsibility with other segments of the agricultural community and all citizens that contribute to nonpoint-source pollution: the pollution that originates from diffuse sources (e.g. agricultural runoff, community stormwater runoff, and erosion). Some segments of the broiler industry, such as processing plants, may also contribute to point-source pollution: the pollution that issues from a known or direct discharge (e.g. stormwater discharged from the end of a pipe or wastewater discharges from processing or treatment plants).

Understanding the complexity of broiler operations can help us address potential environmental quality issues. The industry is separated into hatchery, breeder, broiler, turkey, egg, duck, and live-chicken and egg processing operations. Each of these operations produces dry or liquid co-products and, coupled with death loss, constitutes the main waste streams associated with production. The types of co-products in broiler production include:

1 Litter: a mix of bedding material, manure, feathers, and spilled feed;
2 Hatchery waste: eggshells, infertile eggs, and dead embryos;
3 Carcass waste: dead chickens that need proper disposal; and
4 Processing waste: offal, feathers, wastewater.

Environmental awareness has progressed beyond point sources as with live-chicken processing plants (offal, feathers, and wastewater) to focus on producers that are considered nonpoint-source sites (Gibb, 2023). The shift reflects an increasing awareness of how agricultural runoff affects water quality. It also recognizes that the growth of the industry elevates the need for increased awareness and management of on-farm environmental challenges.

The speed, efficiency, and methods used to produce broiler and broiler products have changed drastically during the last 50 years as mass production techniques have been applied to farming operations and brought the benefits of information technology to the farm. As a result, most broilers are grown in confined operations with a limited environmental footprint. These conditions and changes in scale of production require each producer to dispose of or use large quantities of co-products, which inherently will have an increased footprint as these co-products are used as a fertilizer nutrient source for forages and field crops (Fig. 1).

Any co-product, when improperly managed, can pollute the environment. In this respect, broiler operations are not different from other human activities, and though each operation is unique, most if not all the problems can be prevented or solved through proper co-products disposal or utilization methods and changes in management style or production techniques. While the concepts to reduce environmental impacts are universal, the methodology or onsite strategies are not, as each operation is unique, in a different location, and under diverse management. Still, each one will begin with a model, a technology, or a history of practices that others can apply and adapt to their own circumstances.

It is important for broiler farmers (producers) to know at the beginning of the production cycle: (1) how much co-products will be produced from their operations; (2) how they can measure co-product chemical and physical make up; and (3) how they can account for its potential impact on water quality, the environment, and human health within their community. In addition, producers

**Figure 1** Land application of broiler litter has been the most cost-effective and widely used method for utilizing broiler production co-products for many years. Photo source: USDA-NRCS.

must know (4) how and when such impacts will occur, and (5) what measures they can take to prevent these impacts and manage the co-products in an environmentally safe manner (PEQH, 2025).

The overriding environmental issues facing producers today are to prevent broiler co-products from adversely affecting water and air quality. Potential water pollutants from on-farm broiler operations can be classified as (1) nutrients and salts, (2) organic materials, (3) bacteria, and (4) viruses. Potential air pollutants are dust and volatile organic compounds (gases), and nuisance odors.

These on-farm pollutants may originate in manure, litter, or dead chickens. How such co-products are disposed of, treated, or managed will directly influence the cleanliness and safety of surface and groundwater resources. Air quality may also be affected by improperly handled broiler co-products. Air and soil are less obvious but no less important media for transporting these pollutants into the environment.

Many progressive producers have switched to composting or to rendering mortality as preferred solutions from an environmental and economic viewpoint. A producer must choose a method compatible with his or her individual operation. Improper methods of disposal are unacceptable and cannot be condoned because of the potential for air and water pollution and accompanying lapses in biosecurity.

The magnitude of the problem underscores the advantage to be gained from its alternative: namely, that properly managed broiler co-products such as manure, litter, dead chickens, and wastewater are profitable farm outputs. An effective co-products management plan provides for the proper collection, storage, analysis, and use of broiler co-products (USDA-NRCS, 2010). Materials derived from the co-products will reduce the need for chemical fertilizer, improve soil quality, protect water resources, and support air quality. Effective co-product management will also promote a favorable public attitude toward the industry regarding byproduct and wastewater utilization. There is not a stand-alone best or optimal approach to protect or preserve air and water quality within the environment. It takes the adoption of an array of best management practices that have been approved and implemented to effectively utilize production co-products. Good co-product management practices are essential if the broiler industry is to continue to grow and thrive under today's environmental and societal challenges. It is the intent of this chapter to address the interface between the generation of broiler production co-products and the external environment and to expound upon the challenges and limitations that arise within that interface.

## 2 Nutrient management planning: decades of best management practices

The nutrients and organic materials found in broiler litter are extremely beneficial co-products of broiler production, with broiler litter having both micro- and macronutrients needed for plant growth (Ashworth et al., 2020). Given the size and economic importance of this key agricultural industry, broiler producers must properly use this material to obtain maximum economic value of its fertilizer qualities while assuring protection of the environment. The agronomic and economic requirements of nutrient management planning remain central, but in addition, we must consider the potential impact of these nutrients on environmental quality (Beegle et al., 2000). For many years now, environmental challenges associated with broiler production co-products have been addressed through the process of creating a nutrient management plan (NMP). Identifying best management practices to address each segment of the nutrient management planning process helps to further reduce environmental challenges. A nutrient management plan is essentially a collection of best management practices that assure the appropriate use of animal manures and commercial fertilizers to provide crop nutrient requirements, while simultaneously protecting surface and ground waters from overapplication. In addition, crop production costs can be reduced by effective use of broiler farm co-products as nutrient resources.

Nutrient management planning involves the strategic application of fertilizers and organic amendments to meet crop nutrient needs while minimizing nutrient losses to the environment. Effective nutrient management is critical for improving crop yields, reducing input costs, and protecting water quality. The principles of NMP include (IPNI, 2016):

1 Right source: Selecting the appropriate type of fertilizer or organic amendment based on soil and crop needs;
2 Right rate: Applying the correct amount of nutrients to meet crop requirements without over-application of organic fertilizers;
3 Right time: Timing nutrient applications to coincide with periods of high crop uptake and minimizing seasonal weather influences; and
4 Right place: Placing nutrients where they can be most effectively utilized by crops, while reducing losses to the environment.

Applying these principles through best management practices is the core of NMP. The following are recognized steps for developing an effective nutrient management plan.

## 2.1 Nutrient generation assessment

Broiler litter is a mixture of broiler manure, feathers, wasted feed, and bedding material that contains nitrogen, phosphate, potash, and other nutrients essential for plant growth. Broiler litter is a great natural fertilizer that can boost soil fertility and reduce the use of chemical fertilizers. The primary nutrients in broiler litter are nitrogen and phosphorus. Both nutrients are required as plant food, but they can also cause contamination of surface and ground waters with excessive application. The fertilizer equivalent of broiler litter is approximately 3-3-2 ($N-P_2O_5-K_2O$); however, the actual nutrient content depends on the market weight of broilers raised, bedding materials used, building clean-out frequency, storage and handling practices, use of litter amendments to control ammonia generation, and other factors (Ritz, 2024b). The numbers 3-3-2 represent the percentage by weight of three essential plant nutrients: nitrogen (N), phosphorus (P), and potassium (K). Phosphorus and potassium levels in the litter will continually rise with each flock raised on the same bedding material, as these minerals are in stable form, while the nitrogen level will plateau within the litter after 5–6 flocks on the same bedding material (Ritz, 2024b). Therefore, sampling broiler litter prior to land application to obtain a nutrient content analysis is an important step for managing manure nutrients appropriately.

Broiler litter contains nitrogen in both organic and inorganic forms. Organic nitrogen, primarily from uric acid and undigested proteins, makes up the majority (approximately 90%) of the nitrogen (Chastain et al., 2001). Inorganic nitrogen, in the form of ammonium and a smaller amount of nitrate, is readily available for plant uptake. Organic nitrogen can be converted to inorganic nitrogen by bacteria in the soil and can then be used for plant growth. Excessive organic and ammonium forms of nitrogen can be transformed into nitrate nitrogen, which, if found in high levels within drinking water, can harm human health (USGS, 2018). Excess nitrogen can also be moved from application sites by surface water run-off from rainstorms and by leaching through the soil. This nitrogen can end up in surface or groundwater sources, causing nutrient contamination.

Broiler manure contains phosphorus as a result of natural excretion and from the addition of phosphorus as an ingredient in broiler feeds. Certain forms of phosphorus used in broiler diets are poorly absorbed by the chicken's digestive tract, and this unabsorbed phosphorus is eliminated in the manure. Years of broiler manure applications can result in phosphorus-laden soils. Soils with excessive phosphorus levels, if not managed properly, can be a source of soluble phosphorus contamination via runoff erosion into surface waters (US -EPA.gov). Excessive soluble phosphorus in water can cause increased plant and algal growth that can lead to fish kills due to depleted oxygen levels in the water as the plant material matures and then decays (NOAA.gov).

## 2.2 Soil and litter analysis

Soil testing provides essential information on available nutrient levels, soil fertility potential, and soil pH to help guide fertilizer recommendations and the appropriate use of organic fertilizers.

Laboratory analysis of soil and broiler litter nutrient levels is necessary for determining the appropriate application rates. Soil testing should be conducted regularly to monitor phosphorus and other mineral buildup in the soil. If the soil test for phosphorus is in the high to very high category, local specialists at USDA-NRCS and the Cooperative Extension Service can help determine the phosphorus risk index for crop and forage fields. The phosphorus risk index will indicate if the phosphorus levels in the soil, coupled with co-product management practices, create a significant risk to nearby surface waters.

Obtaining nutrient concentration data for broiler litter is a crucial step in developing and using a nutrient management plan. Measuring the average nutrient concentrations of litter within a broiler house requires sampling procedures that ensure representative sampling. Research shows that spatial variability of nitrogen and phosphorus concentrations can be influenced by conditions such as litter moisture content and waste feed found within broiler production facilities (Ritz, 2024b).

Broiler litter moisture management has perhaps the greatest effect on litter nutrient content. Crusted or caked litter around drinkers and feeders is usually wetter and higher in nitrogen and phosphorus than whole house litter. This caked litter represents approximately 30–35% of the weight of the litter as a whole and typically has different handling characteristics than the rest of the litter material. Broiler litter testing determines the fertilizer value of the litter, which can be used to calculate land application rates or market value if sold off-site.

The attention to detail in the sampling of litter and the soil onto which it is applied will determine how well nutrient applications match the nutrient requirements of the intended crops and forages.

Broiler litter is an unbalanced fertilizer, and repeated land applications based on crop nitrogen needs can cause a build-up of phosphorus and potassium in the soil. Excessively high phosphorus in the erosion and runoff from litter-amended soil can cause water quality issues, such as eutrophication, in nearby surface water bodies. Applying nutrients without a litter analysis could cause crop yield reductions due to lack of adequate nutrients, applying nutrients when they are not needed, or over-applying nutrients that may contribute to water quality impairments (Bolan et al., 2010).

A basic litter sample should be analyzed for nitrogen, phosphorus, potassium, magnesium, manganese, aluminum, zinc, and copper. A pound of manure phosphate or potash has a nutrient value that is equivalent to that of commercial fertilizer. However, only a portion of the nitrogen content is available to crops when first applied. The amount of available nitrogen in the first year of broiler litter application typically ranges from 40% to 60% of the total nitrogen content in the litter, depending on factors like application method, timing, and environmental conditions at the time of application. According to the USDA Agricultural Research Service (ARS), subsurface band application of broiler litter can significantly reduce nitrogen losses through volatilization and runoff, thereby increasing the amount of nitrogen available to crops (Poffenbarger et al., 2015). This method can help synchronize nitrogen release with crop uptake, improving nitrogen use efficiency. A summary of the nitrogen availability from broiler litter in the first year under different application methods is shown in Table 2. These values reflect the proportion of total nitrogen in broiler litter that becomes plant-available in the first growing season. The method of application significantly influences nitrogen retention and crop uptake.

**Table 2** Nitrogen availability from land application of poultry litter during the first growing season under differing application methods (Poffenbarger et al., 2015)

| Application method | Nitrogen availability (%) | Notes |
|---|---|---|
| Surface application | ~40 | Higher ammonia volatilization losses |
| Incorporation | ~50 | Reduces volatilization, improves synchronization with crop uptake |
| Subsurface banding | ~60 | Minimizes losses, enhances nitrogen use efficiency |

Cooperative Extension Service specialists or other qualified professionals can assist with the calculation of manure nutrient availability based on when and how broiler litter is land applied. This information, combined with the soil test report and other crop information, is necessary to develop a nutrient management plan.

The money and effort invested in appropriate litter sampling is small compared to the money that can be saved on commercial fertilizer use. With the increased awareness of potential environmental impacts of land application of manures, broiler producers can more effectively use broiler litter within an NMP through improved sampling procedures (Fig. 2).

## 2.3 Appropriate storage

Improved storage methods for broiler litter are necessary to retain the fertilizer value and preserve the effective use of the nutrients contained in the manure. Appropriate litter storage methods will all have a common purpose: keeping rain and stormwater from entering the stored litter and causing the movement of material and nutrients into surface and groundwater. Appropriate storage and containment are also key to minimizing odors and potential nuisance complaints (Ritz and Cunningham, 2022). Broiler litter storage can be done in a variety of ways with a great range of investment costs and management programs. Storage methods may include: simply leaving the litter in the broiler house until land application occurs; utilizing weighted, removable covers;

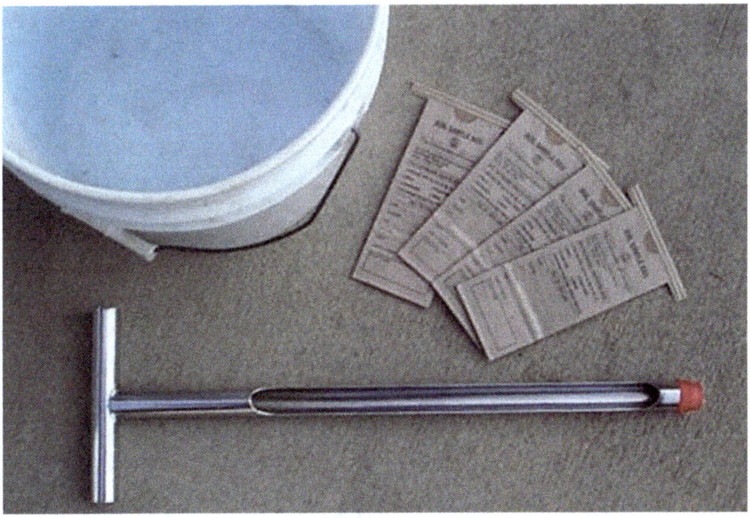

**Figure 2** Soil and manure sampling provides the necessary information on soil fertility and nutrient availability to make informed decisions on the appropriate use of organic fertilizers. Photo source: C.W. Ritz.

or the use of permanent structures. All of these methods can be utilized by producers to store manure fertilizers prior to land application. Broiler litter storage structures are considered a Best Management Practice for protecting the environment and function best at maximizing the fertilizer value of litter. Litter storage structures have become more prevalent and important as the broiler industry continues to implement environmentally sound management programs. Using appropriate methods for storage and application of broiler litter will ensure maximum fertilizer values and environmental protection for the farm and the surrounding rural community (Fig. 3).

Ensuring that stormwater does not enter broiler production facilities or manure storage structures is a vital component of proper litter storage. The unintended release of nutrients from these facilities into surface waters from surging stormwater can create pollution scenarios that we are trying to avoid with proper manure storage methods. Proper siting of the production and manure storage structures on the landscape is a critical component for reducing the potential for stormwater intrusion and subsequent release of nutrients into the environment (Ritz and Cunningham, 2022). Locating structures outside of flood zones and stormwater surge areas, as an example, becomes critical for securing and retaining fertilizer nutrients.

## 2.4 Calibration of spreading equipment

Once the appropriate application rate for broiler litter is determined based on crop nutrient needs and manure analysis, use of spreading equipment

**Figure 3** Litter storage structures or methods that prevent stormwater intrusion into stored broiler litter help to preserve manure nutrients. Photo source: C.W. Ritz.

that is calibrated to deliver this material as accurately as possible is essential. If the spreading equipment is not calibrated, we cannot accurately predict the application rate, nor the ensuing crop yield. We will not know the volume of delivered nutrients and cannot manage or predict results if we have not measured the input components. Overapplication of manure can lead to soil accumulation of nutrients that can then be prone to potential runoff into surface waters during severe weather events. Spreading equipment calibration ensures that application rates are closer to the actual crop fertilizer requirements without over-applying nutrients. Farmers who apply their own organic fertilizer and commercial applicators alike will benefit from effective calibration of manure spreading equipment (Fig. 4).

## 2.5 Land application

To reduce transportation costs and maintain its economic value as a fertilizer, broiler litter is commonly applied to cropland near the farm that generates the litter (Kulesza, 2025). Broiler litter is a good source of plant nutrients and is commonly used as a fertilizer on pastures and hayfields and is much like a complete fertilizer containing not only primary nutrients but secondary and micronutrients.

The inorganic nitrogen (ammonium and nitrate) can be immediately used by plants. Organic nitrogen is not available to plants until it is converted to ammonium or nitrate by microorganisms in the soil. Because this is a biological

**Figure 4** Calibration of spreading equipment promotes more accurate field application of nutrients and more even distribution of litter across the field, resulting in improved crop yield. Photo source: MSU.

process, the rate of conversion depends on soil moisture and temperature. If conditions are extremely dry or cold, little or no nitrogen may be released. One advantage of broiler litter for pastures is that the slow conversion of organic to inorganic nitrogen distributes available nitrogen more evenly over the growing season.

Because there is ammonium in broiler litter, some of the nitrogen can be lost to the atmosphere after the broiler litter is applied. Nitrogen can also be lost through leaching into the soil. Most of the nitrogen not taken up by forages in the first season is either lost to the environment or stabilized as soil organic matter. Little 'carryover' of nitrogen from broiler litter can be expected the second year after application (Harmel et al., 2009).

Applying broiler litter above the recommended rate can result in increased nitrate concentration in forages. Nitrate accumulation is not specific to broiler litter. It occurs when an excess of nitrogen is available and conditions such as drought or cool weather retard forage growth. Producers should be aware of these conditions and use recommended fertilization rates.

Broiler litter is a good source of phosphorus, which is beneficial when soils are low in phosphorus, but can present environmental problems if the soil is already high in phosphorus. Legacy phosphorus or phosphorus that has built up in the soil over time can be a source of surface water pollution if soil runoff from pastures or fields that are historically high in phosphorus results in sedimentation entering surface water.

Since forages normally require three or four times more nitrogen than phosphorus, and broiler litter contains almost equal amounts of both, using broiler litter to meet the nitrogen needs of the forage will cause an overapplication of phosphorus. Phosphorus can quickly accumulate to high levels with overapplication. High rates of broiler litter application can quickly increase the amount of phosphorus in the soil. High phosphorus levels in the soil have been directly linked to water quality problems where there is soil runoff from fields; consequently, using nutrient management planning to carefully supply the nutrient needs of the forage with broiler litter without creating environmental problems is a crucial best management practice.

There are two common management alternatives to reduce the risk of excess phosphorus when applying broiler litter as a fertilizer. One option is to alternate the use of broiler litter with a commercial fertilizer source of nitrogen. Another option is to apply broiler litter to meet the phosphorus needs of forage and crops, and to use commercial fertilizer to meet the nitrogen requirement. Fields located close to surface water sources may need fertilizer application based on crop phosphorus needs versus nitrogen requirement to minimize the phosphorus runoff potential associated with stormwater events. These practices can help reduce the water quality impacts of excess phosphorus in the soil due to over-application of fertilizer sources.

Broiler litter is also a valuable source of potassium. Nearly 100% of the potassium in broiler litter will be available during the growing season when it is applied. Knowing the level of potassium in the litter and soils to which it is applied will ensure that an adequate level of potassium is applied to meet the crop growing needs. Typically, potassium needs will be met when applying litter to meet the plant's need for nitrogen or phosphorus.

Broiler litter is also a source of secondary nutrients such as calcium, magnesium, and sulfur, as well as micronutrients such as copper and zinc. These secondary and micronutrients are an important component of organic fertilizers as they provide mineral needs for crop growth that are not supplied in a standard commercial fertilizer application. Weed species will also germinate or grow rapidly when additional nutrients are applied to pastures. These weeds are noticeable in pastures after broiler litter applications. Studies have shown, however, that properly stored broiler litter does not contain viable weed seeds (Ashabranner, 2023). If broiler litter is stored outside and uncovered, then weed seeds can be introduced from wild birds or other sources.

Producers should use nutrient management planning and recommended land application rates to ensure broiler litter is used in ways that maximize its benefits without harming the environment.

## 2.6 Site-specific conservation and erosion control

Whether siting a new broiler production or manure storage facility, or identifying acreage for manure land application, it is important to include site-specific conservation plans and erosion control practices in order to ensure organic fertilizer nutrients are properly preserved. Utilizing vegetative filter or buffer strips around field edges is an effective practice for reducing soil erosion and with it any applied manure nutrients. Retaining those nutrients in the field not only supports reaching crop yield expectations but also helps to minimize the potential for surface water pollution from fertilizer nutrients associated with stormwater runoff and any accompanying soil erosion (Fig. 5).

## 2.7 Proper dead broiler disposal

Broiler production facilities must deal with the disposal of farm mortalities on a daily basis. Death loss in animal production is an unfortunate reality that requires appropriate handling to prevent the spread of disease, the potential for odor and pest problems, and the possible contamination of surface and ground water. A typical flock of 25 000 broilers averaging 3% total mortality will produce approximately 2000 pounds of carcass during a 6-week growing period. This can represent a significant carcass disposal challenge over the life of the flock. A convenient, sanitary, rapid disposal system is critical for mortality management to be practical, effective, and meet state and local regulations.

**Figure 5** Vegetative filter strips that are established between planted crops and water conveyances help to slow water flow and reduce erosion sedimentation from entering surface water sources. Photo source: UNL.

Regulations governing the disposal of poultry mortality in the United States are primarily outlined by the USDA Natural Resources Conservation Service (NRCS) under the Conservation Practice Standard Code 316: Animal Mortality Facility. This standard provides guidance for the treatment or disposal of routine poultry mortalities on farms (USDA-NRCS).

Typical methods of mortality disposal include burial, incineration, rendering, and composting. Many states have banned the use of burial pits, which historically have been used to dispose of dead birds. Incineration can be costly and raise air quality concerns, and the decreasing number of renderers further complicates disposal (Ritz and Worley, 2022).

Composting is considered a positive alternative method for handling dead birds in an environmentally sound manner. Composting is the most widespread method used in states that have banned pits. It is considered by many as the best alternative for mortality management, although it can be more labor intensive than other methods. The composting process converts dead birds into a useful, inoffensive, stable end-product that can be field-applied for crop use and soil improvement. This relatively inexpensive method of handling dead birds has gained worldwide acceptance throughout the broiler industry. Availability of cost-share funds in the United States to offset composter facility construction expenses has contributed to the increase in the use of this mortality disposal method (Fig. 6).

**Figure 6** Composting daily chicken mortality and manure is a common best management practice for nutrient management planning. Photo source: C.W. Ritz.

Adherence to proper best management practices for mortality disposal is the key to the successful use of this broiler production co-product.

### 2.8 Record keeping

Documentation is critical to demonstrate appropriate nutrient management planning and the use of broiler manure on the farm. The records needed for nutrient management planning include:

- annual manure generation estimation;
- soil and manure nutrient analysis;
- on-farm field application of manure; and
- excess manure removal or off-site transfer.

Accurate and timely recordkeeping assists not only the farmer in making NMP decisions but also helps to protect the environment and minimize the overapplication or misuse of available nutrients. Nutrient management planning practices and use of organic fertilizers can be bolstered with the availability of accurate and up-to-date recordkeeping practices. A record of NMP activities and transactions can assist farmers should their production and co-product management practices come into question (Fig. 7).

## 3 Nutrient management plan challenges

Despite the many benefits, there are farm-level challenges to incorporating an on-farm nutrient management plan. Of concern is the application of expertise and information about nutrients in guiding decisions and actions within a given plan. Farmers make decisions every day about their farming operation, and many of these decisions and subsequent actions influence the cycling and utilization of nutrients on the farm, and nutrient losses from the farm

**Figure 7** Keeping records of NMP best management practices can assist with planning for and reaching environmental stewardship goals. Photo source: C.W. Ritz.

(Oenema and Pietrzak, 2002). Some concerns can be alleviated with additional education or NMP adjustment, while others may require a farmer to consider philosophical or social change in the way that they utilize broiler co-products. Points to consider are as follows:

- phosphorus build-up in soils with nitrogen-based planning;
- reduction in available land if manure is spread at agronomic rates;
- variable and inconsistency of nutrient content of manures;
- minimizing manure transportation and spreading machinery costs;
- land application odors and nuisance complaints;
- assurance of effective manure use practices to alleviate food safety concerns; and
- education on the benefits and practices of using organic fertilizers.

Improper waste management can lead to the proliferation of harmful pathogens such as *Salmonella* and *E. coli*, which can affect both broiler and human health. Consumer safety concerns with the use of organic manure as fertilizer have resulted in changes to the federal standard for use of these materials on crops for human consumption. Under the purview of USDA-FDA and the Food Safety Modernization Act (FSMA), new standards for organic fertilizer use have been put in place to minimize human health concerns associated with organic fertilizers (USDA-AMS). Proper mortality disposal and effective manure use for fertilization of crops for human consumption are best management practices to reduce pathogen proliferation in both broiler and human environments.

## 4 Air quality challenges

The challenges of air quality are twofold: the quality of the air within the broiler production environment, and the ensuing environmental impacts of air emissions from these production facilities. Properly managing the air for optimal broiler health and performance will certainly reduce, but not eliminate, the concentration and volume of the emissions from production facilities. Ammonia emissions from manure can degrade air quality and pose health risks to broiler and farm workers (Ro et al., 2017). Reducing the impact of air emissions on the environment and local rural neighbors and communities can be a function of farm and building siting and other best management practices contained within a nutrient management plan. Utilization of litter treatment products and management strategies to reduce the generation of ammonia during and after production are best management practices that will assist in reducing the impact of ammonia generation and volatilization from production and storage facilities.

Odor is a complex issue that can impact neighbors, and the goal should be to keep odors at non-detectable or non-offensive levels. Proper siting of broiler production facilities and associated litter storage is likely the most important odor control practice that can be implemented. Problems with neighbors can often be avoided or minimized by giving more consideration to the broiler house location. To reduce potential conflicts with neighbors, it makes sense to locate broiler houses as far from property lines as possible, regardless of any existing zoning requirements.

The siting of a broiler production facility is the first step in odor control to minimize impacts on nearby neighbors and public areas. Each broiler farm should have a site-specific odor control plan as part of a nutrient management plan.

Separation distance, predominant wind direction, exposure angle, and terrain are all important considerations when the impact of 'air drainage' is considered when evaluating a potential building site for broiler house construction (Goan, 2002). Topography, local weather patterns, sensitivity of the nearest neighbors, and the characteristics of the animal facility are all factors that should be considered when evaluating the influence of potential separation distances. There are several science-based tools that are available to assist farmers, concerned citizens, and local policy officials in making sound decisions when it comes to establishing separation requirements between animal agriculture and rural communities. An example is the use of wind rose plots. A wind rose can indicate the dominant wind directions and the direction of strongest wind speeds over the course of a designated period of time (Iowa State University) (Fig. 8).

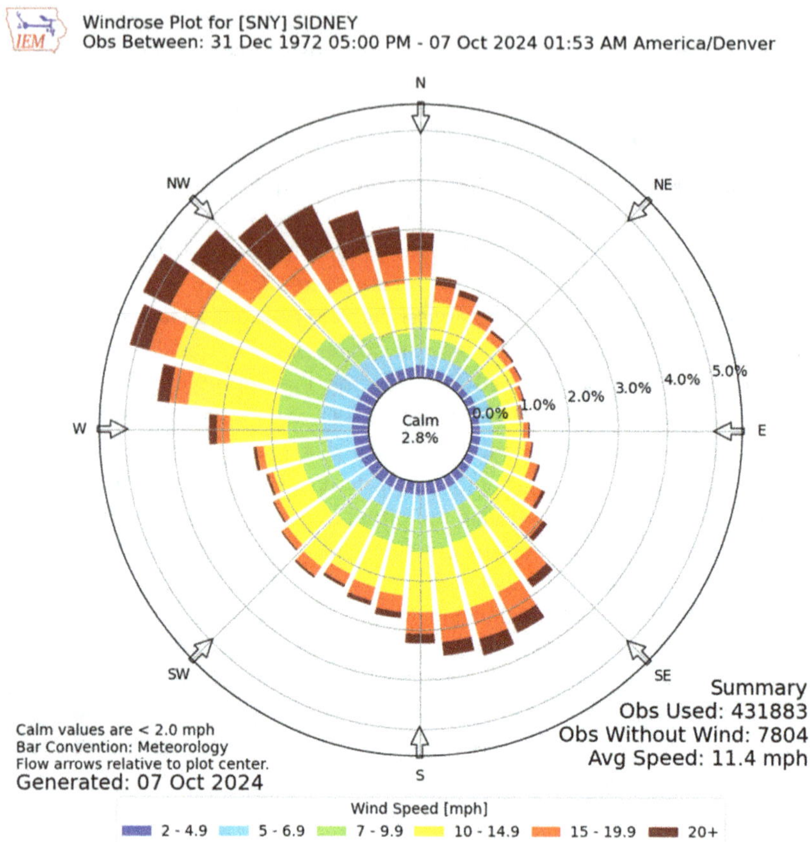


**Figure 8** Wind roses are a plot providing frequencies of wind direction and wind speed. Source: Iowa State University.

Good neighbor best management practices to assist broiler producers with air emissions and odor complaints can include the following:

- Carefully evaluate siting considerations when designing the location of broiler housing and manure storage structures. Pay attention to ventilation airsheds when siting production houses.
- Use natural or well-designed vegetative buffers and visibility screens. Vegetative buffers and windbreaks can filter and dissipate odor.
- Manage mortality disposal appropriately following approved methods.
- Maintain farm sanitation. A clean, appealing farm-scape will go a long way to reduce any perception of mismanagement and disregard for neighbors and the local community.

- Utilize manure spreading strategies to minimize environmental and neighbor impacts. Avoid spreading on weekends and holidays in nice weather when neighbors are likely to be outdoors. Consider local weather conditions when spreading manure – turbulent breezes will dissipate and dilute odors. Hot, humid weather will hold odors closer to the ground.
- When applicable and appropriate based on cropping conditions, incorporate manure into the soil as soon as possible after application. When incorporation is not feasible, pay special attention to where the odor will travel and whom it may impact.

## 5 Sustainability/circularity

The sustainability of agricultural systems is a theme that has become increasingly popular in recent years. The concept of circularity is not new, but has been relegated to small enterprises versus large-scale production. Interest from communities and world governments alike has brought the concept of circularity to the realm of large-scale agricultural production. The Ellen MacArthur Foundation defines a circular economy based on three principles: eliminating waste and pollution, recirculating materials, and regenerating nature.

Agricultural circularity has historically always been present with small farms that raised crops and livestock together so that crops provided food supply for the livestock, which in turn provided manure as fertilizer for the crops. As the intensification of agriculture has evolved, synthetic fertilizers have become mainstream to provide fertilization for crops and in turn produce the greatest yields. Additionally, modern livestock industries have typically become separated from crop production, with the scale of production being a driving factor of enterprise consolidation and separation. As a result, historical and arguably natural nutrient cycles have become disrupted, leading to many of the consequences addressed in this chapter, namely impacts on air, water, and soil quality. The separation of agricultural enterprises also means more transportation is required to utilize manure on cropland that contributes to energy usage and production costs.

Environmental impacts of broiler production include greenhouse gas emissions related to energy use in the production processes, and water quantity and quality impacts from the large quantity of water used in animal processing. Although broiler production has an overall lower footprint than other animal products according to the Food and Agriculture Organization of the United Nations (FAO), broiler processing has been found to be more energy intensive than other meat sectors such as beef and pork, and 41% of emissions within the chicken supply chain are the result of energy use (MacLeod et al., 2013).

Wastewater generated during these activities typically has high biochemical and chemical oxygen demand (BOD and COD) due to the presence of organic materials generated during processing that end up in the wastewater. To address this, broiler processing facilities treat water used in the production of broiler products in accordance with Federal National Pollutant Discharge Elimination System permits, state-issued land application discharge permits, or municipal sewage standards before discharging the water to the environment or a municipal sewer system. The enforceable treatment standards set for a broiler processing facility are dependent on many factors, including the locality, water quality criteria of the receiving stream, the capacity of a municipal wastewater treatment plant, and more. On the farm level, the major environmental concerns are from the greenhouse gas and water quality impacts associated with feed production, energy use on farm, and litter management (PEQH, 2025). According to the FAO, feed production accounts for around 78% of emissions from the global broiler supply chain (MacLeod et al., 2013). Broiler litter can impact environmental quality at the broiler farm itself, as well as on any land to which it is applied as a fertilizer. Litter is a valuable resource, as it can be used in lieu of synthetic fertilizers and provides organic matter to enrich soils. However, nutrient run-off from nearby fields where manure has been applied can negatively affect sensitive ecosystems. As a result, many broiler producers in these sensitive areas haul litter to fields they may have in other locations or sell their litter to farmers in neighboring communities that have fewer challenges with land application. Some states, such as Maryland, have offered cost-share or incentive programs to assist with the transport of broiler litter from nutrient-surplus to nutrient-deficient areas. For example, the state of Maryland's Manure Transport Program provides cost-share assistance to poultry growers to transport excess litter to farms that can use it safely (Maryland.gov). These programs aim to reduce phosphorus buildup in specific watershed areas.

## 6 Improvement opportunities

Implementing production practices that improve energy and water-use efficiency often reduces operating costs and resource waste. At the farm and broiler house level, the major areas of focus are feed production, on-farm equipment, waste and litter management, and water use. Energy and water efficiency upgrades provide a significant opportunity for environmental improvement and cost savings for many farmers. Conducting an energy or water-use audit can assist producers in identifying their biggest opportunity areas for improvement. By analyzing energy consumption patterns, particularly in ventilation systems and other high-demand electrical uses, producers can identify opportunities for energy optimization. Energy-efficient lighting, use of fans to create a 'tunneling effect' to cool chickens, and efficient well-water pumps are just a few areas where energy use can be assessed for efficiency (University of Arkansas).

Broiler producers and poultry companies have an incredible opportunity to address circularity within the broiler production process, from farm to processing facility. While companies may focus first within their own operations, the true opportunity for making an impact on broiler sustainability comes from engaging with the supply chain. Trouble spots with the greatest risk to natural resources in broiler production include air quality, energy consumption, and manure or litter management. Without continued efforts to appropriately utilize broiler production co-products and their environmental impact, pressure to mitigate these areas of concern will only intensify, and so will the environmental, regulatory, and market-based pressures to mitigate them.

The National Chicken Council, a trade association that advocates for the broiler meat industry, issued this statement in their 2020 Sustainability Report regarding the sustainability of the industry, which summarizes the challenges and opportunities when addressing environmental issues:

> Looking ahead, we are focused on sustainable development and the critical role of food systems that include our chicken industry – Opportunities revealed by the described Broiler Production System Life Cycle Assessment: 2020 Update also set the groundwork for next steps for the chicken industry. Based on the data, we know that all five key sustainability intensity metrics {land use, carbon footprint, water consumption, fossil resource use, particulate forming emissions} improved significantly in the past decade. We also know that additional improvements are possible going forward. The research revealed that our continued areas of greatest impact and improvement will come from factors affecting feed consumption and feed conversion ratio. Therefore, further innovations in genetics, feed additives and supplements should be seen as part of our next sustainability frontier – Also based on the 2020 LCA, we learned that external factors associated with increasing crop production, improving fuel efficiency, and increasing adoption of renewable energy sources should become an integral part of our extended purview. Finally, we are mindful of regional differences that affect the opportunities for achieving sustainability progress. Knowing that one-size-does-not-fit-all regarding geography, we will consider regional differences when we advance new solutions... Our chicken industry will continue to innovate as responsible stewards to advance sustainability while feeding the world.
>
> *(NCC, 2020)*

## 7 Conclusion

Adhering to Best Management Practices for storing and applying broiler litter will result in obtaining the maximum fertilizer value of this valuable animal co-product while assuring protection of surface and ground waters. NMPs are essential for the appropriate application of important crop nutrients, but responsible management includes more than just applying the correct amount of broiler litter.

Responsible management also requires consideration of appropriate sites, timing of application, and consideration of neighbors. Maintaining and calibrating spreader equipment on a regular basis will ensure that the calculated rates of application from the NMP are delivered as intended. It is important for all broiler producers to practice Best Management practices for the utilization of their broiler litter and other production co-products. It may take only one poorly managed broiler operation to negatively impact all producers in a community in the court of public opinion.

Environmental challenges facing the broiler industry are evolving as the industry comes into increased competition for resources and space from expanding rural communities. Nutrient management planning is a vital component of sustainable agriculture, offering numerous benefits for crop production, environmental protection, and economic viability. By adopting best nutrient management practices and utilizing modern technologies, producers can achieve efficient and effective on-farm nutrient management and promote improvements in sustainable agricultural efforts.

By addressing environmental challenges with effective strategies and innovative technologies, the broiler production industry can appropriately manage its environmental footprint and contribute to the enhancement of agricultural sustainability both in the United States and on a global scale.

## 8 References

Ashabranner, G. and Fairchild, B., 2023. Does poultry litter contain weed seeds? *University of Georgia Cooperative Extension, Poultry Housing Tip*, 35(9). poultryventilation.com.

Ashworth, A.J., et al., 2020. Nutrient characteristics of poultry manure and litter. In: J.L. Hatfield and B.A. Stewart (eds.), *Animal manure: production, characteristics, environmental concerns, and management*. American Society of Agronomy, ASA Special Publications Book Series.

Beegle, D.B., et al., 2000. Nutrient management planning: justification, theory, practice. *Journal of Environmental Quality*, 29, 72–79.

Bolan, N.S., et al., 2010. Uses and management of poultry litter. *World's Poultry Science Journal*, 66, 673–698.

Chastain, J.P., et al., 2001. Poultry Manure Production and Nutrient Content. In: Clemson Cooperative Extension (ed.), *Confined animal manure managers certification program manual: poultry version*. Clemson University Extension, Clemson, SC. P.3b-1-3b-17.

Ellen McArthur Foundation - Circular Economy.

Gaskin, J.W. and Harris, G.H., 2017. *Poultry litter application on pastures and hayfields*. University of Georgia Extension Publication B-1330.

Gibb, T., 2023. *What's the point and non-point in water quality?* Michigan State University Extension Publication. MSU Water Quality.

Goan, C., 2002. *Site selection factors for new poultry facilities*. Agricultural Extension Service, University of Tennessee, Publication SP-592. UT Site Selection.

Harmel, R.D., et al., 2009. Nitrogen and phosphorus runoff from cropland and pasture fields fertilized with poultry litter. *Journal of Soil and Water Conservation*, 64(6), 400–412.

International Plant Nutrition Institute, 2016. *4R plant nutrition manual: a manual for improving the management of plant nutrition*. IPNI.

Iowa State University ISU Wind Rose.

Kulesza, S., 2025. *Guidelines for the commercial application of poultry litter*. NC State Extension Publications. NCSU Land Application.

MacLeod, M., et al., 2013. *Greenhouse gas emissions from pig and chicken supply chains – A global life cycle assessment*. Food and Agriculture Organization of the United Nations (FAO). https://www.fao.org/4/i3460e/i3460e.pd.

Maryland.gov, 2020. *Maryland's Manure Transport Program provides cost-share assistance to poultry growers to transport excess litter to farms that can use it safely*. Manure Transport Program.

Mississippi State University Extension Service, 2022. *Calibrating poultry litter spreading equipment*. Publication P2495. MSU Calibration.

National Chicken Council Broiler Chicken Industry Sustainability Report, 2020.

NOAA – National Oceanic and Atmospheric Administration. What is eutrophication? https://oceanservice.noaa.gov/facts/eutrophication.htm.

Oenema, O. and Pietrzak, S., 2002. Nutrient management in food production: achieving agronomic and environmental targets. *Ambio*, 31, 159–168.

Poffenbarger, H.J., et al., 2015. Cover crop and poultry litter management influence spatiotemporal availability of topsoil nitrogen. *Soil Science Society of America Journal*, 79, 1660–1673. https://doi.org/10.2136/sssaj2015.03.0134.

Poultry Environmental Quality Handbook, 2025. https://peqh.uga.edu/.

Ritz, C.W., 2024a. *Maximizing poultry manure use through nutrient management planning*. University of Georgia Cooperative Extension, Bulletin 1245.

Ritz, C.W., 2024b. *Poultry litter sampling*. University of Georgia Cooperative Extension, Bulletin 1270. https://extension.uga.edu/publications/detail.html?number=B1270&title=poultry-litter-sampling.

Ritz, C.W., 2025. Personal communication.

Ritz, C.W. and Cunningham, D.L., 2022. *Best management practices for storing and applying poultry litter*. University of Georgia Cooperative Extension Bulletin 1230.

Ritz, C.W. and Worley, J., 2022. *Poultry mortality composting management guide*. University of Georgia Cooperative Extension, Bulleting 1266.

Ro, K.S., et al., 2017. Ammonia and nitrous oxide emissions from broiler houses with downtime windrowed litter. *Journal of Environmental Quality*, 46, 498–504. https://doi.org/10.2134/jeq2016.09.0368.

United Nations - Circular Agriculture for Sustainable Rural Development.

United States Geological Survey, 2018. Nitrogen and Water. USGS Nitrogen and Water.

University of Nebraska-Lincoln, Lesson: Manure Phosphorus and Surface Water Protection III: Transport Factors. https://passel2.unl.edu/view/lesson/4155530d318.

University of Arkansas Agricultural Experiment Station. Chicken House Energy Audit Results in Ideas to Improve Efficiency. https://aaes.uada.edu/research-highlights/chicken-house-electricity-audit/.

USDA-Agricultural Marketing Service. Organic Regulations. https://www.ams.usda.gov/rules-regulations/organic.

USDA-National Agricultural Statistics Service (NASS) 2024. Poultry – production value, 2023, 2023 Summary. https://esmis.nal.usda.gov/sites/default/release-files/m039k491c/b2775j31b/9k4213149/plva0424.pdf.

USDA-Natural Resources Conservation Service (NRCS). 2010. Agricultural Waste Management Field Handbook.

USDA-NRCS, 2015. Conservation Practice Standard – Animal Mortality Facility, Code 316. https://www.nrcs.usda.gov/sites/default/files/2022-08/Animal_Mortality_Facility_316_CPS_9_2105.pdf.

USEPA – United States Environmental Protection Agency. Sources and Solutions: Agriculture. https://www.epa.gov/nutrientpollution/sources-and-solutions-agriculture.

# Chapter 3

## The importance of data sharing between different links in the broiler production chain

*Hasitha Priyashantha, University of Ruhuna, Sri Lanka; Maleeka Nambapana, Uva Wellassa University, Sri Lanka; Ashen Epa Arachchi, University of Peradeniya, Sri Lanka; Shishanthi Jayarathna, University of Ruhuna, Sri Lanka; and Janak K. Vidanarachchi, University of Peradeniya, Sri Lanka*

1 Introduction
2 The broiler chicken supply chain: an overview
3 Levels of integration in the broiler chicken supply chain
4 Data sharing and integration: case studies
5 From data sharing to better products for consumers
6 Future trends and recommendations
7 Conclusion
8 Conflict of interest
9 Acknowledgements
10 Funding
11 Data availability statement
12 References

### 1 Introduction

The world population is growing and income levels are increasing, leading to a surging demand for animal-derived food products such as meat, milk, and eggs. In response, rapid and strategic development of the livestock and poultry sector is often referred to as 'livestock and poultry sector reorganization.' The world population is projected to grow to nine billion by the year 2050. With the growing population, the demand for quality food is expected to rise, along with a growing need for high-quality proteins (Maharjan et al., 2022). In 2050, the demand for meat is expected to increase by 70% compared to 2005 (Yitbarek,

http://dx.doi.org/10.19103/AS.2025.0165.04

2019). The poultry industry, more specifically, the broiler chicken industry, plays a key role in the delivery of high-quality protein and lipids to the world population. The per capita consumption of chicken meat in both developed and developing countries increased significantly from 1995 to 2005 (Scanes, 2007). And over the next decade, per capita chicken meat consumption is expected to rise, particularly in developing countries (Herforth et al., 2020). The rising demand for chicken meat can be attributed to several factors, including its distinction from red meat, the negative effects of red meat on cardiovascular health, fewer religious and cultural restrictions associated with poultry, and the comparatively lower prices of chicken meat and its products compared to beef and pork.

Poultry meat accounted for nearly 40% of global meat production in 2020, following consistent growth over the past three decades. The United States produces 17% of the global poultry production output and is the world's largest poultry meat producer (Lord, 2024). The expansion of the poultry sector is driven by factors such as a growing population, which has led to rising demand for protein, increased economic growth and incomes, urbanization and changing diets, export opportunities, relatively short production cycles, and technological advancements. As a result, the poultry industry plays a vital role in meeting global protein demands amid changing demographics and lifestyles (Mottet and Tempio, 2017; Lord, 2024). In 2022, Asia, Latin America, North America, Europe, the European Union, Africa, and Oceania produced high volumes of chicken meat to meet global chicken meat demand (Table 1). The world's leading chicken meat production companies are also listed in Table 2.

Broiler chicken is one of the most widely consumed meats worldwide, and its production has shown steady growth in recent years. In 2022, global broiler production increased by 9% compared to 2018, reaching a total of 100 thousand tons (Dohlman et al., 2022). During the same year, annual per capita

**Table 1** Chicken meat production per continent

| Region | Chicken meat production - 2022 (Tons) | Share of global chicken meat production 2020 (%) |
|---|---|---|
| World | 119504578 | - |
| Asia | 43491487 | 36.4 |
| Latin America | 26655677 | 22.3 |
| North America | 21794927 | 18.2 |
| Europe | 19684933 | 16.5 |
| European Union | 11036911 | 9.2 |
| Africa | 6378058 | 5.3 |
| Oceania | 1499496 | 1.3 |

*Source*: FAOSTAT (2023).

**Table 2** Top ten global broiler chicken producing companies

| Company | Country | Number of broilers slaughtered annually (millions) |
|---|---|---|
| JBS S.A. | Brazil | 4426 |
| Brazil Tyson Foods Inc. | USA | 1900 |
| BRF USA Brazil | Brazil | 1732 |
| Wen's Food Group | China | 1101 |
| Wellhope Agri-Tech | China | 695 |
| CP Group | Thailand | 685 |
| Koch Foods Inc. | USA | 655 |
| Sanderson Farms Inc. | USA | 635 |
| Industrias Bachoco | Mexico | 627 |
| Cargill | USA | 625 |

*Source*: Clements (2022).

chicken consumption rose to 45.4 kg, surpassing that of beef (25.9 kg) and pork (23.1 kg) (Smith et al., 2023). Geographically, the Americas emerged as the leading producer, contributing 49.5 million tons, or 40.7% of the world's total chicken meat production. Asia followed closely with 44.5 million tons (36.3%), while Europe ranked third with 20 million tons (16.5%). The top ten broiler-producing countries include the United States, Brazil, China, the Russian Federation, India, Indonesia, Mexico, Egypt, Turkey, and Japan (Getiso and Mijena, 2024). Importantly, production growth was observed across all income groups, e.g. low-income, lower-middle-income, upper-middle-income, and even least developed countries, highlighting the global significance and accessibility of chicken meat (Oke et al., 2024).

The broiler production system functions as an interconnected chain involving several key components. The primary stages of the broiler supply chain include input suppliers, breeder farms and hatcheries, broiler farms, processing plants, wholesalers, and retail shops (Blanco et al., 2023). The hierarchy of broiler chicken breeder farms follows a structured pyramid system designed to efficiently multiply and transfer desirable genetic traits for high-quality commercial broiler production (Fig. 1). At the top of the pyramid is the Great-Grandparent (GGP) stock, consisting of pure genetic lines selected for traits like rapid growth, feed efficiency, environmental stress factors, and disease resistance. GGP farms (Aviagen, Cobb-Vantress, Hubbard, Hy-Line International, etc.) are limited in number and operate under strict biosecurity to protect elite genetics. These bird breeders produce the Grandparent (GP) stock, which serves as a bridge between elite genetics and large-scale production, expanding the number of breeders. The GP stock then produces the Parent Stock (PS), commonly known as broiler breeders, which are raised on

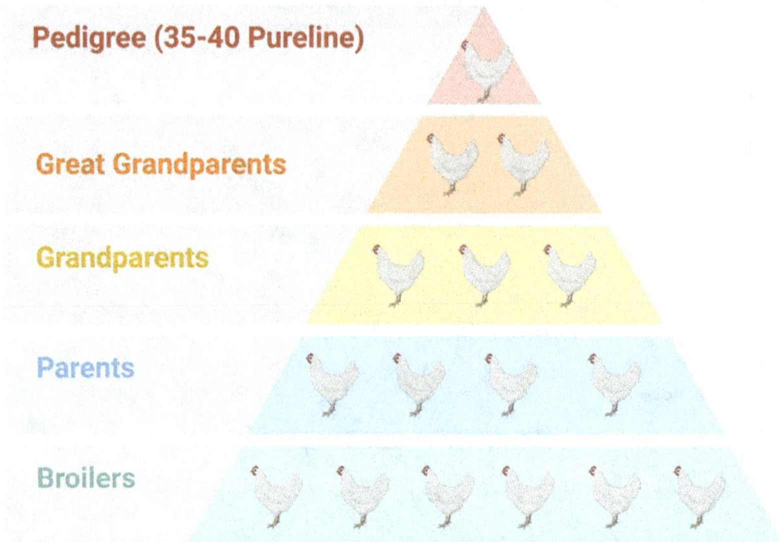


**Figure 1** Expansion of commercial broiler chicken pedigree lines.Source: Pollock (1999). Created in BioRender. Priyashantha, H. (2025) https://BioRender.com/o65w760.

specialized farms to lay fertile eggs. These eggs hatch in specialized hatcheries into commercial broilers, raised specifically for meat production. Commercial broilers are optimized for fast growth and efficient feed conversion and are typically processed within 5–7 weeks. This hierarchical system ensures genetic efficiency, as one GGP breeder can ultimately lead to thousands of commercial broilers, while maintaining biosecurity and consistent broiler chicken quality at scale. Once the chickens reach market weight, they are sent to processing plants that produce chicken meat and related products for distribution through wholesalers and retail outlets.

This interconnected structure is commonly known as the broiler chicken supply chain, which efficiently transforms hatching chicks into high-value meat products by the end of the production cycle. The structure and level of integration within the broiler chicken supply chain vary across countries and companies. According to Narrod et al. (2008), this structure has evolved over time, with a growing preference for large-scale, vertically-integrated supply chain operations. In low- and middle-income countries, however, poultry production is predominantly driven by small- and medium-sized enterprises (SMEs). Blanco et al. (2023) observed a growing trend of collaboration between broiler farms and slaughterhouses to jointly develop and adjust production plans. Enhanced information sharing and coordinated decision-making along the supply chain contribute to greater operational efficiency and improved control over meat product quality and safety.

To meet the growing global demand for chicken meat, effective interaction and information sharing across the interconnected components of the broiler chicken supply chain is essential. Accurate, high-quality data provides valuable insights to reassess past decisions (where these might have been wrong) and support better future decision-making (Davis and Olson, 1984; Redman, 1995; Wang and Strong, 1996). Informed decisions are crucial for accurately forecasting chicken meat production, as well as for optimizing productivity, efficiency, product quality, safety, and profitability throughout the supply chain. Digital tools and emerging AI-based platforms offer promising opportunities to improve communication, integration, and data sharing across the industry. Longer planning horizons are often preferred to take advantage of seasonal demand and better sourcing deals, but these efforts face inherent challenges due to the biological life cycle of broiler chickens.

The broiler chicken industry has increasingly adopted digital technologies not only to enhance food quality, safety, productivity, and animal welfare but also to enable more effective data sharing across the production chain. These technologies serve as tools that collect, analyze, and disseminate data among farmers, integrators, veterinarians, processors, and even consumers, allowing for real-time, informed decision-making and improved coordination across the supply chain.

One such system is the Rapid Alert System for Food and Feed (RASFF), widely used in the European Union. RASFF enables swift communication between producers, government authorities, and retailers regarding food safety risks such as contamination, disease outbreaks, or product recalls by sharing data such as batch numbers, origin, processing history, and distribution records. This system relies heavily on accurate and timely data exchange to ensure traceability and quick responses that protect public health.

Similarly, Precision Livestock Farming (PLF) technologies use sensors, IoT devices, and Artificial Intelligence (AI) to monitor flock health, behavior, environmental conditions, and productivity in real time. These data are continuously shared with farm managers, veterinarians, and integrators, enabling data-driven collaboration and timely decision-making. This supports early disease detection, better welfare practices, and optimized feed use, contributing to a more sustainable and responsive production model (Berckmans, 2017). This information is often uploaded to shared platforms, allowing farm managers, veterinarians, and integrators to access centralized records in real time. By enabling automated monitoring, improved biosecurity, and coordinated decision-making, radio frequency identification (RFID) strengthens data sharing across the entire production chain (Zhang et al., 2016).

Blockchain-based traceability systems further enhance transparency by recording every step of the broiler production process in a secure, tamper-proof ledger. That allows retailers, regulators, and even consumers to access consistent and verified data from farm to fork, ensuring credibility and accountability throughout the supply chain (Kamilaris et al., 2019).

Farm Management Software (FMS) tools consolidate data on feed conversion ratios, growth rates, health indicators, and environmental conditions. This data is often shared among producers, consultants, and supply chain partners to optimize decision-making, adjust strategies, and ensure regulatory compliance. Integration with AI and machine learning algorithms allows these platforms to generate predictions of disease outbreaks and patterns to predict optimal harvest times based on shared historical and real-time data.

In broiler houses, environmental control systems regulate temperature, humidity, and ventilation based on sensor data. When integrated with FMS platforms, these systems can automatically share environmental data with remote farm managers or technical advisors, enabling quicker interventions. Emerging tools such as drones, mobile apps, and virtual reality (VR) also support data collection and training. For example, drones can collect geospatial data on farm layouts, while mobile apps allow workers to input health and production data directly into centralized systems. Finally, quick response (QR) codes on packaging or farm documents are becoming increasingly popular. When scanned, these codes provide immediate access to product histories, including rearing practices, processing dates, and certifications. This enhances consumer trust and enables data transparency between the farm, processing plant, retailer, and end-user. Together, these digital technologies illustrate how data sharing is the backbone of modern broiler chicken production. By enabling real-time collaboration, traceability, and informed decision-making, they help the industry meet growing demands for food safety, efficiency, and sustainability.

Despite the rapid global growth of the broiler chicken meat industry, several challenges persist, particularly concerning sustainability (Castellini et al., 2006; Vaarst et al., 2015; Leinonen and Kyriazakis, 2016; Cui et al., 2021). This chapter focuses on these challenges impacting broiler chicken production and proposes strategies to overcome them. The chapter emphasizes the importance of effective data sharing between different stakeholders to enhance decision-making and streamline operations. The chapter also explores potential future trends, including the role of advanced data analytics and digital platforms in improving productivity and ensuring long-term industry sustainability. The goal is to provide actionable recommendations to support the continuous development of the broiler chicken industry while addressing the growing global demand for high-quality protein sources.

## 2 The broiler chicken supply chain: an overview

### *2.1 Great grandparents, grandparents, parents and hatchery operations*

Artificial selection has been the basis of livestock breeding for millennia, enabling breeders to cultivate desirable traits within animal populations. One prominent example is the broiler chicken, descended from the ancestral *Gallus gallus*. Historically, poultry breeders focused primarily on maximizing live weight as a straightforward metric for success, which significantly enhanced meat yield. As consumer preferences evolved and market demands grew more complex, breeders recognized the economic potential of targeted selection. Consequently, modern breeding programs began incorporating multiple criteria beyond just weight to optimize production outcomes.

Today's selection criteria for broiler chickens encompass growth rate, feed efficiency, carcass quality, disease resistance, environmental adaptability and welfare considerations. These parameters reflect the industry's understanding that efficient production is closely tied to animal health and welfare, and consumer satisfaction. Primary broiler chicken breeding companies serve as the backbone of the chicken industry, providing high-quality genetic material essential for meeting global meat demand across diverse environments and market conditions. According to Pollock (1999), approximately 400 000 pedigree breeders from 35 to 40 distinct pure lines form the genetic foundation for producing around 400 million broiler chickens annually (Fig. 1). Remarkably, just three companies (Cobb-Vantress, Aviagen, and Hubbard) supply 90% of the world's broiler chicks. Intensive selective breeding led to a genetic diversity loss of more than 50% in commercial lines compared to their ancestral breeds (Muir and Cheng, 2014).

The division of breeding populations into male and female lines enables breeders to tailor selection processes to specific traits associated with each sex. Male chickens are typically selected based on growth rate, edible meat yield, and feed conversion ratio, as these traits directly influence production efficiency. In contrast, female chickens are evaluated not only for growth rate and meat yield but also for feed conversion efficiency and egg production potential.

Grandparent (GP) farms play a crucial role in the poultry breeding hierarchy, acting as intermediaries between great-grandparent (GGP) and parent stock (PS) generations. Effective GP farm management is essential for producing genetically superior breeders that drive productivity across the broiler chicken industry (Schippers et al., 2025). The global grandparent generation chicken farming market was valued at USD 1.34 billion in 2023 and is projected to reach USD 2.04 billion by 2031. Key industry players in this sector include Arbor Acres, Cobb-Vantress, Marshall, and Hubbard (Abdullah et al., 2010).

Hatchery management and egg incubation are critical components of the broiler chicken industry, requiring meticulous attention to ensure optimal hatchability rates and high-quality chicks. These processes involve handling, storage, incubation, hatching, and chick handling under carefully controlled conditions. Incubation, whether natural or artificial, relies on maintaining specific temperature and humidity levels to support embryonic development and successful hatching. Single-stage incubation systems generally outperform multi-stage systems in broiler chick production. Factors like hatchery layout, building design, and environmental conditions further influence incubation success (Kuehler, 1993; Underwood et al., 2021). Recent innovations include *in ovo* vaccination, *in ovo* sexing, and targeted incubation strategies that regulate oxygen, temperature, and carbon dioxide levels (Jennison, 2021). Additionally, on-farm hatching has shown potential benefits for broiler chicken welfare by reducing mortality, litter moisture, and footpad lesions, with a negligible impact on body weight and feed conversion ratio at slaughter (De Jong et al., 2012).

## 2.2 Commercial broiler chicken farms

Commercial broiler chicken farms are specialized operations focused on raising broiler chickens for meat production, typically from hatchlings to market weight within 5–7 weeks. These farms use two main housing systems: open houses and closed houses. Open houses (or naturally ventilated houses) rely on natural airflow through sidewall curtains or wire mesh nets and are commonly used in regions with mild and tropical climates. While cost-effective, they offer limited control over temperature and humidity, making broilers more susceptible to environmental stress. In contrast, closed houses (or environmentally controlled houses) use advanced ventilation, cooling, and heating systems to regulate temperature, humidity, and air quality, meant to ensure optimal growing conditions year-round. This system enhances feed conversion, growth rates, and disease control but requires higher initial investment and energy costs. The choice between open and closed housing depends on factors like climate, farm size, and production goals.

## 2.3 Ingredients suppliers, medications, health and veterinary services

The poultry feed industry plays a critical role in global agriculture, specializing in the production of nutritionally balanced diets for species like broiler chickens. Feed mills process raw materials such as grains, cereals, oilseeds, and by-products from the food and agriculture sectors through grinding, mixing, pelleting, crumbling, and packaging steps (Smith et al., 2008). These raw materials are fortified with essential vitamins, minerals, other nutrients, and specific feed additives to create balanced diets that support livestock growth and productivity. The availability of raw materials has increased in recent years, particularly due to enhanced maize cultivation and higher production levels of grains and oilseeds. Gilbert (2010) estimated global animal feed production to be approximately 1000 million tons annually, with 60% being compound feed, the majority of which is allocated to poultry. FAO (2003) projected a significant increase in grain demand, anticipating maize production to rise from 625 to 964 million tons between 2002 and 2030.

Advanced processing techniques in chicken diet manufacturing focus on improving feed quality, digestibility, and safety while maximizing efficiency. The process begins with grinding raw materials to optimize particle size for better nutrient absorption, followed by mixing to ensure uniform distribution of nutrients. Conditioning and pelletizing use steam and pressure to create durable pellets, enhancing feed intake and reducing waste. Sometimes, extrusion technology is applied for high-energy, easily digestible diets, while crumbling and sieving adjust pellet sizes for different broiler rearing stages. Post-pellet liquid application adds heat-sensitive nutrients like exogenous enzymes and probiotics after pelletizing them to maintain their effectiveness.

Extruders in chicken diet conditioning use high temperature, pressure, and mechanical shear to improve feed quality and safety before pelletizing. They enhance starch gelatinization and protein digestibility, leading to better nutrient absorption and improved feed conversion ratios. Extruders also reduce pathogens like *Salmonella*, deactivate anti-nutritional factors, and improve pellet shelf-life. They also allow higher fat incorporation, boosting energy density. While they increase energy costs, the benefits in feed efficiency, broiler performance, and hygiene make extruders cost effective in large-scale broiler chicken production.

Precision formulation software tailor nutrient content based on broiler growth stages and cost efficiency. Sustainability practices, such as using alternative protein sources (e.g. insect meal, algae and microalgae, single-cell proteins, and aquatic plant proteins) and energy-efficient production methods, further enhance modern feed manufacturing, balancing productivity with environmental responsibility. Factors influencing feed composition include raw material availability, price dynamics, regulatory guidelines, and market demand. Global brands, supported by strong research and development capabilities, dominate the feed industry, though regional producers continue to cater to local markets. As the demand for broiler chicken meat grows, the feed industry is expected to continue to expand, with significant investments in more sustainable and high-nutrition feed solutions. In 2023 the poultry feed sector grew by 3.5% with broiler feed accounting for 29.9% of the total global feed tonnage, amounting to 385.04 million metric tons (https://www.alltech.com/agri-food-outlook). As an example of local production, Sri Lanka's poultry feed production was projected to reach 1 220 381 metric tons (MT) in 2024, marking a 5% increase from the 1 165 427 MT produced in 2023.

## 2.4 Chicken meat processing plants

The highest growth in overall meat production is expected to be in broiler chicken meat. Global chicken meat production increased by 9% in 2022 compared to 2018, reaching a total of 100 thousand tons (USDA, 2022). By the end of 2050 the demand for poultry meat is estimated to double compared to the present (Farrell, 2013; Astill et al., 2020). The price of chicken meat is competitive compared to other meats, as raising 1 kg of broiler chicken is less expensive than for beef and pork (USDA, 2022). This is because broiler chicken production produces the highest yield per feed unit among terrestrial animals, making chicken meat a relatively inexpensive source of animal protein. The global poultry industry is projected to grow by 24% from an annual average of 107.6 billion kg of meat in 2012-2014 to 134.8 billion kg in 2024 (OECD, 2015). To meet the demands, the poultry industry has built large, dedicated processing plants with continuously increasing line speeds. Today, a single fast broiler processing line can deal with 15 000 chickens per hour, which is a six-fold increase over the line speed 60 years ago (Barbut, 2020). Other important innovations to increase line speed and efficiency include the conversion of the early batch-type scalding and defeathering process into a continuous line followed by the introduction of automation into the evisceration and cut-up and deboning lines (Fletcher, 2002).

New automated deboning machines have also been developed to produce high-value poultry cuts such as breast and thigh fillets. They incorporate x-ray technology to measure bone length during deboning, allowing about 3500 carcasses to be processed per hour. Breast fillet deboning machines also contain sensors that can determine the size of each breast portion and adjust the cutting parameters accordingly (Barbut, and Leishman, 2022).

It is now possible to use technologies such as ultrasound and computer tomography (CT) to non-destructively scan and measure body composition (including lean meat content) (Dewez et al., 2018). There are also now robotic technologies that can pick and arrange products in a package with a fixed weight, including a computer-controlled blade that can remove excess meat (Barbut and Leishman, 2022). Current grading systems in most countries stipulate mandatory downgrading of poultry due to the presence of feathers on the carcass. According to US regulations e.g., the presence of more than 4 feathers that are equal to or longer than a 1/2 inch change the grade from A to B, while >6 feathers result in a grade C classification (USDA, 2018).

Further processing steps have focused on developing innovative products for increased consumer convenience. Consumers prefer minimal involvement in food preparation and seek convenient, ready-to-eat options that can be consumed anytime and anywhere. They now spend only one-sixth of the time on food preparation compared to consumers 50 years ago (Barbut, 2020).

Chicken fillets are typically perceived by consumers to be a convenient choice due to their versatility, ease of preparation, and fast cooking time. A lean, skinless chicken breast fillet typically contains approximately 74% moisture, 22% protein, and approximately 2% fat. In comparison, chicken thigh meat has approximately 72% moisture, 18% protein, and 8% fat (USDA, 2022). There are some novel techniques to retain moisture in muscle tissues, such as marination of whole muscle meat cuts (Alvarado and McKee, 2007) and straight water addition to ground sausage meat batters (Gómez et al., 2019), where the added moisture can compensate for the losses during cooking. Another approach is to cook meat in liquid such as such as soup or gravy to ensure moisture surrounds the meat cut/portion (Huff-Lonergan and Lonergan, 2005; Warner, 2017).

Modern packaging for processed meat products (cooked, ready-to-eat) is designed to enhance stability by providing an oxygen barrier. This barrier is further strengthened through vacuum or modified atmosphere packaging, which helps slow down spoilage and lipid oxidation (Leory and Degreef, 2015). Most processed poultry products are sold in the form of vacuum-packed items which means consumers rely on visual appearance to guide purchasing decisions. Modern packaging offers re-sealable, microwave safe and family size/single serving options, sometimes with cooking suggestions included for further convenience.

In addition to modern packaging, various techniques are used in the industry to enhance meat preservation and extend shelf life. These include canning, drying, fermentation, salting, high-pressure processing, and smoking (Barbut, 2015; Leory and Degreef, 2015). To improve workforce training and safety, virtual reality (VR) training has been introduced, allowing employees to master complex processing techniques in a controlled, risk-free environment. At the end of processing, the meat is packaged and labeled before being distributed to retailers. Cold chain logistics are crucial here to maintain product freshness and prevent contamination. Recent innovations, such as blockchain technology, have strengthened supply chain transparency, enabling real-time tracking of products from the processing plant to the end consumer. This technology not only enhances traceability but also improves food safety compliance and builds consumer trust. These advancements in processing and distribution ensure the delivery of high-quality broiler chicken meat while meeting modern demands for efficiency, safety, and accountability (Barbut and Leishman, 2022).

## 2.5 Distribution, cold storage facilities and retail and food service operators

Figure 2 illustrates the growing importance of data integration across the broiler chicken supply chain, from breeding to retail. The role of digital platforms and

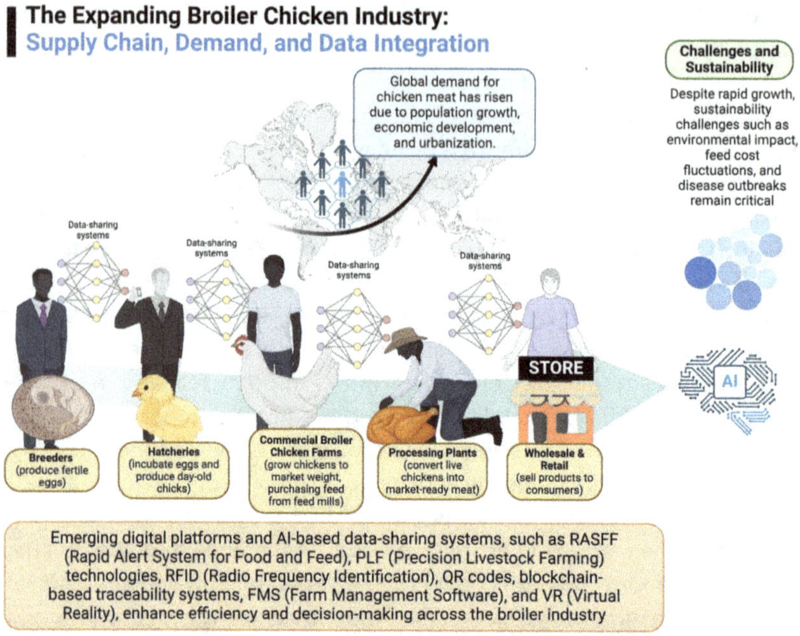


**Figure 2** Expanding broiler chicken industry: supply chain, demand, and data integration. Created in BioRender. Priyashantha, H. https://BioRender.com/p31h574.

AI-based systems such as RASFF, PLF, RFID, blockchain, and FMS is increasingly recognized for their potential to enhance health, welfare, traceability, and efficiency. However, the successful implementation of these systems depends on multiple factors. Currently, data sharing is fragmented, with most producers using isolated management systems and limited interoperability across platforms and stakeholders.

Key challenges include data ownership concerns, lack of standardization, cybersecurity issues, and reluctance to share sensitive operational data due to competition. For data sharing to be successful, it requires trust, robust legal frameworks, stakeholder incentives, and the development of interoperable platforms that can securely exchange and analyze data in real time. Collaborative governance models involving industry, academia, and regulators are essential to build a common infrastructure and culture for open, yet secure, data exchange. Moving forward, success will hinge on how well these systems can balance privacy and transparency, as well as demonstrate clear economic and welfare benefits to all participants in the chain.

Furthermore, the global broiler chicken industry is expanding rapidly due to rising demand for high-quality protein. Its supply chain, from breeders to retailers, relies on data integration and digital platforms to optimize production, forecast demand, disseminate information related to quality and safety, and ensure sustainability.

Once processed, poultry meat is distributed to retailers, markets, hotels, and restaurants through a cold supply chain, utilizing refrigerated containers designed to maintain specific temperature and humidity conditions critical for preserving meat quality and safety. During transportation, chilled poultry meat is maintained at temperatures between 0°C and 4°C, while frozen poultry requires temperatures of −18°C or lower to inhibit microbial growth and enzymatic activity. The duration of transport can range from a few hours for local deliveries to several weeks for international shipments, particularly in the case of frozen products that support year-round availability in global markets.

Maintaining product integrity throughout the supply chain necessitates continuous temperature and humidity monitoring using data loggers and IoT-based tracking systems, ensuring compliance with international food safety standards such as HACCP (Hazard Analysis Critical Control Point) and Codex Alimentarius guidelines. Sanitation protocols for transport containers, including regular disinfection and the use of antimicrobial agents, are essential to prevent cross-contamination. Additionally, proper training of logistics personnel, including transporters and handlers, is critical for ensuring adherence to cold chain protocols and minimizing temperature abuse during loading and unloading processes.

## 2.6 Variability in broiler chicken production across continents and countries

World meat production reached 357 million tons in 2021, up 53% to 124 million tons compared with 2000. There was 4% growth in production between 2020 and 2021, the fastest over the 2000–2021 period. Although many species are raised for meat, three main types (chicken, pork, and beef) accounted for nearly 90% of global production between 2000 and 2021. With 34% of global production in 2021, chicken meat has shown the largest growth in absolute and relative terms since 2000 (107%, or 63 million tons), making it the biggest meat product in 2021 (FAOSTAT, 2023). China and the United States are among the three largest producers of each main meat type. China alone accounted for 45% of world pig meat in 2021. The United States of America produced 17 to 18% of global chicken and cattle meat. The difference between the two countries is that meat production in China is mostly for the domestic market, whereas a significant share of American chicken meat production is exported (FAO, 2022).

The Americas are the center of world chicken production, with production reaching 49.5 million tons or 40.7% of the world's total production of chicken meat. Asia is second with production reaching 44.5 million tons representing around 36.3% of total world production. Third is Europe with production reaching 20 million tons representing 16.5% of total world production.

This global variability in broiler production highlights the increasing complexity of international supply chains and emphasizes the need for effective coordination among stakeholders to balance demand and supply. As production scales, infrastructure, and market priorities differ across countries and regions, data sharing becomes a critical enabler of consistency in quality, traceability, animal health, and welfare standards. Harmonized digital systems can bridge gaps between high- and low-capacity producers by enabling real-time insights, demand forecasting, and disease surveillance across the production chain. Integrated data platforms can help optimize feed usage, reduce production costs, and improve decision-making at farm, processing, and distribution levels. In this context, a globally interconnected data-sharing framework will be essential to ensure sustainability, food safety, and efficiency in the evolving broiler industry.

## 3 Levels of integration in the broiler chicken supply chain

Broiler production is time-efficient and adaptable to market changes given short poultry life cycles. The structure of broiler chicken production varies across countries, with a trend toward large-scale, vertically-integrated operations (Narroad et al., 2007). The broiler chicken production chain operates through multiple levels of integration to optimize efficiency, sustainability, and profitability, as illustrated in Fig. 3. The broiler industry operates through vertical integration, horizontal integration, forward and backward integration, and contract farming (Mottet and Tempio, 2017). Vertical integration streamlines the supply chain by consolidating different production stages under a single entity, reducing costs and enhancing coordination. Horizontal integration allows farms to collaborate and share processing infrastructure, improving resource utilization. Contract farming ensures stable partnerships between commercial broiler chicken farmers and integrators, facilitating market access and resource support. Technological integration, including AI and big data applications, enables data-driven decision-making, precision farming, and long-term sustainability. Together, these integration models contribute to a well-organized, efficient, and resilient broiler chicken production system, ensuring a steady supply of poultry meat to meet global demand.

Vertical integration in the chicken industry refers to a functional technique that helps lower the prices of chicken products, saving on time and cost. It involves combining different biosecurity and sanitation practices, housing technologies, and feeding regimens to improve food safety. Vertical integration in the broiler chicken supply chain involves producers collaborating to supply

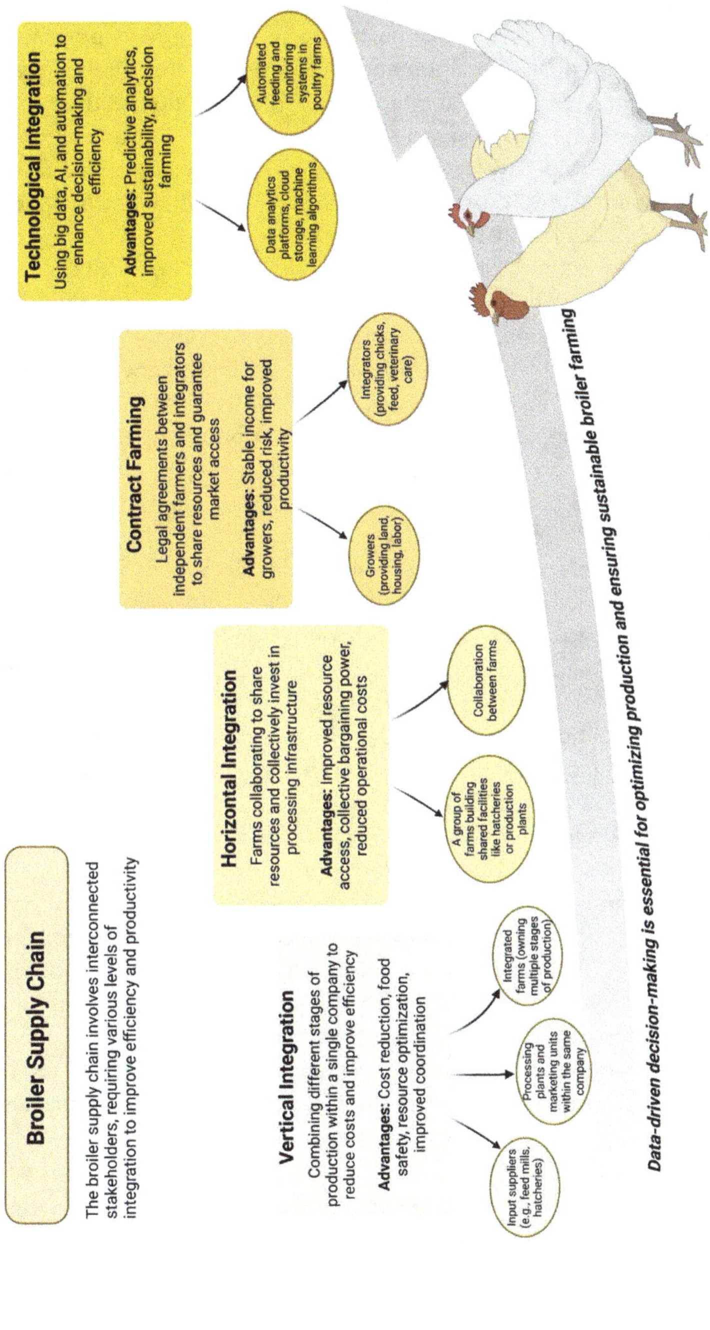


**Figure 3** Levels of integration in the broiler chicken industry. Created in BioRender. Priyashantha, H. https://BioRender.com/e84w007.

various products and services in a coordinated approach to facilitate broiler chicken meat production (National Chicken Council, 2023 available at www.nationalchickencouncil.org/industry-issues/vertical-integration/). It encompasses upstream processes such as day-old chick production, input supply, feed production, slaughterhouse management, processing, transportation, and marketing (Goodwin, 2005). Vertical integration is popular worldwide due to its ability to reduce costs, ensure food safety, and manage large flocks with advanced technology. Moreover, vertical integration enhances coordination among supply chain actors, promoting resource efficiency and sustainable industry growth (Brevik et al., 2020).

A significant part of the global broiler chicken industry operates through contract farming, where integrators and growers collaborate under legal agreements (Sheng and Vukina, 2024; Karaman et al., 2023). Integrators provide resources like chicks, feed, and medications, and technical know-how, while growers contribute land, housing, labor, and utilities (Doye et al., 2017). This collaborative model facilitates efficient chicken meat production, particularly benefiting farmers through repeated production cycles (6–7 times per year) and performance-based bonuses (Grepay, 2009). Contract farming can prove advantageous for developing countries with limited resources, as it supports SMEs (OECD, 2018). Compared to large-scale broiler chicken meat-producing operations, effective broiler meat production requires informed tactical and strategic decisions based on accurate data sharing (Blancoa et al., 2023). Farms must coordinate with hatcheries to determine batch sizes and plan operations efficiently. Feed price forecasting is crucial since broiler chicken rations typically consist of 20–30% soybean meal and 50–60% corn; any disruption in these commodities can significantly impact feed costs and overall production (Sheng and Vukina, 2024; Unveren and Luckstead, 2020).

Horizontal integration occurs when multiple farms jointly establish facilities like hatcheries and processing plants for mutual benefit. Additionally, technological integration offers substantial potential, with modern data collection, storage, and analysis tools enhancing decision-making and operational efficiency. Techniques such as machine learning and deep learning contribute to predictive models that improve profitability, reduce environmental impact, and support animal welfare (Franzo and Legnardi, 2023). Thus, effective integration of broiler chicken supply chain components, supported by advanced technology and data-driven strategies, ensures sustainable, efficient, and responsive production to meet global market demands.

Advanced technologies such as machine learning (ML) and deep learning (DL) are increasingly revolutionizing broiler production across countries like China, Belgium, the United States, and Italy by enabling predictive analytics that improve health monitoring, welfare management, and productivity. In China, AI-enabled IoT systems have been applied to monitor broiler behavior, vocalizations, and environmental factors such as ammonia and humidity levels, facilitating early detection of stress and disease and enabling timely interventions that enhance animal welfare (Ojo et al., 2022). In Canada and Europe, the development of digital phenotyping platforms integrating ML and DL tools supports real-time analysis of broiler growth, behavior, and environment, allowing for optimized feeding strategies, reduced mortality, and more efficient production (Neethirajan, 2023). Visual sensing technologies, particularly thermal and infrared cameras, have also been employed globally. For instance, FLIR Lepton cameras from the United States have been used for temperature monitoring (Bloch et al., 2020), while Dahua IPC-K22A (China) and NHD-818 (USA) cameras have been used to observe drinking and stretching behaviors in pullets and broilers, respectively (Li et al., 2020, 2021). In Italy and China, PRO-1080MSFB and Thermo GEAR-G120 cameras have been used to assess floor distribution and nest occupancy in poultry systems (Guo et al., 2020; Zaninelli et al., 2018). Studies in China using Hikvision cameras have demonstrated effective automation of broiler counting and detection of sick chickens (Cao et al., 2021; Okinda et al., 2019), reducing labor requirements and operational costs. These innovations contribute to the sustainable intensification of poultry farming by improving efficiency, minimizing environmental impact through optimized resource use, and ensuring animal welfare. As the application of ML and DL continues to expand globally, particularly in Asia, North America, and Europe, the importance of robust and interoperable data-sharing systems becomes increasingly evident for integrating insights across all segments of the broiler production chain.

Furthermore, Leat et al. (1998) highlighted that food traceability is vital in reassuring consumers about the origin and safety of food, identifying the source of contaminated or poor-quality products, ensuring compliance with labeling regulations, and aiding in disease control and residue monitoring. The implementation of well-regulated, structured standards has significantly improved food safety practices. In the poultry sector, including broiler chicken meat production, various information tracking and tracing systems have been developed.

Tracking and traceability systems have been established for broiler chicken production and supply chains. In poultry meat production, an integrated supply chain is utilized, with detailed recording and traceability systems implemented at the breeding, hatchery, and meat production stages. These systems focus on animal health and welfare, business performance, and food quality and safety standards (Manning et al., 2006). The types of data recorded and shared include flock health records (e.g. vaccination history, disease status), feed composition and usage, environmental conditions (e.g. temperature, humidity), growth performance metrics (e.g. body weight, feed conversion ratio), slaughter and processing details (e.g. carcass quality, microbiological testing), and transport logistics. Such data are typically shared across stakeholders, including breeders, integrators, processors, and regulators, via digital platforms or centralized databases, enabling real-time decision-making, rapid response to outbreaks, and improved transparency throughout the production chain.

Therefore, tracking and tracing are crucial at each production stage and for all stakeholder interactions. Traceability is the ability to track a product's journey throughout its production process, including the origin and use of raw materials, ingredients, additives, and packaging. It also encompasses critical information such as quality and safety specifications, equipment used, and processing conditions, as illustrated in Fig. 4. While traceability systems enhance transparency, quality control, and consumer trust, their implementation comes with challenges. These may include the complexity of data management, the need for standardized protocols, and the costs associated with advanced tracking technologies.

Guschin et al. (2017) highlighted the risks associated with inefficient data analysis systems, which can negatively impact the broiler chicken meat industry. Fortunately, modern IT-based solutions can mitigate these risks through digital information management. Cloud-based traceability systems offer advantages such as enhanced data reliability, faster response times, protection against data loss, easier maintenance and recovery, and greater system flexibility for future extensions and modifications (Valdokhina and Roiter, 2020). While tracking and traceability systems play a crucial role in ensuring poultry meat safety,

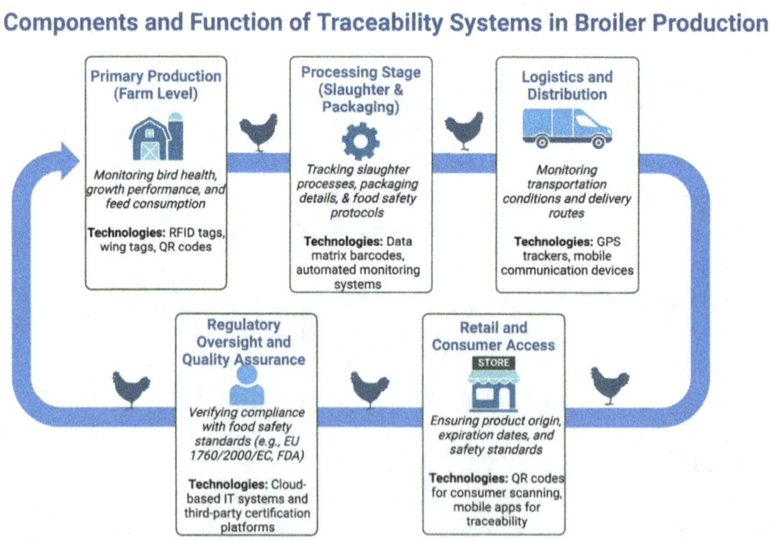


**Figure 4** Components and function of traceability systems in broiler chicken production. Created in BioRender. Priyashantha, H. https://BioRender.com/r01t491.

continuous technological innovation and process improvements are necessary to overcome current limitations and sustain consumer confidence in broiler meat products. Several methods for collecting information along the chicken production chain have emerged, including QR codes, Data Matrix barcodes, identification rings, wing tags, mobile communication technologies, and other IT-based systems. Tarjan et al. (2014) noted that QR codes are particularly well-suited for traceability systems in livestock and poultry products due to their high data storage capacity and flexible usability.

Inerney et al. (2018) has tested the feasibility of using direct laser-printed Data Matrix barcodes on poultry beaks as a practical method for identifying individual chickens, thus enhancing traceability within the food chain. This study found that, while all barcodes were fully readable immediately after printing, readability dropped significantly by day 2 due to loss of contrast caused by natural chick behaviors such as eating. Although applying non-toxic black ink temporarily restored visibility, barcode readability continued to decline over time due to the beak's natural growth and repair processes. Identification performance improved when two barcodes were used (one on each side of the beak), especially in laying hens (though not for broilers). At least one barcode remained readable up to day 7, ensuring reliable identification within that period. For mature laying hens, this extended timeframe for reliable identification could support secure animal movement, enhance producer and consumer confidence in traceability, and ultimately benefit the poultry industry by reinforcing food safety and origin assurance.

With technological advancements, mobile communication tools have become integral to tracking and collecting data from livestock and poultry operations. Personal Digital Assistants (PDAs) are now commonly used for real-time data entry and information sharing (Yu et al., 2018). Furthermore, Chen et al. (2010) developed a standardized system for broiler chicken production, slaughtering, and processing, incorporating third-party certification technology to monitor production activities and ensure product quality. This system also supports a traceability platform that addresses the interests of enterprises, governments, and consumers.

In the twenty-first century, AI technology has gained prominence. Neethirajan (2022) explored its application in identifying individual broiler chickens. AI technology can reorganize traceability processes by automating data collection, analysis, and monitoring tasks across the poultry supply chain. Despite the availability of diverse tracking and tracing methods, some limitations persist. Trienekens et al. (2012) argued that many existing systems fail to maintain the integrity of information and the transparency needed for food quality and safety. To address these shortcomings, Kassahun et al. (2014) proposed a cloud-based service model, suggesting that such systems could improve information sharing and enhance traceability within the food sector.

Updated frameworks like ISO 22000:2018 and HACCP now incorporate real-time monitoring tools, such as IoT sensors and automated data logging to improve accuracy at critical control points in the broiler chicken meat supply chain. Global Food Safety Initiative (GFSI)-recognized standards like Brand Reputation Compliance Global Standards (BRCGS), Food Safety System Certification (FSSC) 22000, and Safe Quality Food (SQF) play a crucial role in ensuring food safety and quality within the broiler chicken industry. These standards emphasize supply chain transparency, enabling full traceability from farm to fork, which is essential in monitoring broiler health, feed quality, and processing conditions. In the broiler chicken sector, allergen management is particularly important during processing stages, where cross-contamination with common allergens can occur in facilities handling multiple products. Additionally, these standards focus on food fraud prevention, safeguarding against issues like mislabeling, adulteration, tampering, or substitution of chicken products, which are critical concerns in global poultry trade. Compliance with GFSI-recognized certifications not only enhances food safety and consumer trust but also facilitates access to international markets by aligning broiler chicken operations with global regulatory requirements.

Blockchain technology has emerged as a key tool for improving traceability in the broiler chicken industry, offering immutable records that enhance transparency, strengthen consumer trust, and simplify recall processes. By recording every stage of the supply chain from breeding, feeding, and vaccination to processing, packaging, and distribution, this would enable real-time tracking of broiler chickens and their products. This ensures that in cases of contamination or disease outbreaks, affected batches can be quickly identified and removed from the market, reducing food safety risks and economic losses. Enhanced frameworks like Threat Assessment and Critical Control Points (TACCP) and Vulnerability Assessment and Critical Control Points (VACC) also address intentional adulteration and supply chain vulnerabilities, ensuring holistic protection across the broiler chicken supply chain.

Figure 4 illustrates the key components and technological functions of traceability systems across the entire broiler production chain, providing a comprehensive view of how data integration supports transparency, food safety, and consumer confidence with respect to broiler production. At the primary production level, novel AI-based technologies ensure accurate on-farm data collection. This data flows into the processing stage, where automated monitoring systems and data matrix barcodes are used to track slaughter conditions, packaging details, and food safety compliance. In the logistics and distribution segment, GPS trackers and mobile communication devices play a critical role in maintaining cold chain integrity and real-time transport monitoring. The system extends to the retail level, where consumers can access product origin, expiration dates, and safety data through QR codes and mobile apps. Importantly, regulatory oversight is maintained through cloud-based IT platforms that verify compliance with national and international food safety standards (e.g. EU 1760/2000/EC, FDA). The closed-loop design of the figure emphasizes how traceability is not a linear process but a continuous cycle that loops back to producers, allowing for rapid response in the event of food safety breaches or quality concerns. This integrated system shows the central role of data sharing and digital technologies in building resilient, transparent, and consumer-focused broiler production chains.

Growing consumer concerns about food-related risks have further heightened the importance of traceability as a tool for building trust (Van et al., 2010). Chemical and physical hazards must also be monitored and traced to ensure food safety, as illustrated in Fig. 5. Chemical contamination risks during broiler chicken meat production may stem from pesticides, biocides, pest

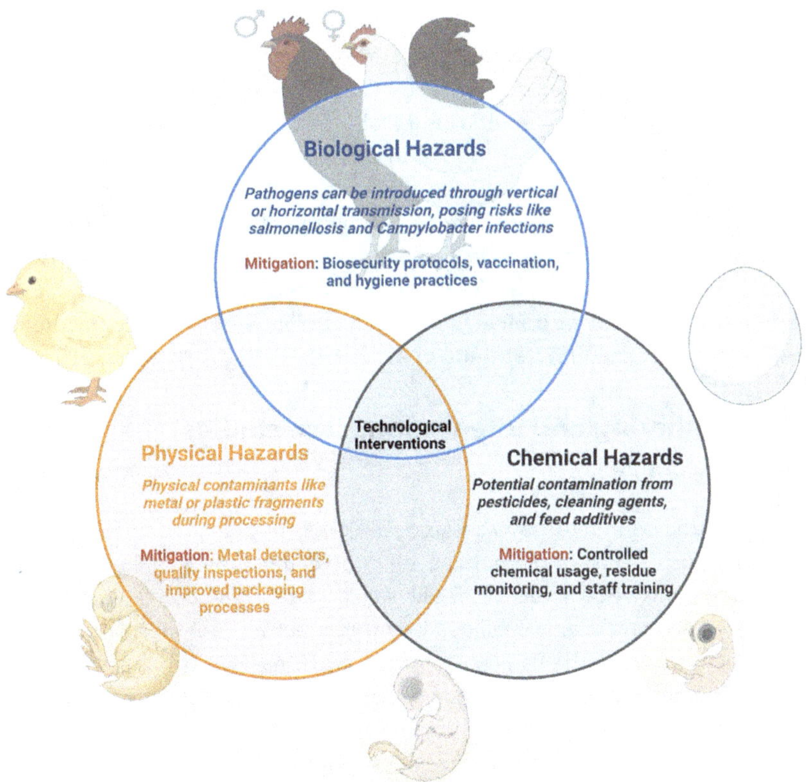


**Figure 5** Food safety hazards and mitigation strategies in broiler production. Advanced technologies like AI-based tracking systems further enhance safety by improving data accuracy and traceability throughout the supply chain. Created in BioRender. Priyashantha, H. https://BioRender.com/t30d135.

control substances such as insecticides and rodenticides, as well as cleaning agents and wood treatment chemicals. Physical hazards can arise at various points in the supply chain and, depending on the processes involved, may be detected and removed during slaughter, processing, or food production stages (Manning et al., 2005). Therefore, implementing necessary measures to minimize contamination risks is crucial, as food safety, particularly meat and egg safety, has become a key factor influencing consumer behavior and purchasing decisions (Viaene and Verbeke, 1998). The sustainability of the broiler meat industry depends heavily on maintaining high-quality standards and ensuring consumer confidence.

Several international organizations and authorities have established standards to ensure the safety and quality of broiler chicken meat as a food product. Producers and sellers are responsible for maintaining these standards to guarantee the quality of the meat they provide for consumers. The original EU food hygiene directive (EU, 1993) was a horizontal regulation that emphasized the use of hazard analysis and risk assessment to manage food safety risks. The guidelines also referred to the Codex Alimentarius General Principles of Food Hygiene, (Codex Alimentarius Commission, 2003) aligning with the international food safety standards established by the Codex Alimentarius Commission to ensure the production of safe and wholesome food (FAO/WHO, 1997). Additionally, the European Union (EU) implemented regulations (1760/2000/EC, 1825/2000/EC) requiring all member states to identify food-producing animals, including both flocks and individual animals (Girish and Barbuddhe, 2020; Zhao et al., 2020). In the United States, the Food and Drug Administration (FDA) introduced similar traceability regulations to ensure food safety (Smith et al., 2005; Dabbene et al., 2014).

## 4 Data sharing and integration: case studies

### *4.1 Case study 1*

In 2003, Buhr conducted a case study on six European organizations in the poultry industry that implemented electronic traceability systems, tracking products from live animals to retail sales. Data collection involved USDA Foreign Agricultural Service offices in the respective countries, supported by organizations such as the Poultry, Livestock, and Meat Board in the Netherlands, various French government agencies, Carrefour Supermarkets in France, and ASDA Supermarkets in the UK.

The study aimed to document supply-chain production protocols, examine governance structures for traceability, and outline traceability methods. Site visits and data collection were conducted at feed mills, farms, processing plants, and retail shops, with interviews held with key personnel at each stage. The findings revealed that each case had unique protocols focused on consumer product attributes, such as organic, group-housing, free-range, or antibiotic-free production to ensure food safety. These protocols specified production inputs like feed, health treatments, and animal-rearing methods, with third-party auditing conducted at all production stages using available records. Monitoring processes included sample collection of feces, feed, and meat, cross-referencing feed delivery times, tracking daily weight gain, and site visits by certified veterinary or farm-management services. The study concluded that all partnerships operated under vertically integrated organizational structures to manage and implement production and data-sharing systems effectively.

## 4.2 Case study 2

Shankar and Thangam (2019) highlighted the need for software-managed data analysis and integration in broiler production systems to facilitate timely actions and accurate predictions. Their findings emphasized that ensuring data quality, preprocessing, and automation could support logical and efficient decision-making. The researchers proposed a framework for poultry data integration to improve prediction, analysis, and decision-making for increased profitability. This framework involves the collection and integration of diverse on-farm data types, including environmental parameters (e.g. temperature, humidity, ammonia levels), production data (e.g. feed intake, body weight, growth rate, mortality), and health and welfare indicators (e.g. behavior patterns, disease occurrence, vaccination status). By consolidating these datasets, the system enables real-time monitoring, early warnings for health issues, and predicted supply and optimized resource management across the poultry production cycle.

Integrated organic farming systems, which incorporate activities like residue recycling, bio-intensive cropping, horticulture, and poultry farming, offer benefits such as cost reduction, improved soil fertility, enhanced food quality, and the development of value-added products (Layek et al., 2023; Agricial, 2023). Yang et al. (2024) noted the growing application of technologies like Meta AI SAM, OpenAI ChatGPT, robotics, and cybernetics to support sustainable poultry production. However, challenges such as resource constraints, inadequate coverage, and service quality issues underscore the need for improved education and extension services for marginalized farmers. Further research is required to integrate these emerging technologies into existing operations to promote animal welfare and environmental sustainability (Bist et al., 2024; Yang et al., 2024; Zhou et al., 2024).

Advanced technologies such as machine learning models and satellite imagery offer promising tools for accurate emissions monitoring and reduction. Predictive models can forecast emissions trends and identify key drivers, providing insights for policymakers and industry stakeholders (Patole, 2021). Various models such as Auto Integrated Moving Average (ARIMA), Long Short-Term Memory (LSTM), and eXtreme Gradient Boosting (XGBoost) have demonstrated effectiveness in predicting time-series data like greenhouse gas emissions. ARIMA models excel at short-term forecasting but struggle with nonlinear relationships and long-term dependencies (Christophorus, 2021). LSTM models, a type of recurrent neural network, can capture long-term dependencies and outperform ARIMA in forecasting complex, nonlinear time-series data (Siami-Namini et al., 2018). XGBoost, with its ensemble learning approach, effectively handles diverse datasets and models complex variable interactions (Sharma, 2024). Integrating machine learning models with satellite platforms like NASA's Giovanni, Sentinel-5P, and OCO-2 enables high-resolution monitoring of methane ($CH_4$) and carbon dioxide ($CO_2$) emissions in broiler chicken production. A study by Jobarteh and Neethirajan (2025) analyzing emissions from 1300 Canadian poultry farms between 2019 and 2023 found significant regional and seasonal variability influenced by farm density and climatic conditions. The LSTM model outperformed ARIMA and XGBoost in forecasting emissions, highlighting the potential of advanced algorithms to track emissions trends accurately. This integration enhances real-time monitoring, supports the development of region-specific mitigation strategies, and contributes to the poultry sector's climate change adaptation and sustainability goals. These systems enable precise tracking of emissions over time and across regions, supporting Canada's climate goals and promoting sustainability in broiler production.

Large-scale broiler chicken meat production operations often adopt vertically integrated models with centralized data management, while medium- and small-scale contract broiler farming relies on collaborative partnerships to access resources and market opportunities (Fig. 6). Effective data sharing across both scales enhances production efficiency, product traceability, and overall sustainability in the broiler chicken industry. In Sri Lanka, the broiler chicken industry has seen remarkable growth in recent years, driven by the adoption of modern farming techniques, including the shift from traditional open-house systems to environmentally controlled closed-house systems, which enhance biosecurity, optimize feed conversion, and improve broiler welfare. According to Priyashantha and Vidanarachchi (2024), this growth is attributed to the increased demand for affordable animal protein and the strategic role of broiler chicken farming in providing technical support, inputs, and guaranteed market access for farmers. The integration of digital tools and real-time data sharing within these contract farming arrangements has further contributed to improving productivity, resource efficiency, and meat quality across the industry.

**Figure 6** (a1) and (a2) represent large-scale broiler meat production operations, while (b1) and (b2) represent medium- and small-scale contract broiler farming, similar to the structure observed in contract grower or buy-back broiler chicken industry, where integrated companies collaborate with contract farmers to meet growing market demands.

## 5 From data sharing to better products for consumers

Vanhonacker et al. (2009, 2012) highlighted that developing an intermediate animal welfare-based certification scheme can contribute to safer poultry products for consumers. Vander Naald and Cameron (2011) observed that consumers are willing to pay a premium for welfare-enhanced meat products, particularly when they trust that 'humanely raised' standards genuinely improve animal welfare. The study further recommended the establishment of an independent certifying body to regulate and enforce welfare standards. These standards should encompass key aspects such as access to veterinary care, nutritionally balanced feed, adequate shelter with proper ventilation, sufficient space for natural behaviors, access to pasture, protection from extreme weather, prohibition of unnecessary physical alterations and electric shocks, maintenance of natural light cycles, and continuous access to food and water.

According to Martinez (2002), contract farming and vertical integration improve production efficiency, ensure consistent product quality, and enable responsiveness to consumer demand for a variety of value-added products. Additionally, the availability of cheaper, more convenient products such as cut-up parts, deboned, pre-cooked, and value-added items benefits consumers.

Countries like Thailand, which were severely affected by Avian Influenza, revitalized their broiler chicken industry by producing heat-treated chicken products and ready-made broiler meals as part of a long-term strategy to ensure high-quality, farm-to-table products (Verbiest and Castillo, 2004). As a key importer and exporter of broiler chicken products, the European Union produces and exports whole chickens, low-value cuts, and mechanically deboned broiler chicken meat while importing prepared and processed poultry products. Countries like Poland, Hungary, and the Czech Republic produce competitively priced, high-quality poultry products due to the low cost of land and labor, along with relatively lower expenses in maintaining environmental and animal welfare standards, which attracts investment to the sector. Additionally, cereal subsidy payments are expected to increase the availability of low-cost feed grains for Hungarian and Polish producers, boosting production. These countries often purchase lower-quality broiler chicken cuts from EU-15 nations for further processing, encouraging industry growth (Smith et al., 2005). Japan predominantly imports high-quality prepared chicken products, whereas Russia mainly imports lower-value whole chickens. China, as a significant importer and exporter, focuses on different product categories. Russia has also developed new markets for value-added fresh and frozen poultry products (O'Keefe, 2004).

## 6 Future trends and recommendations

A framework for the future development of the poultry industry through innovative educational platforms, participatory learning, supportive government policies, targeted training programs, and the integration of advanced technologies is illustrated in Fig. 7. Online tools and experiential learning enhance farmers' knowledge and skills, while government involvement ensures efficient market operations. Educational initiatives covering biosecurity, hygiene, disease control, and welfare improve industry standards. The adoption of robotics and automation technologies further streamlines operations, reduces labor demands, and increases production efficiency, contributing to a more sustainable and competitive poultry industry.

To advance sustainable poultry farming, enhancing accessibility and bridging the knowledge gap are essential. Future education and extension services should prioritize digital platforms (as shown in Fig. 7) to deliver information, training, and interactive communication conveniently and cost-effectively. Participatory and experiential learning approaches, such as field visits, workshops, demonstrations, and farmer-to-farmer networks, can provide practical experiences that develop critical thinking and problem-solving skills. Monitoring tools like surveys, interviews, focus groups, and performance indicators should be employed to assess farmers' and extension agents' knowledge acquisition, behavioral changes, and overall performance.

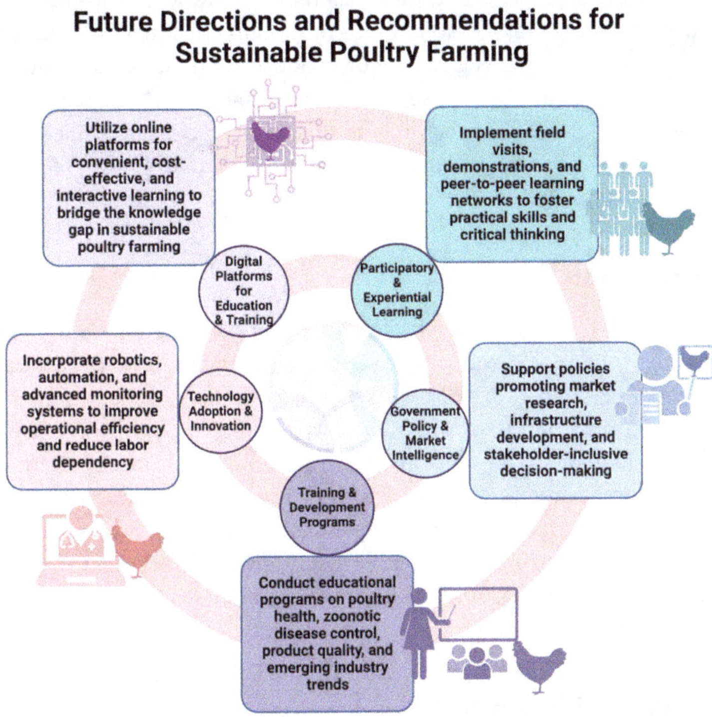


**Figure 7** Future directions and recommendations for sustainable poultry farming. Created in BioRender. Priyashantha, H. (2025) https://BioRender.com/e41t756.

Governments must create an environment that supports productivity improvements across the supply chain. This can be achieved by collecting and disseminating market intelligence, supporting market research, upgrading marketing infrastructure, and reforming input markets (Chang, 2003). Additionally, policy-focused research should adopt a participatory, stakeholder-inclusive approach to build trust and develop integrated, mutually agreed policies.

Educational programs should target farmers, laborers, veterinarians, and agricultural professionals, focusing on:

- biosecurity, hygiene, and disease control;
- genetic and breeding innovations;
- viral diseases and emerging threats;
- zoonotic disease control and farmer health;
- poultry husbandry, behavior, welfare, and ethical farming;
- poultry product quality, hygiene, and industry updates;
- sustainable eco-friendly practices; and
- consumer-driven market shifts.

These programs can strengthen scientific knowledge, refresh existing skills, and promote safer, more efficient poultry production. Technological advancements like robotics and automation such as automated feeding systems, robotic cleaning, and sophisticated monitoring tools can significantly enhance operational efficiency while reducing labor demands (Licardo, 2024). Integrating these innovations will help modernize broiler farming and improve industry sustainability.

## 7 Conclusion

In conclusion, the sustainability of the broiler chicken industry relies on continuous knowledge-sharing, innovative practices, and proactive policy support achieved through good data-sharing practices. Accessible education, digital tools, and hands-on learning opportunities empower stakeholders to adopt efficient, safe, and responsible farming practices. Integrating modern technologies further enhances productivity and competitiveness. A collaborative approach uniting policymakers, industry players, and farming communities will ensure a resilient, sustainable poultry sector capable of meeting future demands while safeguarding animal welfare, human health, and environmental integrity.

## 8 Conflict of interest

The author declares that the article was compiled in the absence of any commercial or financial relationships that could be construed as a potential conflict of interest.

## 9 Acknowledgements

No specific funding, assistance, or collaborations were involved in the preparation of this manuscript.

## 10 Funding

This study did not receive any external grants.

## 11 Data availability statement

The data that support the findings of this study are available from the corresponding authors upon reasonable request.

## 12 References

Abdullah, A.Y. et al. 2010. Growth performance, carcass and meat quality characteristics of different commercial crosses of broiler strains of chicken. *The Journal of Poultry Science*, 47(1), pp.13–21.

Agricial 2023. Organic poultry farming an overview of methods, benefits. Agricial.com https://agricial.com/organic-poultry-farming-methods-benefits/ (Accessed 3 September 2023).

Alvarado, C. and McKee, S. 2007. Marination to improve functional properties and safety of poultry meat. *Poultry Science*, 16, 113–120.

Astill, J., et al. 2020. Smart poultry management: smart sensors, big data, and the internet of things. *Computers and Electronics in Agriculture*, 170, p.105291.

Barbut, S. and Leishman, E.M. 2022. Quality and processability of modern poultry meat. *Animals*, 12(20), p.2766.

Barbut, S. 2015. Chapter 11–Heat Processing, Cooling and Preservation Methods. Free Download. www.poultryandmeatprocessing.com (Accessed 1 January 2022).

Barbut, S. 2020. Meat industry 4.0: A distant future? *Animal Frontiers*, 10, 38–47.

Berckmans, D. 2017. General introduction to precision livestock farming. *Animal Frontiers*, *7*(1), pp.6–11.

Bist, R.B. et al. 2024. Sustainable poultry farming practices: a critical review of current strategies and future prospects. *Poultry Science*, 103, p.104295.

Bloch, V. et al. 2020. Automatic broiler temperature measuring by thermal camera. *Biosystems Engineering*, 199, pp.127–134. https://doi.org/10.1016/j.biosystemseng.2019.08.011.

Brevik, E. et al. 2020. Optimisation of the broiler production supply chain. *International Journal of Production Research*, 58(17), pp.5218–5237.

Brown, S.E. et al. 2018. Evidence for the impacts of agroforestry on agricultural productivity, ecosystem services, and human well-being in high-income countries: a systematic map protocol. *Environmental Evidence*, 7, pp.1–16.

Cao, L. et al. 2021. Automated chicken counting in surveillance camera environments based on the point supervision algorithm: LC-DenseFCN. *Agriculture*, 11(6), 493.

Castellini, C. et al. 2006. Comparison of two chicken genotypes organically reared: oxidative stability and other qualitative traits of the meat. *Italian Journal of Animal Science*, 5(1), pp.29–42.

Chang, H.J. ed. 2003. *Rethinking development economics* (Vol. 1). Anthem Press.

Chang, C.C. et al. 2010. Traceability platform design of production monitoring and products quality for broilers industry technology system. *Chinese Society of Agricultural Machinery*, 41(8), pp.100–106.

Christophorus, B.A.S. 2021. ARIMA model for accurate time series stocks forecasting et al time series analysis and forecasting of coronavirus disease in Indonesia using ARIMA model and PROPHET. *Procedia Computer Science*, 179(12), pp.524–532. https://doi.org/10.1016/j.procs.2021.01.036.

Clements, M. 2022. Top world broiler and egg rankings 2022. WATT Poultry International, October 2022: 6–51. https://www.poultryinternational-digital.com/poultryinternational/October_2022 (Accessed 18 February 2025).

Codex Alimentarius Commission 2003. *General principles of food hygiene (CAC/RCP 1-1969, Rev. 4-2003)*. Rome: Food and Agriculture Organization of the United

Nations and World Health Organization. Available at: https://www.isdi.org/wp-content/uploads/2020/04/CAC-RCP-1-1969.pdf (Accessed 29 March 2026).

Cui, X. et al. 2021. A systematic review and meta-analysis of children with coronavirus disease 2019 (COVID-19). *Journal of Medical Virology*, 93(2), pp.1057–1069.

Dabbene, F. et al. 2014. Traceability issues in food supply chain management: a review. *Biosystems Engineering*, 120, pp.65–80.

Davis, G.B. and Olson, M.H. 1984. *Management information systems: conceptual foundations, structure, and development*. McGraw-Hill, Inc.

Davis, G.S. et al. 2018. Antibiotic-resistant Escherichia coli from retail poultry meat with different antibiotic use claims. *BMC Microbiology*, 18, pp.1–7.

De Jong, I. et al. 2012. Scientific report updating the EFSA opinions on the welfare of broilers and broiler breeders. *EFSA Supporting Publications*, 9(6), p.295E.

Dewez, M.F. et al. 2018. Use of computed tomography to determine body composition of heavy strain turkey hens (Meleagris gallopavo) from rearing to early laying. *Poultry Science*, 97, pp.4093–4106.

Dohlman, E. et al. 2022. *USDA agricultural projections to 2031* (OCE-2022-1). U.S. Department of Agriculture, Office of the Chief Economist. https://www.usda.gov/oce/ag-projections.

Doye, D. et al. 2017. *Broiler production: considerations for potential growers*. Oklahoma State University Extension. https://extension.okstate.edu/fact-sheets/broiler-production-considerations-for-potential-growers.html.

European Council 1993. *Council Directive 93/43/EEC of 14 June 1993 on the hygiene of foodstuffs*. Official Journal of the European Communities, L 175, 19 July, pp.1–11. Available at: http://data.europa.eu/eli/dir/1993/43/oj (Accessed: 29 March 2026).

FAO 2003. World agriculture: towards 2015/30. An FAO perspective, edited by J. Bruinsma. Rome.

FAO 2022. Trade: Crops and livestock products. In: FAOSTAT. [Rome. Cited October 2023]. https://www.fao.org/faostat/en/#data/TCL.

FAOSTAT 2023. Food and Agriculture Organization of the United Nations (FAO), Crops and livestock products. https:// www.fao.org/faostat/en/#data/QCL.

Farrell, D. 2013. The role of poultry in human nutrition. In *Poultry Development Review*; Food and Agriculture Organization: Rome, Italy, pp.2–9.

Fletcher, D.L. 2002. Poultry meat quality. *World's Poultry Science Journal*, 58, pp.131–145.

Food and Agriculture Organization of the United Nations and World Health Organization 1997. *Report of the Twenty-Second Session of the Joint FAO/WHO Codex Alimentarius Commission (ALINORM 97/37), Geneva, 23–28 June 1997*. Rome: FAO/WHO.

Franzo, G. et al. 2023. When everything becomes bigger: big data for big poultry production. *Animals*, 13(11), p.1804.

Getiso, A. and Mijena, D. 2024. Trend Analysis of Poultry production, population growth and distribution in Ethiopia. *Journal of Nutrition and Food Processing*, 7(7), pp.1–8.

Gilbert, C.L. 2010. How to understand high food prices. *Journal of Agricultural Economics*, 61(2), pp.398–425.

Girish, P.S. and Barbuddhe, S.B. 2020. Chapter 10–meat traceability and certification in meat supply chain. In A.K. Biswas & P.K. Mandal (Eds.), *Meat quality analysis* (pp.153–170). Academic Press. https://doi.org/10.1016/B978-0-12-819233-7.00010-0.

Gómez, I. et al. 2019. Physicochemical and sensory properties of sous vide meat and meat analog products marinated and cooked at different temperature-time combinations. *International Journal of Food Properties*, 22, 1693–1708.

Goodwin, H.L. 2005. Location of production and consolidation in the processing industry: the case of poultry. *Journal of Agricultural and Applied Economics*, 37(2), pp.339–346.

Grepay, N. 2009. The main factors affecting poultry production in Libya. *Acta Scientiarum Polonorum. Oeconomia*, 8(4), pp.43–49.

Guo, Y. et al. 2020. A machine vision-based method for monitoring broiler chicken floor distribution. *Sensors*, 20(11), 3179.

Gushchin, A.V. et al. 2017. Use of open electronic courses in educational activity. *European Research Studies Journal*, 20 (Special issue), 541–548.

Herforth, A. et al. 2020. *Cost and affordability of healthy diets across and within countries: Background paper for The State of Food Security and Nutrition in the World 2020*. FAO Agricultural Development Economics Technical Study No. 9 (Vol. 9). Food & Agriculture Org.

Inerney, B. et al. 2011. Preliminary in vivo study on the potential application of a novel method of e-tracking to facilitate traceability in the poultry food chain. *Computers and Electronics in Agriculture*, *77*(1), pp.1–6.

Jennison, R. 2021. 'Hatchery practice', In Barrow, P., Nair, V., Baigent, S., Atterbury, R. and Clark, M. (eds.) *Poultry health: A guide for professionals*. Wallingford: CABI, pp.72–78.

Jobarteh, B. and Neethirajan, S. 2025. Leveraging satellite data for greenhouse gas mitigation in Canadian poultry farming. *Smart Agricultural Technology*, 10, 100704. https://doi.org/10.1016/j.atech.2024.100704.

Kamilaris, A. et al. 2019. The rise of blockchain technology in agriculture and food supply chains. *Trends in food science & technology*, 91, pp.640–652.

Karaman, S. et al. 2023. Profitability and cost analysis for contract broiler production in Turkey. *Animals*, 13(13), p.2072.

Kassahun, A. et al. 2014. Enabling chain-wide transparency in meat supply chains based on the EPCIS global standard and cloud-based services. *Computers and Electronics in Agriculture*, 109, pp.179–190.

Kuehler, C. 1993. Artificial incubation of exotic or "non-domestic" bird eggs. *AFA Watchbird*, 20(2), pp.37–46.

Layek, J. et al. 2023. An integrated organic farming system: innovations for farm diversification, sustainability, and livelihood improvement of hill farmers. *Frontiers in Sustainable Food Systems*, 7, p.1151113.

Leat, P. et al. 1998. Quality assurance and traceability-the Scottish agri-food industry's quest for competitve advantage. *Supply Chain Management: An International Journal*, 3(3), pp.115–117.

Leinonen, I. and Kyriazakis, I. 2016. How can we improve the environmental sustainability of poultry production?. *Proceedings of the Nutrition Society*, 75(3), pp.265–273.

Leroy, F. and Degreef, F. 2015. Convenient meat and meat products. Societal and technological issues. *Appetite*, 94, 40–46.

Li, G. et al. 2020. Assessment of layer pullet drinking behaviors under selectable light colors using convolutional neural network. *Computers and Electronics in Agriculture*, 172, p.105333.

Li, G. et al. 2021. Automated measurement of broiler stretching behaviors under four stocking densities via faster region-based convolutional neural network. *Animal*, 15(1), p.100059.

Licardo, J.T. et al. 2024. Intelligent robotics–a systematic review of emerging technologies and trends. *Electronics*, 13(3), 542. https://doi.org/10.3390/electronics13030542.

Lord, S. 2024. *Data for background report 2024 - FAO the state of food and agriculture*. University of Oxford. https://ora.ox.ac.uk/objects/uuid:cb692ac9-aeb3-49df-98da-187a23f68a85.

Maharjan, P. et al. 2022. Physiological growth trend of current meat broilers, and dietary protein and energy management approaches for sustainable broiler production. *AFMA Matrix*, 31(1), pp.29–35.

Manning, L. et al. 2006. Food safety management in broiler meat production. *British Food Journal*, 108(8), pp.605–621.

Martinez, S. 2002. Vertical coordination of marketing systems: lessons from the poultry, egg, and pork industries. http://www.ers.usda.gov/publications/aer807fm.pdf.

Mottet, A. and Tempio, G. 2017. Global poultry production: current state and future outlook and challenges. *World's Poultry Science Journal*, 73(2), pp.245–256.

Muir, W.M. and Cheng, H.W. 2014. Genetic influences on the behavior of chickens associated with welfare and productivity. In T. Grandin & M. J. Deesing (Eds.), *Genetics and the behavior of domestic animals* (pp.463–506). Academic Press.

Narrod, C. et al. 2008. *Technology transfer, policies, and the role of the private sector in the global poultry revolution*. IFPRI Discussion Paper 00841. The International Food Policy Research Institute (IFPRI).

Neethirajan, S. 2022. Automated tracking systems for the assessment of farmed poultry. *Animals*, 12(3), p.232.

Neethirajan, S. 2023. Digital phenotyping: a game changer for the broiler industry. *Animals*, 13, p.2585. https://doi.org/10.3390/ ani13162585.

OECD 2015. OECD-FAO agricultural outlook 2015. http://dx.doi.org/10.1787/agr_outlook-2015-en.

OECD 2018. OECD statistics. https://stats.oecd.org/.

Ojo, R.O. et al. 2022. Internet of Things and Machine Learning techniques in poultry health and welfare management: A systematic literature review. *Computers and Electronics in Agriculture*, 200, p.107266.

Oke, O.E. et al. 2024. Climate change and broiler production. *Veterinary Medicine and Science*, 10(3), p.e1416.

O'Keefe, T. 2004. Russia's poultry industry is adapting to the free market. *Poultry International*, 43(7), pp.20–28.

Okinda, C. et al. 2019. A machine vision system for early detection and prediction of sick birds: a broiler chicken model. *Biosystems Engineering*, 188, pp.229–242.

Patole, S.C. 2021. Time series forecasting of methane emissions from livestock using machine learning (Doctoral dissertation, Dublin, National College of Ireland). https://norma.ncirl.ie/5210/1/shambhuchandrakantpatole.pdf (Accessed 4 June 2024).

Pollock, D.L. 1999. A geneticist's perspective from within a broiler primary breeder company. *Poultry Science*, 78(3), pp.414–418.

Priyashantha, H. and Vidanarachchi, J.K. 2024. Perspectives on value chain transformation towards resilient animal-source food systems in Sri Lanka. *Cogent Food & Agriculture*, 10(1). https://doi.org/10.1080/23311932.2024.2384329.

Redman, T. 1995. Improve data quality for competitive advantage. *Sloan Management Review*, 36(2), pp.99–107.

Scanes, C.G. 2007. The global importance of poultry. *Poultry Science*, 86(6), p.1057.

Schippers, J. et al. 2025. Bargaining powers in cooperative Carbon Dioxide Removal deployment. *Climate Policy*, 25, pp.1–16.

Shankar, S. and Thangam, S. 2019. Data Integration and Management in Indian Poultry Sector. In *Ubiquitous Communications and Network Computing: Second EAI International Conference, Bangalore, India, February 8–10, 2019, Proceedings 2* (pp.96–102). Springer International Publishing.

Sharma, V. et al. 2024. FL-XGBTC: federated learning inspired with XG-boost tuned classifier for YouTube spam content detection. *International Journal of System Assurance Engineering and Management*, 15(10), pp.4923–4946.

Sheng, Q. and Vukina, T. 2024. Public communication as a mechanism for collusion in the broiler industry. *Review of Industrial Organization*, 64(1), pp.57–91.

Siami-Namini, S., et al., 2018, December. A comparison of ARIMA and LSTM in forecasting time series. In 2018 17th IEEE international conference on machine learning and applications (ICMLA) (pp. 1394–1401). https://doi.org/10.1109/ ICMLA.2018.00227. Orlando, FL, USA.

Smith, G.C. et al. 2005. Traceability from a US perspective. *Meat Science*, 71(1), pp.174–193.

Smith, J. et al. 2023. Fourth report on chicken genes and chromosomes 2022. *Cytogenetic and Genome Research*, 162(8–9), pp.405–528.

Smith, P. et al. 2008. Greenhouse gas mitigation in agriculture. *Philosophical Transactions of the Royal Society B: Biological Sciences*, 363(1492), pp.789–813.

Solano-Blanco, A.L. et al. 2023. Production planning decisions in the broiler chicken supply chain with growth uncertainty. *Operations Research Perspectives*, 10, p.100273.

Tarjan, L. et al. 2014. A readability analysis for QR code application in a traceability system. *Computers and Electronics in Agriculture*, 109, pp.1–11.

Trienekens, J.H. et al. 2012. Transparency in complex dynamic food supply chains. *Advanced Engineering Informatics*, 26(1), pp.55–65.

Underwood, G. et al. 2021. Incubation, hatchery practice and the welfare of layer hens. *Animal Production Science*, 61(10), pp.867–875.

Unveren, H. and Luckstead, J. 2020. Comprehensive broiler supply chain model with vertical and horizontal linkages: Impact of US–China trade war and USMCA. *Journal of Agricultural and Applied Economics*, 52(3), pp.368–384.

USDA Meat Price Spreads. https://www.ers.usda.gov/data-products/meat-price-spreads/ (Accessed 1 January 2022).

USDA 2018. United States Classes, Standards, and Grades for Poultry–AMS 70.200 et seq. Agricultural Marketing Service, Livestock and Poultry Program. https://www.ams.usda.gov/sites/default/files/media/PoultryClassesStandards.pdf.

Vaarst, M. et al. 2015. Sustainable development perspectives of poultry production. *World's Poultry Science Journal*, 71(4), pp.609–620.

Valdokhina, S.I. and Roiter, L.M. 2020. Traceability as a tool for managing the economic viability of a poultry enterprise. In *IOP Conference Series: Earth and Environmental Science* (Vol. 548, No. 2, p. 022032). IOP Publishing.

Van Asselt, E.D. et al. 2010. Selection of critical factors for identifying emerging food safety risks in dynamic food production chains. *Food Control*, 21(6), pp.919–926.

Vander Naald, B. and Cameron, T.A. 2011. Willingness to pay for other species' well-being. *Ecological Economics*, 70(7), pp.1325–1335.

Vanhonacker, F. and Verbeke, W. 2009. Buying higher welfare poultry products? Profiling Flemish consumers who do and do not. *Poultry Science*, 88, pp.2702–2711.

Vanhonacker, F. et al. 2012. The concept of farm animal welfare: citizen perceptions and stakeholder opinion in Flanders, Belgium. *Journal of Agricultural and Environmental Ethics*, 25, pp.79–101.

Verbiest, J.-P.A. and Castillo, C.N. 2004. *Avian flu: An economic assessment for selected developing countries in Asia*. ERD Policy Brief No. 24. Manila: Asian Development Bank. Available at: https://www.adb.org/sites/default/files/publication/28081/pb024.pdf (Accessed 29 March 2026).

Viaene, J. et al. 1998. Quality perception of vegetables by Belgian consumers. In *XXV International Horticultural Congress, Part 14: Horticultural Economics at Micro and Macro Level, International Trade and 524* (pp.89–96).

Wang, R.Y. and Strong, D.M. 1996. Beyond accuracy: what data quality means to data consumers. *Journal of Management Information Systems*, 12(4), pp.5–33.

Warner, R.D. 2017. The eating quality of meat–IV water-holding capacity and juiciness. In F. Toldra (Ed.), *Lawrie's meat science*. Elsevier, pp.419–498.

Yang, X. et al. 2024. Roadmap to achieving sustainable development: does digital economy matter in industrial green transformation? *Sustainable Development*, 32(3), pp.2583–2599.

Yitbarek, M.B. 2019. Livestock and livestock product trends by 2050. *IJAR*, 4, p.30.

Yu, X. et al. 2018. Safety traceability system of livestock and poultry industrial chain. In *Cloud Computing and Security: 4th International Conference, ICCCS 2018, Haikou, China, June 8–10, 2018, Revised Selected Papers, Part VI 4* (pp.3–12). Springer International Publishing.

Zaninelli, M. et al. 2018. Development of a machine vision method for the monitoring of laying hens and detection of multiple nest occupations. *Sensors*, 18, 132.

Zhao, J. et al. 2020. Technologies in individual animal identification and meat products traceability. *Biotechnology & Biotechnological Equipment*, 34(1), pp.48–57. https://doi.org/10.1080/13102818.2019.1711185.

Zhou, H. et al. 2024. Digital technology patents in China: an integrated analysis of patent distribution and transactions. *Technology Analysis & Strategic Management*, 36(12), pp.4281–4297.

# Part 2

## Creating the best day-old broiler chick

# Chapter 4

# Breeding for improved health and welfare in broiler chickens

*Pascal Duenk and Mario Calus, Wageningen University & Research, The Netherlands*

1 Introduction
2 History of broiler breeding
3 Global developments and consequences for broiler welfare
4 Challenges, opportunities and future directions for broiler breeding
5 Conclusion and summary
6 Where to look for further information
7 Acknowledgments
8 References

## 1 Introduction

Genetic selection in broiler chickens has significantly increased growth rates and feed efficiency (Siegel, 2014), resulting in a nutritious animal protein source that is not only affordable but also has a relatively low environmental impact (Poore and Nemecek, 2018). In the past couple of decades, the transition to more balanced breeding goals has helped to partly alleviate several health problems in broilers. Despite these efforts, issues like musculoskeletal deformities and poor foot health persist, and global developments, such as climate change, may introduce new threats to broiler health and welfare in the future.

This chapter aims to (1) provide a concise overview of the broiler breeding history, (2) discuss current and future challenges in broiler breeding with a strong emphasis on improving animal health and welfare, and (3) identify opportunities for industry and academia to further improve the sustainability of broiler breeding. The first section of this chapter provides an overview of the history of broiler breeding, focusing on the development of commercial broiler chickens since the 1940s. In addition, the development of broiler breeding goals over time will be described, highlighting the initial focus on body weight and feed efficiency, and the subsequent shift toward more balanced breeding goals that include traits related to survival, resilience, health, and welfare. In the

http://dx.doi.org/10.19103/AS.2025.0165.05

third section, we will identify three main global developments that could affect the broiler breeding industry during the coming 10 to 15 years. Finally, we will discuss current and future challenges, opportunities, and future directions for improving broiler health and welfare through genetic selection.

## 2 History of broiler breeding

Since the 1940s and 1950s, routine crossbreeding has been used to produce commercial broiler chickens (Tixier-Boichard et al., 2012; Muir and Aggrey, 2003). The commercial broilers are typically a four-way cross, involving crossing four different purebred breeding lines, in a pyramidal breeding structure (Emmerson, 2003). The breeding pyramid contains four layers: (1) the breeding nucleus with the purebred lines, (2) a multiplication layer to increase the number of available purebred animals, being the great grandparents of commercial broilers, (3) a multiplication layer to generate two-way crosses, being the grandparents of commercial broilers, and (4) a multiplication layer to produce the four-way crossbred commercial chickens (Neeteson et al., 2020; Hiemstra and Ten Napel, 2013). There are three main reasons to use crossbred animals in commercial production: (1) to enable using specific favorable characteristics of the involved breeding lines, enabling to select them for at least partly different, but complimentary breeding goals (Smith, 1964), (2) to make use of heterosis (Dickerson, 1973), which is the phenomenon that crossbred animals outperform the average of their purebred parents, and (3) to protect the purebred genetics by not having to distribute purebred animals to primary producers, such that competitors do not get access to it.

Genetic selection in animal breeding is based on the principle that in every generation the best animals, with respect to a certain breeding goal, are selected and used as parents to produce the next generation. The advantage of genetic selection over management improvements is that selection is cumulative, meaning that any genetic improvement that is realized will affect all subsequent generations. In broiler breeding, selection takes place within the four purebred lines at the top of the breeding pyramid (Emmerson, 2003; Hiemstra and Ten Napel, 2013). The power of the breeding pyramid lies in the fact that the genetic superiority of each purebred breeder animal is expressed in millions of commercial great-grand offspring (Hiemstra and Ten Napel, 2013). Traditionally, genetic selection is based on breeding values that are estimated using phenotypic observations of selection candidates and their close relatives, as well as pedigree information. This has also been the method of choice in broilers, until the introduction of genomic prediction in broiler breeding practice, a bit more than a decade ago (Wolc et al., 2016). Using genomic prediction, the transmission of DNA from parents to offspring is modeled across the genome, using so-called single-nucleotide polymorphism

(SNP) markers. In different livestock species, the number of SNPs used in genomic prediction is around 50,000 (Eggen, 2012). Genomic prediction enables to estimate breeding values that differentiate between full sibs before any phenotypes are recorded, while this is not the case if breeding values are estimated using pedigree information. In broilers, the main benefit of genomic prediction comes from increasing accuracy for specific traits that cannot be measured on selection candidates, such as disease resistance and carcass-related traits (that are typically measured on sibs of selection candidates), and reproductive performance traits that cannot be measured at a young age (Wolc et al., 2016). Genomic prediction currently plays an important role in most broiler breeding programs.

## 2.1 Development of broiler breeding goals over time

Initial broiler breeding goals mostly focused on selection for body weight, followed by feed efficiency, which remained the main focus until around 2000 (Siegel, 2014). Measuring individual body weight is relatively straightforward, and the target used to be body weight at 35 or 42 days of age, corresponding to the commercial body weights. This strategy has three main implications. First, body weight can be relatively easily measured on a large number of individuals, enabling to select relatively few individuals for breeding from a large pool of selection candidates, resulting in a high selection intensity. Second, having own observations as well as sibs with observations for a trait like body weight with a moderate to high heritability of ~0.3–0.4 (Mebratie et al., 2019; Mignon-Grasteau, 1999) results in a moderate to high selection accuracy. Third, the body weight of the animals is measured before the animals reach maturity. Broilers reach sexual maturity around 20–27 weeks of age (Lewis and Gous, 2006), which means that the generation interval is relatively short, being 36 to 40 weeks of age, depending on the breeding structure (Avendaño, pers comm). The combination of a high selection intensity, a moderate to high selection accuracy, and a short generation interval has resulted in a spectacular change in body weight and feed efficiency. Indeed, comparisons of performance of broiler genetics from 1957 and 2001, when fed a 1957 diet, showed an increase in body weight at 42 days from 539 g to 2672 g, a reduction in feed conversion ratio from 2.34 to 1.92, and large changes in body composition, but also an increase in mortality (Havenstein et al., 2003a, 2003b). In a similar study that compared broiler genetics from 1957 and 2005, body weight at 42 days increased by over 400% (from 586 g to about 2800 g), feed conversion ratio improved by 50% (from 2.88 to 1.67), and body composition shifted toward much higher breast muscle yield (+79–85%) with less abdominal fat (Zuidhof et al., 2014). However, undesirable correlated responses in some other traits were observed, including increased incidence and severity of skeletal disorders and other leg issues, reduced immunocompetence, and reduced fertility (Hocking, 2014).

Similar to what has been observed in other livestock species, around the year 2000 breeding goals started to become more balanced by including increasingly more traits related to survival, health, and welfare (Neeteson-van Nieuwenhoven et al., 2013; Neeteson et al., 2023). In a balanced breeding goal, the traits of interest are included in a so-called selection index, which is a simple mathematical tool that can be used to optimize selection decisions and thereby maximize the genetic improvement of multiple traits in a population (Hazel, 1943). Each trait included in the index is assigned a weight, which represents the relative importance of that trait to the breeding goal. For production traits, the weights are determined based on their economic value, which reflects the change in farm profit that is realized by increasing the average trait value by one unit. For many welfare-related traits, determining appropriate weights is much more challenging because their economic values are difficult, if not impossible, to derive (Hocking, 2014). This issue is typically solved by the desired gains approach, in which optimal weights are determined based on the desired genetic gain for one or more traits (Nielsen et al., 2014). The framework of the selection index allows breeders to adjust their breeding goals to economic, environmental, and societal developments. Contemporary breeding goals are now mainly directed towards improving traits related to health, welfare, and reproductive fitness. For example, in one of the conventional broiler breeds (Ross 308), these traits collectively account for 59% of the breeding goal, while 25% is reserved for feed efficiency, and 16% for growth and meat yield (Neeteson et al., 2023).

In summary, the broiler breeding industry has seen major developments over the last two to three decades. Initially focused on body weight and feed efficiency, breeding goals have shifted toward a more balanced approach, incorporating traits related to health, welfare, and reproductive fitness. The introduction of genomic prediction has improved the accuracy of breeding values, particularly for traits that are difficult to measure on selection candidates. Contemporary breeding programs now prioritize a combination of production efficiency and animal welfare traits, reflecting broader economic, environmental, and societal developments.

## 3 Global developments and consequences for broiler welfare

### *3.1 Increasing demand for animal protein*

Although this chapter focuses on the requirements for genetic improvement of animal health and welfare, we cannot ignore the fact that the demand for affordable and nutritious animal protein will continue to increase (Henchion et al., 2021). Broiler chickens provide an efficient and sustainable way of producing such protein (Alexander et al., 2016), and they therefore play an important role in providing food security. Worldwide, poultry meat consumption is projected to increase by about 15% between 2023 and 2032, and is expected to account for 41% of the total protein consumed from meat in 2032 (OECD and FAO, 2023). Most of this increase in consumption will probably come from developing and middle-income countries, due to their expected rapid growth in population size and household incomes (OECD and FAO, 2022). In these countries, animal protein needs to be affordable, motivating breeders and farmers to focus on increasing production levels (i.e. growth) and feed efficiency. Global breeding programs will therefore continue to include production and feed efficiency traits in their breeding goals, albeit to a much lesser degree than before.

## 3.2 Climate change

The welfare of broiler chickens is threatened by climate change. In particular, long periods of high ambient temperatures in combination with high humidity (heat waves) are expected to become more frequent, longer-lasting, and more intense in the future (Lau and Nath, 2014). Maintaining thermal balance is especially challenging for broiler chickens because they lack sweat glands and are covered in feathers, relying solely on evaporative heat loss. Additionally, their high growth rate and metabolism make them especially sensitive to heat stress (Nascimento et al., 2017), which triggers a series of physiological responses that negatively impact production efficiency, health, welfare, and mortality rates (Saeed et al., 2019; Wasti et al., 2020; Oluwagbenga and Fraley, 2023). These adverse effects pose a significant threat to broiler welfare in housing systems without effective climate control.

## 3.3 Increasing considerations for animal health and welfare

Consumers and farmers in high-income countries (e.g. in North-West Europe) are increasingly concerned about animal health and welfare (Whitton et al., 2021). This development has (for example) resulted in the abandonment of meat produced by fast-growing broilers in Dutch supermarkets (Schelfaut, 2021), and other countries may follow. In addition to these concerns, the broiler industry poses a risk for human health through bacterial infections transmitted via consumed meat, and the development of resistant pathogen strains following frequent use of antibiotics (Mottet and Tempio, 2017). Part of the solution to this issue is to use less (preventive) antibiotics in livestock (Hedman et al., 2020), which has already shown a declining trend in many parts of the world over the past few decades due to new legislations (World Organisation for Animal Health, 2024). In particular, the use of preventive growth-promoting antibiotics is currently banned in the EU. Although this decrease in antibiotic use could help to prevent the development of resistant pathogen strains, it does not prevent diseases from spreading, and it may actually increase the risk for individual animals to suffer from infection.

The societal concern about animal health and welfare is expected to grow in other parts of the world as well. This highlights the need for a diverse set of breeds that can thrive in multiple challenging environments, and for improving the health and welfare of existing breeds. Improving health and welfare by genetic selection has already been successful for traits that are relatively easy to measure, are moderately to highly heritable, and have relatively low genetic correlations with economically important traits. For example, continued selection on improved leg health has significantly reduced the prevalence of bone deformities, crooked toes, tibial dyschondroplasia, and hock burn (Kapell et al., 2012b). Similarly, genetic selection has made important contributions to the decrease of ascites in the past decades (Neeteson et al., 2023). For other health and welfare problems, improvement by genetic selection may be more difficult, because (1) the pure line selection candidates are typically kept in favorable and bio-secure environments where many health and welfare issues are not observed (Hiemstra and Ten Napel, 2013), (2) it is difficult and costly to obtain many accurate phenotypes from commercial environments where the animals live in very large groups, and (3) some functional traits may have strong unfavorable genetic correlations with production and efficiency traits.

In summary, global demand for affordable chicken meat continues to increase, driving a focus on production efficiency in broilers, particularly in developing countries. Simultaneously, climate change presents a significant welfare challenge for broilers due to their susceptibility to heat stress. Finally, high-income countries show increasing concern for animal health and welfare, leading to reduced antibiotic use and scrutiny of fast-growing breeds. While genetic selection has improved some welfare traits, further improvements face challenges due to data collection difficulties and potential unfavorable genetic correlations with production traits. In the following section, we will discuss these and other challenges in more detail, as well as opportunities and directions for further research that can help overcome these challenges.

## 4 Challenges, opportunities and future directions for broiler breeding

In this section, we will discuss challenges, opportunities, and future directions for broiler breeding, grouped into five main topics: (1) Diversification, (2) Optimizing the breeding goal, (3) Phenotyping, (4) Improving breeding value estimation, and (5) Breeding for disease resistance.

One topic that we will not discuss in detail in this chapter is gene editing. While gene editing technologies like CRISPR-Cas9 are intriguing, their widespread application for improving broiler welfare faces significant hurdles.

First, many welfare-related traits (such as health, robustness, and behavior) are extremely complex, meaning they are influenced by a large number of genes with complex interactions. This makes targeted editing difficult and its outcomes unpredictable (Park et al., 2020). Furthermore, editing one gene might also have unintended and undesirable effects on other characteristics through pleiotropy. Finally, substantial technological challenges remain, particularly regarding the efficiency and precision of gene editing techniques, specifically in avian species, which differ significantly from mammals (Khwatenge and Nahashon, 2021). Although gene editing offers a more precise approach with fewer unintended effects than earlier genetic modification techniques, it remains subject to strict regulations in many regions, especially in the EU (The Nuffield Council on Bioethics, 2021). Together, these factors complicate the development of this technique for commercial use. However, exceptions such as gene editing for specific disease resistance may emerge, as illustrated by the recent approval of gene-edited pigs resistant to PRRS in the United States (American Veterinary Medical Association, 2025).

## *4.1 Diversification*

The global broiler industry is predominantly controlled by a few large multinational companies, offering (at the time of writing) 13 different types of commercial broilers to serve the world market (Cobb-Vantress Inc, 2024; Hubbard Breeders, 2025; Aviagen Group, 2025). These types of broilers include conventional fast-growing and slower-growing genotypes, and breeds that are specifically suitable for small-scale backyard farming systems. However, a single type of broiler may still be used across a wide variety of production environments (i.e. farms), which may differ greatly in terms of climate, housing system, and feed quality. As a result, there may be strong genotype-by-environment interactions for health and welfare traits across commercial production environments. For many traits in poultry, the genetic correlation between nucleus and commercial environments is much smaller than 1 for functional and production traits (Chu et al., 2019; Calus et al., 2023), indicating moderate-to-strong genotype by environment interaction. For example, in the Aviagen breeding lines, the genetic correlation for livability between high- and low-input environments ranged between 0.45 and 0.85 (Neeteson et al., 2023). It is likely that similar genotype by environment interactions exist between commercial environments that differ greatly in quality of climate, management, and feed. Hence, populations that are genetically selected to perform optimally in one environment may not perform optimally in another environment.

Broiler breeding companies typically address genotype-by-environment interactions by including production and survival traits (such as liveweight and livability) as expressed and measured in commercial-like environments in their breeding goals (Avendaño, pers comm). This strategy ensures that genetic selection favors animals with lower environmental sensitivity, but this strategy has several caveats. First, overall genetic improvement may be reduced because of the trade-off between environmental sensitivity and production levels (Rauw and Gomez-Raya, 2015). Second, when many different environments are considered in the breeding goal traits and genetic correlations between environments are relatively low, genetic improvement for each of the environments becomes smaller because of dilution of the breeding goal. Finally, it may be difficult and costly to measure traits across the full range of commercial conditions that the commercial broilers are exposed to. Hence, an alternative (or complementary) strategy to address genotype-by-environment interactions is to further diversify the broiler products that are offered, addressing the needs and desires of farmers and consumers across the world with more specialized products.

One of the most striking differences between environments that broilers are exposed to is the differences in ambient temperature. Due to climate change, the frequency and duration of heat waves continue to increase, especially in oceanic and Mediterranean climates (Lau and Nath, 2014). In a study that compared the Ross 308 (a commercial crossbred) to two local chicken breeds in standard and high temperature environments, there was significant genotype-by-temperature interaction for many production and meat quality traits (Huerta et al., 2023). In a follow-up study, the commercial crossbred was the only breed that was negatively affected by heat stress in terms of growth curve parameters, while the local breeds or their hybrids were not affected (Menchetti et al., 2024). It has indeed been shown that the physiological response to heat stress is less severe in local breeds compared to commercial crosses (Soleimani et al., 2011; Brugaletta et al., 2022). On the other hand, and as shown by these researchers, there is a clear trade-off between heat tolerance and growth performance. These results illustrate that different genotypes respond differently to high ambient temperatures and that some breeds may be better adapted to particular production systems than others.

The likely existence of genotype-by-environment interactions for health and welfare traits in broilers probably requires further diversification to meet the demands of farmers and consumers. One approach for diversification is to use local breeds that are better adapted to the environment in which they are kept. However, the production levels of these local breeds are typically much lower than the commercial breeds (Huerta et al., 2023; Menchetti et al., 2024), which makes it unattractive for farmers to use them. This disadvantage may be partly alleviated by selecting for increased production levels within these breeds, but this requires systematic data recording, which may not be feasible in some extensive production systems. Alternatively, the advantages of local adaptation and high production could be realized by crossing local with commercial (faster-growing) breeds for production environments where adaptability is important, for instance, in small-scale backyard farming. Padhi (2016) reviewed the benefits of such crosses and concluded that the crosses greatly outperformed the local breeds for nearly all production traits. In a study that compared local breeds from Italy (Bionda Piemontese and Robusta Maculata) with their Sasso cross, the cross outperformed the average of the two parental lines for all growth curve parameters, suggesting significant heterosis effects (Mancinelli et al., 2023). Similarly, crosses between Wassachie and Sasso chickens showed higher growth performance and improved carcass and meat quality traits compared to the local purebreds (Dzungwe et al., 2024). Together, these results suggest that in some specific circumstances and production environments, crossing local with commercial breeds may promote the use of local breeds by improving production levels, which has the additional benefit of conserving these local breeds *in vivo* (Woelders et al., 2006).

In addition to improving local adaptability, consumer concern about animal welfare (e.g. in north-western Europe) also warrants the use of different breeds and crossbreds. Research has suggested that slow-growing broilers experience better welfare than fast-growing broilers when the rearing period exceeds six weeks (Wilhelmsson et al., 2019). Dutch supermarkets have responded to this societal concern by abandoning the sale of meat from fast-growing broilers by the end of 2023 (Schelfaut, 2021). It is likely that other countries will follow this development. Although this is an important step towards improved animal welfare, it should be noted that slow-growing hybrids are not completely free from welfare issues either, and fast-growing broilers can have a similar level of welfare as slow-growing broilers when kept in a suitable environment under good management conditions (Neeteson et al., 2023). Nevertheless, the market for poultry meat is expected to diversify further, which could drive further diversification in the broiler breeding industry.

In conclusion, future global market developments may require further diversification within the broiler industry to provide more specialized products tailored to specific environments and consumer needs. We recommend further study into strategies for selecting against environmental sensitivity, and the usefulness of using different breeds and their crosses to ensure high animal welfare standards across the globe.

## 4.2 Optimizing the breeding goal

Breeding companies closely monitor the global developments that impact the broiler industry, and continuously reconsider their breeding goals to serve the needs of the market, farmers, and consumers. The expansion of the breeding goal with new traits is therefore expected to continue in the future (Neeteson et al., 2020). When more and more traits are added to the breeding goal, there is a risk that the selection pressure on each of the traits becomes diluted, reducing genetic gain in each of the considered traits. Breeders should therefore make sure to place enough emphasis on traits that will be important for the broiler of the future, and perhaps even abandon selection for traits that have become irrelevant or that introduce negative welfare consequences. In the following subsections, we will highlight a couple of challenges related to breeding goal optimization and future directions for research.

### 4.2.1 Overcoming trait antagonisms

There may be as many as 50 traits included in a broiler breeding goal (Katanbaf and Hardiman, 2010). These traits are never fully genetically independent of each other, meaning that the genetic correlations between pairs of traits significantly differ from 0. For example, feed intake has a low to moderately positive genetic correlation with growth (Mebratie et al., 2019; Marchesi et al., 2021). The consequence of this positive correlation is that when selection only favors increased growth, feed intake also increases as a correlated response. From the breeder and farmer standpoint, increased appetite (without the equivalent increase in growth) is undesirable, which breeders would solve by balancing selection for growth with selection for feed efficiency.

Many breeding goal traits in broilers have unfavorable genetic correlations, meaning that selection for improvement in one trait has undesirable consequences for another trait. A well-known example in broilers is the strong and negative genetic correlation between growth and reproductive efficiency, which is probably caused by the competition for resources between these two traits. In broilers, this phenomenon is termed the 'broiler-breeder paradox' and may result in major welfare issues in the broiler breeders because of the need for feed restrictions after puberty (Decuypere et al., 2010; Hartcher and Lum, 2020). In some cases, negative welfare consequences that arise through unfavorable genetic correlations can be alleviated through management practices. In the case of the broiler-breeder paradox, alternative feeding strategies such as feed dilution or feeding twice per day may provide a promising solution (Neeteson et al., 2023; van Emous et al., 2024).

Another example of unfavorable genetic correlations is the weak to moderately positive correlation between body weight and several leg health traits, such as long bone deformity, tibial dyschondroplasia, and crooked toes (Johnsson, 2020). Such unfavorable genetic correlations can make it challenging to improve welfare traits without compromising genetic gain in production traits to a degree that is acceptable. However, simultaneous genetic improvement for both traits is still possible (Hill et al., 2016; Bailey et al., 2020), but a reduction in long-term genetic gain for production traits should be accepted. Empirical data have indeed shown that simultaneous improvement of leg health and growth has been realized in the past (Kapell et al., 2012b; Rekaya et al., 2013). In addition, selection has been very effective for other welfare traits as well (leg strength, livability, and cardiovascular function), with limited sacrifice in growth rate improvement (Neeteson et al., 2023). These examples attest to the importance and potential of balanced breeding goals for improving health and welfare traits.

Unfavorable trait antagonisms may be overcome by developing a deeper understanding of the underlying trade-offs and the role of genetic selection. The role of selection could be studied by comparing current commercial crossbreds with unselected ancestral populations, but it remains very difficult to separate the effects of selection from the effects of breed, management, and environmental differences (Pinard-van Der Laan et al., 2020). Fortunately, the continuing developments in genomics and trait recording technologies offer opportunities to further study the ontology of trade-offs and to reduce the impact of the trade-offs between traits by increasing the accuracy of estimated breeding values (Avendaño and Fancher, 2017). For example, functional genomic information such as gene expression and epigenetic profiles could help to identify the genes and biological mechanisms underlying genetic trait correlations. In the area of trait recording technologies, the use of X-ray technology to detect sub-clinical signs of tibial dyschondroplasia has drastically improved the effectiveness of selection against this trait (Kapell et al., 2012b). More recently, the development of computed tomography (CT) and 3D imaging technology has provided new opportunities to uncover detailed information about the broiler's body and generate new phenotypes from live animals. In addition to enabling precise predictions of meat yield, CT facilitates live assessments of skeletal integrity and internal organs (Neeteson et al., 2023).

It has been proposed that non-zero genetic correlations between traits may arise through physiological processes that regulate those traits, instead of pure pleiotropy (Berry et al., 2016). This hypothesis implies that genetic correlations may not be fixed population parameters, but that they can be viewed as non-observed (latent) phenotypes that represent the capacity to regulate the trade-off (Cuyabano et al., 2024). Although a method to estimate such latent traits is not yet available, this different view on genetic correlations can possibly lead to new directions for research that can help to better understand the origin of unfavorable genetic correlations in broiler chicken, and perhaps to develop better selection strategies.

### 4.2.2 Robustness and resilience

The globalization of breeding companies requires that broiler chickens can perform well in a variety of environments. Broiler breeding companies, therefore, aim to improve the *robustness* of their broilers, which is defined as the ability to combine a high production potential with resilience to stressors, allowing for unproblematic expression of a high production potential in a wide variety of environmental conditions (Knap, 2005). In broiler breeding, selection for robustness is realized by measuring production, health, and welfare traits on sibs of selection candidates that are grown in several commercial-like environments with less favorable conditions (low input, non-bio-secure) (Neeteson et al., 2023). These phenotypes are then used in combination with phenotypes from the selection candidates themselves (grown in nucleus environments with high input and bio-security) to estimate breeding values for all breeding goal traits. This ensures that the traits of interest are improved in both the nucleus and a range of commercial environments, increasing robustness of the broiler (Neeteson et al., 2020). It has been shown that collecting data on foot-pad dermatitis in a commercial environment can increase selection response for this trait (Kapell et al., 2012a). More generally, it has been shown that acquiring at least part of the phenotypes from crossbreds in a commercial environment can be beneficial for overall genetic improvement (Chu et al., 2018; Duenk et al., 2021). Although measuring production and welfare-related traits across a variety of environments makes sense, there may be large differences among commercial environments, for example in terms of climate, management, housing, and diet. Continued investment in data collection in a wide variety of environments that accurately reflect the full range of commercial environments is therefore recommended.

In addition to robustness, the intensification of broiler farming requires that broilers be *resilient*. Resilience is defined as the capacity to cope with physical and psychological disturbances, or to rapidly recover from such disturbances (Colditz and Hine, 2016). Although resilience cannot directly be observed in an individual, it can be derived from the deviations of the realized production levels from the expected ones (Berghof et al., 2019b). This method requires phenotypes on production traits that are recorded continuously over time on the same animals (i.e. longitudinal data). For livestock species, such as dairy cattle, such data is easily obtained through the use of milking machines that automatically record milk production. For broilers, obtaining longitudinal data at the individual level is more difficult, but still possible by using identification, such as RFID (Neeteson et al., 2023). Broiler breeding companies currently use these techniques to systematically collect longitudinal data on economically important traits, such as weight and feed intake (Avendaño, pers comm). The future development of new data collection techniques and strategies (such as sensors and computer vision) will eventually make it easier to identify and phenotype animals automatically over time (Brito et al., 2020).

Traditional methods of assessing resilience have primarily focused on deviations in production levels across time, largely because longitudinal production data are relatively easy to obtain. The underlying assumption is that any disturbances, whether physical or psychological, will manifest as drops in production, such as reduced growth rates. While it is suggested that resilience can capture variation in animal welfare (Brito et al., 2020), drops in production level do not fully reflect the current state of an individual's welfare. There is often a delay between the occurrence of a stressor and its impact on production traits, such as growth, making it an imperfect measure of immediate welfare. In addition, drops in growth may not always be related to impaired welfare. To gain a more accurate understanding of welfare over time, it may be beneficial to use phenotypes that respond more directly to environmental stressors, such as feed intake or behavior (Canario et al., 2013; Gorssen et al., 2023). For instance, broilers may exhibit behavioral changes when exposed to stressors, such as high ambient temperatures or aggressive interactions with other broilers. These behavioral changes can provide more immediate and precise insights into the broilers' ability to cope with stressors. Researchers should therefore explore methods of measuring welfare over time, focusing on characteristics that are affected by stressors more directly. Measuring resilience through changes in feed intake might be a feasible option for the near future, considering that feed intake data are routinely being recorded in breeding programs.

### *4.2.3 Heat tolerance*

We have already established that there are significant genotype-by-temperature interactions across different poultry breeds (e.g. Huerta et al. (2023), Menchetti et al. (2024)). It is therefore likely that there is heritable genetic variation in heat tolerance within broiler breeds, presenting an opportunity to improve this trait by selective breeding. The main challenge to improve heat tolerance within breeds is to define the trait and collect high-quality data. Heat tolerance is typically quantified by fitting reaction norms (De Jong, 1995), where performance measures, such as growth, are regressed on temperature or temperature-humidity index (THI) (Ravagnolo and Misztal, 2000). A reaction norm model results in breeding values for the slopes of the reaction norm, which reflect the genetic merit for (in this case) temperature sensitivity based on performance. However, production performance measures may not accurately reflect the current welfare status of the individual, as indicated above. As an alternative, differences in heat tolerance could be measured through behavioral observations among broilers under heat stress. Increased temperatures may affect the daily activity pattern of broilers (Duenk et al., 2024), increase drinking behavior (Li et al., 2015; Cartoni Mancinelli et al., 2023), and decrease eating behavior (Cartoni Mancinelli et al., 2023). Reliable assessment of heat tolerance requires that environmental variables are accurately recorded, and that data are collected across a range of climates (Fodor et al., 2023a).

In practice, breeding companies indirectly select for heat tolerance by selecting for reduced environmental sensitivity based on multi-environmental data on performance and livability (Neeteson et al., 2023). Although directly improving heat tolerance is feasible through the use of specialized lines or selection for genes related to feather cover (e.g. naked neck), these strategies are not widely adopted because of practical difficulties, and consumer preferences for fully feathered birds (Tixier-Boichard, 2020). Further research should focus on how to accurately capture individual differences in heat tolerance at a large scale. The potential welfare benefits of improving heat tolerance in broilers make this a worthwhile pursuit for breeding companies aiming to enhance both the performance and welfare of their broilers in varying climates.

## 4.3 Phenotyping

In the last couple of decades, technologies that capture variation in the genome have rapidly advanced. Still, without phenotypes, breeders cannot effectively capitalize on the vast amounts of genomic data that are produced. A quote that is often reiterated by animal breeders is 'In the age of the genotype, phenotype is king' (Coffey, 2020). Indeed, traits can only be effectively included in a breeding goal when they are measured on selection candidates, or alternatively their sibs, to ensure a reasonable accuracy of the resulting genetic selection. In addition, such measurements should be non-invasive, cost-effective, and standardized. Most importantly, however, they should be measured at a large scale and accurately represent the targeted selection trait (Brito et al., 2020).

In the following paragraphs, we will touch upon some of the main challenges related to measuring health and welfare traits.

Measuring health and welfare traits in broilers can be challenging and costly because they typically require manual measurements or observations. For example, assessment of leg health is currently realized through visual inspection of gait score on individual animals (Avendano, pers comm), which is labor intensive. Technological advancements in the area of sensors, video cameras, and other devices offer unprecedented opportunities for measuring health and welfare traits more efficiently, more accurately, and on a larger scale. One of the main challenges is to translate the data resulting from these devices to phenotypes that accurately reflect the trait of interest. Fodor et al. (2023b) explored the use of automated pose estimation from video recordings to assess walking characteristics associated with lameness. They retrained a deep learning model previously developed by Doornweerd et al. (2021) to detect and track keypoints on broilers' bodies from video recordings. They identified specific walking characteristics, such as hock joint angles and step height, and correlated these characteristics with gait scores. The results showed that broilers with suboptimal gait had sharper hock joint angles, a lower hock-feet distance ratio, and lower step heights compared to those with good gait (Fodor et al., 2023b). These findings show that automated phenotyping of pose can offer an effective and high-throughput way of measuring and monitoring broiler gait. More research is needed before this technology can be applied in a commercial setting where a large number of broilers roam freely.

Machine learning algorithms and computer vision are very promising tools for measuring health and welfare traits, such as locomotion in broilers, but there are a number of challenges that need to be overcome before this can be applied in breeding. First, the algorithms typically need large annotated datasets that are difficult and costly to generate. Technological advancements in data recording devices will probably alleviate this problem in the near future. Second, identifying and re-identifying the same individual over time is extremely difficult (especially in large groups) because broilers have a very similar appearance. This leads to so-called identification switches, where the system mistakenly assigns a different identification number to an individual due to occlusions or proximity to other individuals (Doornweerd et al., 2024). This issue could be addressed by improving the detection algorithms, collecting more annotated data, or combining the video tracking system with other identification systems, such as RFID (e.g. Doornweerd et al., 2023).

Behavioral phenotypes are particularly interesting for studying animal welfare because behavioral traits are often strongly linked to the welfare status of animals (Nawroth et al., 2019), and social interactions can significantly affect animal welfare in large groups (Lee et al., 2022). Developments in video recording and computer vision technologies offer opportunities to measure behavior accurately and at a large scale (Nawroth et al., 2019). It is therefore likely that genetic selection for (or against) certain behaviors becomes feasible in the future. However, before behavioral traits can be included in the breeding goal, it is important to understand the cognitive processes behind behavior (Nawroth et al., 2019; D'Eath et al., 2010). For example, an animal might not express a certain behavior because it was selected to be anxious and docile, and will therefore not move at all. Additionally, it is important to realize that behavior is very context dependent, making it very difficult to define behavioral traits that are relevant across a wide variety of environments. Finally, it is important to carefully consider the ethical implications before selectively breeding for or against specific behaviors. Some may find it unacceptable to reduce the expression of certain behaviors through selective breeding, particularly if it compromises the integrity of the broiler (Coles et al., 2015). For example, breeding to eliminate feather pecking could unintentionally impair the broilers' vision, severely affecting their natural ability to perceive and navigate their environment (Thompson, 2008).

## 4.4 Improving genomic selection

Genomic selection is currently the main method for realizing genetic improvement in broiler breeding programs. With genomic selection, phenotype and genotype marker data are collected from a so-called reference population, which enables the estimation of breeding values for selection candidates that are genotyped (Meuwissen et al., 2001). In practice, breeding companies use reference populations of purebred animals kept in commercial-like environments, accounting for genotype-by-environment interaction between the nucleus and commercial environments. Although genomic selection has already offered benefits for broiler breeding in terms of increased accuracy (Neeteson et al., 2023), there are opportunities for improving this technique even further.

Genomic prediction can be particularly helpful for genetically improving health and welfare traits that occur in commercial broilers, by creating a reference population of crossbred animals that are kept under commercial conditions (Wolc et al., 2016). However, genotyping and phenotyping commercial crossbred animals is not very cost-effective, especially considering that these animals are not used for breeding. In addition, the multiplication steps in a crossbred breeding program result in multiple generations between the commercial crossbreds and the (purebred) selection candidates, limiting the accuracy that can be achieved (Wientjes et al., 2020). One interesting strategy to reduce the costs of creating a crossbred reference population is *DNA pooling*, where individuals with a similar phenotype are genotyped as a single sample. In a study that used data on body weight from about 9,000 crossbred animals, strategies that involved DNA pooling achieved a very reasonable accuracy of breeding values (up to 0.62) compared to genotyping each individual separately (0.88), but for only a fraction (up to 10%) of the total costs (Aldridge et al., 2022). These results warrant further research into the benefits of pooled genotyping in broiler breeding programs, especially for welfare traits.

Genomic selection is typically based on genotype information from a limited set (~50,000) of SNP markers, relying on the idea that these SNPs capture phenotypic variation because they are physically linked to causal variants in (somewhat close) proximity on the genome. It has been proposed that the accuracy of genomic selection could be improved by using all observed variants across the genome (i.e. whole-genome sequence (WGS)). However, the expected benefits of using WGS have not been observed in early studies, probably because genomic prediction models typically assume that each genomic variant has an equal contribution to phenotypic variation. Incorporating prior information, for example from genome annotation, could possibly increase the benefit of using WGS (Clark et al., 2020; Neeteson et al.,

2023). One attractive approach for this is to prioritize genomic variants based on comprehensive impact scores that provide prior information about the likelihood of functional impact of all variants (Bickhart, 2024; Groß et al., 2020).

To date, the benefit of prioritizing genomic variants has shown mixed results for production traits (Bink et al., 2022; Cheruiyot et al., 2022; Raymond et al., 2018). However, it is likely that this technique is more effective for novel welfare and health related traits, because (1) there has been little to no selection for such traits, resulting in a lot of variation at major genes, and (2) prior biological information can be particularly beneficial when the amount of data is relatively limited (as is the case for behavior traits), because in those cases the prior has a relatively strong influence on the posterior estimates (de Los Campos et al., 2013). Although a lot of effort has been put into finding causal genes and variants for complex traits (Alvarenga et al., 2021; Hu et al., 2015), the most effective means of exploiting this information for genetic selection remains unclear. We therefore recommend that breeding companies and researchers should invest in the development of methods that can effectively use genomic information to improve genomic prediction accuracy for health and welfare traits.

In recent years, research on genomic prediction has expanded beyond traditional genotype information to incorporate various omics data, offering new opportunities to enhance prediction accuracy (Martini et al., 2022). One example of such data is transcriptomics, which measures gene expression levels. Several researchers have studied the benefit of using transcriptomic data in genomic prediction in mice and *Drosophila* with varying success (e.g. Perez et al., 2022; Li et al., 2019), but to our knowledge, there are no such studies in broiler chicken. The main challenges of using omics data for genomic prediction are that omics data, such as expression levels, are typically very time and tissue dependent (Perez et al., 2022), and it is very costly to collect such data from all selection candidates. The latter challenge could be overcome by developing cost-effective methods of using omics data to improve genomic prediction, for example by prioritizing genomic variants based on differential gene expression measures in a limited number of individuals.

## 4.5 Breeding for resistance against infectious diseases

The decreasing use of (preventive) antibiotics and increasing concerns about human and animal health require that the broiler of the future be resistant to infectious diseases. Genetic selection could serve as an effective strategy to improve disease resistance in broilers, complementing interventions, such as biosecurity and vaccinations. Most breeding programs indirectly select for general disease resistance by collecting data on sibs of selection candidates that are kept in commercial-like environments that are not bio-secure (Avendaño, pers comm). These sibs are therefore exposed to a general disease challenge, and their measurements on production and functional traits inform about their immunocompetence. However, direct selection for specific disease resistance is rarely used, partly because of the challenges described below (Pinard-van Der Laan et al., 2020).

There have been several studies on the connection between genetics and disease resistance in chickens. Most of these studies focused on identifying genes or genetic markers associated to general or specific disease resistance, for instance by using genome-wide association or differential gene expression studies (Psifidi, 2021; Gul et al., 2022). Together, these studies have shown that disease resistance traits are typically highly polygenic, meaning that there are many genetic variants involved that each have a very small effect (Psifidi, 2021). Furthermore, most causal genetic variants that have been identified are located in non-coding regions of the genome that regulate gene transcription (Pinard-van Der Laan et al., 2020). This polygenic nature of specific disease resistance makes it difficult to select for favorable gene variants.

One of the main issues in selecting for specific disease resistance is the coevolution of pathogens: changes in the broilers' immune function due to genetic selection can lead to the emergence of new strategies in the pathogen that can evade the broilers' immune system. Due to their high mutation rates and replication cycles, pathogens might often have the upper hand in this arms race, which makes it very difficult to breed for specific disease resistance. It has therefore been proposed that the best strategy is to select for high immunocompetence and general disease resistance, rather than specific disease resistance (Pinard-van Der Laan et al., 2020). Selection for general disease resistance may not only improve disease resistance directly, but may also increase the effectiveness of vaccinations. In a long-term experiment on general disease resistance in laying hens, two chicken lines were divergently selected for low and high general disease resistance by measuring the level of natural antibodies. After seven generations of selection, the antibody titer in the high line was 5.2 times higher than in the low line (Bovenhuis et al., 2022), and the high line indeed showed increased resistance to avian pathogenic *Escherichia coli* (Berghof et al., 2019a). To our knowledge, there are no published examples of selection experiments for general disease resistance in broilers. In broilers, directly selecting for high immunocompetence may be challenging, considering that conventional broilers have a maximum lifespan of about six weeks, and the acquired immune system only matures after about three weeks of age (Bar-Shira et al., 2003). Still, genetic selection could effectively improve disease resistance of broilers by strengthening the early-acting innate immune system, which probably requires data from sibs of selection candidates at older ages in commercial-like environments.

## 5 Conclusion and summary

In conclusion, the advancements in broiler breeding and the evolution of breeding goals over the past decades have significantly contributed to the improvement of health and welfare in broiler chickens. The industry can continue to play an important role in improving the sustainability of the broiler industry by integrating modern technologies with genetic selection, and considering global developments, such as climate change and increasing consumer awareness. Future research and development could focus on further diversification of broiler types, optimizing breeding goals by incorporating new traits and deepening our understanding of genetic correlations between traits, and improving methods for phenotyping and genomic selection. By addressing these challenges and opportunities, the broiler breeding industry can contribute to the sustainable and ethical production of animal protein that meets the demands of both farmers and consumers worldwide.

## 6 Where to look for further information

Neeteson et al. (2023) review the key developments in meat poultry breeding and provide an overview of the evolution of breeding goals and what has been achieved through breeding. The book of Aggrey et al. (2020) is a good resource for more information on several advancements in poultry genetics and genomics, with chapters on domestication, genetic background of key traits, breeding for disease resistance, use of omics data, and future challenges in poultry breeding. Finally, Brito et al. (2020) is a good review paper for those interested in large-scale phenotyping to improve animal welfare.

## 7 Acknowledgments

We would like to thank Anne-Marie Neeteson (Aviagen Group), Santiago Avendaño (Aviagen Group), and Rachel Hawken (Cobb-Vantress) for providing useful information and suggestions to an early version of this chapter.

## 8 References

Aggrey, P. S., et al. 2020. *Advances in poultry genetics and genomics*. Cambridge, UK: Burleigh Dodds Science Publishing.

Aldridge, M., et al. 2022. *DNA pooling is a cost effective method of including commercial crossbred data in selection of purebreds*. Proceedings of 12th World Congress on Genetics Applied to Livestock Production (WCGALP), 2022 Rotterdam, the Netherlands. Wageningen Academic Publishers, 781–784.

Alexander, P., et al. 2016. Human appropriation of land for food: the role of diet. *Global Environmental Change*, 41, 88–98.

Alvarenga, A. B., et al. 2021. A systematic review of genomic regions and candidate genes underlying behavioral traits in farmed mammals and their link with human disorders. *Animals*, 11, 715.

American Veterinary Medical Association. 2025. *FDA approves gene-editing tech creating PRRS-resistant pigs* [Online]. Available: https://www.avma.org/news/fda-approves-gene-editing-tech-creating-prrs-resistant-pigs [Accessed 07-05-2025].

Avendaño, S. N. and Fancher, B. 2017. Broiler breeding for sustainability and welfare–are there trade offs? the Poultry Beyond 2023, 6th International Broiler Nutritionists' Conference, 16–20 October 2017 2017 Queenstown, New Zealand, 17.

Aviagen Group. 2025. *Brands* [Online]. Available: https://aviagen.com/eu/brands/ [Accessed 2025-02-25].

Bailey, R. A., et al. 2020. Characterising the influence of genetics on breast muscle myopathies in broiler chickens. *Frontiers in Physiology*, 11, 1041.

Bar-Shira, E., et al. 2003. Establishment of immune competence in the avian GALT during the immediate post-hatch period. *Developmental and Comparative Immunology*, 27, 147–157.

Berghof, T. V., et al. 2019a. Selective breeding for high natural antibody level increases resistance to avian pathogenic Escherichia coli (APEC) in chickens. *Developmental & Comparative Immunology*, 93, 45–57.

Berghof, T. V., et al. 2019b. Opportunities to improve resilience in animal breeding programs. *Frontiers in Genetics*, 9, 692.

Berry, D. P., et al. 2016. Milk production and fertility in Cattle. *Annual Review of Animal Biosciences*, 4, 269–290.

Bickhart, D. M. 2024. *Validation of Including functional genome annotation maps in genomic prediction using FAETH scores*. Plant & Animal Genome Conference 31, San Diego, California.

Bink, M. C. A. M., et al. 2022. Predictive ability of genomic prediction in layers when including CADD scores as genome function information. *Proceedings of 12th World Congress on Genetics Applied to Livestock Production (WCGALP)*. Rotterdam, he Netherlands: Wageningen Academic Publishers.

Bovenhuis, H., et al. 2022. Divergent selection for natural antibodies in poultry in the presence of a major gene. *Genetics Selection Evolution*, 54, 24.

Brito, L. F., et al. 2020. Large-scale phenotyping of livestock welfare in commercial production systems: a new frontier in animal breeding. *Frontiers in Genetics*, 11.

Brugaletta, G., et al. 2022. A review of heat stress in chickens. Part I: insights into physiology and gut health. *Frontiers in physiology*, 13, 934381.

Calus, M., et al. 2023. Animal board invited review: the purebred-crossbred genetic correlation in poultry. *Animal*, 100997.

Canario, L., et al. 2013. Genetics of behavioural adaptation of livestock to farming conditions. *Animal*, 7, 357–377.

Cartoni Mancinelli, A., et al. 2023. Impact of chronic heat stress on behavior, oxidative status and meat quality traits of fast-growing broiler chickens. *Frontiers in Physiology*, 14, 1188175.

Cheruiyot, E. K., et al. 2022. Functionally prioritised whole-genome sequence variants improve the accuracy of genomic prediction for heat tolerance. *Genetics Selection Evolution*, 54, 17.

Chu, T. T., et al. 2018. Benefits of testing in both bio-secure and production environments in genomic selection breeding programs for commercial broiler chicken. *Genetics Selection Evolution*, 50, 52.

Chu, T. T., et al. 2019. Use of genomic information to exploit genotype-by-environment interactions for body weight of broiler chicken in bio-secure and production environments. *Genetics Selection Evolution*, 51, 50.

Clark, E. L., et al. 2020. From FAANG to fork: application of highly annotated genomes to improve farmed animal production. *Genome Biology*, 21, 285.

Cobb-Vantress Inc. 2024. *Products* [Online]. Available: https://www.cobbgenetics.com/products [Accessed 2025-02-25].

Coffey, M. 2020. Dairy cows: in the age of the genotype, #phenotypeisking. *Animal Frontiers*, 10, 19–22.

Colditz, I. G. & Hine, B. C. 2016. Resilience in farm animals: biology, management, breeding and implications for animal welfare. *Animal Production Science*, 56, 1961–1983.

Coles, D., et al. 2015. Ethical issues and potential stakeholder priorities associated with the application of genomic technologies applied to animal production systems. *Journal of Agricultural and Environmental Ethics*, 28, 231–253.

Cuyabano, B., et al. 2024. Genetic correlations: a parameter or a latent phenotype in genetic evaluations? *Interbull Bulletin*, 179–186.

D'Eath, R., et al. 2010. Breeding for behavioural change in farm animals: practical, economic and ethical considerations. *Animal Welfare*, 19, 17–27.

De Jong, G. 1995. Phenotypic plasticity as a product of selection in a variable environment. *The American Naturalist*, 145, 493–512.

de Los Campos, G., et al. 2013. Whole-genome regression and prediction methods applied to plant and animal breeding. *Genetics*, 193, 327–345.

Decuypere, E., et al. 2010. The broiler breeder paradox: ethical, genetic and physiological perspectives, and suggestions for solutions. *British Poultry Science*, 51, 569–579.

Dickerson, G. E. 1973. Inbreeding and heterosis in animals. *Journal of Animal Science*, 1973, 54–77.

Doornweerd, J. E., et al. 2021. Across-species pose estimation in poultry based on images using deep learning. *Frontiers in Animal Science*, 2.

Doornweerd, J. E., et al. 2023. Passive radio frequency identification and video tracking for the determination of location and movement of broilers. *Poultry Science*, 102, 102412.

Doornweerd, J. E., et al. 2024. Tracking individual broilers on video in terms of time and distance. *Poultry Science*, 103, 103185.

Duenk, P., et al. 2021. Review: optimizing genomic selection for crossbred performance by model improvement and data collection. *Journal of Animal Science*, 99.

Duenk, P., et al. 2024. Research note: effects of high barn temperature on group-level dispersion and individual activity in broiler chickens. *Poultry Science*, 103, 103901.

Dzungwe, J. T., et al. 2024. Effect of crossbreeding on growth performance, meat quality, and the economics of production of the pure and reciprocal crosses between the Sasso and Wassachie chickens. *Poultry Science*, 103, 103823.

Eggen, A. 2012. The development and application of genomic selection as a new breeding paradigm. *Animal Frontiers*, 2, 10–15.

Emmerson, D. 2003. Breeding objectives and selection strategies for broiler production. In: Muir, W. M. & Aggrey, S. E. (eds.) *Poultry breeding, genetics and biotechnology*. Wallingford: CABI.

Fodor, I., et al. 2023a. A systematic review of genotype-by-climate interaction studies in cattle, pigs, and chicken. *Frontiers in Animal Science*, 4.

Fodor, I., et al. 2023b. Automated pose estimation reveals walking characteristics associated with lameness in broilers. *Poultry Science*, 102, 102787.

Gorssen, W., et al. 2023. A promising resilience parameter for breeding: the use of weight and feed trajectories in growing pigs. *Journal of Animal Science and Biotechnology*, 14, 101.

Groß, C., et al. 2020. Prioritizing sequence variants in conserved non-coding elements in the chicken genome using chCADD. *PLOS Genetics*, 16, e1009027.

Gul, H., et al. 2022. Genetic resilience in chickens against bacterial, viral and protozoal pathogens. *Frontiers in Veterinary Science*, 9, 1032983.

Hartcher, K. M. & Lum, H. K. 2020. Genetic selection of broilers and welfare consequences: a review. *World's Poultry Science Journal*, 76, 154–167.

Havenstein, G. B., et al. 2003a. Carcass composition and yield of 1957 versus 2001 broilers when fed representative 1957 and 2001 broiler diets. *Poultry Science*, 82, 1509–1518.

Havenstein, G. B., et al. 2003b. Growth, livability, and feed conversion of 1957 versus 2001 broilers when fed representative 1957 and 2001 broiler diets. *Poultry Science*, 82, 1500–1508.

Hazel, L. N. 1943. The genetic basis for constructing selection indexes. *Genetics*, 28, 476–490.

Hedman, H. D., et al. 2020. A review of antimicrobial resistance in poultry farming within low-resource settings. *Animals*, 10, 1264.

Henchion, M., et al. 2021. Review: Trends for meat, milk and egg consumption for the next decades and the role played by livestock systems in the global production of proteins. *Animal*, 15, 100287.

Hiemstra, S. J. & Ten Napel, J. 2013. Study of the impact of genetic selection on the welfare of chickens bred and kept for meat production. *Final report of a project commissioned by the European Commission (DG SANCO 2011/12254).*

Hill, W., et al. 2016. Breeding for sustainability: maintaining and enhancing multi-trait genetic improvement. *CABI*, 193–213.

Hocking, P. M. 2014. Unexpected consequences of genetic selection in broilers and turkeys: problems and solutions. *British Poultry Science*, 55, 1–12.

Hu, Z.-L., et al. 2015. Developmental progress and current status of the Animal QTLdb. *Nucleic Acids Research*, 44, D827–D833.

Hubbard Breeders. 2025. *Hubbard products* [Online]. Available: https://www.hubbardbreeders.com/premium/premium-females/; https://www.hubbardbreeders.com/conventional/conventional-female/ [Accessed 2025-02-25].

Huerta, A., et al. 2023. Resiliency of fast-growing and slow-growing genotypes of broiler chickens submitted to different environmental temperatures: growth performance and meat quality. *Poultry Science*, 102, 103158.

Johnsson, M. 2020. Genetics and genomics of skeletal traits in poultry species. In: Aggrey, P. S., et al. (eds.) *Advances in poultry genetics and genomics.* Cambridge, UK: Burleigh Dodds Science Publishing.

Kapell, D. N., et al. 2012a. Genetic parameters of foot-pad dermatitis and body weight in purebred broiler lines in 2 contrasting environments. *Poultry Science*, 91, 565–574.

Kapell, D. N., et al. 2012b. Twenty-five years of selection for improved leg health in purebred broiler lines and underlying genetic parameters. *Poultry Science*, 91, 3032–343.

Katanbaf, M. N. & Hardiman, J. W. 2010. Primary broiler breeding–Striking a balance between economic and well-being traits1. *Poultry Science*, 89, 822–824.

Khwatenge, C. N. & Nahashon, S. N. 2021. Recent advances in the application of CRISPR/Cas9 gene editing system in poultry species. *Frontiers in Genetics*, 12, 2021.

Knap, P. W. 2005. Breeding robust pigs. *Australian Journal of Experimental Agriculture*, 45, 763–773.

Lau, N.-C. & Nath, M. J. 2014. Model simulation and projection of European heat waves in present-day and future climates. *Journal of Climate*, 27, 3713–3730.

Lee, V. E., et al. 2022. Social behavior in farm animals: applying fundamental theory to improve animal welfare. *Frontiers in Veterinary Science*, 9.

Lewis, P. D. & Gous, R. M. 2006. Effect of final photoperiod and twenty-week body weight on sexual maturity and early egg production in broiler breeders. *Poultry Science*, 85, 377–383.

Li, M., et al. 2015. Effects of heat stress on the daily behavior of wenchang chickens. *Brazilian Journal of Poultry Science*, 17.

Li, Z., et al. 2019. Integrating gene expression data into genomic prediction. *Frontiers in Genetics*, 10, 126.

Mancinelli, A. C., et al. 2023. Crossbreeding to improve local chicken breeds: predicting growth performance of the crosses using the Gompertz model and estimated heterosis. *Poultry Science*, 102, 102783.

Marchesi, J. A. P., et al. 2021. Exploring the genetic architecture of feed efficiency traits in chickens. *Scientific Reports*, 11, 4622.

Martini, J. W. R., et al. 2022. Incorporating omics data in genomic prediction. *Methods Molecular Biology*, 2467, 341–357.

Mebratie, W., et al. 2019. Genetic parameters for body weight and different definitions of residual feed intake in broiler chickens. *Genetics Selection Evolution*, 51, 53.

Menchetti, L., et al. 2024. Effect of genotype and nutritional and environmental challenges on growth curve dynamics of broiler chickens. *Poultry Science*, 103, 104095.

Meuwissen, T. H. E., et al. 2001. Prediction of total genetic value using genome-wide dense marker maps. *Genetics*, 157, 1819–1829.

Mignon-Grasteau, S. 1999. Genetic parameters of growth curve parameters in male and female chickens. *British Poultry Science*, 40, 44–51.

Mottet, A. & Tempio, G. 2017. Global poultry production: current state and future outlook and challenges. *World's Poultry Science Journal*, 73, 245–256.

Muir, W. M. & Aggrey, S. E. 2003. *Poultry genetics, breeding, and biotechnology*. Wallingford: CABI.

Nascimento, S. T., et al. 2017. Metabolic heat production and evaporation of poultry. *Poultry Science*, 96, 2691–2698.

Nawroth, C., et al. 2019. Farm animal cognition–linking behavior, welfare and ethics. *Frontiers in Veterinary Science*, 6, 24.

Neeteson-van Nieuwenhoven, A.-M., et al. 2013. The role of sustainable commercial pig and poultry breeding for food security. *Animal Frontiers*, 3, 52–57.

Neeteson, A.-M., et al. 2020. Poultry breeding for sustainability and welfare. In: Ahmadi, B. V., et al. (eds.) *The economics of farm animal welfare: theory, evidence and policy*. Wallingford: CABI.

Neeteson, A.-M., et al. 2023. Evolutions in commercial meat poultry breeding. *Animals*, 13, 3150.

Nielsen, H. M., et al. 2014. Approaches to formulating practical breeding objectives for animal production systems. *Acta Agriculturae Scandinavica, Section A – Animal Science*, 64, 2–12.

OECD & FAO 2022. Meat. *OECD-FAO Agricultural Outlook 2023–2032*.

OECD & FAO 2023. Meat. *OECD-FAO Agricultural Outlook 2023–2032*.

Oluwagbenga, E. M. & Fraley, G. S. 2023. Heat stress and poultry production: a comprehensive review. *Poultry Science*, 102, 103141.

Padhi, M. K. 2016. Importance of Indigenous breeds of chicken for rural economy and their improvements for higher production performance. *Scientifica*, 2016, 2604685.

Park, J. S., et al. 2020. Precise genome editing in poultry and its application to industries. *Genes*, 11, 1182.

Perez, B. C., et al. 2022. Adding gene transcripts into genomic prediction improves accuracy and reveals sampling time dependence. *G3 Genes|Genomes|Genetics*.

Pinard-van Der Laan, M.-H., et al. 2020. Genetics and genomics of immunity and disease traits in poultry species. In: Aggrey, P. S., et al. (eds.) *Advances in poultry genetics and genomics*. Cambridge, UK: Burleigh Dodds Science Publishing.

Poore, J. & Nemecek, T. 2018. Reducing food's environmental impacts through producers and consumers. *Science*, 360, 987–992.

Psifidi, A. 2021. The genetics of disease resistance in poultry. In: Smith, J. (ed.) *Poultry health: a guide for professionals*. London, UK: CABI Publishing.

Rauw, W. M. & Gomez-Raya, L. 2015. Genotype by environment interaction and breeding for robustness in livestock. *Frontiers in Genetics*, 6, 310.

Ravagnolo, O. & Misztal, I. 2000. Genetic component of heat stress in dairy cattle, parameter estimation. *Journal of Dairy Science*, 83, 2126–2130.

Raymond, B., et al. 2018. Genomic prediction for numerically small breeds, using models with pre-selected and differentially weighted markers. *Genetics Selection Evolution*, 50, 49.

Rekaya, R., et al. 2013. Genetic evaluation for growth, body composition, feed efficiency, and leg soundness. *Poultry Science*, 92, 923–929.

Saeed, M., et al. 2019. Heat stress management in poultry farms: a comprehensive overview. *Journal of Thermal Biology*, 84, 414–425.

Schelfaut, S. 2021. Na bijna 10 jaar strijd is de plofkip zo goed als dood. *Algemeen Dagblad*, 11-08-2021.

Siegel, P. B. 2014. Evolution of the modern broiler and feed efficiency. *Annual Review of Animal Biosciences*, 2, 375–385.

Smith, C. 1964. The use of specialised sire and dam lines in selection for meat production. *Animal Science*, 6, 337–344.

Soleimani, A. F., et al. 2011. Physiological responses of 3 chicken breeds to acute heat stress. *Poultry Science*, 90, 1435–1440.

The Nuffield Council on Bioethics 2021. Genome editing and farmed animal breeding: social and ethical issues. https://www.nuffieldbioethics.org/publications/genome-editing-and-farmed-animal-breeding.

Thompson, P. B. 2008. The opposite of human enhancement: nanotechnology and the blind chicken problem. *Nanoethics*, 2, 305–316.

Tixier-Boichard, M. 2020. From the jungle fowl to highly performing chickens: are we reaching limits? *World's Poultry Science Journal*, 76, 2–17.

Tixier-Boichard, M., et al. 2012. A century of poultry genetics. *World's Poultry Science Journal*, 68, 307–321.

van Emous, R. A., et al. 2024. Effects of different feeding strategies on behavior and performance in broiler breeder pullets. *Poultry Science*, 103, 104336.

Wasti, S., et al. 2020. Impact of heat stress on poultry health and performances, and potential mitigation strategies. *Animals*, 10, 1266.

Whitton, C., et al. 2021. Are we approaching peak meat consumption? Analysis of meat consumption from 2000 to 2019 in 35 countries and its relationship to gross domestic product. *Animals*, 11, 3466.

Wientjes, Y. C. J., et al. 2020. Optimizing genomic reference populations to improve crossbred performance. *Genetics Selection Evolution*, 52, 65.

Wilhelmsson, S., et al. 2019. Welfare Quality® assessment of a fast-growing and a slower-growing broiler hybrid, reared until 10 weeks and fed a low-protein, high-protein or mussel-meal diet. *Livestock Science*, 219, 71–79.

Woelders, H., et al. 2006. Animal genetic resources conservation in the Netherlands and Europe: poultry perspective1. *Poultry Science*, 85, 216–222.

Wolc, A., et al. 2016. Implementation of genomic selection in the poultry industry. *Animal Frontiers*, 6, 23–31.

World Organisation for Animal Health 2024. Annual report on antimicrobial agents intended for use in animals. Paris.

Zuidhof, M. J., et al. 2014. Growth, efficiency, and yield of commercial broilers from 1957, 1978, and 2005. *Poultry Science*, 93, 2970–2982.

# Chapter 5

# Chick quality as affected by the broiler breeder and breeder house management

*Michael J. Wineland, North Carolina State University, USA*

1 Introduction
2 Methods of measuring chick quality typically used
3 Nutrients available to the developing embryo
4 Egg component characteristics
5 Influence of eggshell conductance on chick quality
6 Influence of egg size and breeder age on chick quality
7 Influence of egg storage on chick quality
8 Eggshell quality and its influence on chick quality
9 Egg contamination and egg sweating influence on chick quality
10 Influence of eggs set upside down for incubation and the resulting chick quality
11 Conclusion
12 References

## 1 Introduction

It is often said that you cannot improve chick quality once the broiler chick has hatched. There are many areas of poultry production that impact the resulting chick quality before they reach the broiler house. The maintenance of chick quality is like a chain; if one link is broken in the series, you are unable to produce high-quality chicks. This leaves broiler farms to have to work with the sub-optimal quality that they are given.

Chick quality begins in the breeder house. This includes the genetics and health of the hens laying the eggs, the nutrients that the hen places in the egg for the developing embryo, and the many management procedures such as egg storage conditions and length of storage that are used prior to incubation. It is also important to understand that breeder and egg characteristics can influence incubational processes that can affect chick quality. It has been observed that

http://dx.doi.org/10.19103/AS.2025.0165.06

genetic differences impact how the embryos develop during incubation and then the resulting chick quality. Incubating different breeds whose eggshells possess different qualities in the same incubator is not always advisable but many times unavoidable. Genetic selection has influenced embryonic growth during development which affects how they respond to incubation conditions. Tona et al. (2010) using both Cobb and Ross eggs showed how heat production and growth differences during incubation could impact chick quality if incubated together. Besides the hen's influence, the male also influences how the embryo grows. Van den Brand et al. (2019) showed that besides incubation conditions, the rooster line can influence embryo metabolism and subsequent chick quality.

The eggs produced by the hen can be influenced by many management factors that one finds occurring in the breeder house. Zukiwsky et al. (2021) showed that maternal body weight due to a restricted or relaxed feed regime had little influence on broiler performance. However, chick quality can also be influenced by factors such as social pressure, disease, and heat stress which can impact feed consumption.

The methodology of measuring chick quality is varied, but all procedures are based upon physical criteria of the hatchling. It is most important though to realize that chick quality is dependent upon factors we can't measure in the day-old chicks. Good quality chick has optimal development of its various organ systems to optimize performance. This chapter will discuss how the breeder and its environment influence chick quality and the potential impact on growth in the field.

## 2 Methods of measuring chick quality typically used

Traditionally the weight of the hatched chick had been used as a measure of chick quality (Wolanski et al., 2007). The amount of residual yolk though can vary influencing chick weight depending upon genetic strain (Wolanski et al., 2006), and the relative percent residual yolk can be affected by the age of the breeder hen (Sahan et al. 2014). Research by Joseph et al. (2006) and Wineland et al. (2006a) showed that variations in chick weight can also be attributed to variations in the weight of the residual yolk caused by eggshell conductance and/or incubation conditions. The incubational temperature has been shown to change the expression of yolk sac tissue genes which leads to variation in yolk utilization (Dayan et al. 2020).

Chick length has also been used as a measure of chick quality. Hamidu et al. (2007) showed that chick length increased as the breeder hens became older. Sahan et al. (2014) demonstrated that chick length and chick weight increased as breeders increased in age and as egg size increased. Petek (2010) found a positive correlation between chick weight and chick length and that the longer chicks had better feed efficiency and less mortality. Measuring chick length can sometimes vary when performed by different people because of the procedure used. An alternative to chick length is using shank length which can be measured with less human procedural variation. Wolanski et al. (2007) showed that shank length and chick length correlated more strongly with 14-day body weight than did body weight at hatch.

There have been attempts to use other visual criteria such as hydration status, utilization of residual yolk, alertness, navel condition, incidence of cull chicks, and signs of stressful hatching into a numerical grade. Systems developed at the University of Leuven have the Tona scoring system and Pas Reform having developed the Pasgar scoring system. More recently Cobb-Vantress has developed a similar method using visual characteristics. All criteria used to evaluate chick quality deal with physical attributes. However, when we think of chick quality, what is more important is what we don't see. Most critical is the optimal development of organ systems that will support the growth of the broiler. These organ systems will be influenced by the nutrients available in the egg, eggshell conductance, and incubation conditions. The embryo's organ systems which are developed during incubation and post-hatch are what allow the chick to reach its genetic potential.

## 3 Nutrients available to the developing embryo

The breeder hen is restricted in the amount of feed she receives, so it is critical that diet formulation and feeder management are optimal. The breeder hen must place all the nutrients into the egg that the chick needs to grow as an embryo, to hatch and to do well in the broiler house. Some of the nutrients are found in the yolk, while other nutrients are found in the albumen as well as the shell. How well this occurs can be influenced by many factors. The nutrition available to the breeder hen has vastly improved over the years. However, it is important to keep in mind that the hen may not be receiving the correct nutrients if the feed mill may have made a mistake in mixing the diet. Additionally, it is important that the scales regulating the amount of feed in the breeder house are accurate, that there is proper feed distribution, and that the hen's digestive health is good so that she can assimilate the nutrients she has been provided with. Also important is the condition of the eggshell, as it provides minerals for the developing embryo, it helps to prevent microbial invasion and is important in the movement of gases between the inside and outside of the egg.

Since breeder hens are restricted in feed consumption during their rearing and lay period, there is interest in how the hen responds to feed restriction with egg production and how the restriction influences the chick. A trial which looked at matched or mismatched nutritional environments for hens and their broilers (Van der Waaij et al. 2011), the nutritional environments were either *ad libitum* or restricted. This trial demonstrated that offspring of the hens fed *ad libitum* were heavier than the broilers from the restricted hens. Van Emous et al. (2015) used two different growth patterns of breeder hens to reach body weights of 2200 and 2400 g at 20 weeks of age. Three crude protein intakes were used and were shown not to influence fertility and hatchability, and no difference was noticed other than increased feed consumption of broilers. They suggested that a higher pattern of growth during rearing was effective in fertility and broiler performance of the young breeders.

The protein and energy provided to the hens and placed into the eggs are important for hen productivity and subsequent chick quality. Yadgary et al. (2010) showed that at the time of hatch the amount of carbohydrate, protein, and water in the residual yolk in chicks from old and young breeders was similar. What was of note was that there was less fat relative to yolk-free body weight in chicks from young breeder hens compared to the older breeder hens.

The source of feed energy is important; Oliveira et al. (2010) used a commercial egg laying strain and used four lipid sources in the diet (soybean oil, sunflower oil, linseed oil, and no oil added as the control). They suggest that the fatty acid quantity in the egg yolk could be altered by the source of the lipid in the hen diet. Interestingly there was a decrease in PUFA in the egg yolks as the hens got older. Latour et al. (1998) also examined the type of fat in the diet. They used three different fat sources of corn oil, poultry fat, or lard in breeder diets of hens, which were 38, 51, and 64 weeks of age. They suggest that the age of the breeder can influence the utilization of lipids in the egg by the embryo. Also, they note that the type of fat in the hen diet may also influence the utilization of fat by the embryo. Trials with corn oil and poultry fat (Peebles et al. 1999) demonstrated that energy sources with less saturated fat improved the growout performance of the broilers.

Spratt and Leeson (1987) using maternal diets during lay differing in energy and protein concluded that breeder protein had little or no effect on broiler weight gain and that breeder energy intake had early effects but non-significant effects at market age. Moraes et al. (2014) used different energy and protein diets during rearing and egg production and indicated that hens fed the higher energy diet (2736 versus 2528 kcal/kg) during rearing produced fatter pullets and also produced fatter broilers. A subsequent report by Moraes et al. (2019) using diets during rearing and lay with different protein and energy levels showed transgenerational effects on the broiler, specifically measured by broiler growth. They demonstrated that high crude protein (15.3%) during rearing of the pullets increased broiler body weights. They also showed that during lay the low maternal metabolizable energy (2800 kcal/kg) provided higher body weights in broilers. Lopez and Leeson (1995) utilized hen diets of 16%, 14%, 12%, and 10% crude protein but kept the methionine + cysteine and lysine levels constant. Lower chick weights resulted as protein decreased but had no long-term broiler effects at 48 days with improved feed conversion.

Menge et al. (1964) showed a negative effect on chick growth when breeder diets were deficient in essential fatty acids. Menge and Richardson (1968) demonstrated that a linoleic acid deficiency was found in the plasma, heart, and brain lipids of chicks that hatched from deficient hens. They noted that the deficient chicks were weak and had equilibrium problems and trouble finding feed and water. Hopkins and Nesheim (1962) showed that essential fatty acids–deficient hens produced hatchlings that grew slower than chicks from non-deficient hens. Marion and Edwards (1962) showed that chicks from hens fed a diet adequate in corn oil grew faster than chicks from hens fed a low-fat diet. Noble and Cocchi (1990) indicated that changes in the composition of yolk lipids can be caused by nutritional or environmental stress. Noble et al. (1986) showed that embryos from young parents (no age mentioned in abstract) have trouble transferring yolk lipids causing abnormal tissue development patterns and metabolism which appears to be related to greater embryonic mortality. A trial by Sahan et al. (2014) showed that there were differences in yolk fatty acid composition before incubation and that differences existed at hatch between 36- and 52-week-old breeder chicks. They concluded that the age of the broiler breeder affected yolk absorption and fatty acid concentrations during incubation with resulting differences in chick quality. Oliveira et al. (2010) conclude that the quantity of fatty acids found in the egg yolk can be altered by the source of the lipids in the diet.

Many of the nutrients provided to the hen are assimilated and placed into the egg in proportion to the nutrients in the feed. Vitamin deficiencies of the hen impacting chick quality have been studied extensively. It is important to remember that some vitamins cannot be synthesized by poultry, and thus they must be supplemented in the diet. Studies examining the effects of breeder vitamin supplementation on progeny include Squires et al. (1993) where they demonstrated that depressed concentrations of albumen riboflavin (vitamin B2) increased embryonic developmental problems but also the incidence of club down in chicks. Additionally, they noted decreased egg weights and hen weights with the deficit diet. Patel and McGinnis (1977) and Peterson et al. (1953) showed that vitamin B12 supplied in 6-month-old white leghorn breeder diet was more effective in stimulating the early growth of the chick than supplementation in the chick diet. When hens were fed high levels of vitamin D3, the chicks exhibited body weight improvements, and they suggest that the level of vitamin D3 for obtaining the healthiest chicks is higher than the level needed for optimal reproduction (Atencio et al. 2005). These investigators also noted that the highest levels also resulted in lower incidence of calcium rickets.

Antioxidants in breeder hen diets play an important role in the quality of hatched chicks. Bautista-Ortega et al. (2009) showed that when egg lipids are metabolized by the embryo, long-chain polyunsaturated fatty acids (PUFA) are formed. The high concentrations of PUFA in the cell membranes increase the susceptibility of the tissue to degradation (Selim et al. 2012). Antioxidants can help prevent peroxidation of lipids in the egg yolk when fed to the breeder hen. Compounds with antioxidant activity are selenium and vitamins A, D, and E. Urso et al. (2015) in a study supplementing Vitamin E to breeder hens improved hatchability and the quality of the chicks. Araujo et al. (2019a) showed that providing vitamin E in-ovo improved chick quality and performance by improving the chick's oxidative state. Surai et al. (2019) showed that increased supplementation of vitamin E in the breeder's diet increased the antioxidant defenses and decreased lipid peroxidation in newly hatched chicks.

Vitamin A levels fed to hens come mainly from feed carotenoids which act as antioxidants and help with immune regulation and enhanced antioxidative status and immunity of the chicks (El-Ratel et al. 2024, Wang et al. 2020). Breeders being provided canthaxanthin and 25-OH $D_3$ helped with improved chick quality as both have high antioxidant activity (Araujo et al. 2019b). It was demonstrated by Swaggerty et al. (2023) that the addition of antioxidants and complex biofactors to the diet of breeder hens offered some measure of protection in the chicks from *S.* Enteritidis. Saunders-Blades and Korver (2015) showed that maternal vitamin D increased the chicks' innate immunity towards *E. coli*. The hens' nutritional status because of the nutrients provided to them in the breeder house impacts the embryos' nutritional availability during embryonic growth and subsequent chick quality.

## 4 Egg component characteristics

Gharahveysi et al. (2012) showed that as breeder age increased, the egg components were influenced. It is known that the percentages of yolk and albumen in an egg will vary by hen age. Nangsuay et al. (2015a) comparing broiler and layer type strains showed that broiler eggs have a greater ratio of yolk to albumen than layer eggs and thus the broiler egg had more yolk energy compared to the layer egg. This difference contributed to broilers having greater yolk-free body mass, chick weight, and chick length than in the layers. This also created more metabolic heat production from broiler embryos, which can impact incubation conditions. The interesting part though is that both the broiler and layer embryos were equal in efficiency of converting energy from the egg into yolk-free body mass.

Genetic selection has changed breeder characteristics tremendously over the years, which may influence how the embryo and hatchling respond to incubational and growout conditions. The genetic differences between Cobb 500 and Ross 308 broiler strains (Nangsuay et al. 2015b) demonstrated differences between the two lines in dry matter and crude fat without differences in crude protein. It was also noted that the Cobb 500 chicks hatched about 4 hours earlier than the Ross 308 chicks. Ross 308 was also slightly more efficient in converting energy to yolk-free body mass.

Hamidu et al. (2007) also researched potential differences in egg components between the Ross 308 and Cobb 500 strains. These investigators showed there was no difference in wet or dry yolk percent, wet albumen percent, or wet and dry shell percent. They did note though that the percentage of wet and dry yolk increased while the percentage of wet albumen decreased as the hens aged. The percentage yolk-free body mass calculated as a percentage of original egg weight was higher in young and prime breeders compared to older flocks because of the greater yolk size in older hens (Peebles et al. 2001, Suarez et al. 1997).

Maternal antibodies are placed in the egg by the hen for use by the newly hatched chick until its own immunity system becomes capable (Ismiraj et al. 2019). Hamal et al. (2006) showed that IgY was transferred to the chick at approximately 30% of what was in the hen. A trial by Leandro et al. (2011) suggested that the breeder strain affected the transfer of immunoglobulins to the chicks and had some influence on the development of spleen and bursa. Interestingly, Wang et al. (2004) suggested that the ratio of linolenic acid to alpha linolenic acid of the hens diet had no effect on the hens' humoral response but did influence the passive immunity of their chicks, but further work was needed to understand the mechanism.

Tona et al. (2004) said that day-old chick quality and future performance can be influenced by both incubational factors and breeder flock influences and sometimes an interaction between the two. The egg energy during incubation is portioned between tissue accumulation, maintenance, and growth. These investigators showed that chick quality as measured by physical characteristics of the chick was better from younger breeders than from older breeders.

## 5 Influence of eggshell conductance on chick quality

The influence of egg size and eggshell conductance on chick quality has been investigated by several people. Eggshell conductance is the ability of the eggshell to allow gases $O_2$, $CO_2$, and water vapor) to pass through the pores of the shell by passive diffusion which is important for the proper development of the embryo. During incubation the embryo goes through a period of slight hypoxia when the amount of oxygen passing through the pores has been maximized. The diffusion is limited by the number of eggshell pores, the diameter of the pores, and the length of the pores (eggshell thickness) until the embryo internally pips into the air cell. This period of slight hypoxia is called the plateau stage of oxygen consumption and is a period of development when the embryo will supplement its energy needs using anaerobic sources of energy.

It is important to keep in mind that as egg weight increases, there is not a linear increase in the surface area of the eggshell and thus pore numbers impacting the exchange of gases ($O_2$, $CO_2$, $H_2O$) in and out of the egg. Thus, the less than linear increase in surface area as egg weight increases will affect the respiration of the embryo and the removal of metabolic heat from embryos.

Eggshell conductance has an influence upon organ development during incubation such as intestinal maturation, broiler bone development, and thyroid maturation (Wineland et al. 2006a, 2006b; Oviedo-Rondon et al. 2008). These and other maturational processes are going to impact broiler performance and quality. O'Dea et al. (2004) investigated the effect of eggshell conductance upon chick quality parameters comparing a 1978 line with a 2000's strain (Hubbard). The results showed no difference in eggshell conductance between the strains when eggs were of the same weight but did note a higher conductance as the hens aged. Breeder age effect on conductance was confirmed by Nangsuay et al. (2017). Eggshell conductance studied by Hamidu et al. (2007) using Cobb 500 and Ross 308 when egg weights were the same showed a numerical but not statistical difference with Cobb having a lower eggshell conductance. Tona et al. (2010) did show differences between Cobb and Ross breeder eggs with regard to percent moisture loss in eggs during incubation which would be an indicator that eggshell conductance is different between the two strains. Tona's results were confirmed by Nangsuay et al. (2016, 2017) when they showed Cobb 500 had a significantly lower eggshell conductance than Ross 308. Nangsuay et al. (2016) results also showed significantly less yolk-free body weight for the Cobb chicks. These investigators believe that strain differences in eggshell conductance when incubated using the same eggshell temperature influence chick quality and embryonic growth. Buys et al. (1998) suggest that any differences in eggshell conductance during incubation could predispose chicks to ascites because of altered oxygen availability during embryonic development.

The altitude at which the hens are housed and lay their eggs is important as it can result in altered eggshell conductance. The water vapor conductance of eggs from hens when collected at high altitudes decreases (Carey 1980, Carey et al. 1984, Rahn et al. 1977). Rahn et al. (1982) showed that eggs from hens which had been acclimated at 3800 meters and transferred to 1200 meters changed their eggshell conductance over 2 months. They noted that after the hens were transferred to the lower altitude, the eggshell conductance increased 30%, mainly in cross-sectional areas of individual pores. Hempleman et al. (1993) suggest that the altered conductance due to altitude is in response to hypoxia at high altitudes. Flock age and strain can influence the metabolism of the embryo which in turn could impact chick quality (Hamidu et al. 2007). These investigators showed that oxygen consumption, carbon dioxide production, and metabolic heat production were different between Ross 308 and Cobb 500 during the latter stages of incubation. This suggests that the quality of chicks could be different because of this heat production; if incubation parameters were the same for both strains, they would likely respond differently to the conditions, and this could result in different chick quality. The investigators also demonstrated that metabolic heat production of the embryos was greater in the 55- to 59-week-old flocks compared to 34-, 36-, or 40-week-old flocks, creating chick quality problems.

## 6 Influence of egg size and breeder age on chick quality

The size of the egg has been implicated in chick quality. Wyatt et al. (1985) demonstrated that egg size was important for broiler growth. When comparing eggs between 47–54 g and 58–66 g, the chickens from the larger eggs were heavier at hatch and 91 g heavier at 49 days. A minimum egg weight is regarded as an important criterion when a flock comes into production because it would impact day-old chick weight, chick quality, and subsequent performance. Egg size or egg weight increases as the age of the broiler breeder increases with some variation in size at any one age (McNaughton et al. 1978); therefore, we will discuss egg size and breeder age together. Wolanski et al. (2007) suggest that the strain of breeders can influence average egg weight depending upon the selection criteria of the strain for the broiler. Initial chick weight was found to be positively correlated to egg weight by Sinclair et al. (1990). A lower chick quality from young breeders which tend to have smaller eggs was shown by Peebles et al. (2004). Sinclair et al. (1990) noted that the egg size that is influenced by breeder age influences chick weight and differences in broiler weight at 6 weeks of age. It is well known that within one breeder age you will find variability in egg size. McNaughton et al. (1978) investigated the effects of egg size in breeder flocks that were 29 and 58 weeks of age, where the eggs were separated into two weight groups for each age: 47–54 g and 57–62 g for the 29-week-old breeder and 57–62 g and 67–74 g for the 58-week-old broiler flock. Table 1 shows the market age body weights for each group in the two trials.

The results demonstrated that larger market age broilers were found from the larger eggs for both breeder ages. Interestingly, the 57–62 g eggs in the two ages showed the same final body weights. Damaziak et al. (2018) said that hatching body weight was correlated with egg weight, which has an early effect upon body weight of broilers, but did not influence body weight at 42 days. Iqbal et al. (2017, 2023) (same flock source for eggs) demonstrated in 60-week-old Hubbard Classic that egg weights had less impact on post-hatch performance as the broilers grew. These investigators showed that egg weight affected chick weight and chick length at hatch and performance up to 3 weeks

**Table 1** Egg weight and breeder age effect upon chick body weight.

| Breeder age (weeks) | Egg weight group (g) | Day-old weight (g) | Market age weight |
|---|---|---|---|
| Experiment 1 | | | |
| 29 | 47–54 | 33 | 1598[b] |
| 29 | 57–62 | 37 | 1668[a] |
| 58 | 57–62 | 38 | 1698[a] |
| 58 | 67–74 | 44 | 1717[a] |
| Experiment 2 | | | |
| 29 | 47–54 | 33 | 1480[b] |
| 29 | 57–62 | 36 | 1591[a] |
| 58 | 57–62 | 37 | 1572[a] |
| 58 | 67–74 | 43 | 1591[a] |

McNaughton et al. (1978).

of age but demonstrated no effect on final live body weight, feed conversion, or mortality (60-week-old breeders).

Tufft and Jensen (1991) demonstrated that the size of the egg, rather than age of the hen, determines the early rate of growth of the broiler. The difference in growth they noted though disappeared by 6 weeks of age. Tona et al. (2004) reported that chick hatch weight did not correlate with 42-day broiler weight. This was confirmed by Nasri et al. (2020) when they saw no difference in the final body weight of broilers between 47- and 67-week-old breeders but did see a better feed conversion ratio (FCR) from the 47-week-old breeder broilers. Gardiner (1973) separated eggs by weight from a single flock and showed that by 8 weeks of age the female broilers from larger eggs resulted in heavier female broilers but that the difference that originally was present in males disappeared by 8 weeks.

Franco-Ulmer et al. (2010) used three different egg weights: light, average, and heavy eggs from both 29- and 59-week-old flocks. Greater chick weight was noted from the older hens and for eggs that were heavier egg weights. The 41-day broilers from light eggs were significantly less than the average for the heavy eggs, and the broilers from the older hens were significantly heavier. Tindell et al. (1964) also noted the advantages of higher body weight in the broilers from heavier eggs.

It was discussed by Wolanski et al. (2006) that when looking at the relationship between the growth rate of the broiler and egg weight, you must look at appetite, yolk utilization, and water consumption. Pinchasov (1991) noted that feed intake becomes a better indicator of final body weight than the day-old body weight.

The time of day when oviposition takes place also impacts egg size. Yannakopoulos and Tserveni-Gousi (1987) demonstrated that eggs laid in the afternoon were significantly smaller than those laid in the morning. This was confirmed by Ruiz and Lunam (2002) and Zakaria et al. (2005), showing that eggs laid earlier in the day were heavier than eggs laid later in the day. It was shown by Robinson et al. (1991) that the first egg of the sequence was heavier than subsequent eggs in the sequence.

Nangsuay (2011) demonstrated that egg size influences embryo development when expressed in terms of total and yolk-free body weight and chick length at hatch. Chick length, as mentioned earlier, is one method of evaluating chick quality. Nasri et al. (2020) presented data that showed older hens, 67 weeks of age, produced chicks with larger residual yolk, larger heart and liver as percent of yolk-free body mass compared to chicks from a 47-week-old flock.

Egg weights can also vary by a particular strain, Wolanski et al. (2006) showed that higher egg weights from a strain selected for yield compared to lines selected for multipurpose products. Robinson et al. (1991) using three broiler breeder lines showed that the line that had the greatest egg production had lighter eggs. In a trial by Vieira and Moran (1998a, 1998b) with four different lines, in each instance lighter eggs produced smaller 1-day-old chicks, and after a 49-day growout period, the lighter eggs had significantly lower weight broilers. But of interest was that the lighter eggs had significantly less mortality during growout. Is it possible that the heavier broilers had a greater incidence of metabolic problems?

Ricklefs and Starck (1998) said that egg weight and length of incubation period have the greatest influence on hatch weight. Tona et al. (2010) used Cobb 500 and Ross 308 eggs that were of the same size, where hatch weight was the same, and post-hatch growth and heat production for 7 days post-hatch were elevated for Cobb. Nangsuay et al. (2013) studied the interaction between breeder age (29 versus 53 weeks) and two egg sizes (57–61 g and 66–70 g) on energy expenditure and heat production by embryos. These investigators showed that the efficiency of being able to convert egg energy into chick energy was equal for both egg sizes and both breeder age groups. However, because the eggs were larger in the older flocks, they produced larger yolk-free body mass and had more available energy, and as a result, this was responsible for producing more metabolic heat than from younger flocks.

The resulting chick quality we see in the hatchery and on the broiler farm is obviously influenced by incubation as egg size can impact embryo size and the temperature the embryo develops at, because larger eggs produce more metabolic heat due to embryo size that subsequently needs to be removed as they grow (Lourens et al. 2006). If both egg sizes are in the same incubator, then we will see either too low an incubator temperature for the small eggs or too high an incubator temperature for the large eggs, all of which will impact the proper development of organ systems, influencing post-hatch quality and performance. Air flow is different between many brands of incubators; thus, larger eggs in some incubator types can influence the air flow over the eggs, which helps to remove the excess metabolic heat. Nowak et al. (2019) using eggs from hens 32, 42, and 52 weeks of age examined chicks for developmental defects. They found that the greatest number of cull chicks were showing problems with the abdominal wall and the umbilicus. Additionally, they found the greatest incidence of malformations occurred in the hens of 32 weeks of age and suggested that incubational conditions may need to be altered for these hens' eggs.

Hudson et al. (2004) examined chick quality within a single flock at different ages and felt that some chicks from young breeders were hampered by underdeveloped supply organs. Proper development of organ systems is essential. Maiorka et al. (2016) showed in a study of eggs from 30- to 60-week-old breeders that the intestinal villi were longer with deeper crypts for the embryos from older breeders, suggesting that embryos from older breeders showed greater small intestinal development, thus increasing the potential for better growth. This was confirmed in a work by El Sabry et al. (2013). However, Machado et al. (2020) showed that while chicks from older slow-growing breeders (38-week-old) were heavier at 21 days of age, the chicks from the older hens had taller villi but less crypt depth than chicks from younger hens (51-week-old). It was noted by Sklan et al. (2003) that heavier chicks at hatch had greater satellite cell proliferation and earlier differentiation. The myoblast proliferation was greater in embryos from larger eggs compared to smaller eggs; thus, larger chicks at hatch produce heavier broilers at marketing. Another important factor investigated by Weytjens et al. (1999) discussed development of thermoregulation. These investigators showed that chicks from older breeder flocks were more resilient to cold, while chicks from young breeders were more resistant to heat.

Egg size effects upon processing were examined when eggs were divided into three egg weight groups by Duman and Sekeroglu (2017). These investigators found that the chicks from medium-sized eggs had the best results, expressed as better cooled carcass yield at processing than the other two groups. While the heavier eggs had significantly better 7-day body weights, the differences had disappeared by 39 days.

It is important to realize that the characteristics of the eggs, whether it is egg size, strain, or breeder age, will impact chick quality unless specific incubation conditions are utilized to match the needs of the developing embryo. Also, the length of egg storage, the quality of the eggshell integrity, and cleanliness play an important part. If a poor-quality egg pack is going into the hatchery, one cannot expect the hatchery to produce high-quality chicks.

## 7 Influence of egg storage on chick quality

Egg storage is a necessity in poultry production, as the hens don't stop or slow egg production when a seasonal or holiday cutback in broilers to be processed is needed. When egg storage is necessary, one must consider egg storage temperature. Egg storage beyond seven plus days has been found to alter incubational results but also influence the physiology of the hatched chick. Ruiz and Lunam (2002) describe experiments where they had 1–3-day stored and 9–11-day stored eggs. These investigators showed no differences in chick weight when eggs were stored at 20°C or 16.5°C for short periods, but for the long-stored eggs, the chick weight was better when stored at 10°C versus 16.5°C. Egg storage for 8 days has been found to have a profound effect on chick quality and albumen pH (Lapao et al. 1999) and thus performance. Tona et al. (2003a) demonstrated that 18-day stored eggs had a greater effect, resulting in poorer chick quality than 3-day stored eggs. They also scored chick quality and noted a greater incidence of abnormalities in the chicks from the 18-day stored eggs. Damaziak et al. (2018) used eggs from breeder flocks that were 49, 52, 70, and 73 weeks of age and demonstrated that eggs from older breeders, where the eggs were turned during storage for the 12-day stored eggs, had a higher body weight at 42 days than eggs not turned during storage. This same effect was not seen in eggs from younger hens. Tona et al. (2004) showed that in both 35- and 45-week-old breeders, the eggs stored for 7 days resulted in lower broiler growth to 7 days of age than in fresher eggs. Similarly, Livingston et al. (2019) demonstrated a decrease in weight gain through 42 days in eggs that had been stored for 8–14 days versus eggs stored for 1–7 days (Table 2).

Nasri et al. (2020) also showed that weight gain and feed conversion were poorer in the chicks hatched from 19-day stored eggs. These same investigators showed that at hatch the residual yolk weight was greater and the liver and intestine percentage was less in chicks from longer stored eggs. Shim et al. (2008) demonstrated that when eggs were stored for either 0 days or 10 days, the chick growth was less in the stored eggs for both 27- and 61-week-old breeder flocks. They also noted that 10-day storage exhibited a higher phosphorus rickets score in older flocks (61 weeks old) but not in the younger flock (27 weeks old).

**Table 2** Effect of length of egg storage on body weight gain

| Egg storage length (days) | Body weight 42 days (Kg) |
|---|---|
| 1–7 days | 3.153[a] |
| 8–14 days | 3.056[b] |

Livingston et al. (2019).

It is suggested by Ebeid et al. (2017) that egg storage damaged the antioxidant capacity of hatched chicks, suggesting that during long-term egg storage, there is oxidative stress on eggs. Yang et al. (2021, 2020) demonstrated that prolonged egg storage decreased growth and antioxidant capacity of chicks, but supplementation of hens with vitamin E could alleviate in part the reduction of antioxidant capacity. Yang et al. (2021) showed how long-term storage of eggs reduced the antioxidant capacity of chicks in both the serum and liver. They also noted that while additional maternal vitamin E slightly improved the growth in the first week post hatch, in the stored eggs the chick's growth performance showed no improvement. It was concluded by Whitehead et al. (1985) that after egg storage there was no increase in physical abnormalities in hatched chicks that were typical of vitamin deficiencies. They did note though that chicks from stored eggs did show mild macrocytic anemia characteristic of folic acid (vitamin B9) or cobalamin (vitamin B12) deficiencies.

Ebeid et al. (2017) demonstrated that pre-storage incubation of 14-day stored eggs influenced some of the organ systems of hatched chicks. They showed that pre-incubation before storage for 6 hours improved $T_3$, $T_4$ levels, as well as glutathione peroxidase and IgY concentrations and decreased malondialdehyde levels in newly hatched chicks over those stored eggs not treated. Tona et al. (2003b) demonstrated a higher $T_3/T_4$ ratio and elevated corticosterone levels of newly hatched chicks from 18-day stored eggs versus 3-day stored eggs. Uddin and Hamidu (2014) saw changes in oxygen consumption, carbon dioxide production, and heat production of embryos from stored eggs, which they suggest may attribute to malformation increases and poorer chick quality. Tona et al. (2004) additionally found that 7-day storage eggs affected chick quality from 45-week-old breeders more than from the 35-week-old young breeders. They also showed lower albumen Haugh units with egg storage. One may wish to consider setting eggs from older breeders first and storing the eggs from the younger breeders if eggs must be stored.

Prolonged egg storage has also been suggested to affect the transfer of maternal antibodies of hatched ducklings (El-Fiky et al. 2020). Goliomytis et al. (2015) indicate that prolonged storage of hatching eggs can compromise the hatchlings' immunocompetence. They point out that the effect is on certain aspects of the hatchlings' acquired immunity and their ability to raise protective immunological responses following immunizations. Interestingly, Nilsson et al. (2012) showed that eggs stored at +4.0°C for 4 weeks had no change in antibody titers.

## 8 Eggshell quality and its influence on chick quality

Translucent eggs, sometimes called eggshell mottling, have been a concern in the industry for years, often seen increasing in frequency as the hen ages. Wang et al. (2017) examined both translucent eggs and opaque eggs and believe that both the eggshell and eggshell membranes influence the incidence of translucent eggs. Translucency of the eggshells is due to a disorganization of the mammillary and palisade layers of the shell causing a weaker eggshell. The result of this disorganization of the shell allows water from the egg to accumulate in the eggshell to varying degrees. It is suggested that hairline cracks have a greater opportunity to develop where eggshell mottling exists (Wong et al. 2020). Neto et al. (2024) found no relationship of eggshell translucency to specific gravity, but there was greater egg weight loss due to increased translucency. Severely translucent eggs were found to not only have greater moisture loss during incubation but also a significant reduction in hatchability (Orellana et al. 2023). These investigators also noted that the eggshell was thinner in the more normal eggs and thicker in the more severe translucent eggs. Van den Brand (2023) in trials using organic trace minerals while looking at the severity of translucency concluded that translucency of eggshells is determined by flock characteristics and not necessarily by breeder age.

Specific gravity as a measure of shell quality has always been of interest regarding hatchability and chick quality (McDaniel et al. 1979, 1981, Neto et al., 2024). Lower specific gravity of eggs increased as the flocks aged (McDaniel et al. 1979). The investigators also noted decreased hatch in lower specific gravity eggs. Neto et al. (2024) also noted increased egg contamination with eggs displaying lower specific gravity.

Neto et al. (2024) also examined eggshell color with normal eggshells on hatchability. Comparing light eggshell color with dark eggshell color, they found no difference in hatchability or egg weight loss but did see differences in late embryo mortality (Table 3).

Eggshell cracks obviously are not desired as they can lead to poorer chick quality because of increased incidence of contamination of eggs, the incubators, and excessive moisture loss from the eggs during incubation. Most cracks occur from eggs not being collected often enough, too many eggs being in a nest, or improper collection belts of mechanized nest systems. Most cracks

**Table 3** Effect of eggshell color on hatchability

| Variable | Light eggshell color | Dark eggshell color |
|---|---|---|
| Hatchability | 87.35 ± 1.52 | 87.59 ± 1.87 |
| Egg weight loss % | 10.45 ± 0.27 | 10.02 ± 0.29 |
| Embryonic mortality % 11–18 days | 2.74 ± 0.62[a] | 1.09 ± 0.51[b] |
| Embryonic mortality % 19–21 days | 1.20 ± 0.35[a] | 0.66 ± 0.36[b] |
| Contaminated % | 3.10 ± 0.76 | 0.74± |

Neto et al. (2024)

**Table 4** Effect of eggshell crack type on the hatched chick

| | Yolk-free body mass (g) | Breast weight (g) at hatch | Contamination (%) |
|---|---|---|---|
| Normal eggs | 42.0[a] | 6.29[a] | 2.77[c] |
| Star cracks | 38.7[b] | 4.99[b] | 6.66[b] |
| Hairline cracks | 35.9[c] | 5.062[b] | 10.55[a] |

Khabisi et al. (2012)

of concern to hatchery managers and thus chick quality are hairline cracks or microcracks. Most of these cracks are not seen until the eggs have been setting for a couple of days in storage, and moisture from the albumen has had a chance to move into the crack. Bain et al. (2006) say the microcracks result from the intermammillary crevices on the inner surface of the eggshell. Barnett et al. (2004) compared eggs that were normal or had hairline cracks and showed that eggs with hairline cracks produced lower weight chicks and significantly elevated contamination. Khabisi et al. (2012) compared the effects of normal eggs with star-cracked eggs and hairline-cracked eggs, confirming the work of Barnett et al. (2004). Additionally, Khabisi et al. (2012) found that the yolk-free body mass (YFBM) of chicks from the star-cracked eggs was significantly less than that of the chicks from the intact eggs and that the yolk-free body weight of the chicks from the hairline cracks was worse than from the star-cracked eggs (Table 4). They also noted significant differences in egg contamination.

Many people have looked at ways to mend the cracks in the shell. Peric et al. (2023) used surgical tape to seal cracked eggs and found benefits for hatch parameters with no difference in egg contamination.

## 9 Egg contamination and egg sweating influence on chick quality

Contamination can become a major problem affecting chick quality. Fortunately, the egg is equipped with natural barriers to contamination, namely, the cuticle, eggshell, and shell membranes, but if overloaded with contamination sources, we can see increased chick contamination. Contamination of the eggs can result from hens laying eggs that are smeared with fecal material because of the fact that the reproductive tract and the digestive tract both empty into the cloaca through which the egg passes. Other potential causes are eggs being laid on the floor or the nesting material contaminated with fecal material. If contaminated eggs were to explode in a setter with good-quality eggs, there is an increased risk of additional contamination of chicks. Peric et al. (2022) examined the location of egg laying in the house and its subsequent effect upon chick quality. Their work showed eggs cooling on the floor of the house had a higher risk of contamination compared to those laid in the nest. The clean floor eggs also had a higher level of contamination than nest clean eggs. Van den Brand (2016) compared broiler performance of chicks from clean nest eggs, washed floor eggs, and unwashed floor eggs. They found that floor eggs did have lower hatchability, but the performance of the chicks did not differ until 35 days of age.

Another concern for egg contamination and subsequent contamination of chicks is egg sweating occurring during transport from the breeder farm or in the hatchery prior to setting the eggs. Egg sweating occurs when eggs are moved from a cool environment and into a warmer environment which has elevated relative humidity causing condensation on the eggshell surface. Typically, one sees increased contamination of eggs from the older flocks because of thinner shells when compared to young or prime aged flocks (de Reu et al. 2006). These investigators also saw a non-significant increase of the eggshell penetration from sweating and no difference with regard to the whole egg contamination. Ernst et al. (1998) showed similar results as de Reu et al. (2006), but Fromm and Margolf (1958) did show increased bacterial contamination of egg contents when eggs sweated.

## 10 Influence of eggs set upside down for incubation and the resulting chick quality

A common problem seen in eggs coming to the hatchery is that some eggs in the setter tray are set upside down when the eggs were gathered at the farm. The incidence of upside-down eggs in the setter tray varies with farm, age of breeders, and attitude of farm labor (Bauer et al. 1990). There is a greater chance that eggs set upside down result in embryos oriented with head at the small end of the egg (Waters 1935). Eggs that hatch with the small end up generally result in lethargic chicks that do poorly in the broiler house. Takeshita and McDaniel (1982) showed that eggs that were set upside down resulted in an increase in embryos malpositioned with head in the small end of the egg and lower body weight at hatch when compared to eggs set large end up. They also noted that the malpositioned embryos started to externally pip earlier. When they examined the pip location of eggs set large end up or small end up, they found that not all embryos pipped small end when set upside down. Table 5 shows the results.

Byerly and Olson (1936) showed evidence that eggs set upside hatched 40% less than the eggs set upside down. Talmadge (1977) examined if eggs

**Table 5** Effect of egg orientation on pip location

| Incubation position | Pipped large end | Pipped middle | Pipped small end |
|---|---|---|---|
| Set small end up | 2.9[a] | 27.2[a] | 69.9[a] |
| Set large end up | 86.1[b] | 12.9[b] | 1.0[b] |

Takeshita et al. (1982)

that were originally set upside down but later returned to normal position (large end up) after 8, 9, 10, or 17 days. He found if the eggs set small end up, there was no difference in hatch if moved to large end up on the eighth day of incubation but also noted that as the time of small end up lengthened the incidence of malposition II increased. He also noted a difference related to the breed of parent used; of the embryos in malposition II, 65% of the white leghorns, 83.1% of the sex linked, and 74.2% of the white Plymouth Rock hatched. Bauer et al. (1990) showed that eggs set small end up had a greater incidence of embryos dying after transfer, a greater number of cull chicks, and a lower percentage of grade A chicks. Obviously, if the chicks hatched from the small end of the egg, then they would expend more energy during the hatching process than a chick that had the normal time to rest with beak in air cell, able to evacuate fluids from their lungs, and have increased aerobic respiration before beginning the hatching process.

## 11 Conclusion

During the breeder hen's lifetime she is restricted fed to optimize reproduction. Additionally it is important that during restriction in the breeder house sufficient nutrients are placed in the egg to produce high quality chicks.. The nutrient factors influencing chick quality include the protein and energy intake, vitamin levels, and antioxidant factors. Additionally, the hen's health must be good to assimilate the nutrients and incorporate the nutrients into the egg. Other factors that have impacted chick quality are the breeder age and egg size, where the egg components and thus specific nutrients vary with size and age. The integrity of the shell and eggshell conductance is important not only as a mineral source but as the part they play in the respiration of the developing embryo.

The environment in which the hen is housed has an impact on subsequent chick quality also. The breeder and breeder house management is just the beginning of potential impacts on chick quality. But it must be understood that the hatchery cannot improve the chick quality if the breeder farm supplies a poor-quality egg pack.

## 12 References

Araujo, I.C.S., et al. (2019a) Effect of vitamin E in ovo feeding to broiler embryos on hatchability, oxidative state and performance. *Poultry Science* 98:3652–3661.

Araujo, L.F., et al. (2019b) The dietary supplementation of canthaxanthin in combination with 25-OHD$_3$ results in reproductive, performance and progeny quality gains in broiler breeders. *Poultry Science* 98:5801–5808.

Atencio, A. et al. (2005) *Poultry Science* 84:1058–1068.

Bain, M.M., et al. (2006) Microcracks in eggs. *Poultry Science* 85:2001–2008.

Barnett, D.M., et al. (2004) Hatchability and early chick growth potential of broiler breeder eggs with hairline cracks. *Journal Applied Poultry Research* 13:65–70.

Bauer, F., et al. (1990) Effects of setting eggs small end up on hatchability and post hatch performance of broilers. *British Poultry Science* 31:715–724.

Bautista-Ortega, J., et al. (2009) Egg yolk omega-6 and omega-3 fatty acids modify tissue lipid components, antioxidant status, and ex vivo eicosanoid production in chick cardiac tissue. *Poultry Science* 88:1167–1175.

Buys, N., et al. (1998) Different CO2 levels during incubation interact with hatching time and ascites susceptibility in two broiler lines selected for different growth rate. *Avian Pathology* 27:605–612.

Byerly, T.C. and Olson, M.W. (1936) Lethality of embryo malpositions in the domestic fowl. *Poultry Science* 15:158–162.

Carey, C. (1980) Adaption of the avian egg to high altitude. *American Zoologist* 20:449–459.

Carey, C. et al. (1984) Eggshell conductances of avian eggs at different altitudes in: Seymour, R.S. (eds) *Respiration and metabolism of embryonic vertebrates*. Perspectives in vertebrate science, vol 3. Springer, Dordrecht. https://doi.org/10.1007/978-94-009-6536-2_18.

Damaziak, K., et al. (2018) Short periods of incubation, egg turning during storage and broiler breeder hens age for early development of embryos, hatching results, chick quality and juvenile growth. *Poultry Science* 97:3264–3276.

Dayan, J., et al. (2020) Incubation temperature affects yolk utilization through changes in expression of yolk sac tissue functional genes. *Poultry Science* 99:6128–6138.

De Reu, K., et al. (2006) Influence of eggshell contamination with *Salmonella enterica* Serovar Enteritidis. *Journal of Food Protection* 69:1539–1545.

Duman, M. and Sekeroglu, A. (2017) Effect of egg weights on hatching results, broiler performance and some stress parameters. Brazilian Journal of *Poultry Science* 19:255–262.

Ebeid, T.A., et al. (2017) Influence of pre-storage incubation on hatch traits, thyroid hormones, antioxidative status and immunity of newly hatched chicks at two chicken breeder flock ages. *Animal* 11:1966–1974.

El-Fiky, A.A., et al. (2020) Effect of egg storage period on the immunity and hatched duckling's performance. *Menoufia Journal of Animal, Poultry and Fish Products* 4:25–40.

El-Ratel, I.T., et al. (2024) Effects of supplemental vitamin A on reproduction and antioxidative status of aged laying hens, and growth, blood indices and immunity of their offspring. *Poultry Science* 103:103453.

El Sabry, M.I., et al. (2013) Interaction between breeder age and hatching time affects intestine development and broiler performance. *Livestock Science* 157:612–617.

Ernst, R.A., et al. (1998) Effect of egg sweating on shell penetration of *Salmonella Enteritidis*. *Journal Applied Poultry Research* 7:81–84.

Fromm, D. and Margolf, E.P. (1958) The influence of sweating and washing on weight loss, bacterial contamination and interior physical quality of 12-day old shell eggs. *Poultry Science* 37:1273–1278.

Gardiner, E.E. (1973) Effects of egg weight on post hatching growth rate of broiler chicks. *Canadian Journal of Animal Science* 53:665–668.

Gharahveysi, S., et al. (2012) The effect of broiler breeder ages on the qualitative and quantitative properties of the egg. *American Journal of Animal and Veterinary Science* 7:136–140.

Goliomytis, M., et al. (2015) Effects of egg storage on hatchability, chick quality, performance and immunocompetence parameters of broiler chickens. *Poultry Science* 94:2257–2265.

Hamal, K.R., et al. (2006) Maternal antibody transfer from dams to their egg yolks, egg whites, and chicks in meat lines of chickens. *Poultry Science* 85:1364–1372.

Hamidu, J.A., et al. (2007) The effect of broiler breeder genetic strain and parent flock age on eggshell conductance and embryonic metabolism. *Poultry Science* 86:2420–2432.

Hempleman, S.C., et al. (1993) Oxygen and avian eggshell formation at altitude. *Respiration Physiology* 92:1–12.

Hopkins, D.T. and Nesheim, M.C. (1962) The effect of linoleic acid depletion on performance of hens and their progeny. *Poultry Science* 41:1651.

Hudson, B.P., et al. (2004) Breeder age and zinc source in broiler breeder hen diets on progeny characteristics at hatch. *Journal Applied Poultry Research* 13:55–64.

Iqbal, J., et al. (2017) Effects of egg weight on the egg quality, chick quality and broiler performance at the later stages of production (week 60) in broiler breeders. *Journal Applied Poultry Research* 26:183–191.

Iqbal, J., et al. (2023) Impact of egg size on the quality of eggs, chicks and post hatch overview during the mid stage of yield (45$^{th}$ week) in Hubbard broiler breeders. *Advance Life Science* 10:368–374.

Ismiraj, M.R., et al. (2019) Maternal transfer of natural (auto)antibodies in chickens. *Poultry Science* 98:2380–2391.

Joseph, N.S., et al. (2006) The effects of suboptimal eggshell temperature during incubation on broiler chick quality, live performance and further processing yields. *Poultry Science* 85:932–938.

Khabisi, M.M., et al. (2012) The influence of eggshell crack types on hatchability and chick quality. *Turkish Journal of Veterinary & Animal Sciences* 36:289–295.

Lapao, C., et al. (1999) Effects of broiler breeder age and length of egg storage on albumen characteristics and hatchability. *Poultry Science* 78:640–645.

Latour, M.A., et al. 1998 Broiler breeder age and dietary fat influence the yolk fatty acid profiles of fresh eggs and newly hatched chicks. *Poultry Science* 77:47–53.

Leandro, N.M., et al. (2011) Maternal antibody transfer to broiler progeny varies among strains and is affected by grain source and cage density. *Poultry Science* 90:2730–2739.

Livingston, M.L., et al. (2019) White striping and wooden breast myopathies of broiler breast muscle is affected by time limited feeding, genetic background and egg storage. *Poultry Science* 98:217–226.

Lopez, G. and Leeson, S. (1995) Response of broiler breeders to low protein diets. 2. Offspring performance. *Poultry Science* 74:696–701.

Lourens, A.R.M., et al. (2006) Effect of egg size on heat production and the transition of energy from egg to hatchling. *Poultry Science* 85:770–776.

Machado, J.P., et al. (2020) Effects of breeder age on embryonic development, hatching time, chick quality and growing performance on the slow-growing genotype. *Poultry Science* 99:6697–6704.

Maiorka, A., et al. (2016) Effect of broiler breeder age on the intestinal mucosa development of the embryos at 20 days of incubation. *Brazilian Journal of Poultry Science* 18:689–696.

Marion, J.E. and Edwards H.M. (1962) The response of fat deficient hens to corn oil supplementation. *Poultry Science* 41:1785–1792.

McDaniel, G.R., et al. (1979) The effect of eggshell quality on hatchability and embryonic mortality. *Poultry Science* 58:10–13.

McDaniel, G.R., et al. (1981) Factors affecting broiler breeder performance. 4. The interrelationship of some reproductive traits. *Poultry Science* 60:1792–1797.

McNaughton, J.L., et al. (1978) Effect of age of parents and hatching egg weight on broiler chick mortality. *Poultry Science* 57:38–44.

Menge, H. and Richardson, G.V. (1968) The influence of a linoleic-acid- deficient maternal diet on growth of progeny. *Poultry Science* 47:542–547.

Menge, H., et al. (1964) Effect of a fat free maternal diet on the fatty acid composition of the progeny. *Poultry Science* 43:164–168.

Moraes, T.G.V., et al. (2019) Energy and protein dilution in broiler breeder pullet diets reduced offspring body weight and yield. *Poultry Science* 98:2555–2561.

Moraes, T.G.V., et al. (2014) Effect of maternal dietary energy and protein on live performance and yield dynamics of broiler progeny from young breeders. *Poultry Science* 93:2818–2826.

Nangsuay, A., et al. (2011) Yolk absorption and embryo development of small and large eggs originating from young and old breeder hens. *Poultry Science* 90:2648–2655.

Nangsuay, A., et al. (2013) Energy utilization and heat production of embryos from eggs originating from young and old broiler breeder flocks. *Poultry Science* 92:474–482.

Nangsuay, A., et al. (2015a) Differences in egg nutrient availability, development and nutrient metabolism of broiler and layer embryos. *Poultry Science* 94:415–423.

Nangsuay, A., et al. (2015b) Development and nutrient metabolism of embryos from two modern broiler strains. *Poultry Science* 94:2546–2554.

Nangsuay, A., et al. (2016) Effect of breeder age, broiler strain, and physiological status of embryos and hatchlings. *Poultry Science* 95: 1666–1679.

Nangsuay, A., et al. (2017) Effects of breeder age, strain, and eggshell temperature on nutrient metabolism of broiler embryos. *Poultry Science* 96:1891–1900.

Nasri, H., et al. (2020) Interactions between egg storage duration and broiler breeder age on egg fat content, chicken organ weights and growth performance. *Poultry Science* 99:4607–4615.

Neto, F.L.K., et al. (2024) Eggshell translucency: its relationship with specific gravity and eggshell color and its influence on broiler egg weight loss, hatchability and embryonic mortalities. *Poultry Science* 103:1033528.

Nilsson, E., et al. (2012) IgY stability in eggs stored at room temperature or at +4.0C°. *British Poultry Science* 58: 42–48.

Noble, R.C. and Cocchi, M. (1990) Lipid Metabolism and the neonatal chicken. *Progress in Lipid Research* 29:107–140.

Noble, R.C., et al. (1986) Changes in the lipid metabolism of the chick embryo with parental age. *Poultry Science* 65:409–416.

Nowak, B., et al. (2019) Breeder line and age affect the occurrence of developmental defects, the number of culled one day old broiler chicks and their body mass. *Veterinarni Medicina* 64:323–333.

O'Dea E.E., et al. (2004) Investigating the eggshell conductance and embryonic metabolism of modern and unselected domestic avian genetic strains at two flock ages. *Poultry Science* 83:2059–2070.

Oliveira, D.D., et al. (2010) Effects of lipid sources in the diet of laying hens on the fatty acid profiles of egg yolks. *Poultry Science* 89:2484–2490.

Orellana, L., et al. (2023) Effect of translucency and eggshell color on broiler breeder egg hatchability and hatch chick weight. *Poultry Science* 102:102866.

Oviedo-Rondin, E.O., et al. (2008) Broiler embryo bone development is influenced by incubator temperature, oxygen concentration and eggshell conductance at the plateau stage of oxygen consumption. *British Poultry Science* 49:666–676.

Patel, M.B. and McGinnis, J. (1977) The effect of levels of protein and vitamin $B_{12}$ in hen diets on egg production and hatchability of eggs and on livability and growth of chicks. *Poultry Science* 56:45–53.

Peebles, E.D., et al. (1999) Effects of breeder age and dietary fat on subsequent broiler performance. 1. Growth, Mortality and feed conversion. *Poultry Science* 78:505–511.

Peebles E.D., et al. (2001) Breeder age influences embryogenesis in broiler hatching eggs. *Poultry Science* 80:272–277.

Peebles, E.D., et al. (2004) Relationships among Post Hatch physiological parameters in broiler chicks hatched from young breeder hens and subjected to delayed brooding placement. *International Journal of Poultry Science* 3:578–585.

Peric, L., et al. (2022) Effects of flock age, place of oviposition and cleaning treatments of hatching eggs on hatchability in broiler breeders. *Journal Applied Poultry Research* 31:100279.

Peric, L., et al. (2023) Improving the incubation results by sealing the eggshell of cracked hatching eggs with surgical tape. *Poultry Science* 102:102466.

Petek, M., et al. 2010. Physical chick parameters and effects on growth performance in broiler. *Archiv Tierzucht* 1:108–115.

Peterson, C.F., et al. (1953) Vitamin B12 requirements for hatchability and production of high quality chicks. *Poultry Science* 32:535–542.

Pinchasov, Y. (1991) Relationship between the weight of the hatching eggs and subsequent early performance of broiler chicks. *British Poultry Science* 32:109–115.

Rahn, H., et al. (1977) Reduction of pore area of the avian eggshell as an adaptation to altitude. *Proceedings of the National Academy of Sciences of the United States of America* 74:3095–3098.

Rahn, H., et al. (1982) Changes in eggshell conductance after transfer of hens from altitude of 3,800 to 1,200m. *Journal Applied Physiology: Respiration Environment and Exercise Physiology* 53:1429–1431.

Ricklefs, R.E. and Starck, M.J. (1998) Embryonic growth and development. Pages 31–58 in *Avian growth and development*. Oxford Press, New York, NY.

Robinson, F.E.R.T., et al. (1991) The influence of egg sequence position on fertility, embryo viability and embryo weight in broiler breeders. *Poultry Science* 70:760–765.

Ruiz, J. and Lunam, C.A. (2002) Effect of pre-incubation storage conditions on hatchability, chick weight at hatch and hatching time in broiler breeders. *British Poultry Science*. 43: 374–383.

Sahan, U., et al. (2014) Yolk sac fatty acid composition, yolk absorption, embryo development and chick quality during incubation in eggs from young and old broiler breeders. *Poultry Science* 93:2069–2077.

Saunders-Blades, J.L. and Kover, D.R. (2015) Effect of hen age and maternal vitamin D source on performance, hatchability, bone mineral density and progeny in vitro early innate immune function. *Poultry Science* 94:1233–1246.

Selim, S.A., et al. (2012) Influence of in-ovo administration with vitamin E and ascorbic acid on the performance of Muscovy ducks. *Emirates Journal of Food and Agriculture* 24:264–271.

Shim, M.Y., et al. (2008) The effect of breeder age and egg storage time on the phosphorus utilization by broiler progeny fed a phosphorus deficiency diet with 1α-OH vitamin $D_3$. *Poultry Science* 87:1138–1145.

Sinclair, R.W., et al. (1990) The effects of parent age and posthatch treatment on broiler performance. *Poultry Science* 69:526–534.

Sklan, D., et al. (2003) Heavier chicks at hatch improves marketing body weight by enhancing skeletal muscle growth. *Poultry Science* 8:1778.

Spratt, R.S. and Leeson, S. (1987) Effect of protein and energy intake of broiler breeder hens on performance of broiler chicken offspring. *Poultry Science* 66:1489–1494.99.

Squires, M.W. and Naber, E.C. (1993) Vitamin profiles of eggs as indicators of nutritional status in the laying hen: Riboflavin study. *Poultry Science* 72:483–4942.

Suarez, M.E., et al. (1997) Effect of strain and age of the broiler breeder female on incubation time and chick weight. *Poultry Science* 76:1029–1036.

Surai, P.F., et al. (2019) Nutritional modulation of the antioxidant capacities in poultry: the case of vitamin E. *Poultry Science* 98: 4030–4041.

Swaggerty, C.L., et al. (2023) Addition of a protected complex of biofactors and antioxidants to breeder hen diets confers transgenerational protection against *Salmonella enterica* serovar Enteritidis in progeny chicks. *Poultry Science* 102:102531.

Takeshita, K. and McDaniel, G.R. (1982) Relationship of egg position during incubation on early embryonic growth and hatching of broiler breeder eggs. *Poultry Science* 61:667–672.

Talmadge, D.W. (1977) The effect of incubating eggs narrow end up on malposition II and hatchability. *Poultry Science* 56:1046–1048.

Tindell, D. and Morris, D.R. (1964) The effects of egg weight on subsequent broiler performance. *Poultry Science* 43:534–539.

Tona, K.O.O., et al. (2004) Effects of age of broiler breeders and egg storage on egg Quality, Chick Weight and chick post hatch growth to forty-two days. *Journal Applied Poultry Research* 13:10–18.

Tona, K., et al. (2003a) Effects of egg storage on spread of hatch, chick quality, and chick juvenile growth. *Poultry Science* 82:736–741.

Tona, K., et al. (2003b) Effects of storage time on incubating egg gas pressure, thyroid hormones and corticosterone levels in embryos and on hatching parameters. *Poultry Science* 82:840–845.

Tona, K., et al. (2010) Comparison of Cobb and Ross strains in embryo physiology and chick juvenile growth. *Poultry Science* 89:1677–1683.

Tufft, L.S. and Jensen, L.S. (1991) Effect of age of hen, egg weight and sex on chick performance and lipid retention. *Poultry Science* 70:2411–2418.

Uddin, Z. and Hamidu, J.A. (2014) Prolong egg storage affects broiler breeder embryonic metabolism and chick quality. *Journal Animal Science Advances* 4:973–977.

Ulmer-Franco, A.M., et al. (2010) Hatching egg characteristics, chick quality, and broiler performance at 2 breeder flock ages and 3 egg weights. *Poultry Science* 89:2735–2742.

Urso, U.R., et al. (2015) Vitamin E and selenium in broiler breeder diets. Effect on live performance, hatching process and chick quality. *Poultry Science* 94:976–983.

Van den Brand, H., et al. (2016) Effects of floor eggs on hatchability and later life performance in broiler chickens. *Poultry Science* 95:1025–1032.

Van den Brand, H., et al. (2019) Both the rooster line and incubation temperature affect embryonic metabolism and hatchling quality in laying hen crossbreds. *Poultry Science* 98:2632–2640.

Van den Brand, H., et al. (2023) Effects of trace minerals source in the broiler breeder diet and eggshell translucency on embryonic development of the offspring. *Poultry Science* 102:102455.

Van der Waaij, H., et al. (2011) Effect of match or mismatch of maternal offspring nutritional environment on the development of offspring in broiler chickens. *Animal* 5:741–748.

Van Emous, R.A., et al. (2015) Effects of growth patterns and dietary protein levels during rearing of broiler breeders on fertility, hatchability, embryonic development, and offspring performance. *Poultry Science* 94:681–691.

Vieira, S.L. and Moran, E.T. (1998a) Broiler yields using chicks from egg weight extremes and diverse strains. *Journal Applied Poultry Research* 7:339–346.

Vieira, S.L. and Moran, E.T. (1998b) Broiler chicks hatched from egg weight extremes and diverse breeder strains. *Journal Applied Poultry Research* 7:392–402.

Wang, D.-H., et al. (2017) Traits of eggshells and shell membranes on translucent eggs. *Poultry Science* 96:351–358.

Wang, Y., et al. (2020) Effects of maternal and dietary vitamin A on performance, meat quality, antioxidant status and immune function of offspring broilers. *Poultry Science* 99:3930–3940.

Wang, Y.W., et al. (2004) Maternal dietary ration of linolenic acid to α-linolenic acid affects the passive immunity of hatching chicks. *Poultry Science* 83:2039–2043.

Waters, N.F. (1935) Certain so-called malpositions a natural occurrence in the normal development of the chick embryo. *Poultry Science* 14:208–216.

Weytjens, S., et al. (1999) Thermoregulation of chicks originating from breeder flocks of two different ages. *Journal Applied Poultry Research* 8:139–145.

Whitehead, C.C., et al. (1985) Influence of egg storage on hatchability, embryonic development and vitamin status in hatching broiler chicks. *British Poultry Science* 26:221–228.

Wineland, M.J., et al. (2006a) Incubator temperature and oxygen concentration at the plateau stage in oxygen concentration affects intestinal maturation of broiler chicks. *International Journal of Poultry Science* 5:229–240.

Wineland, M.J., et al. (2006b) Incubator environment interacts with genetic line of broiler at the plateau stage to affect embryo plasma thyroxine and triiodothyronine concentrations. *International Journal of Poultry Science* 5:714–722.

Wolanski, N.J., et al. (2006) Relationship between chick conformation and quality measures with early growth traits in males of eight selected pure or commercial broiler breeder strains. *Poultry Science* 85:1490–1497.

Wolanski, N.J., et al. (2007) Relationships among egg characteristics, chick measurements and early growth traits in ten broiler breeder strains. *Poultry Science* 86:1784–1792.

Wong, H.C., et al. (2020) Classification of inhomogeneous eggshell mottling patterns using pretrained convolutional neural network. *Journal Electron Imaging* 29:1.

Wyatt, C.L., et al. (1985) Influence of egg size, eggshell quality and posthatch holding time on broiler performance. *Poultry Science* 64:2049–2055.

Yadgary, I., et al. 2010. Yolk sac nutrient composition and fat uptake in late-term embryos in eggs from young and old breeder hens. *Poultry Science* 80:2441–2452.

Yang, J., et al. (2020) The effects of broiler breeder dietary vitamin E and egg storage time on the quality of eggs and newly hatched chicks. *Animals* 10:1409.

Yang, J., et al. (2021) The impacts of egg storage time and maternal dietary vitamin E on the growth performance and antioxidant capacity of progeny chicks. *Poultry Science* 100:101142.

Yannakopoulos, A.L. and Tserveni-Gousi, A.S. (1987) Relationship of parents' age, hatching egg weight, and shell quality to day old chick weight as influenced by oviposition time. *Poultry Science* 66:829–833.

Zakaria, A.H., et al. (2005) Oviposition pattern, egg weight, fertility and hatch ability of young and old breeders. *Poultry Science* 84:1505–1509.

Zukiwsky, N.M., et al. (2021) Broiler growth and efficiency in response to relaxed maternal feed restriction. *Poultry Science* 100:100993.

# Chapter 6

## Day-old chick quality

*Itallo Conrado Sousa Araújo, Federal University of Minas Gerais, Brazil; and Roos Molenaar, Aart Lammers and Henry van den Brand, Wageningen University & Research, The Netherlands*

1 Introduction
2 Physical day-old chick quality
3 Immunological day-old chick quality: immunoglobulins
4 Microbiological chick quality
5 Conclusion
6 Where to look for further information
7 References

## 1 Introduction

Hatcheries around the world aim to produce high-quality chicks that can express their production potential, while maintaining their health and welfare (Mesquita et al., 2021). The quality of day-old chicks and the link with post-hatch performance is a widely debated topic and the factors influencing this quality have been extensively studied (Narinç and Aydemir, 2021).

Day-old chick quality can be expressed in different ways. In this book chapter, we aim to discuss day-old chick quality from a physical, immunological, and microbiological perspective, and how these different ways of defining chick quality could predict health and performance until slaughter. The physical quality of chicks refers among others to the visual evaluation of the appearance of newly hatched chicks after hatching. Various scoring systems for this external qualitative evaluation have been developed and are used in hatcheries around the world (Van de Ven et al., 2012; Narinç and Aydemir, 2021). Quantitative criteria, such as chick weight and length, are commonly used to assess the physical quality (Willemsen et al., 2008; Araújo et al., 2016). Additionally, immunological and microbiological aspects can be considered to assess day-old chick quality. These aspects are much less investigated or used in practice but might be valuable parameters that relate to the performance and health of

http://dx.doi.org/10.19103/AS.2025.0165.07

chickens in later life, especially concerning mortality during the first few weeks of life (Chasser et al., 2023).

The relationships between the incubation process, the physical quality of day-old chicks, and health and performance during the rearing phase were frequently investigated a few decades ago, but more recently, less studies have been performed and also questions about the strength of these relationships have been raised. Overall, it was concluded that while qualitative day-old chick quality appears to be a good indicator of the quality of the incubation process, the relationship with later life health and performance is not always strong or even lacking (Decuypere et al., 2001; Tona et al., 2005; Willemsen et al., 2010; Molenaar et al., 2008, 2010; Van den Ven et al., 2012).

## 2 Physical day-old chick quality

Determining the physical quality of newly hatched chicks is widely considered as an essential parameter for chick evaluation at pulling (Hubbard, 2019; Aviagen, 2025b; Cobb-Vantress, 2021). Hatcheries use physical chick quality routinely to assess the quality of the incubation process (Narinç and Aydemir, 2021), but it is not clear to which extent this is done and if the information obtained is also used to adjust hatchery processes. Physical quality can be determined with both quantitative and qualitative methods. In this section, we will discuss several methods to evaluate the physical quality of chicks and their relationship with broiler health and performance during rearing. Moreover, we will evaluate factors that may affect the physical quality of day-old chicks. Finally, we will link the subjective physical quality with quantitative chick quality aspects.

Quantitative methods for analyzing physical quality that are mainly used include chick weight, chick length, yolk free body mass (YFBM), and chick yield (Decuypere et al., 2001; Tona et al., 2005, Willemsen et al., 2008; Molenaar et al., 2010). Qualitative analyses generally involve the Pasgar or Tona scoring system (Boerjan, 2002; Tona et al., 2003) or simplified methodologies developed by commercial companies that are adapted from these systems. Both qualitative evaluations assess several external aspects of the chick, such as activity, eyes, feathers, legs, navel, yolk size, and abdomen filling. With this evaluation, an overall quality score is assigned to the chicks.

### *2.1 Quantitative chick quality*

Hatcheries often choose to weigh chicks after hatching to assess day-old chick quality. This assessment is widely used because it is practical and quick and does not involve euthanizing the chicks. It is also not largely influenced by the person performing the measurement, as it only requires a calibrated and accurate scale and recording sheet.

The relationship between day-old chick weight and broiler performance in later life was evaluated by Sklan et al. (2003). Their findings showed a positive but very weak correlation between body weight (BW) at hatching and body weight at 41 days of age (r = 0.216; $P < 0.01$), indicating that each additional gram of hatching weight contributed to an 8–13 g increase in final body weight. Willemsen et al. (2008) also investigated this relationship and found a better correlation of 0.25–0.33 in Ross breeders aged 39–53 weeks, but not in Cobb breeders of 42 weeks. However, other studies did not find a relationship between day-old chick weight and slaughter weight (Wilson, 1991; Tona et al., 2003, 2005; Molenaar et al., 2008), suggesting that other factors than day-old chick weight are of more importance for slaughter weight. Interfering factors may be egg weight, breeder strain and age, egg storage conditions, and incubation settings.

Body weight includes an unknown amount of residual yolk, meaning that body weight does not accurately reflect the chick's development. While a heavier chick may suggest better development, this is not necessarily the case if the chick contains a large residual yolk and consequently has a low YFBM (van der Wagt et al., 2020).

A more reliable method for assessing physical chick quality seems to be the YFBM, which refers to the weight of the chick after the residual yolk sac has been removed (Cardeal et al., 2022). The YFBM is a direct assessment of how well the chick has absorbed and utilized nutrients from the yolk during incubation and converted these into body tissue (Molenaar et al., 2008; Özlü et al., 2020). Consequently, YFBM is a good reflection of the incubation quality (Molenaar et al., 2008; Araújo et al., 2016). YFBM had a strong correlation with one-day-old body weight (r = 0.73 to 0.80) across breeds and breeder ages (Willemsen et al., 2008). Because the yolk contains essential proteins and immunoglobulins, it can be hypothesized that a better yolk absorption also results in immunological better developed chicks (Cardeal et al., 2021). As a result of a better yolk sac utilization, chicks with a higher YFBM may have a more developed gastrointestinal system and better organ development, allowing them to digest exogenous feed more efficiently post hatch (Willemsen et al., 2008; Cardeal et al., 2022). Whether or not this also means that chicks with a higher YFBM can better cope with (pathogenic) challenges and are thus healthier and more resilient is currently not clear.

YFBM thus seems to be a good indicator of day-old chick quality and may be related to long-term health and resilience. Although this method provides valuable insights into incubation and chick quality, it requires euthanizing the chick, making this method less useful in commercial hatcheries but more applicable in experimental settings.

Another parameter used to assess chick quality is the length of the day-old chick from the chick's beak to the end of its middle toe (Mukhtar et al.,

2013). This parameter can be assessed as an alternative to chick weight, as it provides insight into skeletal growth and general chick vigor (Wolanski et al., 2006). Molenaar et al. (2008) quantified the correlation between hatchling length and later performance. For male broilers, the correlation coefficient between chick length and slaughter weight was 0.36, and for breast meat yield, it was 0.25. This implies a moderate but significant relationship, where every additional millimeter in chick length increased body weight at slaughter age by 24.6 g and breast meat yield by 4.5 g. In contrast, Willemsen et al. (2008) found a low correlation (r = 0.05 to 0.23) between day-old chick length and body weight at day 42 of age in both Ross and Cobb broilers. Differences between studies might be related to the person performing the chick length measurements. Consequently, it is important to standardize the procedure to ensure consistency and reliability across different studies or practical evaluations.

It can be concluded that only a few studies have evaluated quantitative physical day-old chick quality in relation to later life performance. It appears that some day-old chick parameters reflect the quality of incubation (e.g. higher YFBM), but do have a weak relationship with later life performance. Whether physical day-old parameters can be related to later life health, welfare, and resilience is currently unclear.

## 2.2 Qualitative chick quality

The Pasgar and Tona scoring systems are both developed to reduce individual bias and personal influence related to the assessment of visual chick quality (Narinç and Aydemir, 2021). This chick quality assessment can be performed at two different time points: firstly, before selection of second-grade chicks, thereby providing insights primarily into the incubation process quality; or secondly, after selection of second-grade chicks, offering an additional understanding of how the staff hatchery is selecting first-grade chicks.

The Pasgar score evaluates the quality of day-old chicks based on five aspects: activity, navel condition, and the structural quality of legs, beak, and yolk sac (Table 1). For each condition deviating from optimal, one point is subtracted from an initial score of 10. Consequently, a chick of the highest quality scores a 10, while a chick of the lowest quality scores a 5 (Boerjan, 2002). The results of the Pasgar score need to be analyzed as a whole, because the overall score gives insight into the general chick quality of a batch of chicks. However, individual scores of the activity, navel, legs, beak, and belly deviations can help to identify the causes of reduced chick quality. For example, red hocks at the legs may indicate difficulties during hatch from the eggshell. Chicks pushing the hocks against the eggshell during the hatching process (Van den Ven et al., 2012), because of suboptimal hatching conditions, might cause more

**Table 1** Criteria included in the Pasgar score system (Boerjan, 2002)

| Category | Criteria to downgrade by 1 point out of 10 |
|---|---|
| Reflex | The chick is placed on its back on a flat surface and must return to a standing position within 2 s. It is assessed by considering its speed of taking the normal position. |
| Navel | Chick navel inspection, with fans often used to enhance visibility of the navel area. Chick quality is assessed based on whether the navel is clean, closed, and dry. Common issues include a small scab of dried blood, a string of dried yolk sac, a leaky navel, or yolk material outside the abdominal cavity. |
| Hocks | Chick quality is downgraded when there is an abnormality (e.g. swelling, redness, or anatomical anomalies) in the capability of chicks to stand. |
| Beak | The chick needs to have a clean, normal colored nostrils, beak, and comb. Chick quality is downgraded when this deviates (e.g. red dot, swelling, redness, dirty nostrils, or anatomical anomalies). |
| Yolk sac | The chick's belly is examined to ensure it feels soft and smooth. Since this quality is difficult to assess visually, it's important to gently touch and assess the belly. Ideally, there should be no detectable yolk sac, as a large yolk sac (belly distention) can result in a hard belly with tense skin. |

difficulties with hatching, resulting in more chicks with red hocks (Lourens et al., 2005; de Jong et al., 2019).

The Tona score is also a qualitative method, which assesses day-old chicks, using the criteria general activity, outer appearance, presence and amount of residual yolk, condition of the eyes, navel area and legs, presence and amount of the remaining membrane, and the retraction condition of the residual yolk (Table 2; Tona et al., 2003). The maximum score is 100, and for each deviation from optimal for each criterion, the score is lowered, as described in Table 3.

The Tona score is more extensive and in-depth than the Pasgar score and also requires more time to determine. Consequently, the Tona score appears to be more suitable for experimental studies, whereas the Pasgar score is more useful in commercial hatcheries. Largely based on the Pasgar and Tona score, breeding companies (Aviagen, 2020; Cobb-Vantress, 2021) developed their own chick quality scoring systems as well. Cobb-Vantress (2021) categorizes chicks into three groups: good quality, average quality, and cull chicks based in an analysis of reflex, navel, legs, hocks, and defects. In contrast, Aviagen (2020) offered a general guidance for troubleshooting chick quality issues at placement and did not include a scoring system for chick quality in its rearing manual. This includes identifying inactive chicks with slow responses and low energy, as well as evaluating general appearance. Specific concerns included unhealed navels, red hocks or beaks, dark wrinkled legs, and discolored or malodorous yolks or navels.

It can be concluded that although chick scoring systems are a practical tool to assess the day-old chick quality and are used in commercial broiler manuals,

**Table 2** Parameters to determine chick quality according to Tona score (Tona et al., 2003)

| Parameters | Assessment |
|---|---|
| Activity | Activity is assessed by laying the chick on its back and determining how quickly it returns to its feet. A quick jump back onto its feet is regarded as good, but trailing back onto its feet or remaining on its back was assessed as weak. |
| Down and appearance | The chick's body is examined for dry- and cleanness. It is regarded as normal if it is dry and clean. If it is wet or dirty or both (which can be a source of contamination), then it is downgraded. |
| Retracted yolk | The chick is put on its back obliquely on the hand palm until abdominal movement totally stopped. The height of its abdomen is estimated. The consistency of the abdomen to touch is then estimated. If the height of abdomen is estimated to be higher and harder to touch than normal, then yolk retracted is regarded as large and consistent. |
| Eyes | The chick was put on the legs, and its eyes were observed. The state of brightness and wideness of the gape of the eyelids were estimated. |
| Legs | The chick is put on its feet to determine if it remained upright well. The toes are examined for their conformation. If the chick remains upright with difficulty, articulations of the knees are examined to detect signs of inflammation or redness or both. |
| Navel area | Navel and surrounding areas are examined for closure of the navel and its coloration. If the color is different from the skin color of the chick, then it is regarded as bad. |
| Remaining membrane | Observation of the navel area allows estimation of the size of any remaining membrane. The size of any remaining membrane is classified as very large, large, or small. |
| Remaining yolk | Observation of the navel area allows estimation of the size of any remaining yolk. The size of any remaining yolk is classified as very large, large, or small. |

their relationship with health, welfare, and performance in later life seems to be limited (Willemsen et al., 2008).

## 2.3 Pre-hatching factors and their impact on day-old chick physical quality

Several pre-hatching factors, both animal and environmental related, can influence the physical chick quality. Genotype, breeder flock nutrition and age, egg storage duration, and conditions are among other factors that can affect physical chick quality. These factors will shortly be discussed below.

The first animal-related factor affecting physical day-old chick quality is genotype. Different fast-growing broiler strains differed in chick weight and length at pulling when incubated at the same conditions and originating from eggs of the same weight. Nangsuay et al. (2015, 2016) found that at hatch, Ross 308 chicks were 0.2 cm longer and had a 0.6 g heavier YFBM than Cobb 500

**Table 3** Criteria for assessing chick quality using the Tona score method (Tona et al., 2003)

| Parameters | Characteristics | Scores |
|---|---|---|
| Activity | Good | 6 |
| | Weak | 0 |
| Down and appearance | Clean and dry | 10 |
| | Wet | 8 |
| | Dirty and wet | 0 |
| Retracted yolk | Body with normal swallowed yolk | 12 |
| | Body with swallowed large yolk and rather hard to touch | 0 |
| Eyes | Opened and bright | 16 |
| | Opened and not bright | 8 |
| | Closed eyes | 0 |
| Legs | Normal legs and toes | 16 |
| | One infected leg | 8 |
| | Two infected legs | 0 |
| Navel | Completely closed and clean | 12 |
| | Not completely closed and not discolored | 6 |
| | Not closed and discolored | 0 |
| Remaining membrane | No membrane | 12 |
| | Small membrane | 8 |
| | Large membrane | 4 |
| | Very large membrane | 0 |
| Remaining yolk | No yolk | 16 |
| | Small yolk | 12 |
| | Large yolk | 8 |
| | Very large yolk | 0 |

chicks, indicating that day-old chick physical quality differs between strains, even with the same incubation settings, probably as a result of differences in egg composition, eggshell conductance, and embryonic metabolic rates. With a constantly evolving breeding program, physical chick quality can change in time and consequently needs to be evaluated regularly.

A second animal-related factor affecting physical day-old chick quality is breeder age. Older breeders typically lay heavier eggs, resulting in chicks that are both heavier and longer at hatching (Araújo et al., 2016). For example, chicks hatching from 29-week-old fast-growing breeders had lower day-old chick weights (37.3 g) compared to those from 59-week-old breeders (48.9 g) (Ulmer-Franco et al., 2010). Not only is egg weight affected by breeder age but also egg composition. Eggs from older breeders have larger yolk sizes than eggs from younger breeders (Nangsuay et al., 2011, 2013), which results in chicks of different weights, but also in chicks with different body compositions. Eggs of older breeders (53 weeks) resulted in higher day-old chick weight, but those chicks also contained a higher fat and protein content (Nangsuay et al., 2013) than chicks originating from young breeders (29 weeks). The latter might be of importance from a thermoregulation point of view, because Weytjens et al. (1999) and Moreira et al. (2024) demonstrated that chicks originating from older breeders (resulting in heavier chicks containing relatively more body fat and protein) were better able to cope with lower ambient temperatures after placement than chicks from younger breeders (lighter chicks, containing relatively less body fat and protein).

A third important animal-related factor that is related to physical day-old chick quality is egg size. Iqbal et al. (2017) demonstrated in a broiler breeder flock of 60 weeks of age that male chicks from large eggs (74.81±0.11 g) had higher body weights (52.6±0.16 g) and were longer (20.6±0.04 cm) compared to those from medium (48.4±0.16 g; 20.4±0.05 cm) or small (43.3±0.16 g; 19.9±0.05 cm) eggs. However, while egg size influences day-old chick quality characteristics, its effects on feed efficiency (FCR) and mortality rates are inconsistent and limited, with minor differences observed across growth stages and sexes (Iqbal et al., 2017). Because breeder age and egg size are partly entangled, it can be questioned whether breeder age or egg size in itself is most critical in determining physical chick quality.

It can be concluded that animal-related factors, including genotype, breeder age, and egg size, can influence the physical quality of day-old chicks. However, the relationship with later life health, welfare, and performance is largely unknown.

The first environmental factor that might affect physical day-old chick quality is egg storage duration and conditions. Tona et al. (2003) investigated the effects of egg storage duration (1 *vs* 7 days) in 35- and 45-week-old broiler breeders. Chicks from younger breeders consistently showed higher quality ($P < 0.001$), as indicated by the Tona score (97.73%) of chicks achieving a score above 90% after 1 day of storage and 96.92% after 7 days. In contrast, chicks from older breeders (45 weeks) presented reduced quality, with 93.67% of chicks scoring above 90% after 1 day of storage and only 79.36% after 7 days.

This reduction in quality showed a clear interaction between breeder age and storage period ($P < 0.001$), suggesting that prolonged storage negatively affects chick quality more severely in eggs from older breeders (Tona et al., 2003). Although older breeders produced heavier chicks at hatch, this did not translate into higher quality scores. Instead, younger breeders produced chicks of better physiological quality, which may be linked to improved immune function, lower susceptibility to early-life infections, and better early performance, as also suggested in other studies. Additionally, weight gain up to 7 days was highest in chicks from younger breeders after short storage

(96.1 g), suggesting that superior initial quality is associated with higher early growth. Conversely, prolonged storage resulted in lower chick quality and lower early growth, which can predispose chicks to health challenges, such as omphalitis, yolk sac infections, and suboptimal performance (Tona et al., 2003).

In summary, the study highlights that chick quality is not solely dependent on hatch weight but closely linked to maternal age and pre-incubation storage conditions. Tona et al. (2003) used breeder flocks of 35 and 45 weeks, which can both be considered as prime flocks. It can be speculated that the effects obtained in younger and older flocks become larger or not. Additionally, the differences obtained might also be related to the different breeder flocks (location, farm, etc.) rather than only flock age, even when nutrition and management are tried to be the same. Consequently, data of specific treatments in a study should be considered carefully.

The physiological mechanisms underlying chick scoring systems are rarely explored in detail. The Tona score is a subjective evaluation system that assesses physical traits, such as posture and navel condition. However, the physiological basis of these traits, such as their relationships with stress hormone levels, immune competence, or health, remains poorly understood. Similarly, the simplified Pasgar score designed to combine multiple criteria, such as activity, navel quality, and yolk absorption does not address the physiological mechanisms underlying these traits. For example, it remains unclear whether yolk absorption reflects metabolic programming or if chick activity is linked to thermoregulatory capacity or muscle tone development later in life. These questions are often overlooked in the literature but may be relevant for understanding early indicators of performance or health. Additionally, isolated criteria used in scoring systems, such as leg quality or eye condition in the Pasgar or Tona scores, may reflect specific hatchery-related issues, like suboptimal incubation conditions or mechanical failures (Tona et al., 2003). While useful for identifying hatchery problems, these traits are not necessarily reliable predictors of long-term performance, health, or welfare. This underscores the need for further research to clarify the physiological relevance of these characteristics and their implications for later life.

Additionally, it can be questioned whether or not the physical parameters currently used by scoring systems are truly adequate for assessing chick quality, as it is not always clear why different characteristics obtained different weighing factors and how these are related to later life health and performance.

Moreover, it can be questioned whether or not the measurement scoring methods are sensitive enough to detect subtle variations in chick quality. Many of these physical parameters are interconnected, meaning that changes in one can influence others, making it more challenging to interpret the results. For example, a characteristic may appear important simply because it is correlated with another, without necessarily having a direct impact on the chicks' future performance.

At the end of this section about physical day-old chick quality, the following conclusions can be drawn:

- Physical day-old chick quality well reflects (pre)incubation conditions but appears to have a limited relationship with later life performance of broilers.
- The relationship between physical day-old chick and later life health, welfare, and resilience has been hardly investigated.
- Physiological mechanisms explaining the relationship between (pre) incubation conditions and factors and physical day-old chick quality are little investigated.

## 3 Immunological day-old chick quality: immunoglobulins

### 3.1 *Immune system and immunoglobulins*

The immunological status of day-old chicks is crucial for their ability to cope with environmental health challenges (Ulmer-Franco et al., 2010; Alkie et al., 2019; Meijer et al., 2024). While many studies have tried to examine the immune status of newly hatched chicks, few have investigated its relationship with long-term health outcomes, including mortality, welfare, and performance. This section mainly discusses maternal antibodies and the development of immune organs, because the relationships between other immunological aspects and day-old chick quality are less well studied yet. We explore how day-old chick immunoglobulin levels may influence health, welfare, and performance later in life and critically address the scarcity of research.

Large-scale broiler rearing farms, using high stocking densities, foster conditions that promote the transmission of microorganisms among chicks (O'Dea et al., 2006). Because of this high infectious pressure in modern poultry husbandry, day-old chicks should already have an adequate immune competence to resist infections. The main organs of the avian immune system include the Harderian gland, thymus, spleen, bursa of Fabricius, cecal tonsils, Peyer's patch, and the bone marrow (Fig. 1; Kovacs-Nolan and Mine, 2004; Ribatti et al., 2019). Each of these organs plays a role in immune development and functionality of immune responses and consequently in the immunological status of the chick (Ribatti et al., 2019). It is important to highlight that all immune organs are present at hatch, but the immune system is not yet fully functional (Sharma et al., 2024).

The immune system of chicks maintains homeostasis through innate and adaptive responses, including cell-mediated and humoral immunity. Humoral

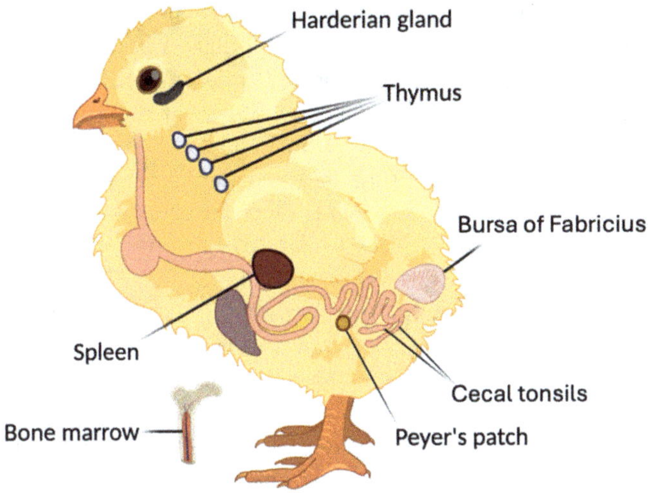


**Figure 1** Chick's main immune system organs. Created with Biorender.com.

immunity involves three immunoglobulin (Ig) classes: IgY, IgM, and IgA. These Ig's or antibodies in day-old chicks originating from the broiler breeders are called maternal antibodies. The antibodies are transferred from the hen to the egg and stored particularly in the yolk (IgY) and albumen (IgM and IgA), providing protection during embryonic development and the early post-hatch period until the chick's own immune system is mature (Carlander et al., 2003; Leandro et al., 2011a, 2011b; Kowalczyk et al., 2019; Cardeal et al., 2021). IgY is the most abundant immunoglobulin, and the transfer from hen to chick begins with the transfer from the hen's bloodstream to the ovarian follicle, and later from the egg yolk to the developing embryo (Ulmer-Franco et al., 2010; Murai et al., 2020).

During the first weeks post hatching, day-old chicks do not have a fully developed immune system to produce their own mature B-lymphocytes (Leandro et al., 2011a), and the chick's humoral immunity depends largely on IgY received from the yolk sac. Maternal antibodies provide essential passive immunity that protects hatchlings from pathogens during the critical initial stages of life, before their adaptive immune system is fully functional. Additionally, it has been hypothesized that these antibodies contribute to immunological imprinting, which influences the development of the endogenous antibody repertoire (Klipper et al., 2004; Berghof et al., 2013). This may implicate that maternal antibodies mediate the transfer of the specific immune system to the offspring. In other words, the hen prepares her offspring for possible future infections.

The newly hatched chick can absorb intact immunoglobulins from the yolk sac during the first days post hatch (Apanius, 1998). The yolk sac is directly connected to the gastrointestinal tract, via the yolk stalk, and is also known for its multiple functions, including its immunological role (van der Wagt et al., 2020; Wong and Uni, 2021). According to Panda et al. (2015) and Wong and Uni (2021), chicks typically absorb the contents of the yolk sac, including immunoglobulins, within the first 24–72 hours post-hatching. Therefore, the first 3 days post hatch are crucial for the transfer of maternal antibodies to ensure the viability and performance of broilers during the first weeks post hatch and possibly also in the long term (Ribatti et al., 2019).

Broiler chicks start producing their own antibodies from approximately 8 to 13 days after hatching onward (Bar-Shira et al., 2003). This means that during the first weeks post hatch, the chicks rely heavily on maternal IgY for immune protection. However, the level of maternal antibodies declines rapidly within 2–3 weeks post hatch (Den Hartog et al., 2016). Gharaibeh and Mahmoud (2013) investigated the half-life of pathogen-specific maternal antibody titers in unvaccinated newly hatched chicks from 1 to 40 days of life. The chicks originated from 40-week-old broiler breeders that had been vaccinated against a range of pathogens. Overall, the half-life of the pathogen-specific maternal antibodies ranged from 3.8 to 7.0 days. The lowest levels of IgY were observed at 21 days post hatch, reflecting the biological half-life of maternally derived IgY, while endogenous production has not yet reached its peak (Simon et al., 2014; Den Hartog et al., 2016). This results in an immunity gap at 2–3 weeks post hatch, during which the chicks are more vulnerable to infectious diseases.

## *3.2 Induction of endogenous IgY in young broiler chicks*

As already mentioned, IgY is poorly produced by the chick in the first two weeks post hatch. Specific IgY in broiler chicks can be either induced by active vaccination of broiler breeders or passively provided by oral administration of specific IgY via the diet.

Hjelm (2018) evaluated IgY levels in breeder serum and egg yolk related with first-week mortality in Ross 308 chicks, studying four different breeder flocks of different ages (33 and 51 weeks). IgY concentrations ranged from 6.12 mg/mL to 12.43 mg/mL in breeder serum and from 3.78 mg/mL to 7.92 mg/mL in egg yolk, with no differences between broiler breeder ages. First-week mortality varied between 2.12% and 4.64%, but no significant correlation was found between maternal IgY levels and chick mortality. These findings showed high individual variation in IgY levels between breeders in serum and egg yolk, but these levels seem not to influence first-week mortality.

Santos et al. (2022) showed that breeder age has a direct impact on passive immune transfer, with older hens depositing higher concentrations of IgY in serum and yolk, resulting in progeny with greater circulating IgY up to 14 days of life. This enhanced early protection, however, contrasts with chicks from younger breeders, which, although receiving lower amounts of maternal IgY, demonstrated a stronger humoral response after vaccination, suggesting a compensatory capacity for endogenous antibody production. These results highlight that maternal age not only determines the initial level of immune protection but also modulates how chicks respond to immunological challenges during the early rearing period. Hjelm (2018) also observed higher IgY concentrations in the yolks of older broiler breeders (52 weeks) compared to eggs from 32- and 42-week-old breeders, suggesting greater prenatal absorption of this antibody in older broiler breeders. This was supported by higher IgY concentrations in the serum of chicks originating from 52-week-old breeders up to 14 days of age. Taken together, these results indicate that higher serum concentrations of IgY in breeder result in higher IgY levels in the offspring. The level of pathogen-specific immunoglobulins in broiler chicks can be enhanced by either transfer from the breeder to the chick by vaccination of the breeders (active route) or by supplementing the chicks with specific immunoglobulins via the broiler diet (passive route). These two aspects will be discussed below.

The antibody repertoire of broiler breeders is a key factor for offspring to receive pathogen-specific antibodies capable of contributing to the defense in the first few days of life (Gharaibeh and Mahmoud, 2013). The pathogen-specific antibody repertoire of broiler breeders can be enhanced by active vaccination procedures. In the current husbandry practice broiler breeders undergo specific vaccination schedules, which vary depending on the region of the world. The main vaccines applied to breeders include Newcastle Disease, Infectious Bursal Disease, Infectious Bronchitis, Avian Encephalomyelitis, Marek's Disease, Reovirus, Fowl Pox, Coccidiosis, and Salmonella (*S. enteritidis*, *S. typhimurium*) (van Emous and Steentjes, 2022). These vaccines help breeders to protect their offspring from a variety of diseases, particularly in the first week after hatch. Additionally, day-old chicks can receive vaccination at the hatchery.

Mahdavi et al. (2010) investigated the effects of IgY supplementation in Ross broiler chicks challenged with *Escherichia coli* O78:K80 at 7-day-old. The challenge was continued for an additional 7 days from day 14 to 21 with 1.0 mL of a late log phase culture (109 CFU/mL) until the level of *E. coli* in feces reached 105 CFU/g. Supplementing the diet with 0.2% or 0.4% of IgY reduced *E. coli* colonization in the ileum by 13.1% and 10.6%, respectively, compared to the control group (without IgY supplementation). Morphological analysis of the broiler chicks' intestines at day 21 to 42 showed that the villus height-to-crypt depth ratio increased from 5.17 in the control group to 6.55 and 6.58 in chicks receiving 0.2% and 0.4% IgY, respectively, indicating improved intestinal health. Additionally, feed conversion efficiency from 21 to 42 days of age decreased from 2.57 (control group) to 2.43 in both IgY-supplemented groups. These findings indicate that while maternal antibodies provide some protection, additional IgY supplementation immediately post-hatch may further enhance gut health and resistance against pathogens.

In conclusion, serum antibodies of day-old chicks appear to play a crucial role in their ability to cope with environmental and pathogenic challenges, especially early in life. Maternal antibodies, particularly IgY, provide passive immunity, protecting chicks during this critical post-hatch period before their immune system fully matures. Higher maternal Ig levels in the day-old chick may contribute to enhanced immune resilience, better health, and improved welfare, potentially leading to better performance outcomes. Although immunological day-old chick quality at hatch may be a key factor for early survival and health, it is currently hardly recognized. Consequently, further research is needed to determine its predictive value for overall health, mortality, welfare, and performance in broilers.

## 4 Microbiological chick quality

To date, it remains unclear what the most optimal microbial composition of day-old chicks should be. Studies evaluating bacterial contamination levels in broiler chicks, setters, and yolk sacs indicate that day-old chicks are minimally colonized or infected with pathogenic microorganisms (Melo et al., 2019; Bain et al., 2019; Wales and Davies, 2020). However, at pulling, using a visual observation it is not possible to determine whether or not chicks are infected, meaning that contaminated chicks could still be classified as first-grade during physical quality assessment. Only a few studies have related the microbiological quality of newly hatched chicks to the success of broiler production.

Vieira et al. (2025) demonstrated that second-grade chicks with incompletely healed navels, residual yolk membrane, or a black button were more likely to show early yolk sac contamination at placement, predominantly with *Enterococcus faecalis*. These defects reflect delayed yolk absorption and inadequate navel closure, which create a microbiological entry point. Nevertheless, by day 5 both groups exhibited colonization by *Escherichia coli* and multiple *Enterococcus* species, indicating substantial environmental or hatchery-derived infection pressure, a pattern consistent with previous reports that highlight the susceptibility of neonatal chicks to opportunistic pathogens during this period.

Despite the initial differences in contamination, mortality was extremely high in both groups, reaching 28.4% in first-grade and 29.4% in second-grade chicks in the first week, with totals of 34.7% and 32.6% at day 14, respectively (Vieira et al., 2025). Under this severe microbial challenge, navel quality did not influence mortality outcomes, suggesting that the magnitude of infection can override any protective effect associated with proper navel healing. Similar findings have been reported in situations where high hatchery contamination masked differences in chick quality. Thus, while navel integrity remains an important indicator of early microbiological vulnerability, the results indicate that mortality is primarily driven by infection pressure rather than the physical navel score when contamination is widespread.

The physical and immunological status of chicks should be closely aligned with the microbiological quality to ensure overall chick health, welfare, performance, and food safety. While chicks may initially appear healthy with good physical characteristics, they can still harbor contaminations with pathogenic microorganisms that may have negative consequences days after placement (Costa et al., 2022; Franco et al., 2023). Microbial contamination of hatching eggs with potential pathogenic bacteria is an important concern for the poultry industry (Wales and Davies, 2020). Such contamination can occur through vertical or horizontal transmission routes. Vertical transmission refers to the direct transfer of pathogens from the breeder to the egg and subsequently to the offspring (Shterzer et al., 2023). Horizontal transmission occurs through contamination from egg-to-egg or from chick to chick (Bain et al., 2019). This section will first explore the microbiological profile of day-old chicks and then examine factors that contribute to horizontal contamination from the time of egg laying to the transportation of chicks to broiler farms.

### 4.1 Early-life microbiota establishment in broiler chicks

The microbiota composition of day-old chicks depends on various factors, including microorganisms present during egg formation, the environment where the egg is laid, and the cooling period immediately after laying, which can all contribute to microorganisms present in the egg (Wu et al., 2024). The microbiota of the chicken's reproductive tract (infundibulum, magnum, isthmus, and uterus) is quite consistent, with a predominance of *Lactobacillus* species (Wen et al., 2021). This oviduct microbiota can be transferred, primarily through the albumen, into the egg. Additionally, microbiota in the cloaca can contaminate the eggshell just before oviposition and subsequently, can pass the eggshell. During incubation, microbiota can be transferred to the embryo's gut, suggesting that both the eggshell and albumen serve as carriers for maternal microorganisms to the developing embryo's intestine (Lee et al., 2019). Certain genera found in the maternal cloaca, eggshell, and egg white, such as *Pseudomonas spp.*, were also abundant in the embryo's intestine at ED18, forming the initial microbiota that will colonize the chick after hatching (Lee et al., 2019).

After oviposition, the bacteria found on the eggshell surface are predominantly Gram-positive, while the bacteria contaminating the egg contents are generally Gram-negative (Legros et al., 2021). Microorganisms present on the eggshell can come into contact with the chick during or after hatching, meaning that day-old chicks can contain both Gram-positive and Gram-negative bacteria. Initially, the intestinal microbiota of chicks is dominated by bacteria from the phyla *Firmicutes* and *Proteobacteria*. This composition undergoes rapid changes in the first days as chicks are exposed to the environment and feed (Ballou et al., 2016; Wu et al., 2024).

During the early post-hatch period, the gastrointestinal tract of chicks undergoes rapid colonization by microbial communities that play a pivotal role in health and development. Initially, the microbiota is dominated by facultative and obligate aerobic bacteria, some of which may include potentially pathogenic taxa, such as *Escherichia coli* and *Enterococcus* spp. (Ballou et al., 2016; Oakley et al., 2014). As the gut environment matures and oxygen levels decrease, a succession occurs favoring the establishment of obligate anaerobes, including members of the *Bacteroidetes* and *Firmicutes* phyla, which are characteristic of a more stable and health-promoting microbiota (Pan and Yu, 2014). This ecological shift is essential for the development of intestinal homeostasis, competitive exclusion of pathogens, and proper immune modulation. Therefore, the transition from an aerobic to anaerobic microbiota is a crucial indicator of microbiological quality and early-life gastrointestinal health in chicks early post-hatch.

The colonization and multiplication of non-pathogenic microorganisms is involved in the initial development of the gastrointestinal tract (Wu et al., 2024), nutrient absorption, and overall performance. According to Melo et al. (2019) and Ballou et al. (2016), commensal species, such as *Escherichia coli*, *Enterobacter cloacae*, *Pantoea agglomerans*, and *Klebsiella pneumoniae*, are part of the normal intestinal microbiota in day-old chicks, aiding in the development of the intestinal microbiome and competing with potentially pathogenic microorganisms. In summary, healthy chicks' intestinal microbiota is initially composed primarily of *Proteobacteria* but quickly shifts to *Firmicutes*, such as *Lactobacillus*, which improves intestinal health and proper development.

## 4.2 Eggshell contamination and bacterial penetration

Initial eggshell and whole egg contamination often occurs shortly after oviposition, with dirty environmental surfaces being the primary contamination source. Hatching eggs, typically laid in breeder farm nests, can become contaminated when bacteria penetrate through the pores of the eggshell (Dolman and Board, 1992; Melo et al., 2019). Cooling of the egg immediately after oviposition creates negative pressure in the egg that draws bacteria through the pores (Bruce and Drysdale, 1994). This means that contaminated eggshells immediately after oviposition (due to contamination in the cloaca or in the nest) increase the risk of penetration of bacteria into the egg. According to Bain et al. (2019), the eggshell cuticle is the first line of defense and plays an important role in protecting the eggshell against pathogenic bacteria, such as pathogenic *E. coli*. An intact cuticle is directly correlated with a lower capacity for bacterial penetration after oviposition (Kulshreshtha et al., 2022). Thus, cuticle deposition is essential for preventing bacterial penetration of eggs. De Reu et al. (2006) studied the penetration of common pathogenic agents through the eggshell and found that shell thickness and number of pores did not influence penetration. Only the cuticle had a protective effect against bacterial penetration, because removal of the cuticle increased penetration by 20–60%. Eggs with poor-quality cuticles were found to have 40% more contamination than eggs with more intact cuticles (26%). Additionally, a correlation was found between the level of contamination of the outside of the eggshell and the degree of penetration, suggesting that outside contaminated eggs are at more risk for penetration than clean eggs. However, it is unclear whether cuticle quality is directly related to day-old chick quality or that other eggshell characteristics are of more importance in this perspective.

For example, non-clustering bacterial strains and those capable of movement like from the genus *Pseudomonas* and *Alcaligenes* were shown to penetrate eggs with poor shell quality more easily and frequently than other species (Sauter and Petersen, 1969; De Reu et al., 2006). Additionally, Sauter and Petersen (1969) discovered that eggs with low specific gravity or poor shell quality were more prone to *Salmonella* penetration. De Reu et al. (2006) reported an increase in eggshell penetration from 12.9% at the start of laying (34- and 46-week-old) to 25.0% at the end of the laying period (69- and 74-week-old) for general bacterial contamination. This was in accordance with Bruce and Johnson (1978) who noted that contamination rates increased as flocks aged. Studying contamination of unhatched eggs, Araújo et al. (2016) also found more contaminated eggs in 59-week-old breeder flocks compared to 29- and 35-week-old breeder flocks (2.23 *vs* 0.54 *vs* 0.54%).

Wales and Davies (2020) emphasized that the primary contamination source in hatcheries often originates from the breeder farm. Fertile eggs, which are the primary source of hatchery contamination, are directly connected to the microbiological quality of the chicks produced (Yoho et al., 2008; Franzo et al., 2020). These eggs can carry contaminants that impair chick hatching and lead to poor physical development, gastrointestinal issues, and performance losses. Ensuring that eggs arrive at hatcheries with minimal pathogenic microbial load is important to allow that chicks start their life in broiler houses with less challenges. Breeder farms must carefully manage litter, laying nests, feeders, drinkers, egg collection, sanitization, and storage.

## 4.3 Chick fungal contamination

Another microbiological issue that can affect hatcheries is fungal contamination, which compromises the microbiological quality of the chicks. Aspergillosis, commonly known as 'brooder pneumonia,' is a fungal infection caused by *Aspergillus fumigatus*, which significantly impacts the microbiological quality and health of day-old chicks (Sharma et al., 2024). This respiratory infection, characterized by labored breathing, white to yellow lesions in the lungs and air sacs, and high mortality rates (5–50%) within the first 3 weeks post-placement, originates from exposure to fungal spores. Hatcheries, with their warm, humid environments (37–45°C), provide ideal conditions for Aspergillus growth, particularly in ventilation systems or on organic matter such as egg yolk residue (Oliveira et al., 2024).

Spores enter eggs via cracks or hairline fractures to reach the yolk, a substrate that promotes fungal proliferation (Sharma et al., 2024). Breeder farms can contribute to contamination, as nesting materials like bedding, manure, and feed can introduce spores, especially to floor eggs, which are more prone to cracks and environmental contamination. Visible mold on the air cell membrane of infected eggs and spikes in embryo mortality around ED16 are early indicators of fungal presence (Jacobsen et al., 2010). If over 0.5% of hatching eggs show signs of Aspergillus infection, there is an increased risk of a fungal outbreak. In all cases of Aspergillus contamination it is important to target the source, followed by effective hygiene practices and biosecurity measures to safeguard the microbiological quality of day-old chicks.

The causes of contamination need to be investigated and addressed, as chicks with poor microbiological quality not only compromise the batch but can also contaminate the hatchery (especially in multi-stage machines where batches are mixed in one machine) and further contaminate production farms when litter is reused for subsequent batches.

## 4.4 Interval between egg collection and eggshell sanitization

Eggs should be collected as soon as possible after laying to minimize contamination from nests, breeders, and the environment. It is recommended that manual or automatic collection occurs about four times a day, and during peak laying periods, around six times to avoid more than three eggs per nest, as well as to reduce breakages and cracks in the eggs, thereby minimizing eggshell contamination (Aviagen, 2025a; Cobb, 2024).

Few scientific studies have evaluated the effect of the interval between egg collection and sanitization on eggshell and the chicks' yolk sac of contamination. Franco et al. (2023) analyzed the collection of clean eggs from both nests and floor, as well as two sanitization times: 30 and 120 minutes after collection. The microorganisms on the eggshell were counted according to the method of sanitization. The results showed no significant differences in eggshell contamination between nest and floor eggs in terms of total aerobic mesophilic microorganisms and *Enterobacteriaceae*. However, more bacteria were observed in the yolk sac of chicks from nest eggs sanitized after 120 minutes, compared to those sanitized after 30 minutes; this number was even higher compared to floor eggs (Table 4).

These findings support theories that microorganisms from the eggshell can penetrate the interior egg despite natural barriers (Berrang et al., 1999). Technologies, such as immediate sanitization in the breeder's house, rather than batch sanitization of all eggs collected throughout the day, could reduce contamination of yolk sacs, improving the microbiological quality of chicks when they are removed from the hatchery.

### 4.5 Floor eggs and microbiological chick quality

Floor eggs are those laid by breeders outside the nests, directly on the litter, which is filled with organic matter, dust, excrement, and other possible contaminants. This type of egg is still incubated in several countries with a high demand for day-old chicks (Costa et al., 2022; Peric et al., 2022). However, the microbiological quality of chicks can be compromised when incubating these floor eggs. Van den Brand et al. (2016) showed that incubation of floor eggs resulted in lower hatchability and contamination of the chick's yolk sac compared to clean nest eggs, which might cause high mortality during the rearing phase. However, no difference in physical quality at hatch was found between chicks that did survive, which was supported by Ahamed et al. (2019).

Costa et al. (2022) found no difference in hatchability and chick quality between clean nest eggs and clean floor eggs that were sanitized with paraformaldehyde 60 minutes after collection. Furthermore, the microbiological quality of the chicks from both groups was the same, highlighting the importance of quick collection and sanitization of this type of egg. Because floor eggs are usually not collected more than twice a day, and thus bacteria can have penetrated the egg, from a biosecurity and day-old chick quality point of view, floor eggs should not be used as hatching egg.

## 5 Conclusion

In conclusion, microbiological quality of day-old chicks is determined by a complex interaction of maternal, environmental, and management-related factors that begin before oviposition and continue until placement in the broiler house. Early-life microbiota establishment plays a central role in gastrointestinal development and immune modulation, but it is highly susceptible to disruptions caused by bacterial and fungal contamination.

Overall, ensuring high microbiological chick quality requires integrated management practices at both the breeder farm and hatchery levels, including egg collection and sanitation protocols, strict biosecurity, and early detection of

**Table 4** Interaction between egg type and sanitization time on counts of different microorganisms ($log_{10}$ CFU/mL) in chick yolk sac content at hatching

| | Sanitization time after collection | | | 30 min |
|---|---|---|---|---|
| | 120 min | Aerobic mesophilic | *Enterobacteriaceae* | Aerobic mesophilic |
| Egg type | *Enterobacteriaceae* | Clean nest | $3.93^{bB}$ | $3.07^{bB}$ |
| $5.11^{aA}$ | $4.73^{aA}$ | Floor | $5.40^{aA}$ | $5.18^{aA}$ |
| $5.30^{aA}$ | $4.69^{aA}$ | | | |

$^{A-B}$ Means within a column lacking a common superscript differ ($P < 0.05$).
$^{a-b}$ Means within a row and species lacking a common superscript differ ($P < 0.05$).
Adapted from Franco et al. (2023).

contamination sources. Further research is needed to better define the optimal microbial profile of day-old chicks and to establish standardized methods for assessing microbiological quality. By reducing early-life microbial challenges, the poultry industry can improve chick robustness, minimize early mortality, and enhance overall flock performance.

## 6 Where to look for further information

Decuypere, E. et al. (2001). The day-old chick: A crucial hinge between breeders and broilers, *World's Poultry Science*. DOI: 10.1079/wps20010010.

Willemsen, H., et al. (2008). Critical assessment of chick quality measurements as an indicator of posthatch performance. *Poultry Science*, *87*(11), 2358–2366.

Vieira, M. C., et al. (2025). Relevance of navel quality of day-old chick on organ development, initial performance, and bacterial presence. *Poultry Science*, *104*(12), 106031.

## 7 References


Ahamed, Y.I., et al. (2019), 'Effects of floor and nest eggs on hatchability and chick quality parameters in broiler breeders', *Poultry Science*, 98(12), 6722–6728.

Alkie, T.N., et al. (2019), 'Development of innate immunity in chicken embryos and newly hatched chicks: a disease control perspective', *Avian Pathology*, 48(4), 288–310.

Apanius, V. (1998), 'Ontogeny of immune function'. In: Starck, J.M. and Ricklefs, R.E. (eds), *Avian growth and development: evolution within the altricial-precocial spectrum*, Oxford University Press, New York, NY, pp. 203–222.

Araújo, I.C.S., et al. (2016), 'Effect of incubator type and broiler breeder age on hatchability and chick quality', *Revista Brasileira de Ciência Avícola*, 18(spe 2), 17–25.

Aviagen (2020), *Management guide: Broiler breeders*. Huntsville: Aviagen.

Aviagen (2025a), *Broiler management handbook 2025*. Huntsville: Aviagen.

Aviagen (2025b), *Hatchery tips 2025*. Huntsville: Aviagen.

Bain, M.M., et al. (2019), 'Cuticle deposition improves the biosecurity of eggs through the laying cycle and can be measured on hatching eggs without compromising embryonic development', *Poultry Science*, 98(4), 1775–1784.

Ballou, A.L., et al. (2016), 'Development of the chick microbiome: how early exposure influences future microbial diversity', *Frontiers in Veterinary Science*, 3, 2.

Bar-Shira, E., et al. (2003), 'Establishment of immune competence in the avian GALT during the immediate post-hatch period', *Developmental & Comparative Immunology*, 27(2), 147–157.

Berghof, T.V.L., et al. (2013), 'Transgenerational epigenetic effects on innate immunity in broilers: an underestimated field to be explored?', *Poultry Science*, 92(11), 2904–2913.

Berrang, M.E., et al. (1999), 'Bacterial penetration of the eggshell and shell membranes of the chicken hatching egg: a review', *Journal of Applied Poultry Research*, 8(4), 499–504.

Boerjan, M. (2002), 'Programs for single stage incubation and chick quality', *Avian and Poultry Biology Reviews*, 13, 237–238.

Bruce, J. and Drysdale, E.M. (1994), 'Trans-shell transmission', In: Starck, J.M. (ed.), *Microbiology of the avian egg*, Springer, Boston, MA, pp. 63–91.

Bruce, J. and Johnson, A.L. (1978), 'The bacterial flora of unhatched eggs'. *British Poultry Science, 19*(5), 681–689.

Cardeal, P.C., et al. (2021), 'Breeder age and post-hatch feed access influence physiology of residual yolk sac, intestinal development and performance of broiler chicks', *Livestock Science*, 251, 104617.

Cardeal, P.C., et al. (2022), 'Effects of breeder age and pre-placement feed on IgY concentration in egg yolk and chick serum', *Journal of Animal Physiology and Animal Nutrition*, 106(3), 561–565.

Carlander, D., et al. (2003), 'Immunoglobulin Y levels in egg yolk from three chicken genotypes', *Food and Agricultural Immunology*, 15(1), 35–40.

Chasser, K.M., et al. (2023), 'Comparison of chick quality, health, and inflammation from two hatchery environments', *Food and Nutrition Sciences*, 14(9), 824–842.

Cobb (2024), *Cobb breeder management guide*, Cobb-Vantress, Siloam Springs, AR.

Cobb-Vantress (2021), *Broiler management guide,* Cobb-Vantress, Siloam Springs, AR.

Costa, B.T.A., et al. (2022), 'Floor eggs: hatchability, microbiological analysis, and effects of post-hatch use of ceftiofur on performance of broiler chicks', *Research in Veterinary Science*, 152, 610–617.

de Jong, I.C., et al. (2019), 'Comparison of performance, health and welfare aspects between commercially housed hatchery-hatched and on-farm hatched broiler flocks', *Animal*, 13(6), 1269–1277.

de Jong, I.C., et al. (2020), 'Effects of on-farm and traditional hatching on welfare, health, and performance of broiler chickens', *Poultry Science*, 99(10), 4662–4671.

De Reu, K., et al. (2006), 'Eggshell factors influencing eggshell penetration and whole egg contamination by different bacteria, including Salmonella enteritidis'. *International journal of food microbiology*, 112(3), 253–260.

Decuypere, E., et al. (2001), 'The day-old chick: a crucial hinge between breeders and broilers', *World's Poultry Science Journal*, 57(2), 127–138.

Den Hartog, G., et al. (2016), 'Intestinal immune maturation is accompanied by temporal changes in the composition of the microbiota', *Beneficial Microbes*, 7(5), 677–686.

Dolman, J. and Board, R.G. (1992), 'The influence of temperature on the behaviour of mixed bacterial contamination of the shell membrane of the hen's egg', *Epidemiology & Infection*, 108(1), 115–121.

Franco, C.M., et al. (2023), 'Sanitization at different times after floor egg collection for hatchability, microbiological quality, and performance of broilers', *Poultry Science*, 102(12), 103156.

Franzo, G., et al. (2020), 'Impact of rotten eggs on hatchery performances: a multicentric study', *Animals*, 10(10), 1725.

Gharaibeh, S. and Mahmoud, K. (2013), 'Decay of maternal antibodies in broiler chickens', *Poultry Science*, 92(9), 2333–2346.

Graham, B.D., et al. (2022), 'Development of an environmental contamination model to simulate the microbial bloom that occurs in commercial hatch cabinets', *Poultry Science*, 101(6), 101890.

Hjelm, C., (2018), 'Identifying factors of importance for chick quality and traits that may predict chick quality', Swedish Univ. Agric. Sci, Uppsala, Sweden (2018) Second cycle, A2E. https://stud.epsilon.slu.se/13865/19/Hjelm_C_180711.pdf (Accessed April 2026).

Hubbard (2019), *Broiler management guide*, Hubbard, Quintin, France.

Iqbal, J., et al. (2017), 'Effects of egg weight on the egg quality, chick quality, and broiler performance at the later stages of production (week 60) in broiler breeders', *Journal of Applied Poultry Research*, 26(2), 183–191.

Jacobsen et al. (2010), 'Embryonated eggs as an alternative infection model to investigate Aspergillus fumigatus virulence'. *Infection and immunity*, 78(7), 2995–3006.

Klipper, E., et al. (2004), 'Maternal antibodies block induction of oral tolerance in newly hatched chicks', *Vaccine*, 22(3–4), 493–502.

Kovacs-Nolan, J. and Mine, Y. (2004), 'Avian egg antibodies: basic and potential applications', *Avian and Poultry Biology Reviews*, 15(1), 25–46.

Kowalczyk, J., et al. (2019), 'Field evaluation of maternal antibody transfer from breeder turkey hens to egg yolks, egg whites, and poults', *Poultry Science*, 98(8), 3150–3157.

Kulshreshtha, G., et al. (2022), 'Properties, genetics and innate immune function of the cuticle in egg-laying species', *Frontiers in Immunology*, 13, 838525.

Leandro, N.M., et al. (2011a), 'Effects of broiler breeder genetic, diet type, and feeding program on maternal antibody transfer and development of lymphoid tissues in chicken progeny', *Journal of Applied Poultry Research*, 20(4), 474–484.

Leandro, N.M., et al. (2011b), 'Maternal antibody transfer to broiler progeny varies among strains and is affected by grain source and cage density', *Poultry Science*, 90(12), 2730–2739.

Lee, S., et al. (2019), 'Characterization of microbial communities in the chicken oviduct and the origin of chicken embryo gut microbiota', *Scientific Reports*, 9(1), 6838.

Legros, J., et al. (2021), 'The role of ovotransferrin in egg-white antimicrobial activity: a review', *Foods*, 10(4), 823.

Lourens, A., et al. (2005), 'Effect of eggshell temperature during incubation on embryo development, hatchability, and posthatch development', *Poultry Science*, 84(6), 914–920.

Mahdavi, A.H., et al. (2010), 'Effects of dietary egg yolk antibody powder on growth performance, intestinal *Escherichia coli* colonization, and immunocompetence of challenged broiler chicks', *Poultry Science*, 89(3), 484–494.

Meijer, M.M.Y., et al. (2024), 'Carvacrol *in ovo* delivery optimization and flow dynamics in broiler chicken eggs', *Poultry Science*, 103(3), 103443.

Melo, E.F., et al. (2019), 'An evaluation of alternative methods for sanitizing hatching eggs', *Poultry Science*, 98(6), 2466–2473.

Mesquita, M.A., et al. (2021), 'Results of hatching and rearing broiler chickens in different incubation systems', *Poultry Science*, 100(1), 94–102.

Molenaar, R., et al. (2008), 'Relationship between hatchling length and weight on later productive performance in broilers', *World's Poultry Science Journal*, 64(4), 599–604.

Molenaar, R., et al. (2010), 'Meeting embryonic requirements of broilers throughout incubation: a review', *Brazilian Journal of Poultry Science*, 12, 137–148.

Moreira, L.M., et al. (2024), 'Early-life cold stress and broiler breeder age: effects on metabolism and performance in broiler chicks', *Animal Production Science*, 64(17), AN24270.

Mukhtar, N., et al. (2013), 'Hatchling length is a potential chick quality parameter in meat type chickens', *World's Poultry Science Journal*, 69(4), 889–896.

Murai, A., et al. (2020), 'Evaluation of a receptor gene responsible for maternal blood IgY transfer into egg yolks using bursectomized IgY-depleted chickens', *Poultry Science*, 99(4), 1914–1920.

Nangsuay, A., et al. (2011), 'Yolk absorption and embryo development of small and large eggs originating from young and old breeder hens', *Poultry Science*, 90(11), 2648–2655.

Nangsuay, A., et al. (2013), 'Energy utilization and heat production of embryos from eggs originating from young and old broiler breeder flocks', *Poultry Science*, 92(2), 474–482.

Nangsuay, A., et al. (2015), 'Development and nutrient metabolism of embryos from two modern broiler strains', *Poultry Science*, 94(10), 2546–2554.

Nangsuay, A., et al. (2016), 'Effects of breeder age, broiler strain, and eggshell temperature on development and physiological status of embryos and hatchlings', *Poultry Science*, 95(7), 1666–1679.

Narinç, D.O. and Aydemir, E. (2021), 'Chick quality: an overview of measurement techniques and influencing factors', *World's Poultry Science Journal*, 77(2), 313–329.

Oakley, B.B., et al. (2014), 'The chicken gastrointestinal microbiome', *FEMS Microbiology Letters*, 360(2), 100–112.

O'Dea, E.E., et al. (2006), 'Investigating the effects of commercial probiotics on broiler chick quality and production efficiency', *Poultry Science*, 85(10), 1855–1863.

Oliveira et al. (2024), 'Obtaining microbiologically safe hatching eggs from hatcheries: using essential oils for integrated sanitization strategies in hatching eggs', poultry houses and poultry. *Pathogens*, 13(3), 260.

Özlü, S.E., et al. (2020), 'Effect of posthatch feed and water access time on residual yolk and broiler live performance', *Poultry Science*, 99(12), 6737–6744.

Pan, D. and Yu, Z. (2014), 'Intestinal microbiome of poultry and its interaction with host and diet', *Gut Microbes*, 5(1), 108–119.

Panda, A.K., et al. (2015), 'Early post hatch nutrition on immune system development and function in broiler chickens', *World's Poultry Science Journal*, 71(2), 285–296.

Peric, L., et al. (2022), 'Effects of flock age, place of oviposition and cleaning treatments of hatching eggs on hatchability in broiler breeders', *Journal of Applied Poultry Research*, 31(3), 100279.

Ribatti, D., et al. (2019), 'The morphological basis of the development of the chick embryo immune system', *Experimental Cell Research*, 381(2), 323–329.

Santos, J.S., et al. (2022), 'Effects of broiler breeder age on immune system development of progeny', *Revista Brasileira de Zootecnia*, 51, e20210127.

Sauter, E.A. and Petersen, C.F. (1969), 'The effect of egg shell quality on penetration by *Pseudomonas fluorescens*', *Poultry Science*, 48(5), 1525–1528.

Sharma, S., et al. (2024), 'In ovo feeding of probiotic lactobacilli differentially alters expression of genes involved in the development and immunological maturation of bursa of Fabricius in pre-hatched chicks', *Poultry Science*, 103(1), 103237.

Shterzer, N., et al. (2023), 'Vertical transmission of gut bacteria in commercial chickens is limited', *Animal Microbiome*, 5(1), 50.

Sklan, D., et al. (2003), 'Ontogeny of brush border carbohydrate digestion and uptake in the chick'. *British Journal of Nutrition*, 89(6), 747–753.

Simon, K., et al. (2014), 'Development of ileal cytokine and immunoglobulin expression levels in response to early feeding in broilers and layers', *Poultry Science*, 93(12), 3017–3027.

Tona, K., et al. (2003), 'Effects of egg storage time on spread of hatch, chick quality, and chick juvenile growth', *Poultry Science*, 82(5), 736–741.

Tona, K., et al. (2005), 'Interrelationships between chick quality parameters and the effect of individual parameter on broiler relative growth to 7 days of age', *European Poultry Science*, 69(2), 67–72.

Ulmer-Franco, A.M., et al. (2010), 'Hatching egg characteristics, chick quality, and broiler performance at 2 breeder flock ages and from 3 egg weights', *Poultry Science*, 89(12), 2735–2742.

Van de Ven, L.J.F., et al. (2012), 'Significance of chick quality score in broiler production', *Animal*, 6(10), 1677–1683.

van der Wagt, I., et al. (2020), 'A review on yolk sac utilization in poultry', *Poultry Science*, 99(4), 2162–2175.

van Emous, R. and Steentjes, A. (2022), 'Optimizing the health of broiler breeder birds'. In: S.de Wit, (ed.), *Optimising poultry flock health*, Burleigh Dodds Science Publishing, Cambridge, UK, pp. 347–386.

Vieira, M.C., et al. (2025), 'Relevance of navel quality of day-old chick on organ development, initial performance, and bacterial presence', *Poultry Science*, 104(12), 106031.

Wales, A. and Davies, R. (2020), 'Review of hatchery transmission of bacteria with focus on *Salmonella*, chick pathogens and antimicrobial resistance', *World's Poultry Science Journal*, 76(3), 517–536.

Wen, C., et al. (2021), 'Microbiota continuum along the chicken oviduct and its association with host genetics and egg formation', *Poultry Science*, 100(7), 101104.

Weytjens, S., et al. (1999), 'Thermoregulation in chicks originating from breeder flocks of two different ages', *Journal of Applied Poultry Research*, 8(2), 139–145.

Willemsen, H., et al. (2010), 'Delay in feed access and spread of hatch: importance of early nutrition', *World's Poultry Science Journal*, 66(2), 177–188.

Willemsen, H., et al. (2008), 'Critical assessment of chick quality measurements as an indicator of posthatch performance', *Poultry Science*, 87(11), 2358–2366.

Wilson, H.R. (1991), 'Interrelationships of egg size, chick size, posthatching growth and hatchability', *World's Poultry Science Journal*, 47(1), 5–20.

Wolanski, N.J., et al. (2006), 'Relationship between chick conformation and quality measures with early growth traits in males of eight selected pure or commercial broiler breeder strains', *Poultry Science*, 85(8), 1490–1497.

Wong, E.A. and Uni, Z. (2021), 'Centennial review: the chicken yolk sac is a multifunctional organ', *Poultry Science*, 100(3), 100821.

Wu, B., et al. (2024), 'Characteristics of gut microbiota in chicken early life', *Journal of Animal and Feed Sciences*, 33(3), 311–320.

Yoho, D.E., et al. (2008), 'Effect of incubating poor quality broiler breeder hatching eggs on overall hatchability and hatch of fertile', *Poultry Science*, 87(Suppl 1), 148.

# Chapter 7

## Early feeding from a broiler perspective

*H. J. Wijnen and C. W. van der Pol, HatchTech, The Netherlands; and R. Molenaar and H. van den Brand, Wageningen University & Research, The Netherlands*

1 Introduction
2 Delayed feeding
3 Residual yolk
4 Growth rate
5 Health
6 Welfare
7 Early feeding case studies
8 Future trends in early feeding
9 Where to look for further information
10 References

## 1 Introduction

Since the second half of the nineteenth century large-scale artificial incubation has grown rapidly (Schippers et al., 2014). Ever since then, batches of hatching eggs from parent flocks are collected on breeder farms and transported to hatcheries, which incubate them and then supply the hatchlings to poultry farmers. Poultry farmers are the first to provide feed and water to these chicks. During the timespan between hatch and farm arrival, chicks do not have access to feed or water. To avert this, alternative hatching systems became available approximately 10 years ago in which nutrients are available directly from hatch onwards, often referred to as 'early feeding'.

Early feeding gained increasing interest since this market introduction and by now, it is at least a known term in much of the global poultry industry. Multiple companies, including key incubator manufacturers, have developed their own early feeding technique for hatcheries. Especially in regions with large-scale broiler production, these alternative hatching systems with early feeding are nowadays often taken into consideration during (new) hatchery constructions. In these regions, early feeding is no longer considered to be an innovation for a niche market, as it was at first during its introduction, but rather as a strategy to

http://dx.doi.org/10.19103/AS.2025.0165.08

increase broiler welfare and yields compared to the current standard practice. In some regions, such as the EU, early feeding is also taken into account by policy makers when revising animal husbandry legislations (EFSA, 2023). The fact that regions with large-scale broiler production are the first to apply these alternative hatching systems does not mean that early feeding is exclusively applicable to this type of poultry production. Some early feeding strategies, such as 'on farm hatching', provide opportunities for smallholder poultry farmers as well. Altogether, this indicates that early feeding is very unlikely to disappear from the poultry market, but instead it is expected that an increasing number of broilers will be raised with early feeding in the coming decades.

It is essential that hatcheries and broiler farmers that consider applying early feeding in their production system are aware of the short- and long-term effects that it can have on broiler needs. Scientific studies revealing such effects will support this awareness. However, although the first commercial early feeding systems were introduced on the poultry market already over 10 years ago, early feeding is still a relatively new concept, and some aspects of it are not yet fully understood or investigated. Current standard management procedures are often accurately formulated for conventional delayed-fed broilers based on decades of research and field experiences, but they may not match the needs of an early fed broiler. It is known that early fed broilers can differ considerably from their delayed fed conspecifics in, for instance, thermoregulation (Van den Brand et al., 2010; Van Roovert-Reijrink et al., 2017; Wijnen et al., 2022b), growth (de Jong et al., 2017), and immunity (Hollemans et al., 2021).

This chapter aims to describe how early feeding may be perceived from a broiler perspective by putting key existing research about early feeding into this context and by pointing out gaps in knowledge. Ultimately, this will help to optimize management of early fed broilers, and consequently, support the global increasing demand for poultry meat in a sustainable and responsible manner in terms of animal health, social acceptance, and the environment, without compromising economic gain for the poultry industry. First, the current practice of delayed feeding and events that may affect its duration will be described. Next, it will be discussed to what extent residual yolk (RY) may fulfill delayed fed broiler's needs. Thereafter, effects of early feeding on broiler growth rate are provided and how this could potentially affect their welfare and nutritional needs. This is followed by effects of feeding strategy on broiler health, focusing on infectious disease responses, and consecutively on broiler welfare linked to affective state, natural behaviour, and physical well-being. Lastly, some new early feeding case studies are provided that discuss chick comfort during transport and future trends.

The current chapter will not include a description of different alternative hatching systems that are commercially available. These have already been described in detail by Van den Brand and Molenaar (2023). However, many early feeding studies are conducted using such alternative hatching systems. Therefore, it is important to recognize that results presented in the current chapter may not always be solely assigned to early or delayed access to feed and water, but perhaps also to other interrelated factors, such as environmental conditions (climate, light, sound) and logistical factors (processing and transport).

## 2 Delayed feeding

Speaking about early feeding, one should first also understand what its counterpart 'delayed feeding' entails for broilers. Delayed feeding can be described as the timespan without feed and water between hatch, which is the moment that a chick emerges from the eggshell, and the first moment of feed and water intake at the broiler farm.

The duration of the delayed feeding timespan that chicks undergo in commercial broiler production depends on various events. Figure 1 shows the most common events that occur in standard practice, along with the estimated times that each event takes. The hatch window comprises the timespan between the first and the last chick that hatch within one batch. It can be seen that in case of short transport (e.g. 2 h), the first and the last chick that hatch have an estimated total delayed feeding timespan between 42 h and 12 h, respectively, whereas in case of long transport (e.g. 12 h), this is estimated to be 52–22 h. This shows that particularly the moment of hatch within the hatch window, as well as the transport duration, contribute to the total delayed feeding timespan that each individual broiler chick experiences.

The hatch window can only be influenced to a limited extent. The moment of hatch varies between individual chicks due to variation in abiotic factors, such as preincubation storage duration of the hatching eggs or temperature variation within an incubator, and biotic factors, such as variation in egg weight and eggshell conductance or sex of the embryo. Variation in abiotic factors can be reduced, but biotic factors will always remain an inherent part of biological systems to some extent, although in the future it might be possible to differentiate, for instance, between eggs of different conductance or sex prior to incubation. In current commercial practice, hatcheries already aim to

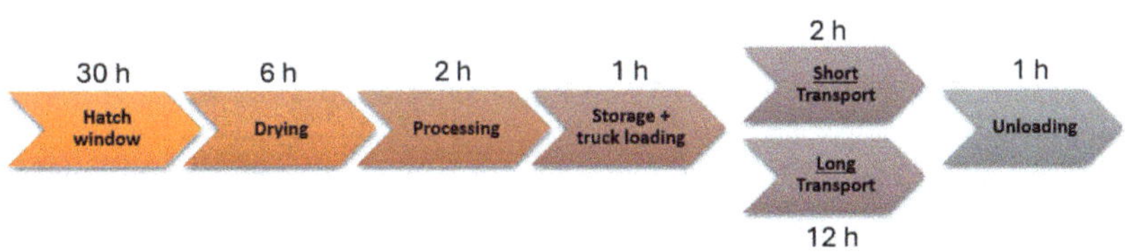


**Figure 1** Events in commercial broiler incubation practice from hatch until farm arrival, including estimated duration of each event.

minimize abiotic and biotic variations. Regardless, the hatch window will very likely often be at least 30 h (Olsen and Winton, 1941; Tona et al., 2003; Careghi et al., 2005; Hollemans et al., 2018; Wijnen et al., 2020, 2022b). Whereas the duration of the hatch window can hardly be influenced, transport duration can in theory be reduced to a minimum through optimizing logistics and transportation methods, and by construction of hatcheries close to areas with local demand for day-old chick to limit chick export. In practice, this is probably more difficult to realize, but realizing that transport duration is important in this perspective is already a step forward.

The longer the timespan without feed and water, the higher the risk of dehydration or malnourishment (Gaglo-Disse et al., 2010; Wang et al., 2014; Khosravinia and Manafi, 2016; Madej et al., 2024), but there is inconsistency amongst studies on the duration at which delayed feeding will have a lasting negative effect on broiler chicks. This inconsistency may partly be explained by the fact that the two key events that comprise the total delayed feeding timespan, hatch moment and transport duration, are not always clearly specified. Consequently, the exact delayed feeding duration is often unclear amongst studies. Additionally, the moment at which feed and water are provided for the first time is often at the same time point for all broilers to synchronize chronological age during grow out, but in such case biological age and delayed feeding duration can vary between individuals because of differences in hatch moment and the duration between feed provision and first feed intake. Some studies try to overcome the variation in biological age by collecting chicks with wet down feathers, indicating that they hatched recently, at a single time point within the hatch window. However, results with this method may not be representative for all chicks since it has been shown that early, mid-term, and late hatching chicks can differ in their response to delayed feeding (Careghi et al., 2005; Lamot et al., 2014; Wang et al., 2014; Deines et al., 2021a & 2021b), although this is still debated (Løtvedt and Jensen, 2014; Boyner et al., 2021).

When effect of delayed feeding are studied, results are generally compared to early feeding, whilst most studies do not include information on the moment at which chicks actually start eating and/or drinking once the feed and water were provided nor the amount that was consumed by the chicks. It can be speculated that early feeding behaviour may differ amongst studies as it likely depends on multiple factors. For example, the amount of artificial light hours and intensity, chick (dis)comfort in (sub)optimal ambient temperature, quality of the water (temperature and disinfectants), feed quality (particle size), the type of alternative hatching system, and quality of the chicks at hatch and especially their gastrointestinal development may play a role in the timing and amount of first feed and/or water intake once provided to the chick. Possibly even transgenerational nutrition affects early feeding behaviour (Roura et al., 2025). Early feeding behaviour has not extensively been studied, but research suggests that >85% of the chicks first ingest a little bit of feed (±0.2 g) within approximately 2.5 h after emergence from the eggshell (Pinchasov and Noy, 1993; Nielsen et al., 2008, 2010; de Jong, unpublished results from 2016; van den Brand et al., 2023) and subsequently ingest a substantial amount of feed filling their crops between 20 h and 40 h after emergence from the eggshell (HatchTech, unpublished results from 2014; Boyner et al., 2021; van den Brand et al., 2023). First water intake occurs a few hours after chicks' first feed intake, at approximately 5–13 h after emergence from the eggshell (Van der Pol et al., 2015; de Jong, unpublished results from 2016). Subsequent drinking behaviour has not been studied yet to the authors knowledge, but it is assumed that, from this point onwards, chicks probably drink every now and then because their total water intake between hatch and pull was found to be 6.4 g on average (HatchTech, unpublished results from 2016). Chick activities in between early feeding behaviour are not studied to our knowledge, but very likely they are mainly resting.

Consistency amongst studies on the duration at which delayed feeding will have a lasting negative effect on broiler chicks are revealed when many studies are combined in a meta-analyses. De Jong et al. (2017) performed such a meta-analysis and summarized existing literature about early feed and water provision published after the year 2000. Based on 75 papers, they concluded that any delayed feeding, regardless of its duration, temporarily impaired broiler (organ) development and permanently lowered body weights throughout the growing period up to 6 weeks of age. The permanent effect of any delayed feeding duration on broiler body weight may be the effect of an earlier onset of growth rather than an altered growth rate, as will be discussed in more detail in Section 4 of this chapter. Additionally, long-delayed feeding durations tended to worsen feed conversion ratio and significantly increased mortality during the total growing period. The exact duration that resulted in these long-term negative effects remained unclear, because studies had to be grouped in delayed feeding durations covering 1 day periods to be able to perform a meta-analyses and besides, studies that reported mortality numbers often did not clearly specify the biological age of the chicken. It was anywhere between 36 h after emergence from the eggshell and 60 h after pull.

To conclude, the exact delayed feeding duration at which broilers are negatively impacted during their remaining life remains unknown to date. Anywhere between 36 h after hatch and 60 h after pull delayed fed chicks will have a lower survivability chance. In commercial practice, chicks that hatch mid-term or late within the hatch window will not probably encounter a delayed feeding timespan of more than 36 h after hatch. However, for early hatching chicks or in case of longer transport durations from the hatchery to the farm, a delayed feeding timespan longer than 36 h seems to be common. A hatch window is unavoidable and not all chicks are mid-term or late hatchers.

Furthermore, unforeseen hold-ups during processing or transport can occur. Moreover, mortality can be regarded as an 'iceberg indicator' which, just as the protruding tip of an iceberg, indicates the submerged bulk beneath the surface (FAWC, 2009; de Jong et al., 2015). This suggest that the higher mortality rate that is found when chicks are delayed fed for longer durations is possibly submerged with other (welfare) measures which may not have been revealed yet. For example, it could not be concluded from the meta-analyses of De Jong et al. (2017) whether or not delayed feeding had any consequences on broiler health or welfare due to the limited amount of studies that were available at that time. This will be elaborated upon in Sections 5 and 6, as additional studies have since been published on this topic. Besides, even the 'average' chick is affected by delayed feeding, regardless of its time span, because delayed feeding durations shorter than 36 h after hatch have also been shown to affect young broiler development. Although these effects were only temporarily, short-term effects can still be relevant when regarded from a broiler's perspective. For example, dehydration, short-term stress, or altered thermoregulation may temporarily affect a broiler's welfare as will be discussed in more detail in the upcoming sections of this chapter.

## 3 Residual yolk

It is important to note that delayed feeding may not be the appropriate term for the delayed access to feed and water between hatch and farm placement, as it implies that chicks do not have access to any nutrients during this period. In fact, this is only true for exogenous sources, whereas chicks possess RY in their abdomen. This RY can be taken up indirectly via the bloodstream or directly into the intestines via the yolk stalk. RY is taken up during the first 5 days post-hatch (reviewed by Van der Wagt et al., 2020), so theoretically, chicks always have access to some kind of nutrients, even in case of long transport duration. It can be questioned though whether uptake of this yolk matches the needs of a newly hatched broiler chick. Approximately, 2 days before the actual moment of hatch, the RY is retracted into the abdominal cavity (Romanoff, 1960). One day later, the chick punctures the membrane of the air chamber and pulmonary respiration is initiated. In the 12 h that follow, various metabolic process start up, while the metabolic rate increases (Vleck et al., 1980) until the chick is able to puncture the outer shell membrane and eggshell with its egg tooth. During the remaining half a day, the chick rotates in the egg while pipping the eggshell and pushing itself upwards until it succeeds to lift the blunt top of the egg. It has been shown that this total process from internal pipping until hatching is energy demanding (De Oliveira et al., 2008; Maatjens, 2016). The nutrients available in the RY may not be sufficient to replenish the energy that was used for these hatching processes, while also matching the nutritional needs for homeostasis and early development, as will be elaborated upon in the next paragraph. Besides, the yolk stalk is attached to the jejenum and RY may therefore not reach anterior parts of the gastrointestinal tract. Likely the foregut, including the crop, remains empty. Actual foraging behaviour and uptake of feed via the beak may be required to fulfill the behavioural needs of a chick, but this remains to be investigated.

In the discussions about early feeding, it is often indicated that chicks at hatch have a kind of 'lunch box' in the form of RY. In the past, it was believed that early feeding would hamper the uptake of RY, leading to digestive troubles (Roberts, 1928). However, to date, not a single study has been published in which early fed chicks showed lower uptake of RY. Later, the opposite effect has been proposed in which early feeding would stimulate the uptake of RY by activating digestive enzymes and/or increasing peristaltic movements (Noy and Sklan, 1998; Sklan and Noy, 2000). Although some studies showed that early feeding indeed resulted in lower weight of the RY compared to delayed feeding (Romanoff, 1944; Bierer and Eleazer, 1965; Moran and Reinhart, 1980; Noy and Sklan, 1998; Noy and Sklan, 2001; El-Husseiny et al., 2008; Bhanja et al., 2009; Cardeal et al., 2021) other studies do not confirm this (Bigot et al., 2003; Gonzales et al., 2003; Franco et al., 2006; Gaglo-Disse et al., 2010; Tabedian et al., 2010; van den Brand et al., 2010; Sarlak et al., 2016; Hollemans et al., 2020; Wijnen et al., 2022b). Possibly early fed chicks take up different nutrients of the RY by utilizing mainly amino acids and minerals for body development, whereas delayed fed chicks mainly utilize lipids and moisture for maintenance and survival (Romanoff, 1944; Moran and Reinhart et al., 1980; Hopcroft et al., 2020).

The amount of nutrients available in the RY for the newly hatched chicks can be of importance for utilization as well. Nangsuay et al. (2013) investigated the nutrient composition of the RY and YFBM of broiler chicks originating from different breeder ages and of different egg sizes. The largest contrast was found between the RY of chicks originating from different egg sizes. These results are provided in Table 1.

Based on Table 1, it can be concluded that broiler chicks at pulling have 1.4–2.1 g of water available in the RY. Assuming that at pulling the average

**Table 1** Effects of broiler breeder eggs size (small; 57-61 g or large; 66-70 g) on residual yolk (RY) weight, DM, protein, fat, and carbohydrate content of chicks at pulling (516-518 h after start of incubation)

| Egg size | RY (g) | DM (%)* (g) | Protein (%)** (g) | Fat (%)** (g) | Carbohydrate (%)** (g) |
|---|---|---|---|---|---|
| Small | 2.7 | 47.3 1.3 | 26.9 0.7 | 13.1 0.3 | 4.2 0.1 |
| Large | 3.9 | 47.2 1.8 | 28.7 1.1 | 11.6 0.5 | 4.4 0.2 |

*Percentage of RY weight. **Percentage of DM.
*Source*: Adapted from Nangsuay et al. (2013).

chick is 24 h of age (Careghi et al., 2005) and chicks use 2.4 g of RY per 24 h in the peri-hatching period (Molenaar et al., 2010), it can be calculated that at the moment of hatching average-sized chicks have a RY weight of 5.0–6.3 g, existing of 2.6–3.3 g water, 1.3–1.8 g protein, 0.6–0.7 g fat, and 0.2–0.3 g carbohydrates. Almeida et al. (2016) found in 'feather-dried' broiler chicks (approximately 3–6 h after hatching) a RY weight of 7.88–9.41 g, depending on incubation temperature, containing 3.5–4.2 g water, 2.8–3.6 g fat, and 0.5–0.7 g protein. With these amounts of nutrients, it can be questioned whether or not a day-old chick can develop optimally in the first 48–72 h post-hatching without the provision of exogenous feed and water. For sure, nutrients including water are available in the chick itself as well and can be used during the delayed feeding period. However, body nutrients are not infinite and using them might decrease developmental opportunities of the chick and can cause dehydration, as will be discussed in the following paragraphs.

### *3.1 Nutrients needed for survival*

For survival probably the most important nutrient is water. After drying, the chick still looses water via evaporation, mainly via respiration. The amount of evaporated water strongly depends on the ambient temperature and the relative humidity the chicks are exposed to. Van der Hel et al. (1991) exposed 2–15 h old broiler chicks to ambient temperatures of 30.8–38.8°C for a period of 24 h. Total body water loss in this 24 h was increased from 3.46 g to 5.73 g (8.2–13.8%) when the temperature increased to 38.8°C. The majority of this water (52–76%) evaporated from the yolk-free body mass (YFBM), meaning that dehydration occurred. More recently, Hamissou Maman et al. (2019) exposed broiler chicks post-pulling (510 h after the start of incubation) to an ambient temperature of 21.7°C (low), 36.2°C (control), or 40.5°C (high) for a period of 12 h. These ambient temperatures aimed to maintain the cloacal temperature at 38.1°C, 40.0°C, or 42.6°C, respectively. Total body weight loss (mostly water loss) during this 12 h exposure period was 1.71 g (3.86%), 1.64 g (3.82%), and 4.43 g (10.20%) for the low, control, and high treatment, respectively. Again, the majority of the water loss originated from the YFBM. Finally, Xin and Harmon (1996) exposed breeder chicks within 5 h after hatching for a period of 50 h to an ambient temperature of 20°C, 25°C, 30°C, or 35°C, paralleled by a decreased in relative humidity from 40% to 17%. An increase from 20°C to 30°C resulted in a slight increase from 3.8 g to 4.1 g water loss.$h^{-1}$.$kg^{-1}$BW (7.37 g to 8.06 g per 50 h). However, an increase of the ambient temperature to 35°C increased the water loss to 4.6 g.$h^{-1}$.$kg^{-1}$BW (9.15 g per 50 h).

Given the 3–5 g of water available in the RY at hatch, and the indicated water losses due to evaporation at normal and particularly to high-ambient temperatures, it is clear that the water stores in the RY are quickly exhausted

(Hamissou Maman et al., 2019), which is reflected in water loss from the YFBM and thus dehydration. This is clearly expressed in higher concentrations of haematological parameters, such as hematocrit and total serum protein in unfed broiler chicks post-hatch compared to immediately fed chicks (Madej et al., 2024). The consequences of dehydration of chicks early post-hatching are hardly investigated, but it can be speculated that particularly at high-ambient temperatures post-hatching (in the hatcher, in the holding room, during transport, during brooding in the broiler farm) dehydration will occur, which might have negative effects on health, welfare, and resilience of the broiler at that moment and in later life. Hammissou Maman et al. (2019) showed that a too high cloacal temperature during 12-h post-hatching had negative effects on performance till slaughter age. Whether or not this is only related to dehydration is disputable, because they also found negative effects immediately after the temperature period on heart, proventriculus, gizzard, and Bursa of Fabricius weight. Based on all these findings, it can be suggested that even at normal temperatures, post-hatching the day-old chick easily runs out of water from the RY, which might have negative consequences for the health and welfare of the broiler.

The importance of the other nutrients (fat, protein, carbohydrates) for survival in the first approximately 72 h post-hatching is much less clear. The meta-analysis from De Jong et al. (2017) found that broiler mortality during the total grow out period increased if birds were deprived of feed and water for 36–60 h after pulling for certain, meaning approximately 60–84 h after hatch, and possibly already at shorter delayed feeding durations. This suggests that fat, protein, and carbohydrates in the RY might not be severely limiting for survival in the first day post-hatching. However, whether or not these nutrients are limiting for more sensitive health and welfare characteristics could not be concluded, because of lack of existing data.

## 3.2 Nutrients needed for development

Nutrients are essential for chick development. Their oxidation generates energy, primarily in the form of ATP. The most efficient pathway to produce ATP is by the Krebs cycle or the citric acid cycle. As the name already suggests, the starting point of the citric acid cycle is citric acid or citrate. To produce citrate, C2 carbon fragments originating from acetyl-CoA are condensed with C3 carbon fragments originating from oxaloacetate. C2 carbon fragments particularly originate from fats and phospholipids, whereas C3 carbon fragments particularly originate from glucose, glycogen, or glucogenic amino acids. So, for a well-running Krebs cycle, besides oxygen, glucose and fats are needed. However, as shown in Table 1, the RY of the day-old broiler chick only contains marginal amounts of carbohydrates and thus glucose. Consequently, in the case the chick is not provided with feed and water post-hatching, amino acids may be used as precursor of oxaloacetate, meaning that protein is used as an energy source and cannot be used as building blocks for body development. This means that sufficient ATP is formed to survive, but less protein is left for development. This might lead to retarded body development in general (Wang et al., 2020) and development of immune related organs (Bursa of Fabricius, caecal tonsils) in particular (Dibner et al., 1998; Madej et al., 2024). Additionally, some of the protein in the RY are immunoglobulins. Indeed, post-hatch feed-deprived chicks have shown to have lower levels of immunoglobulins in their blood (e.g. Dibner et al., 1998; Bar-Shira et al., 2003; Hollemans et al., 2021). It can be speculated that a lower level of IgY in the chick might have negative effects on the health and resilience of broilers post-hatching. Results of Simon et al. (2015) and Wijnen et al. (2021, 2022a) pointed in this direction, indicating that post-hatch feed-deprived broilers tended to be more susceptible for necrotic enteritis and colibacillosis in later life. This will be discussed in more detail in Section 5.

Besides β-oxidation, another metabolic pathway to obtain energy in the post-hatch period is ketogenesis, in which ketone bodies can be used as an energy substrate by various tissues (Beis, 1985). Neonatal ketosis has been found in other species as well during starvation (Inokuchi et al., 2001) and seem to be a way to preserve carbohydrates for other essential biosynthetic processes (Beis, 1985). Higher plasma ketone bodies have been found in chicks at hatch than in adult life, which may indicate that this metabolic pathway is used (Best, 1966; Beis, 1985; Linares et al., 1993). Because of the low availability of oxaloacetate in the liver to start the citric acid cycle, acetyl CoA can be converted to ketone bodies and serve as an energy substrate (Best, 1966; Beis, 1985; Linares et al., 1993). The contribution and consequences of ketone bodies production and utilization in the chick post-hatch are largely unknown (Sato et al., 2006; Khosravinia, 2015).

In the case that feed is provided to the chicks early post-hatching, fat and protein from the RY remain available, but this is complemented with fat, protein, and carbohydrates from the exogenous diet. Fat (from the RY plus exogenous diet) plus carbohydrates (from the exogenous diet only) can be used for ATP production in the Krebs cycle, whereas protein (from the RY plus exogenous diet) can be used for development of the chick. This is supported by the analysis of De Jong et al. (2017), showing advanced body development in early life, mainly expressed by development or organs such as the intestines. Whether this is also revealed in altered growth will be discussed in the following section.

## 4 Growth rate

Decades of genetic selection has resulted in fast-growing meat-type chickens that play a key role in the global protein supply. However, this comes with a negative side, as fast growth rate is associated with health problems such as breast myopathies and leg disorders (Barbut, 2019; Kestin et al., 2001; Lourenço-Silva et al., 2025). This suggests that broilers are performing at their physiological limits, and regardless of economic gain, additional faster growth seems undesirable from a broiler's perspective.

It is often assumed that early feeding increases a broiler's growth rate, but in fact, the opposite is true (Fig. 2). Although early feeding results in significantly higher body weights from start to slaughter age, this is largely attributed to an earlier onset of growth rather than a faster growth rate (Lamot, 2017). The difference in body weight between early- and delayed-fed broilers gets relatively smaller over time. Delayedfed broilers partially catch up in body weight during grow out, which indicates that delayed fed broilers actually have a faster growth rate than early fed broilers.

Accelerated growth after feed restriction, often referred to as 'compensatory growth', is a known phenomenon in animals (Hornick et al., 2000). In broilers, feed restriction between hatch and farm placement also appears to induce compensatory growth (Juul-Madsen et al., 2004; Lamot et al., 2014). Especially during the first 2 weeks of life, delayed fed broilers showed higher average daily weight gain relative to their body weight compared to early fed broilers (Hollemans, 2020). Whether or not this higher relative growth rate in early life of delayed fed broilers coincides with increasing growth-related problems has not been thoroughly investigated to date. The effect of early feeding on breast myopathies has been studied only once to our knowledge, by comparing two alternative hatching systems that included feed and water to a conventional hatcher (Da Silva et al., 2021). In line

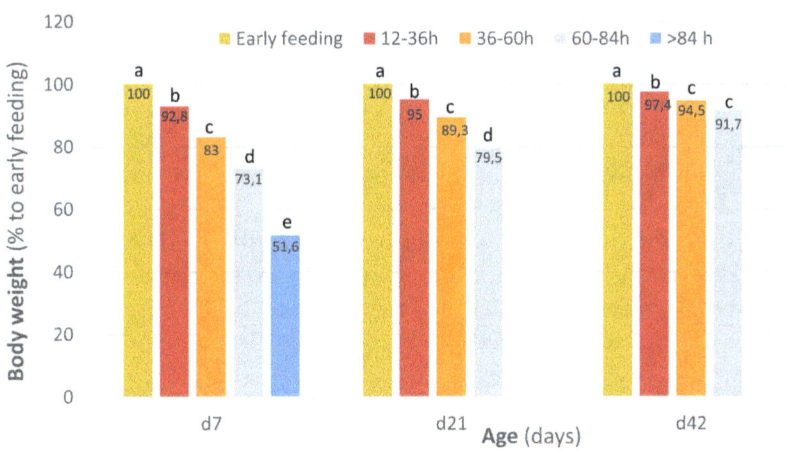


**Figure 2** Body weight at 1, 3, and 6 weeks post-hatch of broilers subjected to different delayed feeding durations at hatch relative to body weight of early fed broilers. Source: Adapted from de Jong et al. (2017).

with previous studies, the delayed fed broilers from the conventional hatcher showed compensatory growth, but regardless of that and contradictory to what was just speculated, a lower incidence of breast myopathies was found at slaughter in the delayed fed broilers compared to the early fed ones. It should be noted though that delayed fed broilers had significantly lower body weights compared to early fed broilers and especially heavier chickens with high breast muscle yield are prone to develop breast myopathies (Kuttappan et al., 2012, 2013, 2017) although growth rate seems to have a larger impact on myopathies and leg disorders than absolute body weight (Meloche et al., 2018a, 2018b, 2018c; Lourenço-Silva et al., 2025). Additionally, only a limited number of studies investigated leg disorders, expressed by alterations in gait score, and did not find differences between alternative hatching systems including early feeding compared to conventional hatching systems including delayed feeding (de Jong et al., 2018; Molenaar et al., 2023; Hanna et al., 2024). Effects of early feeding on other broiler health aspects are discussed in more detail in Section 5.

### 4.1 Body composition

It has been shown in humans and other animals that feed restriction during early life can increase body fat deposition in later life (reviewed by Picó et al., 2012). In older broilers, compensatory growth after feed restriction also seems to affect later body fat deposition (Leeson and Zubair, 1997; Lippens et al., 2002). This has never been investigated yet in young broilers and therefore it remains unknown whether early feeding affects broiler body composition during their remaining life. Findings from a preliminary case study suggest that this may be true. Carcasses (excl. feathers and head) from early and 48 h delayed fed broilers were collected at hatch and day 7, 15, 28, and 35 of age (2 broilers pooled/age/treatment). It was shown that from day 15 of age onwards, early fed broilers consistently had 4–5% lower body fat compared delayed fed broilers (Table 2).

Based on these findings, it can be speculated that early- and delayed-fed broilers may differ in their nutritional requirements, but this requires further investigation. Alternatively, it may also be possible that body composition is altered because current management conditions are suboptimal for early fed broilers compared to delayed fed broilers, such as ambient house temperature, as will be elaborated upon in Section 7.

## 5 Health

Maintaining good health is crucial from a broiler's perspective, because disease often coincides with inflammation, pain, lesions, or even mortality. De Jong et al. (2017) showed that although single studies found no significant

**Table 2** Body composition of early- and 48 h delayed fed broilers at various ages post-hatch (own unpublished results; descriptive statistics)

| | | Day 7 | | Day 15 | | Day 28 | | Day 35 | |
|---|---|---|---|---|---|---|---|---|---|
| | Hatch | Early fed | Delayed fed | Early fed | Delayed fed | Early fed | Delayed fed | Early fed | Delayed fed |
| DM (% of carcass) | 23.2 | 31.3 | 29.0 | 30.6 | 32.7 | 31.7 | 35.0 | 31.6 | 34.6 |
| Crude fat (% of carcass) | 7.1 | 12.6 | 10.6 | 10.6 | 14.6 | 11 | 15.9 | 11.5 | 16.6 |
| Crude protein (% of carcass) | 13.7 | 15.9 | 15.9 | 16.4 | 16.5 | 16.8 | 16.3 | 17 | 16.9 |

effect of early feeding on mortality, once these studies were combined in a meta-analysis, early feeding resulted in lower total mortality during grow out compared to broilers that were delayed fed for 36–60 h after pulling for certain and possibly already at shorter delayed feeding durations. This indicates that early feeding may benefit broiler health. However, the same report also stated that more evidence is needed to draw sound conclusions, because besides the difference in mortality, by the time this report was published only three studies existed that investigated how early- and delayed-fed broilers cope differently once a poultry disease was introduced (Dibner et al., 1998; Yi et al., 2005; Ao et al., 2012). These three studies each challenged broilers through oral *Eimeria* inoculation and found indications that early feeding may support broiler health, expressed by for instance a lower loss in body weight gain, lower mortality rate, and less intestinal damage during disease for early fed broilers compared to delayed fed conspecifics.

The immune system obviously plays a major role in safeguarding broiler health. Numerous studies suggest that early feeding can alter broilers immune responses during grow out (amongst others: Bar-Shira et al., 2005; Simon et al., 2014; Lamot et al., 2016; Sarlak et al., 2016; Hollemans et al., 2020; Hrabia et al., 2022; Martins et al., 2023; Madej et al., 2024; Marcato et al., 2024). However, it should be noted that studies revealing alterations or similarities in immune response between early- and delayed-fed broilers to model antigens can rarely be extrapolated to actual health of broilers. Studying in vivo responses to disease models with infectious pathogens similar to the three previously indicated studies are required to elucidate if early feeding affects broiler health.

Since the De Jong et al (2017) meta-analyses further studies have been published investigating the health of early- and delayed-fed broilers during disease. For instance, Simon et al. (2015) observed that broilers that are challenged with a combination of *E. coli* lipopolysaccharide and human serum albumin at day 28 of age spent more time resting with closed eyes 2 h after administration and reduced their feed intake during 2 days compared to non-challenged conspecifics. These sickness symptoms were lower expressed in early compared to 72 h delayed fed broilers. Additionally, early feeding resulted in a lower variance of body weight deviations and incidence of air sac lesions after inoculation of avian pathogenic *E. coli* (APEC) in the respiratory tract of broilers at day 8 of age (Wijnen et al., 2022a) and tended to lower mortality during necrotic enteritis induction at 3 weeks post-hatch (Wijnen et al., 2021) compared to 48 h delayed fed broilers. Furthermore, Boothe et al. (2024) showed that on farm-hatched broilers, an alternative hatching system that includes early feeding at the broiler house, exhibited significantly higher growth during necrotic enteritis at week 1 of age compared to hatchery-hatched delayed fed broilers. However, on farm-hatched broilers also performed better compared to hatchery-hatched broilers under non-challenged conditions.

Additionally, broilers hatched in alternative hatching systems with early feeding did not differ in their immune response to a live-attenuated infectious bronchitis vaccine virus at day 28 compared to broilers that hatched in a conventional hatcher (Molenaar et al., 2023).

Overall, relatively few studies investigated the effects of early feeding on broiler health using disease models. Therefore, thorough conclusions on this aspect of early feeding cannot be drawn. Results of these studies do point in the direction of a beneficial impact of early feeding on broiler health, and this is supported by the lower overall mortality that was revealed through a meta-analysis by De Jong et al. (2017). Yet, many aspects remain largely unknown to date, such as broiler health during other common diseases like enterococcus, avian influenza, or non-infectious metabolic disorders. Future research observing broiler health during disease models rather than merely observing immune response to vaccinations or model antigens is strongly advised. Furthermore, an increasing number of studies suggest that beneficial alterations in the gut microbiota composition after early feeding potentially explain why broiler health can be improved (Li et al., 2022; Miska et al., 2022; Proszkowiec-Weglarz et al., 2022; Katafuchi et al., 2023; Pottenger et al., 2023; Such et al., 2023; Qiu et al., 2023; Shani and Irani, 2024; Shani et al., 2024; Leterrier et al., 2025). However, more research is required to reveal exact biological mechanisms underlying the potential positive long term effect of early feeding on broiler health.

## 6 Welfare

Early feeding can influence broiler welfare, but it should be noted that effects can be dependent on the type of alternative hatching system as was discussed in the introduction of this chapter. In the current section, the effect of early feeding on several broiler welfare aspects are described, linking to the three welfare pillars commonly used: (1) affective state, (2) physical well-being, and (3) natural behaviour.

### *6.1 Affective state*

Delayed feeding can contribute to early life stress when hunger and thirst are experienced by the broiler chicken, thereby influencing the affective state and behaviour of the broiler chicken. Plasma corticosterone, commonly used as indicator for stress, showed ambiguous results among studies related to delayed feeding (de Jong et al., 2017). Some studies found no difference (Gonzaleset al., 2003; van de Ven et al., 2013; Tong et al., 2015), while others found a higher plasma corticosterone level after delayed feeding (Wijnenet al., 2022b; Witjeset al., 2022; Madejet al., 2024; Broekmeulen et al., 2024). These higher corticosterone levels often disappeared after feed and water were provided (Wijnen et al., 2022b; Broekmeulen et al., 2024) and may indicate that delayed feeding induces a short-term stress response during the fasting period, of which the extent is correlated with the duration. Besides corticosterone level, chick behaviour can also be used as an indicator of early life stress. Khosravinia and Manafi (2016) scored the behaviour of fast- and slow-feathering broiler chicks in boxes during delayed feeding periods up to 48 h and found that escape attempts and active behaviour increased over time. They linked this behaviour to increased frustration the longer the delayed feeding took, and a higher motivation to search for feed and water.

An important welfare indicator is the absence of hunger and thirst. Although there is limited information on how quickly chicks start to eat and drink after hatching, as was discussed in Section 2 'Delayed feeding', prevention of hunger and thirst is one of key aspects of the five animal freedoms (Brambell, 1965). Serum indicators may reflect metabolic changes related to feed and water withdrawal on the one side or early feeding on the other side and this may be used as a reference for the level of hunger and thirst that broiler hatchlings experience. Madej et al. (2024) showed that glucose and triglycerides were significant lower in delayed fed broiler chicks at 1 day after hatching compared to early fed chicks, whereas total serum protein, albumin, and high-density lipids were higher. The lower glucose and triglycerides concentration in blood is probably related to the fact that these nutrients were provided to the early fed chicks, and consequently appeared in the blood. At the same time, the higher serum protein, albumin, and high-density lipids in the delayed fed chicks are probably due to dehydration, meaning that concentrations of these components in blood were increased. Plasma osmolarity, chloride and creatinine, and blood sodium are often used as indicators of feed and water deprivation in later life (Vanderhasselt et al., 2013), but studies measuring this in the post-hatch period are largely lacking.

These ambiguous but rather limited results of early feeding on aspects of affective state in later life may be influenced by on the one hand extrinsic factors, such as duration of delayed feeding, confounding factors of alternative hatching systems, and the chosen behavioral observations (Bergoug et al., 2013; Van den Brand and Molenaar, 2023; Hanna et al., 2024). On the other hand, also intrinsic factors related to strain and the motivation of broilers may explain results found (Giersberg et al., 2020; Hanna et al., 2024).

### 6.2 Physical well-being

The physical well-being of a broiler chicken can be measured by animal-based welfare indicators, and they are mainly scored within studies comparing different alternative hatching systems providing early feed (Van den Brand and Molenaar, 2023). Especially FPD (foot pad dermatitis) tended to be, or was, reduced in alternative hatching systems including early feeding (de Jong et al., 2018; de Jong et al., 2020; Giersberget al., 2021; Lingenset al., 2021; Souza et al., 2023). As litter quality was often not affected in those studies, underlying mechanisms explaining these results are currently largely unclear but may be related to intestinal health and development (Van den Brand and Molenaar, 2023, Akram *et al.*, 2025). Sections 4 and 5 of the current chapter do discuss the effect of early feeding on growth rate related physical well-being and health, respectively.

### 6.3 Natural behaviour

Natural behaviour was investigated in several studies by evaluating the activity and general behaviour of early fed broilers and delayed fed conspecifics, but this was also mainly investigated using alternative hatching systems. Jessen et al. (2021) found indications in slow-growing broilers, and Hanna et al. (2024) in fast-growing broilers that on-farm hatched chicks showed more resting behaviour than hatchery-hatched chicks without early feed and water access. This was especially apparent during the first day of age, possibly as a result of fewer early life stressors, and provision of early feed and water for the on-farm hatched broiler hatchlings. Hanna et al. (2024) showed that play and comfort behaviour, which are often linked to a positive affective state, were not affected in later life between alternative hatching systems (on-farm, hatchery-fed or hatchery-hatched) or different diets (water only, dry or wet feed). Giersberg et al. (2020) also showed that general behaviour and activity did not differ greatly between on-farm or hatchery-hatched broiler chicks in later life. This suggests that behaviour of broilers is mainly affected during the deprivation period and is not largely influenced anymore in later life after providing early feed and water within alternative hatching systems.

Other studies that investigated effects of early feeding on welfare used a variety of different behavioral tests, in which the affective state can be tested in a more challenging environment. Giersberg et al. (2020) found that on-farm hatched fast-growing broilers were more fearful than hatchery-hatched broilers. In contrast, Jessen et al. (2021) found the opposite in slow-growing broilers. However, a subsequent study by Giersberg et al. (2021) found no effect of different alternative hatching systems on several behavioural tests assessing fearfulness, explorational behaviour, and social reinstatement in early or later life. Wijnen et al. (2022b) also found no difference in fearfulness between early fed or 48 h delayed fed fast-growing broilers at day 13 of age. Contrary, Hollemans et al. (2018) found that delayed fed broilers (54 h) were less fearful at 3 days of age, but more fearful at 30 days of age compared to early fed broilers. However, this effect was only found in broilers that were transported for 3 h.

### 6.4 Conclusions

To summarize, effects of early feeding on the affective state, natural behaviour, and physical well-being may result from a combination of different early life events that the chicks are exposed to in the post-hatch period. Underlying mechanisms, such as hormone levels, coping strategies, or brain and cognitive development in early life may explain the results found, but are not intensively studied (Giersberg et al., 2020; Broekmeulen et al., 2024). Therefore, it is important to carefully describe the different procedures surrounding early feed and water provision within studies, and perform more fundamental studies about the relationship between chicken neurobiology and behaviour.

## 7 Early feeding case studies

Numerous scientific studies have investigated the long-term effects of early feeding versus delayed feeding of broilers. However, less attention has been given to the specific requirements and optimal climate conditions once early fed broiler chicks leave the hatchery. Suboptimal conditions during this critical period can have a negative impact on broiler chick thermal comfort, and consequently influence long-term production performance. Therefore, understanding the climate needs of early fed broiler chicks is essential. Do such broiler chicks have different temperature preferences compared to those that are delayed fed? Furthermore, can they tolerate extended transportation times off-feed without adverse effects? Various practical experiments have been conducted to explore the optimal conditions for early fed day-old chicks once they depart from the hatchery. These will be addressed in this section.

Understanding the thermoregulation of day-old chicks is crucial to meeting their post-hatch requirements. A heat production test conducted in climate respiration chambers at Wageningen University & Research demonstrated that chicks with immediate access to feed and water after hatching experience a gradual increase in heat production over time (Van Roovert-Reijrink et al., 2017). Once all chicks had hatched, heat production was approximately 350 mW/chick, which increased to 500 mW/chick at the time chicks would normally be pulled around D21:10H (Fig. 3). At the end of the experiment (14 h after pulling), heat production in early fed chicks was double their heat production at hatch, reaching 700 mW/chick. In contrast, heat production in chicks that did not receive feed and water remained around 350 mW/chick. This means early fed chicks can produce 1.5–2 times more heat than delayed fed chicks before placement in the broiler house. It may therefore be necessary to lower the ambient temperature during transport as well as lowering whole house temperature upon arrival for early fed chicks to offset their higher body heat production. It has been shown for instance that early fed broilers prefer a

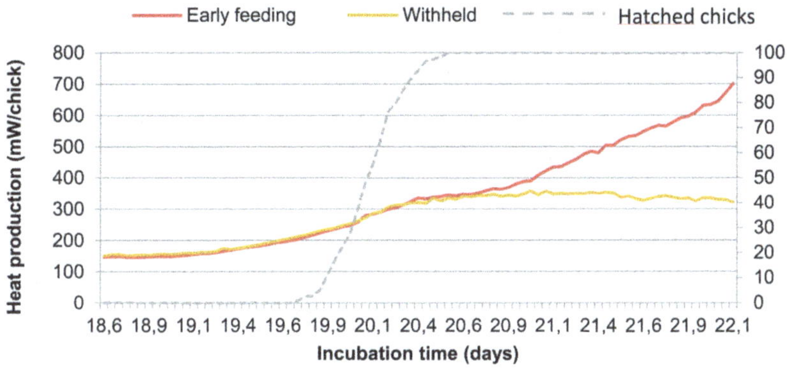


**Figure 3** Heat production and percentage of hatched chicks over incubation time for chicks that received feed and water immediately post-hatch, and chicks for which feed and water were withheld. Source: Adapted from Van Roovert-Reijrink et al. (2017).

~2°C lower ambient temperature upon 12 days post-hatch compared to 48 h delayed fed birds (Wijnen et al., 2022b).

Body temperature regulation is influenced not only by ambient temperature but also by relative humidity and air velocity. Chicks that have immediate access to feed and water have a different hydration status than those that are delayed fed (Gaweł et al., 2023). This could potentially affect their climate preferences, as this may affect their evaporative cooling potential (such as through panting). In an experiment examining the effects of climate conditions, behavioral observations (huddling, panting, and wing spreading) were performed on early- and delayed fed broiler chicks at normal transportation stocking densities under various combinations of air temperature (33–38°C), relative humidity (20–80%), and air velocities (0.5 or 1.0 m/s) (unpublished results). The results showed, on the one hand, that early fed chicks tolerated air temperatures of 35–37°C, provided that relative humidity was low (approximately 40%). Delayed fed chicks, on the other hand, preferred a narrower temperature range of 35°C (at 0.5 m/s) to 36°C (at 1.0 m/s), with little variation due to changes in humidity. Heat loss data indicated that heat loss through evaporation in delayed fed chicks was almost negligible. This was true even at air temperatures above their preferred ambient temperatures (≥37.0°C), which suggests they were less capable of regulating their body temperature through evaporative cooling. Early fed chicks, however, were able to use evaporation as a heat loss mechanism if the air's RH allowed it.

These findings can be used for practical applications for optimizing the post-hatch and transportation climates for both early fed and delayed fed chicks. While early fed chicks produce more heat, they can lose this heat through evaporation as long as relative humidity is maintained at approximately 40% or lower. In hot and humid climates, however, early fed chicks at high stocking densities may be at risk of overheating if environmental conditions such as temperature and humidity are not well-controlled during transportation. This highlights that overheating remains a significant risk for early fed chicks, especially when humidity is high, as this can lead to dehydration and chick quality loss (Van Roovert-Reijrink et al., 2017).

In commercial practice, day-old chicks are transported for periods ranging from 1 h to 4 days, depending on the distance between the hatchery and the poultry house. Concerns have been raised regarding whether early fed chicks can handle prolonged transportation durations as well as delayed fed chicks. Some hatcheries have experienced high first-week mortality in early fed chicks that were transported for extended periods. This could be due to suboptimal climate conditions that led to overheating, as mentioned earlier. Another theory suggests that the gastrointestinal tract of early fed chicks may be more sensitive to being off-feed after having been fed, which could result in worse post-hatch performance. To investigate this, an experiment was conducted using Ross 308 broiler chicks, which were hatched in an alternative hatching system (HatchCare) either given access to feed and water immediately ('Early Fed') or not ('Withheld') from hatch until pull (Van Roovert-Reijrink et al., 2019). These chicks were then transported during 2 h to a grow-out facility, where half of each group received feed and water immediately ('Placement Fed') or after 24 h ('Delayed'), creating four treatment groups in a 2 x 2 factorial design. The 24-h off-feed period was intended to mimic the time chicks might spend off-feed during extended transportation, without the added effects of transportation conditions. To prevent eating of litter, wood shavings were added to the pen only after the delayed period was finished. Results can be seen in Table 3. On day 35 of grow-out (slaughter age), early fed chicks were heavier than Withheld chicks ($P$ = 0.02), regardless of whether they had immediate access to feed and water in the grow-out facility. This demonstrates that providing chicks with feed and water at the hatchery positively influences post-hatch growth, even if they are off-feed for a prolonged period after hatching. Day 35 Placement Fed chicks tended to be heavier than delayed chicks ($P$ = 0.07), who had been off-feed for 24 h after

**Table 3** Body weights from Day 0 (pull at the hatchery) till Day 35 (slaughter age) for broiler chickens that received food and water in the hatchery (Early Fed) or were withheld (Withheld), and then received feed and water upon placement in the broiler farm (Placement Fed) or were delayed for 24 h (Delayed)

| | Body weight (g/chicken) | | | | | |
|---|---|---|---|---|---|---|
| | Day 0 | Day 1 | Day 7 | Day 14 | Day 28 | Day 35 |
| Ealy fed – placement fed | 52.7 | 68.1 | 225 | 593 | 1909 | 2526 |
| Early fed – delayed | 52.7 | 48.3 | 199 | 554 | 1818 | 2436 |
| Withheld – placement fed | 46.2 | 58.5 | 194 | 549 | 1819 | 2420 |
| Withheld – delayed | 46.1 | 43.1 | 178 | 510 | 1775 | 2394 |

arriving at the grow-out facility. However, there was no interaction between the chicks' feeding status at hatch and their growth performance during the first 24 h post-arrival. Furthermore, overall feed-conversion ratio (FCR) and mortality rates did not differ significantly between the groups. These results suggest that concerns on whether early fed chicks can handle prolonged durations off-feed are likely unnecessary if suitable transport conditions are guaranteed. To the authors knowledge there are no sound scientific studies on this topic, except that current results are in line with a study that transported both early and delayed fed chicks during 16 h from the hatchery to the farm and at farm arrival, early fed chicks had advanced gastrointestinal development compared to the delayed fed chicks (Boyner et al., 2023). Furthermore, no differences in mortality or vaccination response were found between both groups and early fed broilers showed higher feed intake and body weight during the first 11 days. On-farm hatching systems, where embryonated eggs at approximately 18 days of incubation are transported directly to the broiler farm, eliminate the need to transport day-old chicks. However, Nordquist et al. (2022) found evidence suggesting that 18-day-old chicken embryos also experience some stress during transport. Therefore, suitable transportation conditions should be ensured for both embryos and chicks of all ages.

In summary, these case studies demonstrate that early fed chicks require specific climate conditions to accommodate their increased heat production. Early fed chicks are more vulnerable to overheating compared to delayed fed chicks due to their higher heat production. A certain period off-feed during transportation appears to have limited effects on early fed chicks.

## 8 Future trends in early feeding

Future research on early feeding will likely focus on understanding the physiological changes that occur in early fed chicks and how these changes interact with the post-hatch environment and handling procedures. One critical area for poultry health is vaccination programs. Hatchery-based vaccination programs, particularly in-ovo vaccination, are expanding, and commercial vaccine programs are typically designed for chicks that have not been fed at the hatchery. A lot is unknown to date on how early feeding influences the immune competence and response of chicks to vaccination. Early fed chicks may require modified vaccination protocols for optimal protection, as they may mount a different immune response or show varying levels of vaccine tolerance. Given that early fed chicks show differences in early gut development, it is also possible that their response to orally administered vaccines, such as spray vaccines, differs from that of delayed fed chicks.

Future research in broiler health would benefit from evaluation of health outcomes within actual disease challenge models. While immune response to vaccination and model antigens provide valuable information on the activation of the immune system, they are not always great predictors of disease resilience or outcome as the mounting of a disease response is complex and multi-factorial.

Another aspect that requires further studies is the potential effect of early feeding on meat quality in general and breast myopathies in particular. In case early or delayed feeding affects the prevalence and/or severity of breast myopathies at slaughter age (all or not related to BW) this might need different growth patterns of the early and delayed fed chicks.

From a welfare perspective, advancing fundamental knowledge in chicken neurobiology and developmental neuroscience is helpful for understanding how early feeding influences brain development and behaviour. A better understanding of the neurobiological mechanisms underlying the effects of early feeding will allow researchers to identify if and how early feeding impacts emotional well-being and adaptability in later life.

Animal welfare concerns have led to a rise in production using slow-growing broiler breeds in western Europe. Research on the early feeding of slow-growing breeds is ongoing, as these breeds have distinct metabolic rates and growth patterns compared to fast-growing counterparts. Factors like feeding timing, diet composition, and environmental conditions during the early stages of life are still being explored. Continued research is necessary to clarify how these factors influence the growth and health of slow-growing broilers and to optimize feeding strategies for these breeds.

## 9 Where to look for further information

A comprehensive description of alternative hatching systems compared to conventional hatching systems can be found in Chapter 16 *Alternative hatching systems for broilers* from the book *Embryo development and hatchery practice in poultry production* edited by Dr Nick French.

Three literature reviews that include early feeding are highly recommended:

- the systematic literature review and meta-analysis *Effects of food and water deprivation in newly hatched chickens* from De Jong et al. (2017) that summarized 75 papers about early feeding published between 2000 and 2016.
- the review from Taha-Abdelaziz et al. (2018) entitled *Effects of early feeding and dietary interventions on development of lymphoid organs and immune competence in neonatal chickens: A review*, which elaborates how early feeding modulates the development and function of the immune system in young chicks.

- the review from Jha et al. (2019) entitled *Early nutrition programming (in ovo and post-hatch feeding) as a strategy to modulate gut health of poultry*, which provides information about the effects of different nutritional compounds in early feed on early broiler development.

Doctoral dissertations:

- *First week nutrition for broiler chickens: Effects on growth, metabolic status, organ development, and carcass composition* published in 2017 by David Lamot that elaborates on individual hatch moment within the hatch window and diet densities. https://edepot.wur.nl/403639.
- *Short and long term effects of early nutrition in broiler chickens* published by Maarten Hollemans in 2020 which focusses, amongst other topics, on immune responses, oral tolerance, and compensatory growth after post-hatch feeding strategies. https://edepot.wur.nl/528572.
- *Route to Resilience: Long term effects of incubation temperature and early feeding in broilers* published in 2022 by Jan Wijnen that focusses on the effects of early feeding and it's interaction with incubation conditions on resilience against infectious diseases. https://edepot.wur.nl/568137.
- *A flying start: Adapted hatching and post-hatch feeding in broiler chickens* published by Malin Boyner in 2023.

The European Union currently revises their animal welfare legislations and requested the European Food Safety Authority Panel for Animal Health And Welfare (EFSA AHaW) their scientific opinion on the welfare of domestic fowl. Amongst others, the welfare consequences of early feeding for broilers are discussed in the report entitled *Welfare of broilers on farm* published in 2023.

## 10 References

Abdulameer, Y.S., et al. (2025). Effect of post-hatch transportation duration on growth performance, immune function, blood profiles and intestinal bacteriology of broiler chicks. *Adv. Anim. Vet. Sci.* 13:909–918.

Akram, M.Z., et al. (2025). Effects of on-farm hatching versus hatchery hatching on growth performance, gut development, and intestinal health and function in broiler chickens. *Poult. Sci.* 104:104770.

Almeida, V.R., et al. (2016). Incubation temperature manipulation during fetal development reduces adiposity of broiler hatchlings. *Poult. Sci.* 95:316–324.

Ao, Z., et al. (2012). Effects of dietary additives and early feeding on performance, gut development and immune status of broiler chickens challenged with Clostridium perfringens. *Asian-Australas. J. Anim. Sci.* 25:541–551.

Barbut, S. (2019). Recent myopathies in broiler's breast meat fillets. *Worlds Poult. Sci. J.* 75:559–581.

Bar-Shira, E., et al. (2003). Establishment of immune competence in the avian GALT during the immediate post-hatch period. *Dev. Comp. Immunol.* 27:147–157.

Beis, A. (1985). Activities of the enzymes of ketone body metabolism in the developing embryo. *Comp. Biochem. Physiol* 81B:671–675.

Bergoug, H., et al. (2013). Effect of transportation duration of 1-day-old chicks on postplacement production performances and pododermatitis of broilers up to slaughter age. *Poult. Sci.* 92:3300–3309.

Best, E.E. (1966). The changes of some blood constituents during the initial post-hatching period in chickens. *Br. Poult. Sci.* 7:23–28.

Bhanja, S.K., et al. (2009). Effect of post hatch feed deprivation on yolk-sac utilization and performance of young broiler chickens. *Asian-Australas. J. Anim. Sci.* 22:1174–1179.

Bierer, B.W. and Eleazer, T.H. (1965). Effect of feed and water deprivation on yolk utilization in chicks. *Poult. Sci.* 44:1608–1609.

Bigot, K., et al. (2003). Effects of delayed feed intake on body, intestine, and muscle development in neonate broilers. *Poult. Sci.* 82:781–788.

Boothe, S.M., et al. (2024). Research note: effects of on-farm and hatchery hatching on broiler performance, intestinal lesions, and immune response during a subclinical necrotic enteritis challenge. *Poult. Sci.* 103(12):104323.

Boyner, M., et al. (2021). Effect of hatching time on time to first feed intake, organ development, enzymatic activity and growth in broiler chicks hatched on-farm. *Animal* 15(2):100083.

Boyner, M., et al. (2023). Effects of access to feed, water, and a competitive exclusion product in the hatcher on some immune traits and gut development in broiler chickens. *Br. Poult. Sci.* 64:330–342.

Brambell, F.R. (1965). *Report of the technical committee to enquire into the welfare of animals kept under intensive livestock husbandry systems*. London: Her Majesty's Stationery Office.

Broekmeulen, C.M.H., et al. (2024). Influence of light exposure during late incubation, early feeding, and litter access posthatch on the stress response and body weight gain in laying hen chicks. *J. Appl. Poult. Res.* 33:100473.

Cardeal, P.C., et al. (2021). Breeder age and post-hatch feed access influence physiology of residual yolk sac, intestinal development and performance of broiler chicks. *Livest. Sci.* 251:104617.

Careghi, C., et al. (2005). The effects of the spread of hatch and interaction with delayed feed access after hatch on broiler performance until seven days of age. *Poult. Sci.* 84:1314–1320.

da Silva, C.S., et al. (2021). Day-old chicken quality and performance of broiler chickens from 3 different hatching systems. *Poult. Sci.* 100(3):100953.

de Jong, I.C., et al. (2015). Implementation of the Welfare Quality® broiler assessment protocol–final report. *Wageningen UR Livest. Res. Rept.* 833:1–57.

de Jong, I.C., et al. (2017). A 'meta-analysis' of effects of post-hatch food and water deprivation on development, performance and welfare of chickens. *PLoS One* 12:e0189350.

de Jong, I.C., et al. (2018). Comparison of performance, health and welfare aspects between commercially housed hatchery-hatched and on-farm hatched broiler flocks. *Animal* 13:1269.

de Jong, I.C., et al. (2020). Effects of on-farm and traditional hatching on welfare, health, and performance of broiler chickens. *Poult. Sci.* 99: 4662–4671.

de Oliveira, J.E., et al. (2008). Important metabolic pathways in poultry embryos prior to hatch. *Worlds Poult. Sci. J.* 64:488–499.

Deines, J.R., et al. (2021a). Effects of hatch window and nutrient access in the hatcher on performance and processing yields of broilers reared with equal hatch window representation, *Animals* 11:1228.

Deines, J.R., et al. (2021b). Effects of hatch window and nutrient access in the hatcher on performance and processing yields of broilers reared according to time of hatch. *Poult. Sci.* 100:101295.

Dibner, J.J., et al. (1998). Early feeding and development of the immune system in neonatal poultry. *J. Appl. Poult. Res.* 7:425–436.

EFSA Panel on Animal Health and Welfare, et al. (2023). Welfare of broilers on farm. *EFSA J.* 21(2):e07788.

El-Husseiny, O.M., et al. (2008). Influence of fasting or early feeding on broiler performance. *Int. J. Poult. Sci.* 7:263–271.

FAWC (2009). *Farm animal welfare in Great Britain: past, present and future*. 57 pp. London: Farm Animal Welfare Council.

Franco, J.R.G., et al. (2006). Influence of delayed placement and dietary lysine levels on small intestine morphometrics and performance of broilers. *Braz. Poult. Sci. J.* 8:233–241.

Gaglo-Disse, A., et al. (2010). Effect of delayed feed access on production and blood parameters of layer-type chicks. *Acta Vet. Hung* 58:211–219.

Gaweł, A., et al. (2023). Impact of early post-hatch nutrition on chickens performance and health. *Abstract XXIInd World Veterinary Poultry Association Congress* 236–236.

Giersberg, M.F., et al. (2020). Comparative assessment of general behaviour and fear-related responses in hatchery-hatched and on-farm hatched broiler chickens. *Appl. Anim. Behav. Sci.* 232: 105100.

Giersberg, M.F., et al. (2021). Effects of hatching system on the welfare of broiler chickens in early and later life. *Poult. Sci.* 100:100946.

Gonzales, E., et al. (2003). Performance and physiological parameters of broiler chickens subjected to fasting on the neonatal period. *Poult. Sci.* 82: 1250–1256.

Hamissou Maman, A., et al. (2019). Effect of chick body temperature during post-hatch handling on broiler live performance. *Poult. Sci.* 98:244–250.

Hanna, H., et al. (2024). Health, welfare and lifetime performance implications of alternative hatching and early life management systems for broiler chickens. *PLoS One* 19:e0303351.

Hollemans, M.S. (2020). *Short and long term effects of early nutrition in broiler chickens* (Doctoral dissertation, Wageningen University and Research).

Hollemans, M.S., et al. (2018). Effects of early nutrition and transport of 1-day-old chickens on production performance and fear response. *Poult. Sci.* 97(7):2534–2542.

Hollemans, M.S., et al. (2020). Effects of early nutrition and sanitary conditions on oral tolerance and antibody responses in broiler chickens. *Vet. Sci.* 7(4):148.

Hollemans, M.S., et al. (2021). Effects of early nutrition and sanitary conditions on antibody levels in early and later life of broiler chickens. *Dev. Comp. Immunol.* 117:103954.

Hopcroft, R.L., et al. (2020). Changes to Cobb 500 chick characteristics, bone ash, and residual yolk mineral reserves during time spent in the hatcher. *Poult. Sci.* 99:2176–2184.

Hornick, J.L., et al. (2000). Mechanisms of reduced and compensatory growth. *Domest. Anim. Endocrinol.* 19:121–132.

Hrabia, A., et al. (2022). Altered gene expression of selected matrix metalloproteinase system proteins in the broiler chicken gastrointestinal tract during post-hatch development and coccidia infection. *Poult. Sci.* 101(8):101915.

Inokuchi, T., et al. (2001). Neonatal ketosis is not rare: experience of neonatal screening using gas chromatography–mass spectrometry. *J. Chromatogr. B Biomed. Sci. Appl.* 758:57–60.

Jessen, C.T., et al. (2021). Effects of hatching on-farm on behaviour, first week performance, fear level and range use of organic broilers. *Appl. Anim. Behav. Sci.* 238:105319.

Jha, R., et al. (2019). Early nutrition programming (in ovo and post-hatch feeding) as a strategy to modulate gut health of poultry. *Front. Vet. Sci.* 6:82.

Juul-Madsen, H.R., et al. (2004). Influence of early or late start of first feeding on growth and immune phenotype of broilers. *Br. Poult. Sci.* 45(2):210–222.

Katafuchi, A., et al. (2023). Effects of delaying post-hatch feeding on the plasma metabolites of broiler chickens revealed by targeted and untargeted metabolomics. *J. Poult. Sci.* 60(2): 2023032.

Kestin, S.C., et al. (2001). Relationships in broiler chickens between lameness, liveweight, growth rate and age. *Vet. Rec.* 148:195–197.

Khosravinia, H. (2015). Effects of intrayolk sac inoculation of olive oil on physiological adaptive responses in newly hatched broiler chicks subjected to neonatal fasting. *J. Poult. Sci.* 52:304–311.

Khosravinia, H. and Manafi, M. (2016). Broiler chicks with slow-feathering (K) or rapid-feathering (k+) genes: effects of environmental stressors on physiological adaptive indicators up to 56 h posthatch. *Poult. Sci.* 95:1719–1725.

Kuttappan, V.A., et al. (2012). Influence of growth rate on the occurrence of white striping in broiler breast fillets. *Poult. Sci.* 91:2677–2685.

Kuttappan, V.A., et al. (2013). Comparison of hematologic and serologic profiles of broiler birds with normal and severe degrees of white striping in breast fillets. *Poult. Sci.* 92:339–345.

Kuttappan, V.A., et al. (2017). Incidence of broiler breast myopathies at 2 different ages and its impact on selected raw meat quality parameters. *Poult. Sci.* 96:3005–3009.

Lamot, D.M. (2017). *First week nutrition for broiler chickens: effects on growth, metabolic status, organ development, and carcass composition* (Doctoral dissertation, Wageningen University and Research).

Lamot, D.M., et al. (2014). Effects of moment of hatch and feed access on chicken development. *Poult. Sci.* 93:2604–2614.

Lamot, D.M., et al. (2016). Effects of feed access after hatch and inclusion of fish oil and medium chain fatty acids in a pre-starter diet on broiler chicken growth performance and humoral immunity. *Animal* 10:1409–1416.

Lamot, D.M., et al. (2017). Diet density during the first week of life: Effects on energy and nitrogen balance characteristics of broiler chickens. *Poult. Sci.* 96:2294–2300.

Leeson, S. and Zubair, A. K. (1997). Nutrition of the broiler chicken around the period of compensatory growth. *Poult. Sci.* 76(7):992–999.

Leterrier, C., et al. (2025). Short-and long-term sex-dependent effects of a nutritional supplement after hatching on growth, metabolism and gut microbiota in broiler chickens. *Br. Poult. Sci.* 24:100343.

Linares, A., et al. (1993). Changes in ketone body utilization by chick liver, duodenal mucosa and kidney during embryonic and postnatal development. *Comp. Biochem. Physiol. B Biochem. Mol. Biol.* 105:277–282.

Lingens, J.B., et al. (2021). Effects of early nutrition of hatched chicks on welfare and growth performance: a pilot study. *Animals* 11:2888.

Lippens, M., et al. (2002). The efficiency of nitrogen retention during compensatory growth of food-restricted broilers. *Br. Poult. Sci.* 43(5):669–676.

Løtvedt, P. and Jensen, P. (2014). Effects of hatching time on behavior and weight development of chickens. *PLoS One* 9(7):e103040.

Li, D.L., et al. (2022). Effects of early post-hatch feeding on the growth performance, hormone secretion, intestinal morphology, and intestinal microbiota structure in broilers. *Poult. Sci.* 101(11):102133.

Lourenço-Silva, M.I., et al. (2025). Fast growth rate is associated with musculoskeletal biomechanical imbalance and dorsal cranial myopathy in broiler chickens. *PLoS One* 20:e0332693.

Maatjens, C. M. (2016). *Effects of temperature and CO2 during late incubation on broiler chicken development* (Doctoral dissertation, Wageningen University and Research).

Madej, J.P., et al. (2024). Impact of early posthatch feeding on the immune system and selected hematological, biochemical, and hormonal parameters in broiler chickens. *Poult. Sci.* 103(3): 103366.

Marcato, F., et al. (2024). Effects of breed and early feeding on intestinal microbiota, inflammation markers, and behavior of broiler chickens. *Front. Vet. Sci.* 11:1492274.

Martins, P.C., et al. (2023). Cellular and humoral immunity of broilers subjected to posthatch fasting and a prestarter diet containing conjugated linoleic acid. *Poult. Sci.* 102(12):103167.

Meloche, K.J., et al. (2018a). Effects of quantitative nutrient allocation on myopathies of the Pectoralis major muscles in broiler chickens at 32, 43, and 50 days of age. *Poult. Sci.* 97:1786–1793.

Meloche, K.J., et al. (2018b). Effects of reduced dietary energy and amino acid density on Pectoralis major myopathies in broiler chickens at 36 and 49 days of age 1. *Poult. Sci.* 97:[[[s-pagenums]]]1794.

Meloche, K.J., et al. (2018c). Effects of reduced digestible lysine density on myopathies of the Pectoralis major muscles in broiler chickens at 48 and 62 days of age. *Poult. Sci.* 97:3311–3324.

Miska, K.B., et al. (2022). Delay of feed post-hatch causes changes in expression of immune-related genes and their correlation with components of gut microbiota, but does not affect protein expression. *Animals* 12(10):1316.

Molenaar, R., et al. (2010). Effect of eggshell temperature and oxygen concentration on survival rate and nutrient utilization in chicken embryos. *Poult. Sci.* 89:2010–2021.

Molenaar, R., et al. (2023). Effects of hatching system on chick quality, welfare and health of young breeder flock offspring. *Poult. Sci.* 102(3):102448.

Moran, E.T. and Reinhart, B.S. (1980). Poult yolk sac amount and composition upon placement: effect of breeder age, egg weight, sex, and subsequent change with feeding or fasting. *Poult. Sci.* 59:1521–1528.

Nangsuay, A., et al. (2013). Energy utilization and heat production of embryos from eggs originating from young and old broiler breeder flocks. *Poult. Sci.* 92:474–482.

Nielsen, B.L., et al. (2008). Ultradian activity rhythms in large groups of newly hatched chicks (Gallus gallus domesticus). *Behav. Process.* 78:408–415.

Nielsen, B.L., et al. (2010). Feeding activity in groups of newly hatched broiler chicks: effects of strain and hatching time. *Poult. Sci.* 89:1336–1344.

Nordquist, R.E., et al. (2022). The effects of transport of 18-day old hatching eggs on physiology and behaviour of slow growing broiler chicken. *Appl. Anim. Behav. Sci.* 257:105789.

Noy, Y. and Sklan, D. (1998). Yolk utilization in the newly hatched poult. *Br. Poult. Sci.* 39:446–451.

Noy, Y. and Sklan, D. (2001). Yolk and exogenous feed utilization in the posthatch chick. *Poult. Sci.* 80:1490–1495.

Olsen, M.W. and Winton, B. (1941). Viability and weight of chicks as affected by shipping and time without feed. *Poult. Sci.* 20:43–250.

Picó, C., et al. (2012). Metabolic programming of obesity by energy restriction during the perinatal period: different outcomes depending on gender and period, type and severity of restriction. *Front. Physiol.* 3: 436.

Pinchasov, Y. and Noy, Y. (1993). Comparison of post-hatch holding time and subsequent early performance of broiler chicks and turkey poults. *Br. Poult. Sci.* 33: 111–120.

Pottenger, S., et al. (2023). Timing and delivery route effects of cecal microbiome transplants on Salmonella Typhimurium infections in chickens: potential for in-hatchery delivery of microbial interventions. *Anim. Microbiome* 5(1):11.

Proszkowiec-Weglarz, M., et al. (2022). Delayed access to feed early post-hatch affects the development and maturation of gastrointestinal tract microbiota in broiler chickens. *BMC Microbiol.* 22(1): 206.

Qiu, K., et al. (2023). Effects of peroral microbiota transplantation on the establishment of intestinal microorganisms in a newly-hatched chick model. *Agriculture* 13(5):1001.

Roberts, R.E. (1928). When should chicks be given first feed?. *Poult. Sci.* 7:220–224.

Romanoff, A.L. (1944). Avian spare yolk and its assimilation. *Auk* 61:235–241.

Romanoff, A.L. (1960). *The avian embryo: structural and function development*. New York: Macmillan Co.

Roura, E., et al. (2025). Developmental programming in chickens including transgenerational nutrition, chemosensory mechanisms and in ovo essential oils, organic acids, and pre-pro-syn-biotics; role of epigenetic modifications. *Poult. Sci.* preproof 105857.

Sarlak, S., et al. (2016). Effects of time of initiation of feeding after hatching and diet composition on performance, carcass characteristics, digestive tract development and immune responses of broilers. *Anim. Product. Sci.* 57(8):1692–1701.

Sato, M., et al. (2006). Heat production and lipid metabolism in broiler and layer chickens during embryonic development. *Comp. Biochem. Physiology* 143A:382–388.

Schippers, H.L., et al. (2014). *Chickens*. Zutphen: Roodbont Publishers.

Shani, M.A. and Irani, M. (2024). Feeding strategy and prebiotic supplementation: Effects on immune responses and gut health in the early life stage of broiler chickens. *Res. Vet. Sci.* 171:105226.

Shani, M.A., et al. (2024). Impact of probiotic supplementation in early or late feeding programs on growth performance and mucin gene expression in broiler chickens. *J. Appl. Poult. Res.* 33(4):100477.

Shira, E.B., et al. (2005). Impaired immune responses in broiler hatchling hindgut following delayed access to feed. *Vet. Immunol. Immunopathol.* 105:33–45.

Simon, K., et al. (2014). Development of ileal cytokine and immunoglobulin expression levels in response to early feeding in broilers and layers. *Poult Sci.* 93:3017–3027.

Simon, K., et al. (2015). Early feeding and early life housing conditions influence the response towards a noninfectious lung challenge in broilers. *Poult Sci.* 94:2041–2048.

Sklan, D., et al. (2000). Decreasing weight loss in the hatchery by feeding chicks and poults in hatching trays. *J. Appl. Poult. Res.* 9:142–148.

Souza, L.F.A., et al. (2023). Effect of hypercapnia during incubation and broiler breeder age on development of the gastrointestinal tract in embryos and hatchlings. *Braz. J.Poult. Sci.* 25.

Such, N., et al. (2023). The hatching time of broiler chickens modifies not only the production traits but also the early bacteriota development of the ceca. *Animals* 13(17):2712.

Tabedian, S.A., et al. (2010). Effect of fasting or post-hatch diet's type on chick development. *J. Anim. Vet. Adv.* 9:406–413.

Taha-Abdelaziz, K., et al. (2018). Effects of early feeding and dietary interventions on development of lymphoid organs and immune competence in neonatal chickens: A review. *Vet. Immun. and Immunopath.* 201:1–11.

Tona, K., et al. (2003). Effects of egg storage time on spread of hatch, chick quality, and chick juvenile growth. *Poult. Sci.* 82:736–741.

Tong, Q., et al. (2015). Physiological status of broiler chicks at pulling time and the relationship to duration of holding period. *Animal* 9:1181–1187.

van de Ven, L.J.F., et al. (2010). Hatching system affects broiler chick behavior. *XIIIth European Poultry Conference, Tours.*

van de Ven, L.J.F., et al. (2013). Perinatal broiler physiology between hatching and chick collection in 2 hatching systems. *Poult. Sci.* 92:1050–1061.

van den Brand, H., et al. (2010). Early feeding affects resistance against cold exposure in young broiler chickens. *Poult. Sci.* 89:716–720.

van den Brand, H., et al. (2023). Alternative hatching systems for broilers. *Embryo development and hatchery practice in poultry production (ed. N French) Burleigh Dodds.*

van der Hel, W., et al. (1991). The upper critical ambient temperature in neonatal chicks. *Poult. Sci.* 70:1882–1887.

van der Pol, C.W., et al. (2015). Effect of feed and water access on hatchling body weight changes between hatch and pull. *20th European Symposium on Poultry Nutrition. Prague, Czech Republic.*

van der Wagt, I., et al. (2020). A review on yolk sac utilization in poultry. *Poult. Sci.* 99:2162–2175.

van Roovert-Reijrink, I., et al. (2017). Heat production of chicks with and without access to feed and water between hatching and pulling. *Eur. Poult. Sci.* 81:30–30.

van Roovert-Reijrink, I., et al. (2019). Effect of feed and water deprivation during transport on post-hatch performance of early fed chicks. *Book of Abstracts Combined workshop of WPSA PDP and IFRG* 75.

Vanderhasselt, R.F., et al. (2013). Dehydration indicators for broiler chickens at slaughter. *Poult. Sci.* 92(3):612–619.

Vleck, C.M., et al. (1980). Patterns of metabolism and growth in avian embryos. *Amer. Zool.* 20 (2):405–416.

Wang, Y., et al. (2014). Spread of hatch and delayed feed access affect post hatch performance of female broiler chicks up to day 5. *Animal.* 8:610–617.

Wang, J.S., et al. (2020). Effects of posthatch feed deprivation on residual yolk absorption, macronutrients synthesis, and organ development in broiler chicks. *Poult. Sci.* 99:5587–5597.

Wijnen, H.J., et al. (2020). Effects of incubation temperature pattern on broiler performance. *Poult. Sci.* 99(8):3897–3907.

Wijnen, H.J., et al. (2021). Low incubation temperature during late incubation and early feeding affect broiler resilience to necrotic enteritis in later life. *Front. Vet. Sci.* 8:784869.

Wijnen, H.J., et al. (2022a). Broiler resilience to colibacillosis is affected by incubation temperature and post-hatch feeding strategy. *Poult. Sci.* 102:102092.

Wijnen, H.J., et al. (2022b). Effects of late incubation temperature and moment of first post-hatch feed access on neonatal broiler development, temperature preference, and stress response. *Poult. Sci.* 101(10):102088.

Witjes, V.L., et al. (2022). Effects of on-farm hatching on short term stress indicators, weight gain, and cognitive ability in layer chicks. *Appl. Anim. Behav. Sci.* 254:105692.

Xin, H. and Harmon, J.D. (1996). Responses of group-housed neonatal chicks to posthatch holding environment. *Transact. ASAE* 39:2249–2254.

Yi, G.F., et al. (2005). Impact of glutamine and oasis hatchling supplement on growth performance, small intestinal morphology, and immune response of broilers vaccinated and challenged with Eimeria maxima. *Poult. Sci.* 84:283–293.

# Chapter 8

## Optimising in ovo delivery of nutrients and bioactive compounds in broiler production

*M. M. Y. Meijer, Wageningen University & Research, The Netherlands and The University of Queensland, Australia; E. Roura, The University of Queensland, Australia; and H. van den Brand, Wageningen University & Research, The Netherlands*

1 Introduction
2 Methodological challenges and variability
3 Case study: in ovo delivery of essential oils
4 Conclusion and future trends
5 Acknowledgements
6 Where to look for further information
7 References

### 1 Introduction

Modern broiler chickens are bred and raised in environments designed to maximise growth rates and production efficiency. As broilers are genetically selected for rapid weight gain, they face a range of health issues that arise from the pressures of intensive production and accelerated growth (Van Der Most et al., 2011; Hafez and Attia, 2020). These issues not only affect the productivity of broiler farming, but also raise substantial welfare and ethical concerns, therefore, addressing them is essential. However, the question remains whether it is possible to adequately influence the health and resilience of broilers during their short lifespan.

Fast-growing broiler chicken strains spend around one-third of their life in embryonic development inside the egg (Fig. 1). This developmental phase is not only critical for establishing foundational systems, including organ structures, immunity, and metabolic processes, but also presents a unique opportunity for early interventions (Romanoff, 1960; De Oliveira et al., 2008; Givisiez et al., 2020; Garcia et al., 2021). Intervening during embryonic development provides the potential to establish a healthier baseline for growth

http://dx.doi.org/10.19103/AS.2025.0165.09

**Figure 1** The average lifespan of a modern broiler chicken is spent roughly ⅓ during embryonic development. Created with BioRender.

and performance immediately after hatching. This approach may also reduce susceptibility to diseases, enhance stress tolerance, and mitigate some of the negative impacts associated with rapid growth, even within the broilers' condensed post-hatch lifespan. As a result, the health and welfare of the broiler chicken might improve.

An interesting approach in early-life interventions is the in ovo delivery of nutrients or bioactive compounds directly into the egg during embryonic development. By targeting this pre-hatch stage, in ovo delivery offers the potential to optimise broiler health and resilience from within the egg, bypassing limitations associated with post-hatching dietary administration, which can be influenced by feed intake, dietary preferences and environmental factors (Khomayezi and Adewole, 2021). Initially developed for vaccination, in ovo delivery has since been expanded into in ovo feeding to optimise embryonic development and resilience (Sharma and Burmester, 1984; Gildersleeve et al., 1993; Uni and Ferket, 2003). By delivering compounds before hatching, this technique may also help mitigate the adverse effects of suboptimal incubation conditions. For example, elevated incubation temperatures during critical developmental stages can lead to oxygen deficits and force the developing embryo to switch from lipid oxidation to glycogen and protein utilisation for energy (De Oliveira et al., 2008; Maatjens et al., 2014). This depletion of essential energy reserves (particularly glycogen) might be restored through in ovo feeding, alleviating some of the negative developmental impacts.

Despite its potential, in ovo delivery for compounds other than vaccines remains largely unused in commercial settings. Evaluating the impact of in ovo delivery remains difficult due to variability in methodologies. Differences in the time of injection, embryonic compartment of administration, and types of compounds used lead to varying results. For example, late-incubation injections (embryonic (E) day 17/18) into the amniotic fluid align with commercial practices, as they can be combined with in ovo vaccination, but mid-incubation injections (E12) into the air cell have been suggested to be preferred for prebiotics due to high vascularisation at this stage (Uni and Ferket, 2004; Siwek et al., 2018). Additionally, early-stage injections into the yolk sac or albumen have been explored for amino acids, carbohydrates and vitamins (Han et al., 2020; Salmanzadeh and Salmanzadeh, 2012; Sgavioli et al., 2019; Zhu et al., 2020) (Fig. 2). These differences in methodologies make it challenging to standardise and compare outcomes.

The aim of this chapter is to provide a practical framework for setting up in ovo delivery experiments, highlighting key factors that should be considered to ensure consistent and meaningful results. These include the dose, volume, timing of administration, injection compartment, compound properties, compatibility with commercial practices and adequate control groups. By clarifying these methodological aspects, this chapter also addresses some of the limitations that may currently hinder broader adoption of in ovo delivery in the field. While this chapter discusses how different types of compounds may require specific formulations or delivery conditions (e.g. solubility, pH, or emulsification), it does not focus on the biological effectiveness of specific compounds or compound groups, as these have been extensively reviewed elsewhere (see Section 6: Where to look for further information). The exception

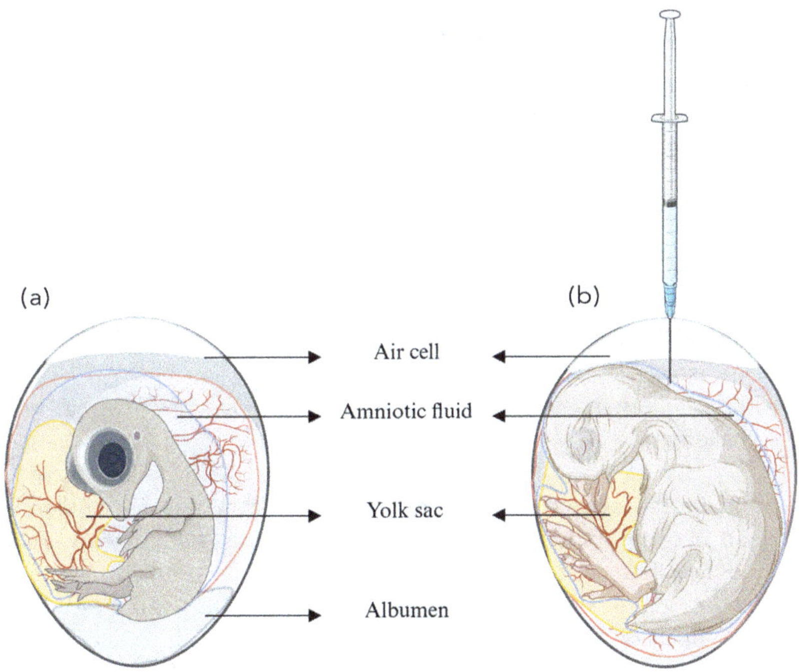


**Figure 2** Different compartments of in ovo delivery of nutrients or biochemical compounds (air cell, amniotic fluid, yolk sac and albumen) at two stages of development; embryonic day 10 (a) and 17/18 (b). Commercial in ovo vaccination occurs at embryonic day 17–18 in the amniotic fluid, as shown in (b). Created with BioRender.

is the case study in Section 3, which illustrates a systematic approach to optimising in ovo delivery of essential oils (EOs) as a model.

Despite promising results, outcomes of in ovo delivery studies remain highly inconsistent. In some cases, benefits are reported for performance, resilience, or physiological mechanisms, while in others, effects are absent or even negative. These discrepancies raise an important question: why do similar interventions yield such different results? Much of this variability can be traced back to differences in methodology. In ovo delivery is not a single technique, but a broad concept with many adjustable variables. Studies differ in injection timing, compartment, dose, compound characteristics, and how or when effects are measured. Often, key methodological details are missing from the published literature, and control treatments are inconsistent or absent, making it difficult to compare results or draw reliable conclusions.

This highlights the need for a structured, hypothesis-driven approach to designing in ovo delivery studies. The following sections provide a practical framework that addresses key methodological considerations and aims to improve consistency, reproducibility, and applicability of findings in both research and practice.

## 2 Methodological challenges and variability

This section aims to highlight the most important methodological factors that influence the outcomes of in ovo delivery, and how these can explain the variation across studies:

- The injected dose and volume;
- The time of injection during embryonic development;
- The embryonic target compartment;
- Properties of the injected compound;
- Viability of the combination with vaccines;
- The use of proper control groups; and
- The circumstances under which effects are quantified – e.g. timing, challenge.

To structure this section and ensure meaningful comparisons, studies were selected based on specific criteria. Only studies that started from a clearly defined hypothesis were included. Effects were evaluated relative to a proper control group, ideally a diluent-injected control, which accounts for the physical effects of injection. If such a control was not included, outcomes were compared to a non-injected group. To maintain clarity and focus, the summary tables in this section focus only on performance parameters (e.g. body weight, feed conversion), as including all physiological or mechanistic outcomes would exceed the scope of this chapter.

By exploring these aspects, this section provides a framework for designing effective, hypothesis-driven in ovo delivery studies, tailored to the compound and the intended application, and ideally suited for translation into commercial practice.

### 2.1 Dose

The first consideration in any in ovo delivery study involving bioactive compounds is the use of the appropriate dose. The optimal dose can significantly impact both the safety and effectiveness of the intervention. If the dose is too low, there may be no measurable effect; if too high, it could be toxic, affecting hatchability and chick quality. This is especially important because many compounds intended for in ovo use are not originally designed for embryonic application, and their efficacy and safety profiles in this context remain uncertain. Consequently, many studies have made dose optimisation a priority before proceeding to broader evaluations (Table 1).

The definition of proper dose may not only depend on the compound, but also on the objective of the study. Not every dose will meet the intended objective, but avoiding harm to hatchability and performance should be considered a minimum standard for quality control. Hence, this highlights the importance of clearly defining objectives and hypotheses before selecting a dose.

The dose range reported across studies is wide and compound-dependent, spanning from less than 1 μg to over 200 mg per egg. In many cases, hatchability and other quality control parameters were unaffected across the tested range. However, some studies showed that certain concentrations, particularly at the higher end, did lead to lower hatchability. For instance, bee pollen extract at 25 mg/egg reduced hatchability by 25%, manganese at 50 μg by 11%, zinc (150–250 μg) by 10–11%, arginine at 12 mg by 9%, and a prebiotic mixture at 7 mg by 32% (Gao et al., 2017; Sun et al., 2018; Geng et al., 2022; Pawłowska et al., 2024). These findings underscore the importance of carefully defining dose thresholds to avoid compromising early viability, especially when working with novel compounds or untested concentrations.

Several studies have demonstrated clear dose-response effects. For example, Xu et al. (2023) reported that 0.8 μg and 0.16 μg calcifediol improved hatchability by 9% and 7% respectively, but 0.8 μg gave the highest BW at d0 (+20%) and d28 (+32.5%). Interestingly, immune-related gene expression (IL-2, IFN-γ, TLR15) was higher in the 0.16 μg group compared to the 0.8 μg group, approximately 1.2–1.5x higher for serum and 1.5–5x higher in the spleen. This suggests that different doses may differentially support either performance or immune activation, highlighting the need to define study objectives when selecting a dose.

Some studies found no negative effects on hatchability, but also no improvements in performance. However, sometimes physiological mechanisms

**Table 1** Overview of effects of dose and volume of in ovo delivery on broiler performance. This table summarises studies investigating the impact of various compound doses and solution volumes on broiler performance following in ovo delivery. Only performance-related outcomes (hatchability, BW-body weight, BWG-body weight gain, FI-feed intake, FCR-feed conversion ratio) are included. When possible, effects are compared to a diluent-injected control to isolate compound-specific effects, or to non-injected controls if diluent data were not available.

| Compound | Dose | Volume (μL) | Day of Injection | Compartment | Control | Time of Measurement[1] | Challenge | Result | Reference |
|---|---|---|---|---|---|---|---|---|---|
| Arginine | 100, 1000, 2500 μg/egg | 100 | E8, E14, E18 | ns | Non, Diluent for each day | d0 | | Higher hatchability with 100 μg at E14 and 2500 μg at E18, lower hatchability with 1000 μg at E8 | (Subramaniyan et al., 2019) |
| Arginine | 0, 3, 6, 12 mg/egg | 600 | E17.5 | Amnion | Non, Diluent | d0, 1, 7, 21 | | Lower hatchability with 12 mg; no differences in BW d0; Highest BW, BWG, FI d21 with 3 or 6 mg; no differences in FCR d21 | (Gao et al., 2017) |
| Threonine | 10.5, 21, 31.5, 42 mg/egg | 600 | E17.5 | Amnion | Diluent | d0, 7, 14, 21 | | No difference in hatchability; Higher BW d0, 14, 21, BWG d0–21 with increasing doses; lower FCR d0–21 with increasing doses | (Moreira Filho et al., 2019) |
| L-carnosine | 25 or 50 μg/egg | ns | E18 | Amnion | Non, Diluent | d0 | | No difference in hatchability, higher BW d0 with 25 and 50 μg | (Keshtkaran et al., 2024) |
| Dextrin | 100, 200 μL/egg | 500 | E18 | Amnion | Diluent | d0 | | Lower hatchability with 100 or 200 μL | (Ghanaatparast-Rashti et al., 2018) |

| | | | | | | | | |
|---|---|---|---|---|---|---|---|---|
| Glycerol | 6.25, 12.5, 18.75, 25.0 nmol/egg | 500 | E17 | Amnion | Non, Diluent | d0 | No difference in hatchability, BW d0 | (Neves et al., 2017) |
| Glucose, fructose, sucrose, maltose, or dextrin | 0.25 g/mL | 100, 400, 700 or 1000 | E18.5 | Amnion | Non, Sham, Diluent | d0 | Lower hatchability with increasing volumes, higher BW d0 with increasing volumes | (Zhai et al., 2011) |
| Ascorbic acid | 3, 6, 12, 36 mg/egg | 100 | E17 | Amnion | Non, Diluent | d0, 14, 28, 40, 45 | No difference in hatchability, BW d0, BW, BWG, FI, FCR d14, 28, 40, 45 | (Zhang et al., 2019) |
| Vitamin D3 | 0.4, 0.6, 0.8 μg/egg | 500 | E18 | Amnion | Diluent | d14, 28, 42 | Highest FI, BWG d0–14 for 0.6 μg; Highest FI d14–28 for 0.6 μg; No differences in FI, BWG, FCR d28–42 | (Abbasi et al., 2017) |
| Calcifediol | 0.8 or 0.16 μg | 100 | E17.5 | Amnion | Non, Diluent | d0, 1, 7, 14, 21, 28 | Higher hatchability with 0.8 and 0.16 μg; Highest BW d1, 7, 14, 21, 28 with 0.8 μg | (Xu et al., 2023) |
| Calcifediol | 0.20, 0.60, 1.80, 5.4 μg/egg | 100 | E18 | Amnion | Non, Diluent | d7, 14, 21, 28, 35, 42 | Higher post-hatch mortality with 5.40 μg; Lower BWG, FI d0–21, no differences in BWG, FI, FCR d21–42 | (Bello et al., 2014) |

*(Continued)*

**Table 1** *Continued*

| Compound | Dose | Volume (μL) | Day of Injection | Compartment | Control | Time of Measurement[1] | Challenge | Result | Reference |
|---|---|---|---|---|---|---|---|---|---|
| Vitamin E | 27.5, 38.5, 49.5, 60.4 IU/egg | 500 | E17.5 | Amnion | Diluent | d0, 7, 14, 21 | | Highest hatchability with 60.4 IU; Higher BW d0 with ≥ 49.5 IU; No differences in BW, BWG, FI, mortality d0–21; Lower FCR d21 for all doses | (Araújo et al., 2019) |
| Vitamin K3 | 2, 6 μg/egg | 500 | E18 | Amnion | Diluent | d14, 28, 42 | | Higher BWG d0-14 for 6 μg; Higher FI d14-28 for both 2 and 6 μg; No differences in FI, BWG, FCR d28–42;6 μg: BWG d0–14 ↑, BWG d14–28 ↓, IgM d42 ↑ | (Abbasi et al., 2017) |
| Copper sulfate, acetate or nanoparticles | 8 or 16 μg/egg | 100 | E10 | Amnion | Non | d0, 21, 35 | | Higher BW d0, BWG d0-35 with 8 or 16 μg; Highest FI d0–35 with 8 μg, lowest FCR d0–21 with 16 μg | (Hassan et al., 2023) |
| Manganese | 6.25, 12.5, 25.0, 50.0 μg/egg | 100 | E9 | ns | Non, Diluent | d0 | Low manganese parents | Lower hatchability with 50 μg; no differences in BW d0 | (Geng et al., 2022) |
| Selenium | 10, 20 μg/egg | 100 | E18 | Amnion | Diluent | d0, 20 | Necrotic enteritis d14–18 | No difference in hatchability, BW d0; Higher BW d20 for 20 μg compared to challenged control, lower lesion score d20 for 10 μg compared to challenged control | (Lee et al., 2014) |

| | | | | | | | | |
|---|---|---|---|---|---|---|---|---|
| Silver nanoparticles | 12.5, 25, 50 μg/egg | 500 | E18 | Amnion/Yolk sac | Non, Diluent | E20, d0, 21, 42 | No differences in BW d0, BWG, FI, FCR d0-42 | (Goel et al., 2017) |
| Zinc | 50, 100, 150, 200, 250 μg/egg | 100 | E9 | Yolk sac | Non, Diluent | d0 | Lower hatchability with 150 or 200 μg; no differences in BW d0 | (Sun et al., 2018) |
| Sodium butyrate | 0.2, 0.6 and 1 μL/egg | 200 | E12 | Air cell | Diluent | d0, 1, 14, 35, 42 | No difference in hatchability, BW, BWG, FI, FCR | (Bełdowska et al., 2024) |
| β-hydroxy β-methylbutyrate | 2.5, 5 μL/egg | 500 | E18 | Amnion | Diluent | d0 | No difference in hatchability | (Ghanaatparast-Rashti et al., 2018) |
| γ-aminobutyric acid | 30, 60, 120 mg/egg | 600 | E17.5 | Amnion | Non, Diluent | d0 | No differences in BW d0 | (Ncho et al., 2021) |
| Astragalus polysaccharide | 1, 2, 4 mg/egg | 500 | E18 | Amnion | Diluent | d0, 1, 7, 14, 21 | No differences in hatchability, BW d1, 7; Higher BW d14, 21 with 2 mg | (Yang et al., 2021) |
| Bee pollen extract | 7, 12, 25 mg/egg | 500 | E18 | Air cell | Diluent | d0, 7, 21, 35, 42 | Lower hatchability with 25 mg; Higher BW d0, d7 with 7 mg; Lower BW d42 with 12 mg; no effect on BWG, FI, FCR, mortality d0–42 | (Pawłowska et al., 2024) |

*(Continued)*

**Table 1** *Continued*

| Compound | Dose | Volume (μL) | Day of Injection | Compartment | Control | Time of Measurement[1] | Challenge | Result | Reference |
|---|---|---|---|---|---|---|---|---|---|
| Black cumin | 2, 4, 6, 8 mg/egg | 500 | E17.5 | Air cell | Non, Diluent | d0, 28, 56 | 39.6 instead of 37.8 E10–E18, 6h/d | Higher hatchability with 6 mg; No differences in BW, chick length d0; Higher BW d28, 56, BWG d28–56 with 6 mg; Lowest FCR d28–56 with 6 mg | (Oke et al., 2021) |
| Grape pomace extract | 2, 4 and 6 mg/egg | 100 | E17.5 | Amnion | Non, Sham, Diluent | d0, 35 | | No difference in hatchability; Higher BW d0 with 4 mg; Higher BW d35 with 4 and 6 mg; No effect on BWG, FI, FCR d0–35 | (Madkour et al., 2024) |
| Resveratrol | 1.5, 3, 4.5 mg/egg | 500 | E7 | Air cell | Non, Diluent | d7, 21, 35 | | No differences in BW, BWG, FI d0-35; Higher FCR d0–35 for 4.5 mg | (El-Fakhrany et al., 2023) |
| CpG-Oligonucleotide | 1, 5, 10, 20, 50 μg/egg | 100 | E18 | Amnion | Diluent | 7d after challenge (d8) | *E.coli* d1 | Higher survival d7 with 20 μg | (Sarfraz et al., 2022) |
| Nucleosides | 25, 50, 100 mg/egg | 100 | E18 | Yolk sac | Diluent | d0, 14, 28, 42 | | No differences in hatchability, BW d0; Higher BW d14, 28, 42 with 50 or 100 mg; Higher FI d28 with 100 mg | (Gopi et al., 2020) |
| Polyinosinic-polycytidylic acid | 5, 10, 20 μg/egg | 100 | E18 | Amnion | Diluent | 7d after challenge (d8) | *E.coli* d1 | Higher survival d7 with 20 μg | (Sarfraz et al., 2022) |

| | | | | | | | | | |
|---|---|---|---|---|---|---|---|---|---|
| *Bacillus subtilis* | $10^5$ and $10^6$ CFU/ egg | 200 | E12.5 | Air cell | Non, Diluent | d0 | | No difference in hatchability, BWd0; Longer chicks d0 with $10^6$ | (Shehata et al., 2024) |
| *Lactobacillus* spp. | $1 \times 10^5$, $10^6$, $10^7$ CFU/ egg | 100 | E18 | Amnion | Non, Diluent | d0, 5, 10 | | No differences in hatchability | (Alizadeh et al., 2020) |
| Multistrain probiotic | $1 \times 10^5$, $10^6$, $10^7$ CFU/ egg | 100 | E18 | Amnion | Non, Diluent | d0, 4, 6, 8, 15, 22 | | No difference in hatchability, BW d0, BW, BWG, FI, FCR d0–22 | (Pender et al., 2017) |
| Raffinose | 2 or 3mg/ egg | 200 | E12.5 | Air cell | Non, Diluent | d0 | | Higher hatchability, BW d0 with 2 or 3 mg | (Shehata et al., 2024) |
| Raffinose | 1.5, 3.0, 4.5 mg / egg | 200 | E12 | Air cell | Non | d0, 7, 14, 21 | | No differences in BW d0, BWG, FI, FCR d0–21 | (Berrocoso et al., 2017) |
| Prebiotic mixture | 0.18, 0.88, 3.5 and 7.0 mg/ egg | 200 | E12 | Air cell | Diluent | d0 | | Lower hatchability with 7 mg | (Bednarczyk et al., 2016) |

Diluent = injected without bioactive compound, but only with its diluent; E = Embryonic day, Non = non-injected; Sham = sham injection, insertion of needle without release; ns = not specified.
[1] Either post-hatching (d) or embryonic day (E).

were still affected. For example, sodium butyrate at 0.2, 0.6, or 1 μL showed no differences in hatchability or production metrics, though the lowest dose did modulate intestinal gene expression. Specifically, 0.2 μL lowered expression of the cytokine IL-4 in the ileum by almost half, while elevating the expression of antimicrobial peptide AvBD1 (four-fold in the ileum), the cytokine IL-8 (five-fold in the cecum), and the mucin gene MUC-2 (four-fold in the ileum) (Bełdowska et al., 2024). These findings suggest that even in the absence of performance effects, in ovo delivery compounds can still influence host physiology in a dose-specific manner.

Another example comes from a disease challenge, where Lee et al. (2014) compared 10 and 20 μg selenium in a necrotic enteritis model. The higher dose improved BW by 20%, while the lower dose reduced lesion scores by 0.75 points on a 0–4 severity scale, compared to the challenged control. The high dose also upregulated IL-1β and IL-6 by approximately five- and two-fold, respectively, showing a trade-off between immune activation and intestinal damage mitigation. This illustrates that outcomes may vary depending on health status or environmental conditions.

These examples demonstrate three key points: optimal dose depends on the objective of the study, physiological responses may not always translate to performance gains, and outcomes may vary depending on environmental and challenge conditions. Together, these findings highlight the importance of preliminary dose screening when literature is lacking for a given compound.

An additional factor to consider is that dose effects may also depend on the timing of injection. Subramaniyan et al. (2019) tested three doses of arginine (100, 1000, and 2500 μg), injected at three different embryonic stages (E8, E14, and E18). Delivery of 100 μg on E14 improved hatchability by ~20% relative to the E14 control, whereas 1000 μg on E8 decreased hatchability by ~5% compared to the E8 control. The study also tested 100 μg on E8, but this dose did not improve hatchability, suggesting that timing, and not just dose, plays a role. These results indicate that even the same dose can have different effects depending on the embryonic stage.

In summary, optimising dose is not only about selecting a non-toxic level, but also about aligning concentration with the study's specific objectives, be it performance, immune modulation, (environmental) conditions or resilience to challenge. Without careful consideration and preliminary screening, studies risk producing undesirable outcomes (i.e. toxicity), or inconclusive or misleading results.

### *2.2 Volume*

Injection volume is another consideration in in ovo delivery, as excessive volumes may influence embryonic development by altering the hydration balance within the egg compartments. Simkiss (1980) described how during incubation, water reserves within the egg compartments are mobilised and redistributed, and that each compartment reaches specific volumes at different stages of development. Thus, injecting additional fluid might disrupt this balance, particularly if excessive volumes alter the physiological functions of these compartments.

Injection volume is particularly relevant in scenarios where in ovo delivery is combined with commercial vaccination practices, which typically inject approximately 50–100 μL (Wakenell et al., 2002; Avakian, 2006; Fatemi et al., 2024a). Most research studies use volumes ranging from 50 μL to 1000 μL (Das et al., 2021). However, few studies have directly compared the effects of varying injection volumes. Zhai et al. (2011) (Table 1) is one of the only studies to test a range of volumes using the same compound: a carbohydrate solution with a fixed concentration of 0.25 g/mL. Thus, a higher injection volume implied higher total carbohydrate doses, confounding volume with dose. In that study, hatchability decreased progressively from 97% at 100 μL to 92% at 400 μL, 88% at 700 μL, and 84% at 1000 μL. Additionally, body weight at hatch increased with higher volumes, likely due to the greater mass of the injected solution.

Interestingly, several other studies have successfully injected volumes up to 1000 μL without reporting negative impacts on hatchability. However, these studies typically did not compare different volumes within the same experiment, making it difficult to assess whether injection volume itself influenced outcomes. This suggests that the hatchability effects observed by Zhai et al. (2011) may have been due to compound overdose rather than volume alone. Moreover, the optimal injection volume likely depends on both the timing and the specific compartment of injection. Embryonic compartments vary in size during incubation, thereby influencing the maximum feasible injection volume. This highlights the importance of carefully considering the time and volume of injection and the specific embryonic target compartment, which will be discussed in the following sections.

### 2.3 Time of injection

A challenge in in ovo delivery research lies in the variability of injection timing and location. Many studies used protocols that do not align with current commercial practices, where injections typically occur in the amnion late in incubation. Research injections vary widely, ranging from before or at the start of incubation up to E19 (Das et al., 2021). This lack of standardisation complicates the interpretation of results and reduces potential commercial applicability. However, the optimal timing of injection may also depend on the properties of the compound, which further highlights the need for both methodological consistency and compound-specific rationale.

The timing and compartment of injection are closely related; mid-incubation injections into the air cell might promote rapid absorption due to higher vascularisation, but such injections are less practical for large-scale commercial operations, which typically use late-incubation amniotic injections for in ovo vaccination, because this moment aligns with candling and transfer of eggs from the setter to the hatcher. Consequently, the commercially viable option currently is limited to late incubation in the amnion, meaning that any compound or intervention delivered in ovo must be demonstrated to be safe and without negative impacts when administered into the amnion during late incubation. However, research aiming at optimisation or future commercialisation can explore alternative days and injection compartments to determine the most effective combinations. Embryonic compartments develop rapidly, thus, the optimal day for injection would be specific for each compartment. For instance, Keshtkaran et al. (2024) reported significantly lower hatchability (13–34% reduction) when injecting L-carnosine into the amnion at E7 compared to E18. Conversely, Ma et al. (2020) observed no negative impact on hatchability when injecting β-hydroxy-β-methylbutyrate into the air cell at E7 and amnion at E18 and found a higher body weight at day 42 (+4%) following E7 air cell injection. These differences highlight that outcomes of early incubation injections may depend strongly on the injection compartment, suggesting that if early incubation injection is desired, the air cell might be a preferable compartment.

Given these findings, careful consideration and standardised approaches to the timing of injections are important, especially when aligning research outcomes with practical, commercial applicability. In line with this, the next section will focus on the embryonic target compartment, as the effectiveness of in ovo delivery is not only influenced by timing, but also by where the compound is delivered.

## *2.4 Embryonic target compartment*

The selection of the injection compartment is critical, as several options exist, each with its own practical and biological implications. These include the air cell, amnion, yolk sac, albumen, allantois, or the embryo itself (Fig. 2). The optimal compartment depends not only on technical feasibility but also on the embryonic stage and compound properties. Early-stage compartments may be more sensitive whereas others can become inaccessible or simply disappear later in development. In addition, some compounds may require direct exposure to specific tissues or fluids to be effective. These decisions may be further complicated by the dynamic nature of embryonic development. Most major compartments form during the early stages of incubation and experience rapid structural changes as development progresses. For example, E12 marks a phase of rapid growth and increased metabolic activity, while the embryo is structurally complete around E17.5 (De Oliveira et al., 2008; Nangsuay et al., 2017). Together, these anatomical and physiological changes highlight the need to match the injection compartment to both the developmental stage and the mode of action of the compound.

Importantly, the intended compartment of injection is not always verified in experimental studies. Some only mention approximate injection sites or needle size, without confirming whether the compound reached its intended location. This lack of precision undermines reproducibility and complicates interpretation. It is important to recognise that the developing embryo and its surrounding structures are not a black box; targeting specific compartments is possible and meaningful.

The amnion, especially during late incubation, appears to be the most practical and effective compartment for in ovo injection, as this is also the standard compartment used in commercial in ovo vaccination practices and consequently, most knowledge of this compartment is available. At this stage, the embryo is in contact with the amniotic fluid, increasing the chance of direct ingestion or absorption of the compound. Additionally, from E12 onwards, the sero-amniotic connection enables albumen to flow into the amnion (Deeming, 1991; Manders et al., 2021), which may facilitate compound transport. Manders et al. (2021) showed high injection success rates in the amnion at several stages, including 95% at E12 (38 mm needle) and 90% at E18 (16 mm needle).

The air cell is easily accessed and frequently used in early-stage research, especially before the embryo has occupied a larger portion of the egg. However, the air cell might not always ensure proper delivery to the embryo, because transport of active ingredients might be limited through the air cell. It does not offer a direct connection to embryonic fluids or tissues, raising concerns about compound absorption.

Yolk sac injections typically require inverting the egg and targeting the yolk at a 45° angle. Success depends on yolk position, which can vary between embryos. While yolk sac injections achieved success rates of up to 100% (Manders et al., 2021), this required matching of needle length to embryonic day, with 38 mm needles used early and 16 mm later. Despite these technical challenges, the yolk sac can still be an effective compartment, especially in early to mid-embryogenesis when the yolk is still the primary nutrient source.

The albumen is only a suitable target during early incubation. Manders et al. (2021) reported reliable targeting (up to 100%) at E6–E8, but almost complete failure at later stages. As development progresses, the albumen is absorbed or displaced, making it a poor target after E10. Additionally, the allantoic cavity is rarely targeted in nutritional or bioactive compound studies and is generally not considered a functional compartment beyond early stages.

Direct embryo injections are typically avoided in nutritional or bioactive compound delivery due to the risk of physical injury and developmental disruption. While used for some DNA vaccines, their application is limited. Manders et al. (2021) found that injections into the embryo itself were only successful from E16 onward, with 100% success at E18 using a 38 mm needle.

Because day and compartment of injection are closely linked, it is essential to account for this when designing studies. Although there are few studies directly comparing injection compartments under standardised conditions, the injection location may have a measurable impact on hatchability and post-hatch performance. Further research in this area would be valuable, particularly studies that systematically compare injection compartments, while controlling for other variables such as day, compound type, and dose.

## 2.5 Compound properties

The characteristics of the injected compound also influence how the in ovo delivery technique must be optimised, in addition to determining the overall suitability of the compound. Factors such as solubility, viscosity, pH, and formulation all influence how the compound behaves within the egg, and whether it reaches the intended target without causing harm to the developing embryo. In addition, highly volatile compounds may pose challenges due to evaporation, premature dispersion, or instability within the egg environment.

Some compounds, such as fat-soluble vitamins, require specific delivery vehicles to improve dispersion and uptake. For instance, vitamin E was successfully delivered into the amniotic fluid using sunflower oil as a diluent (Araújo et al., 2019), but effects were mainly observed at the highest tested dose. This may indicate that the oil did not disperse optimally in the aqueous amniotic fluid, potentially limiting bioavailability at lower concentrations. However, the yolk sac may offer a more compatible environment for delivering lipid-based compounds, as its natural lipid-rich matrix can support solubilisation and may enhance bioavailability without the need for emulsification. In contrast, other hydrophobic substances like EOs have been emulsified to form stable solutions suitable for injection, particularly into aqueous environments such as the amniotic fluid (Meijer et al., 2024a). Emulsifiers themselves could also influence bioavailability and biological activity, making formulation a critical point of consideration (Laothaweerungsawat et al., 2020). Moreover, their potential toxicity or developmental effects in ovo are not always well understood and should be evaluated (Table 2).

Viscosity is another practical consideration, especially for injections using small volumes. High-viscosity solutions can be difficult to administer accurately, as they may clog fine needles (Watt et al., 2019) or disperse unevenly within the egg. This could lead to localised concentrations, reduced efficacy, or tissue

**Table 2** Overview of the effects of embryonic day and injection compartment on broiler performance following in ovo delivery. This table summarises studies comparing the effects of different injection moments and target compartments on broiler performance. Only outcomes related to hatchability and post-hatch performance (BW, BWG, FI, FCR) are included. When possible, results are compared to a non-injected control to exclude procedural effects and isolate the impact of injection timing and location

| Compound | Day of Injection | Compartment | Dose/Volume | Control | Time of Measurement[1] | Challenge | Result | Reference |
|---|---|---|---|---|---|---|---|---|
| Arginine | E8, E14, E18 | ns | 100, 1000, 2500 µg/100 µL | Non, Diluent (all days) | d0 | | Lowest hatchability for E8 (1000 µg) and E18 (diluent, 100, 1000 µg), highest hatchability for E14 (100 µg); lower BW d0 for E18 (100 µg), highest BW d0 for E14 (100 µg) | (Subramaniyan et al., 2019) |
| Amino acid mixture | E7 | CAM, Yolk sac, Extra-embryonic activity, Amnion | 53 mg/500 µL | Non, windowed | d0 | Windowed eggs | Lower or no hatchability for all compartments, including windowed control | (Ohta and Kidd, 2001) |
| L-carnosine | E7, E18 | Amnion | 25 or 50µg/ns | Non, Diluent (both days) | d0 | | Lower hatchability with L-carnosine at E7; higher BW d0 with 25 or 50 µg injected at E18; no impact of diluent on both days | (Keshtkaran et al., 2024) |

(*Continued*)

**Table 2** *Continued*

| Compound | Day of Injection | Compartment | Dose/Volume | Control | Time of Measurement[1] | Challenge | Result | Reference |
|---|---|---|---|---|---|---|---|---|
| Glycerol | E17, E18 | Amnion | 3 mg/500 μL | E17 diluent | d0, 7 | Light eggs | No difference in hatchability, BW d0; Higher BWG, FI d0–7 for E18 | (Dal Pont et al., 2019) |
| Silver nanoparticles | E7, E18 | Extra-embryonic cavity (E7), Amnion/Yolk sac (E18) | 12.5, 25, 50 μg/500 μL | Non | d0 | | No difference in hatchability, BW d0 | (Goel et al., 2017) |
| β-hydroxy-β-methylbutyrate | E7, E18 | Air cell (E7), Amnion (E18) | 1 mg/1000 μL | Non | d0, 7, 21, 42 | | No differences in hatchability, higher BW d0 for both injection days; Higher BWG d0–21, BW d42 for E7 air cell injection | (Ma et al., 2020) |
| Sterilized water | E3, E6, E9 | Yolk sac | Water/100μL | Non | d0 | | Lower hatchability for E3 and E6 | (Sun et al., 2018) |

Diluent = injected without bioactive compound, but only with its diluent; E = Embryonic day, Non = non-injected; Sham = sham injection, insertion of needle without release; ns: not specified.

[1] Post-hatching.

damage. Selecting an appropriate diluent or adjusting concentration and temperature can help mitigate these issues.

The pH of the compound is another important factor in in ovo injection. The internal pH of egg compartments changes throughout incubation. For instance, the albumen has a pH of 7.8 at oviposition, then rises sharply to 9.5 by E2, and gradually decreases to around 7.3 by E15. In contrast, the yolk starts at a lower pH of approximately 6.0 and increases to 8.0 by E15, settling at 7.6 by E19 (Romanoff and Romanoff, 1929). These changes highlight that injected compounds may encounter varying pH environments depending on timing and injection compartment. Although the embryo and the egg have some buffering capacity, particularly in later stages of development, mismatches between the pH of the solution and the local egg environment could lead to metabolic disturbances, such as acidosis or alkalosis.

Compound stability can also be a concern. Some bioactive compounds, such as EOs and vitamins, are prone to degradation through oxidation or hydrolysis. Formulation strategies, like encapsulation or the inclusion of antioxidants, can protect compounds during storage and upon injection, helping to preserve efficacy (Reis et al., 2022). It is also relevant to acknowledge that each compound will behave differently depending on its chemical and physical properties. These include not only solubility and pH, but also volatility, molecular weight, viscosity, and the presence of stabilising agents. These factors may affect how the compound spreads through the egg, whether it reaches the intended compartment, and how it interacts with embryonic tissues.

As mentioned in Section 2.4, compound properties may need to be adapted depending on the injection compartment and timing. Even when a compound is delivered to a specific location within the egg, it does not necessarily remain there. Depending on the properties of the compound, it may remain in the injection compartment (for example, hydrophilic compounds may stay in the amnion), or migrate to other compartments, such as from the amnion to the yolk in the case of more hydrophobic compounds (Meijer et al., 2024a). This highlights the importance of tailoring delivery systems to the specific developmental stage and anatomical context of the embryo, as well as anticipating how compounds may migrate within the egg post-injection.

Given the wide variability in compound types used for in ovo delivery, from nutrients and immune stimulants to nanoparticles, future research should focus on characterising optimal delivery for different classes of compounds. Special attention should be given to how formulation impacts hatchability, absorption, and functional outcomes post-hatch.

## 2.6 Combination with in ovo vaccination

Combining bioactive compound delivery with in ovo vaccination introduces an additional layer of complexity. When considering such combinations, it is crucial that the injected compound does not interfere with the safety or efficacy of the vaccine. On the contrary, certain compounds, particularly immunostimulants, may have the potential to enhance vaccine responses. Any compound co-administered with a vaccine must be compatible with standard in ovo vaccination conditions, including injection day (typically E17–E18), volume (around 50 μL), compartment (amnion), and vaccine diluent.

Some recent studies have begun to explore these combinations (Table 3). For example, Fatemi et al. (2024b) reported that co-injection of Marek's disease vaccine with 1.2 or 2.4 μg of 25-hydroxyvitamin D3 led to higher body weight at hatch (+3% or +4%, respectively) and changed immune gene expression. Specifically, IFN-β expression at hatch was approximately 1.9–2.3 times higher, respectively, compared to the vaccine-only control. The lower dose (0.6 μg) did not show these effects. Similarly, Gaghan et al. (2023) investigated the addition of CpG-oligonucleotides, synthetic DNA sequences that mimic microbial DNA, to in ovo infectious laryngotracheitis (ILT) vaccination. Supplementation with 10 μg, but not 5 μg, resulted in higher expression of IFN-γ (~1.9×), IFN-β (~1.3×), and IL-1β (~1.7×) in the spleen, as well as IFN-γ (~1.9×) and IL-1β (~6×) in the lung compared to the vaccine-only control. Additionally, macrophage numbers in the spleen doubled, indicating more immune activation at hatch. Importantly, neither study observed negative effects on hatchability, suggesting that under the right conditions, co-delivery is feasible and potentially beneficial.

While the available studies are still limited, they provide proof of concept that in ovo vaccination can be safely combined with certain bioactive compounds and may potentially benefit from immunomodulation. Further research is needed to identify which compounds, doses, and formulations are most effective in supporting or amplifying vaccine-induced protection without compromising embryonic development. Further research is also necessary to evaluate how in ovo delivered compounds affect the actual efficacy of the vaccine, including protection levels against disease under field or challenge conditions.

## 2.7 Control groups

The inclusion of control groups in any experimental design is fundamental for drawing reliable conclusions in in ovo research. Given how many variables can affect outcomes, attention to the type and design of control groups is important. Without appropriate controls, observed effects may be incorrectly interpreted.

Many studies rely solely on a non-injected group as a control. While this provides a general reference point, it does not account for the potential effects

**Table 3** Overview of compounds delivered in combination with in ovo vaccination and their effects on broiler performance. This table summarises studies comparing the effects of various compounds co-administered with in ovo vaccines, relative to vaccine-only controls. Only performance-related parameters (hatchability, BW, BWG, FI, FCR) are included.

| Compound | Vaccine | Day of Injection | Compartment | Dose/Volume | Control | Time of Measurement[1] | Result | Reference |
|---|---|---|---|---|---|---|---|---|
| Calcifediol | Marek's disease | E17.5 | Amnion | 1.2 or 2.4 μg/50 μL | Non, Vaccine alone | d0, 14, 28, 40 | No differences in hatchability, BW d0, BW, BWG, FI, FCR d14, 28, 40 | (Fatemi et al., 2024b) |
| Calcifediol | Marek's disease | E17.5 | Amnion | 0.6, 1.2 or 2.4 μg/50 μL | Non, Vaccine alone | d0 | No differences in hatchability. Vaccine + treatment (1.2 and 2.4 μg): BW d0 ↑ | (Fatemi et al., 2024a) |
| CpG-Oligonucleotide | Infectious laryngotracheitis | E18 | Amnion | 5 or 10 μg/100 μL | Diluent, Vaccine only | d0 | No differences in hatchability | (Gaghan et al., 2023) |

CpG = Cytosine-phosphate-Guanine; Diluent = injected without bioactive compound, but only with its diluent; E = Embryonic day, Non = non-injected; Sham = sham injection, insertion of needle without release; ns = not specified.

[1] Post-hatching.

of the injection procedure itself, or the diluent used to deliver a compound. Ideally, studies should include both non-injected and diluent-injected controls to better distinguish procedural effects from those of the test compound. Diluent controls are especially important, since the diluent alone can influence results, for instance by affecting hydration or interacting with embryonic tissues. Zhai et al. (2011) reported that higher injection volumes linearly increased BW at hatch, which may have been due to hydration rather than a biological effect of the compound. In such cases, attributing effects solely to the compound would be misleading. Differences between non-injected and diluent-injected groups may also arise due to the injection process itself. Without a diluent control, the act of injection becomes a confounding effect of the compound tested.

In some studies, control groups with unexpectedly low hatchability are reported, suggesting that incubation conditions or breeder fertility may have been suboptimal. In such cases, observed improvements following in ovo treatment might not reflect a true enhancement, but rather a correction of underlying deficiencies.

To illustrate how the use of different control groups can yield varying outcomes, both in terms of performance and physiological responses, several studies have included more than one control group (Table 4):

- Madkour et al. (2024): A diluent (sterilised DMSO) was compared to a non-injected control and a sham control (punched). No differences in hatchability or BW at d0 were found, but BW at d35 was 4% higher in the diluent group and FCR was 3% lower. The body weight gain (BWG) at d35 was 3–4% higher in both sham and diluent groups. Physiologically, the diluent group showed 7% lower plasma cholesterol, 28% higher liver antioxidant capacity, and 20% lower liver malondialdehyde (MDA) at d0.
- Shehata et al. (2024): Diluent (sterile distilled water) was compared to a non-injected control. No differences were observed in performance, chick length at d0, organ weights, or gene expression linked to gut integrity, nutrient transporters, or immunity.
- Xu et al. (2023): Diluent (sterile water) was compared to a non-injected control. Hatchability and BW were similar up to d14, but BW at d21 was 2% lower in the diluent group. Thymus weight at d14 was 30% higher, with no effect on immune gene expression.
- El-Fakhrany et al. (2023): Diluent (sterile saline) was compared to a non-injected control. While BW and BWG were not affected, FI and FCR were 18% and 30% lower, respectively, at d35. Moreover, the diluent group had lower cholesterol (−16%) and triglycerides (−24%), lower liver GPX and GST (−4–10%), higher liver MDA (+17%), and lower plasma IgY and IgM (−25–30%) at d35.

**Table 4** Overview of control groups used in in ovo delivery studies and their effects on broiler performance. This table summarises studies comparing diluent-injected and/or sham-injected eggs to non-injected controls. Only performance-related parameters (hatchability, BW, BWG, FI, FCR) are included

| Control | Day of Injection | Compartment | Volume | Time of Measurement[1] | Challenge | Result | Reference |
|---|---|---|---|---|---|---|---|
| Non, sham, Diluent | E17.5 | Amnion | 100 μL | d0, 35 | | No differences in hatchability, BW d0; Diluent: BW d35 ↑, FCR ↓; Sham or diluent: BWG d35 ↑ | (Madkour et al., 2024) |
| Non, Diluent each days | E7, E18 | Albumen | Ns | d0 | | No differences in hatchability, BW d0 | (Keshtkaran et al., 2024) |
| Non, Diluent each day | E8, E14, E18 | ns | 100 μL | d0 | | Diluent E18: hatchability ↓; no differences in BW d0 | (Subramaniyan et al., 2019) |
| Non, Diluent each day | E3, E6, E9 | Yolk sac | 100 μL | d0 | | Diluent E3, E6 hatchability ↓ | (Sun et al., 2018) |
| Non, Diluent | E12.5 | Air cell | 200 μL | d0 | | No differences in hatchability, BW d0 | (Shehata et al., 2024) |
| Non, Diluent | E17.5 | Amnion | 100 μL | d0, 1, 7, 14, 21, 28 | | No differences in hatchability, BW d0–14; Diluent: BW d21 ↓ | (Xu et al., 2023) |
| Non, Diluent | E7 | Air cell | 500 μL | d7, 21, 35 | | No differences in BW, BWG d7–35; Diluent; FI, FCR d0–35 ↓ | (El-Fakhrany et al., 2023) |

(*Continued*)

**Table 4** *Continued*

| Control | Day of Injection | Compartment | Volume | Time of Measurement[1] | Challenge | Result | Reference |
|---|---|---|---|---|---|---|---|
| Non, Diluent, Non × HS, Diluent × HS | E9 | ns | 100 μL | d0, 42 | Heat stress | Diluent: hatchability ↓ in 1 Exp; No differences in BW d0, BWG, FI, FCR, mortality; No differences in BWG, FI, FCR, mortality during heat stress | (Geng et al., 2022) |
| Non, Diluent | E10 | Amnion | 500 μL | d0 | Aflatoxin | No differences in hatchability, BW d0 | (Elwan et al., 2022) |
| Non, Diluent | E12 | Air cell | 200 μL | d0 | | No differences in hatchability | (Tavaniello et al., 2022) |
| Non, Diluent, Non × cold, Diluent × cold (only d0) | E17.5 | Amnion | 600 μL | d0 | Low incubation temp | No differences in hatchability, BW d0 after cold temperatures | (Miri et al., 2022) |
| Non, Diluent | E17.5 | Air cell | 500 μL | d0, 28, 56 | High incubation temp | No differences in hatchability, BW d0, BWG, FI, FCR d28–56 | (Oke et al., 2021) |
| Non, Diluent | E17.5 | Amnion | 600 μL | d0 | | No differences in hatchability, BW d0 | (Ncho et al., 2021) |
| Non, Diluent | E18 | Amnion | 100 μL | d0, 5, 10 | | No differences in hatchability | (Alizadeh et al., 2020) |

| | | | | | | | |
|---|---|---|---|---|---|---|---|
| Non, Diluent, Non × HS, Diluent × HS | E12 | Air cell | 200 µL | d42 | Heat stress | No differences in BW, FI, FCR after heat stress | (Slawinska et al., 2020) |
| Non, Diluent | E17 | Amnion | 100 µL | d0, 14, 28, 40, 45 | . | No differences in hatchability, BW d0, BW, BWG, FI, FCR d14–45 | (Zhang et al., 2019) |
| Non, Diluent | E17.5 | Amnion | 600 µL | d0, 1, 7, 21 | . | No differences in hatchability, BW d0, BW, BWG, FI, FCR d0–21 | (Gao et al., 2017) |
| Non, Diluent | E18 | Amnion | 100 µL | d0, 4, 6, 8, 15, 22 | . | No differences in hatchability, BW, BWG, FI, FCR d0–22 | (Pender et al., 2017) |
| Non, Diluent | E18 | Amnion | 100 µL | d7, 14, 21, 28, 35, 42 | . | No differences in BWG, FI, FCR, mortality d0–42 | (Bello et al., 2014) |

Diluent = injected without bioactive compound, but only with its diluent; E = Embryonic day, Non = non-injected; Sham = sham injection, insertion of needle without release; ns = not specified.

[1] Post-hatching.

- Elwan et al. (2022): Diluent (double-distilled water) compared to a non-injected control showed no performance differences, but liver, kidney, and heart tissues had lower GSH (14–32%) and MDA (29%) for the diluent-injected group at d0.
- Zhang et al. (2019) and Gao et al. (2017): Diluent (sterile saline) was compared to a non-injected control. No effects on performance or antioxidant enzymes and digestive parameters were observed.
- Pender et al. (2017): Diluent (sterile water) was compared to a non-injected control. No differences in performance were reported, but reductions in immune gene expression were found in ileum and caecum for the diluent-injected group: TLR2 (2.5–4×), TLR4 (2.5×), IL-4 (3–10×), and IFN-$\gamma$ (3×), measured at d0, d4, and/or d6.

These findings indicate that even when performance is unaffected, underlying physiological or molecular differences may still exist between control groups. Including both non-injected and diluent-injected controls is therefore recommended for accurate interpretation. When the primary goal is to assess the specific contribution of a compound, the diluent control is particularly important, as it isolates effects that occur in addition to those of the vehicle or injection. In all summary tables included in this chapter, the specific control groups used in each study were clearly indicated, allowing readers to assess how results were interpreted based on the chosen controls (Table 5).

## 2.8 Time of measurement and experimental conditions

The discrepancies in the effectiveness of in ovo may also arise from differences in measurement timing and experimental conditions, particularly the presence or absence of an external challenge.

Timing is especially important due to the rapid growth and development of broilers. A single day can make a major difference in measured outcomes. For example, Gunawardana et al. (2019) assessed immune gene expression after in ovo delivery of CpG-oligonucleotides and found differential dynamics depending on the timepoint: expression was measured at 3, 6, 12, 24, 32, 48, 72 hours, and days 5 and 7 post-injection. Results varied depending on the specific gene and time of measurement, demonstrating that timing strongly influences the conclusions drawn. Similarly, Pender et al. (2017) showed that immune gene expression in response to probiotic delivery varied widely depending on when samples were collected. Parameters such as cytokines are highly dynamic, making results difficult to interpret without clearly defined and standardised measurement timepoints.

Environmental stressors or challenges can further influence the effectiveness of in ovo delivery. In many cases, positive effects are only seen under suboptimal or challenged conditions:

**Table 5** Overview of in ovo delivery studies under challenged conditions. This table summarises studies that assessed the effects of in ovo-delivered compounds under environmental or pathogenic stressors. Only results from challenged groups are presented to highlight whether in ovo interventions influenced performance parameters (hatchability, BW, BWG, FI, FCR) or mortality under these conditions

| Compound | Challenge | Timing of Challenge | Day of Injection | Compartment | Dose/Volume | Control | Time of Measurement[1] | Result | Reference |
|---|---|---|---|---|---|---|---|---|---|
| Arginine | Low incubation temp (15°C vs 37.8°C) | E8, 11, 14, 17 for 1h/d | E17.5 | Amnion | 6 mg/600 μL | Non, Non × 15°C, Diluent, Diluent × 15°C (only d0) | d0, 10, 24, 42 | Lower mortality E19–21 with combination 15°C × treatment. No effect of 15°C or treatment on hatchability, BW d0; Higher BW, BWG, FI, lower FCR, mortality d42 with 15°C or treatment | (Miri et al., 2022) |
| Methionine | Aflatoxin | E10 (co-injection) | E10 | Amnion | 5.9 mg/1000 μL | Non, Diluent, only aflatoxin | d0 | Did not look at main effects. Higher mortality with aflatoxin, lower with aflatoxin + treatment. Lower BW d0 with aflatoxin, higher with treatment | (Elwan et al., 2021) |
| Threonine | Salmonella enteritidis | D2 | E17.5 | Amnion | 3.5 μL/1000 μL | Diluent, Diluent × SE | 24, 96, 168 hours post challenge | Lower BW 24 hours post-inoculation for SE, not affected by treatment | (De Moreira Filho et al., 2018) |
| Manganese | Heat stress (34°C vs 22°C) | d29–42 | E9 | ns | 12.5, 25 μg/100 μL | Non, Diluent, Non × 34°C, Diluent × 34°C | d29, 42 | Lower BWG, FI, higher FCR d42 with 34°C; No effect of 34°C or treatment on mortality | (Geng et al., 2022) |

(*Continued*)

**Table 5** *Continued*

| Compound | Challenge | Timing of Challenge | Day of Injection | Compartment | Dose/Volume | Control | Time of Measurement[1] | Result | Reference |
|---|---|---|---|---|---|---|---|---|---|
| γ-aminobutyric acid | High incubation temp (39.6°C vs 37.8°C) | E10–18 6h/d | E17.5 | Amnion | 60 mg/600 μL | Non, Non × 39.6°C | d0, 7, 21, 28 | Did not look at main effects. No effect of 39.6°C or treatment on hatchability, BW d0, BW, BWG, FI d42; Highest FCR d8–21 for combination 39.6°C × treatment | (Ncho et al., 2023) |
| γ-aminobutyric acid | High incubation temp (39.6°C vs 37.8°C) | E10–18 6h/d | E17.5 | Amnion | 60 mg/600 μL | Non, Non × 39.6°C | d0 | Higher BW d0 with 39.6°C or treatment. No effect of 39.6°C or treatment on hatchability | (Goel et al., 2022) |
| γ-aminobutyric acid | Heat stress (38°C vs 29°C) | d10, for 3 hours | E17.5 | Amnion | 60 mg/600 μL | Non, Non × 38°C | d10 | No effects of 38°C or treatment for BW, FI | (Ncho et al., 2021) |
| Bearberry fruit extract | Aflatoxin | E10 (co-injection) | E10 | Amnion | 0.1/500 μL | Non, Diluent, Diluent × aflatoxin | d0 | Did not look at main effects. Lower BW d0 with aflatoxin, higher with aflatoxin + treatment | (Elwan et al., 2022) |
| Probiotic mixture | Eimeria | D3 | E18 | Air cell | 1 × $10^6$/ns | Diluent, Diluent × Eimeria | d9, 15 | Lower BW d9, d14, BWG d3–15, higher lesion score with Eimeria; Higher BW d15, BWG d3–15 with treatment independent of challenge; Lower lesion score for Eimeria in combination with treatment | (Pender et al., 2016) |

| | | | | | | | | | |
|---|---|---|---|---|---|---|---|---|---|
| GOS | Heat stress (30°C vs 20°C) | d36–50 | E12 | Air cell | 3.5 mg/200 μL | Non, Diluent, Non × 30°C | d36, 50 | Lower BW, FI, higher FCR d50 with 30°C; No effect of 30°C or treatment on BWG, mortality | (Tavaniello et al., 2022) |
| GOS | Heat stress (30°C vs 25°C) | d32–42 | E12 | Air cell | 3.5 mg/200 μL | Non, Diluent, Non × 30°C, Diluent × 30°C | d42 | Lower BW, BWG, FI, higher FCR, mortality d42 with 30°C; Higher BW, BWG, FI d42 for treatment independent of challenge; Lower FCR d42 for 30°C in combination with treatment; no effects of 30°C or treatment on mortality | (Slawinska et al., 2020) |
| Postbiotic | Necrotic enteritis | d14 Eimeria, d19/20 *C. perfringens* | E18 | Amnion | 3.1 μL/200 μL | Diluent, Diluent × NE | d14, 21, 28 | Lower BWG, FI, higher FCR with challenge; no effect of treatment | (Dong et al., 2024) |

Diluent = injected without bioactive compound, but only with its diluent; E = Embryonic day, Non = non-injected; Sham = sham injection, insertion of needle without release; ns = not specified.
[1] Post-hatching.

- Elwan et al. (2022): An embryonic aflatoxin challenge at E10 led to a 41% reduction in BW at hatch, ~4× lower serum levels of IgY, IgM, and IgA, and a ~2.5× reduction in total antioxidant activity in serum. When bearberry fruit extract was co-injected, BW at hatch improved by 29%, immunoglobulin levels approximately doubled, and antioxidant activity increased by ~1.75×. Without the challenge, bearberry extract improved BW at hatch by only ~5% and slightly reduced antioxidant activity (~0.9×).
- Elwan et al. (2021): Aflatoxin exposure increased embryonic mortality by ~30%. This higher mortality was reduced by ~21% when the challenge was combined with in ovo delivery of methionine. In the absence of the challenge, methionine had no impact on mortality.
- Dong et al. (2024): Necrotic enteritis negatively affected post-hatch performance, with lower BWG and FI, and higher FCR, and most chickens had jejunal and ileal lesion scores averaging 1 at d21. While in ovo delivery of the postbiotic had no effect on performance, it reduced lesion scores to an average of ~0.3 when co-delivered with the challenge.
- Pender et al. (2016): An Eimeria challenge reduced post-hatch BW by ~25% and increased lesion scores to ~1.8 in both the duodenum and jejunum at d9. Probiotic supplementation did not restore BW, but it did reduce lesion scores to ~1.3 in the duodenum and ~0.6 in the jejunum. In non-challenged chickens, the probiotic did increase BW by ~15%.
- De Moreira Filho et al. (2018): A Salmonella enteritidis challenge at d2 post-hatch led to reduced BW 24 hours post-inoculation and resulted in 15–30% lower ileal villus:crypt ratios. Threonine supplementation did not prevent the drop in BW but improved the villus:crypt ratios by 15–30% at 24 hours and 168 hours post inoculation and reduced Salmonella counts in cecal contents by ~20% at 168 hours post-inoculation.
- Ncho et al. (2021): Acute heat stress (38°C compared to 29°C for 3 hours at d10) increased plasma MDA levels by ~2.5× and hepatic heat shock protein 70 (HSP70) gene expression by ~1.5×. γ-Aminobutyric acid supplementation lowered both markers to the levels of the control, but had no effect when the temperature was 29°C.

These studies highlight the importance of assessing whether in ovo treatments provide true benefits under optimal conditions or only in challenged situations. In some cases, benefits are only observed when baseline performance is reduced, suggesting that the treatments may only be useful when conditions are suboptimal. If so, their relevance under well-managed commercial conditions could be limited.

This again highlights the importance of proper control groups. Including both challenged and non-challenged controls is essential, but it is equally important to verify that non-challenged controls are truly unchallenged and represent normal, healthy performance. In some studies, even when no formal challenge is applied, chickens experience suboptimal conditions, such as poor incubation, low-quality housing, or environmental stress. This can unintentionally act as a challenge, making it difficult to distinguish whether a treatment is genuinely beneficial or simply compensating for these underlying issues. Without confirmation that control animals perform within expected ranges, results may be incorrectly interpreted as evidence that in ovo delivery improves baseline performance. However, if baseline performance is already compromised due to unrecognised suboptimal conditions, such improvements may not reflect a true additive effect, but rather a correction of underlying deficiencies. To avoid this, control group performance could be compared against expected performance objectives to verify that the baseline reflects well-managed commercial or physiological standards.

## 2.9 Other considerations

In addition to the methodological factors discussed previously, this section highlights several additional experimental and reporting aspects worth consideration.

- Hatchability is a critical parameter that reflects the potential toxicity of in ovo interventions. Despite its importance, it is not consistently reported in the literature. Including hatchability of fertile eggs, with a correct assessment of fertility, should be standard practice as a form of quality control, especially when evaluating the safety of a compound or delivery method.
- Diluent type may also influence results. While some diluents appear inert, others have been shown to affect physiological parameters. As discussed in Section 2.7, effects of the diluent alone can range from changes in antioxidant activity to alterations in immune gene expression. Moreover, when in ovo delivery is combined with vaccination, the diluent must be compatible with the vaccine formulation.
- Manual injection techniques vary in how holes are made to access the egg interior: using an egg puncher, needle, rotary tool, tweezers, or other devices. Although this chapter focuses on in ovo delivery rather than commercial injection systems, it is worth noting that manual procedures may also introduce more variability between experiments or operators.
- Sterility of the injection: Contaminants introduced through non-sterile needles, diluents, compounds, or handling procedures may influence embryo viability, immune responses, or post-hatch development. While contamination may not always be visible or reported, ensuring sterile conditions during injection can help minimise confounding effects and enhance reproducibility.

- Environmental conditions during handling and injection may also affect outcomes. Factors such as room temperature, humidity, airflow, and duration of egg exposure outside the incubator may affect embryo viability.
- Injection temperature is another variable that is rarely addressed. Injecting a solution that is much colder than the egg may introduce localised thermal shock, especially during earlier stages of embryonic development. Although this is seldom reported, it could influence embryo viability and should be considered in experimental design.

A number of studies have reported no observable impact of in ovo delivery on either performance or mechanistic parameters, yet these results are often underrepresented in the literature. The studies showing no effects on selected parameters have not been included in the summary tables presented earlier, highlighting a potential bias towards reporting positive findings. Moreover, it is possible that some studies with no significant effects are not published at all, contributing to an incomplete picture of the technique's effectiveness. Examples include experiments where γ-aminobutyric acid failed to mitigate heat stress effects, or selenium and vitamin C did not improve performance during post-hatching heat stress (Sgavioli et al., 2019; Zhao et al., 2023; Ban et al., 2024). Similarly, studies testing pro- and prebiotics have produced mixed results, with some reporting improvements and others finding limited or no effects (Majidi-Mosleh et al., 2017; Singh et al., 2022; Oladokun and Adewole, 2023).

These observations highlight the importance of publishing studies regardless of whether positive effects are observed. Transparent reporting of neutral or negative findings contributes to a more complete understanding of in ovo delivery and prevents overestimation of its effectiveness.

## 3 Case study: in ovo delivery of essential oils

This section presents a case study on the optimisation of in ovo delivery of selected EOs, a group of plant-derived compounds with antimicrobial, antioxidant, and/or anti-inflammatory properties (Brenes and Roura, 2010; Wang et al., 2024). While widely used as feed additives, their application via in ovo delivery remains underexplored in several critical aspects including methodological consistency, compartmental distribution after injection and functional impact in the egg. The overarching hypothesis guiding the research was that EOs could contribute to improved gut health in broiler chickens if delivered before hatch. Specifically, the aim was to investigate whether early-life exposure via in ovo delivery could stimulate gut development or immune responsiveness, potentially leading to improved resilience around hatch and post-hatch performance. This case study describes what initial steps were taken before performing the actual experiments to investigate the effectiveness of EOs after in ovo delivery.

The research initially focused on oregano EO (OEO) and its major bioactive compound, carvacrol. One of the first challenges was its hydrophobicity, which required emulsification for proper solubility in aqueous diluents. Polysorbate 80 was selected after preliminary tests showing effective visual emulsification of OEO at a 1:1 ratio (Meijer et al., 2024a).

The first target was optimising compartment and time of injection. The OEO (0.5 μL) with 0.5 μL polysorbate 80 was diluted in 100 μL 0.9% sterile saline and injected at either E12 or E17.5 into the air cell, yolk, or amniotic fluid (Meijer et al., 2024a). Control groups included non-injected eggs and saline-injected eggs in the air cell at E12 and the amniotic fluid at E17.5, representing the most common compartments used at these respective timepoints. For OEO, injection at E12 led to a 9% lower hatchability than at E17.5, whereas the compartment of injection had no significant effect. Given its compatibility with commercial practice and the absence of adverse effects, we selected injection into the amniotic fluid at E17.5 for subsequent work.

A follow-up experiment was done to assess the effect of the emulsifier, but also to find where carvacrol (OEO's main compound) ends up after in ovo delivery. To test how emulsifier ratio affected compound distribution, three ratios of polysorbate 80 to OEO were compared at 0:1, 0.5:1, or 1:1. The OEO solutions were injected into the amniotic fluid at E17.5 and carvacrol concentrations were measured 9 hours post-injection in yolk, blood, and amniotic fluid. Higher concentrations of carvacrol were found with increased emulsifier ratios across tissues. In a separate time-course experiment, the dynamics of carvacrol distribution over time were examined. Yolk, blood, and amniotic fluid were sampled at 3, 6, and 9 hours post-injection, using a fixed 1:1 ratio of OEO to polysorbate 80. While concentrations declined over time in the amniotic fluid and blood, levels remained consistently high in the yolk. This highlights that despite injection into the amniotic fluid, carvacrol redistributed to the yolk, likely due to its hydrophobicity (Meijer et al., 2024a).

The pH tolerance in injected OEO was also explored. The OEO solutions at pH 4.5 to 9.5 were injected into the amniotic fluid at E17.5, showing no impact on hatchability across treatments. This suggests that the embryo may actually tolerate a range of pH levels, potentially due to its buffering capacity (Khaskheli et al., unpublished data, 2022).

To determine appropriate volume and dose, a fixed amount of OEO (0.5 μL) in increasing saline volumes (100 μL to 1500 μL) was tested. Hatchability remained unaffected across volumes, but BW at hatch increased from 44.1 g (100 μL) to 45.8 g (1500 μL), likely due to the added fluid weight (Santos et al., unpublished data, 2023). A separate dose-response experiment ranging from 5 μL to 50 μL showed that OEO doses above 20 μL reduced hatchability by 66.5% and lowered BW at hatch by 9.3% compared to the control. Even 10 μL showed a hatchability reduction of 6.5%. It was concluded that OEO doses <10 μL per egg were safe for use in further studies (Niknafs et al., 2024).

Regarding controls, non-injected and saline-injected groups were consistently compared during the initial stages of the project. No differences in hatchability or performance were observed; hence, the results were interpreted as a validation of the use of the saline injection as the most appropriate control group in subsequent studies (Meijer et al., 2024a; Niknafs et al., 2024). Additional testing of polysorbate 80 alone (5 μL in 1000 μL saline) compared to saline confirmed its safety (Niknafs et al., unpublished data, 2023).

Subsequently, 27 EOs were screened using the developed method of injecting on E17.5, in the amniotic fluid, with 5 μL EO + 5 μL polysorbate 80 in 1000 μL of saline solution. Several EOs (e.g., thyme, clove, lemongrass) significantly lowered hatchability (13.5–38.5%). Interestingly, OEO reduced BW at hatch by 8.5%, whereas lemon myrtle increased it by 5.5%. In jejunal tissue at d0, some EOs lowered tight junction gene expression (e.g., cassia, eucalyptus for OCLN and sandalwood for ZO-2), and 12 EOs reduced expression of the T-cell marker CD3D (Khaskheli et al., 2025).

In parallel to the EO screening study, additional research was conducted with carvacrol, the main active component of OEO, due to its known antimicrobial and immunostimulatory properties (Liu et al., 2019). Carvacrol was delivered at E17.5 into the amniotic fluid, using a dose of 5 μL combined with 5 μL polysorbate 80 in 1000 μL saline. Hatchability was unaffected, but BW at hatch was 0.8% lower compared to the saline control. Performance until d14 was not impacted by the treatment. Furthermore, immune-related gene expression in jejunum, bursa, and yolk sac was determined at E19.5, d0, and d14. Immune responses were stage-dependent: E19.5 responses were pro-inflammatory (e.g., increased IL-1β, CD3D; reduced IgM), while d14 responses were anti-inflammatory (increased IL-10, reduced IL-6) (Meijer et al., 2024b, 2025b). These findings highlight the importance of time of measurement, as immune-related effects were most pronounced at E19.5 and had shifted by d14, underlining the dynamic nature of immune development. Immune responses can change rapidly within hours, making the exact timing of sampling critical for accurate interpretation of gene expression and functional outcomes.

As a next step, in ovo delivery of carvacrol was tested against post-hatch bacterial LPS (lipopolysaccharide) challenges at d7 and d14. While carvacrol did not affect BW, BWG or FCR, it led to a lower feed intake (~5%) between d0 and d7. The LPS challenge increased the expression of several inflammatory genes, including IL-1β, IL-8, IκB, NF-κB, TNF-α, and TLR4, in the jejunum and spleen. In ovo delivery of carvacrol mitigated these responses, leading to lower expression levels of these genes 6 h post-challenge compared to the challenged control group. In non-challenged chickens, carvacrol increased IFN-$\gamma$ expression, suggesting immunostimulatory effects (Meijer et al., 2025a). To further explore these findings, transcriptomic analyses were performed on splenic tissue 6 h after the second LPS challenge at d14. These analyses revealed enrichment of pathways related to T-cell activation, primarily through upregulation of genes encoding cytokines, chemokines, and their receptors. Interestingly, this enrichment was only observed in challenged chickens treated with carvacrol. In non-challenged chickens, transcriptomic comparison between carvacrol and control groups showed no significant pathway enrichment. This suggests that the immunomodulatory effects of in ovo-delivered carvacrol may only manifest under suboptimal or challenged conditions (Meijer et al., unpublished results, 2025).

Lastly, to assess whether EOs could be combined with existing in ovo vaccination protocols, a preliminary compatibility study was conducted using black pepper EO. In vitro, turkey herpesvirus (HVT), a commonly used proxy for Marek's disease vaccine, was added to cultured chicken embryo fibroblasts to simulate the in ovo vaccination environment. Black pepper EO was added at concentrations ranging from 0 ppm to 400 ppm. A reduction in HVT infection rate was only observed at concentrations above 200 ppm, while lower concentrations did not affect virus infection rates. These findings suggest that co-injection of black pepper EO at ≤200 ppm with Marek's disease vaccine may be feasible without compromising vaccine efficacy (Yang et al., unpublished results, 2025).

This case study illustrates the importance of stepwise optimisation of in ovo delivery methodology, considering compound properties, compartment and timing of injection, controls, and external conditions. While not all potentially influencing factors can be tested in every study, this approach highlights the value of systematic validation to ensure both safety and efficacy of in ovo-delivered compounds. Still, there are opportunities for improvement. While OEO was thoroughly evaluated, other EOs were not tested in a dose-response manner, leaving uncertainty about their optimal inclusion levels. It is also possible that different EOs may require different emulsifiers for proper delivery, which was not explored. Although saline was used as a validated control in key studies, ideally both non-injected and saline-injected controls should be included consistently across all experiments. Furthermore, post-hatching effects were assessed up to d14 only. Follow-up studies extending until slaughter age are necessary to evaluate the long-term relevance and commercial potential of in ovo EO delivery.

## 4 Conclusion and future trends

In ovo delivery applications have shown value in enhancing broiler health, resilience, and performance, but their broader application is limited by variability in both methodology and outcomes. This chapter has highlighted the numerous factors that influence the impact of in ovo interventions, including dose, volume, timing, injection compartment, compound properties, and external conditions. While the concept is appealing and some studies show positive effects, the technique remains context dependent. Some of the most consistent benefits of in ovo applications have been observed in chickens. Benefits may often reflect compensation for suboptimal/stress challenged conditions rather than consistent improvement under commercial settings. Many of these methodological considerations are underreported in the literature, which makes comparison between studies challenging.

One of the most critical steps in designing in ovo delivery studies is to start with a clear, hypothesis-driven objective: what is the physiological mechanism or outcome we aim to improve? Is the goal to enhance resilience under challenge, improve early growth, or modulate immunity? Rather than selecting a compound arbitrarily, studies should begin with a clear rationale linked to a specific problem, pathway, or challenge. Once the goal is defined, the methodology – including the compound, dose, timing, injection compartment and evaluation parameters – should be adapted accordingly. Without a mechanistic rationale, there is a high risk that outcomes will remain exploratory and difficult to interpret.

The case study presented related to in ovo applications of EO showed that simply delivering a compound into the egg does not guarantee that it reaches the intended target compartment. For example, carvacrol redistributed to the yolk despite being injected into the amniotic fluid, demonstrating the importance of testing compound distribution dynamics inside the egg (Meijer et al., 2024a). Timing is equally important, as immune and digestive development and function, including gene expression, can shift rapidly. Our studies showed that carvacrol had opposing immunological effects at E19.5 and d14, highlighting the need for carefully selected timepoints or kinetic studies. Experiments should also assess performance under both optimal and suboptimal conditions to determine whether observed effects represent true enhancements or corrections of underlying deficiencies. Finally, long-term follow-up studies until slaughter age remain rare but are essential for understanding the practical, commercial, and economic payback of potentially costly in ovo interventions.

From a practical and commercial standpoint, the potential of in ovo delivery may be greatest when integrated into existing in ovo vaccination practices. This would ensure compatibility with established infrastructure, while potentially enhancing vaccine efficacy through synergistic immune stimulation. However, upscaling the application to commercial practices, active compound availability, stability and cost, and regulatory requirements remain open challenges.

Future research should therefore aim to clearly define mechanistic hypotheses that guide compound selection and study design. Where available, existing literature can support these decisions and provide a basis for making educated guesses when direct experimental testing is not feasible. Without careful consideration of these elements, in ovo delivery risks becoming a trial-and-error technique rather than a targeted intervention, with unavoidable and undesirable animal welfare implications. Additionally, efforts should focus on optimising and transparently reporting methodological details, including the use of dose-response studies to determine safe and effective delivery levels.

In summary, in ovo delivery offers a promising strategy to improve poultry health and performance, but realising the full potential requires critical, hypothesis-driven study designs, transparent reporting, and greater methodological standardisation. As the field progresses, the success will depend not only on identifying functional compounds of interest (the what), but equally importantly the how, when, and why to inject them.

## 5 Acknowledgements

The case study presented in Section 4 is based on research funded by AgriFutures Australia's Chicken Meat Programme [grant number PRJ-011584]. The authors gratefully acknowledge their financial support, which made this research possible.

## 6 Where to look for further information

- Givisiez et al. (2020): Metabolic and morphological changes during embryonic development are reviewed.
- Jha et al. (2019); Das et al. (2021): The impact of in ovo delivery on performance and gut health.
- Peebles (2018): This review paper focuses on the broader applications of the in ovo technique, also covering vaccines, drugs and hormones.
- Moran (2007): Provides an in-depth look at nutrient metabolism during incubation and early post-hatch, relevant for timing and rationale of in ovo delivery.
- Uni and Ferket (2003): The US patent developed by Uni & Ferket on in ovo feeding.
- Brenes and Roura (2010); Wang et al. (2024): Overview of the use of essential oils in poultry nutrition.

## 7 References

Abbasi, T., et al. 2017. Growth performance parameters, bone calcification and immune response of in ovo injection of 25-hydroxycholecalciferol and vitamin K3 in male Ross 308 broilers. *Theriogenology* 90:260–265.

Alizadeh, M., et al. 2020. Effects of in ovo Inoculation of Multi-Strain Lactobacilli on Cytokine Gene Expression and Antibody-Mediated Immune Responses in Chickens. *Front. Vet. Sci.* 7:105.

Araújo, I. C. S., et al. 2019. Effect of vitamin E in ovo feeding to broiler embryos on hatchability, chick quality, oxidative state, and performance. *Poult. Sci.* 98:3652–3661.

Avakian, A. P. 2006. Understanding in ovo vaccination. *Int. Hatch Prac.* 20:15–17.

Ban, J., et al. 2024. Comparison of selenium-mediated regulation of heat shock protein and inflammation in-vitro and in-ovo for heat resistance enhancement in broiler. *Poult. Sci.* 103:104271.

Bednarczyk, M., et al. 2016. Influence of different prebiotics and mode of their administration on broiler chicken performance. *Animal* 10:1271–1279.

Bełdowska, A., et al. 2024. The effect of sodium butyrate administered in ovo on the health status and intestinal response in broiler chicken. *Poult. Sci.* 103:104108.

Bello, A., et al. 2014. Effects of the commercial in ovo injection of 25-hydroxycholecalciferol on broiler posthatch performance and carcass characteristics. *Poult. Sci.* 93:155–162.

Berrocoso, J. D., et al. 2017. Effect of in ovo injection of raffinose on growth performance and gut health parameters of broiler chicken. *Poult. Sci.* 96:1573–1580.

Brenes, A. and Roura, E. 2010. Essential oils in poultry nutrition: main effects and modes of action. *Anim. Feed Sci. Technol.* 158:1–14.

Dal Pont, G. C., et al. 2019. Glycerol inoculation in eggs of young broiler breeders at different embryonic periods. *Poult. Sci.* 98:3989–3993.

Das, R., et al. 2021. In ovo feeding as a tool for improving performance and gut health of poultry: a review. *Front. Vet. Sci.* 8:754246.

Deeming, D. C. 1991. Reasons for the dichotomy in egg turning in birds and reptiles. In Deeming, D. C. and Ferguson, M.W.J. (eds.), *Egg Incubation: Its Effects on Embryonic Development in Birds and Reptiles*, 307–324. Cambridge: Cambridge University Press.

De Moreira Filho, A. L. B., et al. 2018. Intra-amnionic threonine administered to chicken embryos reduces salmonella enteritidis cecal counts and improves posthatch intestinal development. *J. Immunol. Res.* 2018:9795829.

De Oliveira, J. E., et al. 2008. Important metabolic pathways in poultry embryos prior to hatch. *Worlds Poult. Sci. J* 64:488–499.

Dong, B., et al. 2024. Impact of early postbiotic supplementation on broilers' responses to subclinical necrotic enteritis. *Poult. Sci.* 103:104420.

El-Fakhrany, H. H., et al. 2023. Efficacy of in ovo delivered resveratrol (Trans 3, 4, 5-trihydroxystilbene) on growth, carcass weights, and blood metabolites of broiler chicks. *Anim. Biotechnol.* 34:384–391.

Elwan, H., et al. 2021. Methionine alleviates aflatoxinb1-induced broiler chicks embryotoxicity through inhibition of caspase-dependent apoptosis and enhancement of cellular antioxidant status. *Poult. Sci.* 100:101114.

Elwan, H., et al. 2022. Modulatory effects of arctostaphylos uva-urs extract in ovo injected into broiler embryos contaminated by aflatoxin B1. *Animals* 12:2042.

Fatemi, S. A., et al. 2024a. In ovo administration of the Marek's disease vaccine in conjunction with 25-hydroxyvitamin D3 and its subsequent effects on the performance and immunity-related characteristics of Ross 708 broiler hatchlings. *Poult. Sci.* 103:103199.

Fatemi, S. A., et al. 2024b. Effects of the Marek's disease vaccine on the performance, meat yield, and incidence of woody breast myopathy in Ross 708 broilers when administered alone or in conjunction with in ovo and dietary supplemental 25-hydroxycholecalciferol. *Animals* 14:1308.

Gaghan, C., et al. 2023. Effect of CpG-oligonucleotide in enhancing recombinant herpes virus of turkey-laryngotracheitis vaccine-induced immune responses in one-day-old broiler chickens. *Vaccines* 11:294.

Gao, T., et al. 2017. Effect of in ovo feeding of l-arginine on the hatchability, growth performance, gastrointestinal hormones, and jejunal digestive and absorptive capacity of posthatch broilers. *J. Anim. Sci.* 95:3079–3092.

Garcia, P., et al. 2021. The chicken embryo model: a novel and relevant model for immune-based studies. *Front. Immunol.* 12:791081.

Geng, Y., et al. 2022. Effect of in ovo manganese injection on the embryonic development, antioxidation, hatchability, and performances of offspring broilers under normal and high temperatures. *Poult. Sci.* 101:101936.

Ghanaatparast-Rashti, M., et al. 2018. In ovo feeding of nutrients and its impact on post-hatching water and feed deprivation up to 48 hr, energy status and jejunal morphology of chicks using response surface models. *J. Anim. Physiol. Anim. Nutr.* 102:e806–e817.

Gildersleeve, R. P., et al. 1993. Developmental performance of an egg injection machine for administration of Marek's disease vaccine. *J. Appl. Poult. Res.* 2:337–346.

Givisiez, P. E. N., et al. 2020. Chicken embryo development: metabolic and morphological basis for in ovo feeding technology. *Poult. Sci.* 99:6774–6782.

Goel, A., et al. 2017. In ovo silver nanoparticle supplementation for improving the post-hatch immunity status of broiler chickens. *Arch. Anim. Nutr.* 71:384–394.

Goel, A., et al. 2022. Embryonic thermal manipulation and in ovo gamma-aminobutyric acid supplementation regulating the chick weight and stress-related genes at hatch. *Front. Vet. Sci.* 8:807450.

Gopi, M., et al. 2020. In ovo administration of nucleosides improved the performance, apparent metabolizable energy and gut development in broiler chickens. *Front. Vet. Sci.* 7:583748.

Gunawardana, T., et al. 2019. Synthetic CpG-ODN rapidly enriches immune compartments in neonatal chicks to induce protective immunity against bacterial infections. *Sci. Rep.* 9:341.

Hafez, H. M. and Attia, Y. A. 2020. Challenges to the poultry industry: Current perspectives and strategic future after the COVID-19 outbreak. *Front. Vet. Sci.* 7:516.

Han, G., et al. 2020. Effects of l-leucine in ovo feeding on thermotolerance, growth and amino acid metabolism under heat stress in broilers. *Animal* 14:1701–1709.

Hassan, H. A., et al. 2023. Histological alterations of small intestine and growth performance of broiler chicks after in ovo copper injection at 10 days of embryogenesis period. *Anim. Biotechnol.* 34:585–592.

Jha, R., et al. 2019. Early nutrition programming (in ovo and post-hatch feeding) as a strategy to modulate gut health of poultry. *Front. Vet. Sci.* 6:1–10.

Keshtkaran, M., et al. 2024. Effects of in ovo injection of the L-carnosine on physiological indexes of neonatal broiler chicken. *Poult. Sci.* 103:103380.

Khaskheli, A. A., et al. 2025. The in ovo screening of 27 single essential oils showed selective effects on hatchability, performance and gene expression relevant to gut functions in broilers at hatch. *Poult. Sci.* 104:104670.

Khomayezi, R. and Adewole, D. 2021. Probiotics, prebiotics, and synbiotics: an overview of their delivery routes and effects on growth and health of broiler chickens. *Worlds Poult. Sci. J* 78:57–81.

Laothaweerungsawat, N., et al. 2020. Transdermal delivery enhancement of carvacrol from Origanum vulgare L. essential oil by microemulsion. *Int. J. Pharm.* 579:119052.

Lee, S. H., et al. 2014. Effects of in ovo injection with selenium on immune and antioxidant responses during experimental necrotic enteritis in broiler chickens. *Poult. Sci.* 93:1113–1121.

Liu, S. D., et al. 2019. Effect of carvacrol essential oils on immune response and inflammation-related genes expression in broilers challenged by lipopolysaccharide. *Poult. Sci.* 98:2026–2033.

Ma, Y., et al. 2020. Effect of in ovo feeding of β-hydroxy-β-methylbutyrate on hatchability, muscle growth and performance in prenatal and posthatch broilers. *J. Sci. Food Agric.* 100:755–763.

Maatjens, C. M., et al. 2014. Temperature and CO2 during the hatching phase. II. Effects on chicken embryo physiology. *Poult. Sci.* 93:655–663.

Madkour, M., et al. 2024. Impact of in ovo feeding of grape pomace extract on the growth performance, antioxidant status, and immune response of hatched broilers. *Poult. Sci.* 103:103914.

Majidi-Mosleh, A., et al. 2017. Ileal MUC2 gene expression and microbial population, but not growth performance and immune response, are influenced by in ovo injection of probiotics in broiler chickens. *Br. Poult. Sci.* 58:40–45.

Manders, T. T. M., et al. 2021. Success rates of inoculation of the various compartments of embryonated chicken eggs at different incubation days. *Avian Pathol.* 50:61–77.

Meijer, M. M. Y., et al. 2024a. Carvacrol in ovo delivery optimization and flow dynamics in broiler chicken eggs. *Poult. Sci.* 103:103443.

Meijer, M. M. Y., et al. 2024b. Early-life immunomodulation by carvacrol delivered in ovo in broiler chickens. *Poult. Sci.* 103:104286.

Meijer, M. M. Y., et al. 2025a. The inflammatory response to Escherichia coli lipopolysaccharide is mitigated by in ovo delivery of carvacrol in broiler chicks. *Poult. Sci.* 104:104881.

Meijer, M. M. Y., et al. 2025b. In ovo delivery of carvacrol triggers expression of chemotactic factors, antimicrobial peptides and pro-inflammatory pathways in the yolk sac of broiler chicken embryos. *J. Anim. Sci. Biotechnol.* 16:8.

Miri, B., et al. 2022. Effects of low eggshell temperatures during incubation, in ovo feeding of L-arginine, and post-hatch dietary guanidinoacetic acid on hatching traits, performance, and physiological responses of broilers reared at low ambient temperature. *Poult. Sci.* 101:101548.

Moran, E. T. 2007. Nutrition of the developing embryo and hatchling. *Poult. Sci.* 86:1043–1049.

Moreira Filho, A. L. B., et al. 2019. Enrichment of the amnion with threonine in chicken embryos affects the small intestine development, ileal gene expression and performance of broilers between 1 and 21 days of age. *Poult. Sci.* 98:1363–1370.

Nangsuay, A., et al. 2017. Effects of breeder age, strain, and eggshell temperature on nutrient metabolism of broiler embryos. *Poult. Sci.* 96:1891–1900.

Ncho, C. M., et al. 2023. Effect of in ovo feeding of γ-aminobutyric acid combined with embryonic thermal manipulation on hatchability, growth, and hepatic gene expression in broilers. *Anim. Biosci.* 36:284–294.

Ncho, C. M., et al. 2021. In ovo injection of gaba can help body weight gain at hatch, increase chick weight to egg weight ratio, and improve broiler heat resistance. *Animals* 11:1600.

Neves, D. G., et al. 2017. Effects of in ovo feeding with glycerol for broilers. *J. Anim. Physiol. Anim. Nutr.* 101:434–440.

Niknafs, S., et al. 2024. In ovo delivery of oregano essential oil activated xenobiotic detoxification and lipid metabolism at hatch in broiler chickens. *Poult. Sci.* 103:103321.

Ohta, Y. and Kidd, M. T. 2001. Optimum site for in ovo amino acid injection in broiler breeder eggs. *Poult. Sci.* 80:1425–1429.

Oke, O. E., et al. 2021. In ovo injection of black cumin (Nigella sativa) extract on hatching and post hatch performance of thermally challenged broiler chickens during incubation. *Poult. Sci.* 100:100831.

Oladokun, S. and Adewole, D. 2023. The effect of Bacillus subtilis and its delivery route on hatch and growth performance, blood biochemistry, immune status, gut morphology, and microbiota of broiler chickens. *Poult. Sci.* 102:102473.

Pawłowska, J., et al. 2024. In ovo injection of bee pollen extract on hatchability, chick quality, glycogen reserves and production performance in broiler chickens. *Poult. Sci.* 103:104035.

Peebles, E. D. 2018. In ovo applications in poultry: a review. *Poult. Sci.* 97:2322–2338.

Pender, C. M., et al. 2016. Effects of in ovo supplementation of probiotics on performance and immunocompetence of broiler chicks to an Eimeria challenge. *Benef. Microbes* 7:699–705.

Pender, C. M., et al. 2017. In ovo supplementation of probiotics and its effects on performance and immune-related gene expression in broiler chicks. *Poult. Sci.* 96:1052–1062.

Reis, D. R., et al. 2022. Encapsulated essential oils: a perspective in food preservation. *Future Foods* 5:100126.

Romanoff, A. L. 1960. The Avian Embryo. In Romanoff, A. L. (ed.), *Structural and Functional Development*. Macmillan, New York, 1960. xvi + 1305 pp. Illus. $35.

Romanoff, A. L. and A. J. Romanoff. 1929. Changes in pH of albumen and yolk in the course of embryonic development under natural and artificial incubation. *Biol. Bull.* 57:300–306.

Salmanzadeh, A. M. and Salmanzadeh, M. 2012. The effects of in-ovo injection of glucose on hatchability, hatching weight and subsequent performance of newly-hatched chicks. *Braz. J. Poult. Sci.* 14:137–140.

Sarfraz, M., et al. 2022. Characterization of dosage levels for in ovo administration of innate immune stimulants for prevention of yolk sac infection in chicks. *Vet. Sci.* 9:1–12.

Sgavioli, S., et al. 2019. In ovo injection of ascorbic acid and higher incubation temperature modulate blood parameters in response to heat exposure in broilers. *Br. Poult. Sci.* 60:279–287.

Sharma, J. M. and Burmester, B. R. 1984. Disease control in avian species by embryonal vaccination. *Avian Dis.* 28(3):559–569.

Shehata, A. M., et al. 2024. In-ovo injection of Bacillus subtilis, raffinose, and their combinations enhances hatchability, gut health, nutrient transport- and intestinal function-related genes, and early development of broiler chicks. *Poult. Sci.* 103:104134.

Simkiss, K. 1980. Water and Ionic fluxes inside the egg. *Am. Zoologist* 20:385–393.

Singh, A. K., et al. 2022. Effects of in ovo delivered xylo- and mannan- oligosaccharides on growth performance, intestinal immunity, cecal short-chain fatty acids, and cecal microbiota of broilers. *J. Anim. Sci. Biotechnol.* 13:13.

Siwek, M., et al. 2018. Prebiotics and synbiotics – In ovo delivery for improved lifespan condition in chicken. *BMC Vet. Res.* 14:1–17.

Slawinska, A., et al. 2020. Impact of galactooligosaccharides delivered in ovo on mitigating negative effects of heat stress on performance and welfare of broilers. *Poult. Sci.* 99:407–415.

Subramaniyan, S. A., et al. 2019. Effect of in ovo injection of l-arginine in different chicken embryonic development stages on post-hatchability, immune response, and myo-d and myogenin proteins. *Animals* 9:357.

Sun, X., et al. 2018. Effect of in ovo zinc injection on the embryonic development, tissue zinc contents, antioxidation, and related gene expressions of broiler breeder eggs. *J. Integr. Agric.* 17:648–656.

Tavaniello, S., et al. 2022. Performance and meat quality traits of slow-growing chickens stimulated in ovo with galactooligosaccharides and exposed to heat stress. *Poult. Sci.* 101:101972.

Uni, Z. and Ferket, P. R. 2003. Enhancement of development of oviparous species by in ovo feeding of enteric modulators. North Carolina State University, Raleigh, NC; and Yissum Research Development Company of the Hebrew University of Jerusalem, Jerusalem (Israel), assignees. US Pat. No. 6,592,878.

Uni, Z. and Ferket, P. R. 2004. Methods for early nutrition and their potential. *Worlds Poult. Sci. J.* 60:101–111.

Van Der Most, P. J., et al. 2011. Trade-off between growth and immune function: a meta-analysis of selection experiments. *Funct. Ecol.* 25:74–80.

Wakenell, P. S., et al. 2002. Effect of in ovo vaccine delivery route on herpesvirus of turkeys/SB-1 efficacy and viremia. *Avian Dis.* 46:274–280.

Wang, J., et al. 2024. Phytogenic feed additives as natural antibiotic alternatives in animal health and production: a review of the literature of the last decade. *Anim. Nut.* 17:244–264.

Watt, R. P., et al. 2019. Injectability as a function of viscosity and dosing materials for subcutaneous administration. *Int. J. Pharm.* 554:376–386.

Xu, H., et al. 2023. Improvement in the early growth, immune system and tibia development of broilers in response to the in ovo injection of 25-hydroxyvitamin D3. *J. Appl. Anim. Res.* 51:265–275.

Yang, S. B., et al. 2021. Effects of in ovo Injection of astragalus polysaccharide on the intestinal development and mucosal immunity in broiler chickens. *Front. Vet. Sci.* 8:746757.

Zhai, W., et al. 2011. Effects of commercial in ovo injection of carbohydrates on broiler embryogenesis. *Poult. Sci.* 90:1295–1301.

Zhang, H., et al. 2019. Effects of in ovo injection of L-ascorbic acid on growth performance, carcass composition, plasma antioxidant capacity, and meat quality in broiler chickens. *Poult. Sci.* 98:3617–3625.

Zhao, M., et al. 2023. The effects of in ovo feeding of selenized glucose on selenium concentration and antioxidant capacity of breast muscle in neonatal broilers. *Biol. Trace Elem. Res.* 201:5764–5773.

Zhu, L., et al. 2021. In ovo feeding of vitamin C regulates splenic development through purine nucleotide metabolism and induction of apoptosis in broiler chickens. *Br. J. Nut.* 126:652–662.

# Part 3

## Managing the needs of broilers

# Chapter 9

## Understanding the behavioral needs of broilers

*Leonie Jacobs, Virginia Tech, USA*

1 Introduction
2 Behavior and behavioral needs
3 Normal broiler chicken behavior
4 Abnormal behaviors
5 The impact of genetics on behavior
6 Conclusion
7 Where to look for further information
8 References

### 1 Introduction

Behavioral expression and motivation are valuable indicators for animal welfare. Important aspects to a good life are the ability to perform highly motivated behavior, to have choices, and to have a range of experiences including pleasurable, comfortable ones (Fraser, 2023).

Broiler chicken behavior is discussed mainly for commercially used genetic strains, which are fast-growing, short-living birds housed in large uniform barns. Although slower-growing strains are used in some countries, globally, and in terms of economic value, they still represent the minority. Broilers are generally housed on litter in floor systems, with feeder and drinker lines along the length of the house. Housing conditions may vary slightly, and developments towards broiler chicken welfare are ongoing. For example, companies are now fitting some houses with windows to allow for natural light, companies are recognizing the need for more environmental complexity, and researchers are contributing to the body of work that shows how growth rate impacts broiler chicken welfare (e.g. Bizeray et al., 2000; Kestin et al., 2001; Riber and Wurtz, 2024).

Fast-growing broilers show behavioral motivations similar to their jungle fowl ancestors (Bokkers and Koene, 2004), although the behavioral expression may be modified (Bokkers and Koene, 2003a). A recent scientific opinion listed

http://dx.doi.org/10.19103/AS.2025.0165.10

the behavioral restrictions that broiler chickens may experience in commercial production settings. These included the inability to perform comfort behavior (dustbathing) along with exploratory and foraging behavior, restriction of movement, and problems related to resting (Nielsen et al., 2023). In this chapter, the focus lies on these, and other normal behaviors, with an emphasis on behaviors that are deemed to have a positive valence. In addition, I discuss abnormal behaviors and normal chicken behaviors that are not expressed by broilers. The aim of the chapter is to get a better understanding of broiler chickens' behavioral needs and motivations, even if ability is lacking, with a focus on behaviors that can contribute to good animal welfare.

## 2 Behavior and behavioral needs

Behavioral needs are driven by internal motivations, meaning that the need to express the behavior persists regardless of external factors and environmental design (Duncan, 1998). If the environment allows for the expression of the behavior, the behavior can be performed, and the need met. If the environment does not allow for the expression of the behavior, this may lead to frustration (Duncan, 1998).

Broilers do express some natural behaviors that are driven mainly by external factors, such as responding to a perceived predator. The lack of these external factors in commercial settings (for instance, actual predators other than humans) could mean the behavioral need to respond never occurs. Commonplace things, including humans, can be perceived as a predator (Suarez and Gallup, 1982; Webster, 2002). Objects tossed in the air elicit alarm calls (Webster, 2002), and feather dusters used to clean cages elicit fleeing behavior if possible (Rutter and Duncan, 1991). The presence of these and similar stimuli motivates the chickens to respond, and it is important that the environment facilitates an appropriate response (Duncan, 1998), such as space and shelter to allow chickens to flee to avoid the stimulus.

Bokkers and Koene did excellent work to understand the internal motivations of broiler chickens. Only 20 years ago, a general belief was that fast-growing broilers were not motivated to perform active behaviors besides eating, drinking, and resting, implying that fundamental biological changes took place due to stringent genetic selection for fast growth (Bokkers, 2004). However, we now have a better understanding of their internal needs and observe that many motivations remain, even if ability is lacking.

Ability can be limited by the broiler's physical condition, which can include both clinically healthy broilers and broilers with impairments, such as lameness. Fast-growing broilers have heavy breast muscles that can comprise nearly 20% of their body weight (Xiao et al., 2017). These disproportionately heavy breasts will impact body conformation, center of gravity, and balance, which impact their ability to move. Clinically healthy broilers showed more weight support using both legs rather than one and showed substantial vertical movement of their back when walking compared to jungle fowl (Caplen et al., 2012).

Ability can also be limited by the environment broilers are raised in. For example, consider foraging or perching behavior. If resources that facilitate such behaviors are not provided – such as friable bedding in conventional laying hen cage systems or perching platforms in conventional broiler houses – these behaviors cannot be expressed. When provided with improved environmental conditions, such as platforms, and low stocking densities, broilers demonstrate a willingness to engage in various behaviors (Evans et al., 2023). This indicates an intrinsic motivation that is often limited by their physical constraints and typical commercial rearing environments.

## 3 Normal broiler chicken behavior

In this section, the behavioral repertoire of fast-growing broiler chickens is discussed, with more attention given to behaviors that might be indicative of positive affective state. Behaviors that show evidence of and association with positive affective state are comfort behaviors (dustbathing and preening), perching, and play behaviors (Boissy et al., 2007). Although not discussed in this chapter, vocalizations can also be indicative of emotion or affective states in chickens. For example, specific calls are identified for anxiety (Collins et al., 2024), social isolation (Marx et al., 2001), early-life distress (Herborn et al., 2020), and thermal comfort (Du et al., 2020).

### *3.1 Eating and drinking*

Conventional broiler chickens are selected for fast growth and low feed conversion. This genetic selection has led to a strong motivation to eat (Siegel and Wisman, 1966). Feeding behaviors vary to an extent between genetic lines with divergent selection for growth rates, with differences in meal size, number of meals, and meal duration across lines (Howie et al., 2009). The satiety mechanism dominates the hunger mechanism in broilers, meaning that they eat to their maximum physical capacity, with little influence from the hunger mechanism (Bokkers and Koene, 2003b). It seems that they are not so much driven by hunger, but instead, eat as much as they can until they feel full. This satiety then suppresses their motivation to eat, until this suppression wanes, which will motivate them to eat again.

Unrestricted feeding is conventional in broiler chicken production, and this results in a consistent hourly eating rate if conditions are homogenous, like when they are kept under continuous lighting (Webster, 2002). However, there is a diurnal rhythm to eating behavior. Broilers will eat more at the beginning of the day, which is motivated by hunger because the dark period limits their access to feed. Eating will be limited during the middle of the day, likely because satiety from the morning meal is suppressing motivation to eat and because other motivations are competing. Then, in anticipation of darkness, broilers will eat until satiated depending on crop fill (reviewed by Savory, 1980). Most broilers are grown with at least 4 hours of continuous darkness daily; thus, they will show a diurnal pattern of feeding. Broilers can anticipate the onset of darkness when it coincides with feed removal and will feed more prior to the start of the dark period (May and Lott, 1992).

There is a strong positive relationship between daily water and food intake (Savory, 1978). Broilers drink approximately twice the amount of water as the feed they eat, especially when younger (1.9:1 at 11 days of age), and this decreases as they get older (1.7:1 at 46 days of age; McCreery, 2015). Absolute water intake increases with age, with less frequent drinking bouts as they get older (Jacobs et al., 2019). Broilers will spend more time drinking from drinking nipples compared to bell or cup drinkers due to limited water flow in the former (as reviewed by Savory, 2010). Water intake seems mostly motivated by feed intake when environmental conditions are within the thermal comfort zone (Savory, 2010).

Broilers will work hard (push a weighted door) to access feed (Monckton et al., 2020). They show an even stronger motivation to access feed when feed deprived. However, their ability to access the feed can be hampered by physical ability in older fast-growing broilers, compared to younger fast- and slower-growing broilers (Bokkers and Koene, 2004). These results suggest that feeding is highly rewarding for broilers and may be associated with a positively valenced emotional state. Indeed, there is some evidence that dopamine, a neurotransmitter associated with reward and pleasure, is also involved with satiety in broiler chickens, through its anorectic effect (Foroushan, 2017). Dopamine signaling mediated appetite suppression via histamine in an experiment on young chicks (Foroushan, 2017). Although dopamine was involved, no conclusions were drawn related to the emotional experience. Thus, further research is needed to determine whether this reflects an actual positive experience. If this is the case, ad libitum access to feed might be quite rewarding, despite its detrimental effect on gait and metabolic disorders at higher bodyweights. This may be reflected in the frustration observed in broiler breeders that are kept under severe feed restriction (de Jong et al., 2003; Renema et al., 2007).

### *3.2 Foraging*

Free-roaming chickens can spend approximately half their time foraging (Webster, 2002), which consists of scratching and pecking the ground. However, this behavioral time budget is greatly altered in broiler chicken strains in commercial environments. These broilers spent much of their time inactive, accounting for 61–76% of their time during light hours (Kristensen et al., 2007; Murphy and Preston, 2007; Evans et al., 2023). Foraging, or food-seeking behavior, is considered a highly motivated behavior that broiler chickens will show even when feed is freely available (Bracke and Hopster, 2006; Costa et al., 2012; Jacobs et al., 2021). Foraging behavior has a distinctive behavioral pattern, which includes scratching, stepping backwards, observing the scratched area, and pecking at potential food items. Expression of the behavior can be modified in the fast-growing broiler, where the scratching and stepping backwards sequence is omitted, and rather, broilers sit down while pecking at potential food items within reach. The behavior is also modified by age, with less foraging observed as broilers grow older and heavier (Jacobs et al., 2021; Holt et al., 2023).

Even though the behavioral expression of foraging may be modified in fast-growing broiler chickens compared to free-roaming chickens and conventional laying hens, the motivation to express this behavior is strong. Feed-restricted broiler breeders, with the same genetic background as broilers, would cross a water barrier to access wood shavings, even without feed present. This indicates that they were willing to work for the opportunity to forage even if hunger would not be fulfilled (Dixon et al., 2014). Foraging behavior can be stimulated when preferred resources are provided as a strategy of environmental enrichment (for instance, Bach et al. (2019)). Broilers prefer foraging in a sand substrate compared with other substrates (Arnould et al., 2004; Shields et al., 2005; Toghyani et al., 2010). When broilers had access to seven litter types, they preferred peat and mixed peat for foraging, followed by wood shavings, and fine straw, with the fewest broilers observed in bark, pellets, and course straw (Holt et al., 2023). Similarly, when broilers had access to five litter types, significantly higher levels of foraging were recorded in peat, oat hulls, and wood shavings compared to straw pellets and the litter control (Baxter et al., 2018). Broilers in highly complex environments that included access to a sand-filled dust bath spent more time foraging and had longer bouts of foraging behavior (in the sand) than broilers housed on wood shavings only (Evans et al., 2023).

High stocking density is associated with fewer broilers observed foraging, likely due to the limited space available (van der Eijk et al., 2022; Evans et al., 2023). Broilers at low densities (24, 30, and 36 $kg/m^2$) showed more foraging than broilers at high density (42 $kg/m^2$; van der Eijk et al., 2022). This relationship was linear and negative, with fewer broilers observed foraging with increasing stocking density (van der Eijk et al., 2022). Especially at older ages, stocking density matters (Evans et al., 2023). At 7 weeks of age, broilers were twice as likely to be observed foraging (either standing or sitting) in low-density pens (23.8 $kg/m^2$ at day 50) compared to high-density pens (42 $kg/m^2$ at day 50).

However, there was no difference in foraging frequencies at 2 weeks of age (density of 2.9 vs. 6.2kg/m$^2$; Evans et al., 2023).

The broilers' willingness to work for resources that facilitate foraging and the impacts of environmental modification on its expression indicate that when the right conditions are provided, foraging will be expressed frequently. Thus, broilers have a strong intrinsic motivation to access conditions that allow foraging and spend a substantial amount of time performing this behavior. This 'want' suggests that the act itself is rewarding and may contribute to a positively valenced emotional state during or after its performance.

### *3.3 Dustbathing*

Dustbathing is a complex behavior driven by both internal motivations and external factors (Duncan, 1998). This behavior is an excellent example of a behavioral need that can be satisfied only by performance (Vestergaard et al., 1999). Dustbathing behavior shows a diurnal rhythm, with peak occurrence in the afternoon (in jungle fowl (Hogan and van Boxel, 1993) and in laying hens (Vestergaard, 1982)), and 6–8 hours after the lights come on in Japanese quail (Statkiewicz and Schein, 1980).

Dustbathing is a prolonged behavior (state), and dustbathing bouts last 3–9 minutes in broilers, although bout lengths of 20 minutes were observed (Shields, 2004). Laying hens show substantially longer bouts, lasting 15–30 minutes (van Liere and Wiepkema, 1992), with a duration range between 0.07 and 44 minutes reported (Louton et al., 2016). The behavior is age dependent, observed in broilers by the first week of age, but with frequency (Vestergaard and Sanotra, 1999) and intensity decreasing with age, while bouts get longer (Shields, 2004). Dustbathing was recorded in 2% of the observations in fast-growing broilers between 1 and 6 weeks of age (Bokkers and Koene, 2003a). Although this seems low, it was higher than observations in other studies because of low stocking density and good litter quality. Yet, the prevalence can regularly be underestimated because the behavioral expression depends so strongly on time of day and is easily missed when using scan sampling techniques for behavioral observations.

Three days of deprived access to friable litter resulted in increased frequencies in dustbathing bouts in healthy and lame broilers, compared to broilers that were not deprived of litter access (Vestergaard and Sanotra, 1999a). The behavior is considered compensatory because frequency and durations are increased after deprivation compared to broilers with continuous access to the resource. This compensatory behavior in lame broilers is especially telling of the motivation to perform the behavior, as dustbathing includes vertical wing-shaking that involves forceful leg movements (Vestergaard, 1994), which may be painful for lame broilers (Vestergaard and Sanotra, 1999a).

Most broiler chickens in commercial farms will have access to bedding or litter. Broilers show a preference for fresh, friable, small particle bedding material for dustbathing. When Ross broilers had access to seven types of fresh litter in 1 $m^2$ trays (listed in order of preference), most broilers were observed dustbathing in mixed peat (nine broilers summed over the six scans), peat (eight broilers), and fine straw (seven broilers), compared to wood shavings (three broilers), bark (three broilers), pellets (one broiler), and course straw (two broilers; Holt et al., 2023). In used litter, the broilers seemed to prefer fine straw for dustbathing (five broilers) compared to all other litter types, and fewer broilers dustbathed in used litter compared to fresh litter (one versus four broilers; Holt et al., 2023). Similarly, Baxter et al. (2018) reported that more broilers were observed dustbathing in peat (28% of broilers recorded in the substrate) and oat hulls (19%), compared to straw pellets (2%), or wood shavings (0.5%). When provided the choice of four substrate types, broilers preferred sand for dustbathing (frequency of dustbathing was 3% <21 days of age and 9% >21 days of age) over straw (0%), wood shavings (0.5% or less), and rice hulls (0%; Villagrá et al., 2014). When given the choice between sand and wood shavings, broilers spend more time dustbathing in sand (no means provided per substrate; Shields et al., 2005).

The high motivation to dustbathe is a strong indicator that performance of the behavior induces positive emotions and improves affect. Nicol et al. (2017) argued in their review of farmed bird welfare that the performance of a full sequence of dustbathing returns the birds' motivation to baseline, thus the behavior is a good indicator of positive welfare. Thus, the ability to dustbathe is a need, and once the need is satisfied, it likely leads to a sense of comfort or relaxation. Investigating facial expressions in quail from divergent genetic lines selected for fearfulness, researchers observed increases in crown feather height, throat feather angle, and pupil surface during dustbathing compared to before (control) in the non-fearful line (Bertin et al., 2018). The researchers argued that these changes are indicative of the positive emotions induced by the behavior. When hens were trained to associate a sound cue with a feed reward or dustbathing substrate reward, the hens showed more anticipatory vocalizations after the dustbathing substrate cue was presented compared to the feed reward cues, suggesting that the anticipation of dustbathing substrate induced most arousal (McGrath et al., 2017). Willingness to work for access to a substrate was comparable in deprived hens and hens that had recently dustbathed, which may indicate the behavior leads to a state of pleasure rather than just fulfilling a need (Widowski and Duncan, 2000). Thus, this growing body of work directly assessing poultry emotion and affect supports the interpretation of the behavior as a positive experience.

### 3.4 Preening

Dustbathing and preening are often considered comfort behaviors (Nicol et al., 2009; Trocino et al., 2020; Papageorgiou et al., 2023). For preening, this is because it occurs more frequently when the chickens are free from distress and in a state of relaxation (Spruijt et al., 1992). This behavior involves self-grooming of the feathers with the beak (Bokkers and Koene, 2003a).

Preening was recorded for 6% of the observations between 1 and 6 weeks of age in broilers and became more frequent (9% of observations) for broilers at 7–12 weeks of age (Bokkers and Koene, 2003a). In broilers with differing growth rates, preening was observed once or twice during 5-minute observations across different ages (Dawson et al., 2021).

The anticipation or expectation of a positive stimulus induces a positive affective state (Boissy et al., 2007). Comfort behavior, including preening, was associated with the anticipation of a positive stimulus in laying hens (Zimmerman et al., 2011). Preening, or self-grooming, was associated with relaxation or de-arousal after a high-intensity state (Spruijt et al., 1992).

However, grooming is also observed as displacement behavior, serving as a motivational outlet when two behavioral systems are activated simultaneously (Spruijt et al., 1992). More preening was observed when feeding was thwarted in laying hens (Zimmerman et al., 2000). In thwarting situations, these displacement preening movements were of shorter duration than those in normal situations, thus the behavioral expression differed (Duncan and Wood-Gush, 1972). Preening frequency can increase after heavy-hybrid chickens are handled or injected with adrenocorticotropic hormone (ACTH) to mimic acute distress (Williams and Scampoli, 1984). Increased preening in broilers as they age is theorized to be indicative of frustration originating from a reduced ability to perform active behaviors (Bokkers and Koene, 2003b; Evans et al., 2023).

So, when preening is performed in a state of relaxation, it can be an indicator of positively valenced emotional state. However, care must be taken to interpret the behavior in ambiguous conditions, as it has been observed as a displacement behavior, signaling frustration.

### 3.5 Perching

Chickens, including broiler chickens, are strongly motivated to rest on elevated structures, especially at night (Bailie et al., 2018). This serves as protection from predators (Gottier, 1968). Even without the presence of potential predators in indoor housing, chickens are highly motivated to perch (roost). Laying hens demonstrate continuous use of perches (Duncan et al., 1992), a strong motivation to access perches during dark hours (Duncan et al., 1992; Olsson and Keeling, 2002), and frustration when access is thwarted (Olsson and Keeling, 2000). If not provided with perches, broiler breeders will perch on other furniture, such as on grills above the feeders and on drinking lines (Gebhardt-Henrich et al., 2018).

Perching could play a role in social structures in natural conditions (Gottier, 1968), and the expression depends in part on social status in laying hens (Cordiner and Savory, 2001). Similarly, in commercial broilers, perching seems to be socially facilitated, thus behavioral synchronization can be observed (Bailie and O'Connell, 2015).

Broiler chickens' body confirmation, especially their heavy breast, makes the use of traditional perches difficult as they grow (Bailie and O'Connell, 2015). Broilers used platforms more than perches, and low perches (10 cm) more than high perches (30 cm) at 32 days of age (Norring et al., 2016). Broilers do not readily use traditional perches, but do use platforms, demonstrating that the motivation is still present.

Age impacts perching behavior, with broilers generally perching less frequently as they age (Fig. 1). Fewer broilers were using elevated platforms

**Figure 1** Broiler chickens will perch on elevated structures from an early age. Photo courtesy of Leonie Jacobs.

on day 32 compared to earlier observation days (Norring et al., 2016). Broiler breeders of fast- and slower-growing strains both used traditional perches and wooden bars and perches arranged as a platform, with high usage at night, and overall perching increased with age (Gebhardt-Henrich et al., 2018). Fast-growing broilers consistently used platforms and ramps across ages in an experiment, with 14% of broilers perching on day 8, 18% on day 22, and 15% of broilers on day 29, although the effect of age was not specifically assessed within the strain (Güz et al., 2021). However, most other studies reported a decrease in this behavior as broilers age. A Japanese strain of slower-growing broilers reduced their perching frequency with age, with half as many broilers perching in week 5 and a third as many broilers perching in week 8 compared to week 3 (Ohara et al., 2015). Other studies observed a peak perching frequency (and duration; Bailie and O'Connell (2015)) at 4–5 weeks of age, followed by a drop off throughout the rest of the rearing period (Pettit-Riley and Estevez, 2001; Ventura et al., 2012; Bailie and O'Connell, 2015).

Stocking density also seems to impact perch use. When broilers were housed at a density of 18 broilers/m$^2$ or 13 broilers/m$^2$, 17% or fewer observations of perching were recorded at 4 and 5 weeks of age, compared to approximately 25% of observations at a density of 8 broilers/m$^2$ (Ventura et al., 2012). In contrast, perching frequency was higher at 15 broilers/m$^2$ and 20 broilers/m$^2$ compared to 10 broilers/m$^2$ (Pettit-Riley and Estevez, 2001). In Arbor Acres broilers, perch use depended on perch type (PVC or wooden) and stocking density (Hongchao et al., 2013). For instance, broilers at 38 kg/m$^2$ perched more frequently on the PVC perch than the broilers housed at 29 kg/m$^2$ and 48 kg/m$^2$ (Hongchao et al., 2013).

This general decline in perch use with age, and in some cases with higher stocking density, is theorized to be due to space limitations at later ages (Norring et al., 2016) rather than due to a reduced motivation to perch. To confirm this, it would be valuable to assess whether perching behavior remains stable across ages and densities when ample platform space is provided, so that there is no competition or limited access (although physical limitations associated with genetics could remain).

Perching behavior is argued as an indicator of positive welfare as it promotes a feeling of safety (Nicol et al., 2017; Papageorgiou et al., 2023), which, among others, is reflected in the expression of comfort behaviors when perching (Bach et al., 2019). Due to physical limitations in broilers, the use of traditional perches sharply decreases with age, which is due to their decreased ability to express the behavior rather than due to lower motivation (Norring et al., 2016). Thus, modern broiler strains should have access to elevated platforms with ample space to allow for behavioral synchronization.

### 3.6 Play and exploration

Broiler chickens will play. Play can be defined as 'spontaneous, self-handicapping, non-injurious, non-stereotyped behavior that is energetic but relaxed, occurring in contexts that do not have an immediate survival or reproductive function' (Špinka et al., 2001). Broiler chicken play behavior can be separated into three types. First, object play, where chickens perform worm running (or food running). Second, locomotory play, where chickens will frolic: running with or without wing assistance, wing flap, and jump. Finally, social play, where chickens will spar, or play fight, with their conspecifics (Jacobs et al., 2023). Broilers will spend approximately 1% of their time budget playing at 4 weeks of age (Baxter et al., 2019).

Play is shown in young chickens (Baxter et al., 2019; Rayner et al., 2020) and is theorized to prepare them to deal with 'the unexpected' later in life (Špinka et al., 2001), thus increasing fitness. Fitness can be reflected in growth and survival rates, and indeed, Norwegian commercial flocks that played more had higher body weights and lower mortality rates, although play was negatively impacted by stocking density, but not associated with feed conversion ratio, incidence of lameness, photoperiod schedule, windows, or feed were detected (Newberry et al., 2018).

Play can be more common when broilers are (temporarily or suddenly) provided additional space (Baxter et al., 2019; Vas et al., 2023). This space can be temporarily created when people walk through the house. Disturbing and displacing broilers was effective in stimulating play as open space became available behind the person (Baxter et al., 2019; Liu et al., 2020). The proportion of broilers showing locomotory play increased with greater space availability in general (between 0.06 and 0.075m$^2$ per broiler; Vas et al., 2023) (Fig. 2).

The relationship between play and environmental complexity is inconsistent between strains intended for meat or eggs and when comparing experimental settings. In commercial broiler chicken flocks, social play increased with increasing complexity (for instance, by adding plastic crates, peat bedding, and cardboard boxes), with type of enrichment not included in the analysis (from 0 to 4 types of enrichments; Vas et al., 2023). Pullets housed with multiple resources were more often observed playing than pullets housed with a single resource in week 1–4 of age, with differences disappearing at later ages (Holt et al., 2024). Play did not differ in commercial broilers when housed with no, 1 (platform perches), or 2 enrichments (platform perches and dust baths with peat) between 3 and 5 weeks of age (Baxter et al., 2019). Play was even reduced; broilers were housed with a raised platform, hanging weighing scale, peck stone, and a suet feeder filled with wood shavings (Liu et al., 2020). Type of resources or enrichments, accessibility to these resources or enrichments

**Figure 2** Complex environments with ample space, such as in this experimental pen, can stimulate play and other behaviors. Photo courtesy of Leonie Jacobs.

(number of broilers/resource), duration of exposure (novelty may wear off) may all be factors that impact these inconsistent outcomes.

Play behavior activates dopamine (desire), opioid (reward, mood), and endocannabinoid (pleasure) mechanisms in laboratory rat brains (Siviy and Panksepp, 2011). Although avian brains have differences that have not been studied in the context of play, birds do have dopamine and opioid receptors in equivalent brain regions that were activated by play in rats (Emery and Clayton, 2015). Applying the precautionary principle, it is likely that play in chickens (broilers) may have a similar emotional impact.

In the past, play was referred to as a luxury behavior, because it will only occur when animals have basic survival needs met and are not suffering or in pain (Duncan, 1998). At best, the behavior was considered irrelevant to commercial production, and at worst, something to be discouraged to maintain efficient production (Jacobs et al., 2023). However, the positive emotional impacts contribute to the broilers' quality of life and may help broilers deal with negative stressors (Kaplan, 2020), therefore, play is now understood to be relevant to commercial conditions. Still, a large research gap exists on play behavior in poultry, and it is unclear how it can be used as an indicator for broiler chicken welfare. What is clear, based on research in other species, is that the absence of play indicates an animal welfare deficit (Špinka et al., 2001; Jacobs et al., 2023).

### 3.7 Social behaviors

Broiler chickens are raised in commercial houses with tens of thousands of broilers in a single space. In more natural settings, chickens would live in small groups with a dominant male and fewer than 20 females (Webster, 2002, or see the review by Garnham and Løvlie, 2018). Large groups in the wild may comprise smaller subgroups and synchronize their activities within those groups (Webster, 2002). Synchronization can improve fitness by increasing the probability of finding valuable resources and reducing predation risk (Keeling, Newberry and Estevez, 2017). Synchronization of activities is also observed in commercial facilities, as chickens preen (including allogrooming), dustbathe, rest, and feed at the same time, as observed in laying hens (Keeling et al., 2017). Chickens will develop social hierarchies at 5–6 weeks of age (Guhl, 1968), which is around the same age that fast-growing broiler chickens are sent to slaughter. However, that does not mean there is a lack of sociality among broilers.

Chickens can recognize individuals and prefer the proximity of familiar conspecifics over unfamiliar ones, as tested in genetic strains local to the region, including Padovana, Polverara, and Robusta chickens (Versace et al., 2019). Broilers preferred being in close proximity with their conspecifics even at high stocking densities (Febrer et al., 2006). Chickens show a strong motivation for social reinstatement. In the first week of life, both male and female chickens from an egg-laying strain ran through a runway towards familiar cage mates, although females approached faster than males (Vallortigara and Zanforlin, 1993).

The large numbers of chickens in a broiler house, high stocking density, and limited essential resources (uneven or insufficient resource allocation) may result in a reduction of behavioral synchronization triggered by social facilitation (Keeling et al., 2017). It could lead to an increase in negative social interactions such as competition for resources like platform space or dustbathing material where some broilers may be thwarted access by other broilers (Nielsen et al., 2023). Competition among broilers is not well-studied. One investigation did not observe agonistic interactions associated with competition over resources among broilers (Estevez et al., 1997).

Social distress may negatively impact broiler chicken welfare through the experience of negative affect (pain, fear, frustration; Nielsen et al., 2023). However, there is little research on social stresses in broilers.

One well-studied social stress that broilers experience is disturbed resting due to high stocking density. High density resulted in more frequent disturbances during resting. Broilers performed more resting bouts in higher compared to lower-density conditions, but did not rest longer, suggesting disturbances (Martrenchar et al., 1999; Hall, 2001; Cornetto et al., 2002; Evans et al., 2023). When more barriers were provided, resting improved as these barriers prevented broilers walking over each other to access resources (Ventura et al., 2012). Providing a more complex space with elevated platforms would practically reduce stocking density and provide more opportunities for undisturbed resting, as suggested by Bach et al. (2019). This may reduce some of these disturbances.

### *3.8 Other normal behavior that is not performed*

Some behaviors may be normal for chickens but cannot be performed. Reproductive behaviors are not performed in conventional broiler chickens because standard commercial practices include the harvesting of broiler chickens between 5 and 9 weeks of age (*National Chicken Council*, n.d.; *Raising Broilers | Extension*, n.d.; Junghans et al., 2022). Sexual maturity does not develop until approximately 11 weeks of age in male broilers, and 18–20 weeks of age in female broilers (EFSA, 2023). Therefore, conventional fast-growing broiler chickens do not perform reproductive behaviors. There is an exception. In slow-growing strains, especially males can start exhibiting behaviors associated with reproductive maturity, such as aggressive pecking, kicking, and fighting (Baxter et al., 2019; Nielsen et al., 2023). So, these male broilers can and do perform behaviors associated with reproduction. It is an ethical discussion rather than a scientific one when considering the young age at which standard broiler chickens are processed. Is longevity, including reaching reproductive maturity, an important consideration for broiler chicken welfare?

Another behavior that is not seen in broiler chickens is flying. The ancestors of commercial fast-growing strains can fly quite well, like light-bodied laying hen strains. Yet modern broiler chicken strains are too heavy to fly. Commercial broilers do use their wings when jumping to add thrust, help with balance, or soften a landing (Webster, 2002). If the motivation to fly exists and if it would be associated with positive emotions are unclear. Enjoyment has been suggested in other avian species such as ravens and swans, but empirical evidence is currently lacking (Emery and Clayton, 2015).

## 4 Abnormal behaviors

Broiler chickens are not commonly reported showing abnormal (active) behaviors, although these can be observed in other poultry types. In part that could be due to their short lifespan in commercial production. Some well-documented abnormal behaviors for other poultry include stereotypic pecking, excessive drinking, and pacing in broiler breeders (e.g. Savory et al. (1992) and see Riber et al. (2017) for a review). These behaviors were linked to the restricted feeding and subsequent hunger that occurs in broiler breeder flocks, to ensure body weights conducive to reproductive success (de Jong et al., 2003; Renema et al., 2007). These behaviors do not develop in their offspring. Broilers are in most cases fed ad libitum, so they do not develop these hunger-induced coping mechanisms.

Injurious (conspecific) feather pecking is a common issue in other poultry, especially laying hens (Bestman et al., 2017), and turkeys (Dalton et al., 2018; Erasmus, 2018; Leishman et al., 2022), but also in parent flocks (layer, broiler, turkey) and ducks (Widowski and Rentsch, 2022). Broiler chickens generally do not develop injurious feather pecking behavior, although there may be isolated cases where it does occur. The absence of this abnormal behavior is likely due to a combination of factors. The genetic selection for fast growth rate results in a relatively short growth period, rendering the broilers pre-pubescent when they are sent to processing. A short lifespan also means little time to develop abnormal behaviors. The ethological perspective on injurious pecking in laying hens is contested, but it suggests that a lack of foraging opportunities is a causal factor (van Staaveren et al., 2021). For broilers, the strong motivation to feed is met by their ad libitum access to feed, leaving at least the motivation to feed (rather than feed seeking) fulfilled.

One behavioral trait that is arguably abnormal in broiler chickens is their inactivity (Fig. 3). Broilers will perform behaviors while sitting, such as preening, stretching, and ground pecking (reviewed by Tallentire et al., 2016). Pain

**Figure 3** Broilers spend a large portion of the day inactive. In addition, they have modified behavioral expression of active behaviors, performing them while sitting. Photo courtesy of Leonie Jacobs.

associated with health concerns (bacterial chondronecrosis with osteomyelitis (BCO) or contact dermatitis for instance) may reduce activity and result in reduced expression of behaviors associated with positive affective states. For instance, broilers with impaired gait (gait score 2 on a 0–5 scale) tended to be more inactive, spent more time sitting while feeding, and spent less time performing comfort behavior compared to broilers with healthy gait (Riber et al., 2021).

## 5 The impact of genetics on behavior

Broiler chicken behavior is strongly impacted by genetics. Highly successful genetic selection for fast, efficient growth and breast muscle yield resulted in a 400% growth increase and a 50% reduction in feed conversion ratio from 1957 to 2005 (Zuidhof et al., 2014). This genetic selection has had unintended consequences on broiler chicken welfare, including their behavior, as reviewed by Hartcher and Lum (2020).

Behavioral expression differs depending on the genetic strain that is observed, either because of intensive genetic selection for growth has inadvertently changed behavioral motivations of the chickens, or because it has inadvertently changed the broilers' abilities to express behavioral motivations.

### *5.1 Clinically healthy broilers*

Genetics has direct impacts on broiler behavior. The greatest example of this is seen when comparing a genetic strain selected for egg production with a strain selected for meat production. When directly compared in an experiment, feed intake, rate of consumption, and proportion of time spent drinking were greater in broiler breeder males, while the time spent preening, pecking, and stereotypic pacing were greater in the layer males (Hocking et al., 1997). Both are the same species, and both were housed in cages, yet they greatly differ in behavioral time budget. Behavioral variation is also observed between broiler chicken strains. Fast-growing broilers were more inactive, sat more, and fed more, while spending less time standing, walking, or using enrichments compared to moderate and slow-growing broilers (Dawson et al., 2021). Similarly, fewer fast-growing broilers were perching on platforms, ramps, and barriers, and more fast-growing broilers were found underneath the platforms and ramps than a slow-growing strain, depending on age (Güz et al., 2021). Feeding behavior also differed between genetic strains with fast-growing broilers eating fewer but longer meals, showing almost double the meal size, higher feeding rates (g/min), and more time per day feeding than slower-growing broilers (Howie et al., 2011).

## 5.2 Genetics impact behavior through health modification

Besides the direct impacts, genetics has indirect impacts on broiler behavior through health modification. In a recent review by Riber and Wurtz (2024), the authors presented evidence of the impacts of lameness, high body weights, and altered body conformation of fast-growing broilers on their behavior. These genetically induced changes result in a reduced ability to walk, navigate the environment, access resources, and express natural behavior (Riber and Wurtz, 2024).

Modern broiler strains are prone to musculoskeletal disorders that induce lameness. For example, they are genetically predisposed to BCO (Wideman et al., 2014). When assessing the causes of lameness in fast-growing genotypes raised on litter and wire flooring, all or nearly all incidences were attributed to BCO (Wideman et al., 2014). There were large differences in walking ability among four modern strains, with strong positive correlations between gait score and body weight for all strains (correlation coefficients between 0.43 and 0.53 on day 35 of age), and between gait score and tibial dyschondroplasia scores (a skeletal abnormality characterized by an avascular lesion in the tibial growth plate; Leach and Monsonego-Ornan, 2007) in three out of four strains (correlation coefficients between 0.10 and 0.19; Kestin et al., 1999).

### 5.2.1 Lameness impacts behavior

Lameness (poor walking ability) inhibits the expression of some behaviors, without a decreased motivation to perform them. The lameness-induced behavioral modification is likely attributed to pain (McGeown et al., 1999; Danbury et al., 2000; Weary et al., 2006; Nääs et al., 2009).

Broilers with lameness due to tibial dyschondroplasia dustbathed on fewer days compared to non-lame broilers (Vestergaard and Sanotra, 1999b). Lame broilers reduced their time walking (1.5% in severely lame broilers compared to 3.3%) when at slaughter weight, spent less time standing idle, and spent more time lying with one leg extended. Their feeding behavior was also altered, with fewer feeder visits (30 times per 24 h in the lamest broilers vs. 50 in unaffected broilers; Weeks et al., 2000) and nearly half of feeding bouts were spent sitting down (instead of standing) compared to approximately 1/5th in non-lame broilers (Weeks et al., 2000).

### 5.2.2 Body weight and body confirmation impact behavior

Behavioral differences between genetic strains are often associated with the broilers' genetic potential for growth or body weight. Compared to slower-growing strains at the same age, fast-growing strains were less active, spent more time sitting and feeding, and spent less time standing, walking, and using enrichments (Dawson et al., 2021). Although inactivity for both fast- and slow-growing strains peaked at 78–80% of the day, in fast-growing broilers this peak was reached at an earlier age (Dawson et al., 2021).

Two studies used a controlled approach to compare the impact of weight on broiler chicken walking and activity. In one study, researchers used a harness and a suspension device to reduce broilers' body weight (Djukic-Stojcic and Bessei, 2011). Compared to the control broilers without weight adjustment, the weight-reduced broilers showed higher activity in week 6 of age. Furthermore, the weight-reduced fast-growing broilers showed similar activity to slow-growing broilers without weight adjustment (Djukic-Stojcic and Bessei, 2011). In the other study, weight was reduced by restricting the feed intake (Corr et al., 2003). The feed-restricted broilers walked faster, with higher cadence and took bigger steps. Their walking speed was similar to slow-growing broilers. However, the faster gait could in part be motivated by hunger due to feed restriction. The ad-libitum fed broilers (control group) took wider steps, pointing their toes outwards, and were less active than the slower-growing strain. The authors concluded that gait of the control group was a consequence of the changed body confirmation of the fast-growing broiler, with their center of gravity more forward compared to a broiler with relaxed genetic selection (Corr et al., 2003). They continue that the altered gait would require more energy, explaining the reduced activity in fast-growing broilers (Corr et al., 2003).

## 6 Conclusion

The aim of this chapter was to get a better understanding of broiler chickens' behavioral needs and motivations, focusing on behavior that can contribute to good animal welfare. Broiler chickens do exhibit a range of behaviors that can indicate or are associated with positive affective state, a key aspect of good welfare. These are comfort behaviors (dustbathing and preening), perching (or resting on elevated spaces), play, and exploration. Feeding and foraging can also be associated with positive emotions, based on their motivation to express these behaviors. Some questions do remain about how much genetic selection has impacted behavioral motivations in broiler chickens. Their motivation to eat has changed with genetic selection for growth, and it is clear motivations for most, if not all, normal behaviors remained to an extent, but there is more to understand about changes in motivation for some normal behaviors. Furthermore, it is unclear whether the modified behavioral repertoire provides positive emotional experiences. For instance, broilers spend over half their time inactive. Are they experiencing any enjoyment when being inactive? When gait is impaired, does it feel good to lie down? Dr Bokkers and Dr Koene (Bokkers and Koene, 2003a; Bokkers, 2004) have provided some valuable insights towards these questions. Their findings suggest that all motivations for active behavior remain to some extent, as fast-growing broilers showed all normal behaviors while young and able to perform them (Bokkers and Koene, 2003a). Although physical abilities have changed compared to their ancestors, cognitive abilities of the modern broiler seem very similar to those of the jungle fowl (Garnham and Løvlie, 2018). For instance, (non-broiler) chickens showed self-control in anticipation of receiving a greater reward (Abeyesinghe et al., 2005) and social buffering was observed among mother hens and chicks (Edgar et al., 2015). The lack of research on play warrants more work, as it could be an important aspect of a good life with pleasurable experiences.

Understanding and measuring behavior plays a key part in the investigation of affective states. More research is needed to further unravel the emotions and affect associated with behaviors that broiler chickens show under commercial conditions. With animal welfare research historically focused on the avoidance of harm, a change in direction towards the investigation of a good life will require a better understanding of their experience. Indeed, with more evidence (Marino, 2017) that chickens' cognitive, emotional, and social complexities align with those of other species, including mammals, more behavioral research is needed to understand what chickens want and need to experience a good life.

Recent developments in the quantification of affect and emotion in broiler chickens have contributed to a better understanding of how these broilers feel. The validation (Lourenço da Silva et al., 2024) and application of cognitive bias testing (Košťál et al., 2020; Lourenço-Silva et al., 2023) have served as a valuable tool to gain better understanding of affect. Although some of these studies can have lacking or mixed results, these differences may be due to study design, affect modification approaches, or analysis of the data (Košťál et al., 2020). Further refinement and application of these tests and the application of other methods to determine mental states are needed. Some other promising indicators that can be used to determine the value of behavioral expression may be related to the investigation of neurobiology that is associated with reward or positive feelings. As reviewed for mammals (Coria-Avila et al., 2022), it would be worthwhile to study the neurobiology of chicken behavior more, so we can determine their respective affective states and better understand their behavioral needs.

## 7 Where to look for further information

This review provides a good overview of the impacts of genetics on broiler chicken welfare: Hartcher and Lum (2020).

This book provides comprehensive information on animal behavior across species: Breed and Moore (2022).

This chapter provides solid information on behavior and behavioral needs in poultry: Duncan (1998).

## 8 References


Abeyesinghe, S.M. et al. (2005) 'Can domestic fowl, Gallus gallus domesticus, show self-control?', *Animal Behaviour*, 70(1), pp. 1–11. https://doi.org/10.1016/J.ANBEHAV.2004.10.011.

Arnould, C. et al. (2004) 'Effects of the addition of sand and string to pens on use of space, activity, tarsal angulations and bone composition in broiler chickens', *Animal Welfare*, 13(1), pp. 87–94. https://www.ingentaconnect.com/contentone/ufaw/aw/2004/00000013/00000001/art00013 (Accessed: 6 November 2024).

Bach, M.H. et al. (2019) 'Effects of environmental complexity on behaviour in fast-growing broiler chickens', *Applied Animal Behaviour Science*, 219, p. 104840. https://doi.org/10.1016/J.APPLANIM.2019.104840.

Bailie, C.L. and O'Connell, N.E. (2015) 'The influence of providing perches and string on activity levels, fearfulness and leg health in commercial broiler chickens', *Animal*, 9(4), pp. 660–668. https://doi.org/10.1017/S1751731114002821.

Bailie, C.L. et al. (2018) 'Exploring perch provision options for commercial broiler chickens', *Applied Animal Behaviour Science*, 200, pp. 114–122. https://doi.org/10.1016/j.applanim.2017.12.007.

Baxter, M. et al. (2018) 'An evaluation of potential dustbathing substrates for commercial broiler chickens', *Animal*, 12(9), pp. 1933–1941. https://doi.org/10.1017/S1751731117003408.

Baxter, M. et al. (2019) 'Play behaviour, fear responses and activity levels in commercial broiler chickens provided with preferred environmental enrichments', *Animal*, 13(1), pp. 171–179.

Bertin, A. et al. (2018) 'Are there facial indicators of positive emotions in birds? A first exploration in Japanese quail', *Behavioural processes*, 157, pp. 470–473. https://doi.org/10.1016/J.BEPROC.2018.06.015.

Bestman, M. et al. (2017) 'Feather-pecking and injurious pecking in organic laying hens in 107 flocks from eight European countries', *Animal Welfare*, 26(3), pp. 355–363. https://doi.org/10.7120/09627286.26.3.355.

Bizeray, D. et al. (2000) 'Early locomotor behaviour in genetic stocks of chickens with different growth rates', *Applied Animal Behaviour Science*, 68(3), pp. 231–242. https://doi.org/10.1016/S0168-1591(00)00105-2.

Boissy, A. et al. (2007) 'Assessment of positive emotions in animals to improve their welfare', *Physiology and Behavior*, 92(3), pp. 375–397. https://doi.org/10.1016/j.physbeh.2007.02.003.

Bokkers, E.A.M. (2004) *Behavioural motivations and abilities in broilers*. Wageningen University.

Bokkers, E.A.M. and Koene, P. (2003a) 'Behaviour of fast- and slow growing broilers to 12 weeks of age and the physical consequences', *Applied Animal Behaviour Science*, 81(1), pp. 59–72. https://doi.org/10.1016/S0168-1591(02)00251-4.

Bokkers, E.A.M. and Koene, P. (2003b) 'Eating behaviour, and preprandial and postprandial correlations in male broiler and layer chickens', *British Poultry Science*, 44(4), pp. 538–544. https://doi.org/10.1080/00071660310001616165.

Bokkers, E.A.M. and Koene, P. (2004) 'Motivation and ability to walk for a food reward in fast- and slow-growing broilers to 12 weeks of age', *Behavioural Processes*, 67(2), pp. 121–130. https://doi.org/10.1016/j.beproc.2004.03.015.

Bracke, M.B.M. and Hopster, H. (2006) 'Assessing the importance of natural behavior for animal welfare', *Journal of Agricultural and Environmental Ethics*, 19, pp. 77–89. https://doi.org/10.1007/s10806-005-4493-7.

Breed, M.D. and Moore, J. (2022) *Animal behavior*. 3rd ed. Edited by M.D. Breed and J. Moore. https://books.google.com/books?hl=en&lr=&id=op0tEAAAQBAJ&oi=fnd&pg=PP1&dq=animal+behavior&ots=-WleWIKPyW&sig=bu_-5AzIsYHI8mM5KI7DUVYXNUs#v=onepage&q=animal%20behavior&f=false (Accessed: 20 March 2025).

Caplen, G. et al. (2012) 'Kinematic analysis quantifies gait abnormalities associated with lameness in broiler chickens and identifies evolutionary gait differences', *PLoS One*, 7(7), p. e40800. https://doi.org/10.1371/JOURNAL.PONE.0040800.

Collins, S.A. et al. (2024) 'Do I sound anxious? Emotional arousal is linked to changes in vocalisations in domestic chicks (Gallus gallus dom.)', *Applied Animal Behaviour Science*, 277, p. 106359. https://doi.org/10.1016/J.APPLANIM.2024.106359.

Cordiner, L.S. and Savory, C.J. (2001) 'Use of perches and nestboxes by laying hens in relation to social status, based on examination of consistency of ranking orders and frequency of interaction', *Applied Animal Behaviour Science*, 71(4), pp. 305–317. https://doi.org/10.1016/S0168-1591(00)00186-6.

Coria-Avila, G.A. et al. (2022) 'The neurobiology of behavior and its applicability for animal welfare: a review', *Animals*, 12(7), p. 928. https://doi.org/10.3390/ANI12070928.

Cornetto, T. et al. (2002) 'Using artificial cover to reduce aggression and disturbances in domestic fowl', *Applied Animal Behaviour Science*, 75(4), pp. 325–336. https://doi.org/10.1016/S0168-1591(01)00195-2.

Corr, S. et al. (2003) 'The effect of morphology on walking ability in the modern broiler: a gait analysis study', *Animal Welfare*, 12(2), pp. 159–171. https://doi.org/10.1017/S0962728600025616.

Costa, L.S. et al. (2012) 'Some aspects of chicken behavior and welfare', *Revista Brasileira de Ciencia Avicola*, 14(3), pp. 159–164. https://doi.org/10.1590/S1516-635X2012000300001.

Dalton, H.A. et al. (2018) 'Comparing the behavioural organization of head pecking, severe feather pecking, and gentle feather pecking in domestic turkeys', *Applied Animal Behaviour Science*, 204, pp. 66–71. https://doi.org/10.1016/J.APPLANIM.2018.05.005.

Danbury, T.C. et al. (2000) 'Self-selection of the analgesic drug carprofen by lame broiler chickens', *Veterinary Record*, 146(11), pp. 307–311. https://doi.org/10.1136/VR.146.11.307.

Dawson, L.C. et al. (2021) 'In pursuit of a better broiler: a comparison of the inactivity, behavior, and enrichment use of fast- and slower growing broiler chickens', *Poultry Science*, 100(12), p. 101451. https://doi.org/10.1016/J.PSJ.2021.101451.

de Jong, I.C. et al. (2003) 'Parameters for quantification of hunger in broiler breeders', *Physiology & behavior*, 78(4), pp. 773–783.

Dixon, L.M. et al. (2014) 'Measuring motivation for appetitive behaviour: food-restricted broiler breeder chickens cross a water barrier to forage in an area of wood shavings without food', *PLOS ONE*, 9(7), p. e102322. https://doi.org/10.1371/JOURNAL.PONE.0102322.

Djukic-Stojcic, M. and Bessei, W. (2011) 'The effect of weight load on the legs of broilers behaviour', *Biotechnology in Animal Husbandry*, 27(4), pp. 1667–1671. https://doi.org/10.2298/BAH1104667D.

Du, X. et al. (2020) 'Assessment of laying hens' thermal comfort using sound technology', *Sensors*, 20(2), p. 473. https://doi.org/10.3390/S20020473.

Duncan, E.T. et al. (1992) 'Effect of perches in laying cages on welfare and production of hens', *British Poultry Science*, 33(1), pp. 25–35. https://doi.org/10.1080/00071669208417441.

Duncan, I.J.H. (1998) 'Behavior and behavioral needs', *Poultry Science*, 77(12), pp. 1766–1772. https://doi.org/10.1093/PS/77.12.1766.

Duncan, I.J.H. and Wood-Gush, D.G.M. (1972) 'An analysis of displacement preening in the domestic fowl', *Animal Behaviour*, 20(1), pp. 68–71. https://doi.org/10.1016/S0003-3472(72)80174-X.

Edgar, J. et al. (2015) 'Social buffering in a bird', *Animal Behaviour*, 105, pp. 11–19. https://doi.org/10.1016/J.ANBEHAV.2015.04.007.

EFSA AHAW Panel (EFSA Panel on Animal Health and Welfare). (2023). 'Scientific Opinion: Welfare of broilers on farm', https://doi.org/10.2903/j.efsa.2023.7788.

Emery, N.J. and Clayton, N.S. (2015) 'Do birds have the capacity for fun?', *Current Biology*, 25(1), pp. R16–R20. https://doi.org/10.1016/J.CUB.2014.09.020.

Erasmus, M.A. (2018) 'Welfare issues in turkey production', *Advances in Poultry Welfare*, pp. 263–291. https://doi.org/10.1016/B978-0-08-100915-4.00013-0.

Estevez, I. et al. (1997) 'Broiler chickens: a tolerant social system?', *Etologia*, 5, pp. 19–29.

Evans, L. et al. (2023) 'Environmental complexity and reduced stocking density promote positive behavioral outcomes in broiler chickens', *Animals*, 13(13), p. 2074. https://doi.org/10.3390/ani13132074.

Febrer, K. et al. (2006) 'Forced to crowd or choosing to cluster? Spatial distribution indicates social attraction in broiler chickens', *Animal Behaviour*, 72(6), pp. 1291–1300. https://doi.org/10.1016/J.ANBEHAV.2006.03.019.

Foroushan, G. (2017) 'Interactions between histamine H1 and H3 and dopamine D1 receptors on feeding behavior in chicken', *Iranian Journal of Veterinary Medicine*, 11(1), pp. 63–73.

Fraser, D. (2023) *Understanding animal welfare: the science in its cultural context*. 2nd ed. Wiley Blackwell. https://www.wiley.com/en-us/Understanding+Animal+Welfare%3A+The+Science+in+its+Cultural+Context%2C+2nd+Edition-p-9781119626404 (Accessed: 22 May 2025).

Garnham, L. and Løvlie, H. (2018) 'Sophisticated fowl: the complex behaviour and cognitive skills of chickens and red junglefowl', *Behavioral Sciences*, 8(1), p. 13. https://doi.org/10.3390/BS8010013.

Gebhardt-Henrich, S.G. et al. (2018) 'Use of aerial perches and perches on aviary tiers by broiler breeders', *Applied Animal Behaviour Science*, 203, pp. 24–33. https://doi.org/10.1016/J.APPLANIM.2018.02.013.

Gottier, R.F. (1968) 'The dominance-submission hierarchy in the social behavior of the domestic chicken', *Journal of Genetic Psychology*, 112(2), pp. 205–226. https://doi.org/10.1080/00221325.1968.10533796/ASSET//CMS/ASSET/3AFC164E-5B74-471E-B24D-E4D7E2DE8F22/00221325.1968.10533796.FP.PNG.

Guhl, A.M. (1968) 'Social behavior of the domestic fowl', *Transactions of the Kansas Academy of Science. Kansas Academy of Science*, 71(3), pp. 379–384. https://doi.org/10.2307/3627156.

Güz, B.C. et al. (2021) Effects of pen enrichment on leg health of fast and slower-growing broiler chickens. *PLoS One*, *16*(12), p. e0254462. https://doi.org/10.1371/JOURNAL.PONE.0254462.

Hall, A.L. (2001) 'The effect of stocking density on the welfare and behaviour of broiler chickens reared commercially', *Animal Welfare*, 10(1), pp. 23–40. https://doi.org/10.1017/S096272860002323X.

Hartcher, K.M. and Lum, H.K. (2020) 'Genetic selection of broilers and welfare consequences: a review', *World's Poultry Science Journal*, 76(1), pp. 154–167. https://doi.org/10.1080/00439339.2019.1680025.

Herborn, K.A. et al. (2020) 'Spectral entropy of early-life distress calls as an iceberg indicator of chicken welfare', *Journal of the Royal Society Interface*, 17(167), p. 20200086. https://doi.org/10.1098/RSIF.2020.0086.

Hocking, P.M. et al. (1997) 'Comparison of food intake, rate of consumption, pecking activity and behaviour in layer and broiler breeder males', *British Poultry Science*, 38(3), pp. 237–240. https://doi.org/10.1080/00071669708417978.

Hogan, J.A. and van Boxel, F. (1993) 'Causal factors controlling dustbathing in Burmese red junglefowl: some results and a model', *Animal Behaviour*, 46(4), pp. 627–635. https://doi.org/10.1006/ANBE.1993.1239.

Holt, R.V. et al. (2023) 'A buffet of litters – Broiler chickens behave differently according to litter type and freshness', *Applied Animal Behaviour Science*, 260, p. 105856. https://doi.org/10.1016/J.APPLANIM.2023.105856.

Holt, R.V. et al. (2024) 'Resource choice during ontogeny enhances both the short- and longer-term welfare of laying hen pullets', *Scientific Reports*, 14(1), pp. 1–14. https://doi.org/10.1038/s41598-024-53039-7.

Hongchao, J. et al. (2013) 'Effect of perch type and stocking density on the behaviour and growth of broilers', *Animal Production Science*, 54(7), pp. 930–941. https://doi.org/10.1071/AN13184.

Howie, J.A. et al. (2009) 'The structure of feeding behavior in commercial broiler lines selected for different growth rates', *Poultry Science*, 88(6), pp. 1143–1150. https://doi.org/10.3382/PS.2008-00441.

Howie, J.A. et al. (2011) 'Genetic parameters of feeding behavior traits and their relationship with live performance traits in modern broiler lines', *Poultry Science*, 90(6), pp. 1197–1205. https://doi.org/10.3382/PS.2010-01313.

Jacobs, L. et al. (2019) 'Impact of water sanitation on broiler chicken production and welfare parameters', *Journal of Applied Poultry Research* [Preprint]. https://doi.org/10.1016/j.japr.2019.10.013.

Jacobs, L. et al. (2021) 'Broiler chicken behavior and activity are affected by novel flooring treatments', *Animals*, 11(10), p. 2841. https://doi.org/10.3390/ANI11102841.

Jacobs, L. et al. (2023) 'Enhancing their quality of life: environmental enrichment for poultry', *Poultry Science*, 102(1). https://doi.org/10.1016/j.psj.2022.102233.

Junghans, A. et al. (2022) 'Data evaluation of broiler chicken rearing and slaughter–An exploratory study', *Frontiers in Veterinary Science*, 9. https://doi.org/10.3389/FVETS.2022.957786.

Kaplan, G. (2020) 'Play behaviour, not tool using, relates to brain mass in a sample of birds', *Scientific Reports 2020 10:1*, 10(1), pp. 1–15. https://doi.org/10.1038/s41598-020-76572-7.

Keeling, L.J. et al. (2017) 'Flock size during rearing affects pullet behavioural synchrony and spatial clustering', *Applied Animal Behaviour Science*, 194, pp. 36–41. https://doi.org/10.1016/J.APPLANIM.2017.04.002.

Kestin, S.C. et al. (1999) 'Different commercial broiler crosses have different susceptibilities to leg weakness', *Poultry Science*, 78(8), pp. 1085–1090. https://doi.org/10.1093/PS/78.8.1085.

Kestin, S.C. et al. (2001) 'Relationships in broiler chickens between lameness, liveweight, growth rate and age', *The Veterinary Record*, 148, pp. 195–197. https://search.proquest.com/openview/42dafacb0456f87ae9ced2d58812f47d/1?pq-origsite=gscholar&cbl=2041027 (Accessed: 14 October 2018).

Košťál, Ľ. et al. (2020) 'Use of cognitive bias as a welfare tool in poultry', *Journal of Animal Science*, 98, pp. S63–S79. https://doi.org/10.1093/JAS/SKAA039.

Kristensen, H.H. et al. (2007) 'The behaviour of broiler chickens in different light sources and illuminances', *Applied Animal Behaviour Science*, 103(1–2), pp. 75–89. https://doi.org/10.1016/J.APPLANIM.2006.04.017.

Leach, R.M. and Monsonego-Ornan, E. (2007) 'Tibial dyschondroplasia 40 years later', *Poultry Science*, 86(10), pp. 2053–2058. https://doi.org/10.1093/PS/86.10.2053.

Leishman, E.M. et al. (2022) 'The prevalence of integument injuries and associated risk factors among Canadian turkeys', *Frontiers in Veterinary Science*, 8, p. 757776. https://doi.org/10.3389/FVETS.2021.757776/BIBTEX.

Liu, Z. et al. (2020) 'Play behaviour reduced by environmental enrichment in fast-growing broiler chickens', *Applied Animal Behaviour Science*, 232, p. 105098. https://doi.org/10.1016/J.APPLANIM.2020.105098.

Lourenço da Silva, M.I. et al. (2024) 'Pharmacological validation of an attention bias test for conventional broiler chickens', *PLoS One*, 19(4), p. e0297715. https://doi.org/10.1371/journal.pone.0297715.

Lourenço-Silva, M.I. et al. (2023) 'Social-pair judgment bias testing in slow-growing broiler chickens raised in low- or high-complexity environments', *Scientific Reports*, 13(1), pp. 1–11. https://doi.org/10.1038/s41598-023-36275-1.

Louton, H. et al. (2016) 'Dust-bathing behavior of laying hens in enriched colony housing systems and an aviary system', *Poultry Science*, 95(7), pp. 1482–1491. https://doi.org/10.3382/PS/PEW109.

Marino, L. (2017) 'Thinking chickens: a review of cognition, emotion, and behavior in the domestic chicken', *Animal Cognition*, 20(2), pp. 127–147. https://doi.org/10.1007/S10071-016-1064-4.

Martrenchar, A. et al. (1999) 'Influence of stocking density on behavioural, health and productivity traits of turkeys in large flocks', *British Poultry Science*, 40(3), pp. 323–331. https://doi.org/10.1080/00071669987403.

Marx, G. et al. (2001) 'Vocalisation in chicks (Gallus gallus dom.) during stepwise social isolation', *Applied Animal Behaviour Science*, 75(1), pp. 61–74. https://doi.org/10.1016/S0168-1591(01)00180-0.

May, J.D. and Lott, B.D. (1992) 'Effect of periodic feeding and photoperiod on anticipation of feed withdrawal', *Poultry Science*, 71(6), pp. 951–958. https://doi.org/10.3382/PS.0710951.

McCreery, D. (2015) *Water consumption behavior in broilers*. University of Arkansas. https://www.proquest.com/openview/28d4a3ce4eedef8e28cbb072ff4f1c2a/1?pq-origsite=gscholar&cbl=18750 (Accessed: 29 January 2025).

McGeown, D. et al. (1999) 'Effect of carprofen on lameness in broiler chickens', *Veterinary Record*, 144(24), pp. 668–671. https://doi.org/10.1136/VR.144.24.668.

McGrath, N. et al. (2017) 'Hens vary their vocal repertoire and structure when anticipating different types of reward', *Animal Behaviour*, 130, pp. 79–96. https://doi.org/10.1016/J.ANBEHAV.2017.05.025.

Monckton, V. et al. (2020) 'Broiler chicks' motivation for different wood beddings and amounts of soiling', *Animals*, 10(6), p. 1039. https://doi.org/10.3390/ANI10061039.

Murphy, L.B. and Preston, A.P. (2007) 'Time-budgeting in meat chickens grown commercially', *British Poultry Science*, 29(3), pp. 571–580. https://doi.org/10.1080/00071668808417083.

Nääs, I.A. et al. (2009) 'Impact of lameness on broiler well-being', *Journal of Applied Poultry Research*, 18(3), pp. 432–439. https://doi.org/10.3382/JAPR.2008-00061.

Newberry, R.C. et al. (2018) 'Chickens play in the wake of humans', in *ISAE 2018 Proceedings of the 52nd Congress of the International Society for Applied Ethology*. Wageningen Academic Publishers, p. 199.

Nicol, C.J. et al. (2009) 'Associations between welfare indicators and environmental choice in laying hens', *Animal Behaviour*, 78(2), pp. 413–424. https://doi.org/10.1016/J.ANBEHAV.2009.05.016.

Nicol, C.J. et al. (2017) *Farmed bird welfare science review*. Langford House. https://agriculture.vic.gov.au/__data/assets/pdf_file/0008/529829/Farmed-Bird-Welfare-Science-Review.pdf (Accessed: 19 March 2025).

Nielsen, S.S. et al. (2023) 'Welfare of broilers on farm', *EFSA Journal*, 21(2), p. 7789. https://doi.org/10.2903/j.efsa.2023.7788.

Norring, M. et al. (2016) 'The use of perches and platforms by broiler chickens', *Applied Animal Behaviour Science*, 184, pp. 91–96. https://doi.org/10.1016/J.APPLANIM.2016.07.012.

Ohara, A. et al. (2015) 'Effect of environmental enrichment on the behavior and welfare of Japanese broilers at a commercial farm', *The Journal of Poultry Science*, 52(4), pp. 323–330. https://doi.org/10.2141/JPSA.0150034.

Olsson, I.A.S. and Keeling, L.J. (2000) 'Night-time roosting in laying hens and the effect of thwarting access to perches', *Applied Animal Behaviour Science*, 68(3), pp. 243–256. https://doi.org/10.1016/S0168-1591(00)00097-6.

Olsson, I.A.S. and Keeling, L.J. (2002) 'The push-door for measuring motivation in hens: laying hens are motivated to perch at night', *Animal Welfare*, 11(1), pp. 11–19. https://doi.org/10.1017/S0962728600024283.

Papageorgiou, M. et al. (2023) 'Positive welfare indicators and their association with sustainable management systems in poultry', *Sustainability*, 15(14), p. 10890. https://doi.org/10.3390/SU151410890.

Pettit-Riley, R. and Estevez, I. (2001) 'Effects of density on perching behavior of broiler chickens', *Applied Animal Behaviour Science*, 71(2), pp. 127–140. https://doi.org/10.1016/S0168-1591(00)00174-X.

Raising Broilers | Extension (n.d.) https://extension.unh.edu/resource/raising-broilers (Accessed: 12 September 2024).

Rayner, A.C. et al. (2020) 'Slow-growing broilers are healthier and express more behavioural indicators of positive welfare', *Sci Rep* 10, 15151. https://doi.org/10.1038/s41598-020-72198-x.

Renema, R.A. et al. (2007) 'Implications of changes to commercial broiler and broiler breeder body weight targets over the past 30 years', *World's Poultry Science Journal*, 63(3), pp. 457–472. https://doi.org/10.1017/S0043933907001572.

Riber, A.B. and Wurtz, K.E. (2024) 'Impact of Growth Rate on the Welfare of Broilers', *Animals*, 14(22), p. 3330. https://doi.org/10.3390/ANI14223330.

Riber, A.B. et al. (2017) 'Environmental enrichment for broiler breeders: an undeveloped field', *Frontiers in Veterinary Science*, 4(June), p. 38. https://doi.org/10.3389/fvets.2017.00086.

Riber, A.B. et al. (2021) 'Are changes in behavior of fast-growing broilers with slight gait impairment (GS0-2) related to pain?', *Poultry Science*, 100(3), p. 100948. https://doi.org/10.1016/J.PSJ.2020.12.045.

Rutter, S.M. and Duncan, I.J.H. (1991) 'Shuttle and one-way avoidance as measures of aversion in the domestic fowl', *Applied Animal Behaviour Science*, 30(1–2), pp. 117–124. https://doi.org/10.1016/0168-1591(91)90090-K.

Savory, C.J. (1978) 'The relationship between food and water intake and the effects of water restriction on laying brown leghorn hens', *British Poultry Science*, 19(5), pp. 631–641. https://doi.org/10.1080/00071667808416523.

Savory, C.J. (1980) 'Diurnal feeding patterns in domestic fowls: a review', *Applied Animal Ethology*, 6(1), pp. 71–82. https://doi.org/10.1016/0304-3762(80)90095-4.

Savory, C.J. (2010) 'Nutrition, Feeding and Drinking Behaviour, and Welfare', in I.J.H. Duncan and P. Hawkins (eds) *The welfare of domestic fowl and other captive birds*. Springer Science, pp. 165–187. https://doi.org/10.1007/978-90-481-3650-6_7.

Savory, C.J. et al. (1992) 'Stereotyped behaviour in broiler breeders in relation to husbandry and opioid receptor blockade', *Applied Animal Behaviour Science*, 32(4), pp. 349–360. https://doi.org/10.1016/S0168-1591(05)80027-9.

Shields, S.J. (2004) *Dustbathing by broiler chickens: characteristics, substrate preference, and implications for welfare*. University of California.

Shields, S.J. et al. (2005) 'Effect of sand and wood-shavings bedding on the behavior of broiler chickens', *Poultry Science*, 84(12), pp. 1816–1824. https://doi.org/10.1093/PS/84.12.1816.

Siegel, P.B. and Wisman, E.L. (1966) 'Selection for body weight at eight weeks of age: 6. changes in appetite and feed utilization', *Poultry Science*, 45(6), pp. 1391–1397. https://doi.org/10.3382/PS.0451391.

Siviy, S.M. and Panksepp, J. (2011) 'In search of the neurobiological substrates for social playfulness in mammalian brains', *Neuroscience and Biobehavioral Reviews*, 35(9), pp. 1821–1830. https://doi.org/10.1016/J.NEUBIOREV.2011.03.006.

Špinka, M. et al. (2001) 'Mammalian play: training for the unexpected', *The Quarterly Review of Biology*, 76(2), pp. 141–168. https://doi.org/10.1086/393866.

Spruijt, B.M. et al. (1992) 'Ethology and neurobiology of grooming behavior', *Physiological Reviews*, 72(3), pp. 825–852. https://doi.org/10.1152/PHYSREV.1992.72.3.825.

Statkiewicz, W.R. and Schein, M.W. (1980) 'Variability and periodicity of dustbathing behaviour in Japanese quail (Coturnix coturnix japonica)', *Animal Behaviour*, 28(2), pp. 462–467. https://doi.org/10.1016/S0003-3472(80)80053-4.

Suarez, S.D. and Gallup, G.G. (1982) 'Open-field behavior in chickens: the experimenter is a predator', *Journal of Comparative and Physiological Psychology*, 96(3), pp. 432–439. https://doi.org/10.1037/H0077886.

Tallentire, C.W. et al. (2016) 'Breeding for efficiency in the broiler chicken: a review', *Agronomy for Sustainable Development*, 36(4), pp. 1–16. https://doi.org/10.1007/S13593-016-0398-2/FIGURES/1.

Toghyani, M. et al. (2010) 'Effect of different litter material on performance and behavior of broiler chickens', *Applied Animal Behaviour Science*, 122(1), pp. 48–52. https://doi.org/10.1016/j.applanim.2009.11.008.

Trocino, A. et al. (2020) 'Effect of feed restriction on the behaviour and welfare of broiler chickens', *Animals*, 10(5), p. 830. https://doi.org/10.3390/ANI10050830.

U.S. Broiler Performance - National Chicken Council (n.d.). https://www.nationalchickencouncil.org/statistic/us-broiler-performance/ (Accessed: 12 September 2024).

Vallortigara, G. and Zanforlin, M. (1993) 'Sex differences in social reinstatement motivation of the domestic chick', *Ethology Ecology and Evolution*, 5(3), p. 418. https://doi.org/10.1080/08927014.1993.9523090.

van der Eijk, J.A.J. et al. (2022) 'Reducing stocking density benefits behaviour of fast- and slower-growing broilers', *Applied Animal Behaviour Science*, 257, p. 105754. https://doi.org/10.1016/J.APPLANIM.2022.105754.

van Liere, D.W. and Wiepkema, P.R. (1992) 'Effects of long-term deprivation of sand on dustbathing behaviour in laying hens', *Animal Behaviour*, 43(4), pp. 549–558. https://doi.org/10.1016/S0003-3472(05)81015-2.

van Staaveren, N. et al. (2021) 'A meta-analysis on the effect of environmental enrichment on feather pecking and feather damage in laying hens', *Poultry Science*, 100(2), pp. 397–411. https://doi.org/10.1016/J.PSJ.2020.11.006.

Vas, J. et al. (2023) 'Better welfare for broiler chickens given more types of environmental enrichments and more space to enjoy them', *Applied Animal Behaviour Science*, 261, p. 105901. https://doi.org/10.1016/J.APPLANIM.2023.105901.

Ventura, B.A. et al. (2012) 'Access to Barrier Perches Improves Behavior Repertoire in Broilers', *PLoS One*, 7(1), p. e29826. https://doi.org/10.1371/JOURNAL.PONE.0029826.

Versace, E. et al. (2019) 'Conserved abilities of individual recognition and genetically modulated social responses in young chicks (Gallus gallus)', *bioRxiv*, p. 743765. https://doi.org/10.1101/743765.

Vestergaard, K.S. (1982) 'Dust-bathing in the domestic fowl - diurnal rhythm and dust deprivation', *Applied Animal Ethology*, 8(5), pp. 487–495. https://doi.org/10.1016/0304-3762(82)90061-X.

Vestergaard, K.S. (1994) *Dustbathing and its relation to feather pecking in the fowl: motivational and developmental aspects*. https://www.cabidigitallibrary.org/doi/full/10.5555/19952217118 (Accessed: 17 February 2025).

Vestergaard, K.S. and Sanotra, G.S. (1999a) 'Relationships between leg disorders and changes in the behaviour of broiler chickens', *Veterinary Record*, 144(8), pp. 205–209. https://doi.org/10.1136/vr.144.8.205.

Vestergaard, K.S. and Sanotra, G.S. (1999b) 'Relationships between leg disorders and changes in the behaviour of broiler chickens', *Veterinary Record*, 144(8), pp. 205–209. https://doi.org/10.1136/VR.144.8.205.

Vestergaard, K.S. et al. (1999) 'Regulation of dustbathing in feathered and featherless domestic chicks: the Lorenzian model revisited', *Animal Behaviour*, 58(5), pp. 1017–1025. https://doi.org/10.1006/anbe.1999.1233.

Villagrá, A. et al. (2014) 'Behavior of broiler chickens in four different substrates: a choice test', *Brazilian Journal of Poultry Science*, 16(1), pp. 67–75. https://doi.org/10.1590/S1516-635X2014000100010.

Weary, D.M. et al. (2006) 'Identifying and preventing pain in animals', *Applied Animal Behaviour Science*, 100(1–2), pp. 64–76. https://doi.org/10.1016/j.applanim.2006.04.013.

Webster, A.B. (2002) 'Behavior of chickens', in *Commercial chicken meat and egg production*. Springer, pp. 71–86. https://doi.org/10.1007/978-1-4615-0811-3_6.

Weeks, C.A. et al. (2000) 'The behaviour of broiler chickens and its modification by lameness', *Applied Animal Behaviour Science*, 67(1–2), pp. 111–125. https://doi.org/10.1016/S0168-1591(99)00102-1.

Wideman, R.F. et al. (2014) 'Bacterial chondronecrosis with osteomyelitis in broilers: Influence of sires and straight-run versus sex-separate rearing', *Poultry Science*, 93(7), pp. 1675–1687. https://doi.org/10.3382/PS.2014-03912.

Widowski, T.M. and Duncan, I.J.H. (2000) 'Working for a dustbath: are hens increasing pleasure rather than reducing suffering?', *Applied Animal Behaviour Science*, 68(1), pp. 39–53. https://doi.org/10.1016/S0168-1591(00)00088-5.

Widowski, T.M. and Rentsch, A.K. (2022) 'Farming poultry', in KnightA. , et al. (eds) *Routledge handbook of animal welfare*. 1st edn. Taylor and Francis, pp. 47–63. https://doi.org/10.4324/9781003182351-7/FARMING-POULTRY-TINA-WIDOWSKI-ANA-RENTSCH.

Williams, N.S. and Scampoli, D.L. (1984) 'Handling, ACTH, ACTH1–24, and naloxone effects on preening behavior in domestic chickens', *Pharmacology Biochemistry and Behavior*, 20(5), pp. 681–682. https://doi.org/10.1016/0091-3057(84)90184-9.

Xiao, Y. et al. (2017) 'Association of growth rate with hormone levels and myogenic gene expression profile in broilers', *Journal of Animal Science and Biotechnology*, 8(1), pp. 1–7. https://doi.org/10.1186/S40104-017-0170-8/TABLES/3.

Zimmerman, P.H. et al. (2000) 'The vocal expression of feeding motivation and frustration in the domestic laying hen, Gallus gallus domesticus', *Applied Animal Behaviour Science*, 69(4), pp. 265–273. https://doi.org/10.1016/S0168-1591(00)00136-2.

Zimmerman, P.H. et al. (2011) 'Behaviour of domestic fowl in anticipation of positive and negative stimuli', *Animal Behaviour*, 81(3), pp. 569–577. https://doi.org/10.1016/J.ANBEHAV.2010.11.028.

Zuidhof, M.J. et al. (2014) 'Editor's choice: growth, efficiency, and yield of commercial broilers from 1957, 1978, and 2005', *Poultry Science*, 93(12), p. 2970. https://doi.org/10.3382/PS.2014-04291.

# Chapter 10

## Understanding the nutritional needs of broilers

*E. Delezie and M. Lourenço, Flanders Research Institute for Agriculture, Fisheries and Food (ILVO), Belgium*

1 Introduction
2 Feed formulation and use of raw materials
3 Nutritional requirements of broilers
4 What will broiler diets look like in 10 years?
5 Conclusion
6 Future trends in research
7 Where to look for further information
8 References

## 1 Introduction

There are a number of reasons why so much attention needs to be given to the composition of broiler diets. First of all the feed conversion ratio (FRC) or feed efficiency is a critical measure in broiler farming and represents the amount of feed consumed per unit of meat production. A lower FCR indicates more efficient feed use, leading to improved profitability (Zampiga et al., 2021). Since feed costs account for about 60–70% of the total production costs, improving feed conversion and reducing feed costs can significantly increase the farmer's income (Alhotan, 2021). A high return on investment is warranted but needs to be combined with the delivery of optimal product quality. A properly balanced diet is essential for producing high-quality meat and high slaughter yields (Choi et al., 2023). Nutrients must be present in the diet in the right ratios to ensure high-quality end products. Secondly, diet influences the overall health of broilers (Whitehead, 2002). A balanced diet promotes gut health, helps prevent conditions like osteoporosis (weak bones), and reduces the risk of disorders. Therefore, offering a well-balanced diet is vital to support the broilers' health. Thirdly, the role of the diet in broiler welfare, particularly for broilers, is receiving more attention. In the near future,

http://dx.doi.org/10.19103/AS.2025.0165.11

conditions like footpad dermatitis in broilers, which are linked to poor litter quality, can be exacerbated by an unbalanced diet and high percentages can be penalized (Korver, 2023). Finally, nutrition doesn't only affect the broilers but also the environment. The goal of broiler nutrition is to feed the chickens in a way that the nutrients they ingest are effectively utilized. If excess nutrients are provided in the diet (e.g. safety margins leading to over-supplementation), they may be excreted into the environment, causing pollution. In Europe, the emission of particularly nitrogen, like ammonia, from industrial broiler farming is a growing concern, and legislation is aimed at reducing such emissions (Chalova et al., 2016). Therefore in order to formulate a suitable diet, all these measures need to be taken into account. Key considerations include identifying (new, circular) raw materials that provide the necessary energy, crude protein/amino acids (AA), minerals, and vitamins broilers need. Additionally, it's important to be aware of any anti-nutritional factors (ANFs) that may be present in these raw materials, which can negatively affect broiler welfare and health.

The choice of raw materials and their origin is becoming increasingly important in broiler production. It is not only about nutrient use efficiency, growth, welfare, and emissions but also about contributing to a circular food system and reducing waste. Rather than relying solely on conventional raw materials, nutritionists and the industry as a whole face the challenge of utilizing alternative streams, such as by-products from the food industry or agricultural side-streams. This aligns with the ambition of zero waste and the reduction of the sector's environmental footprint (Puente-Rodriguez et al., 2022).

At the same time, it should be acknowledged that despite ongoing discussions around healthier and more sustainable diets within the EU, global – and also European – demand for broiler meat will continue to rise (Miller et al., 2022). This reinforces the need for innovative solutions that balance economic viability with environmental objectives.

It is also important to recognize that not only nutrition but also factors such as genetics (Marchesi et al., 2021), sex (Tůmová et al., 2021), and feed technology play a crucial role. Smart application of feed technology makes it possible to unlock nutrients from alternative raw materials more efficiently. In combination with targeted genetic selection adapted feeding strategies can lead to more efficient production with lower emissions and reduced waste.

This chapter aims to outline the nutritional needs of broilers, linking feed formulation and ingredient choice to performance, sustainability, and welfare. It addresses nutritional requirements, common and alternative feed ingredients, feed additives, feeding strategies, and particle size. The chapter also explores the future of broiler nutrition, considering circularity, welfare, and how production may evolve over the next decade.

## 2 Feed formulation and use of raw materials

Broiler diets are composed of a variety of raw materials that provide essential nutrients to support the health, welfare, and productivity of broilers. These raw materials are carefully selected and mixed in specific proportions to meet the nutritional requirements of the broilers.

Approximately 95% of the diet is made up of macronutrients, which are the primary sources of energy and protein for the broilers (Swennen et al., 2007). These macronutrients include:

1) carbohydrates providing energy for the broilers' daily activities and growth;
2) crude proteins (CP), which are essential for building and repairing tissues, as well as for growth; and
3) fats also serving as an important energy source and supporting the absorption of fat-soluble vitamins.

The remaining 5% of the diet consists of micronutrients and other functional ingredients, which are added in much smaller quantities but play a critical role in the overall health, welfare, and performance of the broilers (Alagawany et al., 2021; Lauzon, 2006). These micronutrients are:

1) minerals, which are important for bone development, enzyme function, and other physiological processes;
2) organic acids to maintain gut health and support nutrient absorption;
3) synthetic amino acids (AA) crucial for protein synthesis and overall growth;
4) vitamins vital for metabolic processes and overall health; and
5) other micro-ingredients, such as antioxidants, enzymes, and other specialized additives.

The raw materials for broiler diets have different origins, are sourced from various suppliers, and often originate from diverse agricultural and industrial by-products. The goal of feed formulation is to combine these raw materials in an optimal way that ensures both nutritional adequacy and cost-effectiveness, allowing the broilers to thrive, while keeping production costs manageable (Ravindran, 2013).

### 2.1 Common feed ingredients

Energy suppliers for broiler diets are mainly cereals, such as wheat, corn or rye, by-products of cereals (e.g. wheat bran, wheat middlings), and fats or oils, such as animal fat, soy oil, or sunflower oil. Animal fats and vegetable oils are included in broiler diets because of their high energy concentration to support growth performance (Blanch et al., 1996; Ravindran, 2013).

Soybean meal (SBM) is considered the most important plant-based protein source for farm animals. It accounts for about two-thirds of the world's total output of protein feedstuffs, surpassing other major oil meals and fish meal in terms of feeding value. It is used as the standard by which other plant protein sources are measured (Cromwell, 1999). The protein content of soybean meal typically ranges between 47% and 49% for high-protein types (from dehulled seeds), with approximately 3% crude fiber. In comparison, soybean meal offers a much higher protein concentration than soybean hulls (10–13%) which contain less energy due to the presence of fiber and other non-digestible components (Heuzé et al., 2020). The importance of SBM in broiler diets is not only based on its high crude protein content but also on its high and balanced level of both essential and nonessential AA. However, soybeans contain also ANFs that can interfere with digestion and nutrient absorption. Some of the key ANFs are:

- Trypsin inhibitors, which bind with the digestive enzyme trypsin and prevent it from breaking down feed proteins. These inhibitors can also lead to pancreatic enlargement and reduced growth and feed conversion efficiency (Pacheco et al., 2014; Kuenz et al., 2022). However, trypsin inhibitors are heat-labile, which is why heat treatment is necessary before feeding soybean-based products to animals, especially monogastric animals like broilers (Woyengo et al., 2017).
- Lectins are glycoproteins that bind to intestinal villi, reducing nutrient absorption. Like trypsin inhibitors, lectins are also heat-labile (Douglas et al., 1999; Casaubon-Huguenin et al., 2004; Nyachoti et al., 2018).
- Saponins, which are glycosides, can reduce nutrient uptake by making feed less palatable due to their bitter taste, and they can form complexes that irritate or damage the intestinal mucosa, which may cause discomfort and suppress appetite. Their surfactant (soap-like) properties can cause foaming in the gut, interfering with normal digestion and nutrient absorption. Some saponins influence gut hormones and bile salt metabolism, potentially creating a feeling of satiety or reducing feed efficiency (Tadele, 2015; Nyachoti et al., 2018; Ramteke et al., 2019).

Other potential ANFs in soybeans include antivitamins, urease, and plant sterols, which may interfere with the absorption of carotenoids and vitamin E (Kocher et al., 2002; Pope et al., 2023). It is crucial to manage the presence of ANFs in all raw materials used in broiler diets through proper processing methods to ensure the feed's effectiveness and the broilers' health. A failure to do so properly can lead to damage in the intestinal gut wall, leading either to disease or to lower nutrient digestion and absorption. In case of damage to the integrity of the gut wall, either the condition of leaky gut occurs, or opportunistic pathogens will grow in the gut, leading to dysbiosis (Salahi et al., 2025) or even necrotic enteritis (Moore, 2016). At this point translocation of pathogens into the bloodstream is possible, potentially leading to femur head necrosis, for example (Xu et al., 2023; Asnayanti et al., 2024). This will lead to severe pain in the joint, and broilers are unable to walk which in turn leads to stress, discomfort, decreased production (due to decreased feed intake), and eventually to death (McNamee and Smyth, 2000).

Other ANFs, such as non-starch polysaccharides (NSP), will rather lead to increased intestinal digesta viscosity, decreased nutrient absorption, and increased fermentation in the gut which is detrimental to the performance, gut health, and welfare of the broiler (Choct et al., 1996).

When nutrient digestion is affected in the gut, manure consistency will also be affected. When broiler manure has a lower dry matter content, this will contribute to wet litter conditions with increased footpad dermatitis incidence, again leading to discomfort, decreased mobility, and feed intake (Collett, 2012; Taira et al., 2013).

## 2.2 Alternative protein sources

Despite the high quality and use of soybean as protein source in broiler diets, the search for alternative protein sources is ongoing. The fact that in the last years the demand for the most common protein sources used in broiler diets, such as soybean, and eventually fish meal, has increased has led to a mismatch with production capacity of these protein sources, with increases in the price of these feed ingredients (Gale and Arnade, 2015). The diet, as already mentioned, represents 60–70% of the total production costs of broilers. Finding alternative protein sources is vital, not only to reduce feed costs but also to contribute to a more sustainable broiler production and reduce its environmental impact. The production of soybean is connected with deforestation, soil erosion, eutrophication, extensive use of pesticides, loss of biodiversity, and a high $CO_2$ footprint (van Huis, 2015). Due to problems with over-fishing and environmental pollution, marine fish is also regarded as limited resource. There is thus the need to feed these broilers with sustainable ingredients, and hence new protein sources need to be explored.

### 2.2.1 Criteria to be taken into account

When selecting alternative protein sources or feed ingredients for broiler diets, several criteria must be taken into account to ensure that the diet is both effective, sustainable, and cost-efficient. These criteria include: (1) Supply: Reliable availability of raw materials is critical to maintain a steady production cycle and diet composition. (2) Price: The cost of raw materials is a weighing factor in feed formulation. (3) Nutritional aspects: The nutritional content of raw materials is the most important consideration. Diets must provide the right balance of macronutrients and micronutrients to support the health, welfare, and performance of the broilers. (4) Technological aspects: Factors such as mixability, storage stability, and palatability affect the diet's production and quality. The texture of feed ingredients affects how broilers consume and digest the diet. The way raw materials are processed can influence their digestibility, nutrient availability, and overall feed quality. Proper processing techniques (such as heat treatment, milling, or fermentation) are used to enhance the nutritional value and safety of the ingredients.

### 2.2.2 Possible alternative protein sources

The most common possible alternative protein sources to SBM are rapeseed or sunflower meals, or leguminosae. Peas can influence the structure of the diet due to their fibrous outer peel. High levels of pea supplementation can lead to faster fermentation rates, linked to higher gas production and possibility of disbalance in gut microbiome, which can affect overall feed efficiency (David et al., 2024). More recently, attention is being given to leguminosae, such as faba beans and lupins, with its advantages and disadvantages. Next to the traditional plant-based alternatives, attention is being given to algae or even non-plant-based sources such as insects and single cell proteins, meaning microalgae, bacterial, yeast, or fungi, or fermentation of ingredients.

Seaweeds are generally classified into brown, red, and green algae. The chemical composition (crude protein level, amino acid profile) of the algae is quite variable. This variation in crude protein content and AA profile is quite noticeable between species from the same phylum but also within the same species, depending on where the algae have been growing. This can be challenging when looking for an alternative that has to have a relatively stable composition to facilitate feed formulation (Van Nerom et al., 2024).

When comparing to SBM, most of the algae are lower in AA, except for the S-containing ones (Table 1), where we see a limited species difference. Algae have normally a low fat content and can be high in complex carbohydrates and polysaccharides that can limit digestibility and accessibility of other nutrients. How to harvest and process seaweeds to use in broiler diets is an important aspect, as these seaweeds are grown in salt waters and have a high salt content (30–50%) (Abdel-Wareth, 2024; Dewi et al., 2024; Bikker et al., 2020).

Insects have attained great attention as prospective constituents of broiler diets due to their high protein contents (40–60% DM), fats (30–40% DM), vitamins, and minerals (Makkar et al., 2014). Moreover, insects can play a decisive role as a sustainable feed source by efficiently transforming and valorizing bio-wastes

**Table 1** Chemical composition (% DM) and amino acid (AA) profile (g/16gN) of soybean meal, microalgae (Spirulina and Chlorella), black soldier fly insect meal, and bacterial single cell protein (primary strain *Methylococcus*)

| | Soybean meal[1] | Spirulina[2] | Chlorella[2] | BSF meal[3] | Single cell protein[4] |
|---|---|---|---|---|---|
| Chemical composition (% DM) | | | | | |
| Crude protein | 55.2 | 63.8 | 47.1 | 42.1 | 61.8 |
| Crude fat | 1.70 | 6.10 | 11.0 | 26.0 | 7.54 |
| Crude fiber | 4.40 | 2.90 | 2.10 | n.r. | 2.33 |
| Crude ash | 7.30 | 8.90 | 10.3 | n.r. | n.r. |
| Amino acid profile (g/16gN) | | | | | |
| Lysine | 6.20 | 4.40 | 4.40 | 6.60 | 4.92 |
| Methionine | 1.40 | 2.20 | 1.90 | 2.10 | 0.73 |
| Cysteine | 1.60 | 0.90 | 2.30 | 0.10 | 0.45 |
| Threonine | 3.80 | 4.80 | 5.00 | 3.70 | 3.25 |
| Tryptophan | 1.40 | 1.40 | 1.30 | 0.50 | 1.85 |
| Arginine | 7.30 | 4.60 | 4.80 | 5.60 | 3.99 |
| Valine | 4.80 | 6.40 | 5.50 | 8.20 | 3.98 |
| Serine | 4.60 | 4.90 | 6.30 | 3.10 | 2.58 |
| Glycine | 4.20 | 4.80 | 5.60 | 5.70 | 3.91 |

[1]Heuzé et al. (2020); [2]Heuzé et al. (2017); [3]Makkar et al. (2014); [4]Hombegowda et al. (2021); BSF - Black soldier fly; n.r. - not reported

into high-quality feed, thereby exerting minimal pressure on land, water, and energy resources (van Huis, 2015). Furthermore, the exoskeletons of insects are rich in chitin polysaccharides that enhance the palatability of broiler diets. Chitin bioactive compounds can have antimicrobial and immunostimulant effects, making them a viable alternative to antibiotics and beneficial to gut health. Until now, research efforts have focused on the black soldier fly (BSF), housefly (HF), mealworm (MW), crickets, and grasshoppers. However, the cost of production compared to the volumes needed makes this alternative so far not so attractive. Nonetheless, there is a high potential for adoption of insects for broiler diets if they can be produced in sustainable quantities that ensure the viability of broiler farming and feed processing (Sebatta et al., 2018).

Single cell protein production (SCP), protein extracted from cultivated microbial biomass (bacteria, fungi, algae, and yeast), can also be used for protein supplementation (Attia et al., 2003). There are already quite some companies investing in this path. SCP are attractive because they are abundant in proteins, minerals, vitamins, and essential fatty acids with high digestible components (65–95%). Furthermore, the AA profile of the biomass is very similar to the conventional used protein sources (Table 1). In response to protein shortage they present an important potential as they are also sustainable and non-GMO. Microbes can thrive on waste materials, using them as substrates for protein production, which helps minimize environmental pollution and supports recycling. Single-cell protein-producing organisms grow quickly and yield large amounts of protein (Li et al., 2024). However, due to the significant variability in the composition (due to substrate used, production conditions, and strains) and characteristics of the various SCP sources (Nasseri et al., 2011), it is essential to determine appropriate inclusion levels in broiler diets for each source. Currently, information on ileal AA digestibility is lacking, a critical point to assess the nutritional value of these protein sources for broilers. Therefore, additional research is needed before undertaking cost-benefit analyses related to the integration of SCP into broiler diets (Ledbetter, 2025).

## 2.3 Feed additives

### 2.3.1 The use of feed additives

Apart from new ingredients (such as microbial proteins) or nutritional enhancement (e.g. high-oil corn) or reduction in the level of anti-nutritive components in common diet ingredients, growth in using feed additives has been enormous. Additives may not be placed on the market unless authorized after scientific evaluations that show that the additive has no harmful effects on human, animal, and environmental health (Regulation Nr. 1831/2003[1]). EFSA recognizes five categories of feed additives, including technological (organic acids, antioxidants, pellet binders, etc.), sensory (flavors), nutritional (vitamins and amino acids), zootechnical additives (enzymes, probiotics, prebiotics, certain phytogenic), and coccidiostats.

The zootechnical feed additives are a group of products that affect favorably the performance of broilers through improved digestibility or stabilized gut flora, or have a positive impact on the environment. Exogenous enzymes, such as non-starch polysaccharides (NSP's) (cleaving the NSP) or phytase (targeting phytate-complexes in plant-based ingredients), are within the most used and well-accepted feed additives. The use of exogenous enzymes to improve the nutritional value of broiler diets is becoming very common. By adding exogenous NSP enzymes, the anti-nutritive effects of NSP are reduced and an increased use of viscous grains in broiler diets is possible (Choct, 2006). In response to concerns over phosphor pollution, microbial phytase in broiler diets is gaining interest and became one of the most important feed enzymes worldwide (Slominski, 2011). Overall, the supplementation of phytase and xylanase in broiler diets has confirmed their nutritional role enhancing

1 https://eur-lex.europa.eu/eli/reg/2003/1831/oj/eng

nutrient digestibility and growth performance, while also playing functional roles reducing the oxidative stress response, and possibly through the modulation of the mucosal microbiota in the small intestine (Alshelmani et al., 2024). The use of these enzymes allows the use of a wider range of feedstuffs without compromising the optimal performance of the broilers, as well as the environmental concerns associated with a poor hydrolysis of anti-nutritional factors present in the diets, while reducing feed costs (Poernama et al., 2021; Alshelmani et al., 2024).

The interest of plants phenolic compounds as potential natural antioxidant sources has been considered by researchers due to their potential role as feed additives in broiler production. Phenolics are products of plant secondary metabolism. Phenolic compounds can be defined as any compound having a benzene ring with one (phenol) or more (polyphenol) hydroxyl group as esters or methyl esters. Most studies show that dietary polyphenols can be effective as natural feed additives in broilers improving antioxidant capacity, immunity, antimicrobial defense, and production performance. Their effectiveness depends on many factors such as the type of phenolic compound, plant source, maturity, dosage, extraction method, bioavailability, and the specific animal species. Not all phenolic compounds are beneficial, so choosing the right compound at the right level is crucial for achieving positive effects on performance, health, and welfare (Mahfuz et al., 2021).

Health- and performance-promoting effects have been demonstrated for a number of organic acids, including formic, fumaric, citric, and lactic acid and their salts. These organic acids generally lead to a decreased pH in the intestinal tract, promoting the growth of beneficial bacteria with a resulting increased nutrient digestion and immunity (Chukwudi et al., 2024; Du et al., 2024). The supplementation of broiler diets with organic acids can improve the performance of broilers, carcass traits, gut health, the colonization of beneficial bacteria, and the immune response, while reducing the intestinal load of pathogenic bacteria. In this way they are highly valuable as alternative to antibiotics in broiler diets. However, similar to the phenolic compounds, the efficacy of organic acids is influenced by numerous factors, and additional studies are necessary to establish standardized protocols, dosages, and application methods to achieve a more thorough understanding and maximize the effectiveness of organic acids in broiler chicken nutrition (Gomez-Osorio et al., 2021; Ali et al., 2022).

Emulsifiers as feed additives are used to improve fat digestibility and energy efficiency. Gholami et al. (2024) found that emulsifier supplementation compensated for energy reduction (by decreasing corn oil content of the diet) in broiler diets and improved performance, dry matter digestibility, and some intestinal morphology parameters. They suggested using 0.5 g/kg of emulsifier per 100 kcal of ME reduction in broiler diets. Medium-chain fatty acids (MCAs) as feed additive have gained attention due to their potentially favorable antimicrobial effects (Szabó et al., 2023). Another type of feed additive used in broiler diets is prebiotics. Prebiotics are non-digestible food ingredients (readily fermentable sugars) that beneficially affect the host by selectively stimulating the growth and/or activity of one or a limited number of bacteria in the colon (large intestine), and thus improve host health (Patterson and Burkholder, 2003). Most potential prebiotics are carbohydrates (such as oligosaccharides). Some prebiotics that are used in this manner against pathogens are fructo-oligo-saccharides (FOS), xylooligo-saccharides (XOS), mannan-oligo-saccharides (MOS), and galacto-oligo-saccharides (GOS). Finally, probiotic feed additives generally consist of one single strain or a combination of several strains of bacteria, Bacillus spores, or yeast species (multi-strain). A probiotic is defined as a live microbial feed supplement which beneficially affects the host animal by improving its intestinal microbial balance. Authorized for use in animal nutrition in the European Union include different strains of *Enterococcus*, *Bacillus*, *Lactobacillus*, *Pediococcus*, and *Saccharomyces*. The mode of action of probiotic feed additives is mainly based on three principles, including competitive exclusion, bacterial antagonism, and immune modulation (Patterson and Burkholder, 2003). However, studies have reported mixed results regarding the benefits of these additives. Selecting the most suitable pro- or prebiotic for enhancing the performance, health, and welfare of broilers remains challenging, as in practice only a limited number of probiotics meet these criteria (Raza et al., 2024).

## 3 Nutritional requirements of broilers

Nutritional requirements of broilers should not focus only on their performance. Nutritional needs should cover the physiological changes broilers go through during their rearing period. The nutritional needs of broilers in the early stages of life, particularly during the rapid growth phase, play a pivotal role in ensuring optimal growth, health, and welfare. It is known that the growth rate of different organs is different, and development occurs at different life stages (Murawska et al., 2011). In this sense, during the first week of age, broilers will put their energy and nutrients into the development of the gut. A newborn chick is not in a state to digest a complete diet, as the gut is still immature in terms of digestibility and absorption capacity. Indeed, lipid digestion and protein digestion are said to be limiting to the growth of the chick in the first 3–4 days post hatch, as enzymatic secretions are not working to full capacity (Ravindran and Abdollahi, 2021). The same holds for the capacity of the gut to absorb nutrients. After this first week, the broiler will start to divert more nutrients to the skeletal growth and the development of other organs and tissues (immune system, feathers, fat pad, skin). Muscles are also growing, but it is during the last 2 weeks before slaughter that muscle growth increases exponentially (Murawska et al., 2011). The same holds for the capacity of the gut to absorb nutrients.

Unfortunately, accuracy in knowing broiler's specific nutrient requirements on any given day is difficult because nutritional requirements are moving targets influenced by many factors, such as the environment in which the animal is grown, the presence of a disease challenge, and changes in the genetic characteristics of the animal in question (Maharjan et al., 2021; Martínez and Valdivié, 2021). Nevertheless, matching nutrient requirements with feed formulation is essential for the efficiency of nutrient use by the broiler, but also to decrease the environmental impact of broiler production and costs. In Table 2 nutritional recommendations of the most important nutrients in broiler diets are given for different genetic lines.

### *3.1 Amino acid requirements*

Matching the AA requirements with the feed formulation is essential to decrease N in excretions and waste of broilers. Several factors needed to be considered for optimal utilization and retention of dietary N, such as (i) digestible amino acids; (ii) optimal energy intake; (iii) dietary indispensable to dispensable amino acids ratio; and (iv) use or not of exogenous feed additives. Feeding excess N or poor-quality proteins will lead to increased uric acid production, a N waste product. The production of this uric acid is an energy-consuming process that could otherwise be used for growth (Maharjan et al., 2021; Martínez and Valdivié, 2021). Further, protein and amino acid metabolism are heat-producing mechanisms that can lead to reduced feed intake. Free amino acids added to diets allow to lower crude protein requirements in the diet (Heger, 2003; Hilliar et al., 2020). For example, a 1% reduction in crude protein levels through improved amino acid formulations can lead to a 10% reduction in N losses (Rotz, 2004; Nahm, 2007). This means that broilers have more usable energy available for production that otherwise was being used to remove the excess N.

Amino acid requirements can change according to sex, body composition, growth rate, and overall body weight. However, the magnitude of the change in AA requirements is not the same for all individual amino acids. The amino acid profile needed at different ages differs. Amino acid utilization efficiency is 10% higher in the young broiler (10–21 days) than in the older broiler (32–43 days; Maharjan et al., 2021). Needs for tyrosine, glycine, and serine increase with the age of the broiler, whereas the needs of lysine remain similar throughout the rearing period of a broiler, and the needs of the other amino acids decrease with age. This is not strange as tyrosine is needed for the keratinization of feathers and glycine is needed in the uric acid cycle. Serine can be converted

**Table 2** Nutritional requirements of different broiler breeds with different growth rates

| | Ross 308 | Cobb 500 | Redbro Hubbard |
|---|---|---|---|
| **Starter (d1-d10)** | | | |
| Energy (MJ/kg) | 12.4 | 12.1 | 12.6 |
| Protein (%) | 22.0 | 21.0 | 20.0 |
| dLys (%) | 1.26 | 1.26 | 1.08 |
| dMet+Cys (%) | 0.95 | 0.94 | 0.82 |
| dTrp (%) | 0.20 | 0.21 | 0.17 |
| Ca (%) | 0.95 | 0.96 | 0.98 |
| aP (%) | 0.50 | 0.58 | 0.46 |
| **Grower (d10-d28)** | | | |
| Energy (MJ/kg) | 12.9 | 12.3 | 12.8 |
| Protein (%) | 19.0 | 19.0 | 17.8 |
| dLys (%) | 1.06 | 1.16 | 0.96 |
| dMet+Cys (%) | 0.87 | 0.88 | 0.74 |
| dTrp (%) | 0.18 | 0.18 | 0.16 |
| Ca (%) | 0.70 | 0.80 | 0.85 |
| aP (%) | 0.39 | 0.40 | 0.41 |
| **Finisher (d28-d42/d49)** | | | |
| Energy (MJ/kg) | 13.1 | 13.0 | 13.0 |
| Protein (%) | 17.5 | 18.0 | 16.8 |
| dLys (%) | 0.98 | 0.96 | 0.90 |
| dMet+Cys (%) | 0.83 | 0.74 | 0.71 |
| dTrp (%) | 0.17 | 0.17 | 0.16 |
| Ca (%) | 0.60 | 0.72 | 0.70 |
| aP (%) | 0.34 | 0.36 | 0.37 |

dLys - Digestible lysine; dMet+Cys - Digestible Methionine and Cysteine; dTrp - Digestible tryptophan; Ca - Calcium; aP - Available phosphor; Finisher phase for Ross 308 and Cobb 500 is up to 42 days of age (slaughter age), whereas for Redbro Hubbard the finisher phase is up to 49 days of age (slaughter age).

into glycine, so it is normal that the needs of serine also increase with the age of the broiler (Maharjan et al., 2021).

More and more research is showing that amino acid recommendations as requirements might need to be revisited, as requirements have in the past been established based on performance, mainly body weight and FCR, but not the welfare of the broiler, for example. Toghyani et al. (2016) have shown that increasing the dietary levels of methionine to 120% of the established requirement resulted in decreased incidence of footpad dermatitis, hock, and breast burns and led to an overall increase in feather cover. Similarly, Bello et al. (2018) have shown that increasing tryptophan dietary levels from 0.22% to

0.42% resulted in better welfare status (through decreased heat shock protein 70, corticosterone, and serotonin levels) before and after transportation, without negative effects on performance or meat quality. Another indication to revisit broiler amino acid requirements in function of their welfare are the results of Chowdhury et al. (2015), where the authors suggest orally administered citrulline helps in the thermoregulation of the broilers during heat stress by lowering body temperature.

### *3.2 Energy requirements*

As already mentioned, energy requirement is dependent on the physiological stage of the broiler but is also linked to the levels of other nutrient requirements. For example, if excess N is provided by the diet, the energy provided by the diet will be used to excrete the excess N, instead of being used for growth. This could suggest that in this case the need for energy would be higher to cover the needs of excreting excess N and to grow (Martínez and Valdivié, 2021, Ravindran and Abdollahi, 2021).

### *3.3 Factors influencing nutritional needs*

Genetic selection for feed efficiency and for fast-growing broilers has altered chicken's physiology and homeostasis (Maharjan et al., 2021). For example, increasing breast meat production from 15% of total BW in 1994 to >25% of total BW in 2020 has an impact on the nutritional requirements of broilers. At the same time, feed efficiency has also increased in the same period (Maharjan et al., 2021). Broilers have been selected for increased BW in a shorter period of time, to develop lean breast meat, meaning that broilers need to adapt physiologically to a higher protein turnover (McKay et al., 2000), also increasing the nutritional needs of protein and amino acids. Nutritional needs and recommendations have been built over the years based on specific genetic lines that continuously improve. Broiler genotypes have been continually changing since the 1950s (Aftab, 2019). In the last decades genetic lines have been adapted to specific markets, and the performance of the broilers will vary according to the nutritional recommendations used that are either not adapted to the full genetic potential of the broiler and hence not transposable to the housing system (Applegate and Angel, 2014). Indeed, conditions where the recommendations have been developed are not matching the ones where the broiler is being raised in practice. For example, slower-growing broilers need other recommendations than fast-growing broilers (Mohammadigheisar et al., 2021), but recommendations within different breeds of fast-growing broilers are also different. Nutritional needs for Ross 308 or Cobb 500 are not the same (Table 2).

Sex is also an important aspect. Up to week 4 the growth rate seems to be the same between males and females, which would suggest that nutritional needs would be similar (Murawska et al., 2011). However, on the one side, it is well known that generally the males have higher nutritional requirements as they have a higher growth rate than females. On the other side, it is also known that females will develop more feathers and deposit more fat than males (Murawska et al., 2011). Should the broiler industry make the option to rear and formulate different feed diets according to sex? Is this economically feasible? Where is more to gain economically? Formulating and producing different diets for female and male broilers is not the issue, the issue is in the feasibility at the broiler house. At the broiler house this would require male and female broilers to be reared in separate compartments, each with its own feeding system. This is quite an investment for the farmer, but it might be partially gained in reduced emissions, particularly N, on farm. However, looking at the whole production chain, with the feed mill included, the question is whether or not the water and energy used to produce one big batch of broiler diet or two smaller batches is the same or not. On the other side, with the pressure to lower the number of broilers placed in a house due to welfare concerns and with an increased use of slower-growing broilers, producing diets adjusted for male and female broiler requirements is a possibility, similar to what already happens with parent stock and turkey rearing systems.

Next to sex, production system will also have an impact on the nutritional needs of the broilers. Take for example the organic production systems compared to the conventional production systems. Not only is the growth rate of the strains or of the same strain different, but exposure to temperature and space to move is different, and this needs to be taken into account when formulating diets for organic production systems (van Knegsel and van Krimpen, 2008). Broilers will need more energy to cover the higher variations in ambient temperature and movement. This can be partially compensated by a higher feed intake, but this will not be enough. This can be a serious challenge in organic production systems, as the principles of not using exogenous feed enzymes, using less digestible protein ingredients, and not being able to use synthetic free amino acids (Burley et al., 2016; Amato and Castellini, 2022).

As already mentioned, reducing the growth period of the broilers will put even more stress in the first week of life. Not a lot is known on nutritional needs in the first week or days post-hatch. Designing diets for this period is still a challenge, and not a lot is known, or the literature is conflicting. On the one hand, during the first week of age, the microbial load of the gut is lower and the broiler chick could take advantage of this to retrieve more nutrients from the diet, as we are not feeding the microbiome, but only the broiler. At 3 weeks of age a stable microbiome is established, which also consumes a lot of amino acids and energy for growth and maintenance. This needs to be taken into account when formulating diets (Ravindran and Abdollahi, 2021). On the other hand, there is a 300% increase in BW in the first week of life. In this first week a lot of development of the gut, enzymes, and immune, transport, and absorption systems is still going on. The first week of life represents 20–25% of the overall rearing period; thus, its importance cannot be neglected (Ravindran and Abdollahi, 2021). Lipids in the yolk sac are the main energy source for the development of the chicken embryo. Glycogen stored in liver and muscles of the embryo will be used for pipping at hatching. Thus reserves of glycogen are depleted after hatching. Lipids in the residual yolk sac are heavily used in the last week of incubation, but up to 25% of these lipids in the residual yolk sac have not yet been used by the chick after hatching (Van der Wagt et al., 2020). The use of the nutrients in the yolk sac in the first 3 days after hatching is crucial for the chick. Apparently 75% of the residual yolk sac is used by day 3 (Ravindran and Abdollahi, 2021). Residual yolk sac nutrients can be complemented by the nutrients supplied by the feed. This residual yolk sac represents 20% of the BW at hatch and is composed of +/− 50% lipids (Ravindran and Abdollahi, 2021), which can provide the broiler chick with energy immediately after hatching as the glycogen reserves have been depleted during the hatching process. The use of the residual yolk sac's nutrients can happen through direct release in the blood circulation (absorption through the yolk sac membrane) or via expulsion in the gut (Van der Wagt et al., 2020). Some also say that the presence of the residual yolk sac has a positive effect on the utilization of dietary protein and energy, but the exact mechanisms by which this happens are still not clear. Without feed access, use of residual yolk sac nutrients takes place, but this is not an efficient process, as nutrients are valuable for the supply of maternal antibodies, phospholipids, choline, and triglycerides for membrane development. Providing feed will make the use of residual yolk sac nutrients much more efficient (Ravindran and Abdollahi, 2021). Adjusting to a diet based on carbohydrates for energy is vital for the chick, and that is why in the first 4–7 days post-hatch, the priority for the chick is the development of the intestinal tract. There is, however, still discussion if the positive effects of an adapted early feed strategy can still be observed later in life. Effects have been reported to remain up to 3–4 weeks of age, but disappear afterwards, most probably due to compensatory growth of the broilers at a later age (Ravindran and Abdollahi, 2021). This topic will be discussed more in detail in another chapter of this book.

## 3.4 Feeding strategies to meet nutritional needs of broilers

Because of the physiological changes and different nutritional needs at different ages, the broiler industry focuses on providing different diets with different nutritional value, which is known as phase feeding. In these different phases, one tries to match as closely as possible the nutrients offered through the diet to the needs of the broiler. In this way the impact on the environment of excess nutrient excretion can be decreased. Nevertheless, phase feeding will always lead to a limitation at the beginning of the phase and an excess of nutrients at the end of the feeding phase. Typically three or five phase feeding systems are used. Ideally more phases should be used, but this can be a challenge in practice. Buteri et al. (2009) tried to use 28 phases, but these produced similar performance results to a three- or five-phase feeding system.

Another strategy to assure a good gut development with maximum digestibility capacity, which will allow a more efficient use of dietary nutrients, meeting the nutritional needs of the broilers, is to have a basic diet and supplement whole wheat at different inclusion levels according to the age of the broilers. This post-pelleting method involves mixing and pelleting other feed ingredients (pellet concentrate) after which the whole grain is blended with the pelleted concentrate. To achieve positive effects, inclusion should happen after the first week of age as the gastrointestinal tract of the young chickens is not yet fully developed at an earlier age. Furthermore, the increase in wheat inclusion needs to be done gradually so that chickens are able to adapt to it. For example, starting with 5% at 10 days and increasing the supply level towards 35% at slaughter age. By increasing the level of wheat, a dilution of the diet occurs and can be more adapted to the needs of the chickens. However, it is important to notice that not only the inclusion level of wheat will change but also different basic diets will be needed during the growing period of the broilers. By incorporating whole wheat grain into the feed mixture, feed structure is modified. This can have positive effects on the development of the digestive tract, particularly of the gizzard, with increased tonicity (Gabriel et al., 2008) and deliver insoluble fibers that will have a prebiotic effect (Shang et al., 2020). Therefore whole wheat grain is being widely employed in broiler nutrition.

## 3.5 Feed ingredients, particle size and texture

To achieve the maximum genetic potential, diets need to be highly digestible because of the short retention time in the digestive tract. Feed ingredients, particle size, and texture are well-acknowledged factors affecting nutrient digestibility, feed utilization, gut health development, and overall production performance of broiler (Abdollahi et al., 2013).

The technology is rapidly evolving, with new enzymes, enzyme combinations, and novel applications being developed as rapidly as regulatory restrictions allow (Velázquez-De Lucio et al., 2021). However, there is still a great deal of ongoing research into the future enzymes that will be needed when using less conventional raw materials or combinations thereof but also on enzymes that will be developed to improve ingredient quality, food safety, that take into account the broiler's age, but also focusing on the effects of these future enzymes on various side activities and their dosage, leading to a maximization of net income and to a reduction in environmental pollution. Indeed, knowledge will also be needed on the (interactive) effects of combining these future enzymes with the yet to be developed additives (Cowieson, 2010).

To modify the particle size, grinding, the second most expensive process during feed manufacturing (Rubio et al., 2020), is done during processing. It has been shown in literature that changes in the feed ingredients' particle size (length and diameter) can significantly affect broilers' production. The feed form (mash, crumble, pellets) also significantly influences the development of the digestive system and its intestinal microflora, also affecting broilers' overall performance. Providing feed in pelleted form is an advanced form of feeding with many benefits compared to mash, such as improved nutrient intake, feed palatability (reduced dustiness), reduced ingredient segregation and waste, heat-sensitive ANFs are inactivated, and pathogens are destroyed (Abdollahi et al., 2013). However, there are also disadvantages related to pelleting, such as an increased cost, reduced particle properties, and overconsumption, which may result in metabolic disorders (Thaqui et al., 2009; Milanovic, 2018).

Feed structure, alongside the nutritional composition of a feed mixture, is one of the most crucial dietary properties. Chickens select their feed based on visual cues (glossiness) and taste (Niknafs and Roura, 2018). Additionally, despite having relatively few taste buds, they rely on mechanoreceptors in their beaks to determine whether to accept or reject a feed particle (Ferket and Gernat, 2006).

Particle size can have a positive impact on chickens' feed intake. In the case of mash diets, current research suggests that the ideal particle size ranges from 600 to 900 μm to optimize chicken growth (Amerah et al., 2007a). The length and diameter of pellets are also significant factors in enhancing feed intake in broilers. The optimal pellet length and diameter depend on the age of the broilers, being up to 2 mm for chickens until 7 days of age and 3 mm for older chickens. If pellets are smaller than 2 mm, quality and strength increase, resulting in less crumbling during processing (Naderinejad et al., 2016). However, there is a limit of particle size reduction as too fine particles may be difficult to be consumed by chickens. Particle size should ideally be fine to obtain higher pellet quality, but a nutritional conflict arises as coarse particles are required for chickens for sufficient gizzard development and digestive system (Naeem et al., 2024). The composition of the diet may also affect particle size during pelleting due to the high temperature used (level and duration), but also other factors, such as hardness of the main ingredients and the level of added fat, to name a few (Van der Poel et al., 2018; Yameen et al., 2025). In this sense, Amerah et al. (2007a, 2007b) concluded that recommendations concerning optimal particle size are contradictory, because outcomes from different experiments are confounded by factors such as diet complexity, grinding methods, physical feed form, and grain types, among others.

# 4 What will broiler diets look like in 10 years?

## 4.1 Impact of circularity

Over the next decade, broiler diets in countries like Belgium and the Netherlands are expected to shift away from linear, globalized supply chains that heavily rely on imported soy and cereals. This shift will be mainly guided by principles of circular agriculture, environmental sustainability, broiler welfare, and social fairness. This shift is a response to the growing pressure the sector is facing to reduce greenhouse gas emissions, minimize biodiversity loss (Leip et al., 2015), and address the food–feed competition (Mottet, 2017; Trimmer and Guest, 2018). In this way, it is expected that feed production will increasingly focus on locally sourced, circular feed ingredients, making a more efficient use of available biomass while reducing the pressure on natural resources (Puente-Rodríguez et al., 2022).

One of the key pillars of this transition is the choice of feed ingredients that might be less conventional in the broiler sector. For example, unavoidable food waste streams – such as processed vegetable trimmings, bakery residues, or dairy by-products – can be a valuable feed source. Although the collection, processing, and safety regulations regarding the use of these materials remain a challenge, their use can significantly contribute to reducing land use and global feed imports (Seidavi et al., 2021). Additionally, the reintroduction of processed animal proteins (PAPs) – including porcine meal in broiler diets – presents an opportunity to use high-quality protein streams in broiler diets, provided that all safety regulations are met (Pederiva et al., 2022). Other examples of future feed ingredients are insects (Slimen et al., 2023), seaweed (Al-Khalaifah et al., 2024), and microalgae products or side streams (Van Nerom et al., 2024), which, though still in its early production stages, offer opportunities to reduce the environmental impact of broiler feeds (see 3.2.2) (Sajjad et al., 2024). Finally, the development of alternative, locally grown plant-based feed sources – such as field beans, lupins, or biomass from roadside and waterway vegetation – supported by innovations in plant breeding, processing technologies, and market incentives and that could be integrated in poultry diets, would help not only reduce dependency on volatile global markets but also increase regional self-sufficiency and resilience (David et al., 2024).

To be able to incorporate the less conventional feed ingredients in broiler feed, feed technology will also be essential in this transition. To keep broiler production sustainable and responsible, meeting human nutritional needs is not only the choice of feed ingredients that is important, but also the nutritional value of these feed ingredients and how efficient these can be used by the broiler. Before implementation of the above-mentioned future feed ingredients can take place, research is needed not only to evaluate its nutritional value but also on their processing to render the maximum nutrients available to the broilers.

It is not only the choice of feed ingredients and their processing, to help broiler diets to be ecologically responsible, that will be important in the next decades, but the future of broiler diets must also be socially and ethically grounded. To realize this, fair access to feed innovations, economic viability for farmers, and attention to the health and welfare of broilers will be essential. Feed plays a crucial role in broiler health, productivity, and behavior, and a shift towards circular feed systems could help resolve broader broiler welfare challenges inherent to intensive livestock systems (Van Zanten et al., 2018).

In summary, broiler diets in 2035 will be even more diverse in composition, and at the same time more regional in origin, and more tightly integrated into circular, sustainable feed systems. This transition will only be successful if changes in policy, regulations, market structure, and knowledge sharing are allowed to take place. Only then we will have a livestock sector that contributes to climate goals, preserves natural ecosystems, supports farmers, and respects broiler welfare.

## 4.2 Nutrition, welfare and disease control in broiler production

In broiler production, welfare and nutrition are closely intertwined. Nutrition not only influences growth and productivity but also directly affects the physiological status and behavior of the broilers and hence their welfare. Indeed, inadequate or imbalanced diets can lead to welfare problems such as footpad dermatitis, skeletal disorders, poor feathering, and immune suppression, all of which reduce the chickens' ability to cope with environmental or pathogenic stressors. Moreover, high-performing breeds are particularly sensitive to gut disturbances and metabolic imbalances, making the nutritional composition of feed a central component of welfare management (Ducatelle et al., 2023).

One of the main challenges in broiler health today is the high incidence of infectious diseases. Intestinal pathogens, such as *Eimeria* spp. (leading to coccidiosis), *Clostridium perfringens* (leading to necrotic enteritis), and *E. coli*, are responsible for significant economic losses and welfare deterioration due to gut damage, inflammation, and increased mortality. Respiratory pathogens like infectious bronchitis virus (IBV) also compromise broiler health, welfare, and performance. These diseases often have an environmental component linked to management and will strongly interact with the nutritional management of the chickens. These diseases lead to a weakened immune system and poor bird resilience to other stressors, accompanied by an inefficient use of nutrients and drops in performance. To aggravate the situation, early-life stress and poor bird quality can significantly increase vulnerability to disease, emphasizing the need for comprehensive management strategies.

In recent years, nutrition-based interventions have gained attention as alternatives or complements to traditional disease control methods, especially in light of antimicrobial resistance. Among different feed additives, AA seem to play a central role due to their role in immune functions, antioxidant defense, tissue repair, and gut barrier maintenance (Barekatain et al., 2021). For instance, threonine and glutamine support mucus production and intestinal integrity, while tryptophan influences immune modulation and stress responses. Arginine has shown antimicrobial and anti-inflammatory effects, supporting overall disease resilience. Supplementation with specific free AAs has been shown to modulate gut microbiota, reduce inflammation, and enhance the broilers' ability to recover from or resist enteric infections (Bortoluzzi et al., 2018). Targeted amino acid strategies have also shown promising results in mitigating the negative effects of coccidiosis, necrotic enteritis, and other gut-related diseases without relying heavily on antibiotics (Chalvon-Demeersay et al., 2021; Lee et al., 2023), while increasing the welfare of the affected chickens.

*Salmonella* remains a major concern due to its zoonotic potential and public health implications. Nutritional strategies are increasingly being explored to reduce *Salmonella* colonization and shedding in broilers. Organic acids, probiotics, and phytochemicals may contribute to this by strengthening gut barrier function, enhancing mucosal immunity, and promoting beneficial microbial populations that can outcompete pathogens. However, their effectiveness depends on factors such as the broilers' age, health status, environmental conditions, and overall diet composition (Tariq et al., 2022; Naeem and Bourassa, 2024). In addition, feed technology can also play a role in controlling *Salmonella* presence and activity in feed. The balance between the temperature of processing (that needs to be high enough (80°C or higher) to sanitize feed), time at high temperature, and keeping the nutritional value of feed ingredients (e.g. denaturation of protein) is as important. Extrusion processes will normally be good to sanitize feed and control *Salmonella*, but expanding or pelleting is not enough to kill *Salmonella* (Huang et al., 2006). An alternative could be the fermentation of raw materials or of broiler diets to control *Salmonella* contamination (Heres et al., 2003; Jazi et al., 2019).

## 5 Conclusion

In conclusion, improving broiler health and welfare requires an integrated approach that recognizes the complex interactions between nutrition, disease, and physiology. By aligning nutritional interventions with disease prevention and welfare goals, the broiler industry can move towards more sustainable, resilient, and responsible production systems.

## 6 Future trends in research

In the future, feed formulation for broilers will be primarily focused on maximizing profitability which is defined as meeting the broilers' specific nutritional needs, improving their health, and minimizing environmental impact to achieve more sustainable production. By utilizing the above-mentioned innovative ingredients and technologies, feed can be more precisely tailored to meet these specific requirements of broilers. Increasing efforts to enhance the broiler's immunity thereby reducing the need for antibiotics will also be a central focus of future research and feed formulation. In addition, while not discussed in this chapter, the potential role of other nutritional strategies, such as water quality, is gaining increasing attention. The future direction of broiler nutrition will be shaped by ongoing changes in global agriculture and societal concerns. Ultimately, the challenge for the future will be to produce more with fewer resources, while ensuring sustainability and addressing the evolving demands of society.

## 7 Where to look for further information

- Applegate, T. (2024). *Advances in poultry nutrition*. Burleigh Dodds Science Publishing, UK.
- Babinszky, L. and Payan-Carreira, R. (2024). *Feed additives – recent trends in animal nutrition*. IntechOpen, New Delhi. https://doi.org/10.5772/intechopen.111000.
- Blair, R. (2018). *Nutrition and feeding of organic poultry*, 2nd edition, CABI International, Boston.
- Eeckhout, M. and van der Poel, A.F.B. (2024). *Animal feed technology*. Agrimedia – Erling-Verlag, Duitsland.
- FAO (2018). *World livestock: transforming the livestock sector through the sustainable development goals*, FAO, Rome, Italy.
- Moughan, P.J. and Hendriks, W.H. (2018). *Feed evaluation science*. Wageningen Academic Publishers, Wageningen, The Netherlands.

## 8 References

Abdel-Wareth, A.A., et al. (2024). Algae as an alternative source of protein in poultry diets for sustainable production and disease resistance: present status and future considerations. *Frontiers in Veterinary Science*, 11, 1382163.

Abdollahi, M.R., et al. (2013). Influence of pellet diameter and length on the quality of pellets and performance, nutrient utilisation and digestive tract development of broilers fed on wheat-based diets. *British Poultry Science*, 54(3), 337–345.

Aftab, U. (2019). Energy and amino acid requirements of broiler chickens: keeping pace with the genetic progress. *World's Poultry Science Journal*, 75(4), 507–514.

Alagawany, M., et al. (2021). Nutritional significance of amino acids, vitamins and minerals as nutraceuticals in poultry production and health – a comprehensive review. *Veterinary Quarterly*, 41(1), 1–29.

Alhotan, R.A. (2021). Commercial poultry feed formulation: current status, challenges, and future expectations. *World's Poultry Science Journal*, 77(2), 279–299.

Ali, Q., et al. (2022). Microbial short-chain fatty acids: a bridge between dietary fibers and poultry gut health. *Animal Bioscience*, 35(10), 1461–1478.

Al-Khalaifah, H.S. and Al-Nasser, A.Y. (2024). Evaluating the potential of marine algae as sustainable ingredients in poultry feed. *Agriculture*, 14(11), 1889.

Alshelmani, I.M., et al. (2024). *Enzymes in poultry feed*. IntechOpen, New Delhi. https://doi.org/10.5772/intechopen.112927.

Amato, M.G. and Castellini, C. (2022). Adaptability challenges for organic broiler chickens: a commentary. *Animals*, 12, 1354.

Amerah, A.M., et al. (2007a). Feed particle size: implications on the digestion and performance of poultry. *World's Poultry Science Journal*, 63, 439–455.

Amerah, A.M., et al. (2007b). Influence of feed form on gizzard morphology and particle size spectra of duodenal digesta in broiler chickens. *The Journal of Poultry Science*, 44(2), 175–181.

Applegate, T.J. and Angel, R. (2014). Nutrient requirements of poultry publication: history and need for an update. *Journal of Applied Poultry Research*, 23, 567–575.

Asnayanti, A., et al. (2024). Inducing experimental bacterial chondronecrosis with osteomyelitis lameness in broiler chickens using aerosol transmission model. *Poultry Science*, 103(3), 103460.

Attia, Y.A., et al. (2003). Nutritive value of undehulled sunflower meal as affected by multienzymes supplementation to broiler diets. *Archiv Für Geflügelkunde*, 67, 97–106.

Barekatain, R., et al. (2021). Intestinal barrier function and performance of broiler chickens fed additional arginine, combination of arginine and glutamine or an amino acid-based solution. *Animals*, 11(8), 2416.

Bello, A.U., et al. (2018). Gut microbiota and transportation stress response affected by tryptophan supplementation in broiler chickens. *Italian Journal of Animal Science*, 17(1), 107–113.

Bikker, P., et al. (2020). Evaluation of seaweeds from marine waters in Northwestern Europe for application in animal nutrition. *Animal Feed Science and Technology*, 263, 114460.

Blanch, A., et al. (1996). Utilization of different fats and oils by adult chickens as a source of energy, lipid and fatty acids. *Animal Feed Science and Technology*, 61(1–4), 335–342.

Bortoluzzi, C., et al. (2018). Threonine, arginine, and glutamine: influences on intestinal physiology, immunology, and microbiology in broilers. *Poultry Science*, 97(3), 937–945.

Burley, H.K., et al. (2016). Formulation challenges of organic poultry diets with readily available ingredients and limited synthetic methionine. *Journal of Applied Poultry Research*, 25, 443–454.

Buteri, C.B., et al. (2009). Effects of different nutritional plans on broiler performance. *Brazilian Journal of Poultry Science*, 11 (4), 225–234.

Casaubon-Huguenin, M.T., et al. (2004). The effect of raw full-fat soybean and its lectin on the nutrition and pigmentation of broilers. *Journal of Agricultural and Food Chemistry*, 52(18), 5702–5708.

Chalova, V.I., et al. (2016). Reduction of nitrogen excretion and emissions from poultry: a review for conventional poultry. *World's Poultry Science Journal*, 72(3), 509–520.

Chalvon-Demersay, T., et al. (2021). Functional amino acids in pigs and chickens: implication for gut health. *Frontiers in Veterinary Science*, 8, 663727.

Choct, M. (2006). Enzymes for the feed industry: past, present and future. *World's Poultry Science Journal*, 62, 5–16.

Choct, M., et al. (1996). Increased small intestinal fermentation is partly responsible for the anti-nutritive activity of non-starch polysaccharides in chickens. *British Poultry Science*, 37(3), 609–621.

Choi, J., et al. (2023). Nutritional strategies to improve meat quality and composition in the challenging conditions of broiler production: a review. *Animals*, 13(8), 1386.

Chowdhury, V.S., et al. (2015). Oral administration of L-citruline, but not L-arginine or L-Ornithine acts as a hypothermic agent in chicks. *Journal of Poultry Science*, 52, 331–335.

Chukwudi, P., et al. (2024). The effects of organic acids on broiler chicken nutrition: a review. *Animal Research and One Health*, 3, 43–53.

Collett, S.R. (2012). Nutrition and wet litter problems in poultry. *Animal Feed Science and Technology*, 173, 65–75.

Cowieson, A.J., et al. (2010). Interactions between xylanase and glucanase in maize-soy-based diets for broilers. *British Poultry Science*, 51(2), 246–257.

Cromwell, G.L. (1999). Soybean meal - the "gold standard". *The Farmer's Pride, KPPA News*, 11(20), 30.

David, L.S., et al. (2024). Feeding value of lupins, field peas, faba beans and chickpeas for poultry: an overview. *Animals*, 14(4), 619.

Dewi, Y.L., et al. (2024). Processing technology to improve seaweed nutritional quality as a feed for poultry: a review and its implementation. *World's Poultry Science Journal*, 80(1), 207–235.

Douglas, M.W., et al. (1999). Nutritional evaluation of lectin-free soybeans for poultry. *Poultry Science*, 78(1), 91–95.

Du, H., et al. (2024). Organic acids in poultry industry: a review of nutritional advancements and health benefits. *World's Poultry Science Journal*, 80(1), 133–153.

Ducatelle, R., et al. (2023). Poultry gut health and beyond. *Animal Nutrition*, 13, 240–248.

Ferket, P.R. and Gernat, A.G. (2006). Factors that affect feed intake of meat birds: a review. *International Journal of Poultry Science*, 5(10), 905–911.

Gabriel, I., et al. (2008). Effects of whole grain wheat feeding on the development of the digestive tract of broiler chickens. *Animal Feed Science and Technology*, 142(1–2), 144–162.

Gale, F. and Arnade, C. (2015). Effects of rising feed and labor costs on China's chicken price. *International Food and Agribusiness Management*, 18(Special Issue A), 137–150.

Gholami, M., et al. (2024). Effect of emulsifier on growth performance, nutrient digestibility, intestinal morphology, faecal microbiology and blood biochemistry of broiler chickens fed low-energy diets. *Veterinary Medicine and Science*, 10(3), e1437.

Gomez-Osorio, L.M., et al. (2021). Short and medium chain fatty acids and their derivatives as natural strategy in the control of necrotic enteritis and microbial homeostasis in broiler chickens. *Frontiers in Veterinary Sciences*, 8, 773372.

Heger, J. (2003). Essential to non-essential amino acid ratios. In: D'Mello, J.P.F. (ed.), *Amino acids in animal nutrition*. CABI Publishing, Wallingford, pp. 103–124.

Heres, L., et al. (2003). Fermented liquid feed reduces susceptibility of broilers for Salmonella enteritidis. *Poultry Science*, 82(4), 603–611.

Heuzé, V., et al. (2017). Microalgae. Feedipedia, a programme by INRAE, CIRAD, AFZ and FAO.

Heuzé, V., et al. (2020). Soybean meal. Feedipedia, a programme by INRAE, CIRAD, AFZ and FAO.

Hilliar, M., et al. (2020). Using crystalline amino acids to supplement broiler chicken requirements in reduced protein diets. *Poultry Science*, 99(3), 1551–1563.

Hombegowda, G.P., et al. (2021). Growth performance, carcass traits and gut health of broiler chickens fed diets incorporated with single cell protein. *Animal Bioscience*, 34(12), 1951–1962.

Huang, D.S., et al. (2006). Effects of feed particle size and feed form on survival of Salmonella typhimurium in the alimentary tract and cecal S. typhimurium reduction in growing broilers. *Poultry Science*, 85(5), 831–836.

Jazi, V., et al. (2019). Fermented soybean meal ameliorates *Salmonella typhymurium* infection in young broiler chickens. *Poultry Science*, 98(11), 5648–5660.

Kocher, A., et al. (2002). Effects of feed enzymes on nutritive value of soyabean meal fed to broilers. *British Poultry Science*, 43(1), 54–63.

Korver, D.R. (2023). Current challenges in poultry nutrition, health, and welfare. *Animal*, 17, 100755.

Kuenz, S., et al. (2022). Effects of gradual differences in trypsin inhibitor activity on the estimation of digestible amino acids in soybean expellers for broiler chickens. *Poultry Science*, 101(4), 101740.

Lauzon, D.A. (2006). *The effects of micronutrients on pullets and broilers*. Louisiana State University Master's Theses, 4151.

Ledbetter, F. (2025). Single-Cell protein for feed and food. *The EuroBiotech Journal*, 9(1), 117–127.

Lee, J.T., et al. (2023). Functional properties of amino acids: improve health status and sustainability. *Poultry Science*, 102(1), 102288.

Leip, A., et al. (2015). Impacts of European livestock production: nitrogen, sulphur, phosphorus and greenhouse gas emissions, land-use, water eutrophication and biodiversity. *Environmental Research Letters*, 10, 115004.

Li, Y.P., et al. (2024). Recent advances and challenges in single cell protein (SCP) technologies for food and feed production. *NPJ Science of Food*, 8(1), 66.

Maharjan, P., et al. (2021). Review: Physiological growth trend of current meat broilers and dietary protein and energy management approaches for sustainable broiler production. *Animal*, 15, 100284.

Mahfuz, S., et al. (2021). Phenolic compounds as natural feed additives in poultry and swine diets: a review. *Journal of Animal Science and Biotechnology*, 12, 1–18.

Makkar H.P.S., et al. (2014). State of the art on use of insects as animal feed. *Animal Feed Science and Technology*, 197, 1–33.

Marchesi, J.A.P., et al. (2021). Exploring the genetic architecture of feed efficiency traits in chickens. *Scientific Reports*, 11(1), 4622.

Martínez, Y. and Valdivié, M. (2021). Efficiency of Ross 308 broilers under different nutritional requirements. *Journal of Applied Poultry Research*, 30, 100140.

McKay, J.C., et al. (2000). *Broiler production around the world*. XXI. In: Proceedings of the Worlds Poultry Congress, 20–24 August, 2000, Montreal, Canada, pp. 20–24.

McNamee, P.T. and Smyth, J.A. (2000). Bacterial chondronecrosis with osteomyelitis ('femoral head necrosis') of broiler chickens: a review. *Avian Pathology*, 29, 253–270.

Milanovic, S. (2018). *Literature review on the influence of milling and pelleting on nutritional quality, physical characteristics, and production cost of pelleted poultry feed*. Master Thesis, Norwegian University of Life Sciences, p. 47.

Miller, M., et al. (2022). *Poultry expected to continue leading global meal imports as demand rises*. The Economics of Food, Farming, Natural Resources, and Rural America, Amber Waves.

Mohammadigheisar, M., et al. (2021). Growth performance, organ attributes, nutrient and caloric utilization in broiler chickens differing in growth rates when fed a corn-soybean meal diet with multienzyme supplement containing phytase, protease and fiber degrading enzymes. *Poultry Science*, 100, 101362.

Moore, R.J. (2016). Necrotic enteritis predisposing factors in broiler chickens. *Avian Pathology*, 45(3), 275–281.

Mottet, A., et al. (2017). Livestock: on our plates or eating at our table? a new analysis of the feed/food debate. *Global Food Security*, 14, 1–8.

Murawska, D., et al. (2011). Age-related changes in the percentage content of edible and non-edible components in broiler chickens. *Asian-Australasian Journal of Animal Science*, 24(4), 532–539.

Naderinejad, S., et al. (2016). Influence of feed form and particle size on performance, nutrient utilisation, and gastrointestinal tract development and morphometry in broiler starters fed maize-based diets. *Animal Feed Science and Technology*, 215, 92–104.

Naeem, M. and Bourassa, D. (2024). Optimizing poultry nutrition to Combat Salmonella: insights from the literature. *Microorganisms*, 12(12), 2612.

Naeem, M., et al. (2024). Investigating the effect of pelleting on the particle size of commercial poultry feeds: a case study. *Journal of Applied Poultry Research*, 33, 100390.

Nahm, K.H. (2007). Feed formulations to reduce N excretion and ammonia emission from poultry manure. *Bioresource Technology*, 98(12), 2282–2300.

Nasseri, A.T., et al. (2011). Single cell protein: production and process. *American Journal of Food Technology*, 6(2), 103–116.

Niknafs, S. and Roura, E. (2018). Nutrient sensing, taste and feed intake in avian species. *Nutrition Research Reviews*, 31, 256–266.

Nyachoti, C.M., et al. (2018). Plant secondary compounds and antinutritional factors; In: Moughan, P.J. and Hendriks, W.H. (eds), *Feed evaluation science*. Wageningen Academic Publishers, Wageningen, pp. 145–172.

Pacheco, W.J., et al. (2014). Effects of trypsin inhibitor and particle size of expeller-extracted soybean meal on broiler live performance and weight of gizzard and pancreas. *Poultry Science*, 93(9), 2245–2252.

Patterson, J.A. and Burkholder, K.M. (2003). Application of prebiotics and probiotics in poultry production. *Poultry Science*, 82(4), 627–631.

Pederiva, S., et al. (2022). Processed animal proteins (PAPs) in animal nutrition: assessment of the chemical risk of essential and non-essential elements. *Journal of Trace Elements in Medicine and Biology*, 71, 126959.

Poernama, F., et al. (2021). The effect of feeding phytase alone or in combination with non-starch polysaccharides-degrading enzymes on broiler performance, bone mineralization, and carcass traits. *Journal of Applied Poultry Research*, 30(1), 100134.

Pope, M., et al. (2023). Quantifying the value of soybean meal in poultry and swine diets. *Journal of Applied Poultry Research*, 32(2), 100337.

Puente-Rodríguez, D., et al. (2022). A circularity evaluation of new feed categories in the Netherlands–squaring the circle: a review. *Sustainability*, 14(4), 2352.

Ramteke, R., et al. (2019). Antinutritional factors of feed and fodder used for livestock and poultry feeding. *Acta Scientific Nutritional Health*, 3(5), 39–48.

Ravindran, V. (2013). Main ingredients used in poultry feed formulations. In: *Poultry development review*, FAO, pp. 67–69. ISBN 978-92-5-108067-2.

Ravindran, V. and Abdollahi, M.R. (2021). Nutrition and digestive physiology of the broilers chick: state of the art and outlook. *Animals*, 11, 2795.

Raza, A., et al. (2024). Role of probiotics in increasing meat and egg production in poultry: a review. *Kafkas Universitesi Veteriner Fakultesi Dergisi*, 30(6), 753–760.

Rotz, C.A. (2004). Management to reduce nitrogen losses in animal production. *Journal of Animal Science*, 82(13), E119–E137.

Rubio, A.A., et al. (2020). Effects of corn particle size on broiler performance during the starter, grower and finisher periods. *Journal of Applied Poultry Research*, 29(2), 352–361.

Sajjad, M., et al. (2024). Insect larvae as an alternate protein source in poultry feed improve the performance and meat quality of broilers. *Animals*, 14(14), 2053.

Salahi, A., et al. (2025). Gut dysbiosis: nutritional causes and risk prevention in poultry, with reference to other animals. *South African Journal of Animal Science*, 55(2), 33–63.

Sebatta, C., et al. (2018). Farmers' acceptance of insects as an alternative protein source in poultry feeds. *International Journal of Agricultural Research, Innovation and Technology*, 8(2), 32–41.

Seidavi, A., et al. (2021). Practical application of some common agro-processing wastes in poultry diets. *World's Poultry Science Journal*, 77(4), 913–927.

Shang, Q., et al. (2020). The impact of wheat bran on the morphology and physiology of the gastrointestinal tract in broiler chickens. *Animals*, 10(10), 1831.

Slimen, I.B., et al. (2023). Insects as an alternative protein source for poultry nutrition: a review. *Frontiers in Veterinary Science*, 10, 1200031.

Slominski, B.A. (2011). Recent advances in research on enzymes for poultry diets. *Poultry Science*, 90(9), 2013–2023.

Swennen, Q., et al. (2007). Implications of dietary macronutrients for growth and metabolism in broiler chickens. *World's Poultry Science Journal*, 63(4), 541–556.

Szabó, R.T., et al. (2023). Effects of short-and medium-chain fatty acids on production, meat quality, and microbial attributes–a review. *Molecules*, 28(13), 4956.

Tadele, Y. (2015) Important anti-nutritional substances and inherent toxicants of feeds. *Food Science and Quality Management*, 36, 40–47.

Taira, K., et al. (2013). Effect of litter moisture on the development of footpad dermatitis in broiler chickens. *Journal of Veterinary Medical Sciences*, 76(4), 583–586.

Tariq, S., et al. (2022). Salmonella in poultry; an overview. *International Journal of Multidisciplinary Sciences and Arts*, 1(1), 80–84.

Thaqui, M., et al. (2009). *Testing of pelleting efficiency, quality and effect of pelleted feed on broiler performance*. Proceedings 37th International Symposium 'Actual Tasks on agricultural engineering', Opatija, Croatia, 10–13 Veljače, pp. 527–537.

Toghyani, M., et al. (2016). Effect of stocking density and dietary sulfur amino acids on welfare indices of broiler chicks. *Journal of Advanced Agricultural Technologies*, 3(2), 124–127.

Trimmer, J.T. and Guest, J.S. (2018). Recirculation of human-derived nutrients from cities to agriculture across six continents. *Nature Sustainability*, 1, 427–435.

Tůmová, E., et al. (2021). Research note: The effects of genotype, sex, and feeding regime on performance, carcasses characteristic, and microbiota in chickens. *Poultry Science*, 100(2), 760–764.

van der Poel, A.F.B., et al. (2018). Feed processing. In: Moughan, P.J. and Hendriks, W.H. (eds), *Feed evaluation science*. Wageningen Academic Publishers, Wageningen, pp. 295–336.

Van der Wagt, I., et al. (2020). A review on yolk sac utilization in poultry. *Poultry Science*, 99(4), 2162–2175.

van Huis A. (2015). Edible insects contributing to food security? *Agriculture and Food Security*, 4, 20.

van Knegsel, A.T.M. and van Krimpen, M.M. (2008). *Energy and protein requirements of organic housed poultry*. Wageningen University, Rapport.

Van Nerom, S., et al. (2024). Microalgae as feed additives in poultry: a review on the health-promoting effects. *Algal Research*, 83, 103733.

Van Zanten, H.H.E., et al. (2018). Defining a land boundary for sustainable livestock consumption. *Global Change Biology*, 24, 4185–4194.

Velázquez-De Lucio, B.S., et al. (2021). Exogenous enzymes as zootechnical additives in animal feed: a review. *Catalysts*, 11(7), 851.

Whitehead, C.C. (2002). Nutrition and poultry welfare. *World's Poultry Science Journal*, 58(3), 349–356.

Woyengo, T.A., et al. (2017). Effects of anti-nutritional factors of oilseed co-products on feed intake of pigs and poultry. *Animal Feed Science and Technology*, 233, 76–86.

Xu, T.T., et al. (2023). Gut microbiome dysregulation drives bone damage in broiler tibial dyschondroplasia by disrupting glucose homeostatsis. *Biofilms and Microbiomes*, 9, 1.

Yameen, R.K., et al. (2025). A review on feed particle size and form: implications on the performance and gut health of poultry. *Poultry Science Journal*, 13(1), 1–15.

Zampiga, M., et al. (2021). Importance of feed efficiency for sustainable intensification of chicken meat production: implications and role for amino acids, feed enzymes and organic trace minerals. *World's Poultry Science Journal*, 77(3), 639–659.

# Chapter 11

## Understanding the environmental needs of broilers

*Edgar Orlando Oviedo-Rondón, North Carolina State University, USA*


1 Introduction
2 Temperature
3 Ventilation
4 Air quality
5 Variability in environmental parameters
6 Flooring systems
7 Lighting
8 Space management
9 Modern technologies to achieve environmental needs
10 Conclusion
11 References


### 1 Introduction

Environments that support broiler homeostasis are essential for their proper metabolism, growth, and development, which, in turn, affect health and welfare, as is the case with any living organism (Tao and Xin, 2003; Mesa et al., 2017; Lima et al., 2021). Environmental needs can be grouped into interconnected dimensions. These dimensions include temperature, humidity, air quality, air velocity, lighting, and space management, such as stocking density and environmental enrichment.

Optimizing the housing environment according to age, growth rate, and stocking density can improve welfare conditions, maintain health, and promote high productivity in broiler chicken flocks, while maintaining energy efficiency and minimizing environmental impact, thereby making it a sustainable (Li et al., 2022) and profitable practice (Trentin et al., 2025). Broiler productivity, low morbidity, and low mortality have been identified as primary measures for assessing the effectiveness of housing conditions. However, welfare parameters, such as the incidence and severity of footpad dermatitis and hock

http://dx.doi.org/10.19103/AS.2025.0165.12

burns, as well as behavioral parameters, need to be considered nowadays, given the increasing pressure on welfare from consumers and, in some areas of the world, government regulations.

The building itself, the drinking, feeding, and ventilation equipment, the litter material, the chickens housed, the feed nutrient composition, and the interactions among these factors, and the outdoor climate influence the final set of environmental variables (Abeyesinghe et al., 2001; McLean et al., 2002; Freire et al., 2003; Lima et al., 2021). There are numerous complex interactions among these factors; the environmental data exhibit substantial dispersion and multidimensionality, making it challenging to optimize variables dynamically as the broilers grow and age (Benvenga et al., 2020; Martinez et al., 2021; Li et al., 2022). Currently, most broiler houses worldwide are equipped with electrical devices, electronic sensors, controllers, and monitoring equipment for environmental management. The appropriate management, settings, and maintenance of these devices determine the actual accuracy in obtaining the desired parameters and evaluating the results for future improvement.

Broiler production is optimized when the rearing environment is nearly ideal (Lima et al., 2011; Chou et al., 2004; Stamp et al., 2004; Aradas et al., 2005; Trentin et al., 2025). This can occur when broilers are housed in a thermoneutral environment that varies dynamically with broiler age, stocking density, air speed, external temperature, moisture, and atmospheric pressure (Pecoraro et al., 2023; Omomowo et al., 2021; Ferreira et al., 2024). However, these conditions are difficult to predict and adapt to due to their variability. Despite advances in electronic sensors, house controllers, and predictive algorithms (Abbas et al., 2022; George and George, 2023; Brassó et al., 2025), many broiler houses do not achieve optimal conditions at certain times during the grow-out period. Monitoring and detecting environmental inefficiencies and variability in flock performance and welfare are necessary to improve broiler housing.

Broiler welfare, health, and performance may be compromised by the physiological adjustments required by a harsh housing environment (Tao and Xin, 2003; Nawab et al., 2018), especially under hot weather. The hot ambient air affects indoor air quality, moisture levels, and litter quality. The performance and welfare of broilers housed under intensive conditions are significantly affected by heat accumulation and its effects on air quality within the house, rather than by chicken stocking density (Dawkins et al., 2004; Nawab et al., 2018).

Air quality is critical for broiler production (Soliman et al., 2020; Van Limbergen et al., 2020; Barbosa et al., 2024). Effective ventilation is essential for dissipating excess heat, moisture, gases from broilers and waste breakdown products, and dust particles released indoors. Suboptimal ventilation can increase dust, noxious gases, and water vapor, which may contribute to the incidence of broiler respiratory diseases (Menegali et al., 2012; Mesa et al., 2017; Liu et al., 2021; Vissers et al., 2021; Barbosa et al., 2024).

There is considerable variation in housing conditions across geographic locations, broiler rearing characteristics, flooring or litter type, and the use of different cooling and ventilation systems (Qaid et al., 2023; Honig et al., 2024). Consequently, it is difficult to generalize these concepts, and each type of housing and equipment may have different optimal parameters for achieving the best possible results. On the other hand, over the past decades, climate change has made weather conditions more variable and extreme, making it more difficult to maintain optimal environmental conditions in broiler houses.

Numerous authors have demonstrated that if environmental conditions are not met in broiler houses, the zoo-hygienic recommended standards are not achieved, increasing pathogen proliferation in the house environment that may affect health (Özkan et al., 2003; Tuyttens et al., 2005; Skomorucha et al., 2007, 2010). Consequently, another critical goal in poultry housing is to minimize conditions that increase microbial proliferation.

The objective of this chapter is to discuss the critical environmental parameters that should be controlled to guarantee conditions for optimum broiler growth, welfare, and health. In the following sections, the main dimensions of the broiler house environment will be discussed separately, with their interactions with other dimensions highlighted. The development of new technologies for controlling and monitoring environmental parameters will be reviewed.

## 2 Temperature

House temperature is one of the basic decisions that broiler growers must make. However, air temperature is not the only factor to consider in broiler thermal comfort. Optimum temperatures have been extensively researched for several decades and vary depending on genetics, growth rate as broilers age, body mass, stocking density, litter material, type of heating source (Charles, 1986; Donkoh, 1989; Nawab et al., 2018), air velocity (Akter et al., 2022), and air relative humidity (Omomowo, 2021). The temperature that broilers sense, or effective temperature, may depend on the interactions among all these factors, and, in general, the relationship is simplified by thermal indices (Del Valle et al., 2021; Ferreira et al., 2024). Among the factors that may affect broiler production and efficiency, the effective temperature as the flock ages is probably the most relevant (Baracho et al., 2019).

The optimal temperatures at each age and condition vary as genetic selection for rapid growth advances and metabolic heat production increases (Gous and Morris, 2005). The heat production of broilers is related to their growth rates, and the comfort temperatures, or thermoneutral zones, decrease as heat production increases (Table 1). It can be observed from recommendations from genetic companies over the past 20 years that optimal growth temperatures have decreased, based on expected body weight gains. The air relative humidity also affects the optimum temperature for each and average flock weight (Table 2).

Temperature affects broiler development from the embryonic period, depending on incubation conditions, and has effects in the post-hatch life (Maatjens et al., 2016; Wijnen et al., 2020; Tona et al., 2022; Wang et al., 2025; Meteyake et al., 2023). However, this chapter primarily addresses the multiple effects of post-hatch brooding temperature on broiler behavior, welfare, and performance (Deaton et al., 1996; Henriksen et al., 2016). Nevertheless, it is important to consider that incubation temperature may affect the optimum brooding temperature (Bruzual et al., 2000) or the temperature preference of chicks (Morita et al., 2016; Wijnen et al., 2022). Incubation temperatures and early post-hatch thermal conditioning can also be used to increase thermotolerance and enhance the performance and health of broilers under cold or heat stress (Loyau et al., 2015; Zaboli et al., 2017; Meteyake et al., 2023).

During the first week of life, the underdeveloped thermoregulatory system of young chicks makes them extremely sensitive to temperature changes. This constraint renders them functionally poikilothermic, necessitating external heat sources to sustain optimal body temperature. The optimal internal body temperature for broiler chickens increases with age. It is approximately 39.5–40.1°C at hatch; it reaches 40.0–40.6°C during the first week of life, increasing to 41.1°C by the second week of life (11–14 days), and generally reaches 41.2–41.8°C in 5–8-week broilers under thermoneutral conditions.

Even short periods of exposure to suboptimal temperatures can significantly affect the growth, welfare, and performance of chicks (Quintana-Ospina et al., 2023). Chicks exhibit high metabolic plasticity during the embryonic period and the first days of life; consequently, environmental factors can permanently modify their phenotype (Naraballobh et al., 2016; Weeks et al., 2022).

Maman et al. (2019) demonstrated that chicks exposed to 40.48°C ± 0.21°C had cloacal temperatures significantly higher (42.60°C ± 0.12°C) than those of control chicks (39.96°C ± 0.12°C), which were maintained at 36.19°C ± 0.21°C for 12 h. Chicks exposed to high temperatures had lower body weight at placement, lower feed intake (−3.1%) and body weight (−4.3%), and higher feed conversion ratio (+2.74%) and mortality (5.3% vs 1.78%) than the control group up to 35 days of age. In contrast, another group exposed to low temperatures (21.68°C ± 0.21°C) for 12 h showed performance no different from that of the control group. However, low temperatures raise the risk of hypothermia, which can lead to pulmonary hypertension syndrome (ascites), according to studies by Hasanpur et al. (2015), Kamely et al. (2015), and others. High temperatures cause atrophy of lymphoid organs, affecting immunological

**Table 1** Heat production in male and female broilers, estimated from observed body weights and the estimated comfort temperature[#]

| Age days | Heat production, kJ/d | | | | Comfort temperature, °C | | | |
|---|---|---|---|---|---|---|---|---|
| | 1970 | | 2004 | | 1970 | | 2004 | |
| | Males | Females | Males | Females | Males | Females | Males | Females |
| 7 | 180 | 180 | 204 | 200 | 32.0 | 32.0 | 29.0 | 29.0 |
| 14 | 410 | 350 | 468 | 458 | 30.0 | 29.5 | 25.0 | 25.5 |
| 21 | 760 | 620 | 845 | 843 | 28.0 | 27.0 | 20.0 | 21.0 |
| 28 | 1030 | 866 | 1260 | 1250 | 25.5 | 24.0 | 15.5 | 17.0 |
| 35 | 1444 | 1030 | 1545 | 1600 | 23.5 | 21.0 | 12.0 | 14.5 |
| 42 | 1650 | 1165 | 1785 | 1840 | 21.5 | 18.5 | 11.5 | 15.0 |

[#]Calculated as the average of the lower and the upper critical temperature within the thermoneutral range.
*Source*: Adapted from Gous and Morris, 2005.

**Table 2** Temperatures recommended (°C) for broilers Ross 308 and Ross 308 AP according to relative humidity

| Age (days) | 2006 | | | | | 2025 | | | | |
|---|---|---|---|---|---|---|---|---|---|---|
| | Body weight (g) | Relative humidity (%) | | | | Body weight (g) | Relative humidity (%) | | | |
| | | 50 | 60 | 70 | 80 | | 50 | 60 | 70 | 80 |
| 0 | 42 | 33.5 | 30.5 | 28.6 | 27.5 | 44 | 33.2 | 30.8 | 29.2 | 28.5 |
| 3 | 76 | 32.0 | 29.5 | 27.6 | 27.2 | 103 | 31.2 | 28.9 | 27.3 | 26.8 |
| 6 | 129 | 31.0 | 28.5 | 26.6 | 26.3 | 185 | 29.9 | 27.7 | 26.0 | 25.8 |
| 9 | 206 | 29.7 | 27.5 | 25.6 | 25.3 | 290 | 28.6 | 26.7 | 25.0 | 24.9 |
| 12 | 309 | 27.2 | 25.0 | 23.8 | 22.9 | 425 | 27.8 | 25.7 | 24.0 | 23.8 |
| 15 | 440 | 26.2 | 24.0 | 22.5 | 21.8 | 590 | 26.8 | 24.8 | 23.0 | 22.9 |
| 18 | 599 | 25.0 | 23.0 | 21.5 | 20.0 | 790 | 25.5 | 23.6 | 21.9 | 21.8 |
| 21 | 785 | 24.0 | 22.0 | 20.5 | 19.8 | 1015 | 24.7 | 22.7 | 21.3 | 21.2 |
| 24 | 992 | 23.0 | 21.0 | 19.5 | 18.9 | 1260 | 23.5 | 21.7 | 20.2 | 20.1 |
| 27 | 1217 | 23.0 | 21.0 | 19.5 | 18.6 | 1530 | 22.7 | 20.7 | 19.3 | 19.2 |

*Source*: Aviagen Ross recommendations 2006 and 2025.

responses to pathogens (Honda et al., 2015; Hirakawa et al., 2020). In addition, heat stress induces higher body temperature, faster respiratory rate, metabolic acidosis, and dehydration (Farag and Alagawany, 2018).

Optimal brooding temperatures can vary with broiler breeder age due to egg and chick size, incubation temperature (Leksrisompong et al., 2007, 2009; Morita et al., 2016), and relative humidity (Bruzual et al., 2000). The effects can be observed in thermal preferences and responses to heat stress throughout the rearing phase (Piestun et al., 2011; Akşit et al., 2013; Wijnen et al., 2022; Meteyake et al., 2023).

In older broilers, Hamrita and Conway (2017) demonstrated that deep body temperatures (DBT) of 8.6-, 9.0-, and 9.4-week-old broilers rose from 1.5°C to 3°C (average 1.2°C to 1.9°C) from 41.2°C to 43.8°C when environmental temperature increased from 20°C to 33°C during 4 h to 5 h, and no ventilation was used. These authors used biotelemetry electronic sensors to demonstrate that air speeds of 3.6 m/s, normally used in tunnel ventilation, helped to minimize this change in DBT, and only 0.5°C to 1.0°C higher temperatures were observed. Tunnel ventilation (3.6 m/s) maintained body weight gain in heavy broilers between 118 g/day and 187 g/day, whereas control groups without ventilation experienced body weight loss (maximum of 50 g) to 167 g/day.

## 2.1 Methods to measure thermal needs

Thermal comfort indices, such as the temperature-humidity index (THI), integrate the effects of temperature, humidity, and, in some cases, air velocity (Tao and Xin, 2003; Akter et al., 2022) to predict the impact of thermal conditions on performance. However, the THI does not consider thermal radiation (solar and longwave from heaters), duration of exposure, or differences in age, sex, or genotype. Several THI have been generated or adapted for broilers (Table 3). These THI vary with genetics, growth rates, and environmental conditions during development (Table 4).

**Table 3** Temperature-humidity indexes (THI) for broilers

| Equation | Reference |
|---|---|
| $(1.8 \times T + 32) - [(0.55 - 0.0055 \times RH) \times (1.8 \times T - 26.8)]$ | NRC (1971) |
| $T + 0.35\,dpT + 41.5$ | Thom (1979) |
| $0.6 \times T_{DB} + 0.4 \times T_{WB}$ | DeShazer and Beck (1988) |
| $0.85 \times T_{DB} + 0.15 \times T_{WB}(0.85 \times T_{DB} + 0.15 \times T_{WB}) \times V^{-0.058}$ | Tao and Xin (2003) |
| 3 ~ 4 weeks: $0.62 \times T_{DB} + 0.38 \times T_{WB}$5 ~ 6 weeks: $0.71 \times T_{DB} + 0.29 \times T_{WB}$ | Chepete et al. (2005) |
| $0.85 \times T_{DB} + (RH \times (T_{DB} - 14.3)/100) + 46.3$ | De Moraes et al. (2008) |

Note: dpT = Dew point temperature; DB = Dry bulb; WB = Wet bulb; T = Temperature; RH = Relative humidity.

**Table 4** Comfort range for temperature heat index (THI), black globe and humidity index (BGHI), and enthalpy (H) for broilers between 1 and 6 weeks of age

| Week | THI range[1] | BGHI range[2] | H range[3] |
|---|---|---|---|
| 1 | 73–81 | 77–82 | 76–87 |
| 2 | 69–77 | 74–77 | 70–78 |
| 3 | 65–73 | 69–71 | 60–70 |
| 4 | 62–70 | 68–70 | 52–60 |
| 5 | 56–64 | 68–70 | 40–55 |
| 6 | 54–62 | 68–70 | 37–52 |

[1]Thom (1979);
[2]Buffington et al. (1981);
[3]Rodrigues et al. (2010).

The black globe temperature and humidity index (BGHI) incorporates thermal radiation, providing a more comprehensive assessment of the environment as perceived by broilers, replacing the dry-bulb temperature in the THI with the black globe temperature (Buffington et al., 1981; Ponciano et al., 2013; Ferreira et al., 2024). The comfortable BGHI values have been considered to be between 77 and 81.3 in the first week and between 74.5 and 77.0 in the second week (Table 4).

Another method to estimate the need for specific temperature is to use enthalpy (H), a thermodynamic property. It estimates the thermal energy that must be removed to ensure adequate comfort or survival conditions within the facility. This is calculated using the following equation (Rodrigues et al., 2010):

$$H = 1.006 \times t + RH/BP \times 10^{(7.5 \times t/237.3 + t)} \times (71.28 + 0.052 \times t)$$

where

H = enthalpy (kJ kg of dry air$^{-1}$);

t = air temperature (°C);

RH = relative humidity (%);

BP = local barometric pressure (mmHg).

The THI, BGHI, and H could be used in electronic controllers to determine dynamically comfort ranges for broilers to integrate multiple pieces of information from sensors located in the broiler house.

Broilers generate metabolic heat proportional to their growth rate, body mass, and body composition (Martinez et al., 2022). Core broiler body temperature can be influenced by environmental factors (Han et al., 2021), especially in young chicks (Tzschentke, 2007). These effects can be measured in the animal or predicted using equations. Body temperature can be measured with cloacal thermometers. However, this invasive procedure causes stress-induced hyperthermia. Other methodologies currently used to study body temperature include surgically implanted radio-telemetry data loggers

(Ahmmed et al., 2024), telemetry devices, or data loggers that can be placed in broiler crops (Iyasere et al., 2017; Hamrita and Conway, 2017) to measure DBT.

An additional method is infrared thermography using pictures and video (Hernández-Sánchez et al., 2024; Vercellino et al., 2025). Featherless body areas are adequate for predicting variations in body temperature across different environmental conditions. Target body areas can be age-specific, but comb and wattles (Vercellino et al., 2025) and facial skin (Yehia et al., 2025) are among the most reliable. Bloch et al. (2020) developed an automated system for measuring broiler temperatures using thermal cameras. The predicted body temperature was compared with the actual core body temperature measured by temperature loggers implanted in the abdominal cavity. The accuracy was ±0.27°C, measuring one broiler every 16 min on average.

According to Yehia et al. (2025), skin temperature measured with infrared cameras at the face, eye, and breast from day 28 to 33 post-hatch (Fig. 1) was correlated with cloacal temperature ($r$ = 0.88, 0.89, and 0.92, respectively). This recent study also concluded that cloacal temperature is more variable under heat stress than at thermoneutral conditions, and that skin temperature is preferred for evaluating body temperature under heat stress. The heat is transferred from the core to the eyes at a significantly higher rate than to the skin of the face and breast. However, the skin of the breast dissipates more heat.

## 2.2 Thermal stress

Thermal stress is caused by adverse combinations of inappropriate temperature, relative humidity, and air ventilation in the micro-climate surrounding the broiler.

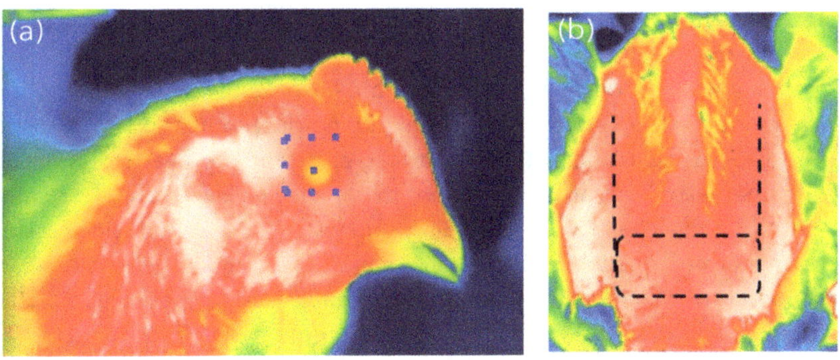


**Figure 1** Infrared thermal images illustrating the anatomical regions where facial, eye (a), and breast skin (b) temperatures were recorded in broilers between 28 and 33 days of age. The facial temperature was recorded along the three dotted lines around the eye, and the average of the three lines was taken as the final facial temperature. The eye temperature was measured at the exact center of the eye. The breast temperature was measured in the dashed box and along the two dashed lines, and the average of these readings was taken as the final temperature of the breast skin. Source: Yehia et al., 2025.

The thermoneutral zone, or range of environmental temperature at which a stable core body temperature can be maintained with minimal physiological adaptation, varies with age. Cold stress has a more negative effect on young chicks. It can be observed when brooding temperatures are below 25–27°C, or unconditioned cold air enters the chicken house, causing condensation and cooling chicks during the first week of life.

Litter moisture condensation is a good marker of potential cold stress during the brooding phase. Moisture condenses at the dew point temperature when external cold air is not directed to the ceiling to mix with the hottest house air. In a broiler house with a set point of 29°C and a relative humidity of 45%, incoming air below 15.8°C will cause condensation and cool chicks.

Nevertheless, under commercial conditions it is more common to observe heat stress due to the heating conditions, high stocking density (>30 chicks/$m^2$) during brooding, and the housing facilities normally used (Mesa et al., 2017). Heat stress occurs when chickens cannot maintain a balance between metabolic heat production and environmental heat loss, usually triggered by temperatures above 25°C in chickens older than 4 weeks of age, but observed at temperatures above 34°C in chicks during the first week of life. It causes hyperthermia, leading to behavioral changes such as wing extension, panting, reduced feed intake, impaired growth performance, and even increased mortality.

The hyperthermia level at which sufficient recovery occurs was found to be between 44°C and 44.5°C. Above this level, broilers can suffer a heat stroke (Yahav et al., 2009). The impacts of heat stress on physiology, health, welfare, behaviors, and performance have been summarized by Liu et al. (2020), Onagbesan et al. (2023), and Apalowo et al. (2024). Heat stress also affects meat quality (Song and King, 2015; Nawaz et al., 2021; Bejaoui et al., 2023) and gut health (Brugaletta et al., 2022).

Heat stress can occur at any time in broiler production, including during chick transportation, brooding, any week during grow-out, or during transport to the processing plant. Mitigation strategies to reduce heat stress include environmental or housing modifications to enhance ceiling and wall insulation, air cooling and air velocity, or providing cold water (<20°C), reducing stocking density, thermal conditioning during incubation and early life, genetic selection, and nutritional manipulations or restricted feeding (Onagbesan et al., 2023; Mangan and Siwek, 2024). Among all measurements, the most effective has been to increase air velocity, since high relative humidity may limit the use of evaporative cooling systems and foggers, which increase moisture in the hot air even further.

Housing aspects to mitigate heat stress include proper ceiling and wall insulation, airflow, and cooling systems. Roofing materials and ceiling insulation are the most critical for avoiding heat stress, since solar radiation has a greater impact on increasing house temperatures. Mesa et al. (2017) reported lower mortality in broiler houses with clay roofs than in those with concrete roofs in 977 different farms in Brazil. Akinwemoye et al. (2025) detected differences in broiler performance and hematological parameters between corrugated-iron-roofed houses, houses with asbestos-cement roofs, and houses with tarpaulin roofs. In this case, houses with corrugated-iron roofs had broilers with the worst performance and blood parameters indicative of greater stress. Hematological analysis showed higher heterophil, basophil, and heterophil-to-lymphocyte (H/L) ratios in broilers housed in corrugated-iron-roofed houses, indicating increased heat stress. In contrast, broilers on asbestos-cement and tarpaulin roofs exhibited higher lymphocyte, monocyte, hemoglobin, and hematocrit levels. Due to their high thermal conductivity, corrugated-iron roofs resulted in heat stress, reduced growth, and impaired hematological parameters. Conversely, asbestos cement and tarpaulin roofs provided better thermal insulation, enhancing growth performance and health.

## 3 Ventilation

Ventilation is key to counterbalance the negative effects of high stocking density by improving air quality and minimizing moisture and heat stress. The main types of ventilation are natural, mechanical, and combined systems. In mechanical ventilation, broiler houses can be under negative or positive pressure. Negative ventilation can generally be designed with tunnel ventilation or cross ventilation, with a capacity that increases air velocity as needed. In commercial broiler houses, as the flock ages, ventilation transitions from minimum to transitional and then to tunnel ventilation.

A negative pressure ventilation system includes fans, inlets, and controllers. Fans can be exhaust, roof, or circulation fans to keep the air moving. Fans are only the air pump. Broiler house ventilation can be improved by optimizing the airflow direction and length in the flow-guiding device at the inlet, thereby improving environmental factors such as temperature, wind speed, and carbon dioxide distributions. Inlet size can increase the velocity of the air emerging from the opening. There are multiple types of inlets that can be installed on the walls at different heights or on the roof. A common rule of thumb is to size air inlets with 0.2 $m^2$ of inlet area per 1000 L/s of exhausted air (1 $ft^2$/1000 CFM). The size of the air inlet opening can be calculated with this formula: $A = Q/1000\ V$

Where A = cross-sectional area of inlet in square meters ($ft^2$)

Q = fan exhaust capacity in L/s (CFM)

V = inlet velocity in meters per second (ft/min)

Note: value 1000 does not occur in the formula with imperial units.

It is desirable to place inlets across the ceiling to avoid cold drafts on the birds. In any broiler house, incoming air should be warmed by warm room air before it reaches the area where the birds are located. Also, the air velocity at bird level should be within acceptable limits to avoid drafty conditions. The air inlet must be adjustable to maintain a high inlet air velocity. When cold air enters a warm barn, it is heavier than the surrounding warm air and sinks rapidly to the floor, causing cold drafts. To solve this problem, the cold air must be directed along the ceiling and jetted forcefully (4–5 m/s or 800–1000 ft/min) into the room to promote rapid air mixing and prevent sinking. Since a constant incoming air speed is desired, a steady static pressure must be maintained regardless of the number of fans operating. This static pressure should be kept in the range of 1.0–2.5 mm (0.04–0.10 inches) of water gauge. During cold weather, it is important to keep the static pressure at the high end of this range and lower during warm weather (Park et al., 2018). Controllers can be electrical or electronic. Electronic controllers can integrate information from temperature, humidity, $CO_2$, and ammonia sensors to regulate heaters, and the number and speed of fans based on target indoor and external temperatures (Elwakeel, 2025).

Every broiler ventilation system is energy-dependent, and the choice of ventilation and cooling equipment affects power consumption and the efficiency of the poultry house (Costantino et al., 2020; Saleeva et al., 2020). Energy conservation is becoming a priority in broiler housing, given the significant share of energy costs in the cost of poultry products (9–14%) and rising energy prices.

### 3.1 Minimum ventilation during brooding

During the brooding phase, which comprises the first two weeks of life for broilers, ventilation focuses on replenishing air, removing moisture and noxious gases ($CO_2$, CO, ammonia, and volatile organic compounds), and conserving energy for the heating system (Menegali et al., 2012). Minimizing $CO_2$ and litter moisture accumulation during this phase significantly affects broiler viability and performance (Soliman et al., 2021). The goal is to maintain $CO_2$ concentration below 3000 ppm. The $CO_2$ emissions from a chicken house can be as high as 247 kg/h.

Another aspect of ventilation during the brooding phase is to minimize relative humidity and litter moisture and enhance temperature uniformity. High litter moisture (>25%) strongly correlates with higher ammonia and increased footpad lesions. Ventilation to maintain relative humidity at 60% or lower, using circulation fans, and proper drinker management are the preferred methods to control litter moisture. Mou et al. (2025) showed that circulation fans at 150 feet/minute or 0.76 m/s were able to homogenize floor temperature, reduce litter moisture from 28–30% to 18% at 14–29 days of brooding, maintain, and minimize footpad lesions to 10% compared to 29% in the control house evaluated in the third week. McMillan et al. (2025) also observed that circulation fans significantly reduced foot pad dermatitis (32% with no lesions vs 22%), hock burns (18% vs 32%), and feather cleanliness (8% completely clean compared to 2%).

Ventilation rates are set for minimum ventilation during brooding. To calculate these rates, it is necessary to know the number of birds, the ventilation rate per bird ($m^3$/h/bird), the ventilation fan capacity ($m^3$/h), the number of fans available, and the cycle time in seconds (300 for 5-min cycles). Calculations are made in two steps:

1. The minimum ventilation requirement = number of birds in the house × ventilation rate per bird (Table 5).
2. The On and Off times are calculated using this formula.

ON Time = ventilation required/(minimum ventilation fan capacity × number of fans) × cycle time (300 s).

However, these minimum ventilation rates are frequently adjusted based on broiler stocking density to maintain relative humidity below 60%, carbon dioxide below 3000 ppm, and ammonia below 5 ppm. Weather conditions, growth rates, water and feed intake, litter moisture, and previous use of the litter

**Table 5** Minimum ventilation rates for broilers according to average body weight

| **Body weight (kg)** | **Minimum ventilation rates ($m^3$/h)** | **Body weight (kg)** | **Minimum ventilation rates ($m^3$/h)** | **Body weight (kg)** | **Minimum ventilation rates ($m^3$/h)** |
|---|---|---|---|---|---|
| 0.05 | 0.080 | 0.65 | 0.625 | 1.50 | 1.171 |
| 0.10 | 0.141 | 0.70 | 0.661 | 1.60 | 1.229 |
| 0.15 | 0.208 | 0.75 | 0.696 | 1.70 | 1.286 |
| 0.20 | 0.258 | 0.80 | 0.731 | 1.80 | 1.343 |
| 0.25 | 0.305 | 0.85 | 0.765 | 1.90 | 1.398 |
| 0.30 | 0.350 | 0.90 | 0.798 | 2.00 | 1.453 |
| 0.35 | 0.393 | 0.95 | 0.831 | 2.20 | 1.561 |
| 0.40 | 0.435 | 1.00 | 0.864 | 2.40 | 1.666 |
| 0.45 | 0.475 | 1.10 | 0.928 | 2.60 | 1.769 |
| 0.50 | 0.514 | 1.20 | 0.991 | 2.80 | 1.87 |
| 0.55 | 0.552 | 1.30 | 1.052 | 3.00 | 1.969 |
| 0.60 | 0.589 | 1.40 | 1.112 | 3.20 | 2.067 |

*Source*: Aviagen, 2018.

or litter treatments can affect minimum ventilation rates. These adjustments are commonly made manually every day, but monitoring with electronic sensors and advanced algorithms could make them automatic in the near future.

### 3.2 Tunnel ventilation

Tunnel fan ventilation is set to increase air speed and cool broilers. Calculations require the house's cross-sectional area to determine the required fan capacity ($m^3$/h). The number of fans required is determined by dividing the required fan capacity by the capacity per fan at the assumed pressure. Generally, tunnel ventilation also requires sufficient cooling pad area for adequate air intake. This area ($m^2$) is calculated with this formula:

- = total operating fan capacity ($m^3$/h)/design air speed through cooling pads (m/s)/3600 (s)

In broiler houses, the greatest heat losses occur through exhaust ventilation, reaching up to 29 000 gigacalories in poultry farms during the heating season. Fans alone account for 64% of the total power consumption (Menegali et al., 2012). Ventilation systems and lighting account for 85% of the total power consumption in a broiler house. Recently, Liang and Costello (2024) found that ventilation fan electricity consumption accounted for 88% of the total electric load in a commercial broiler house in Arkansas, USA. The average electric energy consumption per metric ton of live weight is 43 kWh in a winter flock and 184 kWh in a summer flock. Even with substantial electricity use, studies have shown that achieving ventilation efficacy in broiler houses remains challenging, despite advancements in cooling systems (Casey et al., 2010; Calvet et al., 2010). Ventilation performance changes when different equipment and settings are used or when the equipment's maintenance varies. Due to differences in ventilation systems, it is challenging to compare houses in terms of temperature, air speed, levels of harmful gases, litter quality, and how these factors affect flock productivity (Miragliotta et al., 2006).

Enhancing air velocity may improve litter quality and reduce ammonia concentrations, improving broiler performance and leg health (West et al., 2024) and meat quality (Leishman et al., 2021). West et al. (2024) reported that heavy broilers from 42 to 61 days of age, with a final stocking density of more than 40 kg/$m^2$ exposed to high air speeds that vary between 1 m/s and 4.5 m/s depending on environmental temperatures had better body weights (4890 g vs 4520 g), lower foot pad dermatitis (2.01 vs 2.45), and hock burn (1.05 vs 1.30) severity at 61 days of age than broilers maintained in chambers with a 20% lower ventilation rate. Optimum ventilation is necessary to guarantee the performance and welfare of heavy broilers under summer conditions.

Ventilation is critical to prevent broiler heat stress and to maintain homeostasis, feed intake, and growth. Hamrita and Conway (2017) conducted 3 consecutive experiments with 8.6-, 9.0-, and 9.4-week-old broilers, each spanning approximately 12 h, to measure DBT using biotelemetry electronic sensors. The DBT responses of broilers under no ventilation increased by as much as 3°C in response to step increases in ambient temperature from 20°C to 31°C, 32°C, or 33°C for 5 h. Broilers exposed to tunnel ventilation (3.6 m/s) maintained a lower DBT by up to 0.9°C. During the experiment days, broilers exposed to tunnel ventilation consistently gained weight with a percentage weight gain ranging from 1% to 11%. Broilers not exposed to tunnel ventilation showed less consistent responses, with some gaining as much as 14%, while others losing as much as 9%.

Heavy broilers with body weights above 3.2 kg typically require higher ventilation rates to maintain thermal comfort and production efficiency compared to smaller chickens. Akter et al. (2022a) observed that surface temperatures of heavy broilers were lower under high air velocity (4–4.3 m/s) and low humidity (<55%). High air velocity increased the number of chickens feeding, standing, and walking (Akter et al., 2022b). These heavy chickens, between 35 and 61 days of age, pant in mild-to-moderate thermal conditions (28°C to 33°C and 50% to 70% RH), indicating that upper critical temperatures may be lower at higher body weights.

## 4 Air quality

Air quality is measured by how clean or polluted the air is. This includes the presence of harmful particles and gases. Maintaining optimal air quality is crucial for broiler welfare, with elevated gas levels aimed at limiting respiratory distress, eye irritation, and immune system impairment. While ammonia is the primary concern, carbon dioxide, carbon monoxide, and particle matter (PM) must also be managed (Table 6).

**Table 6** Maximum recommended levels, welfare standards, and short-term exposure to gases

| Air factor | Maximum recommended level | Welfare standard | Short-term exposure |
|---|---|---|---|
| Ammonia ($NH_3$) | 25 ppm | 15 ppm | 35 ppm for 10 min |
| Carbon dioxide ($CO_2$) | 2500 ppm | | 4000–50000 ppm |
| Carbon monoxide (CO) | 0 ppm | | |
| Particle matter | 5 mg/m$^3$ | | 8 mg/m$^3$ |

### 4.1 Ammonia

Ammonia ($NH_3$) is the most important gas emitted in broiler houses due to the degradation of litter, excreta, and feed residues (Vilela et al., 2020). $NH_3$ negatively affects broiler performance and health (Miles et al., 2004). This gas can have toxic effects in the respiratory tract (Liu et al., 2020), eyes (Miles et al., 2006), brain (Dasarathy et al., 2017), spleen (Zhao et al., 2020), liver (Sa et al., 2018), intestine (Wang et al., 2019), and even gut microflora (Han et al., 2021). Many reports indicate that $NH_3$ negatively affects broiler immune function (Wei et al., 2015; Shah et al., 2020; Zhou et al., 2020). However, Aston et al. (2019) concluded that ambient $NH_3$ (30 ppm to 60 ppm) during the 14 days of life did not impact vaccine or challenge virus replication and clearance, ciliostasis, or tracheal histopathology scores.

Ammonia emissions can be mitigated by litter amendments, biofiltration, acid scrubbers, dietary manipulations, and bedding materials. The mitigation efficiency has been determined to be 50% for litter amendments and 95% for multi-stage acid scrubbers (Bist et al., 2023).

### 4.2 Carbon dioxide management

Carbon dioxide concentrations in broiler houses fluctuate with multiple factors, typically increasing during the brooding period due to the operation of gas burners. Primary influences include the number and size of birds, ventilation rates, heating system usage, and individual bird characteristics such as feed consumption, diet composition, activity level, and age.

While young chicks tolerate $CO_2$ levels up to 4000 ppm without adverse effects, much higher concentrations become dangerous. At 118 000 ppm, birds exhibit gasping and respiratory distress, while 174 000 ppm proves lethal. Excessive $CO_2$ levels alter metabolism by reducing plasma thyroxine and glycogen concentrations while affecting bird activity, behavior, and production performance. These findings underscore the critical importance of maintaining appropriate $CO_2$ levels for optimal bird health, welfare, and productivity (Burggren et al., 2015).

### 4.3 Particle matter

The PM in broiler houses originates from multiple sources, with manure comprising the largest fraction at 72.1%, followed by feathers at 21.3%, wood shavings at 5.8%, and ambient particles at 0.7% (Wicklin and Czarick, 1997). The PM inside broiler houses can be 10–100 times higher than outdoors and higher than in other livestock facilities. Particles are classified by their average size. Those with a diameter of less than 10 μm ($PM_{10}$) can reach the upper respiratory tract of broilers, while those with less than 2.5 μm ($PM_{2.5}$) can enter the lungs (Wang et al., 2022). Air in poultry houses should have less than 5 $mg/m^3$ dust at the broiler level. Dust levels of 8 $mg/m^3$ can be tolerated if the broilers are not stressed by ammonia, heat, or respiratory disease agents.

The PM can irritate the respiratory tract, decrease immune resistance, and cause respiratory disease by inhaling pathogenic and non-pathogenic microorganisms. The PM can also contain endotoxins from gram-negative bacteria, lipopolysaccharides, and mycotoxins produced by fungi. Zhou et al. (2023) observed that broilers exposed to $PM_{2.5}$ for 21 days exhibited significant disruption of lung histological structures, chronic inflammation, and higher expression of interleukins IL-6, TNFα, and IFNγ, with decreased IL-10 expression. $PM_{2.5}$ significantly altered the α- and β-diversity of the pulmonary microbiota. At the phylum level, $PM_{2.5}$ exposure decreased Firmicutes abundance and increased Actinobacteria and Proteobacteria abundances. At the genus level, $PM_{2.5}$ exposure significantly increased the abundance of *Rhodococcus*, *Achromobacter*, *Pseudomonas*, and *Ochrobactrum*. These authors observed positive associations between the above altered genera and lung TNFα and IFNγ expression, as well as chronic lung inflammation.

The concentration of PM in broiler houses depends on several factors, including age, stocking density, ventilation, humidity, season, and sampling time. The most important factors affecting total suspended PM are house temperature and relative humidity. Dry environments normally contain more PM. Dust can be mitigated through frequent house cleaning, optimal lighting, liquid spraying, bedding management, and air filtration systems (Ullman et al., 2004; Bist et al., 2022; Guo et al., 2022).

Spraying water, acidic water, or oil-water combinations can reduce $PM_{10}$ and $PM_{2.5}$ emissions by 18% to 64%. However, proper dosage and pH control are crucial, as excessive water application may inadvertently increase ammonia emissions (Winkel et al., 2016). The selection of appropriate bedding materials substantially impacts particle generation. Alternatives such as peat or clay pellets produce less dust than traditional wood shavings or straw. While reusing litter may increase PM emissions, topping existing litter with fresh bedding can reduce concentration by up to 40% (van Harn et al., 2012). Sophisticated filtration systems, including dry filters, water filters, and biofilters, demonstrate high efficiency in PM removal. Acid scrubbers show particularly impressive results, reducing $PM_{10}$ by 61–93% and $PM_{2.5}$ by 47–90% (Patterson, 2005). This technology charges airborne particles, causing them to precipitate onto surfaces. Studies demonstrate reductions of up to 94% for PM and 96% for airborne bacteria (Winkel et al., 2016). Proper ventilation remains fundamental for diluting and removing particles. High ventilation rates, especially during warmer periods, significantly reduce PM concentrations. Additionally, adjusting lighting schedules to include dark periods decreases bird activity and associated dust production.

## 5 Variability in environmental parameters

Broiler houses generally exhibit spatial heterogeneity in temperature, humidity, air velocity, $NH_3$, $CO_2$, and illuminance distributions (Ferraz et al., 2019, Ferreira et al., 2024; Çayli, 2025). To provide suitable environmental conditions in the indoor sections, especially on extremely hot days, it is vital to determine the system capacity correctly and take the necessary measures to ensure more efficient operation and uniformity. Although the internal conditions of the broiler house may be ideal for production, the airflow pattern of the tunnel ventilation system should be optimized to minimize temperature differences across certain areas of the house, and the efficiency of the evaporative cooling pads should be improved. Avoiding microenvironments in the house is critical to guarantee uniformity in flock results and welfare for all birds.

Despite the efforts to maintain uniformity in broiler houses, variability occurs. Geostatistical techniques can aid in quantifying spatial variability and detecting problem areas, which can be visualized with kriging maps. de Sans et al. (2021) described the in-barn heterogeneity of broiler chicken welfare in 4 closed-sided and 13 open-sided houses in 2 seasons in the southern Brazilian subtropical climate. These researchers observed a pattern for the spatial distribution, heading from the house center to the west end of both house types, with worse results for three environmental indicators: higher temperature, $NH_3$, and $CO_2$ concentrations (Fig. 2) and three animal-welfare indicators: higher prevalence of hock burn, bird soiling, and footpad dermatitis (Fig. 3). In close-sided houses, illuminance was very low (4.4 lx to 6.7 lx) compared to open-sided houses (119.8 lx to 145.3 lx); in both house types, the prevalence of lameness was high (50.9% to 78.0%), even though both illuminance and lameness prevalence were evenly distributed across all houses. The kriging maps allowed for the identification of worse welfare problems in the west direction, which, in closed-sided houses, means near exhaust fans and in open-sided houses, the direction of positive-pressure mechanical ventilation by fans. Understanding the variability within each housing type and environmental condition is critical for determining improvements needed to enhance flock performance and welfare.

Broiler chicken production involves a dynamic cycle, in which environmental variables are regularly adjusted to meet the animals' needs at different stages of growth. The technologies used in dark-house facilities normally create a stable, comfortable environment that primarily mitigates heat stress, which directly improves productivity. However, most broiler facilities have limitations in moving air and moisture from certain areas of the house, which can increase $NH_3$, $CO_2$, and other gases that affect chicken health and welfare. Circulating fans and air deflectors or baffles can help optimize aerodynamics and improve

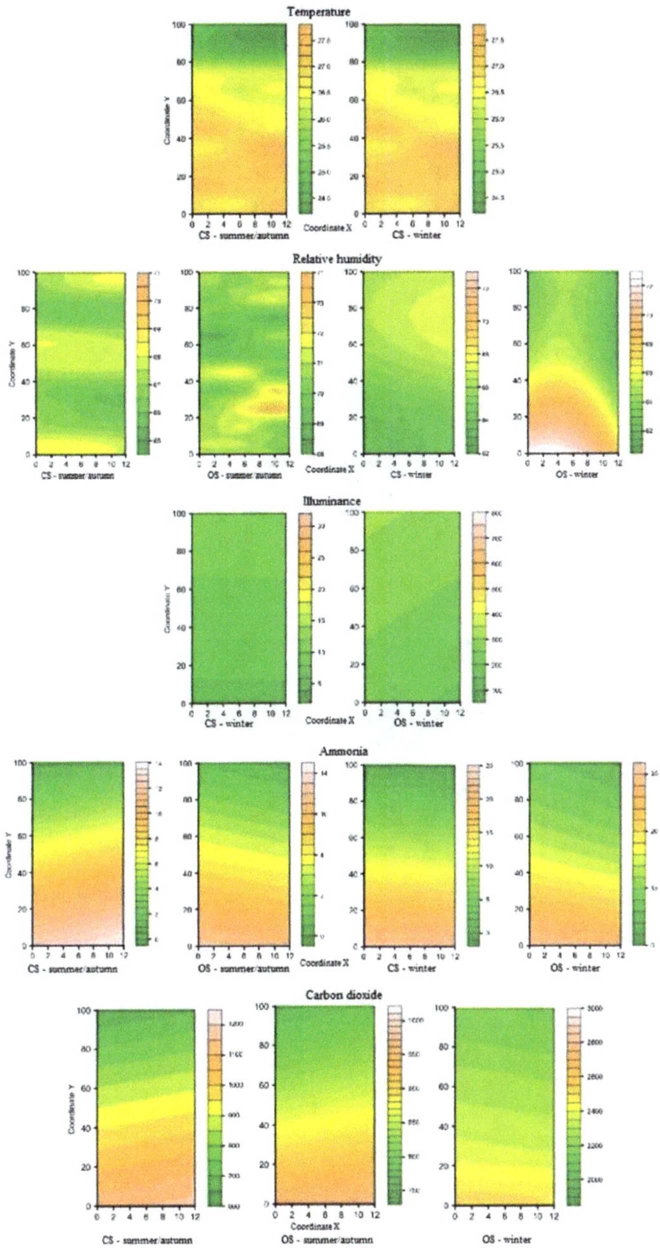


**Figure 2** Spatial distribution of temperature (T °C), relative humidity (%), illuminance (lx), ammonia (ppm), and carbon dioxide concentration assessed in four closed-sided and 13 open-sided houses, during summer/autumn and winter seasons 2019, in the subtropical climate. Source: Sans et al., 2021.

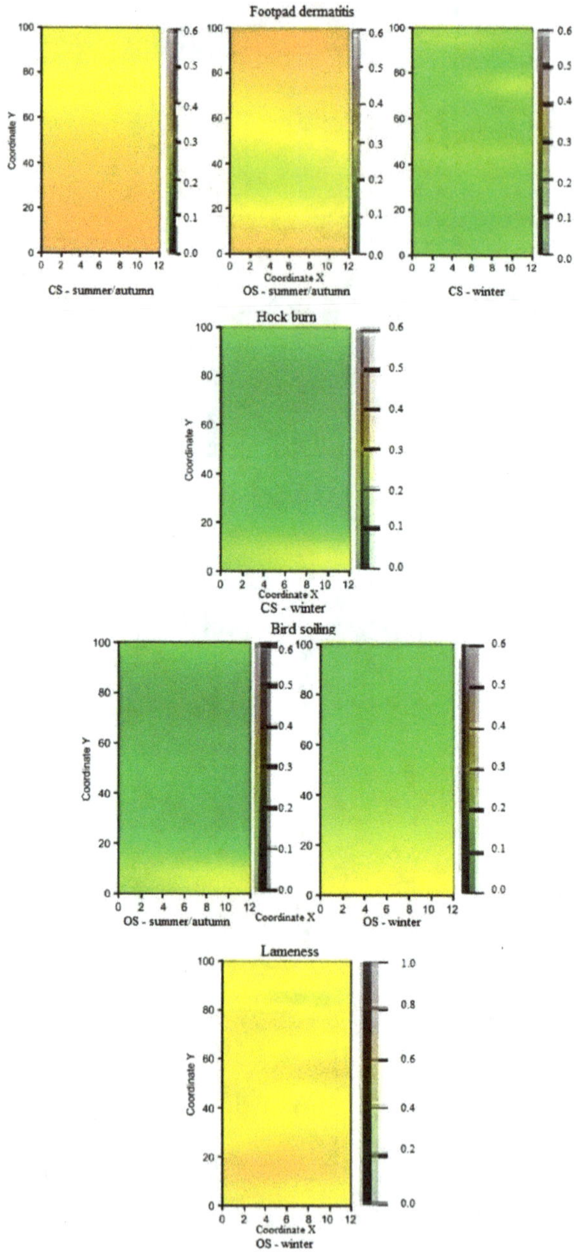


**Figure 3** Spatial distribution of the absence or presence of footpad dermatitis, hock burn, bird soiling, and lameness assessed in four closed-sided and 13 open-sided houses, during summer/autumn and winter seasons 2019, in the subtropical climate. Source: Sans et al., 2021.

ventilation. Minimizing microenvironments within broiler houses is a critical area to improve environmental conditions.

## 6 Flooring systems

The selection of a flooring system impacts performance, welfare, and the environmental impact of broiler farms (Li et al., 2017; Mesa et al., 2017; Soliman et al., 2020; Topal et al., 2021). Mesa et al. (2017) observed that concrete flooring significantly improved feed conversion in broilers compared to hard-packed dirt floors after analyzing more than 3000 flocks. However, flocks raised on dirt floors had significantly lower mortality. Litter or bedding materials are used in broiler houses in most parts of the world, while some countries use plastic and wire slats or cages. The selection of bedding materials is critical for welfare and health (Almeida Paz et al., 2010; Diarra et al., 2021).

Ideal broiler litter material must be absorbent, non-toxic, lightweight, and cost-effective, with key considerations being moisture control (15–30% range), particle size, and, critically, the ability to release moisture to prevent caking. Proper litter reduces ammonia emissions, prevents breast blisters, hock burns, and, ultimately, supports better bird welfare, footpad health, and carcass quality. Bedding materials include wood shavings, sawdust, wood chips, rice hulls, straw, hay, corn cobs, sugar cane bagasse, sand, shredded newspapers, peanut hulls, and gypsum (Diarra et al., 2021).

Depending on the material selected, the smaller size has a higher specific surface area and promotes water loss. In many countries, litter is reused due to a lack of new bedding material at times and to environmental regulations that prevent the spread or use of litter. Other areas of the world prohibit or seek to avoid the reuse of litter due to food safety regulations aimed at controlling foodborne pathogens, such as Salmonella spp. (Dunn et al., 2022).

Reuse of litter is an established practice in broiler production. Generally, the live performance of broilers grown on reused litter is not affected when compared with those on new litter. However, unsuitable use of recycled litter can lead to many problems, such as increased $NH_3$ emissions into the atmosphere, which adversely affect bird performance and health, as well as workers' health, leading to loss of profits. Also, unsuitable use of recycled litter could be increasing coliform levels and coccidia outbreaks in poultry flocks and increasing disease transmission within the farm from flock to flock.

Under conditions of reused litter rice hulls, litter has proven to be better than reused wood shavings substrate due to low moisture content, pH, and ammonia concentration. The pH of the litter usually exceeds 8; litter acidification plays a significant role in minimizing ammonia volatilization, since alkaline pH favors microbial activity that increases $NH_3$ production. Litter acidification can be achieved by applying products that contain sodium bisulfate, aluminum sulfate, and acidic clays. Application is generally done before chick placement.

Litter with moisture below 35% is generally more friable and retains the ability to release and absorb moisture. However, most broiler litter reaches 40% moisture near drinkers or cooling pads. A critical housing factor for controlling litter is relative humidity and temperature. Higher flock stocking density also increases litter moisture (Dunlop et al., 2016). Continuous air movement over the litter using circulation fan systems helps distribute heat, achieve uniform floor temperatures, and maintain drier litter (Mou et al., 2025).

## 7 Lighting

The light photoperiod, intensity, and the light spectrum or wavelengths affect broiler growth rate, behavior, diurnal rhythms, activity, welfare, metabolism, and tissue development (Alvino et al., 2009a, 2009b; Collins et al., 2011; Deep et al., 2012; Wu et al., 2022). Chicken physiology can be affected by its impact on the eyes, pineal gland, and pituitary gland, which are located near the hypothalamus. Light affects the neuroendocrine system, especially the hypothalamic-pituitary-gonadal axis, influencing circadian rhythms and various physiological activities, including the release of gonadotropin-releasing hormone, growth hormone, and thyroid hormones (Kuenzel et al., 2015).

Optimum light periods vary with age. The light photoperiod, or the number of consecutive hours of light in a 24-h period, can affect broiler growth and development from the embryonic phase (Van der Pol et al., 2019; Riaz et al., 2021).

The illumination can be provided continuously or intermittently. Continuous exposure to varying light intensities had a minor effect on blood physiological variables in broiler chickens, whereas a short photoperiod had a marked effect on most blood physiological variables without inducing physiological stress in broilers (Olanrewaju et al., 2013). In contrast, the intermittent lighting program promotes yields of drumsticks and thighs since broilers are stimulated to get up more frequently (Abreu et al., 2011). When intermittent lighting (3 L:1 D) was applied, feed efficiency, innate immunity, and oxidative status were enhanced compared with continuous lighting programs in broilers (Ghanima et al., 2020). Then, near-continuous lighting seems less advisable for broilers, particularly at older marketing ages.

A photoperiod of 16 h to 24 h had no effect on thyroid gland development or function, as measured by both biochemical and morphometric parameters, in broilers (Ozkanlar et al., 2021), suggesting that the photoperiod could be slightly shortened. Furthermore, the intermittent lighting program did not hinder broiler performance and promoted energy efficiency (Manfio et al., 2019). Although long photoperiod programs may reduce mortality and improve feed efficiency, negative impacts on body weight at young marketing ages cannot be ignored. Based on this, a 16 L:8 D light duration was suggested for broiler production. However, it should be adjusted according to seasonal changes if light is not completely controlled and the different opening and lighting degrees of different chicken house types.

Light sources may differ in the intensity that they can provide, their color, and wavelength. Light sources commonly used in the broiler industry include incandescent (ICD), compact fluorescent (CF), cold cathode fluorescent (CCF), and light-emitting diode (LED) lamps. Consequently, differences in broiler performance and behavior have been reported (Kristensen et al., 2007; Wu et al., 2022). The CF and LED lights in broiler houses were found to have better energy utilization than ICD lights (Olanrewaju et al., 2018; Bennato et al., 2021). The CCF is more energy-efficient and has a longer working lifespan than ICD light, but is less efficient than LED light (Alberts et al., 2010). Besides, the body weight, feed conversion, and mortality rate of broilers did not differ significantly when raised under either ICD or LED light (Olanrewaju et al., 2015, 2016), but broilers reared under CCF light exhibited lower body weight and higher heterophil-to-lymphocyte (H:L) ratios, indicating the occurrence of chronic stress (Rogers et al., 2015a, 2015b).

The color of light is determined by the different wavelengths in the visible spectrum. White light is composed of all these wavelengths; however, its color temperature is contingent upon the spectral characteristics, which are the power emitted by each wavelength. Color temperature, therefore, indicates the overall color appearance of white light, with warmer colors associated with lower temperatures. In contrast to humans, chickens can also detect UV-A light, which lies immediately below the visible light spectrum.

The use of LEDs has become popular for saving energy, reducing the costs of bulb replacements, and influencing chicken behavior and tissue development (Senaratna et al., 2012; Sultana et al., 2013; Riber, 2015). The LED can emit white light with a range of color temperatures or monochromatic light. Light color is measured in terms of correlated color temperature (CCT) in Kelvin (K). Light appears warm or more yellow or red/orange with a low K (2700 K), or cool with a K of 5000 or 6500 K, producing a light that is more blue/white. Riber (2015) investigated broilers' preference for neutral-white (4100 K) and cold-white (6065K). Chickens preferred the 6065 K at 16, 28, and 34 days of age. Chickens exposed to this cold-white light were 67.4 g ± 19.2 g heavier at 35 days and had heavier *Pectoralis minor*. However, no differences in welfare parameters were observed between the two light colors.

Research has indicated that broilers prefer monochromatic blue light when they are reared in green, red, and white light, whereas those reared in blue light prefer green light (Prayitno et al., 1997). Xie et al. (2008) reported that blue light may alleviate the stress response in broilers by reducing serum IL-1β levels.

Blue and green light have been shown to positively affect growth in broilers (Rozenboim et al., 1999, 2004).

The effects of light intensity, brightness, or light strength on broiler growth, health, immunity, and, especially, welfare have been investigated. Regarding the minimum light intensity, 5 lux has been considered the standard to ensure the productivity and welfare of broiler chickens. However, improved growth, breast meat yield, and broiler welfare were observed with increased (5–10 lx) light intensity (Deep et al., 2012; Yang et al., 2018). In contrast, broilers reared at 20 lx were more active with slower growth and lighter eye weight, while the welfare parameters or leg health were not affected (Rault et al., 2017). Broilers exposed to a lower light intensity threshold of 1 lx showed more rest and less preening, with increased ulcerative footpad lesions and eye size, and were associated with reduced welfare state (Deep et al., 2012; Deep et al., 2012). In addition, preferences for different intensities vary with age, diurnal variation, and certain behaviors (mainly walking and lying) of broilers. Older broilers were characterized as less active, with more lying and sleeping behavior under dim light (5 lx), but had heavier body weights compared to 20 lx and 320 lx (Senaratna et al., 2012).

# 8 Space management

## *8.1 Stocking density*

Stocking densities vary worldwide and even among companies and farms in similar regions. In the United States, the National Chicken Council specifies stocking density allowances by broiler weight, ranging from 32 kg/m$^2$ to 44 kg/m$^2$. In Canada, the required stocking density should not exceed 31 kg/m$^2$ unless certain conditions are met, which allows a maximum of 38 kg/m$^2$. In the European Union, the maximum stocking density is 33 kg/m$^2$, but it can be increased to 39 kg/m$^2$ or 42 kg/m$^2$ if the producers comply with specific requirements.

Due to house dimensions, market demands, and economic constraints, the stocking density varies. High stocking density (near 40 kg/m$^2$) can be detrimental, especially during the finishing phase. High stocking density reduces feed intake, growth rates, slaughter weight, and increases feed conversion ratio (Averós and Estevez, 2018; Bergeron et al., 2020; Shynkaruk et al., 2023).

Optimum stocking densities depend mainly on the ability to provide an environment that dissipates heat and moisture and renovates air, maintaining good litter conditions. However, at very high density, the air cannot move between the broilers. Recent studies have determined the optimal stocking densities for broilers. Thema et al. (2022) determined the optimal stocking density to be 15 birds/m$^2$ (27.27 kg/m$^2$), considering welfare parameters.

Shynkaruk et al. (2023) observed a linear decrease in feed intake and body weight as stocking density varied from 31 $kg/m^2$ to 41.5 $kg/m^2$. However, no effects were detected on feed conversion ratio, flock uniformity, or carcass condemnations.

The effects of stocking density on broiler welfare are more consistent. Footpad dermatitis severity and lameness increased with greater densities of 34 $kg/m^2$ and 36 $kg/m^2$ compared to 30 $kg/m^2$ (Bailie et al., 2018). Broilers housed at stocking densities of 24 $kg/m^2$ and 30 $kg/m^2$ showed greater comfort, foraging, frolicking, running, and sparring than those housed at 36 $kg/m^2$ and 42 $kg/m^2$ (van der Eijk, 2022).

## 8.2 Environmental enrichment for broilers

Scientists have expressed concerns about the high level of inactivity in modern broilers, which may be linked to a range of welfare issues. Compared with lines unselected for fast growth, broilers spent 60–80% of their daytime sitting and resting (Bizeray et al., 2002a; McLean et al., 2002). Long-term contact with litter increases the risk of contact dermatitis (Bessei, 2006). Low activity and fast growth increase the incidence and severity of leg problems (Reiter and Bessei, 2009).

Broilers raised at high stocking density (42 $kg/m^2$) may exhibit decreased activity, and consequently, reducing stocking density is frequently considered an intervention with greater potential to improve welfare (Tahamtani et al., 2020). However, this approach obviously increases production costs. It has been proposed that environmental enrichment may promote more activity. Environmental enrichment is defined as adaptations in the rearing environment that increase behavioral opportunities, thereby improving biological responses, such as physical activity (Abdallah et al., 2024).

Pederson and Forkman (2019) reviewed 62 studies to determine whether environmental enrichment can improve leg health in broilers. They included 56 randomized trials and 6 cross-sectional studies that evaluated light programs, light intensity, stocking density, perches, straw bales, and resource separation. These authors concluded that optimal enrichment strategies will reduce leg problems and increase mobility, thereby improving broiler welfare. Providing perches and increasing light intensity showed limited effectiveness in improving leg health, primarily affecting contact dermatitis. In contrast, evidence suggested that reduced stocking density and a dark/light schedule could improve leg health.

Activity levels and behavior can be improved, at least temporarily, by environmental modifications that increase environmental complexity. However, more research is needed on the impacts of enrichments on broiler resilience, on broilers in commercial conditions, and on slower-growing strains. Incorporating farmers' expertise can greatly benefit the design and implementation of enrichment on commercial farms (Jacobs et al., 2023).

Examples of environmental enrichments include providing perches or elevated resting places, panels and barriers, placing hay or straw bales in housing, introducing sand trays, providing covered verandas and outdoor ranges, vegetable materials, hanging objects, dustbathing substrates, and mirrors. Introducing these materials or equipment changes the environment and incurs additional costs. However, enrichment should be both practical and profitable, and it should significantly improve welfare or health. Otherwise, it may not be implemented by farmers (Riber et al., 2018).

In numerous studies, tests of environmental enrichment yield conflicting results (Riber et al., 2018; Güz et al., 2021; Adballah et al., 2024). For example, Ohara et al. (2015) found that the provision of hay bales and perches increased broiler activity levels (more standing and moving) in both male and female broilers and reduced the severity of footpad dermatitis (FPD) in female broilers. Güz et al. (2021) evaluated enrichments such as barrier perches, angular ramps, horizontal platforms, a large distance between feed and water, and the provision of live Black Soldier fly larvae in a dustbathing area. In this study, pen enrichment increased broiler activity and improved bone development in both fast- and slow-growing chickens. However, environmental enrichment resulted in lower body weight gain. On the other hand, Bailie et al. (2013) found no effect of either straw bales or natural light on FPD, while de Sans et al. (2012) found no effect on FPD of providing perches, a sand tray, kale, green cabbage, or hanging objects.

Recently, the economic aspects of environmental enrichment have been addressed (Riber et al., 2018). Jones et al. (2020) evaluated eight types of environmental enrichments that included roughage, vertical panels, straw bales, elevated platforms (5 cm and 30 cm), increased distances between feed and water (7 m and 3.5 m), and stocking density reduced from 40 kg/m$^2$ to 34 kg/m$^2$, plus a control group. Of all these interventions, only increasing the distance between feed and water to 3.5 m improved the flock's financial performance. The authors concluded that most environmental enrichment interventions would have a negative financial impact due to additional costs and that implementation would be feasible only if consumers were willing to pay a premium for the chickens produced under these conditions.

Consequently, the evidence indicates that environmental enrichment may stimulate locomotion and other activities, improve bone development, and even minimize footpad dermatitis or hock burns, but generally reduces growth and profitability. The benefits of environmental enrichment are variable and not completely proven, and its adoption reduces the profitability of the production system. However, environmental enrichment may gain relevance due to welfare regulations and consumer pressure.

## 9 Modern technologies to achieve environmental needs

Current broiler housing can control temperature, humidity, lighting, and airflow, and minimize $NH_3$ and $CO_2$ emissions. However, responses to external weather variations are not always appropriate, and adjustments to chicken weight gain, drinking behavior, and stocking densities are manual and often subjective. This combination of technology and art leads to inconsistencies in achieving optimums. As the number of personnel available to manage broiler houses declines, automated and early-warning systems are becoming more important.

In enclosed broiler houses, equipment malfunctions in ventilation systems lead to mass mortality events. Dust accumulation around ventilation devices and the release of gases such as ammonia and hydrogen sulfide can lead to device malfunctions. There are several methods for detecting and notifying of device malfunctions (Kim and Lee, 2022).

Intelligent systems offer real-time tracking to automatically adjust temperature, lighting, ventilation, and feeding. Sensors can monitor air quality, movement, vocalizations, and other factors to assess bird health and well-being (George and George, 2023). Computer vision and artificial intelligence (AI) can detect injuries, illnesses, and behavioral issues, enabling early intervention and preventing welfare issues. Automated poultry farming processes will improve standardization. Portable devices, wireless sensors, and Internet of Things (IoT) are gaining increasing significance and attention in broiler housing, so web-based, remote monitoring of facilities using mobile applications may be the future of broiler housing (Brassó et al., 2025).

Precision tools can offer promising strategies for alleviating heat stress by continuously measuring animals' physiological parameters or optimizing feed intake to reduce metabolic heat production. Stress and health problems can be monitored and detected at an early stage, enabling preventive treatments, reducing costs, and production losses. Despite these benefits, implementing these technologies poses challenges, including user knowledge of how to use and maintain the devices, internet connectivity, and initial investment costs. Correct positioning and calibration of electronic sensors are critical to represent housing conditions (Curi et al., 2017).

## 10 Conclusion

Broilers are fast-growing animals that require optimal environmental conditions for health, welfare, and performance. These optimal conditions should consider factors such as temperature, humidity, ventilation, air speed, lighting, air and litter quality, stocking density, and environmental enrichment. All these factors interact, and consequently, their interactions should always be included when deciding on the parameters to use. Achieving optimal temperatures and avoiding heat stress are the most impactful factors. However, effective temperature depends on relative humidity and air speed and can be affected by stocking density. Temperature-humidity indices are more appropriate for determining comfort in broiler housing. Lighting may affect broiler activity, metabolism, heat production, and behavior. Environmental enrichment has been promoted to enhance broiler welfare; however, implementing any option does not guarantee improved welfare or a good return on investment unless customers are willing to pay for the additional costs. Reducing stocking density has the most significant effect on broiler welfare; however, it also reduces system profitability. Electronic sensors, controllers, the Internet of Things, artificial intelligence, and smart systems are becoming more frequent in broiler production worldwide, and their widespread adoption can help improve environmental control accuracy, detect equipment malfunctions, and alert farmers to make timely interventions and prevent welfare issues or catastrophic losses. Broiler housing will continue to advance in technology and precision, energy efficiency, and will also be driven by animal welfare perceptions and regulations.

## 11 References

Abbas, G., et al. (2022), Prospects and challenges of adopting and implementing smart technologies in poultry production. *Pak. J. Sci.* 74(2): 108–127.

Abdallah, N., et al. (2024), Environmental enrichment/improvement: effect on performance of commercial broiler strains. *Braz. J. Poult. Sci.* 26(3): eRBCA-2023.

Abeyesinghe, S. et al. (2001), The aversion of broiler chickens to concurrent vibrational and thermal stressors. *Applied Animal Behaviour Science* 73(3): 199-215.

Abreu, V. M. N., et al. (2011), Curtain color and lighting program in broiler production: II. carcass and parts yield and abdominal fat deposition. *R. Bras. Zootec.* 40: 2035–2038.

Ahmmed, P., et al. (2024), A Subcutaneously injectable implant for multimodal physiological monitoring in animals. *IEEE Sensors J.* 24(7): 11205–11216.

Akinwemoye, A., et al. (2025), Effects of different roofing materials on the growth performance, haematological profile and welfare of broiler chickens. *Direct Res. J. Agric. Food Sci.* 13(1): 36–39.

Akşit, M., et al. (2013), Broilers respond to cooler ambient temperatures after temperature acclimation during incubation and early postnatal age. *J. Appl. Poult. Res.* 22(2): 298–307.

Akter, S., et al. (2022a), Impacts of air velocity treatments under summer condition: part I–Heavy broiler's surface temperature response. *Animals* 12(3): 328.

Akter, S., et al. (2022b), Impacts of air velocity treatments under summer conditions: part II–Heavy broiler's behavioral response. *Animals* 12(9): 1050.

Alberts, I. L., et al. (2010), Hollow cathode effect in cold cathode fluorescent lamps: a review. *J. Display Technol.* 6(2): 52–59.

Almeida Paz, I. C. L., et al. (2010), Selecting appropriate bedding to reduce locomotion problems in broilers. *Braz. J. Poult. Sci.* 12: 189–195.

Alvino, G. M., et al. (2009a), Behavioural time budgets of broiler chickens reared in varying light intensities. *Appl. Anim. Behav. Sci.* 118(1–2): 54–61.

Alvino, G. M., et al. (2009b), Light intensity during rearing affects the behavioural synchrony and resting patterns of broiler chickens. *Br. Poult. Sci.* 50(3): 275–283.

Apalowo, O. O., et al. (2024), Impact of heat stress on broiler chicken production. *Poultry* 3(2): 107–128.

Aradas, M. E. C., et al. (2005), Comparing the thermal environmental in broiler housing using two bird's densities under tropical conditions. *Agric. Eng. Int.: CIGR Journal.*

Aston, E. J., et al. (2019), Ambient ammonia does not appear to inhibit the immune response to infectious bronchitis virus vaccination and protection from homologous challenge in broiler chickens. *Vet. Immunol. Immunopathol.* 217: 109932.

Averós, X. and Estevez, I. (2018), Meta-analysis of the effects of intensive rearing environments on the performance and welfare of broiler chickens. *Poult. Sci. 97*(11): 3767–3785.

Bessei, W. (2006), Welfare of broilers: a review. World's Poultry Science Journal 62(3): 455–466.

Bailie, C. L., et al. (2013), Influence of the provision of natural light and straw bales on activity levels and leg health in commercial broiler chickens. *Animal* 7: 618–626.

Bailie, C. L., et al. (2018), Effects of stocking density and string provision on welfare-related measures in commercial broiler chickens in windowed houses. *Poult. Sci.* 97(5): 1503–1510.

Baracho, M. S. et al. (2019), Factors affecting broiler production: a meta-analysis. *Brazilian Journal of Poultry Science* 21(3): eRBCA-2019-1052.

Barbosa, L. V., et al. (2024), Predicting risk of ammonia exposure in broiler housing: correlation with incidence of health issues. *Animals* 14(4): 615.

Bejaoui, B. et al. (2023), Physicochemical properties, antioxidant markers, and meat quality as affected by heat stress: a review. *Molecules* 28(8): article 3332.

Bennato, F., et al. (2021), Evaluation of chemical composition and meat quality of breast muscle in broilers reared under light-emitting diode. *Animals* 11(6): 1505.

Benvenga, M. A. C., et al. (2020), Optimization algorithm applied to environmental control in broiler houses. *Braz. J. Poult. Sci.* 22: eRBCA-2020.

Bergeron, S., et al. (2020), Commercial poultry production stocking density influence on bird health and performance indicators. *Animals* 10(8): 1253.

Bist, R. B. and Chai, L. (2022), Advanced strategies for mitigating particulate matter generation in poultry houses. *Appl. Sci.* 12(22): 11323.

Bist, R. B., et al. (2023), Ammonia emissions, impacts, and mitigation strategies for poultry production: A critical review. *J. Environ. Manag.* 328: 116919.

Bizeray, D. et al. (2002a), Effects of increasing environmental complexity on the physical activity of broiler chickens. *Appl. Anim. Behav. Sci.* 79(1): 27-41.

Bloch, V. et al. (2020), Automatic broiler temperature measuring by thermal camera. *Biosystems Engineering* 199: 127-134.

Brassó, L. D., et al. (2025), Modern technologies for improving broiler production and welfare: a review. *Animals* 15(4): 493.

Brugaletta, G., et al. (2022). A review of heat stress in chickens. Part I: insights into physiology and gut health. *Front. Physiol.* 13: 934381.

Bruzual, J. J., et al. (2000), Effects of relative humidity during the last five days of incubation and brooding temperature on performance of broiler chicks from young broiler breeders. *Poult. Sci.* 79(10): 1385–1391.

Buffington, D. E., et al. (1981), Black Globe-Humidity Index (BGHI) as comfort equation for dairy cows. *Trans. ASAE* 24: 711–714.

Burggren, W. W., et al. (2015), Hypercapnic thresholds for embryonic acid–base metabolic compensation and hematological regulation during $CO_2$ challenges in layer and broiler chicken strains. *Resp. Physiol. Neurobiol.* 215: 1–12.

Calvet, S., et al. (2010), Ventilation rates in mechanically-ventilated commercial poultry buildings in Southern Europe: measurement system development and uncertainty analysis. *Biosyst. Eng.* 106(4): 423–432.

Casey, K. D., et al. (2010), Ammonia emissions from a US broiler house–comparison of concurrent measurements using three different technologies. *J. Air Waste Manag. Assoc.* 60(8): 939–948.

Çayli, A. (2025), Investigation of thermal environment uniformity in a tunnel-ventilated broiler house in the summer. *Appl. Ecol. Environ. Res.* 23(2): 3589–3602.

Costantino, C. et al. (2020), The reduction of gas concentrations in broiler houses through ventilation: Assessment of the thermal and electrical energy consumption. *Biosystems Engineering* 199: 135-148.

Charles, D. R. (1986), Temperature for broilers. *World's Poult. Sci. J.* 42(3): 249–258.

Chepete, J., et al. (2005), Production performance and temperature-humidity index of Cobb 500 broilers reared in open-sided naturally ventilated houses in Botswana. In Proceedings of the Seventh International Symposium, Livestock Environment VII, Beijing, China, 18–20 May 2005.

Chou, C. C., et al. (2004), Risk factors for cumulative mortality in broiler chicken flocks in the first week of life in Taiwan. *Br. Poult. Sci.* 45(5): 573–577.

Collins, S., et al. (2011), Investigating the importance of vision in poultry: comparing the behaviour of blind and sighted chickens. *Appl. Anim. Behav. Sci.* 133: 60–69.

Curi, T. M. R. D. C., et al. (2017), Positioning of sensors for control of ventilation systems in broiler houses: a case study. *Sci. Agric.* 74: 101–109.

Dasarathy, S., et al. (2017), Ammonia toxicity: From head to toe? *Metab. Brain. Dis.* 32: 529–538.

Dawkins, M. S. et al. (2004), Chicken welfare is influenced more by housing conditions than by stocking density. *Nature* 427(6972): 342-354.

De Moraes, S. R. P., et al. (2008), Classification of the temperature and humidity index (THI), aptitude of the region, and conditions of comfort for broilers and layer hens in Brazil. In Proceedings of the International Conference of Agricultural Engineering, XXXVLL Brazilian Congress of Agricultural Engineering, International Livestock Environment Symposium-ILES VIII, Iguassu Falls City, Brazil, 31 August–4 September 2008.

de Sans, E. C. O., et al. (2012), Effects of environmental enrichment on the welfare of industrial broiler chickens, *World's Poult. Sci. J.* 1: 64–67.

Deaton, J. W., et al. (1996), The effect of brooding temperature on broiler performance. *Poult. Sci.* 75(10): 1217–1220.

Deep, A., et al. (2012), Effect of light intensity on broiler behaviour and diurnal rhythms. *Appl. Anim. Behav. Sci.* 136(1): 50–56.

Del Valle, J. E., et al. (2021), Unrest index for estimating thermal comfort of poultry birds (*Gallus gallus domesticus*) using computer vision techniques. *Biosyst. Eng.* 206: 123–134.

DeShazer, J. A. and Beck, M. M. (1988), *University of Nebraska Report for Northeast Regional Poultry Project NE-127*. Agricultural Research Division, University of Nebraska: Lincoln, NE, USA.

Diarra, S., et al. (2021), Alternative bedding materials for poultry: availability, efficacy, and major constraints. *Front. Vet. Sci.* 8: 669504.

Donkoh, A. (1989), Ambient temperature: a factor affecting performance and physiological response of broiler chickens. *Int. J. Biometeorol.* 33(4): 259–265.

Dunlop, M. W., et al. (2016), Water activity of poultry litter: Relationship to moisture content during a grow-out. *J. Environ. Manag.* 172: 201–206.

Dunn, L. L., et al. (2022), The prevalence and concentration of *Salmonella enterica* in poultry litter in the southern United States. *PLoS One* 17(5): e0268231.

Elwakeel, A. E. (2025), A smart automatic control and monitoring system for environmental control in poultry houses integrated with earlier warning system. *Sci. Rep.* 15(1): 31630.

Farag, M. R. and Alagawany, M. (2018), Physiological alterations of poultry to the high environmental temperature. *J. Therm. Biol.* 76: 101–106.

Ferraz, P. F. P. et al. (2019), Spatial variability of litter temperature, relative air humidity and skin temperature of chicks in a commercial broiler house. *Agronomy Research* 17(2): 408-417.

Ferreira, J. C., et al. (2024), Dynamics of the thermal environment in climate-controlled poultry houses for broiler chickens. *Ag. Eng.* 6(4): 3891–3911.

Freire, R. et al. (2003), Behaviour and welfare of individual laying hens in a non-cage system. *British Poultry Science* 44: 22-29.

George, A. S. and George, A. H. (2023), Optimizing poultry production through advanced monitoring and control systems. *Partners Universal Int. Innovation J.* 1(5), 77–97.

Ghanima, M. M. A., et al. (2020), Growth, carcass traits, immunity and oxidative status of broilers exposed to continuous or intermittent lighting programs. *Anim. Biosci.* 34(7): 1243.

Gous, R. M. and Morris, T. R. (2005), Nutritional interventions inalleviating the effects of high temperatures in broiler production. *World's Poult. Sci. J.* 61: 463–475.

Guo, L., et al. (2022), Mitigation strategies of air pollutants for mechanical ventilated livestock and poultry housing–a review. *Atmosphere* 13(3): 452.

Güz, B. C., et al. (2021), Effects of pen enrichment on leg health of fast and slower-growing broiler chickens. *PLoS One* 16(12): e0254462.

Hamrita, T. K. and Conway, R. H. (2017), Effect of air velocity on deep body temperature and weight gain in the broiler chicken. *J. Appl. Poult. Res.* 26(1): 111–121.

Han, H., et al. (2021), Effects of ammonia on gut microbiota and growth performance of broiler chickens. *Animals 11*(6): 1716.

Hasanpur, K., et al. (2015), Influence of ascites syndrome on growth pattern of chickens reared at normal or cold ambient temperature. *Ann. Animal Sci.* 15(2): 373–385.

Henriksen, S., et al. (2016), Effects of post-hatch brooding temperature on broiler behavior, welfare, and growth. *Poult. Sci.* 95(10): 2235–2243.

Hernández-Sánchez, R. C., et al. (2024), Systematic review and meta-analysis of thermal stress assessment in poultry using infrared thermography in specific body areas. *Animals* 14(22): 3171.

Hirakawa, R., et al. (2020), Heat stress causes immune abnormalities via massive damage to effect proliferation and differentiation of lymphocytes in broiler chickens. *Front. Vet. Sci.* 7: 46.

Honda, B. T. B., et al. (2015), Effects of heat stress on peripheral T and B lymphocyte profiles and IgG and IgM serum levels in broiler chickens vaccinated for Newcastle disease virus. *Poult. Sci.* 94(10): 2375–2381.

Honig, H., et al. (2024), Comparative analysis of broiler housing systems: Implications for production and wellbeing. *Animals* 14(11): 1665.

Iyasere, O. S., et al. (2017), Validation of an intramuscularly-implanted microchip and a surface infrared thermometer to estimate core body temperature in broiler chickens exposed to heat stress. *Comp. Electr. Agric.* 133: 1–8.

Jacobs, L., et al. (2023), Enhancing their quality of life: environmental enrichment for poultry. *Poult. Sci.* 102(1): 102233.

Jones, P. J., et al. (2020), The productivity and financial impacts of eight types of environmental enrichment for broiler chickens. *Animals* 10(3): 378.

Kamely, M., et al. (2015), Incidence of ascites syndrome and related hematological response in short-term feed-restricted broilers raised at low ambient temperature. *Poult. Sci.* 94(9): 2247–2256.

Kim, S. J. and Lee, M. H. (2022), Design and implementation of a malfunction detection system for livestock ventilation devices in smart poultry farms. *Agriculture* 12(12): 2150.

Kristensen, H. H., et al. (2007), The behaviour of broiler chickens in different light sources and illuminances. *Appl. Anim. Behav. Sci.* 103(1–2): 75–89.

Kuenzel, W. J., et al. (2015), Exploring avian deep-brain photoreceptors and their role in activating the neuroendocrine regulation of gonadal development. *Poult. Sci.* 94(4): 786–798.

Leishman, E. M., et al. (2021), Meta-analysis to predict the effects of temperature stress on meat quality of poultry. *Poult Sci.* 100(11): 101471. https://doi.org/10.1016/j.psj.2021.101471.

Leksrisompong, N., et al. (2007), Broiler incubation. 1. Effect of elevated temperature during late incubation on body weight and organs of chicks. *Poult. Sci.* 86(12): 2685–2691.

Leksrisompong, N., et al. (2009), Broiler incubation. 2. Interaction of incubation and brooding temperatures on broiler chick feed consumption and growth. *Poult. Sci.* 88(6): 1321–1329.

Li, H., et al. (2017), Effects of two different broiler flooring systems on production performances, welfare, and environment under commercial production conditions. *Poult. Sci.* 96(5): 1108–1119.

Li, Y., et al. (2022), Design considerations for net zero energy buildings for intensive, confined poultry production: A review of current insights, knowledge gaps, and future directions. *Renew. Sustain. Energy Rev.* 154: 11874.

Liang, Y. and Costello, T. (2024), Measurement of dynamic electric consumption trend in a broiler house in Arkansas. *Appl. Eng. Agric.* 40(1): 143–150.

Lima, K. A. O., et al. (2011), Ammonia emissions in tunnel-ventilated broiler houses. *Braz. J. Poult. Sci.* 13: 265–270.

Lima, K. A., et al. (2021), Applying multi-criteria analysis to select the most appropriate broiler rearing environment. *Inf. Process. Agric.* 8(2): 205–218.

Liu, L., et al. (2020), Heat stress impacts on broiler performance: a systematic review and meta-analysis. *Poult. Sci.* 99(11): 6205.

Liu, Q. X., et al. (2021), Broilers' head behavior as an early warning index of production and lung health under ammonia exposure. *Poult. Sci.* 100(3): 100814.

Loyau, T., et al. (2015), Cyclic variations in incubation conditions induce adaptive responses to later heat exposure in chickens: a review. *Animal* 9(1): 76–85.

Maatjens, C. M., et al. (2016), Temperature during the last week of incubation. I. Effects on hatching pattern and broiler chicken embryonic organ development. *Poult. Sci.* 95(4): 956–965.

Maman, A. H., et al. (2019), Effect of chick body temperature during post-hatch handling on broiler live performance. *Poult. Sci.* 98(1): 244–250.

Manfio, E. S., et al. (2019), Intermittent lighting program does not hinder the performance of broiler chickens and promotes energy economy. *Can. J. Anim. Sci.* 100(2): 228–233.

Mangan, M. and Siwek, M. (2024), Strategies to combat heat stress in poultry production—a review. *J. Anim. Physiol. and Anim. Nutr.* 108(3): 576–595.

Martinez, A. A. G., et al. (2021), A Heuristic and data mining model for predicting broiler house environment suitability. *Animals* 11(10): 2780.

Martinez, D. A., et al. (2022), The fasting heat production of broilers is a function of their body composition. *Animal-Open Space* 1(1): 100029.

McMillian, Z., et al. (2025), Effects of circulation fans on broiler welfare indicators in commercial houses during cold seasons. *Journal of Applied Animal Welfare Science* 2025: 1–14.

McLean J. A., et al. (2002), Welfare of male and female broiler chickens in relation to stocking density, as indicated by performance, health and behaviour. *Animal Welfare* 11(1): 55-73.

Menegali, I., et al. (2012), Influence of different systems of minimum ventilation on air quality in broiler houses. *Eng. Agric.* 32: 1024–1033.

Mesa, D., et al. (2017), Broiler-housing conditions affect the performance. *Braz. J. Poult. Sci.* 19(2): 263–272.

Meteyake, H. T., et al. (2023), Thermal manipulation during incubation: effects on embryo development, production performance, meat quality, and thermal tolerance of broiler chickens. *J. World's Poult. Res.* 13(1): 29–40.

Miles, D. M., et al. (2004), Atmospheric ammonia is detrimental to the performance of modern commercial broilers. *Poult. Sci.* 83: 1650–1654.

Miles, D. M., et al. (2006), Ocular responses to ammonia in broiler chickens. *Avian Dis.* 50: 45–49.

Miragliotta, M. Y., et al. (2006), Spatial analysis of stress conditions inside broiler house under tunnel ventilation. *Sci. Agric.* 63: 426–432.

Morita, V. S., et al. (2016), Incubation temperature alters thermal preference and response to heat stress of broiler chickens along the rearing phase. *Poult. Sci.* 95(8): 1795–1804.

Mou, C. T., et al. (2025), Evaluation of the effects of maintaining a moderate humidity (50–60%) and increased air movement on litter moisture and footpad health in a commercial broiler house. *J. Appl. Poult. Res.* 34(1): 100476.

Naraballobh, W., et al. (2016), Transient shifts of incubation temperature reveal immediate and long-term transcriptional response in chicken breast muscle underpinning resilience and phenotypic plasticity. *PLoS One* 11(9): e0162485.

National Research Council (NRC). (1971), *A Guide to Environmental Research on Animals.* National Academic Science, Washington, DC, USA.

Nawab, A., et al. (2018), Heat stress in poultry production: Mitigation strategies to overcome the future challenges facing the global poultry industry. *J. Thermal Biol.* 78: 131–139.

Nawaz, A. H., et al. (2021), Poultry response to heat stress: Its physiological, metabolic, and genetic implications on meat production and quality including strategies to improve broiler production in a warming world. *Front. Vet. Sci.* 8: 699081.

Ohara, A., et al. (2015), Effect of environmental enrichment on the behaviour and welfare of Japanese broilers at a commercial farm. *J. Poult. Sci.* 52: 323–330.

Olanrewaju, H. A., et al. (2013), Interactive effects of photoperiod and light intensity on blood physiological and biochemical reactions of broilers grown to heavy weights. *Poult. Sci.* 92(4): 1029–1039.

Olanrewaju, H. A., et al. (2015), Effects of color temperatures (kelvin) of LED bulbs on growth performance, carcass characteristics, and ocular development indices of broilers grown to heavy weights. *Poult. Sci.* 94(3): 338–344.

Olanrewaju, H. A., et al. (2016), Effects of light sources and intensity on broilers grown to heavy weights. Part 1: growth performance, carcass characteristics, and welfare indices. *Poult. Sci.* 95(4): 727–735.

Olanrewaju, H. A., et al. (2018), Influence of light sources and photoperiod on growth performance, carcass characteristics, and health indices of broilers grown to heavy weights. *Poult. Sci.* 97(4): 1109–1116.

Omomowo, O. O. (2021), Temperature-humidity index and thermal comfort of broilers in humid tropics. *Agric. Eng. Int.* 23(3): 101–110.

Onagbesan, O. M., et al. (2023), Alleviating heat stress effects in poultry: updates on methods and mechanisms of actions. *Front. Vet. Sci.* 10: 1255520.

Özkan, S. et al. (2003), The effect of short-term fasting on performance traits and rectal temperature of broilers during the summer season. *British Poultry Science* 44(1): 88–95.

Ozkanlar, S., et al. (2021), Effects of photoperiod on thyroid gland development and function in growing chicks: a biochemical and morphometric study. *Anim. Prod. Sci.* 61(16): 1652–1658.

Park, G., et al. (2018), Ventilation rate formula for mechanically ventilated broiler houses considering aerodynamics and ventilation operating conditions. *Biosyst. Eng.* 175: 82–95.

Patterson, P. H. (2005), Management strategies to reduce air emissions: emphasis–Dust and ammonia. *J. Appl. Poult. Res.* 14(3): 638–650.

Pecoraro, C. A., et al. (2023), Enthalpy as a thermal comfort index in broiler poultry production. *Rev. Bras. Eng. Agric. Ambient.* 28(1): e270399.

Pederson, I. J. and Forkman, B. (2019), Improving leg health in broiler chickens: a systematic review of the effect of environmental enrichment. *Animal Welfare* 28(2): 215–230.

Piestun, Y., et al. (2011), Thermal manipulations during broiler embryogenesis improves post-hatch performance under hot conditions. *J. Therm. Biol.* 36(7). 469–474.

Ponciano, P. F., et al. (2013), Spatial variability of air-dry bulb temperature and black globe humidity index in a broiler house during the heating phase. *Eng. Agríc.* 33: 433–444.

Prayitno, D. S. et al. (1997), The effects of color of lighting on the behavior and production of meat chickens. *Poultry Science*. 76 (3): 452-457.

Qaid, M. M., et al. (2023), Effect of housing system and housing density on performance, viability, and gastrointestinal tract growth of broiler chicks during the first 2 wk of age. *Poult. Sci.* 102(7): 102752.

Quintana-Ospina, G. A., et al. (2023), Effect of environmental and farm-associated factors on live performance parameters of broilers raised under commercial tropical conditions. *Animals* 13(21): 3312. https://doi.org/10.3390/ani13213312.

Rault, J-L., et al. (2017), Light intensity of 5 or 20 lux on broiler behavior, welfare and productivity. *Poultry Science* 96(4): 779-787.

Reiter, K and Bessei, W. (2009), Effect of locomotor activity on leg disorders in fattening chickens. Berl. Munch. *Tierarztl. Wochenschur*. 122: 264-270.

Riaz, M. F., et al. (2021), Impact of light stimulation during incubation on hatching traits and post-hatch performance of commercial broilers. *Trop. Anim. Health Prod.* 53(1): 107.

Riber, A. B. (2015), Effects of color of light on preferences, performance, and welfare in broilers. *Poult. Sci.* 94(8): 1767–1775.

Riber, A. B., et al. (2018), Review of environmental enrichment for broiler chickens. *Poult. Sci.* 97(2): 378–396.

Rodrigues, V. C., et al. (2010), A correct enthalpy relationship as thermal comfort index for livestock. *Int. J. Biometeorol.* 55: 455–459.

Rogers, A. G., et al. (2015a), I. Evaluation of the impact of alternative light technology on male broiler chicken growth, feed conversion, and allometric characteristics. *Poult. Sci.* 94(3): 408–414.

Rogers, A. G., et al. (2015b), II. Evaluation of the impact of alternative light technology on male broiler chicken stress. *Poult. Sci.* 94(3): 331–337.

Rozenboim, I., et al. (1999), The effect of monochromatic light on broiler growth and development. *Poult. Sci.* 78: 135–138.

Rozenboim, I., et al. (2004), The effect of a green and blue monochromatic light combination on broiler growth and development. *Poult. Sci.* 83: 842–845.

Sa, R. N., et al. (2018), Atmospheric ammonia alters lipid metabolism-related genes in the livers of broilers (*Gallus gallus*). *J. Anim. Physiol. Anim. Nutr.* 102: e941–e947.

Saleeva, I. et al. (2020), Efficiency of poultry heating and ventilation. IOP Conference Series Earth and Environmental Science 433(1): *article* 012041.

Senaratna, D., et al. (2012), Preference of broiler chicken for different light colors in relation to age, session of the day and behaviour. *Tropical Agric. Res.* 23: 193–203.

Shah, S. W. A., et al. (2020), Ammonia inhalation impaired immune function and mitochondrial integrity in the broilers bursa of fabricius: implication of oxidative stress and apoptosis. *Ecotoxicol. Environ. Saf.* 190: 110078.

Shynkaruk, T., et al. (2023), Impact of stocking density on the welfare and productivity of broiler chickens reared to 34 d of age. *J. Appl. Poult. Res.* 32(2): 100344.

Skomorucha, I. et al. (2007), Effect of restricted access to water on broiler house climate and productivity of broiler chickens. *Paper at ISAH 2007 Conference, Tartu, Estonia.*

Soliman, E. S. and Hassan, R. A. (2020), Influence of housing floor on air quality, growth traits, and immunity in broiler chicken farms. *Adv. Anim. Vet. Sci.* 8(9): 997–1008.

Soliman, E. S., et al. (2021), Impact of heating systems on air and litter quality in broiler houses, performance, behavior, and immunity in broiler chickens. *Adv. Anim. Vet. Sci.* 9(2): 301–314.

Song, D. J. and King, A. J. (2015), Effects of heat stress on broiler meat quality. *World's Poult. Sci. J.* 71(4): 701–709.

Sosnówka-Czajka, E. et al. (2010), Effect of different housing systems on productivity and welfare of laying hens. *Ann. Anim. Sci.* 10(4): 349-360.

Stamp Dawkins, M., et al. (2004), Chicken welfare is influenced more by housing conditions than by stocking density. *Nature* 427(6972): 342–344.

Sultana, S., M. R., et al. (2013), The effect of monochromatic and mixed LED light colour on the behaviour and fear responses of broiler chicken. *Avian Biol. Res.* 6: 207–214.

Tahamtani, F. M., et al. (2020), Effects of environmental complexity on welfare indicators of fast-growing broiler chickens. *Poult Sci.* 99(1): 21–29.

Tao, X. and Xin, H. (2003), Acute synergistic effects of air temperature, humidity, and velocity on homeostasis of market-size broilers. *Trans. ASAE* 46(2): 491.

Thom, E. C. (1979), The discomfort index. *Weatherwise* 12: 57–59.

Tona, K., et al. (2022), Chicken incubation conditions: role in embryo development, physiology and adaptation to the post-hatch environment. *Front. Physiol.* 13: 895854.

Topal, E. and Petek, M. E. T. İ. N. (2021), Effects of fully or partially slatted flooring designs on the performance, welfare and carcass characteristics of broiler chickens. *Br. Poult. Sci.* 62(6): 804–809.

Trentin, A., et al. (2025), Technical and economic performance favours fully automated climate control broiler housing. *Br. Poult. Sci.* 66(1): 63–70.

Tuyttens, F. (2005), Comparison of broiler chicken health and welfare in organic versus traditional production systems. *Animal Science Papers and Reports* (Polish Academy of Sciences) 23 (Suppl. 1): 217–222.

Thema, K. K., et al. (2022), Stocking density-induced changes in growth performance, blood parameters, meat quality traits, and welfare of broiler chickens reared under semi-arid subtropical conditions. *PLoS One* 17(10): e0275811.

Tzschentke, B. (2007), Attainment of thermoregulation as affected by environmental factors. *Poult. Sci.* 86(5), 1025–1036.

Ullman, J. L., et al. (2004), A review of literature concerning odors, ammonia, and dust from broiler production facilities: 4. Remedial management practices. *J. Appl. Poult. Res.* 13(3): 521–531.

van der Eijk, J. A., et al. (2022), Reducing stocking density benefits behaviour of fast-and slower-growing broilers. *Appl. Anim. Behav. Sci.* 257: 105754.

van der Pol, C. W., et al. (2019), Light-dark rhythms during incubation of broiler chicken embryos and their effects on embryonic and post hatch leg bone development. *PLoS One* 14(1): e0210886.

van Harn, J. et al. (2012), Effect of bedding material on dust and ammonia emissions from broiler houses. *Transactions of the ASABE* 55(1): 219-226.

Van Limbergen, T., et al. (2020), Risk factors for poor health and performance in European broiler production systems. *BMC Vet. Res.* 16(1): 287.

Vercellino, R. D. A., et al. (2025), Tracking heat stress in broilers: a thermographic analysis of anatomical sensitivity across growth stages. *Animals* 15(15): 2233.

Vilela, M. D. O., et al. (2020), Nitrogen transformation stages into ammonia in broiler production: sources, deposition, transformation, and emission into the environment. *Dyna* 87(214): 221–228.

Vissers, L. S., et al. (2021), Exploring the performance of system changes in Dutch broiler production to balance animal welfare, ammonia emissions and particulate matter emissions with farm profitability. *Agric. Syst.* 193: 103217.

Wang, G. et al. (2025), Impact of chronic heat stress in broiler breeders on egg quality, hatchability and day-old chick quality. *Poultry Science* 104: article 105452.

Wang, S., et al. (2019), Application of transcriptome analysis: Oxidative stress, inflammation and microtubule activity disorder caused by ammonia exposure may be the primary factors of intestinal microvilli deficiency in chicken. *Sci. Total Environ.* 696: 134035.

Wang, G., et al. (2022). Effects of different ammonia concentrations on pulmonary microbial flora, lung tissue mucosal morphology, inflammatory cytokines, and neurotransmitters of broilers. *Animals* 12(3): article 261.

Weeks, B. C., et al. (2022), Temperature, size and developmental plasticity in birds. *Biol. Lett.* 18(12): p.20220357.

Wei, F. X., et al. (2015), Ammonia concentration and relative humidity in poultry houses affect the immune response of broilers. *Genet. Mol. Res.* 14(2): 3160–3169.

West, D., et al. (2024), Impacts of air velocity treatments under summer conditions: Part III–Litter characteristics, ammonia emissions, and leg health of heavy broilers. *Animals* 14(23): 3525.

Wicklin, G. and Czarick, M. (1997), Particulate emissions from poultry housing. In Proceedings of the ASAE Annual International Meeting, Minneapolis Convention Center, Minneapolis, MN, USA, 10–14 August 1997; pp. 10–14.

Wijnen, H. J., et al. (2022), Effects of late incubation temperature and moment of first post-hatch feed access on neonatal broiler development, temperature preference, and stress response. *Poult. Sci.* 101(10): 102088.

Wijnen, H. J., et al. (2020), Effects of incubation temperature pattern on broiler performance. *Poult. Sci.* 99(8): 3897–3907.

Winkel, A., et al. (2016), Evaluation of oil spraying systems and air ionisation systems for abatement of particulate matter emission in commercial poultry houses. *Biosyst. Eng.* 150: 104–122.

Wu, Y., et al. (2022), Light regimen on health and growth of broilers: an update review. *Poultry Science* 101(1): article 101545.

Xie, D., et al. (2008), Effects of monochromatic light on immune response of broilers. *Poultry Science* 87 (8): 1535-1539.

Yahav, S. (2009), Alleviating heat stress in domestic fowl: different strategies. *World's Poultry Science Journal* 65: 719-732.

Yang, Y., et al. (2018), The quantitative models for broiler chicken response to monochromatic, combined, and mixed light-emitting diode light: a meta-analysis. *Poult. Sci.* 97(6): 1980–1989.

Yehia, M., et al. (2025), Rectal temperature and heat transfer dynamics in the eye, face, and breast of broiler chickens exposed to moderate heat stress. *Poult. Sci.* 104(2): 104748.

Zaboli, G. R., et al. (2017), Thermal manipulation during Pre- and Post-Hatch on thermotolerance of male broiler chickens exposed to chronic heat stress. *Poult. Sci.* 96(2): 478–485.

Zhao, F., et al. (2020), The imbalance of Th1/Th2 triggers an inflammatory response in chicken spleens after ammonia exposure. *Poult. Sci.* 99: 3817–3822.

Zhou, Y., et al. (2020), Effects of ammonia exposure on growth performance and cytokines in the serum, trachea, and ileum of broilers. *Poult. Sci.* 99(5): 2485–2493.

Zhou, Y., et al. (2023), Fine particulate matter perturbs the pulmonary microbiota in broiler chickens. *Animals*, 13(18): 2862.

# Part 4

## Challenges in broiler production

# Chapter 12

## Challenges in achieving antibiotic-free broiler production

*Elijah G. Kiarie, University of Guelph, Canada and North-West University, South Africa; Emily Kim, New-Life Mills – Division of Parrish & Heimbecker Ltd., Canada; Aizwarya Thanabalan, Jones Feed Mill, Canada; Munene Kithama, Pwani University, Kenya; and Lisa Bean-Hodgins, BSC Animal Nutrition Inc., Canada*

1 Introduction
2 Antibiotic use in broiler production: a brief history
3 Genesis of antibiotic-free broiler chicken production
4 State of antibiotic-free broiler production
5 Challenges for achieving antibiotic-free broiler production
6 Opportunities for achieving antibiotic-free broiler production
7 Conclusion and future perspectives
8 Author contributions
9 Conflict of interest statement
10 Acknowledgments
11 Where to look for further information
12 References

## 1 Introduction

Advances in genetics, nutrition, health, management, and digital technology have collectively contributed to the greater biological and economic performance seen in modern broiler production value chain (Gous, 2010, Havenstein et al., 2003, Krupa et al., 2017, Merks, 2018). However, the industry is under pressure to produce food products in ways that are ethical, welfare-friendly, environmentally sustainable, and wholesome. For example, there are growing concerns around the world on indiscriminate use of antibiotics and nexus with the burgeoning antimicrobial resistance (AMR) in human and veterinary pathogens (Bean-Hodgins and Kiarie, 2021, Emes et al., 2024, Mak et al., 2022). Defined as the survival capacity of pathogenic bacteria in

http://dx.doi.org/10.19103/AS.2025.0165.14

the presence of previously effective single or multiple antibiotics, the AMR is a serious threat to humanity, as drug-resistant pathogens of common human infections are currently claiming over 1 million lives annually and projected to rise to 10 million and attendant economic loss of in excess of US$100 trillion by 2050 (Kirwin et al., 2019, Ahmad et al., 2017, Murray et al., 2022). Although scientists do not know the precise origin of AMR, indiscriminate antimicrobial use (AMU) in food animal production has been pinpointed to be a major contributing factor (PHAC, 2016, Landers et al., 2012, Emes et al., 2024, Van Boeckel et al., 2017, Khan et al., 2024). In this context, food animal production chains are faced with regulatory, industry, and consumer pressure to ban and restrict AMU to stem AMR (Laxminarayan et al., 2015, Van Boeckel et al., 2017). This chapter will attempt to highlight emerging challenges and opportunities for achieving antibiotic-free broiler (AFB) production.

## 2 Antibiotic use in broiler production: a brief history

The discovery of penicillin in the early 1920s paved the way for the development of other antibiotic molecules and their subsequent mass production for use in human and veterinary medicine. By late 1940, a serendipitous discovery showed that feeding sub-therapeutic doses of antibiotics to healthy broiler chickens promoted faster growth with tremendous feed conversion efficiency (Moore et al., 1946). This discovery became the basis of the widespread practice of AMU for growth promotion as it permitted more rapid and cost-effective chicken production. Nexus to growth promotion is metaphylactic administration to control disease from spreading to healthy chickens in a flock with some diseased birds, or when outbreaks have been detected in nearby houses or farms. This application may also protect already infected, but asymptomatic chickens. Prophylactic administratton, on the other hand, entails the preventative administration of antibiotics to the entire healthy flock due to the high likelihood of disease outbreak. Approval of using antibiotics as feed additives without veterinary prescription for prophylactic and growth promotion in the 1950s heralded a new era of food animal production in the EU and US (Castanon, 2007, Laxminarayan et al., 2015). It has been estimated that on standardized basis, the AMU in humans (118 mg/kg) is less than in food animals (133 mg/kg of animal product) (Van Boeckel et al., 2017). However, AMU in animals is far greater in terms of quantity and diversity of the molecules because of the massive biomass of food-producing animals and distinct patterns of application for prophylactic and production purposes (Van Boeckel et al., 2017, van Bunnik and Woolhouse, 2017, Khan et al., 2024). The global AMU in food animal production was projected to reach over 200,000 tons by 2030, representing a 50% increase of the consumption levels recorded for 2013 (Van Boeckel et al., 2017). Arguably further expansion of AMU in food animal production is expected, because of anticipated growth in global demand for animal proteins (Mottet and Tempio, 2017).

## 3 Genesis of antibiotic-free broiler chicken production

The widespread AMU has, over the years, raised concerns about increased selection pressure, the emergence, and spread of AMR pathogens of human and veterinary importance. For example, in 1970, the UK banned, and the US proposed the withdrawal of penicillin and tetracycline from use as growth promoters in food animal production. However, global awareness of AMR hit the center stage in early 2000 with the World Health Organization highlighting the risks associated with indiscriminate use of antibiotics in human medicine and agriculture (Laxminarayan et al., 2015, WHO, 2021). The circularity of resistant microbes and resistance genes between humans, animals, food, water, and the environment posits a deeper link between AMU in food animals and human health risks (Van Boeckel et al., 2017, van Bunnik and Woolhouse, 2017, Khan et al., 2024). The antibiotic molecules used in food animal production and human medicine are structurally and functionally similar. Studies in Canada showed that *in-ovo* use of ceftiofur to control omphalitis was linked to resistant *Salmonella enterica* serovar Heidelberg (Dutil et al., 2010). Furthermore, AMR in zoonotic pathogens including to the last resort drugs, such as colistin, generated serious concerns in human medicine linked to the possibility of untreatable infections (Liu et al., 2016). However, much stronger evidence for AMR in common foodborne pathogens such as *Campylobacter*, *Escherichia coli*, and *Salmonella* linked to AMU for production purposes in broiler chickens highlighted the unassailable risk (Emes et al., 2024, Khan et al., 2024). Arguably, the AMR also threatens the effectiveness of treating infectious diseases in poultry and as such risking the food security and livelihoods of millions of primary producers and consumers.

Public policy discussions have focused on the inappropriate AMU in humans and food animal production. A scenario that has led to banning or restricting AMU in food animal production through legislative interventions and voluntary actions by processors, retailers, and consumers. Following an early ban in Sweden and Denmark, the EU banned AMU for growth promotion in 2006. Other countries, including the US and Canada, followed with similar regulations over the next decade (Bean-Hodgins and Kiarie, 2021, Mak et al., 2022). Other countries in the Organisation for Economic Co-operation and Development (OECD) block, such as Australia, Mexico, South Korea, and New Zealand, have implemented ban on AMU for growth promotion, whereas other OECD countries, such as Japan, still permit AGP (Laxminarayan et al., 2015). Brazil, China, India, and much of South Asian countries, including Vietnam, Bangladesh, Bhutan, Indonesia, Myanmar, Nepal, Sri Lanka, and Thailand, have announced some degree of banning and restrictions (Maria Cardinal et al., 2019). Instructively, AMU for any purpose in animal production continuum is not currently banned in many other countries. A common goal of banning AMU in food animal production is to reduce the AMR gene pool in pathogens and normal gastrointestinal microbiota, to reduce the risk of transmission of antibiotic residues and resistant bacteria to humans. In this context, the World Health Organization has been pressing for global-scale banning because of the potential spread of AMR organisms from one country to another through trade, travel, and migration (WHO, 2021).

## 4 State of antibiotic-free broiler production

Ever since the EU implemented ban on AMU for growth promotion in 2006, AFB production has been defining the industry trends for marketing, consumer preferences, and regulatory initiatives (Bean-Hodgins and Kiarie, 2021, Smith, 2019). Estimating the global broiler production under AFB is difficult in large part because of data availability. However, AFB production is a significant portion of broiler production in the EU (Augère-Granier, 2019) and has been reported to constitute 50–60% of total broiler production in the US (Cervantes and McDougald, 2023). Approximately 40% of broiler flocks in Canada are reared without medically important antibiotics in categories I, II, and III (Hein, 2023). Where Category I is of very high importance in human medicine, for which there are no alternatives. Category II is of high importance in human medicine, with other options generally available. Category III is of medium importance in human medicine, with alternatives readily available. There are indications of a growing trend for AFB in other countries as legislative and consumer pressure lump up (Laxminarayan et al., 2015, Mak et al., 2022). For example, projections indicated increased AMU in food animal production in the coming years, particularly in Asia, Africa, and Latin America, where further growth in meat production is expected (Van Boeckel et al., 2015, 2017). Moreover, broiler meat is one of the most traded animal proteins with the EU importing significant quantities from countries such as Brazil, Thailand, Ukraine, Chile, and China (Augère-Granier, 2019). It is plausible such products must adhere to strict EU regulations on AMU. Considering that the risk of AMR is associated with the quantity and frequency of AMU, it is imperative for coordinated and collaborative global efforts. For example, producers must adhere to strict regulations and standards for AFB labeling, which can vary between countries. Ensuring compliance requires careful record-keeping and management practices. Moreover, consumer understanding of what AFB means can be limited or distorted by marketing messaging (Laxminarayan et al., 2015). This can create confusion and affect purchasing decisions. Surveys in the US, Canada, and the EU indicated that consumers would place a high value on AFB products and would be willing to pay more for such products (Barrett et al., 2021). However, most countries do not have the capability or resources for inspecting farms or traceability programs (Haque et al., 2020). Moreover, it is worth noting the diversity of integration of broiler production systems for economic efficiency. For example, according to the National Chicken Council, over 90% of the broiler chicken production chain in the US is vertically integrated with feed mills, hatcheries, and processing plants with contract grower operations. The level of integration has a significant impact on the ease with which policies can be implemented. Integration means fewer decision-makers influence a larger share of production, which can institute changes more easily than thousands of entities covering small shares of production. This is particularly true in developing countries, where broiler production is often organized in separate entities, such as feed mills, hatcheries, farms, and processors. As such, ensuring that the entire supply chain, from hatcheries to feed suppliers, is aligned with antibiotic-free standards can be challenging.

## 5 Challenges for achieving antibiotic-free broiler production

The largest cost of raising broiler chickens is feed; therefore, maintaining chickens in an environment that supports proper nutrition is essential for productivity and profitability. The primary challenges with AFB production at the farm level are high burden of neonatal infections, incidences of intestinal health problems, impaired nutrient digestion, and absorption, ultimately leading to poor feed conversion ratio (FCR), increased mortalities, and condemnation at the processing plant (Bean-Hodgins and Kiarie, 2021, Fernández et al., 2024, Kogut et al., 2024). In this context, strategies for supporting functional and healthy gastrointestinal (GIT) are seen critical for lifetime survival, and resiliency in broilers raised without antibiotics (Klasing, 2007, Kogut and Klasing, 2009). Consequently, managing gut health is an emerging frontier for preventing enteric diseases and bolstering immunity for productivity in AFB. Tremendous efforts have been directed towards research and development of dietary strategies for modulating GIT health and function. For example, organic acids (Dai et al., 2021), prebiotics and probiotics (Tomczyk et al., 2024), phytogenic compounds (Yang et al., 2015, Kithama et al., 2021b), and immune modulators (Kim and Lillehoj, 2019), among others. We have run several studies to test such dietary strategies, for example, epidermal growth factor, probiotics, seaweed, fruit pomaces, insect protein, functional amino acids (glutamine and glutamic acid), and a combination of probiotics, organic acids, yeast, and essential oils as alternatives. A meta-analysis of our studies showing magnitude of growth performance responses against antibiotics/ionophores (AGP) is illustrated in Fig. 1. The mean effect size of average daily gain (g/d) was 0.567 in favor

(a)

| Study name | Std diff in means | Standard error | Variance | Lower limit | Upper limit | Z-Value | p-Value |
|---|---|---|---|---|---|---|---|
| Mbao et al., 2020a | -2.043 | 0.617 | 0.380 | -3.251 | -0.834 | -3.312 | 0.001 |
| Mbao et al., 2020b | -0.035 | 0.577 | 0.333 | -1.167 | 1.097 | -0.060 | 0.952 |
| Kithama et al., 2020 | -0.414 | 0.505 | 0.255 | -1.404 | 0.576 | -0.819 | 0.413 |
| Mohammadigheisar et al., 2020 | 1.244 | 0.546 | 0.298 | 0.174 | 2.315 | 2.278 | 0.023 |
| Bean-Hodgins et al. 2022a | 0.000 | 0.354 | 0.125 | -0.693 | 0.693 | 0.000 | 1.000 |
| Bean-Hodgins et al. 2022b | 0.063 | 0.354 | 0.125 | -0.630 | 0.756 | 0.178 | 0.859 |
| Fruci et al., 2023 | -0.892 | 0.524 | 0.275 | -1.919 | 0.136 | -1.701 | 0.089 |
| Kim 2023a | 0.734 | 0.596 | 0.356 | -0.435 | 1.903 | 1.231 | 0.218 |
| Kim 2023b | 1.690 | 0.673 | 0.452 | 0.372 | 3.008 | 2.512 | 0.012 |
| Kim 2023c | 0.916 | 0.607 | 0.368 | -0.273 | 2.106 | 1.510 | 0.131 |
| Kim 2023d | 1.833 | 0.688 | 0.473 | 0.484 | 3.181 | 2.664 | 0.008 |
| Kithama et al., 2023a | 0.920 | 0.397 | 0.158 | 0.141 | 1.699 | 2.314 | 0.021 |
| Kithama et al., 2023b | 0.947 | 0.399 | 0.159 | 0.165 | 1.728 | 2.375 | 0.018 |
| Kithama et al., 2023c | 1.444 | 0.424 | 0.180 | 0.612 | 2.276 | 3.402 | 0.001 |
| Kithama et al., 2023d | 1.460 | 0.425 | 0.181 | 0.626 | 2.294 | 3.432 | 0.001 |
| Facey et al., 2023 | 1.822 | 0.595 | 0.354 | 0.657 | 2.988 | 3.064 | 0.002 |
| Tsementzis 2023 | 0.142 | 0.501 | 0.251 | -0.839 | 1.123 | 0.284 | 0.777 |
| Pooled | 0.567 | 0.230 | 0.053 | 0.117 | 1.017 | 2.468 | 0.014 |
| Prediction Interval | 0.567 | | | -1.195 | 2.329 | | |

Q-test for heterogeneity 58.7; P<0.001, I-squared 73%; Tau-squared 0.63, SD=0.79

(b)

| Study name | Std diff in means | Standard error | Variance | Lower limit | Upper limit | Z-Value | p-Value |
|---|---|---|---|---|---|---|---|
| Mbao et al., 2020a | -0.667 | 0.514 | 0.264 | -1.674 | 0.340 | -1.298 | 0.194 |
| Mbao et al., 2020b | 0.500 | 0.586 | 0.344 | -0.649 | 1.649 | 0.853 | 0.394 |
| Kithama et al., 2020 | 0.000 | 0.500 | 0.250 | -0.980 | 0.980 | 0.000 | 1.000 |
| Mohammadigheisar et al., 2020 | -1.200 | 0.543 | 0.295 | -2.265 | -0.135 | -2.209 | 0.027 |
| Bean-Hodgins et al. 2022a | -0.250 | 0.355 | 0.126 | -0.946 | 0.446 | -0.704 | 0.481 |
| Bean-Hodgins et al. 2022b | -0.500 | 0.359 | 0.129 | -1.204 | 0.204 | -1.393 | 0.164 |
| Fruci et al., 2023 | 0.625 | 0.512 | 0.262 | -0.379 | 1.629 | 1.221 | 0.222 |
| Kim 2023a | -0.667 | 0.593 | 0.352 | -1.829 | 0.496 | -1.124 | 0.261 |
| Kim 2023b | 0.333 | 0.581 | 0.338 | -0.806 | 1.473 | 0.573 | 0.566 |
| Kim 2023c | 0.667 | 0.593 | 0.352 | -0.496 | 1.829 | 1.124 | 0.261 |
| Kim 2023d | 0.000 | 0.577 | 0.333 | -1.132 | 1.132 | 0.000 | 1.000 |
| Kithama et al., 2023a | -1.000 | 0.401 | 0.161 | -1.786 | -0.214 | -2.494 | 0.013 |
| Kithama et al., 2023b | -1.250 | 0.413 | 0.171 | -2.060 | -0.440 | -3.025 | 0.002 |
| Kithama et al., 2023c | -2.000 | 0.463 | 0.214 | -2.907 | -1.093 | -4.320 | 0.000 |
| Kithama et al., 2023d | -2.250 | 0.483 | 0.233 | -3.197 | -1.303 | -4.659 | 0.000 |
| Facey et al., 2023 | -0.667 | 0.514 | 0.264 | -1.674 | 0.340 | -1.298 | 0.194 |
| Tsementzis 2023 | -0.333 | 0.503 | 0.253 | -1.320 | 0.653 | -0.662 | 0.508 |
| Pooled | -0.545 | 0.200 | 0.040 | -0.937 | -0.154 | -2.731 | 0.006 |
| Prediction Interval | -0.545 | | | -2.008 | 0.918 | | |

Q-test for heterogeneity 45.8; P<0.001, I-squared 65%; Tau-squared 0.43, SD=0.66

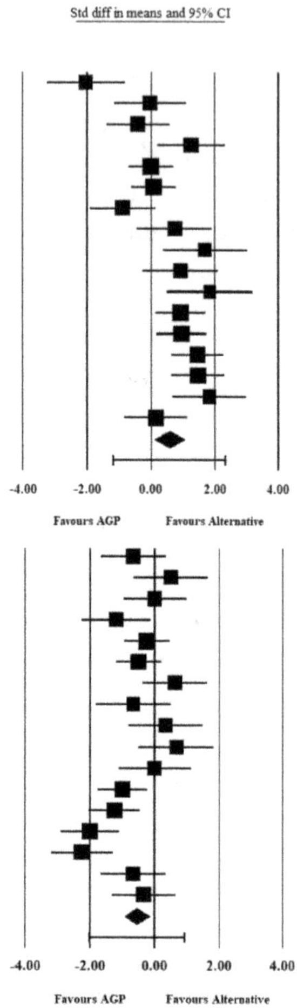


**Figure 1** Forest plot on standardized difference in means of average daily gain (ADG, (a)) and feed conversion ratio (FCR, (b)) effects of antibiotic growth promoters (AGP) vs. alternatives in 17 broiler chicken studies. Meta-analyses conducted using Comprehensive Meta-Analysis Version 4 (Borenstein, 2022).

(*P*=0.014) of the alternatives relative to AGP with a 95% confidence interval of +0.117 to +1.017, whereas the mean effect of FCR was −0.545 with 95% confidence interval of −0.937 to −0.154 in favor (*P*=0.006) of alternatives. However, the Q-value was 58.7 ($P < 0.001$) for average daily gain (ADG) and 45.8 ($P<0.001$) for FCR, indicating the response depended on the type of dietary strategy linked to the mode of action. Our studies extended the body of data demonstrating promising but variable responses of alternative feed additives and feeding strategies as effective replacements for AGP, for example (Moore, 2024). Indeed, a recent survey of 300 global poultry and feed industry stakeholders (nutritionists, veterinarians, consultants, purchasing agents, producers, retailers, feed mills, etc.) by WATT Global Media indicated that over 90% of respondents have implemented programs to eliminate or reduce AMU for production purposes (Roembke, 2025). However, the respondents indicated that the biggest challenges for AGP-free production were losses associated with diseases linked to the inconsistent responses of alternative feed additives. In this context, there is room to explore strategies to mitigate suboptimal health and performance in AFB. Subsequent sections will highlight how adjustments in management, nutrition, and vaccines are the most promising strategies for making AFB production successful.

## 6 Opportunities for achieving antibiotic-free broiler production

### 6.1 Management

Using the Danish experience of banning AMU for growth promotion, it has been reported that the loss in performance in pigs could be compensated for by improving animal management practices and biosecurity (Aarestrup et al., 2010, Aarestrup et al., 2008). Poor management practices have been associated with increased AMU linked to bacterial dysbiosis, immunocompromised chick, and abundant and persistent pathogens with resultant dissemination of AMR bacteria (Khan et al., 2024). Management decisions have a huge influence on growth performance as the chicken environment, husbandry practices, sanitary conditions, and feeding contribute to resilience and immunocompetence (Kogut et al. 2024). Considering the breeders, the egg and hatching conditions, the chick that shall grow to market weight, and the general housing and welfare conditions, broiler management considerations would entail the summations below (Fig. 2).

#### 6.1.1 The breeder

The goal of a broiler breeder farm is to produce as many good-quality fertile eggs as possible to maximize hatchability and ensure that the chicks can tolerate field placement challenges and achieve performance targets. The quality of the newly hatched chick is a major factor in determining its livability, growth, and health and thus relevant in AFB production endeavors. The ancestors of domestic chickens inhabited and were adapted to a living environment different from that used for current commercial production. Different husbandry measures for broiler breeders and chicks may result in a mismatch between generations, resulting in impaired disease resistance in the neonate (Godfrey et al., 2007). Broiler breeder houses must be managed to produce nest-clean hatching eggs. These could be achieved through optimization of lighting, ventilation, litter quality, nest space utilization, egg collection, and storage

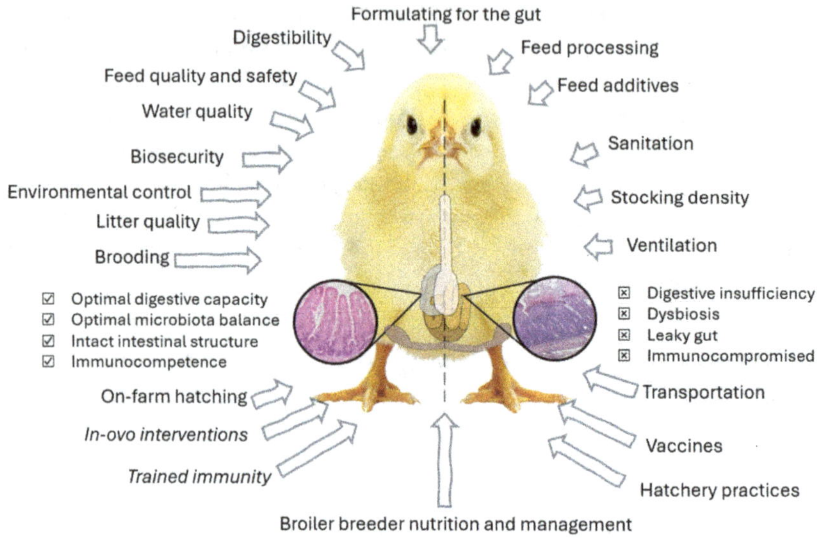


**Figure 2** Pathways for achieving antibiotic-free broiler production.

practices (Kogut et al., 2024). Evolutionally, the hatching eggs are enriched with integrated structural barriers and innate immune components critical for nourishment and protection during embryogenesis through to early post-hatch period (Hincke et al., 2019). Freshly laid egg is an enclosure that must remain free of any microorganisms despite the surface microbiota of the eggshell and function as transgenerational carrier of the maternal nutrients, microbiome, and functional molecules to sustain embryo and neonate (Hincke et al., 2019, Friedman et al., 2012). The broiler breeder age, management, and nutrition prior to sexual maturity and during egg production affect hen gut microbiome, hormonal status, and egg attributes (Thanabalan et al., 2024, Thanabalan et al., 2022). The broiler breeder house sanitary conditions, heat stress, vaccination programs, immunocompetence, and pathogen exposure have implications on hatching egg attributes and subsequent impact on the progeny (Kogut et al., 2024, Yalçin et al., 2001). Collectively, integrating broiler breeder nutrition and management in gut health management and overall, the well-being of broiler chickens will become increasingly important for the success of AFB production.

### *6.1.2 The egg*

Manipulation of hatching eggs from oviposition in broiler breeder farms through to placement in the hatchery could affect embryo metabolism, tissue development, and yolk utilization. For example, egg storage conditions have been shown to affect hatchability, chick quality at hatch and post-hatch intestinal function, and immunocompetence (Goliomytis et al., 2015, Yalçin et al., 2017). Conditions in the incubator such as temperature, humidity, egg turning, ventilation, and lighting have been reported to affect GIT development and immunity (Oznurlu et al., 2010). Incubation temperature regimen impacts hatch time and hatch window (Agyekum et al., 2022), leading to delayed placement at the farms and prolonged period without feed and water. Collectively, delayed feeding post-hatch results in slow yolk sac absorption (van der Wagt et al., 2020), maturation of the gut-associated lymphoid tissue (Bar-Shira and Friedman, 2005, Bar-Shira et al., 2005), damage to intestinal integrity (Proszkowiec-Weglarz et al., 2020), and increased risk of exposure to pathogenic bacteria (Proszkowiec-Weglarz et al., 2022, de Jong et al., 2022). The eggshell surface and hatchery environment are the major sources of pioneer colonizers (Chasser et al., 2021). Contaminated embryos, or non-viable embryonated eggs, have the potential to explode during incubation due to microbial overgrowth and may harbor pathogens (Karunarathna et al., 2022). Chicks hatch at different time points (spanning up to 72 hours) in commercial hatcheries, and early hatchers start eating whatever soft and edibles present in the hatcher (e.g. shell membranes, meconium, and egg content from exploded eggs) (Uni, 1998, Uni et al., 2003). Subsequently, timely access to feed and water is critical for maintaining intestinal mucosa development, morphology, and immune response with lasting effects on growth performance and welfare of the chickens. To overcome issues with delayed feed and water, an emerging concept is that of hatching chicks at the farm. Briefly, eggs are transferred to the placement farm between 17 and 19 days of incubation for hatching in the presence of feed and water. Appraisal of this concept has demonstrated benefits to early growth performance and welfare compared to conventional hatcheries (De Jong et al., 2019, Giersberg et al., 2021). Relevant to antibiotic-free production is a study that showed commercial flocks from on-farm hatching had less AMU and had greater proportion of flocks being AFB in full cycle production than conventionally hatched flocks (Jerab et al., 2023). This suggests that hatching chicks at the farm to allow immediate access to feed and water could be a relevant approach to supporting chicken health. However, widespread acceptance and application of on-farm hatching is still out of reach in many operations due to many biological and technical factors as discussed recently (Kogut et al., 2024). For more information about early provision of feed and water after hatching, including related hatching systems, please see elsewhere in this book.

### 6.1.3 The chick

An emerging and pressing issue in the broiler industry is persistent poor chick quality as evidenced by burgeoning embryonic and first week mortality and morbidity (early life mortality, ELM) with significant economic losses

(Karunarathna et al., 2022, Karunarathna et al., 2017, Swelum et al., 2021, Olsen et al., 2012, Yerpes et al., 2020). Although ELM has been linked to various factors, bacterial infections are recognized as the leading cause (Waliaula et al., 2024). Pivotally, bacterial infections during embryogenesis induce ELM, unthrifty chicks, and leaky navels in survivor chickens and increase susceptibility to other infections later in life (Nguyen et al., 2022). A survey indicated that 56% of ELM in commercial hatcheries in western Canada were due to co-infection of *Enterococcus* species and Avian pathogenic *E. coli* (APEC) (Karunarathna et al., 2017). Since *E. coli* is a ubiquitously distributed facultative anaerobe, it easily survives in the environment, and the likelihood that *E. coli* will encounter newly hatched chicks is high. Asymptomatic chick carriers of pathogenic *E. coli* strains spread it within the flock through fecal-oral or aerosol transmission (Karunarathna et al., 2022, Karunarathna et al., 2017). In the context of *necrotic enteritis*, chicks exposed to *E. coli* during incubation were determined to have gut conditions that favored clostridial growth, highlighting the need to control inflammation-causing bacteria pre-hatch (Chasser et al., 2021). There is no simple non-antibiotic solution for ELM (Waliaula et al., 2024). However, with respect to broiler breeder-chick quality interface, several factors could affect the chick quality, and some complex interactions are apparent, including hen/rooster physiological status, hen/rooster nutrition, diet formulation, farm and hatchery management, as well as transportation and brooding management (Chang et al., 2016, Leeson and Summers, 2005, Smith, 2019).

### 6.1.4 Broiler chicken houses

Management is core in planning, implementing, organizing, and operating the farm for optimal and profitable commercial broiler operation (Khan et al., 2024). Management decisions on husbandry practices, temperature control, lighting program, ventilation, biosecurity, and sanitary have an impact on GIT development, microbiome, and immunity (Wickramasuriya et al., 2022, Khan et al., 2024). This premise is becoming more critical in managing gut health in the context of AFB production (Oviedo-Rondón, 2019, Kogut et al., 2024). Among the management aspects that are more impactful are those associated with starting the chicks right. Transportation conditions from the hatchery to farm, brooding, ventilation, water temperature, and lighting regimen can influence feed and water consumption with lifetime impact on growth and health of chickens. The quality of the litter is also a focal point as its microbial load greatly impacts microbial succession and health (Schokker et al., 2021, Bindari and Gerber, 2022). Litter moisture management is even more difficult when AFB production is coupled with an all-vegetable feeding program because of increased excretion of undigested proteins (Smith, 2019). High litter moisture and wet litter promote foot and hock burns, skin lesions, poor air quality (and attendant respiratory diseases), and parasitic and bacterial infections (Kogut et al., 2024, Hermans and Morgan, 2007). As such, AFB production will require consideration on aspects of litter quality. Research has indicated that ventilation, the type of materials, frequency of cleaning between flocks, and management practices during grow out and between flocks have impact on the nature of microorganisms in contact with broiler chickens (Bucher et al., 2020, de Toledo et al., 2020, Díaz Carrasco et al., 2022). Water quality attributes, such as temperature, and physicochemical properties are essential for optimal intake. The variation in water quality within farms, between farms and regions in a country can be huge, and as such, water treatment methods vary. Water lines are also used for medication and certain nutritional interventions. These practices influence microbiological, physical, and chemical properties of the water. Biosecurity is an integral part of farm management and entails implementation of measures to minimize introduction of pathogens to the establishment and dissemination within or between the farms (Dhaka et al., 2023). In this context, management is pivotal for maintaining gut health and preventing proliferation of pathogens of import to chicken health and food safety. Indeed gut health management is much easier if biosecurity measures are in place and well maintained (Kogut et al., 2024). Stocking density impacts growth performance, GIT health, and chicken welfare. Although stock density is regulated and aligned with code practices in some countries, it might not be the case in many countries. Moreover, there are considerable variations in recommended density reflective of environmental control capabilities and welfare standards.

### *6.2 Nutrition*

There is recognition that a healthy and functional GIT is important for effective immune function and digestive capacity. A functional gut is a result of the status of many interrelated elements such as villi architecture, gut-associated immune system, and microbiome (Kiarie et al., 2013). However, microbial status is very critical to the overall health and functionality of the gut. The gut microbes also facilitate IgA responses, which in turn regulate the composition and function of the gut microbiota (Friedman et al., 2012). Thus, the reciprocal regulation of the gut microbiota and the host immune system may influence the balance between homeostasis and enteric disease (Klasing, 2007, Friedman et al., 2012). Development and application of molecular-based tools in poultry research have made immense contributions to the understanding of intestinal microbial profile and functionality (Mak et al., 2022). This has enabled development of feed additives for modulating gut health and immunity such as probiotics, prebiotics, and feed enzymes. However, as previously discussed, inconsistent responses and cost are significant barriers for their application. The reasons for lack of consistency in alternatives to AMU have been reported elsewhere, for example, Abd El-Hack et al. (2022), Moore (2024), and Bean-Hodgins and Kiarie (2021). Some level of efficacy of gut health additives has been demonstrated in trials conducted at research institutions. However, the farm conditions are replete with variables that can obscure efficacy (Bean-Hodgins and Kiarie, 2021). For example, targeting gut microbial ecology and function is the overriding concept behind numerous alternative feeding strategies to replace AMU in broiler chicken production (Kiarie et al., 2013). However, many factors influence the gut environment of the broiler, including the type of feed ingredients and digestibility of components, water quality, resident microbiota, and metabolites (Stanley et al., 2014). Broilers under farm conditions may have discordant gut ecology compared to experimental flocks. The development and application of feed additives for modulating intestinal health and immunity have great promise in advancing AFB production. However, such products ought to be tested in diverse production environments to elucidate their overall effectiveness. For example, understanding the roles of microbiota in feed efficiency, productivity, and health of chickens will be critical, specifically, the quantitative insights into the interactions between microbiome, immunity, and bird performance under diverse production settings. Recent evidence has shown that artificial intelligence and machine learning algorithms can relate excreta microbiota status and biomarkers for flock performance (Kuttappan et al., 2024). Further development of such tools will ease farm-level decision-making on the right dietary intervention. With respect to nutritional interventions, the current application is justified mainly on account of demonstrable benefits on production metrics such as feed efficiency. Finding the right dietary strategy or combination of strategies and how they influence the broiler physiological response to deliver consistent, measurable return to investment is the next frontier for ABF production. Such an approach would require shifting away from evaluation of feed additives on account of impact on feed cost to animal response premised on mitigating enteric pathogens, enhancing digestive capacity, intestinal integrity, and immunocompetence.

### 6.2.1 Digestive capacity

Although broiler chickens are highly efficient in converting feed to food products, they still excrete significant amounts of undigested nutrients (Ravindran, 2012). The peculiarity is that feedstuffs contain anti-nutritional factors and components that are not sufficiently degraded in the GIT (Kiarie et al., 2013, Kiarie et al., 2016). Microbiota derive nourishment from endogenous and undigested dietary components (Kiarie et al., 2013). Therefore, enhancing digestion is a critical frontier for managing GIT microbiota. This inherent digestive inefficiency in broiler chickens is the basis of development and application of exogenous feed enzymes technology (Kiarie et al., 2013, Munyaka et al., 2016, Thanabalan et al., 2021, Mohammadigheisar et al., 2021). We have demonstrated that feed enzymes not only improve nutritive value of feed but also impact GIT microbial ecology by reducing undigested substrates (Kiarie et al., 2019, Kiarie et al., 2013, Kiarie et al., 2022b). We have also investigated numerous functional ingredients to augment digestive capacity and resilience to enteric pathogens, for example, epidermal growth factor (Kim et al., 2017, Kim et al., 2020) and yeast metabolites (Kiarie et al., 2019, Leung et al., 2019, Leung et al., 2018, Maina et al., 2022, Kiarie et al., 2022a). Feed processing for optimal particle size can influence gizzard function, protein digestibility, and subsequently modulate the gut ecology (Kiarie and Mills, 2019, Kiarie et al., 2021). Feed quality, in particular, presence of mycotoxins, can damage the intestinal tract providing entry points for enteric pathogens. Therefore, mycotoxin management strategies are pivotal for enhancing digestive capacity and boosting intestinal health (Xu et al., 2022, Mwaniki et al., 2021).

### *6.2.2 Mitigating enteric pathogens*

The emphasis on mitigating enteric pathogens is to shift from fighting the unfavorable organisms to nurturing the favorable organisms, i.e. to ensure a favorable and stable intestinal ecology (Fig. 2). Three approaches have been suggested: (1) 'seeding' with favorable organisms, (2) 'feeding' the favorable microorganisms, and (3) 'weeding' out the unfavorable organisms. Extensive research has been reported as to the utility of probiotics, for example (Neijat et al., 2019, Mohammadigheisar et al., 2019, Gelinas et al., 2024) Mak et al., 2025. Probiotic microbes exploit a wide variety of mechanisms to benefit their survival while simultaneously benefiting poultry by suppressing pathogen invasion via competitive exclusion and antimicrobial peptides (Rantala and Nurmi, 1973). It is plausible that pre-hatch exposure to probiotics could be a potential strategy for modulating the prevalence of pathogenic bacteria during late embryogenesis and in the first week of life. *In-ovo* feeding is one of the pre-hatch applications of probiotics that has been evaluated (Das et al., 2021, Oladokun and Adewole, 2020). However, *in-ovo* feeding is still work in progress and is not commercially feasible (Peebles, 2018, Das et al., 2021, Thøfner et al., 2021). Since eggshell and hatchery contamination are considered major sources of pioneer colonizers, studying vertical and horizontal exposure of probiotics can provide insights into early microbial influence. Provisioning breeders with probiotics will be beneficial to broiler breeders (through immune modulation) and progeny.

Competitive exclusion generally refers to a reduction in colonization by a pathogen due to several possible mechanisms: physical occupation of a site, resource competition in a physical or chemical niche, or direct physical or chemical insult to the potential colonist (Oakley et al., 2014). Emerging strategies have capitalized on increased understanding of the molecular basis that pathogens use to attach to the mucosa for colonization. Microbe attachment to host cell docking sites on the intestinal epithelium is dependent on surface molecule structure, and this is the pivotal first step in the colonization and infection (Giron et al., 2002). For example, several bacteria exhibit a binding effect specific for the sugar mannose (Mirelman et al., 1980). Mannose in the cell wall may cause the yeast or its residue to act as a decoy for the attachment of bacteria to the intestinal wall, and this has been the basis of commercial success of many yeast-based products (Kiarie et al., 2010a, Kiarie et al., 2010b). Feed enzymes can impact gut microbial ecology by producing oligosaccharides *in situ* from complex dietary carbohydrates with potential prebiotic effects (Kiarie et al., 2007, 2008, 2022b, Sanchez et al., 2019, Rho et al., 2020). We have demonstrated that enzyme hydrolysis products could control enterotoxigenic *E. coli* diarrhea in a piglet model (Kiarie et al., 2008).

In other endeavors, dietary protein supply may have a significant impact on the intestinal microbiota, both qualitatively and quantitatively. High protein diets increase the concentrations of proteolytic bacteria, especially *Clostridia* and *E. coli* (Parenteau et al., 2020, Kiarie et al., 2021, Tsementzis et al. 2025). Adjusting protein supply and amino acid profiles can be considered essential to achieve optimal performance and to control the intestinal formation of metabolites such as ammonia and biogenic amines from protein fermentation generally considered detrimental. The use of supplemental amino acids would offset or minimize the need to use some expensive animal proteins, which could reduce the cost of feeds. Furthermore, extensive use of supplemental amino acids would allow them to more precisely meet the animal dietary requirements while reducing dietary crude protein. This change in formulation can positively impact gut health and the environment by reducing environmental excretion of nitrogen and reduce metabolic stress of detoxifying N-catabolites (Parenteau et al., 2020).

### 6.2.3 Intestinal integrity

It is prudent to expect that poultry raised without antimicrobial growth promoters have a higher chance to experience prolonged intestinal inflammation. The metabolic changes induced by inflammation are homeostatic in nature, and thus nutrients that would have been utilized for growth and skeletal muscle accretion are diverted to support host defense systems (Klasing, 2007). Specifically, nutrient absorption and gut adaptation to luminal inflammatory stress is a challenge in AFB as the pathogens compromise epithelial permeability and intestinal barrier function (Awad et al., 2017). The implications are the high metabolic rate and cellular turnover of gut tissues, resulting in substantial first-pass utilization of dietary nutrients, especially amino acids, to maintain gut function. Thus, unraveling the host-microflora interactions that influence gut function will be critical for developing dietary strategies to increase the efficient use of ingested nutrients for meat production in AFB. It is also known that many amino acids are used to synthesize proteins involved in the immune response and may be required in higher amounts in challenge conditions and may be beneficial for robustness and resilience in antibiotic-free production (Iseri and Klasing, 2014). Our research has focused on characterizing the impact of enteric pathogens on digestibility of amino acids (Kim et al., 2022a, 2022b). We have demonstrated that the digestibility of certain amino acids was more affected than others by coccidiosis challenge, suggesting the need to reconsider further supplementation to support gastrointestinal health. In further experimentation, supplementation with functional amino acids (glutamine and glutamic acid) increased growth performance in AFB, but had no effect on growth performance, plasma AA concentration, histomorphology, and breast weight when broiler chickens were exposed to coccidiosis (Kim, 2023). We have also explored the effects of feeding natural sources of bioactives such as fruit pomaces (Kithama et al., 2021, 2023a, 2023b) and seaweed (Mohammadigheisar et al., 2020). The results inbroiler studies showed the potential of the fruit pomaces, such as cranberry pomace in increasing the population of beneficial Lactobacillus and Bacteroides species while reducing the abundances of pathogenic bacteria and resistance to lincosamide and macrolide based antibiotics (Kithama et al., 2025).

### *6.2.4 Immunocompetence*

The protective role of maternal immunoglobulins (IgY and IgA) is of particular interest due to the precocial nature of chicken (Friedman et al., 2012). These maternal antibodies are provided during the process of egg formation and continue to function in the hatchling until its own immune response can take over (Friedman et al., 2012). In broiler chickens, the nutrients and factors deposited in the hatching egg by the hen are the only source of nutrients available to the embryo (Ferket, 2012, Uni et al., 2005, Uni and Ferket, 2004). Maternally derived factors (nutrients, immunoglobulins, microbiota) have the potential to improve disease resistance of neonatal chicks in a nonspecific manner when the adaptive immune system has not yet fully developed. However, there is limited exploration of whether immune and antimicrobial proteins in hatching eggs are amenable to breeder physiological status, nutritional, and microbial exposure. The transmission of immunoglobulins to progeny can be manipulated through dam nutrition, in a concept called transgenerational immunity. For example, dietary components such as cereal grain, vitamin E, trace minerals, and polyunsaturated fatty acids in broiler breeder diets increased antibody transfer and humoral response in chicks (Wang et al., 2004, Leandro et al., 2011, Oviedo-Rondón et al., 2013, Boa-Amponsem et al., 2001). Our research has focused on feeding broiler breeders omega-3 fatty acids (Thanabalan and Kiarie, 2021, 2022, Thanabalan et al., 2020), yeast metabolites (Lu et al., 2019, Maina et al., 2025), and trace minerals (copper, manganese, and zinc) (Bakhshalinejad et al., 2024b, Bakhshalinejad et al., 2025). Although results depended on the tested concept and response criteria, the broiler chickens hatched from these breeders showed better growth performance, skeletal integrity, breast meat yield, and resilience to enteric challenge (Maina et al., 2022, Thanabalan et al., 2024, Bakhshalinejad et al., 2024a).

### 6.3 Vaccines

Healthy chickens with functioning immune systems are remarkably resistant to stressors and fighting off diseases (Linden, 2015, Olsen et al., 2012). The chicken industry has a long history of application of vaccine technology for bolstering the immune system against specific pathogens (Marangon and Busani, 2007, Abdelaziz et al., 2024, Shamriz et al., 2026). Vaccination has benefited the industry in terms of maintaining flock health and productivity through prevention and control of contagious and infectious diseases. Vaccines are available to target various diseases, including new castle disease, infectious bronchitis, and fowl pox (Abdelaziz et al., 2024). According to the field survey, coccidiosis, necrotic enteritis, and colibacillosis are the most prevalent enteric diseases in AFB production (Roembke, 2025). Coccidiosis management through effective application of vaccination and permitted drugs is achievable in the field (Chapman and Jeffers, 2014, Chapman, 2014). There have been initiatives to develop vaccines for *Clostridium perfringens*, the causative agent for necrotic enteritis (Moore, 2024) and avian pathogenic *E. coli*, the causative agent for colibacillosis (Waliaula et al., 2024). However, antigenic variability of these pathogens has made it difficult to develop effective vaccines. The quest for effective vaccines for necrotic enteritis and colibacillosis is ongoing. For example, colibacillosis prototypes with multiple antigens or vectors encoding virulence genes targeting different stages of disease pathogenesis have shown promise (Yuan et al., 2022, Wilde et al., 2019, Lynne et al., 2012, Chaudhari et al., 2012). With respect to necrotic enteritis, it afflicts broilers at 21 days of age, presenting limited opportunity for mass vaccine delivery through injection route (Moore, 2024). Consequently, current innovations of necrotic enteritis vaccine target developing effective application hatchery without the need for field booster application (Gautam et al., 2025). Designing effective vaccines against necrotic enteritis and colibacillosis will be pivotal for mainstreaming AFB production.

## 7 Conclusion and future perspectives

The discourse on AMU in broiler chicken production is binary, whereas the farmer desires to produce food in the most profitable and sustainable manner, the consumers desire affordable, wholesome, safe, and ethically produced food. Progress towards AFB production will reduce abundance of resistant strains making it easier to treat flocks in the event of disease outbreaks. Adjustments in farm management practices and diet manipulation have enabled the industry to move towards achieving AFB production; however, persistent intestinal health challenges and higher cost of production are significant barriers. Further investments in research and development will be required for innovative, efficacious, and consistent feed additives. With respect to application in the field, the focus on production metrics has hindered implementation of feed additives on account of justification for the added cost. Finding the right dietary strategies and how they influence the broiler physiological response to deliver consistent, measurable return on investment is the next frontier for ABF production. A potential consideration is co-application of distinct dietary strategies aligned with their mode of action on intestinal functionality. Such an approach would require shifting away from evaluation of feed additives on account of the impact on feed cost to animal response. Moreover, integrating breeder and hatchery interface in the management of broiler gut health will become increasingly important in antibiotic-free broiler chicken production. Collectively, management and nutrition strategies along with effective vaccines for necrotic enteritis and colibacillosis will be pivotal for advanced antibiotic-free broiler production.

## 8 Author contributions

EM, AT, MK, and LH were PhD students under EGK tutelage and conducted foundational research that informed several concepts discussed in the chapter. They also contributed to writing and reviewing the chapter. EGK conducted the literature search, wrote substantial portions of the chapter, and had the overall conceptual and editorial responsibility.

## 9 Conflict of interest statement

The authors declare that the research was conducted in the absence of any commercial or financial relationships that could be construed as a potential conflict of interest.

## 10 Acknowledgments

Contributions of the current and former graduate students and research associates of EGK Monogastric Nutrition Research Laboratory at the University of Guelph are appreciated. Gratitude to Veronica Cheng for the technical support.

This chapter presents research carried out in the EGK Monogastric Nutrition laboratory at the University of Guelph from 2016 to 2025. The work described here was made possible through the generous support provided by several organizations and funding bodies. Financial assistance was received from Canadian government programs, including NSERC, MITACS, and the AAFC Science Cluster. Additional funding was provided by Ontario government programs such as the Ontario Agri-Food Alliance and the Ontario Centers of Excellence. Producer programs contributed to this research, including Egg Farmers of Canada, Egg Farmers of Ontario (LRIC), and the Canadian Poultry Research Council. The allied feed industry, both Canadian and international, also supported these efforts. The contributions of all these organizations are gratefully acknowledged.

## 11 Where to look for further information

For readers interested in exploring the topics discussed in this chapter in greater depth, the following publications offer comprehensive background, methodological insights, and updated perspectives across antimicrobial resistance, poultry production systems, gut health, and nutritional strategies:

- Antimicrobial use and resistance in poultry systems: Mak, P. H. W., et al. 2022. Production systems and important antimicrobial resistant-pathogenic bacteria in poultry: a review. *Journal of Animal Science and Biotechnology,* 13(1):148. https://doi.org/10.1186/s40104-022-00786-0. Provide an in-depth review of antimicrobial-resistant pathogens in poultry production systems and the underlying drivers of AMR emergence at the animal–environment–human interface. This article offers an excellent foundation for understanding the broader public health context of antibiotic-free production.
- Regulatory changes and their implications: Bean-Hodgins, L. and Kiarie, E. G. 2021. Mandated restrictions on the use of medically important antibiotics in broiler chicken production in Canada: implications, emerging challenges, and opportunities for bolstering gastrointestinal function and health: a review. *Canadian Journal of Animal Science,* 101(4):602–629. https://doi.org/10.1139/cjas-2021-0015. Discusses mandated restrictions on medically important antibiotics in Canada and detail the resulting challenges and opportunities for enhancing gastrointestinal function. Readers seeking policy context and its biological implications will find this review highly relevant.

- Economic and policy dimensions of AMU reduction: Laxminarayan, R., et al. 2015. The economic costs of withdrawing antimicrobial growth promoters from the livestock sector. https://doi.org/10.1787/5js64kst5wvl-en. Examines the economic consequences of withdrawing antimicrobial growth promoters from livestock systems, offering insights into global policy pressures and the cost–benefit considerations of transitioning to antibiotic-free production.
- Historical context for antibiotic use in poultry: Castanon, J. I. R. 2007. History of the use of antibiotic as growthpromoters in European poultry feeds. *Poultry Science,* 86(11):2466–2471. https://doi.org/10.3382/ps.2007-00249. Provides a historical overview of antibiotic use as growth promoters in European poultry feed, helping readers understand how past practices shaped today's production paradigms.

Together, these resources present a complementary viewpoint: from molecular and microbiological mechanisms to policy, economics, and production practice and serve as useful starting points for readers who wish to investigate specific aspects of antibiotic-free broiler production more deeply.

## 12 References

Aarestrup, F. M., et al. 2008. Resistance in bacteria of the food chain: epidemiology and control strategies. *Expert Review of Anti-infective Therapy,* 6, 733–750.

Aarestrup, F. M., et al. 2010. Changes in the use of antimicrobials and the effects on productivity of swine farms in Denmark. *American Journal of Veterinary Research,* 71, 726–733.

Abd El-Hack, M. E., et al. 2022. Alternatives to antibiotics for organic poultry production: types, modes of action and impacts on bird's health and production. *Poultry Science,* 101, 101696.

Abdelaziz, K., et al. 2024. Advances in poultry vaccines: leveraging biotechnology for improving vaccine development, stability, and delivery. *Vaccines (Basel),* 12, 134.

Agyekum, G., et al. 2022. Impact of incubation temperature profile on chick quality, bone, and immune system during the late period of incubation of Cobb 500 broiler strain. *Poultry Science,* 101,10199.

Ahmad, A., et al. 2017. Pharmaceutical waste and antimicrobial resistance. *The Lancet Infectious Diseases,* 17, 578–579.

Augère-Granier, M.-L. 2019. *The EU poultry meat and egg sector – Main features, challenges and prospects – In-depth analysis.* Publications Office [Online]. Available: https://www.europarl.europa.eu/thinktank/en/document/EPRS_IDA(2019)644195 [Accessed Novermber, 2025].

Awad, W. A., et al. 2017. Enteric pathogens and their toxin-induced disruption of the intestinal barrier through alteration of tight junctions in chickens. *Toxins (Basel),* 9, 60.

Bakhshalinejad, R., et al. 2024a. *Effects of feeding hydroxychloride or organic sources of zinc, copper, and manganese to broiler breeders on growth performance of offsprings when fed hydroxychloride or inorganic sources of the same minerals*. Animal Nutrition Conference of Canada (ANCC) May 14 to 16 2024 Winnipeg, MB. Ottawa: Animal Nutrition Associationn of Canada, 183–184.

Bakhshalinejad, R., et al. 2024b. Comparative efficacy of hydroxychloride and organic sources of zinc, copper, and manganese on hatching eggs, embryo and hatchlings attributes1. *Poultry Science*, 104, 104728.

Bakhshalinejad, R., et al. 2025. Comparative efficacy of hydroxychloride and organic sources of zinc, copper, and manganese on hatching eggs, embryo and hatchlings attributes. *Poultry Science,* 104, 104728.

Bar-Shira, E. & Friedman, A. 2005. Expression of innate immune functions in developing broiler gut associated lymphoid tissue in the immediate pre and posthatch period. *Poultry Science,* 84, 27–27.

Bar-Shira, E., et al. 2005. Impaired immune responses in broiler hatchling hindgut following delayed access to feed. *Veterinary Immunology and Immunopathology,* 105, 33–45.

Barrett, J. R., et al. 2021. Consumer perceptions of antimicrobial use in animal husbandry: a scoping review. *PLoS One,* 16, e0261010.

Bean-Hodgins, L. & Kiarie, E. G. 2021. Mandated restrictions on the use of medically important antibiotics in broiler chicken production in Canada: implications, emerging challenges, and opportunities for bolstering gastrointestinal function and health – a review. *Canadian Journal of Animal Science,* 101, 602–629.

Bindari, Y. R. & Gerber, P. F. 2022. Centennial review: factors affecting the chicken gastrointestinal microbial composition and their association with gut health and productive performance. *Poultry Science,* 101, 101612.

Boa-Amponsem, K., et al. 2001. Antibody responses of hens fed vitamin E and passively acquired antibodies of their chicks. *Avian Diseases,* 45, 122–127.

Borenstein, M. (2022). Comprehensive Meta-Analysis Software. In *Systematic Reviews in Health Research* (eds M. Egger, J.P.T. Higgins and G. Davey Smith). https://doi.org/10.1002/9781119099369.ch27.

Bucher, M. G., et al. 2020. Reused poultry litter microbiome with competitive exclusion potential against Salmonella Heidelberg. *Journal of Environmental Quality,* 49, 869–881.

Castanon, J. I. R. 2007. History of the use of antibiotic as growth promoters in European poultry feeds. *Poultry Science,* 86, 2466–2471.

Cervantes, H. M. & McDougald, L. R. 2023. Raising broiler chickens without ionophore anticoccidials. *Journal of Applied Poultry Research,* 32, 100347.

Chang, A., et al. 2016. Can feeding the broiler breeder improve chick quality and offspring performance? *Animal Production Science,* 56, 1254–1262.

Chapman, H. D. 2014. Milestones in avian coccidiosis research: a review. *Poultry Science,* 93, 501–511.

Chapman, H. D. & Jeffers, T. K. 2014. Vaccination of chickens against coccidiosis ameliorates drug resistance in commercial poultry production. *International Journal for Parasitology-Drugs and Drug Resistance,* 4, 214–217.

Chasser, K. M., et al. 2021. Enteric permeability and inflammation associated with day of hatch Enterobacteriaceae inoculation. *Poultry Science,* 100, 101298.

Chaudhari, A. A., et al. 2012. Construction of an attenuated salmonella delivery system harboring genes encoding various virulence factors of avian pathogenic escherichia coli and its potential as a candidate vaccine for chicken colibacillosis. *Avian Diseases,* 57, 88–96.

Dai, D., et al. 2021. Organic acids as alternatives for antibiotic growth promoters alter the intestinal structure and microbiota and improve the growth performance in broilers. *Frontiers in Microbiology,* 11.

Das, R., et al. 2021. In ovo feeding as a tool for improving performance and gut health of poultry: a review. *Front Veterinary Science,* 8, 754246.

de Jong, I., et al. 2019. Comparison of performance, health and welfare aspects between commercially housed hatchery-hatched and on-farm hatched broiler flocks. *Animal,* 13, 1269–1277.

de Jong, I., et al. 2022. Early life environment affects behavior, welfare, gut microbiome composition, and diversity in broiler chickens. *Frontiers in Veterinary Science,* 9, 977359–977359.

de Toledo, T. D. S., et al. 2020. An assessment of the impacts of litter treatments on the litter quality and broiler performance: a systematic review and meta-analysis. *PLoS One,* 15, e0232853.

Dhaka, P., et al. 2023. Can improved farm biosecurity reduce the need for antimicrobials in food animals? A scoping review. *Antibiotics,* 12, 893.

Díaz Carrasco, J. M., et al. 2022. The role of farm environment and management in shaping the gut microbiota of poultry. In M. Kogut and G. Zhang (Eds.), *Gut microbiota, immunity, and health in production animals (Series: The microbiomes of humans, animals,plants, and the environment, Volume 4),* Springer. Pages 193-224.

Dutil, L., et al. 2010. Ceftiofur resistance in Salmonella enterica serovar Heidelberg from chicken meat and humans, Canada. *Emerging Infectious Diseases,* 16, 48–54.

Emes, E., et al. 2024. The contribution of animal antibiotic use to antibiotic resistance in human infections: Panel evidence from Denmark. *One Health,* 19, 100856.

Ferket, P. R. 2012. Embryo epigenomic response to breeder management and nutrition. *XXIV World's Poultry Congress.* Salvador, Bahia, Brazil.

Fernández, M. M. E., et al. 2024. How did antibiotic growth promoters increase growth and feed efficiency in poultry? *Poultry Science,* 103, 103278.

Friedman, A., et al. 2012. The gut associated lymphoid system in the post-hatch chick: dynamics of maternal IgA. *Israel Journal of Veterinary Medicine,* 67, 75–81.

Gautam, H., et al. 2025. Immunoprotection of broiler chickens against necrotic enteritis by oral delivery of a live clostridium perfringens vaccine adjuvanted with cholera toxin at hatch. *Avian Diseases.* 69, 91-101.

Gelinas, A., et al. 2024. Growth performance, organs weight, intestinal histomorphology, and oocyst shedding in broiler chickens offered novel single strain Bacillus subtilis (BS-9) isolated from camel dung and challenged with Eimeria. *Poultry Science*, 103, 103519.

Giersberg, M. F., et al. 2021. Effects of hatching system on the welfare of broiler chickens in early and later life. *Poultry Science,* 100.

Giron, J. A., et al. 2002. The flagella of enteropathogenic Escherichia coli mediate adherence to epithelial cells. *Molecular Microbiology,* 44, 361–379.

Godfrey, K. M., et al. 2007. Epigenetic mechanisms and the mismatch concept of the developmental origins of health and disease. *Pediatric Research,* 61, 5–10.

Goliomytis, M., et al. 2015. Effects of egg storage on hatchability, chick quality, performance and immunocompetence parameters of broiler chickens. *Poultry Science,* 94, 2257–2265.

Gous, R. M. 2010. Nutritional limitations on growth and development in poultry. *Livestock Science,* 130, 25–32.

Haque, M. H., et al. 2020. Sustainable antibiotic-free broiler meat production: current trends, challenges, and possibilities in a developing country perspective. *Biology*, 9, 411.

Havenstein, G. B., et al. 2003. Growth, livability, and feed conversion of 1957 versus 2001 broilers when fed representative 1957 and 2001 broiler diets. *Poultry Science*, 82, 1500–1508.

Hein, T. 2023. Broiler antibotics update. *Canadian Poultry Magazine [Online]*.. Available: https://www.canadianpoultrymag.com/broiler-antibiotics-update/ [Accessed August, 31 2024].

Hermans, P. G. & Morgan, K. L. 2007. Prevalence and associated risk factors of necrotic enteritis on broiler farms in the United Kingdom; a cross-sectional survey. *Avian Pathology*, 36, 43–51.

Hincke, M. T., et al. 2019. Dynamics of structural barriers and innate immune components during incubation of the avian egg: critical interplay between autonomous embryonic development and maternal anticipation. *Journal of Innate Immunity*, 11, 111–124.

Iseri, V. J. & Klasing, K. C. 2014. Changes in the amount of lysine in protective proteins and immune cells after a systemic response to dead escherichia coli: implications for the nutritional costs of immunity. *Integrative and Comparative Biology*, 54, 922–930.

Jerab, J. G., et al. 2023. Antimicrobial use in on-farm hatching systems vs. traditional hatching systems: a case study. *Animals*, 13, 3270.

Karunarathna, R., et al. 2017. Increased incidence of enterococcal infection in nonviable broiler chicken embryos in Western Canadian hatcheries as detected by matrix-assisted laser desorption/ionization-time-of-flight mass spectrometry. *Avian Diseases*, 61, 472–480.

Karunarathna, R., et al. 2022. Exposure of embryonating eggs to Enterococcus faecalis and Escherichia coli potentiates E. coli pathogenicity and increases mortality of neonatal chickens. *Poultry Science*, 101, 101983.

Khan, Y. R., et al. 2024. Exposure factors associated with antimicrobial resistance and identification of management practices for preharvest mitigation along the broiler production systems: a systematic review. *Journal of Global Antimicrobial Resistance*. 9, 212-223.

Kiarie, E. G. & Mills, A. 2019. Role of feed processing on gut health and function in pigs and poultry: conundrum of optimal particle size and hydrothermal regimens. *Frontiers in Veterinary Science*, 6, 19.

Kiarie, E. G., et al. 2007. Growth performance, gastrointestinal microbial activity, and nutrient digestibility in early-weaned pigs fed diets containing flaxseed and carbohydrase enzyme. *Journal of Animal Science*, 85, 2982–2993.

Kiarie, E. G., et al. 2008. Nonstarch polysaccharide hydrolysis products of soybean and canola meal protect against enterotoxigenic Escherichia coli in piglets. *Journal of Nutrition*, 138, 502–508.

Kiarie, E. G., et al. 2010a. Weaned piglet responses to Escherichia coli K88+ oral challenge when fed diets containing a S. cerevisiae fermentation product with or without in-feed antibiotics. *Journal of Animal Science*, 88(E-Suppl. 2), 653.

Kiarie, E. G., et al. 2010b. Effect of products derived from hydrolysis of wheat and flaxseed non starch polysaccharides by carbohydrase enzymes on net absorption in enterotoxigenic Escherichia coli (K88) challenged piglet jejunal segments. *Journal of Animal Science*, 81, 63–71.

Kiarie, E. G., et al. 2013. The role of added feed enzymes in promoting gut health in swine and poultry. *Nutrition Research Reviews,* 26, 71–88.

Kiarie, E. G., et al. 2016. Performance, digestive function, and mucosal responses to selected feed additives for pigs. *Journal of Animal Science,* 94, 169–180.

Kiarie, E. G., et al. 2019. Utility of Feed Enzymes and Yeast Derivatives in Ameliorating Deleterious Effects of Coccidiosis on Intestinal Health and Function in Broiler Chickens. *Frontiers in Veterinary Science,* 6, 473.

Kiarie, E. G., et al. 2021. Impact of feeding modified soy protein concentrate in the starter phase on growth performance and gastrointestinal responses in broiler chickens through to day 42 of age. *Poultry Science,* 100, 101147.

Kiarie, E. G., et al. 2022a. Effects of early feeding of enzymatically treated yeast on growth performance, organ weights, intestinal histomorphology, and ceca microbial metabolites in broiler chickens subjected to Eimeria challenge. *Poultry Science,* 101, 101967.

Kiarie, E. G., et al. 2022b. Does supplementing β-mannanase modulate the feed-induced immune response and gastrointestinal ecology in poultry and pigs? An appraisal. *Frontiers in Animal Science,* 3, 875095.

Kim, E. 2023. *Investigating the effects of coccidiosis on amino acid digestibility and utilization in broiler chickens raised without antibiotics.* PhD, University of Guelph, 183 pages.

Kim, W. H. & Lillehoj, H. S. 2019. Immunity, immunomodulation, and antibiotic alternatives to maximize the genetic potential of poultry for growth and disease response. *Animal Feed Science and Technology,* 250, 41–50.

Kim, E., et al. 2017. Growth performance and gastrointestinal responses of broiler chickens fed corn-soybean meal diet without or with exogenous epidermal growth factor upon challenge with Eimeria. *Poultry Science,* 96, 3676–3686.

Kim, E., et al. 2020. In ovo feeding of epidermal growth factor (EGF): embryonic expression of intestinal EGF receptor and post-hatch growth performance and intestinal development in broiler chickens1. *Poultry Science, 99, 5736-5743.*

Kim, E., et al. 2022a. Standardized ileal digestibility of amino acids in broiler chickens fed single or mixture of feed ingredients-based diets with or without Eimeria challenge. *Poultry Science,* 101, 101839.

Kim, E., et al. 2022b. Centennial review: a meta-analysis of the significance of Eimeria infection on apparent ileal amino acid digestibility in broiler chickens. *Poultry Science,* 101, 101625.

Kirwin, E., et al. 2019. Comparing methods to estimate incremental inpatient costs and length of stay due to methicillin-resistant Staphylococcus aureus in Alberta, Canada. *BMC Health Services Research,* 19, 743.

Kithama, M., et al. 2021. The enzymatic digestion of pomaces from some fruits for value-added feed applications in animal production. *Frontiers in Sustainable Food Systems,* 5, 611259.

Kithama, M., et al. 2023a. Growth performance, organ weight, and plasma metabolites in broiler chickens fed corn-soybean meal diet containing berry pomaces and fed without or with multienzymes supplement. *Poultry Science,* 102, 102544.

Kithama, M., et al. 2023b. Utilization of grape (Vitis vinifera), cranberry (Vaccinium macrocarpon), wild blueberry (Vaccinium angustifolium), and apple (Malus pumila/domestica) pomaces in broiler chickens when fed without or with multi-enzyme supplement. *Canadian Journal of Animal Science,* 103, 15–25.

Kithama et al. 2025.Modulation of cecal microbiota and resistome of broiler chicken fed berry pomace alone or in combination with a multienzymes mixture. Microorganisms, 13, 1044. https://doi.org/10.3390/microorganisms13051044.

Klasing, K. C. 2007. Nutrition and the immune system. *British Poultry Science,* 48, 525–537.

Kogut, M., et al. 2024. *Environmental effects on gut health in production animals.* Wageningen Academic.

Kogut, M. H. & Klasing, K. 2009. An immunologist's perspective on nutrition, immunity, and infectious diseases: Introduction and overview. *Journal of Applied Poultry Research,* 18, 103–110.

Krupa, E., et al. 2017. Estimation of economic values for traits of pig breeds in different breeding systems: II. Model application to a three-way crossing system. *Livestock Science,* 205, 70–78.

Kuttappan, V. K., et al. 2024. *Impact of a novel combination of postbiotic and essential oil compounds to support gut microbiome maturation and performance in broilers.* 113th PSA Annual Meeting, July 2024 2024 Louisville, Kentucky. Poultry Science Association, 141.

Landers, T. F., et al. 2012. A review of antibiotic use in food animals: perspective, policy, and potential. *Public Health Reports (Washington, D.C.: 1974),* 127, 4–22.

Laxminarayan, R., et al. 2015. The Economic Costs of Withdrawing Antimicrobial Growth Promoters from the Livestock Sector. *OECD Food, Agriculture and Fisheries Papers,* No. 78, OECD Publishing, Paris [Online]. Available: https://doi.org/10.1787/5js64kst5wvl-en. [Accessed September 15, 2024).

Leandro, N. M., et al. 2011. Maternal antibody transfer to broiler progeny varies among strains and is affected by grain source and cage density. *Poultry Science,* 90, 2730–2739.

Leeson, S. & Summers, J. D. 2005. *Commercial poultry nutrition.* University Books.

Leung, H., et al. 2019. Nucleotide-rich yeast extract fed to broiler chickens challenged with Eimeria: impact on growth performance, jejunal histomorphology, immune system, and apparent retention of dietary components and caloric efficiency1. *Poultry Science,* 98, 4375–4383.

Leung, H., et al. 2018. Responses of broiler chickens to Eimeria challenge when fed a nucleotide-rich yeast extract1. *Poultry Science,* 98, 1622–1633.

Linden, J. 2015. *Colibacillosis in layers: an overview* [Online]. The Poultry Site. Available: https://thepoultrysite.com/articles/colibacillosis-in-layers-an-overview [Accessed April 25, 2020].

Liu, Y. Y., et al. 2016. Emergence of plasmid-mediated colistin resistance mechanism MCR-1 in animals and human beings in China: a microbiological and molecular biological study. *Lancet Infectious Diseases,* 16, 161–168.

Lu, Z., et al. 2019. The effects of feeding yeast bioactives to broiler breeders and/or their offspring on growth performance, gut development, and immune function in broiler chickens challenged with Eimeria. *Poultry Science,* 98, 6411–6421.

Lynne, A. M., et al. 2012. Recombinant Iss as a potential vaccine for avian colibacillosis. *Avian Diseases,* 56, 192–199.

Maina, A. N., et al. 2022. Enzymatically treated yeast bolstered growth performance of broiler chicks from young broiler breeders linked to improved indices of intestinal function, integrity, and immunity. *Poultry Science,* 101, 102175.

Maina, A. N., et al. 2025. Effects of lifetime feeding of hydrolyzed yeast to broiler breeders on egg production, quality, and hatchling attributes. *Poultry Science*, 104, 104826.

Mak, P. H. W., et al. 2022. Production systems and important antimicrobial resistant-pathogenic bacteria in poultry: a review. *Journal of Animal Science and Biotechnology*, 13, 148.

Mak, P. H. W., et el. 2025. Growth performance and body composition of *Eimeria*-infected broiler fed berry pomaces and a *Limosilactobacillus reuteri* isolate. *Poultry Science*, 104, 105812.

Marangon, S. & Busani, L. 2007. The use of vaccination in poultry production. *Rev Sci Tech*, 26, 265–274.

Maria Cardinal, K., et al. 2019. Withdrawal of antibiotic growth promoters from broiler diets: performance indexes and economic impact. *Poultry Science*, 98, 6659–6667.

Merks, J. W. M. 2018. One century of genetic changes in pigs and the future needs. *BSAP Occasional Publication*, 27, 8–19.

Mirelman, D., et al. 1980. Screening of bacterial isolates for mannose-specific lectin activity by agglutination of yeasts. *Journal of Clinical Microbiology*, 11, 328–331.

Mohammadigheisar, M., et al. 2019. Growth performance and gastrointestinal responses in heavy Tom turkeys fed antibiotic free corn–soybean meal diets supplemented with multiple doses of a single strain Bacillus subtilis probiotic (DSM29784). *Poultry Science*, 98, 5541–5550.

Mohammadigheisar, M., et al. 2020. Growth performance, breast yield, gastrointestinal ecology and plasma biochemical profile in broiler chickens fed multiple doses of a blend of red, brown and green seaweeds. *British Poultry Science*, 61, 590–598.

Mohammadigheisar, M., et al. 2021. Growth performance, organ attributes, nutrient and caloric utilization in broiler chickens differing in growth rates when fed a corn-soybean meal diet with multienzyme supplement containing phytase, protease and fiber degrading enzymes. *Poultry Science*, 100, 101362.

Moore, P. R., et al. 1946. Use of sulfasuxidine, streptothricin, and streptomycin in nutritional studies with the chick. *Journal of Biological Chemistry*, 165, 437–441.

Moore, R. J. 2024. Necrotic enteritis and antibiotic-free production of broiler chickens: Challenges in testing and using alternative products. *Animal Nutrition*, 16, 288–298.

Mottet, A. & Tempio, G. 2017. Global poultry production: current state and future outlook and challenges. *World's Poultry Science Journal*, 73, 245–256.

Munyaka, P. M., et al. 2016. Impact of combined beta-glucanase and xylanase enzymes on growth performance, nutrients utilization and gut microbiota in broiler chickens fed corn or wheat-based diets. *Poultry Science*, 95, 528–540.

Murray, C. J., et al. 2022. Global burden of bacterial antimicrobial resistance in 2019: a systematic analysis. *The Lancet*, 399, 629–655.

Mwaniki, A. W., et al. 2021. Comparative efficacy of commercially available deoxynivalenol detoxifying feed additives on growth performance, total tract digestibility of components, and physiological responses in nursery pigs fed diets formulated with naturally contaminated corn1. *Translational Animal Science*, 5, txab050.

Neijat, M., et al. 2019. Bacillus subtilis strain DSM 29784 modulates the cecal microbiome, concentration of short-chain fatty acids, and apparent retention of dietary components in shaver white chickens during grower, developer, and laying phases. *Applied and Environmental Microbiology*, 85, e00402-19.

Nguyen, T. T. T., et al. 2022. Enhancement of protective efficacy of innate immunostimulant based formulations against yolk sac infection in young chicks. *Poultry Science*, 102119.

Oakley, B. B., et al. 2014. The chicken gastrointestinal microbiome. *FEMS Microbiol Lett*, 360, 100–112.

Oladokun, S. & Adewole, D. I. 2020. In ovo delivery of bioactive substances: an alternative to the use of antibiotic growth promoters in poultry production–a review. *Journal of Applied Poultry Research*, 29, 744–763.

Olsen, R. H., et al. 2012. An investigation on first-week mortality in layers. *Avian Diseases*, 56, 51–57, 7.

Oviedo-Rondón, E. O. 2019. Holistic view of intestinal health in poultry. *Animal Feed Science and Technology*, 250, 1–8.

Oviedo-Rondón, E. O., et al. 2013. Broiler breeder feeding programs and trace minerals on maternal antibody transfer and broiler humoral immune response1 1Mention of trade names or commercial products in this publication is solely for the purpose of providing specific information and does not imply recommendation or endorsement by North Carolina State University. *Journal of Applied Poultry Research*, 22, 499–510.

Oznurlu, Y., et al. 2010. Histochemical and histological evaluations of the effects of high incubation temperature on embryonic development of thymus and bursa of Fabricius in broiler chickens. *British Poultry Science*, 51, 43–51.

Parenteau, I. A., et al. 2020. Egg production and quality responses to increasing isoleucine supplementation in Shaver white hens fed a low crude protein corn-soybean meal diet fortified with synthetic amino acids between 20 and 46 weeks of age. *Poultry Science*, 99, 1444–1453.

Peebles, E. D. 2018. In ovo applications in poultry: a review. *Poultry Science*, 97, 2322–2338.

PHAC 2016. Canadian antimicrobial resistance surveillance system –report 2016. Ottawa, Ontario.

Proszkowiec-Weglarz, M., et al. 2020. Effect of delayed feeding post-hatch on expression of tight junction- and gut barrier-related genes in the small intestine of broiler chickens during neonatal development. *Poultry Science*, 99, 4714–4729.

Proszkowiec-Weglarz, M., et al. 2022. Delayed access to feed early post-hatch affects the development and maturation of gastrointestinal tract microbiota in broiler chickens. *BMC Microbiology*, 22, 206.

Rantala, M. & Nurmi, E. 1973. Prevention of the growth of Salmonella infantis in chicks by the flora of the alimentary tract of chickens. *British Poultry Science*, 14, 627–630.

Ravindran, V. 2012. Advances and future directions in poultry nutrition: an overview. *Korean Journal of Poultry Science*, 39, 53–62.

Rho, Y., et al. 2020. Fiber degrading enzymes increased monosaccharides release and fermentation in corn distillers dried grains with solubles and wheat middlings steeped without or with protease. *Translational Animal Science*, 4, txaa153.

Roembke, J. 2025. Poultry nutrition and feed survey. In *Feed strategy*. WATT Global Meadia [Online]. Available https://www.feedstrategy.com/ [Accessed, July 5, 2025].

Sanchez, J., et al. 2019. Growth performance, gastrointestinal weight, microbial metabolites and apparent retention of components in broiler chickens fed up to 11% rice bran in a corn-soybean meal diet without or with a multi-enzyme supplement. *Animal Nutrition*, 5, 41–48.

Schokker, D., et al. 2021. Factors influencing the succession of the fecal microbiome in broilers. *Livestock Science*, 247, 104486.

Shamriz, S., et al., 2026. Protein nanoparticle-based vaccine candidate produced in *Nicotiana benthaminana* againt non-typhoidal *Salmonella enterica* in poultry. Frontier Plant Science. 17, 1761000.

Smith, J. A. 2019. Broiler production without antibiotics: United States field perspectives. *Animal Feed Science and Technology,* 250, 93–98.

Stanley, D., et al. 2014. Highly variable microbiota development in the chicken gastrointestinal tract. *PLoS One,* 8, e84290.

Swelum, A. A., et al. 2021. Ways to minimize bacterial infections, with special reference to Escherichia coli, to cope with the first-week mortality in chicks: an updated overview. *Poultry Science,* 100, 101039.

Tsementzis, A. K., et al. 2025. Comparative impact of partial replacement of soybean meal with select specialty protein ingredients in broiler chicken starter feeding program on growth, organ, intestinal, plasma, and litter attributes to 49 days of age. *Poultry Science*, 104, 105408.

Thanabalan, A. & Kiarie, E. G. 2021. Influence of feeding omega-3 polyunsaturated fatty acids to broiler breeders on indices of immunocompetence, gastrointestinal, and skeletal development in broiler chickens. *Frontiers in Veterinary Science,* 8, 653152.

Thanabalan, A. & Kiarie, E. G. 2022. Body weight, organ development and jejunal histomorphology in broiler breeder pullets fed n-3 fatty acids enriched diets from hatch through to 22 weeks of age. *Poultry Science,* 101, 101514.

Thanabalan, A., et al. 2020. Effects of feeding broiler breeder hens a coextruded full-fat flaxseed and pulses mixture without or with multienzyme supplement. *Poultry Science,* 99, 2616–2623.

Thanabalan, A., et al. 2021. Amino acids and energy digestibility in extruded or roasted full fat soybean fed to broiler chickens without or with multienzyme supplement containing protease, phytase, and fiber degrading enzymes. *Poultry Science,* 100, 101511.

Thanabalan, A., et al. 2022. A meta-analysis on the significance of dietary omega-3 fatty acids on bone development and quality in egg- and meat-type chickens. *Frontiers in Animal Science,* 3, 875944.

Thanabalan, A., et al. 2024. Significance of successive feeding of sources of n-3 fatty acids to broiler breeders and their progeny on growth performance, intestinal lesion scores, lymphoid organs weight and plasma immunoglobulin A in broiler chickens challenged with Eimeria. *Poultry Science,* 103, 103796.

Thøfner, I., et al. 2021. Recovery of probiotic bacteria from the intestinal tract of broilers after noninvasive pre-hatch application. *Beneficial Microbes,* 12, 467–477.

Tomczyk, G., et al. 2024. Probiotic supplementation as an alternative to antibiotics in broiler chickens. *J Vet Res,* 68, 147–154.

Uni, Z. 1998. Impact of early nutrition on poultry: review of presentations. *Journal of Applied Poultry Research,* 7, 452–455.

Uni, Z. & Ferket, R. P. 2004. Methods for early nutrition and their potential. *Worlds Poultry Science Journal,* 60, 101–111.

Uni, Z., et al. 2003. Pre- and posthatch development of goblet cells in the broiler small intestine: effect of delayed access to feed. *Poultry Science,* 82, 320–327.

Uni, Z., et al. 2005. In ovo feeding improves energy status of late-term chicken embryos. *Poultry Science,* 84, 764–770.

Van Boeckel, T. P., et al. 2015. Global trends in antimicrobial use in food animals. *Proceedings of the National Academy of Sciences,* 112, 5649–5654.

Van Boeckel, T. P., et al. 2017. Reducing antimicrobial use in food animals. *Science,* 357, 1350–1352.

van Bunnik, B. A. D. & Woolhouse, M. E. J. 2017. Modelling the impact of curtailing antibiotic usage in food animals on antibiotic resistance in humans. *Royal Society Open Science,* 4, 161067.

van der Wagt, I., et al. 2020. A review on yolk sac utilization in poultry. *Poultry Science,* 99, 2162–2175.

Waliaula, P. K., et al. 2024. Predisposition factors and control strategies of avian pathogenic Escherichia coli in laying hens. *Frontiers in Veterinary Science,* 11.

Wang, Y. W., et al. 2004. Maternal dietary ratio of linoleic acid to alpha-linolenic acid affects the passive immunity of hatching chicks. *Poultry Science,* 83, 2039–2043.

WHO 2021. *Antimicrobial resistance and the United Nations sustainable development cooperation framework: Guidance for United Nations Country teams*. World Health Organization.

Wickramasuriya, S. S., et al. 2022. Role of physiology, immunity, microbiota, and infectious diseases in the gut health of poultry. *Vaccines,* 10, 172.

Wilde, S., et al. 2019. Salmonella-vectored vaccine delivering three Clostridium perfringens antigens protects poultry against necrotic enteritis. *PLoS One,* 14, e0197721.

Xu, R., et al. 2022. Nutritional impact of mycotoxins in food animal production and strategies for mitigation. *Journal of Animal Science and Biotechnology,* 13, 69.

Yalçin, S., et al. 2001. Effects of strain, maternal age and sex on morphological characteristics and composition of tibial bone in broilers. *British Poultry Science,* 42, 184–190.

Yalçin, S., et al. 2017. Effect of egg storage duration and brooding temperatures on chick growth, intestine morphology and nutrient transporters. *Animal,* 11, 1791–1797.

Yang, C., et al. 2015. Phytogenic compounds as alternatives to in-feed antibiotics: potentials and challenges in application. *Pathogens,* 4, 137–156.

Yerpes, M., et al. 2020. Factors associated with cumulative first-week mortality in broiler chicks. *Animals (Basel),* 10, 310.

Yuan, B., et al. 2022. Immunization with pooled antigens for clostridium perfringens conferred partial protection against experimental necrotic enteritis in broiler chickens. *Vaccines,* 10, 979.

# Chapter 13

## Concepts of circular broiler production

*Ollie van Hal, Ben van Selm, Asaah Ndambi and Simon Oosting, Wageningen University & Research, The Netherlands*

1 Introduction
2 Circularity indicators and circularity of broiler production
3 Strategies for more circular broiler production
4 The potential of broilers in a more circular food system
5 Conclusion
6 Where to look for further information
7 References

## 1 Introduction

The global food system is a complex web of interconnected food production and supply chains that provide nutrition to the global human population. As this population is expected to grow from 8 billion in 2023 to 9.7 billion in 2050, the demand for food will increase (United Nations, 2024). Simultaneously, as people become more prosperous, they consume more animal source foods that have a high resource use and environmental impact compared to plant source foods (Marques et al., 2018). This while the resources needed for food production, such as arable land, are limited and food production already contributes significantly to global environmental impacts such as global warming and biodiversity loss, which already surpass environmental boundaries (Crippa et al., 2021). It is increasingly acknowledged that the high environmental impact of our food system is largely due to our disintegrated linear production systems that extract, use, and discard resources at a high rate that is no longer aligned with the rate at which nature generates resources and neutralizes wastes (Giampietro, 2019).

For a sustainable food future, the food requirements of the world population should be met while staying within the carrying capacity of our planet. To this end, a transition to a more circular food system is increasingly proposed by scientists (Hamam et al., 2021) and governments (European Commission, 2020). The circular food system concept is based on principles of the circular

http://dx.doi.org/10.19103/AS.2025.0165.15

bioeconomy and has various definitions. A narrow definition assumes that the circular food system aims to reduce resource use and environmental impacts by closing the loop of materials and substances biomass (Ghisellini et al., 2016; Jurgilevich et al., 2016). Solutions to improve circularity then focus on enhancing the reuse and recycling of resources and nutrients (Kirchher et al., 2017). Under a broader definition, a circular food system ought to minimize the depletion of finite resources (e.g. phosphate rock, fossil fuel), encourage regenerative practices (e.g. restore fish stocks, soil carbon), prevent the loss of natural resources (e.g. carbon, nutrients, water), and stimulate reuse and recycling of inevitable losses and by-products to their highest utility (Muscat et al., 2021). To achieve that, natural and agricultural ecosystems should be safeguarded by using resources within the regenerative and absorptive capacity. This requires avoiding the use of non-essential products and the loss of essential ones; prioritizing basic human needs; using biomass effectively to meet these needs; and recycling unavoidable losses, waste, and by-products at their highest utility.

Assuming this broader, holistic view, a circular food system uses resources efficiently to provide adequate nutrition to all people, within the environmental boundaries of our planet. Spatial agricultural resources, such as land and waterbodies, are used most effectively when used for direct food production (de Boer and van Ittersum, 2018). The alternative, using these resources to produce feed for animals, is less efficient due to the losses when animals convert feed into animal source food (Wilkinson, 2011). In a fully circular food system, agricultural land is used to produce food crops for human consumption, where possible (de Boer and van Ittersum, 2018). Similarly, all edible fish captured through sustainable fisheries are used as food. Producing, processing, and consuming this food, however, results in unavoidable by-products and waste that people cannot or do not want to eat. Animals can contribute to the resource use efficiency of the food system by upcycling these food leftovers and grass produced on marginal lands into animal source food (van Kernebeek et al., 2016). The role of animals in a circular food system is thus to upcycle so-called low opportunity-cost biomass (LOCB) into nutrient-dense foods (Fig. 1). The amount of animal source food in the diet is then limited by the availability of LOCB and estimated to be about 30 g animal protein per capita per day globally, much less than currently consumed in high-income countries (van Zanten et al., 2018; van Selm et al., 2023). This implies that a more circular food system in such countries requires changes in both what food is consumed and how this is produced. It is important to note that a food system will never be fully circular, and making the food system as circular as possible is not the aim. Improving circularity is a means to get our food system to function within a safe operating space where nutritional demands are met without exceeding the planetary boundaries.

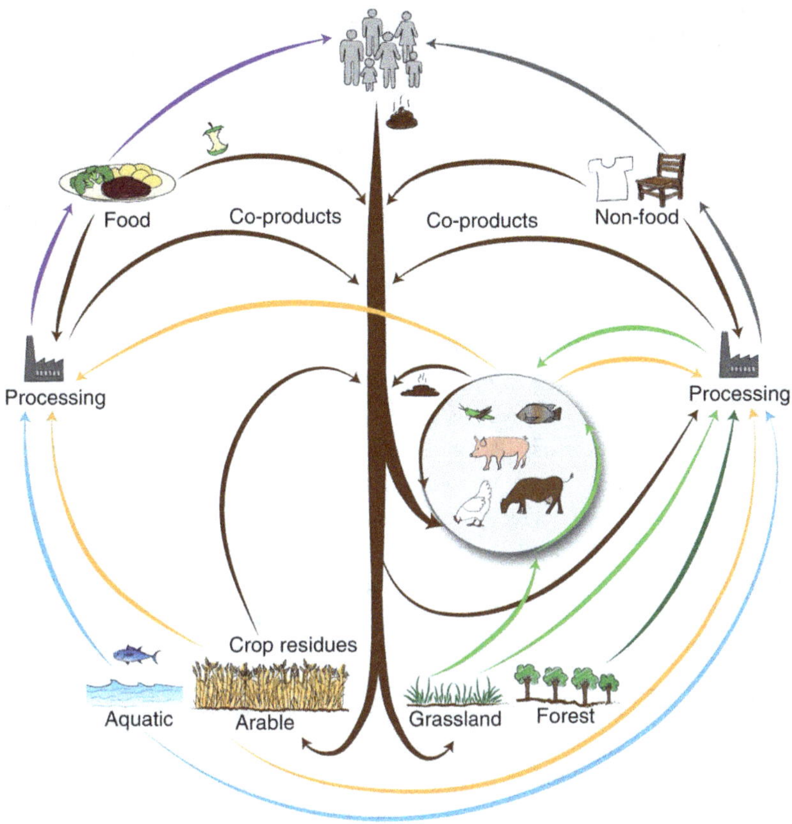


**Figure 1** Concept of a fully circular food system and bioeconomy (broad definition) where animals are fed solely on low opportunity cost biomass (Muscat et al., 2021).

This chapter aims to explore the circularity of current broiler systems and the potential and options to improve it. Section 2 outlines how circularity of livestock systems can be assessed and the circularity performance of broiler production based on various indicators. Section 3 explores various options to improve the circularity of broiler production. Finally, the Dutch case study in Section 4 illustrates the role of broiler production in two scenarios for a sustainable future food system.

## 2 Circularity indicators and circularity of broiler production

Indicators in general have proved to be highly valuable in improving the sustainability of how food is produced and consumed (Saidani et al., 2019). This is because they enable both farmers and consumers to make informed decisions to produce and/or consume food with a limited environmental impact. Furthermore, they can be used in policy tools to illustrate the contribution of farms to goals set at various organizational scales (De Oliveira et al., 2021). To stimulate a transition to more circular food systems, where livestock contributes to its resource use efficiency, farm and product-level (e.g. per kg meat) indicators that reflect the circularity of livestock systems are needed (Saidani et al., 2019; Aznar-sánchez et al., 2020). In recent years, a variety of circularity assessment methods and indicators have been proposed and applied at various organizational (e.g. product, sector, food system) or geographical (e.g. city, region, country) scales (Parchomenko et al., 2019; Poponi et al., 2022). They often propose to include such circularity indicators within a broader set of environmental impact indicators, for example on global warming potential, to make a holistic assessment (Poponi et al., 2022). Improving circularity, however, is not a goal but a means to achieve more sustainable food systems; ultimately circularity should be accounted for in product-level environmental impact assessment (Harris et al., 2021). Common environmental impact assessment methods, such as life cycle assessment (LCA), however, oversimplify interconnections in the food system needed to consistently account for changes in stocks of resources and are therefore unable to reflect circularity (Peña et al., 2021; van Hal et al., 2019a). Accounting for circularity when assessing product-level environmental impacts is complex, and no indicators are available for practical application.

The best currently possible is to enrich conventional environmental impact assessments with indicators that assess the nutrient or material circularity of a product, industry, or geographical area (Harris et al., 2021) if following the narrow interpretation of circularity. Following the broader interpretation of circularity, supplementary circularity indicators should indicate the contribution of a livestock product to the resource use efficiency of the food system (van Hal et al., 2019b). This chapter explains such indicators one by one and shows the performance of industrial broiler production on these indicators. While there is a variation of industrial broiler production systems (e.g. conventional, free range, organic), they are similar around the world. In smallholder farms that provide food for the maturity of people, however, indigenous chicken farming is a more common means to the same goal of producing chicken meat (Padhi, 2016) and is therefore discussed as a contrast.

## 2.1 Nutrient and material circularity indicators

The narrow interpretation of circularity considers circularity a means to achieve a higher resource use efficiency and lower environmental impact, due to a higher reuse of flows and matter and a lower dependency on external inputs. Studies taking this interpretation have developed and used nutrient or material circularity indicators related to recycling and self-sufficiency of biomass or specific nutrients such as nitrogen (N) and phosphorus (P), which are both essential for agricultural production and detrimental when excessively present in nature (Koppelmaki et al., 2021, Harchaoui et al., 2023, Tang et al., 2025). They are often combined with indicators on nutrient use efficiency and loss, which are commonly included in LCA.

Recycling indicators focus on the share of system inputs that are of recycled origin, the share of waste (unintended system outputs) that is recycled, or a combination of the two. On the input side, the secondary to total input ratio (Papangelou et al., 2021) and the more specific secondary feed or fertilizer ratios (Tang et al., 2025) indicate how much of the inputs of a system are of recycled origin, while the virgin feedstock rate (Gallo et al., 2023) is the invert of that. For livestock production recycled inputs can either be the feeds themselves or the fertilizers used to produce feeds. Harchaoui et al. (2023) found that in France, 27% of the N in broiler feed is of recycled origin. Gallo et al. (2023) found that in Italy, conventional broiler production used 2% recycled inputs, while organic broiler production used 94% recycled inputs. In smallholder indigenous chicken farming, feed is largely composed of food leftovers, crop residues, and scavenged insects, worms, and grass (Magothe et al., 2012). It could, therefore, be considered mainly recycled (Wilson et al., 2022).

On the output side, the recycling rate indicates the share of unrecoverable waste that is recycled (Papangelou et al., 2021; Tang et al., 2025), while the unrecovered waste rate (Gallo et al., 2023) indicates the invert. In livestock production, waste outputs are manure and slaughter waste, which can be recycled as feed or fertilizer. Harchaoui et al. (2023) found that in France 54% of N in broiler manure is recycled into the cropping system, while the remainder is lost during manure storage and application. Gallo et al. (2023) found that in Italy, in conventional broiler production, 79% of the wastes are recycled, and in organic broiler production, 84% of the wastes are recycled. Due to a large manure surplus in the Netherlands, about 430 kt of poultry manure is incinerated for energy production (bmc Moerdijk, 2024). While this is a form of recycling and even called circular in politics, during incineration all nitrogen and carbon components are lost. In smallholder indigenous chicken farming a much larger share of manure is used, mainly as fertilizer, but also in dairy feed (Carron et al., 2017). Gallo et al. (2023) finally combined the virgin feedstock rate and unrecovered waste rate into the material circularity indicator (MCI), a value between 0 (fully linear) and 1 (fully circular). They find an MCI of 0.33 for conventional broiler meat, making it a linear system, and while with an MCI of 0.89, organic broiler meat production was much more circular. This is mainly due to the use of only organic fertilizer in feed production.

Self-sufficiency indicators are included to reflect the dependency of a production system on external inputs, especially imported ones. This is highly relevant for industrial livestock production, because they are highly connected to national and global food systems through the international feed trade. Thereby livestock production has been decoupled from the production of feed, which has resulted in imbalanced nutrient flows between regions and countries (Koppelmaki et al., 2021). This in turn causes an accumulation of excess nutrients, in manure and emissions, in net feed-importing countries, while soils are degraded in feed-exporting countries, resulting in high use of artificial fertilizer or expansion of the agricultural area (deforestation) (Koppelmaki et al., 2021). This can be avoided by reducing feed and food trade and decreasing the dependency on imports. Harchaoui et al. (2023) found that in France, the feed self-sufficiency of the predominantly industrial poultry production is 10%, indicating the high dependence on feed import and thereby contributing to nutrient imbalances. Smallholder indigenous chickens, in contrast, mainly scavenge their feed, supplemented with homemade rations, including food scraps and by-products, making them less dependent on feed imports (Wilson et al., 2022).

Efficiency indicators express how much of the inputs to a system are captured in the intended output (e.g. broiler meat) and are generally expressed as a fraction or percentage. Loss indicators show the amount of material or nutrient lost per hectare and thus express how much a production system affects its environment. A study that compared the N use efficiency (NUE) of the five most common livestock species in four western European countries found the NUE of the predominantly industrial broiler production ranged between 48% and 52% and was consistently higher than for any other species considered (Groenestein et al., 2019). In line with that broiler production in France has an NUE of 49%, higher than those of laying hen production (28%) and duck and geese production (38), and thus the efficiency of the overall poultry sector (44%) in 2020 that has a domestic N loss rate of 12 kg hectare per year (Harchaoui et al., 2023). In smallholder indigenous chicken farming the nitrogen efficiency likely varies with conditions and is challenging to assess as the actual feed intake is often unknown. In Indonesian indigenous chicken farming, the NUE was estimated to be 15%, much lower than in industrial broiler farms, due to a lower production efficiency (Hantoro et al., 2024). The number of these animals kept per acre, however, is much lower, resulting in a much lower nitrogen loss per hectare (Oosting et al., 2021).

## 2.2 Indicators on contribution to resource use efficiency of the food system

Industrial broiler production is highly appreciated for its high production efficiency, which is reflected in a low feed conversion ratio (FCR). With the FCR for conventional industrial production averaging between 1.7 and 2.0, reflecting the amount of kg feed needed to produce 1 kg meat (Fry et al., 2018), broilers produce meat with lower environmental impact per kg protein than other terrestrial livestock (de Vries and de Boer, 2015). Only milk and eggs outperform broiler meat regarding environmental impacts per kg protein, because producing a daily product is more efficient than slaughtering and regrowing an animal (de Vries and de Boer, 2015). This has led many to believe that if one wants to eat meat, eating broiler meat is most environmentally sound, with the co-benefit of boiler meat healthier than red, fatty, and processed meats (Hallström et al., 2015). While the environmental impact of the individual diet would indeed reduce from shifting from beef to boiler meat, applying this strategy at large will result in adverse consequences most studies fail to account for. To achieve its high production efficiency, industrial broilers eat diets of a much higher quality, containing much more human edible ingredients than their beef counterparts. Replacing beef with broilers requires more feed produced on arable land, reducing the land use efficiency of the food system, while land use is critical in many environmental impacts. In other words, the high production efficiency of broiler production may come at a cost for circularity, which is not accounted for in the above-described indicators, because they do not consider whether the inputs used in a production system compete with food production. Similarly, when implementing strategies that improve circularity or reduce environmental impacts per kg livestock product, this does not necessarily reduce the environmental impact of the food system (Tukker, 2015; van Hal et al., 2019a). Therefore, studies that assume the broader definition of circularity have developed indicators that express the contribution of a livestock system to the resource use efficiency of the food system. Such indicators indicate to what extent a livestock system competes with food production directly (by feeding human edible products) or indirectly (by feeding on products unsuitable for human consumption, but grown on land suited for food crop production). These indicators tend to focus on protein as this is the most valued macronutrient provided by animal products.

Many similar direct feed-food competition indicators have been introduced with slightly different names, such as the edible feed conversion ratio (eFCR; Wilkinson, 2011), the human edible protein conversion ratio (HEPCR; van Riel et al., 2023), and the edible protein efficiency (de Cuyper et al., 2022). The HEPCR, for example expresses the amount of human edible protein in the feed needed to produce 1 kg of human edible protein of the livestock product (van Riel et al., 2023). First the edible protein content per kg feed is calculated by summing the inclusion rate (kg/kg) multiplied with an edibility coefficient and the protein content of each edible feed ingredient. This edible protein content is then multiplied with the FCR to calculate the amount of edible feed protein used to produce one kg of animal product. To calculate the HEPR this value is then divided by the human edible protein content of 1 kg animal product which is calculated by multiplying the edible yield of the animal product (e.g.

edible share of the LW) with its protein content. If the value is below 1, this implies that the production system is a net supplier of human edible protein.

For the UK, Wilkinson et al. (2011) found a HEPCR of 2.1, implying a broiler consumes twice the amount of human edible protein it supplies. This is mainly because 75% of the ingredients in the broiler diet were considered human edible, the highest for all conventional livestock species. In the Netherlands industrial broiler feeds contain between 21% and 24% human edible protein ingredients (te Pas et al., 2021). Assuming a human edible protein content per kg broiler life weight of 191 g (Wilkinson, 2011), this results in a HEPCR between 1.85 and 2.13 (te Pas et al., 2021). The edible protein efficiency is the invert of the HEPCR, where a score below 1 implies a food system is a net edible protein consumer. De Cuyper et al. (2022) show that in Belgium, broilers have an edible protein efficiency of 0.61 and are thus net consumers of protein. De Cuyper et al. (2022) further more illustrate the large effect of different assumptions in human edibility, had wheat grains not been considered human edible (as none of the wheat cultivated in Belgium is of commercial baking quality), the edible protein efficiency increases to 0.96 for broilers. The indicator is highly sensitive to the assumed human edibility of feed, while assumptions and the rationale behind them vary between studies. Some studies assume a feed ingredient is either edible or not, thus using binary edibility coefficients, based on whether it is currently used as food to any extent (van Riel et al., 2023). Other studies base the edibility coefficients on the fraction that is currently consumed in a specific region, or is expected to be consumed with social and technological development (Laisse et al., 2019). So far, the human edible protein ration has not been calculated for indigenous poultry farming, but since there are hardly any edible products fed and a large share of the meat is consumed, it is likely much lower than for industrial broiler farming.

The only known indicator that also consistently considers indirect feed food competition is the land use ratio (LUR) first introduced by van Zanten et al. (2016). This indicator illustrates the amount of human edible protein that could have been produced on the land needed to produce one kg of edible protein in livestock products. So far, the land use ratio has not been calculated for broilers, but studies that calculate both the HEPR and LUR of a production system find that the LUR is consistently higher than the HEPR (Hennessy et al., 2021). As poultry and, especially, broilers have the highest share of food competing feedstuff (68%) in their feed compared to other conventional livestock species, they likely have the highest LUR as well (Sandström et al., 2022). This is also reflected in an earlier indicator, the land use efficiency (LUE), which shows that with 3.09 $m^2$ arable land needed to produce 100 g animal protein, broiler production is the least land use efficient among conventional livestock systems in the US when excluding pastures (Peters et al., 2014). This indicator, however, does indicate whether using the land for food crops would be more efficient. The study, furthermore, considered all concentrate feed to be grains and soybean, both with a high LU, which underestimates the share LOCB.

## 3 Strategies for more circular broiler production

From the circularity assessments of broiler production, it can be derived that the production efficiency in industrial broiler production is high but requires high-quality feed with high feed-food competition. As broiler production largely relies on international or interstate feed imports, it is decoupled from the land used for this feed production, hampering effective recycling of nutrients (manure). To improve the circularity of broiler production, and thereby its contribution to the resource use efficiency of the food system, this chapter considers three types of strategies: improved efficiency, improved recycling, and food system redesign. These are based on the efficiency, circularity (narrow definition), and sufficiency strategies of Spiller et al. (2024). First, improving the production efficiency of broiler production is considered, followed by improving the recycling of broiler production, considering both the inputs (feed) and outputs (manure and waste), in line with the narrow interpretation of circularity. Finally, in line with the broad interpretation of circularity, restructuring or redesigning the food system and the role of broilers therein is considered.

### *3.1 Improved production efficiency*

As a circular food system intends to use resources efficiently, one way for broiler production to become more circular is to improve the production efficiency of the broiler production system itself. Higher production efficiency implies that less resource input is needed per output produced. In livestock this is generally realized through improved management (e.g. less diseases), breeding, or feeding strategies. Note that improved production efficiency can only improve circularity if it does not require an increased demand for high-quality feed, as this will increase global feed-food competition.

As indicated in Section 2.2 conventional industrial broiler production has a remarkably high production efficiency, which has increased significantly over the last decades by focused breeding efforts combined with management and feed adjustments (Zuidhof et al., 2014). As high growth rates can contribute to severe health issues (Hartcher and Lum, 2020), further increases can be deemed infeasible (Harchaoui et al., 2023). Often even a reduction in growth rate is deemed necessary to meet societal animal welfare demands in the Netherlands resulting in the slower growing broiler (van der Eijk et al., 2023), known to have higher environmental impact per kg meat (Chan et al., 2022). Production efficiency, however, might be improved in other parts of the supply chain such as in feed production, slaughter, and manure management.

Harchaoui et al. (2023) indicate that such improved efficiency can improve the N use efficiency to 51% (+7%), reduce N loss to 103 kg N/ha (−23 kg n/ha), increase the N recycling rate to 29% (+4%) but does not affect self-sufficiency.

Smallholder indigenous chicken farming has the theoretical efficiency benefit that the poultry is kept until it stops laying eggs, after which they are slaughtered, and a high proportion of their carcass is consumed including the offal and other parts that are only used for pet food production in high-income countries. In practice, their potential to increase production efficiency is larger than in industrial production as feed availability and time of slaughter are often (periodically) suboptimal (Wilson et al., 2022). While the indigenous breeds are more resistant to diseases and rough conditions, they are also much more exposed by free roaming and scavenging (Dessie et al., 2001). These inefficiencies can partly be overcome by improved management which requires knowledge transfer, improved infrastructure, and improved economic conditions for the farmers (Wilson et al., 2022). The indigenous breeds, furthermore, genetically have a lower production potential for meat and a lower feed use efficiency (Dessie et al., 2011). To address this inefficiency as well as the increasing demand for poultry meat, some farmers switch to improved indigenous breeds or dual-purpose breeds that grow and lay more efficiently. While these breeds can be of economic benefit to smallholder indigenous chicken farmers, they require more financial investment and experience in management than indigenous chickens. To achieve their production potential these breeds require higher quality feed, resulting in more external inputs of feed that is potentially of food quality (Wilson et al., 2022; Oosting et al., 2021).

## 3.2 Improved recycling

In a circular food system, animals play a key role in recycling losses in the food system into valuable food. On the one hand, increasing the share of LOCB in broiler feed would reduce their feed-food competition and thus make them more circular. On the other end, improved circularity can be achieved through better recycling of the unintended outputs of broiler production, namely, manure and slaughter wastes.

### 3.2.1 Feeding low opportunity-cost biomass (LOCB)

There are various types of LOCB that could be used in broiler diets. A large share of available LCOB considers grass from marginal land, which appears unsuited for broiler production. Producing novel ingredients with such resources and other resources unsuited for food production can, however, improve their suitability as broiler feed. Another large share of LOCB is leftovers from the food system which can be classified based on where in the food supply chain they are generated, which is usually a good indicator of their quality and how complicated it is to safely use them as feed. Food supply chains run from primary production (e.g. livestock and crop farming), through food processing (e.g. animal slaughter and wheat milling), food manufacturing (e.g. bread baking, cheese production), food distribution and retail (e.g. (whole)sales) to final food preparation (e.g. cooking) and consumption or discard. In each stage by-product losses and/or wastes are generated that people cannot or do not want to eat (see Fig. 2). By-products are defined as the unintended outputs of processes with multiple outputs. Wheat bran, for example is a by-product from processing wheat grains into wheat flour. Food loss is the biomass in a shape not yet intended for consumption, that does not move on to the next stage of the supply chain, for example the whole grains lost on the field at harvest. Food waste is biomass in the shape intended for consumption but not consumed, like food scraps or supermarket waste. The nutritional quality of food loss and waste is generally higher than that of by-products as it is, or contains, the component. intended for human consumption. Along the supply chain the nutritional quality leftovers generated appears to increase, as they come closer to food quality. Simultaneously, feed safety becomes more difficult to guarantee when collecting leftovers at the end of the supply chain. Collecting household kitchen waste, for example has a considerable risk of contamination with pathogens, but also inedible components. Especially for feeding food waste, risks have been thoroughly reviewed, but the same risks also apply to other food system leftovers.

**Figure 2** Loss, waste, and by-products generated along the food supply chain, and their value and collectability as livestock feed. In gray, losses that cannot be collected. Adapted from van Hal (2020).

### 3.2.1.1 Risks of feeding LOCB

Risks when feeding food leftovers can relate to characteristics of the leftover itself or contaminations or spoiling it may have endured along the supply chain. The following risks of feeding food system leftovers hold for all livestock species and thus also for broilers. Due to the young age of broilers, toxins may, for example already be problematic at lower concentrations. Compounds that become problematic as they accumulate in the body may, however, be less relevant due to the short lifespan of broilers.

Contamination with pathogenic microorganisms can be effectively controlled by sterilization (Chen et al., 2015), but this does not affect other risks related to contamination or spoilage such as mycotoxin, plastics, heavy metals, biogenic amines, oxidized lipids, and anti-nutritional factors (Simmons et al., 2023). Mycotoxins are toxins produced when a product is contaminated with specific microorganisms and commonly occur in cereal grains and legumes. Aflatoxin, the most potent mycotoxin, can cause kidney and liver damage and poor feed efficiency (Murugesan et al., 2015). Methods to remove mycotoxins from feed are in early development; thus, for now, inclusion of contaminated ingredients should be restricted. A study that produced and examined pelleted food waste found mycotoxin and heavy metal levels were below the EU regulation level for feed (Castrica et al., 2018), making them especially suitable for broiler production as their short lifespan does not allow these compounds to accumulate. Larger non-organic contaminations such as macroplastics can cause gastrointestinal obstruction or damage, especially in young animals, but can be removed by sieving (Meng et al., 2022). Much harder to address is contamination with microplastics, which are shown to cause intestinal damage and reduce growth and increasingly occur also in conventional feed ingredients through contaminated soils (Meng et al., 2022; Walkinshaw et al., 2022). Spoiling of food leftovers can result in both oxidized lipids and biogenic amines. Fats in precooked food and food exposure to air may rancidify, resulting in oxidized lipids which when consumed excessively may cause nutritional encephalomalacia (Mezes et al., 1997), to which especially young chicks appear susceptible (Budowski et al., 1987). Biogenic amines can be beneficial in small amounts, but when excessive, for example in decaying/rotting foods, they are toxic (Simmons et al., 2023). So far, no removal technology is available for oxidized lipids and biogenic amines, so rotting and decay should be avoided (Feddern et al., 2019). Fermentation is expected to have potential in avoiding accumulation of biogenic amines and mycotoxins (Feddern et al., 2019). Anti-nutritional factors generally relate to naturally occurring substances in feed. Lectins, for example are commonly found in legumes and may inhibit growth (Tufarelli et al., 2018). Alkaloids are found in many plants with high concentrations in some legumes. They are found to reduce feed intake (Tufarelli et al., 2018), but some may also improve feed conversion ratios (Cottrell et al., 2022). Phytate, found in cereal brans and oilseed meals, reduces protein and nutrient digestion (Meng et al., 2022; Walkinshaw et al., 2022). Tannins, found in many ingredients but in excessive amounts in coffee by-products and rapeseed meal, reduced nutrient utilization and production (Redondo et al., 2014). Soaking, fermenting, and milling, as well as supplementation with neutralizing compounds such as phytase may reduce anti-nutrtitional factors (Simmons et al., 2023). Due to their young age, broiler chicks may be more sensitive to such anti-nutritional factors.

### *3.2.1.2 Potential of feeding LOCB*

Multiple studies have explored including more LOCB in the diets of industrially produced broilers. To avoid tradeoffs with other sustainability issues they aim to maintain high N retention efficiency while replacing raw ingredients with alternative (protein) ingredients that do not compete with human consumption (Harchaoui et al., 2023). This requires that the appropriate ratio of protein to calories is maintained, and protein quality is sufficient so that broiler growth rate and FCR can be maintained. In laying hens, feeding a diet consisting only of LOCB while maintaining production was found feasible and reduced the LUR even below 1, depending on the LOCB fed (van Hal et al., 2019a). The following section explores the suitability of various LOCB in broiler feeds.

Regarding food processing by-products, many of them are already commonly used in broiler feed, but research has shown potential for less conventional by-products. No negative effects on growth and feed conversion ratio were found when including up to 16% dried distillers grains (Damasceno et al., 2020), 7.5% almond hulls (Wang et al., 2021), 8% modified meat meal (Navidshad et al., 2009), 5% citrus pulp (Mourao et al., 2008), 10% dried apple pomace Matoo et al. 2001), 50% fermented apple pomace (Joshi et al., 2000), 15% of dried tomato pomace (Kavitha et al., 2004), 1% oyster mushroom by-product (Fard et al., 2014), and 24% fresh sake lees (Ito et al., 2022), while positive effects were sometimes found at lower inclusion levels. Limitations were generally due to high fiber or moisture contents (Wadhwa et al., 2013). At low inclusion levels, apple, orange, and wine by-products have been shown to also improve broiler health and meat quality, mainly because they provide antioxidants (Georganas et al., 2023).

Regarding food waste, the potential varies for distinct types of food waste. During processing, retail, and distribution, food waste can be collected per food group, resulting in a constant and homogenous waste stream. Unsold bread and bakery products, for example are increasingly collected and used as feed. Bread waste can replace up to 100% of the grain in the broiler diet (Abido et al., 2021; Chauhan et al., 2024). Fruit and vegetable waste can contain significant amounts of sugar and protein, but the main challenge in incorporating them in broiler diets is the high fiber and moisture contents (90%) (Bakshi et al., 2016). In broiler production wasted peas could be included in the diet up to 20% without affecting production of feed efficiency (Paliwal et al., 1980); the same holds for 3% of dried wasted tomatoes (Barroso et al., 2008) and 20% of dried wasted apples (Matoo et al., 2001). In retail different wasted vegetables and fruits are collected together; their inclusion potential in broiler diets, to our knowledge, remains unstudied (Bakshi et al., 2016).

Post consumer food waste (PCFW) considers food scraps, which vary a lot in characteristics. In general, PCFW is fatter (15%) and more fibrous (8%) than common broiler diet ingredients such as maize and soybean meal. The protein content highly varies (range 1–40%), but is on average (19%) higher than maize, lower than soybean meal, and similar to common broiler feed (20%) (Simmons et al., 2023). Research has shown that mixed food wastes can be included in broiler diets, but effects vary due to the variation in composition. As the feed intake capacity of broilers is limited, most studies focused on the inclusion of dried food waste. Most studies show that with inclusion levels up to 20% productivity can be improved or maintained, while higher levels reduce productivity (Cho et al., 2004; Hossein and Dahlan, 2015; Gelan and Aberra 2021; Sani et al., 2021). Others found that inclusions of 10% to 20% may already reduce growth (Giamouri et al., 2021; Johnson et al., 2022). Fiber content is likely a limiting factor; while small amounts of fiber enhance intestinal health and thereby nutrient digestion and utilization, excessive fibers may negatively affect digestion and energy and other nutrients and consequently limit growth (Tejeda and Kim, 2021; Fung et al., 2019; Wang et al., 2021; Boumans et al., 2022). As broilers become better at digesting fiber over time, food waste may be more suitable to feed in later stages (e.g. 4–8 weeks; Chen et al., 2007; Mateos et al., 2012; Jha and Mishra, 2021). In many regions, such as the EU and the US, however, feeding PCFW is prohibited by food safety legislation in response to major disease outbreaks (Leib et al., 2016).

Regarding novel feed ingredients, several recent studies have illustrated the potential of especially novel protein sources such as aquatic proteins (e.g. algae and duckweed), insect proteins, and green leaf protein (Bikker and Jansman, 2023). For these ingredients to be circular they should not be suited for direct human consumption nor should the resources (including land and waterbodies) they use. Green leaf protein concentrates, concentrated protein obtained from, for example grass, clover, or lucerne, are promising due to their high protein content but require significant input for protein extraction. In these proteins, lysine and methionine contents are limited (Stødkilde et al., 2019). Insects are a natural food source for chickens; many species are of high protein quality, and they can be produced on wastes and manure that are unsuitable as food or livestock feed (REF Parodi). The latter is, however, not allowed in some regions due to feed safety legislation, restricting them to conventional livestock feed sources reducing the net food supply when insects are used as intermediary (Veldkamp and Bosch, 2015). Inclusion of spirulina, a microalgae, up to 16% increased growth performance of industrial broilers, while higher levels reduced production due to a lower feed intake (Evans et al., 2015). Macro algae, such as duckweed, had no effect on growth performance up to 25% inclusion in dried form; higher inclusion rates had negative effects (Demann et al., 2022).

#### *3.2.1.3 Increasing the potential of LCOB as broiler feed*

In conclusion, many unconventional food leftovers can be included in broiler diets while maintaining production and feed efficiency. The amount that can be included of each product is, however, quite limited due to high moisture, fat, and fiber contents, which then also holds for combinations of products with the same components. To enable higher inclusion levels, leftovers can be dried, as already assumed for PCFW, but this requires energy. The high fiber content may, furthermore, be overcome by ensiling or fermentation, which also improves preservation (Jinno et al., 2018; King, 2013). To achieve such feeding strategies in practice, feed formulations should not be guided only by market prices, but by the origin, feed-food competition, and environmental impacts of feed ingredients as well (Garcia-Launay et al., 2018). This could be stimulated by appreciating farmers for their environmental performance, which drives them to better value sustainable feeds. This requires a consistent application of environmental and circularity indicators in the feed industry, that is under development in the Netherlands.

When trying to meet the nutritional value (energy and protein) of conventional broiler diets solely with LOCB, this requires large amounts of high-quality co-products such as biscuit waste, oilseed meals, gluten feed, and animal fat (Bikker and Jansman, 2023). This results in a diet much lower in starch and much higher in fat, sugar, and fiber than conventional broiler diet. This provides a limited perspective for broiler production as high-quality co-products are only limitedly available, while high fat, sugar, and fiber contents would likely negatively affect production (Bikker and Jansman, 2023). The potential of feeding broilers with LOCB could be improved by integrating the concept of resource use efficiency, thus the ability to feed on LCOB, in breeding programs (van Hal et al., 2019b). This is especially relevant as studies that look at which animals can best utilize LOCB showed dairy to be preferred over pig and poultry, and poultry was only able to feed on food waste if dried, while pigs could also use it as a wet feed (Karlsson et al., 2019; van Hal et al., 2019b; van Hal, 2020).

Alternatively, the potential of using LOCB in broiler diets could be enhanced by allowing for a reduction in growth rate. If broilers are provided with a feed with a lower energy and or protein content, they first adapt by increasing their feed intake. If they cannot meet the nutrient requirement for their high growth within their feed intake capacity, their growth will reduce (Bos et al., 2023). This would allow a larger share of LOCB but may increase the environmental impact per kg meat (Bos et al., 2023). This is, however, largely due to our environmental impact assessment methods, and methods that better account for the environmental benefits of feeding LOCB are needed (van Hal et al., 2019a). Under the broad definition of circularity, an increase in environmental impact of a specific product is not undesired if it reduces the environmental impact of the food system as a whole. Smallholder indigenous poultry production has a lower productivity and generally has a larger share of LOCB in their diet, if not fully (Wilson et al., 2022). As traditional breeds are adapted to a more natural system, they respond better to feeding of live insects that are a natural feed source compared to fast-growing broiler breeds (Dessie et al., 2011). In Kenya, where feeding manure-fed insects to poultry is allowed, integrated insect poultry systems developed where insects are grown on poultry manure in crates above the broiler coop and fall down when they are large and crawl out. Additionally, smallholder farmers with improved indigenous breeds venture into growing algae on farm, especially duck weeds, to feed their chickens, with no negative effects observed on laying and growth performance. For the slower-growing naked neck breed (1.7 kg in 17 weeks), where up to 100% of wet food waste (35% DM) in the diet did not affect growth performance or meat quality (Bughio et al., 2021).

### *3.2.2 Valorizing manure and slaughter wastes*

At the other end, recycling of the outputs of broiler production can also be enhanced. Besides the intended broiler meat, broiler production results in slaughter by-products and manure. Slaughter by-products include cuts and organs people cannot or do not want to eat, bones, and feathers. The share of the chicken that is consumed varies per region; this share is very limited in Europa and the USA (below 50% of slaughter weight; Laisse et al., 2019) compared to, for example many places in Africa (up to 80%, Carron et al., 2017; Wilson et al., 2022), and increasing this share would be better than improving recycling of wasted by-products.

Originally, generating manure was an important purpose of keeping livestock; before artificial fertilizers came into play, crop production depended on manure for nutrients in an available form. Currently, in many major poultry production areas there is a surplus of manure, the manure is no longer used to produce food or feed and is often even incinerated. As mentioned, such incineration can politically be incentivized to address local nitrogen surplus. In the Netherlands, it is defined as manure treatment, which is considered desirable, and is valued as it generates energy. Valuable organically bound carbon and nitrogen are, however, lost from the food system, while these are highly needed feed exporting regions that currently depend on artificial fertilizers and are prone to soil erosion and degradation. In smallholder indigenous chicken farming, poultry manure is fully used as fertilizer or integrated into cattle diets (Carron et al., 2017).

Chiarelotto et al. (2021) explored composting of all broiler manure, waste, and by-products in Brazil and found it could be a valuable way to return nutrients back to crop production, but it does generate greenhouse gas emissions. Furthermore, extra carbon sources, for example crop residues, should be added to have an appropriate C to N ratio for composting. Harchaoui et al. (2023) found that improving recovery of poultry manure and by-products back into agriculture in France would increase NUE from 44% to 51% and recycling rate from 24% to 29% and decrease N loss from 125 kg N ha/yr to 102 kg N ha/yr.

## *3.3 Restructure livestock production: reintegrating and recoupling*

The major circularity issue with industrial livestock production is that the (spatial) decoupling of livestock and feed production results in mining of nutrients where feed is produced, while they accumulate in regions dense with industrial livestock production (van Selm et al., 2023). While the efficiency and recycling solutions discussed above may reduce the dependency on imported feed and reduce manure surplus, they do not assure that nutrients are returned to crop production in a balanced way. To this aim, studies propose systems where crop and livestock production are reintegrated (van Selm et al., 2023). Studies that strive for sustainable food systems (e.g. that operate within the planetary boundaries) take a broader perspective on such redesign and propose to recouple livestock with feed production, for example on a national level.

### *3.3.1 Integrated crop-broiler production systems*

Integrated crop-livestock systems are production systems that combine crop farming and livestock farming in the same area or enterprise, allowing for the direct recycling of nutrients in manure as fertilizer and crops or crop residue as feed. Of common livestock species, broilers are not the most common to consider in integrated crop-livestock systems in industrialized areas. This is likely because the high production efficiency broilers are valued for requires a highly controlled environment and feeding, to which concessions may be needed when integrating systems. Systems can, however, be integrated to various extents, which determines how much the management of each system is affected.

He et al. (2023) explored the environmental benefit of better connecting crop-laying hen production in China, while not housing the hens in the crop fields and thereby limiting management changes. They assume this connection to enhance the use of poultry manure on crops, but also assumed a lower protein feed, feed additives, and optimized fertilization. Compared to separate conventional broiler and crop systems, this integrated system increased NUE from 33% to 50% and reduced loss of reactive forms of N with 55% and greenhouse gas (GHG) emissions by 35%, while increasing economic benefit by 21%.

A system that pastured laying hens in arable cropping of sunflowers and beans produced higher crop yields and had improved soil fertility throughout the field (Hilimire et al., 2013). Similar results were found when pasturing broilers in orchards, but soil fertility improvements were reduced as broiler manure had less of a liming effect than that of laying hens, which are fed more calcium additives. Poultry could thus effectively fertilize crops directly over time and are furthermore effective in controlling weeds and pests (Hilimire, 2011). By feeding on these weeds and pests, part of the feed requirement is fulfilled, demanding less feed input. This integration requires severe management adaptations to both crop and poultry farming. Pasturing poultry in orchards or vineyards, a form of agroforestry, likely requires less management changes. This integration has the additional benefit that the upper story vegetation provides better shelter and leaves more space for the animals, resulting in less plant damage (Hilimire, 2011). In organic systems, orchards have potential in providing the mandatory outdoor access. So far, they are especially popular for laying hens, but in the EU, land-use legislation that demands agricultural land can only have one function is limiting practical implications. Rauw et al. (2022) proposed a novel integrated system of legumes, poultry, and bees. They proposed increased use of legume crops in European crop rotations, to reduce dependency on soy in poultry production, while also reducing artificial fertilizer use for the next crop in rotation. Proper integration with beekeeping enhances legume yields, while providing a rich foraging habitat for honey production and wild pollinators. Bees in turn provide valuable products for both human and poultry nutrition and health. Bee pollen and bee bread for example are high protein foods with minerals and vitamins, and royal jelly, honey, and propolis are energy dense, and both have antioxidant, anti-microbial, and anti-inflammatory properties (Mahmoud et al., 2016). This system, however, is not found in practice yet.

Smallholder indigenous farming in both Asia and sub-Saharan Africa commonly integrates crop and livestock farming, and, thereby, recycles better (Herrero et al., 2010). However, integrating dairy cattle in these farming systems appeared to be financially more beneficial than integrating poultry (Udo et al., 2011). Regarding poultry, integration with on-farm insect or algae production, which is used as feed, is upcoming. Poultry manure is used as a fertilizer for these feed sources.

### 3.3.2 Recoupling livestock with feed production

While integrated crop-livestock systems ensure that nutrients are balanced within the system, another option is to ensure this balance at a larger scale, a strategy called recoupling livestock with feed production. Recoupling creates a natural ceiling for livestock numbers by allowing the feed production capacity of a region to determine the size of the local livestock population. By no longer importing nutrients through feed, this strategy reduces the accumulation of manure in livestock-dense areas. It, however, requires either a reduction in animal numbers or an increase in the amount of domestic arable land dedicated to feed production. Partially recoupling broiler production in France, reducing feed imports by 50%, while maintaining broiler numbers, increased domestic arable land use with 0.23 Mha. As such land is not available, it would come at the cost of food production or feed for other livestock species. Nevertheless, Harchaoui et al. (2023) estimated this would increase the NUE from 44% to 46%, the N self-sufficiency from 10% to 15%, and the recycling rate from 24% to 27% and a slight decrease of N loss from 125 kg N ha/yr to 119 kg N ha/yr. Circularity improvement would likely have been more pronounced if the recoupling was paired with better recycling of feed resources, or if broiler numbers would have reduced.

In food systems redesign, recoupling is posed as a solution towards a sustainable food system that meets human food demand or requirement, while operating within the planetary boundaries. Exploring the full potential of recoupling livestock should, therefore, also explore reductions in animal numbers and what animals are most suited to produce in a region. The case study below (Section 4) explores what broiler production would look like if livestock were recoupled with feed production in the Netherlands. As recoupling livestock still allows for feeding crops grown on arable land that would be used more efficiently for food crop production, which goes against circularity principles, the role of broiler production, if additionally avoiding all feed-food competition, was assessed as well.

## 4 The potential of broilers in a more circular food system

This case study, based on van Selm et al. (2023), uses the food system model FOODSOM to explore how the role of broilers changes in circular food systems with different feed availabilities. FOODSOM is uniquely suited for this assessment, as it captures inter-linkages between food production and food consumption and includes high detail on animal production systems.

FOODSOM is an agro-ecological food system optimization model designed to minimize environmental objectives (e.g. GHG emissions) or maximize productivity objectives (e.g. protein production). This example takes the Netherlands as a case study. The Netherlands is a food trading nation; large quantities of animal feed are imported to sustain the domestic livestock population and the export of animal products. The Netherlands has the highest livestock density in Europe, which has led to local environmental issues, such as eutrophication and acidification, also referred to as the nitrogen crisis. Van Selm et al. (2023) identified recoupling livestock and feed production in the Netherlands as a promising strategy to enhance food self-sufficiency and reduce environmental impacts. How does broiler production change in the Netherlands when recoupling livestock with domestic feed production (Recouple scenario: no feed import allowed) and avoiding all feed-food competition (Recouple NFFC scenario: no feed food competition allowed)?

---

Agricultural land and marine fisheries form the basis of FOODSOM. Within the model 1.7 million hectares of agricultural land, 49 representative crops (one productivity level, based on current management) can be grown to produce food for humans and feed for animals, while marine fisheries only provide a source of food for humans. To make some of these crops and marine fish suitable for human consumption, processing (e.g. wheat into flour) is required, which produces co-products (e.g. wheat bran) to be used as animal feed or to be composted and returned to the soil as a source of nutrients. Moreover, food is lost (e.g. during storage or during processing), and food is wasted (e.g. during consumption or in the supermarket). These losses and wastes are suitable for animal feed or can be used as a source of nutrients for the (e.g. compost). Seven animal systems (dairy cattle, beef cattle, broiler chickens, laying hens, pigs, dairy goats, and meat sheep) including parent (e.g. sows in a pig system) and reproductive stocks (e.g. dairy heifers in a dairy cattle system) can consume a selected number of crops, co-products, and food losses/waste; in turn, these animals produce food suitable for human consumption, as well as animal manure to fertilize crops, and co-products (e.g. blood and bone meal) to be fed back to animals or composted and applied to the soil as fertilizer. Food from crops and animals is consumed by humans to satisfy their dietary and nutritional requirements (27 nutrients). The model can be found at https://git.wur.nl/aps/foodsom-public.

---

The Reference scenario represents the current situation in the Netherlands (2017/18). The broiler system includes 49 million broiler chickens largely fed on wheat and soybean meal (Fig. 3). The soybeans are imported from abroad, processed into soybean meal and soybean oil, the oil is exported, and the soybean meal remains in the Netherlands to be fed to livestock. The primary driver for importing soybeans is to produce soybean meal which creates an abundance of co-products for the domestic livestock population.

In the Recouple scenario feed imports and exports are prohibited; this includes soybeans, as they are mainly imported for soybean meal as livestock feed. Livestock in the Netherlands are, then, solely fed with feed produced in the Netherlands, which reduces feed availability and livestock numbers. This resulted in a substantial reduction in animal numbers, most pronounced

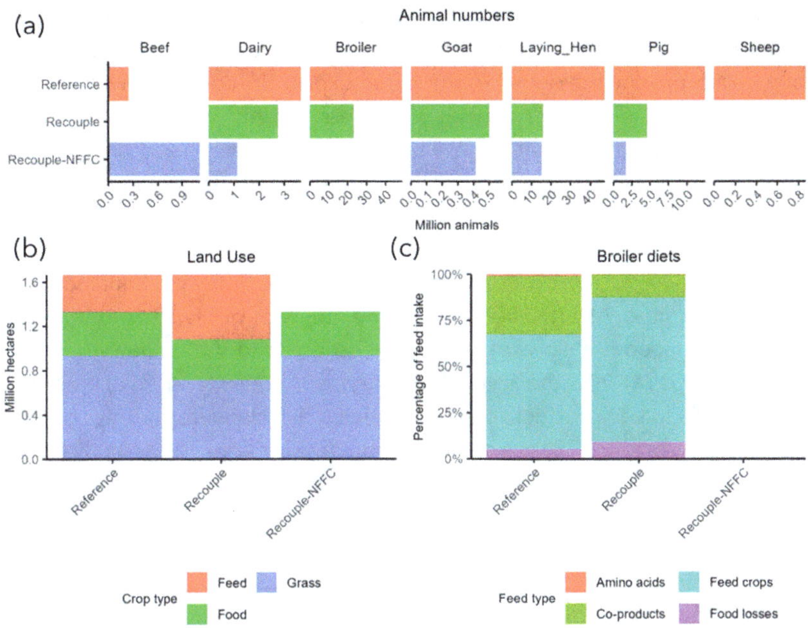


**Figure 3** Broiler production currently and when recoupling (Recouple) and avoiding feed-food competition (Recouple NFFC): (a) animal numbers (million), (b) land use (million hectares), (c) broiler diets (% of total feed intake).

in monogastrics, including broiler chickens (−57%; Fig. 3). The quantity of co-products decreased in broiler diets reduced from 32% in the reference scenario to 10% in the recouple scenario (Fig. 3), because the supply of high-quality co-products (soybean meal) was dependent on the import of oil crops, which was prohibited in this scenario. Instead, the smaller population of broiler chickens was fed more feed crops (75%) or food losses (14%). The reduction in animal numbers allowed the share of food losses in animal diets to increase, as the availability of food losses was similar across all scenarios. The reduction in animal numbers reduced animal source food exports, especially for pig and poultry products that were needed to maintain the current diet of the local population (see van Selm et al., 2023, for further information). In countries with a lower population density, more animal source food may be available for export.

If circular food system principles are followed, livestock should be fed solely with LOCB and arable land should be used to produce food for humans instead of feed for animals. In the Recouple-NFFC scenario, where using arable land to produce livestock feed is prohibited, broiler chickens did not play a role. The limited co-products and food losses available were prioritized for laying hens and pigs (Fig. 3). Pigs require a less nutrient-dense feed, allowing them to consume low-quality feeds, especially co-products and food losses. Laying hens were selected to produce eggs for the Dutch diet, while meat consumption was partially made up by beef cattle. Beef cattle could utilize permanent grassland to produce meat for the Dutch diet, at the expense of dairy cattle producing dairy products for export. In countries with relatively more grassland, this consequence may be less severe.

These results show that the availability and quality of feed products have an influence on the role broilers can play in the food system. When significant quantities of high-quality co-products or feed crops are available, broilers are selected as efficient meat producers for the human diet. However, if strictly following circularity principles and only feeding LOCB, the potential for broiler chickens is limited. As indicated before, the ability of broilers to valorize LOCB may be enhanced if a slower growth rate is accepted. An earlier study of van Hal et al. (2019a) into the role of livestock productivity in valorizing LOCB showed that lower productive dairy cattle and pigs were indeed preferred when feeding only LOCB, but not low productive broilers. This is likely because the high production efficiency is the main competitive advantage of broilers.

## 5 Conclusion

Industrial broiler production is highly efficient; it requires less feed to produce meat than any other terrestrial livestock species. To achieve this high production efficiency, they do, however, require feed of a high quality, of which a large share is imported. Thereby, broiler production scores lower on recycling and self-sufficiency than other terrestrial livestock species. They, furthermore, compete more for resources with food production directly (using human edible feed) and indirectly (using feed produced on arable land). Enhancing the circularity of broiler production by increasing the share of non-competing feed ingredients, so-called low opportunity cost biomass (LOCB) has limited potential when maintaining the current production efficiency. Due to high feed quality requirements, only high-quality LOCB (e.g. soybean meal) can be fed to conventional broilers, which is available in limited quantities. Reducing the productivity of broilers could enhance their ability to utilize LOCB, but this also reduces their competitive advantage over other livestock species. Therefore, it is concluded that the circularity of broilers is and will likely remain limited compared to other livestock species. They can, however, play a key role in a sustainable food system if one wishes to maintain substantial meat production and consumption.

## 6 Where to look for further information

### Resources focused on circularity principles

- Muscat, A., et al. (2021), Principles, drivers and opportunities of a circular bioeconomy. *Nat. Food* 2(8), 561–566. https://doi.org/10.1038/s43016-021-00340-7.

### Resources focused on circularity indicators

- Harchaoui, S., et al. (2023), Challenges and opportunities for improving circularity in the poultry meat and egg sector: the case of France. *Resour. Conserv. Recycl.* 193, 106963. https://doi.org/10.1016/j.resconrec.2023.106963.
- Hennessy, D.P., et al. (2021), The net contribution of livestock to the supply of human edible protein: the case of Ireland. *J. Agric. Sci.* 159(5–6), 463–471. https://doi.org/10.1017/S0021859621000064.

### Resources focused on feeding low-opportunity-cost biomass

- Boumans, I.J.M.M., et al. (2022), Feeding food losses and waste to pigs and poultry: implications for feed quality and production. *J. Clean. Prod.* 378, 134623.
- te Pas, M.F.W., et al. (2021), Adaptation of livestock to new diets using feed components without competition with human edible protein sources–a review of the possibilities and recommendations. *Animals* 11, 2293. https://doi.org/10.3390/ani11082293.

### Resources focused on food systems modeling and the role of animals therein

- van Hal, O. (2020), *Upcycling biomass in a circular food system – the role of livestock and fish*. Dissertation Wageningen University. https://doi.org/10.18174/524412.
- van Selm, B. (2024), *Future scenarios for more circular food systems: insights from the Netherlands & Europe*. Dissertation Wageningen University. https://doi.org/10.18174/638762.

## 7 References

Abido, A.M.M., et al. (2021), Evaluation of using some untraditional energy sources in broiler chicken diets on performance, carcass characteristics and economic efficiency: 1 – Cake by-product. *EJNF* 24(2), 129–135.

Aznar-Sánchez, J.A., et al. (2020), Indicators for circular economy in the agri-food sector. *Resour. Conserv. Recycl.* 163, 105028.

Bakshi, M.P.S., et al. (2016), Waste to Worth: Vegetable Wastes as Animal Feed, published in CAB Reviews, 11, 1–26.

Barroso F.G., et al. (2008), Alimentación de ovejascon ensilados basados en el fruto del tomate. *Albeitar* 120, 6–9.

Bikker, P., and Jansman, A.J.M. (2023), Review: Composition and utilisation of feed by monogastric animals in the context of circular food production systems. *Animal*, 17(7), 100892.

Bmc Moerdijk (2024), https://www.bmcmoerdijk.nl/nl/.

Bos, A.P., et al. (2023), Welzijn, milieu en economie van vleeskuikens: Een overzicht van de inzichten uit het project Greenwell. Wageningen Livestock Research. https://edepot.wur.nl/591335.

Boumans, I.J.M.M., et al. (2022), Feeding food losses and waste to pigs and poultry: implications for feed quality and production. *J. Clean. Prod.* 378, 134623.

Budowski, P., et al. (1987), Nutritional encephalomalacia in the chick: an exposure of the vulnerable period for cerebellar development and the possible need for both omega 6- and omega 3-fatty acids. *Br. J. Nutr.* 58(3), 511–520.

Bughio, E., et al. (2021), Effects of production system and feeding regimen on carcass and meat quality traits of Naked Neck chicken. *S. Afr. J. Anim. Sci.* 51, 2.

Carron, M., et al. (2017), The broiler meat system in Nairobi, Kenya: Using a value chain framework to understand animal and product flows, governance and sanitary risks. *Prev. Vet. Med.* 147(1), 90–99.

Castrica, M., et al. (2018), Pet food as the most concrete strategy for using food waste as feedstuff within the European context: a feasibility study. *Sustainability* 10(6), 2035.

Chan, I., et al. (2022), The 'sustainability gap' of US broiler chicken production: trade-offs between welfare, land use and consumption. *R. Soc. Open Sci.* 9(6), 210478.

Chauhan, K., et al. (2024), Bakery waste as a potential unconventional feed ingredient in broiler ration: a review. *IJAN* 41(1), 1–8. https://epubs.icar.org.in/index.php/IJAN/article/view/142182.

Chen, K.L., et al. (2007), Effect of dietary inclusion of dehydrated food waste products on taiwan native chicken (Taishi No. 13). *AJAS* 20(5), 754–760.

Chen, T., et al. (2015), A safety analysis of food waste-derived animal feeds from three typical conversion techniques in China. *Waste Manag.* 45, 42–50.

Chiarelotto, M., et al. (2021), Composting organic waste from the broiler production chain: a perspective for the circular economy. *J. Clean Prod.* 329, 129717.

Cho, Y.M., et al. (2004), Effects of feeding dried leftover food on growth and body composition of broiler chicks. *AJAS* 17(3), 386–393.

Cottrell, J.J., et al. (2022), Recent advances in the use of phytochemicals to manage gastrointestinal oxidative stress in poultry and pigs. *Anim. Prod. Sci.* 62(12), 1140–1146.

Crippa, M., et al. (2021), Food systems are responsible for a third of global anthropogenic GHG emissions. *Nat. Food* 2, 198–209.

Damasceno, J.L., et al. (2020), Effect of feeding dried distillers' grains with solubles to broiler chickens from day 0 to 21. *Livest. Sci.* 241, 104235.

de Boer, I.J. and van Ittersum, M.K. (2018), *Circularity in agricultural production*. Wageningen University. https://www.nutrientplatform.org/wp-content/uploads/2019/06/3939_Circularity-in-agricultural-production-012019final.pdf.

De Cuyper, C., et al. (2022), Total versus human-edible protein efficiency of Belgian pigs and poultry. In *Book of Abstracts of the 73rd Annual Meeting of the European Federation of Animal Science* (p. 365).

De Oliveira, C.T., et al. (2021), Nano and micro level circular economy indicators: Assisting decision-makers in circularity assessments. *Sustain. Prod. Consum.* 26, 455–468.

de Vries, M., et al. (2015), Comparing environmental impacts of beef production systems: a review of life cycle assessments. *Livest. Sci.* 178, 279–288.

Demann, J., et al. (2022), Nutritional value of duckweed as protein feed for broiler chickens–digestibility of crude protein, amino acids and phosphorus. *Animals* 13(1), 130.

Dessie T., et al. (2011), Current state of knowledge on phenotypic characteristics of indigenous chickens in the tropics. *World. Poult. Sci. J.* 67(3), 507–516.

European Commission (2020), Farm to fork strategy. https://ec.europa.eu/info/strategy/priorities-2019-2024/european-green-deal/actions-being-taken-eu/farm-fork_en (accessed 3.8.24).

Evans, A.M., et al. (2015), Effects of algae incorporation into broiler starter diet formulations on nutrient digestibility and 3 to 21 d bird performance. *J. Appl. Poult. Res.* 24, 206–214.

Fard, S., et al. (2014), Effect of oyster mushroom wastes on performance, immune responses and intestinal morphology of broiler chickens. *Int. J. Recycl. Org. Waste. Agric.* 3, 141–146.

Feddern, V., et al. (2019), A review on biogenic amines in food and feed: toxicological aspects, impact on health and control measures. *Anim. Prod. Sci.* 59(4), 608.

Fry, J.P., et al. (2018), Feed conversion efficiency in aquaculture: do we measure it correctly?. *ERL* 13(2), 024017.

Fung, L., et al. (2019), Estimated energy and nutrient composition of different sources of food waste and their potential for use in sustainable swine feeding programs. *Transl. Anim. Sci.* 3(1), 359–368.

Gallo, F., et al. (2023), Integration of a circular economy metric with life cycle assessment: methodological proposal of compared agri-food products. *Int. J. Life Cycle Assess.* 29, 1359–1379.

Garcia-Launay, F., et al. (2018), Multiobjective formulation is an effective method to reduce environmental impacts of livestock feeds. *Br. J. Nutr.* 120(11), 1298–1309.

Gelan, O. and Melesse, A. (2021), Significance of cafeteria food leftover as alternative cheap feed resource in poultry nutrition: a viable strategy to reduce its polution impact on the environment. In *Proceedings of the Ahi Evran International Conference on Scientific Research*. Turkey.

Georganas, A., et al. (2023), Utilization of agro-industrial by-products for sustainable poultry production. *Sustainability 15*(4), 3679.

Ghisellini, P., et al. (2016), A review on circular economy: the expected transition to a balanced interplay of environmental and economic systems. *J. Clean Prod.* 114, 11–32.

Giamouri, E., et al. (2021), The food for feed concept. Performance of broilers fed hotel food residues. *Br. Poult. Sci.* 62(3), 452–458.

Giampietro, M. (2019), On the circular bioeconomy and decoupling: implications for sustainable growth. *Ecol. Econ.* 162, 143–156.

Groenestein, C.M., et al. (2019), Comparison of ammonia emissions related to nitrogen use efficiency of livestock production in Europe. *J. Clean Prod.* 211, 1162–1170.

Hallström, E., et al. (2015), Environmental impact of dietary change: a systematic review. *J. Clean. Prod.* 91, 1–11.

Hamam, M., et al. (2021), Circular economy models in agro-food systems: a review. *Sustainability* 13(6), 3453.

Hantoro, F.R.P., et al. (2024), The effect of stocking density and crude protein level on performance, carcass characteristics, nitrogen use, and ammonia emissions of the Indonesian native chickens. *J. Indonesian Trop. Anim. Agric.* 49(2), 157–168.

Harchaoui, S., et al. (2023), Challenges and opportunities for improving circularity in the poultry meat and egg sector: the case of France. *Resour. Conserv. Recycl.* 193, 106963.

Harris, S., et al. (2021), Circularity for circularity's sake? scoping review of assessment methods for environmental performance in the circular economy. *Sustain. Prod. Consump.* 26, 172–186.

Hartcher, K.M. and Lum, H.K. (2020), Genetic selection of broilers and welfare consequences: a review. *World Poult. Sci. J.* 76(1), 154–167.

He, Z., et al. (2023), Mitigation of nitrogen losses and greenhouse gas emissions in a more circular cropping-poultry production system. *Resour. Conserv. Recycl.* 189, 106739.

Hennessy, D.P., et al. (2021), The net contribution of livestock to the supply of human edible protein: the case of Ireland. *J. Agric. Sci.* 159(5–6), 463–471.

Herrero, M.T., et al. (2010), Drivers of change in crop-livestock systems and their potential impacts on agro-ecosystems services and human well-being to 2030. Presentation at the Nairobi Forum, ILRI, Nairobi, 21 September 2010. Nairobi, Kenya, ILRI.

Hilimire, K. (2011), Integrated crop/livestock agriculture in the United States: a review. *J. Sustain. Agric.* 35(4), 376–393.

Hilimire, K., et al. (2013), Soil fertility and crop growth under poultry/crop integration. *Renew. Agr. Food Syst.* 28(2), 173–182.

Hossein, S. and Dahlan, I. (2015), Growth performance of free-range village chickens fed dehydrated processed food waste. *Mal. J. Anim. Sci.* 18(1), 77–86.

Ito, K.R., et al. (2022), Utilization of Sake lees as Broiler Feedstuff and its Effects on Growth Performance and Intestinal Immunity. *J. Poult. Sci.* 59(3), 247–259.

Jha, R. and Mishra, P. (2021), Dietary fiber in poultry nutrition and their effects on nutrient utilization, performance, gut health, and on the environment: a review. *JASB* 12(1), 51.

Jinno, C., et al. (2018), Enzymatic digestion turns food waste into feed for growing pigs. *AFST* 242, 48–58.

Johnson, J.T., et al. (2022), Formulation and nutrient assessment of poultry feed from domestic waste and its effect on the growth of poultry birds. *IJAT* 2(1), 1–7.

Joshi, V., et al. (2000), Production and evaluation of fermented apple pomace in the feed of broilers. *J. Food Sci. Technol.* 37(6), 609–612.

Jurgilevich, A., et al. (2016), Transition towards circular economy in the food system. *Sustainability* 8(1), 69.

Karlsson, J.O. and Röös, E. (2019), Resource-efficient use of land and animals–Environmental impacts of food systems based on organic cropping and avoided food-feed competition. *Land Use Policy* 85, 63–72.

Kavitha, P., et al. (2004), Nutrient utilization in broilers fed dried tomatopomace with or without enzyme supplementation. *IJAN* 21, 17–21.

King, A. (2013), Removal of excess cellulose and associated polysaccharides in fruit and vegetable by-products – implication for use in feed for monogastric farm animals. In van de ven, T. and Godbout, L. (Eds), *Cellulose - Fundamental aspects*. IntechOpen.

Kirchherr, J., et al. (2017), Conceptualizing the circular economy: an analysis of 114 definitions. *Resour. Conserv. Recycl.* 127, 221–232.

Koppelmaki, K., et al. (2021), Nested circularity in food systems: a Nordic case study on connecting biomass, nutrient and energy flows from field scale to continent. *Resour. Conserv. Recycl.* 164, 105218.

Laisse, S., et al. (2019), L'efficience nette de conversion des aliments par les animaux d'élevage: une nouvelle approche pour évaluer la contribution de l'élevage à l'alimentation humaine. *INRA Prod. Anim.* 31, 269–288.

Leib, E., et al. (2016), *Leftovers for livestock: a legal guide for using excess food as animal feed*. The Harvard Food Law and Policy Clinic and the Food Recovery Project at the University of Arkansas School of Law. Leftovers-for-Livestock_A-Legal-Guide_August-2016.pdf.

Magothe, T.M., et al. (2012), Indigenous chicken production in Kenya: I. Current status. *World. Poultry Sci. J.* 68(1), 119–132.

Mahmoud, U.T., et al. (2016), Functions of propolis as a natural feed additive in poultry. *World. Poultry Sci. J.* 72, 37–48.

Marques, A.C., et al. (2018), Economic growth, sustainable development and food consumption: evidence across different income groups of countries. *J. Clean. Prod.* 196, 245–258.

Mateos, G.G., et al. (2012), Poultry response to high levels of dietary fiber sources varying in physical and chemical characteristics. *JAPR* 21(1), 156–174.

Matoo, F.A., et al. (2001), Performance of broiler fed onapple pomace diets supplemented with enzymes. *IJAN* 18, 349–352.

Meng, X., et al. (2022), Polystyrene microplastics induced oxidative stress, inflammation and necroptosis via NF-κB and RIP1/RIP3/MLKL pathway in chicken kidney. *Toxicol.* 478, 153296.

Mezes, M., et al. (1997), Nutritional metabolic diseases of poultry and disorders of the biological oxidant defense system. *Acta Vet. Hung.* 45, 349–360.

Mourao, J.L., et al. (2008), Effect of dietary dehydrated pasture and citrus pulp on the performance and meat quality of broiler chickens. *Poult. Sci.* 87, 733–743.

Murugesan, G.R., et al. (2015). Prevalence and effects of mycotoxins on poultry health and performance and recent development in mycotoxin counteracting strategies. *Poultry Science*, 94(6), 1215–1232.

Muscat, A., et al. (2021), Principles, drivers and opportunities of a circular bioeconomy. *Nat. Food* 2(8), 561–566.

Navidshad, B., et al. (2009), Effect of dietary levels of a modified meat meal on performance and small intestinal morphology of broiler chicken. *Afr. J. Biotechnol.* 8, 5620–5626.

Oosting, S., et al. (2021), Farmed animal production in tropical circular food systems. *Food Secur.* 14, 273–292.

Padhi, M.K. (2016), Importance of indigenous breeds of chicken for rural economy and their improvements for higher production performance. *Scientifica* 2016(1), 2604685.

Paliwal, V.K., et al. (1980), Utilization of agroindustrial byproducts by livestock and poultry. II. Effect of higher level for pea waste on broiler chicks. *HAU J. Res.* 10, 283.

Papangelou, A. and Mathijs, E. (2021), Assessing agro-food system circularity using nutrient flows and budgets. *Environ. Manage.* 288, 112383.

Parchomenko, A., et al. (2019), Measuring the circular economy-a multiple correspondence analysis of 63 metrics. *J. Clean. Prod.* 210, 200–216.

Peña, C., et al. (2021), Using life cycle assessment to achieve a circular economy. *Int. J. Life Cycle Assess.* 26, 215–220.

Peters, J.P., et al. (2014), Feed conversions, ration compositions, and land use efficiencies of major livestock products in U.S. agricultural systems. *Agric. Syst.* 130, 35–43.

Poponi, S., et al. (2022), Evaluating the transition to the circular economy in the agri-food sector: Selection of indicators. *Resour. Conserv. Recycl.* 176, 105916.

Redondo, L.M., et al. (2014), Perspectives in the use of tannins as alternative to antimicrobial growth promoter factors in poultry. *Front. Microbiol.* 5, 118.

Saidani, M., et al. (2019), A taxonomy of circular economy indicators. *J. Clean. Prod.* 207, 542–559.

Sandström, V., et al. (2022), Food system by-products upcycled in livestock and aquaculture feeds can increase global food supply. *Nat. Food* 3, 729–740.

Sani, U.M., et al. (2021), Growth response, nutrient digestibility and carcass characteristics of broiler chickens fed dried kitchen waste meal as a partial replacement for corn. *IOP Conf. Ser.: Earth Environ. Sci.* 761, 012094.

Simmons, C.W., et al. (2023), Incorporation of recovered food and food scraps into poultry feed: a systematic review. *CABI Rev.* 18, 1.

Spiller, M., et al. (2024), Beyond circularity! Integration of circularity, efficiency, and sufficiency for nutrient management in agri-food systems. *Nutr. Cycl. Agroecosyst.* 129(3), 287–297.

Stødkilde, L., et al. (2019), Digestibility of fractionated green biomass as protein source for monogastric animals. *Animal* 13, 1817–1825.

Tang, C., et al. (2025), Assessing nitrogen circularity in food systems in the North China Plain. *Resour. Conserv. Recycl.* 213, 108015.

Tejeda, J.O. and Kim, K.W. (2021), Role of dietary fiber in poultry nutrition. *Animals* 11(2), 461.

te Pas, M.F.W., et al. (2021), Adaptation of livestock to new diets using feed components without competition with human edible protein sources–a review of the possibilities and recommendations. *Animals* 11, 2293.

Tufarelli, V., et al. (2018), Feeding forage in poultry: a promising alternative for the future of production systems. *Agriculture* 8(6), 81.

Tukker, A. (2015), Product services for a resource-efficient and circular economy – a review. *J. Clean. Prod.* 97, 76–91.

Udo, H.M.J., et al. (2011), Impact of intensification of different types of livestock production in smallholder crop-livestock systems. *Livest. Sci.* 139(1–2), 22–29.

United Nations (2024), *World population prospects 2024: summary of results*. UN DESA/POP/2024/TR/NO. 9. United Nations.

van der Eijk, J.A.J., et al. (2023), Fast- and slower-growing broilers respond similarly to a reduction in stocking density with regard to gait, hock burn, skin lesions, cleanliness, and performance. *Poult. Sci.* 102(5), 102603.

van Hal, O. (2020), *Upcycling biomass in a circular food system: The role of livestock and fish*. Doctoral Dissertation, Wageningen Universtity and Research. Promotors, Imke de Boer and Hannah van Zanten. https://doi.org/10.18174/524412.

van Hal, O., et al. (2019a), Accounting for feed-food competition in environmental impact assessment: Towards a resource efficient food-system. *J. Clean. Prod.* 240, 118241.

van Hal, O., et al. (2019b), Upcycling food leftovers and grass resources through livestock: impact of livestock system and productivity. *J. Clean. Prod.* 219, 485–496.

van Kernebeek, H.R.J., et al. (2016), Saving land to feed a growing population: consequences for consumption of crop and livestock products. *Int. J. Life Cycle Asses.* 21(5), 677–687.

van Riel, A-J., et al. (2023), Feed-food competition in global aquaculture: current trends and prospects. *Rev. Aquacult.* 15(3), 1142–1158.

van Selm, B., et al. (2023), Recoupling livestock and feed production in the Netherlands to reduce environmental impacts. *STOTEN* 870, 165540.

van Zanten, H.H.E., et al. (2016), Global food supply: land use efficiency of livestock systems. *Int. J. Life Cycle Assess.* 21(5), 747–758.

van Zanten, H.H.E., et al. (2018), Defining a land boundary for sustainable livestock consumption. *Glob. Change Biol. 24*(9), 4185–4194.

Veldkamp, T. and Bosch, G., (2015), Insects: a protein-rich feed ingredient in pig and poultry diets. *AF* 5, 45–50.

Wadhwa, M. and Bakshi, M. (2013), *Utilization of fruit and vegetable wastes as livestock feed and as substrates for generation of other value-added products*. FAO, Bangkok.

Walkinshaw, C., et al. (2022), Detection and characterisation of microplastics and microfibres in fishmeal and soybean meal. *Mar. Pollut. Bull.* 185, 114189.

Wang, J., et al. (2021), Effect of almond hulls on the growth performance, body composition, digestive tract weight, and liver antioxidant capacity of broilers. *JAPR* 30, 2.

Wilkinson, J.M. (2011), Re-defining efficiency of feed use by livestock. *Animal* 5(7), 1014–1022.

Wilson C.W., et al. (2022), The diversity of smallholder chicken farming in the Southern Highlands of Tanzania reveals a range of underlying production constraints. *Poult. Sci.* 101(10), 102062.

Zuidhof, M.J., et al. (2014), Growth, efficiency, and yield of commercial broilers from 1957, 1978, and 2005. *Poult. Sci.* 93(12), 2970–2982.

# Chapter 14

## Challenges in achieving low-input, extensive broiler production

*Claire Bonnefous and Veronika Maurer, Research Institute of Organic Agriculture (FiBL), Switzerland; and Elisabeth Duval and Anne Collin, Biologie des Oiseaux et Aviculture – INRAE and Université de Tours, France*

1 Introduction
2 Outdoor ranges: characteristics and use by chickens
3 Providing outdoor range access: risks and potential solutions
4 Providing outdoor range access: welfare issues
5 Impacts of outdoor rearing systems on the environment
6 Conclusion
7 References

### 1 Introduction

There are hundreds of different poultry production systems around the world. These range from large-scale, intensive, indoor (including cage) systems with fast-growing broilers reaching 3 kg in 30 days (Xu et al., 2025) to slow-growing, backyard broilers reared outdoors in smallholdings, reflecting a broad range of histories and cultures. An example is the change in broiler rearing systems in Europe over the last century. While slow-growing backyard rearing used to be common, food shortages after two world wars led to intensification of production. Improvements in nutrition formulation and genetic selection contributed to developing fast-growing broilers with a dressing percentage (a measure of the amount of edible meat post slaughter) of up to 75% (Sosnówka-Czajka et al., 2023). However, problems such as the extensive use of antibiotics in intensive production, health issues in broilers such as reduced mobility and lameness, and the limited possibilities for birds to express natural behaviours, has, in recent decades, led consumers to favour more extensive methods of production such as organic or free-range systems.

http://dx.doi.org/10.19103/AS.2025.0165.16

This chapter focuses on challenges and potential solutions in optimising rearing of meat-type chickens specifically in low-input and extensive production systems. Parr et al. (1990) defined such systems as:

'Low Input Farming Systems (LIFS) [which] seek to optimise the management and use of internal production inputs (i.e. on-farm resources) and to minimise the use of production inputs (i.e. off-farm resources), such as purchased fertilisers and pesticides, wherever and whenever feasible and practicable, to lower production costs, to avoid pollution of surface and groundwater, to reduce pesticide residues in food, to reduce a farmer's overall risk, and to increase both short- and long-term farm profitability'.

This definition, which is still used by the Food and Agriculture Organization (FAO), highlights three main aspects:

- Use of on-farm resources, which in this chapter refers to optimisation of eating grass as a key source of nutrition. This extends to the impact of minimising reliance on bought-in feed (typically derived from plants such as soya) in reducing competition with crop production for human food.
- Limited use of off-farm resources, including limiting the use of antibiotics in favour of curative veterinary products derived e.g. from herbs and other plants.
- Limited environmental impact, related to the impact of rearing chickens outdoors.

Extensive broiler production also implies not confining birds indoors, the provision of outdoor, free-range conditions and lower densities of birds.

The aim of this chapter is to highlight the challenges of rearing broilers in a low-input system and to summarise a range of solutions that are adaptable to different settings or environments. Within this chapter, we specifically discuss the challenges in producing chickens with outdoor access and what solutions there are within the scope of low-input production systems. Finally, we consider the impact of producing chickens outdoors on the environment. While some challenges are specific to broilers, others are common for meat and egg production chickens, hence the discussion and references including broilers and laying hens where appropriate.

## 2 Outdoor ranges: characteristics and use by chickens

Providing outdoor access is the first step. However, the use of an outdoor range by broilers is variable both at the group level (Dawkins et al., 2003) and at the individual level (Bonnefous et al., 2023; Ferreira et al., 2019).

## 2.1 Why a group of chickens might or might not use an outdoor range

Chicken thermoregulation capacity develops during incubation (Loyau et al., 2014) and over the first 10 days of life (Nichelmann and Tzschentke, 2002). Outdoor range access in Europe is therefore provided after 21 days (Biosuisse, 2025; Council of the European Union and European Parliament, 2018; INAO, 2009). The first day of trap opening is therefore accompanied by the discovery of a new environment. The use by chickens of this new environment, the outdoor range, increases day by day for slow-growing broiler production systems (Collet et al., 2024; Stadig et al., 2017a), with age being one of the main factors in increasing outdoor range use (Taylor et al., 2017a). In increasing the use of the outdoor range, allowing chickens to view it before access (Fig. 1a) might reduce the fear of novelty associated with the discovery of a new environment (Taylor et al., 2023).

The outdoor environment can be more or less attractive for chickens depending on weather conditions and the outdoor layout. The influence of these environmental factors on range use has been well-documented. Among climatic factors, rainfall, UV radiation and wind speed contribute negatively to chickens' range use (Collet et al., 2024; Stadig et al., 2017a; Taylor et al., 2017a), while an increase of temperature from 15°C to 25°C leads to increased range use (Stadig et al., 2017b; Rodriguez-Aurrekoetxea et al., 2014). To limit the negative impact of rainfall and wind, the design of the outdoor range can be adjusted. However, variables such as temperature, wind, rainfall and species make it difficult to quantify the effects of enrichment.

Meadows without the provision of any cover and shade are less used by chickens compared to outdoor ranges with artificial or natural covers (Fig. 1b) such as trees (Stadig et al., 2017b; Dawkins et al., 2003). Range use was increased by 5% when the range was enriched by short rotation coppice instead of an artificial structure (Stadig et al., 2017b). Individual range use by S757N broilers

(a) (b)

**Figure 1** (a) Visual access provided to 1-day-old chicks (optimising ranging behaviour by free-range meat chickens, 2023). (b) Artificial structure for broilers in outdoor areas (Stadig et al., 2017b).

was around 1 m per observation in a meadow range (Ferreira et al., 2019) and around 3.5 m per observation in a tree-covered range (Bonnefous et al., 2023). Cover provides protection from rainfall, UV radiation, wind and also from birds of prey. Among covers, trees and short rotation grass seem to be preferred compared to artificial structures: in these conditions broilers used significantly more of the range further away than 5m from the broiler house (Stadig et al., 2017b; Dawkins et al., 2003).

Last but not least, genetic background also impacts range use at the strain level (Castellini et al., 2016). Local, slow-growing strains use the range significantly more than fast-growing strains both at group and individual level (Collet et al., 2024).

## 2.2 Why a specific individual might or might not use an outdoor range

Even under optimised free-range conditions, some chickens never use the outdoor range. Individual range use behaviour has been repeatedly shown to be consistent from 21 to above 80 days of life and across multiple environmental conditions (Collet et al., 2024; Bonnefous et al., 2023; Ferreira et al., 2019; Taylor et al., 2017b). As 'behavioural differences between individuals that are consistent overtime and across situations' (Réale et al., 2010), range use may be considered as a personality trait.

Multiple behaviours relate to range use. Chickens' attitudes to foraging, desire for locomotion, social behaviour or even cognition influence their use of the range (Bonnefous et al., n.d.a). As gregarious animals, chickens may be driven by social behaviour to follow certain chickens who either range outside or stay inside (Bonnefous et al., 2023; Gómez et al., 2022; Ferreira et al., 2020a, 2020b, 2020c). Additionally, more active chickens, expressing more foraging and locomotor behaviours, seem to be more prone to use the range (Ferreira et al., 2021, 2022; Mattioli et al., 2021; Cartoni Mancinelli et al., 2020). Understanding chicken behaviour is a challenge, and most behavioural studies suffer from limitations as direct observations are time-consuming and observer-dependent. Introducing high-throughput technology contributes to bridge these gaps (Collet et al., 2024; Gómez et al., 2022), but improvements to capture precise behavioural variation are still needed.

Measures promoting foraging and locomotor behaviours in chickens, such as enrichment and genetic selection, have been proposed as means to increase range use. Enrichment measures such as perches and hay bales within the barn before opening of traps to allow range access did not have a significant influence on the outdoor range use (Taylor et al., 2023). While different enrichment systems such as laser beams (Meyer et al., 2021) may enhance chickens' locomotor behaviour in the barn, their efficacy in the case of slow-growing chickens and consequences for outdoor range use are still unknown. Genetic selection is an option when precise, reliable and repeatable indicators can be used. Radio Frequency Identification (RFID) technologies have been shown to track range use in commercial chicken strains (Collet et al., 2024). Studying a pedigree line should make it possible to calculate the heritability of these indicators and determine their potential for further genetic selection to promote range use.

## 3 Providing outdoor range access: risks and potential solutions

Details on the risks from outdoor range are reviewed in this section. Fig. 2 summarises risks and potential solutions (Fig. 2).

### *3.1 Risk of disease is increased by outdoor access*

As chickens roam in the outdoor area, they eat grass (Dal Bosco et al., 2014), increasing the risk of an endo-parasitic infestation (Lolli et al., 2019). Van De Weerd et al. (2009) did not identify endo-parasites as a major health issue for organic broilers. However, Souillard et al. (2019) reported a huge increase in helminth infection after outdoor access in broilers: helminths were noted in only one farm before range use but in more than 12 out of 15 independent farms and in 38 out of 70 affiliated farms (i.e. farms linked to a company). An efficient solution against endo-parasites is the use of anthelminthics. However, within a low-input broiler production system, one aim is to limit the use of such synthetic medicinal products. To meet this objective without impairing

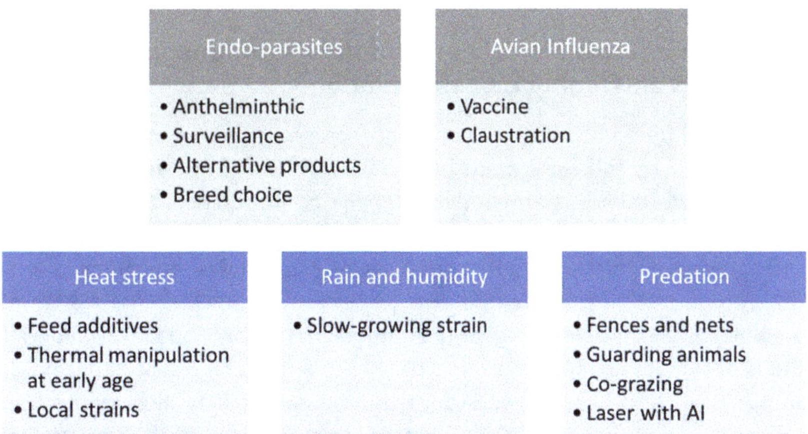


**Figure 2** Challenges due to limited control of the environment and levers of action to produce low-input, extensive broilers with outdoor access.

the health of broilers, three types of action can be implemented: surveillance, limitation of infestation and choice of an appropriate strain. First, monitoring potential infestation based on faecal egg count (Cain et al., 2024; Daş et al., 2011) limits excessive anthelminthic use by targeting the precise point when anthelminthics are needed. Second, management of the outdoor space by rotation of chickens in differing outdoor areas and by regular change of litter in a winter garden (where provided) helps contain parasitic infestation (Maurer et al., 2013). Third is the use of alternative medicines, such as plant extracts in the form of essential oils or directly planted as vegetation in the outdoor range, which may play a role in managing endo-parasitic infestation. *In vitro* tests (Poulopoulou et al., 2023) and experimental tests (Rodenbücher et al., 2023) show promising results, but on-farm tests show the difficulty in finding the right plant extract and the right dosage to show efficacy (Bonnefous et al., 2025). Finally, slow-growing strains seem to cope better with nematode infections than fast-growing strains (Stehr et al., 2019).

Providing outdoor access also facilitates contact between production broilers and wild birds, therefore increasing the risk of avian influenza. Avian influenza is typically monitored via a national surveillance programme (Aznar et al., 2023). While there are only seasonal outbreaks in some countries, outbreak risk remains high in others. Vaccination restricts the transmission rate of avian influenza in layer chickens, in addition to reducing numbers of dead and recovered animals following an influenza episode (Nielsen et al., 2024; Capua and Alexander, 2008). The main current mitigation strategy within low-input outdoor and organic broiler farms is based on confinement (Germain et al., 2024), though vaccination has been proposed as a national strategy in duck farms in France since 2023 (Guinat et al., 2025).

## 3.2 Outdoor access limits control of the environment: weather impacts and higher risk of predation

Climate change, including a general increase in global temperatures, is accompanied by more frequent heat waves. However, outdoor temperature cannot be controlled using ventilation or air conditioning as in conventional rearing in buildings. Heat stress may become more and more common for outdoor-reared chickens. Consequences of heat stress are numerous and compromise the health, welfare and performance of chickens (Akinyemi & Adewole, 2021). While mitigation strategies can contribute to limiting the effects of climate change, it is also possible to adapt chickens better to higher temperatures and heat waves. Adaptation strategies may include the use of feed additives (Bora et al., 2024; Fatima et al., 2024), incubation strategies (Meteyake et al., 2023; Loyau et al., 2014) or the use of better adapted strains such as local slow-growing crossbreeds (Menchetti et al., 2024) or naked-neck strains better able to lose heat (Meteyake et al., 2023).

The consequences of extreme or excessive rain extend from limiting range use to foot health problems. An increase in litter moisture from 25% to 31% is positively related to a 50% increase in the prevalence of foot pad dermatitis (Mou et al., 2025). However, as range use leads to less excretion in the poultry barn and consequently potentially drier litter, the prevalence of pododermatitis may be reduced. However, pododermatitis prevalence has been sometimes demonstrated as higher in organic production than in conventional broiler systems (Van De Weerd et al., 2009). A Barred Plymouth Rock crossed with a Ross 308 strain suffered from twice as much pododermatitis with outdoor access compared to indoor access only, though this was not the case with a Barred Plymouth Rock crossed twice with a Ross 308 (Sarica et al., 2014; Van De Weerd et al., 2009). Two reasons for this result may be: (1) the weight of different strains and (2) free range conditions. Pododermatitis is positively correlated to weight even in free-range rearing as higher incidence of foot pad dermatitis was observed in fast-growing strains compared to slow-growing strains in an organic system (Cartoni Mancinelli et al., 2020; Sarica et al., 2014). In slow-growing strains, no difference in the number of cases of pododermatitis between chickens raised in free range and indoor situations was observed even in rainy conditions (Bonnefous et al., 2024; Stadig et al., 2017a; Fanatico et al., 2008). However, in one study a higher incidence of pododermatitis was observed for groups using an outdoor range more compared to a group using the range less in winter (Dal Bosco et al., 2014). This indicates that chickens walking more in a muddy area may then increase the moisture percentage in litter indoors, hence the need for improved management of the poultry house.

Beyond weather, strain and free-range conditions, the outdoor area in an extensive system also exposes broilers to the risk of predation from the ground by foxes with 2% mortality due to foxes in free-range laying hens (Moberly et al., 2004) and from the air, notably by birds of prey, with up to 4% mortality in free-range laying hens (Bestman & Bikker-Ouwejan, 2020). To protect chickens from ground predation, electric fences and guard animals (such as dogs) with a separate range surrounding the outdoor range for chickens have been proven to be efficient and long-lasting measures (McKellar et al., 2023; Zingaro et al., 2018; Hennovation, 2017). Similarly, nets contribute to limiting predation by birds of prey, but those may be very difficult to install in a larger free-range area with trees and need a lot of maintenance. Birds of prey may be deterred thanks to acoustic or visual deterrents (e.g. scarecrows) but birds do become habituated to these devices. Two approaches have been developed to tackle predation by birds of prey. Co-grazing of cattle and chickens has helped to protect the chickens from goshawks in particular

(Hübner et al., 2024). A laser beam combined with a camera vision system and artificial intelligence for data analysis can drive away predators perched on fences (Chen et al., 2024; Elbers and Gonzales, 2021). However, the latter device is not able to detect and track predators in the air, which is a necessity for birds of prey such as goshawks.

## 4 Providing outdoor range access: welfare issues

### 4.1 Consequences for farmers and consumers

The livelihood of some farmers depends on rearing broilers under low-input extensive systems. Beyond chicken growth performance, farmer income also depends on consumer perception of meat quality. Socio-economic research has shown that better knowledge of the production system involved can promote greater consumption of produce from responsible farming systems (Harju et al., 2024). Range use has been positively and significantly correlated to levels of chicken activity (Collet et al., 2024; Cartoni Mancinelli et al., 2020). Enhanced movement in low-input and extensive systems, together with less access to energy-rich feed concentrates, may lead to lower live weight. However, results may depend on the daily weight gain of a particilar strain. While weight in fast-growing strains may predispose chickens to greater use of a range (Taylor et al., 2018), a negative trade-off between growth performances and range use has been observed after range access for strains, with an average daily weight gain between 20 g/d and 40 g/d (Bonnefous et al., 2024; Marchewka et al., 2020; Stadig et al., 2016). Range use had no significant impact on live weight of slow-growing strains with growth rates under 20 g/d (Bonnefous et al., 2024; Marchewka et al., 2020). Local strains even displayed greater body weight at slaughter under a free-range system compared to an indoor system (Menchetti et al., 2024). The negative growth performance consequences of rearing broilers under an outdoor extensive system may therefore be avoided by selecting an adapted local strain.

Chickens roaming in an extensive system forage and eat grass in particular (Ferreira et al., 2022; Dal Bosco et al., 2014). Omega-3 polyunsaturated fatty acids and carotenes were documented as being 7 to almost 200 times more concentrated in grass than in concentrate feed (Mattioli et al., 2022; Dal Bosco et al., 2016; Lopez-Bote et al., 1998). The consequences for meat are both nutritional and visual, with yellower meat from higher forage intake (Mattioli et al., 2022) and indirectly from higher range use (Bonnefous et al., 2024). As colour appears to be a key element in consumer choice (Altmann et al., 2023), marketing to explain the distinct colour of meat from low-input, extensive systems is key to farmers selling their products at a premium price.

### 4.2 Broiler welfare

The definition of animal welfare has expanded from just health parameters to encompass an animal's mental as well as physical needs (ANSES, 2018). Chickens under an extensive system allocate more of their time to active natural behaviours. This includes dustbathing, expressed five times more with outdoor provision compared to indoor rearing; foraging or locomotion, as chickens spent 2–20% more time moving when provided with outdoor access compared to chickens kept indoors (Zhao et al., 2014; Branciari et al., 2009; Sosnówka-Czajka et al., 2007). Broilers provided with an outdoor run also demonstrated significantly less negative behaviours such as feather pecking events compared to chickens reared indoors (Zhao et al., 2014; Bestman and Wagenaar, 2003). Broilers with low density (2.5 $m^2$ per bird) given outdoor access expressed less aggressive behaviour than broilers with a high density (0.5 $m^2$ per bird) with outdoor access (Phillips and Heins, 2021). The evidence suggests abnormal behaviours occur less in extensive outdoor systems than in intensive indoor systems.

However, in the event of confinement (e.g. as a biosecurity measure due to an avian influenza outbreak), organic and free-range broilers are reared under similar conditions to conventional indoor broilers. While we are not aware of any research on broiler behaviour during confinement, we can hypothesise that behaviour may be similar to chickens only reared indoors, or that negative behaviours such as feather pecking may even be worse among chickens used to outdoor access.

## 5 Impacts of outdoor rearing systems on the environment

While environmental conditions have a major impact on range use by chickens, chickens also have a significant impact on their environment.

As chickens do not have a sphincter, their excretion is distributed according to their location and state of their digestive tract. Outdoor excretion has been found to range between 2% and 44% of total excretion (Meda et al., 2015; Bonnefous et al., n.d.a). This substantial variation depends particularly on degree of use of the outdoor range. Environmental factors, such as increased rainfall and lower temperature in winter, may contribute to lower outdoor use and consequently to 50-100% lower outdoor excretion in winter compared to the spring season in French conditions (Meda et al., 2015). Additionally, the presence of trees seemed to influence outdoor excretion, especially in winter, with a significant disparity in outdoor excretion from 2% in a meadow to 20% in a pasture with trees (Meda et al., 2015). In the case of one outdoor range, outdoor excretion varied significantly with higher excretion and resulting emissions closer to the poultry house (Meda et al., 2012), where chickens tend to spend most of their time when outdoors (Collet et al., 2024). As the amount of nitrogen (N) and phosphorus (P) excreted close to the poultry house (Meda et al., 2015; Bonnefous et al., n.d.b) was substantially above the threshold set by the European Council (1991), improving outdoor use to better distribute excretion over the whole range is a challenge that needs to be tackled.

Chickens using an outdoor range the most are also chickens who forage the most (Ferreira et al., 2022; Mattioli et al., 2022; Dal Bosco et al., 2014). Under a global context of limited food resources, feeding animals with ingredients that could be used for human food is becoming a pressing societal issue. Providing chickens with roughage may provide a solution to feed-food competition and enable more self-sufficient and resilient farming systems. In one trial the percentage of grass intake was around 20% with a dual-purpose strain and a laying hen strain (Leiber et al., 2024, 2025) but was lower in a slow-growing broiler strain compared to a medium-growing strain (Bonnefous et al., n.d.b). However, providing roughage did not limit the consumption of commercial feed with medium-growing chickens, and the breast yield was unaffected by the type of roughage provided (low or high crude protein) (Leiber et al., 2025). Limiting the protein content in commercial feed and planting the range with high-protein content plants such as alfalfa could contribute to improving range use and solve the feed-food competition issue (Singh and Cowieson, 2013). Initial results have shown that protein consumption from range plants represented up to 9% of total protein consumption (Goujon et al., 2024). The growing performance of local strains and crossbreeds from local strains (measured by the Gompertz curve, body weight and feed intake) were less affected by diminished crude protein content in commercial feed and replacement of soybean with local ingredients compared to a fast-growing strain (Menchetti et al., 2024; Fanatico et al., 2008). Chickens are also omnivorous and eat larvae and other insects (Lomu et al., 2004). The benefits of rearing chickens under tree cover may extend to controlling pests in orchards. The potential of silvopastoral agroforestry systems using chickens needs further study to identify their benefits (Bosshardt et al., 2022).

## 6 Conclusion

Rearing low-input, extensive meat-type chickens brings unique challenges. Outdoor access increases the risk of diseases from wild birds, outdoor parasites and the risk of predation. Multiple solutions are available. However, these solutions have potential disadvantages in terms of cost, whether financial or in terms of the labour and time involved for the farmer, and may also affect animal welfare. The use of outdoor space has consequences for both the animal and the environment. A key element for successfully rearing broilers in low-input, extensive systems is to use a resilient local strain that will be able to cope with the local climate, be equipped with a better immune system, limit potential loss of growth due to range use or even enhance growth by fully exploiting outdoor resources. Selling an extensively-reared meat product at the right price is a further challenge (Göransson and Lundmark Hedman, 2024). Support from regulation and policies is an essential factor in the success of this type of chicken production system. Deployment of marketing campaign to raise awareness of these system among customers, as well as the introduction of financial support for farmers, may help to expand low-input extensive farming.

## 7 References

Akinyemi, F. and Adewole, D., 2021. Environmental stress in chickens and the potential effectiveness of dietary vitamin supplementation. *Frontiers in Animal Science*. https://doi.org/10.3389/fanim.2021.775311.

Altmann, B.A., et al., 2023. Consumer preferences for the color of unprocessed animal foods. *Journal of Food Science*. https://doi.org/10.1111/1750-3841.16485.

ANSES, 2018. Avis de l'Agence nationale de sécurité sanitaire de l'alimentation, de l'environnement et du travail relatif au « *Bien-être animal: contexte, définition et évaluation* » 33, 1–34.

Aznar, I., et al., 2023. Annual report on surveillance for avian influenza in poultry and wild birds in Member States of the European Union in 2022. *EFSA Journal* 21. https://doi.org/10.2903/j.efsa.2023.8480.

Bestman, M. and Bikker-Ouwejan, J., 2020. Predation in organic and free-range egg production. *Animals* 10. https://doi.org/10.3390/ani10020177.

Bestman, M. and Wagenaar, J.P., 2003. Farm level factors associated with feather pecking in organic laying hens. *Livestock Production Science* 80, 133–140. https://doi.org/10.1016/S0301-6226(02)00314-7.

Biosuisse, 2025. Cahier des charges.

Bonnefous, C., et al., 2023. Behavioural indicators of range use in four broiler strains. *Applied Animal Behaviour Science* 260. https://doi.org/10.1016/j.applanim.2023.105870.

Bonnefous, C., et al., 2024. Performance, meat quality and blood parameters in four strains of organic broilers differ according to range use. *Scientific Reports* 14, 30854. https://doi.org/10.1038/s41598-024-81672-9.

Bonnefous, C., et al., 2025. Limiting ascarid infection of organic laying hens with vinegar: an on-farm trial. In *76th Annual Meeting of the European Federation of Animal Science*, p. 274.

Bonnefous, C., et al., n.d.a. Research note: Range use associates with grass intake but not outdoor excretion in free-range organic broilers. In preparation.

Bonnefous, C., et al., n.d.b. Individual outdoor use in free-range broilers: from associations with behaviours to physiological predictors. Submitted.

Bora, S., et al., 2024. Effect of Vitamin E and selenium supplementation on growth performance and heat shock protein 70 levels in broiler chickens exposed to summer heat stress. *Indian Journal of Animal Research*. https://doi.org/10.18805/ijar.b-5261.

Bosshardt, S., et al., 2022. Changing perspectives on chicken-pastured orchards for action: a review based on a heuristic model. *Agricultural Systems* 196, 103335. https://doi.org/10.1016/j.agsy.2021.103335.

Branciari, R., et al., 2009. Effect of genotype and rearing system on chicken behavior and muscle fiber characteristics. *Journal of Animal Science* 87, 4109–4117. https://doi.org/10.2527/jas.2009-209.

Cain, J.L., et al., 2024. An automated faecal egg count system for detection of Ascaridia galli ova in chickens. *Journal of Helminthology* 98. https://doi.org/10.1017/S0022149X24000373.

Capua, I. and Alexander, D.J., 2008. Avian influenza vaccines and vaccination in birds. *Vaccine* 26. https://doi.org/10.1016/j.vaccine.2008.07.044.

Cartoni Mancinelli, A., et al., 2020. Performance, behavior, and welfare status of six different organically reared poultry genotypes. *Animals* 10, 1–12. https://doi.org/10.3390/ani10040550.

Castellini, C., et al., 2016. Adaptation to organic rearing system of eight different chicken genotypes: Behaviour, welfare and performance. *Italian Journal of Animal Science* 15, 37–46. https://doi.org/10.1080/1828051X.2015.1131893.

Chen, Y.C., et al., 2024. Automatic wild bird repellent system that is based on deep-learning-based wild bird detection and integrated with a laser rotation mechanism. *Scientific Reports* 14. https://doi.org/10.1038/s41598-024-66920-2.

Collet, J.M., et al., 2024. High-throughput phenotyping to characterise range use behaviour in broiler chickens. *Animal* 18. https://doi.org/10.1016/j.animal.2024.101099.

Council of the European Union, European Parliament, 2018. Regulation (EU) 2018/848 of the European Parliament and of the Council of 30 May 2018 on organic production and labelling of organic products and repealing Council Regulaiton (EC) No 834/2007. *Official Journal of the European Union* 2018, 1–92.

Dal Bosco, A., et al., 2016. Transfer of bioactive compounds from pasture to meat in organic free-range chickens. *Poultry Science* 95, 2464–2471. https://doi.org/10.3382/ps/pev383.

Dal Bosco, A., et al., 2014. Effect of range enrichment on performance, behavior, and forage intake of free-range chickens. *Journal of Applied Poultry Research* 23, 137–145. https://doi.org/10.3382/japr.2013-00814.

Daş, G., et al., 2011. Precision, repeatability and representative ability of faecal egg counts in Heterakis gallinarum infected chickens. *Veterinary Parasitology* 183, 87–94. https://doi.org/10.1016/j.vetpar.2011.07.005.

Dawkins, M.S., et al., 2003. What makes free-range broiler chickens range? In situ measurement of habitat preference. *Animal Behaviour* 66, 151–160. https://doi.org/10.1006/anbe.2003.2172.

Elbers, A.R.W. and Gonzales, J.L., 2021. Efficacy of an automated laser for reducing wild bird visits to the free range area of a poultry farm. *Scientific Reports* 11, 1–11. https://doi.org/10.1038/s41598-021-92267-z.

European Council, 1991. Directive 91/676/EEC of 12 December 1991–concerning the protection of waters against pollution caused by nitrates from agricultural sources.

Fanatico, A.C., et al., 2008. Performance, livability, and carcass yield of slow- and fast-growing chicken genotypes fed low-nutrient or standard diets and raised indoors or with outdoor access. *Poultry Science* 87, 1012–1021. https://doi.org/10.3382/ps.2006-00424.

Fatima, N., et al., 2024. Role of feed additives in mitigating the impact of heat stress on poultry gut health and productivity. *World's Poultry Science Journal*. https://doi.org/10.1080/00439339.2024.2410877.

Ferreira, V.H.B., et al., 2019. Relationship between ranging behavior and spatial memory of free-range chickens. *Behavioural Processes* 166, 1–8. https://doi.org/10.1016/j.beproc.2019.103888.

Ferreira, V.H.B., et al., 2020a. Social motivation and the use of distal, but not local, featural cues are related to ranging behavior in free-range chickens (Gallus gallus domesticus). *Animal Cognition* 23, 769–780. https://doi.org/10.1007/s10071-020-01389-w.

Ferreira, V.H.B., et al., 2020b. Range use is related to free-range broiler chickens' behavioral responses during food and social conditioned place preference tests. *Applied Animal Behaviour Science* 230, 1–8. https://doi.org/10.1016/j.applanim.2020.105083.

Ferreira, V.H.B., et al., 2020c. Uninhibited chickens: Ranging behaviour impacts motor self-regulation in free-range broiler chickens (Gallus gallus domesticus). *Biology Letters* 16, 1–5. https://doi.org/10.1098/rsbl.2019.0721.

Ferreira, V.H.B., et al., 2021. Working for food is related to range use in free-range broiler chickens. *Scientific Reports* 11, 6253. https://doi.org/10.1038/s41598-021-85867-2.

Ferreira, V.H.B., et al., 2022. Foraging behavior shows individual-consistency over time, and predicts range use in slow-growing free-range male broiler chickens. *Frontiers in Veterinary Science* 9, 1–12. https://doi.org/10.3389/fvets.2022.814054.

Germain, K., et al., 2024. L'élevage de poulets biologiques: les verrous à lever et les opportunités offertes par cette production. *INRAE Productions Animales* 37, 8237. https://doi.org/10.20870/productions-animales.2024.37.2.8237.

Gómez, Y., et al., 2022. Similarity in temporal movement patterns in laying hens increases with time and social association. *Animals* 12, 1–18. https://doi.org/10.3390/ani12050555.

Göransson, L. and Lundmark Hedman, F., 2024. The perks of being an organic chicken – animal welfare science on the key features of organic poultry production. *Frontiers in Animal Science* 5, 1–30. https://doi.org/10.3389/fanim.2024.1400384.

Goujon, M., et al., 2024. Le parcours, une obligation réglementaire ou un atout pour l'autonomie alimentaire des volailles biologiques? *Fourrages* 259, 79. https://doi.org/10.64256/FOU2426.

Guinat, C., et al., 2025. Promising effects of duck vaccination against highly pathogenic avian influenza, France 2023–2024. *Emerging Infectious Diseases* 31, 1468–1471. https://doi.org/10.3201/eid3107.241445.

Harju, C., et al., 2024. Consumers' views on egg quality and preferences for responsible production – results from nine European countries. *British Food Journal* 126, 623–644. https://doi.org/10.1108/BFJ-03-2023-0183.

Hennovation, 2017. *Recommendation for using alpacas as guardians of free range hens*. Hennovation.

Hübner, S., et al., 2024. Mix it–cograzing with cattle reduces broiler losses and increases broiler range use. *Poultry Science* 103, 103906. https://doi.org/10.1016/j.psj.2024.103906.

INAO, 2009. Cahier des charges Label Rouge - LA 17/87 - Poulet blanc fermier-élevé en plein air entier et découpes, frais ou surgelés 1–16.

Leiber, F., et al., 2024. Hat der Raufutteranteil eine besondere Bedeutung für Zweinutzungshennen? *Agrarforschung Schweiz* 15, 89–97. https://doi.org/10.34776/afs15-89.

Leiber, F., et al., 2025. Effects of dietary forage on feed efficiency of poultry from a slow-growing and a dual-purpose strain for organic fattening systems. *Biological Agriculture & Horticulture* 1–16. https://doi.org/10.1080/01448765.2025.2483212.

Lolli, S., et al., 2019. Effect of range use on endo- and ectoparasite infestation in Italian organic egg production. *Italian Journal of Animal Science* 18, 690–695. https://doi.org/10.1080/1828051X.2018.1564377.

Lomu, M.A., et al., 2004. Metabolizable energy of crop contents in free-range hens. *International Journal of Poultry Science* 3, 728–732. https://doi.org/10.3923/ijps.2004.728.732.

Lopez-Bote, C.J., et al., 1998. Effect of free-range feeding on n–3 fatty acid and α-tocopherol content and oxidative stability of eggs. *Animal Feed Science and Technology* 72, 33–40. https://doi.org/10.1016/S0377-8401(97)00180-6.

Loyau, T., et al., 2014. Cyclic variations in incubation conditions induce adaptive responses to later heat exposure in chickens: a review. *Animal* 9, 76–85. https://doi.org/10.1017/S1751731114001931.

Marchewka, J., et al., 2020. Associations between welfare and ranging profile in free-range commercial and heritage meat-purpose chickens (Gallus gallus domesticus). *Poultry Science* 99, 4141–4152. https://doi.org/10.1016/j.psj.2020.05.044.

Mattioli, S., et al., 2022. Intake of nutrients (polyunsaturated fatty acids, tocols, and carotenes) and storage efficiency in different slow-growing chickens genotypes reared in extensive systems. *PLoS One* 17(11). https://doi.org/10.1371/journal.pone.0275527.

Mattioli, S., et al., 2021. How the kinetic behavior of organic chickens affects productive performance and blood and meat oxidative status: a study of six poultry genotypes. *Poultry Science* 100, 101297. https://doi.org/10.1016/j.psj.2021.101297.

Maurer, V., et al., 2013. Effects of paddock management on vegetation, nutrient accumulation, and internal parasites in laying hens. *Journal of Applied Poultry Research* 22, 334–343. https://doi.org/10.3382/japr.2012-00586.

McKellar, R.A., et al., 2023. Chicken's best friend- Livestock guardian dog bonding with free-ranging chickens. *Translational Animal Science* 7, 1–12. https://doi.org/10.1093/tas/txad014.

Meda, B., et al., 2012. Greenhouse gas emissions from the grassy outdoor run of organic broilers. *Biogeosciences* 9, 1493–1508. https://doi.org/10.5194/bg-9-1493-2012.

Meda, B., et al., 2015. Influence of season and outdoor run characteristics on excretion behaviour of organic broilers and gaseous emissions. *Biosystems Engineering* 139, 35–47. https://doi.org/10.1016/j.biosystemseng.2015.08.001.

Menchetti, L., et al., 2024. Effect of genotype and nutritional and environmental challenges on growth curve dynamics of broiler chickens. *Poultry Science* 103, 104095. https://doi.org/10.1016/j.psj.2024.104095.

Meteyake, H.T., et al., 2023. Naked neck gene and intermittent thermal manipulations during embryogenesis improve posthatch performance and thermotolerance in slow-growing chickens under tropical climates. *Poultry Science* 102, 102912. https://doi.org/10.1016/j.psj.2023.102912.

Meyer, M.M., et al., 2021. Laser enrichment device stimulates broiler laser-following behavior while increasing individual bird locomotion and pen-wide movement. *Frontiers in Animal Science* 2, 784408. https://doi.org/10.3389/fanim.2021.784408.

Moberly, R.L., et al., 2004. Mortality due to fox predation in free-range poultry flocks in Britain. *Veterinary Record* 155, 48–52. https://doi.org/10.1136/vr.155.2.48.

Mou, C.T., et al., 2025. Evaluation of the effects of maintaining a moderate humidity (50–60%) and increased air movement on litter moisture and footpad health in a commercial broiler house. *Journal of Applied Poultry Research* 34, 100476. https://doi.org/10.1016/j.japr.2024.100476.

Nichelmann, M. and Tzschentke, B., 2002. Ontogeny of thermoregulation in precocial birds. *Comparative Biochemistry and Physiology Part A: Molecular & Integrative Physiology*.

Nielsen, S.S., et al., 2024. Vaccination of poultry against highly pathogenic avian influenza – Part 2. Surveillance and mitigation measures. *EFSA Journal* 22, 1–70. https://doi.org/10.2903/j.efsa.2024.8755.

Parr, J.F., et al., 1990. Sustainable agriculture in the United States. In *Sustainable agricultural systems*, pp. 50–67.

Phillips, H.N. and Heins, B.J., 2021. Effects of outdoor stocking density on growth, feather damage and behavior of slow-growing free-range broilers. *Animals* 11, 1–16. https://doi.org/10.3390/ani11030688.

Poulopoulou, I., et al., 2023. In vitro evaluation of the effects of methanolic plant extracts on the embryonation rate of Ascaridia galli eggs. *Veterinary Research Communications* 47, 409–419. https://doi.org/10.1007/s11259-022-09958-9.

Réale, D., et al., 2010. Evolutionary and ecological approaches to the study of personality. *Philosophical Transactions of the Royal Society B: Biological Sciences* 365, 3937–3946. https://doi.org/10.1098/rstb.2010.0222.

Rodenbücher, A.L., et al., 2023. Pumpkin seeds, lemongrass essential oil and ripleaf leaves as feed additives for Ascaridia galli infected laying hens. *Veterinary Research Communications* 47, 817–832. https://doi.org/10.1007/s11259-022-10042-5.

Rodriguez-Aurrekoetxea, A., et al., 2014. Environmental complexity and use of space in slow growing free range chickens. *Applied Animal Behaviour Science* 161, 86–94. https://doi.org/10.1016/j.applanim.2014.09.014.

Sarica, M., et al., 2014. Einfluss des produktionssystems auf das vorkommen von Fußballen-Dermatiden (FPD) bei langsam, mittelschnell und schnell wachsenden broilern. *European Poultry Science* 78, 1–10. https://doi.org/10.1399/eps.2014.52.

Singh, M. and Cowieson, A.J., 2013. Range use and pasture consumption in free-range poultry production. *Animal Production Science* 53, 1202–1208. https://doi.org/10.1071/AN13199.

Sosnówka-Czajka, E., et al., 2007. Effect of management system and flock size on the behaviour of broiler chickens. *Annals of Animal Science* 7, 329–335.

Sosnówka-Czajka, E., et al., 2023. Performance and meat quality of broiler chickens fed with the addition of dried fruit pomace. *Poultry Science* 102, 1–9. https://doi.org/10.1016/j.psj.2023.102631.

Souillard, R., et al., 2019. Husbandry practices, health, and welfare status of organic broilers in France. *Animals* 9, 1–12. https://doi.org/10.3390/ani9030097.

Stadig, L.M., et al., 2016. Effects of free-range access on production parameters and meat quality, composition and taste in slow-growing broiler chickens. *Poultry Science* 95, 2971–2978. https://doi.org/10.3382/ps/pew226.

Stadig, L.M., et al., 2017a. Effects of shelter type, early environmental enrichment and weather conditions on free-range behaviour of slow-growing broiler chickens. *Animal* 11, 1046–1053. https://doi.org/10.1017/S1751731116002172.

Stadig, L.M., et al., 2017b. Effect of free-range access, shelter type and weather conditions on free-range use and welfare of slow-growing broiler chickens. *Applied Animal Behaviour Science* 192, 15–23. https://doi.org/10.1016/j.applanim.2016.11.008.

Stehr, M., et al., 2019. Resistance and tolerance to mixed nematode infections in chicken genotypes with extremely different growth rates. *International Journal for Parasitology* 49, 579–591. https://doi.org/10.1016/j.ijpara.2019.03.001.

Taylor, P.S., et al., 2017a. Ranging behaviour of commercial free-range broiler chickens 1: factors related to flock variability. *Animals* 7, 1–14. https://doi.org/10.3390/ani7070054.

Taylor, P.S., et al., 2017b. Ranging behaviour of commercial free-range broiler chickens 2: individual variation. *Animals* 7, 1–9. https://doi.org/10.3390/ani7070055.

Taylor, P.S., et al., 2018. Ranging behavior relates to welfare indicators pre- and post-range access in commercial free-range broilers. *Poultry Science* 97, 1861–1871. https://doi.org/10.3382/ps/pey060.

Taylor, P.S., et al., 2023. Visual access to an outdoor range early in life, but not environmental complexity, increases meat chicken ranging behavior. *Poultry Science* 102, 1–27. https://doi.org/10.1016/j.psj.2023.103079.

Van De Weerd, H.A., et al., 2009. A review of key health-related welfare issues in organic poultry production. *World's Poultry Science Journal* 65, 649–684. https://doi.org/10.1017/S0043933909000464.

Xu, H., et al., 2025. Effects of tannic acid on growth performance, intestinal health, and tolerance in broiler chickens. *Poultry Science* 104. https://doi.org/10.1016/j.psj.2024.104676.

Zhao, Z.G., et al., 2014. Effects of housing systems on behaviour, performance and welfare of fast-growing broilers. *Asian-Australasian Journal of Animal Sciences* 27, 140–146. https://doi.org/10.5713/ajas.2013.13167.

Zingaro, M., et al., 2018. Are the livestock guarding dogs where they are supposed to be? *Applied Animal Behaviour Science* 198, 89–94. https://doi.org/10.1016/j.applanim.2017.10.002.

# Chapter 15

## Heat stress in broiler production

*Victoria Anthony Uyanga, Lincoln University of Missouri, USA; Semiu Folaniyi Bello, Agriculture and Agri-Food Canada, Canada; and Folasade Olukemi Oke, Okanlawon Mohammed Onagbesan and Oyegunle Emmanuel Oke, Federal University of Agriculture, Nigeria*

1 Introduction
2 Understanding heat stress in broiler production
3 Metabolic and physiological changes during heat stress
4 Genetic basis of heat resistance in broilers
5 Impacts of heat stress on broiler's growth and productivity
6 Impacts of heat stress on broiler's behavior and welfare
7 Conclusion and future trends
8 Where to look for further information
9 References

### 1 Introduction

Globally, the poultry sector has rapidly grown from 9 to 133 million tons between 1961 and 2020, owing to increasing human population, industrialization, and the drive for urbanization in various world regions. As of 2020, poultry meat contributed up to 40% of global meat production, and the largest producers are the United States of America (supplying 17% of global output), followed by China and Brazil (FAO, 2024). The U.S. poultry industry is significant enough that total production in 2023 was $67.1 billion. Of this, broiler production accounted for $42.6 billion (64%) with 59.7 billion pounds produced as live-weight broilers (USDA-NASS, 2024).

Over the past few decades, intensive genetic selection and nutritional advancements have led to broiler strains that grow four to five times faster and reach market weight in less than half the time compared to broilers from the 1950s (Havenstein et al., 2003; Sarica et al., 2014). For example, while broilers in the 1950s required around 70 days to reach 1.5 kg, modern strains can achieve over 2.5 kg in just 35–42 days. The birds now have a higher metabolic rate and feed efficiency (Yahav, 2009). Broilers are homoeothermic animals that can only

http://dx.doi.org/10.19103/AS.2025.0165.18

maintain their body temperature efficiently when ambient temperatures are within certain limits (Bell and Weaver, 2002; Lourençoni *et al.*, 2019). Decades of selective breeding programs have shown that commercial broilers produce more body heat, yet have an impaired capacity to dissipate it, making them highly susceptible to high environmental temperatures (Lara and Rostagno, 2013; Tickle et al., 2018).

With the incidence of climate change, global temperature is rising, along with increasing variability in precipitation, causing climatic extremes to occur more frequently. The Earth's mean surface temperature increased by 1.09°C between 2011 and 2020 compared to the periods between 1850 and 1900 (IPCC, 2021). Environmental temperature, precipitation patterns, and atmospheric $CO_2$ are all shifting due to global climate change. These variabilities in ambient temperatures, unpredictable rainfall, long-lasting droughts, etc., upset the delicate ecosystemic balance, adversely affecting animal and food production, thus posing a major threat to food security (Sundström et al. 2014). Due to increasing changes in environmental temperatures and varied ecosystems, the tropics are especially vulnerable to the consequences of climate change (Oke et al., 2021a, 2021b, 2022). Importantly, achieving ideal broiler performance is hindered by climate change since it upsets the delicate equilibrium of the thermoneutral environment required to attain the genetic potential of broilers (Oke et al., 2024a).

Several noteworthy reviews have been published on the impacts of heat stress on poultry production (Saeed et al., 2019; Abd El-Hack et al., 2020; Abdel-Moneim et al., 2021; Kpomasse et al., 2021; Oke et al., 2024). A major obstacle to broiler production is heat stress, which has been aggravated by rising global temperatures occasioned by climate change, deteriorating the health and productivity of poultry, particularly in tropical and subtropical climates. Due to their limited sweat glands, feather cover, rapid growth, and high metabolic rates, broilers are especially susceptible to heat stress, which can seriously impair their performance and well-being (Zhang et al., 2017a; Oke et al., 2017; Kpomasse et al., 2023). Heat stress is implicated in several physiological dysfunctions, including respiratory alkalosis, neuroendocrine disorders, metabolic conditions, immune dysregulation, and electrolyte imbalances in chickens (Lara and Rostagno, 2013; Uyanga et al., 2022a).

Studies have revealed that the average daily feed intake is typically decreased by 15–30 g/bird/day under chronic heat stress conditions (32–35°C), resulting in a 10–20% decrease compared to birds housed under thermoneutral conditions (~22–24°C) (Lara and Rostagno, 2013; Sohail et al., 2012). According to Sohail et al. (2012), after 42 days of age, broilers exposed to prolonged heat stress had a greater feed conversion ratio (+25.6%), lower body weight (−32.6%), and decreased feed intake (−16.4%). Furthermore, broilers subjected to a cyclic temperature of 35°C (8 h per day) for 7 days had lower body weight, average daily gain, and average feed intake than those raised in thermoneutral conditions (23°C) (Shao et al., 2019). The authors found that body weight gain decreased, ranging from Δ = 150 to 250 g/bird, which was directly attributed to reduced feed intake. Similar to this, heat stress caused the feed conversion ratio (FCR) to decline, increasing by Δ = 0.10 to 0.25 points compared to controls, suggesting less effective nutrient utilization (Quinteiro-Filho et al., 2010). Overall, the physiology, health, welfare, and productivity of broilers are adversely impacted by heat stress, resulting in significant financial losses for the poultry industry. This necessitates a comprehensive and in-depth understanding of the thermotolerance of broilers, the predisposing factors to thermal insults, the impacts of heat stress on broilers, and efficient ways to mitigate heat stress in broiler production. This chapter explores the physiological, metabolic, and performance-related impacts of heat stress on broilers.

## 2 Understanding heat stress in broiler production

Heat stress occurs when there is an imbalance in heat production and heat loss between the animal and its environment. Heat is generated during physical activity, muscular contractions, and metabolic processes. With increasing environmental temperature, the core body temperature and heat production from metabolic processes are elevated when the upper critical temperature is exceeded. Heat can be lost through water evaporation during panting, which aids in shedding surplus heat from the respiratory tract of chickens, and through conduction, convection, and radiation to expel excess body heat. Given that broilers have a limited capacity for heat dissipation, the core body temperature gradually increases beyond the zone of thermal comfort, causing the animal to succumb to hyperthermia and heat stress (Gonzalez-Rivas et al., 2020). When the amount of heat lost is not proportional to the heat generated, the thermoregulatory equilibrium is disturbed, resulting in the onset of heat stress. Heat stress can be categorized as acute, moderate, or chronic depending on a combination of factors, including environmental conditions (ambient temperature, humidity, ventilation), individual conditions (breed, genetics, metabolic rate, age, thermoregulatory ability), and stress conditions (intensity and duration). An improved understanding of heat stress in the broiler industry will enhance health, welfare, and productivity and reduce economic losses, thereby improving the available protein supply to meet the growing human population and rising standard of living.

### *2.1 Susceptibility of broilers to heat stress*

Heat stress remains a global challenge for the poultry industry, since chickens are required to maintain a constant body temperature through internal homeostatic mechanisms, ranging from 40.0 to 42.2°C. As endothermic homeotherms, chickens lack sweat glands, and their body surface is primarily insulated with feathers. These chickens are susceptible to heat stress since they largely depend on evaporative heat losses for thermoregulation, and thus cannot efficiently dissipate body heat during exposure to high environmental temperatures (Uyanga et al., 2022a).

Comparative studies on breed's susceptibility to heat stress revealed variations in physiological traits between the genetically distinct lines of Leghorn and Fayoumi chickens (Wang et al., 2018). Fayoumi chickens are native to Egypt (largely sub-tropical climate) and have acquired heat tolerance following years of inbreeding and natural selection. The leghorn chickens have been selectively bred to maximize their feed efficiency and productivity, resulting in varying thermotolerance. Under heat stress exposure (35–38°C; 50% humidity), both breeds respond differently, with the leghorn chickens exhibiting an acute response characterized by electrolyte imbalance and metabolic alkalosis. In contrast, Fayoumi chickens maintained a relatively balanced electrolyte status, demonstrating their heat resilience and ability to thrive under high environmental temperatures. In another study with two broiler breeds, normal-growth White Rocks and dwarf-growth White Rock with varying feathering speeds were exposed to high environmental temperature (32 ± 2.5°C) for 8 weeks (Amad et al., 1992). The dwarf broilers managed under high temperatures had lower live weights than normal broilers at the eighth week. More so, dwarfed broilers and animals under normal temperatures possessed a smaller slaughtering proportion than normal-growth broilers under a higher temperature.

Based on the above findings, the genetic disposition contributes significantly to the broiler's ability and capacity for heat stress response. Exposure to heat stress can lead to the development of heat-tolerant traits in broilers, which may enhance long-term productivity of chickens. This adaptation may enhance their ability to cope with high temperatures in the future, potentially leading to improved survival rates in hot environments. Interestingly, these challenges posed by heat stress have prompted research into genetic improvement and selection for heat-resistant breeds (Juiputta et al., 2023).

## 2.2 Factors contributing to heat stress in broiler production

High environmental temperatures in combination with high humidity impose detrimental effects on chickens. Other environmental factors that may influence the vulnerability of chickens under high temperatures are solar radiation, wind velocity, and previous exposure to heat stress and acclimatization of the chickens. The temperature and latency of heat stress exposure are important factors when considering the impact of heat stress. At different ages, broilers perform optimally within their zone of thermal comfort (thermoneutral temperature) ranging from 34°C to 32°C (week 1), 32–28°C (week 2), 28–26°C (week 3), 26.24°C (week 4), and 18–24°C (week 5 to 6) (Cassuce et al., 2013). From a series of studies, chronic heat stress ranges from 28°C to 38°C for more than 7 days (typically 1–3 weeks); acute heat stress ranges from 30°C to 38°C for 4–24 h, while moderate heat stress lasts less than 7 days (Liu et al., 2020). Chickens exposed to a constant (24 h per day) high temperature of 30–35°C would suffer greater deleterious effects than those encountering cyclic or alternating temperatures (typically 6–12 h per day) of 30–35°C during daylight and 18–25°C by nighttime (Daghir, 2008). Several factors affect body temperature regulation in chickens, and their predisposition to heat stress is discussed below.

### 2.2.1 Age of broilers

Broilers at different ages are bound to respond differentially to thermal insults. Comparing the impacts of acute heat exposure of 32°C, 75% RH for 2 h on 35 and 63-day-old broilers revealed that the extent of the hyperthermia and alterations to acid/base balance and muscle myopathy were more severe in older broilers than younger ones (Sandercock et al., 2001). Given the close association between heat stress and oxidative stress, the age-differential responses of broilers to heat stress may be related to their responses to oxidative stress. Younger broilers exhibit better antioxidant responses due to increased expression/activity of antioxidant enzymes, which help combat the production of ROS during heat stress (Del Vesco et al., 2017). In addition, subjecting young broilers to early-age heat exposure would effectively condition their thermoregulatory system to adapt to subsequent heat stress, improving their thermotolerance (Liu et al., 2024). The age of broilers will greatly contribute to their ability to adapt to heat stress, and younger broilers resist better due to high expression of antioxidant enzymes.

### 2.2.2 Sex

Male and female broilers perform differently in their growth performance, carcass weights, and nutrient requirements (England et al., 2023). Male broilers are more profitable since they weigh more than the females at market age. Hence, their increased growth rate is associated with higher metabolic demand and increased metabolic heat production (Dozier et al., 2011), consequently causing them to generate a higher body temperature than females. The effects of heat stress are primarily observed in male broilers because of their faster growth rate than females.

### 2.2.3 Physical activity

To improve animal welfare, perches are utilized to increase movement and walking ability and reduce leg weakness in chickens. Chickens engaged in increased mobility or activity would tend to have a higher body temperature due to increased metabolic rates and muscular activity (Saeed et al., 2019). As an environmental enrichment during hot summer conditions, cooled perches are introduced in rearing houses to mitigate heat stress and decrease stress levels in broilers (Karaarslan et al., 2024).

### 2.2.4 Body size and weight

There exists a negative relationship between body size and heat tolerance in chickens (Nawaz et al., 2023). As such, chickens with heavier body weight would tend to be more vulnerable to heat stress. Commercial broilers are more susceptible to heat stress than indigenous chickens due to their heavier size than indigenous chickens (Soleimani et al., 2011).

### 2.2.5 Breed

Generally, indigenous and slow-growing breeds of chickens that are native to the tropical regions are hardy and can withstand the effects of higher environmental temperatures better than the fast-growing strains of commercial broilers. Breed comparison between red jungle fowl, village fowl, and commercial broilers revealed that the domestication process and selective breeding for phenotypic traits have altered the physiological traits of chickens such as the body temperature, heterophil: lymphocyte ratio, and plasma corticosterone concentration, producing chickens that are more susceptible to heat stress (Soleimani et al., 2011). The ability of broilers to withstand heat stress greatly depends on their genetic makeup and predisposition. The impacts of heat stress conditions on broiler genetics are discussed in Section 3.

### 2.2.6 Stocking density

High stocking density and high temperature create a poor microenvironment that allows for the accumulation of noxious gases, promotes oxidative stress, and impairs the growth performance and productivity of chickens. Stocking broilers at a higher density of 26 chickens/m$^2$ under high temperatures (33°C, 24 h for 7 days) decreased the body weight and increased rectal temperature, respiratory rate, and corticosterone levels relative to broilers housed under lower stocking densities (16 and 18 birds/m$^2$) (Song et al., 2013). Additionally, the blood superoxide dismutase activity was decreased, whereas liver malondialdehyde was increased, suggesting the incidence of oxidative stress (Song et al., 2013).

Altogether, these factors can directly and/or indirectly influence the predisposition, adaptability, resilience, and responsiveness of broilers to heat stress. Importantly, the physiological responses to heat stress can be dynamic, changing over time and depending on the duration and severity of the stress.

## *2.3 Economic implications of heat stress to the poultry industry*

Although information on the economic impacts of climate change on heat stress in poultry, particularly in the tropics and subtropics, is still scarce, climate change is known to have a far-reaching impact on animal production through its effects on crop yield and productivity, which makes up a crucial input for animal feed. Indirect effects of climate change on livestock productivity include lower-quality feed output and outright feed shortages (Vandana et al., 2021). As such, both crops and livestock subsectors have suffered the highest losses (both around 50%) from weather anomalies and extreme events relative to other agricultural sectors between the periods of 2007 and 2022 (FAO, 2023). Data from the National Oceanic and Atmospheric Administration–National Centers for Environmental Information (NOAA-NCEI) provides that from 1980 to 2024, there have been 403 confirmed weather/climate disaster events with losses exceeding $1 billion each affecting the United States. These events have included drought, floods, freeze, severe storms, tropical cyclones, wildfires, and winters (NOAA-NCEI, 2025). Hatfield et al. (2014) reported that the economic cost of animal production losses due to high temperatures in the United States was higher than $1 billion in 2011.

High environmental temperatures cause the U.S. livestock production business to suffer severe losses of $1.69 to $2.36 billion, of which the poultry industry accounts for $128 to $165 million, with broiler production contributing $51.8 million in annual costs (St-Pierre et al., 2003). Liverpool-Tasie et al. (2019) documented that broilers are the most susceptible to rising environmental temperatures among the breeds of chicken, since they are raised primarily for meat production and have a high feed-to-meat conversion rate. Furthermore, the proportion of losses from heat stress in different livestock sectors has been reported as 16–22% in beef, 13–18% in swine, and 7–8% in poultry (Key et al., 2014). These economic consequences emphasize how important it is to manage heat stress to improve the long-term health and profitability of broiler production. They are projected to increase even more in the upcoming years as global temperatures rise, impacting the productivity, efficiency, and profitability of the poultry industry, and the agricultural sector as a whole.

## 3 Metabolic and physiological changes during heat stress

Heat stress contributes to a cascade of metabolic and physiological disturbances in the body, as extensively revealed by studies that have reviewed the impacts of heat stress on poultry physiology (Lara and Rostagno, 2013; Wasti et al., 2020; Uyanga et al., 2022a). Some of these physiological changes triggered by heat stress may enhance certain metabolic pathways and neuroendocrine pathways to support the energy and thermoregulatory needs of chickens during periods of high metabolic demand. Importantly, the changes in the levels of blood metabolites and hormones in the body may serve as indicators for evaluating the well-being and health status of broilers. Broilers are homeotherms, and exposure to high environmental temperatures overcomes the thermoregulatory capacity of these birds, creating a plethora of events as summarized in Fig. 1.

Detailed discussion into these events are explained in the following paragraphs.

### *3.1 Changes in body temperature*

Determining the body temperature is an important index when examining heat stress in broilers. This may be performed as measurements of the cloacal, core body, or surface temperature. The cloacal temperature can be examined by inserting the probe of the cloacal thermometer 2–3 cm into the vent and reading the resulting outcomes. These are easy to insert and use, but may need secure handling due to the chicken's movement (Brown-Brandl et al., 2003). With the use of electronic data loggers, continuous measurements of body temperature can be taken from broilers. This involves a minor surgical protocol to place the transmitter inside the peritoneal cavity of chickens, and corresponding readings for core body temperature are obtained from the datalogger (Uyanga et al., 2021b). With recent advances in technology, infrared thermal imaging can be utilized to detect the surface temperature of chickens (Uyanga et al., 2022b). Given the close relationship between skin surface temperature and core body temperature in broilers ($R^2$ values of 0.75 to 0.89), this provides a non-invasive and non-contact method for examining body temperature in animals and thus improving the diagnosis of environmental stress (Giloh et al., 2012). The main areas of heat loss in chickens are the head (comb and barbs), the shank, and to a lesser extent, the wings (Giloh et al., 2012). During the rearing period of 1 to 6 weeks, Moghbeli et al. (2018) identified that the warmest areas were the broiler's eyes and underwings, probably due to the lack of feather coverage in these areas, which could serve as insulation. Loyau et al. (2016) also demonstrated that the feathering score for the belly and back areas was significantly correlated with the shank (−0.73 and −0.48) and comb (−0.44 and −0.58) temperatures of 28-week-old hens housed at thermoneutrality

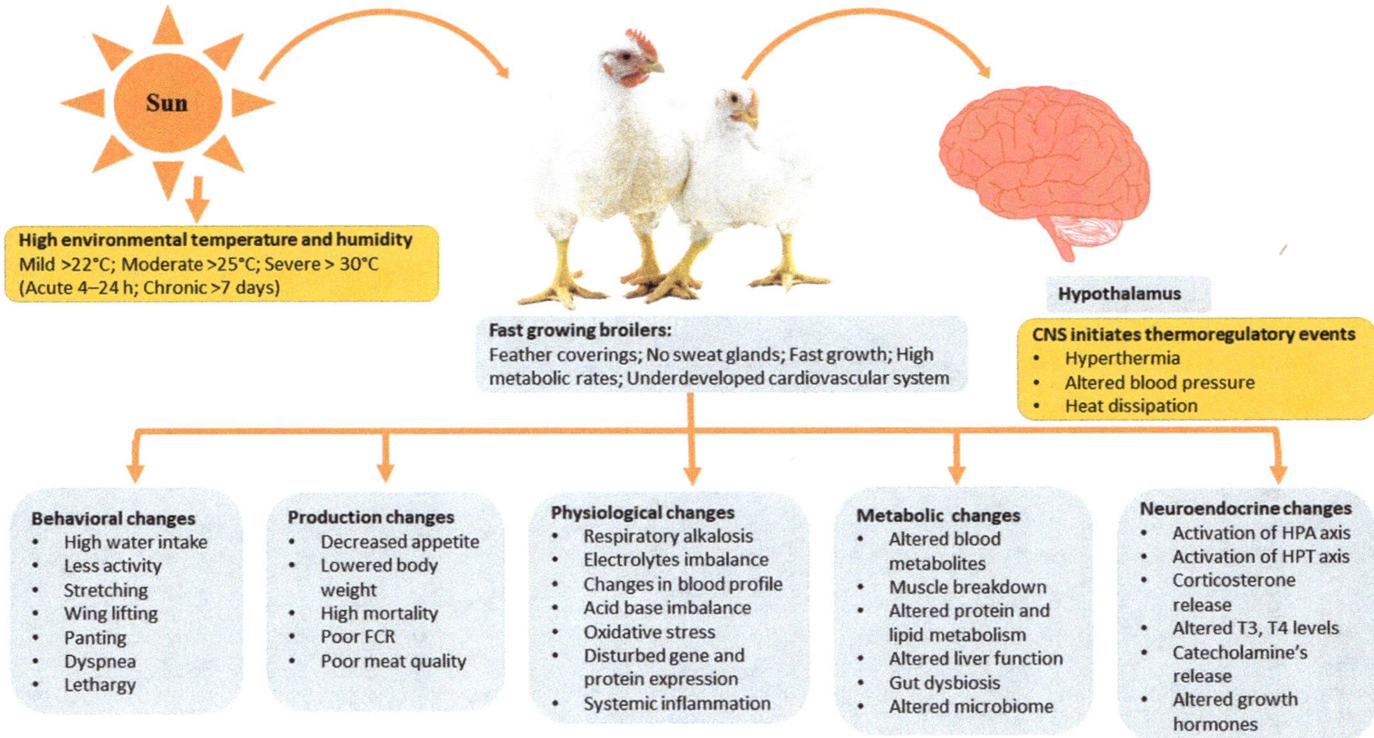


**Figure 1** Schematic diagram showing the behavioral, production, physiological, metabolic, and neuroendocrine responses in broilers during exposure to heat stress conditions.

(18–22°C), suggesting that better plumage quality was associated with higher body temperatures.

### 3.2 Changes in blood components and metabolites

Exposure to heat stress induces compositional changes in the blood such that the lymphocytes and monocytes content are decreased, whereas the heterophils and basophils count increases (Altan et al., 2003). This increases the heterophil/lymphocyte (H/L) in heat-stressed broilers, which is a widely acceptable and reliable index for predicting stress in poultry (Xu et al., 2018). Acute heat stress primarily disturbs the levels of blood metabolites responsible for energy supply and reestablishing homeostasis. In contrast, chronic heat stress causes tissue damage, as reflected by alterations in biomarkers associated with organ function, such as lactate dehydrogenase, glutamic oxaloacetic transaminase, and creatine kinase (Xie et al., 2015). Exposure to acute heat stress at 32 ± 1°C for 24 h increased the circulating concentrations of blood metabolites, including glucose and cholesterol levels, while triglyceride was decreased (He et al., 2020). Subjecting 38-day-old broilers to moderate heat stress of 32 ± 1°C for 3 days elevated the serum levels of glucose, blood urea nitrogen, low-density lipoprotein cholesterol, and the protein carbonyl group (Jinxue et al., 2023). Therefore, heat stress is accompanied by disturbances in blood nutrient concentrations and their metabolic products. Circulating aspartate aminotransferase (AST) and alanine aminotransferase (ALT) enzyme activities are often measured as markers of heat stress-induced liver damage. Both parameters have been found to respond to change or remain unaffected in heat-stressed broilers (Del Vesco et al., 2017). The variations in these blood metabolites and their subsequent recovery to afford thermotolerance may be influenced by the temporal effects of heat stress conditions.

### 3.3 Neuroendocrine changes

In response to heat stress insult, distinct neuroendocrine pathways are activated to release hormones that will engage in the regulation of various physiological functions, including body temperature, heart rate, respiratory rate, blood pressure, and glucose metabolism. The activation of the sympathetic adreno-medullary (SAM) pathway controls the fight or flight response and allows for the release of the catecholamines (epinephrine and norepinephrine), which affect muscle activity. The hypothalamic–pituitary–adrenal (HPA) axis is also activated for the release of the glucocorticoid, corticosterone, which acts as the major stress hormone in chickens. The release of corticosterone can exert either positive or negative impacts on the body depending on the duration and latency of heat exposure. Corticosterone improves heat stress resistance by facilitating the breakdown of energy substrates from muscle protein and adipose tissue via gluconeogenesis to provide sufficient energy for the body to resist stress (Xu et al., 2018). It can promote immune response, alertness, and vigilance during acute stress, but its sustained release during chronic stress impairs growth, immunity, and reproduction (Oluwagbenga and Fraley, 2023). Some studies have also reported that heat stress conditions did not exert any significant impact on circulating corticosterone levels, probably owing to variabilities in heat stress conditions, pulsatile hormonal activity, tissue-specificity, and species of animal studied (Lin et al., 2006; Bai et al., 2020; Uyanga et al., 2021a; Oni et al., 2024).

Hormones secreted via the hypothalamic-pituitary-thyroid (HPT) axis, including triiodothyronine (T3) and thyroxine (T4), are crucial for regulating metabolic heat production and body temperature. The levels of these hormones are altered during heat stress, and a decline in T3 levels has been attributed to reduced peripheral deiodination of T4 to T3 (Wasti et al., 2020). Heat stress decreases the size and activity of the thyroid gland, reducing the concentrations of both T3 and T4 hormones as an adaptive response to minimize heat load (Oluwagbenga and Fraley, 2023).

### 3.4 Oxidative stress

Reactive oxygen species (ROS) are primarily free radicals and peroxides generated during cellular metabolism (Oke et al., 2024b). They are essential for several biochemical processes, such as transcription, immunomodulation, and ion transport. However, increasing levels of ROS may cause damage to biological macromolecules, including proteins, carbohydrates, lipids, and nucleic acids. Heat stress leads to the production of ROS, and their accumulation overcomes the antioxidant capacity of cells, resulting in oxidative stress. Heat stress alters the activities of antioxidant enzymes (catalase, superoxide dismutase, and glutathione peroxidase) and increases protein and lipid peroxidation markers, such as malondialdehyde (Uyanga et al., 2022a). The activation of the antioxidant defense system is necessary to protect cells during high temperatures since heat stress is closely associated with oxidative stress. Broilers subjected to 38°C for 24 h at 21 days had higher hepatic levels of thioredoxin reductase 1, glutathione synthetase, and superoxide dismutase than at 42 days old (Del Vesco et al., 2017). Additionally, the 42-day-old broiler generated higher levels of hydrogen peroxide, revealing that older chickens may be more susceptible to heat and oxidative stress than younger chickens.

### 3.5 Immune system

Heat stress affects the immune system by suppressing innate and adaptive immune responses (Bagath et al., 2019). Heat stress in chickens reduced antibody levels after vaccination and total blood cell volume (Maxwell et al., 1992). It also increased their susceptibility to infectious diseases, thereby affecting poultry health and welfare (Monson et al., 2018). Heat stress affects the number of T cells, secretion of antibodies and cytokines, immunoglobulin concentrations, and proliferation of lymphocytes. Both the innate and adaptive immunity are compromised during heat stress exposure, with the innate immunity serving as the first line of defense during thermal insults, while the adaptive immunity mediates the actions of lymphocytes and reactivity. Activation of both the HPA and SAM axes during heat stress elicits changes in the immune response. The release of corticosterone acts as a stimulus for the activation of the immune system during acute stress; however, this can result in immune suppression during chronic stress, causing infiltration of cells from the adaptive immune system (Cantet et al., 2021). Heat stress initiates a strong inflammatory response in the broiler spleen via significant upregulation of interleukins – IL-1β, IL-4, IL-6, and NF-κβ molecule expression (Medzhitov and Horng, 2009). Chronic heat stress triggers the atrophy of immune organs, such as the bursa of Fabricius, thymus, and spleen, leading to impaired functioning in broilers (Quinteiro-Filho et al., 2010; Jahanian and Rasouli, 2015). Damage to the spleen tissue induces the activation of innate immunity and inflammatory responses by enhancing the mRNA expression of genes (TLR2, TLR4, and DAI), transcriptional factors (NF-κB, AP-1, and IRF3), and concentration of downstream inflammatory cytokines in broilers (Ma et al., 2024). Additionally, heat stress reduces the total level of circulating IgM and IgG antibodies during primary and secondary immune responses (Niu et al., 2009). Detailed discussions on the impacts of heat stress on the immune system and the corresponding events have been extensively reviewed (Cantet et al., 2021; Nagai and Iriki, 2001; Zmrhal et al., 2023).

## *3.6 Metabolic processes*

Hyperventilation during panting in heat-stressed broilers leads to respiratory alkalosis. This distorts the mineral balance and increases the amount of sodium and potassium ions broilers excrete, resulting in acid-base imbalance (Ajakaiye et al., 2011; Onagbesan et al., 2023). There is a reduction in blood pH resulting from urinary excretion of bicarbonate ions, leading to metabolic acidosis (Gonzalez-Rivas et al., 2020). Furthermore, heat-stressed chickens have decreased protein deposition and reduced body fat mobilization, with increased protein breakdown via gluconeogenesis to promote glucose synthesis as a major energy source during heat stress.

Heat stress impairs intestinal barrier function and alters the activity of digestive enzymes and nutrient transporters, compromising nutrient digestibility, absorption, and assimilation. A comparison of broilers under different heat stress models revealed that nutrient digestibility and metabolism were negatively impacted during constant heat stress (35°C for 24 h), relative to acute (35°C for 4 h) and cyclic heat stress (12 h at 24°C and 12 h at 35°C) (Teyssier et al., 2023). More so, constant heat stress decreased the apparent ileal digestibility for nitrogen, fat, and total amino acids by 6.26%, 2.60%, and 5.05%, respectively, compared to the thermoneutral group (24°C for 24 h), altering post-absorptive metabolism and nutrient utilization. This alteration in nutrient metabolism aggravates the deleterious effects of heat stress in degrading the performance and productivity of chickens (Uyanga et al., 2022a, 2022b).

### 3.7 Transcriptional changes

Heat shock proteins (HSPs) function as chaperones that bind other proteins after heat stress damage to preserve their structure, manage their migration across membranes or organelles, or control receptor availability or specific enzyme functionality (Kampinga and Craig, 2010). HSPs are rapidly induced and their contents are increased, which rapidly enhances the heat tolerance capacity in broilers subjected to heat stress (Das et al., 2010; Uyanga et al., 2021b). HSP70 and HSP90 are members of the most conserved and well-studied HSP family which serve as markers for heat stress in chickens (Rivera et al., 2005; Xie et al., 2014). The HSP70 gene plays a key role in protecting the body from the deleterious effects of oxidative stress via serving as a molecular chaperone that helps in the proper folding and assembly of proteins (Favatier et al., 1997), while HSP90 interacts with client proteins during the later stages of folding and modifies their configuration (Wegele et al., 2004).

Importantly, the expression of the HSP-related genes (HSP70 and HSP790 genes) and antioxidant enzymes are associated with heat preconditioning (Hisaeda et al., 1997). Broilers possess altered expression profiles of these stress-associated genes, such as the HSF3, HSP70, and HSP90 when exposed to heat stress (Cedraz et al., 2017; Al-Zghoul et al., 2019; Uyanga et al., 2021b). The expression levels of HSP70 in three Egyptian local breeds (Fayoumi, Dandarawi, and Sinai) were investigated with and without the naked neck (Na) gene under normal and heat stress conditions (Galal et al., 2019). Higher HSP70 expression levels in crossbreeds with the Na gene were discovered, suggesting that the Na gene is responsible for the up-regulation of HSP70 expression and has a positive impact on heat stress adaptation not only by reducing feather cover (Galal et al., 2019).

## 4 Genetic basis of heat resistance in broilers

Genetic background plays a crucial role in the animal's response to heat stress. Local breeds of chicken have developed unique traits adapted to their respective environments, featuring exceptional ecological adaptability traits, including disease resistance, resilience to poor nutritional conditions, and tolerance to extreme weather conditions. More so, these breeds have proven to be superior counterparts for crossbreeding compared to exotic varieties, playing an essential function in improving environmental adaptability within species, improving production, and enabling the selection for more adaptable commercial lines (Xie et al., 2024). Local breeds have proven to be highly effective for crossbreeding, significantly enhancing the overall sustainability of animal farming practices and fostering the production of high-quality outputs (Soglia et al., 2020). It is important to understand the genetic background of the broiler as this significantly determines their susceptibility to harsh environmental conditions.

### 4.1 Selective breeding for heat tolerance

In livestock breeding, genomic selection is a major factor in selecting the fittest animals (Ibtisham et al., 2018). Selecting animals for progeny improvement based on their ability to tolerate heat and improve performance metrics is desired. The most significant factor is genetic selection for heat-tolerant chickens which is predicated on identifying and choosing genes linked to disease resistance and heat tolerance (Ranjan et al., 2019). The broilers' intense genetic selection for rapid growth rates caused significant challenges for heat stress adaptation (Yahav, 2000). Furthermore, the selection of chickens for production qualities has made them more vulnerable to heat stress (DeSmit et al., 2005), which raises the risk of oxidative stress and mitochondrial dysfunction in poultry (Akbarian et al., 2016; Oke et al., 2024b). The ability of chickens to adapt to heat can be enhanced by selecting the traits in high-temperature environments (Mathur and Horst, 1994). Identifying genetic markers related to heat tolerance makes it possible to perform selective breeding for heat resilience in chickens. Characterization of genetically controlled physiological attributes that change in response to heat stress will lay the foundation for utilizing these attributes as potential markers to select for heat stress tolerance (Mack et al., 2013). Heat tolerance, growth performance, and reproductive traits in chickens may be improved by using potentially heat-resistant genes, such as naked neck (Na), dwarf (Dw), frizzle (F), and slow feathering (K) (Fathi et al., 2022; Onagbesan et al., 2023).

### 4.2 Role of genomics and biotechnology

The management of desired traits, including heat resistance, has resulted in the development of novel technologies to investigate genetic relations with these phenotypes. An excellent approach to detecting the functional genes and polymorphic variants associated with heat stress resilience is through the use of high-throughput screening technologies, such as genome-wide analyses of genetic variations (Hoffmann et al., 2011; Bush et al., 2017). In addition, new molecular techniques have been developed, including RNA-Sequencing,

Single Nucleotide Polymorphisms (SNPs), and bioinformatics approaches, such as Genome-Wide Association Study (GWAS). Figure 2 shows the different genomics and biotechnology approaches applied in the study of heat stress and their related genomic mechanisms in poultry species.

### 4.2.1 Single nucleotide polymorphisms

Single nucleotide polymorphisms (SNPs) are a type of genetic markers used in the selection of chromosomal regions of poultry that are resistant to heat stress (Pawar et al., 2016). Nucleotide polymorphisms in the gene expression of some heat shock proteins (such as HSP70 and HSP90) have been identified as the genetic basis for heat stress adaptation in chickens (Budi et al., 2024). Comparison between native Chinese breeds, Dwarf White Chicken, and Red Jungle fowl uncovered sequence variations within and between chicken species. Genetic sequencing identified 34 SNPs and 2 indel mutations in the coding regions of the chicken HSP70 gene (Gan et al., 2015). Studies on genetic diversity in different chicken breeds including local chicken breeds and red jungle fowl, are performed using metabarcoding and microsatellite genotyping data to examine the diversity of neutral genetic markers and examine the inherent variations between breeds. An understanding of these genetic variations and their conservation will contribute to the development of selection and breeding programs targeted at improving thermotolerance in poultry.

### 4.2.2 Quantitative trait loci

Quantitative trait loci (QTL) fine-mapping allows for the screening of high-heat stress-tolerant genes, which can serve as selected genomic markers for heat tolerance in poultry (Van Goor et al., 2015). Genetics has contributed to the

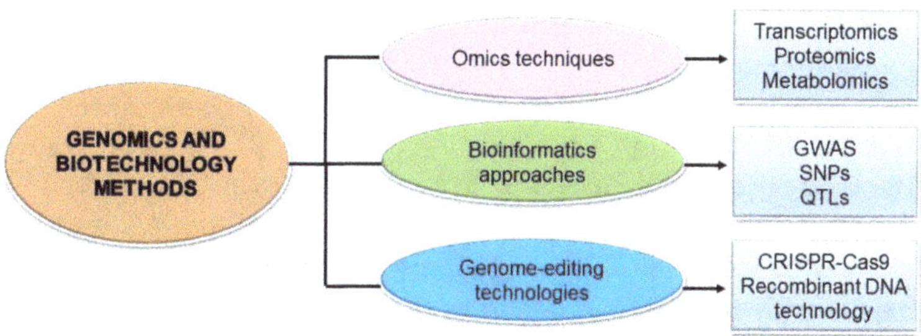


**Figure 2** Different genomics and biotechnology approaches used to investigate heat stress tolerance in poultry. Genome-Wide Association Study (GWAS); Single Nucleotide Polymorphisms (SNPs); Quantitative trait loci (QTL); Clustered regularly interspaced short palindromic repeats (CRISPR).

understanding of heat stress and Newcastle disease virus (NDV) infection in diverse chicken lines and breeds by demonstrating the potential for genetic improvement of heat tolerance and disease resistance in chickens. Van Goor et al. identified QTLs for body temperature, body weight, and breast muscle yield in a Fayoumi x broiler advanced intercross population. In another study, 61 QTLs were identified to reveal the candidate genes responsible for the chickens' response to heat stress. Functional analysis of these genes identified the angiopoietin pathway's involvement in the chicken's response to heat stress and the heritabilities of some blood parameters under thermoneutral and heat stress conditions (Van Goor et al., 2016).

### 4.2.3 Genome-wide association study

Genome-Wide Association Study (GWAS) was developed to unveil the genetic background of phenotypic traits based on the hypothesis of links between genetic variants and different phenotypic traits (Gebreselassie et al., 2020). A GWAS study unveiled genomic regions associated with mortality from heat stress, body temperature, cholesterol levels, glucose levels, bacterial and viral infection in a layer line (Wolc et al., 2019). GWAS was applied to study the body temperature changes in roosters under acute heat stress. Several SNPs were identified as candidate genes that differed with thermotolerance, and they were mainly associated with apoptosis or responses to external stimuli (Zhuang et al., 2019). GWAS has also been used to detect QTLs associated with viral load, antibody response, and body weight change in Hy-Line Brown (HLB) chicks subjected to heat stress and NDV infection (Saelao et al., 2019).

### 4.2.4 Next-generation sequencing

A high-throughput screening technology largely used is RNA sequencing (RNA-Seq). RNA molecules carry the transcribed information encoded in selected genes, which can be translated into proteins or used to check gene expression directly (i.e. miRNA) or indirectly (i.e. heterogeneous nuclear ribonucleoproteins – hnRNPs). Therefore, the rate of RNAs expressed in a known circumstance shows the cell state and can disclose stress mechanisms (Perini et al., 2021).

The capacity of local chickens to rapidly adapt to a hotter environment was analyzed using the transcriptome from various tissues (heart, breast muscle, and spleen) in Ethiopian lowland (adapted to lowland thermal conditions) and highland chickens (non-adapted to lowland environments). The highland chickens responded rapidly to heat stress, and their muscle tissue had higher HSP regulation, suggesting that the muscle tissue is particularly vulnerable to heat stress than other tissue types (te Pas et al., 2019). Following a whole-genome resequencing study of 186 chickens from various regions of Asia, the genetic structure analysis revealed high levels of inbreeding among some chicken breeds due to artificial selection and low genetic diversity of tropical chickens. Sixty-nine differentially expressed genes were identified between tropical and cold-resistant chickens, including heat adaptation genes, and their biological and metabolic functions in establishing thermotolerance in chickens were established (Bai et al., 2024).

### *4.2.5 Genome editing*

Genome editing can precisely modify targeted genes' DNA sequences to improve livestock productivity or disease resistance without introducing any undesired abnormalities. However, traditional selective breeding relies on naturally occurring genetic variations, which take longer to gain desirable attributes (Park et al., 2020). Contrarily, it is possible to introduce desirable genetic variations quickly using genome-editing technology. Thus, genome editing represents an effective method that can easily be combined with traditional breeding. Different studies have adopted genome editing to improve livestock productivity or disease resistance (Ricroch, 2019; Koslová et al., 2020). Genome-editing technologies such as the CRISPR/Cas9 system have been developed, and these methods have been applied to the production of a variety of genome-edited animals, including livestock (Ricroch, 2019). This approach has not yet been deeply utilized in modifying the targeted genes' DNA sequences of heat stress in broilers. The key findings of some studies that adopted genomic and biotechnology techniques in their understanding of thermotolerance in heat-stressed broilers are summarized in Table 1.

## 5 Impacts of heat stress on broiler's growth and productivity

Broilers undergo physiological, neuroendocrine, and behavioral changes when exposed to ambient temperatures above the thermoneutral zone, negatively impacting their growth, development, and productivity.

### *5.1 Growth rate and performance*

As an adaptive response to heat stress, broilers reduce their feed consumption to minimize metabolic heat production (Nawab et al., 2018). Heat stress directly alters the feed intake, body weight gain, and productivity of broilers, consequently, elevating the feed conversion ratios and mortality of chickens (Uyanga et al., 2022c; Onagbesan et al., 2023). Consequently, this reduces the body weight gain and feed efficiency. Reduction in feed intake directly depreciates growth performance, since dietary nutrients are unavailable or insufficient from less feed consumed, which also limits the amount of

**Table 1** Genomic and biotechnology techniques applied in the understanding of thermotolerance in heat-stressed broilers

| Broiler breeds and age | Heat conditions | Technique(s) adopted | Considered trait(s) | Key findings | References |
|---|---|---|---|---|---|
| Lingnan yellow 14-day-old female | 34–38°C for 28 days | 16S rRNA high-throughput sequence analyses | Gut microbiota structure | 1. Heat stress causes profound changes in the composition of gut microbiota influencing the abundance and diversity of species.<br>2. Increase in Phyla *Firmicutes*, *Tenericutes*, and *Proteobacteria* and the genera *Anaeroplasma* and *Lactobacillus*.<br>3. Decrease in Phyla *Bacteroidetes* and *Cyanobacteria* and the genera *Bacteroides*, *Oscillospira*, *Faecalibacterium*, and *Dorea*. | (Shi et al., 2019) |
| Indigenous Huaixing yellow-feather 8-week-old | 32.5 ± 1.4°C for 8 h/day from 9:00 am to 17:00 pm | 16S rRNA biodiversity sequencing analysis | Ileal microbiome | 1. No significant effects on the alpha diversity index in ileal microflora of heat stress broilers.<br>2. Heat stress significantly reduced the relative abundance of *Campylobacter* and increased the abundance of *Delftia*.<br>3. Heat stress significantly enhanced the abundance of the microflora related to lipid metabolism, carbohydrate metabolism, and xenobiotics biodegradation and metabolism.<br>4. Heat stress reduced the abundance of the microflora related to nucleotide metabolism and amino acid metabolism. | (Jin et al., 2022) |
| Arbor Acres 28-day-old | 32°C, 55–60% for 14 days | Proteomics | Meat quality and textural attributes | 1. Chronic heat stress impaired the breast quality.<br>2. Fluctuation of muscular carbonylation degree at amino acid level in heat stress broilers.<br>3. The carbonylated proteins were mainly annotated to glycolytic process, calcium homeostasis, apoptosis, and collagen fibril organization. | (Liu et al., 2023) |

| | | | | | |
|---|---|---|---|---|---|
| Ross-308<br>20-day-old | 34°C from days 21 to 35 | Proteomics | Liver component | 1. The expression of heat shock protein 70 significantly increased in the heat-stressed with low body weight (HL) group.<br>2. High temperature aggravated oxidative stress in broilers.<br>3. MRP-126, fatty acid binding protein, ferritin heavy chain, glutathione S-transferase, agmatinase; mitochondrial, alpha-enolase, 60 kDa heat shock protein; mitochondrial, and tubulin beta-7 chain were revealed. | (Park et al., 2022) |
| Huaixiang<br>8-week-old | 32.5 ± 1.4°C, 8 h/day, from 9:00 am to 17:00 pm | 16S rRNA | Production performance, redox status, and intestinal morphology | 1. Broilers in heat stress group had higher drip loss of breast muscle than control group.<br>2. Lower total antioxidant capacity in serum and jejunum of HS broilers.<br>3. Malondialdehyde contents in breast muscle, duodenum, and jejunum were elevated by heat stress exposure.<br>4. Relative mRNA expression of *Occludin* and *ZO-1* in the duodenum, *Occludin*, *Claudin-1*, *Claudin-4*, *ZO-1*, *Mucin-2* in the jejunum, and the *Claudin-4* and *Mucin-2* in the ileum was down-regulated by heat stress.<br>5. Decline in production performance, reduced antioxidant capacity, disrupted intestinal barrier function, and negatively affected cecal microbiota in heat stress broilers. | (Liu et al., 2022) |

(*Continued*)

**Table 1** *Continued*

| | | | | | |
|---|---|---|---|---|---|
| Broiler-Type Strain of Taiwan Country Chickens<br>47-week-old roosters | 38°C for 4 h | Microarray and quantitative real-time polymerase chain reaction (qRT-PCR) analyses | Testes | 1. The body temperature and respiratory rate increased significantly in heat stress broilers.<br>2. 163 genes including those involved in the response to stimulus, protein metabolism, signal transduction, cell adhesion, transcription, and apoptosis were found to be expressed significantly different in the testes of heat stress chickens.<br>3. Numerous transcripts in the testes exhibited distinct expressions between the heat-stressed broiler-type. | (Wang et al., 2015) |
| Ross 708<br>21 days posthatch | 35–37°C for 8 h/day | RNA-seq analysis | Heart development | 1. Body weight and normalized heart weight were significantly reduced in heat stress broilers.<br>2. Decreased cell cycle activity and increased apoptosis in heat stress broilers. | (Zhang et al., 2017a) |
| Generations F18 and F19 of a broiler (heat-susceptible) × Fayoumi (heat-resistant) advanced intercross line (AIL)<br>17-day-old | 35°C for 7 h/day | Quantitative trait loci (QTL) | Body temperature (BT), body weight (BW), breast yield, and digestibility | 1. QTLs for BT on *Gallus gallus* chromosome (GGA) 14, 15, 26, and 27 were identified.<br>2. QTLs on BW on GGA1 to 8, 10, 14, and 21.<br>3. Dry matter digestibility on GGA19, 20, and 21.<br>4. QTL of very large effect for breast muscle yield on GGA1, 15, and 22 with a single 1-Mb window on GGA1 explaining more than 15% of the genetic variation. | (Van Goor et al., 2015) |
| Ross 308 broilers22-day-old | 35°C (LCT-HS) from 22 to 42 days of age | RNA-seq analysis | Thermotolerance and nutrient absorption | 1. Genes such as *EMX2*, *WFIKKN2*, *SLC6A4*, *WNT10A*, and *PHOX2B* related to central nervous system development2. Genes including *NPY*, *AgRP*, *GAL*, *POMC*, and *NMU* responsible for the regulation of feed intake in chickens were identified. | (Uyanga et al., 2023) |

metabolizable energy available to sustain average output (Geraert et al., 1996). Furthermore, heat stress impairs development and production by altering hormonal balance, suppressing immunological function, and upsetting nutrient metabolism (Yahav et al., 2005). Exposure of chickens to extreme environmental temperature (35°C) at 2 weeks of age for 2 weeks significantly reduced feed intake (29%) and resulted in a lower growth rate (37%) accompanied by increased in vivo uptake of galactose and methionine (Mitchell and Carlisle, 1992). Wang et al. (2021) documented that the average daily growth and feed intake in heat-stressed broilers were lower than under thermoneutral conditions. Zeferino et al. (2016) also reported a 32% reduction in daily weight gain, a 20% decrease in feed intake, and a 16% rise in feed conversion ratio in heat-stressed broilers. Typically, broilers experience a 15–30% reduction in average daily feed intake at temperatures greater than 30°C from 2 weeks of age onward.

## 5.2 Meat quality

Heat stress results in conditions like pale, soft, and exudative meat, which lowers consumer acceptance and market value. Elevated temperatures leading up to slaughter can worsen these consequences, resulting in lower-quality carcasses and a lower meat yield that can cause financial losses (Vandana et al., 2021). The meat color is greatly affected in heat stress chickens prior to slaughter. Heat-stressed chicken meat has a drop in redness and an increase in lightness and yellowness (Suman et al., 2016). Lu et al. (2007) found that prolonged heat stress causes meat to become lighter overall, with varying effects on its yellowness and redness. The reduced redness value of meat from heat-stressed chickens suggests an excess of oxidized myoglobin (Mancini and Hunt, 2005). Gregory (2010) revealed that heat stress may impact the ultimate pH and the rate at which postmortem muscle pH declines. Glycogen levels and the rate at which glycolysis occurs influence postmortem muscle pH, due to the influence of preslaughter stress and muscle temperature, respectively (Chauhan and England, 2018). However, the effects of heat stress on meat quality vary among different animal species or breeds, possibly due to differences in thermotolerance. Whereas Zeferino et al. (2016) found no discernible impact of heat stress on meat quality, some studies have documented a detrimental effect (Zhang et al. 2012a; Kadim et al. 2014). The varying impacts of heat stress on meat quality attributes may be explained by the different physiological and biochemical mechanisms triggered by elevated temperatures. These include altered muscle metabolism, reduced muscle glycogen stores, leading to higher ultimate pH, oxidative stress that damages muscle proteins and lipids, and impaired post-mortem proteolysis. Additionally, heat stress can influence blood flow distribution and carcass composition, ultimately affecting traits such as water-holding capacity, tenderness, color, and shelf-life stability (Prates, 2025).

## 5.3 Carcass yield

Commercial broilers often achieve a slaughter weight of 2.5 ~ 3.0 kg between 4 and 7 weeks of age. Under heat-stressed conditions, both live body weight and carcass yield of broilers are decreased (Toghyani et al., 2012). In two distinct experiments, Liu et al. (2019) demonstrated decreased carcass production in heat-stressed chickens. High ambient temperatures negatively affect chickens, causing low slaughter weight, increased mortality, and ultimately lower meat yield and quality in broilers. Zhang et al. (2012b) reported a loss in breast muscle weight and an increase in thigh muscle, whereas Zeferino et al. (2016) documented a drop in breast weight, while thigh weight remained the same. Additional research confirmed the prior findings of Zhang et al. (2012b), which indicated a decrease in breast muscle weight and an increase in thigh muscle (Liu et al., 2019). The differences in myofibril metabolism between the thigh and breast muscles may be attributed to weight variations. Red slow-contracting oxidative fiber is abundant in the leg, while white fast-contracting glycolytic fiber, dependent on glycogen supply for metabolism and hypertrophy, is abundant in breast muscle, as further documented by Ismail and Joo (2017). Reduction in feed intake also causes the body to use less glycogen and produces less protein in the breast muscle (Temim et al., 2000). The extent of changes in carcass characteristics may vary between species, but overall, heat stress–induced carcass yield loss has been estimated to cause extensive economic losses to livestock industries (St Pierre et al., 2003).

Heat stress in broilers triggers a cascade of physiological responses. The mechanisms involve complex thermoregulatory adaptations, metabolic shifts, and stress-related hormonal changes that significantly affect both the overall health of the bird and the quality of meat produced (Nawaz et al., 2021). Panting, the bird's primary cooling mechanism, leads to respiratory alkalosis, disrupting acid-base balance and muscle function (Nawaz et al., 2021; Kpomasse et al., 2021). Blood is redirected from the viscera to the skin to facilitate heat dissipation, reducing oxygen and nutrient supply to muscle tissue. Simultaneously, reduced gut motility and compromised nutrient absorption contribute to lower muscle glycogen stores, limiting post-mortem lactic acid production (Prates, 2025). This results in a higher ultimate pH, a darker meat color, and an increased risk of producing dark, firm, and dry meat. These effects are further compounded by dehydration and electrolyte imbalances, which interfere with muscle relaxation and can negatively influence meat tenderness and juiciness (Liu et al., 2020; Onagbesan et al., 2023).

In addition, heat stress induces oxidative stress through excessive production of reactive oxygen species, leading to lipid and protein oxidation in muscle tissue. This damages the cell membrane, reduces the water-holding capacity, and compromises meat flavor and shelf life (Prates, 2025). Mitochondrial dysfunction may also impair cellular energy production, affecting post-mortem muscle metabolism (Nawaz et al., 2021). The activation of the hypothalamic-pituitary-adrenal axis increases corticosterone secretion, altering protein turnover and suppressing immune function. Furthermore, the production of inflammatory cytokines can promote muscle protein degradation. Reduced blood perfusion to muscles under heat stress exacerbates hypoxia and accelerates metabolic fatigue, all of which contribute to inconsistent and often undesirable meat quality outcomes in broiler production (Lara and Rostagno, 2013).

## 6 Impacts of heat stress on broiler's behavior and welfare

### *6.1 Altered feeding behavior*

Heat stress results in adjustments in feed consumption, thereby negatively impacting the behavior, physiology, productive capacity, and well-being of chickens (Renaudeau et al., 2012; Rostagno, 2020; Uyanga et al., 2022c). When exposed to high temperatures, chickens will often spend less time feeding and walking and more time drinking, resting, and panting (Mack et al., 2013). Chickens exposed to heat stress have shorter behavioral responses, like lying down and eating (Branco et al., 2021). Under certain instances, exposure to acute heat stress did not impact the feed intake and feeding frequency of broilers, probably due to the highly integrated thermoregulatory mechanisms (Rajaei-Sharifabadi et al., 2017). It is understood that the control of appetite in chickens is regulated by both peripheral and central mechanisms, which may respond differentially depending on stress stimulus. Overall, broilers cultivate voluntary decrement in feed intake to lower the thermogenesis associated with digestion, absorption, and nutrient utilization (Baumgard and Rhoads, 2013). Evidently, the lowered productivity of heat-stressed chickens is commonly related to decline in feed consumption (Brugaletta et al., 2021).

Nutritional approaches can be employed to mitigate the deleterious impacts of heat stress on broilers, including the use of feed additives such as probiotics, enzymes, antioxidants, phytobiotics, electrolytes, and mineral have been reviewed (Madkour et al., 2022; Abdel-Moneim et al., 2023). Nutraceuticals and medicinal plants contain bioactive compounds that possess antioxidant, antimicrobial, anti-inflammatory, and immunomodulatory effects, which may prove beneficial as thermoregulatory agents in poultry diets (Abd El-Hack et al., 2020; Mahasneh et al., 2024). Table 2 summarizes recent reports

**Table 2** Summary of the effects of dietary interventions and heat stress condition on the responses of broilers

| Broiler breeds | Dietary supplements | Experimental conditions | Impacts on performance and physiology | References |
|---|---|---|---|---|
| Cobb day-old chicks | Microalgae (MAG) | HS and HS + 3% MAG (HS+MAG) broilers subjected to cyclic HS from d 22 to 35 (32–33°C for 8 h). | 1. Improvements in FCR in HS+MAG compared to HS broilers; similar BW until 21 days<br>2. Increased villus surface area; increase in ileal antioxidant (GPX3), immune-related (IL4), and tight-junction (CLDN2) genes | (Chaudhary et al., 2023) |
| Day-old broiler chicks | Dried plum (DP) | HS and HS + DP groups were subjected to cyclic HS conditions (33 °C for 8 h during daytime) for 3 weeks. | 1. BW and AVG decreased in HS broilers from 28 days; FCR was higher in HS broilers<br>2. Increase in total VFAs in HS + DP; increase in HSPs, antioxidant-related, tight-junction, and immune-related genes in ileal of HS + DP | (Wasti et al., 2020) |
| One-day-old male Ross broilers | Dimethyl itaconate (DI) | HS group, HS + 50 mg/kg DI group, HS + 150 mg/kg DI group, and HS + 200 mg/kg DI group. HS: 32 ± 1°C for 8 h daily. | 1. Declines in serum T-AOC and activities of peroxidase, SOD, and CAT | (Wang et al., 2023) |
| Ross 308 broilers | Steam-exploded pine particle (SPP) | 0%, 1%, and 2% SPP and two temperature conditions (NT: 21°C; CHS: 31°C) for 7 days. | 1. Decrease in BW in SPP groups; decrease in the spleen and plasma total protein in CHS<br>2. Increase in total protein levels of CHS | (Goel et al., 2022) |
| Day-old Cobb male broiler chicks | Vitamin E and organic Selenium (Se) | Vitamin E and organic Se at the rate of 250 mg/kg and 1 mg/kg, respectively. HS: 35 ± 1°C from 10:00 am to 2:00 pm and reduced to 25°C ± 1 daily. | 1. No significant impact of dietary treatment on immunity in jejunal tissues at days 28 and 35<br>2. Low mortality rate in broilers fed with Se<br>3. Decrease in broiler carcass, tissue, and lean weights on 35 days | (Calik et al., 2022) |

| | | | | |
|---|---|---|---|---|
| One-day-old Ross 308 chicks | Betaine (BT) and organic minerals (OM) | Diets containing BT (0 and 2,000 ppm) and OMs (0, 250, and 500 ppm), singly or in combination. HS: 35°C ± 2°C for 6 h daily. | 1. Elevated serum total protein and globulin concentrations in BT group<br>2. Decrease in final BW, feed intake and FCR<br>3. Increase in crude protein digestibility in OM group | (Saleh et al., 2023) |
| Eight-day-old Ross 308 broiler chicks | Solubles from shredded, steam-exploded pine particles (SSPP) | 0%, 0.1%, and 0.4% SSPP was added to basal diet from 8 to 36th day old. HS (up to 31°C) was introduced from 29th day. | 1. Increase in HSPs 70 and 90 in 0% SSPP HS2. Decrease in BW and level of triglyceride of HS broilers | (Goel et al., 2023) |
| Arbor Acres male broilers | Selenomethionine (SeMet) | Basal diet with 0.0 (CHS- 33 ± 2°C), 0.2 (CHS + 0.2 Se), 0.4 (CHS + 0.4 Se), and 0.6 (CHS + 0.6 Se) mg Se/kg SeMet. | 1. Increase in hepatic selenium concentration<br>2. Higher serum ALT, GLU, TBA, and TG<br>3. Decrease in relative weight and increase in HSP70 of liver | (Jing et al., 2024) |
| One-day-old unsexed Arbor Acres broilers | Biogenic selenium nanoparticles using Alginate Oligosaccharides (SeNPs-AOS) | SeNPs-AOS (5 mg/kg feed vs. no inclusion) with HS (33 ± 2°C for 10 h/day within 21 days. | 1. Oxidative damage and metabolite changes in the breast muscle<br>2. Decrease in shear force, cooking loss, and freezing loss<br>3. Decrease in level of T-AOC, T-SOD, CAT, and GSH-Px enzymes; increase in MDA | (Yang et al., 2023) |
| One-day-old Ross 308 broilers | Resveratrol-loaded liposomal nanocarriers (Resv-Lipo NPs) | Basal diet supplemented with Resv-Lipo NPs I, II, and III at the levels of 0, 50, 100, and 150 mg/kg diet. HS: 6 h/day at 36 ± 2°C. | 1. Improved lipid profile – serum triglyceride, cholesterol, LDL, and VLDL levels were significantly reduced in the groups that received a higher dosage of Resv-Lipo NPs, regardless of HS<br>2. Significant low body weight and FCR<br>3. Low total protein and serum concentration in HS broilers | (Kishawy et al., 2023) |

(Continued)

**Table 2** *Continued*

| Broiler breeds | Dietary supplements | Experimental conditions | Impacts on performance and physiology | References |
|---|---|---|---|---|
| One-day-old broiler chicks (Ross 308) | Probiotics and clove essential oils (CEO) | Clove essential oils (300 mg/kg), PRO – CON with probiotics (2 g/kg), and PC – CON with probiotics and clove essential oils. HS: 35 °C for 4 h/ 5days/week until 35 days old. | 1. Higher BW gain in PRO and PC groups<br>2. Improved FCR in all groups except control<br>3. Reduced nutrient digestibility in control<br>4. Significant decrease in serum MDA levels | (Elbaz et al., 2023) |
| Ross-308 broiler chicks | *Spirulina platensis* (SP) and/or garlic powder (GP) | Basal diet enriched with SP (1 g/kg diet), GP (200 mg/kg diet), or SP/GP (1 g SP/kg + 200 mg GP/kg diet), respectively. HS: 35 ± 1°C for 9 h/day | 1. A significant elevation in triglycerides, cholesterol, and LDL levels; decrease in HDL levels impacted the intestinal morphology2. SP and GP impacted the negative effects of HS on IgM, IgA, and IgY, and antibodies against NDV and IBD | (Attia et al., 2011) |

Notes: GSH-Px: Glutathione peroxidase; VFA: Volatile fatty acids; MDA: Malondialdehyde; T-AOC: Total antioxidant capacity; SOD: Superoxide dismutase; CAT: Catalase; BW: body weight; ALT: Alanine transaminase; GLU: Glucose; TBA: Total bile acids; TG: Triglyceride; CHS: Chronic heat stress; HSP: Heat Shock proteins; LDL: Low-density lipoprotein; HDL: High-density lipoprotein; IBD: Infectious Bursal disease; NDV: Newcastle disease investigating the impacts of heat stress and nutritional interventions that have been studied on the performance, productive outcomes, and physiological responses of broilers.

## 6.2 Changes in drinking patterns

Exposure to high temperatures drastically affects the drinking pattern, reduces mobility, and increases panting in heat-stressed chickens (Lara and Rostagno, 2013). For instance, broilers exposed to 32–35°C can exhibit a 2–3-fold increase in water intake compared to those kept at thermoneutral conditions (21–24°C), while feed intake may decrease by up to 30–40% (Syafwan et al., 2012; Lara and Rostagno, 2013). Panting frequency can rise from a normal 20–30 breaths/min to over 200 breaths/min under acute heat stress, accompanied by a notable reduction in voluntary movement as birds attempt to conserve energy and limit metabolic heat production (Onagbesan et al., 2023). Consequently, there is a high water loss via the respiratory tract and increased water intake to replenish and restore the thermoregulatory balance. Exposing broiler chickens to high temperatures (35°C) for 4 h at 45 days of age led to increased water loss by 64% compared with birds housed at 24°C (Belay and Teeter, 1992). The differences between the temperature of consumed water and body temperature is increased, enhancing heat dissipation and cold water intake under heat stress conditions (Fairchild and Ritz, 2015). High temperature increased water loss by evaporation (via panting) or excretion to reestablish body homeostasis (Braun, 2015). In a recent study, the impacts of heat stress on water consumption, water conversion ratio, and water/feed ratio in broiler lines divergently selected for growth rate were examined (Greene et al., 2024). More so, an adequate supply of cold water after feed withdrawal in the afternoon could help decrease body temperature (Onagbesan et al., 2023). In addition, water consumption helps enhance the rapid uptake of essential nutrients with a better absorption rate in the gut during exposure to heat stress in broilers (Kolluri et al., 2022).

## 6.3 Changes in activity patterns

Behavioral assessments can provide valuable insights into the welfare status of an animal, and it can be combined with other physiological assessments to better understand an animal's welfare status (Oluwagbenga and Fraley, 2023). Changes in environmental temperature are correlated to group-level behaviors, such as clustering and increased locomotor activity, escape attempts, frequent postural changes, and other signs of general unrest – a term that encompasses agitation, restlessness, and elevated alertness among the flock (Pereira et al., 2020; Del Valle et al., 2021) and the frequencies of individual-level behaviors, such as drinking, eating, panting, and roosting (Cartoni Mancinelli et al., 2023). Alterations in behavior are the first visible indicator of heat stress. Broilers respond physiologically and behaviorally when facing the negative influences of heat stress by attempting to reestablish homeostasis (Ratriyanto and Mosenthin, 2018). Heat-stressed chickens express behavioral responses targeted at dissipating heat such as panting, elevating wings, roosting, stretching, decreased foraging, less walking, increased drinking, and decreased feeding (Khattak et al., 2012; Haldar et al., 2015; Cartoni Mancinelli et al., 2023).

Fearfulness is one of the abnormal behaviors used as an indicator for the evaluation of chickens' adaptability to physical, physiological, and psychological stressors (Campbell et al., 2019). Fearfulness tests, such as the novel object test, touch test, isolation test, and tonic immobility test, are regularly adopted in psychological investigations in humans and various animals, including chickens (Dennis et al., 2013). Studies have shown the relationship between fearful reactions and neuroendocrine pathways, including the adrenergic and serotonergic systems. For instance, exposing broilers to 34 ± 1°C for 3 h on day 41 of age increased circulating corticosterone and HSP70 concentration in high-fear-response broilers with enhanced tonic immobility reactions (Zulkifli et al., 2009).

## 7 Conclusion and future trends

Heat stress is a recurring problem in broiler production as global temperatures rise, necessitating flexible solutions to preserve animal welfare and operational effectiveness. The genetic composition of broilers significantly determines their capacity to tolerate heat stress conditions. Additionally, age, sex, and diet determine how heat-stressed broilers express their genetic potentials under harsh environmental conditions. Several genetic approaches have been adopted to examine the molecular mechanisms, including genes and pathways involved in the thermotolerance of broilers under diverse environmental conditions. An in-depth research of identified candidate genes and associated pathways needs to be explored, as this could provide better knowledge and understanding on the susceptibility, responses, and adaptability of the different broiler breeds. Future developments are anticipated to maximize broilers' physiological tolerance to high temperatures by utilizing genetic selection for heat-resistant strains. Innovations in feed formulations, including immune modulators, antioxidants, and plant-based additives, can boost feed efficiency and growth rates by reducing oxidative stress brought on by high temperatures. Heat stress management is also changing due to the incorporation of precision livestock farming technologies. Proactive, customized responses to temperature variations are made possible by AI-driven analytics, real-time monitoring systems, and sensor based technologies. Automated changes are made to environmental controls, such as lighting, ventilation, and cooling systems. Stress will be further reduced with emerging technological designs that use thermal insulation, energy-efficient cooling, and movable ventilation to better regulate the microclimate. These proactive methods point to a dynamic, multidimensional approach that blends infrastructure, technology, nutrition, and genetics, providing a model for a robust and climate-smart poultry sector.

## 8 Where to look for further information

A standard introduction to broiler production and the physiological responses of birds under stress is the Poultry Science textbook by Scanes and Christensen (See references for full details). For information on the impacts of heat stress in poultry systems, Poultry Production in Hot Climates, by Daghir, 2008 provides a comprehensive knowledge of these subjects.

Other sites that include relevant information and resources include:

1. Food and Agriculture Organization (FAO), https://www.fao.org.
2. USDA Agricultural Research Service (USDA- ARS, https://www.ars.usda.gov/oc/webinars/livestock-heat-stress-series/).
3. Department for Environment, Food and Rural Affairs (DEFRA, www.defra.gov.uk).
4. The International Livestock Research Institute (ILRI) (https://www.ilri.org/).

There are several current research projects aimed at understanding heat stress in poultry, its impacts, mitigation strategies, and the application of advanced biotechnological tools to develop heat tolerance in birds. These publications can be assessed from periodicals, including:

1. Poultry Science Journal (https://www.sciencedirect.com/journal/poultry-science).
2. World Poultry Science Journal (https://www.tandfonline.com/journals/twps20).
3. Journal of Applied Poultry Research (https://www.sciencedirect.com/journal/journal-of-applied-poultry-research).
4. Journal of Thermal Biology (https://www.sciencedirect.com/journal/journal-of-thermal-biology).

## 9 References

Abd El-Hack, M.E., et al. (2019), Managerial and nutritional trends to mitigate heat stress risks in poultry farms. *Handb. Environ. Chem.*, 77, 325–38. https://doi.org/10.1007/698_2018_290.

Abd El-Hack, M.E., et al. (2020), Herbs as thermoregulatory agents in poultry: an overview. *Sci. Total Environ.*, 703, 134399. https://doi.org/10.1016/j.scitotenv.2019.134399.

Abdel-Moneim, A.E., et al. (2021), Nutritional manipulation to combat heat stress in poultry – A comprehensive review. *J. Therm. Biol.*, 98, 102915. https://doi.org/10.1016/j.jtherbio.2021.102915.

Ajakaiye, J.J., et al. (2011), Impact of heat stress on egg quality in layer hens supplemented with l-ascorbic acid and dl-tocopherol acetate. *Vet. Arhiv.*, 81, 119–132.

Akbarian, A., et al. (2016), Association between heat stress and oxidative stress in poultry; mitochondrial dysfunction and dietary interventions with phytochemicals. *J. Anim. Sci. Biotechnol.*, 7, 1–14.

Akosile, O.A., et al. (2023), Potential implications of in ovo feeding of phytogenics in poultry production. *TAS*, 7, 1–12. https://doi.org/10.1093/tas/txad094.

Al Amaz, S., et al. (2024), Pre-hatch thermal manipulation of embryos and post-hatch baicalein supplementation mitigated heat stress in broilers. *J. Anim. Sci. Biotechnol.*, 15(1), 1–15. https://doi.org/10.1186/s40104-023-00966-6.

Altan, Ö., et al. (2003), Effect of heat stress on oxidative stress, lipid peroxidation and some stress parameters in broilers. *Br. Poult. Sci.*, 44, 545–550.

Al-Zghoul, M.B., et al. (2019), Effects of pre-hatch thermal manipulation and post-hatch acute heat stress on the mRNA expression of interleukin-6 and genes involved in its induction pathways in 2 broiler chicken breeds *Poult. Sci.*, 98(4), 1805–1819. https://doi.org/10.3382/ps/pey499.

Amad, A., et al. (1992), Differentiation of fattening and slaughtering performance and carcass quality of broiler genotypes with and without the dwarf factor and fast or slow feathering under heat stress. *Beiträge zur tropischen Landwirtschaft und Veterinärmedizin*, 30(4), 407–425.

Attia, Y.A., et al. (2011), Effect of ascorbic acid or increasing metabolizable energy level with or without supplementation of some essential amino acids on productive and physiological traits of slow-growing chicks exposed to chronic heat stress. *J. Anim. Physiol. Anim. Nutr.*, 95(6), 444–755.

Bagath, M., et al. (2019), The impact of heat stress on the immune system in dairy cattle: a review. *Res. Vet. Sci.*, 126, 94–102. https://doi.org/10.1016/j.rvsc.2019.08.011.

Bai, H., et al. (2024), Whole-genome resequencing identifies candidate genes associated with heat adaptation in chickens. *Poult. Sci.*, 103, 104139. https://doi.org/10.1016/j.psj.2024.104139.

Bai, S., et al. (2020), Effects of dietary inclusion of Radix Bupleuri and Radix Astragali extracts on the performance, intestinal inflammatory cytokines expression, and hepatic antioxidant capacity in broilers exposed to high temperature. *Anim. Feed Sci. Technol.*, 259, 114288. https://doi.org/10.1016/j.anifeedsci.2019.114288.

Baumgard, L.H. and Rhoads, R.P. (2013), Effects of heat stress on postabsorptive metabolism and energetics. *Annu. Rev. Anim. Biosci.*, 1, 311–317. https://doi.org/10.1146/annurev-animal-031412-103644.

Belay, T. and Teeter, R. (1992), Broiler water balance and thermobalance during thermoneutral and high ambient temperature exposure. *Poult. Sci.*, 72(1), 116–124. https://doi.org/10.3382/ps.0720116.

Bell, D.D. and Weaver, W.D. (2002), Commercial chicken meat and egg production: 5th edition. *J. Appl. Poult. Res.*, 11(2), 224–225. https://doi.org/10.1093/japr/11.2.224.

Branco, T., et al. (2021), The sequential behavior pattern analysis of broiler chickens exposed to heat stress. *Agric. Eng.*, 3, 447–457. https://doi.org/10.3390/agriengineering3030030.

Braun, E.J. (2015), Osmoregulatory systems of birds. In Scanes, G.C. (Eds.), *Sturkie's Avian Physiology: Sixth Edition*. Academic Press, pp. 285–300.

Brown-Brandl, T., et al. (2003), A new telemetry system for measuring Core body temperature in livestock and poultry. *Appl. Eng. Agric.*, 19, 583–589.

Brugaletta, G., et al. (2021), Effect of cyclic heat stress on feeding-related hypothalamic neuropeptides of three broiler populations and their ancestor jungle fowl. *Front. Physiol.*, 12, 809341. https://doi.org/10.3389/fphys.2021.809341.

Budi, T., et al. (2024), Thai local chicken breeds, chee fah and fah luang, originated from chinese black-boned chicken with introgression of Red junglefowl and domestic chicken breeds. *Sustainability*, 15(8), 6678. https://doi.org/10.3390/su15086878.

Bush, E.C., et al. (2017), PLATE-Seq for genome-wide regulatory network analysis of high-throughput screens. *Nat. Commun.*, 8(1), 105. https://doi.org/10.1038/s41467-017-00136-z.

Calik, A., et al. (2022), Influence of dietary vitamin E and selenium supplementation on broilers subjected to heat stress, Part II: oxidative stress, immune response, gut integrity, and intestinal microbiota. *Poult. Sci.*, 101(6), 101858. https://doi.org/10.1016/j.psj.2022.101858.

Campbell, D.L.M., et al. (2019), Application of open field, tonic immobility, and attention bias tests to hens with different ranging patterns. *Peer. J.*, 11, e8122. https://doi.org/10.7717/peerj.8122.

Cantet, J.M., et al. (2021). Heat stress-mediated activation of immune-inflammatory pathways. *Antibiotics (Basel, Switzerland)*, 10, 1285. https://doi.org/10.3390/antibiotics10111285.

Cartoni Mancinelli, A., et al. (2023), Impact of chronic heat stress on behavior, oxidative status and meat quality traits of fast-growing broiler chickens. *Front. Physiol.*, 14, 1242094. https://doi.org/10.3389/fphys.2023.1242094.

Cassuce, D., et al. (2013), Thermal comfort temperature update for broiler chickens up to 21 days of age. *Engenharia Agrícola*, 33, 28–36. https://doi.org/10.1590/S0100-69162013000100004.

Cedraz, H., et al. (2017), Heat stress induces expression of HSP genes in genetically divergent chickens. *PLoS One*, 12(10), e0186083. https://doi.org/10.1371/journal.pone.0186083.

Chaudhary, A., et al. (2023), Dietary supplementation of microalgae mitigates the negative effects of heat stress in broilers. *Poult. Sci.*, 102(10), 102958. https://doi.org/10.1016/j.psj.2023.102958.

Chauhan, S.S. and England, E.M. (2018), Postmortem glycolysis and glycogenolysis: insights from species comparisons. *Meat. Sci.*, 144, 118–126. https://doi.org/10.1016/j.meatsci.2018.06.021.

Daghir, N.J. (2008), Nutrient requirements of poultry at high temperatures. In: Gowe, R., et al. (eds) *Poultry Production in Hot Climates: Second Edition*. CAB International, pp. 132–159.

Das, S., et al. (2010), Stress inducible heat shock protein 70: a potent molecular and toxicological signature in arsenic exposed broiler chickens. *Mol. Biol. Rep.*, 37(7), 3151–3155. https://doi.org/10.1007/s11033-009-9894-7.

Del Valle, J.E., et al. (2021), Unrest index for estimating thermal comfort of poultry birds (*Gallus gallus domesticus*) using computer vision techniques. *Biosyst. Eng.*, 206, 123–134. https://doi.org/10.1016/j.biosystemseng.2021.03.018.

Del Vesco, A.P., et al. (2017), Age-related oxidative stress and antioxidant capacity in heat-stressed broilers. *Animals*, 11, 1783–1790.

Dennis, R.L., et al. (2013), Effects of early serotonin programming on behavior and central monoamine concentrations in an avian model. *Behav. Brain Res.*, 253, 290–296. https://doi.org/10.1016/j.bbr.2013.07.043.

DeSmit, L., et al. (2005), Comparison of three lines of broilers differing in ascites susceptibility or growth rate. 2. Egg weight loss, gas pressures, embryonic heat production, and physiological hormone levels. *Poult. Sci.*, 84, 1446–1452. https://doi.org/10.1093/ps/84.9.1446.

Dozier, W.A., et al. (2011), Apparent metabolizable energy needs of male and female broilers from 36 to 47 days of age1. *Poult. Sci.*, 90, 804–814. https://doi.org/10.3382/ps.2010-01132.

Elbaz, A.M., et al. (2023), Effectiveness of probiotics and clove essential oils in improving growth performance, immuno-antioxidant status, ileum morphometric, and microbial community structure for heat-stressed broilers. *Sci. Rep.*, 13(1), 18846. https://doi.org/10.1038/s41598-023-45868-9.

England, A., et al. (2023), Influence of sex and rearing method on performance and flock uniformity in broilers-implications for research settings. *Anim. Nutr.*, 12, 276–283.

Fairchild, B.D. and Ritz, C.W. (2015), *Poultry Drinking Water Primer*. University of Georgia, Cooperative Extension Bulletin No. 1301.

FAO. (2023), *The Impact of Disasters on Agriculture and Food Security 2023 – Avoiding and Reducing Losses Through Investment in Resilience*. Food and Agriculture Organization of the United Nations. https://doi.org/10.4060/cc7900en:1-168. (accessed 30 December 2022).

FAO. (2024), *Production-Gateway to Poultry Production and Products*. Food and Agriculture Organization of the United Nations. https://www.fao.org/poultry-production-products/en/.

Fathi, M.M., et al. (2022), Using major genes to mitigate the deleterious effects of heat stress in poultry: An updated review. *Poult. Sci.*, 101, 102157.

Favatier, F., et al. (1997), Variation in hsp gene expression and Hsp polymorphism: do they contribute to differential disease susceptibility and stress tolerance? *Cell Stress Chaperones*, 2(3), 141–155. https://doi.org/10.1379/1466–1268.

Galal, A., et al. (2019), Expression levels of HSP70 and CPT-1 in three local breeds of chickens reared under normal or heat stress conditions after the introduction of the naked neck gene. *J. Therm Biol.*, 80, 113–118.

Gan, J.K., et al. (2015), Analysis of genetic diversity of the heat shock protein 70 gene on the basis of abundant sequence polymorphisms in chicken breeds. *Genet. Mol. Res.*, 14, 1538–1545.

Gebreselassie, G., et al. (2020), Review on genomic regions and candidate genes associated with economically important production and reproduction traits in sheep (Ovies aries). *Animals*, 10(1), 33. https://doi.org/10.3390/ani10010033.

Geraert, P.A., et al. (1996), Metabolic and endocrine changes induced by chronic heat exposure in broiler chickens: growth performance, body composition and energy retention. *Br. J. Nutr.*, 75(2), 195–204.

Giloh, M., et al. (2012), Skin surface temperature of broiler chickens is correlated to body core temperature and is indicative of their thermoregulatory status, *Poult. Sci.*, 91, 175–188.

Goel, A., et al. (2022), Dietary shredded steam-exploded pine particle supplementation as a strategy to mitigate chronic cyclic heat stress by modulating gut microbiota in broilers. *Sci. Rep.*, 12(1), 19704. https://doi.org/10.1038/s41598-022-24031-w.

Goel, A., et al. (2023), Dietary supplementation of solubles from shredded, steam-exploded pine particles modifies gut length and cecum microbiota in cyclic heat-stressed broilers. *Poult. Sci.*, 102(4), 102498. https://doi.org/10.1016/j.psj.2023.102498.

Gonzalez-Rivas, P.A., et al. (2020), Effects of heat stress on animal physiology, metabolism, and meat quality: a review. *Meat Sci.*, 162, 108025.

Greene, E.S., et al. (2024), Effect of heat stress on the hypothalamic expression of water channel- and noncoding RNA biogenesis-related genes in modern broilers and their ancestor red jungle fowl. *Brain Res.*, 1830, 148810. https://doi.org/10.1016/j.brainres.2024.148810.

Gregory, N.G. (2010), How climatic changes could affect meat quality. *Food Res. Int.*, 43(7), 1866–1873. https://doi.org/10.1016/j.foodres.2009.05.018.

Haldar, S., et al. (2015), Effects of supplementation of betaine hydrochloride on physiological performances of broilers exposed to thermal stress. *Open Access Anim. Physiol.*, 7, 111–120.

Hatfield, J., et al. (2014), Ch. 6: Agriculture. In: Melillo, J. M., et al. (eds) *Climate change impacts in the United States: The Third National Climate Assessment*. U.S. Global Change Research Program, pp. 150–174. https://doi.org/10.7930/J02Z13FR.

Havenstein, G., et al. (2003), Growth, livability, and feed conversion of 1957 versus 2001 broilers when fed representative 1957 and 2001 broiler diets. *Poult. Sci.*, 82(10), 1500–1508. https://doi.org/10.1093/ps/82.10.1500.

He, S., et al. (2020), Characterization of heat stress affecting the growth performance, blood biochemical profile, and redox status in male and female broilers at market age. *Trop. Anim. Health Prod.*, 52, 3833–3841.

Hisaeda, H., et al. (1997), Heat shock protein 65 induced by gammadelta T cells prevents apoptosis of macrophages and contributes to host defense in mice infected with Toxoplasma gondii. *J. Immunol.*, 159(5), 2375–2381.

Hoffmann, T.J., et al. (2011), Next generation genome-wide association tool: Design and coverage of a high-throughput European-optimized SNP array. *Genomics*, 98(2), 79–89.

Ibtisham, F., et al. (2018), Effect of nutrition on reproductive efficiency of dairy animals. *Med. Weter.*, 74, 356–361.

Intergovernmental Panel on Climate Change (IPCC) (2021), Summary for policymakers. In: V. Masson-Delmotte, et al. (eds) *Climate Change 2021: The Physical Science Basis. Contribution of Working Group I to the Sixth Assessment Report of the Intergovernmental Panel on Climate Change*. Cambridge University Press, pp. 3–32.

Ismail, I. and Joo, S.T. (2017), Poultry meat quality in relation to muscle growth and muscle fiber characteristics. *Korean J. Food Sci. Anim. Resour.*, 37, 873–883. https://doi.org/10.5851/kosfa.2017.37.6.87.

Jahanian, R. and Rasouli, E. et al. (2015), Dietarychromium methionine supplementation could alleviate immunosuppressive effects of heat stress in broiler chicks. *Journal of Animal Science*, 93(7).

Jin, Y.Y., et al. (2022), Effect of heat stress on ileal microbial community of indigenous yellow-feather broilers based on 16S rRNA gene sequencing. *Vet. Med. Sci.*, 8(2), 642–653.

Jing, J., et al. (2024), Selenomethionine alleviates chronic heat stress-induced breast muscle injury and poor meat quality in broilers via relieving mitochondrial dysfunction and endoplasmic reticulum stress. *Anim Nutr.* 16, 363–375. https://doi.org/10.1016/j.aninu.2023.12.008.

Jinxue, D., et al. (2023), Sex-based responses of heat stress and subsequent recovery on the growth performance, metabolic changes, and redox status of broilers at market age. *Int. J. Biometeorol.*, 67, 1669–1677.

Juiputta, J., et al. (2023), Appropriate genetic approaches for heat tolerance and maintaining good productivity in tropical poultry production: a review. *Vet. Sci.*, 10(10), 591. https://doi.org/10.3390/vetsci10100591.

Ju, X., et al. (2023), DNA methylation in poultry: a review. *J. Anim. Sci. Biotechnol.*, 14(1), 138. https://doi.org/10.1186/s40104-023-00939-9.

Kadim I.T., et al. (2014), Effects of the transportation during hot season and electrical stimulation on meat quality characteristics of goat Longissimus dorsi muscle. *Small Rumin. Res.*, 121(1), 120–124. https://doi.org/10.1016/j.smallrumres.2014.01.010.

Kampinga, H.H. and Craig, E.A. (2010), The HSP70 chaperone machinery: proteins as drivers of functional specificity. *Nat. Rev. Mol. Cell Biol.*, 11(8), 579–592.

Karaarslan, S., et al. (2024), Effects of cooled perch and different floor types on animal- and management-based welfare indicators, heterophil/lymphocyte ratio, and perching behavior in broiler chickens reared at high ambient temperature. *Res. Vet. Sci.*, 180, 105433.

Key, N., et al. (2014), Climate change, heat stress, and U.S. Dairy Production. U.*S. Dept. Agric. Econom. Res. Ser.*, 175, 1–45.

Khattak, F.M., et al. (2012), Comparative efficacy of different supplements used to reduce heat stress in broilers. *Pakistan J. Zool.*, 44(1), 31–41.

Kishawy, A.T.Y., et al. (2023), Impact of resveratrol-loaded liposomal nanocarriers on heat-stressed broiler chickens: Effects on performance, sirtuin expression, oxidative stress regulators, and muscle building factors. *Front. Vet. Sci.*,10, 1137896. https://doi.org/10.3389/fvets.2023.1137896.

Kolluri, G., et al. (2022), Effects of spirulina (Arthrospira platensis) as a drinking water supplement during cyclical chronic heat stress on broiler chickens: Assessing algal composition, production, stress, health and immune-biochemical indices. *J. Therm. Biol.*, 103, 103100. https://doi.org/10.1016/j.jtherbio.2021.103100.

Koslová, A., et al. (2020), Precise CRISPR/Cas9 editing of the NHE1 gene renders chickens resistant to the J subgroup of avian leukosis virus. *PNAS*, 117(4), 2108–2112. https://doi.org/10.1073/pnas.1913827117.

Kpomasse, C.C., et al. (2023). Juvenile growth, thermotolerance and gut histomorphology of broiler chickens fed curcuma longa under hot-humid environments. Heliyon, 9, e13060. https://doi.org/10.1016/j.heliyon.2023.e13060.

Kpomasse, C.C., et al. (2021), Broiler production challenges in the tropics: a review. *Vet. Med. Sci.*, 7, 831–842.

Lara, L.J. and Rostagno, M.H. (2013), Impact of heat stress on poultry production. *Animals*, 3(2), 356–369. https://doi.org/10.3390/ani3020356.

Lin, H., et al. (2006), Strategies for preventing heat stress in poultry. *World's Poult. Sci. J.*, 61(1), 71–85.

Liu, L., et al. (2020), Heat stress impacts on broiler performance: a systematic review and meta-analysis. *Poult. Sci.*, 99, 6205–6211.

Liu, W., et al. (2019), Effects of dietary betaine on growth performance, digestive function, carcass traits, and meat quality in indigenous yellow-feathered broilers under long-term heat stress. *Animals*, 9(8), 506. https://doi.org/10.3390/ani9080506.

Liu, W.C., et al. (2022), Effects of heat stress on production performance, redox status, intestinal morphology and barrier-related gene expression, cecal microbiome, and metabolome in indigenous broiler chickens. *Front. Physiol.*, 13, 890520.

Liu, Y., et al. (2023), Insight on the meat quality and carbonylation profile of breast muscle of broilers in response to chronic heat stress: a proteomic research. *Food Chem.*, 423, 136437.

Liu, Y.S., et al. (2024), Early-age heat exposure improved growth performance and heat tolerance in broilers. *Poult. Sci.*, 103, 104254. https://doi.org/10.1016/j.psj.2024.104254.

Liverpool-Tasie, L.S.O., et al. (2019), Climate change adaptation among poultry farmers: evidence from Nigeria. *Clim. Change.*, 157(3), 527–544. https://doi.org/10.1007/s10584-019-02574-8.

Lourençoni, D., et al. (2019), Productive responses from broiler chickens raised in different commercial production system - part II: Impact of climate change. *Eng. Agric.*, 39(1), 11–17.

Loyau, T., et al. (2016), Heritability of body surface temperature in hens estimated by infrared thermography at normal or hot temperatures and genetic correlations with egg and feather quality. *Animal*, 10, 1594–1601. https://doi.org/10.1017/S1751731116000616.

Lu, Q., et al. (2007), Effect of chronic heat exposure on fat deposition and meat quality in two genetic types of chicken. *Poult. Sci.*, 86(6), 1059–1064. https://doi.org/10.1093/ps/86.6.1059.

Ma, D., et al. (2024), Dietary peppermint extract inhibits chronic heat stress-induced activation of innate immunity and inflammatory response in the spleen of broiler chickens. *Animals*, 14(8), 1157.

Mack, L.A., et al. (2013), Genetic variations alter production and behavioral responses following heat stress in 2 strains of laying hens. *Poult. Sci.*, 92(2), 285–294.

Madkour, M., et al. (2022), Mitigating the detrimental effects of heat stress in poultry through thermal conditioning and nutritional manipulation. *J. Therm. Biol.*, 103, 103169. https://doi.org/:10.1016/j.jtherbio.2021.103169.

Mahasneh, Z.M.H., et al. (2024), Effects of medical plants on alleviating the effects of heat stress on chickens. *Poult. Sci.*, 103, 103391. https://doi.org/10.1016/j.psj.2023.103391.

Mancini, R.A. and Hunt, M. (2005), Current research in meat color. *Meat Sci.*, 71, 100–121. https://doi.org/10.1016/j.meatsci.2005.03.003.

Mathur, P.K. and Horst, P., (1994), Genotype by environment interactions in laying hens based on relationship between breeding values of sires in temperate and tropical environments. *Poult. Sci.*, 73, 1777–1784.

Maxwell, M.H., et al. (1992), The fine structure of broiler chicken blood cells, with particular reference to basophils, after severe heat stress. *Comp. Haematol. Int.*, 2, 190–200.

Medzhitov, R. and Horng, T. (2009), Transcriptional control of the inflammatory response. *Nat. Rev. Immunol.*, 10, 692–703.

Mitchell, M.A. and Carlisle, A.J. (1992), The effects of chronic exposure to elevated environmental temperature on intestinal morphology and nutrient absorption in the domestic fowl (Gallus domesticus). *Comp. Biochem. Physiol. A.*, 101(1), 137–142.

Moghbeli, M., et al. (2018), Evaluation of body surface temperature in broiler chickens during the rearing period based on age, air temperature and feather condition. *Iranian J. Appl. Anim. Sci.*, 8, 499–504.

Monson, M.S., et al. (2018), Immunomodulatory effects of heat stress and lipopolysaccharide on the bursal transcriptome in two distinct chicken lines. *BMC Genom.*, 19(1), 1–15.

Nagai, M. and Iriki, M. (2001), Changes in immune activities by heat stress. In: Kosaka, M., et al. (eds) *Thermotherapy for Neoplasia, Inflammation, and Pain*. Springer.

Nawaz, A.H., et al. (2021), Poultry response to heat stress: its physiological, metabolic, and genetic implications on meat production and quality including strategies to improve broiler production in a warming world. *Front. Vet. Sci.*, 8, 699081. https://doi.org/10.3389/fvets.2021.699081.

Nawaz, A.H., et al. (2023), Investigating the heat tolerance and production performance in local chicken breed having normal and dwarf size. *Animal* 17, 100707. https://doi.org/:10.1016/j.animal.2023.100707.

Niu, Z.Y., et al. (2009), Effects of different levels of vitamin E on growth performance and immune responses of broilers under heat stress. *Poult. Sci.*, 88(10), 2101–2107.

NOAA National Centers for Environmental Information (NCEI) U.S. Billion-Dollar Weather and Climate Disasters (2025), https://www.ncei.noaa.gov/access/billions/.

Oke, O.E., et al. (2017), Physiological responses and performance of broiler chickens offered olive leaf extract under hot humid tropical climate. *J. App. Poult. Res.*, 26(3), 376–382. https://doi.org/10.3382/japr/pfx005.

Oke, O.E., et al. (2021a), In ovo injection of black cumin (Nigella Sativa) extract on hatching and post hatch performance of thermally challenged broiler chickens during incubation, *Poult. Sci.*, 100, 100831. https://doi.org/10.1016/j.psj.2020.10.072.

Oke, O.E., et al. (2021b), Environmental stress and livestock productivity under hot-humid tropics: alleviation and future perspectives. *J. Therm. Biol.*, 100, 103077. https://doi.org/10.1016/j.jtherbio.2021.103077.

Oke, O.E., et al. (2022), Adaptation of the white fulani cattle to the tropical environment. *J. Therm. Biol.*, 110, 103372. https://doi.org/10.1016/j.jtherbio.2022.103372.

Oke, O.E., et al. (2024a), Climate change and broiler production. *Vet. Med. Sci.*, 10, e1416. https://doi.org/10.1002/vms3.1416.

Oke, O.E., et al. (2024b), Oxidative stress in poultry. *Poult. Sci.*, 104003. https://doi.org/10.1016/j.psj.2024.104003.

Oluwagbenga, E.M. and Fraley, G.S. (2023), Heat stress and poultry production: a comprehensive review. *Poult. Sci.*, 102, 103141.

Onagbesan, O.M., et al. (2023), Alleviating heat stress effects in poultry: updates on methods and mechanisms of actions. *Front. Vet. Sci.*, 26(10), 1255520.

Oni, A.I., et al. (2024), The roles of phytogenic feed additives in stress mitigation in broiler chickens. *J. Anim. Physiol. Anim. Nutr.*, 108(1), 81–98. https://doi.org/10.1111/jpn.13869.

Park, J.S., et al. (2022), Proteomic changes in broiler liver by body weight differences under chronic heat stress. *Poult. Sci.*, 101(5), 101794.

Park, J.S., et al. (2020), Precise genome editing in poultry and its application to industries. *Genes*, 11, 1182. https://doi.org/10.3390/genes11101182.

Pawar, S., et al. (2016), Assessing and mitigating the impact of heat stress in poultry. *Adv. Anim. Vet. Sci.*, 4, 332–341.

Pereira, D.F., et al. (2020), Cluster index for estimating thermal poultry stress (*Gallus gallus domesticus*). *Comput. Electron. Agric.*, 177, 105704.

Perini, F., et al. (2021), Emerging genetic tools to investigate molecular pathways related to heat stress in chickens: a review. *Animals*, 11, 46.

Prates, J.A. (2025), Impact of heat stress on carcass traits, meat quality, and nutritional value in monogastric animals: Underlying mechanisms and nutritional mitigation strategies. *Foods*, 14(9), 1612. https://doi.org/10.3390/foods14091612.

Quinteiro-Filho, W.M., et al. (2010), Heat stress impairs performance parameters, induces intestinal injury, and decreases macrophage activity in broiler chickens. *Poult. Sci.*, 89(9), 1905–1914. https://doi.org/10.3382/ps.2010-00812.

Rajaei-Sharifabadi, H., et al. (2017), Surface wetting strategy prevents acute heat exposure-induced alterations of hypothalamic stress- and metabolic-related genes in broiler chickens. *J. Anim. Sci.*, 95, 1132–1143.

Ranjan, A., et al. (2019), Effect of heat stress on poultry production and their managemental approaches. *Int. J. Curr. Microbiol. App. Sci.*, 8(2), 1548–1555.

Ratriyanto, A. and Mosenthin, R. (2018), Osmoregulatory function of betaine in alleviating heat stress in poultry. *J. Anim. Physiol. Anim. Nutr.*, 102(6), 1634–1650.

Renaudeau, D., et al. (2012), Adaptation to hot climate and strategies to alleviate heat stress in livestock production. *Animal*, 6, 707–728.

Ricroch, A., et al. (2019), Global developments ofgenome editing in agriculture. *Transgenic Research,* 28.

Rivera, R.E., et al. (2005), Influence of selenium on heat shock protein 70 expression in heat stressed turkey embryos (*Meleagris gallopavo*). *Comp. Biochem. Physiol. A Mol. Integr. Physiol.*, 142(4), 427–432.

Rostagno, M.H. (2020), Effects of heat stress on the gut health of poultry. *J. Anim. Sci.*, 98(4), skaa090. https://doi.org/10.1093/jas/skaa090.

Saeed, M., et al. (2019), Heat stress management in poultry farms: a comprehensive overview. *J. Therm. Biol.*, 84, 414–425. https://doi.org/10.1016/j.jtherbio.2019.07.025.

Saelao, P., et al. (2019), Genetics and genomic regions affecting response to newcastle disease virus infection under heat stress in layer chickens. *Genes*, 10(1), 61. https://doi.org/10.3390/genes10010061.

Saleh, A.A., et al. (2023), Effect of dietary supplementation of betaine and organic minerals on growth performance, serum biochemical parameters, nutrients digestibility, and growth-related genes in broilers under heat stress. *Poult Sci.*, 102(11), 103051. https://doi.org/10.1016/j.psj.2023.103051.

Sandercock, D.A., et al. (2001), Acute heat stress-induced alterations in blood acid-base status and skeletal muscle membrane integrity in broilers at two ages: implications for Meat Quality. *Poult. Sci.*, 80, 418–425.

Sarica, M., et al. (2014), Comparing slow-growing chickens produced by two- and three-way crossings with commercial genotypes. 2. Carcass quality and blood parameters. *Eur. Poult. Sci.*, 78, 1–12.

Scanes, C.G. and Christensen, K.D. (2020), *Poultry Science*. 5th Edition. Waveland press.

Shao, D., et al. (2019), Effects of cyclic heat stress on the phenotypic response, meat quality and muscle glycolysis of breasts and thighs of yellow-feather broilers. *Ital. J. Anim. Sci.*, 18(1), 301–308. https://doi.org/10.1080/1828051X.2018.15200.

Shi, D., et al. (2019), Impact of gut microbiota structure in heat-stressed broilers. *Poult. Sci.*, 98, 2405–2413.

Soglia, D., et al. (2020), Growth performance analysis of two italian slow-growing chicken breeds: Bianca di saluzzo and Bionda piemontese. *Animals*, 10(6), 969.

Sohail, M.U., et al. (2012), Effect of supplementation of prebiotic mannan-oligosaccharides and probiotic mixture on growth performance of broilers subjected to chronic heat stress. *Poult. Sci.*, 91, 2235–2240.

Soleimani, A.F., et al. (2011), Physiological responses of 3 chicken breeds to acute heat stress. *Poult. Sci.*, 90, 1435–1440.

Song, J., et al. (2013), Cello-oligosaccharide ameliorates heat stress-induced impairment of intestinal microflora, morphology and barrier integrity in broilers. *Anim. Feed Sci. Technol.*, 185, 175–181. https://doi.org/10.1016/j.anifeedsci.2013.08.001.

St-Pierre, N.R., et al. (2003), Economic losses from heat stress by US livestock industries. *J. Dairy Sci.*, 86(SUPPL. 1), E52–E77.

Suman, S.P., et al. (2016), Factors influencing internal color of cooked meats. *Meat Sci.*, 120, 133–144. https://doi.org/10.1016/j.meatsci.2016.04.006.

Sundström, J.F., et al. (2014), Future threats to agricultural food production posed by environmental degradation, climate change, and animal and plant diseases–a risk analysis in three economic and climate settings. *Food Sec.*, 6, 201–215.

Syafwan, S., et al. (2012). Dietary self-selection by broilers at normal and high temperature changes feed intake behavior, nutrient intake, and performance. *Poult. Sci.*, 91(3), 537–549. https://doi.org/10.3382/ps.2011-01559.

Te Pas, M.F.W., et al. (2019), Transcriptomic profiles of muscle, heart, and spleen in reaction to circadian heat stress in Ethiopian highland and lowland male chicken. *Cell Stress Chap.*, 24, 175–194.

Temim, S., et al. (2000), Chronic heat exposure alters protein turnover of three different skeletal muscles in finishing broiler chickens fed 20 or 25% protein diets. *J. Nutr.*, 130, 813–819. https://doi.org/10.1093/jn/130.4.813.

Teyssier, J.R., et al. (2023), Influence of different heat stress models on nutrient digestibility and markers of stress, inflammation, lipid, and protein metabolism in broilers. *Poult. Sci.*, 102, 103048.

Tickle, P.G., et al. (2018), Energy allocation and behaviour in the growing broiler chicken. *Sci. Rep.*, 8(1), 4562. https://doi.org/10.1038/s41598-018-22604-2.

Toghyani, M., et al. (2012), Chromium supplementation can alleviate the negative effects of heat stress on growth performance, carcass traits, and meat lipid oxidation of broiler chicks without any adverse impacts on blood constituents. *Biol. Trace Elem. Res.*, 146, 171–180. https://doi.org/10.1007/s12011-011-9234-3.

USDA-NASS. (2024), *Poultry - Production and Value 2023 Summary (April 2024)*. United States Department of Agriculture-National Agricultural Statistics Service: 1–14.

Uyanga, V.A., et al. (2021a), Potential implications of citrulline and quercetin on gut functioning of monogastric animals and humans: a comprehensive review. *Nutri.*, 25(13), 3782.

Uyanga, V. A., et al. (2021b), L-Citrulline influences the body temperature, heat shock response and nitric oxide regeneration of broilers under thermoneutral and heat stress conditions. *Front. Physiol.*, 12, 671691.

Uyanga, V.A., et al. (2022a), Functional roles of taurine, L-theanine, L-citrulline, and betaine during heat stress in poultry. *J. Anim. Sci. Biotechnol.*, 13, 23.

Uyanga, V.A., et al. (2022b), Dietary L-citrulline influences body temperature and inflammatory responses during nitric oxide synthase inhibition and endotoxin challenge in chickens. *Stress*, 25, 74–86.

Uyanga, V.A., et al. (2022c), Dietary L-citrulline modulates the growth performance, amino acid profile, and the growth hormone/insulin-like growth factor axis in broilers exposed to high temperature. *Front. Physiol.*, 13, 937443.

Uyanga, V.A., et al. (2023), Transcriptomics analysis unveils key potential genes associated with brain development and feeding behavior in the hypothalamus of L-citrulline-fed broiler chicken. *Poult. Sci.*, 102(12), 103136.

Van Goor, A., et al. (2016), Quantitative trait loci identified for blood chemistry components of an advanced intercross line of chickens under heat stress. *BMC Genom.*, 17(1), 287. https://doi.org/10.1186/s12864-016-2601-x.

Van Goor, A., et al. (2015), Identification of quantitative trait loci for body temperature, body weight, breast yield, and digestibility in an advanced intercross line of chickens under heat stress. *Genetics Selct. Evol.*, 47(1), 96. https://doi.org/10.1186/s12711-015-0176-7.

Vandana, G.D., et al. (2021), Heat stress and poultry production: impact and amelioration. *Int. J. Biomet.*, 65, 163–179.

Wang, G., et al. (2021), Effects of heat stress on gut-microbial metabolites, gastrointestinal peptides, glycolipid metabolism, and performance of broilers. *Animals*, 11, 1286. https://doi.org/10.3168/jds.S0022-0302(03)74043-0.

Wang, H., et al. (2023), Protective effects of dietary dimethyl itaconate supplementation on oxidative stress, inflammation, and apoptosis in broilers under chronic heat stress. *J. Anim. Sci.*, 101, 356. https://doi.org/10.1093/jas/skad356.

Wang Y., et al. (2018), Physiological responses to heat stress in two genetically distinctchicken inbred lines. *Poultry Science,* 97(3).

Wang, S.H., et al. (2015), Acute heat stress induces differential gene expressions in the testes of a broiler-type strain of Taiwan country chickens. *PLoS One*, 10(5), e0125816. https://doi.org/10.1371/journal.pone.0125816.

Wasti, S., et al. (2020), Impact of heat stress on poultry health and performances, and potential mitigation strategies. *Animals*, 10(8), 1266.

Wegele, H., et al. (2004), Hsp70 and Hsp90--a relay team for protein folding. *Rev. Physiol. Biochem. Pharmacol.*, 151, 1–44. https://doi.org/10.1007/s10254-003-0021-1.

Wolc, A., et al. (2019), Genome wide association study for heat stress induced mortality in a white egg layer line. *Poult. Sci.*, 98(1), 92–96. https://doi.org/10.3382/ps/pey403.

Xie, J., et al. (2014), Differential expression of heat shock transcription factors and heat shock proteins after acute and chronic heat stress in laying chickens (*Gallus gallus*). *PLoS One*, 9(7), e102204. https://doi.org/10.1371/journal.pone.0102204.

Xie, J., et al. (2015), Effects of acute and chronic heat stress on plasma metabolites, hormones and oxidant status in restrictedly fed broiler breeders. *Poult. Sci.*, 94, 1635–1644.

Xie, X.F., et al. (2024), Genome-wide scans for selection signatures in indigenous chickens reveal candidate genes associated with local adaptation. *Animal*, 18(5), 101151.

Xu, Y., et al. (2018), Effect of chronic heat stress on some physiological and immunological parameters in different breed of broilers. *Poult. Sci.*, 97, 4073–4082.

Yahav, S. (2000), Relative humidity at moderate ambient temperatures: its effect on male broiler chickens and turkeys. *Br. Poult. Sci.*, 41, 94–100.

Yahav, S. (2009), Alleviating heat stress in domestic fowl: Different strategies. *World's Poult. Sci. J.*, 65, 719–732.

Yahav, S., et al. (2005), Air velocity alters broiler performance under harsh environmental conditions. *Poult. Sci.*, 83(7), 1210–1220.

Yang, Y.-Y., et al. (2023), Biogenic selenium nanoparticles synthesized using alginate oligosaccharides attenuate heat stress-induced impairment of breast meat quality via regulating oxidative stress, metabolome and ferroptosis in broilers. *Antioxidants*, 12(12), 2032. https://doi.org/10.3390/antiox12122032.

Zeferino, C.P., et al. (2016), Carcass and meat quality traits of chickens fed diets concurrently supplemented with vitamins C and E under constant heat stress. *Animal*, 10, 163–171.

Zhang, C., et al. (2017a), Dietary resveratrol supplementation prevents transport-stress-impaired meat quality of broilers through maintaining muscle energy metabolism and antioxidant status. *Poult. Sci.*, 96(7), 2219–2225.

Zhang, J., et al. (2017b), Transcriptome analysis reveals potential mechanisms underlying differential heart development in fast-and slow-growing broilers under heat stress. *BMC Genom.*, 18(1), 295. https://doi.org/10.1186/s12864-017-3675-9.

Zhang, Z.F., et al. (2012a), Effects of B-glucan and Bacillus subtilis on growth performance, blood profiles, relative organ weight and meat quality in broilers fed maize-soybean meal based diets. *Livest. Sci.*, 150(1–3), 419–424.

Zhang, Z.Y., et al. (2012b), Effects of constant and cyclic heat stress on muscle metabolism and meat quality of broiler breast fillet and thigh meat. *Poult. Sci.*, 91(11), 2931–2937.

Zhuang, Z.-X., et al. (2019), Genome-wide association study on the body temperature changes of a broiler-type strain Taiwan country chickens under acute heat stress. *J. Therm. Biol.*, 82, 33–42.

Zmrhal, V., et al. (2023), The influence of heat stress on chicken immune system and mitigation of negative impacts by baicalin and baicalein. *Animals*, 13, 2564. https://doi.org/10.3390/ani13162564.

Zulkifli, I., et al. (2009), Crating and heat stress influence blood parameters and heat shock protein 70 expression in broiler chickens showing short or long tonic immobility reactions. *Poult. Sci.*, 88(3), 471–476.

# Part 5

## Precision farming technologies in broiler production

# Chapter 16

## Sensor technologies for monitoring broiler welfare indicators

*Malou van der Sluis, Wageningen University & Research, The Netherlands*

1 Introduction
2 Types of sensor technologies
3 Implementing sensors for recording welfare indicators
4 Case studies
5 Conclusion and future trends
6 Where to look for further information
7 References

## 1 Introduction

The welfare of farm animals in production systems, including meat chickens (or broilers), is considered more and more important by society, as shown for example in a recent survey by the European Commission. According to the survey, 91% of respondents believe it is important to protect the welfare of farmed animals, and 84% indicated they believe the welfare of farmed animals should be better protected than was done at that point in time (in 2023; European Commission, 2023).

Monitoring broiler welfare is challenging for several reasons. First, welfare is an individual experience (Broom, 2010), meaning that resource-based or flock-level observations may be insufficient to obtain a full picture of the welfare of all individual broilers in the flock. Furthermore, the welfare of broilers is not a static state, but may constantly change. Therefore, frequent monitoring is required, ideally at the individual level.

Manually recording components of welfare (e.g. good walking ability, thermal comfort or absence of disease) is possible, for example using the Welfare Quality® protocol for poultry (Welfare Quality® Consortium, 2009). However, this is time-consuming to conduct and may be subjective. Moreover, it requires people to visit farms to do the scoring, which may be undesirable for biosecurity reasons.

http://dx.doi.org/10.19103/AS.2025.0165.19

Automated approaches for recording broiler welfare may provide solutions for the drawbacks of recording manually. Generally, it is not possible to record welfare directly or as a whole using sensors (similarly for manual assessment). However, different aspects of – or proxies for – welfare may be feasible to monitor, such as physical aspects of the broilers, resource use, activity, walking ability, vocalizations, or specific behaviours. Automated approaches are becoming increasingly popular in the field of broiler research (Fig. 1) and, even though most research is still in the experimental or small-scale phase, there appears to be potential for wider implementation in practice in the future.

Sensor technologies for monitoring welfare indicators can have applications in different settings. In research settings, these technologies can be used to assess the welfare impact of different interventions, for example when evaluating different housing systems for broilers. In breeding settings, sensor technologies can contribute to automated phenotyping for different components of welfare, which can subsequently be taken into account in broiler breeding programs. Finally, there is potential to use sensor technologies to monitor welfare on farms in practice, allowing for quick detection and intervention when welfare issues arise.

This chapter aims to provide an overview of how existing sensor technologies can be used to automatically monitor welfare indicators in broilers. To this end, different types of sensor technologies for recording animal-based indicators of welfare are discussed, including examples of existing implementations (Section 2). Subsequently, it is discussed what different types of recordings can be made with these sensors in different environments or situations and how to move from sensor data to implementation in practice (Section 3). Two case studies will provide more insight into broiler monitoring options (Section 4),

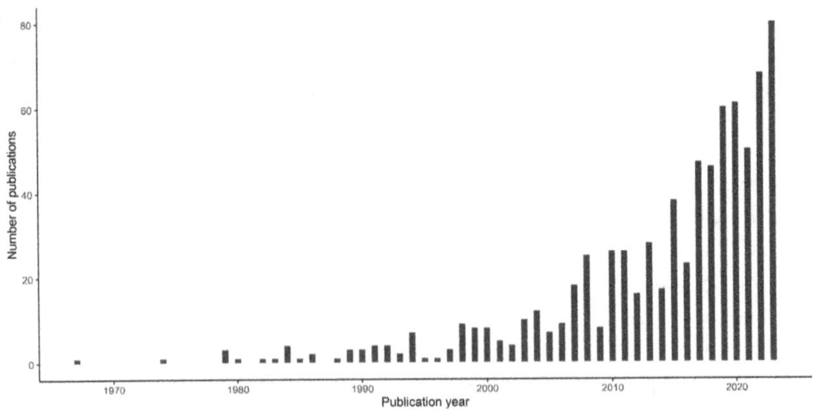


**Figure 1** Number of publications in the past decades (up to 2023) mentioning broiler and sensor, broiler and track* or broiler and automat*. Derived from Scopus, on the 6th of May 2024.

before an overall summary and a discussion of future trends will be presented (Section 5). Moreover, an overview table is presented that highlights examples of implementation of the different sensors that are discussed in this chapter.

## 2 Types of sensor technologies

Many sensor technologies exist that could be implemented to monitor welfare in broiler chickens. A general distinction can be made into body-worn sensor technologies, which make use of sensors attached to individual broilers, and remote technologies (Ellen et al., 2019), which make use of a sensor that is placed in the external environment in which the broilers are housed without attaching a device to individual broilers. In this section, these two general classes of sensor technologies are described in more detail, including examples of existing implementations, and the pros and cons of both will be discussed. Furthermore, the potential of combining body-worn and remote sensors will be discussed.

### *2.1 Body-worn sensor technologies*

A variety of body-worn sensors exist, that have been implemented for broilers mainly in research settings (Table 1). In this section, three main technologies will be discussed in more detail: accelerometers, radio frequency identification (RFID) systems and ultra-wideband (UWB) systems.

#### *2.1.1 Accelerometers*

Accelerometers are electromechanical devices that can detect acceleration forces (Chapa et al., 2020). Microscopic crystals within the accelerometer are stressed by movement, which results in a voltage (Benjamin and Yik, 2019). The size of this voltage is subsequently used to determine the orientation and velocity of the movement (Chapa et al., 2020). Tri-axial accelerometers gather information in the x-, y- and z-directions, i.e. they collect three-dimensional information, and furthermore, record the gravitational pull based on the angle of the device (Benjamin and Yik, 2019). In this way, accelerometers record movement data. Interpretation of the movements is part of the post-processing of the data (see Section 3).

Depending on the type of accelerometer, the data generated by the accelerometer can either be stored on the device itself, which then needs to be (periodically) retrieved from the animal to access the data, or can be downloaded using remote data transfer technology (Brown et al., 2013). As mentioned earlier, the data that accelerometers generate are generally not readily interpretable and require subsequent processing to obtain insight into displacement and velocity, and the underlying specific behaviour patterns (Benjamin and Yik, 2019; Chapa et al., 2020).

**Table 1** Examples of the use of sensor technologies for monitoring potential welfare indicators in broilers. Note that some indicators shown in the column of examples can be recorded in multiple ways (i.e. with different sensors)

| Category | Type of sensor | Subcategory | Recorded variable | Level of measurement[1] | Examples of implementations |
|---|---|---|---|---|---|
| Body-worn sensors | Accelerometer | | Acceleration forces (i.e. movement) | Individual | Classification of behaviours (Yang et al., 2021b); disease detection based on behaviour (Mei et al., 2023) |
| | Radio frequency identification | | Presence at RFID antenna | Individual | Feeding and drinking (Li et al., 2018); locomotor activity (van der Sluis et al., 2020; Doornweerd et al., 2023); range use (Collet et al., 2024; Taylor et al., 2017; Taylor et al., 2020) |
| | Ultra-wideband | | Location | Individual | Locomotor activity (van der Sluis et al., 2019; Giersberg et al., 2023); locomotor activity as proxy for gait (van der Sluis et al., 2021); range use (Stadig et al., 2018); tracking and movement patterns (Baxter and O'Connell, 2020; Baxter and O'Connell, 2023) |
| Remote sensors | Cameras | Optical flow/ pixel changes | Pixel changes | Group | Activity levels (van Leuffen et al., 2023); deviations in activity levels (Kristensen and Cornou, 2011); gait score prediction (Van Hertem et al., 2018; Silvera et al., 2017; Aydin et al., 2010); gait score distribution (van der Eijk et al., 2023); hock burn and mortality (Dawkins et al., 2012); footpad dermatitis (Dawkins et al., 2017); unrest index as indicator for thermal comfort (Del Valle et al., 2021); predicting equipment malfunction (e.g. of feeding, drinking, heating or ventilation systems; Kashiha et al., 2013) |

| | | | | |
|---|---|---|---|---|
| | Broiler detection and/or classification | (number of) broilers in an image | Individual | Detecting birds on floor (Guo et al., 2023); resource use (van der Eijk et al., 2022); detecting sick birds (Zhuang and Zhang, 2019); detecting dead birds (Hao et al., 2022); spatial distribution (Guo et al., 2020); body weight prediction (Amraei et al., 2018); comb colour detection as health indicator (Wei et al., 2024); detection over time to assess movement patterns (Doornweerd et al., 2023); displacement speed (Pereira et al., 2021); lying behaviour (Aydin et al., 2023); walking features (Aydin, 2017a); stretching behaviour (Li et al., 2021c); nine behaviours in broiler breeders (Pereira et al., 2013) |
| | Tracking | Location over time | Individual | Classification of activity levels (Campbell et al., 2024); mobility assessment (Jaihuni et al., 2023); tracking head at feeder as feeding indicator (Nasiri et al., 2023); tracking location and movement (Doornweerd et al., 2024); movement speed (Yang et al., 2024); walking around feeder (Li et al., 2021b); stretching and preening behaviour (Nasiri et al., 2024b); drinking behaviour (Nasiri et al., 2024a) |
| | Pose estimation | Relative position of body parts | Individual | Postures (Fang et al., 2022); beak and head motion during feeding (Abdanan Mehdizadeh et al., 2015); walking characteristics (Fodor et al., 2023); lameness recognition (Nasiri et al., 2022); behaviour classification (Fang et al., 2021) |
| | Infrared thermography | Temperatures in an image | Individual | Estimating body temperature (Bloch et al., 2020); head surface temperature as fever indicator (Xiong et al., 2019) |
| | Depth (i.e. 3D information) | Depth information in an image | Individual | Weight prediction (Mortensen et al., 2016); detecting lying events as indicator for gait scores (Aydin, 2017b); locomotion metrics (Li et al., 2023) |
| Force plates | | Forces | Individual/Group | Pecking patterns at feeder (Seber et al., 2022); gait characterization (Nääs et al., 2010) |

[1] Individual level does not necessarily mean 'identified' – it can be individual level without knowing which specific broiler over time.

The location of placement of the accelerometer on the animal's body may affect the precision of detection of specific movements, as some behaviours result in clear movement patterns in only certain parts of the body (Chapa et al., 2020; Brown et al., 2013). Moreover, secure, stable placement of the accelerometer is important, as the interpretation of the three-dimensional data may change when the tag shifts position relative to the position of the animal (Brown et al., 2013).

While location data cannot be obtained directly from accelerometers, these can be approximated through a process called 'dead-reckoning' or path integration when speed (derived from acceleration), heading and change in depth or altitude are available (Brown et al., 2013). However, the resulting location estimates are often not reliable, as there may be some general noise in the accelerometer data, and these small errors in movement detection may accumulate over time (Brown et al., 2013).

Existing implementations of accelerometers for monitoring welfare indicators in broilers include using the accelerometer data to extract information on, and classify, behaviour and to detect potential disease based on extracted behaviour (Table 1).

### 2.1.2 Radio frequency identification (RFID) and ultra-wideband (UWB)

The term RFID refers to systems that use radio frequency fields for wireless communication (Debouzy and Perrin, 2012). RFID systems typically consist of three components: tags, readers and a host system (Roberts, 2006). The tags can be attached to the animal of interest for tracking. Each tag includes an electronic microchip for identification and an antenna (Debouzy and Perrin, 2012). The readers can communicate with the tags via an antenna that works on the same radio frequency, and subsequently send the information that is retrieved from the tags to a host system (i.e. a computer) (Ngai et al., 2008). The resulting data generally consist of a log file with tag ID code, time of registration and location (i.e. antenna number) of registration.

There are many different types of RFID systems that vary in their specifications, affecting their respective suitability for different environments and applications. A first distinction can be made in the type of tag used. RFID tags can broadly be categorized as passive or active (Roberts, 2006). Passive tags, on the one hand, do not contain their own power supply and thus need to obtain power from the field of the reader to allow communication with the reader (Roberts, 2006). Active tags, on the other hand, contain a battery and can actively broadcast a signal to the readers (Roberts, 2006).

Whether a tag is passive or active is important for its potential implementation. On the one hand, passive tags can be smaller and more lightweight (Roberts, 2006) and are therefore more suitable for use on small chickens than the larger and heavier active tags. On the other hand, active tags generally have longer communication ranges (Roberts, 2006), meaning that the distance between the tag and reader can be much larger. What communication ranges can be achieved with the system depends on the operating frequency (see further on), but ranges up to 30 m may be achieved (Ruiz-Garcia and Lunadei, 2011). However, it is important to keep in mind that batteries might run low at some point, so these may need to be checked and replaced periodically.

A second distinction between different RFID systems can be made in the operating frequency. Generally, four main operating frequencies are distinguished: low frequency (LF, commonly operating at 125–134 KHz), high frequency (HF, commonly operating at 13.56 MHz), ultra-high frequency (UHF, commonly operating at 868–928 MHz) and microwave (commonly operating at 2.4 GHz) (Nekoogar and Dowla, 2011). However, microwave frequencies are not commonly used for animal tracking (Ellen et al., 2019). The operating frequency can affect the communication range of the RFID system as well as its sensitivity to interference from the environment. Generally speaking, with higher operating frequencies, the communication can be faster and over longer ranges (Ruiz-Garcia and Lunadei, 2011). At the same time, higher frequency systems are more prone to interference from liquids and metals than lower frequency systems (Ruiz-Garcia and Lunadei, 2011).

An exception to this general distinction in operating frequency are ultra-wideband systems, which do not use one specific radio frequency to communicate, but instead make use of low-power pulses across a wide portion of the frequency spectrum (Alarifi et al., 2016; Nekoogar and Dowla, 2011). This generally makes UWB systems less susceptible to signal interference in environments containing a lot of metal (Nekoogar and Dowla, 2011), which often applies to broiler housing.

Examples of implementation of RFID systems for monitoring indicators of broiler welfare include using the RFID detections to assess feeding and drinking, locomotor activity and range use (Table 1). Examples of UWB implementation for broilers include using the resulting location information to assess locomotor activity in itself or as a proxy for gait, range use, or to track broilers and assess movement patterns (Table 1).

To summarize, body-worn sensor technologies make use of a sensor attached to the animal of interest, and the most common body-worn sensors for broilers are accelerometers, RFID and UWB systems. A lot of variation exists in RFID systems in terms of power supply and operating frequency. In the case studies discussed in Section 4, examples of implementation of a passive, high-frequency RFID system and a UWB system are presented.

### 2.2 Remote sensor technologies

Remote sensor technologies make use of sensors that are placed in the environment in which the broilers are housed (Ellen et al., 2019). A variety of remote sensors have been implemented in broiler research (Table 1). Here, two main categories of remote technologies will be discussed in more detail: cameras (including thermal and depth cameras) and force plates, and examples of implementations of both technologies for monitoring broiler welfare will be given. Microphones and olfactory sensors are discussed elsewhere in this book.

#### 2.2.1 Cameras

Cameras form the main input for computer vision systems (Li et al., 2021a). Cameras are often installed to provide a top-view of the broilers, but can also be used to assess certain aspects from the front, side or back of the broilers (Li et al., 2021a). The cameras can be regular (RGB) cameras, but may also be other types of cameras, such as thermal or depth (i.e. 3D) cameras. The cameras send their output to recording units that collect the images or videos and save or transfer these for further processing (Li et al., 2021a). Video data can be analyzed in many ways, and a general distinction can be made in whether overall pixel changes are studied or (individual) chickens are recognized and assessed.

Optical flow is a method that assesses the change in pixels between consecutive frames as an indication of movement in the field of view (Li et al., 2020). It does not distinguish between different chickens that create the movement, but instead only records overall pixel changes. Subsequent analyses of optical flow data can determine, for example, the mean flow rate (i.e. the overall level of movement across a specified period of time), the variance in movement, the skew of movement and the kurtosis of movement (Dawkins et al., 2012). Such movement patterns have been shown to be linked to, for example, hock burn, foot pad dermatitis and gait scores of broiler flocks (Table 1).

Other analytical approaches focus on detecting and analyzing the objects of interest (in this case: chickens) in the images. Several main computer vision tasks can be defined: image classification, object detection, semantic or instance segmentation, pose estimation and tracking (reviewed by Li et al., 2021a; Fig. 2). Image classification concerns classifying an image as a whole, e.g. the image is classified as 'chickens'. Object detection consists of detecting and locating individual objects of interest (i.e. chickens) and adding bounding boxes around them. Segmentation tasks segment the objects of interest from a background, determining for each pixel whether or not it belongs to an object of interest. As opposed to object detection, it does not draw a bounding box around the chicken, but shows its full outlined shape. Pose estimation tasks determine key points and skeletons of the objects of interest and use

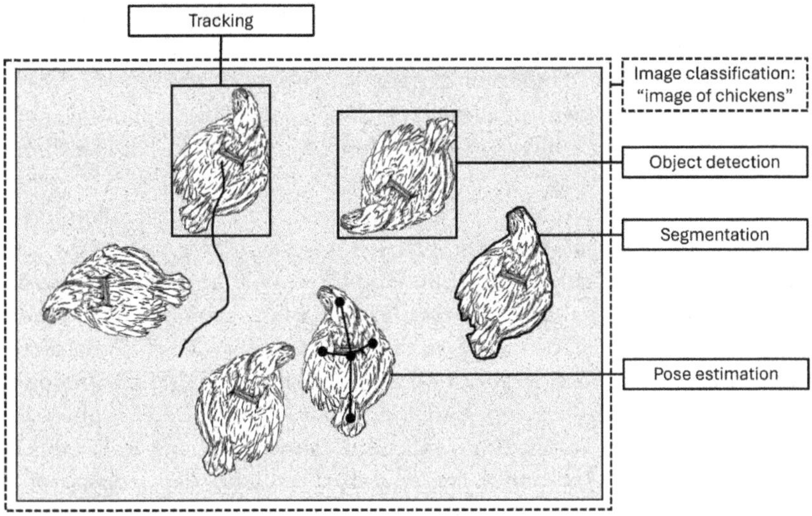


**Figure 2** Schematic overview of different computer vision tasks. Image classification, object detection, segmentation, pose estimation and tracking approaches are indicated.

these to estimate poses (Li et al., 2021a). Tracking tasks consist of detecting objects within video frames and linking these detections between consecutive frames (Doornweerd et al., 2024). These types of computer vision tasks require extensive training. Generally, between 1000 and 5000 images are labelled manually for training (Li et al., 2021a).

As presented in Table 1, object detection and/or classification approaches have for example been implemented for broilers to assess resource use (detecting how many broilers are present at a resource), spatial distribution, or to detect stretching behaviour. Tracking approaches have been implemented to, for example, assess mobility, location and movement of broilers. Pose estimation approaches, recording the relative position of body parts, have been used to classify different behaviours or to assess walking ability.

Specific types of cameras can be used to collect additional or alternative information. Infrared thermography can be used to capture the infrared radiation from a chicken and convert this to a temperature value that is displayed visually on a colour-scaled image (Gunaratnam et al., 2024). Infrared thermography can only measure surface temperatures, and areas of the body that are not covered in feathers provide a closer approximation of the body temperature than areas with feather coverage (Gunaratnam et al., 2024; Bloch et al., 2020). It is, however, important to note that the precision and outcomes of infrared thermography recordings depend on many factors, including environmental factors, whether the animal's body is clean and the animal's level of stress or discomfort (Gunaratnam et al., 2024), making it difficult to compare recordings made in different situations. Implementations of infrared thermography for monitoring broiler welfare indicators mainly focus on recording or estimating body temperature as an indicator of potential disease (Table 1).

Depth cameras can provide three-dimensional information, generally in one of three common ways (Brown-Brandl and Condotta, 2023): through stereoscopy, structured light or time-of-flight. In stereoscopy, two cameras are used to capture the same scene from two perspectives that differ slightly, which creates a measurable difference in the position of objects in the scene that can be used to calculate depth. Structured light approaches consist of projecting a known pattern of infrared dots on the scene and subsequently assessing the pattern distortion as an indicator for depth. Time-of-flight cameras use the time of flight of an artificial light signal to determine the distance between the camera and the scene it captures. The phase shift between the light signal and the reflection is recorded and translated into depth information. For depth cameras, the resulting data consist of the distance for every pixel in the image in a 2D array, forming a depth map of 3D points (Brown-Brandl and Condotta, 2023). The benefit of depth cameras is that they provide information in a third dimension, as opposed to regular cameras. Depth cameras have been implemented for broilers to, for example, predict body weight or detect lying events as an indicator for gait scores (Table 1).

### 2.2.2 Force plates

Force plate sensors are sensors that can measure ground reaction forces (Azarpajouh et al., 2022). Generally, these systems consist of two components: a hardware component, i.e. the force plate sensor, that records the force data, and a software component that processes, analyses and visualizes the force data (Nääs et al., 2010). The force sensor is made up of load cells, that contain piezoelectric elements, strain gauges or beam load cells (Lamkin-Kennard and Popovic, 2019). These distort in response to a force being applied on them, which causes a measurable change in voltage (Lamkin-Kennard and Popovic, 2019) that can be used for further analysis of the forces. Examples of the use of force plates for monitoring broiler welfare indicators include the detection of pecking patterns at the feeder or to characterize gait (Table 1).

To summarize, remote sensors are placed in the environment of the broilers, instead of being attached to the broilers themselves. Especially computer vision approaches offer many opportunities for assessing different behaviours or other welfare indicators. In Section 4, a case study using cameras for assessing walking ability in broilers is presented, in which a pose estimation approach is implemented.

## 2.3 Pros and cons of body-worn and remote sensors

Both body-worn and remote sensors come with their own benefits and downsides for implementation in broilers. In this section, some general benefits and downsides of both are discussed.

Main benefits of body-worn sensors include its reliable individual identification (one sensor is attached to one specific broiler) and its ability to function without a direct line of sight, both of which are convenient in broiler barns with many similar looking individuals. However, there are also downsides to body-worn sensors. First, they need to be attached to the broiler of interest. In this, there is a risk that the sensor itself may affect the chicken and its behaviour, albeit maybe only for a short acclimatization period (Anderson et al., 2023). Moreover, one sensor per broiler is required. This means that if you want to monitor a large flock, many sensors are needed, which may be expensive and time-consuming to fit to the broilers (Ellen et al., 2019). Furthermore, there may be a trade-off between, on the one hand, the size and weight of the sensor and, on the other hand, the battery life, sampling frequency and data storage capacities. Higher sampling frequencies may decrease the battery life, which results in the need to have either larger batteries, making the sensor larger and heavier, or to change the battery more often, which means more (and potentially stressful) interaction with the animal to demount and remount the sensor (Chapa et al., 2020).

Within the body-worn sensors for location tracking in broilers (UWB and RFID) further pros and cons can be observed. UWB tags can provide detailed location information due to its triangulation-based location determination, but have relatively heavy tags (due to the batteries) that make implementation of this system not feasible for very young broilers (although it must be noted that with the recent technological advancements, these tags will likely continue to become smaller and smaller). RFID tags are much lighter and can therefore already be implemented from hatching onwards. However, the level of detail that passive RFID can provide is more constrained, depending on the number of antennas, their size and their configuration.

Regarding remote sensors, main benefits include that they can often monitor multiple broilers at the same time, and therefore fewer sensors are needed to cover a larger group of broilers. However, remote sensors do pose a challenge in terms of individual identification, i.e. which specific individual you see, hear or detect with the sensor, and often require a direct line of sight to the animals of interest. Specifically for computer vision approaches, there are challenges with the farm environment, generalizability across different environments, time-consuming training and data storage. The farm environment can be challenging for computer vision tasks: chickens may occlude each other, there is variation in light conditions and it can be difficult to separate chickens from the background of the image (Okinda et al., 2020). Computer vision algorithms often generalize poorly across animals in different environments (Li et al., 2021a), resulting in additional (re)training work being required when implementing cameras in a novel setting. Moreover, recording video in high resolution and with high frame rates may result in very large datasets that require substantial storage capacity.

Overall, both body-worn and remote sensors come with their benefits and challenges. The main benefit of body-worn sensors is that they provide individual-level data and the identity of the broilers is known, given that a sensor is attached to a specific broiler. However, a main downside is that a sensor is needed for every broiler of interest. The main benefit of remote sensors is that they can record data on multiple broilers simultaneously, but individual identification of broilers is challenging.

### 2.4 Combining sensors

One way to overcome some of the challenges of body-worn and remote sensors is to implement a sensor-fusion approach, in which different sensors provide different aspects of the required information (Ellen et al., 2019). For example, computer vision and RFID can be combined, in which computer vision is used for the main behaviour quantification or tracking and RFID is used to support identification of individual chickens (e.g. Nakarmi et al. (2014) in laying hens; Doornweerd et al. (*in preparation*) in broilers), or in which thermal cameras are used to record body temperatures and RFID provides the chickens' identities (Bloch et al., 2020). Other examples include a combination of a rod platform weighing system and a camera system, where the camera system can be used to detect the number of broilers on the rod (Wang et al., 2018), or a combination of an electronic platform weigher and a depth camera, in which the depth camera system can classify the sex of the broiler based on its comb size (Liu et al., 2021). In this way, limitations of one sensor can be solved by using another, additional sensor. However, it does require two sets of equipment (body-worn and remote), and may thus require a larger (financial) investment. Moreover, the sensors need to be adequately synchronized in time to allow successful linking of the resulting data. In other words, both sensors should record the same timestamp for an event that they record simultaneously.

To summarize, a combination of sensors can be used to provide different aspects of the information one wants to collect. However, this requires the sensors to be synchronized well, to be able to link the resulting data.

## 3 Implementing sensors for recording welfare indicators

Animal welfare is difficult to capture as a whole at once, but it is possible to assess different components or aspects of welfare, using sensor technology.

For example, sensor technologies can help to record physical aspects of the chickens, resource use, activity, walking ability, and specific behaviour, or can help to detect potential diseases (Table 1).

In general, it is important to make a distinction between resource-based versus animal-based indicators. Given the aim of this chapter to provide insight into sensor technologies that can be used to monitor welfare indicators, the examples presented in Table 1 focus solely on animal-based indicators, as animal-based indicators are aimed at directly reflecting how well animals cope with their environment (Winckler, 2023). Moreover, the examples are limited to live, on-farm recording and do not include measurements made at slaughterhouses, because of the focus of this chapter on recording welfare indicators for potential interventions. Recording welfare indicators at slaughter is of importance for reflection on the welfare of chickens during their lives and can help to improve the welfare of future chickens, but does not allow for improving the welfare of these individuals themselves anymore.

Another important distinction lies in individual-level versus group-level recordings. Whether it is highly relevant to have information on the welfare state of all individual animals or only on a group of animals as a whole depends on the purpose of the welfare recordings (Winckler, 2023). Overall, animal welfare is a characteristic of the individual animal (Broom, 2010), and therefore, individual-level recordings of welfare aspects may be preferable (Winckler, 2023). Ideally, such individual-level recordings are made while broilers are housed in their common production context, to best reflect their welfare in daily life. With group-level recordings it may not be clear how well a measurement reflects all animals in the group equally, which sparks the question whether the group mean or the worst experience of an individual is what is most important for welfare assessment (Winckler, 2023). In practice, the decision to record at the individual or group level may depend on multiple factors (discussed in Section 3.1), including costs and feasibility of fitting all individuals with a sensor, the intended purpose of the collected data and whether it is feasible to intervene at the individual level if welfare concerns are detected.

In any case, it is important to realize that sensor technologies for monitoring broiler welfare aspects are generally not plug-and-play: there are several steps to take in the process from sensor data collection to being able to reliably assess indicators of broiler welfare. The generally required steps are discussed in Section 3.2.

### *3.1 Deciding on what sensor to use in what context*

As can be seen in Table 1, different sensors (e.g. RFID versus cameras) can be used to assess similar behaviours (e.g. locomotor activity). Which sensor technology is most suitable for assessing aspects of broiler welfare depends on the goal of the implementation.

In research settings, sensor technologies for recording animal welfare may be implemented for two main purposes: to assess the impact of different interventions (e.g. environmental enrichment) on welfare aspects as a research question in itself or to monitor welfare during animal studies as a form of checking that scientific welfare standards are being continuously met. Automated welfare recordings in research may be at the individual or group level, depending on the research question (e.g. depending on whether the intervention that is studied is implemented at the individual or group level). Research settings are in general highly controlled environments, in which sensors have a bit less to endure from the environment in terms of, for example, fluctuating temperatures, dust and dirt than in practical production settings. That means that more sensitive sensors may be selected for research settings than for other settings.

In breeding settings, individual-level data are of great importance to allow inclusion of the data in breeding programs (Ellen et al., 2019). Therefore, sensors that can record data at the individual level and link these to specific chickens are highly preferable in breeding. Sensor technologies that can contribute to automated phenotyping of welfare components could be very interesting in relation to breeding for improved welfare. For example, leg health problems are a concern for broiler welfare (Knowles et al., 2008). It has been shown that breeding for improved leg health aspects is feasible in broilers (Kapell et al., 2012). However, leg health aspects are currently difficult and time-consuming to collect, commonly requiring manual observations (Kapell et al., 2012). Therefore, the implementation of sensor technology for frequent recording of leg health at the individual level in broilers could facilitate inclusion of this phenotype in breeding goals (see also the second case study in Section 4).

In broiler production practice, i.e. on farms, sensor technologies can be implemented to monitor welfare and allow for quick detection and intervention when welfare issues arise. Given that in current broiler production interventions often occur at the group level (e.g. application of medicine, changes in diet composition, vaccinations) and not at the individual level, group-level recordings will likely be sufficient in these settings (Dawkins, 2023). It is important that the sensors are able to withstand a farm environment (e.g. in terms of dust, ammonia, flies). Moreover, it is important that the sensor technologies provide clear perspectives for action. For example, if a sensor technology can help to quickly detect defects in water lines, this can help the farmer to intervene quickly and fix the water lines in question. However, if a sensor technology would detect that there are a few lame chickens in the flock – but not many – there might not be a very clear direct action a farmer can take to (1) find the affected chickens and (2) treat the lameness. Therefore, the second example might be less suitable for broiler production practice at this stage.

To summarize, which sensor is most suitable for recording broiler welfare depends not only on the welfare indicator of interest and how that can be measured but also on the research question or aimed-for objective of the implementation.

### 3.2 From recording data to assessing welfare

Sensor technologies for monitoring broiler welfare aspects are generally not plug-and-play. Instead, there are several steps to take in the process from sensor data collection to being able to reliably assess indicators of broiler welfare. These steps include data collection, data processing, validation of the recordings and (if applicable) setting thresholds.

Implementing sensors to monitor welfare indicators starts with the implementation of a sensor to record data on broilers. Already in the initial setup of the sensor, important decisions need to be made. For body-worn sensors, one needs to consider how the sensor can be attached in such a way that it does not affect the chickens too much. For remote sensors, one needs to consider where in the environment to place the sensors. For microphones and cameras, one does not want the chickens to be in direct contact with the sensor, as they could start pecking at the sensor which might cause the sensor to shift position or to be broken. Another important consideration in the data collection is how frequently one wants to sample data. In Section 4, Case study 1, it is discussed how the sampling frequency of a UWB system may affect the under- or overestimation of distances moved in broilers. Also data storage is important to consider, especially in the case of video data, where large amounts of data are collected, and substantial digital storage space is required for the raw data.

Data processing consists of moving from raw data to something that can be interpreted as a component of broiler welfare. Often, one starts with examining the raw data, for example in terms of average values in the dataset, variation in the values over time, presence of patterns in the data, and presence of outliers. It is important to consider whether these values (broadly) make sense. For example, if one records activity levels, are the recorded activity levels realistic for broilers and do you see the likely expected difference in activity between day and night? If the raw data are not directly insightful or are not already processed into something tangible using algorithms in pre-existing software, relevant information must be extracted by performing additional analytical steps. An example of this is the use of accelerometers, that provide data on acceleration forces (Chapa et al., 2020) but not directly on behaviour of the broilers. Therefore, this type of data requires subsequent processing to obtain insight into the underlying specific behaviour patterns (Benjamin and

Yik, 2019; Chapa et al., 2020). This can for example be achieved by live or video scoring of the behaviour of the broilers at specific time points and linking this to the accelerometer data of the same broilers at that same point in time (see e.g. Yang et al., 2021b).

To validate the recordings from sensors, it is important to implement the sensor in a setting in which an alternative form of recording is implemented simultaneously, to correlate the outcomes of the two types of recordings. Ideally, manual recordings are used as a gold standard, as alternative forms of automated recordings may come with their own inaccuracies or limitations. How good is 'good enough' in terms of correlation between the sensor data and the gold standard data depends on whether it is okay for the intended application to sometimes incorrectly classify broilers as having good or poor welfare. The first case study presented in Section 4 describes an example of sensor validation. In that case study, a UWB system and an RFID system were validated for recording distances moved in broilers by comparing them to each other and to (partially manual) tracking on video, when these three methods were implemented simultaneously for the same group of broilers.

Once it is confirmed that a sensor can reliably record a specific behavioural, physical or other aspect of broiler welfare, it is important to consider when this indicator – or lack thereof or change therein – is an indication of improved or reduced welfare. Unless presence of the indicator in itself already indicates overall poor or good welfare, thresholds need to be defined for the indicator to be able to assess when a welfare impairment or improvement is present. Such thresholds may for example be based on existing literature or on comparisons between individuals with and without a known welfare impairment.

Ideally, one uses the broilers as their own reference (preferably at the individual level) by examining changes in the level of the welfare indicator over time. In this way, individual variation in the baseline level of the indicator of interest can be accounted for. For example, if some broilers are always more active than some of their pen mates, a drop in activity may not be detected when only comparing to a set threshold defined on average group-level activity, whereas the relative drop in activity may indicate reduced welfare of that individual.

To summarize, several steps need to be taken in the process from sensor data collection to being able to reliably assess indicators of broiler welfare, including data collection, data processing, validation of the recordings and (when applicable) setting thresholds.

## 4 Case studies

In this section, two case studies will be presented as examples of the use of sensors to record broiler welfare indicators. The first case study focusses on using body-worn sensors for locomotor activity tracking, and the second case study focusses on monitoring walking ability using remote sensing.

## 4.1 Case study 1: Locomotor activity tracking using body-worn sensors

Locomotor activity appears to be a key trait to monitor in broilers, given its relationship with a range of health and welfare aspects (Dawkins, 2023). Two body-worn sensors for recording locomotor activity at the individual level in small groups of broilers were tested: UWB (van der Sluis et al., 2019) and passive RFID (van der Sluis et al., 2020).

For the UWB tracking, two-week-old broilers were fitted with battery-powered UWB tags. These tags had a size of approximately 3.8 cm by 3.9 cm and a weight of around 25 g. The tags were fitted to the broilers' backs, using elastic bands around their wing base. These tags sent out a signal approximately every 7 s, to four receivers in the environment (Fig. 3). Using triangulation, the location of the tags – and thus of the broilers – was recorded over time. The resulting data consisted of a log of the tag ID, the date and time and the x- and y-coordinates of the position of the broiler. Distances moved by the broilers were automatically determined using TrackLab software (Noldus Information Technology, Wageningen, The Netherlands).

For the passive RFID tracking, a small group of day-old broilers was fitted with tags on one of their legs using rubber bands. These tags had a size of approximately 15 mm by 3.7 mm and weighed less than 1 g. The pen in which the broilers were housed was fitted with a grid of 30 RFID antennas (41 cm by 32 cm each) underneath a false floor. Every second, these antennas would detect the presence of RFID tags within their range (Fig. 4). The resulting log file consisted of the tag ID code, the date and time and the location of detection (i.e. pre-assigned antenna number). From this log file, distances moved were calculated using the shortest-line distance between centre points of consecutively registered antenna locations.

When comparing the locomotor activity records from both approaches to (partly manual) video tracking as a gold standard, moderately strong correlations were observed. For UWB tracking, a correlation of 0.71 (95%-CI 0.64–0.77, $p < 0.001$) with video was observed, based on one-hour samples per day, for day 15 to 33 of life. However, it was observed that at low distances moved on video (<15 m in 1 h), the UWB system on average overestimated the distance moved by 40%. Likely, this is linked to the noise in the detection (i.e. the triangulation-based location of the tag may deviate slightly from the actual location and this deviation may occur in different directions over time): when broilers move very little, the detection noise can make up a relatively large part of the recorded distance. At high distances moved on video (>30 m

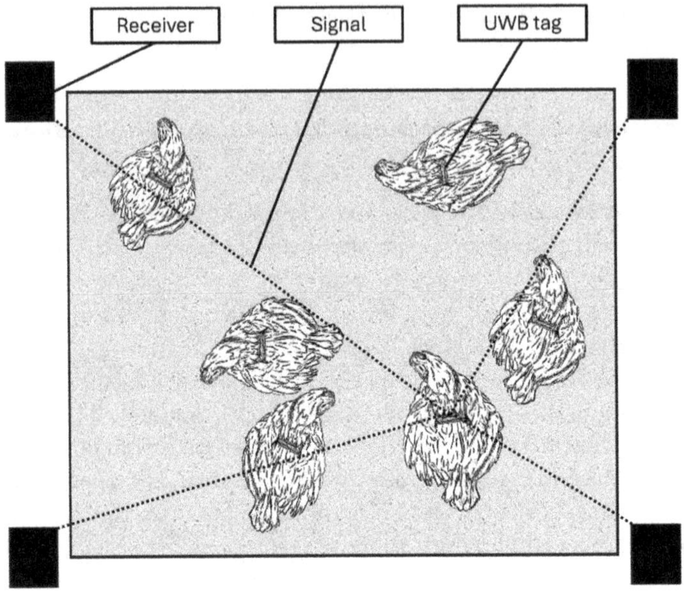


**Figure 3** Schematic visualization of the UWB system (not to scale). Four receivers in the environment are shown, and broilers are wearing UWB tags on their back.

in 1 h), the UWB system on average underestimated the distance moved by 15%. Due to working with a sampling rate of approximately 7 s, it could occur that a broiler moves from position A to B and back to A within 7 s and then the full distance covered between A and B (back and forth) is not registered, as the broiler will be detected near position A at both timepoints. Nonetheless, the relative recorded distances (i.e. the ranking of broilers in terms of distances moved) correlated well with the video observations.

For passive RFID tracking, a rank correlation of 0.82 (95%-CI 0.61–0.92, $p < 0.001$) with video was observed, using one-hour samples from days 25, 26 and 28 to 33 of life for four colour-marked broilers. However, the RFID system consistently underestimated the distance moved according to the video. This is likely the result of the RFID system only being able to detect movement between antenna grid cells, but not within antenna grid cells (i.e. a change in location would only be registered if a broiler moved from one antenna to another).

The correlation between RFID and UWB was also assessed when both systems were implemented simultaneously (van der Sluis et al., 2020). A rank correlation of 0.71 (95%-CI 0.65–0.75, $p < 0.001$) was observed when assessing the average distance moved per hour, calculated per day and broiler, for the period from 15 to 33 days old. The UWB tracking consistently indicated larger distances moved than the RFID system, likely again due to the fact that the

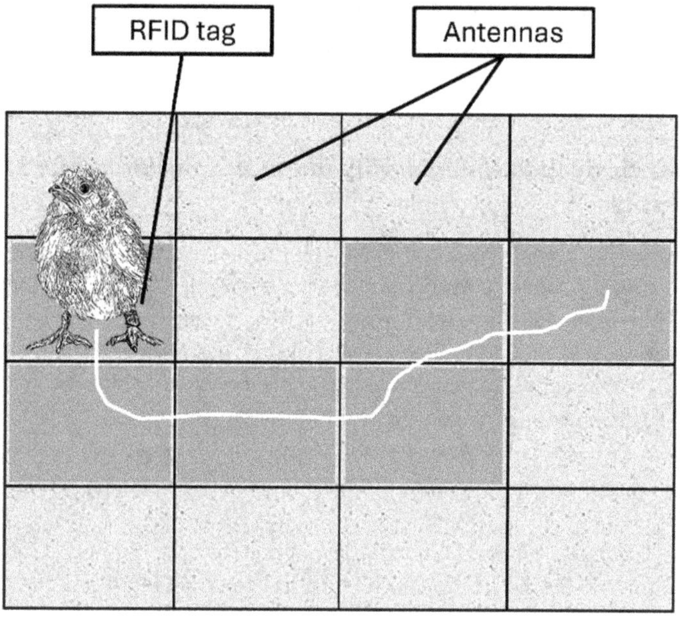


**Figure 4** Schematic visualization of the passive RFID system (not to scale). The different grid cells indicate RFID antennas underneath a false floor. The white line shows the fictional trajectory of a broiler, with the dark colour antenna grid cells indicating registrations at RFID antennas over time. The broiler wears an RFID leg tag.

RFID system could only detect movement between antenna grid cells, and not within, while the UWB system could detect every locational change. When looking within days (i.e. accounting for repeated measurements across days), a correlation of 0.65 (95%-CI 0.59–0.70, $p$ <0.001) was observed, and the patterns looked similar across days and thus ages of the broilers.

Regardless of the under- or overestimations of the true metrical distances moved, the moderately strong correlations between UWB, RFID and video indicated that relative locomotor activity levels (i.e. which broilers are more or less active than their pen mates) could be recorded well.

Using the UWB tracking system described here, it has been shown that locomotor activity decreases with age in broilers and that broilers with a higher body weight tend to be less active than broilers with lower body weights (van der Sluis et al., 2019). Moreover, it has been shown that broilers with a poorer gait score demonstrate lower locomotor activity levels than broilers with a good gait score, although the inclusion of body weight in the analyses revealed that this difference was only visible in light-weight broilers and not in heavy-weight broilers (van der Sluis et al., 2021). Using the RFID system, it was furthermore observed that broilers that were more variable in their activity levels in the first two weeks of life had a lower average daily body weight gain (van der Sluis et al., 2022). Overall, it appears that sensor recordings of locomotor activity levels can provide interesting insights into broiler welfare (see also Table 1).

### 4.2 Case study 2: Walking ability monitoring using remote sensors

Impaired walking ability is a common welfare issue in broilers (Knowles et al., 2008) that might cause pain and can affect the broilers' behaviour, for example resulting in more inactivity and less perching behaviour (Riber et al., 2021; Abeyesinghe et al., 2021). Walking ability is often assessed manually (Wurtz and Riber, 2024), for example using the gait scoring system of Kestin et al. (1992). However, this can be very time-consuming and subjective, as assigned scores may differ between observers (Fodor et al., 2023). Therefore, it was studied whether computer vision could be implemented to assess broiler walking ability.

Two different computer vision approaches were implemented: back-view pose estimation (Fodor et al., 2023) and top-view tracking (Fodor et al., 2024). In both studies, broilers of 33 (Fodor et al., 2023) or 34 (Fodor et al., 2024) days old were placed one by one in a walkway that ended in their home pen, and their gait was manually scored while videos were collected either from the back at broiler-height or from the top for subsequent analyses. The manually determined gait scores could range from a score of 0, representing broilers that walk very well, to a score of 5, representing broilers that can barely walk and also use their wings for support during walking (the scoring system is described in more detail in van der Sluis et al. (2021) and is similar to the well-known scoring system of Kestin et al. (1992)).

For the back-view pose estimation, a deep learning model was used to detect and track eight key points on the broilers in the videos: the head, the neck, and the left and right knees, hocks and feet. From the leg key points, seven pose features were derived (illustrated in Fig. 5): the hock joint lateral angles, the shank-versus-floor angle, the tibiotarsus length, the shank length, the hock-knee distance ratio, the hock-feet distance ratio and the step height. The tibiotarsus length, shank length and the step height were calculated in relation to the vertical difference between the highest knee and lowest foot as a normalization factor. Based on the manually determined gait scores, broilers were subsequently categorized as having a good gait (gait scores ≤ 2) or a suboptimal gait (gait scores > 2), and it was assessed whether the derived pose features differed between these two categories of broilers.

It was observed that broilers with a suboptimal gait had sharper hock joint angles, indicating a more sagging posture, than broilers with a good gait. Moreover, these broilers had lower hock-feet distance ratios, i.e. the feet of

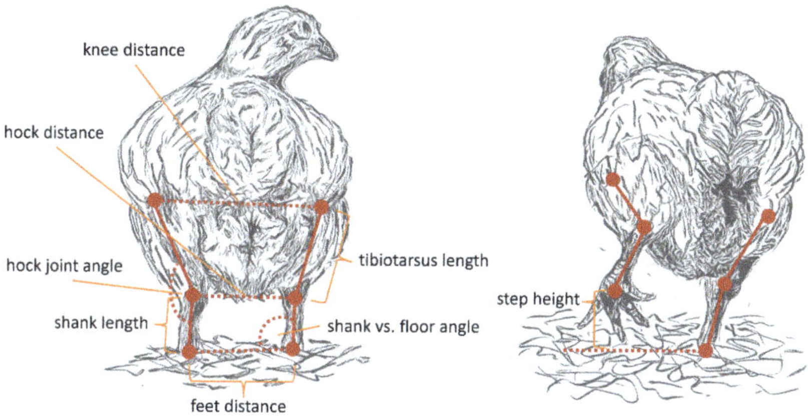


**Figure 5** Illustration of the seven pose features (the hock joint lateral angles, the shank-versus-floor angle, the tibiotarsus length, the shank length, the hock-knee distance ratio, the hock-feet distance ratio and the step height) that were derived from back-view videos of broilers walking, for the double-support stance phase (left) and the leg lift phase (right) of walking.

these broilers were relatively more spread out than the hocks, compared to the broilers with a good gait. During steps, it was observed that the relative step height (i.e. how high they lift their feet during the maximum leg lift) was lower in the broilers with a suboptimal gait. Overall, it appears that back-view pose estimation can provide insight into the walking characteristics of broilers.

However, a major challenge for the implementation of back-view pose estimation is that it requires a clear back-view of broilers walking. This would be challenging to obtain under commercial conditions, where large groups of broilers are kept in a single area, and they may occlude each other. Therefore, it was subsequently studied whether proxies for walking ability can also be derived using top-view videos, which would be easier to obtain in practice.

For the top-view tracking, a deep learning model was developed to detect the broiler in the walkway as well as the location of its head, as an indicator of the direction the broiler was facing. Four features were derived from the detections of the broiler over time: (1) the lateral body oscillation, calculated as deviations from the movement trajectory of the broiler, (2) step count, (3) completion time of the walkway and (4) the length-width ratio of the broilers while walking as an indicator of wing support. Subsequently, it was examined how these features were related to the broilers' manually determined gait scores.

It was observed that broilers that were manually categorized as having worse gait scores had longer estimated completion times and higher step counts, and furthermore showed a statistical trend for higher lateral body oscillation levels in the walkway setup compared to broilers having a better gait. The length-width ratio of the broilers during walking was found not to be a suitable proxy for wing support. Overall, the outcomes of this study indicate that top-view derived walking characteristics have the potential to contribute to future automated gait scoring in broilers.

## 5 Conclusion and future trends

Attention for broiler welfare is increasing, and sensor technologies appear to have great potential for recording different components of welfare (e.g. good walking ability or absence of disease) automatically. There is a wide variety of sensor technologies that have been successfully tested in research settings, but each sensor comes with its own challenges. Using a combination of sensors may help to (partly) overcome these. Nonetheless, implementation in practice is still limited due to the challenging broiler production environment with large numbers of highly similar looking individuals housed together. As recently stated by Dawkins (2023), this technical difficulty of large numbers of highly similar looking individuals, combined with a lack of agreement on how to measure broiler welfare, may underlie the yet incompletely fulfilled potential of sensor technologies for improving broiler welfare.

Yet, a lot of progress has been – and is being – made in developing smaller and more lightweight body-worn sensors and in more advanced (computer vision) algorithms for remote sensors. Likely, this progress will result in more applications of sensor technologies for monitoring broiler welfare in the future. Especially computer vision seems to hold promise, given that this technology allows for remote monitoring of multiple broilers simultaneously and different potential indicators of welfare (e.g. activity levels, resource use, specific behaviours; see Table 1) can be assessed from the same videos. However, for applications in which individual-level data with known identities of broilers is required, individual identification of the broilers in the video remains a hurdle that requires further attention in the coming years.

Capturing welfare as a whole is challenging and therefore most existing studies focussed on recording specific aspects of welfare. With the increasing attention of consumers for animal welfare, it is likely that more welfare aspects will become relevant in the future, also regarding positive welfare (i.e. the animal flourishing through the experience of predominantly positive mental states and the development of competence and resilience (Rault et al., 2025)). Novel (implementations of existing) sensors are needed to capture potential proxies for positive welfare, such as play behaviour, too.

Overall, by combining sensors and recordings of different aspects of welfare over time, we will hopefully be able in the future to monitor broiler welfare continuously at the individual level and in this way ensure a sustainable broiler production in which the chickens experience good welfare.

## 6 Where to look for further information

Further reading:

- A review and guidance on computer vision systems for monitoring farm animals is provided by Li et al. (2021a) in 'Practices and applications of convolutional neural network-based computer vision systems in animal farming: a review'.
- Guidance on implementing RFID for monitoring poultry can be found in 'Setting the stage to tag "n" track: a guideline for implementing, validating and reporting a radio frequency identification system for monitoring resource visit behavior in poultry' by Alindekon et al. (2023).

Scientific conferences on this topic include:

- European Conference on Precision Livestock Farming (ECPLF).
- European Symposium on Poultry Welfare (ESPW).
- International Conference on the Welfare Assessment of Animals at Farm Level (WAFL).

## 7 References

Abdanan Mehdizadeh S, et al. (2015) Image analysis method to evaluate beak and head motion of broiler chickens during feeding. *Computers and Electronics in Agriculture* 114: 88–95.

Abeyesinghe SM, et al. (2021) Associations between behaviour and health outcomes in conventional and slow-growing breeds of broiler chicken. *Animal* 15(7): 100261.

Alarifi A, et al. (2016) Ultra wideband indoor positioning technologies: analysis and recent advances. *Sensors (Basel)* 16(5): 707.

Alindekon S, et al. (2023) Setting the stage to tag "n" track: a guideline for implementing, validating and reporting a radio frequency identification system for monitoring resource visit behavior in poultry. *Poultry Science* 102(8): 102799.

Amraei S, et al. (2018) Development of a transfer function for weight prediction of live broiler chicken using machine vision. *Engenharia Agrícola* 38(5): 776–782.

Anderson G, et al. (2023) Impact of body-worn sensors on broiler chicken behavior and agonistic interactions. *Journal of Applied Animal Welfare Science* 6: 1–10.

Aydin A, et al. (2010) Application of a fully automatic analysis tool to assess the activity of broiler chickens with different gait scores. *Computers and Electronics in Agriculture* 73(2): 194–199.

Aydin A (2017a) Development of an early detection system for lameness of broilers using computer vision. *Computers and Electronics in Agriculture* 136: 140–146.

Aydin A (2017b) Using 3D vision camera system to automatically assess the level of inactivity in broiler chickens. *Computers and Electronics in Agriculture* 135: 4–10.

Aydin A, et al. (2023) Automatic classification of measures of lying to assess the lameness of broilers. *Animal Welfare* 24(3): 335–343.

Azarpajouh S, et al. (2022) Smart farming: a review of animal-based measuring technologies for broiler welfare assessment. *CABI Reviews* 17: 1–14.

Baxter M and O'Connell NE (2020) Testing ultra-wideband technology as a method of tracking fast-growing broilers under commercial conditions. *Applied Animal Behaviour Science* 233: 105150.

Baxter M and O'Connell NE (2023) Large variation in the movement of individual broiler chickens tracked in a commercial house using ultra-wideband backpacks. *Scientific Reports* 13(1): 7634.

Benjamin M and Yik S (2019) Precision livestock farming in swine welfare: a review for swine practitioners. *Animals (Basel)* 9(4): 133.

Bloch V, et al. (2020) Automatic broiler temperature measuring by thermal camera. *Biosystems Engineering* 199: 127–134.

Broom DM (2010) Animal welfare: an aspect of care, sustainability, and food quality required by the public. *Journal of Veterinary Medical Education* 37(1): 83–88.

Brown DD, et al. (2013) Observing the unwatchable through acceleration logging of animal behavior. *Animal Biotelemetry* 1: 20.

Brown-Brandl T and Condotta I (2023) Depth cameras for animal monitoring. In: Zhang Q (ed) *Encyclopedia of Digital Agricultural Technologies*. Cham: Springer Nature, pp. 294–306.

Campbell M, et al. (2024) A computer vision approach to monitor activity in commercial broiler chickens using trajectory-based clustering analysis. *Computers and Electronics in Agriculture* 217: 108591.

Chapa JM, et al. (2020) Accelerometer systems as tools for health and welfare assessment in cattle and pigs - A review. *Behavioural Processes* 181: 104262.

Collet JM, et al. (2024) High-throughput phenotyping to characterise range use behaviour in broiler chickens. *Animal* 18(3): 101099.

Dawkins MS (2023) Active walking in broiler chickens: a flagship for good welfare, a goal for smart farming and a practical starting point for automated welfare recognition. *Frontiers in Veterinary Science* 10: 1345216.

Dawkins MS, et al. (2012) Optical flow, flock behaviour and chicken welfare. *Animal Behaviour* 84(1): 219–223.

Dawkins MS, et al. (2017) Early warning of footpad dermatitis and hockburn in broiler chicken flocks using optical flow, bodyweight and water consumption. *Veterinary Record* 180(20): 499.

Debouzy J-C and Perrin A (2012) RFID. In: Perrin A and Souques M (eds) *Electromagnetic Fields, Environment and Health*. Paris, France: Springer, pp.81–87.

Del Valle JE, et al. (2021) Unrest index for estimating thermal comfort of poultry birds (Gallus gallus domesticus) using computer vision techniques. *Biosystems Engineering* 206: 123–134.

Doornweerd JE, et al. (2023) Passive radio frequency identification and video tracking for the determination of location and movement of broilers. *Poultry Science* 102(3): 102412.

Doornweerd JE, et al. (2024) Tracking individual broilers on video in terms of time and distance. *Poultry Science* 103(1): 103185.

Ellen ED, et al. (2019) Review of sensor technologies in animal breeding: phenotyping behaviors of laying hens to select against feather pecking. *Animals (Basel)* 9(3): 108.

European Commission (2023) Special Eurobarometer 533 – attitudes of Europeans towards animal welfare. DOI: 10.2875/872312.

Fang C, et al. (2021) Pose estimation and behavior classification of broiler chickens based on deep neural networks. *Computers and Electronics in Agriculture* 180: 105863.

Fang C, et al. (2022) Study on poultry pose estimation based on multi-parts detection. *Animals (Basel)* 12(10): 1322.

Fodor I, et al. (2023) Automated pose estimation reveals walking characteristics associated with lameness in broilers. *Poultry Science* 102(8): 102787.

Fodor I, et al. (2024) Top-view characterization of broiler walking ability and leg health using computer vision. *Poultry Science* 104(2): 104724.

Giersberg MF, et al. (2023) Group level and individual activity of broiler chickens hatched in 3 different systems. *Poultry Science* 102(7): 102706.

Gunaratnam A, et al. (2024) Computer vision in livestock management and production. *Engineering Applications in Livestock Production*. pp.93–128.

Guo Y, et al. (2020) A machine vision-based method for monitoring broiler chicken floor distribution. *Sensors (Basel)* 20(11): 3179.

Guo Y, et al. (2023) Detecting broiler chickens on litter floor with the YOLOv5-CBAM deep learning model. *Artificial Intelligence in Agriculture* 9: 36–45.

Hao H, et al. (2022) A dead broiler inspection system for large-scale breeding farms based on deep learning. *Agriculture* 12(8): 1176.

Jaihuni M, et al. (2023) Broiler mobility assessment via a semi-supervised deep learning model and neo-deep sort algorithm. *Animals (Basel)* 13(17): 2719.

Kapell DN, et al. (2012) Twenty-five years of selection for improved leg health in purebred broiler lines and underlying genetic parameters. *Poultry Science* 91(12): 3032–3043.

Kashiha M, et al. (2013) Development of an early warning system for a broiler house using computer vision. *Biosystems Engineering* 116(1): 36–45.

Kestin SC, et al. (1992) Prevalence of leg weakness in broiler chickens and its relationship with genotype. *Veterinary Record* 131(9): 190–194.

Knowles TG, et al. (2008) Leg disorders in broiler chickens: prevalence, risk factors and prevention. *PLoS One* 3(2): e1545.

Kristensen HH and Cornou C (2011) Automatic detection of deviations in activity levels in groups of broiler chickens: a pilot study. *Biosystems Engineering* 109(4): 369–376.

Lamkin-Kennard KA and Popovic MB (2019) Sensors: natural and synthetic sensors. In: Popovic MB (ed) *Biomechatronics*. New York: Academic Press, pp. 81–107.

Li G, et al. (2018) Radio-frequency identification (RFID) system for monitoring specific behaviors of group housed broilers. In: *10th International Livestock Environment Symposium (ILES X)*.

Li G, et al. (2021a) Practices and applications of convolutional neural network-based computer vision systems in animal farming: a review. *Sensors (Basel)* 21(4): 1492.

Li G, et al. (2021b) Development and evaluation of a method to detect broilers continuously walking around feeder as an indication of restricted feeding behaviors. *Computers and Electronics in Agriculture* 181: 105982.

Li G, et al. (2021c) Automated measurement of broiler stretching behaviors under four stocking densities via faster region-based convolutional neural network. *Animal* 15(1): 100059.

Li N, et al. (2020) Review: automated techniques for monitoring the behaviour and welfare of broilers and laying hens: towards the goal of precision livestock farming. *Animal* 14(3): 617–625.

Li G, et al. (2023) Tracking and characterizing spatiotemporal and three-dimensional locomotive behaviors of individual broilers in the three-point gait-scoring system. *Animals (Basel)* 13(4): 717.

Liu D, et al. (2021) Separate weighing of male and female broiler breeders by electronic platform weigher using camera technologies. *Computers and Electronics in Agriculture* 182: 106009.

Mei W, et al. (2023) Identification of aflatoxin-poisoned broilers based on accelerometer and machine learning. *Biosystems Engineering* 227: 107–116.

Mortensen AK, et al. (2016) Weight prediction of broiler chickens using 3D computer vision. *Computers and Electronics in Agriculture* 123: 319–326.

Nääs IdA, et al. (2010) Assessing locomotion deficiency in broiler chicken. *Scientia Agricola* 67(2): 129–135.

Nakarmi AD, et al. (2014) Automated tracking and behavior quantification of laying hens using 3D computer vision and radio frequency identification technologies. *Transactions of the ASABE* 57(5): 1455–1472.

Nasiri A, et al. (2022) Pose estimation-based lameness recognition in broiler using CNN-LSTM network. *Computers and Electronics in Agriculture* 197: 106931.

Nasiri A, et al. (2023) Estimating the feeding time of individual broilers via convolutional neural network and image processing. *Animals (Basel)* 13(15): 2428.

Nasiri A, et al. (2024a) An automated video action recognition-based system for drinking time estimation of individual broilers. *Smart Agricultural Technology* 7: 100409.

Nasiri A, et al. (2024b) Automated detection and counting of broiler behaviors using a video recognition system. *Computers and Electronics in Agriculture* 221: 108930.

Nekoogar F and Dowla F (2011) *Ultra-wideband radio frequency identification systems.* Berlin/Heidelberg, Germany: Springer Science & business Media.

Ngai EWT, et al. (2008) RFID research: An academic literature review (1995–2005) and future research directions. *International Journal of Production Economics* 112(2): 510–520.

Okinda C, et al. (2020) A review on computer vision systems in monitoring of poultry: a welfare perspective. *Artificial Intelligence in Agriculture* 4: 184–208.

Pereira DF, et al. (2013) Machine vision to identify broiler breeder behavior. *Computers and Electronics in Agriculture* 99: 194–199.

Pereira DF, et al. (2021) Movement analysis to associate broiler walking ability with gait scoring. *AgriEngineering* 3(2): 394–402.

Rault JL, et al. (2025) A consensus on the definition of positive animal welfare. *Biology Letters* 21(1): 20240382.

Riber AB, et al. (2021) Are changes in behavior of fast-growing broilers with slight gait impairment (GS0-2) related to pain? *Poultry Science* 100(3): 100948.

Roberts CM (2006) Radio frequency identification (RFID). *Computers & Security* 25(1): 18–26.

Ruiz-Garcia L and Lunadei L (2011) The role of RFID in agriculture: applications, limitations and challenges. *Computers and Electronics in Agriculture* 79(1): 42–50.

Seber RT, et al. (2022) Classifier's performance for detecting the pecking pattern of broilers during feeding. *AgriEngineering* 4(3): 789–800.

Silvera AM, et al. (2017) Lameness assessment with automatic monitoring of activity in commercial broiler flocks. *Poultry Science* 96(7): 2013–2017.

Stadig LM, et al. (2018) An automated positioning system for monitoring chickens' location: Accuracy and registration success in a free-range area. *Applied Animal Behaviour Science* 201: 31–39.

Taylor PS, et al. (2017) Ranging behaviour of commercial free-range broiler chickens 1: factors related to flock variability. *Animals (Basel)* 7(7): 54.

Taylor PS, et al. (2020) Frequent range visits further from the shed relate positively to free-range broiler chicken welfare. *Animal* 14(1): 138–149.

van der Eijk JAJ, et al. (2022) Seeing is caring – automated assessment of resource use of broilers with computer vision techniques. *Frontiers in Animal Science* 3: 945534.

van der Eijk JAJ, et al. (2023) Individuality of a group: detailed walking ability analysis of broiler flocks using optical flow approach. *Smart Agricultural Technology* 5: 100298.

van der Sluis M, et al. (2019) Validation of an ultra-wideband tracking system for recording individual levels of activity in broilers. *Animals (Basel)* 9(8): 580.

van der Sluis M, et al. (2020) Assessing the activity of individual group-housed broilers throughout life using a passive radio frequency identification system: a validation study. *Sensors (Basel)* 20(13): 3612.

van der Sluis M, et al. (2021) The relationship between gait and automated recordings of individual broiler activity levels. *Poultry Science* 100(9): 101300.

van der Sluis M, et al. (2022) Early locomotor activity in broilers and the relationship with body weight gain. *Poultry Science* 101(10): 102086.

van Hertem T, et al. (2018) Predicting broiler gait scores from activity monitoring and flock data. *Biosystems Engineering* 173: 93–102.

van Leuffen N, et al. (2023) Comparing pixel changes and manual observations for mapping broiler activity during dried black soldier fly larvae (Hermetia illucens) provisioning. *Animals (Basel)* 13(13): 2200.

Wang K, et al. (2018) An image-assisted rod-platform weighing system for weight information sampling of broilers. *Transactions of the ASABE* 61(2): 631–640.

Wei TY, et al. (2024) Comb color analysis of broilers through the video surveillance system of a poultry house. *Brazilian Journal of Poultry Science* 26(1): 1–8.

Welfare Quality® Consortium (2009) Welfare Quality assessment protocol for poultry (broilers, laying hens).

Winckler C (2023) Assessing animal welfare at the farm level: do we care sufficiently about the individual? *Animal Welfare* 28(1): 77–82.

Wurtz KE and Riber AB (2024) Overview of the various methods used to assess walking ability in broiler chickens. *Veterinary Record* 195(4): e4398.

Xiong X, et al. (2019) An automatic head surface temperature extraction method for top-view thermal image with individual broiler. *Sensors (Basel)* 19(23): 5286.

Yang X, et al. (2021a) Characterizing sounds of different sources in a commercial broiler house. *Animals* 11(3): 916.

Yang X, et al. (2021b) Classification of broiler behaviours using triaxial accelerometer and machine learning. *Animal* 15(7): 100269.

Yang X, et al. (2024) Deep learning methods for tracking the locomotion of individual chickens. *Animals (Basel)* 14(6): 911.

Zhuang X and Zhang T (2019) Detection of sick broilers by digital image processing and deep learning. *Biosystems Engineering* 179: 106–116.

# Chapter 17

# Use of vocalisation technology for early detection of challenges in broiler production

*Gerardo José Ginovart-Panisello, CEALVET S.L.U. and La Salle Ramon Llull University, Spain; Silvia Riva and Tesa Panisello Monjo, CEALVET S.L.U., Spain; and Rosa Ma Alsina-Pagès, CEALVET S.L.U. and La Salle Ramon Llull University, Spain*

1 Introduction
2 Acoustic features of broilers
3 Animal disease detector
4 Tracking of animal behaviour
5 Case study
6 Conclusion and future trends
7 Where to look for further information
8 References

## 1 Introduction

Chickens are among the most vocal of all animal species and constitute the largest population among domestic animals, with numbers exceeding 26 billion in 2022 (Statista 2024). Animal bioacoustics covers all aspects of sound production, transmission, and reception and, in the case of birds, consists of the analysis of their vocalisations (Mellinger 2011). A trend in the use of acoustics in domestic poultry (*Gallus gallus, Gallus gallus domesticus*) is to monitor the condition of the animals and identify their behaviour through their vocal signals (Laurijs et al. 2021). Bird calls provide information about their sex (Dale et al. 2022), age (Cordeiro et al. 2018), emotional state (Neethirajan, Reimert, and Kemp 2021; De Lanerolle and OM 1978; Briefer and Comber 2012), growth rate (Fontana et al. 2017), and many other aspects and behaviours. Vocalisations can therefore be an expression of the degree of well-being of animals, which can be influenced by both internal and external factors (Manteuffel and Schön 2004). It is widely acknowledged that animal vocalisation sounds strongly

http://dx.doi.org/10.19103/AS.2025.0165.20

correlate with welfare conditions (Laurijs et al. 2021). For example, vocalisations such as screams are often associated with pain or stressful conditions.

The monitoring of vocalisations through sound is considered a minimally invasive process, as it does not interfere with animal activities or farm management practices (Li et al. 2020). An additional advantage is that installation and maintenance can be carried out without the presence of animals, thus preserving the biosecurity chain (Arzey and Littleton 2007).

Acoustic monitoring requires technology to collect data, detect, isolate, and process vocalisations. In recent years, the combination of technology for processing large amounts of data, together with precision engineering, has led to the development of specific technologies called Precision Livestock Farming (PLF), used to monitor animals in real time with the aim of ensuring their health and well-being (Berckmans 2017), leading to higher production rates and cost reductions, while ensuring animal health (Rowe, Dawkins, and Gebhardt-Henrich 2019).

Currently, several sensors are available on the market for farmers to measure different farm parameters. The most widely used sensors in the livestock industry include those that measure temperature, humidity, carbon dioxide, ammonia levels, food and water supply, and ventilation rate. New sensors that use Artificial Intelligence (AI) data processing to monitor animals have recently been introduced. These devices incorporate technologies such as microphones, cameras, and other sensors (e.g. body temperature and skin conductivity sensors) (Halachmi et al. 2019).

This chapter aims to define the relationships between animal vocalisations and chicken diseases and welfare, while highlighting the novel technologies for their detection. It is structured as follows. Section 1 is a general introduction to the use of PLF technologies, and in particular bioacoustics, for monitoring the health status of broilers through the analysis of their vocalisations. The most common acoustic features of broiler chickens to describe their calls are detailed in Section 2. Call detectors can be used for disease detection as detailed in Section 3 or for tracking animal behaviour in Section 4. Section 5 shows a case study on detecting prolonged fasting vocalisations in newly hatched chickens. A summary and future trends can be found in Section 6.

## 2 Acoustic features of broilers

Broiler chicken vocalisations are generated by a specialised organ called the syrinx, which is located between the bronchi and trachea and acts as an air valve (Beddard 1898). At the syrinx, the trachea divides into two bronchi, which lead to the lungs. The air flows from the lungs, through the bronchi, syrinx, trachea, larynx, and is finally released through the beak or nasal cavity. The vocalisation generated by the syrinx is modulated in the vocal tract (trachea and bronchi), which acts as a resonating tube. The variety of frequencies of bird calls depends on the control of airflow through the expansion and contraction of the syrinx (Suthers 1990). By controlling airflow, birds can produce different frequencies of calls that encode their health status (Collins et al. 2024). Therefore, calls can be used as biomarkers.

Bioacoustic monitoring is an innovative and emerging technology for Precision Livestock Farming. Until recently, the cost of electronics and storage to apply real-time algorithms was very high due to the large amount of data to be processed. Consequently, it has not been economically feasible to continuously monitor birds on a large scale. As Fontana et al. (2016) explained, broiler chickens have a maximum peak frequency above 4 kHz and a vocal range that includes harmonics up to about 10 kHz. To record the entire call range, a minimum sampling rate of 20 kHz, with a minimum resolution of 8 bits, is required. With these values, one day of continuous recording without compression is equivalent to 13.82 GB, and a 45-day production cycle is equivalent to 621 GB. The raw data is useful for listening to birdsong after recording and for applying post-processing techniques to extract acoustic features related to bird vocalisations.

By isolating animal vocalisations to extract acoustic features, it is possible to filter background noises, such as fans, feeders, etc., and extract acoustic features from the vocalisations that can be used as biomarkers, such as pitch frequency.

Extracting animal vocalisations can be done with two different approaches: using signal processing techniques to filter out non-animal sounds, or using pattern recognition algorithms to identify vocal regions of an audio file. Some signal processing algorithms used for broiler calls identification are (i) maximal overlap discrete wavelet transform (MODWT) (Yang et al. 2021), (ii) filter the frequency bands that do not contain vocalisations with band pass filters or spectral subtraction (Ginovart-Panisello et al. 2024), (iii) delete samples with predominant noise and filtering (Ginovart-Panisello et al. 2022). Some pattern recognition algorithms used in broilers to identify calls: (i) Hidden Markov models (Kogan and Margoliash 1998), (ii) neural networks with acoustic features or spectrum (Thomas et al. 2023; Mao et al. 2022), (iii) distinguish vocal syllables (Bardeli et al. 2010; Mahdavian et al. 2021). Vocalisation detection algorithms are generally more computationally complex to extract vocal features than algorithms that simply filter the sound before extracting vocal features, but they offer greater accuracy in ensuring that the extracted acoustic features actually come from a vocalisation.

A wide range of temporal and frequency features can be extracted from each animal vocalisation using audio samples. These features represent the characteristics of the animal's call that can be used to detect challenges during broiler production. This section provides a detailed description of the most common acoustic features of poultry found in the literature.

Temporal Features include various time-dependent aspects of sound, such as duration, rhythm, repetition rates, and temporal patterns, such as pauses or intervals between calls.

- Duration: measures the duration, usually in milliseconds, of the animal's call.
- Vocalisation rate: the number of vocalisations in a given time interval. The same feature is also called VocalNum or number of vocalisations (Curtin, Daley, and Anderson 2014).
- Sound Pressure Level (SPL): the sound pressure level of the sound source, expressed in decibels.
- Energy: measures the energy of a sound source. It is a simple and effective classifying parameter for voiced and unvoiced segments (Yang et al. 2010).
- Zero Crossing Rate (ZCR): calculates the number of times the wave crosses the zero amplitude of an audio time series (Rabiner and Schafer 2010).
- Absolute extremum: measures the absolute value of the maximum amplitude of the sound.

Frequency Features include various frequency-dependent aspects of sound, such as fundamental frequency, harmonics, bandwidth, and frequency modulation patterns, among others. Frequency parameters help identify species-specific calls and emotional states.

- F0 or pitch: indicates the fundamental frequency at which the birds vocalise. It can be extracted using a method based on auto-correlation (Rabiner 1977).
- F1 to F4: the frequency range of each formant, hence the difference between the maximum and minimum frequency of that formant. The feature estimate can be calculated from the LPC analysis data (Snell and Milinazzo 1993).
- Spectral centroid: the frequency that identifies the centre of mass of the spectrum (Klapuri and Davy 2007).
- Spectral bandwidth: the difference between the highest and lowest observed frequencies (Hz). It provides information about the distribution of energy across different frequency bins. A wider bandwidth suggests that the sound contains a wider range of frequencies, while a narrower bandwidth suggests that the sound is concentrated in a smaller range of frequencies. (Klapuri and Davy 2007).
- Spectral entropy: a measure of its spectral power distribution. The Shannon entropy is calculated from the signal's normalised power distribution in the frequency domain as a probability distribution (Misra et al. 2004).

- Spectral flatness: a measure to quantify how similar a sound is to noise. It is also known as the tonality coefficient. A value of spectral flatness closer to 1 indicates that the spectrum is similar to white noise, while a value closer to 0 is similar to tones (Dubnov 2004).
- Spectral rolloff: defined as the frequency of a spectral bin, such that at least the rolloff value (usually 0.85) of the spectral energy is contained in that bin and the bins below it (Scheirer and Slaney 1997).
- Mel Frequency Cepstral Coefficients (MFCC): a group of features based on the human auditory system. Each coefficient is a sum of various spectrum bins to calculate the energy compressed in a frequency region following the Mel filter bank. The first filter is very narrow and indicates the energy near 0 Hz. As the frequency increases, the filter becomes wider. Since most of the acoustic information is below 20 kHz, 12 cepstral coefficients are sufficient to modulate the sound signal (Rabiner and Schafer 2010).
- Sound Pressure Measurement (SPL) of frequency bands: indicates the sound pressure after applying a weighting filter (Z/A/C) for a frequency bandwidth as octaves or 1/3 octave (Harris and Tocci 1993). The measurement shows the sound pressure as a function of the obtained frequency and the acoustic spectrum.

Spectrotemporal Features combine both frequency and time-related aspects to provide a complete understanding of acoustic signals. These features include dynamic changes in frequency over time, amplitude modulation, and structure of harmonic patterns.

- Spectrogram: a two-dimensional visual representation where the x-axis represents time, while frequency is shown on the y-axis, and the colour of each dot represents the amplitude of energy at a specific time (Yu 1967). Frequency can be converted to different scales for display: linear, logarithmic, Mel scale, constant-Q, variable-Q, and chromatic. Figure 1 shows a Mel spectrogram.
- Tempograms: a two-dimensional visual representation of time-tempo that encodes the local tempo of a sound signal over time (Grosche, Müller, and Kurth 2010). See Fig. 2. Tempo is the speed at which a sound is played, measured in beats per minute (BPM). The more vocalisations, the faster the tempo.

The features described above are typically extracted from short audio windows, from milliseconds to a few seconds. Each feature has its own configuration parameters, and different results can be obtained depending on the configuration. Most projects start by calculating a large number of features to

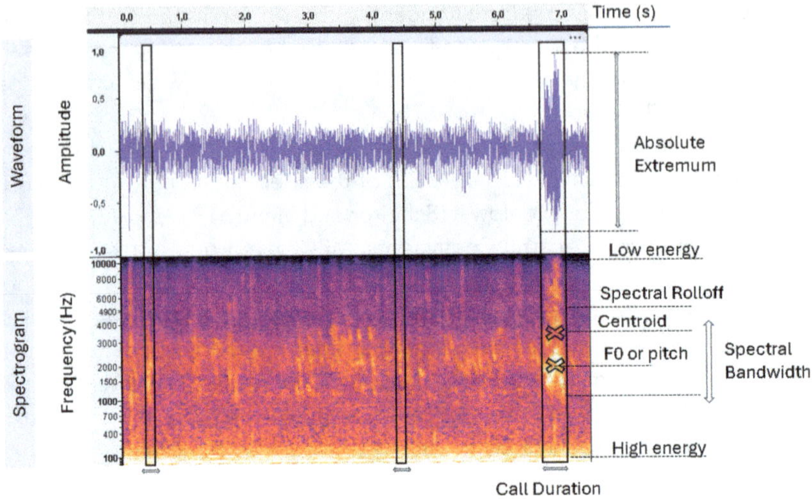


**Figure 1** Representation of acoustical features in waveform and spectrogram graphics.

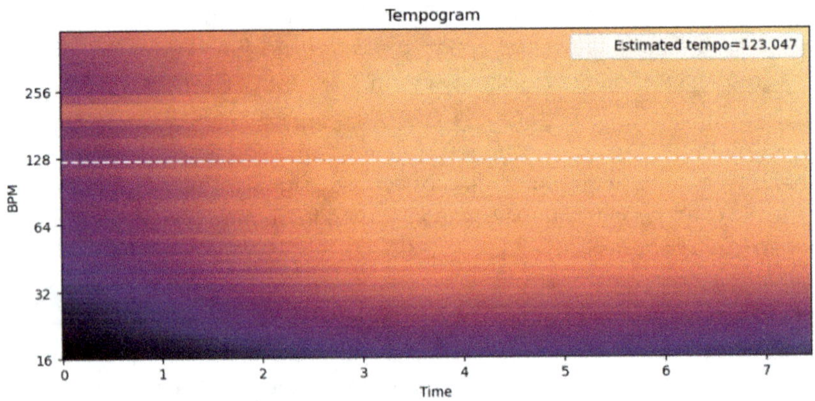


**Figure 2** Tempogram representation of a 7 second sample of chicken vocalisation.

discard the less relevant parameters. Depending on the desired output to be detected or predicted, the most important features may vary.

A fine-tuning process based on the most important features is needed to adapt the setup to the data recording and achieve the desired vocalisation characteristic. Furthermore, this data can be presented statistically for larger time windows. For example, the mean, standard deviation, and minimum and maximum values can be calculated. The statistical data is calculated on the vector obtained after calculating the feature in all the short time windows included in the large time windows.

The acoustic features of vocalisations can be converted into indicators or serve as a warning signal if relationships are found between the acoustic characteristics and the health and welfare status of animals (diseases, stress, zootechnical parameters, etc.). The most commonly used algorithms in the literature are:

- Feature Selection: a machine learning technique that ranks features by importance to indicate their contribution to a prediction or output (Zien et al. 2009). This method identifies non-linear relationship between features and outputs. The most important features highlight the most correlated features that help predict the output. While the less important features can be eliminated without significantly affecting the accuracy of the model. It also reduces the amount of data and computing power needed to train the model. These algorithms are useful for finding relationships between input and output signals, as performed by Ginovart-Panisello et al. (2022), and for reducing the dimensionality of the data.
- Dimensionality Reduction: the transformation of high-dimensional data into a meaningful representation of reduced dimension (Van Der Maaten et al. 2009). Audio signals usually have a large data size, often causing redundancy. Similar data information in the dataset requires more computational resources and complex models. Therefore, dimensionality reduction facilitates, among other things, data classification, visualisation, and compression. The most common techniques to reduce dimensionality are Principal Components Analysis (PCA) (Kurita 2019), factor analysis (Garson 2022), and Fisher Discriminant Analysis (FDA) (Wang et al. 2013), among others. Feature importance selection is also a non-linear dimensionality reduction method.
- Statistical Analysis: a scientific method developed to interpret and draw conclusions from data (Anderson and Finn 2012). It analyses data and draws inferences from samples to the entire population (Winters, Winters, and Amedee 2010). Acoustic features obtained from animal calls can be used in statistical analyses to determine whether a specific treatment (e.g. vaccination, product administration, environmental conditions, etc.) alters animal vocalisations, and hence their health and welfare, compared to a control group. This analysis requires the use of t-test (Posten 1984) and ANOVA (Ståhle et al. 1989) to prove the treatment hypothesis (Ginovart-Panisello et al. 2024). In case the acoustic features data are not normally distributed, data transformation might be necessary before the analysis.
- Automatic Detection: a technology that enables computers and digital devices to learn, analyse, and perform classification tasks commonly associated with intelligent beings (Garg 2021). Subdisciplines of AI are machine learning and deep learning algorithms, which learn and adapt from training data to predict an outcome. Animal health and welfare can be classified according to different parameters: positive/negative

emotions, stress, disease, among others. An example of machine learning algorithms to monitor poultry health through acoustic features extraction is the work done by Mahdavian et al. (2021) for the diagnosis of infectious bronchitis and Newcastle disease. The classification results, achieving an accuracy of over 80%, underscored the strong relationship between animal vocalisations and their health status. These findings not only validated the significance of vocalisation patterns in assessing animal well-being but also enhanced the efficiency of automated health monitoring systems. The most common machine learning algorithms used in literature for poultry studies are: Support Vector Machine (SVM) (Mammone, Turchi, and Cristianini 2009), K-nearest neighbour (KNN) (Mucherino et al. 2009), Classification and regression trees (CART) (Berk 2008), Bootstrap aggregating (Bagging) (Breiman 1996), Random Forest (RF) (Parmar, Katariya, and Patel 2019). Meanwhile, the most common deep learning algorithms used in poultry are: Deep neural networks (DNN) (Cichy and Kaiser 2019) and Convolutional neural networks (CNNS) (Ketkar et al. 2021), using spectrum images.

## 3 Animal disease detector

As defined by the World Organisation for Animal Health, an animal is in a good state of welfare if it is healthy, comfortable, well-nourished, safe, capable of expressing innate behaviour, and if it does not suffer from unpleasant states, such as pain, fear, and distress (Commission 2008). However, in practice, it is difficult to verify and indicate the state of welfare of an animal. Most welfare definitions are based on anthropomorphic views of the animal. Humans attribute human emotional and behavioural characteristics to non-human animals (Mota-Rojas et al. 2021). However, some anthropomorphic perceptions of animal condition, both physical (e.g. movement, weight, nutrition) and emotional (e.g. stress, fear, anxiety, and aggression), may not represent the true level of animal welfare. Furthermore, quantifying animal welfare on the basis of subjective parameters, such as stress or aggression makes it even more difficult to obtain a measure that is independent of the assessor (Urquiza-Haas and Kotrschal 2015).

Vocalisations, especially the wide variety of those of birds, can be used to infer whether animals are experiencing positive or negative feelings, indicating a quantitative level of animal welfare. Therefore, new devices based on animal sound and movement are being developed to identify incorrect or inadequate management practices, or to detect behavioural or health problems, helping to improve livestock rearing (Ben Sassi, Averós, and Estevez 2016). In recent years, detection technology has become increasingly diverse, accurate, and accessible, thus starting to find application in farm management practices, simplifying the work of farmers. In particular, non-invasive acoustic sensors can be used to study reliable indicators of stress and therefore represent an interesting approach when trying to assess animal welfare (ibid.).

Nowadays, certain diseases such as necrotic enteritis caused by *Clostridium perfringens*, Newcastle disease, Infectious Bronchitis, and Avian Influenza are particularly concerning due to the significant economic losses they can cause to the poultry industry (Riva and Monjo 2021; Swayne 2007). Vocalisations provide important information about the health status of animals (Manteuffel and Sch¨on 2004); there is a difference between the sound of a healthy chicken and that of a sick one (Li et al. 2020). Several vocalisation studies that calculated the relationship with diseases are detailed below.

Rizwan et al. (2016) proposed an algorithm based on Extreme Learning Machine (ELM) and Support Vector Machine (SVM) classifiers to detect wheezing in chickens suffering from respiratory diseases. The experiment involved two groups of newborn chickens: a control group and an experimental group, each comprising six chickens. The birds were housed in two Horsfall isolators equipped with microphones, cameras, and weatherboards. Continuous audio recordings were captured at a 48 kHz sampling rate throughout the 25-day experiment and stored as minute-long FLAC files. On day 15, the experimental group was exposed to the infectious bronchitis virus, developing noticeable symptoms of the disease between days 20 and 24. In contrast, the control group remained healthy throughout the experiment. A 20-min segment of data containing numerous rales was selected for training and testing the algorithms. The labelled audio samples were converted into Mel-scale spectral features, and, after training, the models achieved an accuracy of 97%, demonstrating the ability to detect wheezes in poultry.

Based on the vocalisations emitted, an Artificial Intelligence method was developed to identify and classify broiler chickens affected by *Clostridium perfringens* type A (Sadeghi et al. 2015). The study, conducted in February and March 2014 at the Faculty of Agriculture, Tarbiat Modares University, involved 30 14-day-old Ross chickens divided into two groups of 15. One group served as healthy controls with no exposure to *Clostridium perfringens*, while the other was inoculated with the bacterium on day 14. Both groups were housed separately to prevent cross-contamination, and all chickens were vaccinated against common diseases. From days 16 to 22, the birds were removed from their cages daily at 10:00 am and placed individually in 50 $cm^3$ boxes. Recordings began 3 min later to mitigate handling stress. Results showed that healthy chickens exhibited higher vocalisation intensity and greater uniformity compared to infected ones. Infected birds displayed lower sound intensity but higher frequency in the main low-frequency range. Analysis of maximum values, root mean square, standard deviation (SD), and crest factor further confirmed these findings. Healthy birds had more intense and consistent vocalisations, while infected birds exhibited less uniform and more irregular vocal signals, supporting the feature extraction and signal analysis results. Neural network pattern recognition was used to distinguish healthy chickens from infected ones. With an accuracy of 66.6% after the first 2 days and 100% after 8 days post-infection, the results demonstrated the effectiveness of an intelligent method to determine the health status of animals, in this case before the onset of disease symptoms (Sadeghi et al. 2015).

A prediction model based on acoustic signals was developed to diagnose avian diseases by using Data mining methods, SVM and Dempster-Shafer evidence theory (Banakar, Mohammad Sadeghi, and Shushtari 2016). The study was conducted at the Agricultural School of Tarbiat Modares University in Tehran, focusing on a brood of chickens. Fourteen-day-old Ross 308 chickens were divided into four groups of 60, each housed in a separate room. The first group served as the control. The second, third, and fourth groups were infected with Newcastle Disease Virus, Infectious Bronchitis Virus, and Avian Influenza Virus, respectively, administered via eye drops (0.1 cc per eye). A total of 720 sound signals were collected from 240 samples over 2 consecutive days following the virus infection. To achieve the information from chicken's sound signals, 25 statistical features from frequency domains and 75 statistical features from time-frequency domains were extracted and processed using Fast Fourier Transform and Discrete Wavelet Transform (Banakar et al. 2016). The accuracy of the method based on Dempster-Shafer infusion of sound data reached 91.15% for the diagnosis of the three diseases (Banakar et al. 2016).

Mahdavian et al. (2021) used five acoustic features to determine the health conditions of three groups of broiler chickens, two of which were affected by Infectious Bronchitis Virus or Newcastle disease. The wavelet entropy (WET) was the feature with the best performance, with an accuracy of 83% to detect Infectious Bronchitis after 3 days from inoculation. In the case of Newcastle disease the wavelet entropy and the Mel cepstral coefficients exhibited similar accuracy, 80% and 78%, respectively, on the fourth day after inoculation, but WET was more reliable in detecting healthy chickens, while the MFCC performed better in detecting chickens with health problems. Huang, Wang, and Zhang (2019) and Cuan et al. (2020) also developed a method to automatically detect Avian Influenza based on MFCC and other poultry vocal features. Despite the accuracy of both studies (around 90%), the development of algorithms for predicting poultry health was based on a small number of animals. Therefore, as promising as they are, these experimental studies remain the first step towards a more detailed and precise use of bioacoustics as a PLF technology (Coutant, Villain, and Briefer 2024). To test and refine these algorithms, studies with a much larger number of animals are needed, such as that of Liu et al. (2020), who developed a method to detect abnormal sounds in a chicken house involving a group of 20,000 Ross 308 white-feather broilers, aged 30 days, over a 1-week period starting on March 29, 2019. The broilers, weighing between 1.6 and 2.2

kg, were reared on the floor in a controlled environment with a temperature of 20–22°C and humidity levels of 46–50%. They were provided ad libitum access to compound feed via hoppers and natural drinking water. The building was illuminated for 21 h a day throughout the experiment. The study focused on detecting three types of broiler sounds: cough, snore, and interfering sounds. Hidden Markov Models (HMM) were trained using various input features, including MFCCs, WMFCCs, PWMFCCs, CWMFCCs, and weighted PWMFCCs. The results demonstrated that the algorithm effectively identified abnormal broiler sounds. Among the inputs, weighted CWMFCCs achieved the best performance, with an average classification accuracy of 93.8%, precision of 94.4%, recall of 94.1%, and F1-score of 94.2%. This highlights the effectiveness of HMM with weighted CWMFCCs for broiler sound classification.

Deep Learning techniques have also been tested to detect diseases. Cuan et al. (2022) presented an automatic detection method of the early symptoms of Newcastle disease. The accuracy within the first, second, third, and fourth days after infection was 82.15%, 90.00%, 93.60%, and 98.50%, respectively.

Non-invasive acoustic detection was used to detect the reaction of hens after administration of the vaccine against infectious laryngotracheitis (ILT) virus (Ginovart-Panisello et al. 2024). The hens were divided into two groups, both vaccinated, and only the treatment group was administered a natural anti-inflammatory product to prevent or mitigate the reaction to the vaccine. The research was based on the study of the acoustic features most correlated with stress: the centroid, the number of vocalisations (VocalNum), and the MFCC. Furthermore, it was divided into phases: the pre-reaction phase (days before vaccination), the reaction phase (days with vaccine reaction symptoms), and the post-reaction phase (days after reaction symptoms). The control group was used to model the animals' reaction to the vaccine and its side effects based on the vocalisations emitted. The acoustic results obtained confirmed the hypothesis: the administration of the eye-drop vaccine against ILT was a stressful procedure, which can be detected and quantified through acoustic parameters. The study highlighted stress-related acoustic features like the centroid and VocalNum. These features showed marked increases during the reaction period, particularly in the first 3–4 days after vaccination. These findings suggested that manual administration of the vaccine via eye drops caused stress in the hens, accompanied by significant side effects. Data collected during nighttime intervals proved to be more reliable due to lower environmental noise interference. The nighttime analysis revealed statistically significant differences between the treated and control groups. The centroid and VocalNum features, in particular, had *P*-values as low as $2.41 \times 10^{-14}$ and $1.42 \times 10^{-6}$, respectively, indicating the treatment's effectiveness in reducing stress responses. Moreover, the study explores the potential for automating vaccine reaction detection. Among the algorithms tested, the Gaussian Naive

Bayes (GNB) model delivered the best performance, achieving an F1 score and accuracy above 0.80. The centroid and MFCC4 were identified as the most critical features for detection, aligning with previous studies on poultry behaviour, including their use as indicators of food intake and environmental $CO_2$ levels. Finally, the study assessed the daily detection rates of vaccine reactions, which peaked at over 75% on day three after vaccination. By day five, reactions had significantly diminished, although 30% of side effects were still detectable. Validation data further confirmed the efficacy of the treatment, with reduced reaction percentages compared to test predictions. (Ginovart-Panisello et al. 2024). These results may encourage the use of bioacoustics as an effective and non-invasive method to verify the efficacy and effects of vaccine administration and other products on animal health. For each challenge, the relevant acoustic features may vary, but among the most common are pitch, frequency, and number of vocalisations.

## 4 Tracking of animal behaviour

In the animal world, some behaviours can be associated with certain sounds or vocalisations, and bioacoustics can be a valid tool for detecting the execution of behaviours that may be of interest for monitoring the health status of farm animals.

Monitoring feeding behaviour can be an interesting aspect to study to control feed intake of large groups of animals (Coutant, Villain, and Briefer 2024). Huang et al. (2021) proposed a four-threshold automatic selection method based on the short time energy and the short time zero crossing rates of poultry vocalisations and environmental sounds previously analysed (ibid.). In the experiment, vocalisations of 18 chickens were collected and characterised according to time variation. The sounds were classified into three networks (PV-net 1, PV-net 2, and PV-net 3) based on normal vocalisations and those emitted during feeding. The results showed that the recognition rates of eating of PV-net1, PV-net2, and PV-net3 were 93.5%, 94.5%, and 96%, respectively; their sensitivity was 96%, 97%, and 96%, respectively; and their specificities were 91%, 92%, and 96%, respectively.

- Aydin Bahr et al. (2014) investigated the relation between pecking sounds and the amount of feed intake by developing an algorithm to identify individual beak sounds. Based on the measurements of feed consumption and waste, the feed intake was calculated and used for validation of the proposed algorithm. The experiments were conducted using 28-day-old Ross 308 broilers, consisting of 12 males. Recordings were performed over 3 consecutive days. Each bird participated in three experimental trials, resulting in a total of 36 experiments. Prior to each trial, individual chickens were housed in separate cages without access to food or water for 4 h to stimulate pecking behaviour at the beginning of the experiment. Each trial lasted 15 min, during which the bird was placed in an isolated cage. Pecking sounds were generated as the chickens' beaks touched and dipped into the feed in the feeder. These sounds were continuously recorded and subsequently analysed. The results showed that 93% of the pecking sounds were clearly identified by the algorithm, while the 7% were false positives. Feed uptake was measured using a weighing system, and these measurements were correlated with the results of the sound algorithm. On average, feed intake per pecking was determined to be 0.025 g. Before estimating the absolute feed intake of chickens based on the pecking sound algorithm, the relationship between the number of peckings and feed uptake was assessed. A linear relationship was identified between these variables. Linear regression analysis revealed a high coefficient of determination ($R^2$ = 0.995), indicating a strong relationship. Moreover, 90% of feed intake events were accurately monitored through sound analysis. As depicted in Fig. 9, the number of peckings was found to be highly correlated with the feed intake of broiler chickens. These results suggested the potential of this non-invasive and non-intrusive system. This last algorithm was used for a further study, in which, in addition to the pecking sounds of 10 male, 39-day-old broilers, video recordings were obtained around the feeders, and automatic systems were installed for weighing feed uptake (Aydin and Berckmans 2016). The relationship between these three surveillance systems demonstrated a strong positive correlation. Additionally, 90% accuracy was achieved in meal size, 95% in meal duration, 94% in number of meals per day, and 89% in feeding rate. These results showed that this automatic, non-invasive, and non-intrusive system could be used as a tool to monitor in real time the short-term feeding behaviour of a flock of chickens throughout their life span (Aydin and Berckmans 2016).

The use of bioacoustics as a monitoring system has not been limited only to broilers but has ranged to other breeds and species of birds. Du, Lao, and Teng (2018) developed a surveillance system using microphone arrays of Kinects to automatically monitor laying hens' abnormal vocalisations (2018). By monitoring the number of vocalisations and acoustic features, the model detected anomalous states during the night for specific areas of the pen. The model may detect an increase in the number of nocturnal vocalisations in otherwise relatively quiet flocks, which could indicate a lack of food or water. Even increased peaking sounds for water intake can, for example, be associated with a lack of solid feed (2018).

Although very promising, these studies to monitor real-time feeding intake or feeding behaviour of birds need to be validated in a husbandry context involving larger numbers of birds and environmental noise (Coutant, Villain, and Briefer 2024).

Animal behaviour can be influenced by environmental conditions. Ginovart-Panisello et al. (2022) studied vocalisations as a behavioural response of chickens to changes in environmental $CO_2$ levels. $CO_2$ is naturally produced by animals, as a result of their metabolism and their litter, and it is therefore used as a tracer gas in order to determine the ventilation rates in animal houses. Moreover, it is an anaesthetic acid gas, which causes unpleasant pain sensations in animals. The increase of carbon dioxide inside the farm due to poor ventilation is directly related to the increase of other noxious gases, which together can cause the chicken to rapidly lose brain functions, leading to convulsions and death. The study collected data from two commercial broiler farms in north-west Spain between May and December 2021, both operating under similar weather conditions. The farms featured highly automated management systems to minimise noise from workers. Farm A housed 42,000 broilers, with data collected over three production cycles, while Farm B had 40,000 broilers, with one cycle recorded. Both farms followed similar flock management practices and diets, adhering to industrial standards. Each production cycle lasted 42–44 days, depending on target weight and dietary requirements. The study focused on the first 40 days of the chickens' lives, during which they had unrestricted access to feed, water, and movement. The results obtained demonstrated a relationship between the acoustic parameters MFCC 6, 9, 4, and 3 with the environmental levels of $CO_2$ between 400 and 4000 ppm.

Among the most significant features, the MFCC 9 was the one that best isolated vocalisations from environmental sounds, with similar performance and representativeness, describing with greater accuracy the behaviour of chickens with respect to the concentration of environmental carbon dioxide (Ginovart-Panisello et al. 2022).

Ginovart-Panisello et al. (2024) focused on chicks' vocalisations as reliable indicators of changes in stress, anxiety, and well-being. An acoustic recording campaign was conducted at a Mediterranean chicken farm in Jesús, southern Catalonia (Spain). The farm comprised two identical houses, vertically stacked, each housing 45,000 layers of Hy-Line Brown genetics. The hens were raised for 18 weeks, starting from day one, under artificial light from 5:00 am to 10:00 p.m. At 10 weeks of age, the birds were manually vaccinated with the ILT eye drop vaccine, a process lasting 3 days. The study aimed to record bird vocalisations before, during, and after vaccination to analyse changes in their vocal behaviour. Data collection occurred between 28 November and 10 December 2022, using two identical non-invasive sound sensors placed in each house to ensure measurement consistency. Acoustic data were processed to isolate hen calls, extracting 13 Mel-frequency cepstral coefficients (MFCCs), the spectral centroid, and the number of vocalisations. In the control group, a Gaussian Naive Bayes model achieved a balanced accuracy of 80% in distinguishing between reactions and non-reactions to the ILT vaccine. The spectral centroid and the fourth MFCC (MFCC4) were identified as key features in detecting vaccine reactions. Applying this model to the treatment group, 99% of vocalisations were classified as non-reactions, suggesting that the anti-inflammatory product effectively mitigated vaccine-related side effects. This non-invasive method demonstrates potential for real-time monitoring of poultry health and welfare through vocalisation analysis. (2024).

Du et al. (2020) developed a hen vocalisation detection method to assess their thermal comfort condition. The study focused on quantitatively measuring frustration-related vocalisations to understand if they could be useful indicators for evaluating animal thermal comfort condition and, consequently, their well-being. The experiments were conducted at the Shangzhuang experimental station of China Agricultural University in Beijing, using 100 Jingfen hens aged 18–20 weeks. The hens were housed in a perch husbandry system with *ad libitum* access to food and water and a timer-controlled light schedule (8:00 am to 6:00 pm). The experiments were carried out in a temperature-controlled chamber to monitor and record hens' vocalisations under varying thermal conditions, classified using the Temperature-Humidity Index (THI). THI levels were recorded every 5 min and categorised into four zones: comfort (¡ 70), alert (70–75), danger (76–81), and emergency (¿ 81). Vocalisations were classified into four types: gakel, alarm, squawk, and others. Using a machine learning-based approach, researchers developed a detection method that analysed nine temporal and spectral features based on the source-filter theory, employing a Support Vector Machine (SVM) classifier. The optimised SVM model achieved a sensitivity of 95.1 ± 4.3% and a precision of 97.6 ± 1.9%. Key findings revealed significant correlations between specific vocalisation types and the Temperature-Humidity Index (THI): alarm calls showed a negative correlation ($R = -0.414$, $P = 0.01$), while squawk calls displayed a positive correlation ($R = 0.594$, $P = 0.01$) (ibid.). These results suggested that monitoring vocalisations can provide an effective and non-invasive method to assess thermal comfort of laying hens, offering valuable insights to improve poultry welfare and management practices.

## 5 Case study

Most domestic birds are transported as day-old chickens from hatcheries to production sites. As reported by the European Food Safety Authority (EFSA), in 2020, the European Union exported over 200 million day-old chickens to non-EU countries, primarily by truck, though air transport is also commonly used

(Nielsen et al. 2022). Broilers usually hatch over a 3-day period, with a "hatching window" of 24–48 h, during which most eggs in a batch hatch. At hatching, the chickens weigh around 45 g, with an 8 g yolk sac providing their only nutrients until food is introduced (Ginovart-Panisello et al. 2024). This usually happens after the chickens arrive at the farm, remaining fasted for the entire time of transport, reaching a maximum of 72 h without water and food. This prolonged fasting can jeopardise their health and proper development (2024).

A recent approach to address this issue is the use of nutrient-enriched hydrated gels, which provide essential nutrients during the chickens' first critical hours, avoiding prolonged fasting (Riva and Monjo 2020). These gels contain easily digestible vitamins, minerals, prebiotics, and amino acids and are administered directly in the incubation room in the form of droplets, for rapid ingestion (Prabakar et al. 2016). For this study, the product Licuicel ® Complex, a hydrated gel enriched with high-quality nutrients, from the company CEALVET S.L.U., was used. This research is based on Ginovart-Panisello et al.'s (2024) publication, where the reader can find more details.

Acoustic data collection was conducted continuously for 16 months, from February 2023 to May 2024, in a Spanish commercial hatchery. The study aimed to verify the potential health benefits associated with the administration of the product Licuicel ® Complex. The animals were divided into two groups: a control group, whose vocalisations were recorded during the odd weeks, and a treatment group, whose vocalisations were recorded during the even weeks. All newly hatched broiler chickens in the shipping room, where the acoustic sensor was installed, were monitored during hatching on Monday, Tuesday, and Thursday, for 7 h, from the time the first chicken was stored until the time before transport to the truck.

To create the acoustic dataset, a class 1 sound level meter and recorder were placed in the centre of the shipping room, 0.5 m from the ceiling. Figure 3 shows an example of where the microphone was installed. The sensor consisted of a 1/2" condenser microphone connected to an ADC and a processing unit that continuously recorded data with a sampling frequency of 48 kHz, a resolution of 24 bits, and a sensitivity of 25 mV/Pa. It also had an IP65 protection rating for the entire sensor, so it did not need to be removed when the room was cleaned (2024).

The acoustic recording data mainly contained raw vocalisations of newborn chickens isolated in the shipping room, and some noise detected from other sources, such as ventilation, or from adjacent rooms. To eliminate background noise, the following two filters were applied: a 10th-order Butterworth low-pass filter with a cut-off frequency of 1 kHz and a 10th-order Butterworth high-pass filter with a cut-off frequency of 6 kHz (2024). The acoustic data were recorded continuously and split into 1-min files with FLAC compression.

**Figure 3** Microphone installed in a commercial hatchery to detect chicken vocalisation.

In this study, the frequency of the chickens' vocalisations was analysed, as it is a parameter that can be correlated with animal distress. The acoustic data was filtered starting 4 h after the administration of hydrated gel (HG), as the product required a certain amount of time to take effect. According to the Shapiro-Wilk test (Shapiro and Wilk 1965), the data were not derived from populations with a normal distribution. Since the normality criterion was not met, Levene's test was performed to find significant evidence (for alpha = 0.05) that the variances were different between the two populations (Schultz 1985). Finally, a t-test with Welch's correction was performed to determine whether or not the mean of acoustic features between the groups differed.

Data were subjected to statistical analysis using an independent sample t-test (Posten 1984) to determine whether the two groups differed from each other, with a *P*-value < 0.05 to validate the hypothesis of a significant difference.

The results obtained, reported in Fig. 4 and Table 1, showed that the vocalisations of the treatment group had a lower mean vocalisation frequency, and this difference was statistically significant at 1.95e-12. The reduction was also observed in the minimum, maximum, 25%, 50%, and 75% values of the data analysed. This reduction was attributed to the gel administration, being the only difference between the two groups. Additionally, the standard deviation of the treated group (HG) was lower than that of the control group, indicating that the data were less scattered around the mean, suggesting more similar vocalisation frequencies.

A detailed study of these data was also carried out by filtering the data according to the day of the week of birth, in Fig. 5 and Table 2.

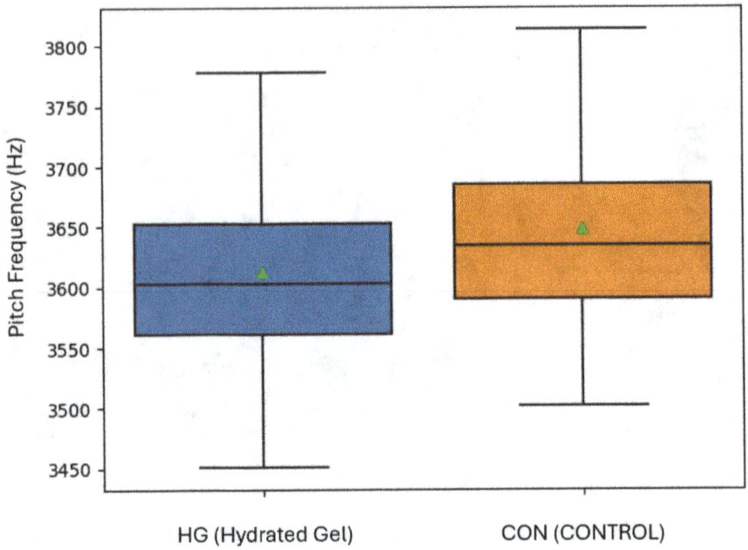


**Figure 4** Pitch frequency values per minute shown in Box plot by treatment group.

**Table 1** Pitch frequency values per minute (mean, minimum, maximum, 25%, 50%, 75%) by treatment group

| Group | Mean | Std | Min | 25% | 50% | 75% | Max | *P*-val |
|---|---|---|---|---|---|---|---|---|
| CON | 3648 | 90 | 3435 | 3590 | 3634 | 3684 | 4226 | 1.95e-12 |
| HG | 3612 | 73 | 3406 | 3561 | 3603 | 3652 | 4145 | |

The results by day of the week showed that there was a statistically significant difference for the 3 days of hatching, with a *P*-value < 0.05 (Table 2). Monday was the day with the lowest mean (3585 Hz), indicating that it was the day with the least stress for the treated animals. Tuesday was the day with the largest difference between the two groups, giving the highest *P*-value. Differences between days could be due to different flocks, ages, and days of egg transfer.

In conclusion, it can be deduced that, through the use of bioacoustics, it was possible to detect and measure how the administration of the product Licuicel® Complex to newly hatched chickens can attenuate the effects of prolonged fasting and, consequently, reduce physical stress, significantly reducing the frequency of calls.

## 6 Conclusion and future trends

Over the last decade, the price of computing resources has dropped significantly, new IoT devices and companies have entered the market,

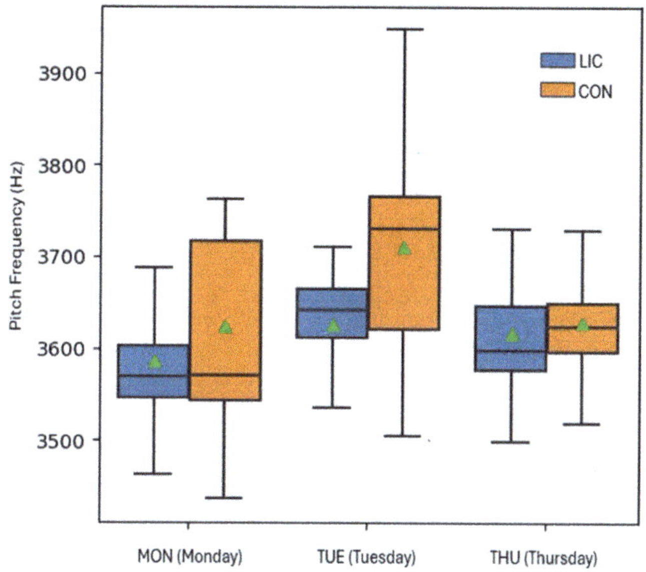


**Figure 5** Pitch frequency Box plot by day of week of birth and treatment group.

and standard algorithms have become available in open-source libraries. All of these technological advances, combined with society's demand for environmentally friendly and animal-friendly production, have accelerated research into Precision Livestock Farming.

As the demand for poultry products has increased, the size of poultry farms has also increased. Intensive production of meat for human consumption can sometimes create unpleasant conditions for the animals, such as pain, fear, and distress, which negatively affect their welfare. Traditionally, livestock monitoring was done by operators, but the increase in animal number has complicated this task.

Acoustic PLF technologies in poultry farming focus on monitoring animals with AI models to detect diseases and control their behaviour.

Many studies are beginning to detect through acoustic systems and the use of algorithms, some of the most widespread diseases in poultry, such as necrotic enteritis, Newcastle disease, bronchitis, Avian Influenza, and side effects of vaccine administration. The accuracy detection rates of diseases or vaccine reactions depend on the study and vary between 80% and 97%. Another focus of research is the study of animal behaviour based on the detection of abnormal vocalisations under certain stress factors, such as lack of food or water, high $CO_2$ concentrations, prolonged fasting in newly hatched chickens, and thermal stress. Animal behaviour is one of the first indicators of potential health problems.

**Table 2** Pitch frequency values per minute (mean, minimum, maximum, 25%, 50%, 75%) by day of week of birth and treatment group

| Day | Group | Mean | Std | Min | 25% | 50% | 75% | Max | *P*-val |
|---|---|---|---|---|---|---|---|---|---|
| Mon | CON | 3625 | 120 | 3436 | 3543 | 3571 | 3716 | 4109 | 0.008 |
| | HG | 3585 | 79 | 3407 | 3542 | 3570 | 3603 | 4146 | |
| Tue | CON | 3710 | 111 | 3505 | 3622 | 3731 | 3766 | 4096 | 2.7e-15 |
| | HG | 3626 | 55 | 3451 | 3612 | 3643 | 3665 | 3778 | |
| Thu | CON | 3629 | 54 | 3509 | 3596 | 3624 | 3649 | 3835 | 0.036 |
| | HG | 3616 | 65 | 3499 | 3577 | 3597 | 3646 | 3973 | |

Although the results are promising, most studies have been conducted in experimental farms with a small population of birds and a controlled environment. The challenge for future research will be to obtain the same results in farms and hatcheries with background noise independent of the type of farm, machinery, and management.

The industry will adopt this new technology when real-time detection will have an accuracy of over 90%, thus improving animal health, increasing productivity or reducing production costs, applying treatments only when necessary, and anticipating possible medical complications. Regulatory bodies may recommend the use of this technology for animal health certification, as it is a non-invasive system that provides a quantifiable and replicable level of welfare, independent of the assessor.

In the coming decades, future trends will be related to technological innovation, resources, and communication protocols. The industry will continue to automate based on AI models to reduce human interaction with animals, maximising their welfare. New electronic resources will allow the use of more complex algorithms and larger data sets, leading to better disease and behaviour detectors. Advances in IoT technology and data interrelationship between sensor types will lead to new applications and equipment for monitoring poultry production.

## 7 Where to look for further information

Further information on bioacoustic approaches in poultry production can be found within the broader literature on Precision Livestock Farming (PLF), animal bioacoustics, and sensor-based welfare assessment. Readers are encouraged to consult peer-reviewed journals in animal behaviour, animal welfare, biosystems engineering, and applied signal processing, where poultry-specific studies on vocalisation analysis, health monitoring, and stress detection are increasingly reported. Edited books and review papers on PLF provide an integrated perspective on bioacoustics alongside other sensing technologies. In addition, proceedings from international PLF and animal welfare conferences, as well as open-access repositories hosting audio datasets and signal-analysis tools, offer practical and methodological resources for researchers and practitioners.

## 8 References

Anderson, T.W. and Finn, J.D. (2012). *The new statistical analysis of data*. Springer, New York.

Arzey, G. and Littleton, I. (2007). *NSW biosecurity guidelines for free range poultry farms*. NSW Department of Primary Industries, Orange, NSW.

Aydin, A. and Berckmans, D. (2016). "Using sound technology to automatically detect the short-term feeding behaviours of broiler chickens". *Computers and Electronics in Agriculture* 121, pp. 25–31.

Aydin Bahr, C. et al. (2014). "A novel method to automatically measure the feed intake of broiler chickens by sound technology". *Computers and Electronics in Agriculture* 101, pp. 17–23.

Banakar, A. et al. (2016). "An intelligent device for diagnosing avian diseases: newcastle, infectious bronchitis, avian influenza". *Computers and Electronics in Agriculture* 127, pp. 744–753.

Bardeli, R. et al. (2010). "Detecting bird sounds in a complex acoustic environment and application to bioacoustic monitoring". *Pattern Recognition Letters* 31.12, pp. 1524–1534.

Beddard, F.E. (1898). *The structure and classification of birds*. Longmans, Green, and Company, London.

Ben Sassi, N. et al. (2016). "Technology and poultry welfare". *Animals* 6.10, p. 62.

Berckmans, D. (2017). "General introduction to precision livestock farming". *Animal Frontiers* 7.1, pp. 6–11.

Berk, R.A. (2008). "Classification and regression trees (CART)". In: Berk, R.A. (ed.) *Statistical learning from a regression perspective*. Springer, Cham, pp. 1–65.

Breiman, Leo (1996). "Bagging predictors". *Machine Learning* 24, pp. 123–140.

Briefer, E. and Comber, S. (Sept. 2012). "Vocal expression of emotions in mammals: Mechanisms of production and evidence". *Journal of Zoology* 288, pp. 1–20. https://doi.org/10.1111/j.1469-7998.2012.00920.x.

Cichy, R.M. and Kaiser, D. (2019). "Deep neural networks as scientific models". *Trends in Cognitive Sciences* 23.4, pp. 305–317.

Collins, S.A. et al. (2024). "Do I sound anxious? Emotional arousal is linked to changes in vocalisations in domestic chicks (Gallus gallus dom.)." *Applied Animal Behaviour Science* 277, p. 106359.

Commission, Terrestrial Animal Health Standards (2008). "Report of the meeting of the oie terrestrial animal health standards commission". In: *OIE Terrestrial Animal Health*.

Cordeiro, A.F. et al. (2018). "Use of vocalisation to identify sex, age, and distress in pig production". *Biosystems Engineering* 173, pp. 57–63.

Coutant, M. et al. (2024). "A scoping review of the use of bioacoustics to assess various components of farm animal welfare". *Applied Animal Behaviour Science* 300, p. 106286.

Cuan, K. et al. (2020). "Detection of avian influenza-infected chickens based on a chicken sound convolutional neural network". *Computers and Electronics in Agriculture* 178, p. 105688.

Cuan, K. et al. (2022). "Automatic Newcastle disease detection using sound technology and deep learning method". *Computers and Electronics in Agriculture* 194, p. 106740.

Curtin, R.R. et al. (2014). "Classifying broiler chicken condition using audio data". In: *2014 IEEE Global Conference on Signal and Information Processing (GlobalSIP)*. IEEE, pp. 1141–1144.

Dale, S.S. et al. (2022). "Distinguishing sex of northern spotted owls with passive acoustic monitoring". *Journal of Raptor Research* 56.3, pp. 287–299.

De Lanerolle, N.C. and Youngren, O.M. (1978). "Chick vocalization and emotional behavior influenced by apomorphine". *Journal of Comparative and Physiological Psychology* 92.3, p. 416.

Du, X. et al. (2020). "Assessment of laying hens' thermal comfort using sound technology". *Sensors* 20.2, p. 473.

Du, X. et al. (2018). "A sound source localisation analytical method for monitoring the abnormal night vocalisations of poultry". *Sensors* 18.9, p. 2906.

Dubnov, S. (2004). "Generalization of spectral flatness measure for nongaussian linear processes". *IEEE Signal Processing Letters* 11.8, pp. 698–701.

Fontana, E.T. et al. (2016). "Vocalisation sound pattern identification in young broiler chickens". *Animal* 10.9, pp. 1567–1574.

Fontana, E.T. et al. (2017). "Sound analysis to model weight of broiler chickens". *Poultry Science* 96.11, pp. 3938–3943.

Garg, P.K. (2021). "Overview of artificial intelligence". In: Jokanović, V.R. (ed.) *Artificial intelligence*. Chapman and Hall/CRC, Boca Raton, pp. 3–18.

Garson, G.D. (2022). *Factor analysis and dimension reduction in R: a social scientist's toolkit*. Routledge, New York.

Ginovart-Panisello, G. et al. (2022). "Trend and Representativeness of Acoustic Features of Broiler Chicken Vocalisations Related to CO2". *Applied Sciences* 12.20, p. 10480.

Ginovart-Panisello, G. et al. (2024). "Acoustic detection of vaccine reactions in hens for assessing anti-inflammatory product efficacy". *Applied Sciences* 14.5.

Grosche, P. et al. (2010). "Cyclic tempogram–a mid-level tempo representation for musicsignals". In: *2010 IEEE International Conference on Acoustics, Speech and Signal Processing*. IEEE, pp. 5522–5525.

Halachmi, I. et al. (2019). "Smart animal agriculture: application of real-time sensors to improve animal well-being and production". *Annual Review of Animal Biosciences* 7, pp. 403–425.

Harris, C.M. and Tocci, G.C. (1993). "Handbook of acoustical measurements and noise control". *The Journal of the Acoustical Society of America* 93.6, pp. 3532–3532.

Huang, J. et al. (2019). "Method for detecting avian influenza disease of chickens based on sound analysis". *Biosystems Engineering* 180, pp. 16–24.

Huang, J. et al. (2021). "An intelligent method for detecting poultry eating behaviour based on vocalization signals". *Computers and Electronics in Agriculture* 180, p. 105884.

Ketkar, N. et al. (2021). "Convolutional neural networks". In: Apress Berkeley, CA. (ed) Springer (Deep Learning with Python: A Hands-on Introduction | Springer Nature Link, pp. 197–242. (https://link.springer.com/book/10.1007/978-1-4842-2766-4).

Klapuri, A. and Davy, M. (2007). *Signal processing methods for music transcription*. Springer Science & Business Media. Chap. 5.

Kogan, J.A. and Margoliash, D. (1998). "Automated recognition of bird song elements from continuous recordings using dynamic time warping and hidden Markov models: a comparative study". *The Journal of the Acoustical Society of America* 103.4, pp. 2185–2196.

Kurita, T. (2019). "Principal component analysis (PCA)". In: Springer, Cham (ed.). *Computer vision: a reference guide*, pp. 1–4. https://link.springer.com/rwe/10.1007/978-3-030-63416-2_649.

Laurijs, K. A. et al. (2021). "Vocalisations in farm animals: a step towards positive welfare assessment". *Applied Animal Behaviour Science* 236, p. 105264.

Li, N. et al. (2020). "Automated techniques for monitoring the behaviour and welfare of broilers and laying hens: towards the goal of precision livestock farming". *Animal* 14.3, pp. 617–625.

Liu, L. et al. (2020). "A novel method for broiler abnormal sound detection using WMFCC and HMM". *Journal of Sensors* 2020.1, p. 2985478.

Mahdavian, A. et al. (2021). "Acoustic features of vocalization signal in poultry health monitoring". *Applied Acoustics* 175, p. 107756.

Mammone, A. et al. (2009). "Support vector machines". *Wiley Interdisciplinary Reviews: Computational Statistics* 1.3, pp. 283–289.

Manteuffel, G. et al. (2004). "Vocalization of farm animals as a measure of welfare". *Applied Animal Behaviour Science* 88.1, pp. 163–182.

Mao, A. et al. (2022). "Automated identification of chicken distress vocalizations using deep learning models". *Journal of the Royal Society Interface* 19.191, p. 20210921.

Mellinger, D.K. (2011). "Introduction to Animal Bioacoustics". *The Journal of the Acoustical Society of America* 129.4, pp. 2406–2406.

Misra, H. et al. (2004). "Spectral entropy based feature for robust ASR". In: *2004 IEEE International Conference on Acoustics, Speech, and Signal Processing*. Vol. 1. IEEE, pp. I–193.

Mota-Rojas, D. et al. (2021). "Anthropomorphism and its adverse effects on the distress and welfare of companion animals". *Animals* 11.11, p. 3263.

Mucherino, A. et al. (2009). "K-nearest neighbor classification". *Data mining in agriculture*, Springer, New York, NY. pp. 83–106. https://link.springer.com/chapter/10.1007/978-0-387-88615-2_4.

Neethirajan, S., Inonge, R. Reimert, I., and Kemp, B. (2021). "Measuring farm animal emotions Sensor based approaches". *Sensors* 21.2, p. 553.

Nielsen, S.S. et al. (2022). "Welfare of domestic birds and rabbits transported in containers". *EFSA Journal* 20.9, e07441.

Parmar, A. et al. (2019). "A review on random forest: An ensemble classifier". In: *International conference on intelligent data communication technologies and internet of things (ICICI) 2018*. Springer, pp. 758–763.

Posten, H.O. (1984). "Robustness of the two-sample t-test". In: Rasch, D. and Tiku, M.L. (eds) *Robustness of statistical methods and nonparametric statistics*. Springer, Cham, pp. 92–99.

Prabakar, G. et al. (2016). "Early nutrition and its importance in poultry: a review". *Indian Journal of Animal Nutrition* 33.3, pp. 245–252.

Rabiner, L. (1977). "On the use of autocorrelation analysis for pitch detection". *IEEE Transactions on Acoustics, Speech, and Signal Processing* 25.1, pp. 24–33.

Rabiner, L. and Schafer, R. (2010). *Theory and applications of digital speech processing*. Prentice Hall Press, Upper Saddle River.

Riva, S. and Monjo, T.P. (2020). "The importance of early nutrition in broiler chickens: Hydrated gels enriched with nutrients, an innovative feeding system". In: *Animal husbandry, dairy and veterinary science*. Academic Press, London, pp. 45–60.

Riva, S. and Monjo, T.P (2021). "The major predisposing factors for necrotic enteritis in broiler chickens and the use of probiotics as new strategy to prevent the disease". *Journal of Veterinary Medicine and Animal Sciences* 4, p. 1051.

Rizwan, M. et al. (2016). "Identifying rale sounds in chickens using audio signals for early disease detection in poultry". In: *2016 IEEE Global Conference on Signal and Information Processing (GlobalSIP)*, pp. 55–59.

Rowe, E. et al. (2019). "A systematic review of precision livestock farming in the poultry sector: Is technology focussed on improving bird welfare?" *Animals* 9.9, p. 614.

Sadeghi, M. et al. (2015). "An intelligent procedure for the detection and classification of chickens infected by clostridium perfringens based on their vocalization". *Brazilian Journal of Poultry Science* 17, pp. 537–544.

Scheirer, E. and Slaney, M. (1997). "Construction and evaluation of a robust multifeature speech/music discriminator". In: *1997 IEEE international conference on acoustics, speech, and signal processing*. Vol. 2. IEEE, pp. 1331–1334.

Schultz, B.B. (1985). "Levene's Test for Relative Variation". *Systematic Biology* 34.4, pp. 449–456.

Shapiro, S.S. and Wilk, M.B. (1965). "An analysis of variance test for normality (Complete samples)". *Biometrika* 52.3/4, pp. 591–611.

Snell, R.C. and Milinazzo, F. (1993). "Formant location from LPC analysis data". *IEEE Transactions on Speech and Audio Processing* 1.2, pp. 129–134.

Ståhle, L. et al. (1989). "Analysis of variance (ANOVA)". *Chemometrics and Intelligent Laboratory Systems* 6.4, pp. 259–272.

Statista (2024). *Number of chickens worldwide from 1990 to 2022*. https://www.statista.com/statistics/263962/number-of-chickensworldwide-since-1990/. Accessed: 2024-04-26.

Suthers, R.A (1990). "Contributions to birdsong from the left and right sides of the intact syrinx". *Nature* 347.6292, pp. 473–477.

Swayne, D.E. (2007). "Report of the subcommittee on avian influenza and Newcastle disease". In: *One hundred and sixteenth anual meeting*, p. 465.

Thomas, P. et al. (2023). "Using a neural network based vocalization detector for broiler welfare monitoring". In: *10th Convention of the European Acoustics Association*.

Urquiza-Haas, E.G. and Kotrschal, K. (2015). "The mind behind anthropomorphic thinking: Attribution of mental states to other species". *Animal Behaviour* 109, pp. 167–176.

Van Der Maaten, L. et al. (2009). "Dimensionality reduction: a comparative". *Journal of Machine Learning Research* 10, pp. 66–71.

Wang, H. et al. (2013). "Fisher discriminant analysis with L1-norm". *IEEE Transactions on Cybernetics* 44.6, pp. 828–842.

Winters, R. et al. (2010). "Statistics: a brief overview". *Ochsner Journal* 10.3, pp. 213–216.

Yang, X. et al. (2021). "Characterizing sounds of different sources in a commercial broiler house". *Animals* 11.3.

Yang, X. et al. (2010). "Comparative study on voice activity detection algorithm". In: *2010 International Conference on Electrical and Control Engineering*. IEEE, pp. 599–602.

Yu, T.S. (1967). "Information content of a sound spectrogram". *Journal of the Audio Engineering Society* 15.4, pp. 407–413.

Zien, A. et al. (2009). "The feature importance ranking measure". In: *Machine Learning and Knowledge Discovery in Databases: European Conference, ECML PKDD 2009, Bled, Slovenia, September 7-11, 2009, Proceedings, Part II 20*. Springer, pp. 694–709.

# Chapter 18

# The potential of using volatile organic compounds (VOCs) to monitor health issues in broiler production

*Nelle Lauwers, Maarten Hertog, Nadia Everaert and Matthias Corion, Katholieke University of Leuven, Belgium*

1 Introduction
2 Volatile organic compounds: primary metabolic pathways
3 Volatile organic compounds: secondary metabolic pathways
4 Sampling volatile organic compounds in chickens
5 Pathogen influence on volatile organic compounds
6 General challenges in identifying reliable pathogen biomarkers
7 Internal factors influencing volatile organic compound fingerprints
8 External factors influencing volatile organic compound fingerprints
9 Volatile organic compound measurement and analysis
10 Conclusion and future trends
11 Where to look for further information
12 References

## 1 Introduction

Volatile organic compounds (VOCs) have recently gained attention as a novel monitoring tool in broiler production. These small organic molecules with a high vapor pressure at ambient temperature and standard atmospheric pressure can evaporate easily and disperse over long distances in both gas and liquid phases (Dell'Olio et al., 2024; Epping & Koch, 2023). They can be emitted by the birds themselves (Hankel et al., 2022) as well as external sources such as feed (Recharla et al., 2017), manure (Ranadheera et al., 2017), and the surrounding environment (Huanhong et al., 2023). Within poultry barns, VOCs exist in the ambient air and can vary significantly based on diet (David et al., 2014), infections (Hankel et al., 2022), and environmental conditions (Liao et al., 2001). These variations make VOCs promising biomarkers for assessing animal

http://dx.doi.org/10.19103/AS.2025.0165.21

health, welfare, air quality, and overall production efficiency, though further research and validation are needed for routine application. This chapter focuses on organic volatiles, while also considering certain inorganic volatiles, such as ammonia or hydrogen sulfide, due to their relevance in the broiler industry.

The identification and characterization of VOCs fall under the domain of volatilomics, a branch of metabolomics (Rodríguez-Hernández et al., 2022), just as lipidomics is another branch of metabolomics (Ferreira et al., 2010). Volatilomics complements other 'omics' disciplines, such as genomics (DNA), transcriptomics (RNA), and proteomics (proteins), providing insights into host-microbiota interactions and metabolic processes (Fig. 1).

Earlier studies on VOCs in livestock primarily examined their impact on air quality in animal housing. In broiler production, for instance, the accumulation of manure and litter is a significant source of VOC emissions, contributing to ground-level ozone formation more than other livestock sectors (Yao et al., 2018). Addressing these emissions is crucial for improving air quality, and VOC dispersion models have been developed to determine optimal farm-to-residence distances (Guo and Yuan, 2011).

More recently, the focus has shifted toward VOCs as biomarkers for respiratory and intestinal health in poultry. Specific VOC signatures have been linked to microbial infections in the respiratory and gastrointestinal tracts, distinguishing between bacterial and viral infections (Kamal et al., 2021). For example, VOC profiling has successfully identified *Campylobacter* infections in broilers (Hankel et al., 2022). The noninvasive nature of VOC analysis makes it a promising diagnostic tool, allowing repeated assessments without compromising animal welfare.

Integrating VOC monitoring in broiler production offers multiple benefits, from early disease detection and antibiotic reduction (Grilli et al., 2018) to improved feed formulation and waste management (Urrego et al., 2021; Yan et al., 2021). Besides air sampling, VOCs can be noninvasively collected from various sources, including feathers, skin, breath, and uropygial secretions (Campagna et al., 2012), providing a comprehensive profile of bird health. By leveraging these insights, broiler production can adopt more sustainable and data-driven approaches to health management and air quality control.

The aim of this chapter is to provide a comprehensive overview of the current knowledge on VOCs as potential biomarkers in broiler production. We discuss the metabolic and microbial origins of VOCs, their links with broiler health and disease, relevant sample types, and analytical techniques. Throughout the chapter, we highlight both opportunities and challenges, offering practical insights and a critical perspective to guide future research and application in broiler health monitoring.

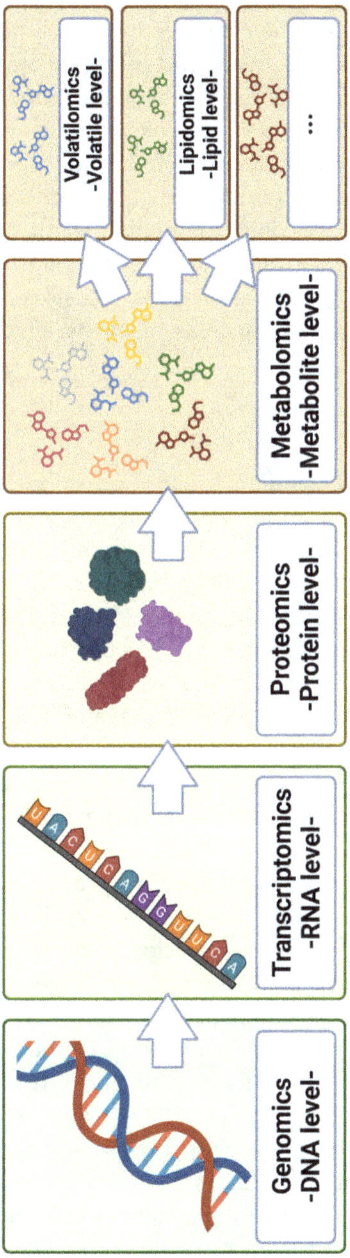


**Figure 1** Overview of the different 'omics' domains. This figure illustrates the 'omics' disciplines from genomics to volatilomics, with the flow of biological information going from DNA to volatiles (Figure created with BioRender.com).

## 2 Volatile organic compounds: primary metabolic pathways

The production of VOCs in broilers arises from biochemical reactions involving various metabolic pathways in the animal. Proteins, lipids, and carbohydrates act as key substrates for VOC production (Oliphant and Allen-Vercoe, 2019). These macronutrients are metabolized within intestinal cells, producing primary metabolites, some of which directly contribute to VOC production. Subsequently, interactions between these metabolites and the gut microbiota can activate secondary metabolic pathways, leading to the formation of specific VOCs (van Vorstenbosch et al., 2023). A minimal understanding of these underlying processes is essential for using VOCs as effective indicators of broiler health, particularly when potential disruptions, such as infections, affect VOC production.

### *2.1 Protein metabolism*

Protein metabolism includes various pathways leading to the production of VOCs. Proteins are first hydrolyzed into peptides and subsequently into amino acids, which can then be degraded through processes such as transamination (Bender, 2007), deamination (Lammers-Jannink et al., 2022), decarboxylation (Feddern et al., 2019), Strecker degradation (Belitz, Grosch and Schieberle, 2009), and amino acid oxidation (Jaimes-Mogollón et al., 2021). Under normal physiological conditions, these metabolic routes produce a characteristic pattern of VOCs. However, this balance can be disrupted, for example, by oxidative stress, which leads to the generation of reactive oxygen species (ROS). An imbalance between free radicals and antioxidant defenses can lead to tissue damage, where ROS attack amino acids, resulting in volatile compounds as by-products (Kessler and Remmer, 1990; Oyerinde et al., 2023).

Figure 2 provides an overview of these metabolic processes. Decarboxylation and deamination of amino acids yield amines, and the remaining carbon skeletons, often referred to as keto acids, can either undergo transamination, enter the gluconeogenesis cycle, or serve as precursors for fatty acid synthesis via acetyl-CoA (Bender, 2007). Additionally, the Strecker degradation, a reaction involving amino acids and dicarbonyl compounds, produces aldehydes with the release of ammonia and carbon dioxide (Belitz, Grosch and Schieberle, 2009; Luo et al., 2022). While this reaction is primarily recognized in the context of food processing, it is noteworthy for its significance in the formation of VOCs that contribute to the flavor profile of chicken meat (Luo et al., 2022). Furthermore, metabolic amino acid oxidation contributes to the generation of diverse volatile compounds, arising either through cytochrome P450-mediated catabolic pathways or via oxidative stress mechanisms involving ROS (Jaimes-Mogollón et al., 2021;

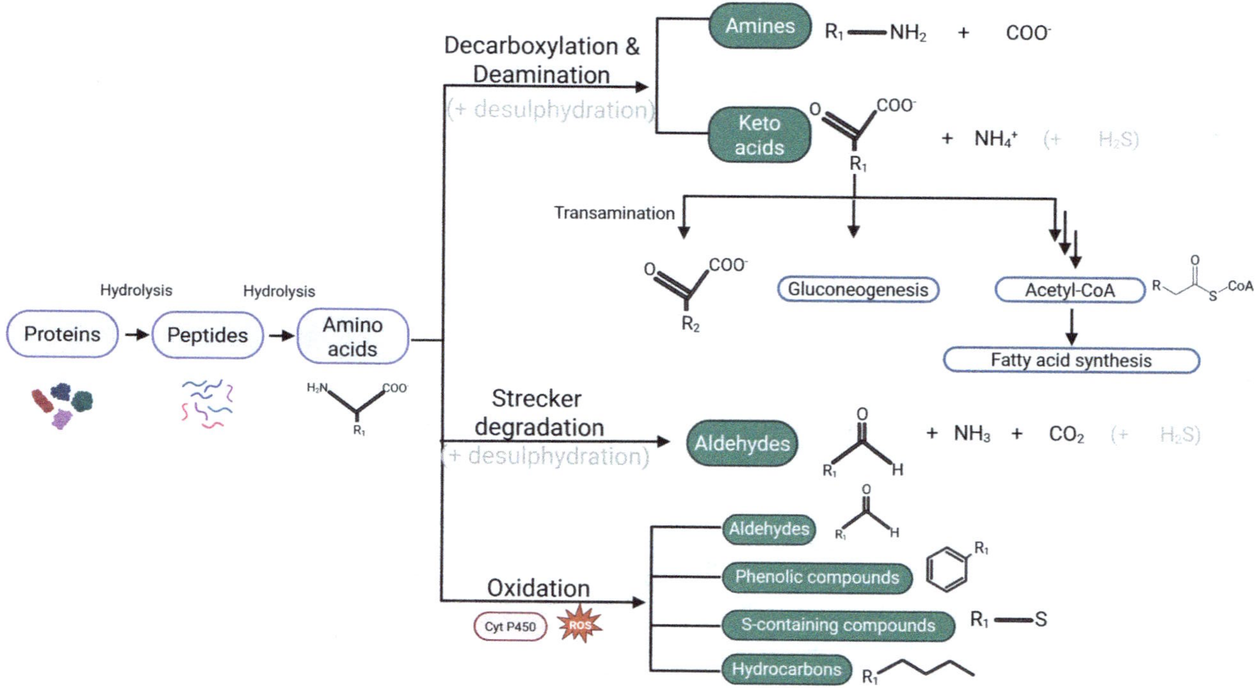


**Figure 2** Pathways of VOC production from proteins. Proteins are initially hydrolyzed into peptides and subsequently into amino acids, which undergo various catabolic processes, including decarboxylation, producing amines and keto acids with the release of ammonia. The keto acids can be further metabolized through gluconeogenesis or fatty acid synthesis. Strecker degradation generates aldehydes alongside ammonia and carbon dioxide, while amino acid oxidation, mediated by cytochrome P450 enzymes or oxidative stress via reactive oxygen species (ROS), leads to the formation of diverse VOCs. Sulfur-containing amino acids additionally release hydrogen sulfide via desulfhydration. Substrates are shown in blue, while the resulting VOCs are highlighted in green (Figure created with BioRender.com).

Oyerinde et al., 2023). The specific volatile compounds formed depend on the chemical nature of the amino acids. Overall, sulfur-containing amino acids also release hydrogen sulfide (Lammers-Jannink et al., 2022). Table 1 provides an additional overview, including examples of produced VOCs for each pathway.

## 2.2 Fat metabolism

The fat metabolism is illustrated in Fig. 3 and arises from both dietary sources and endogenous pathways such as *de novo* lipogenesis in the liver, which mainly results in the synthesis of monounsaturated fatty acids (MUFAs). In contrast, dietary intake of polyunsaturated fatty acids (PUFAs) stimulates the β-oxidation and can decrease the enzyme activity responsible for the *de novo* synthesis of MUFAs in the liver (Kouba and Mourot, 1998; Sanz et al., 2000). However, in commercial broiler production, lipid-poor diets are preferred, making the *de novo* lipogenesis the primary source of fat accumulation (Dridi et al., 2022). Fat metabolism in general involves the hydrolysis of triglycerides into fatty acids and esters. The fatty acids are further metabolized by the β-oxidation pathway in the mitochondria to form acetyl-CoA, which is subsequently used for energy production through the citric acid cycle or converted into cholesterol (Bender, 2007). The cholesterol metabolism yields hydrocarbons, mostly alkenes (Jaimes-Mogollón et al., 2021). The main volatile products produced by β-oxidation are ketones (Jaimes-Mogollón et al., 2021). Similar to proteins, potential disruptions of these pathways can cause oxidative stress, which significantly influences VOC production from fats by increasing the peroxidation of fatty acids, particularly polyunsaturated fatty acids (PUFAs), ω3-, and ω6-fatty acids (Oyerinde et al., 2023). This results in the production of hydrocarbons and ketones (Jaimes-Mogollón et al., 2021). Furthermore, the secondary products include alcohols and aldehydes, where aldehydes can either arise directly from hydrocarbons or via alcohol metabolism (Oyerinde et al., 2023). Examples of these VOC classes are provided in Table 1.

## 2.3 Carbohydrate metabolism

Carbohydrates are initially hydrolyzed into small oligosaccharides, which are further broken down into monosaccharides and disaccharides for absorption by enterocytes in the small intestine. The metabolism of these saccharides involves several pathways, collectively known as energy-yielding metabolism, and is shown in Fig. 4 (Bender, 2007). One primary pathway is glycolysis, a series of enzymatic reactions that convert glucose into pyruvate. According to Zimmermann et al. (2007), this pathway is also associated with benzaldehyde production.

Published by Burleigh Dodds Science Publishing Limited, 2026.

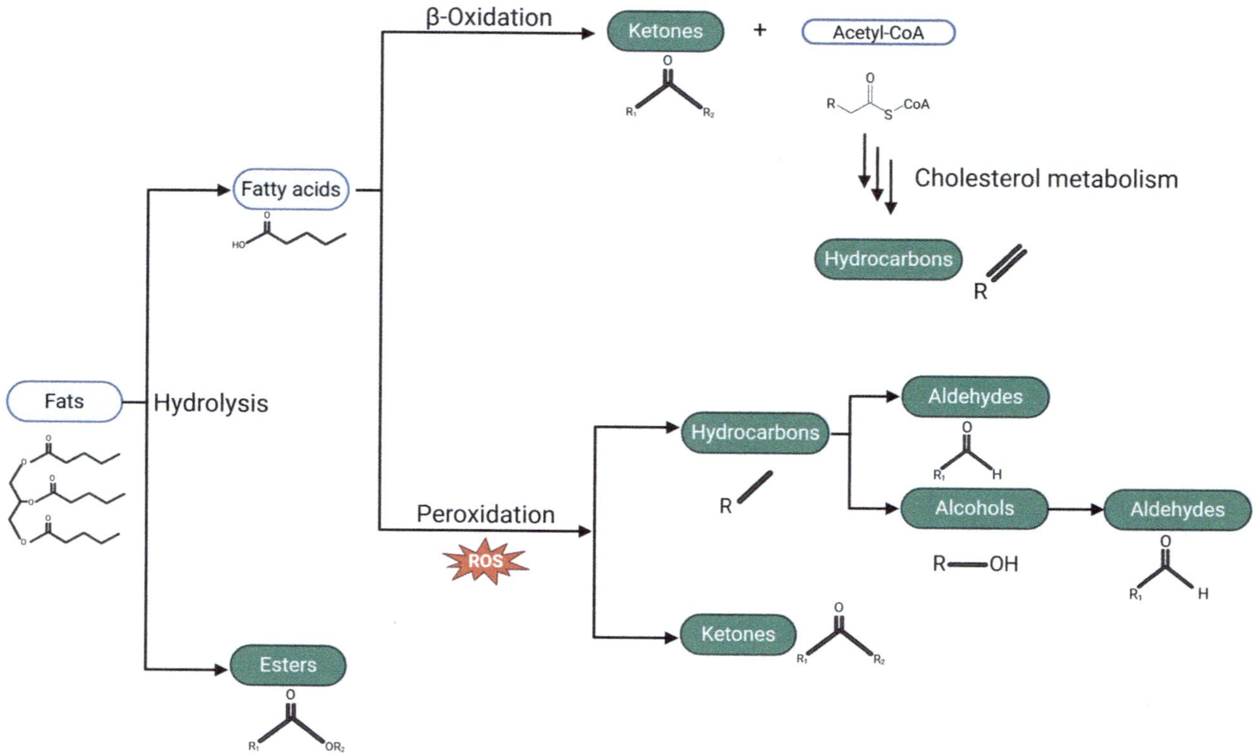


**Figure 3** Pathways of VOC production from fats. Fats are hydrolyzed into esters or fatty acids, which can undergo β-oxidation, producing ketones and acetyl-CoA. Acetyl-CoA contributes to cholesterol synthesis, releasing hydrocarbons (alkenes) during metabolism. Fatty acids may also undergo peroxidation via reactive oxygen species (ROS), generating hydrocarbons and ketones as primary products. Secondary VOCs include aldehydes and alcohols, with alcohols further breaking down into aldehydes. Substrates are shown in blue, while the resulting VOCs are highlighted in green (Figure created with BioRender.com).

**Table 1** Overview of products from various primary metabolic pathways. This table summarizes the metabolic pathways of proteins, fats, and carbohydrates, highlighting the classes and examples of VOCs they produce

| Nutrient | Metabolism pathway | Class of VOCs* | Examples of VOCs | Reference |
|---|---|---|---|---|
| Proteins | Deamination and decarboxylation | Amines | Histamine, putrescine, cadaverine | Feddern et al. (2019) |
| | | N- and S-containing compounds | Ammonia, hydrogen sulfide | Lammers-Jannink et al. (2022) |
| | | Keto acids | Pyruvic acid | Lammers-Jannink et al. (2022) |
| | Strecker degradation | Aldehydes | Benzaldehyde | Luo et al. (2022) |
| | | N- and S-containing compounds | Ammonia, hydrogen sulfide | Belitz, Grosch and Schieberle (2009) |
| | Amino acid oxidation | Aldehydes | Hexanal, nonanal, 2-methylbutanal | Fischer et al. (2015) |
| | | Aromatic compounds | Phenol, 3-methylindole, *1H*-pyrrole | Garner et al. (2007) |
| | | S-containing compounds | 2-methoxy-5-methylthiophene | Jaimes-Mogollón et al. (2021) |
| | | Hydrocarbons | Ethene, propane, ethane | Kessler and Remmer (1990) |
| Fats | β-oxidation | Ketones | 2-Pentanone | Jaimes-Mogollón et al. (2021) |
| | Peroxidation | Hydrocarbons | Dodecane, pentane, ethane | Janfaza et al. (2019) |
| | | Aldehydes | Nonanal, decanal, butanal, pentanal | Oyerinde et al. (2023) |
| | | Alcohols | Cyclohexanol, propan-1-ol | Oyerinde et al. (2023) |
| | | Ketones | Heptan-2-one, pentan-2-one | Oyerinde et al. (2023) |

Published by Burleigh Dodds Science Publishing Limited, 2026.

| | | | | |
|---|---|---|---|---|
| | Cholesterol metabolism | Hydrocarbons | 2-Methylbuta-1,3-diene | Fischer et al. (2015) |
| | Lipid hydrolysis | Esters | Isopropylmyristate | Oyerinde et al. (2023) |
| Carbohydrates | Glycolysis | Aldehydes | Benzaldehyde | Zimmermann et al. (2007) |
| | Pyruvate oxidative decarboxylation | Carboxylic acids | Acetate | Bender (2007) |
| | Ketogenesis | Ketones | Acetone, acetoacetate, 3-hydroxybutanoate | Bender (2007) |

*Abbreviations: N- and S-containing compounds, nitrogen- and sulfur-containing compounds.

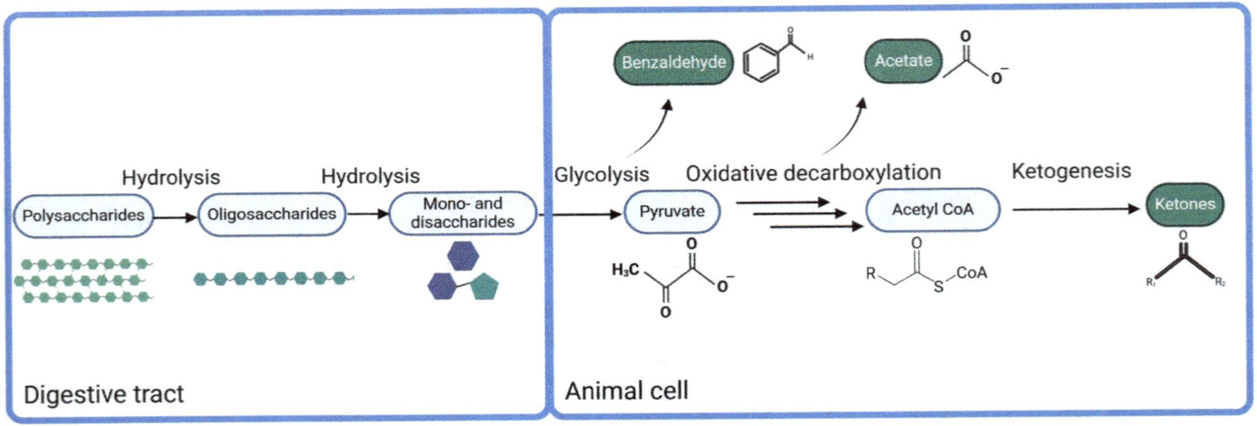


**Figure 4** Pathways of VOC production from carbohydrates. Carbohydrates are initially hydrolyzed in the digestive tract into oligosaccharides, which are further broken down into mono- and disaccharides. On cell level, their metabolism follows energy-yielding pathways, including glycolysis, which converts glucose into pyruvate and is associated with benzaldehyde production. Pyruvate undergoes oxidative decarboxylation to release acetate and produce acetyl-CoA. Acetyl-CoA is, under fasting conditions, redirected toward ketogenesis, leading to the production of ketone bodies. Substrates are shown in blue, while the resulting VOCs are highlighted in green (Figure created with BioRender.com).

Pyruvate can subsequently undergo oxidative decarboxylation to form acetate, which can be esterified to coenzyme A to form acetyl-CoA. This molecule enters the citric acid cycle for further oxidation. This cycle catalyzes the oxidation of acetyl-CoA derived not only from carbohydrates, but also from other sources, including the β-oxidation of fatty acids, catabolism of ketone bodies, ethanol metabolism, and the degradation of certain amino acids (Bender, 2007).

Besides acetate, produced during pyruvate oxidation, carbohydrate metabolism does not directly yield VOCs under normal conditions. Under fasting conditions, acetyl-CoA generated through the citric acid cycle or fatty acid oxidation is redirected toward ketogenesis. This process results in the production of ketone bodies (Bender, 2007). However, the upregulation of genes involved in ketogenesis typically occurs after approximately 16 h of fasting, a duration not commonly encountered under standard commercial broiler practices (Désert et al., 2008). Consequently, the endogenous production of ketone bodies under normal conditions is expected to be minimal. Nonetheless, the presence of such metabolites may serve as a potential biomarker for reduced feed intake, reflecting compromised health status in broilers. Table 1 provides several examples of these chemical classes.

## 3 Volatile organic compounds: secondary metabolic pathways

The animal microbiome comprises unicellular microorganisms, including bacteria, fungi, archaea, viruses, and protozoa (Bindari and Gerber, 2022). These microorganisms engage in complex interactions with their host, which can be detrimental (pathogenic microbiota) or beneficial (symbiotic microbiota) (Aruwa et al., 2021). Microbiota inhabit various regions of the body, such as the skin (Cholewińska et al., 2021), feathers (Jacob et al., 2018), and the respiratory tract (Ivulic et al., 2022), but the highest density is found in the gastrointestinal tract (Aruwa et al., 2021). Microbial VOCs are produced by the metabolic activity of microbiota within the host, in contrast to the host's own metabolic processes discussed in the previous section. To differentiate these from host-derived VOCs, *in vitro* headspace experiments can be conducted to characterize the VOC profiles of microbial cultures. However, discrepancies between *in vivo* and *in vitro* VOC profiles may arise, underscoring the need to establish VOC signatures for each bacterial genus and species, using standardized instrumentation and normalization methods (Belizário, Faintuch and Malpartida, 2021). Table 2 provides a detailed overview of the chemical classes produced by various fermentation processes of the macronutrients by microbiota, along with some common examples of these classes.

An extensive review from Belizário, Faintuch, and Malpartida (2021) about microbial VOCs in exhaled breath of humans reported that the identified

**Table 2** Overview of products from various secondary metabolic pathways. This table summarizes the metabolic pathways of proteins, fats, and carbohydrates, highlighting the classes and examples of microbial VOCs that are produced

| Metabolism | Chemical class* | VOCs | References |
|---|---|---|---|
| Protein metabolism | SCFAs | 3-MethylbutanoateAcetate | Portune et al. (2016)Chen et al. (2024) |
| | Aromatic compounds | 1*H*-indolePhenol4-Methylphenol | Portune et al. (2016) |
| | S-containing compounds | MethanethiolHydrogen sulfide | Portune et al. (2016)Chen et al. (2024) |
| Fat metabolism | Aldehydes | AcetaldehydeProp-2-enal3-Hydroxypropanal | Chen et al. (2024)De Weirdt et al. (2010) |
| | Alcohols | Propane-1,3-diol | De Weirdt et al. (2010) |
| | Amines | Dimethylamine | Fennema, Phillips and Shephard (2016) |
| Carbohydrate metabolism | SCFAs | AcetatePropanoateButanoate | Bindari and Gerber (2022) |
| | Alcohols | Ethanol | Józefiak, Rutkowski and Martin (2004) |
| | Ketones | Acetoin | Kimball et al. (2013) |

* Abbreviations: SCFAs, short-chain fatty acids

Published by Burleigh Dodds Science Publishing Limited, 2026.

volatiles could mainly be classified as aldehydes, phenols, fatty acids, and ketones. However, research on microbial VOCs in poultry is mostly focused on the gastrointestinal tract. Those microbiota serve as a major source of substrates within the body by metabolizing undigested carbohydrates, proteins, and fats from feed, producing vitamins, short-chain fatty acids (SCFAs), organic acids, and antimicrobial compounds (Shang et al., 2018; Chen et al., 2024). Additionally, ammonia is generated by cecal bacteria through the metabolism of nitrogenous host compounds, such as uric acid (Shang et al., 2018). The VOCs produced are thus metabolic by-products of catabolic reactions, facilitating energy and nutrient production. However, these compounds can also contribute to mediating long-range chemical intra- and inter-kingdom interactions (Weisskopf, Schulz and Garbeva, 2021).

## 4 Sampling volatile organic compounds in chickens

VOCs are found in various biological samples and excretions of chickens or animals in general (Fig. 5). These sources include feces and gut content (Garner et al., 2008), breath (Nol et al., 2020), blood (Shirasu and Touhara, 2011), and feathers and skin (Bernier et al., 2008). Depending on the type of diet, infection, environmental conditions, or disruption in endogenous metabolic pathways, the VOC profiles of these sources can be affected (Liao et al., 2001; David et al., 2014; Ellis et al., 2014). For accurate disease detection, the choice of biological

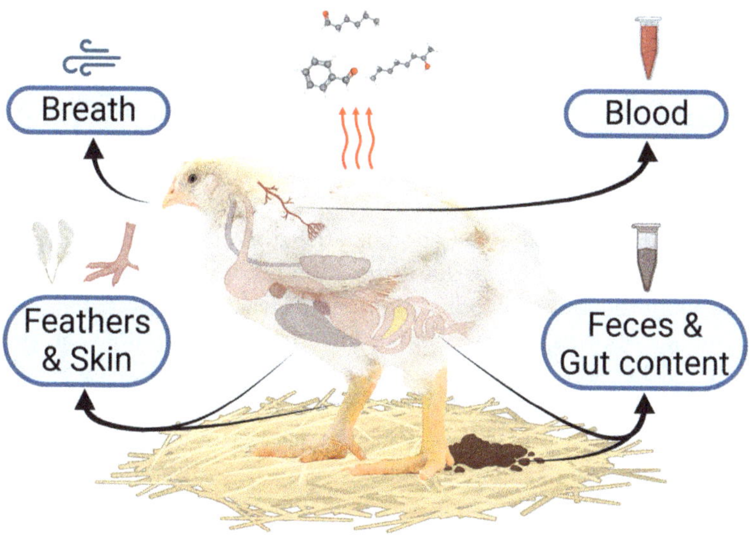


**Figure 5** Types of biological samples in chickens: This figure highlights the various types of biological samples, including feces and gut content, breath, feathers and skin, and blood, from which VOCs can be measured for disease diagnosis (Figure created with BioRender.com).

sample for VOC analysis should align with the study's aim and consider whether the relevant VOCs are produced and released locally or systemically. Additionally, physiological processes from growth and metabolism will cause background variability in the *in vivo* VOCs (Fischer et al., 2015). Therefore, it is essential to understand the characteristics and various types of biological samples available for assessment. An overview of these sample types is provided in Table 3 and is discussed in more detail as follows.

## 4.1 Feces and gut content

Feces and gut content are valuable resources for studying gut health, microbial fermentation, and diet-related metabolic changes (Rodríguez-Hernández et al., 2022). Among these, feces stand out as the most accessible biological sample in chickens, serving as a primary route for eliminating metabolic by-products. This accessibility, combined with their practicality and noninvasive nature, makes feces particularly well-suited for VOC measurements. They hold substantial potential to reveal detailed insights into the animal's internal physiology (Karthikeyan et al., 2013) and offer a particularly representative matrix for the distal colon environment (Jones et al., 2021b). While small intestine contents provide a more precise reflection of the upper gut environment, their invasive collection methods are less practical, highlighting the need for further research to evaluate whether feces reliably serve as a proxy for these upper regions in the gastrointestinal tract (Yan et al., 2019). Additionally, as no research was found concerning VOCs from gut content samples, we focus more on feces in this section.

For accurate fecal VOC profiling, proper sample handling is crucial. Methods to preserve VOCs include immediate freezing, typically at −20°C or −80°C, or storing at 4°C overnight (Jaimes-Mogollón et al., 2021). Subsequently, headspace analysis of the samples can be performed, a process that requires minimal sample handling (Garner et al., 2008). Interpreting fecal volatilomics presents significant challenges due to the inherent heterogeneity of the samples, which include both endogenous components (such as microbiota, epithelial cells, and metabolites) and exogenous materials (like feed) (Beauclercq et al., 2019). In chicken feces, this complexity is further heightened by the presence of urine in the form of uric acid (Beauclercq et al., 2019). Moreover, fecal sample composition can vary due to factors such as differences between the inner and outer portions of the sample, storage conditions, and sampling time during the day (Jones et al., 2021a). Moisture content adds another layer of difficulty, as it influences aerobic microbial activity and metabolite production due to limited oxygen availability in the fecal matrix (Su et al., 2022). Addressing these challenges requires careful management of storage conditions and the standardization of sampling protocols to ensure reliable VOC analysis.

Published by Burleigh Dodds Science Publishing Limited, 2026.

**Table 3** Overview table of different types of biological samples for VOC analysis in broilers. This table summarizes the main biological sample types used in VOC research, detailing the information each provides, associated challenges, storage and handling considerations, and relevant infections or disorders that can be detected

| Sample type | Information provided | Challenges | Storage/handling | Detectable infections/ disorders |
|---|---|---|---|---|
| Feces and gut content | • Gut health<br>• Microbial fermentation<br>• Diet impact | • Interpretation of VOC profile<br>• Sample heterogeneity<br>• High moisture variability<br>• Gut content: invasive sampling | • Storage at 4°C, −20°C, or −80°C | • Digestive disorders<br>• Stress and welfare status<br>• Intestinal infections |
| Breath | • Respiratory health<br>• Real-time metabolic status<br>• VOC composition in the blood (noninvasively) | • Complex sampling<br>• Individual handling<br>• Dead space air | • Immediate analysis or sorbent materials can be stored at −80°C or at 4°C | • Respiratory diseases<br>• Metabolic disorders |
| Feathers and skin | • External contaminants or infections<br>• Microbial and fungal activity | • Environmental VOC contamination | • Swab of the feather or skin surface<br>• Storage at −80°C or 4°C | • Skin infections<br>• External parasites |
| Blood | • Systemic health status<br>• Immune responses<br>• Nutritional and metabolic status | • Invasive sampling<br>• Stressful sampling<br>• Time-consuming sample handling | • Immediate use of whole blood or serum separation<br>• Storage at −80°C or −20°C | • Systemic infections<br>• Metabolic disorders |

Research has increasingly focused on the use of fecal VOCs as biomarkers for the detection of various pathogens and disorders. For instance, Garner et al. (2008) identified a *Campylobacter*-specific VOC signature from six compounds extracted from broiler feces collected on farm sites, achieving an accuracy of over 95%. Additionally, Beauclercq et al. (2019) identified VOCs as early markers for 12-day-old broilers as a response to negative postnatal events, such as a delay in introducing broilers in the barn or transport without feed access. These markers could contribute to monitoring and defining broilers' stress and welfare status (Pérez-Calvo et al., 2019).

### 4.2 Breath

Breath represents another promising biological sample for detecting VOC biomarkers, as it contains volatiles representing respiratory health and internal organ systems with their microbiomes (Belizário, Faintuch and Malpartida, 2021). Most volatiles released by the breath originate from the blood, diffusing across the pulmonary alveolar membrane (Sethi, Nanda and Chakraborty, 2013; Brebu et al., 2023). In addition to endogenous sources, exogenous volatiles can also contribute to breath profiles after inhalation, due to distribution in the chicken's body or exhalation without prior intake in the body (Brebu et al., 2023).

Breath samples offer the advantage of noninvasive collection (Brebu et al., 2023). Once collected on a sorbent tube, typically Tenax, samples can be stored at −80°C (Ellis et al., 2014) or at 4°C (Nol et al., 2020) until analysis, or they can be analyzed immediately (Kamal et al., 2021). However, collecting breath samples from broilers presents practical challenges. Unlike larger animals, like goats (Fischer et al., 2015) or cattle (Ellis et al., 2014), poultry anatomy complicates the use of face masks, which must fit tightly to prevent leakage and collect sufficient breath. A solution was proposed by Hughes, Tivey, and Butler (2008) with helmets for breath collection, but these require individual handling, which is time-consuming, stressful for the animals, and reduces feasibility in large-scale settings. Previous alternatives, such as tracheal cannulas or gluing tubes into chickens' nostrils (Scheid and Piiper, 1969), raise even more significant animal welfare concerns. Additionally, the mixing of dead-space air, which does not participate in gas exchange, with alveolar air in exhaled breath is a challenge across species. Dead-space air primarily reflects inhaled air rather than systemic VOCs (Boots et al., 2012), making $CO_2$-controlled sampling necessary to ensure that VOC profiles represent alveolar gas and accurately reflect blood VOC composition (Fischer et al., 2015). Despite these challenges, VOC analysis from breath has been successfully applied in other species, such as cattle (Ellis et al., 2014) and humans (Kamal et al., 2021), yet, to our knowledge, not in birds.

### 4.3 Feathers and skin

VOCs from feathers and skin originate from various sources, including secretions from the uropygial gland spread across their surfaces (Campagna et al., 2012). Traditionally studied in the context of avian body scent and chemical communication, such as pheromones (Hirao, Aoyama and Sugita, 2009; Campagna et al., 2012), these VOCs also reflect environmental odors, bacterial enzymatic activity, and chemical transformations, making them relevant for health monitoring (Campagna et al., 2012). Feathers and skin exhibit notable antibacterial activity, partly due to the presence of specific VOCs (Rajchard, 2010). For instance, Soini et al. (2006) reported elevated levels of linear *n*-alcohols in dark-eyed juncos during breeding conditions. These alcohols exhibit inherent antimicrobial and antifungal properties, and their increased concentrations may indicate increased microbial activity at elevated ambient temperatures. Such findings highlight the dual role of VOCs as both indicators of microbial dynamics and potential biomarkers for health-related changes.

To facilitate VOC collection from feathers and skin, Soini et al. (2006) introduced an in situ rolling-stir bar method to collect VOCs originating from uropygial gland secretions of the avian species *Junco hyemalis*. This noninvasive technique involves rolling the stir bar over the feather or skin surface. Samples collected using this method can be stored at 4°C or –80°C, which is also used to collect volatiles from the human upper back (Oxner et al., 2023). However, one significant challenge in working with feathers and skin is the potential for environmental VOC contamination, which can complicate data interpretation (Oxner et al., 2023).

The application of VOCs from skin samples for diagnostic purposes has already been established in cattle (Brebu et al., 2023) and humans (Shirasu and Touhara, 2011). These studies indicate that most VOCs in skin samples are primarily derived from symbiotic bacteria on the skin surface, though they may also originate from endogenous metabolic or hormonal processes. Brebu et al. (2023) reported that numerous VOCs excreted from the skin are products of physiological and metabolic responses to infections, which can lead to oxidative stress, lipid peroxidation, and cytochrome P450-mediated alcohol metabolism. Consequently, skin-excreted compounds metabolized into volatiles by bacteria hold potential as VOC biomarkers for infectious diseases, such as bovine tuberculosis.

### 4.4 Blood

Currently, no published research specifically addresses the capture of VOCs from blood or blood serum samples in poultry. However, this approach could offer a direct and reliable method for assessing a chicken's health status. Blood serum, a complex matrix rich in content, transports a wide range of compounds throughout the organism, facilitating exchanges with key organs, such as the intestines and lungs (Rodríguez-Hernández et al., 2021; Liu et al., 2024). As such, blood-derived samples provide insights into nutritional, metabolic, and immune status (Shirasu and Touhara, 2011), reflecting the internal physiology of the organism. Variations in the VOC profile of these samples may indicate infections, as they closely mirror cellular metabolic alterations (Shirasu and Touhara, 2011; Liu et al., 2024). As mentioned earlier, VOCs from the blood diffuse across the pulmonary alveolar membrane, meaning breath and blood samples offer similar insights into endogenous biochemical processes (O'Hara et al., 2009; Sethi, Nanda and Chakraborty, 2013).

Blood sample collection is a valuable method for assessing systemic health status, but involves individual immobilization, which can induce stress reactions in chickens (Rodríguez-Hernández et al., 2021). Additionally, sample handling requires careful attention, as serum separation is necessary to prevent red blood cell lysis when stored at freezing temperatures, which could alter the matrix equilibrium and affect analysis (Ashley et al., 1996; Shirasu and Touhara, 2011). However, serum samples can be reliably stored at –40°C or –80°C (Jaimes-Mogollón et al., 2021). Although these factors present challenges for large-scale applications, they do not rule out the potential of blood-derived samples for health monitoring in broilers via VOCs.

Infections caused by *Mycobacterium bovis*, *Mycobacterium avium* subspecies *paratuberculosis*, and *Brucella* spp. have been studied through VOC analysis of serum samples in various animal species, including cattle, badgers, white-tailed deer, and wild boars (Rodríguez-Hernández et al., 2021). Similarly, blood samples from rats and mice have been utilized for VOCs analysis to investigate neurological disorders, such as Parkinson's disease and amyotrophic lateral sclerosis (Jaimes-Mogollón et al., 2021). In humans, blood samples are considered an ideal diagnostic medium, being the second most commonly employed biological matrix for metabolomics after urine (Liu et al., 2024). Therefore, research on VOC analysis from human blood samples exists and focuses for example on identifying biomarkers for early cancer diagnosis, including serum VOCs for esophageal (Liu et al., 2024) and colorectal cancer (Oxner et al., 2023).

## 5 Pathogen influence on volatile organic compounds

VOC levels in healthy animals are influenced by various factors, including age, species, circadian rhythm, diet, and many other variables (Fischer et al., 2015; Recharla et al., 2017; Jones et al., 2021a; Weng et al., 2021). However, the onset of diseases caused by pathogens can significantly alter VOC profiles (Wilson and Forse, 2023). Pathogens such as bacteria, viruses, parasites, and fungi disrupt host and microbiota metabolic pathways, leading to biochemical alterations associated with infections. These disruptions result in the production of new VOCs or alterations in existing VOC ratios, creating distinct VOC fingerprints that reflect the nature of the disease or the causative pathogen (Shirasu and Touhara, 2011; Grilli et al., 2018; Wilson and Forse, 2023). The earliest reported time point at which VOCs could be distinguished between infected and noninfected samples was 1 h post-infection by avian influenza virus, as determined using selected ion flow tube mass spectrometers (SIFT-MS). However, this finding was based on *in vitro* studies using chicken cells rather than *in vivo* experiments (Filaire et al., 2024). In general, most studies have reported the detection of infection-related VOCs approximately 36 to 48 h post-infection. Table 4 provides an overview of pathogens relevant to broilers and their associated VOCs, which will be further discussed.

## 5.1 Bacteria

Bacteria are a major source of VOCs in biological systems, with many compounds being secondary metabolites produced during aerobic or anaerobic fermentation (Żuchowska and Filipiak, 2024)). In broilers, the predominant pathogenic bacterial species include *Campylobacter spp.*, *Salmonella spp.*, *Enterococcus faecalis*, *Staphylococcus spp.*, *Escherichia coli*, and *Clostridium perfringens* type A (Belizário, Faintuch and Malpartida, 2021). The VOCs associated with these bacteria span various chemical classes, including carbohydrates, aldehydes, ketones, alcohols, acids, isocyanates, amines, terpenes, and sulfides. Some VOC patterns are pathogen-specific, providing potential biomarkers for diseases caused by these pathogens (Żuchowska and Filipiak, 2024).

For broilers, VOC analyses have primarily focused on *Campylobacter jejuni* and *Salmonella* infections. Garner et al. (2008) developed a statistically significant model to identify *C. jejuni*-positive fecal samples from chickens, using six key VOCs: hexanal, E-2-octenal, pyrrole, ethyl ethanoate, methyl alcohol, and 2-heptanone. Hankel et al. (2022) also detected *C. jejuni*-positive broilers by analyzing air samples with sensors sensitive to 1-pentanol, 1-octanol, 1-nonanol, acetoin, dimethyl disulfide, and isovaleraldehyde. Regarding *Salmonella*, Wang et al. (2024) identified VOCs associated with *Salmonella* contamination in chicken meat, while van Veen et al. (2025) observed that *Salmonella* vaccination in 3-week-old pullets led to consistent changes in volatile peak intensities. For example, they found that (Z)-6-tridecene consistently decreased across different pullet flocks, following either *Salmonella* or a viral/bacterial cocktail vaccination, regardless of age or housing conditions. The authors hypothesized that (Z)-6-tridecene may originate from membrane lipid peroxidation and suggested that vaccination could reduce oxidative stress, thereby lowering its production.

**Table 4** Pathogen-associated VOCs in broilers: This table provides an overview of VOCs linked to common broiler pathogens across various biological samples. Pathogens are categorized by type (bacteria, viruses, parasites, fungi), with details on associated broiler diseases, studied sample sources (including non-chicken species to give extra examples or if no chicken data is available), and VOC chemical classes

| Pathogen type | Pathogen | Pathology | Species and biological sample | VOC chemical class | VOC example | Reference |
|---|---|---|---|---|---|---|
| Bacteria | *Campylobacter* spp. | Acute gastroenteritis | Air samples, feces | Alcohols | 1-pentanol, 2-propyl-1-pentanol, 1-heptadecanol, methyl alcohol | (Probert, et al., 2004; Garner et al., 2008; Hankel et al., 2022; Santos, et al., 2024) |
| | | | | Ketones | Acetoin, isophorone, 2-heptanone | |
| | | | | S-containing compounds | Dimethyl disulfide | |
| | | | | Aldehydes | Isovaleraldehyde, hexanal, (E)-2-octenal | |
| | | | | Aromatic compounds | Phenol, indole, pyrrole | |
| | | | | Esters | Ethyl ethanoate | |
| | *Salmonella* spp. | Extraintestinal infections | Feces, inoculated chicken meat | (Long-chain) alcohols | 1-decanol, 1-propanol 1-dodecanol | (Audrain et al., 2015; Belizário, et al., 2021; Wang et al., 2024; van Veen et al., 2025) |
| | | | | Carboxylic acids | Acetic acid, pentanoic acid | |
| | | | | Esters | Methyl ester | |
| | | | | Hydrocarbons | Dodecane, hexadecane, (Z)-6-tridecene | |

| | | | | | |
|---|---|---|---|---|---|
| *Enterococcus faecalis* | Intestinal dysbiosis, infections | Human breath, *E. faecalis* culture, urine, exudates | Alcohols | 1-butanol, 1-pentanol | (Bos, et al., 2013; Ratiu et al., 2020; Pennington, et al., 2024) |
| | | | Ketones | 2-aminoacetophenone, acetophenone, 2-pentanone | |
| | | | Aldehydes | Benzaldehyde, Formaldehyde | |
| | | | Hydrocarbons | Dodecane | |
| | | | Aromatic compounds | Methylpyrzaine | |
| | | | S-containing compounds | Hydrogen sulfide | |
| *Staphylococcus spp.* | Respiratory infections | Human breath | Aldehydes | Acetaldehyde, 2-methyl-butanal, 2-methy-propanal | (Boots et al., 2012; Bos, et al., 2013; Belizário, et al., 2021; Jenkins and Bean, 2023) |
| | | | Alcohols | Ethanol, butanol | |
| | | | Ketones | 2-nonanone, acetoin, 2-aminoacetophenone | |
| | | | Carboxylic acids | Isovaleric acid, acetic acid | |
| | | | S-containing compounds | Dimethyl sulfide, hydrogen sulfide | |
| *Escherichia coli* | Avian colibacillosis (respiratory chronic respiratory disease complex or enteritis) | Exhaled air of human | Aromatic compounds | Indole | (Bos, et al., 2013; Ratiu et al., 2020; Belizário, et al., 2021) |

*(Continued)*

**Table 4** Continued

| Pathogen type | Pathogen | Pathology | Species and biological sample | VOC chemical class | VOC example | Reference |
|---|---|---|---|---|---|---|
| | | | | Alcohols | Methanol, pentanol | |
| | | | | Esters | Ethyl acetate | |
| | | | | Carboxylic acid | Propionic acid | |
| | *Clostridium perfringens* type A | Necrotic enteritis, diarrhea | Animal and human breath samples, stool samples | Carboxylic acids | Butyric acid, 2-methylbutanoic acid, methylpropionic acid, isocaproic acid | (Garner et al., 2007; Belizário, et al., 2021) |
| | | | | Alcohols | Ethanol, isopropanol, butanol | |
| Viruses | Influenza virus | Respiratory infections | Breath, cell cultures, chicken cells, duck feces, swine breath | Alcohols | 2-methoxyethanol, 3,7-dimethyloctan-3-ol | (Kimball et al., 2013; Aksenov et al., 2014; Traxler et al., 2018; Filaire et al., 2024) |
| | | | | Carboxylic acids | Propanoic acid, butanoic acid | |
| | | | | Ketones | Octan-2-one, acetoin | |
| | | | | Aldehydes | Decanal, acetaldehyde, propanal | |
| | | | | Esters | n-propyl acetate, methyl methacrylate | |
| | | | | Ethers | 1-dipropoxypropane | |
| | | | | Aromatic compounds | Styrene | |
| | Coronaviridae | Infectious bronchitis | Air samples, human breath | Alcohols | Methanol, ethanol, propanol | (Wilson and Forse, 2023) |

| | | | | | | |
|---|---|---|---|---|---|---|
| | | | | Aldehydes | Acetaldehyde, heptanal, octanal | |
| | | | | Carboxylic acid | Propionic acid | |
| | | | | Hydrocarbon | Butane, dodecane | |
| | Rotavirus | Diarrhea and enteritis in young birds | Children's stool | Carboxylic acids | Ethanoic acid | (Al-Kateb et al., 2012) |
| | | | | Ketones | 2,3-butanedione | |
| | | | | Aldehydes | | |
| | Rhinovirus | Respiratory infections, asthma | Children's breath | Long-chain alkanes | 2,4,6-trimethyl-decane | (Van De Kant et al., 2013; Gould et al., 2020) |
| | | | | Alkenes | 2-methyl-1-butene | |
| | | | | Ketones | 6-methyl-2-heptanone | |
| | | | | Alcohols | 1,2-propanediol | |
| | | | | Carboxylic acid | Pentanoic acid | |
| | | | | Aldehydes | 4-methyl-2-pentenal | |
| Parasites | Eimeria (Coccidiosis) | Enteric diseases (hemorrhagic diarrhea, malabsorption) | Air samples from chicken barns | VOCs were unidentified as they were measured with an E-Nose | / | (Grilli et al., 2018; Borgonovo et al., 2020, 2024) |
| | *Dermanyssus gallinae* (poultry red mite) | Skin irritations, anemia, vascular issues, ... | Headspace above mites | Hydrocarbons | 1-heptadecene, heptadecane, octadecane | (Gay et al., 2020) |
| | | | | Aldehydes | (E)-non-2-enal, octanal | |
| | | | | Carboxylic acids | Nonanoic acid | |

(*Continued*)

**Table 4** Continued

| Pathogen type | Pathogen | Pathology | Species and biological sample | VOC chemical class | VOC example | Reference |
|---|---|---|---|---|---|---|
| | | | | Esters | Ethyl acetate, (Z)-ligustilide (repellent) | |
| | | | | Alcohols | 9-octadecene-ol, oct-1-en-3-ol | |
| Fungi | Aspergillus | Brooder pneumonia | Feed crops | Aldehydes | Hexanal, benzaldehyde, octanal | (Kamiński et al., 1972; Afsah-Hejri et al., 2023; Okechukwu et al., 2024) |
| | | | | Alcohols | 1-octanol, 2-pentanol, 3-methylbutanol, 3-octanol | |
| | | | | Carboxylic acids | Butyric acid | |
| | | | | Ethers | 1,4-benzodioxin | |
| | | | | Esters | Ethyl acetate, methyl formate | |
| | | | | Hydrocarbons | Tridecane, 1,6-octadiene | |
| | | | | Ketones | benzophenone, 3-octanone | |
| | | | | N-containing compounds | 1-cyclopentyl, acetonitrile, proline | |
| | Candida albicans | Lesions in oropharynx and crop, thickened mucosa, enteritis | C.albicans culture | Alcohols | Longifolol, β-santalol, 2-methyl-1-propanol | (Karami et al., 2017) |
| | | | | Hydrocarbons | 1,3-butadiene | |
| | | | | N-containing compounds | Carbamic acid, 4-t-butyl-2-(1-methyl-2-nitroethyl)cyclohexane | |
| | | | | S-containing compounds | 2-methyl-2-undecanethiol | |

Published by Burleigh Dodds Science Publishing Limited, 2026.

Research on other common bacteria in broilers is limited, but studies on VOC detection from human breath, urine, and exudates indicate potential applications for broilers. Table 4 provides an overview of these pathogens and the VOC analyses conducted so far.

### *5.2 Viruses*

Numerous viral infections in broilers, such as those caused by *Influenza virus*, *Coronaviridae*, *Rotavirus*, and *Rhinovirus*, significantly alter host cell activity and metabolism, leading to detectable changes in VOC profiles (Wernery, 2016; Traxler et al., 2018). These infections often trigger inflammatory processes marked by elevated VOCs, particularly aldehydes, such as acetaldehyde (Traxler et al., 2018; Wilson and Forse, 2023).

For instance, VOC analysis of duck feces showed that influenza infection decreased 1-octen-3-ol and increased acetoin concentrations, which are compounds linked to respiratory tract infections (Kimball et al., 2013). This rise in acetoin levels likely results from viral impacts on the pyruvate pathway, where *Bacillus* bacteria convert pyruvate to acetoin under glucose-rich conditions. Viral infections boost glycolytic flux, leading to pyruvate accumulation and subsequent acetoin production, marking it as a potential biomarker for influenza (Spizizen, 1957; Kimball et al., 2013).

Research on influenza is more advanced than on other viruses. For example, Filaire et al. (2024) successfully distinguished healthy from infected chicken cells and identified high- and low-pathogenicity strains in an *in vitro* experiment. In swine, influenza-inoculated animals were distinguishable from healthy ones by day 4 post-infection, based on breath VOCs, like acetaldehyde, propanal, *n*-propyl acetate, methyl 2-methylprop-2-enoate, styrene, and 1,1-dipropoxypropane (Traxler et al., 2018). Acetaldehyde was linked to airway inflammation, while propanal was associated with $\omega$-3 lipid peroxidation. Furthermore, viral interactions with epithelial cells altered metabolic pathways, leading to increased n-propyl acetate levels (Traxler et al., 2018).

In contrast, VOC research on *Coronaviridae*, *Rotaviruses*, and *Rhinoviruses* is less developed in broilers, but has shown promise in human studies, using breath and stool samples. Relevant findings are summarized in Table 4. Despite progress in some areas, there remains a gap in robust animal models for viral VOC diagnostics, especially compared to those used for parasitic or oncological conditions. This gap highlights significant opportunities for further research and development in viral VOC profiling (Kimball et al., 2013).

### 5.3 Parasites

Concerning parasites in poultry, *Eimeria*, the causative agent of coccidiosis, and the poultry red mite, *Dermanyssus gallinae*, are among the most significant threats to flock health and productivity (Wernery, 2016). Research on these pathogens is summarized in Table 4.

For coccidiosis, a pioneering approach utilized an electronic nose (E-nose) equipped with a sensor array to detect VOC patterns specific to the infection (Grilli et al., 2018). The specific *Eimeria* strains were not reported, but air samples were collected using Nalophan bags in the barn, and the samples were measured afterwards on the E-nose. Although such a device does not identify individual VOCs, it provides qualitative data and concentration estimates (Sohn et al., 2008). Results demonstrated that the E-nose used in this study could reliably differentiate between healthy and infected broilers, classified through oocyst counting, presenting a promising tool for precision livestock farming (Grilli et al., 2018).

VOCs linked to poultry red mite infestations have not been extensively explored. Koenraadt and Dicke (2010), Gay et al. (2020), and Auffray et al. (2023) have investigated attractant and repellent VOCs to manage mite populations. These findings suggest that attract-and-kill strategies, successfully used for pests, like fruit flies and codling moths in other agricultural systems (Charmillot, Hofer and Pasquier, 2000; Vargas, Miller and Stark, 2003), could be adapted to control red mites in poultry. However, further research into VOCs emitted by chickens in response to poultry red mite infestation may offer valuable insights into infestation severity and host health. As poultry red mites are known to cause significant stress through anemia, and stress can alter VOC profiles, volatile analysis from biological chicken samples shows promise as a noninvasive diagnostic tool.

### 5.4 Fungi

Concerning fungal infections in broilers, aspergillosis and candidiasis are the most prevalent (Wernery, 2016). Aspergillosis, caused by fungi from the genus *Aspergillus*, primarily affects the respiratory system, often resulting in brooder pneumonia (Wernery, 2016; Hauck, Cray and França, 2020). Although no studies have yet investigated VOC detection for diagnosing *Aspergillus* infections in animals, research has focused on VOC profiles from *Aspergillus flavus* in feed crops to enhance food safety and security (Okechukwu et al., 2024). These studies revealed significant differences in VOC profiles, as summarized in Table 4.

Candidiasis, caused by *Candida albicans*, is another common fungal disease in chickens. It leads to lesions in the oropharynx and crop, mucosal thickening, and occasionally enteritis (Wernery, 2016). Karami et al. (2017) cultured *C. albicans* and identified several key VOCs, also summarized in Table 4. These findings emphasize the potential of VOC analysis as a diagnostic tool for fungal infections in broilers.

## 6 General challenges in identifying reliable pathogen biomarkers

Identifying reliable biomarkers for pathogens is challenging due to the complex and dynamic nature of VOC production. Various factors influence the presence and concentration of VOCs in biological samples, complicating the identification of consistent biomarkers (Ratiu et al., 2020).

One challenge is the lack of specificity of single VOCs for a particular pathogen. For example, Żuchowska and Filipiak (2024) found that indole is produced by multiple bacterial species, including *E. coli*, *S. aureus*, and *K. pneumoniae*. This overlap necessitates the use of VOC fingerprints rather than individual compounds for accurate detection (Palma et al., 2018).

Another critical factor is the stability of VOCs during sample collection and handling. For instance, 2-acetoaminophenone, associated with *Pseudomonas aeruginosa*, remains stable in glass bulbs, but degrades in Tedlar polycarbonate sampling bags (Ratiu et al., 2020). Therefore, selecting suitable materials for sample collection is crucial. Additionally, differences in the developmental stages and subtypes of pathogens may introduce genetic variability that influences the enzymatic activity involved in VOC production (Bos, Sterk and Schultz, 2013).

Another layer of complexity involves co-infections of multiple pathogens that alter the host metabolism through pathogen-pathogen interactions, complicating VOC pattern interpretations (Ratiu et al., 2020). Differentiating between bacterial and viral infections is essential for guiding treatment and minimizing unnecessary antibiotic use (Gould et al., 2020). Abd El Qader et al. (2015) demonstrated that hexan-1-ol serves as a biomarker for viral infections, while heptadec-1-ene indicates bacterial ones, enabling accurate distinction between infection types. Pathogen-immune interactions and antibiotic treatments can further influence VOC patterns (Hankel et al., 2022).

VOC kinetics depend on bacterial metabolic activity or total bacterial load. VOC concentrations generally correlate with bacterial cell numbers, but may decrease if host-related VOCS are used as energy or substrates in bacterial metabolism. This indicates an inverse relationship with bacterial load. Additionally, metabolic consumption within the body complicates the accurate determination of VOC production levels. Antibiotics further influence these dynamics by targeting bacterial cells (Żuchowska and Filipiak, 2024).

## 7 Internal factors influencing volatile organic compound fingerprints

When using VOC analysis as a diagnostic tool in broilers, it is essential to account for factors that might influence its reliability and accuracy. VOC patterns can be affected not only by the health status of the birds, but also by individual variability among healthy birds. Figure 6 illustrates the significant influence of internal and external factors on these profiles. To ensure effective diagnostics, these biological and environmental influences must be carefully considered and addressed (Shirasu and Touhara, 2011). This section focuses on internal factors, with the following section focusing on external factors.

### 7.1 Age

Age significantly influences VOC profiles as metabolic and physiological processes evolve throughout an organism's life. For example, Corion et al. (2024a) measured VOCs from chicken eggs during embryonic development. Changes in the relative abundances of chemical classes were observed over 12 days of incubation, with carboxylic acids showing elevated levels in fertilized eggs toward the 12th day. This could be attributed to oxidative stress induced by hypercapnia and hypoxia conditions, which leads to increased lipid peroxidation. Similarly, Fischer et al. (2015) investigated this factor by tracking VOC emissions in goats over one year during standardized ambient conditions, and using early castration to eliminate the influence of sex pheromones. They observed notable changes in VOC concentrations in both feces and breath samples.

A potential confounding factor when studying age is diet, as dietary composition typically changes over time. Weng et al. (2021) found age-related VOC differences in pigs and chickens, with nursery houses emitting higher levels of combustible gases compared to general housing. Similarly, Fischer et al. (2015) observed that dietary changes in goats altered VOC profiles due to shifts in energy metabolism. In contrast, Dragonieri et al. (2016) reported that age-related differences in breath VOC profiles were not detectable in humans using E-nose analysis, suggesting that the influence of age on VOCs may vary across species.

### 7.2 Breeds

Inter-breed differences in chicken eggs have been observed across various breeds. Corion et al. (2024c) found that (E)-undec-2-enal was more abundant in the Dekalb breed compared to Ross on incubation days 2 and 8. This variation may have a genetic basis, as (E)-undec-2-enal is likely a product of oleic acid autoxidation, and oleic acid concentrations can differ between breeds, even when fed similar diets (Bunea et al., 2017). Additionally, the

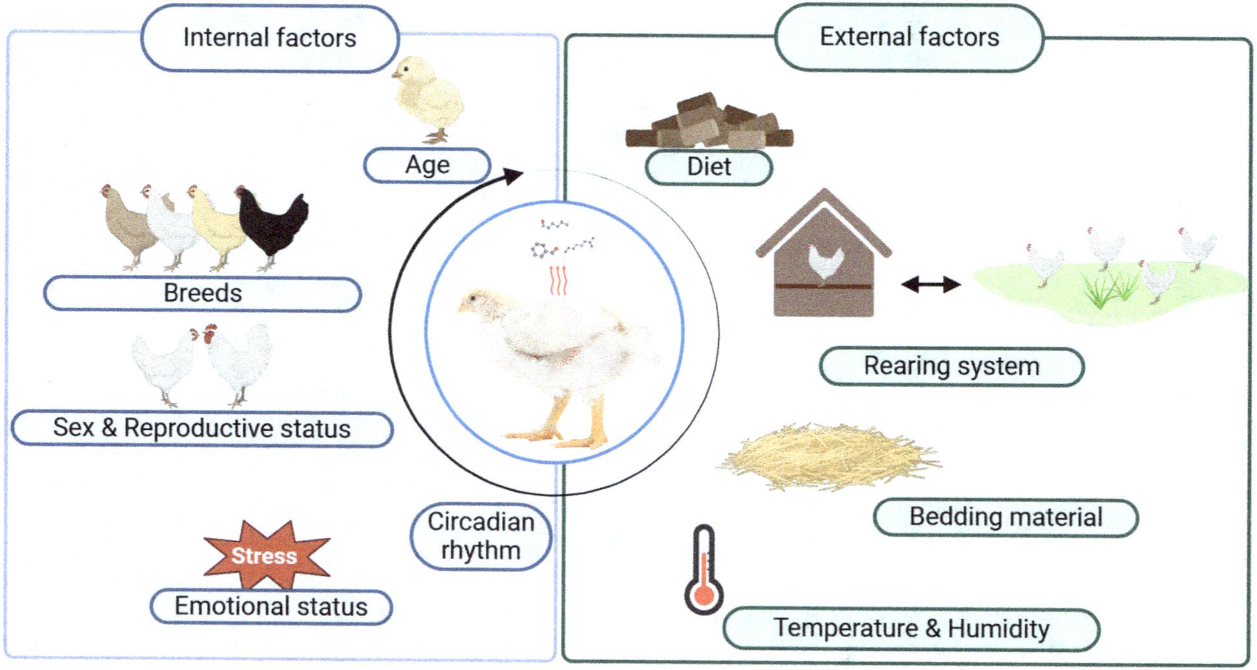


**Figure 6** Overview of other influencing factors, besides pathogens, on VOC fingerprints in broilers: The blue box represents internal factors, including age, species and breed, sex and reproductive status, emotional state, and circadian rhythm. The green box illustrates external factors, such as diet, rearing system, bedding material, and environmental conditions like temperature and humidity (Figure created with BioRender.c om).

natural compounds tetradecanal and tridecan-1-ol were present at higher concentrations in Shaver compared to Dekalb. These differences may reflect breed-specific metabolic activity; however, it cannot be excluded that they are partially influenced by differences in optimal incubation conditions between breeds. Supporting this, Yalçin et al. (2008) demonstrated that incubation temperature can affect the oxidation of specific fatty acids. The potential of VOCs to discriminate between breeds was further confirmed by Li et al. (2024), who identified 20 VOCs in meat from different Chinese breeds as candidate markers for breed differentiation.

## 7.3 Sex and reproductive status

Sex and reproductive status also influence VOC emissions. In layers, sex-specific VOC differences are detectable even during early incubation, with different VOC emission profiles (Corion et al., 2024b). Another sex-associated source of VOCs is the uropygial gland of chicken, as females use its volatile secretions for mate selection and social interactions (Hirao, Aoyama and Sugita, 2009).

Similar patterns are observed across species. For example, Jones et al. (2021b) found higher VOC abundances in male wolf urine compared to females. Liang, Bushman, and FitzGerald (2015) reported greater fecal VOC variability in female mice due to more pronounced gut microbiota oscillations. Karthikeyan et al. (2013) identified distinct fecal VOCs, such as 4-methylphenol and trans-verbenol, in water buffaloes during estrus. These findings suggest the potential for using VOCs to enhance reproductive management in livestock.

## 7.4 Emotional state

The emotional state in broilers can also influence VOC emissions. Bombail et al. (2018) reported that stress induced the production of specific volatile compounds recognizable by animals such as rats. These rats could distinguish between feces from stressed and non-stressed chickens, suggesting that emotional states shape VOC profiles. Moreover, stress-related odors remained detectable even two weeks after chickens had experienced stress. In humans, VOC profiles vary significantly between fearful and neutral or pleasant emotional states. For example, Smeets et al. (2020) found that fear was associated with elevated levels of linear aldehydes and ketones, whereas neutral states were linked to esters and cyclic compounds. These findings highlight the role of VOCs in intra- and interspecies communication and their potential for assessing emotional states to improve animal health and productivity.

### *7.5 Circadian rhythm*

The timing of sample collection during the day significantly influences the VOC composition detected during analysis. Studies have shown that morning fecal samples exhibit the highest VOC levels for both chickens and pigs (Weng et al., 2021). Similarly, Hankel et al. (2022) found that morning air samples from chicken cages were more effective in distinguishing between *C. jejuni*-infected and noninfected birds. This is likely because chickens follow a diurnal feeding rhythm, starting to eat almost immediately after the dark period, leading to a high volume of fresh excreta in the morning.

Additionally, ongoing bacterial fermentation makes fecal composition highly unstable over time when stored at room temperature (Gratton et al., 2016). Metabolite levels also exhibit diurnal variations, driven by the host's circadian clock, which regulates microbiome structure and activity rhythms (Murakami and Tognini, 2020). Supporting this, Jiang et al. (2020) found that artificial light at night – known to disrupt circadian rhythms – significantly altered the taxonomic composition, species diversity, and community structure of the intestinal microbiota in Eurasian tree sparrows.

To improve the accuracy of VOC analysis for broiler health, it may be beneficial to standardize sample collection in the morning, as current evidence suggests that VOC concentrations are consistently higher at that time. However, further research is needed to confirm this approach.

## 8 External factors influencing volatile organic compound fingerprints

### *8.1 Diet*

Diet influences VOC emissions in broilers through variations in feed additives, protein levels, and fiber content. Du et al. (2000) investigated the effect of dietary conjugated linoleic acid as a feed additive on the VOC fingerprint of raw chicken meat, reporting significant changes in VOC composition. Similarly, Shu et al. (2024) found that curcumin supplementation increased aldehyde concentrations, identifying them as the most abundant chemical group in chicken meat. Beyond feed additives, crude protein levels also play a role. Excess crude protein can increase nitrogen excretion and odor emissions, as undigested protein is excreted in feces (Heo et al., 2023). Additionally, higher dietary fiber levels lead to increased production of volatile SCFAs by gut microbiota (Jha and Mishra, 2021). While diet is often considered a key driver of VOC emissions due to its role in providing metabolic substrates, it does not necessarily overrule the influence of other factors such as sex, climate, mental state, or rearing system. Future research should aim to integrate these factors for a more comprehensive understanding and management of VOC production in broiler systems.

### 8.2 Rearing system

Comparative studies on different rearing systems reveal significant differences in VOC profiles. As reviewed by Campagna et al. (2012), free-living birds generally exhibit higher concentrations of carboxylic acids and benzaldehyde in their uropygial gland secretions compared to captive ones. In chicken farming, Huanhong et al. (2023) summarized evidence indicating that while free-range systems improve welfare and sustainability, they also introduce greater variability in VOC emissions due to inconsistent feed intake. This variability arises from a combination of outdoor foraging and supplementary feeding.

### 8.3 Bedding material

Additional factors, such as bedding material, manure management, dust generation, and ventilation systems, further influence VOC profiles in free-range and closed housing systems. Emission patterns also vary between floor housing and traditional cage systems for laying hens (Nimmermark and Gustafsson, 2005). Bedding materials, when exposed to moisture and microbial activity, undergo decomposition as a result of fermentation, releasing additional VOCs, with excess moisture particularly linked to elevated ammonia levels (Huanhong et al., 2023). Spiehs et al. (2013) compared various bedding materials for beef cattle, finding that corn cobs produced the highest VOC emissions, whereas wood shavings produced the lowest. This discrepancy likely stems from the antimicrobial properties of essential oils in wood shavings, which inhibit fermentation, whereas the high fermentability of corn cobs promotes rapid VOC production.

Emissions from bedding material can contribute to background signals in VOC measurements within housing systems. Moreover, the inclusion of bedding material during fecal sampling may introduce additional volatile compounds, altering the overall VOC profile. A background measurement of the bedding material alone could serve as a reference for distinguishing its contribution to VOC emissions.

### 8.4 Temperature and humidity

Temperature and humidity are other crucial factors influencing VOC emissions. While chicken house temperatures are typically optimized for growth and energy efficiency, fluctuations can impact both thermoregulation in chickens and VOC release. On the one hand, rising temperatures may trigger behavioral adaptations, such as wing lifting to enhance cooling through airflow (Gerken, Afnan and Dörl, 2006). Combined with temperature gradients that drive

Published by Burleigh Dodds Science Publishing Limited, 2026.

buoyant convection (Angle et al., 2016), this could redistribute VOCs from the birds within the environment. On the other hand, increased temperature and humidity can directly affect VOC emissions from the environment. Nimmermark and Gustafsson (2005) observed that warmer, more humid conditions in floor housing systems for laying hens correlated with higher VOC emissions, likely due to enhanced volatilization and stimulated microbial activity.

## 9 Volatile organic compound measurement and analysis

Advancements in VOC capturing and analysis are enhancing our understanding of their relationship with animal health. As shown in Fig. 7, VOCs can be captured in three main ways: direct measurement on the instrument (Sharma et al., 2017), headspace collection in a container (Yao et al., 2018), or enrichment on a sorbent material for later desorption and analysis (Corion et al., 2024a). Enrichment involves exposing a sorbent material to the sample's headspace for a set incubation time and temperature. Sampling can be active (e.g. using a sorbent tube) or passive (e.g. with a fiber or sorbent stir bar). Once enriched, the sorbent material can be stored and later desorbed using heat for analysis on a VOC measurement device.

The choice of VOC capture method is typically aligned with the analytical technique employed, although there are exceptions. For example, a gas chromatography-mass spectrometer (GC-MS) necessitates VOC extraction and pre-concentration on a sorbent material prior to column analysis, whereas an E-nose allows direct measurement of VOCs in the headspace (Epping and Koch, 2023). Additionally, VOCs can be enriched on a sorbent material and subsequently desorbed onto an E-nose for analysis (Hankel et al., 2022), demonstrating the versatility of this approach.

In some cases, a 0.2 μm Teflon filter can be used to remove suspended particles from the air stream before measuring or storing the headspace (Trabue et al., 2010), avoiding contaminations with for instance microorganisms. Storage containers, such as bags or canisters, can also be reused through cleaning procedures using nitrogen flushing (Ho et al., 2017) or heat treatment (Filipiak et al., 2012), reducing contamination risks. Table 5 provides a comprehensive overview of the various VOC capture methods, detailing their respective advantages and disadvantages to guide the selection of the most suitable approach for specific analytical needs.

After VOCs are captured, they can be analyzed using various measurement devices. These devices can measure VOCs directly from the headspace or after capturing them using one of the previously mentioned methods. Figure 7 and Table 6 summarize the most commonly used VOC analysis instruments in research, ranked from most to least portable, along with their applications and

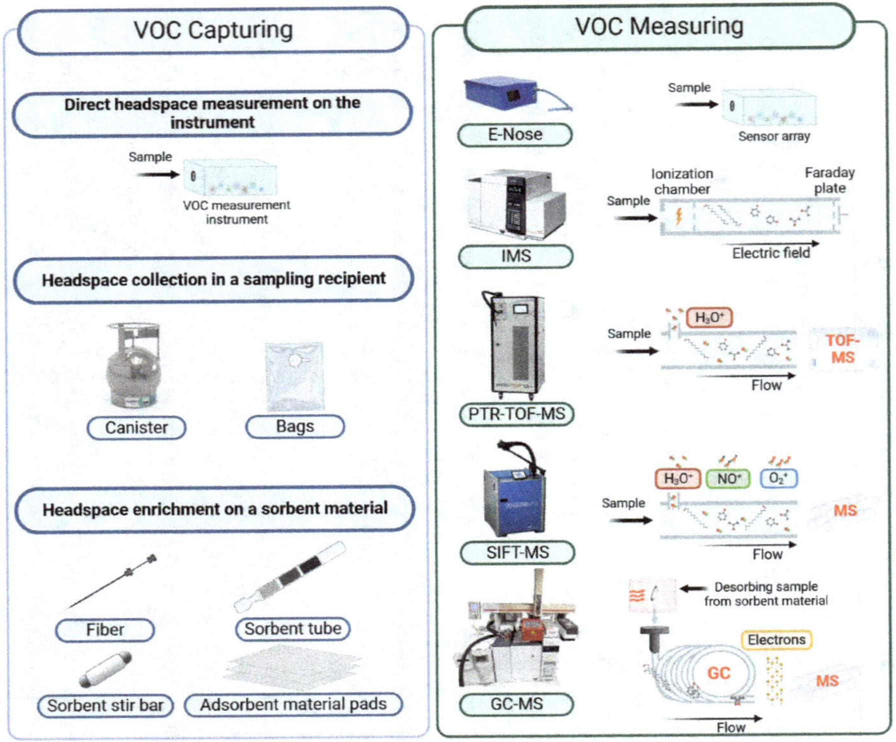


**Figure 7** Overview of VOC capturing methods and measurement devices. This figure illustrates the various methods for capturing VOCs (left) and the principal measurement devices used for their analysis (right). The VOC capturing methods range from direct measurement to headspace collection in a container, as well as enrichment on sorbent materials for subsequent desorption and analysis. On the right-hand side, the main VOC measurement devices are displayed alongside their working principles. Abbreviations: E-nose, electronic nose; IMS, ion mobility spectrometer; PTR-TOF-MS, proton transfer reaction-time-of-flight-mass spectrometer; SIFT-MS, selected ion flow tube-mass spectrometer; GC-MS, gas chromatography-mass spectrometer (Figure created with BioRender.com).

**Table 5** Overview of the different capture methods for VOC measurements

| VOC capture method | Advantages | Disadvantages | Application |
|---|---|---|---|
| Direct on the instrument | • Real-time VOC analysis and no sample loss or degradation due to storage | • The instrument must be available on-site | Bacterial culture (Sharma et al., 2017) |
| Canister (glass or fused silica lined, e.g. 1 L) | • Two weeks of storage<br>• Suitable for slow filling | • Slow filling<br>• Water condensation may lead to component loss | Broiler barn air (Yao et al., 2018) |
| Bags (Nalophan, Tedlar, Teflon, Stainless steel, e.g. 1–25 L) | • Portable and easy to use | • Limited VOC storage due to the porosity of some materials (Nalophan holds for the shortest duration) | Broiler barn air (Grilli et al., 2018) |
| SPME fiber | • Different coating materials allow for various affinities for molecular masses and polarities | • Limited sample volume capacity | Equine feces (Hough, Archer and Probert, 2018) |
| Sorbent Stir Bar | • Allows both passive and active VOC extraction<br>• Less sensitive to humid samples<br>• More sensitive than SPME fibers due to the large surface area | • No major disadvantages | Egg VOCs (Corion et al., 2024a) |
| Sorbent tube (Carbotrap, Tenax) | • Highly sensitive due to the large adsorption area | • Sensitive to humid environments (excess water affects performance) | Layer feces (van Veen et al., 2025) |
| Adsorbent material pads | • Can enrich VOCs from large air volumes (e.g. 92 $m^3/h$) | • Depending on the material, the pad may produce background VOC signals | Broiler cage air (Hankel et al., 2022) |

Abbreviations: SPME, solid phase microextraction fiber.

**Table 6** Comparison of VOC measurement devices (Sources: Arce et al., 2014; Epping and Koch, 2023; Arce et al., 2014; Epping and Koch, 2023; Corion et al., 2024c)

| Device | Principle of operation | Sensitivity | Selectivity | Speed | Portability | Application |
|---|---|---|---|---|---|---|
| E-nose | A sensor array producing unique patterns for VOCs | Moderate (ppb) | Low – Enhanced by combining more sensors on the array | Real-time | High (portable) | Broiler barn air (Sohn et al., 2008; Grilli et al., 2018) |
| IMS | Ionizes VOCs; measures drift time in electric field | High (ppt to ppb) | Moderate – limited compound differentiation | Fast (seconds) | Moderate (portable) | Pig feces (Pérez-Calvo et al., 2019), Human serum (Liu et al., 2024) |
| PTR-TOF-MS | Proton transfer ionization ($H_3O^+$) and mass spectrometry | Very high (ppt) | Very high – Precise mass spectra for selective analysis | Real-time | Moderate (ambient air as carrier gas) | Human breath (Liangou et al., 2021), Chicken eggs (Corion et al., 2024c) |
| SIFT-MS | Selective ionization (e.g. $H_3O^+$, $NO^+$, or $O_2^+$) followed by mass spectrometry | Very high (ppt) | High – Different reagent ions improve selectivity | Real-time | Low (helium carrier gas) | Broiler litter headspace (Sharma et al., 2018), Bacterial culture (Sharma et al., 2017) |
| GC-MS | Separates VOCs via GC; identifies and quantifies using MS | Very high (ppt) | Highest – Column separation facilitates the identification of compound-specific mass spectra | Slow (minutes to hours) | Lowest (lab-based) | Poultry feces (Garner et al., 2008; van Veen et al., 2025) Bacterial culture (Karami et al., 2017) |

Abbreviations: E-nose, electronic nose; IMS, ion mobility spectrometer; PTR-MS, proton transfer reaction-mass spectrometer; SIFT-MS, selected ion flow tube-mass spectrometer; GC-MS, gas chromatography-mass spectrometry.

sample types. VOC measurement devices can be applied in both untargeted approaches, where VOCs of interest are not known beforehand, and targeted approaches, where specific VOCs are analyzed.

An E-nose consists of an array of sensors selected to increase device specificity. The sensors react to VOC adsorption on the surface, but an E-nose cannot identify specific compounds. However, the sensor's known specificities indicate the types of molecules it can detect (Epping and Koch, 2023). Given that an E-nose is restricted to its selection of sensors, it is considered more as a targeted approach. The ion mobility spectrometer (IMS) is a more complex technique that ionizes VOCs at ambient pressure and temperature, rapidly separating analytes in a drift tube. Selectivity can be enhanced by coupling it with a capillary column (GC) (Arce et al., 2014). Proton transfer reaction-time-of-flight-mass spectrometers (PTR-TOF-MS) and selected ion flow tube mass spectrometers (SIFT-MS) are direct trace gas mass spectrometers using soft chemical ionization to ionize sample analytes, producing product ions separated by a mass spectrometer (Corion et al., 2024c). Finally, the GC-MS is considered the gold standard lab-based analytical system for identifying VOCs through separation and mass detection. For this reason of identification, the GC-MS is also the preferred system for untargeted VOC analysis (Corion et al., 2024c).

Each instrument has its unique advantages in animal health and VOC analysis. The E-nose is highly portable and adaptable for on-site applications (Hankel et al., 2022), whereas GC-MS remains the gold standard for VOC identification due to its analytical precision (Epping and Koch, 2023). However, a key limitation of E-nose technology is its reliability and stability, which remain challenging (Liu et al., 2023). A crucial factor in improving E-nose performance is selecting appropriate sensors for target VOC recognition, a process that can be guided by preliminary GC-MS analysis. Complementarity between VOC analysis devices is also evident in the use of PTR-MS and GC-MS. For instance, PTR-MS, particularly when coupled with a time-of-flight mass spectrometer, excels at detecting smaller molecules, while GC-MS provides greater sensitivity for heavier compounds (Corion et al., 2024c).

VOC measurements generate high-dimensional datasets that require advanced analytical approaches. GC-MS yields data in the form of VOC peak areas, whereas E-noses produce sensor response signals (Hankel et al., 2022). In contrast, datasets from real-time VOC analyzers such as PTR-TOF-MS, SIFT-MS, and IMS consist of variables representing product ion signals (Pérez-Calvo et al., 2019; Corion et al., 2024c). To manage the complexity of such datasets, a combination of univariate and multivariate data analysis techniques is typically employed.

Univariate statistical methods, including t-tests, ANOVA, Kruskal–Wallis tests, and fold change analysis, are commonly used to identify significant differences between groups (e.g. healthy vs. diseased or vaccination status), typically following normality checks and applying false discovery rate corrections (Peters et al., 2017; van Veen et al., 2025). Correlation analyses can further link VOC levels to clinical response variables or pathogen load (Traxler et al., 2018), enhancing the interpretation of biological relevance.

Multivariate approaches are essential for pattern recognition, dimensionality reduction, and classification. Principal component analysis (PCA) and hierarchical clustering analysis (HCA) are frequently applied for unsupervised data exploration, enabling visualization of sample groupings and relationships between VOCs (Yongwei et al., 2009; Corion et al., 2024a). For supervised modeling, partial least squares discriminant analysis (PLS-DA) and partial least squares regression (PLS-R) are widely used, often combined with variable selection methods such as variable importance in projection (VIP) scores to identify discriminative VOCs or predict response variables (Corion et al., 2024b; Li et al., 2024; van Veen et al., 2025).

In recent years, nonlinear machine learning algorithms such as Random Forests (RF) and Support Vector Machines (SVM) have gained traction for VOC classification tasks (Palma et al., 2018; Liu et al., 2024). These models enhance robustness and generalizability when validated through techniques like cross-validation. As VOC datasets continue to grow in size and complexity, the integration of advanced multivariate modeling and machine learning is becoming increasingly valuable for identifying reliable volatile biomarkers.

## 10 Conclusion and future trends

VOCs hold promising potential as noninvasive biomarkers for enhancing broiler production by monitoring animal health, welfare, and productivity. By leveraging volatilomics, this approach provides insights into host-microbiota interactions, disease detection, and metabolic efficiency. Although in the early stages of development, VOC analysis could contribute to more climate-smart broiler production by facilitating early disease detection, reducing antibiotic use, and optimizing feed formulations. This may enhance productivity while minimizing environmental impacts, such as ammonia emissions and odor pollution. Additionally, pathogen-specific VOC profiles could improve targeted disease management and biosecurity.

However, realizing this potential requires addressing several key challenges. Standardized protocols for VOC sampling, handling, and analysis are needed to ensure reproducibility and enable meaningful cross-study comparisons. Future research must systematically assess the effects of storage duration and temperature on VOC stability to establish reliable biomarkers. Comprehensive animal trials are also essential to understand the influence of internal (e.g. age, sex, emotional state) and external factors (e.g. diet, housing systems) on VOC emissions, facilitating field validation.

Pathogen-induced VOC profiling remains underexplored despite its relevance for early disease detection. Future studies should prioritize profiling for key broiler pathogens, including *Campylobacter*, *Salmonella*, *Influenza*, and *Aspergillus*, to enhance diagnostic capabilities. Understanding the metabolic pathways driving pathogen-specific VOC emissions will be crucial for advancing diagnostic techniques.

Emerging advancements in data analytics, particularly through machine learning and advanced multivariate modeling, offer significant opportunities to improve pattern recognition and classification (Palma et al., 2018). These approaches can improve differentiation between processes of infected and noninfected animals. Integrating VOC analysis with other omics technologies, such as metabolomics, can further deepen insights into host-pathogen interactions. Recently applied in plant sciences, this integrative approach enables the tracing of volatile origins through correlation and network analysis, providing a more comprehensive understanding of metabolic pathways (Li et al., 2023; Bartels et al., 2024). Applying this methodology in broiler research could significantly optimize disease monitoring and health management.

While the application of VOCs in broiler production shows great promise, it is still in an exploratory phase. Continued research is necessary to validate diagnostic accuracy and practical application. If successful, VOC analysis could revolutionize broiler health monitoring, supporting more sustainable, resource-efficient, and climate-smart farming practices.

## 11 Where to look for further information

For comprehensive insights into the use of VOCs as diagnostic tools, we refer to two recent review articles:

- Rodríguez-Hernández, P. et al. (2022) 'Analytical tools for disease diagnosis in animals via fecal volatilome', *Critical Reviews in Analytical Chemistry*, 52(5), pp. 917–932. Available at: https://doi.org/10.1080/10408347.2020.1843130.
- Jaimes-Mogollón, A.L. et al. (2021) 'Review on volatolomic studies as a frontier approach in animal research', *Advanced Biology*, 5(6), pp. 1–27. Available at: https://doi.org/10.1002/adbi.202000397.

Additionally, the following papers present interesting insights into advanced VOC analytical techniques:

- Corion, M., et al. (2024) 'Non-destructive egg breed separation using advanced VOC analytical techniques HSSE-GC-MS, PTR-TOF-MS, and SIFT-MS: assessment of performance and systems' complementarity', *Food Research International*, 176, Article 113802. Available at: https://doi.org/10.1016/j.foodres.2023.113802.
- Żuchowska, K. and Filipiak, W. (2024) 'Modern approaches for detection of volatile organic compounds in metabolic studies focusing on pathogenic bacteria: current state of the art', *Journal of Pharmaceutical Analysis*, 14(4), Article 100898. Available at: https://doi.org/10.1016/j.jpha.2023.11.005.

## 12 References


Abd El Qader, A., et al. (2015) 'Volatile organic compounds generated by cultures of bacteria and viruses associated with respiratory infections', *Biomedical Chromatography*, 29(12), pp. 1783–1790. Available at: https://doi.org/10.1002/bmc.3494.

Afsah-Hejri, L., et al. (2023) 'Identification of volatile organic compounds (VOCs) by SPME-GC-MS to detect Aspergillus flavus infection in pistachios', *Food Control*, 154, Article 110033. Available at: https://doi.org/10.1016/j.foodcont.2023.110033.

Aksenov, A.A., et al. (2014) 'Cellular scent of influenza virus infection', *ChemBioChem*, 15(7), pp. 1040–1048. Available at: https://doi.org/10.1002/cbic.201300695.

Angle, C., et al. (2016) 'Canine detection of the volatilome: a review of implications for pathogen and disease detection', *Frontiers in Veterinary Science*, 3(JUN), Article 47. Available at: https://doi.org/10.3389/fvets.2016.00047.

Arce, L., et al. (2014) 'Ion mobility spectrometry a versatile analytical tool for metabolomics applications in food science', *Current Metabolomics*, 2(4), pp. 264–271. Available at: https://doi.org/10.2174/2213235x03999150212102944.

Aruwa, C.E., et al. (2021) 'Poultry gut health – microbiome functions, environmental impacts, microbiome engineering and advancements in characterization technologies', *Journal of Animal Science and Biotechnology*. 12, Article 119. Available at: https://doi.org/10.1186/s40104-021-00640-9.

Ashley, D.L., et al. (1996) 'Measurement of volatile organic compounds in human blood', *Environmental Health Perspectives*, 104(suppl 5), pp. 871–877. Available at: https://doi.org/10.1289/ehp.96104s5871.

Audrain, B., et al. (2015) 'Role of bacterial volatile compounds in bacterial biology', *FEMS Microbiology Reviews*, 39(4), pp. 222–233. Available at: https://doi.org/10.1093/femsre/fuu013.

Auffray, T., et al. (2023) 'Attractant activity of host-related chemical blends on the poultry red mite at different spatial scales', *Journal of Chemical Ecology*, 49(1–2), pp. 18–35. Available at: https://doi.org/10.1007/s10886-022-01399-7.

Bartels, N., et al. (2024) 'Paired metabolomics and volatilomics provides insight into transient high light stress response mechanisms of the coral Montipora mollis', *Metabolomics*, 20(4), Article 36. Available at: https://doi.org/10.1007/s11306-024-02136-9.

Beauclercq, S., et al. (2019) 'A multiplatform metabolomic approach to characterize fecal signatures of negative postnatal events in chicks: a pilot study', *Journal of Animal Science and Biotechnology*, 10(1), Article 21. Available at: https://doi.org/10.1186/s40104-019-0335-8.

Belitz, H.D., et al. (2009) *Food chemistry, food chemistry*. Springer Berlin Heidelberg. Available at: https://doi.org/10.1007/978-3-540-69934-7.

Belizário, J.E., et al. (2021) 'Breath biopsy and discovery of exclusive volatile organic compounds for diagnosis of infectious diseases', *Frontiers in Cellular and Infection Microbiology*, 10, Article 564194. Available at: https://doi.org/10.3389/fcimb.2020.564194.

Bender, D.A. (2007) *Introduction to nutrition and metabolism*, 4th ed. CRC Press.

Bernier, U.R., et al. (2008) 'Volatile compounds from the integument of White Leghorn Chickens (Gallus gallus domesticus L.): candidate attractants of ornithophilic mosquito species', *Journal of Separation Science*, 31(6–7), pp. 1092–1099. Available at: https://doi.org/10.1002/jssc.200700434.

Bindari, Y.R. and Gerber, P.F. (2022) 'Centennial review: factors affecting the chicken gastrointestinal microbial composition and their association with gut health and productive performance', *Poultry Science*, 101(2), Article 101612. Available at: https://doi.org/10.1016/j.psj.2021.101612.

Bombail, V., et al. (2018) 'In search of stress odours across species: Behavioural responses of rats to faeces from chickens and rats subjected to various types of stressful events', *Applied Animal Behaviour Science*, 205(May 2017), pp. 216–226. Available at: https://doi.org/10.1016/j.applanim.2017.10.013.

Boots, A.W., et al. (2012) 'The versatile use of exhaled volatile organic compounds in human health and disease', *Journal of Breath Research*. 6(2), Article 027108. Available at: https://doi.org/10.1088/1752-7155/6/2/027108.

Borgonovo, F., et al. (2020) 'A Data-Driven Prediction Method for an Early Warning of Coccidiosis in Intensive Livestock Systems: A Preliminary Study', *Animals*, 10(4), Article 747. Available at: https://doi.org/10.3390/ani10040747.

Borgonovo, F., et al. (2024) 'An innovative approach for analysing and evaluating enteric diseases in poultry farm', *Acta IMEKO*, 13(1), pp. 1–5. Available at: https://doi.org/10.21014/actaimeko.v13i1.1627.

Bos, L.D.J., et al. (2013) 'Volatile metabolites of pathogens: a systematic review', *PLoS Pathogens*. 9(5), Article e1003311. Available at: https://doi.org/10.1371/journal.ppat.1003311.

Brebu, M., et al. (2023) 'Putative volatile biomarkers of bovine tuberculosis infection in breath, skin and feces of cattle', *Molecular and Cellular Biochemistry*, 478(11), pp. 2473–2480. Available at: https://doi.org/10.1007/s11010-023-04676-5.

Bunea, A., et al. (2017) 'Chromatographic analysis of lypophilic compounds in eggs from organically fed hens', *Journal of Applied Poultry Research*, 26(4), pp. 498–508. Available at: https://doi.org/10.3382/japr/pfx022.

Campagna, S., et al. (2012) 'Potential semiochemical molecules from birds: a practical and comprehensive compilation of the last 20 years studies', *Chemical Senses*, 37(1), pp. 3–25. Available at: https://doi.org/10.1093/chemse/bjr067.

Charmillot, P.J., et al. (2000) 'Attract and kill: a new method for control of the codling moth Cydia pomonella', *Entomologia Experimentalis et Applicata*, 94(2), pp. 211–216. Available at: https://doi.org/10.1046/j.1570-7458.2000.00621.x.

Chen, L., et al. (2024) 'Firmicutes primarily drive odor emission profiles in poultry manure treatments', *Poultry Science*, 103(11). Article 104250. Available at: https://doi.org/10.1016/j.psj.2024.104250.

Cholewińska, P., et al. (2021) 'The level of selected bacterial phyla on the skin surface of small ruminants according to the breed and species', *Animals*, 11(9). Article 2734. Available at: https://doi.org/10.3390/ani11092734.

Corion, M., et al. (2024a) 'Noninvasive assessment of chicken egg fertility during incubation using HSSE-GC-MS VOC profiling', *Journal of Agricultural and Food Chemistry*, 72(15), pp. 8434–8443. Available at: https://doi.org/10.1021/acs.jafc.3c09824.

Corion, M., et al. (2024b) 'In ovo sexing of chickens: evaluating volatile organic compounds analysis techniques and daily prediction performance from the onset of incubation', *Biosystems Engineering*, 247(March), pp. 109–118. Available at: https://doi.org/10.1016/j.biosystemseng.2024.08.013.

Corion, M., et al. (2024c) 'Non-destructive egg breed separation using advanced VOC analytical techniques HSSE-GC-MS, PTR-TOF-MS, and SIFT-MS: assessment of performance and systems' complementarity', *Food Research International*, 176, Article 113802. Available at: https://doi.org/10.1016/j.foodres.2023.113802.

David, L.A., et al. (2014) 'Diet rapidly and reproducibly alters the human gut microbiome', *Nature*, 505(7484), pp. 559–563. Available at: https://doi.org/10.1038/nature12820.

Dell'Olio, A., et al. (2024) 'Non-invasive VOCs detection to monitor the gut microbiota metabolism in-vitro', *Scientific Reports*, 14(1). Article 17386. Available at: https://doi.org/10.1038/s41598-024-66303-7.

Désert, C., et al. (2008) 'Transcriptome profiling of the feeding-to-fasting transition in chicken liver', *BMC Genomics*, 9. Article 611. Available at: https://doi.org/10.1186/1471-2164-9-611.

De Weirdt, R., et al. (2010) 'Human faecal microbiota display variable patterns of glycerol metabolism', *FEMS Microbiology Ecology*, 74(3), pp. 601–611. Available at: https://doi.org/10.1111/j.1574-6941.2010.00974.x.

Dragonieri, S., et al. (2016) 'Influence of age and gender on the profile of exhaled volatile organic compounds analyzed by an electronic nose', *Jornal Brasileiro de Pneumologia*, 42(2), pp. 143–145. Available at: https://doi.org/10.1590/S1806-37562015000000195.

Dridi, S., et al. (2022) 'Editorial: Fat Metabolism and Deposition in Poultry: Physiology, Genetics, Nutrition and Interdisciplinary Research, Volume I', *Frontiers in Physiology*. 13, Article 937081. Available at: https://doi.org/10.3389/fphys.2022.937081.

Du, M., et al. (2000) 'Influence of dietary conjugated linoleic acid on volatile profiles, color and lipid oxidation of irradiated raw chicken meat', *Meat Science*, 56(4), pp. 387–395. Available at: https://doi.org/10.1016/S0309-1740(00)00067-X.

Ellis, C.K., et al. (2014) 'A pilot study exploring the use of breath analysis to differentiate healthy cattle from cattle experimentally infected with Mycobacterium bovis', *PLoS One*, 9(2). Article e89280. Available at: https://doi.org/10.1371/journal.pone.0089280.

Epping, R. and Koch, M. (2023) 'On-site detection of volatile organic compounds (VOCs)', *Molecules*, 28(4), Article 1598. Available at: https://doi.org/10.3390/molecules28041598.

Feddern, V., et al. (2019) 'A review on biogenic amines in food and feed: toxicological aspects, impact on health and control measures', *Animal Production Science*, pp. 608–618. Available at: https://doi.org/10.1071/AN18076.

Fennema, D., et al. (2016) 'Trimethylamine and trimethylamine N-oxide, a flavin-containing monooxygenase 3 (FMO3)-mediated host-microbiome metabolic axis implicated in health and disease', *Drug Metabolism and Disposition*, pp. 1839–1850. Available at: https://doi.org/10.1124/dmd.116.070615.

Ferreira, C.R., et al. (2010) 'Proteomics, metabolomics and lipidomics in reproductive biotechnologies: the MS solutions', *Acta Scientiae Veterinariae*, 38(2), pp. 591–603. Available at: https://pesquisa.bvsalud.org/portal/resource/pt/biblio-1411904 (Accessed: 14 March 2025).

Filaire, F., et al. (2024) 'High and low pathogenicity avian influenza virus discrimination and prediction based on volatile organic compounds signature by SIFT-MS: a proof-of-concept study', *Scientific Reports*, 14(1), Article 18432. Available at: https://doi.org/10.1038/s41598-024-67219-y.

Filipiak, W., et al. (2012) 'Molecular analysis of volatile metabolites released specifically by staphylococcus aureus and pseudomonas aeruginosa', *BMC Microbiology*, 12(1), Article 113. Available at: https://doi.org/10.1186/1471-2180-12-113.

Fischer, S., et al. (2015) 'Physiological variability in volatile organic compounds (VOCs) in exhaled breath and released from faeces due to nutrition and somatic growth in a standardized caprine animal model', *Journal of Breath Research*, 9(2), Article 027108. Available at: https://doi.org/10.1088/1752-7155/9/2/027108.

Garner, C.E., et al. (2007) 'Volatile organic compounds from feces and their potential for diagnosis of gastrointestinal disease', *The FASEB Journal*, 21(8), pp. 1675–1688. Available at: https://doi.org/10.1096/fj.06-6927com.

Garner, C.E., et al. (2008) 'Identification of Campylobacter infection in chickens from volatile faecal emissions', *Biomarkers*, 13(4), pp. 413–421. Available at: https://doi.org/10.1080/13547500801966443.

Gay, M., et al. (2020) 'Control of Dermanyssus gallinae (De Geer 1778) and other mites with volatile organic compounds, a review', *Parasitology*. Cambridge University Press, pp. 731–739. Available at: https://doi.org/10.1017/S0031182020000530.

Gerken, M., et al. (2006) 'Adaptive behaviour in chickens in relation to thermoregulation', *Archiv für Geflügelkunde*, 70(5), pp. 199–207.

Gould, O., et al. (2020) 'Breath analysis for detection of viral infection, the current position of the field', *Journal of Breath Research*, 14(4), Article 041001. Available at: https://doi.org/10.1088/1752-7163/ab9c32.

Gratton, J., et al. (2016) 'Optimized Sample Handling Strategy for Metabolic Profiling of Human Feces', *Analytical Chemistry*, 88(9), pp. 4661–4668. Available at: https://doi.org/10.1021/acs.analchem.5b04159.

Grilli, G., et al. (2018) 'A pilot study to detect coccidiosis in poultry farms at early stage from air analysis', *Biosystems Engineering*, 173, pp. 64–70. Available at: https://doi.org/10.1016/j.biosystemseng.2018.02.004.

Guo, H. and Yuan, Y. (2011) 'Annual variations of odor concentrations and emissions from swine gestation, farrowing, and nursery buildings', *Journal of the Air and Waste Management Association*, 61(12), pp. 1361–1368. Available at: https://doi.org/10.1080/10473289.2011.623636.

Hankel, J., et al. (2022) 'Monitoring of Campylobacter jejuni in a chicken infection model by measuring specific volatile organic compounds and by qPCR', *Scientific*

*Reports*, 12(1), Article 11725. Available at: https://doi.org/10.1038/s41598-022-15863-7.

Hauck, R., et al. (2020) 'Spotlight on avian pathology: aspergillosis', *Avian Pathology*. Taylor and Francis Ltd., pp. 115–118. Available at: https://doi.org/10.1080/03079457.2019.1696946.

Heo, Y.-J., et al. (2023) 'Effects of dietary protein levels on performance, nitrogen excretion, and odor emission of growing pullets and laying hens', *Poultry Science*, 102(8), Article 102798. Available at: https://doi.org/10.1016/j.psj.2023.102798.

Hirao, A., et al. (2009) 'The role of uropygial gland on sexual behavior in domestic chicken Gallus gallus domesticus', *Behavioural Processes*, 80(2), pp. 115–120. Available at: https://doi.org/10.1016/j.beproc.2008.10.006.

Ho, S.S.H., et al. (2017) 'Influences of relative humidities and temperatures on the collection of C2-C5 aliphatic hydrocarbons with multi-bed (Tenax TA, Carbograph 1TD, Carboxen 1003) sorbent tube method', *Atmospheric Environment*, 151, pp. 45–51. Available at: https://doi.org/10.1016/j.atmosenv.2016.12.007.

Hough, R., et al. (2018) 'A comparison of sample preparation methods for extracting volatile organic compounds (VOCs) from equine faeces using HS-SPME', *Metabolomics*, 14(2), Article 19. Available at: https://doi.org/10.1007/s11306-017-1315-7.

Huanhong, K., et al. (2023) 'Volatile organic compound emissions in free-range chicken production: Impacts on environment, welfare and sustainability', *AIMS Agriculture and Food*, 8(4), pp. 1071–1091. Available at: https://doi.org/10.3934/agrfood.2023058.

Hughes, R.J., et al. (2008) 'A Method for Collecting Breath Samples from Individual Chickens for Analysis of 13 CO2, H2, and CH4', *Poultry Science*, 87(9), pp. 1804–1809. Available at: https://doi.org/10.3382/ps.2007-00420.

Ivulic, D., et al. (2022) 'Litter Management Strategies and Their Impact on the Environmental and Respiratory Microbiome Might Influence Health in Poultry', *Microorganisms*, 10(5), Article 878. Available at: https://doi.org/10.3390/microorganisms10050878.

Jacob, S., et al. (2018) 'Chemical regulation of body feather microbiota in a wild bird', *Molecular Ecology*, 27(7), pp. 1727–1738. Available at: https://doi.org/10.1111/mec.14551.

Jaimes-Mogollón, A.L., et al. (2021) 'Review on Volatolomic Studies as a Frontier Approach in Animal Research', *Advanced Biology*, 5(6), pp. 1–27. Available at: https://doi.org/10.1002/adbi.202000397.

Janfaza, S., et al. (2019) 'Digging deeper into volatile organic compounds associated with cancer', *Biology Methods and Protocols*. 4(1), Article bpz014. Available at: https://doi.org/10.1093/biomethods/bpz014.

Jenkins, C.L. and Bean, H.D. (2023) 'Current Limitations of Staph Infection Diagnostics, and the Role for VOCs in Achieving Culture-Independent Detection', *Pathogens*. 12(2), Article 181. Available at: https://doi.org/10.3390/pathogens12020181.

Jha, R. and Mishra, P. (2021) 'Dietary fiber in poultry nutrition and their effects on nutrient utilization, performance, gut health, and on the environment: a review', *Journal of Animal Science and Biotechnology*, 12(1), Article 51. Available at: https://doi.org/10.1186/s40104-021-00576-0.

Jiang, J., et al. (2020) 'The effects of artificial light at night on Eurasian tree sparrow (Passer montanus): Behavioral rhythm disruption, melatonin suppression and intestinal microbiota alterations', *Ecological Indicators*, 108, Article 105702. Available at: https://doi.org/10.1016/j.ecolind.2019.105702.

Jones, J., et al. (2021a) 'Fecal sample collection methods and time of day impact microbiome composition and short chain fatty acid concentrations', *Scientific Reports*, 11(1), Article 13964. Available at: https://doi.org/10.1038/s41598-021-93031-z.

Jones, M., et al. (2021b) 'Differential expression of urinary volatile organic compounds by sex, male reproductive status, and pairing status in the maned wolf (Chrysocyon brachyurus)', *PLOS ONE*. Edited by B. Senthilkumaran, 16(8), Article e0256388. Available at: https://doi.org/10.1371/journal.pone.0256388.

Józefiak, D., et al. (2004) 'Carbohydrate fermentation in the avian ceca: a review', *Animal Feed Science and Technology*. Elsevier, pp. 1–15. Available at: https://doi.org/10.1016/j.anifeedsci.2003.09.007.

Kamal, F., et al. (2021) 'Virus-induced volatile organic compounds are detectable in exhaled breath during pulmonary infection', *American Journal of Respiratory and Critical Care Medicine*, 204(9), pp. 1075–1085. Available at: https://doi.org/10.1164/rccm.202103-0660OC.

Kamiński, E., et al. (1972) 'Identification of the predominant volatile compounds produced by *Aspergillus flavus*', *Applied Microbiology*, 24(5), pp. 721–726. Available at: https://doi.org/10.1128/am.24.5.721-726.1972.

Karami, N., et al. (2017) 'Diagnosis of three different pathogenic microorganisms by gas chromatography-mass spectrometry', *F1000Research*, 6, Article 1415. Available at: https://doi.org/10.12688/f1000research.12003.1.

Karthikeyan, K., et al. (2013) 'Faecal chemical cues in water buffalo that facilitate estrus detection', *Animal Reproduction Science*, 138(3–4), pp. 163–167. Available at: https://doi.org/10.1016/j.anireprosci.2013.02.017.

Kessler, W. and Remmer, H. (1990) 'Generation of volatile hydrocarbons from amino acids and proteins by an iron/ascorbate/GSH system', *Biochemical Pharmacology*, 39(8), pp. 1347–1351. Available at: https://doi.org/10.1016/0006-2952(90)90011-9.

Kimball, B.A., et al. (2013) 'Avian influenza infection alters fecal odor in Mallards', *PLoS One*, 8(10), Article e75411. Available at: https://doi.org/10.1371/journal.pone.0075411.

Koenraadt, C.J.M. and Dicke, M. (2010) 'The role of volatiles in aggregation and host-seeking of the haematophagous poultry red mite Dermanyssus gallinae (Acari: Dermanyssidae)', *Experimental and Applied Acarology*, 50(3), pp. 191–199. Available at: https://doi.org/10.1007/s10493-009-9305-8.

Kouba, M. and Mourot, J. (1998) 'Effect of a high linoleic acid diet on delta 9–desaturase activity, lipogenesis and lipid composition of pig subcutaneous adipose tissue', *Reproduction Nutrition Development*, 38(1), pp. 31–37. https://doi.org/10.1051/rnd:19980103.

Lammers-Jannink, K.C.M., et al. (2022) 'Microbial protein metabolism in the monogastric gastrointestinal tract: a review', pp. 435–466. Available at: https://doi.org/10.19103/as.2021.0089.23.

Li, H., et al. (2024) 'GC-IMS and multivariate analyses of volatile organic components in different Chinese breeds of chickens', *Heliyon*, 10(8), Article e29664. Available at: https://doi.org/10.1016/j.heliyon.2024.e29664.

Li, J., et al. (2023) 'Integration of volatilomics and metabolomics unveils key flavor-related biological pathways in different carambola cultivars', *Journal of Agricultural and Food Chemistry*, 71(28), pp. 10850–10862. Available at: https://doi.org/10.1021/acs.jafc.3c02015.

Liang, X., et al. (2015) 'Rhythmicity of the intestinal microbiota is regulated by gender and the host circadian clock', *Proceedings of the National Academy of Sciences of the United States of America*, 112(33), pp. 10479–10484. Available at: https://doi.org/10.1073/pnas.1501305112.

Liangou, A., et al. (2021) 'A method for the identification of COVID-19 biomarkers in human breath using proton transfer reaction time-of-flight mass spectrometry', *eClinicalMedicine*, 42, Article 101207. Available at: https://doi.org/10.1016/j.eclinm.2021.101207.

Liao, C.M., et al. (2001) 'An inhalation dose model for assessing dust-borne VOC-Odor exposure from feeding in swine buildings', *Transactions of the ASAE*, 44(6), pp. 1813–1824. Available at: https://doi.org/10.13031/2013.7019.

Liu, Q., et al. (2024) 'Serum-volatile organic compounds in the diagnostics of esophageal cancer', *Scientific Reports*, 14(1), Article 17722. Available at: https://doi.org/10.1038/s41598-024-67818-9.

Liu, T., et al. (2023) 'Review on algorithm design in electronic noses: challenges, status, and trends', *Intelligent Computing*, 2, Article 0012. Available at: https://doi.org/10.34133/icomputing.0012.

Luo, N., et al. (2022) 'A comparison of different tissues identifies the main precursors of volatile substances in chicken meat', *Frontiers in Physiology*, 13, Article 927618. Available at: https://doi.org/10.3389/fphys.2022.927618.

Murakami, M. and Tognini, P. (2020) 'The circadian clock as an essential molecular link between host physiology and microorganisms', *Frontiers in Cellular and Infection Microbiology*. Available at: https://doi.org/10.3389/fcimb.2019.00469.

Nimmermark, S. and Gustafsson, G. (2005) 'Influence of temperature, humidity and ventilation rate on the release of odour and ammonia in a floor housing system for laying hens', *The CIGR Journal of Scientific Research and Development*, pp. 1–16. Available at: https://www.researchgate.net/publication/228510976.

Nol, P., et al. (2020) 'Evaluation of volatile organic compounds obtained from breath and feces to detect mycobacterium tuberculosis complex in wild boar (Sus scrofa) in Doñana National Park, Spain', *Pathogens*, 9(5), Article 346. Available at: https://doi.org/10.3390/pathogens9050346.

O'Hara, M.E., et al. (2009) 'Endogenous volatile organic compounds in breath and blood of healthy volunteers: examining breath analysis as a surrogate for blood measurements', *Journal of Breath Research*, 3(2), Article 027005. Available at: https://doi.org/10.1088/1752-7155/3/2/027005.

Okechukwu, V.O., et al. (2024) 'Room temperature detection of aspergillus flavus volatile organic compounds (VOCs) under simulated conditions using graphene oxide and tin oxide Nanorods (SnO2 NRs-GO)', *Food Chemistry*, 456, Article 140068. Available at: https://doi.org/10.1016/j.foodchem.2024.140068.

Oliphant, K. and Allen-Vercoe, E. (2019) 'Macronutrient metabolism by the human gut microbiome: major fermentation by-products and their impact on host health', *Microbiome*, 7(1), Article 91. Available at: https://doi.org/10.1186/s40168-019-0704-8.

Oxner, M., et al. (2023) 'The versatility and diagnostic potential of VOC profiling for noninfectious diseases', *BME Frontiers*, 4, Article 0002. American Association for the Advancement of Science. Available at: https://doi.org/10.34133/bmef.0002.

Oyerinde, A.S., et al. (2023) 'Potential role of oxidative stress in the production of volatile organic compounds in obesity', *Antioxidants*, 12(1), Article 129. MDPI. Available at: https://doi.org/10.3390/antiox12010129.

Palma, S.I.C.J., et al. (2018) 'Machine learning for the meta-analyses of microbial pathogens' volatile signatures', *Scientific Reports*, 8(1), Article 3360. Available at: https://doi.org/10.1038/s41598-018-21544-1.

Pennington, T., et al. (2024) 'Identification of volatile metabolites produced from levodopa metabolism by different bacteria strains of the gut microbiome', *BMC Microbiology*, 24(1), Article 236. Available at: https://doi.org/10.1186/s12866-024-03373-7.

Pérez-Calvo, E., et al. (2019) 'The measurement of volatile organic compounds in faeces of piglets as a tool to assess gastrointestinal functionality', *Biosystems Engineering*, 184, pp. 122–129. Available at: https://doi.org/10.1016/j.biosystemseng.2019.06.005.

Peters, A.L., et al. (2017) 'Volatile organic compounds in exhaled breath are independent of systemic inflammatory syndrome caused by intravenous lipopolysaccharide infusion in humans: Results from an experiment in healthy volunteers', *Journal of Breath Research*, 11(2), Article 026003. Available at: https://doi.org/10.1088/1752-7163/aa6545.

Portune, K.J., et al. (2016) 'Gut microbiota role in dietary protein metabolism and health-related outcomes: The two sides of the coin', *Trends in Food Science and Technology*, 57, pp. 213–232. Available at: https://doi.org/10.1016/j.tifs.2016.08.011.

Probert, C.S.J., et al. (2004) 'A novel method for rapidly diagnosing the causes of diarrhoea', *Gut*, 53(1), pp. 58–61. Available at: https://doi.org/10.1136/gut.53.1.58.

Rajchard, J. (2010) 'Biologically active substances of bird skin: a review', *Veterinární medicína*, 55(9), pp. 413–421. Available at: https://doi.org/10.17221/2981-VETMED.

Ranadheera, C.S., et al. (2017) 'Strategies for eliminating chicken manure odour in horticultural applications', *World's Poultry Science Journal*, 73(2), pp. 365–378. Available at: https://doi.org/10.1017/S0043933917000083.

Ratiu, I.A., et al. (2020) 'An optimistic vision of future: diagnosis of bacterial infections by sensing their associated volatile organic compounds', *Critical Reviews in Analytical Chemistry*. Taylor and Francis Ltd., pp. 501–512. Available at: https://doi.org/10.1080/10408347.2019.1663147.

Recharla, N., et al. (2017) 'Effects of amino acid composition in pig diet on odorous compounds and microbial characteristics of swine excreta', *Journal of Animal Science and Technology*, 59(1), Article 28. Available at: https://doi.org/10.1186/s40781-017-0153-5.

Rodríguez-Hernández, P., et al. (2021) 'Application of volatilome analysis to the diagnosis of mycobacteria infection in livestock', *Frontiers in Veterinary Science*, 8, Article 635155. Available at: https://doi.org/10.3389/fvets.2021.635155.

Rodríguez-Hernández, P., et al. (2022) 'Analytical tools for disease diagnosis in animals via fecal volatilome', *Critical Reviews in Analytical Chemistry*, 52(5), pp. 917–932. Available at: https://doi.org/10.1080/10408347.2020.1843130.

Santos, D.F.M.S., et al. (2024) 'Volatile organic compounds produced by thermophilic and non-thermophilic Campylobacter spp.: influence of growth phase and nutrient composition', *International Food Research Journal*, 31(3), pp. 551–566. Available at: https://doi.org/10.47836/ifrj.31.3.02.

Sanz, M., et al. (2000) 'Nutrient metabolism-research communication abdominal fat deposition and fatty acid synthesis are lower and-oxidation is higher in broiler chickens fed diets containing unsaturated rather than saturated fat', *The Journal of Nutrition*, 130(12), pp. 3034–3037. Available at: https://doi.org/10.1093/jn/130.12.3034.

Scheid, P. and Piiper, J. (1969) 'Volume, ventilation and compliance of the respiratory system in the domestic fowl', *Respiration Physiology*, 6(3), pp. 298–308. Available at: https://doi.org/10.1016/0034-5687(69)90029-2.

Sethi, S., et al. (2013) 'Clinical application of volatile organic compound analysis for detecting infectious diseases', *Clinical Microbiology Reviews*, 26(3), pp. 462–475. Available at: https://doi.org/10.1128/CMR.00020-13.

Shang, Y., et al. (2018) 'Chicken gut microbiota: importance and detection technology', *Frontiers in Veterinary Science*, 5, Article 254. Available at: https://doi.org/10.3389/fvets.2018.00254.

Sharma, N.K., et al. (2017) 'Emissions of volatile odorous metabolites by Clostridium perfringens - in vitro study using two broth cultures', *Poultry Science*, 96(9), pp. 3291–3297. Available at: https://doi.org/10.3382/ps/pex129.

Sharma, N.K., et al. (2018) 'Necrotic enteritis challenge and high dietary sodium level affect odorant composition or emission from broilers', *Poultry Science*, 97(1), pp. 39–46. Available at: https://doi.org/10.3382/ps/pex257.

Shirasu, M. and Touhara, K. (2011) 'The scent of disease: volatile organic compounds of the human body related to disease and disorder', *Journal of Biochemistry*, pp. 257–266. Available at: https://doi.org/10.1093/jb/mvr090.

Shu, Y., et al. (2024) 'Dietary effect of curcumin on amino acid, fatty acid, and volatile compound profiles of chicken meat', *Foods*, 13(14), Article 2230. Available at: https://doi.org/10.3390/foods13142230.

Smeets, M.A.M., et al. (2020) 'Chemical fingerprints of emotional body odor', *Metabolites*, 10(3), Article 84. Available at: https://doi.org/10.3390/metabo10030084.

Sohn, J.H., et al. (2008) 'Implementation of an electronic nose for continuous odour monitoring in a poultry shed', *Sensors and Actuators, B: Chemical*, 133(1), pp. 60–69. Available at: https://doi.org/10.1016/j.snb.2008.01.053.

Soini, H.A., et al. (2006) 'Seasonal variation in volatile compound profiles of preen gland secretions of the Dark-eyed Junco (Junco hyemalis)', *Journal of Chemical Ecology*, 33(1), pp. 183–198. Available at: https://doi.org/10.1007/s10886-006-9210-0.

Spiehs, M.J., et al. (2013) 'Effect of bedding materials on concentration of odorous compounds and escherichia coli in beef cattle bedded manure packs', *Journal of Environmental Quality*, 42(1), pp. 65–75. Available at: https://doi.org/10.2134/jeq2012.0251.

Spizizen, J. (1957) 'The effect of virus infection on pyruvate metabolism', *Biochimica et Biophysica Acta*, 23(3), pp. 333–341. Available at: https://doi.org/10.1016/0006-3002(57)90336-0.

Su, W., et al. (2022) 'Standards for collection, preservation, and transportation of fecal samples in TCM clinical trials', *Frontiers in Cellular and Infection Microbiology*, 12, Article 783682. Available at: https://doi.org/10.3389/fcimb.2022.783682.

Trabue, S., et al. (2010) 'Speciation of volatile organic compounds from poultry production', *Atmospheric Environment*, 44(29), pp. 3538–3546. Available at: https://doi.org/10.1016/j.atmosenv.2010.06.009.

Traxler, S., et al. (2018) 'VOC breath profile in spontaneously breathing awake swine during Influenza A infection', *Scientific Reports*, 8(1), Article 14857. Available at: https://doi.org/10.1038/s41598-018-33061-2.

Urrego, M.I.G., et al. (2021) 'Dietary protein sources and their effects on faecal odour and the composition of volatile organic compounds in faeces of French Bulldogs',

*Journal of Animal Physiology and Animal Nutrition*, 105(S1), pp. 65–75. Available at: https://doi.org/10.1111/jpn.13605.

Van De Kant, K.D.G., et al. (2013) 'Exhaled breath profiling in diagnosing wheezy preschool children', *European Respiratory Journal*, 41(1), pp. 183–188. Available at: https://doi.org/10.1183/09031936.00122411.

van Veen, L.A., et al. (2025) 'Manure odor profiling for flock-level monitoring on commercial layer pullet farms: vaccination events as a model stressor', *Poultry Science*, 104(2), Article 104681. Available at: https://doi.org/10.1016/j.psj.2024.104681.

van Vorstenbosch, R., et al. (2023) 'Systematic review: contribution of the gut microbiome to the volatile metabolic fingerprint of colorectal neoplasia', *Metabolites*. MDPI, 13(1), Article 55. Available at: https://doi.org/10.3390/metabo13010055.

Vargas, R.I., et al. (2003) 'Field trials of spinosad as a replacement for naled, DDVP, and malathion in methyl eugenol and cue-lure bucket traps to attract and kill male oriental fruit flies and melon flies (Diptera: Tephritidae) in Hawaii', *Journal of Economic Entomology*, 96(6), pp. 1780–1785. Available at: https://doi.org/10.1093/jee/96.6.1780.

Wang, Y., et al. (2024) 'Identification and biotransformation analysis of volatile markers during the early stage of Salmonella contamination in chicken', *Food Chemistry*, 431, Article 137130. Available at: https://doi.org/10.1016/j.foodchem.2023.137130.

Weisskopf, L., et al. (2021) 'Microbial volatile organic compounds in intra-kingdom and inter-kingdom interactions', *Nature Reviews Microbiology*. Nature Research, pp. 391–404. Available at: https://doi.org/10.1038/s41579-020-00508-1.

Weng, X., et al. (2021) 'Detection of volatile organic compounds (VOCs) in livestock houses based on electronic nose', *Applied Sciences*, 11(5), Article 2337. Available at: https://doi.org/10.3390/app11052337.

Wernery, U. (2016) 'Infectious Diseases', in *Avian Medicine*. Elsevier, pp. 434–521. Available at: https://doi.org/10.1016/B978-0-7234-3832-8.00014-6.

Wilson, A.D. and Forse, L.B. (2023) 'Potential for early noninvasive COVID-19 detection using electronic-nose technologies and disease-specific VOC metabolic biomarkers', *Sensors*, 23(6), Article 2887. Available at: https://doi.org/10.3390/s23062887.

Yalçin, S., et al. (2008) 'Acclimation to heat during incubation. 2. Embryo composition and residual egg yolk sac fatty acid profiles in chicks', *Poultry Science*, 87(6), pp. 1229–1236. Available at: https://doi.org/10.3382/ps.2007-00436.

Yan, G., et al. (2021) 'Incorporating VOC emission control in China's Hazardous waste regulatory system', In *Environmental Science and Technology*. American Chemical Society, pp. 15569–15571. Available at: https://doi.org/10.1021/acs.est.1c03207.

Yan, W., et al. (2019) 'Efficacy of fecal sampling as a gut proxy in the study of chicken gut microbiota', *Frontiers in Microbiology*, 10, Article 2126. Available at: https://doi.org/10.3389/fmicb.2019.02126.

Yao, Q., et al. (2018) 'Using a vegetative environmental buffer to reduce the concentrations of volatile organic compounds in poultry-house atmospheric emissions', *Journal of Agricultural and Food Chemistry*, 66(31), pp. 8231–8236. Available at: https://doi.org/10.1021/acs.jafc.8b00088.

Yongwei, W., et al. (2009) 'Monitoring storage time and quality attribute of egg based on electronic nose', *Analytica Chimica Acta*, 650(2), pp. 183–188. Available at: https://doi.org/10.1016/j.aca.2009.07.049.

Zimmermann, D., et al. (2007) 'Determination of volatile products of human colon cell line metabolism by GC/MS analysis', *Metabolomics*, 3(1), pp. 13–17. Available at: https://doi.org/10.1007/s11306-006-0038-y.

Żuchowska, K. and Filipiak, W. (2024) 'Modern approaches for detection of volatile organic compounds in metabolic studies focusing on pathogenic bacteria: current state of the art', *Journal of Pharmaceutical Analysis*, 14(4), Article 100898. Available at: https://doi.org/10.1016/j.jpha.2023.11.005.

# Part 6

## Maintaining the health of broilers

# Chapter 19

## Vaccination strategies in broilers

*Alessandra Piccirillo, University of Padua, Italy; Giacomo Berto, Veterinary Services Local Health Unit (ULSS) 8 "Berica" – Vicenza, Italy; Davide Giovanardi, Tre Valli Laboratory, Italy; Luca Bano, Experimental Zooprophylactic Institute of the Venezie, Italy; and Andrea Laconi, University of Padua, Italy*

1 Introduction
2 The immune system in broilers
3 Vaccination strategies in broiler production
4 Types of vaccines used in broilers
5 Vaccine delivery methods
6 Factors influencing vaccination in broilers
7 New developments and future perspectives
8 Conclusion
9 Where to look for further information
10 References

## 1 Introduction

Broiler production faces multiple challenges that impact productivity and sustainability. These challenges range from economic pressures and climate change to welfare concerns and disease management (Castro et al., 2023; Korver, 2023; Oke et al., 2024). While genetic selection and modern farming systems have significantly improved production efficiency, they have also introduced new risks related to animal health (Julian, 2005; Hossain and Akter, 2022; Liu et al., 2023a). Modern broilers are increasingly affected by metabolic, developmental, and growth-related disorders due to their accelerated metabolism and rapid growth rate (Korver, 2023; Hartcher and Lum, 2020). These physiological imbalances can also weaken immune response, making broilers more susceptible to infections (Bessei, 2006; Hill and Kirkpatrick, 2019). Additionally, intensive broiler farming provides an ideal environment for the rapid spread of infectious agents. As a result, infectious diseases remain one of the most pressing concerns, highlighting the need for effective prevention and

http://dx.doi.org/10.19103/AS.2025.0165.24

control strategies (Collett, 2020; Hafez and Attia, 2020). The infectious disease issues affecting broilers include bacterial, viral, and parasitic infections, which can lead to high mortality rates, reduced performance, and increased antibiotic use (Oviedo-Rondón, 2022; Yehia et al., 2023). However, many broiler diseases are multifactorial, arising from interactions between infectious agents, nutrition, and environmental factors, with a pathogenesis often not fully understood (Maes et al., 2021; Mench et al., 2021). Additionally, immunosuppression resulting from infectious or non-infectious factors further complicates the health threat by increasing susceptibility to disease or lack of responsiveness to vaccination (Hoerr, 2010; Schat and Skinner, 2022). The prevention and control of infectious diseases in broiler production requires a multifactorial approach. This includes several key factors such as genetic selection for disease-resistant broilers; rearing of breeder flocks free from key vertically transmitted pathogens (e.g. *Salmonella pullorum*, *Mycoplasma* spp.) and adequate maternal immunity, implementing stringent biosecurity measures to limit pathogen exposure; adopting improved management practices to reduce stress and enhance resilience; and applying vaccination programmes targeting prevalent pathogens (Collett, 2020; Hafez and Attia, 2020). Although genetic selection has increased resistance to diseases such as Marek's disease, its effectiveness in modern broiler farming remains limited due to the complexity of field infections, where broilers are exposed to multiple pathogens simultaneously (Saif and Swayne, 2020; Psifidi, 2021). Disease prevention and control primarily rely on biosecurity, which encompasses all procedures aimed at reducing infection risks and managing outbreaks (Jennison, 2021; Vaillancourt et al., 2022). However, ensuring strict biosecurity compliance remains a challenge due to factors such as economic constraints, lack of awareness, and inconsistent implementation (Vaillancourt et al., 2022; Pao et al., 2023). Therefore, biosecurity alone may not be sufficient and should be complemented with effective management strategies and targeted vaccination programmes.

Vaccination is a cornerstone of disease prevention in the broiler industry, serving as a key tool to protect against infectious diseases and ensure optimal flock health and productivity. In high-density poultry farming, the risk of pathogen transmission is significant, making well-structured and adaptable vaccination programmes essential (Marangon and Busani, 2006; Collett, 2020). A variety of infectious agents pose a continuous threat to broiler health; therefore, vaccination is essential to prevent health and welfare issues and major economic losses. Vaccines targeting prevalent poultry pathogens, such as Marek's disease virus (MDV), Newcastle disease virus (NDV), infectious bronchitis virus (IBV), infectious bursal disease virus (IBDV), and coccidia, are extensively employed in commercial broiler production. Administration of vaccines against these pathogens typically begins at hatch or, in some cases, during incubation via *in ovo* vaccination. To ensure efficacy, vaccination programmes must be carefully tailored to the specific epidemiological context and disease pressures present in each production system or geographic region (Sharma, 1999; Cserep, 2009; Schat, 2014). Given the growing concern over antimicrobial resistance and the increasing pressure to reduce antibiotic use in animal agriculture, vaccination has become an essential tool in promoting antibiotic-free poultry farming. By preventing both viral and bacterial infections, and reducing the need for therapeutic antibiotic use, vaccination supports a more sustainable approach to poultry health management (Smith et al., 2020; Hoelzer et al., 2018a; Hoelzer et al., 2018b; Islam and Rahman, 2023). Moreover, vaccination significantly enhances broiler welfare by reducing mortality and stress while promoting a more stable, ethical production system aligned with societal expectations (Schat, 2014; Bist et al., 2024). Furthermore, innovative methods, such as *in ovo* vaccination, also contribute to broiler welfare as they enable efficient mass immunization, while minimizing handling and stress (Li et al., 2024; Abd El-Ghany, 2025). Ultimately, vaccination should be part of a comprehensive disease management strategy that includes biosecurity, good husbandry and management practices, and continuous disease monitoring and surveillance (Marangon and Busani, 2006; Collett, 2020). This integrated approach supports improved animal health and welfare, reduced economic losses, responsible antimicrobial use, and better public health outcomes, while contributing to improved productivity and the long-term economic viability of poultry farming.

The aim of this chapter is to provide a comprehensive overview of vaccination strategies in broiler production, focusing on the factors that influence their implementation and effectiveness. To this end, it integrates current knowledge on the development of the avian immune system, pathogen epidemiology, vaccine technologies, and management practices. The chapter begins by examining the ontogeny and maturation of the avian immune system, from embryogenesis through the early post-hatch period. It then explores the principal infectious diseases targeted by vaccination and reviews immunization practices and approaches used in both fast- and slow-growing broilers, including those raised under alternative production systems, such as on-farm hatching. A detailed description of conventional and new generation vaccines is provided, along with the main delivery methods used in commercial settings. The chapter also discusses the multifactorial determinants of vaccine performance – including host-, pathogen-, vaccine-, and environment/management-related factors – and highlights key issues, such as immunosuppression, vaccine interference, and administration errors. Although aspects such as host genetic resistance and regulatory frameworks are acknowledged, they fall outside the scope of this chapter and are addressed in greater detail in scientific literature and official guidelines (some suggestions are provided in the section 'Where to look for further information').

## 2 The immune system in broilers

The avian immune system is a highly specialized and efficient defence mechanism that enables birds to recognize and neutralize pathogens, ensuring their survival in environments with high microbial exposure, such as intensive farming systems (Schat, 2014; Rautenschlein, 2020a). It is a well-coordinated network, consisting of two interconnected branches: innate and adaptive immunity, which work in synergy to protect birds against infections (Kaiser, 2010; Kaiser and Balic, 2014). Innate immunity serves as the first line of defence, providing a rapid and non-specific response, while also activating adaptive immunity. This initial response relies on physical barriers (e.g. skin, mucosal surfaces) that prevent pathogen entry; molecular mediators (e.g. cytokines, chemokines, host defence peptides, acute phase proteins) that regulate immune signalling and pathogen elimination; and cellular components (e.g. granulocytes, macrophages, dendritic cells, natural killer (NK) cells, thrombocytes) that detect and neutralize threats (Gobel and Smith, 2022; Yoshimura et al., 2024). Adaptive immunity, in contrast, provides specific and long-term protection. It functions through humoral immunity, where B lymphocytes produce IgM and IgY antibodies to target extracellular pathogens, and cell-mediated immunity, mediated by cytotoxic CD8+ T cells, which eliminate infected cells, and helper CD4+ T cells, which coordinate immune responses, enhancing protection against both intracellular and extracellular pathogens (Ratcliffe, 2006; Ratcliffe and Hartle, 2022; Smith and Gobel, 2022). Additionally, mucosal immunity protects critical pathogen entry points, such as the gastrointestinal and respiratory tracts, through secretory IgA antibodies, which prevent pathogen adhesion and neutralize infections at their initial site of entry (Smiałek et al., 2011; Kaspers and Schat, 2022).

Although vaccination stimulates both humoral and cell-mediated immunity, research has mainly focused on antibody production, leaving cell-mediated immunity comparatively less explored. However, effective immune protection is achieved only through the coordination between innate and adaptive responses, including mucosal immunity (Hao et al., 2021; Yoshimura et al., 2024). Upon administration, vaccines stimulate pathogen-associated molecular patterns (PAMPs), which are recognized by pattern recognition receptors (PRRs), leading to the activation of innate immune pathways. This triggers cytokine secretion, enhancing antigen presentation and facilitating the activation of B and T lymphocytes, which are essential for adaptive immunity (Schat, 2014; Rautenschlein and Schat, 2023; Yoshimura et al., 2024). The type of vaccine and its route of administration significantly influence the immune response. Live attenuated vaccines, which replicate within the host, induce a strong cell-mediated immune response, along with humoral immunity (Sharma, 1999; Schijns et al., 2014). These vaccines, when applied via mucosal routes, stimulate local IgA production at pathogen entry sites, providing broad protection (Al-Garib et al., 2003; Chhabra et al., 2015). In contrast, inactivated vaccines, administered parenterally, primarily stimulate humoral immunity with high antibody titres, but elicit only limited cell-mediated immunity (Sharma, 1999; Schijns et al., 2014). Moreover, adjuvants enhance vaccine efficacy by stimulating innate immune receptors, prolonging antigen presentation, and promoting stronger proliferation of B and T lymphocytes (Bodman-Harris et al., 2024).

Following vaccination, the first antibody to be produced is IgM, providing an initial but short-lived response. In adult chickens with a functional immune system, IgM can be detected as early as 3 days after a single vaccination, generally peaking between 4 and 8 days before rapidly declining (Benedict et al., 1963; Higgins, 1975). Although usually short-lived, IgM responses may persist in the case of chronic bacterial infections (Suresh et al., 1994). This early response is followed by an isotype switch to IgA at mucosal surfaces or to IgY in the systemic circulation, depending on the site of antigen exposure (Schijns et al., 2014; Rautenschlein and Schat, 2023). Live vaccines, commonly administered via ocular or conjunctival routes (e.g. spray or drinking water), mimic natural infection and promote the production of secretory IgA, which is the predominant antibody at mucosal sites, such as the respiratory tract (Lillehoj, 2018). These surfaces often represent the initial sites of pathogen invasion. Consequently, IgA is first detected locally (e.g. in tears) and later appears in the systemic circulation. For instance, after vaccination with live attenuated IBV, a marked increase in IgA antibodies can be observed in tears by day 3 post-vaccination, although no rise is seen in plasma at this stage (Orr-Burks et al., 2014).

Immune priming initiates immune activation by exposing naïve B and T lymphocytes to an antigen, establishing immunological memory. Booster immunization enhances this response, leading to stronger and prolonged immunity. This strategy increases IgY levels in circulation for systemic protection and boosts IgA production at mucosal surfaces, reinforcing the first line of defence. The prime-boost approach, combining live attenuated and inactivated vaccines, ensures broader and long-lasting immunity, particularly in long-lived birds like breeders and layers (Schijns et al., 2014; de Wit and Montiel, 2022; Rautenschlein and Schat, 2023).

## *2.1 Ontogeny and maturation of the immune system*

The development of the avian immune system begins during early embryogenesis and continues after hatching, gradually acquiring full functionality. Figure 1 illustrates the timeline of key developmental events during the ontogeny of the avian immune system, from embryonic day (ED)

**Figure 1** Timeline of key events occurring during the ontogeny of the avian immune system.

1 to hatching. During embryogenesis, hematopoiesis initiates in the yolk sac around ED 2 to 4, producing the first immune cells, including macrophages and heterophils. By ED 6, hematopoietic stem cells migrate from the aorta-gonad-mesonephros region to the embryonic liver, which temporarily becomes the primary site of blood and immune cell production. As the embryo develops, the bone marrow assumes its role as the primary hematopoietic organ around ED 14, laying the basis for a functional immune system (McGrath and Palis, 2005; Tavares Guedes et al., 2014; Yvernogeau et al., 2022).

The primary lymphoid organs (PLOs), the thymus and bursa of Fabricius, develop early in embryogenesis. The thymus forms around ED 3–4 and is colonized by T cell precursors during three major waves at ED 6.5, 12, and 18. These cells undergo differentiation into functional T lymphocytes, including cytotoxic (CD8+) and helper (CD4+) subtypes. Similarly, the bursa of Fabricius, becoming visible by ED 10, is colonized by B cell precursors around ED 7. These cells start expressing surface IgM by ED 10, followed by IgY at ED 14 and IgA at ED 16, with full antibody repertoire diversification occurring closer to hatching (Ko et al., 2018; Alkie et al., 2019; Nagy et al., 2022). By ED 12–14, chicken embryos can mount adaptive immune responses to antigens, although immune tolerance can also be induced depending on the nature of the antigen and the timing of exposure (Schijns et al., 2014; de Wit and Montiel, 2022). Unlike adaptive immunity, innate immunity develops earlier and is functional at hatch. By ED 18, key innate immune cells, such as macrophages, heterophils, and dendritic cells, are capable of pathogen recognition via PRRs, including Toll-like receptors (TLRs). These cells coordinate an early defence, supported by cytokines and host defence peptides (HDPs), such as defensins (Alkie et al., 2019; Gobel and Smith, 2022). In contrast, adaptive immunity requires several weeks post-hatch to achieve full functionality. While T and B lymphocytes are present before hatching, their activity remains limited (Kaiser and Balic, 2014; Yvernogeau et al., 2022). Furthermore, the rapid growth rate of broilers imposes a competition for resources between immune system maturation and tissue development, potentially affecting the efficiency of early immune responses (Scheele, 1999; Riber and Wurtz, 2024).

During late embryonic development, the immune system begins to respond to antigens, as demonstrated by *in ovo* vaccination. Vaccines administered between ED 17 and 19 prime the immune system before hatch. When delivered into the amniotic cavity or the body of the embryo, these vaccines are ingested and absorbed by the embryo, stimulating both mucosal and systemic immune activation. Additionally, early immunization contributes to the development of immune memory, enhancing protection against pathogens during early life (Li et al., 2024; Abd El-Ghany, 2025). Vaccines administered *in ovo*, particularly those targeting MDV and IBDV, singly or in combination, have demonstrated the embryo's ability to generate protective immune responses, often superior to those elicited by post-hatch vaccination (Gagic, et al., 1999; Sharma et al., 2002; Gimeno et al., 2015; de Wit et al., 2021). However, for other pathogens, such as NDV and IBV, the immune response induced by *in ovo* vaccination may remain suboptimal compared to those generated by a fully mature immune system (Kapczynski et al., 2012; Saiada et al., 2019).

After hatching, the immune system undergoes rapid maturation. Innate immune cells, such as heterophils and macrophages, improve in efficiency, enhancing their phagocytic and microbial-killing functions. B lymphocytes continue antibody repertoire diversification in the bursa, while T lymphocytes migrate to secondary lymphoid organs (SLOs), such as the spleen and mucosa-associated lymphoid tissues (MALT), where they refine their immunological functions (Kaiser and Balic, 2014; Yvernogeau et al., 2022). By the end of the first two weeks post-hatch, the chicken's immune system reaches a greater level of functional competence, allowing for stronger immune responses against pathogens. Although day-one vaccination does not immediately stimulate antibody production due to the incomplete development of SLOs in newly hatched chickens, it has been shown to effectively stimulate both innate and adaptive immunity (de Wit and Montiel, 2022; Ceccopieri and Madej, 2024). Day-old vaccination against NDV and IBV, as well as against other key pathogens, such as MDV, has been demonstrated to provide early protection. However, some interference from maternal antibodies (MDAs) may occur with live vaccines (Sharma and Witter, 1983; Chansiripornchai and Sasipreeyajan, 2006; Awad et al., 2016; Ike et al., 2021). These antibodies are essential for passive immunity, providing protection to chickens during the first weeks of life, while the immune system matures (Isham et al., 2023; Wein et al., 2024). In contrast, the efficacy of day-old vaccination against IBDV is more severely affected by high MDA titres, which can neutralize the vaccine virus and impair immunogenicity (Block et al., 2007). Conversely, live *Eimeria* spp. vaccines are not impacted by the presence of MDAs, allowing effective immunization from the first days of life (Albanese et al., 2018).

During early growth, immune strength is influenced by multiple factors, including genetics, environmental and nutritional conditions, social stress, intestinal microbiota composition, MDA transfer, and vaccination. However, the synergy between MDAs and the microbiota is essential for maintaining broiler health in early life (Broom and Kogut, 2018; Gong et al., 2023). Newly hatched chickens are highly vulnerable to microbial infections, and the interaction between maternal immunity and microbiota significantly enhances their survival in environments with high pathogen loads, such as intensive production settings.

## 2.2 Maternal antibody transfer

Maternal immunity transfer is the passive immunity established through the deposition of MDAs into the egg (Hincke et al., 2019; Hartle et al., 2022). These antibodies are produced by the hen in response to environmental pathogens, making them highly specific to the microbial challenges in her environment. Antigen-specific IgY antibodies are transferred from the hen's bloodstream to the egg yolk. These antibodies become detectable in the egg approximately 5–6 days after appearing in the hen's serum, with their concentration in the yolk generally mirroring that in the maternal bloodstream. However, not all MDAs are transferred equally to the yolk. IgY is selectively transported through specialized receptor-mediated transcytosis, whereas IgM and IgA, which are incorporated into the egg albumen via oviductal secretions, are present in much smaller quantities (Tesar et al., 2008; Tian and Zhang, 2012; Ulmer-Franco, 2012). The yolk sac membrane mediates IgY transfer to the embryonic circulation, initiating around ED 7 and peaking during the final three days before hatch, ensuring sufficient systemic immunity at hatching. However, only about 10% of the total IgY present in the yolk is transferred into the embryo's bloodstream, leading to neonatal serum antibody titres that are approximately eight times lower than those found in the yolk (Kowalczyk et al., 1985; Murai, 2013). In addition, egg yolk infection may impair the transfer of antibodies to the embryonic bloodstream (Sander et al., 1998). IgM and IgA, initially secreted into the oviduct and incorporated into the egg albumen, are absorbed by the embryo through the ingestion of amniotic fluid around ED 10–12. These antibodies provide localized protection to the intestinal mucosa at hatching and continue to support gut immunity as the yolk sac is gradually absorbed during the first few days of life (Ulmer-Franco, 2012; Hartle et al., 2022; Gong et al., 2023).

At hatching, chickens benefit from systemic immunity, provided by circulating IgY, and localized gut protection from IgM and IgA. MDAs provide temporary protection, reaching peak levels within 1–2 days post-hatch before gradually declining over 2–3 weeks. The *de novo* synthesis of IgY by the chicken's immune system begins around day 10 post-hatch, marking the transition from maternal immunity to active endogenous immune function (Murai, 2013; Rautenschlein, 2020a; Hartle et al., 2022). In addition to antibody-mediated protection, newly hatched chickens also absorb immune cells from the residual yolk sac, such as macrophages and monocytes, which provide temporary immune support, particularly in the early days post-hatch before the chicken's own immune system is fully functional (Alkie et al., 2019; Rautenschlein, 2020a; Hartle et al., 2022). This period of declining MDAs represents a critical window in which the risk of infection significantly increases if exposure to pathogens occurs before the chicken's immune system becomes fully competent. Studies suggest that this period may differ between fast- and slow-growing broilers, with slow-growing chickens tending to maintain higher levels of maternal IgY in early life and exhibiting a more gradual and potentially more effective transition to active immunity (Cheema et al., 2003; Akyüz et al., 2022). To overcome the challenge posed by this critical window, vaccination schedules must be strategically planned to align with the decline of MDAs, a principle routinely applied in the scheduling of IBDV vaccination, and the gradual maturation of the chicken's adaptive immune system (Dey et al., 2019).

Maternal immunity can also be strategically boosted through targeted breeder vaccination programmes. Breeders are commonly vaccinated at key stages to maximize antibody levels during laying, ensuring effective transfer of immunity to their offspring. Point-of-lay pullets receive booster vaccinations to raise circulating antibodies, strengthening maternal immunity for chickens (Cserep, 2008; van Emous, 2022). By vaccinating hens against specific pathogens or key diseases, such as reovirus, avian encephalomyelitis virus (AEV), and chicken anaemia virus (CIAV), high levels of protective antibodies are produced and subsequently passed to the eggs (Calnek, 1998; Jones, 2000; Fatoba and Adeleke, 2019). This approach ensures that chickens hatch with passive immunity, providing early protection against these pathogens in their farm environment. However, while MDAs provide crucial early protection, they can also interfere with early vaccination by neutralizing live vaccine antigens, as described later in this chapter.

### 2.3 Role of microbiota in immune development

The microbiota plays a crucial role in the immune development of chickens, particularly in shaping their innate and adaptive immune systems during early life (Rodrigues et al., 2021; Kogut et al., 2020). Specifically, gut-associated lymphoid tissue (GALT), a key component of the avian mucosal immune system, undergoes significant development in response to microbial colonization of the gut, playing a pivotal role in immune maturation. This interaction is vital for ensuring robust immune responses, as it helps to mature the immune system and create a balanced defence against pathogens, while maintaining tolerance to harmless antigens (Bouwens and Savelkoul, 2019; Kallam and Sejian, 2021).

In natural environments, chickens acquire maternal microbiota through direct contact with the hen's skin, feathers, droppings, and nesting material, as well as through ingestion of environmental microbes. This exposure begins immediately after hatching and plays a significant role in seeding the chicken's microbiota. While most of the chicken's microbiota is acquired horizontally post-hatch, there is limited but biologically relevant vertical microbiota transfer via the egg, particularly through internal egg contamination and maternal cloacal microbiota (Lee et al., 2019; Wen et al., 2021). The diverse microbial community introduced in this way stimulates the immune system, leading to the early development of GALT, particularly in the caecal tonsils and Peyer's patches (Ding et al., 2017; Shehata et al., 2021; Gong et al., 2023). However, in commercial hatcheries, chickens hatch in highly sanitized environments, often in incubators devoid of natural microbiota. The absence of maternal microbiota in these settings may delay colonization of the gut and slow the activation and maturation of the immune system (Shterzer et al., 2023; Ding et al., 2024). This delay can increase the vulnerability of chickens to infections during the critical early weeks of life. Moreover, the lack of microbial exposure may reduce the diversity of the microbiota, which is associated with weaker immune responses and delayed development of immune tolerance (Shehata et al., 2021; Zenner et al., 2021; Rautenschlein and Schat, 2023).

Microbial products, such as lipopolysaccharides (LPS), peptidoglycans, and short-chain fatty acids (SCFAs), interact with PRRs, particularly TLRs, on intestinal epithelial and immune cells, initiating innate immune responses. This stimulation triggers cytokine production and antimicrobial peptide secretion, helping to protect the chicken against early infections (Broom, 2019; Liu et al., 2021; Smith et al., 2022). In addition, the microbiota supports immune homeostasis by promoting regulatory T cell (Treg) development, which is crucial for maintaining immune tolerance to non-pathogenic antigens and preventing excessive inflammation. The microbiota also plays a key role in maintaining the integrity of the intestinal barrier (Kogut, 2022; Dascalu et al., 2023).

Microbial exposure also drives the development of adaptive immunity. Dendritic cells in the gut capture microbial antigens and present them to naïve T and B lymphocytes in GALT, stimulating antigen-specific immune responses and immunological memory. For instance, commensal bacteria promote the production of secretory IgA, which plays a key role in mucosal immune defence by neutralizing pathogens and preventing their adhesion to epithelial cells (Kogut et al., 2020; Smith et al., 2022). In addition to stimulating local immune responses, the microbiota influences systemic immunity. Studies in germ-free chickens have shown delayed development of both cell-mediated immunity and antibody-mediated responses when microbial exposure is absent (Kogut et al., 2020). This highlights the microbiota's importance in priming the immune system for both localized and systemic pathogen defence, which should be considered when vaccinating chickens *in ovo* or at hatch.

In poultry, the role of commensal microbiota in vaccine efficacy remains largely unexplored. However, recent studies have specifically demonstrated its contribution to improving immune responses to anticoccidial vaccines, thereby enhancing protection against coccidiosis (Cai et al., 2023; Liu et al., 2024). Modulating the microbiota through probiotics or prebiotics has been shown to stimulate both innate and adaptive immunity, further supporting the effectiveness of vaccination strategies against *Eimeria* spp. (Lee et al., 2022).

## 2.4 Trained immunity

Innate immunity has long been considered incapable of generating immunological memory; however, recent studies have challenged this view, demonstrating that innate immunity can develop memory responses, known as 'trained immunity'. This process enables the innate immune system to exhibit an enhanced and more efficient response upon subsequent exposure to pathogens or immune stimulants, despite lacking antigen-specific memory (Netea et al., 2016; Gobel and Smith, 2022). Unlike adaptive immunity, which depends on antigen-specific T and B lymphocytes, trained immunity is primarily driven by innate immune cells such as monocytes, macrophages, and NK cells, which undergo metabolic and epigenetic reprogramming to mount a heightened immune response upon re-exposure. Neutrophils and dendritic cells may also contribute to this phenomenon by improving pathogen recognition and amplifying cytokine signalling (Bouwens and Savelkoul, 2019; Rautenschlein and Schat, 2023; Yoshimura et al., 2024). Although most research on trained immunity has been conducted in mammals, there is growing evidence that poultry can also develop innate immune memory (Rautenschlein and Schat, 2023; Yoshimura et al., 2024).

Several strategies have been explored to induce trained immunity in chickens, including vaccination, feed additives, and modulation of the gut microbiota (Verwoolde et al., 2020; Verwoolde et al., 2021; Yoshimura et al., 2024). Live attenuated vaccines, such as NDV and IBV vaccines, have been shown to enhance innate immune responses by upregulating TLRs and stimulating cytokine production (Shimizu et al., 2020; Yoshimura et al., 2023). Additionally, epigenetic modifications in innate immune cells, such as histone methylation and acetylation, are fundamental in sustaining long-term changes in immune responsiveness (Kogut, 2022). Certain dietary components, such as β-glucans, interact with innate immune receptors like dectin-1, triggering a more robust immune response upon subsequent pathogen encounters. These interactions lead to metabolic shifts, such as increased glycolysis and glutaminolysis, which sustain trained immunity by providing rapid energy for immune activation (Guo et al., 2003a; Cox et al., 2010; Bouwens and Savelkoul, 2019). The composition of the gut microbiota influences immune training. Some commensal bacteria produce metabolites, such as butyrate and propionate, that affect epigenetic regulation in immune cells, enhancing their ability to mount trained immune responses. The presence of beneficial microbes and prebiotic supplementation favours the differentiation and activation of monocytes and macrophages, promoting a more effective trained immunity response and improving resistance to infections (Kogut, 2022).

In summary, the avian immune system is a highly dynamic and complex network that begins developing during embryogenesis and continues to mature after hatching. In broilers, live attenuated vaccines are predominantly used as they induce rapid immune responses that are crucial within the species' short lifespan. The possibility of vaccinating broilers *in ovo* enables very early immunological priming and has proven particularly effective, even though the immune system is still maturing at this stage. Passive immunity via MDAs provides vital protection during the first days of life but may also interfere with certain live vaccines. The microbiota also plays a central role in shaping both innate and adaptive immunity during early life, and its influence on vaccine responses and immune resilience is increasingly recognized. Additionally, the emerging concept of trained immunity highlights the functional plasticity of the innate immune system and suggests new opportunities for early-life immunomodulation through vaccination, nutrition, and microbiota-targeted interventions. Advancing our understanding of immune ontogeny and its modulation is key to designing integrated health programmes that optimize vaccine timing, enhance maternal immunity, support microbiota development, and strengthen effective management practices to improve disease resistance and reduce reliance on antibiotics in broiler production.

## 3 Vaccination strategies in broiler production

Vaccination plays a central role in preventing infectious diseases in intensive broiler production. The high stocking density in commercial farms increases the risk of disease transmission, making structured vaccination programmes essential for maintaining flock health and ensuring optimal productivity. By stimulating the immune system, these programmes help reduce infections and antibiotic use, enhance broiler welfare, and contribute to a stable, profitable production system (Ravikumar et al., 2022; Islam and Rahman, 2023). Over the years, advancements in vaccine technology have significantly enhanced immunization efficiency and coverage. One major innovation is *in ovo* vaccination, which enables chickens to prime immunity before hatching, thereby reducing susceptibility to early-life infections and improving vaccine uniformity. Similarly, day-of-hatch vaccination techniques, such as spray, gel, and other delivery systems, allow efficient mass immunization, while minimizing labour costs and human errors (Abdul-Cader et al., 2018; Abd El-Ghany, 2025). Another notable advancement is the use of recombinant vaccines, which employ viral vectors or genetic engineering to stimulate immune responses in a timely, efficient, and safe manner, making them particularly suited for large-scale operations. Moreover, these vaccines provide broader protection against multiple virus strains, making them a valuable addition to vaccination strategies (Hein et al., 2021; Wang et al., 2024). Despite the availability of diverse vaccine types and delivery methods, only a subset is widely used in intensive broiler production worldwide.

Developing a vaccination programme for broilers requires careful consideration of multiple factors to ensure optimal disease prevention and maintain production efficiency. These include the local epidemiological context, the immune status of broilers at the time of vaccination, the vaccine characteristics, and the administration method. Company strategy and national regulations may also influence the overall approach. Main elements are summarized below and discussed in detail in the following sections:

- Disease risk profile varies depending on geographic location, farm management practices, and biosecurity measures. Understanding the distribution of diseases in a specific region is essential for determining which vaccines should be included in the programme (Marangon and Busani, 2006; Collett, 2020).
- Precise timing is critical due to the short rearing period of broilers, which limits the development of protective immunity. Vaccination schedules must be aligned with the kinetics of MDA's decline and the risk of early pathogen exposure (Cserep, 2009; de Wit and Montiel, 2022). Administering vaccines too early (when MDAs are still high or the immune system is immature) may compromise efficacy, while delaying vaccination increases the risk of pathogen exposure before protection is established (Rautenschlein, 2020a; de Wit and Montiel, 2022).
- Vaccine type should match production dynamics. Live and recombinant vaccines are routinely used. While some provide effective protection after a single dose, others require a prime-boost strategy (Sharma, 1999; Schat, 2014; Rautenschlein, 2020a). Inactivated vaccines are routinely used in breeders to ensure adequate transfer of MDAs. However, high MDA levels can interfere with early live vaccination in broilers (de Wit and Montiel, 2022; Ganapathy and Parthiban, 2023).
- Method of administration is crucial for ensuring effective and uniform immunity. Hatchery vaccination is increasingly preferred for its efficiency and uniformity. *In ovo* vaccination provides early protection, while subcutaneous vaccination at day-old is standard for some viral diseases. Spray vaccination is widely used for respiratory diseases in the first weeks of life (Abdul-Cader et al., 2018; Ganapathy and Parthiban, 2023).
- Storage and handling are essential to preserve vaccine efficacy. Live vaccines require cold chain management and must be reconstituted and administered promptly. Water-based vaccines need dechlorinated water to prevent inactivation, and spray vaccinations require proper droplet size calibration for respiratory uptake (Cargill, 1999; Cserep, 2008, 2009; Guerin et al., 2023).
- Vaccine interference is a frequent challenge in broilers due to their short lifespan and the need to administer multiple vaccines in a limited

timeframe. Immune suppression or antigenic competition can reduce vaccine efficacy. Well-designed schedules and field monitoring (e.g. serology) are key to managing interference risks (Marangon and Busani; 2006; de Wit and Montiel, 2022).

- Environmental and management factors affect vaccine responsiveness. Stress conditions, such as temperature extremes, transport, or feed transitions, can impair immune responses. Proper ventilation, optimal stocking densities, and well-balanced nutrition are essential to support effective vaccination (Surai and Fisinin, 2016a, 2016b; Abo-Al-Ela et al., 2021).
- Economic and logistical feasibility must be taken into account. While broiler production operates on tight profit margins, omitting essential vaccines can lead to economic losses from mortality, poor growth, and high medication costs (Landman, 2012; Lemiere, 2013; Jones et al., 2018).
- Post-vaccination monitoring is essential to assess programme effectiveness. Serological testing, flock performance metrics, and monitoring for clinical signs help identify protection gaps and guide adjustments in future protocols (Marangon and Busani, 2006; Collett, 2020).

### 3.1 Vaccination against major broiler diseases

Globally, the major disease threats requiring vaccination in broiler production include respiratory pathogens, such as NDV and IBV, immunosuppressive agents, such as MDV and IBDV, and protozoal infections, such as coccidiosis. These diseases are considered endemic in many regions with intensive broiler farming and can severely impair flock health and welfare, performance, and productivity (Jackwood, 2017; Reddy et al., 2017; Toro et al., 2009; Zaheer et al., 2022; Shafaati et al., 2024). Respiratory viruses spread rapidly through flocks, contributing to elevated morbidity and mortality; immunosuppressive agents weaken the host immune system, increasing vulnerability to secondary infections and undermining the effectiveness of other vaccines; and enteric pathogens, particularly *Eimeria* spp., compromise intestinal integrity, reduce nutrient absorption, and negatively impact feed conversion efficiency and growth (Cargill, 1999; Collett, 2020; Ganapathy and Parthiban, 2023). In specific epidemiological contexts, infectious laryngotracheitis (ILT), avian metapneumovirus (aMPV), and avian influenza (AI) may also assume significant epidemiological relevance and necessitate vaccination in broilers (Garcia and Zavala, 2019; Salles et al., 2023; Studniski et al., 2023). In contrast, certain vertically transmitted or early-life immunosuppressive diseases such as reovirus infections, AE, and CIA are primarily managed by vaccinating breeders (Calnek, 1998; Fatoba and Adeleke, 2019; Markis, 2022).

Among respiratory diseases, NDV remains a primary concern due to its high transmissibility and potential to cause systemic disease affecting the respiratory, digestive, and nervous systems, often culminating in high flock mortality (up to 100% in naïve flocks). Vaccination strategies involve a combination of live attenuated and recombinant vaccines, adapted to MDA levels and regional epidemiological pressure. These vaccines aim to prevent clinical disease, reduce mortality, and limit virus shedding and spreading. Under experimental conditions, both live and recombinant vaccines can confer nearly 100% protection against morbidity and mortality caused by highly virulent (velogenic) NDV strains. However, infection and shedding of the challenge virus are not fully controlled, particularly with live vaccines. In contrast, rHVT-ND vaccines achieve superior reduction of virus shedding, even in the presence of MDAs (Bell et al., 1991; Kapczynski and King, 2005; Tatár-Kis et al., 2020; Palya et al., 2021; Dewidar et al., 2022). Recombinant vaccines, administered *in ovo* or at hatch, act as priming immunization and can provide life-long protection within 3–5 weeks. Nevertheless, booster vaccination with live vaccines after 2–3 weeks may be implemented in high-risk conditions to ensure adequate flock-level protection. Depending on the local epidemiological context and the level of MDAs provided by the breeder flocks, broiler vaccination programmes against ND may include either a single administration of live attenuated asymptomatic/lentogenic strains via spray at the hatchery, or a sequential administration of lentogenic or mesogenic (not permitted in some countries) strains via drinking water within 21 days. In high-risk areas with elevated NDV challenge pressure or in production cycles of slow-growing broilers, an additional booster via drinking water may be required to ensure adequate and sustained protection (Dimitrov et al., 2017; Miller and Koch, 2020; Hu et al., 2022).

IBV is another economically significant respiratory pathogen, causing lesions in the respiratory and renal systems, with wide variations in clinical outcomes. Although overall mortality is generally low, it can still be significant, particularly in cases involving nephropathogenic strains, with reported rates as high as 20–30%. Respiratory infections are frequently exacerbated by secondary bacterial infections, resulting in increased antibiotic use and higher condemnation rates at slaughter. Vaccination against IBV is particularly challenging due to the virus's high antigenic variability and the continuous emergence of novel variants. Nonetheless, vaccination with homologous or heterologous strains significantly reduces disease severity and mortality in both experimental and field conditions, with full protection often achieved using homologous strains (Awad et al., 2016; Sultan et al., 2019; Kutle et al., 2020). Furthermore, omitting day-old vaccination can lead to small-scale outbreaks of the disease in broiler flocks (Toffan et al., 2013). Infected, vaccinated chickens also shed significantly less homologous virus for shorter durations, thereby reducing environmental contamination and transmission risk (de Wit et al.,

1998). Protection strategies rely on early immunization with live attenuated vaccines delivered at hatch, via coarse spray or gel, using strains representing prevalent serotypes or protectotypes (e.g. Mass vaccines). A second dose (including also variant strains, such as 793/B) is routinely administered within 21 days of age. Within this 'protectotype' strategy, a broader cross-protection is reached when vaccines are administered in sequence. However, due to the relatively short duration of immunity conferred by live vaccines, generally lasting only a few weeks, a further booster may be required to maintain protective immunity in slow-growing broiler production cycles (Jordan, 2017; Jackwood and de Wit, 2020; Abozeid, 2023).

Among immunosuppressive diseases, MD remains a major concern due to its oncogenic effects, manifesting as cutaneous or visceral tumours leading to condemnations at slaughter, and its immunosuppressive effect, especially early in life. Although vaccination does not prevent infection or virus shedding, it effectively protects against mortality and tumour development. Before vaccination was introduced in the early 1970s, MDV caused 30–60% mortality and up to 10% condemnation rates in broilers. The introduction of vaccines dramatically reduced these losses, making MD a largely controllable disease in modern poultry production. However, MDV still circulates in the field, and more virulent strains have emerged, likely due to the non-sterilizing nature of vaccines. This has led to the widespread use of stronger vaccines, such as CVI988/Rispens and bivalent/trivalent combinations (e.g. HVT + Rispens ± SB-1), especially in areas with high challenge pressure. Vaccination is routinely performed *in ovo* or at hatch using live attenuated or recombinant HVT vaccines, which generally do not require boosting. These vaccines induce early immunity, detectable approximately one week post-vaccination, and provide long-lasting protection against field strains (Reddy et al., 2017; Nair et al., 2020).

IBDV causes significant B cell immunosuppression by targeting the bursa of Fabricius, thereby compromising humoral immunity and responses to other co-administered vaccines. In unvaccinated flocks, the disease can cause near 100% morbidity, with mortality ranging from negligible to over 90%, particularly in cases involving very virulent (vv) IBDV strains. Beyond mortality, IBDV is responsible for subclinical infections that result in serious immunosuppression, adversely affecting health, welfare, and causing economic losses. Although vaccination remains an effective control measure, outbreaks may still occur due to the emergence of antigenic variants. Protection rates following IBDV vaccination in broilers range from 60% to 100%, depending on vaccine type, bird age, and levels of MDAs (Haddad et al., 1997; Rautenschlein et al., 2005; Chansiripornchai and Sasipreeyajan, 2009; Sultan et al., 2012; Gelb et al., 2016; Sedeik et al., 2019). Importantly, the immunosuppressive potential of live vaccine strains, particularly intermediate-plus/hot live vaccines, must also be considered when designing vaccination programmes. In breeder flocks, a combination of live and inactivated IBDV vaccines is used to ensure adequate MDA transfer to progeny. Immune complex and rHVT-IBD vaccines, administered *in ovo* or at hatch, are especially effective in the presence of high MDAs or where uniform early protection is required. They induce long-lasting immunity starting at 3–4 weeks of age without the need for booster doses. Live vaccines are typically administered via drinking water between 7 and 21 days of age, when MDA titres have declined sufficiently, as determined by serological monitoring, to allow optimal vaccine take. Due to the several variables involved (e.g. MDA levels, vaccine strain type, and host immune response), the onset of immunity following IBDV vaccination can vary considerably. Protective immunity may develop as early as 1–2 weeks when intermediate plus vaccines are used, whereas lower attenuation strains typically require 3–4 weeks to achieve full protective efficacy (Muller et al., 2012; Jackwood, 2017; Eterradossi and Saif, 2020).

Among enteric diseases, coccidiosis, caused by various *Eimeria* spp., remains the most costly protozoal disease in broiler production, with global losses estimated around USD 15 billion (Blake et al., 2020). The disease leads to intestinal epithelial damage, impaired nutrient absorption, poor feed conversion, and, in severe cases, high mortality. *E. tenella* in particular is associated with the highest clinical and sub-clinical disease burden. While chemoprophylaxis has historically been the primary control strategy, increasing drug resistance and regulatory restrictions have promoted the use of live coccidial vaccines. These vaccines effectively prevent clinical coccidiosis in broilers by reducing mortality, intestinal lesions, and pathogen load. Field trials have demonstrated equal or superior performance compared to anticoccidial drugs, with additional gut health benefits (Williams et al., 1999; Suo et al., 2006; Jansson et al., 2022). Vaccination also results in lower oocyst shedding and induces robust protective immunity (Gaghan et al., 2022). Coccidial vaccines contain oocysts from multiple *Eimeria* spp. and are administered orally (e.g. gel droplets or coarse spray) at hatch or within the first few days of life at the farm. They rely on controlled oocyst cycling in the litter rather than booster doses. Following initial exposure, immunity begins to develop after the first oocyst cycle (approximately 7–10 days) and is generally complete by 21–28 days of age, after two to three cycles (Sharman et al., 2010; Cervantes et al., 2020; Hauck and Macklin, 2023).

Other diseases may require targeted vaccination in broilers under specific epidemiological conditions, notably ILTV, aMPV, and AI. ILTV, which causes severe upper respiratory disease and increased mortality (ranging from 5% to 20%), particularly in slow-growing broilers, is managed using live attenuated vaccines and, increasingly, recombinant vector vaccines. Due to the virus's capacity to establish latency and reactivate under stress, effective control relies on strict biosecurity and timely vaccination, especially in endemic areas

(Garcia and Zavala, 2019). aMPV, another important respiratory pathogen, may require vaccination in broilers under specific epidemiological conditions, such as high virus circulation, proximity to turkey farms, and multi-age production settings. The virus causes upper respiratory tract infections, typically associated with low mortality (around 2%), but can predispose broilers to secondary bacterial infections, leading to airsacculitis and reduced performance (Rautenschlein, 2020b). Like ILTV, aMPV is controlled through live vaccines typically administered by spray or drinking water at day-old and/or during the first week of life. Vaccination may be particularly relevant in slow-growing broiler production systems or multi-age environments (Salles et al., 2023). Highly pathogenic avian influenza (HPAI) represents a global concern due to its zoonotic potential and devastating economic impact. In endemic regions, inactivated and recombinant vector vaccines (e.g. HVT-AI) are used to limit clinical disease and viral shedding. However, vaccination must be carefully coordinated with surveillance systems, as it may hinder outbreak detection, interfere with serological monitoring, and promote antigenic drift under immune pressure. For this reason, AIV vaccination programmes are regulated under strict national and international frameworks (Studniski et al., 2023).

Finally, certain pathogens, such as avian reovirus, CIAV, and AEV, are managed through immunization of breeders rather than direct broiler vaccination, which provides passive protection during the critical early post-hatch period. Reovirus causes viral arthritis and growth retardation; CIAV induces anaemia and immunosuppression; AEV is neurotropic and can result in neurologic signs and mortality. In all cases, vaccination of breeder flocks with live and/or inactivated vaccines is essential, as early post-hatch vaccination is either ineffective or contraindicated due to immune immaturity and interference from MDAs (Cserep 2008, 2009; Ganapathy and Parthiban, 2023).

The design and timing of vaccination programmes must also consider the specific production system adopted, particularly the immunological and management differences between fast- and slow-growing broilers, as discussed in the following section. An illustrative example of a broiler vaccination programme is provided in Fig. 2, summarizing the key components such as timing, vaccine type, route of administration, and the corresponding target diseases throughout the production cycle.

### 3.2 Vaccination in fast- vs. slow-growing broilers

Vaccination strategies in broiler production vary between fast-growing and slow-growing broilers due to differences in growth rate, metabolic demands, immune system development, disease risks, and management practices (Bessei, 2006; Gul et al., 2022; Rautenschlein and Schat, 2023). These factors influence how vaccines are applied and the extent to which they contribute to the overall flock

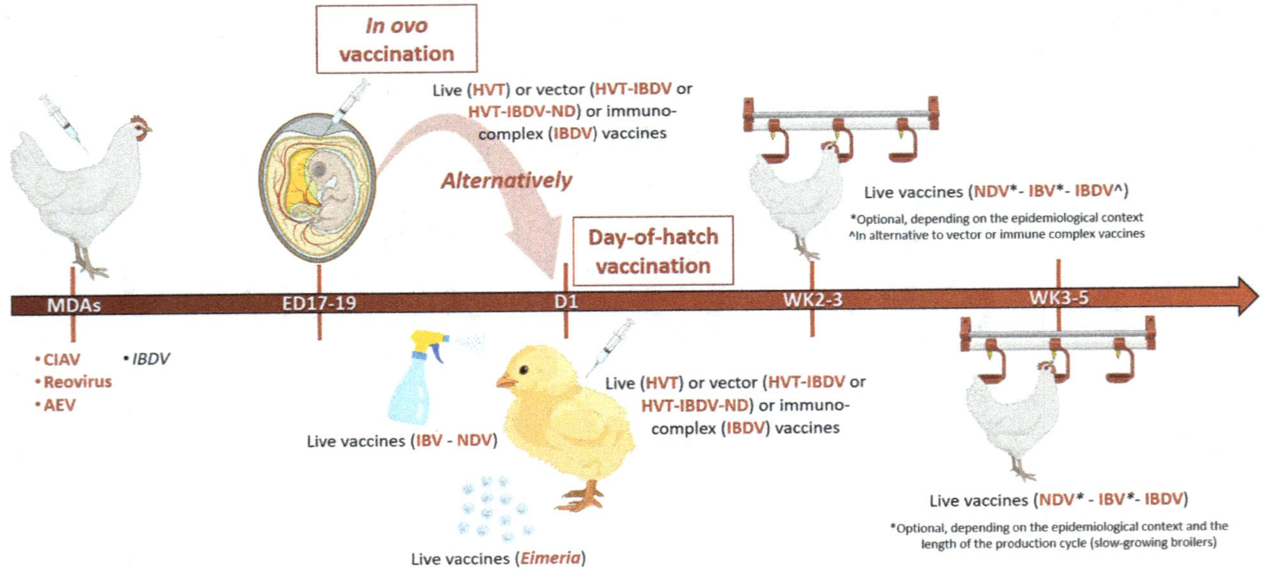


**Figure 2** Schematic example of a broiler vaccination program, illustrating the timing, vaccine types, delivery methods, and targeted diseases throughout the production cycle.

health and welfare. The short life span of fast-growing broilers creates unique challenges, requiring a highly targeted and efficient vaccination programme that provides rapid protection. In contrast, slow-growing broilers benefit from longer-lasting immunity and greater flexibility for booster vaccinations, which can enhance their resistance to a broader range of pathogens (Swaggerty et al., 2009; Kjærup et al., 2017; Tellis et al., 2024). Understanding these differences is crucial for optimizing vaccination strategies, improving broiler health and welfare, and ensuring sustainable poultry production.

Fast-growing broilers are genetically selected for rapid weight gain and efficient feed conversion, reaching market weight in a few weeks. However, this rapid growth challenges their immune system, as immune organ development lags behind muscular and skeletal growth. The immune system of broilers matures more slowly than their skeletal and muscular structures, making them particularly vulnerable to infections during their early life stages (Martínez et al., 2021; Song et al., 2021). Since these broilers live for only 35–50 days, there is limited time for natural immunity to develop, meaning that vaccination must be precisely timed to ensure effective protection. Additionally, the short production cycle limits booster effectiveness and the presence of MDAs can interfere with live vaccines, requiring precise timing of immunization (de Wit, 2000; Block et al., 2007). Many diseases affecting broilers, such as ND, IB, and IBD, are highly contagious and can spread rapidly in intensive production settings (Glisson, 1998; Villegas, 1998). Without vaccination, disease outbreaks can cause high mortality rates, severe production losses, and increased reliance on antibiotics to manage secondary bacterial infections (Alkie and Rautenschlein, 2016; Bhuiyan et al., 2021; Hu et al., 2022). Given the high density in which these broilers are raised, vaccines are crucial for preventing pathogen circulation and mitigating outbreak risk. The advantages of vaccination in fast-growing broilers are well established. Besides reducing mortality, vaccination improves feed efficiency and meat quality. For example, ND+IB and IBD vaccination using live vaccines at different ages showed different growth performance, with cumulative feed conversion ratio (FCR) ranging from 1.84 to 1.98; improvement in carcass characteristics, including breast meat yield (556.33 g to 569.00 g); and enhanced meat quality, such as reduced drip loss (5.68% to 7.71%) and improved water-holding capacity (Chung et al., 2020). In another study (Wegner et al., 2023), differences in carcass yield (77.1% vs. 74.7%) and breast meat percentage (31.7% vs. 29.5%) were observed between two rHVT-IBD-based vaccination programmes administered *in ovo*. Fast-growing broilers typically receive most of their vaccines at the hatchery through *in ovo* or at-hatch injection, spray, gel and other administration methods depending on the vaccine type. These include live attenuated, immune complex and recombinant vaccines targeting pathogens, such as MDV, IBDV, NDV, IBV and *Eimeria* spp. *In ovo* and at-hatch vaccination guarantees early immune stimulation, while reducing handling stress and ensuring uniform coverage. During the first weeks of life, live vaccines may be administered via drinking water or aerosol spray in the poultry house to reinforce immunity (Alkie and Rautenschlein, 2016; Reddy et al., 2017; Bhuiyan et al., 2021; Hu et al., 2022; Zaheer et al., 2022).

In contrast, slow-growing broilers, such as those raised in free-range, organic, or specialty systems, have a longer production cycle, typically being slaughtered between 8 and 12 weeks, and in some systems, even beyond. This extended lifespan means these broilers encounter different disease risks and require a different vaccination approach. One of the key differences in slow-growing broilers is that their immune system has more time to mature, resulting in stronger and more sustained vaccine-induced responses (Hofmann et al., 2020; Khazali et al., 2025). However, they are also more exposed to environmental pathogens, particularly in outdoor systems where contact with wild birds, contaminated soil, and water sources increases the risk of diseases such as ND, AI, *Salmonella*, and coccidiosis (Koch and Elbers, 2006; Wierup et al., 2017; McMullin, 2022; Munoz-Gomez et al., 2024). This makes vaccination critically important for minimizing disease outbreaks and maintaining flock health. Since slow-growing broilers live longer, vaccines administered in early life have sufficient time to develop full protective immunity before exposure to field challenges. This extended window allows for the implementation of more complex immunization schedules, including multiple booster doses, which are often not feasible in fast-growing broilers. Booster vaccinations, particularly against NDV and IBV, are commonly applied at 3–5 weeks of age and repeated if the production cycle extends beyond 8–9 weeks. These revaccinations help maintain adequate immunity levels throughout the extended fattening period and reduce the risk of clinical disease outbreaks (Gallili and Ben-Nathan, 1998; Cook et al., 1999). However, multiple booster doses add labour costs and logistical challenges, particularly in outdoor settings where broilers' handling is more complex and stressful. Additionally, unlike fast-growing, slow-growing broilers may require vaccines against bacterial (e.g. *E. coli*, *Pasteurella*) and parasitic infections due to prolonged exposure to environmental challenges and the higher risk of contact with contaminated substrates (McMullin, 2022; Hauck and Macklin, 2023). Many organic and premium-label production systems also restrict or prohibit the use of antibiotics, making preventive vaccination strategies essential to maintain health, welfare, and performance without relying on therapeutic interventions (Smith et al., 2020; Slegers et al., 2024).

### *3.3 Vaccination in on-farm hatching systems*

In recent years, on-farm hatching has gained attention as an alternative to traditional hatchery practices, aiming to enhance chicken welfare, immune system development, and early growth performance. In this system, fertilized eggs are transferred to the farm at 18 days of incubation, allowing chickens to hatch directly in their production environment (de Jong et al., 2020; EFSA, 2023). This method eliminates the stress associated with transport from hatcheries, particularly the dehydration and handling stress that occur when day-old chickens are moved to rearing facilities (Giersberg et al., 2021; Jessen et al., 2021a).

As a result, vaccination strategies must be adapted to align with the physiological and environmental differences related to on-farm hatching, ensuring optimal immune protection from the earliest stages. In conventional hatcheries, vaccination schedules are well-established, benefiting from automated methods for both *in ovo* and post-hatch immunization. Traditional hatcheries operate under controlled conditions, where *in ovo* vaccination is widely used to protect against diseases such as MD and IBD, ensuring early immune stimulation before hatching. Additionally, post-hatch vaccinations for diseases such as ND, IB, and coccidiosis are typically administered through automated spray systems, providing precise dosing with minimal handling stress (Williams, 2023; Bublot, 2023; Abd El-Ghany, 2025). In contrast, on-farm hatching requires a different approach to immunization that combines *in ovo* vaccination with targeted post-hatch interventions (Jessen et al., 2021b; EFSA, 2023). Therefore, a key component of this strategy is *in ovo* vaccination, which can take place before eggs are transferred to the farm. Vaccines against MD, ND, and IBD can be injected into the embryo, ensuring early protection and reducing the need for additional handling after hatch. However, *in ovo* vaccination does not cover all necessary immunizations, and certain vaccines, such as those for IB and coccidiosis, must still be given within the first few days of life. On-farm vaccines can be administered via commonly used mass delivery methods, such as spray or drinking water application, which minimize handling stress (Abdul-Cader et al., 2018; Bublot, 2023; Guerin et al., 2023). However, the effectiveness of these methods depends on uniform vaccine uptake, which can be influenced by chicken behaviour, environmental conditions (e.g. temperature and humidity), and water quality. Ensuring consistent and adequate vaccine intake is essential, as poor distribution or uneven consumption can lead to gaps in immunity (de Wit and Montiel, 2022; Guerin et al., 2023).

Additionally, early access to feed plays a crucial role in supporting immune function and vaccine response. Research suggests that early feeding enhances metabolic balance, supports gut microbiota development, and improves nutrient absorption, all of which contribute to a stronger immune system (Madej et al., 2024; Akram et al., 2025). This accelerated immune maturation may improve vaccine efficacy, particularly for oral vaccines targeting the GALT, which plays a fundamental role in mucosal immunity. Furthermore, energy availability from early feeding supports antibody production, potentially leading to a more robust immune response following vaccination (Simon et al., 2014; Kogut et al., 2020). However, delays in vaccination can create a critical window of susceptibility to common poultry diseases, making precise scheduling essential to ensure timely protection.

On-farm vaccination also introduces biosecurity considerations. Unlike hatchery vaccination, which takes place in highly controlled environments with strict sanitation protocols, on-farm hatching exposes chickens to farm-specific microbial populations from the moment they hatch. While this early exposure may contribute to microbiota diversity and immune system priming, it also increases the risk of disease transmission if proper biosecurity measures are not maintained (Guilloteau et al., 2024; Kustra et al., 2024). Rigorous hygiene protocols and farm-level disease prevention plans are essential to mitigate these risks. Despite these challenges, studies suggest that on-farm hatched chickens may develop stronger immunity and exhibit lower mortality rates compared to conventionally hatched and transported chickens, potentially decreasing the need for antibiotic interventions later in life (de Jong et al., 2019; de Jong et al., 2020; Jerab et al., 2023).

In summary, vaccination is a fundamental pillar in broiler health management, especially in the context of high-density, intensive production systems. The design of effective immunization programmes must consider not only the major pathogens affecting broilers but also the specific challenges posed by different production models. Fast-growing broilers require early, efficient, and minimally invasive vaccine delivery strategies, often relying on hatchery-based *in ovo* or day-of-hatch administration, due to their short lifespan and high production turnover. In contrast, slow-growing broilers, typically reared in alternative systems, face longer exposure periods and broader pathogen spectra, thus benefiting from more flexible schedules, including booster doses and expanded antigenic coverage. Moreover, innovations such as recombinant vaccines and on-farm hatching systems are reshaping vaccination approaches. While recombinant and immune complex vaccines have improved protection in the presence of MDAs, on-farm hatching introduces both new opportunities and risks. This approach enhances early immune maturation and welfare but requires tailored post-hatch vaccination protocols and strict farm-level biosecurity. Across all systems, the success of vaccination depends not only on the choice of antigens or delivery methods but on precise timing, environmental and management conditions, and continuous monitoring of vaccine performance. Ultimately, vaccination strategies must be part of an integrated health management plan that balances efficacy, logistics, welfare, and sustainability, ensuring that broilers are adequately protected throughout their production cycle, regardless of the system in which they are raised.

## 4 Types of vaccines used in broilers

A wide range of vaccines is available for the prevention and control of viral, bacterial, and parasitic diseases in poultry. However, in broiler production, only a limited number is routinely used due to several factors, including the broilers' short lifespan, the prevalence and impact of specific diseases, and both practical and economic considerations. When selecting vaccines for broilers, it is crucial to consider their specific immune response dynamics, which differ from those of long-lived poultry, such as layers or breeders (Koenen et al., 2002). Moreover, practical aspects such as ease of administration, cost-effectiveness, and the ability to vaccinate large flocks quickly and uniformly are also central to vaccine selection. Additionally, vaccine availability on the market and regulatory approvals further influence the choice of products to be used (Bublot, 2023; Schrader and Smith, 2023; Werner et al., 2023). Vaccines can be broadly categorized into two main categories: conventional vaccines, including live attenuated and inactivated vaccines, and new-generation vaccines, which include vector, subunit, and nucleic acid vaccines. In broiler production, live vaccines, whether conventional live attenuated or those developed using new technologies, are the most commonly used. Their widespread application is due to their ability to induce rapid and robust immunity, as well as their suitability for mass vaccination using automated systems, such as *in ovo* administration. These methods not only enhance vaccine coverage but also reduce labour costs, making live vaccines a highly practical and cost-efficient option for broiler farms (Bublot, 2023; Ganapathy and Parthiban, 2023; Williams, 2023).

### 4.1 Conventional vaccines

Conventional poultry vaccines are categorized into live attenuated and inactivated vaccines, both of which leverage distinct immunological mechanisms to confer protection against infectious agents.

Live attenuated vaccines are derived from strains that are either naturally non- or low-pathogenic or have been artificially attenuated through controlled methods (e.g. serial passages in tissue cultures, embryonated eggs, selective media), depending on whether the agent is viral, bacterial, or parasitic. These vaccines retain the ability to replicate in the host, thereby closely mimicking natural infection and eliciting robust mucosal and systemic (innate and adaptive) immune responses. Their key advantages include the early onset of protection (within 2–3 weeks) and the suitability for mass administration (e.g. spray, drinking water), making them a practical and cost-effective solution for large-scale broiler operations (Cserep, 2008, 2009; de Wit and Montiel, 2022). While live vaccines are typically associated with early protection, their immunity is generally short-lived following primary vaccination, especially for NDV and

IBV (Corbanie et al., 2008; de Wit et al., 2010), often requiring booster doses or combination strategies with inactivated vaccines, especially in long-lived birds. In the case of IBDV, live attenuated vaccines are administered following the decline of MDAs and may also require boosting depending on field pressure and vaccine strain virulence (Block et al., 2007; Ramon et al., 2022). By contrast, live MD vaccines are administered individually *in ovo* or on the day of hatch via subcutaneous injection and provide durable, long-term immunity (Sharma, 1999; Bublot and Sharma, 2004). Similarly, live vaccines containing *Eimeria* spp. do not require boosters, as protective immunity is achieved through controlled oocyst cycling in the litter (Chapman et al., 2002; Albanese et al., 2018). In broilers, live vaccines are widely used against key pathogens. NDV vaccines are available in asymptomatic/lentogenic forms (e.g. LaSota, Hitchner B1, C2 or V4, NDW 1–2), which are safer but may require multiple doses or booster vaccinations, and in mesogenic strains (e.g. Roakin, Komarov or Mukteswar), which induce stronger immunity, but carry an increased risk of post-vaccinal reactions (Senne et al., 2004; Afonso and Miller, 2013; Dimitrov et al., 2017). Live IBV vaccines are commonly based on the Massachusetts strain, which often requires combination with other serotypes (e.g. D274, 793B, QX, Var-2, or other variants) to achieve broad cross-protection due to the high antigenic variability among circulating strains (Bru et al., 2017; Smialek et al., 2017; Keep et al., 2020a). The most commonly used MDV vaccines can be based on the CVI988 (or Rispens) strain, a serotype 1 vaccine that has demonstrated superior protection compared to HVT (serotype 3; FC126 strain most widely used) and MDV 2 (serotype 2) strains. CVI988 is effective against very virulent (vvMDV) and very virulent plus (vv+MDV) strains, which are prevalent in the field (Bublot and Sharma, 2004; Nair, 2004). For IBDV, live vaccines are based on serotype 1 and are categorized into mild, intermediate, and intermediate-plus/hot strains depending on their attenuation level. The main challenge is balancing safety with immunogenicity, as more attenuated IBDV strains are safer but may not confer adequate protection, while more virulent ones offer stronger protection but increase the risk of immunosuppression (Muller et al., 2003; Muller et al., 2012; Smialek et al., 2024). Live coccidiosis vaccines, containing attenuated *Eimeria* spp. (e.g. *E. acervulina*, *E. maxima*, and *E. tenella*), are also widely used to stimulate immunity against this protozoal disease. These vaccines, administered at the hatchery, stimulate protective intestinal immunity through ingestion and replication in the gut (Williams, 2002; Dasgupta and Lee, 2000; Liu et al., 2023b). Despite their advantages, live vaccines have certain limitations, including interference by MDAs or by co-administration of multiple vaccines, or adverse reactions, as discussed later.

Immune complex vaccines consist of a virus bound to specific antibodies, forming immune complexes that are less susceptible to early neutralization by MDAs. This technology has been particularly successful for IBDV, where conventional live vaccines are often ineffective in the presence of strong maternal immunity. Typically administered both *in ovo* or at hatch, the immune complex protects the virus during early life and allows for a delayed release as MDAs wane, ensuring timely and effective immune stimulation even in chickens with high passive immunity (Whitfill et al., 1995; Jeurissen et al., 1998; Ramon et al., 2022).

Inactivated vaccines are produced by rendering the pathogen non-viable through chemical agents (e.g. formaldehyde, β-propiolactone, binary ethyleneimine) or heat/radiation inactivation, while preserving its antigenic structure to maintain immunogenicity. As these vaccines do not replicate in the host, they require the inclusion of adjuvants, such as oil-in-water emulsions or alum-based formulations, to potentiate and prolong the immune response (Cserep, 2008, 2009; de Wit and Montiel, 2022). Recent research is investigating novel adjuvants based on vegetable, fish and animal oils, which are showing promising immunostimulatory potential and may offer improved safety and efficacy profiles (Abdelaziz et al., 2024; Bodman-Harris et al., 2024). Inactivated vaccines offer several advantages, including the ability to induce a strong, long-lasting humoral immune response without the risk of reversion to virulence. However, their application in broilers is limited by several factors. These include high production and formulation costs, the necessity of adjuvants to generate a robust immune response, and the requirement for individual administration via parenteral injection, which is labour-intensive and logistically challenging for large-scale production (Collett, 2020; de Wit and Montiel, 2022). Another important limitation is their reduced efficacy in very young chickens, whose adaptive immune systems are not yet fully developed. As a result, inactivated vaccines may fail to induce a sufficient antibody response if administered too early (Mast and Goddeeris, 1999; Wlazlak et al., 2023). Furthermore, because they are administered systemically and not at mucosal surfaces, they do not elicit strong mucosal immunity, reducing their effectiveness against, for instance, respiratory pathogens, which enter primarily via the respiratory tract (Rautenschlein and Schat, 2023; Akter et al., 2024). To enhance protection, inactivated vaccines are often used in prime-boost strategies, where an initial live vaccine primes the immune system, and a subsequent inactivated vaccine boosts the response, achieving broader and more durable immunity (de Wit and Montiel, 2022; Rautenschlein and Schat, 2023).

Several inactivated vaccines are commercially available for poultry. However, they are primarily administered in breeder flocks as booster immunizations and to ensure the transfer of high levels of MDAs to their progeny (Cserep, 2008, 2009; Ganapathy and Parthiban, 2023). With the exception of CIA, for which only live vaccines are currently available and no inactivated vaccines are commercially available (Fatoba and Adeleke, 2019), breeders are routinely vaccinated with live/inactivated vaccines against reovirus (Wood et al., 1986; French, 2022) and AE (Calnek, 1998) to ensure efficient MDA transfer to broilers.

## 4.2 New-generation vaccines

The advent of new generation vaccines has revolutionized disease control in poultry production by addressing many of the limitations associated with conventional live and inactivated vaccines. These innovative platforms, which rely on recombinant DNA technology and reverse genetics to genetically modify viruses or bacteria, are designed to enhance antigen presentation while reducing pathogenicity, thereby improving efficacy, safety, and delivery efficiency (Bublot, 2023; Ravikumar et al., 2022). Among the various technologies developed, vector-based vaccines have been the most successfully implemented in broiler production for the prevention of major infectious diseases. In contrast, other new generation vaccines, such as subunit and nucleic acid (DNA and mRNA) vaccines, are still largely in the experimental or early commercialization phases and have not yet been widely adopted in broiler production (Bublot, 2023; Wang et al., 2024).

Vector-based vaccines employ non-pathogenic viruses or bacteria as vectors to deliver genes encoding protective antigens from target pathogens. An example of how recombinant HVT-vectored vaccines are constructed is provided in Fig. 3, showing rHVT vaccines expressing protective antigens from IBDV, NDV, or a combination of both. This allows the host to express the antigen *in vivo* and elicit both humoral and cell-mediated immune responses (Romanutti et al., 2020; Hein et al., 2021). A major advantage of vector vaccines is their ability to avoid risks associated with live pathogenic microorganisms and to circumvent interference of MDAs, which often limits the effectiveness of conventional live vaccines in young chickens. Additionally, they eliminate the risk of reversion to virulence, a concern with some live attenuated vaccines, and are considered highly safe and stable, making them ideal for large-scale broiler operations. They also reduce the need for multiple booster doses and prevent vaccine virus shedding, minimizing the risk of environmental contamination or pathogen reversion (Williams and Hopkins, 2011; de Wit and Montiel, 2022). By leveraging modern biotechnological techniques, vector vaccines provide stronger and longer-lasting protection and are well-suited to mass vaccination strategies, such as *in ovo* and hatchery-based immunization (i.e. subcutaneous route in day-old chickens), which are essential for achieving early and uniform flock protection. These platforms offer direct protection against infections and serve as versatile tools for multivalent vaccine development, potentially targeting multiple pathogens through a single application (Peebles, 2018; Francis, 2021; Li et al., 2024). Their use therefore contributes to improved biosecurity and animal welfare by minimizing post-hatch handling.

Recombinant vector vaccines have been developed against several viral pathogens affecting broilers, including IBDV and NDV, for which they have demonstrated high efficacy (Ravikumar et al., 2022; Abdelaziz et al., 2024).

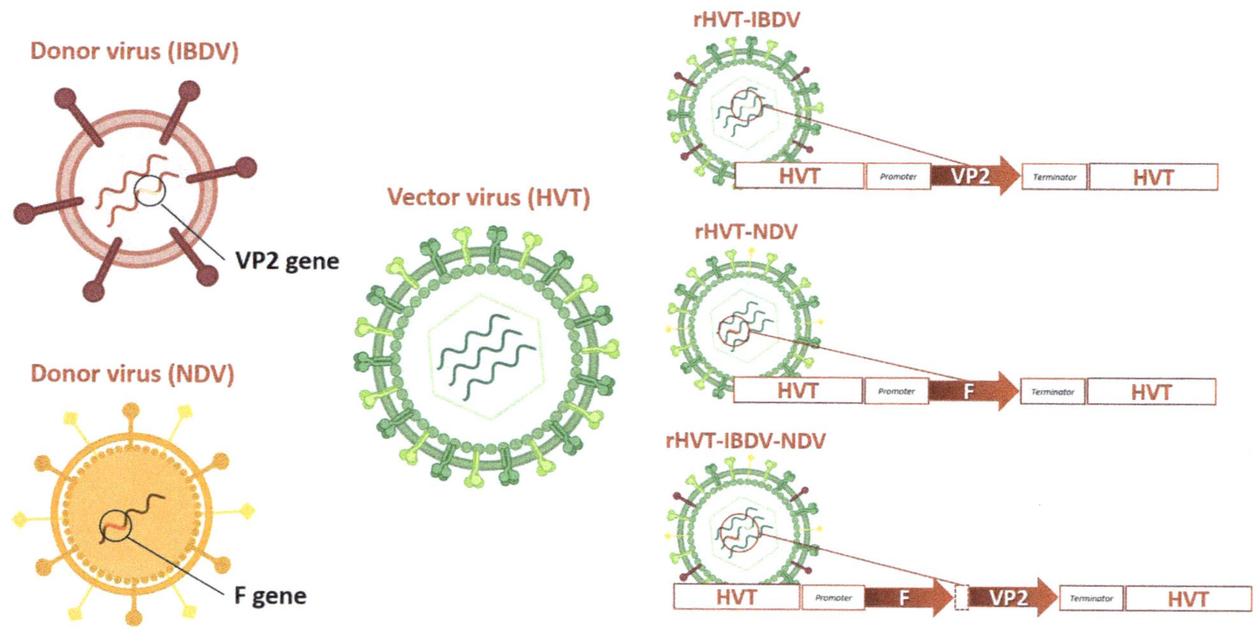


**Figure 3** Schematic example of recombinant herpesvirus of turkey (rHVT) vector vaccines expressing antigens from Infectious bursal disease virus (IBDV) and/or Newcastle disease virus (NDV).

Among the most widely used platforms are recombinant herpesvirus of turkey (rHVT), fowlpox virus (FPV), and NDV, each serving as a delivery vehicle for protective antigens. However, *in ovo* administration is currently applicable primarily to vaccines using the rHVT vector, while other vectors like FPV are generally used post-hatch, e.g. in breeder vaccination against several diseases including AE (Sondermeijer et al., 1993; Nakaya et al., 2001; Skinner et al., 2005; Bublot, 2023).

The use of rHVT as a vaccine backbone offers several immunological and practical advantages. As a cell-associated virus, HVT establishes persistent infection, ensuring long-term antigen expression and early replication. rHVT stimulates both humoral and cell-mediated immunity with a single administration, providing broad protection even in the presence of MDAs. Additionally, rHVT vaccines administered *in ovo* do not induce immunological tolerance (Tsukamoto et al., 2002; Zhang and Sharma, 2003; Iqbal, 2012; Dimitrov et al., 2021). In contrast, mucosal immunity is not stimulated and the onset of immunity is slower than that induced by conventional vaccines, with protective effects becoming evident between the second and fourth week of age (Ingrao et al., 2017; Ingrao et al., 2018; Dunn et al., 2019; Ravikumar et al., 2022). For example, rHVT-IBD vaccines express the VP2 gene of IBDV, the primary antigen responsible for virus-neutralizing antibodies. Unlike live IBD vaccines, they do not replicate in the bursa and are effective even in the presence of MDAs. This allows for early administration, including *in ovo*, without the risk of immunosuppression or bursal damage (Tsukamoto et al., 2002; Bublot et al., 2007; Perozo et al., 2009). Similarly, rHVT-ND vaccines express the F (fusion) protein gene of NDV, the main immunogen responsible for inducing protective antibodies. These vaccines provide robust and long-lasting immunity, even in the presence of MDAs, which often interfere with conventional live ND vaccines. Their use in *in ovo* or day-old vaccination protocols enables early protection; however, live attenuated vaccines are still needed post-hatch to stimulate mucosal immunity (Rauw et al., 2010; Palya et al., 2012; Rauw et al., 2014).

More recently, bivalent rHVT vector vaccines have been developed to provide simultaneous protection against two pathogens. Examples include rHVT-ND-IBD (Ferreira et al., 2020) and vHVT-IBD-AI and vHVT-ND-AI (Criado et al., 2023), combining protection against multiple diseases in a single dose and optimizing hatchery vaccination strategies. Recombinant vectored vaccines against IBV, CIAV, and reovirus are not yet available, although CIAV vector vaccines remain in the experimental stage (Chellappa et al., 2022). Beyond viral diseases, vector vaccines are also being explored for bacterial and parasitic pathogens, such as *Salmonella* spp., *Escherichia coli*, *Clostridium perfringens*, and *Mycoplasma gallisepticum*, as well as *Eimeria* spp.; however, their use in broilers remains uncommon (Deslauriers and Boulianne, 2023; Hauck and Macklin, 2023).

In summary, understanding vaccines' immunological properties and practical applications is essential for optimizing flock health. Conventional vaccines have long formed the foundation of broiler immunization programmes. Live vaccines are widely used for their ability to induce rapid, robust immune responses and are suitable for mass administration, but require precise timing and are sensitive to MDA interference. The emergence of new generation vaccines, particularly recombinant vector vaccines, has expanded the immunological toolbox by offering safer, more targeted, and MDA-resistant alternatives. These platforms enable early immunization (even *in ovo*), reduce handling stress, and support multivalent protection, crucial advantages in both intensive and alternative production systems. However, they may not yet fully replace conventional vaccines due to limitations in mucosal stimulation and slower onset of protection, often requiring combination strategies.

## 5 Vaccine delivery methods

Over the years, various vaccine delivery methods have been developed to meet the evolving needs of poultry production, aiming to ensure efficient and uniform immunization across large flocks in a cost- and labour-efficient manner. The selection of a delivery system depends not only on the nature of the vaccine formulation but also on the specific pathogen targeted, the desired immune response, and the stage of production at which vaccination occurs. The effectiveness of a vaccine is closely linked to the ability to deliver the full dose to all broilers in a flock, as well as to the precision, consistency, and overall suitability of the method of administration itself (Cserep, 2008, 2009; Collett, 2020; Cargill, 2021). For this reason, the optimization of delivery strategies remains a critical component of vaccination planning, particularly in broilers, where the vaccination method must balance immunological efficacy with practical feasibility. Vaccine administration can be broadly categorized into mass application methods, including spray and drinking water vaccination, and individual application methods, such as *in ovo* vaccination, eye-drop or intranasal delivery, and parenteral injection.

Mass vaccination techniques are widely used in broiler production, primarily due to their ability to deliver vaccines to large numbers of broilers quickly and uniformly. These approaches are particularly effective for live attenuated vaccines, which can replicate in the host and elicit both mucosal and systemic immune responses. These methods align well with the operational needs of high-density production systems, where speed, uniformity, and labour efficiency are essential. Furthermore, mass application is associated with reduced stress on the broilers, as it avoids the need for individual capture and restraint.

In contrast, individual vaccination involves administering the vaccine directly to each chicken. While more precise, this method is labour-intensive and time-consuming, making it impractical for large-scale broiler flocks. It is, therefore, predominantly employed in breeder and layer systems, where the higher economic value and longer productive lifespan of the chickens justify the increased resource investment. Individual vaccination is also indispensable when using inactivated or adjuvanted vaccines, which cannot be delivered effectively by mass methods and require parenteral administration to stimulate a strong, durable immune response (de Wit and Montiel, 2022; Rautenschlein and Schat, 2023; Guerin et al., 2023; Bodman-Harris et al., 2024).

## 5.1 Mass application methods

One of the most widely adopted methods in broiler vaccination is spray vaccination, particularly for the administration of live attenuated vaccines against respiratory agents, such as IBV and NDV, which typically induce a protective immune response within 2–3 weeks post-vaccination (Corbanie et al., 2008; Landman et al., 2015; Roh et al., 2015; Al-Rasheed et al., 2023). This method can be applied both at the hatchery and in the poultry house, depending on the immunization strategy and disease challenge. In both hatchery and farm settings, the methodology involves dispersing droplets of a specific size over the broilers, ensuring they are inhaled or settle on the mucous membranes.

Spray vaccination is highly sensitive to droplet size, sprayer calibration and flock positioning (Degefa et al., 2004; de Wit, 2013). In particular, droplet size is critical (ranging from 70 μm to 150 μm): if too fine (less than 20 μm), droplets may be exhaled or penetrate too deeply into the lower respiratory tract, while droplets that are too large may not reach the intended mucosal sites and instead settle on feathers or the environment, reducing efficacy (Cargill, 1999; Guerin et al., 2023). At the hatchery, spray vaccination is performed on day-old broilers, delivering vaccine-laden droplets over the chickens in a controlled environment (e.g. spray cabinets). This allows for early mucosal immune stimulation with minimal handling, ensuring rapid and uniform coverage. In this setting, coarse spray is preferred to facilitate deposition on the upper respiratory tract, where it stimulates local immune tissues such as the Harderian gland and respiratory-associated lymphoid tissues (Corbanie et al., 2006; Tell et al., 2012; Espejo et al., 2024). In the poultry house, spray vaccination is commonly used as a booster or follow-up dose, particularly in flocks at risk of secondary respiratory challenges or in regions with endemic circulation of viruses. Field application allows for revaccination under real production conditions, but it requires careful attention to ensure effective coverage and uptake. Environmental conditions such as temperature, humidity, and ventilation must also be managed during spray application, as they directly influence droplet behaviour and vaccine absorption (de Wit et al., 2010; de Wit, 2013). To ensure vaccine uptake, it is common practice to expose broilers to the spray for 5–10 seconds and stop ventilation fans for approximately 15 minutes following vaccination (Cserep, 2008, 2009; Guerin et al., 2023).

Spray vaccination is also widely used to provide early protection against coccidiosis, though through a different mechanism of action, in which direct ingestion of the vaccine is critical. In this case, live attenuated vaccines containing oocysts of *Eimeria* spp. are applied as a coarse spray over the chickens, primarily targeting the feathering on the head and back. The oocysts are then ingested by the chickens during natural preening, ensuring direct and efficient uptake with minimal stress and initiating a protective immune response in the intestinal tract (Chapman et al., 2002; Albanese et al., 2018). To enhance this process, colouring agents and gel-based carriers are often added to the vaccine formulation, making it more visible and palatable, thus promoting oral uptake (Danforth et al., 1997; Tensa and Jordan, 2019). This method has become increasingly popular in large commercial hatcheries due to its ease of application, uniform coverage, and reduced vaccine wastage. More recently, it has also been successfully applied to other pathogens, such as IBV (Legnardi et al., 2020; Al-Rasheed et al., 2023). Unlike conventional spray methods, which may result in uneven coverage or loss due to evaporation, gel-based systems provide a more controlled and visible delivery, enhancing consistency across the flock (Albanese et al., 2018).

Drinking water delivery of vaccines is particularly employed for live attenuated vaccines targeting diseases, such as IBD and various respiratory infections (e.g. NDV and IBV) (Ganapathy and Parthiban, 2023; Guerin et al., 2023). The simplicity, speed, and minimal chicken handling associated with this approach make it a highly cost-effective solution for broiler farms. By leveraging the existing drinking infrastructure on the farm, it enables rapid immunization of thousands of broilers without direct physical contact. In this method, the vaccine is diluted into the drinking water and distributed via the water lines (Cargill, 1999; Guerin et al., 2023).

Although it offers logistical advantages, the success of water vaccination depends on several factors, including the uniformity of water intake across broilers. Natural variability in drinking behaviour can lead to inconsistent vaccine exposure and, consequently, uneven immunization within the flock. Moreover, improper preparation or administration may not only reduce efficacy but, in some cases, may even trigger risks such as reversion to virulence (Cserep, 2008, 2009; Guerin et al., 2023). To mitigate these issues, careful planning and supervision during administration are essential. Standard practice includes withholding water for 1 to 2 h before vaccination to increase thirst and stimulate synchronous drinking. The water containing the vaccine is then offered for a period of 2 to 4 h to ensure sufficient vaccine uptake (Cserep, 2008, 2009; Guerin et al., 2023).

Beyond behavioural factors, water quality, waterline sanitation, or inadequate mixing can inactivate live vaccine viruses and result in non-uniform distribution across the flock. Residual chlorine, heavy metals (e.g. iron or copper), and other dissolved contaminants can rapidly inactivate live vaccine viruses, undermining the immunization process (Gentry and Braune, 1972; Saleh et al., 2023). To address this, it is recommended to use dechlorinating agents, like sodium thiosulfate, sodium ascorbate, or even skim milk powder, before vaccine preparation and delivery (Woodward and Tudor, 1975). Additionally, incorporating stabilizers, such as commercial vaccine protectants, can help preserve viral integrity during transit through the water system (Kamau et al., 2010). Cleaning water lines beforehand to eliminate biofilm and ensuring even water pressure throughout the house further supports consistent vaccine distribution and uptake (Cserep, 2009; Munster and Kemper, 2024).

## 5.2 Individual vaccination methods

One of the most significant innovations in broiler vaccination has been the development and widespread adoption of *in ovo* vaccination, a technique in which vaccines are delivered directly into the amniotic fluid or embryonic tissue of developing chickens using an automated injection machine, typically around ED 18 (range 17.5 to 19.2), when eggs are transferred from the incubator to the hatcher (Ricks et al., 1999; Williams, 2023). Over 90% of broiler hatcheries in the US are now equipped with *in ovo* vaccination machines (Peebles, 2018; Abd El-Ghany, 2025). Although applied on a large scale, *in ovo* vaccination is an individual technique: each embryo receives a precise, controlled dose, combining the accuracy of single-chicken injection with the efficiency of mass application (Peebles, 2018; Guerin et al., 2023).

Besides optimal embryo vitality, timing is critical: both premature and delayed delivery may compromise vaccine uptake and immunogenicity (Williams and Hopkins, 2011; Fernandes et al., 2016; Johnston et al., 1997). ED 18 has been identified as the optimal timepoint for *in ovo* administration, as the embryonic immune system is sufficiently developed to mount a response without compromising chick viability (Gagic, et al., 1999; de Wit et al., 2021). This approach has revolutionized poultry production by replacing post-hatch injection and enabling immune system priming prior to hatch, thereby offering early and consistent protection from the moment chickens are exposed to environmental pathogens (Gildersleeve et al., 1993; Zhang and Sharma, 2001; Negash et al., 2004; Okwor et al., 2014). Despite the initial investment required for automation and training, *in ovo* vaccination has proven to be one of the most economically advantageous and efficient mass vaccination tools in the poultry industry. Its high precision, scalability, and automation provide significant benefits, including reduced chick handling, improved hatchery workflow, and uniform vaccine distribution. In addition to early protection, it enhances the uniformity of the immune response across the flock, minimizing individual variation and improving overall vaccine efficacy (Li et al., 2024; Abd El-Ghany, 2025). However, successful implementation requires specialized hatchery equipment and strict adherence to hygiene protocols, including the use of sterile single-use needles or properly disinfected multi-use injectors, precise machine calibration, and consistent sanitation to prevent bacterial and fungal contamination and ensure dose integrity (Williams et al., 1994; Williams and Zedek, 2010; Castaneda et al., 2019).

Initially developed for MD vaccination, and available for all the three MDV serotypes (Sharma and Burmester, 1982; Sharma and Witter, 1983), *in ovo* administration is now widely adopted in commercial broilers worldwide, driven by the availability of newer vaccine formulations, such as the IBDV immune complex vaccine (Jeurissen et al., 1998) and rHVT-based vector vaccines (Tsukamoto et al., 2002; Rauw et al., 2010), which are currently the only vaccines demonstrated to be safe, stable, and compatible with the embryonic environment. rHVT-vectored vaccines, such as rHVT-IBD and rHVT-ND, are routinely administered via this route, enabling protection against multiple pathogens with a single dose. Recent developments include rHVT constructs capable of vectoring two antigens simultaneously, offering broader protection within a single immunization (Ge et al., 2014; Ferreira et al., 2020; Criado et al., 2023). Beyond these, *in ovo* vaccination is increasingly being explored for other pathogens, including AIV, ILTV, avian metapneumovirus (aMPV) and reovirus, all of which may negatively impact broiler health in the post-hatch period (Guo et al., 2003b; Hess et al., 2004; Johnson et al., 2010; Astill et al., 2018). Experimental and commercial trials have also assessed the feasibility of *in ovo* delivering of coccidiosis vaccines (Sokale et al. 2017) and bacterial antigens such as those from *E. coli* and *Mycoplasma* (Elliott et al., 2018; Fatemi et al., 2023), with promising results in terms of immunogenicity and integration into hatchery routines.

Although eye-drop and intranasal administration are among the most immunologically effective routes for delivering live respiratory vaccines, particularly for pathogens such as NDV and IBV (Eidson et al., 1982; Devlin et al., 2008; Al-Rasheed et al., 2023), thanks to their ability to directly stimulate mucosal immunity via lymphoid tissues such as the Harderian gland, their use in broiler production is not routinely implemented. These routes elicit stronger and more consistent immune responses than mass application methods like spray or drinking water, offering superior protection under field conditions (Beltran et al., 2019; Espejo et al., 2024). However, their practicality is limited by the need for individual chicken handling, which increases labour, costs, and stress, raising both logistical and animal welfare concerns (de Wit and Montiel, 2022; Guerin et al., 2023).

Parenteral injection (i.e. subcutaneous [SC] or intramuscular [IM]) of MD vaccines at one day of age remains an essential and highly effective practice, balancing the need for individual administration with the operational demands of high-volume poultry production (Ricks et al., 1999; Abdul-Cader, 2018). The vaccine, which is often based on live MDV, is cell-associated and must be delivered via injection to remain effective, as it cannot be administered through mass methods, like spray or drinking water (Bublot and Sharma, 2004). This process is typically automated in modern hatcheries, using high-throughput injection systems that allow for accurate, sterile, and rapid parenteral delivery to tens of thousands of chickens within a narrow time window. Although it technically qualifies as individual vaccination, its integration into automated hatchery workflows enables efficient large-scale implementation, making it an exception among injection-based approaches in broiler systems (Ricks et al., 1999; Guerin et al., 2023).

Nevertheless, the success of day-old MD vaccination depends on several critical factors, including precise dosing, rigorous equipment maintenance, and strict hygiene protocols to avoid contamination and ensure consistent protection across the flock (Geerligs and Hoogendam, 2007). Furthermore, despite offering precise dosing, it increases labour requirements and chick handling stress (Peebles et al., 2017; Abdul-Cader et al., 2018). Despite the limited use of this delivery method in broilers, it is widely used in breeders to inoculate inactivated vaccines for booster immunization against IBDV and reovirus, primarily to ensure transfer of specific MDAs to offspring for early protection (Al-Natour et al., 2004; Gharaibeh et al., 2008).

In summary, the effectiveness of vaccination in broiler production relies not only on the antigen formulation but also on the method of delivery. A well-designed delivery strategy ensures that each broiler receives the correct dose at the optimal time, thereby maximizing protection and minimizing health, welfare, and economic risks. Mass application methods are widely adopted due to their scalability, speed, and reduced handling stress. These approaches are particularly well-suited to live attenuated vaccines and meet the operational demands of large-scale production. However, their success depends on the precise control of technical parameters to ensure uniform vaccine uptake and immune response. In contrast, individual vaccination methods, particularly *in ovo* administration and automated subcutaneous injection at day-old, have transformed broiler immunization. These techniques allow early, precise, and standardized delivery, striking an effective balance between accuracy and throughput, especially in hatchery-based programmes. Although eye-drop and intranasal routes offer superior immunogenicity, they remain impractical for broiler production due to labour intensity and animal welfare concerns.

## 6 Factors influencing vaccination in broilers

Vaccination failures in broiler production represent a significant challenge in disease prevention and control, often resulting in outbreaks despite rigorous immunization programmes. The success of a vaccination programme depends on a complex interplay of factors related to the immune status of the broilers, characteristics of the pathogen, vaccine selection and administration, and environmental and management conditions. These failures are frequently associated with immunosuppression or other host-related factors that prevent broilers from mounting an adequate immune response, even when vaccines are administered correctly. As a result, vaccination failures can occur, leading to increased disease incidence, higher mortality rates, compromised animal welfare, reduced productivity, and substantial economic losses (de Wit and Montiel, 2022; Schat and Skinner, 2022). Figure 4 provides a schematic overview of the multifactorial elements influencing vaccination effectiveness in broilers, which are explored in detail in the following paragraphs.

### *6.1 Factors related to the host*

The success of a vaccination programme is closely tied to the health status and immunocompetence of broilers at the time of vaccine administration. Broilers that are clinically ill, physiologically weak, or immunosuppressed are less likely to mount an effective immune response, resulting in suboptimal or delayed protection against target pathogens. Moreover, broilers are typically vaccinated at a very young age, when the immune system is not yet fully developed, which can further compromise vaccine efficacy. In addition to these, several intrinsic host-related parameters may influence vaccine responsiveness. These include the level of MDAs, which can interfere with the replication and effectiveness of live vaccines; the genetic background of the broilers, which significantly modulates the immune response; flock uniformity, as variability in body weight, immune maturity, and health status within a flock can lead to inconsistent vaccine uptake and coverage (Marangon and Busani, 2006; de Wit and Montiel, 2022).

MDA interference primarily results from the neutralization of live vaccine viruses before they can replicate and elicit an effective immune response. While MDAs offer crucial passive protection during the early post-hatch period, their persistence can impair the replication and immunogenicity of live attenuated vaccines if these are administered before antibody titres have sufficiently declined. This effect is especially marked for IBDV, and to a lesser extent for NDV, whereas the impact on MDV, IBV, and coccidia vaccines is relatively limited (Reddy et al., 2017; Hu et al., 2022; Gao et al., 2023; Hauck and Macklin, 2023; Isham et al., 2023). MD vaccines, delivered *in ovo* or at day-old, generally provide early protection without significant MDA neutralization; however, reduced efficacy has been reported for cell-free HVT formulations

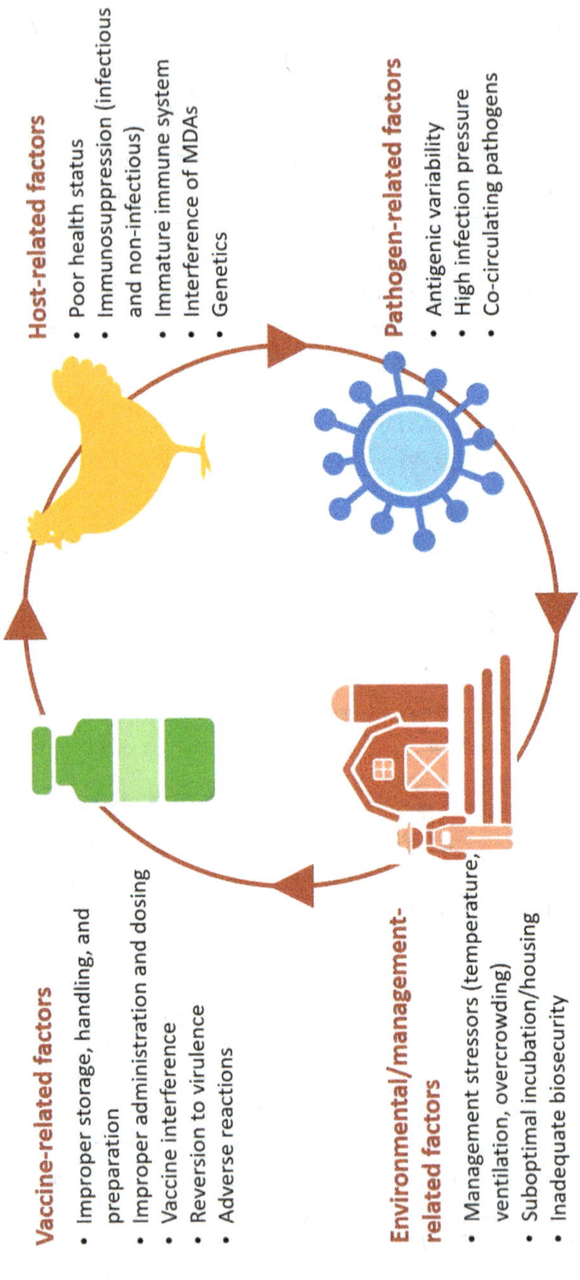


**Figure 4** Schematic overview of the main factors negatively influencing vaccination effectiveness in broilers.

(Witter and Burmester, 1979). To mitigate MDA interference, some protocols, particularly for IBDV, delay live vaccine administration until antibody titres decline. Alternatively, immune complex or recombinant vector-based vaccines are used, which are less susceptible to MDA neutralization (de Herdt et al., 2005; Block et al., 2007; Ramon et al., 2022). Nevertheless, live IBV and NDV vaccines are routinely administered at day-old, as they can confer clinical protection even in the presence of MDAs. In these cases, broilers may remain seronegative, highlighting the crucial role of mucosal and cell-mediated immunity in early protection (Davelaar and Kouwenhoven, 1981; de Wit, 2000; Rauw et al., 2009; Yosipovich et al. 2015).

Immunosuppression represents one of the most prevalent and multifactorial challenges in modern broiler production. By compromising immune function, it increases susceptibility to infectious diseases and significantly reduces vaccine effectiveness. Even when vaccination protocols are correctly followed, immunosuppressed chickens may fail to develop protective immunity, leading to vaccine failure and disease outbreaks. In addition to elevating the risk of infection, immunosuppression often exacerbates disease severity, delays recovery, and contributes to extended viral shedding, thereby facilitating transmission within flocks (de Wit and Montiel, 2022; Schat and Skinner, 2022). In broilers, immunosuppression may arise from a wide range of infectious and non-infectious factors, including chronic stress, exposure to mycotoxins, and physiological or management-related stressors (Hoerr, 2010; Gimeno and Schat, 2018). Among infectious agents, IBDV, MDV, and CIAV are recognized as the primary viral pathogens contributing to immunosuppression due to their targeted effects on lymphoid organs and impairment of both humoral and cellular immunity (Hoerr, 2010; Gimeno and Schat, 2018). IBDV specifically targets the bursa of Fabricius, resulting in B cell depletion and a severe compromise in antibody production (Zhang et al., 2024). Chickens infected at an early age are more susceptible to infectious and parasitic infections in diseases and often fail to develop effective immunity following vaccination against a wide range of key broiler pathogens (Pejkovski et al., 1979; Okoye and Shoyinka, 1983; Sharma, 1984; Rosenberger and Cloud, 1989; Faiz et al., 2016). Similarly, CIAV and MDV contribute to immunosuppression through depletion and dysfunction of T cell populations (Otaki et al., 1988; Markowski and Schat, 2002; Zhu et al., 2024). CIAV infection, in particular, is associated with increased mortality and reduced responsiveness to vaccines, such as those against NDV. For example, in chickens vaccinated with live NDV vaccine at 1 day of age, co-infection with CIAV led to severe respiratory signs, over 30% mortality, and significant growth retardation, although humoral response to NDV was not impaired (de Boer et al., 1994). In chickens vaccinated against MD, CIAV co-infection caused a reduction in bursa/body weight ratio, increased mortality (26.7%), and reduced NDV antibody titres (Zhang et al., 2017). This effect is frequently compounded in co-infections with IBDV, which further impairs immune function and compromises vaccine-induced protection (Cloud et al., 1992; Toro et al., 2009). Moreover, CIAV-induced immunosuppression may cause persistence of IBV vaccine strains that undergo genetic changes during serial passages in vaccinated flocks (Gallardo et al., 2012). Other viral agents, such as adenoviruses and reoviruses, can also contribute to immunosuppression by damaging lymphoid tissues and suppressing immune cell function (Neelima et al., 2003; Niu et al., 2017). Beyond pathogens, non-infectious factors can significantly affect immune competence. Among these, stress plays a central role. Activation of the hypothalamic-pituitary-adrenal (HPA) axis during stress leads to corticosterone (CORT) release, which directly modulates immune cells through its receptors. Chronic CORT elevation induces lymphocyte apoptosis, suppresses cytokine production, and increases vulnerability to infections (Surai and Fisinin, 2016a, 2016b; Dalgaard et al., 2022; Oluwagbenga and Fraley, 2023). In broilers, the combination of rapid growth and environmental, management, or nutritional stressors amplifies this effect, further impairing immune responsiveness (Quinteiro-Filho et al., 2012; Gogoi et al., 2021). Mycotoxins, including aflatoxins, fumonisins, deoxynivalenol (DON), T-2 toxin, ochratoxins, and zearalenone, are potent immunosuppressive agents. These toxins damage lymphoid organs, thereby impairing humoral and cell-mediated immunity, and reducing vaccine efficacy (Batra et al., 1991; Elaroussi et al., 2006; Yunus et al., 2012; Awad et al., 2013; Kufuor-Mensah et al., 2016; Okasha et al., 2024; Shanmugasundaram et al., 2025).

Vaccination itself can also contribute to transient immunosuppression under certain conditions. This is particularly evident when live vaccines are improperly scheduled or when multiple immunosuppressive vaccines are administered simultaneously (Hoerr, 2010; Gimeno and Schat, 2018). For example, intermediate-plus/hot IBDV vaccine strains can cause bursal atrophy, thereby interfering with the immune response to subsequent vaccinations (Ezeokoli et al., 1990; Killian et al., 2017; Bashir et al., 2019). Studies have shown impaired responses to NDV and *Salmonella* vaccines and reduced growth performance following administration of these IBDV strains (Guittet et al., 1992; Rautenschlein et al., 2007; Arafat et al., 2017). Likewise, live MDV, CIAV, and reovirus vaccines may replicate in lymphoid tissues and, if not properly scheduled, may compromise the broiler's ability to respond effectively to later immunizations (Friedman et al., 1992; Guo et al., 2004; Vaziry et al., 2011).

### 6.2 Factors related to pathogens

Pathogen-related factors include a variety of characteristics that critically influence vaccine performance, particularly the selection of appropriate vaccine strains based on circulating field variants. Antigenic variation between vaccine and field strains, particularly among RNA viruses characterized by high mutation rates and recombination events, can lead to reduced protection. Vaccine efficacy is also affected by the level of environmental infection pressure, the route of natural exposure, and interactions with co-circulating pathogens, which can further modulate host susceptibility and immune response. These elements influence not only the clinical manifestation of disease but also define the threshold of immune protection required for effective disease control (Marangon and Busani, 2006; de Wit and Montiel, 2022).

The selection of appropriate vaccine strains is a key determinant of vaccination success and must be tailored to the virulence and antigenic properties of circulating field viruses. Vaccination programmes employing strains of differing pathogenicity and immunogenicity should be designed according to the degree of virulence and the prevailing epidemiological context. When vaccine antigens do not match those of circulating field strains, even well-structured immunization protocols may fail to confer adequate protection (Marangon and Busani, 2006). Several avian viruses, including MDV, NDV, IBV, and IBDV, are prone to genetic and antigenic variability, which complicates vaccination strategies. Both vaccine-derived and field strains may mutate and recombine, giving rise to novel variants that escape existing immunity. In the case of MDV, the emergence of more virulent pathotypes (such as very virulent (vv) and very virulent plus (vv+) strains) has progressively reduced the protective efficacy of older vaccines based on HVT alone (Witter, 1997; Read et al., 2015). This has led to the adoption of more robust vaccination strategies, such as the use of bivalent combinations (HVT + SB-1) or the incorporation of CVI988/Rispens strains, which offer superior protection against these highly pathogenic field isolates (Nair et al., 2020). In areas where velogenic NDV strains are endemic, the use of lentogenic vaccines may be insufficient. These vaccines often provide only partial protection, fail to prevent viral shedding, and may even contribute to the persistence and evolution of virulent field viruses (Miller et al., 2007; Miller et al., 2009). This phenomenon is linked to a positive correlation between the pathogenicity of the vaccine strain and the magnitude of the immune response it elicits. To achieve adequate protection without causing clinical disease, vaccination strategies should include the sequential use of live vaccines with increasing virulence (Miller and Koch, 2020). Similar challenges are also observed with IBV, where high antigenic variability frequently leads to mismatches between vaccine and field strains. Frequent point mutations and recombination events in IBV contribute to the emergence of new antigenic variants capable of partially or fully evading vaccine-induced protection. Even multivalent vaccines targeting multiple IBV serotypes may fail to prevent outbreaks caused by variant strains. Clinical cases have been reported in flocks vaccinated with heterologous serotypes that provided only partial or no cross-protection (de Wit et al., 1998; Cook et al., 1999; Jackwood and Lee,

2017). This highlights the need for regionally adapted multivalent vaccines or tailored combinations of serotype-specific formulations, as well as periodic re-evaluation and updating of vaccine compositions based on current regional surveillance data (Jackwood and de Wit, 2020). A comparable challenge has been described for IBDV, particularly following the emergence of vvIBDV and variant strains that differ significantly from classical vaccine strains. These strains have been shown to cause immunosuppression and mortality even in flocks vaccinated with mild or intermediate vaccines (Rosenberger et al., 1987; Ismail and Saif, 1991). In such cases, vaccine failure is often linked to low antigenic homology and insufficient immunogenicity. These differences may lead to reduced cross-protection and vaccination failure, even when appropriate vaccination schedules are followed. Continuous molecular surveillance of circulating IBDV strains is therefore essential to ensure compatibility between field and vaccine strains, thereby maintaining vaccine efficacy (Eterradossi and Saif, 2020).

Finally, both the environmental infection pressure and the route of pathogen exposure (whether respiratory, enteric, or systemic) are factors that may influence vaccine efficacy. Airborne viruses such as NDV and IBV can rapidly spread in high-density broiler production systems, potentially overwhelming vaccine-induced protection, particularly when flock immunity is uneven (de Wit et al., 2010; Wiseman and Berman, 2017). Environmentally persistent pathogens, such as IBDV or MDV, are highly resistant under field conditions and often infect chickens before adequate immune protection is established. Early challenge with IBDV has been associated with reduced protective efficacy of vaccines against several pathogens (e.g. MDV, NDV, and IBV), as well as prolonged persistence of NDV and IBV in the trachea (Faragher et al., 1974; Jen and Cho, 1980; van Ginkel et al., 2008; Dey et al., 2019).

### 6.3 Factors related to vaccines

A successful vaccination programme relies on the use of high-quality, properly stored, correctly prepared and administered vaccines. Several interdependent factors contribute to vaccine efficacy, including the intrinsic quality of the antigen and strict adherence to transport, storage, preparation, and administration protocols. Each of these steps directly influences the magnitude, duration, and uniformity of the immune response, ultimately determining the level of protection conferred to the flock (Breytenbach, 2007; Cserep, 2008, 2009).

The method, timing, and precision of vaccine administration significantly influence the outcome of vaccination, including the severity of post-vaccination reactions. Improper delivery is among the leading causes of vaccination failure. Each administration route presents specific technical and biological challenges that have been already discussed in Section 5. The stability, potency, and immunogenic integrity of vaccines are highly dependent on compliance with proper storage conditions, mainly cold chain requirements. Live vaccines are particularly sensitive to thermal stress and must be stored consistently at 2–8°C to avoid degradation. Some formulations are also photosensitive and require protection from light. Failure to adhere to cold chain protocols, such as exposure to freezing or high ambient temperatures, can result in substantial potency loss (Cargill, 2021; Abdelaziz et al., 2024). For instance, lyophilized HVT vaccines remained stable at 4°C for 1 year, but lost ~50% infectivity at 24°C in 4 months and 90–99.7% within 4 weeks at 37–45°C. After reconstitution, the half-life was 1–5 h at 4°C and ~1 h at 45°C (Siegmann et al., 1980). IBDV remained infective up to 42°C, became non-infective above 72°C, and showed minimal sensitivity to pH variations (Rani and Kumar, 2015). Live NDV vaccines also retained potency at 4°C but degraded significantly after 1 week at 37°C (Wood et al., 1980). IBV vaccine studies showed that viral RNA dropped ~100-fold at 41°C vs. 37°C (Keep et al., 2020b). Most live vaccines used in broiler production are freeze-dried or frozen. They must be transported and stored under refrigeration, except for cell-associated MDV and HVT-vectored vaccines, which require storage in liquid nitrogen at ultra-low temperatures, making them particularly vulnerable to inactivation if exposed to temperatures above −150°C, with significant losses occurring even at −70°C, and complete inactivation within a few days at 4°C (Garrido et al., 1972; Hein et al., 2021). Vaccine preparation immediately prior to administration is another critical point; in particular, freeze-dried vaccines demand meticulous reconstitution and rapid use. Even high-quality vaccines may fail if improperly reconstituted or diluted. Common errors include the use of contaminated or chlorinated water, inappropriate diluents or volumes, inadequate mixing, and delayed administration post-reconstitution. Inadequate mixing of reconstituted vaccines may lead to dosage inconsistencies, with some chickens receiving insufficient antigen and others excessive amounts, both scenarios compromising the desired immune response (Cserep, 2009; Cargill, 2021). Live vaccines, particularly those used against MDV, are highly sensitive to improper handling, with a significant risk of viral inactivation before administration. For example, they may lose up to 95% of their potency if thawed and refrozen (Halvorson and Mitchel, 1979; Geerligs and Hoogendam, 2007). Accurate dosing is equally critical: underdosing may result in inadequate immune stimulation, while overdosing can increase the likelihood of vaccine reactions without providing additional benefit. This issue is particularly relevant in mass application methods, such as drinking water or spray, where intra-flock variation in vaccine uptake can lead to non-uniform immunization and protection gaps (Gentry and Braune, 1972; Breytenbach, 2007; Cserep, 2009; Cargill, 2021). For example, spray or drinking water administration has been reported to reduce protective antibody levels by approximately 47% in broilers, compared to ocular administration, which achieved up to 93% seroconversion following ND vaccination (Degefa et al., 2004).

Vaccine interference refers to immunological interactions that may compromise the host response to one or more vaccine components, thereby reducing their efficacy. This phenomenon can result from several mechanisms, including neutralization of vaccine viruses by MDAs before viral replication occur (discussed in detail in Section 6.1), competition between live vaccines for replication in specific target tissues (e.g. respiratory mucosa) or immune modulation, in which the response induced by one vaccine suppresses or alters the response to another (Schijns et al., 2014; de Wit and Montiel, 2022). Beyond MDA interference, direct competition or mutual suppression between live vaccines represents an important mechanism of interference. A well-documented example is the co-administration of live NDV and IBV vaccines, especially when administered via coarse spray at hatch, which compete for replication sites in the respiratory mucosa, delaying or impairing immune responses and reducing vaccine efficacy (Thornton et al., 1973; Thornton et al., 1975; Winterfield, 1984; Gelb et al., 2007). For example, in chickens vaccinated simultaneously with monovalent NDV and IBV live vaccines, IBV significantly interfered with NDV replication and reduced protection against NDV challenge to 36%, compared to >90% in other vaccination groups (Gelb et al., 2007). Similar antagonistic effects have been observed when NDV and IBV vaccines are applied concurrently with live ILT vaccines, decreasing replication and seroconversion (Vagnozzi et al., 2010), or with live AMPV vaccines, resulting in temporary suppression of AMPV replication and protective effect (Cook et al., 2001; Ganapathy et al., 2005). An additional example includes co-administration of live IBDV and CIAV vaccines, which may induce transient immunosuppression and blunt responses to other vaccines (Cloud et al., 1992). Furthermore, interference has also been documented with the use of multiple rHVT-vectored vaccines, where mutual suppression between constructs can diminish the efficacy of one or both components (Dunn et al., 2019). Infection with *Eimeria* spp. has also been shown to impair the immunological response to NDV vaccination, suggesting that pre-existing enteric infections may act as non-specific modulators of vaccine efficacy (Akhtar et al., 2015).

Live attenuated vaccines also carry an inherent risk of reversion to virulence. Under certain circumstances, such as suboptimal vaccine administration, prolonged chicken-to-chicken transmission, or host immunosuppression, live vaccine viruses may revert to a virulent form. In particular, improper vaccine application, including hatchery-based mass administration and inaccurate dosing, can result in heterogeneous uptake, leaving part of the flock non- or insufficiently immunized. In these conditions, vaccine strains may circulate extensively among susceptible broilers, triggering the so-called 'rolling reactions' within the flock. This phenomenon has been well-documented for

NDV and IBV live vaccines, where point mutations or recombination events may result in the emergence of pathogenic revertants. Field isolates genetically related to vaccine strains have been associated with clinical disease (Bande et al., 2015; Dimitrov et al., 2017). In some cases, the reverted strains have produced pathological lesions indistinguishable from those caused by wild-type viruses (Eidson et al., 1977; Hopkins and Yoder, 1986).

Although most vaccines are safe when correctly administered, improper technique or suboptimal application conditions can lead to adverse reactions, particularly with live respiratory vaccines, such as those against NDV and IBV (Cavanagh, 2003; Landman et al., 2015). When administered via spray with inadequate droplet size or distribution, these vaccines can cause epithelial damage or penetrate too deeply into the respiratory tract, causing transient signs like sneezing, conjunctivitis, and mild respiratory distress. These effects are exacerbated in immunosuppressed broilers or in those coinfected with *Mycoplasma* spp. or *E. coli*, especially under poor ventilation or high ammonia levels (Villegas et al., 1976; Nakamura et al., 1992; Matthijs et al., 2005). In the case of NDV, insufficient flock coverage resulting from improper administration of live vaccines may lead to suboptimal immunity, enabling extended viral shedding from vaccinated to unvaccinated broilers. This prolonged circulation within the flock can trigger post-vaccinal respiratory signs and negatively impact growth performance and feed conversion efficiency (van Boven et al., 2008; Wang et al., 2015; Wiseman and Berman, 2017). Localized post-injection reactions can also be observed following parenteral administration of MDV vaccines at hatch, with reports of localized inflammation and tissue damage at the injection site (Gustafson et al., 2003).

Finally, given the high number and diversity of vaccines, and the limited timeframe in which broilers are vaccinated, the immature immune system may become overstimulated. This may lead to immune fatigue, delayed or inconsistent seroconversion, and transient immunosuppression, particularly in the presence of concurrent environmental or physiological stressors (Hoerr, 2010; Schijns et al., 2014; Landman, 2015; Abo-Al-Ela et al., 2021).

## 6.4 Environmental and management factors

Beyond host-, pathogen-, and vaccine-related factors, environmental and managerial conditions play a critical role in shaping vaccine outcomes. Although often underestimated, these factors are fundamental in determining both the broilers' immune competence and the overall success of vaccination programmes. Environmental variables such as ambient temperature, ventilation and air quality, litter condition, stocking density, and nutritional status are strongly associated with stress responses in broilers, which can disrupt immune function and impair vaccine responsiveness (Dalgaard et al., 2022; Wlazlak et al., 2023).

The incubation, hatching, and immediate post-hatch phases constitute highly sensitive periods in broiler development during which environmental and management-related stressors can exert negative effects on immune system maturation (Surai and Fisinin, 2016a). Suboptimal incubation conditions (e.g. inappropriate temperature, humidity, or gas exchange), together with inadequate hatchery practices (e.g. prolonged hatch windows, delayed chick retrieval, and intensive handling procedures), have been shown to exacerbate physiological stress responses (Decuypere and Bruggeman, 2007; Tona et al., 2007; Willemsen et al., 2010; Noiva et al., 2014). These stressors negatively affect the development of PLOs, which are essential for the establishment of adaptive immune competence (Panda et al., 2015; Wijnen et al., 2020; Madej et al., 2024). Chickens exposed to thermal or metabolic stress during this critical period frequently exhibit delayed lymphocyte differentiation, reduced proliferation of B and T lymphocytes, and diminished antibody responses (McFarlane and Curtis, 1989; Hulet et al., 2007). These effects are further exacerbated by suboptimal brooding conditions, such as delayed access to feed and water or insufficient thermal comfort, which impair the early development of GALT and mucosal immunity (Bar Shira et al., 2005; Panda et al., 2015). These early-life stressors compromise not only baseline immune function but also the capacity to elicit robust and protective vaccine-induced immune responses. As a result, the immunogenicity and overall efficacy of several vaccines administered during the incubation and the first days of life may be significantly diminished under such conditions. For instance, the performance of *in ovo*-administered vaccines against MDV has been reported to decline when administered under suboptimal incubation or hatchery management conditions (Williams and Hopkins, 2011; Fernandes et al., 2016). The timing of administration is particularly critical: vaccination is effective only when performed between 17.6 and 18.5 days of incubation, as administering it earlier can reduce hatchability by up to 4.68% (Fernandes et al., 2016). Delaying vaccination, on the other hand, increases the risk of incorrect anatomical placement, which drastically affects efficacy. Protection is null when the vaccine is injected into the air cell, limited to 28.3% in the allantoic fluid, but reaches 94.4% into the amniotic fluid and 93.9% when delivered directly into the embryo body (Wakenell et al., 2002). Similarly, spray-applied vaccines targeting NDV and IBV have demonstrated reduced efficacy in thermally stressed or immunologically immature chickens, with impaired mucosal immune activation and lower seroconversion rates (van Ginkel et al., 2015; Honda et al., 2015). Live *Eimeria* spp. vaccines, administered orally in the first days of life, rely heavily on the early development of gut-associated immune structures. Delayed access to feed has been shown to impair the development of GALT, reduce intestinal IgA secretion, and compromise the protective response to coccidial challenge (Price et al., 2015). Moreover, brooding-related stressors such as heat stress or dehydration can disrupt gut integrity and immune homeostasis, interfering with normal oocyst cycling and significantly reducing the efficacy of live coccidial vaccines (Ghareeb et al., 2022).

Environmental and management stressors commonly encountered in the poultry house, such as inadequate temperature, humidity, and ventilation (including high ammonia exposure), as well as overcrowding, are well-recognized contributors to chronic stress and consequent immunosuppression, thereby increasing disease susceptibility and severity, as well as reducing the effectiveness of vaccination (Surai and Fisinin, 2016a; Dalgaard et al., 2022; Wlazlak et al., 2023). Among these, heat stress is particularly detrimental. In broilers older than 2 weeks of age, exposure to temperatures exceeding 28–29°C (or above 27°C under high humidity) induces oxidative stress and respiratory alkalosis due to hyperventilation. Studies have shown that at around 35°C, it also leads to regression of key lymphoid organs, including the thymus, spleen, and bursa of Fabricius (Hirakawa, 2020; Wang et al., 2023). Heat stress also raises CORT levels, suppressing lymphocyte proliferation and antibody production (IgM, IgG), thereby dysregulating immune responses and impairing vaccine-induced immunity (Nawaz et al., 2021; Oke et al., 2024). These alterations have been linked not only to reduced efficacy of vaccines against NDV, as demonstrated by lower antibody titres and delayed seroconversion in heat stressed broilers (Beard and Mitchell, 1987; Honda et al., 2015), but also to an increased incidence of bacterial infections affecting the respiratory and gastrointestinal tracts, including *S. enterica*, *E. coli*, and *Clostridium perfringens* (Quinteiro-Filho et al., 2012; Nanto-Hara et al., 2020; Wickramasuriya et al., 2022).

Ammonia exposure, typically resulting from poor litter management and inadequate ventilation, is known to cause epithelial damage and ciliostasis in the respiratory tract, impairing mucociliary clearance. Compromised respiratory health both increases the risk of secondary viral and bacterial infections and negatively affects the performance of respiratory vaccines (Kristensen and Wathes, 2000; Hofmann et al., 2020). A recent study indicated that ammonia exposure does not impair the immune response to IBV vaccines in chickens, although it may increase the incidence of airsacculitis in unvaccinated broilers (Aston et al., 2019).

High stocking densities and social stress further contribute to immunosuppression by elevating CORT levels, impairing lymphocyte activity, and suppressing antibody responses (Yanai et al., 2018; Hofmann and Stefanski, 2025). In addition, feed restriction or dietary imbalances during early growth represent another major non-infectious factor affecting immune cell proliferation and cytokine production, leading to incomplete or delayed responses to vaccination. Deficiencies in critical micronutrients, such as vitamins A, E, and selenium, are known to impair antioxidant defences, reduce lymphoid organ integrity, and diminish both humoral and cellular responses (Klasing, 1998; Dalgaard et al., 2022). Experimental studies have shown that vitamin E or selenium deficiency reduces antibody production following NDV and IBV vaccination (Singh et al., 2006; Ma et al., 2019). Additionally, broilers subjected to restricted feeding during brooding exhibit reduced protection against *Eimeria* spp. and decreased response to oral and drinking-water-administered vaccines (Rasheed and Matsler, 2020).

Additionally, biosecurity compliance is a crucial variable influencing vaccination success. Inadequate biosecurity may increase pathogen circulation pressure, reduce the effectiveness of vaccination by promoting early exposure before immunity develops, and facilitate the introduction of immunosuppressive co-infections. Therefore, vaccination programmes should be designed and implemented as part of a comprehensive health management plan that includes strict and measurable biosecurity practices (Van Limbergen et al., 2018).

Finally, systematic monitoring of vaccine performance through post-vaccination monitoring and serological testing is essential (Marangon and Busani, 2006; Breytenbach, 2007; de Wit and Montiel, 2022). These tools enable the early detection of vaccine failures, which may result from inaccurate administration, breaches in the cold chain, or antigenic drift in circulating field strains. Continuous evaluation and refinement of vaccination protocols, based on farm-specific disease pressure, flock performance, and diagnostic data, are critical to maintaining long-term efficacy and flock protection. To optimize vaccine application, poultry producers should ensure comprehensive training of personnel in vaccination techniques, regularly calibrate and maintain vaccination equipment, and monitor vaccine uptake across the flock, with particular attention to uniformity and early identification of potential adverse reactions.

In summary, the efficacy of vaccination in broiler production depends not only on the quality of vaccines or the immunological principles underlying their action but also on a complex network of biological, environmental, and managerial factors. Host-related variables such as immunosuppression, the presence of MDAs, and flock uniformity can profoundly influence vaccine uptake and response. Likewise, pathogen characteristics, including antigenic variability and virulence, require constant monitoring and strain matching to avoid immune escape. Even the most potent vaccines may fail if compromised by improper storage, preparation, or administration, or if delivered using suboptimal methods. In parallel, environmental stressors, especially heat, overcrowding, poor ventilation, or nutritional deficits, can undermine immune function and reduce vaccine responsiveness. These elements highlight the importance of integrating vaccination into a broader health and management framework that includes biosecurity, environmental control, staff training, and post-vaccination monitoring. Only through a holistic approach can broiler vaccination programmes deliver consistent, flock-wide protection in the face of evolving disease and production pressures.

## 7 New developments and future perspectives

**Innovation in vaccine platforms:** The continuous evolution of poultry production systems, the increasing restrictions on antimicrobial use, and the emergence of novel pathogen variants are key drivers of innovation in broiler vaccination. Advances in molecular virology, reverse genetics, vector engineering, and automation technologies (e.g. *in ovo* vaccination) have led to the development of recombinant and multivalent vaccines. These formulations can elicit broader and more durable immune responses, even in the presence of passive immunity, supporting early immune activation and more consistent flock-wide protection. Platforms such as immune complex and rHVT vector vaccines represent important milestones and are expected to expand to additional pathogens.

**Development of bacterial vaccines:** Considerable efforts are underway to develop effective bacterial vaccines targeting *Escherichia coli*, *Clostridium perfringens*, and *Salmonella enterica*. These pathogens are major contributors to broiler morbidity, antimicrobial use, and economic losses. Innovations in adjuvant formulations, antigen delivery systems (e.g. micro- and nanoparticles), and mucosal immunization strategies are expected to improve the efficacy and field applicability of bacterial vaccines.

**Emerging vaccine technologies:** New vaccine platforms, including mRNA-based vaccines and nanoparticle-based delivery systems, are being actively explored for their ability to induce targeted and potent immune responses, particularly at mucosal surfaces. These approaches hold promise for rapid-response vaccination strategies and the development of multivalent formulations capable of addressing complex and evolving disease challenges.

**Precision vaccination and digital tools:** Precision vaccination is gaining traction with the goal of adapting immunization protocols in real time. Technologies such as sensor-based devices, immunological biomarkers, and digital decision-support tools may soon enable personalized or flock-level adjustments of vaccine schedules based on dynamic monitoring of immune status, vaccine uptake, and population variability.

**Integrative health approaches:** Future strategies will increasingly account for the interaction between early immune imprinting, gut microbiota composition, and environmental factors influencing vaccine efficacy. This points to the development of integrated approaches combining vaccination with nutritional immunomodulation, microbiota-targeted interventions (e.g. probiotics and postbiotics), and improved hatchery practices such as on-farm hatching and early feeding.

**Towards sustainable broiler production:** Recent trends in vaccinology indicate a convergence of immune modulation, genetic selection, and microbiota management aimed at promoting disease resistance and sustainability. Greater automation, improved antigen design, and reduced antibiotic reliance will continue to shape disease control strategies and enhance production efficiency. Ultimately, next-generation vaccination programmes must balance immunological robustness, operational feasibility, and sustainability goals, supporting animal resilience and public health in a post-antibiotic era.

## 8 Conclusion

Effective vaccination is a critical pillar of broiler health management, enabling the control of major infectious diseases while reducing reliance on antibiotics. In the context of intensive poultry production, the success of immunization programmes depends not only on the immunological properties of vaccines but also on a wide range of biological, environmental, and operational factors. Understanding the ontogeny of the avian immune system, the role of maternal antibodies, and the influence of the microbiota is essential for designing immunization schedules that provide early and robust protection. Likewise, the choice of vaccine type, namely conventional or new generation, and delivery method must be adapted to the broiler production system, considering the short lifespan and rapid growth of modern broilers. Advances in recombinant technologies, immune complex platforms, and *in ovo* administration have expanded the possibilities for early, safe, and uniform immunization. However, vaccine efficacy can be compromised by host immunosuppression, antigenic mismatch, incorrect application, or environmental stress.

These challenges highlight the need for an integrated approach to vaccination – one that encompasses precise timing, optimized delivery, strict biosecurity, and continuous monitoring. Looking ahead, innovation in broiler vaccinology is moving towards greater precision, automation, and immunological targeting. As pressure mounts to reduce antimicrobial use and improve sustainability, vaccination strategies must evolve to support immune resilience while maintaining operational feasibility. The integration of advanced vaccines with microbiota modulation, nutritional immunology, and digital health tools will be key to building robust, adaptable, and future-ready broiler production systems.

## 9 Where to look for further information

### *Key books*

- Swayne, D. (2020), *Diseases of poultry*, John Wiley & Sons, Hoboken, USA.

- Kaspers, B., et al. (2022), *Avian immunology*, Academic Press, London, UK.
- Samal, S.K. (2019), *Avian virology - current research and future trends*, Caister Academic Press, Norfolk, UK.

### Key book chapters/reviews

### Genetic resistance

- Gul, H., et al. (2022), 'Genetic resilience in chickens against bacterial, viral and protozoal pathogens', *Front. Vet. Sci.*, 9, 1032983.
- Pinard-van der Laan, M.H., et al. (2020), 'Genetics and genomics of immunity and disease traits in poultry species', In S. Aggrey, et al. (Eds), *Advances in poultry genetics and genomics*, Burleigh Dodds Science Publishing, Cambridge, UK, pp. 263–305.
- Zekarias, Z., et al. (2002), 'Immunological basis of differences in disease resistance in the chicken', *Vet. Res.*, 33, 109–125.

### Regulatory frameworks

- Schrader, J.S. and Smith, H.E. (2023), 'Licensing requirements for poultry veterinary biologics in the United States', *Avian Dis.*, 67, 467–473.
- Werner, E., et al. (2023), 'Licensing requirements for avian vaccines within the European Union', *Avian Dis.*, 67, 474–479.
- World Organization for Animal Health (OIE) (2024), 'Vaccination'. OIE, Paris, France. Available at: https://www.woah.org/en/what-we-do/standards/codes-and-manuals/; accessed: 27th June, 2025.
- World Organization for Animal Health (OIE) (2024), 'Veterinary vaccines'. OIE, Paris, France. Available at: https://www.woah.org/en/what-we-do/standards/codes-and-manuals/; accessed: 27th June, 2025.

## 10 References


Abd El-Ghany, W.A. (2025), '*In ovo* vaccination technology: an alternative approach to post-hatch vaccination in modern poultry operations', *Microbiol. Res.*, 16, 7.

Abdelaziz, K., et al. (2024), 'Advances in poultry vaccines: leveraging biotechnology for improving vaccine development, stability, and delivery', *Vaccines*, 12, 134.

Abdul-Cader, M.S., et al. (2018), 'Hatchery vaccination against poultry viral diseases: potential mechanisms and limitations', *Viral Immunol.*, 31, 23–33.

Abo-Al-Ela, H.G., et al. (2021), 'Stress and immunity in poultry: light management and nanotechnology as effective immune enhancers to fight stress', *Cell Stress Chaperones*, 26, 457–472.

Abozeid, H.H. (2023), 'Global emergence of infectious bronchitis virus variants: evolution, immunity, and vaccination challenges', *Transbound. Emerg. Dis.*, 29, 1144924.

Afonso, C.L. and Miller, P.J. (2013), 'Newcastle disease: progress and gaps in the development of vaccines and diagnostic tools', *Dev. Biol. (Basel)*, 135, 95–106.

Akhtar, M., et al. (2015), 'The effect of infection with mixed *Eimeria* species on hematology and immune responses following Newcastle disease and infectious bursal disease booster vaccination in broilers', *Vet Q.*, 35, 21–26.

Akram, M.Z., et al. (2025), 'Effects of on-farm hatching versus hatchery hatching on growth performance, gut development, and intestinal health and function in broiler chickens', *Poult. Sci.*, 104, 104770.

Akter, R., et al. (2024), 'Clarification of a unique mucosal vaccination route for improved systemic and mucosal immune response in broiler', *Heliyon*, 10, e39621.

Akyüz, H.C., et al. (2022), 'Age and sex related changes in fattening performance, dermatitis, intestinal histomorphology, and serum IgG level of slow- and fast-growing broilers under the intensive system', *Trop. Anim. Health Prod.*, 54, 312.

Albanese, G.A., et al. (2018), 'Evaluation of a coccidia vaccine using spray and gel applications', *Poult. Sci.*, 97, 1544–1553.

Al-Garib, S.O., et al. (2003), 'Review of Newcastle disease virus with particular references to immunity and vaccination', *World's Poul. Sci. J.*, 59, 185–200.

Alkie, T.N. and Rautenschlein, S. (2016), 'Infectious bursal disease virus in poultry: current status and future prospects', *Vet. Med. Res. Rep.*, 7, 9–18.

Alkie, T.N., et al. (2019), 'Development of innate immunity in chicken embryos and newly hatched chicks: a disease control perspective', *Avian Pathol.*, 48, 288–310.

Al-Natour, M.Q., et al. (2004), 'Effect of different levels of maternally derived antibodies on protection against infectious bursal disease virus', *Avian Dis.*, 48, 177–182.

Al-Rasheed, M., et al. (2023), 'Evaluation of protection and immunity induced by infectious bronchitis vaccines administered by oculonasal, spray or gel routes in commercial broiler chicks', *Vaccine*, 41, 4508–4524.

Arafat, N., et al. (2017), 'Effect of infectious bursal disease (IBD) vaccine on *Salmonella* Enteritidis infected chickens', *Vaccine*, 35, 3682–3689.

Astill, J., et al. (2018), 'Induction of immune response in chickens primed *in ovo* with an inactivated H9N2 avian influenza virus vaccine', *BMC Res. Notes*, 11, 428.

Aston, E.J., et al. (2019), 'Ambient ammonia does not appear to inhibit the immune response to infectious bronchitis virus vaccination and protection from homologous challenge in broiler chickens', *Vet. Immunol. Immunopathol.*, 217, 109932.

Awad, W., et al. (2013), 'The toxicological impacts of the fusarium mycotoxin, deoxynivalenol, in poultry flocks with special reference to immunotoxicity', *Toxins (Basel)*, 5, 912–925.

Awad, F., et al. (2016), 'Heterologous live infectious bronchitis virus vaccination in day-old commercial broiler chicks: clinical signs, ciliary health, immune responses and protection against variant infectious bronchitis viruses', *Avian Pathol.*, 45, 169–177.

Bande, F., et al. (2015), 'Progress and challenges toward the development of vaccines against avian infectious bronchitis', *J. Immunol. Res.*, 2015, 424860.

Bar Shira, E., et al. (2005), 'Impaired immune responses in broiler hatchling hindgut following delayed access to feed', *Vet. Immunol. Immunopathol.*, 105, 33–45.

Bashir, K., et al. (2019), 'Combination of TLR2 and TLR3 agonists derepress infectious bursal disease virus vaccine-induced immunosuppression in the chicken', *Sci. Rep.*, 9, 8197.

Batra, P., et al. (1991), 'Effect of aflatoxin B1 on the efficacy of turkey herpesvirus vaccine against Marek's disease', *Res. Vet. Sci.*, 51, 115–119.

Beard, C.W. and Mitchell, B.W. (1987), 'Influence of environmental temperatures on the serologic responses of broiler chickens to inactivated and viable Newcastle disease vaccines', *Avian Dis.*, 31, 321–326.

Bell, I., et al. (1991), 'The resistance of meat chickens vaccinated by aerosol with a live V4 Newcastle disease virus vaccine in the field to challenge with a velogenic Newcastle disease virus', *Aust. Vet. J.*, 68, 97–101.

Beltran, G., et al. (2019), 'Immune responses in the eye-associated lymphoid tissues of chickens after ocular inoculation with vaccine and virulent strains of the respiratory infectious laryngotracheitis virus (ILTV)', *Viruses*, 11, 635.

Benedict, A.A., et al. (1963), 'The temporal synthesis and some chromatographic and ultracentrifugal characteristics of chicken antibodies', *J. Immunol.*, 90, 399–411.

Bessei, W. (2006), 'Welfare of broilers: a review', *World's Poult. Sci. J.*, 62, 455–466.

Bhuiyan, M.S.A., et al. (2021), 'Infectious Bronchitis virus (Gammacoronavirus) in poultry farming: vaccination, immune response and measures for mitigation', *Vet. Sci.*, 8, 273.

Bist, R.B., et al. (2024), 'Sustainable poultry farming practices: a critical review of current strategies and future prospects', *Poult. Sci.*, 103, 104295.

Blake, D.P., et al. (2020), 'Re-calculating the cost of coccidiosis in chickens', *Vet. Res.*, 51, 115.

Block, H., et al. (2007), 'A field study on the significance of vaccination against infectious bursal disease virus (IBDV) at the optimal time point in broiler flock with maternally derived IBDV antibodies', *Avian Pathol.* 36, 401–409.

Bodman-Harris, O., et al. (2024), 'Approaches to enhance the potency of vaccines in chickens', *Vaccines*, 12, 1337.

Bouwens, M. and Savelkoul, H.F.J. (2019), 'Animal nutrition and immunity in pigs and poultry', In W.H. Hendriks, et al. (Eds), *Poultry and pig nutrition: challenges of the 21st century*, Wageningen Academic Publishers, Wageningen, The Netherlands, pp. 105–127.

Breytenbach, J.H. (2007), 'Correct and effective vaccination of poultry', *Int. Poult. Prod.*, 12, 17–21.

Broom, L.J. and Kogut, M.H. (2018), 'The role of the gut microbiome in shaping the immune system of chickens', *Vet. Immunol. Immunopathol.*, 204, 44–51.

Broom, L.J. (2019), 'Host-microbe interactions and gut health in poultry - focus on innate responses', *Microorganisms*, 7, 139.

Bru, T., et al. (2017), 'Protection of chickens vaccinated with combinations of commercial live infectious bronchitis vaccines containing Massachusetts, Dutch and QX-like serotypes against challenge with virulent infectious bronchitis viruses 793B and IS/1494/06 Israel variant', *Avian Pathol.*, 46, 52–58.

Bublot, M. and Sharma, J.M. (2004), 'Vaccination against Marek's disease', In F. Davison and V. Nair (Eds), *Marek's disease. An evolving problem*, Elsevier Academic Press, London, England, pp. 268–285.

Bublot, M., et al. (2007), 'Use of a vectored vaccine against infectious bursal disease of chickens in the face of high-titred maternally derived antibody', *J. Comp. Pathol.*, 137, S81-S84.

Bublot, M. (2023), 'Poultry vaccine technology platforms', *Avian Dis.*, 67, 1–22.

Cai, H., et al. (2023), 'Effects of a complex probiotic preparation, Fengqiang Shengtai and coccidiosis vaccine on the performance and intestinal microbiota of broilers challenged with *Eimeria* spp.', *Parasites Vectors*, 16, 253.

Calnek, B.W. (1998), 'Control of Avian Encephalomyelitis: a historical account', *Avian Dis.*, 42, 632–647.

Cargill, P. (1999), 'Vaccine administration in poultry', *In Practice*, 6, 323–328.

Cargill, P. (2021), 'Vaccination strategies and application of vaccines', In P. Barrow, et al. (Eds), *Poultry health - a guide for professionals*, CABI International, Wallingford, Oxfordshire, UK.

Castaneda, C.D., et al. (2019), 'Evaluating bacterial colonization of a developing broiler embryo after *in ovo* injection with a bioluminescent bacteria', *Poult. Sci.*, 98, 2997–3006.

Castro, F.L.S., et al. (2023), 'Poultry industry paradigms: connecting the dots', *J. Appl. Poult. Res.*, 32, 100310.

Cavanagh, D. (2003), 'Severe acute respiratory syndrome vaccine development: experiences of vaccination against avian infectious bronchitis coronavirus', *Avian Pathol.*, 32, 567–582.

Ceccopieri, C. and Madej J.P. (2024), 'Chicken secondary lymphoid tissues-structure and relevance in immunological research', *Animals*, 14, 2439.

Cervantes, H.M., et al. (2020), 'Coccidiosis', In D.E. Swayne (Ed), *Diseases of poultry*, John Wiley & Sons, Hoboken, NJ, USA, pp. 1193–1217.

Chhabra, R., et al. (2015), 'Immune responses to virulent and vaccine strains of infectious bronchitis viruses in chickens', *Viral Immunol.*, 28, 478–488.

Chansiripornchai, N. and Sasipreeyajan, J. (2006), 'Efficacy of live B1 or Ulster 2C Newcastle disease vaccines simultaneously vaccinated with inactivated oil adjuvant vaccine for protection of Newcastle disease virus in broiler chickens', *Acta Vet. Scand.*, 48, 2.

Chansiripornchai, N. and Sasipreeyajan, J. (2009), 'Comparison of the efficacy of the immune complex and conventionally live vaccine in broilers against Infectious Bursal Disease infection', *Thai J. Vet. Med.*, 39, 115–120.

Chapman, H.D., et al. (2002), 'Sustainable coccidiosis control in poultry production: the role of live vaccines', *Int. J. Parasitol.*, 32, 617–629.

Cheema, M.A., et al. (2003)., 'A Comparison of the immune response of a 2001 commercial broiler with a 1957 randombred broiler strain when fed representative 1957 and 2001 broiler diets', *Poult. Sci.*, 82, 1519–1529.

Chellappa, M.M., et al. (2022), 'Newcastle disease virus vectored chicken infectious anaemia vaccine induces robust immune response in chickens', *Viruses*, 13, 1985.

Chung, E.L.T., et al. (2020), 'Do different vaccination regimes affect the growth performance, immune status, carcase characteristics and meat quality of broilers?', *Br. Poult. Sci.*, 62, 32–37.

Cloud, S.S., et al. (1992), 'Immune dysfunction following infection with chicken anemia agent and infectious bursal disease virus. II. Alterations of *in vitro* lymphoproliferation and in vivo immune responses', *Vet. Immunol. Immunopathol.*, 34, 353–366.

Collett, S.R. (2020), 'Principles of disease prevention, diagnosis, and control', In D.E. Swayne (Ed), *Diseases of Poultry*, John Wiley & Sons, Hoboken, NJ, USA, pp. 3–78.

Cook, J.K.A., et al. (1999), 'Breadth of protection of the respiratory tract provided by different live-attenuated infectious bronchitis vaccines against challenge with infectious bronchitis viruses of heterologous serotypes', *Avian Pathol.*, 28, 477–485.

Cook, J.K.A., et al. (2001), 'Infectious bronchitis virus vaccine interferes with the replication of avian pneumovirus vaccine in domestic fowl', *Avian Pathol.*, 30, 233–242.

Corbanie, E.A., et al. (2006), 'Deposition of differently sized airborne micro-spheres in the respiratory tract of chickens', *Avian Pathol.*, 35, 475–485.

Corbanie, E.A., et al. (2008), 'Vaccination of broiler chickens with dispersed dry powder vaccines as an alternative for liquid spray and aerosol vaccination', *Vaccine*, 26, 4469–4476.

Cox, C.M., et al. (2010), 'Immune responses to dietary β-glucan in broiler chicks during an *Eimeria* challenge', *Poult. Sci.*, 89, 2597–2607.

Criado, M.F., et al. (2023), 'Efficacy of multivalent recombinant herpesvirus of turkey vaccines against high pathogenicity avian influenza, infectious bursal disease, and Newcastle disease viruses', *Vaccine*, 41, 2893–2904.

Cserep, T. (2008), 'Vaccines and vaccination', In M. Pattison, et al. (Eds), *Poultry diseases*, Elsevier Limited, Philadelphia, PA, USA, pp. 66–81.

Cserep, T. (2009), 'Vaccination: theory and practice', In P. Hocking (Ed), *Biology of breeding poultry*, CAB International, Wallingford, UK, pp. 377–390.

Dalgaard, T.S., et al. (2022), 'Factors modulating the avian immune system', In B. Kaspers, et al. (Eds), *Avian immunology*, Academic Press, London, UK, pp. 419–435.

Danforth, H.D., et al. (1997), 'Evaluation of a gel-immunization technique used with two different Immucox vaccine formulations in battery and floor-pen trials with broiler chickens', *Parasitol. Res.*, 83, 445–451.

Dascalu, S., et al. (2023), 'The influences of microbial colonisation and germ-free status on the chicken TCRb repertoire', *Front. Immunol.*, 13, 1052297.

Dasgupta, T. and Lee, E.H. (2000), 'A gel delivery system for coccidiosis vaccine: uniformity of distribution of oocysts', *Can. Vet. J.*, 41, 613–616.

Davelaar, F.G. and Kouwenhoven, B. (1981) Study on the local effect of eyedropvaccination against infectious bronchitis in 1-day-old chicks with maternal antibodies,Avian Pathology, 10:1, 83-90, DOI: 10.1080/03079458108418460.

de Boer, G.F., et al. (1994), 'Interaction between chicken anaemia virus and live Newcastle disease vaccine', *Avian Pathol.*, 23, 263–275.

Decuypere, E. and Bruggeman, V. (2007), 'The endocrine interface of environmental and egg factors affecting chick quality', *Poult. Sci.*, 86, 1037–1042.

de Herdt, P., et al. (2005), 'Evaluation of the enzyme-linked immunosorbent assay for the detection of antibodies against infectious bursal disease virus (IBDV) and the estimation of the optimal age for IBDV vaccination in broilers', *Avian Pathol.*, 34, 501–504.

de Jong, I.C., et al. (2019), 'Comparison of performance, health and welfare aspects between commercially housed hatchery-hatched and on-farm hatched broiler flocks', *Animal*, 13, 1269–1277.

de Jong, I.C., et al. (2020), 'Effects of on-farm and traditional hatching on welfare, health, and performance of broiler chickens', *Poult. Sci.*, 99, 4662–4671.

Degefa, T., et al. (2004), 'Technical and economic evaluation of different methods of Newcastle disease vaccine administration', *J. Vet. Med. A.*, 51, 365–369.

Deslauriers, N. and Boulianne, M. (2023), 'Evolution of bacterial vaccines: from Pasteur to genomics', *Avian Dis.*, 67, 421–426.

Devlin, J.M., et al. (2008), 'Comparison of the safety and protective efficacy of vaccination with glycoprotein-G-deficient infectious laryngotracheitis virus delivered via eye-drop, drinking water or aerosol', *Avian Pathol.*, 37, 83–88.

Dewidar, A.A.A., et al. (2022), 'Genotype VII.1.1-based Newcastle disease virus vaccines afford better protection against field isolates in commercial broiler chickens', *Animals*, 12, 1696.

de Wit, J.J., et al. (1998), 'Transmission of infectious bronchitis virus within vaccinated and unvaccinated groups of chickens', *Avian Pathol.*, 27, 464–471.

de Wit, J.J. (2000), 'Detection of infectious bronchitis virus', *Avian Pathol.*, 29, 71–93.

de Wit, J.J. (2013), 'Underestimation of the difficulties of vaccination against viral respiratory diseases by mass application methods', *XVIIIth International Congress of the World Veterinary Poultry Association*, France, 63–67.

de Wit, J.J., et al. (2010), 'Efficacy of infectious bronchitis virus vaccinations in the field: association between the alpha-IBV IgM response, protection and vaccine application parameters', *Avian Pathol.*, 39, 123–131.

de Wit, J.J., et al. (2021), '*In ovo* application of a live infectious bursal disease vaccine to commercial broilers confers proper immunity', *Avian Pathol.*, 50, 531–539.

de Wit, J.J. and Montiel, E. (2022), 'Practical aspects of poultry vaccination', In B. Kaspers, et al. (Eds), *Avian immunology*, Academic Press, London, UK, pp. 469–488.

Dey, S., et al. (2019), 'Infectious bursal disease virus in chickens: prevalence, impact, and management strategies', *Vet. Med. Res.*, 10, 85–97.

Dimitrov, K.M., et al. (2017), 'Newcastle disease vaccines: a solved problem or a continuous challenge?', *Vet. Microbiol.*, 206, 126–136.

Dimitrov, K.M., et al. (2021), 'Novel recombinant Newcastle disease virus-based *in ovo* vaccines bypass maternal immunity to provide full protection from early virulent challenge', *Vaccines*, 9, 1189.

Ding, J., et al. (2017), 'Inheritance and establishment of gut microbiota in chickens', *Front. Microbiol.*, 8, 1967.

Ding, P., et al. (2024), 'Effects of surface sterilization of fertile eggs on the yolk microbiota during the chicken embryo development', *Front. Vet. Sci.*, 11, 1493415.

Dunn, J.R., et al. (2019), 'Evaluation of protective efficacy when combining turkey herpesvirus-vector vaccines', *Avian Dis.*, 63, 75–83.

EFSA Panel on Animal Health and Animal Welfare / EFSA AHAW Panel, et al. (2023), 'Scientific Opinion on the welfare of broilers on farm', *EFSA Journal 2023*, 21(2), 7788. https://doi.org/10.2903/j.efsa.2023.7788.

Eidson, C.S., et al. (1977), 'Comparison of the efficacy and transmissibility of conventional NDV vaccines and vaccines prepared after back-passage through chickens', *Poult Sci.*, 56, 19–25.

Eidson, C.S., et al. (1982), 'Vaccination of broiler chicks from breeder flocks immunized with a live or inactivated oil emulsion Newcastle disease vaccine', *Poult Sci.*, 61, 1621–1629.

Elaroussi, M.A., et al. (2006), 'Experimental ochratoxicosis in broiler chickens', *Avian Pathol.*, 35, 263–269.

Elliott, K.E.C., et al. (2018), 'Early post-hatch survival and humoral immune response of layer chickens when *in ovo* vaccinated with strain F *Mycoplasma gallisepticum*', *Poult. Sci.* 97, 3860–3869.

Espejo, R., et al. (2024), 'Immune responses in the Harderian gland after newcastle disease vaccination in chickens with maternal antibodies', *Avian Dis.*, 68, 192–201.

Eterradossi, N. and Saif, Y.M. (2020), 'Infectious bursal disease', In D.E. Swayne (Ed), *Diseases of poultry*, John Wiley & Sons, Hoboken, NJ, USA, pp. 257–283.

Ezeokoli, C.D., et al. (1990), 'Immunosuppression and histopathological changes in the bursa of Fabricius associated with infectious bursal disease vaccination in chicken', *Comp. Immunol. Microbiol. Infect. Dis.*, 13, 181–188.

Faiz, N., et al. (2016), 'Early infection with Marek's disease virus can jeopardize protection conferred by laryngotracheitis vaccines: a method to study MDV-induced immunosuppression', *Avian Pathol.*, 45, 606–615.

Faragher, J.T., et al. (1974), 'Immunosuppressive effect of infectious bursal agent on vaccination against Newcastle disease', *Vet. Rec.*, 95, 385–388.

Fatemi, S.A., et al. (2023), 'Effects of the *in ovo* injection of an *Escherichia coli* vaccine on the hatchability and quality characteristics of commercial layer hatchlings', *Poult Sci.*, 102, 103057.

Fatoba, A.J. and Adeleke, M.A. (2019), 'Chicken anemia virus: a deadly pathogen of poultry', *Acta Virol.*, 63, 19–25.

Fernandes, J.I.M.I., et al. (2016), 'Evaluation of incubation yield, vaccine response, and performance of broilers submitted to *in-ovo* vaccination at different embryonic ages', *Rev. Bras. Cienc. Avic.*, 18 (S2), 55–63.

Ferreira, H.L., et al. (2020), 'Protection conferred by commercial NDV live attenuated and double recombinant HVT vaccines against virulent California 2018 Newcastle disease virus (NDV) in chickens', *Vaccine*, 38, 5507–5515.

Francis, M.J. (2021), 'Spotlight on avian pathology: the importance of recombinant vector platform technologies in poultry vaccination', *Avian Pathol.*, 50, 109–111.

French, D. (2022), 'Incidence and economic impact of reovirus in the poultry industries in the United States', *Avian Dis.*, 66, 432–434.

Friedman, A., et al. (1992), 'Marek's disease vaccines cause temporary B-lymphocyte dysfunction and reduced resistance to infection in chicks', *Avian Pathol.*, 21, 621–631.

Gaghan, C., et al. (2022), 'Characterization of vaccine-induced immune responses against coccidiosis in broiler chickens', *Vaccine*, 40, 3893–3902.

Gagic, M., et al. (1999), '*In ovo* vaccination of specific-pathogen-free chickens with vaccines containing multiple agents', *Avian Dis.*, 43, 293–301.

Gallardo, R.A., et al. (2012), 'Effects of chicken anaemia virus and infectious bursal disease virus-induced immunodeficiency on infectious bronchitis virus replication and genotypic drift', *Avian Pathol.*, 41, 451–458.

Gallili, G.E. and Ben-Nathan, D. (1998), 'Newcastle disease vaccines', *Biotechnol. Adv.*, 16, 343–366.

Ganapathy, K., et al. (2005), 'Interaction between live avian pneumovirus and Newcastle disease virus vaccines in specific pathogen free chickens', *Avian Pathol.*, 34, 297–302.

Ganapathy, K. and Parthiban, S. (2023), 'Pros and cons on use of live viral vaccines in commercial chicken flocks', *Avian Dis.*, 67, 410–420.

Gao, H., et al. (2023), 'Genetic insight into the interaction of IBDV with host - a clue to the development of novel IBDV vaccines', *Int. J. Mol. Sci.*, 24, 8255.

Garcia, M. and Zavala, G. (2019), 'Commercial vaccines and vaccination strategies against infectious laryngotracheitis: what we have learned and knowledge gaps that remain. *Avian Dis.*, 63, 325–334.

Garrido, C., et al. (1972), 'Storage of a cell-associated herpes-virus of turkeys (HVT strain FC126) vaccine', *Avian Dis.*, 16, 45–51.

Ge, J., et al. (2014), 'Novel in-ovo chimeric recombinant Newcastle disease vaccine protects against both Newcastle disease and infectious bursal disease', *Vaccine*, 32, 1514–1521.

Geerligs, H.J. and Hoogendam, A. (2007), 'Determination of optimal conditions for thawing and diluting cell-bound CVI 988 Marek's disease vaccine and stability of the diluted vaccine', *Avian Dis.*, 51, 969–973.

Gelb, J., et al. (2007), 'Evaluating viral interference between infectious bronchitis virus and Newcastle disease virus vaccine strains using quantitative reverse transcription-polymerase chain reaction', *Avian Dis.*, 51, 924–934.

Gelb, J., et al. (2016), 'Efficacy of recombinant HVT-IBD vaccines administered to broiler chicks from a single breeder flock at 30 and 60 weeks of age', *Avian Dis.*, 60, 603–612.

Gentry, R.F. and Braune, M.O. (1972), 'Prevention of virus inactivation during drinking water vaccination of poultry', *Poult. Sci.*, 51, 1450–1456.

Gharaibeh, S., et al. (2008), 'Field evaluation of maternal antibody transfer to a group of pathogens in meat-type chickens', *Poult. Sci.*, 87, 1550–1555.

Ghareeb, A.F.A., et al. (2022), 'Heat stress modulates the disruptive effects of *Eimeria maxima* infection on the ileum nutrient digestibility, molecular transporters, and tissue morphology in meat-type chickens', *PLoS One*, 17, e0269131.

Giersberg, M.F., et al. (2021), 'Effects of hatching system on the welfare of broiler chickens in early and later life', *Poult. Sci.*, 100, 100946.

Gildersleeve, R.P., et al. (1993), 'Developmental performance of an egg injection machine for administration of Marek's disease vaccine', *J. Appl. Poult. Res.*, 2, 337–346.

Gimeno, I.M., et al. (2015), '*In ovo* vaccination with turkey herpesvirus hastens maturation of chicken embryo immune responses in specific-pathogen-free chickens', *Avian Dis.*, 59, 375–383.

Gimeno, I.M. and Schat, K.A. (2018), 'Virus-induced immunosuppression', *Avian Dis.*, 62, 272–285.

Glisson, J.R. (1998), 'Bacterial respiratory disease of poultry', *Poult. Sci.*, 77, 1139–1142.

Gobel, T.W. and Smith, A.L., (2022), 'Introduction to the avian innate immune system; properties, effects, and integration with other parts of the immune system', In B. Kaspers, et al. (Eds), *Avian immunology*, Academic Press, London, UK, pp. 163–166.

Gogoi, S., et al. (2021), 'Impact of heat stress on broilers with varying body weights: Elucidating their interactive role through physiological signatures', *J. Therm. Biol.*, 97, 102840.

Gong, H., et al. (2023), 'Maternal effects drive intestinal development beginning in the embryonic period on the basis of maternal immune and microbial transfer in chickens', *Microbiome*, 11, 41.

Guerin, J.L., t al. (2023), 'Vaccination technology in poultry: principles of vaccine administration', *Avian Dis.*, 67, 489–494.

Guilloteau, L.A., et al. (2024), 'On-farm hatching and contact with adult hen post hatch induce sex-dependent effects on performance, health and robustness in broiler chickens', *Peer Comm. J.*, 4, e20.

Guittet, M., et al. (1992), 'Safety of infectious bursal disease vaccines: assessment of an acceptability threshold', *Dev. Biol. Stand.*, 79, 147–152.

Gul, H., et al. (2022), 'Genetic resilience in chickens against bacterial, viral and protozoal pathogens', *Front. Vet. Sci.*, 9, 1032983.

Guo, Y., et al. (2003a), 'The influence of β-glucan on immune responses in broiler chicks', *Immunopharmacol. Immunotoxicol.*, 25, 461–472.

Guo, Z.Y., et al. (2003b), 'Influence of a reovirus-antibody complex vaccine on efficacy of Marek's disease vaccine administered *in ovo*', *Avian Dis.*, 47, 1362–1367.

Guo, Z.Y., et al. (2004), 'Effect of *in ovo* administered reovirus vaccines on immune responses of specific-pathogen-free chickens', *Avian Dis.*, 48, 224–228.

Gustafson, C.R., et al. (2003), 'Postural and neurologic deficits in broiler chicks after cervical vaccination with live vaccine', *J. Vet. Diagn. Invest.*, 15, 361–364.

Haddad, E., et al. (1997), 'Efficacy of a novel infectious bursal disease virus immune complex vaccine in broiler chickens', *Avian Dis.*, 41 4, 882–889.

Hafez, H.M. and Attia, Y.A. (2020), 'Challenges to the poultry industry: current perspectives and strategic future after the COVID-19 outbreak', *Front. Vet. Sci.*, 26, 516.

Halvorson, D.A, and Mitchel, D.O. (1979), 'Loss of cell-associated Marek's disease vaccine titer during thawing, reconstitution and use', *Avian Dis.*, 23, 848–853.

Hao, X., et al. (2021), 'The evaluation of cellular immunity to avian viral diseases: methods, applications, and challenges', *Front. Microbiol.* 12, 794514.

Hartcher, K.M. and Lum, H.K. (2020), 'Genetic selection of broilers and welfare consequences: a review', *Worlds Poult. Sci. J.*, 76, 154–167.

Hartle, S., et al. (2022), 'Structure and evolution of avian immunoglobulins', In B. Kaspers, et al. (Eds), *Avian immunology*, Academic Press, London, UK, pp. 101–120.

Hauck, R. and Macklin, K.S. (2023), 'Vaccination against poultry parasites', *Avian Dis.*, 67, 441–449.

Hein, R., et al. (2021), 'Review of poultry recombinant vector vaccines', *Avian Dis.*, 65, 438–452.

Hess, M., et al. (2004), 'Hatchability, serology and virus excretion following *in ovo* vaccination of chickens with an avian metapneumovirus vaccine', *Avian Pathol.*, 33, 576–580.

Higgins, D.A. (1975), 'Physical and chemical properties of fowl immunoglobulins. *Vet. Bull.*, 45, 139–154.

Hill, W.G. and Kirkpatrick, M. (2019), 'What animal breeding has taught us about evolution', *Annu. Rev. Ecol. Evol. Syst.*, 41, 1–19.

Hincke, M.T., et al. (2019), 'Dynamics of structural barriers and innate immune components during incubation of the avian egg: critical interplay between autonomous embryonic development and maternal anticipation', *J. Innate Immun.*, 11, 111–124.

Hirakawa, R., et al. (2020), 'Heat stress causes immune abnormalities via massive damage to effect proliferation and differentiation of lymphocytes in broiler chickens', *Front. Vet. Sci.*, 7, 46.

Hoelzer, K., et al. (2018a), 'Vaccines as alternatives to antibiotics for food producing animals. Part 1: challenges and needs', *Vet. Res.*, 49, 64.

Hoelzer, K., et al. (2018b), 'Vaccines as alternatives to antibiotics for food producing animals. Part 2: new approaches and potential solutions', *Vet. Res.*, 49, 70.

Hoerr, F.J. (2010), 'Clinical aspects of immunosuppression in poultry', *Avian Dis.*, 54, 2–15.

Hofmann, T., et al. (2020), 'Impact of housing environment on the immune system in chickens: a review', *Animals*, 10, 1138.

Hofmann, T. and Stefanski, V. (2025), 'Keep the distance - social stress in chickens and its consequences for the immune system', *World's Poult. Sci. J.*, 81, 1, 65–86.

Honda, B.T.B., et al. (2015), 'Effects of heat stress on peripheral T and B lymphocyte profiles and IgG and IgM serum levels in broiler chickens vaccinated for Newcastle disease virus', *Poult. Sci.*, 94, 2375–2381.

Hopkins, S.R. and Yoder, H.W. (1986), 'Reversion to virulence of chicken-passaged infectious bronchitis vaccine virus', *Avian Dis.*, 30, 221–223.

Hossain, M.E. and Akter, N. (2022), 'Further insights into the prevention of pulmonary hypertension syndrome (ascites) in broiler: a 65-year review', *World's Poult. Sci. J.*, 78(3), 641–688.

Hu, Z., et al. (2022), 'Current situation and future direction of Newcastle disease vaccines', *Vet. Res.*, 53, 99.

Hulet, R., et al. (2007), 'Influence of egg shell embryonic incubation temperature and broiler breeder flock age on posthatch growth performance and carcass characteristics. *Poult. Sci.*, 86, 408–412.

Ike, A.C., et al. (2021), 'Towards improved use of vaccination in the control of Infectious bronchitis and Newcastle disease in poultry: understanding the immunological mechanisms', *Vaccines*, 9, 20.

Ingrao, F., et al. (2017), 'Characterization of two recombinant HVT-IBD vaccines by VP2 insert detection and cell-mediated immunity after vaccination of specific pathogen-free chickens', *Avian Pathol.*, 46, 289–299.

Ingrao, F., et al. (2018), 'Early immune responses and profiling of cell-mediated immunity-associated gene expression in response to rHVT-IBD vaccination', *Vaccine*, 36, 615–623.

Iqbal, M. (2012), 'Progress toward the development of polyvalent vaccination strategies against multiple viral infections in chickens using herpesvirus of turkeys as vector', *Bioengineered*, 3, 222–226.

Isham, I.M., et al. (2023), 'Impact of maternal antibodies on infectious bronchitis virus (IBV) infection in primary and secondary lymphoid organs of chickens', *Vaccines*, 11, 1216.

Islam, M.S. and Rahman, M.T. (2023), 'A comprehensive review on bacterial vaccines combating antimicrobial resistance in poultry', *Vaccines*, 11, 616.

Ismail, N.M. and Saif, Y.M. (1991), 'Immunogenicity of infectious bursal disease viruses in chickens', *Avian Dis.*, 35, 460–469.

Jackwood, D.J. (2017), 'Advances in vaccine research against economically important viral diseases of food animals: Infectious bursal disease virus', *Vet. Mic.*, 206, 121–125.

Jackwood, M.W. and Lee, D.H. (2017), 'Different evolutionary trajectories of vaccine-controlled and non-controlled avian infectious bronchitis viruses in commercial poultry', *PLoS One*, 12, e0176709.

Jackwood, M.W. and de Wit, S. (2020), 'Infectious bronchitis', In D.E. Swayne (Ed), *Diseases of poultry*, John Wiley & Sons, Hoboken, NJ, USA, pp. 167–188.

Jansson, D., et al. (2022), 'Anticoccidial vaccination is associated with improved intestinal health in organic chickens', *Vet. Sci.*, 9, 347.

Jen, L.W. and Cho, B.R. (1980), 'Effects of infectious bursal disease on Marek's disease vaccination: suppression of antiviral immune response', *Avian Dis.*, 24, 896–907.

Jennison, R. (2021), 'Biosecurity', In P. Barrow, et al. (Eds), *Poultry :health a guide for professionals*, CABI International, Wallingford, Oxfordshire, UK.

Jerab, J.G., et al. (2023), 'Antimicrobial use in on-farm hatching systems vs. traditional hatching systems: a case study', *Animals*, 13, 3270.

Jessen, C.T., et al. (2021a), 'Effects of hatching on-farm on performance and welfare of organic broilers', *Poult. Sci.*, 100, 101292.

Jessen, C.T., et al. (2021b), 'Effects of hatching on-farm on behaviour, first week performance, fear level and range use of organic broilers', *Appl. Anim. Behav. Sci.*, 238, 105319.

Jeurissen, S.H., et al. (1998), 'The working mechanism of an immune complex vaccine that protects chickens against infectious bursal disease', *Immunology*, 95, 494–500.

Johnson, D.I., et al. (2010), 'Protection against infectious laryngotracheitis by *in ovo* vaccination with commercially available viral vector recombinant vaccines', *Avian Dis.*, 54, 1251–1259.

Johnston, P.A., et al. (1997), 'Applications in *in ovo* technology', *Poult. Sci.*, 76, 165–178.

Jones, R.C. (2000), 'Avian reovirus infections', *Rev. Sci. Tech. Off. Int. Epiz.*, 19, 614–625.

Jones, P.J., et al. (2018), 'A review of the financial impact of production diseases in poultry production systems', *Anim. Prod. Sci.*, 59, 1585–1597.

Jordan, B. (2017), 'Vaccination against infectious bronchitis virus: a continuous challenge', *Vet. Microbiol.*, 206, 137–143.

Julian, R.J. (2005), 'Production and growth related disorders and other metabolic diseases of poultry-a review', *Vet. J.*, 169, 350–369.

Kaiser, P. (2010), 'Advances in avian immunology-prospects for disease control: a review', *Avian Pathol.*, 39, 309–324.

Kaiser, P. and Balic A. (2014), 'The avian immune system', In C.G. Scanes (Ed), *Avian immunology*, Academic Press, London, UK, pp. 392–407.

Kallam, N.R.K. and Sejian, V. (2021), 'Gut health and immunity in improving poultry production', In A.K. Patra (Ed), *Advances in poultry nutrition research*, IntechOpen, London, UK, pp. 1–21.

Kamau, N., et al. (2010), 'Effect of Vac-Pac Plus on the viability of a live infectious bronchitis vaccine', *J. Appl. Poult. Res.*, 19, 152–156.

Kapczynski, D. and King, D. (2005), 'Protection of chickens against overt clinical disease and determination of viral shedding following vaccination with commercially available Newcastle disease virus vaccines upon challenge with highly virulent virus from the California 2002 exotic Newcastle disease outbreak', *Vaccine*, 26, 3424–3433.

Kapczynski, D.R., et al. (2012), 'Protection from clinical disease against three highly virulent strains of Newcastle disease virus after *in ovo* application of an antibody–antigen complex vaccine in maternal antibody positive chickens', *Avian Dis.*, 56, 555–560.

Kaspers, K.A. and Schat, K.A. (2022), 'The mucosal immune system', In B. Kaspers, et al. (Eds), *Avian immunology*, Academic Press, London, UK, pp. 299–301.

Keep, S., et al. (2020a), 'Temperature sensitivity: a potential method for the generation of vaccines against the avian coronavirus infectious bronchitis virus', *Viruses*, 12, 754.

Keep, S., et al. (2020b), 'Limited cross-protection against infectious bronchitis provided by recombinant infectious bronchitis viruses expressing heterologous spike glycoproteins', *Vaccines*, 8, 330.

Khazali, D., et al. (2025), 'Comparison of immune-related responses of several slow-growing indigenous chickens with a fast-growing broiler in Iran', *Vet. Med. Sci.*, 11, e70288.

Killian, M.P., et al. (2017), 'Induction of apoptosis in the bursa of Fabricius by vaccination against Gumboro disease', *Avian Pathol.*, 46, 526–534.

Kjærup, R.B., et al. (2017), 'Comparison of growth performance and immune parameters of three commercial chicken lines used in organic production', *Vet. Immunol. Immunopathol.* 187, 69–79.

Klasing, K.C. (1998), 'Nutritional modulation of resistance to infectious diseases. *Poult. Sci.*, 77, 1119–1125.

Ko, K.H., et al. (2018), 'Changes in bursal B cells in chicken during embryonic development and early life after hatching', *Sci. Rep.*, 8, 16905.

Koch, G. and Elbers, A.R.W. (2006), 'Outdoor ranging of poultry: a major risk factor for the introduction and development of high-pathogenicity avian influenza', *NJAS-Wagen. J. Life Sci.*, 54, 179–194.

Koenen, M.E., et al. (2002), 'Immunological differences between layer- and broiler-type chickens', *Vet. Immunol. Immunopathol.*, 89, 47–56.

Kogut, M.H., et al. (2020), 'Microbiome and pathogen interaction with the immune system', *Poult. Sci.*, 99, 1906–1913.

Kogut, M.H. (2022), 'Impact of the gut microbiota on the immune system', In B. Kaspers, et al. (Eds), *Avian immunology*, Academic Press, London, UK, pp. 353–364.

Korver, D.R. (2023), 'Current challenges in poultry nutrition, health, and welfare', *Animal*, 17, 100755.

Kowalczyk, K., et al. (1985), 'Quantitation of maternal-fetal IgG transport in the chicken', *Immunology*, 54, 755–762.

Kristensen, H.H. and Wathes, C.M. (2000), 'Ammonia and poultry welfare: a review', *World's Poult. Sci. J.*, 56, 235–245.

Kufuor-Mensah, E., et al. (2016), 'Effects of T-2 toxin on turkey herpesvirus-induced vaccinal immunity against Marek's disease', *Avian Dis.*, 60, 56–62.

Kustra, K., et al. (2024), 'A conventional hatchery vs "on-farm" hatching of broiler chickens in terms of microbiological and microclimatic conditions', *Animal*, 18, 101223.

Kutle, L., et al. (2020), 'Efficacy of Infectious Bronchitis GI-13 (793B) vaccine candidate tested according to the current european union requirements and for cross-protection against heterologous QX-Like challenge', *Viral. Immunol.*, 33, 555–564.

Landman, W.J.M. (2012), 'The downside of broiler vaccination', *Vet. Q.*, 32, 121–122.

Landman, W.J.M., et al. (2015), 'Comparison of Newcastle disease vaccine administered as powder or liquid in relation to the serum antibody response and adverse vaccinal reactions in broilers', *Avian Pathol.*, 44, 114–123.

Lee, S., et al. (2019), 'Characterization of microbial communities in the chicken oviduct and the origin of chicken embryo gut microbiota', *Sci. Rep.*, 9, 6838.

Lee, Y., et al. (2022), 'Coccidiosis: recent progress in host immunity and alternatives to antibiotic strategies', *Vaccines*, 10, 215.

Legnardi, M., et al. (2020), 'Infectious bronchitis hatchery vaccination: comparison between traditional spray administration and a newly developed gel delivery system in field conditions', *Vet. Sci.*, 8, 145.

Lemiere, S. (2013), 'The cost benefits of vaccination in poultry production', *Int. Poult. Prod.*, 21, 19–21.

Li, X., et al. (2024), 'How to break through the bottlenecks of *in ovo* vaccination in poultry farming', *Vaccines*, 12, 48.

Lillehoj, H.S. (2018), 'Cell-mediated immunity in parasitic and bacterial diseases', In J.M. Sharma (Ed), *Avian cellular immunology*, Routledge, New York, USA, pp. 155–182.

Liu, L., et al. (2021), 'Biological function of short-chain fatty acids and its regulation on intestinal health of poultry', *Front. Vet. Sci.*, 8, 736739.

Liu, K.L., et al. (2023a), 'Leg disorders in broiler chickens: a review of current knowledge.', *Anim. Biotechnol.*, 34, 5124–5138.

Liu, Q., et al. (2023b), 'Live attenuated anticoccidial vaccines for chickens', *Trends Parasitol.*, 39, 1087–1099.

Liu, P.Y., et al. (2024), 'Multi-omics analysis reveals regime shifts in the gastrointestinal ecosystem in chickens following anticoccidial vaccination and *Eimeria tenella* challenge', *mSystems*, 9, e00947-24.

Ma, X., et al. (2019), 'Combined adjuvant effect of ginseng stem-leaf saponins and selenium on immune responses to a live bivalent vaccine of Newcastle disease virus and infectious bronchitis virus in chickens', *Poult. Sci.*, 98, 3548–3556.

Madej, J.P., et al. (2024), 'Impact of early posthatch feeding on the immune system and selected hematological, biochemical, and hormonal parameters in broiler chickens', *Poult. Sci.*, 103, 103366.

Maes, D., et al. (2021), 'Bacterial diseases in pigs and poultry: occurrence, epidemiology, and biosecurity measures', In N. Foster, et al. (Eds), *Advancements and technologies in pig and poultry bacterial disease control*, Academic Press, London, UK, pp. 25–51.

Marangon, S. and Busani, L. (2006), 'The use of vaccination in poultry production', *Rev. Sci. Tech. Off. Int. Epiz.*, 26, 265–274.

Markowski-Grimsrud, C.J. and Schat, K.A. (2002), 'Immunity to chicken infectious anemia virus', *Dev. Comp. Immunol.*, 26, 209–223.

Martínez, Y., et al. (2021), 'Effect of age on the immune and visceral organ weights and cecal traits in modern broilers', *Animals*, 11, 845.

Mast, J. and Goddeeris, B.M. (1999), 'Development of immunocompetence of broiler chickens', *Vet. Immunol. Immunopathol.*, 93, 245–256.

Matthijs, M.G., et al. (2005), 'Effect of IBV-H120 vaccination in broilers on colibacillosis susceptibility after infection with a virulent Massachusetts-type IBV strain', *Avian Dis.*, 49, 540–545.

McFarlane, J.M. and Curtis S.E. (1989), 'Multiple concurrent stressors in chicks. 3. Effects on plasma corticosterone and the heterophil:lymphocyte ratio', *Poult. Sci.*, 68, 522–527.

McGrath, K.E. and Palis, J. (2005), 'Hematopoiesis in the yolk sac: more than meets the eye', *Exp. Hematol.*, 33, 1021–1028.

McMullin, P. (2022), 'Infectious diseases in free-range compared to conventional poultry production', *Avian Pathol.*, 51, 424–434.

Mench, J., et al. (2021), 'Influences of housing and management on broiler welfare', In A. Butterworth, et al. (Eds), *Broiler chickens: welfare in practice*, 5m Books, Lings, Great Easton, UK, pp. 43–64.

Miller, P.J., et al. (2007), 'Antigenic differences among Newcastle disease virus strains of different genotypes used in vaccine formulation affect viral shedding after a virulent challenge', *Vaccine*, 25, 7238–7246.

Miller, P.J., et al. (2009), 'Comparison of viral shedding following vaccination with inactivated and live Newcastle disease vaccines formulated with wild-type and recombinant viruses', *Avian Dis.*, 53, 39–49.

Miller, P.J. and Koch, G. (2020), 'Newcastle disease', In D.E. Swayne (Ed), *Diseases of poultry*, John Wiley & Sons, Hoboken, NJ, USA, pp. 112–129.

Markis, M. (2022), 'Evaluation of pathogenicity and antigenicity of avian reoviruses and disease control through vaccination', *Avian Dis.*, 66, 435–442.

Muller, H., et al. (2003), 'Research on infectious bursal disease: the past, the present and the future', *Vet. Microbiol.*, 97, 153–165.

Muller, H., et al. (2012), 'Current status of vaccines against infectious bursal disease', *Avian Pathol.*, 41, 133–139.

Munoz-Gomez, V., et al. (2024), 'Global and regional prediction of coccidiosis and ascaridiosis prevalence in extensive backyard chickens in low-income and middle-income countries', *Vet. Parasitol.*, 331, 110268.

Munster, P. and Kemper, N. (2024), Long-term analysis of drinking water quality in poultry and pig farms in Northwest Germany', *Front. Anim. Sci.*, 5, 1467287.

Murai, A. (2013), 'Maternal transfer of immunoglobulins into egg yolks of birds', *J. Poul. Sci.*, 50, 185–193.

Nagy, N., et al. (2022), 'Structure of the avian lymphoid system', In B. Kaspers, et al. (Eds), *Avian immunology*, Academic Press, London, UK, pp. 11–44.

Nair, V. (2004), 'Successful control of Marek's disease by vaccination', *Dev. Biol.*, 119, 147–154.

Nair, V., et al. (2020), 'Marek's disease', In D.E. Swayne (Ed), *Diseases of poultry*, John Wiley & Sons, Hoboken, NJ, USA, pp. 550–585.

Nakamura, K., et al. (1992), 'The effect of mixed live vaccines of Newcastle disease and infectious bronchitis on the chicken respiratory tract', *J. Comp. Pathol.*, 106, 341–350.

Nakaya, T., et al. (2001), 'Recombinant Newcastle disease virus as a vaccine vector', *J. Virol.*, 75, 11868–11873.

Nanto-Hara, F., et al. (2020), 'Heat stress directly affects intestinal integrity in broiler chickens', *J. Poult. Sci.*, 57, 284–290.

Nawaz, A.H., et al. (2021), 'Poultry response to heat stress: Its physiological, metabolic, and genetic implications on meat production and quality including strategies to improve broiler production in a warming world', *Front. Anim. Sci.*, 8, 699081.

Neelima, S., et al. (2003), 'Avian reovirus induces an inhibitory effect on lymphoproliferation in chickens', *Vet. Res. Commun.*, 27, 73–85.

Negash, T., et al. (2004), 'Comparison of *in-ovo* and post-hatch vaccination with particular reference to infectious bursal disease: a review', *Vet. Q.*, 26, 76–87.

Netea, M.G., et al. (2016), 'Trained immunity: a program of innate immune memory in health and disease', *Science*, 352, aaf1098.

Niu, Y., et al. (2017), 'Pathogenicity and immunosuppressive potential of fowl adenovirus in specific pathogen free chickens', *Poult. Sci.*, 96, 3885–3892.

Noiva, R.M., et al. (2014), 'Influence of temperature and humidity manipulation on chicken embryonic development', *BMC Vet. Res.*, 10, 234.

Okasha, H., et al. (2024), 'Hidden hazards revealed: mycotoxins and their masked forms in poultry', *Toxins*, 16, 137.

Oke, O.E., et al. (2024), 'Climate change and broiler production', *Vet. Med. Sci.*, 10, e1416.

Okoye, J.O. and Shoyinka, S.V. (1983), 'Newcastle disease in a vaccinated flock which had experienced subclinical infectious bursal disease', *Trop. Anim. Health Prod.*, 15, 221–225.

Okwor, G.O., et al. (2014), 'Profile of maternally derived antibody in broiler chicks and in-ovo vaccination of chick embryo against Newcastle disease', *World J. Vaccines*, 4, 72–80.

Oluwagbenga, E.M. and Fraley, G.S. (2023), 'Heat stress and poultry production: a comprehensive review', *Poult. Sci.*, 102, 103141.

Orr-Burks, N., et al. (2014). 'Immunoglobulin A as an early humoral responder after mucosal avian coronavirus vaccination', *Avian Dis.*, 58, 279–286.

Otaki, Y., et al. (1988), 'Immune response of chicks inoculated with chicken anemia agent alone or in combination with Marek's disease virus or turkey herpesvirus', *Nihon Juigaku Zasshi*, 50, 1040–1047.

Oviedo-Rondón, E.O. (2022), 'Optimizing the health of broilers', In S. de Wit (Ed), *Optimizing poultry flock health*, Burleigh Dodds Science Publishing, Sawston, Cambridge, UK, pp. 249–287.

Palya, V., et al. (2012), 'Advancement in vaccination against Newcastle disease: recombinant HVT NDV provides high clinical protection and reduces challenge virus shedding with the absence of vaccine reactions', *Avian Dis.*, 56, 282–287.

Palya, V., et al. (2021), 'Efficacy of a turkey herpesvirus vectored Newcastle disease vaccine against genotype VII.1.1 virus: challenge route affects shedding pattern', *Vaccines*, 9, 37.

Panda, A.K., et al. (2015), 'Early post hatch nutrition on immune system development and function in broiler chickens', *Worlds Poult. Sci. J.*, 71, 285–295.

Pao, H.N., et al. (2023), 'The attitude-behaviour gap in biosecurity: applying social theories to understand the relationships between commercial chicken farmers' attitudes and behaviours', *Front. Vet. Sci.*, 9, 1070482.

Peebles, E.D., et al. (2017), 'Comparative effects of *in ovo* versus subcutaneous administration of the Marek's disease vaccine and pre-placement holding time on the post-hatch performance of Ross 708 broilers', *Poult. Sci.*, 96, 1071–1077.

Peebles, E.D. (2018), '*In ovo* applications in poultry: a review', *Poult. Sci.*, 97, 2322–2338.

Pejkovski, C., et al. (1979), 'Immunosuppressive effect of infectious bursal disease virus on vaccination against infectious bronchitis', *Avian Path.*, 8, 95–106.

Perozo, F., et al. (2009), 'Efficacy of single dose recombinant herpesvirus of turkey infectious bursal disease virus (IBDV) vaccination against a variant IBDV strain', *Avian Dis.*, 53, 624–628.

Price, K.R., et al. (2015), 'Shedding of live *Eimeria* vaccine progeny is delayed in chicks with delayed access to feed after vaccination', *Vet. Parasitol.*, 208, 242–245.

Psifidi, A. (2021), 'The genetics of disease resistance in poultry', In P. Barrow, et al. (Eds), *Poultry health: a guide for professionals*, CABI International, Wallingford, Oxfordshire, UK.

Quinteiro-Filho, W.M., et al. (2012), 'Acute heat stress impairs performance parameters and induces mild intestinal enteritis in broiler chickens: role of acute hypothalamic-pituitary-adrenal axis activation', *J. Anim. Sci.*, 90, 1986–1994.

Ramon, G., et al. (2022), 'Efficacy of live attenuated, vector and immune complex infectious bursal disease virus (IBDV) vaccines in preventing field strain bursa colonization: a European multicentric study', *Front Vet. Sci.*, 9, 978901.

Rani, S. and Kumar, S. (2015), 'Evaluation of infectious bursal disease virus stability at different conditions of temperature and pH', *Biologicals*, 43, 515–518.

Rasheed, M. and Matsler, P.L. (2020), 'Assessment of protection against *Eimeria tenella* in broiler breeders conferred by a live anticoccidial vaccine and effects of vaccination on early pullet growth', *J. Appl. Poult. Res.*, 29, 447–454.

Ratcliffe, M.J. (2006) 'Antibodies, immunoglobulin genes and the bursa of Fabricius in chicken B cell development', *Dev. Comp. Immunol.*, 30, 101–118.

Ratcliffe, M. and Hartle, S. (2022), 'B cells, the bursa of Fabricius, and the generation of antibody repertoires', In B. Kaspers, et al. (Eds), *Avian immunology*, Academic Press, London, UK, pp. 71–100.

Rautenschlein, S., et al. (2005), 'Protective efficacy of intermediate and intermediate plus Infectious Bursal Disease Virus (IBDV) vaccines against very virulent IBDV in commercial broilers', *Avian Dis.*, 49, 231–237.

Rautenschlein, S., et al. (2007), 'Bilateral effects of vaccination against infectious bursal disease and Newcastle disease in specific-pathogen free layers and commercial broiler chickens', *Avian Dis.*, 51, 14–20.

Rautenschlein, S. (2020a), 'Host factors for disease resistance', In D.E. Swayne (Ed), *Diseases of poultry*, John Wiley & Sons, Hoboken, NJ, USA, pp. 79–108.

Rautenschlein, S. (2020b), 'Avian Metapneumovirus', In D.E. Swayne (Ed), *Diseases of poultry*, John Wiley & Sons, Hoboken, NJ, USA, pp. 135–143.

Rautenschlein, S. and Schat, K.A. (2023), 'The immunological basis for vaccination', *Avian Dis.*, 67, 366–379.

Rauw, F., et al. (2009), 'Humoral, cell-mediated and mucosal immunity induced by oculo-nasal vaccination of one-day-old SPF and conventional layer chicks with two different live Newcastle disease vaccines', *Vaccine*, 27, 3631–3642.

Rauw, F., et al. (2010), 'Improved vaccination against Newcastle disease by an *in ovo* recombinant HVT-ND combined with an adjuvanted live vaccine at day-old', *Vaccine*, 28, 823–833.

Rauw, F., et al. (2014), 'The combination of attenuated Newcastle disease (ND) vaccine with rHVT-ND vaccine at 1 day old is more protective against ND virus challenge than when combined with inactivated ND vaccine', *Avian Pathol.*, 43, 26–36.

Ravikumar, R., et al. (2022), 'Vaccines against major poultry viral diseases: strategies to improve the breadth and protective efficacy', *Viruses*, 14, 1195.

Read, A.F., et al. (2015)', Imperfect vaccination can enhance the transmission of highly virulent pathogens', *PLoS Biol.*, 13, e1002198.

Reddy, S.M., et al. (2017), 'Marek's disease vaccines: current status, and strategies for improvement and development of vector vaccines', *Vet. Microbiol.*, 206, 113–120.

Riber, A.B. and Wurtz, K.E. (2024), 'Impact of growth rate on the welfare of broilers', *Animals*, 14, 3330.

Ricks, C.A., et al. (1999), '*In ovo* vaccination technology', *Adv. Vet. Med.*, 41, 495–515.

Rodrigues, D.R., et al. (2021), 'Proper immune response depends on early exposure to gut microbiota in broiler chicks', *Front. Physiol.*, 12, 758183.

Roh, H.J., et al. (2015), 'Hatchery spray cabinet administration does not damage avian coronavirus infectious bronchitis virus vaccine based on analysis by electron microscopy and virus titration', *Avian Dis.*, 59, 149–152.

Romanutti, C., et al. (2020), 'Current status of virus-vectored vaccines against pathogens that affect poultry', *Vaccine*, 38, 6990–7001.

Rosenberger, J., et al. (1987), 'Use of infectious bursal disease virus variant vaccines in broilers and broiler breeders', *36th Western Poultry Disease Conference*, California (USA), 105–109.

Rosenberger, J.K. and Cloud, S.S. (1989), 'The effects of age, route of exposure, and coinfection with infectious bursal disease virus on the pathogenicity and transmissibility of chicken anemia agent (CAA)', *Avian Dis.*, 33, 753–759.

Saiada, F., et al. (2019), 'Early vaccination of chickens induces suboptimal immunity against infectious bronchitis virus', *Avian Dis.*, 63, 38–47.

Saif, Y.M. and Swayne, D.E. (2020), 'Emerging diseases and diseases of complex or unknown etiology', In D.E. Swayne (Ed), *Diseases of poultry*, John Wiley & Sons, Hoboken, NJ, USA, pp. 1385–1386.

Saleh, N.A., et al. (2023), 'Influence of water quality and pollution on broiler's performance, vaccine and antibiotic efficiencies: a review', *Ann. Anim. Sci.*, 23, 1021–1036.

Salles, G.B.C., et al. (2023), 'Trends and challenges in the surveillance and control of Avian Metapneumovirus', *Viruses*, 15, 1960.

Sander, J.E., et al. (1998), 'The effect of inoculating *Enterococcus faecalis* into the yolk sac on chick quality and maternal antibody absorption', *Avian Dis.*, 42, 359–363.

Schat, K.A. (2014), 'Vaccines and vaccination practices: key to sustainable animal production', In N. Van Alfen (Ed), *Encyclopedia of agriculture and food systems*, Academic Press, London, UK, pp. 315–332.

Schat, K.A. and Skinner, M.A. (2022), 'Avian immunosuppressive diseases and immune evasion', In B. Kaspers, et al. (Eds), *Avian immunology*, Academic Press, London, UK, pp. 387–417.

Scheele, C.W. (1999), 'Pathological changes in metabolism of poultry related to increasing production levels', *Vet. Q.*, 19, 127–130.

Schijns, V.E.J.C., et al. (2014), 'Practical aspect of poultry vaccination', In C.G. Scanes (Ed), *Avian immunology*, Academic Press, London, UK, pp. 345–362.

Schrader, J.S. and Smith, H.E. (2023), 'Licensing requirements for poultry veterinary biologics in the United States', *Avian Dis.*, 67, 467–473.

Sedeik, M., et al. (2019), 'Comparative evaluation of HVT-IBD vector, immune complex, and live IBD vaccines against vvIBDV in commercial broiler chickens with high maternally derived antibodies', *Animals*, 9, 72.

Senne, D.A., et al. (2004), 'Control of Newcastle disease by vaccination', *Dev. Biol.*, 119, 165–170.

Shafaati, M., et al. (2024), 'An overview of progress in the development of Newcastle disease vaccines, from empirical to rational design in modern vaccine development', *World's Poult. Sci. J.*, 80, 299–327.

Shanmugasundaram, R., et al. (2025), 'Exposure to subclinical doses of fumonisins, deoxynivalenol, and zearalenone affects immune response, amino acid digestibility, and intestinal morphology in broiler chickens', *Toxins*, 17, 16.

Sharma, J.M. and Burmester, B.R. (1982), 'Resistance to Marek's disease at hatching in chickens vaccinated as embryos with the turkey herpesvirus', *Avian Dis.*, 26, 134–149.

Sharma, J.M. and Witter, R.L. (1983), 'Embryo vaccination against Marek's disease with serotypes 1, 2 and 3 vaccines administered singly or in combination', *Avian Dis.*, 27, 453–463.

Sharma, J.M. (1984), 'Effect of infectious bursal disease virus on protection against Marek's disease by turkey herpesvirus vaccine', *Avian Dis.*, 28, 629–640.

Sharma, J.M. (1999), 'Introduction to poultry vaccines and immunity', *Adv. Vet. Med.*, 41, 481–522.

Sharma, J.M., et al. (2002), 'Field trial in commercial broilers with a multivalent *in ovo* vaccine comprising a mixture of live viral vaccines against Marek's disease, infectious bursal disease, Newcastle disease, and fowl pox', *Avian Dis.*, 46, 613–622.

Sharman, P.A., et al. (2010), 'Chasing the golden egg: vaccination against poultry coccidiosis', *Parasite Immunol.*, 32, 590–598.

Shehata, A.M., et al. (2021), 'Managing gut microbiota through *in ovo* nutrition influences early-life programming in broiler chickens', *Animals*, 11, 3491.

Shimizu, M., et al. (2020), 'Effects of avian infectious bronchitis with Newcastle disease and Marek's disease vaccinations on the expression of toll-like receptors and avian β-defensins in the kidneys of broiler chicks', *Poult. Sci.*, 99, 7092–7100.

Shterzer, N., et al. (2023), 'Vertical transmission of gut bacteria in commercial chickens is limited', *Anim. Microbiome*, 5, 50.

Siegmann, O., et al. (1980), 'Short- and long-term stability studies on four lyophilised and one cell-associated turkey herpesvirus vaccines against Marek's disease of chickens', *Avian Pathol.*, 9, 21–32.

Simon, K., et al. (2014), 'Development of ileal cytokine and immunoglobulin expression levels in response to early feeding in broilers and layers', *Poult. Sci.*, 93, 3017–3027.

Singh, H., et al. (2006), 'Effects of dietary supplements of selenium, vitamin E or combinations of the two on antibody responses of broilers', *Br. Poult. Sci.*, 47, 714–719.

Skinner, M.A., et al. (2005), 'Fowlpox virus as a recombinant vaccine vector for use in mammals and poultry', *Expert Rev. Vaccines*, 4, 63–76.

Slegers, Y., et al. (2024), 'Broiler flocks in production systems with slower-growing breeds and reduced stocking density receive fewer antibiotic treatments and have lower mortality', *Poult. Sci.*, 103, 104197.

Smiałek, M., et al. (2011), 'Local immunity of the respiratory mucosal system in chickens and turkeys', *Pol. J. Vet. Sci.*, 14, 291–297.

Smialek, M., et al. (2017), 'Immunological aspects of the efficiency of protectotype vaccination strategy against chicken infectious bronchitis', *BMC Vet. Res.*, 13, 44.

Smialek, M., et al. (2024), 'Comparison of three different IBD vaccination protocols in broiler chicken - efficacy, serological baselines and histo-pathological lesions in the bursa of Fabricius', *Animals*, 14, 3116.

Smith, J.A., et al. (2020), 'Disease prevention and control in antibiotic-free production', In D.E. Swayne (Ed), *Diseases of poultry*, John Wiley & Sons, Hoboken, NJ, USA, pp. 40–51.

Smith, A.L. and Gobel, T.W. (2022), 'Avian T cells: antigen recognition and lineages', In B. Kaspers, et al. (Eds), *Avian immunology*, Academic Press, London, UK, pp. 121–134.

Smith, A.L., et al. (2022), 'The avian enteric immune system in health and disease', In B. Kaspers, et al. (Eds), *Avian immunology*, Academic Press, London, UK, pp. 303–326.

Sokale, A.O., et al. (2017), 'Effects of coccidiosis vaccination administered by *in ovo* injection on the hatchability and hatching chick quality of broilers', *Poult. Sci.*, 96, 541–547.

Sondermeijer, P.J., et al. (1993), 'Avian herpesvirus as a live viral vector for the expression of heterologous antigens', *Vaccine*, 11, 349–358.

Song, B., et al. (2021), 'Effects of age on immune function in broiler chickens', *J. Animal Sci. Biotechnol.*, 12, 42.

Studniski, M., et al. (2023), 'Vaccination of poultry against influenza', *Avian Dis.*, 67, 402–409.

Sultan, H., et al. (2012), 'Efficacy of HVT-IBDV vector vaccine against recent Egyptian vvIBDV in commercial broiler chickens', *Int. J. Poult. Sci.*, 11, 710–717.

Sultan, H.A., et al. (2019), 'Protective efficacy of different live attenuated Infectious Bronchitis virus vaccination regimes against challenge with IBV variant-2 circulating in the Middle East', *Front. Vet. Sci.*, 6, 341.

Suo, X., et al. (2006), 'The efficacy and economic benefits of Supercox, a live anticoccidial vaccine in a commercial trial in broiler chickens in China', *Vet. Parasitol.*, 142, 63–70.

Surai, P.F. and Fisinin, V.I. (2016a), 'Vitagenes in poultry production: part 1. Technological and environmental stresses', *World's Poult. Sci. J.*, 72, 721–734.

Surai, P.F. and Fisinin, V.I. (2016b), 'Vitagenes in poultry production: part 2. Nutritional and internal stresses', *World's Poult. Sci. J.*, 72, 761–772.

Suresh, P., et al. (1994), 'Mucosal and systemic humoral immune response to *Bordetella avium* in experimentally infected turkeys', *Avian Dis.*, 38, 225–230.

Swaggerty, C.L., et al. (2009), 'Selection of broilers with improved innate immune responsiveness to reduce on-farm infection by foodborne pathogens', *Foodborne Pathog Dis.*, 6, 777–783.

Tatár-Kis, T., et al. (2020), 'A herpesvirus of turkey-based vector vaccine reduces transmission of Newcastle disease virus in commercial broiler chickens with maternally derived antibodies', *Vaccines*, 8, 226.

Tavares Guedes, P., et al. (2014), 'Histological analyses demonstrate the temporary contribution of yolk sac, liver, and bone marrow to hematopoiesis during chicken development', *PLoS One*, 9, e90975.

Tell, L.A., et al. (2012), 'Study of nebulization delivery of aerosolized fluorescent microspheres to the avian respiratory tract', *Avian Dis.*, 56, 381–386.

Tellis, C., et al. (2024), 'Pasture vs. coop: biomarker insights into free-range and conventional broilers', *Animals*, 14, 3070.

Tensa, L.R. and Jordan B.J. (2019), 'Comparison of the application parameters of coccidia vaccines by gel and spray', *Poult. Sci.*, 98, 634–641.

Tesar, D.B., et al. (2008), 'The chicken yolk sac IgY receptor, a mammalian mannose receptor family member, transcytoses IgY across polarized epithelial cells', *Mol. Biol. Cell*, 19, 1587–1593.

Thornton, D.H. and Muskett, J.C. (1973), 'Comparison of immunity to Newcastle disease after vaccination with Newcastle disease vaccine given alone or together with infectious bronchitis vaccine', *Vet. Rec.*, 92, 373–374.

Thornton, D.H. and Muskett, J.C. (1975), 'Effect of infectious bronchitis vaccination on the performance of live Newcastle disease vaccine', *Vet. Rec.*, 96, 467–468.

Tian, Z. and Zhang, X. (2012), 'Progress on research of chicken IgY antibody-FcRY receptor combination and transfer', *J. Recept. Signal Transduct.*, 32, 231–237.

Toffan, A., et al. (2013), 'Diagnostic and clinical observation on the infectious bronchitis virus strain Q1 in Italy', *Vet. Ital.*, 49, 347–355.

Tona, K., et al. (2007), 'Non-ventilation during early incubation in combination with dexamethasone administration during late incubation: 1. Effects on physiological hormone levels, incubation duration and hatching events', *Domest. Anim. Endocrinol.*, 33, 32–46.

Toro, H., et al. (2009), 'Effects of chicken anemia virus and infectious bursal disease virus in commercial chickens', *Avian Dis.* 53, 94–102.

Toro, H. (2021), 'Global control of infectious bronchitis requires replacing live attenuated vaccines by alternative technologies', *Avian Dis.*, 65, 637–642.

Tsukamoto, K., et al. (2002), 'Complete, long-lasting protection against lethal infectious bursal disease virus challenge by a single vaccination with an avian herpesvirus vector expressing VP2 antigens', *J. Virol.*, 76, 5637–5645.

Ulmer-Franco, A.M. (2012), 'Transfer of chicken immunoglobulin Y (IgY) from the hen to the Chick', *Avian Biol. Res.*, 5, 81–87.

Vagnozzi, A., et al. (2010), 'Protection induced by infectious laryngotracheitis virus vaccines alone and combined with Newcastle disease virus and/or infectious bronchitis virus vaccines', *Avian Dis.*, 54, 1210–1219.

Vaillancourt, J.P., et al. (2022), 'Improving biosecurity in poultry flocks', In S. de Wit (Ed), *Optimizing poultry flock health*, Burleigh Dodds Science Publishing, Cambridge, UK, pp. 107–148.

van Boven, M., et al. (2008), 'Herd immunity to Newcastle disease virus in poultry by vaccination', *Avian Pathol.*, 37, 1–5.

van Emous, R. (2022), 'Optimizing the health of broiler breeder birds', In S. de Wit (Ed), *Optimizing poultry flock health*. Burleigh Dodds Science Publishing, Sawston, Cambridge, UK, pp. 347–385.

van Ginkel, F.W., et al. (2008), 'Infectious bronchitis virus in the chicken Harderian gland and lachrymal fluid: viral load, infectivity, immune cell responses, and effects of viral immunodeficiency', *Avian Dis.*, 52, 608–617.

van Ginkel, F.W., et al. (2015), 'Age-dependent immune responses and immune protection after avian coronavirus vaccination', *Vaccine*, 33, 2655–2661.

Van Limbergen, T., et al. (2018), 'Scoring biosecurity in European conventional broiler production', *Poult. Sci.*, 97, 74–83.

Vaziry, A., et al. (2011), 'Chicken infectious anaemia vaccinal strain persists in the spleen and thymus of young chicks and induces thymic lymphoid cell disorders', *Avian Pathol.*, 40, 377–385.

Verwoolde, M.B., et al. (2020), 'Training of primary chicken monocytes results in enhanced pro-inflammatory responses', *Vet. Sci.*, 7, 115.

Verwoolde, M.B., et al. (2021), 'Innate immune training and metabolic reprogramming in primary monocytes of broiler and laying hens', *Dev. Comp. Immunol.*, 114, 103811.

Villegas, P., et al. (1976), 'Effect of route of Newcastle disease vaccination on the incidence of airsacculitis in chickens infected with *Mycoplasma synoviae*', *Avian Dis.*, 20, 395–400.

Villegas, P. (1998), 'Viral diseases of the respiratory system', *Poult. Sci.*, 77, 1143–1145.

Wakenell, P.S., et al. (2002), 'Effect of in ovo vaccine delivery route on herpesvirus of Turkeys/SB-1 efficacy and viremia', *Poult. Sci.*, 46, 274–280.

Wang, X., et al. (2015), 'Effect of different doses of Newcastle disease vaccine immunization on growth performance, plasma variables, and immune response of broilers', *J. Anim. Sci. Biotechnol.*, 6, 20.

Wang, Y., et al. (2023), 'Heat stress affects immune and oxidative stress indices of the immune organs of broilers by changing the expressions of adenosine triphosphate-binding cassette subfamily G member 2, sodium-dependent vitamin C transporter-2, and mitochondrial calcium uniporter', *Poult. Sci.*, 102, 102814.

Wang, H., et al. (2024), 'Current status of poultry recombinant virus vector vaccine development', *Vaccines*, 12, 630.

Wegner, M., et al. (2023), 'Effect of different vaccination programs on production parameters, carcass, leg bones, and digestive system characteristics of broilers', *Poult. Sci.*, 102, 102668.

Wein, Y., et al. (2024), 'Immunization and host responses to MB-1, a live hatchery vaccine against infectious bursal disease', *Vaccines*, 11, 1316.

Wen, C., et al. (2021), 'Microbiota continuum along the chicken oviduct and its association with host genetics and egg formation', *Poult. Sci.*, 100, 101104.

Werner, E., et al. (2023), 'Licensing requirements for avian vaccines within the European Union', *Avian Dis.*, 67, 474–479.

Whitfill, C.E., et al. (1995), 'Determination of optimum formulation of a novel infectious bursal disease virus (IBDV) vaccine constructed by mixing bursal disease antibody with IBDV', *Avian Dis.*, 39, 687–699.

Wickramasuriya, S.S., et al. (2022), 'Role of physiology, immunity, microbiota, and infectious diseases in the gut health of poultry', *Vaccines*, 10(2), 172.

Wierup, M., et al. (2017), 'Occurrence of *Salmonella* spp.: a comparison between indoor and outdoor housing of broilers and laying hens', *Acta Vet. Scand.*, 59, 13.

Wijnen, H., et al. (2020), 'Effects of eggshell temperature pattern during incubation on primary immune organ development and broiler immune response in later life', *Poult. Sci.*, 99, 6619–6629.

Willemsen, H., et al. (2010), 'Delay in feed access and spread of hatch: importance of early nutrition', *Worlds Poult. Sci. J.*, 66, 177–188.

Williams, C., et al. (1994), 'A microbiological survey of broiler hatcheries in North America - the prevalence of aspergillus and other deuteromycetes', *Poult. Sci.*, 73, 166.

Williams, R.B., et al. (1999), 'The efficacy and economic benefits of Paracox, a live attenuated anticoccidial vaccine in commercial trials with standard broiler chickens in the United Kingdom', *Int. J. Parasitol.*, 29, 341–355.

Williams, R.B. (2002), 'Anticoccidial vaccines for broiler chickens: pathways to success', *Avian Pathol.*, 31, 317–353.

Williams, C.J. and Zedek, A.S. (2010), 'Comparative field evaluation of *in-ovo* applied technology', *Poult. Sci.*, 89, 189–193.

Williams, C.J. and Hopkins, B.A. (2011), 'Field evaluation of the accuracy of vaccine deposition by two different commercially available *in ovo* injection systems', *Poult Sci.*, 90, 223–226.

Williams, C.J. (2023), '*In ovo* vaccination of chicken eggs', In N. French (Ed), *Embryo development and hatchery practice in poultry production*, Burleigh Dodds Science Publishing, Cambridge, UK, pp. 1–34.

Winterfield, R.W. (1984), 'Vaccination of chickens with Newcastle disease and infectious bronchitis vaccines administered singly and in combination', *Poult Sci.*, 63, 182–184.

Wiseman, A. and Berman, E.M. (2017), 'Herd immunity to Newcastle disease virus in broiler flocks in Israel', *Avian Pathol.*, 46, 396–402.

Witter, R.L, and Burmester, B.R. (1979), 'Differential effect of maternal antibodies on efficacy of cellular and cell-free Marek's disease vaccines', *Avian Pathol.*, 8, 145–156.

Witter, R.L. (1997), 'Increased virulence of Marek's disease virus field isolates', *Avian Dis.*, 41, 149–163.

Wlazlak, S., et al. (2023), 'Modulation of the immune system of chickens a key factor in maintaining poultry production-a review', *Poult. Sci.*, 102, 102785.

Wood, G.W., et al. (1980), 'The stability of live Newcastle disease vaccines', *J. Biol. Stand.*, 8, 157–163.

Wood, G.W., et al. (1986), 'Observations on the ability of avian reovirus vaccination of hens to protect their progeny against the effects of challenge with homologous and heterologous strains', *J. Comp. Pathol.*, 96, 125–129.

Woodward, H.L. and Tudor D.C. (1975), 'A skim milk stabilized water vaccine for Newcastle disease (B1-type LaSota): its effectiveness under modern commercial cage layer methods of delivery', *Poult. Sci.*, 54, 866–871.

Yanai, Y., et al. (2018), 'Effect of stocking density on productive performance, behaviour, and histopathology of the lymphoid organs in broiler chickens', *Europ. Poult. Sci.*, 82, 1–15.

Yehia, N., et al. (2023), 'Common viral and bacterial avian respiratory infections: an updated review', *Poult. Sci.*, 102, 102553.

Yoshimura, Y., et al. (2023), 'Effects of Newcastle disease/infectious bronchitis vaccine and feeding yeast products on the innate immune system in the proventriculus and ileum of broiler chicks', *J. Poult. Sci.*, 60, 2023005.

Yoshimura, Y., et al. (2024), 'Innate immune training in chickens for improved defense against pathogens: a review', *J. Poult. Sci.*, 61, 20204008.

Yosipovich, R., et al. (2015), 'Overcoming the susceptibility gap between maternal antibody disappearance and auto-antibody production', *Vaccine*, 33, 472–478.

Yunus, A.W., et al. (2012), 'Deoxynivalenol as a contaminant of broiler feed: effects on bird performance and response to common vaccines', *Poult. Sci.*, 91, 844–851.

Yvernogeau, L., et al. (2022), 'Development of the avian hematopoietic and immune systems', In B. Kaspers, et al. (Eds), *Avian immunology*, Academic Press, London, UK, pp. 45–69.

Zaheer, T., et al. (2022), 'Vaccines against chicken coccidiosis with particular reference to previous decade: progress, challenges, and opportunities', *Parasitol. Res.*, 121, 2749–2763.

Zenner, C., et al. (2021), 'Early-life immune system maturation in chickens using a synthetic community of cultured gut bacteria', *mSystems*, 6, e01300-20.

Zhang, Y. and Sharma, J.M. (2001), 'Early posthatch protection against Marek's disease in chickens vaccinated *in ovo* with a CVI988 serotype 1 vaccine', *Avian Dis.*, 45, 639–645.

Zhang, Y. and Sharma J.M. (2003), 'Immunological tolerance in chickens hatching from eggs injected with cell-associated herpesvirus of turkey (HVT)', *Dev. Comp. Immunol.*, 27, 431–438.

Zhang, Y., et al. (2017), 'Depression of vaccinal immunity to Marek's disease by infection with chicken infectious anemia virus', *Front. Microbiol.*, 8, 1863.

Zhang, T., et al. (2024), 'Advances on adaptive immune responses affected by infectious bursal disease virus in chicken', *Front. Immunol.*, 14, 1330576.

Zhu, Z.J., et al. (2024), 'Immune escape of avian oncogenic Marek's disease herpesvirus and antagonistic host immune responses', *Vaccines*, 9, 109.

# Chapter 20

## Managing health issues in broilers: first-week mortality

*Serdar Özlü, Ankara University, Turkey*


1 Introduction
2 Mortality patterns
3 Causes of first-week mortality: animal-related factors
4 Causes of first-week mortality: egg and incubation-related factors
5 Causes of first-week mortality: external factors
6 Conclusion
7 Where to look for future information
8 References


## 1 Introduction

First-week mortality in the life of broilers is a significant issue which is widely monitored in poultry production and which can serve as an indicator of the health status and performance of the broiler flock during the rearing period. According to reports based on information from commercial production systems across Europe, mean first-week mortality ranges from 0.59% to 2.1% between studies (Baxter et al., 2021; Forseth et al., 2024; Heier et al., 2002; Junghans et al., 2022; Rayner et al., 2020; Van Limbergen et al., 2020; Yassin et al., 2009; Yerpes et al., 2020). However, in individual cases, much higher mortalities have been demonstrated, indicating that the first-week mortality is a big issue in broiler production.

Various factors, including housing conditions, broiler flock size, biosecurity routines, climatic control, season, stocking density, nutrition, growth rate, genetic susceptibility, and general management routines, have been shown to affect first-week mortality (Heier et al., 2002; de Jong and van Riel, 2020; Van Limbergen et al., 2020).

During the first week of rearing, risk factors for mortality also include animal-related factors, environmental factors, incubation-related factors, and management factors (Yerpes et al., 2020; Yassin et al., 2009). This chapter

http://dx.doi.org/10.19103/AS.2025.0165.25

discusses these potential factors, and how they are related to subsequent broiler health status and performance.

The aim of this chapter is:

- to discuss first-week mortality patterns of broilers;
- to discuss potential animal-related factors, egg and incubation-related factors, and external factors that affect first-week mortality of broilers.

## 2 Mortality patterns

Generally, a high first-week mortality can be a signal of a problem. Mortality in the first 1 to 3 days after placement might be related to poor incubation conditions or poor early post-hatching management. Increased mortality a couple of days later (day 4–5) is often caused by a bacterial infection. Higher mortality at the end of the first week (day 6–7) is considered a signal of poor placement and management conditions at the broiler farm (Lourens et al., 2021).

A first-week mortality rate up to 1% is unavoidable. Vaccination reactions can occur in cases where a live vaccine has been administered, and this can lead to higher mortality during the first week (Eidson et al., 1982). The peak in mortality at day 3 to 4 may coincide with the disappearance of the yolk sac in the intestine of chicks (Lourens et al., 2021). Chicks may survive the first few days with the yolk sac alone, but once this food source is absorbed, the chick will start to perish. While chicks having a good start are active, avidly eating feed and move away quickly when approached, cull chicks will often stand by themselves, chirp, and refuse to move away as the farmer comes near. When cull chicks are found, they should be immediately removed and humanely killed by an approved method (Watkins, 2003). Removing cull chicks at this early stage will prevent chick suffering and will improve broiler flock uniformity, making management of feeder and drinker height much easier as the flock ages.

Gillingham (2023) has suggested examining dead chicks that die for signs that could be related back to broiler farm conditions and the peak in mortality by day 4 (Fig. 1).

Tabler et al. (2004) investigated the relationship between early mortality and late mortality. Broiler flocks that lost the most chicks early (first week) tended to lose the most chicks late. In addition, when first-week mortality was high, broiler flock uniformity in later life was often lower, with poorer feed conversion ratios (Fig. 2).

Lourens et al. (2021) reported that higher first-week mortality is often a signal of a problem (Fig. 3). Immediate post-hatch mortality often indicates poor pre-hatching conditions, as reflected by a peak in mortality in the first few days post-hatching. High mortality a couple of days later is often caused by bacterial infection (with a mortality peak at day 4–5 post-hatching). Mortality at

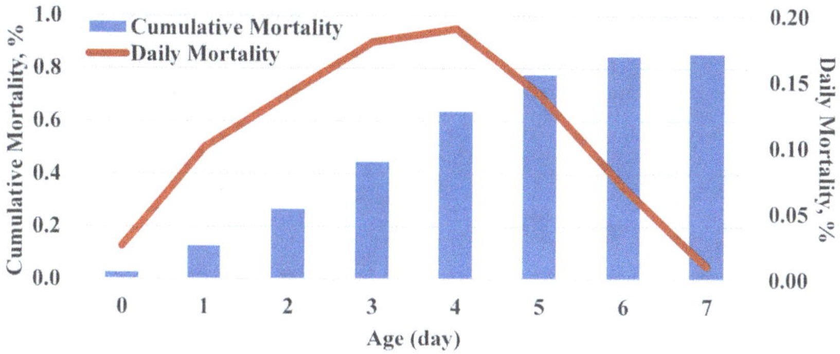


**Figure 1** Regular pattern of the first-week mortality in broiler flocks. Source: Adapted from Gillingham (2023).

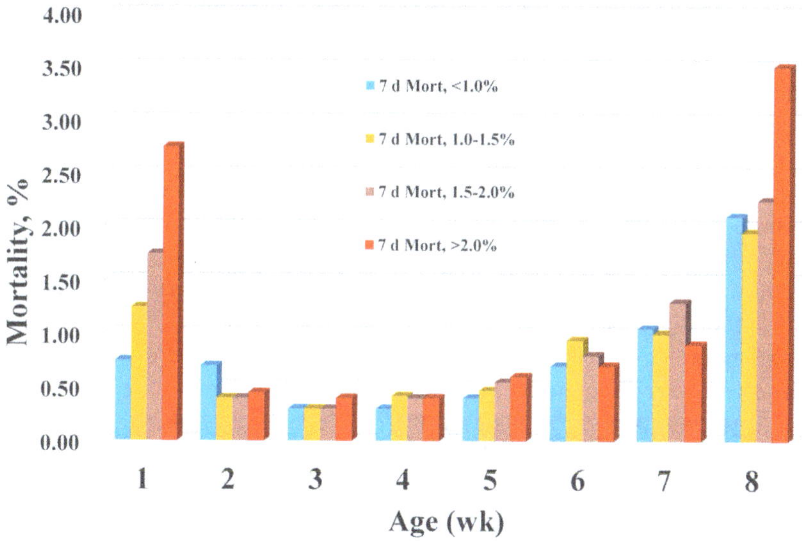


**Figure 2** First-week mortality in broiler chicks in relationship to later life mortality. Source: Adapted from Tabler et al. (2004).

the end of the first week is often a signal of poor placement conditions at the broiler farm (higher mortality from day 5 onwards).

## 3 Causes of first-week mortality: animal-related factors

Risk factors for first-week mortality can be divided into three main categories:

- animal-related factors
- egg and incubation factors
- external factors

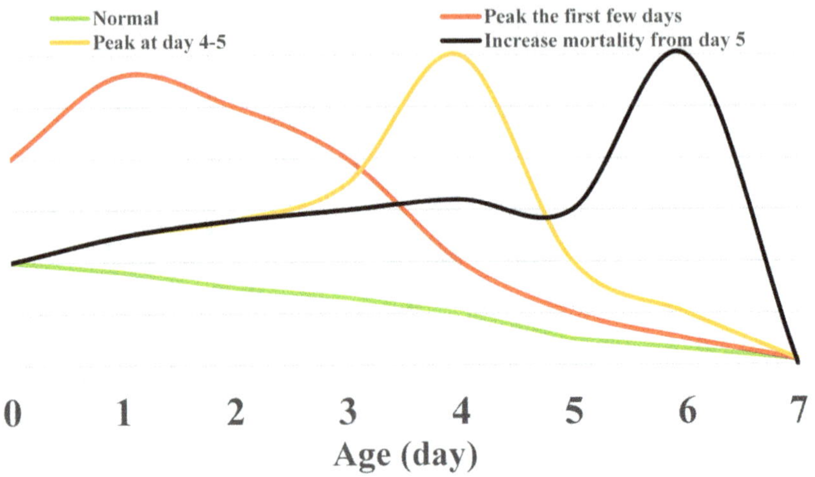


**Figure 3** Potential patterns in first-week mortality of broiler flock, which might be related to different causes of mortality. Source: Adapted from Lourens et al. (2021).

Factors include sex, chick quality, strain, breeder flock age, levels of maternal antibodies in breeders, egg weight, egg quality, egg storage duration, egg shell contamination, incubation environmental conditions, and hatchery routines. These differ from factors in the later period such as feed availability in the hatcher or hatch facility on farm or longer post-hatch transportation (duration, environmental conditions, availability of feed and/or water or gel). There are also more general factors affecting first-week mortality, such as cold stress, gastroenteric disorders and other infectious diseases, feed and water availability, handling stress, and dehydration.

Animal-related factors affecting first-week mortality are strain (Fast growing: Cobb, Hubbard, Ross; Slow growing: Hubbard, Redbro), sex (male, female, and mixed), breeder flock age, categorized according to Hamidu et al. (2007) (Young: 25–30 weeks, Peak: 31–36 weeks, Post Peak (Prime): 36–44 weeks, Mature: 45–54 weeks, Old: 55–58 weeks, and Very Old: 59–68 weeks). These animal-related factors were chosen due to the relationship detected in other studies with the viability and chick quality at the moment of hatching or pulling (Elibol and Brake, 2003; McNaughton et al., 1978; Peebles et al., 2004; Suarez et al., 1997; Van den Brand et al., 2016; Vieira and Moran, 1999b; Yassin et al., 2009; Yerpes et al., 2020).

### 3.1 Strain

Genetic selection for fast growth is a risk factor for mortality in broiler chickens (Dixon, 2020; Havenstein et al., 2003). Fast growth has been negatively associated with adaptive immune system function (Cheema et al., 2003) and metabolic function (Lubritz et al., 1995), which might explain a higher mortality in fast-growing broilers.

Chicks from Ross strain had a significantly higher first-week mortality, and a higher probability of increasing mortalities of the whole rearing period compared to Cobb and Hubbard breed chicks (Forseth et al., 2024; Yassin et al., 2009; Awobajo et al., 2007; Yerpes et al., 2020). However, the breeds used to carry out the comparison were different in all the studies. Some breed-related factors influencing chick viability and quality and consequently first-week mortality are the difference in egg weight (Vieira and Moran, 1998a) and the metabolic rate of the embryo during incubation (Hamidu et al., 2007). In addition to numerous external factors, the health and robustness of the chicks is also influenced by genetics (EFSA Panel on Animal Health Welfare, 2011). However, studies comparing mortality between hybrids of different growth rates are scarce; two studies reported lower first-week mortality for slower-growing broilers than for fast-growing broilers (Baxter et al., 2021; Rayner et al., 2020). Whether or not these lower first-week mortalities are really due to strain effects or related to for example different incubation conditions or breeder flock characteristics remains unclear. For example, feeding strategies and nutrient levels in breeder nutrition can influence livability and immunity of progeny (Chang et al., 2016; Leandro et al., 2011a; Leandro et al., 2011b; Oviedo-Rondon et al., 2013) and consequently might be a factor in first-week mortality.

### 3.2 Sex

The growth rate, hatch time, and metabolic functions of broilers are affected by chick sex even if the rearing period is relatively short.

In general female chicks showed a lower first-week mortality and a higher probability that this rate would decrease, compared to male and mixed chicks (Leitner et al., 1989; Van Emous et al., 2015). Male and female broilers differ in their development of immune response, making males more susceptible to pathogens, and consequently, they have a higher first-week mortality (Yerpes et al., 2020). On the other hand, the male chicks have a longer incubation period and may be placed on a broiler farm immediately after hatch, when they are not ready to eat and drink. This means late hatched male chicks are at higher risk of mortality after placement if environmental conditions are poor (Shiranjang, 2015).

### 3.3 Breeder flock age

The age of the breeder flock has been found to be a significant factor in first-week mortality, as has been found in numerous studies (Grochowska et al., 2019; Hays, 1955; Ipek and Sozcu, 2015; Junghans et al., 2022; Lourens and Kuijpers,

2002; McNaughton et al., 1978; óDea et al., 2004; Pedroso et al., 2005; Peebles et al., 1999; Peebles et al., 2004; Suarez et al., 1997; Wilson, 1991; Yassin et al., 2009; Yerpes et al., 2020; Sedeik et al., 2019; de Jong and van Riel, 2020). As an example, de Jong and van Riel (2020) reported that the highest first-week mortality was found in young breeders, with a fast decline toward prime flock breeders and a slight increase again in older breeders (Fig. 4).

Although some of them found that chicks hatched from eggs collected from older breeder flocks had a higher first-week mortality than those of young and prime breeder flocks, some of them reported that the highest rate of first-week mortality occurs in chicks from young breeders, followed by chicks from old breeders. These differences might be because most of the studies only used two of the five breeder flock age categories. For example, Wilson (1991) used chicks from young and old breeders, and Suarez et al. (1997) used chicks from young and prime breeders. Ipek et al. (2014) reported a higher first-week mortality (3.61%) in a 62-week-old breeder flock compared to a 33-week-old breeder flock (0.83%), but McNaughton et al. (1978) found the opposite comparing first-week mortality of chicks originating from a 29-week-old breeder flock (3.87%) with a 58-week-old breeder flock (1.57%). The ambiguous results might be related to the fact that comparing first-week mortality of chicks originating from different breeder flock ages mostly means that different flocks are compared, each flock might differ from another flock in several aspects, and consequently, breeder flock age is often confounded with other factors, such as breeder nutrition, egg storage duration, and incubation conditions.

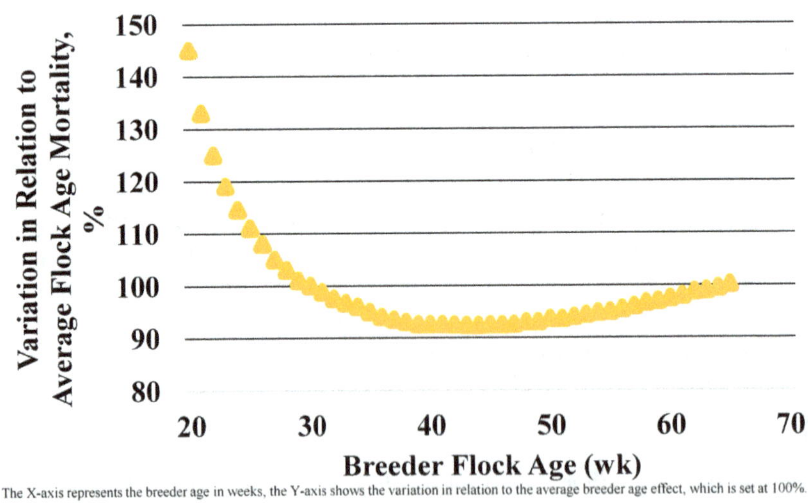


**Figure 4** Effect of broiler breeder flock age on the first-week mortality of the offspring. Source: Adapted from De Jong and van Riel (2020).

A possible reason for the relationship between breeder flock age and the first-week mortality could be the egg size and, consequently, the incubation duration (Pedroso et al., 2005; Suarez et al., 1997; Wilson, 1991; Sedeik et al., 2019). Another possible reason is that chicks from young or old breeders have lower chick quality and viability than chicks from prime breeders and a worse healed navel, increasing the probability of suffering diverse pathologies and infections of the yolk sac, which might increase first-week mortalities (Joseph and Moran, 2005; Pedroso et al., 2005; Tona et al., 2004; Willemsen et al., 2008; Yassin et al., 2009).

When broiler breeders age increases, this is related to changes in internal egg composition, larger egg weight, lower shell quality, and increased embryo mortality (Elibol and Brake, 2003; Peebles et al., 1998; Tona et al., 2004; Nangsuay et al., 2011; Weytjens et al., 1999). Besides, Wilson (1991) explained that the relatively low egg weights in young breeders might lead to less available nutrients, as well as a reduced mobilization of energy from the yolk sac. These also could affect first-week mortality.

## 4 Causes of first-week mortality: egg and incubation-related factors

First-week mortality could be affected by egg source (clean, nest, floor, or washed eggs), egg weight, egg storage duration, egg shell conditions (dirty, stained, clean), and incubation environmental properties.

Embryos utilize egg components for development from the early stages of embryo formation until hatching (Romanoff and Romanoff, 1967). Yolk nutrients are transported through the yolk sac membrane and the surrounding vascular system (Noble and Cocchi, 1990). This process accelerates after approximately 12 d of incubation, at which time a rapid transfer of lipids from the yolk to the embryo begins (Speake et al., 1998). The yolk sac is drawn into the body cavity on the 19th day of incubation, and the entire yolk residue is located within the abdomen at the end of day 20. At hatch, approximately 30% of the original nutrients in the yolk are still present in the yolk residue, and these nutrients function as a major source of energy for the hatchling during the first few days of life (Etches, 1998; Noy and Sklan, 2001). Because of the details explained above, the nutrition of the yolk sac needs to be absorbed by chicks to survive in the first-week period.

### *4.1 Egg weight*

Skoglund and Tomhave (1949), Wiley (1950), and Tindell and Morris (1964) found no relationship between egg weight and first-week mortality. Skoglund et al. (1952) only showed a trend toward higher mortality in chicks hatched from small eggs. McNaughton et al. (1978) reported that higher first-week mortality occurred when chicks hatched from eggs weighing 47–54 g compared to either 57–62 g or 67–74 g hatching eggs. Particularly the increase from 47–54 g to 57–62 g within the same young breeder flock showed clear effects on the first-week mortality (Table 1).

Several studies have demonstrated a strong relation between egg size and chick weight (Vieira and Moran, 1998a; Vieira and Moran, 1998b; Wiley, 1950; Wyatt et al., 1985). This relationship was observed in experiments where an increase in egg weight was associated with an increase in breeder flock age (Vieira and Moran, 1998a; Suarez et al., 1997) as well as an increase in egg size at the same hen age (Vieira and Moran, 1998b; Lourens et al., 2006). Smaller chicks from young hens have been shown to have higher mortality after placement compared to prime or older breeder flock chicks (Sedeik et al., 2019; Shanawany, 1987; Vermeersch and Vanschoubroek, 1968; Washburn and Guill, 1974).

## 4.2 Egg storage duration

Prolonged egg storage durations before incubation are associated with lower hatchability, longer incubation durations (Fasenko, 2007; Fasenko et al., 2001; Özlü, 2021), and lower chick quality (Suarez et al., 1997; Tona et al., 2003; Tona et al., 2004). It has been suggested that this prolonged storage of the egg may induce embryonic stress, resulting in irreparable damage to the embryo, increase in embryonic mortality, and decrease in performance (Fasenko, 2007; Hamidu et al., 2011). Moreover, Özlü and Elibol (2023) reported that cooling the EST to 24°C within 45 min (rapid cooling) compared to 6 h (control) after

**Table 1** Effect of breeder age and egg weight (within same flock) on the first-week mortality of broiler chicks

| Experiment | Breeder age, wk | Egg weight, g | First-week mortality, % |
|---|---|---|---|
| 1 | 29 | 47–54 | $3.35^{a}$ |
| | | 57–62 | $2.49^{ab}$ |
| | 58 | 57–62 | $1.51^{b}$ |
| | | 67–74 | $1.62^{b}$ |
| 2 | 29 | 47–54 | $5.40^{x}$ |
| | | 57–62 | $4.22^{x}$ |
| | 58 | 57–62 | $1.51^{y}$ |
| | | 67–74 | $1.62^{y}$ |

$^{a,b}$ Values within a column lacking a common superscript differ ($P \leq 0.05$).
$^{x,y}$ Values within a column lacking a common superscript differ ($P \leq 0.01$).
*Source*: Adapted from McNaughton et al. (1978).

laying retarded the blastoderm developmental stage and hatch time of eggs from both young and old broiler breeder flocks. The blastoderm developmental stage directly affects incubation duration and chick quality of the broiler in their first week of life.

Grochowska et al. (2019) observed that selected periods of storage time at hatchery, i.e. 4, 6, and 11 d, linearly affected (increased) the first-week mortality. These findings were partly consistent with Yassin et al. (2009) and Merritt (1963), who showed that increased egg storage periods before incubation was positively associated with the first-week mortality. However, Yerpes et al. (2020) found that chicks from eggs stored for less than seven days had a higher first-week mortality (1.94% vs. 1.76%) and also a higher mortality risk during the first week of life.

## 4.3 Egg shell contamination

When a hatching egg is laid, it is warm. The cuticle is still wet and can be penetrated by bacteria (De Reu et al., 2006). During the cooling process, bacteria on the shell surface can be sucked into the pores of the egg shell. An egg laid on the floor or in a dirty laying nest can never be made 'clean', because the dirt on an egg is full of bacteria and molds (Table 2). Even clean floor eggs had a significantly higher percentage of contaminated eggs compared to nest eggs (Perić et al., 2022).

It appears that hatchability is lower in floor eggs and washed eggs than in clean nest eggs, but once the chicken has hatched from floor and washed eggs, the quality is not different from that of chickens originating from clean nest eggs (Van den Brand et al., 2016). Comparable results were found by Buhr et al. (1994) who found that sanitized or not sanitized dirty hatching eggs had a 5.6% to 7.0% lower hatchability than clean nest eggs, but no significant difference in the first-week mortality was found.

**Table 2** Effect of egg shell cleanliness on bacterial load on the egg shell of broiler breeder eggs and on mortality during the first two weeks of live in the offspring

| Egg source | Bacterial load (count per mm$^2$) | Egg shell condition (count per mm$^2$) | Bacterial load | Of which coli-type bacteria (%) | Mortality within 14 days |
|---|---|---|---|---|---|
| Fresh Egg[1] | 300 | Nest Egg (Clean) | 600 | 123 | 0.9 |
| Clean Egg | 3000 | Nest Egg (Stained) | 20 000 | 904 | 2.3 |
| Floor Egg | 25 000 | Floor Egg (Dirty) | 80 000 | 1307 | 4.1 |
| Dirty Egg | 400 000 | | | | |

[1]Fresh egg: Newly laid egg.
*Source*: Adapted from Lourens et al., (2021).

### 4.4 Incubation conditions

It is well known that even when all conditions for good quality hatching eggs are fulfilled, egg storage, incubation, and hatching conditions and post-hatch handling are crucial to produce good quality day-old chicks (Bergoug et al., 2013a; de Jong et al., 2017; Decuypere et al., 2001; Willemsen et al., 2010).

First-week mortality and first-week culled chicks are characteristics that are influenced by incubation temperature (Collin et al., 2007; Joseph et al., 2006; Lourens et al., 2005; Tzschentke and Halle, 2009). For example, Maatjens et al. (2016) reported that the first-week mortality for egg shell temperature (EST) after 15–19 d of incubation on 35.6°C, 36.7°C, 37.8°C, and 38.9°C were 1.1%, 0.0%, 0.0%, and 2.5%, respectively, suggesting that a too high EST in the last phase of incubation might negatively affect the first-week mortality. Avşar et al. (2022) found no significant difference in the first-week mortality when the eggs were incubated at high (38.6°C) or control (37.8°C) EST during the first week of incubation. Moreover, there was no significant difference in the first-week mortality between eggs incubated at low (33.3–36.7°C), control (37.8–38.2°C) or high (38.9–40.0°C) EST from day 10 to day 18 of the incubation period (Ipek et al., 2014).

Bruzual et al. (2000b) reported that the relative humidity (RH) level of the incubation period did not alter the first 12-day mortality of the chicks. Moreover, the same researchers found that the interaction of RH × brooding temperature in an experiment that approached significance suggested that the chicks coming from the lower RH treatment (43%) and brooded at the cool temperature (26°C) exhibited four times higher mortality. This effect was not seen when the chicks were brooded under warm conditions (30°C).

Egg position or turning of the eggs during the incubation period is an environmental factor to achieve the best result of the hatchability. Moreover, Uçar et al. (2021) determined that chicks hatched from the egg position SEU (Small End Up) and pipped through the small end of the egg, not through the air cell (SSE), had a numerically higher first-week mortality when compared to two groups, which were chicks hatched from the egg position LEU (Large End Up) and pipped through the air cell (LAC) and hatched from the egg position SEU and pipped through the air cell (SAC),

First-week mortality might also be affected by hatching systems, which are often related to the immediate or delayed provision of feed and water. Van den Brand et al. (2023) reviewed alternative hatching systems for broilers in two categories where the first category is chickens hatch in a hatchery and have feed (and water) available in the hatcher (hatchery fed), and the second category consists of on-farm hatching systems in which, after candling at day 18 of incubation, fertile eggs are transported to the rearing farm where they can hatch.

Lingens et al. (2021) found that the mortality rate in the first week of life was 1.81% and 1.24%, in trials 1 and 2, respectively, for the group of delayed nutrition on-hatchery hatched (control group) vs. 1.28% and 0.87% for the group of early nutrition on-farm hatched (experimental group) in trials 1 and 2, respectively. This difference might be related to the early provision of feed and water post-hatching (Van de Ven et al., 2009), but it might also be related to the lack of transportation of on-farm hatched chicks.

The first-week mortality was higher in chicks hatched at a hatchery and transported as day-old to the farm than in chicks hatched on-farm (Jessen et al., 2021a, 2021b).

## 4.5 Day-old chick quality and hatching time

First-week mortality is a common parameter used to assess chick quality. Chick quality can be defined as chicks that are optimally developed during incubation and show a high performance potential and survivability (Tona et al., 2005a). The quality of day-old chicks has been associated with the first-week mortality (Yassin et al., 2009; Goodhope, 1991).

Several studies reported that the day-old chick quality depends on the genetic line of the breeders, breeder age, egg weight, egg storage conditions and duration, and incubation conditions such as temperature, humidity, gas levels, and altitude (Decuypere and Bruggeman, 2007; Decuypere et al., 2001; Fasenko, 2007; Peebles et al., 1999; Tona et al., 2004, 2005b; Vieira and Moran, 1999b; Wilson, 1991).

In a study conducted by Ebeid et al. (2017), the percentage of good-quality chicks was greater for eggs stored for 4 d than for those stored for 14 d, and Elibol et al. (2002) reported that eggs subjected to extended storage (14 d vs. 1 d) resulted in a greater number of poor-quality chicks.

Özlü and Elibol (2023) reported that the number of second-grade chicks from eggs from the young breeder flock was lower than that from eggs from the old breeder flock (0.85 vs. 2.33%). This was in agreement with the study by Özlü (2021).

Hamdy et al. (1991) found that late hatched chicks had a significantly higher body weight at day 0, but the body weight gain and feed consumption were similar as their earlier. However, due to the fact that the early hatched chicks spent more time in the hatcher than those of the later hatched, their mortality could be higher, especially in high temperature conditions.

Eggs within a batch do not hatch all at the same moment, but within a certain hatch window. It has been shown that late-hatched chicks had a significantly higher first-week mortality (Table 3) compared with early or middle-hatched chicks (Dişa et al., 2022; Özlü et al., 2018; Shiranjang et al., 2018). A possible

**Table 3** Effect of hatch time of broiler chicks on the first-week mortality

| Hatch time | 0–7 d mortality, % | Hatch time | 0–7 d mortality, % | Hatch time | 0–7 d mortality, % |
|---|---|---|---|---|---|
| Early (470–488 h) | 0.0$^{b}$ | Early (471–477 h) | 0.7$^{b}$ | Early (478–490 h) | 1.1$^{b}$ |
| Middle (488–492 h) | 0.5$^{ab}$ | Middle (480–486 h) | 0.5$^{b}$ | Middle (490–496 h) | 1.1$^{b}$ |
| Late (492–510 h) | 1.8$^{a}$ | Late (494–510 h) | 2.2$^{a}$ | Late (496–510 h) | 1.7$^{a}$ |

*Source*: Adapted from Shiranjang et al. (2018).
*Source*: Adapted from Özlü et al. (2018).
*Source*: Adapted from Dişa et al. (2022).
$^{a-b}$Values within a column lacking a common superscript differ ($P \leq 0.05$).

reason is that the late-hatched chicks had a significantly higher residual yolk sac, and they did not start eating immediately after placement compared to early and middle-hatched chicks (Özlü et al., 2018).

Starting from a conditioned life in the hatchery, they move on to an independent life in a barn where they have to adjust to new feed, water, thermoregulation, competition, and fight infections at the same time (Vieira and Moran, 1999b).

Although the first few days of the life of a chick are known to be crucial to later performance (Bruzual et al., 2000a; Tona et al., 2005b), it is questionable whether the physiological requirements of the hatchlings can be met with current incubation systems and hatchery management procedures. Moreover, the first-week mortality is a good indicator of chick quality.

The delayed placement resulting from early hatch also increased mortality during the first week of the rearing period as did holding chicks in boxes in excess of 30 h (Fanguy et al., 1980). Studies have reported unhealed navels and enlarged yolk sacs to be the abnormalities most frequently registered during chick quality assessment of both on-farm and hatchery-hatched chicks, but with a higher prevalence in chicks hatched on-farm (Jessen et al., 2021a; Van de Ven et al., 2012; de Jong and van Riel, 2020). For example, Jessen et al. (2021a) reported unhealed navels to occur in 8.2% of the chicks hatched on farm, whereas the prevalence was 5.3% for the hatchery-hatched chicks. The severity of unhealed navels of newly hatched chicks can be considered high, as it may result in mortality, mainly during the first week of life, which is likely due to the open navel being the entry point of pathogenic bacteria leading to infection (Fasenko and O'Dea, 2008).

Fasenko and O'Dea (2008) showed that the presence of a leaky or button navel did not negatively influence chick body weight. However, body weight gains during the first week were significantly lower in chicks with navel buttons compared with chicks with leaky navels or normal navels. Moreover, the navel

**Table 4** Effect navel condition of broiler chicks at pull time on the first-week mortality

| Navel condition[1] | 0–7 d mortality (%) |
|---|---|
| Button | 12.7[a] |
| Leaky | 10.1[ab] |
| Normal | 5.7[b] |
| SEM | 2.7 |

[a-b] Means within a column lacking a common superscript differ ($P \leq 0.05$).
[1]Button = Chicks having a scab of blood less than 3 mm in diameter formed over the navel. Leaky = Chicks with moist down feathers or discolored down feathers surrounding the navel. Normal = Chicks with well-healed navel that the abdomen sealed nicely after the yolk was retracted into the body.
*Source*: Adapted from Fasenko and O'Dea (2008).

condition affected of the first-week mortality of the chicks; each deviation from normal resulted in a higher mortality (Table 4).

## 5 Causes of first-week mortality: external factors

Besides animal-related factors, egg and incubation-related factors, there are also external factors that can influence the first-week mortality, such as transportation conditions, preplacement holding time, season, broiler flock size, stocking density, type of ventilation, litter quality, and type of drinkers. These factors will be discussed below.

### *5.1 Chick transportation conditions*

After hatching, the thermoregulatory system of chickens is still immature (Nichelmann and Tzschentke, 2002), and warmth is a critical need for young chicks. Young chickens behave in the first days as poikilotherms, which means that they are dependent on the external conditions to keep their body temperature at the desired level (Nichelmann and Tzschentke, 2002). Appropriate effective temperature (Nielsen, 2020) is therefore required to keep the body temperature of chickens at the desired level (40.0–40.5°C) (Hamissou Maman et al., 2019; Mujahid and Furuse, 2009). If the body temperature of a chicken drops, it will become inactive (lethargic) and lie down, which further accelerates the undercooling (Mujahid and Furuse, 2009).

Early mortality in chickens has been related to suboptimal truck temperatures and longer duration of transport from the hatchery to the farm (Carver et al., 2002; Chou et al., 2004). Yerpes et al. (2021) investigated transport-related variables in their effects on chick weight and first-week mortality. They reported that the journey duration, height inside the trolley, month of the year, number of stops during transport, and time of the day to start the journey did not have a significant effect on first-week mortality.

However, temperature during transportation might have an effect on first-week mortality. The modern breed of commercial chicks has a much higher metabolic rate than older breeds (EFSA, 2004). It is important to avoid high temperatures and keep proper ventilation during transportation of the chicks in order to reduce first-week mortality and to reach a better performance (Joseph and Moran, 2005).

In fact, first-week mortality can reflect the stress of the transportation process (Bayliss and Hinton, 1990; Yassin et al., 2009). The transportation of chicks may cause stress and adversely affect chick development (Valros et al., 2008). Chick mortality occurs mostly during the first week of the rearing period (Heier et al., 2002) due to the combined stress of post-hatch handling in hatcheries, such as counting, sexing, vaccination, transportation, and poor adaptation to rearing conditions (Bayliss and Hinton, 1990). When chick transportation was increased from 50 to 300 km, first-week mortality also increased from 1.2% to 1.7% (Chou et al., 2004). However, Bergoug et al. (2013b) reported that transportation of 1-day-old chicks for 4 h or 10 h compared with those moved directly to the farm (less than 5 min) had no effect on the first-week mortality.

During post-hatch handling, temperature is the main environmental condition affecting the quality of day-old chicks and their subsequent broiler performance (Decuypere et al., 2001). In practice, most hatcheries recommend that environmental temperatures for chick rooms or transportation of the boxes containing chicks remain between 24 and 26°C (Mitchell et al., 1996; Mitchell, 2009) to maintain the optimal chick body temperature (40.0 to 40.5°C), When above the ideal range, the air temperature may induce hyperthermia and dehydration in the chicks, leading to increased mortality (Joseph and Moran, 2005; Xin and Harmon, 1996; Xin and Rieger, 1995). During the first days of life, a chick still functions as a poikilotherm and cannot adjust its heat production sufficiently to ambient temperatures (Weytjens et al., 1999). Therefore, cold stress may be another problem in day-old chicks that could affect muscle growth and development (Davison and Lickiss, 1979; Sagher, 1975).

Chick body temperature, one of the important indicators of chick welfare, affects the quality of day-old chicks and their subsequent broiler performance (Decuypere et al., 2001; Hamissou Maman et al., 2019). The optimum day-old chick vent temperature should be 39.4°C to 40.5°C during the post-hatch holding period (Aviagen, 2021).

Hamissou Maman et al. (2019) reported there were no deaths in any of the groups during the 12 h post-hatch handling period (low body temperature = 38.1°C, control body temperature = 40.0°C, and high body temperature = 42.6°C), and there was one dead chick in the high body temperature group in the first week. Moreover, mortality at 35 d of age in the low, control, and high body temperature groups was 2.68%, 1.79%, and 5.36%, respectively, which was numerically higher in the high body temperature group than in the other two groups.

Several studies have shown that provision of hydrogels during the post-hatch holding period improves growth performance of broilers, poults, and emus (Batal and Parsons, 2002; Dibner et al., 1998; Mozdziak et al., 2002; Noy and Sklan, 1999) but has no effect on mortality rates (Batal and Parsons, 2002; Henderson et al., 2008; Lowman and Parkhurst, 2014; Noy and Sklan, 1999). Moreover, a recent study showed a lack of effects of provision of hydrogels in the chick boxes during the preplacement holding period on the yolk-free body mass or residual yolk sac of broilers at placement, the state of crop filling 3 h after placement in the barn or on the body weight gain and cumulative mortality after the first week (Özlü et al., 2022).

In poultry production, 1-day-old chicks are delivered from the hatchery to the broiler farm, and heat stress can occur when hot climates and long distances are involved (Ernst et al., 1984).

Ambient temperatures greater than 32°C caused body-weight loss and increased mortality of neonatal chicks due to dehydration (Daghir and Jones, 2008; Deaton et al., 1996; Van der Hel et al., 1991).

## 5.2 Pre-placement holding time (post-hatch food and water deprivation)

After chick collection from the hatcher, further hatchery procedures, such as sexing, vaccination, packaging, and transportation, increase the time until placement in the broiler house and thus first feed and water intake, for part of the flock by up to 50 h or more (Özlü, 2016; Sklan et al., 2000; Careghi et al., 2005). Suboptimal conditions during transport and a delay in the moment of placement and the first feed and water intake are associated with higher early mortality in chicks and poults (Kingston, 1979; Bergoug et al., 2013a; Carver et al., 2002; Chou et al., 2004; de Jong et al., 2017).

The time until the first feed and water intake 'holding period' may take up to 72 h when long transportation distances are involved (Van de Ven et al., 2009). The long transportation process could exacerbate the depletion of reserves and dehydration through excessive thermoregulatory demands and stress, thus possibly affecting chicks' body weight and mortality rates (de Jong et al., 2017; Jacobs et al., 2017). The EFSA Panel on Animal Health Welfare (2011) specifies that newly hatched broiler chicks can be transported in the first 72 h after hatching, including a maximum transport time of 24 h without feed and water. This recommendation is based on the fact that the chick's metabolic reserves stored in the yolk sac last up to 3 d.

Xin and Rieger (1995) found that up to 72 h pre-placement time affected the dead on arrival percentage and it reached 11%, and the first-week mortality reached to 50%. Similarly, Fanguy et al. (1980) found that the first-week mortality had been increased when placement was delayed 72 h, with another significant increase resulting from holding chicks 100 h. Moreover, Pinchasov and Noy (1993) conducted several studies showing that the first-week mortality could be increased after 24 h to 48 h pre-placement holding, especially with heat stress conditions.

In general, the immune system of early-fed broilers seems to be ahead in development compared to delayed-fed broilers, especially during the first weeks of life (Dibner et al., 1998; Hollemans et al., 2020; Panda et al., 2010; Bar Shira et al., 2005). At first sight, a head start of approximately 48 h in development might seem limited. However, the broiler immune system is immature at the moment of hatch, and it develops rapidly during the first week of age (Mast and Goddeeris, 1999; Bar Shira et al., 2005). At later ages, early fed broilers showed enhanced growth performance, lower mortality rate, and different immune responses compared to delayed-fed broilers (de Jong et al., 2017; Juul-Madsen et al., 2004; Lamot et al., 2016; Bar Shira et al., 2005).

It cannot be ruled out that other factors such as chick handling (Knowles et al., 2004), dust, pathogens in the hatchery (de Gouw et al., 2017) and stress during transport (Mitchell, 2009) may also contribute to the increased mortality of on-hatchery-hatched chicks subjected to delayed nutrition. On the other hand, *in ovo* feeding of specific micronutrients to embryonated eggs could enhance the health status of the chicks as recently reported by Gouda et al. (2022).

First-week mortality was higher in chickens after 84 h post-hatch feed and water deprivation than in early-fed chickens (de Jong et al., 2017). However, it should be noted that only a small number of scientific papers included such long durations of food and water deprivation as the number of database records for food and water deprivations of more than 84 h was relatively low (Table 5).

Feed and water are typically either both absent or both present in the hatcheries and during the holding time. Thus, if prolonged hunger occurs, then prolonged thirst is likely to co-occur and vice versa. When a newly hatched chick experiences prolonged hunger and thirst, body weight loss will occur. A

**Table 5** Relative value after food and water deprivation

| | **Relative value after food and water deprivation** | | | | |
|---|---|---|---|---|---|
| | **0 h (0–12 h)** | **24 h (12–36 h)** | **48 h (36–60 h)** | **72 h (60–84 h)** | **>84 h** |
| | | | **(%)** | | |
| Mortality day 0–7 | 100$^{bc}$ | 81.4$^{c}$ | 143$^{bc}$ | 226$^{b}$ | 827$^{a}$ |

$^{a\text{–}c}$ Values within a row lacking a common superscript differ ($P \leq 0.05$).
*Source*: Adapted from de Jong et al. (2017).

post-hatch feed deprivation period of on average 48 h (between 36 h and 60 h) or longer resulted in increased mortality and impaired performance even if the yolk sac may provide some nutrients for up to 72 h post-hatch (de Jong et al., 2017). However, there is a high uncertainty of the exact threshold for post-hatch food and water deprivation to result in increased mortality, as studies varied in whether they assessed these effects post-pull or post-hatch, in the hatching and housing conditions, and the exact durations of post-hatch feed and water deprivation. Moreover, only mortality and performance effects could be analyzed in the study of de Jong et al. (2017), as studies including other indicators of prolonged hunger, such as latency to feed or total feed intake, were lacking.

The reduction in growth and increased post-hatch mortality were primarily due to inaccessibility of feed and water, resulting in dehydration and a shortage of available energy (Vieira and Moran, 1999a). However, this was not the case in Özlü et al.'s (2020) study, which clearly demonstrated that up to 40 h of delayed access to feed and water under optimum chick holding environmental conditions did not affect mortality compared with the immediate feed (2 h) group in both experiments.

Elibol et al. (2023) found that incubation time had no impact on mortality at 7 d after placement. However, mortality was affected by preplacement holding time after pull time and reached 1.82% in the 72 h group at 7 d, which was significantly higher than that in the other preplacement holding time groups (Table 6). Özlü et al. (2022) also reported similar findings, where holding chicks for 72 h increased 7 d mortality more than those held for 6–60 h preplacement.

Özlü et al. (2022) reported that mortality was 1.4% in the 72 h group at 3 d, which was more than double that in the other preplacement holding time (6 h to 60 h) groups. During the rest of the first week (4–7 d), mortality was similar among the groups (0.3–0.5%), However, the 72 h holding time group still had significantly higher mortality for the whole 7 d period (0–7 d) compared to the other pre-placement holding time groups (Table 6).

**Table 6** Effect of preplacement holding time on dead on arrival and the first-week mortality

| Preplacement holding time, h | Dead on arrival, % | 0–7 mortality, % | Dead on arrival, % | 0–7 d mortality, % |
|---|---|---|---|---|
| 6 | 0.020[a] | 1.02[b] | 0.005[c] | 0.57[b] |
| 24 | 0.039[b] | 0.99[b] | 0.024[bc] | 0.52[b] |
| 48 | 0.081[ab] | 0.94[b] | 0.032[bc] | 0.89[b] |
| 60 | 0.119[ab] | 0.81[b] | 0.057[ab] | 1.15[b] |
| 72 | 0.244[a] | 1.64[a] | 0.107[a] | 1.82[a] |
| SEM | - | 0.189 | | 0.311 |

[a-c]Means within a column lacking a common superscript differ ($P \leq 0.05$).
*Source*: Adapted from Özlü et al. (2022) and Elibol et al. (2023), respectively.

Given that optimal house brooding temperatures are some 6–8°C warmer than those normally used in chick delivery vehicles, these chicks almost certainly overheated, which could be expected to increase early mortality (Hamissou Maman et al., 2019). On the other hand, in a much larger-scale experiment, Dişa et al. (2022) examined the interaction effect of hatching time and pull time on broiler live performance. Chicks were held in the hatcher for 7, 17, 26, 31, 41, or 50 h after hatching, and the highest mortality at 41 d was found in chicks that were held in the hatcher for the shortest time (7 h),

### *5.3 Season*

Seasonal variations in first-week mortality have been described in several studies (Forseth et al., 2024; Junghans et al., 2022; Yassin et al., 2009; Yerpes et al., 2020), but the causal factors behind seasonal differences are not fully understood. Figure 5 shows that the first-week mortality was the largest in autumn, but this was only one broiler strain, whereas in another strain no effect of season was found (Forseth et al., 2024).

Yerpes et al. (2020) found that the highest first-week mortality was detected in autumn and winter, whereas Yassin et al. (2009) observed higher first-week mortality during the spring and lower rates during autumn. Moreover, Junghans et al. (2022) suggested that first-week mortality is lowest in the

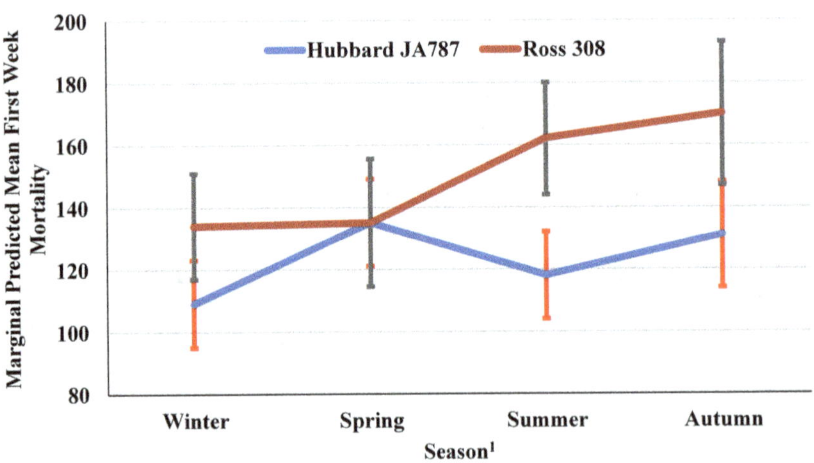


**Figure 5** Effect of season and strain on first-week mortality of two broiler strains. Source: Adapted from Forseth et al. (2024).

[1]Season: Winter = December, January, and February, Spring = March, April, and May, Summer = June, July, and August, Autumn = September, October, and November.

autumn. These ambiguous results among different studies might be related to the geographical location in which the study was executed, but might also be related to animal-related aspects, incubation-related aspects, or environmental aspects, as discussed in this chapter.

## 5.4 Environmental factors in the rearing period

During the first week of life, chicks will experience cold stress when the ambient temperature is below 30°C – the lower limit of their thermal comfort zone (30–35°C) – at the hatchery or at placement (EFSA Ahaw Panel et al., 2023). Chicks that are too cold will clump together and are less likely to go to feed and water, which will lead to dehydration and starvation. In addition, these chickens will also be more susceptible to infections, and with prolonged exposure out of the thermal comfort zone, the risk of mortality increases, particularly during the first week of age (Heier et al., 2002). As the mortality will be the result of the presence of *E. coli* and the health status and resistance, low body temperatures at placement will not help the chicks (Lourens and Kuijpers, 2002).

Ambient and litter temperature at placement have already received significant attention in previous reports (Bruzual et al., 2000b; Cobb, 2008; Deaton et al., 1996; Harris et al., 1975; Lourens and Kuijpers, 2002; Renwick and Washburn, 1982), and if chicks are exposed to low ambient temperatures due to improper post-hatch handling, the chances of increasing first-week mortality are up to three times greater (Yerpes et al., 2019). Shiranjang et al. (2018) reported that the greatest mortality occurred in late hatching chicks from younger breeders when subjected to cool litter temperature (Hot: 34°C vs. Cool: 28°C) during the first 2 d of brooding. These findings are in general agreement with the results of Butcher and Nilipour (2002), Ates et al. (2004), Tona et al. (2004), and Yalcin et al. (2017).

Van Limbergen et al. (2020) indicated potential factors that can influence first-week mortality in broilers, which are floor quality, ventilation type, and the presence of other professional activities of the farmer. However, the fresh air requirement for a day-old chick is only 0.08 $m^3$/h, and there is no significant accumulation of ammonia concentration in the house during the first week (Chou et al., 2004). Moreover, Olanrewaju et al. (2008) conducted a study where there were four treatmentss which were 0, 3000, 6000, and 9000 ppm $CO_2$ during the first 2 weeks of the rearing period. The researchers did not find any significant difference in the first 2 weeks of mortalities among treatments.

The relative humidity during the brooding period can affect the litter quality, but as a matter of fact, mortality during the first days and litter quality were not related, as found by Ritz et al. (2009) and Toghyani et al. (2010).

### 5.5 Broiler flock size and stocking density

Heier et al. (2002) described a relationship between stocking density and first-week mortality, whereby broiler flocks with higher stocking density had lower mortality in the first week after placement. However, in a study of Junghans et al. (2022), the stocking density did not affect first-week mortality, but the broiler flock size did. The highest first-week mortality was observed in flocks categorized as medium, while the lowest was found in flocks categorized as small.

### 5.6 Drinker system and water quality

Despite the important management reasons for choosing drinker types in the farms, nipple drinkers uses for easy and protection of the litter quality, but this inconsistency with their natural behavior (spooning), Moreover, Stamps and Andrews (1995) reported that there is no significant difference in mortality with the three different types of waterers (satellite waterer, plasson, plasson with ring). These findings were partly consistent with reports of Heier et al. (2002).

Carpenter et al. (1990) compared performance indices in chicks raised with nipple drinkers or with nipple drinkers supported by supplementary drinkers (plate type), They detected that mortality decreased in the latter case. On the other hand, Quilumba et al. (2015) did not find significant differences for first-week mortality among the different flow rates of the water in the nipple drinkers (Table 7).

Determine of effects of as a water ingredient carbohydrate had been conducted by McNaughton et al. (1978) who reported a significant decrease in broiler mortality when chicks were given a sucrose solution during the 15-h period following placement.

### 5.7 Vaccination and diseases

Broiler breeder management, nutrition, and vaccination play a large role in controlling diseases of the breeders and the offspring (Oviedo-Rondón,

**Table 7** Effect of drinking system and flow rate of the water on the first-week mortality

| Drinking system | 0–7 mortality, % | Flow rate, mL/min | 0–7 d mortality, % |
|---|---|---|---|
| Drinking cups | 1.3 | 50 | 0.335 |
| Nipple drinkers | 1.5 | 75 | 0.335 |
| Round-drinkers | 1.6 | 100 | 0.445 |
| | | 120 | 0.000 |

*Source*: Adapted from Heier et al. (2002) and Quilumba et al. (2015).

2022). The first-week mortality in broilers partly depends on the effective transfer of antibodies from immunized hens to their progeny (Kariyawasam et al., 2004). However, no significant differences in the first-week mortality were observed between flocks originating from vaccinated and non-vaccinated parents (Gregersen et al., 2010). The bacterium is predominantly held to be responsible for respiratory disease and lameness resulting from lesions of tracheitis, pneumonia, airsacculitis, arthritis, and osteomyelitis (De Herdt et al., 2012). Vaccination reactions can occur in cases where a live vaccine has been administered, and this can lead to higher mortality during the first week.

Junghans et al. (2022) found that the first-week mortality was numerically higher in flocks without reported antibiotic treatment (0.67%) than in flocks reported to have been treated with antibiotics (0.64%). As a major cause of high first-week mortality, *E. coli* infections lead to various disease syndromes in chicks, including yolk sac infection or omphalitis, enteritis, swollen head, respiratory tract infection, cellulitis, and septicemia (Dho-Moulin and Fairbrother, 1999; Lutful Kabir, 2010; Nolan et al., 2020; Noy et al., 1996; Shahjada et al., 2017).

Similar to *E. coli*, paratyphoid *Salmonella* species can cause yolk sac infection and septicemia in newly hatched chicks, which may result in increased first-week mortality (Kemmett et al., 2014; Yassin et al., 2009).

Bacterial infections in chicken eggs often cause mortality of embryos and clinical consequences in chicks, but the pathological mechanism is unclear (Rezaee et al., 2021). Minimizing bacterial infections, especially *E. coli*, during the first week of age in chicks is very critical. Colibacillosis in poultry is associated with different lesion patterns, including airsacculitis, salpingitis, peritonitis, and septicemia (Barnes et al., 2008). Colibacillosis causes decreased performance and higher first-week morbidity and mortality (Fancher et al., 2021; Guabiraba and Schouler, 2015). Kemmett et al. (2014) reported a broiler flock mortality in the first week as 1.03% and 0.44% for the first 72 h. Lower standards of hygiene and biosecurity may result in deterioration in chick quality, manifested by omphalitis, aspergillosis, and colibacillosis (Shane, 1999). The biosecurity, selection of good quality chicks, and immune modulation remain as the main factors to achieve the best results (Swelum et al., 2021).

## 6 Conclusion

The first-week mortality has been shown to be an indicator of the broiler flock health and is affected by animal-related factors, egg and incubation-related factors, and external environmental factors. The first-week mortality can provide information not only about farm environmental conditions but also about sources of the eggs or the chicks besides infections.

Rapidly growing strains tend to have higher first-week mortality due to their egg source and specific requirements when compared to slow-growing strains.

Additionally, females exhibit lower first-week mortality compared to males. Flock age directly impacts egg weight, chick weight, and maternal antibodies passed to the progeny. However, these effects vary between flocks and across different ages within the same category of breeder age. First-week mortality also increases when eggs are stored for more than 7 days. Bacterial infections from contaminated eggs can contribute to first-week mortality. Incubation conditions, particularly eggshell temperature, influence the prevalence of second-grade chicks and affect first-week mortality during the rearing period. The likelihood of first-week mortality is directly linked to the quality of day-old chicks. Late-hatched chicks, which are not immediately ready to eat after placement, experience significantly higher first-week mortality. Additionally, navel conditions also play a role in early chick mortality.

The chick mortality during first week of life can reflect the stress of the transportation process, such as environmental characteristics, journey duration, load percentage, and number of stops. Transport conditions and a delay in the moment of placement and thus first feed and water intake are not related to first-week mortality when the chicks were not dehydrated. However, high temperatures during pre-placement holding and transport (it needs to be checked by chick body temperature) are expected to increase early mortality.

Young chickens behave in the first days as poikilotherms, which means that they are dependent on the external conditions to keep their body temperature at the desired level, meaning that the temperature at placement is very important for chicks. The other factors during the rearing period do not consistently affect the first-week mortality of broilers.

Broiler breeder management, nutrition, and immunization play a big role in the occurrence of diseases and the prevalence of first-week mortality in broilers, because of the transfer of antibodies from immunized hens to their progeny.

In conclusion, the first-week mortality in the life of broilers is a significant factor, which is widely assessed in poultry production and can serve as an indicator of the health status and performance of the broiler flock during the fattening period. Therefore, there is a need to put more effort into the first 7 days of the chicks' life to make sure they develop good drinking and feeding behavior and thus can cope with potential challenges quickly.

## 7 Where to look for future information

Readers interested in learning more about all infectious diseases of broilers can refer to the book *Diseases of Poultry* (eds D. E. Swayne, M. Boulianne, C. M. Logue, L. R. McDougald, V. Nair, D. L. Suarez, S. Wit, T. Grimes, D. Johnson, M. Kromm, T. Y. Prajitno, I. Rubinoff, and G. Zavala).

The following sources are also useful in learning more about animal welfare and transportation:

- EFSA Ahaw Panel, et al. (2023), Welfare of broilers on farm. *EFSA J*, 21(2): p. e07788.
- Health, E.P.o.A., et al. (2022), Welfare of domestic birds and rabbits transported in containers. *EFSA J*, 20(9): p. e07441.

## 8 References


Ates C, et al. (2004) The effect of storage period of eggs on hatching time and broiler performance. *Proc. of XXII World's Poult. Congr., Istanbul, Turkey*. 209.

Aviagen (2021) *Factors affecting chick comfort and liveability from hatcher to brooding house*. Available at: https://www.thepoultrysite.com/articles/factors-affecting-chick-comfort-and-liveability-from-hatcher-to-brooding-house (accessed November 2021).

Avşar KO, et al. (2022) Effect of high eggshell temperature during the early period of incubation on hatchability, hatch time, residual yolk, and first-week broiler performance1. *Journal of Applied Poultry Research* 31(1): 100197.

Awobajo O, et al. (2007) The mortality rate of two different breeds of broilers after brooding stage to maturity. *Middle-East Journal of Scientific Research* 2(1): 37–42.

Bar Shira E, et al. (2005) Impaired immune responses in broiler hatchling hindgut following delayed access to feed. *Veterinary Immunology and Immunopathology* 105(1–2): 33–45.

Barnes HJ, et al. (2008) Colibacillosis. 12th Ed, In: *Diseases of poultry (YM Saif et al)* Wiley Blackwell Publishing, USA, pp: 631–652.

Batal A and Parsons C (2002) Effect of fasting versus feeding oasis after hatching on nutrient utilization in chicks. *Poultry Science* 81(6): 853–859.

Baxter M, et al. (2021) A comparison of fast growing broiler chickens with a slower-growing breed type reared on Higher Welfare commercial farms. *PLoS One* 16(11): e0259333.

Bayliss P and Hinton M (1990) Transportation of broilers with special reference to mortality rates. *Applied Animal Behaviour Science* 28(1–2): 93–118.

Bergoug H, et al. (2013a) Effect of pre-incubation and incubation conditions on hatchability, hatch time and hatch window, and effect of post-hatch handling on chick quality at placement. *World's Poultry Science Journal* 69(2): 313–334.

Bergoug H, et al. (2013b) Effect of transportation duration of 1-day-old chicks on postplacement production performances and pododermatitis of broilers up to slaughter age. *Poultry Science* 92(12): 3300–3309.

Bruzual JJ, et al. (2000a) Effects of relative humidity during incubation on hatchability and body weight of broiler chicks from young breeder flocks1. *Poultry Science* 79(6): 827–830.

Bruzual JJ, et al. (2000b) Effects of relative humidity during the last five days of incubation and brooding temperature on performance of broiler chicks from young broiler breeders1. *Poultry Science* 79(10): 1385–1391.

Buhr RJ, et al. (1994) Automated spray sanitizing of broiler hatching eggs 2. Hatchability of nest clean and dirty eggs. *Journal of Applied Poultry Research* 3(3): 226–233.

Butcher GD and Nilipour AH (2002) *Management of hatching eggs and broiler performance*. University of Florida Cooperative Extension Service, Institute of Food and Agricultural Sciences, EDIS.

Careghi C, et al. (2005) The effects of the spread of hatch and interaction with delayed feed access after hatch on broiler performance until seven days of age. *Poult. Sci.* 84(8): 1314–1320.

Carpenter G, et al. (1990) Effects of the presence or absence of satellite chick-waterers in conjunction with nipple-drinkers on the mortality and productive performance of broiler chicks from young and old dams. *Poultry Science* 69(1): 45–47.

Carver D, et al. (2002) Hatchery and transportation factors associated with early poult mortality in commercial turkey flocks. *Poultry Science* 81(12): 1818–1825.

Chang A, et al. (2016) Can feeding the broiler breeder improve chick quality and offspring performance? *Animal Production Science* 56(8): 1254.

Cheema M, et al. (2003) A comparison of the immune response of a 2001 commercial broiler with a 1957 randombred broiler strain when fed representative 1957 and 2001 broiler diets. *Poultry Science* 82(10): 1519–1529.

Chou CC, et al. (2004) Risk factors for cumulative mortality in broiler chicken flocks in the first week of life in Taiwan. *British Poultry Science* 45(5): 573–577.

Cobb (2008) *Hatchery management*. Cobb-Vantress.

Collin A, et al. (2007) Effects of thermal manipulation during early and late embryogenesis on thermotolerance and breast muscle characteristics in broiler chickens. *Poultry Science* 86(5): 795–800.

Daghir N and Jones R (2008) Breeder and hatchery management in hot climates. In *Poultry production in hot climates*. CABI, pp. 294–329.

Davison T and Lickiss P (1979) The effect of cold stress on the fasted, water-deprived, neonate chicken (Gallus domesticus). *Journal of Thermal Biology* 4(2): 113–120.

de Gouw P, et al. (2017) Effects of dust, formaldehyde and delayed feeding on early postnatal development of broiler chickens. *Research in Veterinary Science* 112: 201–207.

De Herdt P, et al. (2012) Improved broiler performance associated with Ornithobacterium rhinotracheale vaccination in breeders. *Avian Diseases* 56(2): 365–368.

de Jong IC and van Riel JW (2020) Relative contribution of production chain phases to health and performance of broiler chickens: a field study. *Poultry Science* 99(1): 179–188.

de Jong IC, et al. (2017) A'meta-analysis' of effects of post-hatch food and water deprivation on development, performance and welfare of chickens. *PLoS One* 12(12): e0189350.

De Reu K, et al. (2006) Eggshell factors influencing eggshell penetration and whole egg contamination by different bacteria, including Salmonella enteritidis. *International journal of food microbiology* 112(3): 253–260.

Deaton JW, et al. (1996) The effect of brooding temperature on broiler performance. *Poultry Science* 75(10): 1217–1220.

Decuypere E and Bruggeman V (2007) The endocrine interface of environmental and egg factors affecting chick quality. *Poultry Science* 86(5): 1037–1042.

Decuypere E, et al. (2001) The day-old chick: a crucial hinge between breeders and broilers. *World's Poultry Science Journal* 57(02): 127–138.

Dho-Moulin M and Fairbrother JM (1999) Avian pathogenic Escherichia coli (APEC). *Veterinary Research* 30(2–3): 299–316.

Dibner JJ, et al. (1998) Early feeding and development of the immune system in neonatal poultry. *Journal of Applied Poultry Research* 7(4): 425–436.
Dişa R, et al. (2022) Research note: interaction between hatching time and chick pull time affects broiler live performance. *Poultry Science* 101(6): 101845.
Dixon LM (2020) Slow and steady wins the race: the behaviour and welfare of commercial faster growing broiler breeds compared to a commercial slower growing breed. *PLoS One* 15(4): e0231006.
Ebeid TA, et al. (2017) Influence of pre-storage incubation on hatchability traits, thyroid hormones, antioxidative status and immunity of newly hatched chicks at two chicken breeder flock ages. *Animal* 11(11): 1966–1974.
EFSA Ahaw Panel, et al. (2023) Welfare of broilers on farm. *EFSA J* 21(2): e07788.
EFSA (2004) Opinion of the Scientific Panel on Animal Health and Welfare (AHAW) on a request from the Commission related to the welfare of animals during transport. *EFSA Journal* 2(5): 44.
EFSA Panel on Animal Health Welfare (2011) Scientific opinion concerning the welfare of animals during transport. *EFSA Journal* 9(1), 1–36.
Eidson C, et al. (1982) Vaccination of broiler chicks from breeder flocks immunized with a live or inactivated oil emulsion Newcastle disease vaccine. *Poultry Science* 61(8): 1621–1629.
Elibol O and Brake J (2003) Effect of frequency of turning from three to eleven days of incubation on hatchability of broiler hatching eggs. *Poultry Science* 82(3): 357–359.
Elibol O, et al. (2002) The effect of storage and pre-warming periods on hatch time and hatchability. *Avian and Poultry Biology Reviews* 13(4): 243–244.
Elibol O, et al. (2023) Effects and interactions of incubation time and preplacement holding time on mortality at placement, yolk sac utilization, early feeding behavior and broiler live performance. *Animals* 13(24): 3827.
Ernst R, et al. (1984) Effects of heat stress on day-old broiler chicks. *Poultry Science* 63(9): 1719–1721.
Etches RJ (1998) *Reproduction in poultry*. CABI Publishing.
Fancher CA, et al. (2021) Prevalence and molecular characteristics of avian pathogenic escherichia coli in "No Antibiotics Ever" Broiler Farms. *Microbiology Spectrum* 9(3): e00834–00821.
Fanguy R, et al. (1980) Effect of delayed placement on mortality and growth performance of commercial broilers. *Poultry Science* 59(6): 1215–1220.
Fasenko G and O'Dea E (2008) Evaluating broiler growth and mortality in chicks with minor navel conditions at hatching. *Poultry Science* 87(3): 594–597.
Fasenko GM (2007) Egg storage and the embryo. *Poultry Science* 86(5): 1020–1024.
Fasenko GM, et al. (2001) Prestorage incubation of long-term stored broiler breeder eggs: 1. Effects on hatchability. *Poultry Science* 80(10): 1406–1411.
Forseth M, et al. (2024) Mortality risk on farm and during transport: a comparison of 2 broiler hybrids with different growth rates. *Poultry Science* 103(3): 103395.
Gillingham S (2023) *Broiler brooding best management practices*. Available at: https://thebigbookproject.org/agri/poultry-farming/broiler-brooding-best-management-practices/ (accessed August, 2023).
Goodhope R (1991) *First week broiler mortality–Influence on production*. Second Western Meeting of Poultry Clinicians and Pathologists. http://www.westvet.com/1st_week_mortality.htm Accessed November 2007.

Gouda A, et al. (2022) Influences of vitamin A, L-carnitine, and folic acid in ovo feeding on embryo and hatchling characteristics and general health status in ducks. *Animal Biotechnology* 33(1): 150–158.

Gregersen RH, et al. (2010) Impact of Escherichia coli vaccine on parent stock mortality, first week mortality of broilers and population diversity of E. coli in vaccinated flocks. *Avian Pathology* 39(4): 287–295.

Grochowska E, et al. (2019) Field study on the factors affecting egg weight loss, early embryonic mortality, hatchability, and chick mortality with the use of classification tree technique. *Poultry Science* 98(9): 3626-3636.

Guabiraba R and Schouler C (2015) Avian colibacillosis: still many black holes. *FEMS Microbiology Letters* 362(15), 118.

Hamdy AMM, et al. (1991) Effects of incubation humidity and hatching time on heat tolerance of neonatal chicks: growth performance after heat exposure. *Poultry Science* 70(7): 1507–1515.

Hamidu JA, et al. (2007) The effect of broiler breeder genetic strain and parent flock age on eggshell conductance and embryonic metabolism. *Poultry Science* 86(11): 2420–2432.

Hamidu JA, et al. (2011) Broiler egg storage induces cell death and influences embryo quality. *Poultry Science* 90(8): 1749–1757.

Hamissou Maman A, et al. (2019) Effect of chick body temperature during post-hatch handling on broiler live performance1. *Poultry Science* 98(1): 244–250.

Harris G, et al. (1975) The influence of air temperature during brooding on broiler performance. *Poultry Science* 54(2): 571–577.

Havenstein G, et al. (2003) Growth, livability, and feed conversion of 1957 versus 2001 broilers when fed representative 1957 and 2001 broiler diets. *Poultry Science* 82(10): 1500–1508.

Hays F (1955) Age of parents and mortality in offspring: A. During Growing Period B. during the first laying year. *Poultry Science* 34(2): 472–474.

Heier B, et al. (2002) Factors associated with mortality in Norwegian broiler flocks. *Preventive Veterinary Medicine* 53(1–2): 147–158.

Henderson SN, et al. (2008) Effect of an early nutritional supplement on broiler performance. *International Journal of Poultry Science* 7(3): 211–214.

Hollemans MS, et al. (2020) Effects of early nutrition and sanitary conditions on oral tolerance and antibody responses in broiler chickens. *Veterinary Science* 7(4): 148.

Ipek A and Sozcu A (2015) The effects of broiler breeder age on intestinal development during hatch window, chick quality and first week broiler performance. *Journal of Applied Animal Research* 43(4): 402–408.

Ipek A, et al. (2014) The effects of different eggshell temperatures on embryonic development, hatchability, chick quality, and first-week broiler performance. *Poultry Science* 93(2): 464–472.

Jacobs L, et al. (2017) Impact of transportation duration on stress responses in day-old chicks from young and old breeders. *Research in Veterinary Science* 112(Supplement C): 172–176.

Jessen CT, et al. (2021a) Effects of hatching on-farm on behaviour, first week performance, fear level and range use of organic broilers. *Applied Animal Behaviour Science* 238: 105319.

Jessen CT, et al. (2021b) Effects of hatching on-farm on performance and welfare of organic broilers. *Poultry Science* 100(9): 101292.

Joseph N, et al. (2006) The effects of suboptimal eggshell temperature during incubation on broiler chick quality, live performance, and further processing yield. *Poultry Science* 85(5): 932–938.
Joseph NS and Moran ET (2005) Effect of flock age and postemergent holding in the hatcher on broiler live performance and further-processing yield. *Journal of Applied Poultry Research* 14(3): 512–520.
Junghans A, et al. (2022) Data evaluation of broiler chicken rearing and slaughter–an exploratory study. *Frontiers in Veterinary Science* 9: 957786.
Juul-Madsen HR, et al. (2004) Influence of early or late start of first feeding on growth and immune phenotype of broilers. *British Poultry Science* 45(2): 210–222.
Kariyawasam S, et al. (2004) Resistance of broiler chickens to Escherichia coli respiratory tract infection induced by passively transferred egg-yolk antibodies. *Veterinary Microbiology* 98(3): 273–284.
Kemmett K, et al. (2014) The contribution of systemic Escherichia coli infection to the early mortalities of commercial broiler chickens. *Avian Pathology* 43(1): 37–42.
Kingston DJ (1979) Some hatchery factors involved in early chick mortality. *Australian Veterinary Journal* 55(9): 418–421.
Knowles T, et al. (2004) Welfare aspects of chick handling in broiler and laying hen hatcheries. *Animal Welfare* 13(4): 409–418.
Lamot DM, et al. (2016) Effects of feed access after hatch and inclusion of fish oil and medium chain fatty acids in a pre-starter diet on broiler chicken growth performance and humoral immunity. *Animal* 10(09): 1409–1416.
Leandro N, et al. (2011a) Maternal antibody transfer to broiler progeny varies among strains and is affected by grain source and cage density. *Poultry Science* 90(12): 2730–2739.
Leandro N, et al. (2011b) Effects of broiler breeder genetic, diet type, and feeding program on maternal antibody transfer and development of lymphoid tissues in chicken progeny. *Journal of Applied Poultry Research* 20(4): 474–484.
Leitner G, et al. (1989) Sex-related differences in immune response and survival rate of broiler chickens. *Veterinary Immunology and Immunopathology* 21(3–4): 249–260.
Lingens JB, et al. (2021) Effects of early nutrition of hatched chicks on welfare and growth performance: a pilot study. *Animals* 11(10): 2888.
Lourens A and Kuijpers M (2002) Control temperature of young chicks to reduce mortality. *World Poultry* 18(11): 24–26.
Lourens A, et al. (2005) Effect of eggshell temperature during incubation on embryo development, hatchability, and posthatch development. *Poultry Science* 84(6): 914–920.
Lourens A, et al. (2006) Effect of egg size on heat production and the transition of energy from egg to hatchling. *Poultry Science* 85(4): 770–776.
Lourens S, et al. (2021) *Hatchery signals: a practical guide to improving hatching results.* Agricultural Publishers.
Lowman Z and Parkhurst C (2014) The effect of feeding Hydrogel-95 to emu chicks at hatch. *Journal of Applied Poultry Research* 23(1): 129–131.
Lubritz D, et al. (1995) Heritability of ascites and the ratio of right to total ventricle weight in broiler breeder male lines. *Poultry Science* 74(7): 1237–1241.
Lutful Kabir S (2010) Avian colibacillosis and salmonellosis: a closer look at epidemiology, pathogenesis, diagnosis, control and public health concerns. *International Journal of Environmental Research and Public Health* 7(1): 89–114.

Maatjens CM, et al. (2016) Temperature during the last week of incubation. II. Effects on first week broiler development and performance. *Poultry Science* 95(9): 2136–2144.

Mast J and Goddeeris B (1999) Development of immunocompetence of broiler chickens. *Veterinary immunology and immunopathology* 70(3–4): 245–256.

McNaughton J, et al. (1978) Effect of age of parents and hatching egg weight on broiler chick mortality. *Poultry Science* 57(1): 38–44.

Merritt E (1963) Pre-incubation storage effects on subsequent performance of chickens 1. *British Poultry Science* 5(1): 67–73.

Mitchell M, et al. (1996) Defining the optimum thermal environment for the transportation of 1-d-old chicks: physiological responses during transport simulations. *British Poultry Science* 37: S89.

Mitchell MA (2009) Chick transport and welfare. *Avian Biology Research* 2(1–2): 99.

Mozdziak P, et al. (2002) The effect of early posthatch nutrition on satellite cell mitotic activity. *Poultry Science* 81(11): 1703–1708.

Mujahid A and Furuse M (2009) Behavioral responses of neonatal chicks exposed to low environmental temperature. *Poultry Science* 88(5): 917–922.

Nangsuay A, et al. (2011) Yolk absorption and embryo development of small and large eggs originating from young and old breeder hens. *Poultry Science* 90(11): 2648–2655.

Nichelmann M and Tzschentke B (2002) Ontogeny of thermoregulation in precocial birds. *Comparative Biochemistry and Physiology Part A: Molecular & Integrative Physiology* 131(4): 751–763.

Nielsen BL (2020) Understanding the sensory perception of chickens. In *Understanding the behaviour and improving the welfare of chickens*. Burleigh Dodds Science Publishing, pp. 37–58.

Noble R and Cocchi M (1990) Lipid metabolism and the neonatal chicken. *Progress in lLpid Research* 29(2): 107–140.

Nolan L, et al. (2020) Colibacillosis. 14th Ed, In: *Diseases of Poultry (DE Swayne et al) Wiley Blackwell Publishing, USA, pp:* 770–830.

Noy Y and Sklan D (1999) Energy utilization in newly hatched chicks. *Poultry Science* 78(12): 1750–1756.

Noy Y and Sklan D (2001) Yolk and exogenous feed utilization in the Posthatch chick. *Poultry Science* 80(10): 1490–1495.

Noy Y, et al. (1996) Routes of yolk utilisation in the newly-hatched chick. *British Poultry Science* 37(5): 987–996.

óDea EE, et al. (2004) Investigating the eggshell conductance and embryonic metabolism of modern and unselected domestic avian genetic strains at two flock ages. *Poultry Science* 83(12): 2059–2070.

Olanrewaju H, et al. (2008) Growth performance and physiological variables for broiler chickens subjected to short-term elevated carbon dioxide concentrations. *International Journal of Poultry Science* 7(8): 738–742.

Oviedo-Rondón EO (2022) *Optimizing the health of broilers*. North Carolina State University.

Oviedo-Rondon EO, et al. (2013) Broiler breeder feeding programs and trace minerals on maternal antibody transfer and broiler humoral immune response. *Journal of Applied Poultry Research* 22(3): 499–510.

Özlü S (2016) *Effects of hatch time and holding time in hatcher on broiler performance*. Ankara University, Graduate School of Natural and Applied Science.

Özlü S (2021) Research Note: Storage period and prewarming temperature effects on synchronous egg hatching from broiler breeder flocks during the early laying period. *Poultry Science* 100(3): 100918.

Özlü S and Elibol O (2023) Rapid egg cooling rate after oviposition influences the embryonic development, hatchability, and hatch time of young and old broiler hatching eggs. *Poultry Science* 102(12): 103083.

Özlü S, et al. (2018) Effect of hatching time on yolk sac percentage and broiler live performance. *Brazilian Journal of Poultry Science* 20: 231–236.

Özlü S, et al. (2020) Effect of posthatch feed and water access time on residual yolk and broiler live performance1. *Poultry Science* 99(12): 6737–6744.

Özlü S, et al. (2022) Influence of the preplacement holding time and feeding hydration supplementation before placement on yolk sac utilization, the crop filling rate, feeding behavior and first-week broiler performance. *Poultry Science* 101(10): 102056.

Panda A, et al. (2010) Effect of post-hatch feed deprivation on growth, immune organ development and immune competence in broiler chickens. *Animal Nutrition and Feed Technology* 10(1): 9–17.

Pedroso AA, et al. (2005) Fertility and hatchability of eggs laid in the pullet-to-breeder transition period and in the initial production period. *Animal Reproduction Science* 90(3–4): 355–364.

Peebles ED, et al. (1998) Effects of dietary fat and eggshell cuticle removal on egg water loss and embryo growth in broiler hatching eggs. *Poultry Science* 77(10): 1522–1530.

Peebles ED, et al. (1999) Effects of breeder age and dietary fat on subsequent broiler performance. 1. Growth, mortality, and feed conversion. *Poultry Science* 78(4): 505–511.

Peebles ED, et al. (2004) Relationships among post-hatch physiological parameters in broiler chicks hatched from young breeder hens and subjected to delayed brooding placement. *International Journal of Poultry Science* 3(9): 578–585.

Perić L, et al. (2022) Effects of flock age, place of oviposition and cleaning treatments of hatching eggs on hatchability in broiler breeders. *Journal of Applied Poultry Research* 31(3): 100279.

Pinchasov Y and Noy Y (1993) Comparison of post-hatch holding time and subsequent early performance of broiler chicks and Turkey poults. *British Poultry Science* 34(1): 111–120.

Quilumba C, et al. (2015) Evaluation of different water flow rates of nipple drinkers on broiler productivity. *Journal of Applied Poultry Research* 24(1): 58–65.

Rayner AC, et al. (2020) Slow-growing broilers are healthier and express more behavioural indicators of positive welfare. *Scientific Reports* 10(1): 15151.

Renwick G and Washburn K (1982) Adaptation of chickens to cool temperature brooding. *Poultry Science* 61(7): 1279–1289.

Rezaee MS, et al. (2021) Bacterial infection in chicken embryos and consequences of yolk sac constitution for embryo survival. *Veterinary Pathology* 58(1): 71–79.

Ritz CW, et al. (2009) Litter quality and broiler performance. *World's Poultry Science Journal* 65(2): 265–277.

Romanoff AL and Romanoff AJ (1967) *Biochemistry of the avian embryo: a quantitative analysis of prenatal development*. Interscience Publishers.

Sagher B (1975) The effect of cold stress on muscle growth in young chicks. *Growth* 39(2): 281–288.

Sedeik ME, et al. (2019) Comparative evaluation of HVT-IBD vector, immune complex, and live IBD vaccines against vvIBDV in commercial broiler chickens with high maternally derived antibodies. *Animals* 9(3): 72.

Shahjada Z, et al. (2017) Bacteria causing omphalitis in newly hatched chicks from broiler and layer flocks and their antibiotic profiles. *International Journal of Natural and Social Sciences* 4(2): 73–81.

Shanawany M (1987) Hatching weight in relation to egg weight in domestic birds. *World's Poultry Science Journal* 43(02): 107–115.

Shane S (1999) Promoting chick quality and livability. *Zootecnica International* 22: 46–49.

Shiranjang R (2015) *Genç Damızlıklarda Kuluçka ve Broyler Performansının Artırılmasına Yönelik Uygulamalar*. Ankara University, Fenbilimleri Enstitüsü Zootekni ABD.

Shiranjang R, et al. (2018) The effect of egg storage period, hatching time, and initial brooding litter temperature on performance of chicks from young breeder. In *International poultry science congress of WPSA Turkish Branch'2018*. Cappadocia-Nigde, pp. 145–150.

Sklan D, et al. (2000) Decreasing weight loss in the hatchery by feeding chicks and poults in hatching trays. *Journal of Applied Poultry Research* 9(2): 142–148.

Skoglund W, et al. (1952) Growth of broiler chicks hatched from various sized eggs when reared in competition with each other. *Poultry Science* 31(5): 796–799.

Skoglund WC and Tomhave AE (1949) Relationship between egg weight, initial chick weight and subsequent broiler growth. *Poultry Science* 28(4): 500–510.

Speake BK, et al. (1998) The utilization of yolk lipids by the chick embryo. *World's Poultry Science Journal* 54(04): 319–334.

Stamps LK and Andrews LD 1995. "Effects of Delayed Housing of Broiler Chicks and Three Different Types of Waterers on Broiler Performance", Poultry Science, 74 (12),1935-1941.

Suarez ME, et al. (1997) Effect of strain and age of the broiler breeder female on incubation time and chick weight. *Poultry Science* 76(7): 1029–1036.

Swelum AA, et al. (2021) Ways to minimize bacterial infections, with special reference to Escherichia coli, to cope with the first-week mortality in chicks: an updated overview. *Poultry Science* 100(5): 101039.

Tabler G, et al. (2004) Mortality patterns associated with commercial broiler production. *Avian Advice* 6(1): 1–3.

Tindell D and Morris D (1964) The effects of egg weight on subsequent broiler performance. *Poultry Science* 43(3): 534–539.

Toghyani M, et al. (2010) Effect of different litter material on performance and behavior of broiler chickens. *Applied Animal Behaviour Science* 122(1): 48–52.

Tona K, et al. (2003) Effects of egg storage time on spread of hatch, chick quality, and chick juvenile growth. *Poultry Science* 82(5): 736–741.

Tona K, et al. (2004) Effects of age of broiler breeders and egg storage on egg quality, hatchability, chick quality, chick weight, and chick posthatch growth to forty-two days. *The Journal of Applied Poultry Research* 13(1): 10–18.

Tona K, et al. (2005a) Day-old chick quality: Relationship to hatching egg quality, adequate incubation practice and prediction of broiler performance. *Avian Poult. Biol. Rev* 16(2): 109–119.

Tona K, et al. (2005b) Interrelationships between chick quality parameters and the effect of individual parameter on broiler relative growth to 7 days of age. *Avian and Poultry Biology Reviews* 16(1): 71–71.

Tzschentke B and Halle I (2009) Influence of temperature stimulation during the last 4 days of incubation on secondary sex ratio and later performance in male and female broiler chicks. *British Poultry Science* 50(5): 634–640.

Uçar A, et al. (2021) Research note: the effects of chick pipping location on broiler live performance. *Poultry Science* 100(9): 101381.

Valros A, et al. (2008) Effect of simulated long transport on behavioural characteristics in two strains of laying hen chicks. *Applied Animal Behaviour Science* 109(1): 58–67.

Van de Ven LJF, et al. (2009) Effects of a combined hatching and brooding system on hatchability, chick weight, and mortality in broilers. *Poultry Science* 88(11): 2273–2279.

Van de Ven LJF, et al. (2012) Significance of chick quality score in broiler production. *Animal* 6(10): 1677–1683.

Van den Brand H, et al. (2016) Effects of floor eggs on hatchability and later life performance in broiler chickens. *Poultry Science*. https://doi.org/10.3382/ps/pew008.

Van den Brand H, et al. (2023) Alternative hatching systems for broilers. In *Embryo development and hatchery practice in poultry production*. Burleigh Dodds Science Publishing, pp. 401–424.

Van der Hel W, et al. (1991) The upper critical ambient temperature in neonatal chicks. *Poultry Science* 70(9): 1882–1887.

Van Emous RA, et al. (2015) Effects of growth patterns and dietary protein levels during rearing of broiler breeders on fertility, hatchability, embryonic mortality, and offspring performance. *Poultry Science*. https://doi.org/10.3382/ps/pev024.

Van Limbergen T, et al. (2020) Risk factors for poor health and performance in European broiler production systems. *BMC Veterinary Research* 16(1): 1–13.

Vermeersch G and Vanschoubroek F (1968) The quantification of the effect of increasing levels of various fats on body weight gain, efficiency of food conversion and food intake of growing chicks. *British Poultry Science* 9(1): 13–30.

Vieira S and Moran E (1998a) Broiler chicks hatched from egg weight extremes and diverse breeder strains. *The Journal of Applied Poultry Research* 7(4): 392–402.

Vieira S and Moran E (1998b) Eggs and chicks from broiler breeders of extremely different age. *The Journal of Applied Poultry Research* 7(4): 372–376.

Vieira S and Moran E (1999a) Effects of delayed placement and used litter on broiler yields. *Journal of Applied Poultry Research* 8(1): 75–81.

Vieira S and Moran E (1999b) Effects of egg of origin and chick post-hatch nutrition on broiler live performance and meat yields. *World's Poultry Science Journal* 55(02): 125–142.

Washburn K and Guill R (1974) Relationship of embryo weight as a percent of egg weight to efficiency of feed utilization in the hatched chick. *Poultry Science* 53(2): 766–769.

Watkins, S. 2003. "Animal welfare audits: What to expect and how to be prepared", Avian Advice, 5 (4),6-8.

Weytjens S, et al. (1999) Thermoregulation in chicks originating from breeder flocks of two different ages. *The Journal of Applied Poultry Research* 8(2): 139–145.

Wiley WH (1950) The Influence of Egg Weight on the Pre-Hatching and Post-Hatching Growth Rate in the FowlII. Egg Weight-Chick Weight Ratios1,2. *Poultry Science* 29(4): 595–604.

Willemsen H, et al. (2008) Critical assessment of chick quality measurements as an indicator of posthatch performance. *Poultry Science* 87(11): 2358–2366.

Willemsen H, et al. (2010) Delay in feed access and spread of hatch: importance of early nutrition. *World's Poultry Science Journal* 66(2): 177–188.

Wilson H (1991) Interrelationships of egg size, chick size, posthatching growth and hatchability. *World's Poultry Science Journal* 47(1): 5–20.

Wyatt C, et al. (1985) Influence of egg size, eggshell quality, and posthatch holding time on broiler performance. *Poultry Science* 64(11): 2049–2055.

Xin H and Harmon JD (1996) Responses of group-housed neonatal chicks to posthatch holding environment. *Transactions of the ASAE* 39(6): 2249–2254.

Xin H and Rieger SR (1995) Physical conditions and mortalities associated with international air transport of young chicks. *Transactions of the ASABE* 38(6): 1863.

Yalcin S, et al. (2017) Effect of egg storage duration and brooding temperatures on chick growth, intestine morphology and nutrient transporters. *Animal* 11(10): 1791–1797.

Yassin H, et al. (2009) Field study on broilers' first-week mortality. *Poultry Science* 88(4): 798–804.

Yerpes M, et al. (2019) Thermal stress in day-old chicks: risk factors and effects on mortality. *Animal Welfare* 28: 123–132.

Yerpes M, et al. (2020) Factors associated with cumulative first-week mortality in broiler chicks. *Animals* 10(2): 310.

Yerpes M, et al. (2021) Effect of environmental conditions during transport on chick weight loss and mortality. *Poultry Science* 100: 129–137.

# Chapter 21

## Managing health issues in broilers: leg health and leg problems

*Wendy I. Muir and Peter J. Groves, The University of Sydney, Australia; and Rodney L. Reece, Veterinary Consultant, Australia*

1 Introduction
2 Broiler leg problems
3 Assessment of leg problems
4 Improving leg health and reducing leg problems
5 Future trends in research
6 Conclusion
7 Where to look for further information
8 References

### 1 Introduction

Chicken meat is one of the most commonly consumed animal foods across the globe (Mottet and Tempio, 2017). More than 90% of chicken meat is produced in specialised conventional systems where broilers are housed on litter bedding with mechanised provision of water and age-related formulated feed. Under these conditions modern meat chickens or broilers have a high growth rate, reaching slaughter weight of 2.3 kg by 35 days of age and consuming approximately 3.2 kg feed (Anon, 2024b), achieving a feed conversion ratio (FCR) of approximately 1.39 (Anon, 2022).

The broilers' outstanding FCR has been attained predominantly through selective breeding for body weight gain (Havenstein et al., 1994; Havenstein et al., 2003), fine-tuned nutrition, improvements in chicken management including broiler husbandry and health (Anon, 2024b), and advances in equipment within the poultry house, especially shed ventilation (Barger-Weathers, 2025). Together these have generated a 400% increase in broiler growth rate in 60 years (Santos et al., 2022). When compared to other meat production systems, for example beef cattle and sheep meat, chicken meat is produced with relatively low resource usage and cost (Phibbs et al., 2021). Its

http://dx.doi.org/10.19103/AS.2025.0165.26

comparative affordability and perception as a healthy alternative to red meat, ease of preparation, low carbon footprint, and appeal across a diverse range of religions and cultures (Mottet and Tempio, 2017) have driven its increasing popularity. In Australia, the consumption of chicken meat has doubled in the last 30 years, now being integral to the main meal 2–3 times each week (Anon, 2024a). Globally chicken meat consumption is expected to grow by 15% between 2023 and 2032, when it is predicted to contribute 41% of the total meat protein consumed (OECD/FAO, 2023).

However, the very rapid growth of broilers, and in particular, their fast accretion of muscle mass, is assembled onto a skeleton with relatively immature bone development (Santos et al., 2022). Broiler market weight (~2.3 kg) is typically reached when the broiler is 5 weeks old while maturation of the skeleton occurs closer to 25 weeks of age (Rath et al., 2000). This mismatch in muscle deposition and skeletal development has resulted in alterations in broiler locomotion and mobility which compromise broiler welfare (Berg and Butterworth, 2021). Reduced broiler welfare due to leg problems and poorer mobility is evident in several contexts including: pain induced by the condition (McGeown, 1999; Cook, 2000; Tahamtani et al., 2021), injury (Kestin et al., 1992), which may also be acquired from scratches and skin damage, potentially causing cellulitis, when walked over by flock mates (Groves and Muir, 2017); an inability of the broiler to access feed and water, which can lead to dehydration and starvation (Butterworth et al., 2002); and, a change in broiler behaviour, such as reduced frequency of dustbathing (Vestergaard and Sanotra, 1999) and greater time spent sitting or lying versus walking (Weeks et al., 2000). Given the welfare implications of leg problems, the low value of the individual broiler and their unlikely recovery, lame broilers are typically culled (Berg and Butterworth, 2021).

Frequently referenced estimates of the impact of leg disorders in broiler populations are from some time ago (Sanotra et al., 2003; Knowles et al., 2008). A survey of 176 flocks in the United Kingdom identified poor mobility in >27% of broilers with ~ 3.3% almost unable to walk (Knowles et al., 2008) when close to slaughter size (40 days old). More recently Kittelsen et al. (2017) assessed 59 commercial Ross 308 broiler flocks in Norway, identifying that, when 4 weeks old, ~ 25% of the broilers had a moderate to severe defect in their walking ability (i.e. gait score ≥3). For several decades broiler breeder companies have selected for improved growth rate while also removing broilers with leg and bone defects from their breeding programmes (Neeteson et al., 2023; Barger-Weathers, 2025). This has contributed to a reduction in leg problems (Angel, 2007), but fast-growing broilers continue to have gait scores indicative of compromised locomotion, walking ability, and leg problems (Riber and Wurtz, 2024). Further, given the ~4-year delay in observing the impact of selection decisions on commercial broiler flocks (Barger-Weathers, 2025), current data on leg problems is required to garner an accurate reflection of current prevalence and aetiology.

Throughout the literature leg disorders are referred to in several ways, including leg weakness (Kestin et al., 1992), leg strength (Muir and Groves, 2019), leg health (Lister, 2011; Phibbs et al., 2021), leg problems (Sanotra et al., 2003), leg disorders (Knowles et al., 2008; Liu et al., 2023), and lameness (McGeown, 1999; Nääs et al., 2009). Overall, these terms all refer to the reduced agility and alterations to the gait and time budget of fast-growing (FG) broilers (Bessei, 2006; Dawson et al., 2021). Within this chapter the broad variety of leg disorders will generally be referred to as leg problems.

Leg problems are caused by a variety of conditions, and they may be classified by a number of criteria including non-infectious and infectious factors (Kierończyk et al., 2017) or as degenerative, developmental, and infectious causation (Bradshaw, 2002). However, these are not mutually exclusive. The aim of this chapter is to appraise the causes and consequences of leg problems in broilers and to evaluate approaches that may reduce their incidence and impact, thereby improving broiler welfare and the sustainability of the chicken meat sector. To provide a framework for the discussion on leg problems broiler embryology and bone development are briefly addressed, establishing what is normal for the broiler. This is then contrasted with leg problems which are considered in terms of metabolic, infectious, developmental, and traumatic aetiologies. Techniques for assessing leg integrity are presented in tabular form, followed by a review of several strategies that have been evaluated as potential mechanisms to improve broiler leg health and to reduce leg problems. These include breeding programmes, aspects of the parent flock, fertile egg incubation, and broiler management. Focal areas for future research are briefly considered, and finally suggested sources for further information on broiler leg problems are listed.

## 2 Broiler leg problems

### 2.1 Introduction

We will present a brief overview of the embryology and development of the legs of the chicken to provide a scaffolding for contextualisation of the potential problems associated with the legs of broilers. Each of these leg problems is potentially serious but may be absent, intermittent, or sporadic. They can affect productivity and have consequences for animal welfare. We will address nutritional and metabolic disorders represented by osteodystrophies including rickets; the infectious diseases, bacterial osteomyelitis, femoral head necrosis and associated phenomena, and pododermatitis; tibial dyschondroplasia as an example of a developmental problem; and rupture of the gastrocnemius tendon due to trauma. Meaningful data on the flock

Published by Burleigh Dodds Science Publishing Limited, 2026.

prevalence, let alone incidence, and industry or company occurrence and impact, of these conditions, locally, nationally, and internationally, is difficult to come by, as such information is mostly treated as confidential in-house. Observations derived from laboratory submissions are, by their very nature, biased towards particular cases of morbidity or mortality and must be handled and interpreted accordingly. However, Reece et al. (1985), besides documenting the diagnoses in 1402 broiler flock submissions, also included data on significance within the affected flocks. For example, there were 116 cases (~8% of problem flocks submitted) of osteodystrophy with up to 40% flock prevalence, 37 flocks with tibial dyschondroplasia affecting from 0.5% to 70% of the flock, 7 flocks with ruptured tendons affecting 2–5%, and 2 flocks with Staphylococcal osteomyelitis affecting about 1%.

## 2.2 Broiler embryology and development

By 12 days of incubation in the chicken embryo, the avascular cartilaginous limb buds have a trabeculated cortex; at hatch this is still in a parallel herringbone-like arrangement (Fig. 1a), and in layer-type chickens this feature remains for about 1 week (R.L. Reece, unpublished observations). Bone mineralisation in the late embryo is facilitated by mobilisation of minerals from the inner layer of the eggshell via the chorioallantoic membrane (Halgrain et al., 2022) and plateaus just before hatch. It then accelerates a few days post-hatch (Yair et al., 2012), which coincides with a feed-derived nutrient surge when the chicken is placed in the grower house. The long bones are still pliable until about 10 days of age (Sanchez-Rodriguez et al., 2019). The diaphyseal cortex of the broiler femur and tibiotarsus are circular in the embryo but become elliptical by 1 week post-hatch (Yair et al., 2012). This feature requires consideration when assessing bone breaking strength as that will vary depending upon point of pressure application. The outer diaphyseal cortex expands by intramembranous ossification involving periosteal osteoblasts and becomes lamellated, while the inner aspect is being eroded continuously during growth by osteoclasts: thus, the cortex has a relatively short period to mature before it is effectively remodelled and removed (Fig. 1b). In slower-growing (SG) broilers, cortical osteoblasts proliferate more rapidly and have lower alkaline phosphatase activity, than in fast-growing (FG) broilers, and the osteons in FG broilers do not fill out as completely, probably associated with the more rapid radial expansion (Rawlinson et al., 2009). The vascular spaces in the cortex, the Haversian or osteon canals, extend longitudinally via extensions of the nutrient artery and are prominent in broilers, but rapidly growing broilers also have abundant large radial vessels (Pratt and Cooper 2018), probably contributing to cortical porosity. Compared to the diaphyseal cortex, the metaphyseal cortex is somewhat trabeculated and porous rather than extensively lamellated,

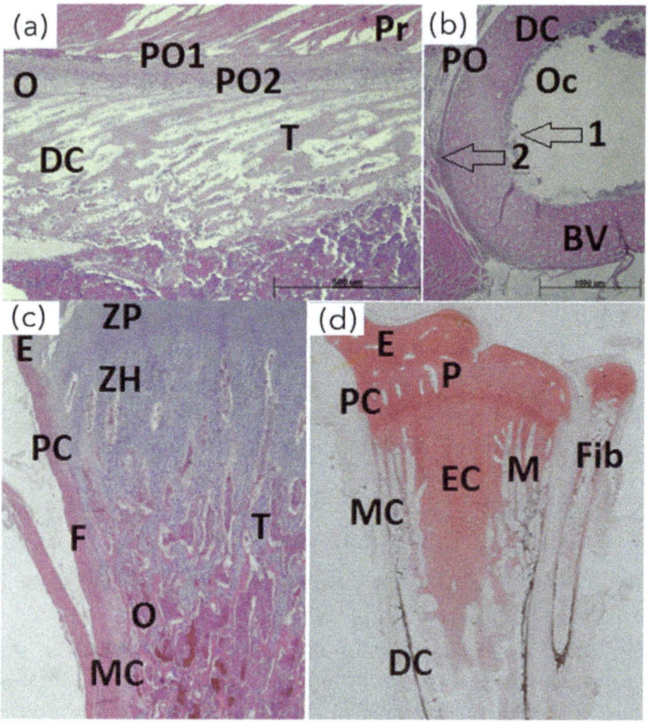


**Figure 1** (a) Femur of hatchling broiler. Longitudinal section. Demineralised. HEx100. PO1, outer periosteum of connective tissue. PO2, inner periosteum of osteoblasts and precursors. O, prominent focus of intramembranous ossification (this section is from the diaphyseal cortex). DC, diaphyseal cortex. Pr, proximal. T, well-formed trabeculae arranged in a 'herringbone' fashion. (b) Tibiotarsus, 14-day-old broiler. Midshaft diaphysis. Transverse section cut with fine-toothed hand-saw, trimmed, fixed, demineralised. HEx40. PO, periosteum separated away from cortex (in the upper aspect) is a post-mortem artefact. DC, diaphyseal cortex of lamellar bone formed by intramembranous ossification. Oc, osteoclasts, lining and rapidly eroding, the inner layer of cortex. Arrow 1 indicates the direction of radial vessels derived from nutrient artery entering cortex from medulla. Arrow 2 indicates the direction of peripheral cortical expansion. BV, blood vessels in broiler cortex are large and prominent and include both longitudinal Haversian or osteonic canals, and radial vessels, thus creating significant porosity. (c) Proximal tibiotarsus, 3-week-old chicken. HEx40. E epiphysis. ZP, zone of proliferation. ZH, zone of hypertrophy. PC, physeal cortex. F, thick fibrous outer periosteum. T, ossifying trabeculae are part of the endochondral ossification process. O, cortex is formed by intramembranous ossification, and the inner layer is not fully lamellated at this stage. MC, metaphyseal cortex. (d) Tibiotarsus, 4-day-old chicken. Longitudinal medio-lateral section; fixed but not demineralised. Von Kossa stain reveals calcium (black), with a red/pink counterstain, x40. E, epiphysis. P, physis. PC, physeal cortex. M, metaphysis. MC, metaphyseal cortex. Note that PC and MC are not mineralised and therefore much weaker than the mineralised DC. EC, embryonic cartilaginous cone. This is a normal feature in long bones of recently hatched chickens and distinct from tibial dyschondroplasia. DC, mineralised diaphyseal cortex. Fib, fibula.

as it is also rapidly extending in length (Fig. 1c). The cortex at the physis is a fibro-cartilaginous collar and a site of weakness. A large wedge of embryonic cartilage is still present at hatch (Fig. 1d). Longitudinal growth of the bones comes from the physis which is sharply demarcated from the epiphysis by a thin line of resting chondrocytes. These multiply, creating neat stacks of flattened chondrocytes (designated as zone of proliferation, ZP), richly supplied by an arcade of non-anastomosing, thin-walled, fenestrated epiphyseal vascular channels, each ending as a capillary plexus (Wideman and Prisby, 2013). In birds, it is the depth of the ZP, characterised by flattened chondrocytes, that determines the growth rate of the long bones (Kember et al., 1990). There is a narrow band of relatively inactive prehypertrophic chondrocytes embedded in avascular cartilage (zone of prehypertrophy, ZpH). As the chondrocytes develop and hypertrophy (by 5–10x), they still present as columns continuous with the proliferating stacks (zone of hypertrophy, ZH) and secrete cartilaginous matrix vesicles, and Ca and P, which crystallise as hydroxyapatite around collagen fibres. This provisional mineralisation (zone of provisional mineralisation, ZPM) leads to apoptosis of chondrocytes; the cartilage is eroded by incoming osteoclasts associated with thick-walled metaphyseal arteries; capillaries penetrate into the vacated lacunae; and osteoblasts enter, mature into osteoblasts, and remain there secreting bone (zone of ossification, ZO) and forming trabeculae, which are modified to follow lines of stress or tension (Figs. 2, 3, 4a and 4b). The sequential changes from chondrocyte proliferation to hypertrophy and subsequent differentiation are tightly controlled by many growth factors activating chondrocyte-specific transcriptions (Adams et al., 2007). The chondrocytes in the developing physis multiply and mature, but they do not migrate: it is their relationship to surrounding structures and blood vessels which changes. As the epiphyseal vascular channels withdraw, they leave behind fibrinoid debris, and as the metaphyseal arteries and capillaries advance, there may be partial luminal constriction due to cartilage protrusion from activated hypertrophic chondrocytes, and vascular debris is common. Contrary to some earlier reports, patent transphyseal vessels are occasionally encountered. A secondary centre of endochondral ossification develops in many epiphyses.

The chicken is bipedal, but during walking, the femur remains relatively immobile, and it is the tibiotarsus which rotates in a semi-arc to keep the centre of gravity over the leading leg (Paxton et al., 2010). The modern broiler with well-developed breast muscle has a forward-placed centre of gravity. In a healthy 'normal' chicken the femur and tibiotarsus have an outward shaft rotation and a bowing along their length, with the medial diaphyseal cortex somewhat thicker than the remainder. Applegate and Lilburn (2002) showed that the pattern of development of femur was different to that of the tibiotarsus in broilers, with the femur being more sensitive to nutritional changes and its

Published by Burleigh Dodds Science Publishing Limited, 2026.

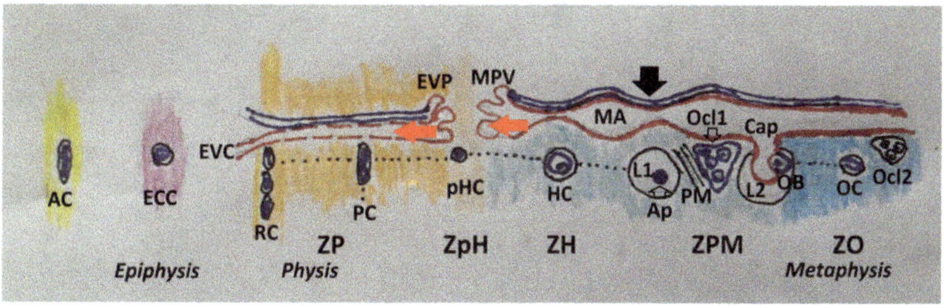


**Figure 2** A diagrammatic representation so that the reader becomes more familiar with the major events and processes involved in endochondral ossification of a long bone in a chicken. EVC, epiphyseal vascular channels (thin-walled and fenestrated). EVP, epiphyseal vessel plexus. MVP, metaphyseal vascular plexus. MA, metaphyseal artery (thick-walled with continuous endothelium). Black arrow: compression restriction of metaphyseal artery lumen because abutting hypertrophic chondrocytes produce abundant cartilage matrix. Red arrows: direction of growth of blood vessels as the bone elongates. L1 and L2, lacunae in which chondrocytes reside. AC, articular cartilage and chondrocyte. ECC, epiphyseal chondrocyte and hyaline cartilage. RC, resting chondrocyte. PC, proliferating chondrocyte. pHC, prehypertrophic chondrocyte. HC, hypertrophic chondrocyte. Ap, apoptotic chondrocyte. PM, provisional mineralisation of cartilage. Ocl1, osteoclast removing hypertrophic cartilage scaffold and apoptotic chondrocyte. OB, osteoblast. OC, osteocyte. ZP, zone of proliferation. ZpH, zone of prehypertrophy. ZH, zone of hypertrophy. ZPM, zone of provisional mineralisation. ZO, zone of ossification forming bony trabeculae. Ocl2, osteoclast remodelling the formed bony trabecula.

length plateauing earlier, while also having to accommodate for rapid changes in weight and conformation (increasing breast muscle development with age) and thus being more susceptible to skeletal abnormalities. The tibiotarsus in a broiler grows almost twice as fast proximally as distally, and between 11- and 18-days post-hatch may increase by up to 2.4 mm per day (0.1 mm per hour) (Riddell, 1975; Williams et al, 2000), but that extension may be pulsatile rather than uniform. This rapid elongation of the long bones will cause stretching of the ligaments, tendons, and muscles, and perhaps affects their attachments, leading to some degree of discomfort, pain, and/or vulnerability similar to growing pains and Osgood-Schlatter syndrome in rapidly growing children (Vaishya et al., 2016), but this has not been adequately addressed in broilers. In broilers, right-sided leg dominance is most prevalent (Duff and Thorp, 1985), but the reason for this, and its consequences, particularly how it may or may not impact performance and evaluation of gait and movement studies, has only been tentatively addressed (Paxton et al., 2013).

The modern broiler is derived and selected from various English game fowl including Old English Game and cross-bred in nineteenth-century Cornwall, UK, with Asian game fowl (particularly Malay; and Indian Red Assel) to create 'Indian Game' (UK) or 'Cornish' (USA), and later crossed with Plymouth Rock (USA), Light Sussex (UK), or other heavy breeds, and is therefore different in many features from more conventional breeds with which we are familiar.

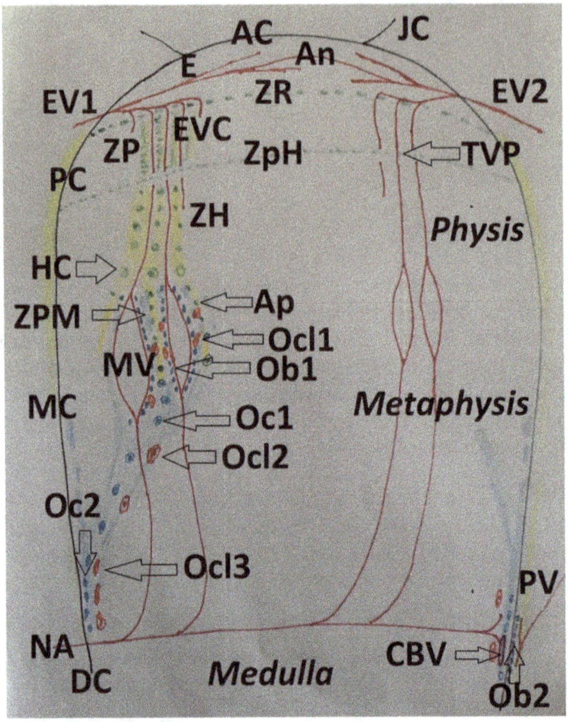


**Figure 3** Diagram labelled with components of the growth plate of a chicken, so the reader becomes more familiar with terminology. AC, articular cartilage. JC, joint capsule. E, epiphysis. EV1 and EV2, epiphyseal vessels (often more than one enters epiphysis below the JC and above the physis. An, anastomoses between epiphyseal vessels occur, but not common. ZR, resting chondrocytes of physis sharply demarcated from epiphysis. EVC, narrow epiphyseal vascular channels penetrate between columns of proliferating chondrocytes. ZP, physeal zone of chondrocyte proliferation. ZpH, zone of prehypertrophy. TVP, transphyseal vessels being anastomosed between epiphyseal and metaphyseal vessels are rare. ZH, zone of hypertrophy where chondrocytes gradually increase in size and secrete more cartilaginous matrix. PC, physeal cortex of thick outer periosteal fibrous tissue and inner thin rim of cartilage. HC, hypertrophic chondrocytes. Ap, apoptotic chondrocytes. ZPM, zone provisional mineralisation. MV, metaphyseal vessel. Ocl1, osteoclasts accompanying incoming MV remove apoptotic chondrocytes and cartilage. Ob1, osteoblasts accompany MV and move into vacated lacunae. Oc1, osteocytes mature and produce bony trabeculae as part of endochondral ossification process. Ocl2, osteoclasts remodelling trabeculae. MC, metaphyseal cortex. Ocl3, osteoclasts removing inner cortex to allow expansion outwards. Oc2, osteocytes in the inner layer of cortex. NA, nutrient artery enters laterally in mid-diaphysis. DC diaphyseal cortex is tubular, solid, and distal to metaphysis. Medulla has sparse trabeculae. PV, periosteal vessels supply outer periosteum. CBV, cortical blood vessels. Ob2, osteoblasts of the inner layer of periosteum form the cortex by intramembranous ossification.

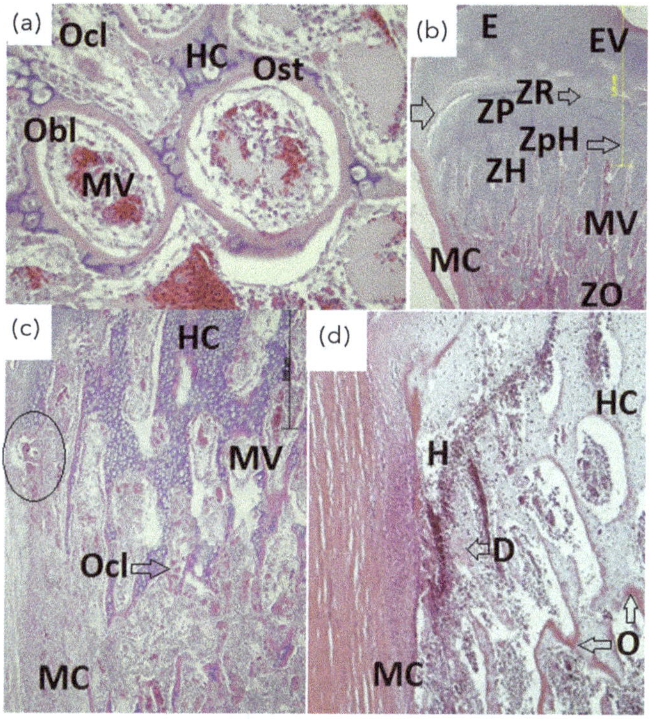


**Figure 4** (a) Proximal tibiotarsus, 4-week-old broiler. Cross-section of metaphysis. HEx400. Ocl, osteoclast. HC, cross-section of trabeculae of hypertrophic cartilage (resemble islands). Ost, rim of osteoid. Obl, circle of osteoblasts accompanying incoming blood vessels. MV, metaphyseal vessels. (b) Proximal tibiotarsus, 19-day-old broiler. HEx40. E, epiphysis. EV, epiphyseal vessels. Large arrow: a dilated vessel near the epiphyseal/physeal boundary, at least partially artefactual from processing. This is a well-recognised, but not common, finding at this site, and must be interpreted cautiously. ZR, narrow linear zone of resting chondrocytes. ZP, zone of proliferation. ZpH, narrow, relatively indistinct at this magnification, zone of prehypertrophy. ZH, zone of hypertrophy. MV, metaphyseal vessels (ascending artery and descending venules). MC, metaphyseal cortex. ZO, zone of ossification developing from provisional mineralisation. (c) Proximal tibiotarsus, 14-day-old broiler. HEx100. HC, hypertrophic cartilage. MV, metaphyseal vessels. Ocl, small cluster of osteoclasts. MC, metaphyseal cortex. Circle encloses small focus of debris adjacent to metaphyseal cortex which has necrotic cartilage and osteoid, free erythrocytes (mild focal haemorrhage), and quite a few adjacent osteoclasts. This is probably focal blunt trauma impact. (d) Proximal tibiotarsus, 6-week-old broiler. HEx100. H, line of recent haemorrhage towards the periphery of the physis in the upper zone of hypertrophy. This is probably a linear shearing tear at the edge of the non-mineralised physeal cortex, which is a vulnerable site. There is haemorrhage, only a few infiltrating inflammatory cells (not obvious at this magnification), and early cartilage degeneration – so this in recently ante-mortem, not immediately peri-mortem. It is perhaps a traumatic consequence of catching this broiler by one leg, and upending it while it was moved to transport container, a few hours prior to submission. HC, hypertrophic cartilage. MC, metaphyseal cortex. D, slightly eosinophilic, homogenous, degenerating cartilage. O, thin rim of early ossification (homogenous and eosinophilic) at the edge of hypertrophic cartilage (much paler) trabecula.

Compared to leghorn chickens, the ZP are deeper and more disorderly, incoming metaphyseal vessels are larger, and ZH and ZPM are less orderly (Reiland et al., 1978). The diaphyseal cortex of rapidly growing broilers is relatively more porous than in SG broilers (Rawlinson et al., 2009) and perhaps less robust, and that may predispose them to bone breakage during catching, transport, and handling and contribute to a weaker framework to support extra muscle (Williams et al., 2004). Broilers may spend up to 80% of their time lying (Weeks et al., 2000), but forcing them to exercise does not consistently improve walking ability, nor bone strength (Bizeray-Filoche et al., 2004; Foutz et al., 2007a; Ruiz-Feria et al., 2014). Exercise is known to increase bone mass, but that takes time to come into effect. Movement itself may cause microfractures, which require remodelling (Allan et al., 2014), and 'normal' minor traumatic events of life may result in small transverse tears in the physes adjacent to the non-mineralised fibro-cartilaginous cortex collar (Fig. 4c and d).

In summary, bone development and maintenance is a complex process involving various physiological and biochemical factors, but we have not attempted to detail the metabolic pathways, genes involved, nor the chemical signalling between cells that occurs. Broilers grow rapidly because of genetic selection, combined with optimal nutrition and management, and disease control. They have been selected for muscle growth, and that creates an increased demand upon skeletal integrity (Williams et al., 2000). The features of bone growth outlined above are 'normal' in broilers and other chickens and will also be encountered, or perhaps accentuated, in other conditions, pathological or otherwise: some of these 'normal' features have been reported in detail, supposedly as a component of significant other pathologies. Bones are relatively solid, and preparation for examination can result in damage and artefacts; histology processing requires aggressive demineralisation. Some features require appropriate presentation and alignment to be fully appreciated. The reader is advised to be cautious and not to over-interpret what they observe when examining broiler leg disorders.

## 2.3 Metabolic and nutritional problems: osteodystrophies, including rickets

The modern poultry industry became possible because of the addition of vitamin $D_3$ ($VD_3$) to their diet (then as cod-liver oil; currently, commercial $VD_3$ is mainly derived from sheep's wool lanolin) in the 1920s (Norris, 1958; Torella, 2024). $VD_3$ has to be absorbed, then metabolised in the liver and kidney to become active: among other activities it controls calcium and phosphorus absorption and deposition, whereas parathyroid hormone and calcitonin control circulating concentrations. Calcium and phosphorus requirements are greatest during rapid growth, but plant phytate (inositol hexaphosphate), which can account for up to 80% of plant phosphorus in grains, is unavailable to the chicken and also reduces calcium bioavailability; therefore, phytase is routinely added to the feed. Even though apparently adequate amounts of $VD_3$ are present, absorption can be reduced in infectious stunting syndrome and related diseases, and/or impeded by competition at the micelle level with other fat-soluble ingredients, such as excess carotenoids added for carcass colouration.

In calcium and/or $VD_3$, absolute or relative, deficient diets ('rickets'), there is inhibition of chondrogenesis and failure of hypertrophic chondrocytes to undergo apoptosis because of reduced provisional mineralisation, among other derangements, resulting in accumulated non-mineralised cartilage. The ZP is disrupted, disorganised, and deeper, and often wider than the underlying cortex, due to accumulation rather than increased proliferation (Fig. 5a). There is also an apparent expansion of the ZpH. The epiphyseal vascular channels are relatively normal, but of increased length, while the metaphyseal vessels are blunted. In relative or absolute P deficiency (often designated as 'osteodystrophy'), there is apparent increase in depth of ZP due to accelerated

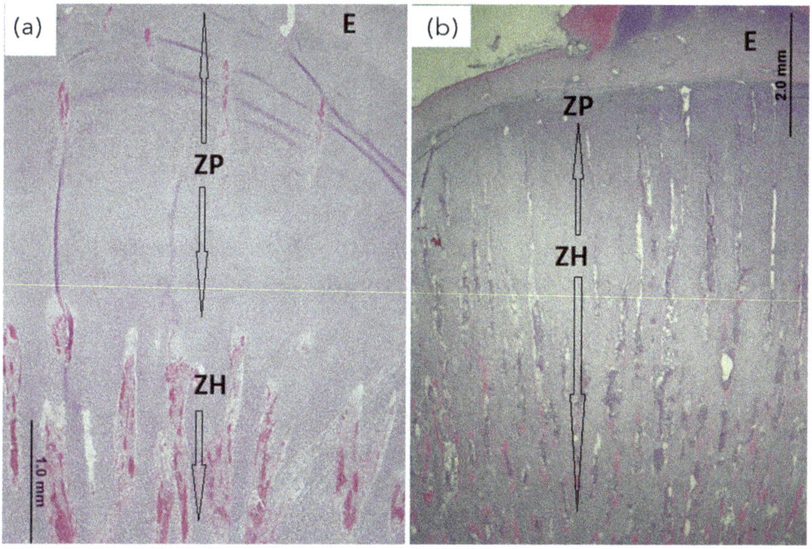


**Figure 5** (a) Proximal tibiotarsus, 14-day-old broiler. HEx40. E, epiphysis. ZP, extremely deep zone of proliferation, in which blood vessels are narrow and difficult to discern, and the stacks of proliferating chondrocytes are a little disorderly – this is consistent with vitamin $D_3$/Ca-deficient rickets. ZH, zone of hypertrophy is also somewhat deepened. The physis in this case slightly bulges out beyond the diaphyseal cortex. (b) Proximal tibiotarsus, 19-day-old broiler. HEx40. E, epiphysis. ZP, shallow zone of proliferation. ZH, deep zone of hypertrophy with long and wide columns of poorly mineralising chondrocytes, and wide and prominent metaphyseal vessels – this is consistent with hypophosphataemic osteodystrophy. In this case, the physis is aligned with the cortex.

chondrogenesis, but failure of transition to hypertrophy, although the many hypertrophied chondrocytes accumulate leading to relatively increased depth of ZH (Fig. 5b), and there is excessive persistence of islands of non-mineralised cartilage within trabeculae (Lacey and Huffer, 1982; Dinev, 2012a). The difference is often over-simplified and depicted as an increase in depth of ZP in calcium/ $VD_3$ deficiency and an increase in ZH in P deficiency. The poor mineralisation of bones, and resorption of minerals, especially from the diaphyseal cortex, in an attempt to maintain homeostasis, results in weak, soft, and rubbery bones, with resultant fractures and collapse of the rib cage. Chickens with such potentially serious problems are encountered often in the field, but the presenting pathologies may overlap, thus obscuring a precise diagnosis (Randall, 1991) and leading to broad treatment. Outward bowing of the legs takes time, and parathyroid hypertrophy is only observed after several weeks of marked calcium or $VD_3$ deficiency. These metabolically induced conditions are distinct from valgus and varus (angular leg and bone) deformities which have different causes and manifestation (van den Brand et al., 2022).

In summary, the rapid growth of broilers makes them vulnerable to metabolic disorders hindering mineral deposition in the skeleton, thus potentially leading to significant bone disorders, particularly during periods of great proportional growth, such as the third week post-hatch (see Broiler embryology and development, Section 2.2). Affected broilers are lame and have reduced feed intake, further exacerbating the problem. In the event of such metabolic or nutritional osteodystophies, appropriate treatment can lead to rapid response, although with residual lost growth.

## 2.4 Infectious problems: bacterial osteomyelitis, femoral head necrosis and pododermatitis and related problems

### 2.4.1 Bacterial osteomyelitis

Chickens are reared on litter, comprising low-cost bedding material such as wood shavings, to which becomes added droppings, feathers, spilt feed, etc., to absorb moisture, provide environmental insulation, and allow some scratching behaviour. Litter develops a rich microbiome, including a low proportion of potential pathogenic bacteria, such as coliforms and staphylococci, and can produce significant amounts of ammonia from degradation of excreta and nitrogenous waste. The chicken mostly inhales air that is less than 30 cm above the litter surface.

Bacterial osteomyelitis in chickens was developed as an animal disease model, and much is known of the sequential pathogenesis (Emslie et al., 1983). Bacteria such as *Escherichia coli* and *Staphylococcus aureus* enter the circulation intermittently or as septicaemias, from compromised respiratory and intestinal tract, and/or damaged integument. These opportunistic bacteria then lodge in vessels, particularly the growing ends of metaphyseal vessels (Emslie and Nade, 1983) where circulation is often sluggish due to constriction of vessel lumina by adjacent hypertrophic cartilage (Fig. 2); they infiltrate and multiply in the adjacent interstitium and colonise lacunae of the ZH. The ensuing inflammatory process of vascular changes, congestion, oedema, infiltration of heterophils and macrophages, and fibrin coagulation can lead to infarctive necrosis of a wedge of the physis forming a sequestra (Figs. 6a and b). Heterophilic granulomas develop rapidly with their wall of giant cells, typical of avian inflammation (Fig. 6c).

Such lesions are usually unilateral, and flock prevalence is usually low. Nairn and Watson (1972) mention that clinical incidence in some meat chicken flocks may reach 50% associated with a relatively low mortality rate of 5%, but no further details were supplied, and this figure is often cited in reviews. Bacterial osteomyelitis results in lameness, and affected birds are often moribund. As such infections are secondary to compromised respiratory, alimentary, and/or integumentary systems, treatment of the flock with appropriate antibiotics may reduce the likelihood of other chickens developing infection leading to osteomyelitis but will not address the dilemma of the necrotic sequestra. The predisposing problems, which are probably multifactorial, require attention!

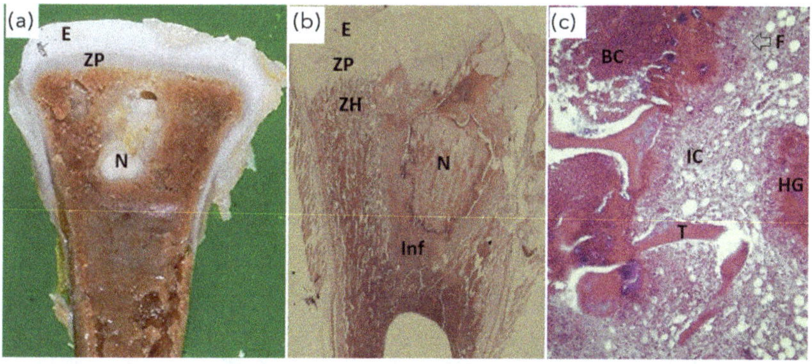


**Figure 6** (a) Proximal tibiotarsus, 3-week-old broiler. E, epiphysis. ZP, zone of proliferation. N, fibrino-necrotic osteomyelitis due to *Escherichia coli*. (Birds react to infection by producing a fibrinous exudation, which is soft and relatively homogenous, not fluid 'pus'). (b) Proximal tibiotarsus from broiler above. Sectioned and fixed. HEx20. E, epiphysis. ZP, zone of proliferation. ZH, zone of hypertrophy. N, sequestra of necrotic cartilage surrounded by fibrin and inflammatory cell infiltrate (Inf), with significant distortion of the adjacent physeal structures. (c) Femoral metaphysis, 6-week-old broiler. HEx200. F, fibrin-rich plasma. IC, mixed inflammatory cell infiltrate (details not discernible at this magnification, but comprises heterophils, macrophages, some lymphocytes, and fibroblasts). BC, colonies of bacterial coliforms. HG, heterophilic granuloma with central eosinophilic debris from degranulated heterophils, and bacterial colonies, surrounded by a wall of giant cell macrophages, then a thin layer of lymphocytes, etc. T, bony trabecula. (This is a coliform osteomyelitis of 5–7 days duration.)

Published by Burleigh Dodds Science Publishing Limited, 2026.

### 2.4.2 'Femoral head necrosis'

'Femoral head necrosis' is a broad descriptor for a number of different phenomena which blur into each other, including femoral head separation, femoral head degeneration, femoral head haemorrhage, and bacterial chondronecrosis with osteomyelitis. It is a well-recognised, but of variable reported incidence, cause of lameness in broilers: McNamee and Smyth (2000) in Northern Ireland found that 'femoral head necrosis' peaked at about 5 weeks of age and was present in about 20% of lame broilers, thus affecting <1% of broilers placed. Even so, this still has potentially significant economic and welfare implications.

The femoral head in chickens has several sources of blood supply, more so than in mammals (Xu et al., 2010). Osteochondrotic clefts at the epiphyseal-physeal boundary and within the physis are often observed histologically in broilers at post-mortem, but unless they include haemorrhage, degeneration, or necrosis, they cannot be proven to be ante-mortem lesions (Fig. 7a and b), although they do indicate structural fragility (Wideman and Prisby, 2013). In live chickens, 'femoral head separation' is a detachment of epiphyses from off the femur, and similar lesions can occur in the proximal tibiotarsus, resulting in

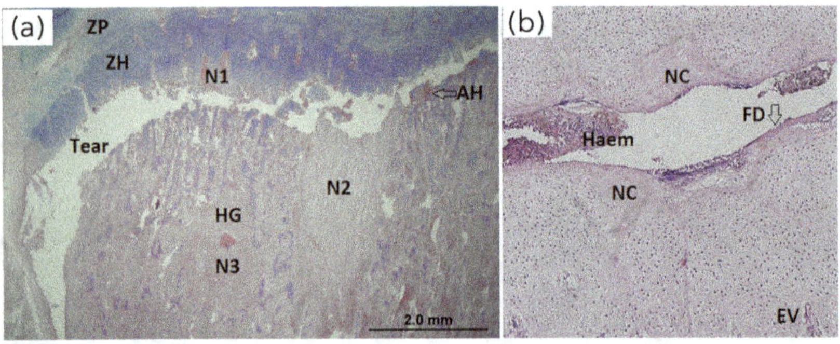


**Figure 7** (a) Physis of proximal femur, 5-week-old broiler. HEx20. ZP, zone of proliferation. ZH, zone of hypertrophy. AH, acid haematin is from the reaction of haemoglobin with demineralisation fluid, at the site of free erythrocytes, indicating small focal haemorrhage, i.e. ante-mortem lesion). Tear: transverse tear through the metaphysis but not adjacent cortex (there is a little frank haemorrhage further along); the bulk of this tear is artefactual (post-mortem) from processing, but it is following along a plane of weak tissue with focal pre-existing haemorrhage and necrosis. N1, necrotic cartilage. N2, large sequestra of necrotic cartilage. N3, large sequestra of necrotic cartilage which appears continuous with N1. HG heterophilic granuloma. *Escherichia coli* was isolated from this lesion (coliform osteomyelitis of femoral head). (b) Epiphysis of tibiotarsus, 6-week-old broiler. HEx100. Haem: haemorrhage in expanded plane. NC, necrotic cartilage (homogenous and slightly eosinophilic). FD, fibrinoid deposit (homogenous, dense, and intensely eosinophilic); probably vascular wall remnants. EV, epiphyseal vessel. (This is an ante-mortem linear tear lesion with haemorrhage and necrosis of epiphyseal cartilage.)

micro-haemorrhage, focal necrosis, and cartilaginous atrophy. There is a down-regulation of certain genes and an enrichment of other biological processes involved in altering osteogenesis, chondrogenesis, and angiogenesis, in affected chickens (Ibelli et al., 2022), but whether these are consequential responses or reactions to the already present significant pathology, or are predisposing and/or causal, awaits clarification. The potential role and activities of many molecular mechanisms have been described (Choppa and Kim, 2023). The true incidence of minor or partial femoral head separation is unknown as affected chickens may not be lame, and the aetiology is still undetermined (Packialakshmi et al., 2015). Thorp et al. (1993b) observed many non-bacterial pathologies in the proximal femurs of non-lame 'control' broilers, but these changes were also seen in lame chickens and probably provided foci for the establishment of bacterial infection. Femoral head bacterial osteochondronecrosis (Fig. 8a and b) is similar to bacterial osteomyelitis as described above but is increased in broilers by rearing on wire, which amplifies torque and shear stressors, resulting in minor damage (Wideman and Prisby, 2013). Such low viability tissues are vulnerable to haematogenously spread opportunistic bacteria. Investigation of blood-borne biota has revealed a wide array of potential bacteria that could be involved in femoral head chondronecrosis and osteomyelitis (Mandal et al., 2016). The potential role in these conditions by air sac diverticula associated with the femoral head (Fig. 8c) is rarely considered.

Separation of the cartilaginous femoral epiphysis from off the underlying bony metaphysis (avulsion) and haemorrhage is an acute traumatic episode associated with capture and handling (Duff and Randall, 1986) when heavy broilers are caught by one leg and upended by their captor until crated (so this lesion may be occurring several hours ante-mortem; not immediately peri-mortem). Separation of the femoral head at the physes during necropsy (Fig. 8d) and fissures and artefactual tears created during histological processing are post-mortem changes. These must be differentiated from ante-mortem femoral head separation and osteomyelitis.

In summary, femoral head lesions are usually of an intermittent, low, and sporadic incidence but are sometimes a serious problem of broilers. Wideman and Prisby (2013) noted that the highest incidence was in the fastest-growing flocks. Many affected broilers appear to have had normal feed intake and growth until the relatively recent or sudden onset of femoral head lesions, and they may not be obviously lame. Much about this condition is still not fully understood. There are probably anatomical, developmental, and/or biomechanical predispositions which are aggravated by environmental factors and become sites of bacterial infection leading to osteomyelitis.

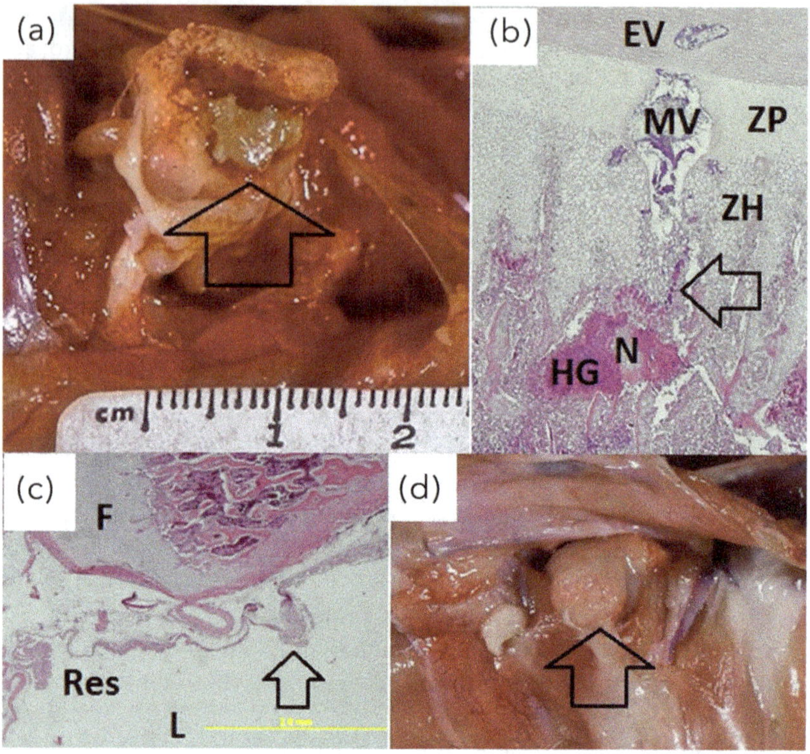


**Figure 8** (a) Exposed femoral head, 33-day-old broiler. Arrow: fibrino-necrotic exudate in proximal femur. *Escherichia coli* isolated from lesion. This is coliform, femoral head osteomyelitis. (b) Proximal femur, 6-week-old broiler. HEx100. EV, prominent epiphyseal vessel. MV, greatly distended ascending metaphyseal vessel with peripheral bacterial colonies. ZP, zone of proliferation. ZH, zone of hypertrophy. Arrow: bacterial colonies (*Staphylococcus aureus*) occupying cartilage lacunae. N, sequestra of necrotic hypertrophic cartilage. HG, early stage of a heterophilic granuloma with central degranulated heterophil debris, but wall of macrophages just forming (i.e. this lesion is only a few days old). (c) Proximal femoral region, 6-week-old broiler. HEx40. F, proximal femur. Res: tuft of ciliated respiratory epithelium. L, lumen of femoral diverticulum extending from the posterior abdominal air sac. Arrow: small focus of sloughed epithelium and debris with underlying mild mixed inflammatory cell infiltrate in propria. (This chicken has infectious laryngotracheitis and significant secondary bacterial coliform respiratory disease, but it is uncertain as to the exact cause of this small lesion.) (d) Proximal femur, 6-week-old 'healthy' broiler. Arrow: exposed, deep metaphyseal bony trabeculae. (During dis-articulation of the femur while performing necropsy of this broiler, the femoral head itself has remained in the acetabulum: this is the 'naked', deep metaphyseal bony trabeculae from below the zone of pre-mineralisation. This is a post-mortem avulsion. The broiler does not have any femoral head lesions.)

### 2.4.3 Pododermatitis, 'scald' (hock burn) and breast blisters

The litter produces ammonia from the degradation of excreted uric acid and accumulated other nitrogenous wastes: reused litter tends to release more. Moisture content of litter depends upon the nature, quality and depth of the bedding material, the age, weight and stocking density of the chickens, their diet, and environmental conditions, especially humidity, inclement weather, temperature, air circulation, drainage, and attention to litter maintenance (Dunlop et al., 2016). For turkeys, >30% litter moisture is associated with significant health problems (Wu and Hocking, 2011). Damp litter cakes over, becoming slippery. When the litter no longer adequately controls moisture or ammonia, the chickens' foot pads (from walking and standing; in severe cases this becomes 'bumble-foot'), and in-contact skin when sitting (skin covering hock and underlying structures, tendons, etc. hence 'scald') and lying (the skin covering the ventral keel and the underlying sterna synovial bursa, hence 'breast blister'), become macerated, inflamed, and ulcerated with subcutaneous and synovial infection and subsequent fibrosis.

Affected chickens are in pain (so this becomes a welfare issue) and change their behaviour, reduce feed intake, and may incur systemic infections entering via damaged integument. The affected carcasses are downgraded at abattoirs, and there can also be staffing issues associated with pick-up and collection. Bilgili et al. (2006), using experiments to imitate field conditions, found the incidence of severe pododermatitis to be 16–27% and noted that such lesions excluded chicken feet or 'paws' from niche markets.

In summary, the incidence of pododermatitis and related skin damage varies greatly but can be largely controlled by appropriate management practices.

## 2.5 Developmental problems: tibial dyschondroplasia

Tibial dyschondroplasia (TD) is a well-recognised problem in broilers, but not other breeds, and also occurs in rapidly growing meat turkeys and ducks (Orth and Cook, 1994). Mild to moderate cases do not present with any significant clinical signs. Diagnosis requires radiology or necropsy, and minimal lesions (Fig. 9a) may be overlooked or dismissed, so cumulative incidence is usually unknown. Severe cases may be associated with lameness and secondary complications, such as deformity, fractures, and/or osteomyelitis, but these broilers have usually grown well until then.

TD is a focal or extensive expansion of the ZpH, especially in the caudo-medial proximal tibiotarsus (Fig. 9b): other physes can also be affected (Farquharson and Jeffries, 2000) (Fig. 9c). Grossly, there is an avascular glistening cartilagenous plug (Fig. 9d). The overlying ZP appears 'normal', but the underlying and surrounding ZH is disorderly, partly due to intrusion of this space-occupying

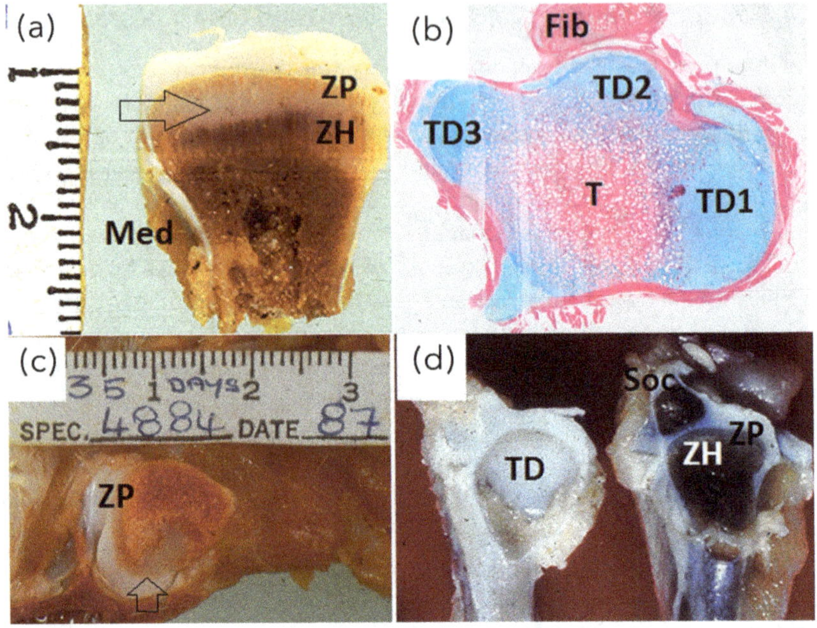


**Figure 9** (a) Proximal tibiotarsus, 4-week-old broiler. ZP, zone of proliferation. ZH, zone of hypertrophy. Arrow: relatively narrow transverse lesion of tibial dyschondroplasia. Med: TD lesion is more prominent and deeper medially, and there is a slight medial bulge (such lesions can be easily missed or overlooked). (b) Proximal tibiotarsus sectioned transversely just below physis, 6-week-old broiler. Alcian blue/HE x20. Fib: the fibula is attached to the lateral aspect of the tibiotarsus, somewhat cranially, and articulates with the lateral femoral condyle. The fibula tapers off, moving caudally, towards the lower third of the tibiotarsus. T, metaphyseal trabeculae. TD1, large medial plug of tibial dyschondroplasia. TD2, small caudo-lateral TD lesion. TD3, smaller lateral lesion. Alcian blue stains acid mucopolysaccharides, such as aminoglycans in cartilage, but the TD lesions are relatively pale, confirming TD lesions' paucity of such components. (c) Proximal femur, 5-week-old broiler. ZP, zone of proliferation. Arrow: tibial dyschondroplastic lesion extending down from the ZP. (d) Proximal tibiotarsii, 6-week-old broilers. Sectioned cranio-caudal. Left: TD large, glistening, wedge-shaped, tibial dyschondroplastic lesion in physis (cut somewhat medially). Right: non-TD affected, flock mate (sectioned more centrally). Soc: secondary centre of ossification towards the cranial aspect. ZP, zone of proliferation. ZH, zone of hypertrophy.

lesion, and partly due to the irregular transition at the leading edge. TD is distinct from osteodystrophy (Fig. 10a). The lesion is bilaterally symmetrical. Wong-Valle et al. (1993) examined different lines of broilers at 7 weeks (by which time many minor cases would have resolved), and the prevalence of TD varied from 3.5% to 61.9%. Burton et al. (1981) studied three separate flocks and found prevalence of TD varied from 14% and 35%. Dinev et al. (2012b) examined broilers by necropsy from 20 commercial flocks (>300,000 broilers) and found prevalence of TD in three different strains was 24–28%, with an apparently earlier than anticipated

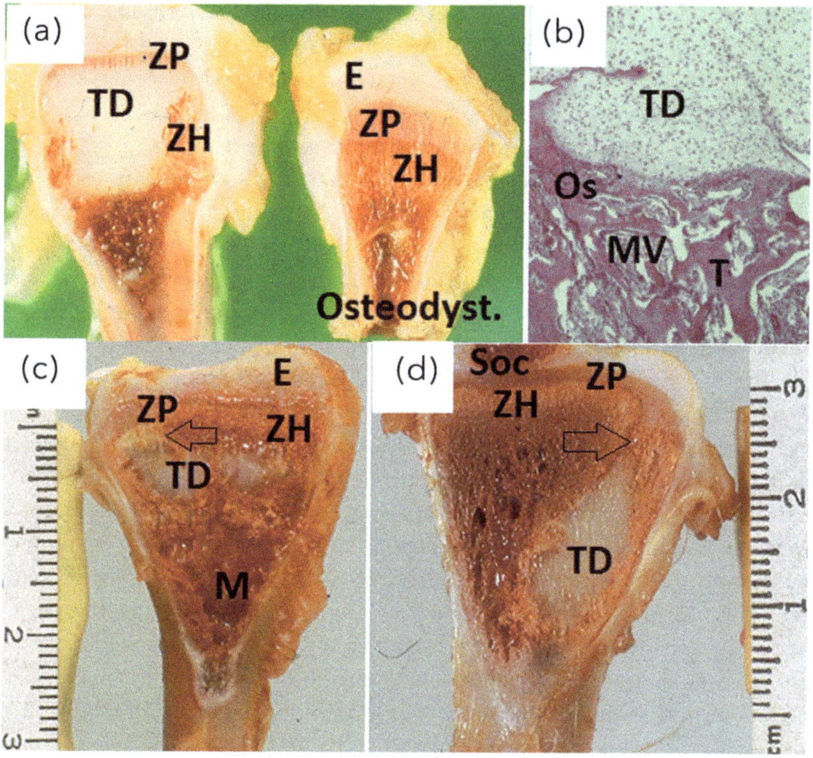


**Figure 10** (a) Proximal tibiotarsus, 4-week-old broilers. Left: TD, large opaque plug of tibial dyschondroplasia causing some medial bulging and displacement of physis. ZP, shallow, uniform zone of proliferation. ZH, there are blocks of zone of hypertrophy at the edge of the TD lesion. Right (one of many lighter cohorts from the same flock): E, epiphysis. ZP, zone of proliferation is deep and almost the same depth as ZH. ZH, zone of hypertrophy. Osteodyst., This is an osteodystrophy with a deep ZP suggestive of marginal Vitamin $D_3$/Ca deficiency. (b) Proximal tibiotarsus, 26-day-old broiler. HEx20. TD, edge of tibial dyschondroplastic lesion merges abruptly into ossified trabeculae. Os, endochondral ossification in the metaphysis. T, trabeculae. MV, abundant large metaphyseal vessels appear 'normal', but there are no blood vessels penetrating into the periphery of the TD lesion. (c) Proximal tibiotarsus, 7-week-old broiler. E, epiphysis. ZP, zone of proliferation is glistening and deeper towards one side (medial?), than on the other, and associated with some apparent outward bulging. ZH, zone of hypertrophy. TD, tibial dyschondroplastic lesion is largely separated from the ZP and divided. Arrow: small focus of necrotic metaphyseal cartilage (these complications are usually found medially). M, medullary bone marrow. (d) Proximal tibiotarsus, 8-week-old broiler. Cranio-caudal section. Soc, secondary ossification centre at cranial aspect. ZP, zone of proliferation. ZH, zone of hypertrophy. TD, large opaque plug of tibial dyschondroplastic cartilage. Arrow: shows that although most of it is detached from ZP and resolving, there is still a narrow bridge connecting it to lower ZP.

onset at 16 days in some strains, but other strains did not have first cases until almost 5 weeks of age; only a low proportion (<10%) were severe cases with haemorrhage, necrosis, and/or fractures thus potentially causing lameness: however, the criteria for selection for examination of these ~3,000 necropsied broilers is unclear and may only include those considered lame, rather than indicating true flock prevalence. Field broilers reared under comparable conditions have very different prevalence and severity of TD, implying multifactorial involvement, such as genetics, gender, growth rate, nutrition, etc., and therefore a precise time to determine flock prevalence by non-invasive procedures, or post-mortem examination, cannot be given: although some can be detected at about 2 weeks, most occur between 3 and 8 weeks of age.

The TD microanatomical features and biological function are consistent with TD being derived from ZpH: the chondrocytes are small, relatively inactive, and produce sparse proteoglycans but of normal composition (Sauveur and Mongin, 1978), the collagen is more cross-linked than expected (Farquharson and Jeffries, 2000), probably related to an extended sojourn at this phase. The avascular cartilage plug is sited between the epiphyseal vascular channels in the ZP and the metaphyseal arteries in ZH. The involved chondrocytes have been described as apoptotic or necrotic (Rath et al., 1998), but this may be autophagy, a survival process in stressed cells (Leach and Monsonego-Ornan, 2007). Vascularisation and progression through hypertrophy to ossification at the leading edge is narrow but otherwise normal (Gay et al., 2007) (Fig. 10b); so the chondrocytes must remain viable and able to be aroused (Farquharson et al., 1995). The chondrocytes in TD lesions, as expected, differ significantly in many ways from active chondrocytes in ZH. Chondrocytes from TD-affected areas fail to excrete collagen X matrix and lack transforming growth factor-β and *c-myc*, both involved in cartilage differentiation, but these changes may be consequential, not causal (Whitehead, 1997). Shi et al. (2024) analysed the function of many hundreds of genes in the growth plate of TD compared to non-thiram treated, non-TD-affected, control, Chinese yellow broilers and found, among many other changes, that in TD affected physes, those genes involved in pathways of cartilage development were changed and those involved in forming the alpha chain of type XII collagen and calcium signalling were enriched. Studies comparing the morphology and function of TD-involved chondrocytes to 'normal' prehypertrophic chondrocytes are sparse. Whether such changes are causal, consequential, or co-occurring is hard to deduce. Whatever causes the retardation or inhibition of transition in development of the chondrocytes wanes with time, and the lesions resolve completely with the mass ultimately undergoing standard endochondral ossification *in situ* or after it detaches into the metaphysis (Fig. 10d). Thorp et al. (1993a) advised caution in interpreting changes observed in established lesions as they may be obscuring what occurs at initiation.

Published by Burleigh Dodds Science Publishing Limited, 2026.

Spontaneous TD in chickens is restricted to rapidly growing broilers, confirming that it is associated with growth rate in susceptible chickens, not genetics alone (Williams et al., 2004). Males are inconsistently reported as more affected than females. Incubation procedures can affect the growth rate of broilers, and thence the prevalence of TD (Yalçin et al., 2007; Groves and Muir, 2017, Section 4.6).

Various experimental models have been developed, but they all require TD-prone, rapidly growing broilers, subjected to significant dietary alterations or toxins. These procedures increase the prevalence and severity of TD, but too often have other untoward features. Thiram, a carbamate commonly used to treat grain, when given at high doses, can increase the prevalence of TD, but the chickens do not grow well, mortality is significant, and there is considerable derangement of liver function (Kang et al., 2020; Iqbal et al., 2023) compared to 'controls', indicating this is probably due to thiram, not directly a consequence of TD. Farquharson and Jeffries (2000) wondered if distinct and separate mechanisms could result in the occurrence of similar lesions.

Within chickens, TD occurs only in broilers, and the incidence can be increased or decreased by selection (Wong-Valle et al., 1993), although no particular gene has been identified (Praul et al., 2000). The incidence of TD is probably being considered in recent breeding programmes, and therefore declining. Current field prevalence varies, but true incidence is difficult to quantify. The prevalence and severity can be reduced by slowing the growth rate, but that is generally unacceptable (Edwards and Sorensen, 1987); slower-growing (SG) broilers are still affected, but apparently more mildly (Santos et al., 2022). Feeding activated $VD_3$ appears to minimise TD, but only in otherwise imbalanced diets (Whitehead et al., 2004). Mild to moderate TD is not associated with overt lameness, nor does it appear to adversely affect growth, so these manifestations are probably not animal welfare problems: severe cases are a different matter. Although a reduction in incidence can be achieved, eradication is probably not a realistic option. The actual cause of TD has not been clearly elucidated.

## 2.6 Traumatic problems: gastrocnemius tendon rupture

Rupture of the gastrocnemius tendon is encountered from time to time in well-grown, heavy broilers and breeders. It is usually unilateral, and there are strong indications that this is usually a sudden traumatic event (Hill et al., 1988). It tends to affect the tendon as it begins to arc above the hock near insertion into the distal tibiotarsus, where there is a large hypovascular zone with an intratendinous cartilaginous plaque (Duff and Anderson, 1986). Affected broilers or breeders are severely lame and have a dropped hock, and their carcasses are downgraded. There is extensive haemorrhage; if the rupture occurred a few days previously, early resolution of the haematoma results in green discolouration, hence 'green leg' (Fig. 11a). Besides haemorrhage and clotting, there may be oedema, infiltration of macrophages and heterophils, haemociderin, fibroplasia, and vascular proliferation; repair of non-aligned fragments is disorderly (Fig. 11b). If an infectious agent such as reovirus has caused inflammation and weakening of tendons, and underlies the rupture, there ought to be numerous lymphoid germinal centres and plasma cells (Fig. 11c), although these affected chickens tend to be older (Duff and Randall,

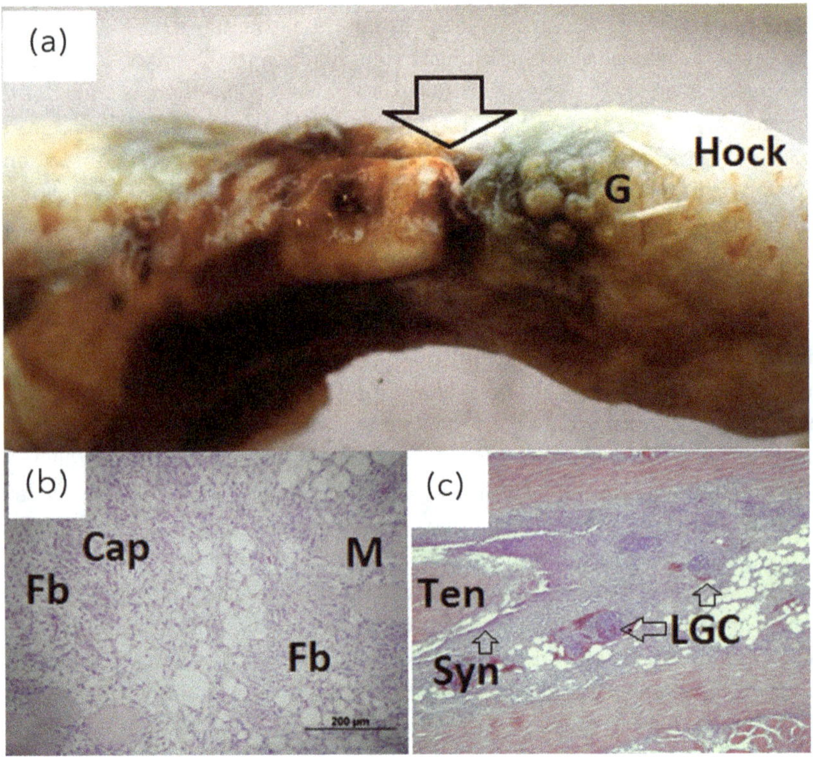


**Figure 11** (a) Ruptured gastrocnemius tendon, broiler. Arrow: free end of ruptured tendon. G, 'green' discolouration due to breakdown of haemoglobin from extravascular erythrocytes released by haemorrhage (hence 'green-leg') – this is several days after the initial event. Hock joint indicated. (b) Ruptured gastrocnemius tendon, 8-week-old broiler. HEx200. M, muscle fibre. Cap: thick-walled, reactive, capillaries. Fb, foci of fibroblasts which are not aligned in any particular pattern because the ruptured ends are loose. (There are small numbers of scattered macrophages not readily discernible at this magnification. This section is slightly distant from the site of frank haemorrhage. Lesion is 4–5 days old.) (c) Gastrocnemius tendon sheath, 8-week-old broiler. HEx100. Ten, tendon. Syn, synovium lining tendon sheath. LGC, lymphoid germinal centres with dense lymphoplasmacytic and macrophage infiltrate surrounding the tendon sheath. (LGC take about 2 weeks to form up like this). (Neither *Mycoplasma* spp. nor standard bacteria were isolated from this tissue, so it was probably of viral aetiology.)

1986). Histological examination of tendons collected from the processing line after scalding can be difficult to interpret.

Assemblage and maturation of tendon collagen is sequential, and development of full strength is beyond the life span of the broiler. Enforced exercise did not improve the biomechanical properties of broiler gastrocnemius tendons in the time-frames available (Foutz et al., 2007b; Ruiz-Feria et al., 2014). The tendons of chickens, and other birds, have cartilaginous foci in many well-recognised sites as a 'normal' feature (Van Der Berge, 1979) and those should not be over-interpreted.

## 3 Assessment of leg problems

There are several methods of assessing leg problems and/or broiler mobility that are referred to in this chapter. For succinctness, in Table 1 the methods are categorised for the leg problem they assess, which includes walking ability, specific leg conditions, leg strength, and bone characteristics. Each assessment method is identified as being either impartial or subjective, with a brief suggestion on the scenarios they are most suited to, tools or equipment they require, comments and recommendations when undertaking the assessment, and accompanying references.

## 4 Improving leg health and reducing leg problems

Continual increases in broiler body weight for age have been achieved through ongoing genetic selection for rapid growth and high metabolic rate, coupled with optimisation of the broiler's nutritional needs and computerised feed formulation (Havenstein et al., 1994; Maharjan et al., 2021). However, the broiler's skeletal system has not developed at a comparable rate, being less able to support the weight-bearing load required with such rapid muscle accretion (Santos et al., 2022; Riber and Wurtz, 2024). Slower-growing (SG) broilers with lower rate of growth have lower incidence of leg problems (Section 4.2), but this also entails additional costs due to lower feed efficiency and an extended grow-out (Nicol et al., 2024), thereby reducing sustainability. Considerable research is being invested into SG broilers (Willemsen, 2022). Currently they are grown in areas in North-West Europe, including the Netherlands, Germany, Denmark (van der Eijk et al., 2023), France, the United Kingdom (Willemsen, 2022), and the USA (Hartmann, 2022), being marketed as a higher quality product with a higher price tag (Hartmann, 2022). While the shift to SG broilers is building momentum, currently FG broilers are the dominant source of chicken meat. Hence the industry continues to search for mechanisms to improve their bone integrity and decrease the incidence of leg problems. To this end, aside from reducing their growth rate, other mechanisms include selecting parent breeders without any leg defects (Duggan, 2024; Barger-Weathers, 2025),

**Table 1** Assessment of leg problems and mobility in broilers

| Aspect of leg problems | Impartial or subjective assessment | Applications: on-farm (OF), processing plant (PP), research (R) | Requirements* | Comments/recommendations | References |
|---|---|---|---|---|---|
| **Walking ability** | | | | | |
| Gait scoring | Subjective | OF, R | Assessor training for consistency | Time consuming<br>[1]Assess 250 broilers selected at random/flock<br>[2]Assess 150 broilers from random areas of the house<br>[3]Modified gait scoring system to improve objectivity | Kestin et al. (1992)<br>Butterworth and Haslam (2009)[1]<br>Welfare Quality® (2009)[2]<br>Garner et al. (2002)[3] |
| Optical flow | Impartial | OF, R | Recording equipment (phone) and algorithm platform | Aligns with gait scoring<br>Longitudinal assessment of walking ability | Dawkins et al. (2009) |
| Force plate pressure | Impartial | R | Force plate apparatus | Assess 4–7-week-old broilers<br>Aligns with gait scoring | Nääs et al. (2009) |
| Runway +/– obstacle timed test | Impartial | R | Runway/obstacles set-up | Measure the time taken to complete the course<br>Concurrently observe gait parameters | McGeown (1999)<br>Caplen et al. (2014)<br>Tahamtani et al. (2021) |
| **Specific leg conditions** | | | | | |
| Angulation (varus-valgus) | Subjective | OF, PP | One assessor for comparison of results | Degree of angulation scored<br>[4]Fix the legs at femorotibiotarsal joint. | Kestin et al. (1999)<br>van den Brand et al. (2022)[4] |
| Tibial dyschondroplasia | Subjective | OF (culls or mortalities), R | One assessor for comparison of results | Score at post-mortem<br>[5]Use imaging technologies, e.g. X-ray or computer tomography on live broilers | Kaup (1918)<br>Groves and Muir (2017)<br>Neeteson et al. (2023)[5] |

| | | | | | |
|---|---|---|---|---|---|
| Femoral head necrosis | Impartial | OF (culls or mortalities), R | | Yes or no outcome | Packialakshmi et al. (2015) |
| Contact dermatitis: hock burn; foot pad dermatitis (FPD) | Subjective | OF, PP, R | One assessor for comparison of results | [6]Score based, several different score ranges have been used<br>[7,8]Preliminary assessment to identify early signs of lesions through:<br>[7]optical flow<br>[8]thermal imaging<br>[9]FPD assessment recommended OF or post-processing, not on processing line | Butterworth and Haslam (2009)[6]<br>Ekstrand et al. (1998)[6]<br>Kestin et al. (1999)[6]<br>Sherlock et al. (2010)[6]<br>Welfare Quality® (2009)[6]<br>Dawkins et al. (2017)[7]<br>Jacob et al. (2016)[8]<br>Hunter et al. (2017)[9] |
| **Leg strength** | | | | | |
| Latency-to-lie (LTL) down (using aversion to water) | Impartial | R | Water-tight pen or tubs, tepid water, stopwatch<br>Broiler contained in a box for the assessment | [10]Group testing in a pen with water added, requiring a settling down period – time-consuming. Broilers 5–6 weeks old<br>[11]Modified LTL test on individual, 5-week-old broilers, out of the sight of other broilers, in tubs prefilled with water<br>[10,11]Strong negative relationship with gait scoring<br>[12]Assess 5-week-old male broilers | Weeks et al. (2002)[10]<br>Berg and Sanotra (2003)[11]<br>Muir and Groves (2019)[12] |
| LTL down (without water) | Impartial | R<br>OF | | LTL without water<br>[13]Reduce stress for the broiler<br>[14]Evaluating feasibility for OF assessment | Bailie et al. (2013)[13]<br>Wurtz et al. (2025)[14] |
| Bone breaking strength | Impartial | R | Texture analyser | Requires consistent bone placement for comparison of results | Duff and Thorp (1985)<br>Santos et al. (2022) |

(*Continued*)

**Table 1** *Continued*

| Aspect of leg problems | Impartial or subjective assessment | Applications: on-farm (OF), processing plant (PP), research (R) | Requirements* | Comments/recommendations | References |
|---|---|---|---|---|---|
| **Bone characteristics** | | | | | |
| Bone porosity/density | Impartial | R | Histology resources | Assessed through NIH Image (ImageJ) | Williams et al. (2000) |
| Bone ash | Impartial | R | Muffle furnace | | Williams et al. (2000) |
| Bone mineral density | Impartial | R | Dual energy X-ray absorptiometry (DEXA) | [15]For comparison bones should be in the same condition (dry or wet) | Aguado et al. (2015) Orban et al. (1993)[15] |

* Always record broiler age at time of assessment.

[1-15] Numbered superscripts relate specific comments to the source publication

Published by Burleigh Dodds Science Publishing Limited, 2026.

parent flock age and diet composition (Favero et al., 2013; Torres and Korver, 2018), *in-ovo* administration of nutrients (Kpodo and Proszkowiec-Weglarz, 2023), management of incubation conditions (Groves and Muir, 2014; Muir and Groves, 2019), broiler nutrition (Whitehead, 2002), enrichment of the broiler housing environment (Riber et al., 2018), and deepening our understanding of the implications of immune responses on bone resorption (Sharma et al., 2023). Each of these areas is discussed in the following sections.

### 4.1 Selection against leg problems

The fine details of the breeding programmes used by chicken meat breeding companies to select the parent stock of commercial lines are commercially sensitive and retained in-house. However, leg health is a core trait, with continued selection of favourable skeletal characteristics including straight legs and walking ability, while also selecting against contact dermatitis, foot pad dermatitis, hock lesions, toe defects, and TD (Neeteson et al., 2023; Barger-Weathers, 2025). Despite these programmes, leg problems continue to be a challenge in commercial FG broiler production systems, which has driven the search for ways to reduce their occurrence and impact.

### 4.2 Growth rate and stocking density

There is agreeance that FG broilers have poorer leg health than SG broilers (Dawson et al., 2021; Santos et al., 2022). Following a systematic literature review Nicol et al. (2024) reported gait scores that were 0.65 points higher and a 15–25% higher incidence of contact dermatitis in FG compared to SG broilers. Additionally, FG broilers had poorer tibia mineral content, density, and strength (Güz et al., 2022) and lower tibia osseous and total volume (Güz et al., 2021) at a comparable weight compared to SG broilers. Interestingly, Santos et al. (2022) proposed that morphometric features, such as bone thickness and length, may affect bone function in FG compared to SG broilers at a similar weight, more so than bone strength and mineralisation. Both Singh et al. (2021) and Baxter et al. (2021) compared the LTL time of FG and SG broilers, the former when the broilers were of similar weight but different age (2.0–2.2 kg, attained at 4.5 and 8 weeks of age respectively), the latter when the broilers were the same age (4–7 weeks old), and presumably SG broilers were of lighter body weight. In both cases the LTL time of the SG broilers was longer than the FG broilers, indicating better mobility.

Following a comprehensive review of growth rate on walking ability Riber and Wurtz (2024) suggested that improvements in leg health and gait scores can be achieved with even small decreases, for example 4–5 g/day, in growth rate. However, as previously mentioned, we need to be mindful of the consequences of the SG broiler, as its longer growing period will require additional resources, in particular feed, for each batch of broilers. This will increase the cost of chicken meat production and subsequently its cost to the consumer (Dixon, 2020; Phibbs et al., 2021). Consumer acceptance of the higher price may require greater education about chicken meat production systems, the advantages and disadvantages of SG broilers, and development of branding to distinguish SG chicken meat from other products. Depending on the consumer's acceptance of and ability to pay the higher price (Phibbs et al., 2021), this may risk the sourcing of cheaper chicken meat grown under poorer welfare standards (Butterworth and Haslam, 2009), including poorer leg health.

Reducing broiler stocking density can improve their leg health (Shynkaruk et al., 2023), which has been observed in broilers of different growth rates (Guinebretière et al., 2024). A slower growth rate appears to favour improved leg health more so than a lower stocking density (Rayner et al., 2020), but the two could be used in unison (Guinebretière et al., 2024). Further, the role of the environment in which the broilers are housed, including the quality of the stockperson, on broiler leg health should not be underestimated (Dawkins et al., 2004; Shynkaruk et al., 2023).

As early bone development and mineralisation occur *in-ovo*, characteristics of the eggs, including egg size, may influence chick skeletal development (Yair et al., 2017; Varol Avcilar et al., 2023), with the latter specifically exploring the effects of differences in egg size mediated by hen age. Notably the mechanical properties of the bones of FG embryos that hatch from relatively small eggs may be compromised and have a greater risk of bone pathology compared to a SG embryo hatching from a similarly small egg (Yair et al., 2017). Hence selecting parent flocks that produce larger eggs, especially in FG lines, may benefit the skeletal properties of their offspring. Further, when breeder diets are supplemented with nutrients, a larger egg may accommodate larger quantities of nutrient uptake which is then available to the developing embryo (Sections 4.3 and 4.5).

### 4.3 Breeder diet

Kidd (2003) provided a comprehensive overview of the effect that breeder hen nutrition may have on their progeny. The more recent review of Torres and Korver (2018) affords a deep dive into the role of trace minerals, together with breeder flock age, on the skeletal integrity of the offspring.

The nutrients required for embryonic bone development are included in the egg during its progress along the oviduct (Dibner et al., 2007). Hence the nutrient composition of the diet of the parent flock will influence the availability of nutrients to the developing embryo, its skeletal growth and development, and the bone integrity of the broiler. While boosting levels of minerals in the breeder diet had limited effect on their concentration in the egg (Angel, 2007), a more successful approach has been to enrich the maternal diet with organic minerals including zinc (Kidd et al., 1992) and zinc, manganese, and copper (Favero et al., 2013). This increased embryonic skeletal size: width of the femur and tibia, their calcification (Favero et al., 2013), and dry bone weight (Kidd et al., 1992). This is in part due to the increased bioavailability, chemical stability and reduced interference from antinutritional factors found in the gastrointestinal tract, on organic minerals (Yi et al., 2007; Dobrzanski et al., 2008). However, van den Brand et al. (2023) reported no effect on tibial ossification of embryos or chicks hatched from eggs of Ross 308 breeders fed a diet with partial replacement of inorganic trace minerals (zinc, manganese, copper, and selenium) with organic forms. Interestingly, when comparing SG and FG lines, Güz et al. (2022) identified greater tibia total and osseous volume and mineral density in older progeny (when 2.6 kg body weight) of hens of SG line fed organic minerals, but not in the FG line. There were no differences in these features between the progeny of the FG and SG lines at lighter bodyweight (1.7 kg). The reason for these different findings warrants further investigation to fully appreciate the circumstances in which feeding organic minerals to parent flocks may benefit the bone integrity of their offspring.

In addition to minerals, other nutritional components of the breeder diet may also impact bone development in the progeny. For example, including dietary omega-3 polyunsaturated fatty acids (n-3 PUFA) in the breeder diet can enrich the egg with n-3 PUFA, which appears to favour the translocation of stem cells into osteoblasts, supporting embryonic bone formation and mineralisation. Further, the anti-inflammatory properties of n-3 PUFA can reduce the activity of bone-resorbing osteoclasts (Thanabalan and Kiarie, 2021).

Supplementing breeder diets that are deficient in vitamin D with vitamin D or 25-hydroxycholecalciferol (25(OH)$D_3$), the latter being a vitamin D metabolite involved with the mobilisation of calcium and mineralisation of bone, can improve chick bone ash at hatch (Atencio et al., 2005). However, when breeders fed diets with adequate levels of vitamin D received 25(OH)$D_3$ supplementation through the drinking water, there was no effect on the bone mineral density of their progeny when they were 6 weeks of age (Saunders-Blades and Korver, 2014). Hence it appears that when dietary vitamin D levels are adequate, further enrichment of the parent diet may not impact progeny bone density.

In a study reported by Driver et al. (2006) breeder hens were fed diets of either lower or higher levels of vitamin $D_3$ ($LD_3$ and $HD_3$, respectively). All chicks from the former and half of the chicks from the latter were then fed a corn-soybean-based diet low in calcium (to induce TD), and the other chicks from the latter received an adequate diet. It was concluded that the incidence and severity of TD in the offspring were related to hen age. In younger hens (39 and 44 weeks of age) the TD scores and severity of TD were similarly high for chicks from hens that received both the $LD_3$ and $HD_3$ diet. However, when older hens (53 and 64 weeks old) were fed the $HD_3$ diet, the incidence and severity of TD in their offspring were lower than in chicks from hens fed the $LD_3$ diet. It is proposed that as the hen ages and egg production declines, the $HD_3$ may support greater deposition of vitamin $D_3$ into the egg, thereby provisioning adequate vitamin $D_3$ and reducing TD compared to hens fed the $LD_3$ diet. This is indicative of the potential role of hen age on bone features of their progeny when supplementing the maternal diet with nutrients for translocation to the embryo.

### 4.4 Breeder age

As hens age, egg production declines and egg size increases, and the proportion of yolk within the egg also increases (Nangsuay et al., 2016). This enhances the supply of yolk-derived nutrients (Lourens et al., 2006), including copper, zinc (Torres and Korver, 2018), and manganese (Favero et al., 2013) to the embryo. Hence, the larger eggs produced by older hens are likely to support their larger embryos through higher levels of bone mineralisation compared to embryos from smaller eggs typical of younger flocks (Torres and Korver, 2018). Further, the age of the parent flock can impact offspring bone strength, with the bones of progeny from a 65-week-old flock being more resistant to breakage compared to bones from progeny of a 25-week-old breeder flock (Shaw et al., 2010). However, maternal age was not found to impact the presence of TD (Nelson et al., 1992). Alfonso-Torres (2009) compared offspring from breeder flocks of either 38 or 60 weeks of age, and while the offspring from the older breeders had the heavier and wider (absolute measure) femur bones, the ossification of cartilage to bone and bone length did not differ.

It should be noted that when evaluating the role that breeder age may have on the bone characteristics of their progeny, there may be variation in the outcomes due to the feature being assessed and the age of the progeny at the time of the assessment. For example, in studies by Yalçin et al. (2001), the effect of maternal age on bone weight and ash per cent was evident only when the progeny were day-old, whereas bone width was altered in progeny up to 5 weeks, but not when they were 7 weeks old. The breaking strength (both maximal force and elastic force) of the tibia of the progeny of Dwarf breeder hens increased with parental age and dietary phosphorus intake. Progeny bone ash was also higher with phosphorus supplementation of the parent diet (Triyuwanta and Nys, 1992), but this was only observed in young (1-day-old) chicks and was not evident when the broilers were 7 weeks of age.

## 4.5 In-ovo *administration of nutrients*

Minerals required for embryonic bone development, including phosphorus, zinc, manganese, copper, and calcium, come from the yolk, and after approximately day 10 incubation, calcium is also sourced from the eggshell (Ono and Tuan, 1991; Yair and Uni, 2011). The amount of each mineral within the yolk tends to decrease throughout incubation, but reports do differ (Yair and Uni, 2011; Hopcroft et al., 2019). The former found low levels of all minerals apart from calcium, in the yolk, by hatch, while the latter identified particularly low levels of phosphorus, copper, and zinc, and an increase in calcium at 17.5 days embryonation compared to day 0 embryonation. To bolster mineral reserves and their availability for embryonic development Yair and Uni (2011) delivered them *in-ovo* at day 17 embryonation. This enhanced levels of copper, iron, manganese, and zinc in the yolk, but not calcium nor phosphorus. *In-ovo* administration of organic minerals increased bone ash at hatch (Oliveira et al., 2015b), but *in-ovo* administration of organic minerals and vitamin $D_3$ did not alter latency-to-lie (LTL) in 5-week-old broilers (Yair et al., 2015).

Kpodo and Proszkowiec-Weglarz (2023) reported conflicting outcomes in chick leg integrity following *in-ovo* administration of minerals and vitamins. The different observations are likely due to differences in the experimental design, the chemical form of the additives, their targeted delivery point, and the age of the embryo and chick at sampling. For example Oliveira et al. (2015b) delivered organic zinc, manganese, and copper *in-ovo*. While an increase in bone ash was observed at hatch, bone mineralisation was variable during the 3 weeks post-hatch (Oliveira et al., 2015a). In a series of reports Bello and co-authors (Bello et al., 2014a, 2014b, 2014c) described the outcomes of *in-ovo* delivery of 25(OH)$D_3$ on bone development and mineralisation in the broilers post-hatch. When they were 3 weeks old, measures of tibial bone ash suggested that a higher dose of 25(OH)$D_a$ (i.e. >1.20 ug) (Bello et al., 2014a) may improve bone mineralisation. However, curiously it was the vaccine diluent (carrier control) that generated higher bone breaking strength in 2-week-old male broilers, which was not observed when they were older (4 weeks of age).

As previously mentioned, vitamin $D_3$ regulates the availability of calcium from the yolk to the embryo, which, from day 10 of incubation, is supplemented with calcium from the eggshell, the latter being regulated by vitamin K (Ono and Tuan, 1991). In light of this Abbasi et al. (2017) compared the *in-ovo* delivery of different quantities of both vitamins administered either alone or together on leg characteristics. Their combined administration at 0.4 μg vitamin $D_3$ and 6 μg vitamin $K_3$ reduced calcium levels in the eggshell (indicating calcium translocation to the embryo), and generated the highest tibial strength, diameter, and length, when the broilers were 6 weeks old.

## 4.6 Associations between fertile egg incubation and broiler leg weakness

The rapid growth rate of broilers has increased their weight for age (Zuidhof et al., 2014) and reduced their age at market weight (currently 30–35 days), while the duration of egg incubation (21.5 days) has remained constant (Uni et al., 2012). Incubation now accounts for over 37% of a broiler's entire lifespan, and the conditions experienced by the developing embryo are likely to have an ever-increasing influence on its life post-hatch. Selection for FG genotypes has also impacted embryonic metabolism (Tona et al., 2004), in addition to chicken growth and development after hatch.

Incubation conditions influence chick quality and will affect subsequent growth and performance (Decuypere et al., 2001). As described in Section 2.2, embryonic bone development occurs in the later stages of incubation (Yair et al., 2012), thereby providing opportunity for incubation conditions to impact bone development and the occurrence of leg problems. In broilers, the development of splayed legs, crooked toes, valgus deformity, TD, and muscle or tendon weaknesses has been implicated with embryo incubation (Oviedo-Rondón et al., 2009b; Shim and Pesti, 2011; Groves and Muir, 2017). However, studies aiming to link varying incubation conditions with subsequent leg weakness may also suffer from differences in other components of the study design, and hence published results vary. The line of broiler being studied and its continual genetic selection (Neeteson et al., 2023; Barger-Weathers, 2025) may be central to some of the variation. Additionally, other factors not related to incubation, including genetic growth rate, nutrition, infectious causes, and litter conditions, can also influence broiler leg strength and potentially contribute to disparate outcomes between studies.

### 4.6.1 Effects of incubation temperature on broiler leg strength

Numerous studies have searched for associations between incubation conditions and leg integrity. Several aspects of incubation have been alleged to affect bone health and weakness post-hatch. In particular higher temperatures at various stages of incubation have featured as a candidate contributor to leg health issues (Oviedo-Rondón et al., 2009b).

Towards the end of incubation, the development of the thyroid gland leads to large increases in the level of embryonic thyroid hormones which play a major role in chicken embryonic development (Grommen et al., 2011). This includes bone growth and chondrocyte differentiation (Oviedo-Rondón et al., 2009a), yolk sac absorption, pulmonary respiration, and the hatching process (Grommen et al., 2011). Incubation temperatures above those recommended affect the thyroid-IGF I-GH hormonal axis, thereby altering bone, tendon, and muscle development and thyroid metabolism (Grommen et al., 2011).

Published by Burleigh Dodds Science Publishing Limited, 2026.

High incubation temperature can also depress the expression of collagen type X and transforming growth factor-β which regulate bone mineralisation (Oviedo-Rondón et al., 2008;Lopez and Hergott, 2015). Further, the alignment of collagen fibres, which is central to proper embryonic bone development, may be adversely affected by variations in incubation temperature (lower early, higher in later incubation) (Oviedo-Rondón and Wineland, 2011).

The heart of the chicken embryo is subject to circadian rhythm regulation with the density of myocytes being higher in the scotoperiod compared to the photoperiod (Ko et al., 2010). It has been postulated that this is mediated through enzymes that are responsive to circadian regulation (Ko et al., 2010) and that the lack of a circadian light pattern could influence metabolic dysfunction. A circadian effect on embryo development during incubation has not been explored in depth, but limited studies have suggested that, as commercial fertile eggs are largely incubated in total darkness, the absence of a periodic light regime during incubation may lead to skeletal inadequacies (van der Pol et al., 2019). Additionally, the photoperiod during incubation could affect bone development for the life of the broiler (van der Pol et al., 2019). The latter study involved Ross 308 eggs exposed to either continuous darkness, continuous light, or a 12-h dark: 12-h light cycle throughout incubation. Embryos exposed to continuous light hatched with lower bone ossification than those exposed to a 12-h circadian rhythm, while growth hormone levels were higher in embryos experiencing continuous darkness. Effects on bone growth throughout the subsequent broiler grow-out period were maintained in terms of bone size and strength. Continuous light during incubation resulted in a higher prevalence of TD compared to the circadian rhythm and continuous darkness exposure. It was concluded that some darkness is necessary for proper embryonic bone development and that continuous light is detrimental.

There is a marked difference in embryonic growth between some of the major broiler lines (Tona et al., 2010), with the Cobb chicken developing earlier than the Ross broiler. It has been shown that what was regarded as the 'ideal' incubation temperature for broiler eggs (Lourens et al., 2005) can compromise skeletal integrity in the Cobb broiler (Groves and Muir, 2014). In the latter, Cobb broilers hatched 18–24 h earlier than Ross when incubated at the defined optimal eggshell temperature (EST) of 37.8°C. This earlier hatching time can affect bone mineralisation when chicks are pulled from the incubators, which can lead to later skeletal problems (Groves and Muir, 2017). An analysis across several studies (Groves and Muir, 2014) indicated a positive correlation of bone ash at hatch and male broiler LTL at 5 weeks of age. Further studies comparing incubation temperature profiles also showed differences between Cobb and Ross broilers in bone mineralisation at hatch (Hopcroft et al., 2020) and subsequent variations in LTL at 5–6 weeks of age (Groves and Muir, 2017).

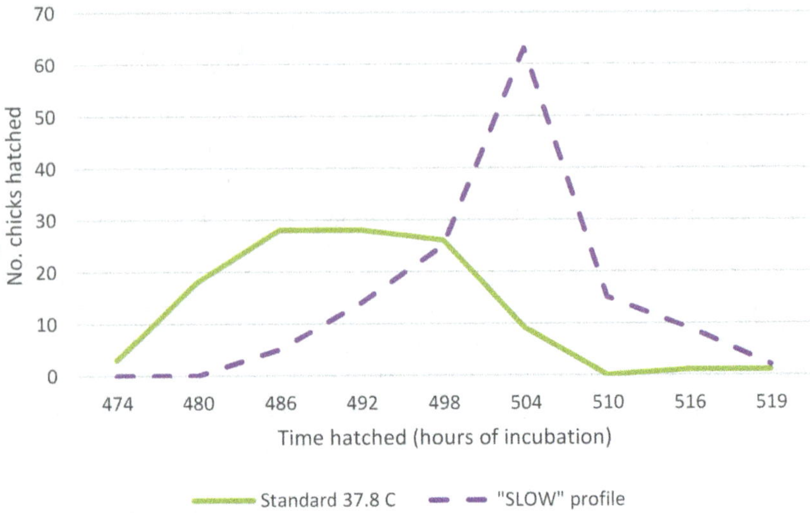


**Figure 12** Hatch time distribution of Cobb chicks incubated under a Standard eggshell temperature (EST) profile (continuous 37.8°C) and a 'Slow-start' EST profile (37.3°C at embryonic day (ED)1 rising to 37.8°C by ED16). Mean hatch time for the Standard profile was 488 h, and this was extended to 500 h by the Slow-start profile ($P < 0.001$).

Differences in incubation pattern between the broiler lines, Cobb and Ross, have already been mentioned (Tona et al., 2010). The Cobb line follows faster embryological development when exposed to the same EST, producing more heat, and hatching earlier with a faster early growth rate. Groves and Muir (2017) manipulated incubation temperature programmes where a Slow-start incubation, i.e. 37.3°C EST at embryonic day (ED)1 rising to 37.8°C EST by ED16, slowed the Cobb embryo development, with average hatch time of 500 h of incubation (Fig. 12), which more closely matched the later, 504 h of incubation, hatching time of the Ross embryo.

At the time of chick pulling, 516 h of incubation, the earlier hatching chicks from the Standard incubation temperature profile were lower in bodyweight than chicks from the Slow-start incubation (39.2 g compared to 41.4 g, respectively), the difference being due to a lower retained yolk weight in the earlier hatched chicks. Further, the earlier hatching chicks were slightly longer (from beak to top of middle toenail; 18.8 cm versus 18.5 cm), indicating continued bone growth after hatching without being fed, but supported through utilisation of yolk reserves. These observations were attributed to the length of time the chick spent outside the egg while still in the hatcher (referred to as its 'sojourn'). A longer sojourn resulted in a lower total chick weight, increased chick length, and a resulting decreased yolk weight. Additionally, during the longer sojourn, serum calcium increased while serum phosphorus decreased. A subset of chicks was grown out, and the earlier hatched chickens gained more weight in the first

7 days than the later hatching Slow-start group (123 g gain compared to 112 g: $P < 0.05$). This faster early growth may have been the result of hunger in chicks with a long sojourn, as their feed intake on the first day was significantly higher following their protracted sojourn. Their first week weight advantage was not maintained until 6 weeks of age. Further, the hatch-time delay was not thought to be fully effective and may have a greater impact if closer to the desired mean hatch-time of 504 h incubation.

In a further experiment (Muir and Groves, 2018) Cobb embryos started incubation at 37.2°C EST which was gradually raised to 37.8°C EST by ED13. The reduced temperature at the start of incubation increased average time to hatch by 11 h. Some of the early hatched chicks (hatched <492 h) were sampled at 492 h (group EE), while others were immediately fed (group EF) and sampled at 516 h from the start of incubation. Another group remained in the hatcher for an extended sojourn (group EL). At pulling (516 h after the start of incubation), late hatching chicks (>492 h; group LL) were also sampled. At 516 h after the start of incubation the weight of LL chicks was similar to EE chicks at the time of hatch but were greater than EL chicks which experienced a 24 h sojourn. Chicks that were fed (EF) were significantly heavier than the other groups at 516 h. EE and LL chicks were the shortest, but LL had significantly better bone mineralisation (bone ash%) than those that hatched early (EE or EL).

Measurements of chick serum and yolk calcium and phosphorus, yolk weight and femur ash are presented in Table 2. Early hatched chicks that experienced a long sojourn (EL) showed a decline in yolk weight and in the total amount of calcium and phosphorus in the residual yolk compared to EE chicks. The EL chicks displayed higher serum calcium and lower serum phosphorus compared to chicks hatched early without a sojourn (EE), indicating they were growing while drawing on minerals from the residual yolk during their longer time in the hatcher. A subset of these groups of chickens were also grown out. The EL chickens had a faster growth rate over their first 7 days compared to the later hatched chickens (LL) and early hatched chickens which were immediately fed (EF). This suggests that these sojourned chicks were hungry and consumed feed quickly at their first opportunity.

The longer post-hatch sojourn may explain associations drawn in other studies. In some circumstances chick length has been adopted as a measure of chick quality (Maatjens et al., 2009; Petek et al., 2010) as longer chicks are deemed to have utilised their yolk reserves more effectively and tend to grow more rapidly in the first 1 to 2 weeks after hatch. However, as early hatched chicks with a long hatcher sojourn continue to grow while absorbing their yolk sac, a longer chick may be one which has just hatched early. These chicks have a ravenous appetite when they finally reach the grower house and hence grow relatively quickly over the first 3 days. This may give them the appearance of thriving, but they may have weaker bones (as seen in lower bone ash at hatch)

**Table 2** Serum and yolk calcium and phosphorus levels, yolk weight, and femur ash in Cobb chicks from different hatch windows and post-hatch management

| Hatch window and post-hatch management | n | Serum Ca (mmol/L) | Serum P (mmol/L) | Yolk weight (gm) | Total yolk Ca (mg) | Total yolk P (mg) | Femur ash (%) |
|---|---|---|---|---|---|---|---|
| EE | 39 | $2.34^{C}$ | $1.46^{A}$ | $5.67^{A}$ | $2.25^{A}$ | $0.105^{A}$ | $22.8^{B}$ |
| EL | 41 | $2.46^{B}$ | $1.22^{C}$ | $3.34^{C}$ | $1.74^{B}$ | $0.062^{C}$ | $23.4^{B}$ |
| LL | 38 | $2.40^{C}$ | $1.21^{C}$ | $4.74^{B}$ | $1.80^{B}$ | $0.080^{B}$ | $24.9^{A}$ |
| EF | 37 | $2.76^{A}$ | $1.31^{BC}$ | $2.96^{C}$ | $2.40^{A}$ | $0.079^{B}$ | $26.3^{A}$ |
| *P* | | *<0.001* | *<0.001* | *<0.001* | *<0.001* | *<0.001* | *<0.0001* |

EE: chicks hatched before 492 h, pulled at 492 h, sampled at 492 h.
EL: chicks hatched before 492 h, remained in the incubator until 516 h, sampled at 516 h.
LL: chicks hatched after 492 h, pulled at 516 h, sampled at 516 h.
EF: chicks hatched before 492 h, pulled at 492 h and placed on feed for 24 h, sampled at 516 h.
n: number of chicks.
$^{ABC}$ Means with different letter superscripts in the same column differ at $P < 0.001$.

and be more likely to have locomotory problems later in life, due to bone growth throughout the sojourn with limited mineral supply from the yolk for bone mineralisation.

In conclusion this series of studies provide evidence that altering the incubation and EST profile of the Cobb line may have positive consequences including improved leg integrity. Hatchability is improved, and bone mineralisation is higher at pulling (Muir and Groves, 2018) which correlated with better leg strength later in the grow-out (Groves and Muir, 2014; Muir and Groves, 2019). This series of studies was funded by AgriFutures Australia – Chicken Meat Program which is gratefully acknowledged.

Several leg problems have been investigated in relation to incubation conditions. Of leg conditions observed in broilers, putative aetiological associations with incubation have been inferred for splayed legs at hatch, crooked toes, varus-valgus deformity, TD, and some musculo-tendon weaknesses. The relationships between these conditions and incubation will now be considered in more detail.

### 4.6.2 Splayed legs

Chicks with splayed legs or 'spraddle legs' are often observed in hatcheries. The condition can be unilateral or bilateral. It is commonly associated with high humidity in the hatcher phase of incubation or with slippery surfaces in the hatchery (Crespo, 2020). However, Oviedo-Rondón and Wineland (2011) explained that leg asymmetry (limbs of differing lengths) can provoke splayed legs. Their studies suggested that higher temperatures in the final incubation phase contribute to the incidence of splayed legs, as exposure to higher temperatures can lead to lower muscle glycogen levels and reduced muscle size, tendon development, and thyroid metabolism. Further, higher temperature throughout incubation may lead to greater anaerobic glycogen metabolism resulting in the accumulation of lactate. This can hinder muscle function, making the chickens more prone to splayed legs (Lopez and Hergott, 2015).

### 4.6.3 Crooked toes

Crooked toes are seen relatively often in broilers, but usually less than 3% of the flock is affected (Oviedo-Rondón et al., 2009a). Low incubation temperatures or inadequate preheating can predispose to toe deformities and leg weaknesses, and high temperatures in later incubation slow skeletal development and increase leg asymmetry (Liu et al., 2023). Although genetics, rapid growth rate, nutritional factors, lighting programmes, and other stressors could influence crooked toes, Oviedo-Rondón et al. (2009a) also incriminated high temperatures during the later phase of incubation in affecting leg symmetry and skeletal development. Outcomes of a large field trial demonstrated that commercial single-stage incubation produced better gait scores and fewer crooked toes than did multi-stage incubation techniques (Oviedo-Rondón et al., 2009b).

### 4.6.4 Valgus deformity

Varus-valgus deformity (VVD), or 'slipped tendon', is a deformity which may affect one or both legs. Varus refers to medial deviation of the tibia, while valgus refers to lateral deviation. Valgus is much more common (Van Wettere, 2025) and may predispose to lateral slipping of the gastrocnemius tendon off the distal condyle of the tibia. Males are more commonly affected (Bradshaw et al., 2002; Van Wettere, 2025). Although signs can be seen at hatch (Bradshaw et al., 2002), the condition usually becomes obvious after 2 weeks and is most often observed from 4 weeks of age (Guevara-Torres et al., 2021; Van Wettere, 2025). In a study revealing an overall prevalence of 2% of chickens with locomotory problems, it was estimated that ~20% of these were due to valgus (Guevara-Torres et al., 2021).

The aetiology of VVD in chickens has been associated with genetics, rapid growth, and restriction of exercise (Bradshaw et al., 2002). Bone mineralisation in the first weeks of broiler life is comparatively low. Hence, rapidly growing bones, such as the tibiotarsus, are susceptible to deformation from increased weight bearing on soft bones (Bradshaw et al., 2002). Further, rotated tibia and valgus can be associated with earlier rickets or other causes of poor bone mineralisation (Van Wettere, 2025).

Precise influences of incubation conditions on valgus deformity have not been described, but faster growth has been associated with its increased severities and TD (Shim et al., 2012). Hence, it follows that incubation conditions which encourage faster early growth may be implicated with valgus deformity.

### 4.6.5 Tibial dyschondroplasia

TD is well recognised as a developmental abnormality in the long bones of broiler chickens (Section 2.5), and rapid early growth may contribute to its occurrence (Orth and Cook, 1994; Bradshaw et al., 2002; Shim et al., 2012; Van Wettere, 2025). The rate of bone growth in the first few weeks exceeds that at older ages with TD reaching a maximum between 3 and 8 weeks of age.

While Bradshaw et al. (2002) did not implicate incubation conditions with TD, subsequent studies made that connection. Oviedo-Rondón and Wineland (2011) commented that incubation could influence the occurrence of TD, mediated by effects on chondrocyte differentiation during incubation through the thyroid-IGF1-GH hormonal axis. Studying Ross embryos Yalçin et al. (2007) were able to vary TD expression through changes in early or late incubation temperature (either above or below ideal) and described their relationship with tibial bone ash at hatch.

Incubating Cobb broiler eggs under the Standard temperature profile of 37.8°C was more likely to predispose them to TD during grow out when they were fed a diet conducive to TD, compared to incubation under the previously described Slower-start temperature regime that was found to be more appropriate for Cobb (Groves and Muir, 2017). As previously outlined, the Slow-start regime resulted in chicks hatching later with higher bone ash and subsequently lower TD scores when 5 weeks old, compared to earlier hatching chicks from Standard temperature profile. In contrast, Shim and Pesti (2011), while finding that variations in incubation temperature could modify hatching time and chick weight, found no association with subsequent TD or rickets in Cobb chickens fed three different diets designated as being TD and ricket inducive through either calcium or phosphorus deficiency.

van der Pol et al. (2019) identified higher TD prevalence in chickens incubated under continuous light compared to the circadian rhythm and continuous darkness exposure. The effect of varying photoperiod during incubation is an area of research needing further evaluation.

### 4.6.6 Muscle/tendon weakness

Muscle cell numbers increase progressively throughout embryonic development and is almost complete at hatch (Velleman, 2007), with post-hatch muscle growth occurring mostly by fibre enlargement. The health of muscle cells at hatch can impact early locomotion, and defects caused during incubation can have an ongoing impact.

Al-Musawi et al. (2012) explored differences in incubation temperature profiles for muscle growth in layer (WL) and broiler (Ross) lines. Raising incubation temperature from 37.5°C to 38.5°C between days 4 and 7 led to breed-based differential changes in the gastrocnemius muscle. The layers experienced increased myogenesis, whereas muscle reduction and lengthening of leg bones and their cross-sectional area was observed in broilers. Compared to the layers, broilers showed no stimulation of osteogenesis.

Elevated incubation temperatures can also affect tendon and muscle development (Lopez and Hergott, 2015). As the rate of oxygen passage through the eggshell becomes limiting late in incubation, and oxidation of the yolk lipids is somewhat decreased, the embryo relies on anaerobic metabolism of glycogen from the embryonic liver and muscle stores, especially the breast muscle, to supply its energy needs. This is accentuated by higher incubation temperatures. Lactate is produced and the resultant acidity hinders muscle function, which could result in muscle weakness and poor chicken growth.

### 4.6.7 Conclusions: the effect of incubation conditions on leg weakness

Incubation conditions can have notable effects on the subsequent leg strength of broilers. Many studies have implicated the exposure of embryos to higher than recommended temperatures during incubation with detrimental effects on bones, muscle, and tendons throughout the broiler's short lifespan.

During incubation attention must be paid to the following to achieve optimal leg strength in broilers:

- Utilise appropriate incubation temperatures and especially avoid temperatures higher than recommended at all stages of incubation;
- Follow preferred incubation temperature profiles recommended for different broiler lines;
- Avoid chicks spending a lengthy sojourn in the hatcher after hatching which can lead to very rapid growth in the first week of life; and
- Avoid continuous light during incubation and evaluate whether a circadian rhythm can enhance subsequent leg strength.

## 4.7 Broiler diet

Minerals make up 70% of bones with calcium and phosphorus being predominant (Kierończyk et al., 2017). Typically an approximate ratio of 2:1 calcium and phosphorus is recommended in broiler starter diets (Whitehead, 2002). While infrequently observed in broiler diets, calcium deficiency can stimulate bone resorption and mineral release, mediated through parathyroid hormone and vitamin D, which can, in turn, decrease bone strength (Rath et al., 2000).

Discrepancies in phosphorus metabolism can also impact bone strength. As mentioned in Section 2.3, the exogenous enzyme phytase is routinely included in poultry diets to release phytate-bound phosphorus found in plant feed sources (Kierończyk et al., 2017). Broilers fed diets low in calcium and phosphorus have improved bone strength, tibia ash, and calcium and phosphorus following phytase supplementation. Notably, benefits in bone strength and tibia ash were achieved within the starter phase (Nuamah et al., 2024).

Replacing inorganic mineral sources with organic sources has been observed to improve tibia ash content, mineral density, and strength (Ao et al., 2019; Güz et al., 2019). Dietary supplementation with vitamin D or its metabolites can reduce TD (Whitehead et al., 2004) and improve bone calcification (Gomez-Verduzco et al., 2012) and strength (Suarez et al., 2023). However, the inclusion of calcium, phosphorus, and vitamin D in the diet needs to be carefully managed as their complex interactions can have detrimental effects on bone integrity (Rath et al., 2000).

Published by Burleigh Dodds Science Publishing Limited, 2026.

In addition to vitamin D, vitamins C, K, and B, in particular $B_6$ (pyridoxine), are also critical for bone strength, directing the production of the non-mineral matrix including collagen and collagen cross-links (Massé et al., 1996; Weber, 1999). Further, Thanabalan and Kiarie (2021) proposed n-3 PUFA supplementation of the broiler diet may reduce intestinal inflammation and the opportunity for bacterial-translocation from the alimentary tract into the bloodstream. Once in the bloodstream bacteria may reside in areas of stress and fracture in the tibia, causing leg problems of an infectious nature, for example bacterial osteomyelitis or femoral head necrosis (Section 2.4). However, partial replacement of dietary soybean and palm oil with fish oil, a source of n-6 PUFA, did not benefit tibial characteristics nor gait score, when compared to the control diet (Güz et al., 2019)

### 4.8 Enriching the housing environment

Enriching an animal's environment is intended to have positive outcomes, including expanding their behavioural repertoire and increasing their activity, thereby improving their physiology and welfare (Newberry, 1995), including leg health. Enrichment may involve introducing additional structures, for example straw bales or perches/platforms, to increase the environmental complexity or through alterations within the broiler house, such as stocking density, distance between feeding and water lines, and type of litter (Reiter and Bessei, 2009; Ventura et al., 2010; Riber et al., 2018; Tahamtani et al., 2020; Güz et al, 2021; van der Eijk et al., 2023). Interestingly changes in light intensity (Sherlock et al., 2010) did not affect broiler activity or leg health nor did the introduction of various substrates, including sand, kale, cabbage, hanging objects, and perches (Su et al., 2000; de Oliveira Sans et al., 2014). Some benefits in leg health were reported with floor substrate of either zeolite litter (Khan et al., 2023) or wood shavings (Su et al., 2000; Almeida Paz et al., 2010).

Adjustments to stocking density and presence/height and design of raised platforms had varying impacts on broiler walking ability and foot pad dermatitis (Ventura et al., 2010; Tahamtani et al., 2020) and require further evaluation. Assessing the impact of enrichments including platforms, ramps, barrier perches, and the distance between the feed trough and water line (Güz et al., 2021) on leg characteristics and the behaviour of FG and SG broilers, found improvements in tibia ossification and mineralisation, and higher bird activity, but lower body weight in both the FG and SG broilers. In Australia, perches are an increasingly common inclusion in broiler sheds, being required for accreditation by the Royal Society for the Prevention of Cruelty to Animals (RSPCA Australia, 2020). However, research assessing the role of perches in improving leg health has met with varying outcomes (Phibbs et al., 2021).

These varying results are indicative of the need for further research to ascertain the potential for environmental enrichment to reduce broiler leg problems.

Riber et al. (2018) provide an overview of environmental enrichments and their impact on broiler behaviour, activity, leg health, and welfare. They conclude that an accurate evaluation of environmental enrichment needs to encompass both its impact on the animal's physiology and the economic consequences for the farming operation. As with the introduction of SG broilers, if environmental enrichment increases the cost of broiler production, consumers may need to be informed of the reasons for the higher prices.

### 4.9 Broiler health

Maintaining chicken health and minimising disease is likely to also support bone health. Pathogen incursion stimulates innate immunity, including inflammation, during which proinflammatory cytokines can induce osteoclastic activity and bone resorption (Manolagas, 1998; Roux and Orcel, 2000). For example, exposing broilers to *E. coli* lipopolysaccharide induced losses in tibia calcium, weight, length, and breaking strength (Mireles et al., 2005; Tong et al., 2020). Bone homeostasis was also altered following an *Eimeria* challenge characterised by reduced bone growth and increased bone resorption, which led to losses in bone mineral density and mineral content (Tompkins et al., 2022, 2023).

Given the intestinal damage caused by *Eimeria* oocysts, compromised bone integrity in broilers infected with coccidiosis has traditionally been attributed to poor absorption of calcium, phosphorus and trace minerals (Turk and Stephens, 1969; Turk, 1973). However, studies on restricted pair feeding that matched the reduced feed intake observed with *Eimeria* infection did not result in a comparable loss in bone integrity, indicating a role of the inflammatory response in altering bone homeostasis (Sharma et al., 2023). Specifically, pro-inflammatory cytokines, including IL-17, may mediate bone loss through changes in the expression of receptor activator of nuclear factor kappa-B ligand (RANKL). RANKL is antagonised by osteoblast-derived osteoprotegerin (OPG) (Tong et al., 2020), and osteogenesis is regulated through the relative expression of RANKL and OPG (Tong et al., 2020, Sharma et al., 2023). There is more to understand in this area, and the mediation of pathogen-induced immune responses on bone loss requires further elucidation. This may then inform the development of strategies to induce selective immune activation against pathogens with minimal disruption to bone homeostasis.

## 5 Future trends in research

As breeder companies acknowledge the importance of selecting against leg problems (Neeteson et al., 2023; Barger-Weathers, 2025), it is anticipated this will continue to be a priority. Due to commercial sensitivity, details of the breeding programmes are confidential, but it is hoped that overall progress in reducing leg problems will continue to be shared. It is also important to acknowledge that over time, selection against leg problems at the breeder level is likely to introduce variation in the outcomes of research on leg health involving commercial broiler lines.

There is an increasing amount of research exploring the benefits of SG broilers for broiler leg health and welfare (Dixon, 2020; Baxter et al., 2021; Güz et al., 2021, 2022; Santos et al., 2022; van der Eijk et al., 2023). Further, even small reductions in growth rate may be combined with other features to bring valuable improvements in broiler mobility and leg quality (Guinebretière et al., 2024). The impact of combined treatments on leg problems also requires more evaluation. As SG broilers are becoming more common in some regions, and in light of their welfare benefits, including improvements in leg health, further research on SG broilers as an alternative to FG broilers in commercial settings will continue (Hartmann, 2022; Willemsen, 2022). However, as previously mentioned, this should be completed hand in hand with an evaluation of their economic viability in different markets and, consumer education about the products, their systems of production and overall sustainability.

Further research on lowering EST during specific phases of incubation as a mechanism to reduce the incidence of leg problems is recommended. As the broiler genotypes continue to evolve, the ideal incubation EST for improvements in leg strength post-hatch may also evolve. Importantly management of EST to achieve shorter sojourn times of the chick in the hatcher to increase leg strength has been verified (Groves and Muir, 2014; Groves and Muir, 2017; Muir and Groves, 2018, 2019).

Identifying effective environmental interventions that reduce leg problems in FG broilers, together with the cost of their implementation (Riber et al., 2018; Khan et al., 2023), is also worthy of further research. This may achieve benefits not only for broiler leg health but in other areas of broiler physiology and welfare.

As our understanding of the impact that infectious disease and the associated activation of the immune system have on bone integrity is expanded (Tong et al., 2020; Sharma et al., 2023; Tompkins et al., 2023), our ability to strategically manage these processes is likely to also develop. It is critical to identify poultry health programmes that induce protective immunity, minimise inflammation, and sustain bone integrity.

## 6 Conclusion

Throughout this chapter the normal process of leg bone development in broilers is described. The details of some leg problems that commonly arise in fast-growing broilers, reducing their locomotion and mobility with implications for their welfare, are outlined. Potential causes, the impacts of leg problems, and some procedures that may improve leg health and decrease the incidence and severity of leg problems are discussed. Attention must be given to the correct incubation profile being used for the particular broiler strain to ensure proper musculoskeletal development during incubation and immediately post-hatching. Focal areas for ongoing research to improve leg strength and broiler mobility are also highlighted. Ultimately reducing broiler leg problems will improve broiler welfare and the sustainability of the chicken meat sector.

## 7 Where to look for further information

The following articles provide an overview of leg health and leg problems in broilers:


- Angel, R. 2007. Metabolic Disorders: Limitations to growth of and mineral deposition into the broiler skeleton after hatch and potential implications for leg problems. *Journal of Applied Poultry Research,* 16, 138–149.
- Butterworth, A. & Haslam, S. M. 2009. A lameness control strategy for broiler fowl. *In:* Miele, M. & Roex, J. (ed.) *Welfare Quality Reports.* Cardiff: Cardiff University.
- Bradshaw, R., et al. 2002. A review of the aetiology and pathology of leg weakness in relation to welfare *Avian and Poultry Biology Reviews*, 13, 45–103.
- Groves, P. J. & Muir, W. I. 2014. A meta-analysis of experiments linking incubation conditions with subsequent leg weakness in broiler chickens. *PLoS* One, 9. https://doi.org/10.1371/journal.pone.0102682.
- Muir, W. I. & Groves, P. J. 2019. The leg strength of two commercial strains of meat chicken subjected to different incubation profiles. *Animal*, 13, 1489–1497.
- Nicol, C. J., et al. 2024. An analysis of the welfare of fast-growing and slower-growing strains of broiler chicken. *Frontiers in Animal Science*, 5. https://doi.org/10.3389/fanim.2024.1374609.
- Rath, N. C., et al. 2000. Factors regulating bone maturity and strength in poultry. *Poultry Science*, 79, 1024–1032.
- Riber, A. B. & Wurtz, K. E. 2024. Impact of growth rate on the welfare of broilers. *Animals* [Online], 14. https://doi.org/10.3390/ani14223330.
- Welfare Quality® 2009. Welfare Quality® assessment protocol for poultry (broilers, laying hens). Welfare Quality® Consortium, Lelystad, Netherlands.

# 8 References


Abbasi, T., et al. 2017. Growth performance parameters, bone calcification and immune response of *in-ovo* injection of 25-hydroxycholecalciferol and vitamin K3 in male Ross 308 broilers. *Theriogenology*, 90, 260–265.

Adams, S. L., et al. 2007. Integration of signaling pathways regulating chondrocyte differentiation during endochondral bone formation. *Journal Cellular Physiology*, 213, 635–641.

Aguado, E., et al. 2015. Bone mass and bone quality are altered by hypoactivity in the chicken. *PLoS One*, 10. https://doi.org/10.1371/journal.pone.0116763.

Al-Musawi, S. L., et al. 2012. *In ovo* temperature manipulation differentially influences limb musculoskeletal development in two lines of chick embryos selected for divergent growth rates. *The Journal of Experimental Biology*, 215, 1594–1604. https://doi.org/10.1242/jeb.068791

Alfonso-Torres, K. A., et al. 2009. Breeder age and bone development in broiler chicken embryos. *Arquivi Brasileiro de Medicina Veterinaria e Zootecnia*, 61, 219–226.

Allan, G. H., et al. 2014. Blood flow for bone remodelling correlates with locomotion in living and extinct birds. *Journal Experimental Biology*, 217, 2956–2962.

Almeida Paz, I., et al. 2010. Selecting appropriate bedding to reduce locomotion problems in broilers *Brazilian Journal of Poultry Science*, 12, 189–195.

Angel, R. 2007. Metabolic disorders: limitations to growth of and mineral deposition into the broiler skeleton after hatch and potential implications for leg problems. *Journal of Applied Poultry Research*, 16, 138–149.

Anon. 2022. *Aviagen Ross 308 Ross 308 FF Performance Objectives 2022* [Online]. Available: https://aviagen.com/assets/Tech_Center/Ross_Broiler/RossxRoss308-BroilerPerformanceObjectives2022-EN.pdf [Accessed 26 November 2024].

Anon. 2024a. *Australian Chicken Meat Federation: Facts and Figures* [Online]. Australian Chicken Meat Federation. Available: https://chicken.org.au/our-product/facts-and-figures/ [Accessed 26 November 2024].

Anon. 2024b. *Meat Chicken (broiler) Industry* [Online]. Poultry Hub, Australia. Available: https://www.poultryhub.org/production/meat-chicken-broiler-industry [Accessed 26 November 2024].

Ao, T., et al. 2019. Growth performance and bone characteristics of broiler chickens fed corn-soy diet supplemented with different levels of vitamin premix and sources of mineral premix. *Journal of Applied Animal Nutrition*, 7. https://doi.org/10.1017/jan.2019.4.

Applegate, T. J. & Lilburn, M. S. 2002. Growth of the femur and tibia of a commercial broiler line. *Poultry Science*, 81, 1289–1294.

Atencio, A., et al. 2005. Twenty-five hydroxycholecalciferol as a cholecalciferol substitute in broiler breeder hen diets and its effect on the performance and general health of the progeny. *Poultry Science*, 84, 1277–1285.

Bailie, C. L., et al. 2013. Influence of the provision of natural light and straw bales on activity levels and leg health in commercial broiler chickens. *Animal*, 7, 618–626.

Barger-Weathers, K. 2025. *The Challenge and Responsibility of Using Genetics to Improve Poultry Health and Welfare* [Online]. Cobb-Vantress Inc. Available: https://www.cobbgenetics.com/en_US/articles/the-challenge-and-responsibility-of-using-genetics-to-improve-poultry-health-and-welfare [Accessed 1 February 2025].

Baxter, M., et al. 2021. A comparison of fast growing broiler chickens with a slower-growing breed type reared on higher welfare commercial farms. *PLoS One*, 16. https://doi.org/10.1371/journal.pone.0259333

Bello, A., et al. 2014a. Effects of commercial *in-ovo* injection of 25-hydroxycholecalciferol on broiler bone development and mineralization on days 0 and 21 posthatch. *Poultry Science*, 93, 1053–1058.

Bello, A., et al. 2014b. Effects of commercial *in-ovo* injection of 25-hydroxycholecalciferol on bone development and mineralization in male and female broilers. *Poultry Science*, 93, 2734–2739.

Bello, A., et al. 2014c. Effects of the commercial *in-ovo* injection of 25-hydroxycholecalciferol on broiler posthatch performance and carcass characteristics. *Poultry Science*, 93, 155–162.

Berg, L. & Butterworth, A. 2021. *Health and Disease Impacts on Broiler Welfare Broiler Chickens: Welfare in Practice*. Great Easton, United Kingdom: 5m Publishing Ltd.

Berg, C. & Sanotra, G. S. 2003. Can a modified latency-to-lie test be used to validate gait-scoring results in commercial broiler flocks? *Animal Welfare*, 12, 655–659.

Bessei, W. 2006. Welfare of broilers: a review. *World's Poultry Science Journal*, 62, 455–466.

Bilgili, S. F., et al. 2006. Influence of age and sex on foot pad quality and yield – broiler chickens reared on low and high density diets. *Journal Applied Poultry Research*, 15, 433–441.

Bizeray-Filoche, D., et al. 2004. Making broilers walk: what for and how? *Production Animales, Institut Nationale de la Recherche Agronomique*, 17(1), 45–57.

Bradshaw, R., et al. 2002. A review of the aetiology and pathology of leg weakness in relation to welfare. *Avian and Poultry Biology Reviews*, 13, 45–103.

Burton, R. W., et al. 1981. The incidence and importance of tibial dyschondroplasia to the commercial broiler industry in Australia. *British Poultry Science*, 22(3), 153–160.

Butterworth, A., et al. 2002. Dehydration and lameness in a broiler flock. *Animal Welfare*, 11, 89–84.

Butterworth, A. & Haslam, S. M. 2009. A lameness control strategy for broiler fowl. *In:* MIELE, M. & ROEX, J. (eds.) *Welfare Quality Reports*. Cardiff, Wales: Cardiff University.

Caplen, G., et al. 2014. Lameness is consistently better at predicting broiler chicken performance in mobility tests than other broiler characteristics. *Animal Welfare*, 23, 179–187.

Choppa, V. S. R. & Kim, W. K. 2023. A review of pathophysiology and molecular mechanisms of bacterial chondronecrosis and osteomyelitis in commercial broilers. *Biomolecules*, 13(7), 1032–1048.

Cook, M. E. 2000. Skeletal deformities and their causes: Introduction. *Poultry Science*, 79, 982–984.

Crespo, R. 2020. Developmental, metabolic, and other noninfectious disorders. *In: Diseases of Poultry,* D.E. Swayne (ed.) 14th Ed. Chapter 30. Hoboken, NJ: Wiley & Sons Inc.

Dawkins, M. S., et al. 2004. Chicken welfare is influenced more by housing conditions than by stocking density. *Nature*, 427, 342–344.

Dawkins, M. S., et al. 2009. Optical flow patterns in broiler chicken flocks as automated measures of behaviour and gait. *Applied Animal Behaviour Science*, 119, 203–209.

Dawkins, M. S., et al. 2017. Early warning of footpad dermatitis and hockburn in broiler chicken flocks using optical flow, bodyweight and water consumption. *Veterinary Record*, 180, 499–499.

Dawson, L. C., et al. 2021. In pursuit of a better broiler: a comparison of the inactivity, behavior, and enrichment use of fast- and slower growing broiler chickens. *Poultry Science*, 100. https://doi.org/10.1016/j.psj.2021.101451.

de Oliveira Sans, E., et al. 2014. Environmental enrichment on the welfare of industrial broiler chickens. *XXIV World's Poultry Science Congress 2012, Book of Abstracts*, 64–67.

Decuypere, E., et al. 2001. The day-old chick: a crucial hinge between breeders and broilers. *World's Poultry Science Journal*; 57, 127–138.

Dibner, J. J., et al. 2007. Metabolic challenges and early bone development. *Journal of Applied Poultry Research*, 16, 126–137.

Dinev, I. 2012a. Clinical and morphological investigations on the incidence of forms of rickets and their association with other pathological states in broiler chickens. *Research in Veterinary Science*, 92, 273–277.

Dinev, I., et al. 2012b. Comparative clinical and morphological studies on the incidence of tibial dyschondroplasia as a cause of lameness in three commercial lines of broiler chickens. *Journal of Applied Poultry Research*, 21(3), 637–644.

Dixon, L. M. 2020. Slow and steady wins the race: The behaviour and welfare of commercial faster growing broiler breeds compared to a commercial slower growing breed. *PLoS One*, 15. https://doi.org/10.1371/journal.pone.0231006.

Dobrzanski, Z., et al. 2008. Influence of organic forms of copper, manganese and iron on bioaccumulation of these metals and zinc in laying hens. *Journal of Elementology*, 13(3), 309–313.

Driver, J. P., et al. 2006. The effect of maternal dietary vitamin D3 supplementation on performance and tibial dyschondroplasia of broiler chicks. *Poultry Science*, 85, 39–47.

Duff, S. R. & Anderson, I. A. 1986. The gastrocnemius tendon in domestic fowl: histological findings in different strains. *Research in Veterinary Science*, 41, 402–409.

Duff, S. R. & Randall, C. J. 1986. Tendon lesions in broiler fowls. *Research in Veterinary Science*, 40, 333–338.

Duff, S. R. & Thorp, B. H. 1985. Patterns of physiological bone torsion in the pelvic appendicular skeleton of domestic fowl. *Research in Veterinary Science*, 39(3), 307–312.

Duggan, B., et al. 2024. *Decades of Breeding for Welfare and Sustainability* [Online]. Aviagen Group. Available: https://aviagen.com/assets/Welfare/AviagenDecades-of-WelfareReport_2023.pdf [Accessed 26 November 2024].

Dunlop, M. W., et al. 2016. Water activity in poultry litter: relationship to moisture content during a grow-out. *Journal of Environmental Management*, 172, 201–206.

Edwards, H.M. Jr & Sorensen, P. 1987. Effect of short fasts on the development of tibial dyschondroplasia in chickens. *Journal of Nutrition*, 117, 197–200.

Ekstrand, C., et al. 1998. Prevalence and control of foot-pad dermatitis in broilers in Sweden. *British Poultry Science*, 39, 318–324.

Emslie, K. R. & Nade, S. 1983. Acute hematogenous staphylococcal osteomyelitis. A description of the natural history in an avian model. *American Journal of Pathology*, 110(3), 333–345.

Emslie, K. R., et al. 1983. Acute haematogenous osteomyelitis: an experimental model. *Journal of Pathology*, 141, 151–167.

Farquharson, C. & Jeffries, D. 2000. Chondrocytes and longitudinal bone growth: the development of tibial dyschondroplasia. *Poultry Science*, 79, 994–1004.

Farquharson, C., et al. 1995. Regulation of chondrocyte differentiation in tibial dyschondroplasia: an in vivo and in vitro study. *Bone*, 17, 279–286.

Favero, A., et al. 2013. Development of bone in chick embryos from Cobb 500 breeder hens fed diets supplemented with zinc, manganese, and copper from inorganic and amino acid-complexed sources. *Poultry Science*, 92, 402–411.

Foutz, T.L., et al. 2007a. Effects of increased physical activity on juvenile avian bone. *Transactions of the American Society of Agricultural and Biological Engineers*, 50(1), 213–219.

Foutz, T.L., et al. 2007b. Effects of activity on avian gastrocnemius tendon. *Poultry Science*, 86, 211–218.

Garner, J. P., et al. 2002. Reliability and validity of a modified gait scoring system and its use in assessing tibial dyschondroplasia in broilers. *British Poultry Science*, 43, 355–363.

Gay, C.V., et al. 2007. Immunolocalisation of vascularisation factors in normal, tibial dyschondroplasia and rachitic cartilage. *Avian Pathology*, 36(6), 445–451.

Gomez-Verduzco, G., et al. 2012. Use of 25-hydroxycholecalciferol in diets of broiler chickens: effects on growth performance, immunity and bone calcification. *The Journal of Poultry Science*, 50, 60–64.

Grommen, S. V. H., et al. 2011. Ontogenic expression profiles of thyroid-specific genes in embryonic and hatching chicks. *Domestic Animal Endocrinology*, 40, 10–18.

Groves, P. J. & Muir, W. I. 2014. A meta-analysis of experiments linking incubation conditions with subsequent leg weakness in broiler chickens. *PLoS One*, 9. https://doi.org/10.1371/journal.pone.0102682

Groves, P. J. & Muir, W. I. 2017. Earlier hatching time predisposes Cobb broiler chickens to tibial dyschondroplasia. *Animal*, 11, 112–120.

Guevara-Torres, B. L., et al. 2021. Comparison of locomotion problems and its economic impact on Cobb and Ross broiler strains. *Veterinary Science Research*, 3, 40–46. https://doi.org/10.30564/vsr.v3i2.4126.

Guinebretière, M., et al. 2024. Effects of strain and stocking density on leg health, activity, and use of enrichments in conventional broiler chicken production. *Poultry Science*, 103. https://doi.org/10.1016/j.psj.2024.103993.

Güz, B. C., et al. 2019. Effects of dietary organic minerals, fish oil, and hydrolyzed collagen on growth performance and tibia characteristics of broiler chickens. *Poultry Science*, 98, 6552–6563. https://doi.org/10.3382/ps/pez427.

Güz, B. C., et al. 2021. Effects of pen enrichment on leg health of fast and slow-growing broiler chickens. *PLoS One*, 16(12), e0254462. https://doi.org/10.1371/journalpone.0254462.

Güz, B. C., et al. 2022. Effects of organic macro and trace minerals in fast and slower growing broiler breeders' diet on offspring growth performance and tibia characteristics. *Poultry Science*, 101. https://doi.org/10.1016/j.psj.2021.101647.

Halgrain, M., et al. 2022. Egg shell decalcification and mineralisation during chicken embryonic development: defining candidate genes in the chorioallantoic membrane. *Poultry Science*, 101(2). https://doi.org/10.1016/j.psj.2021.101622.

Hartmann, F. 2022. *Market Trends: Slow-Growing Breeds and the Impact on Efficiency* [Online]. The Poultry Site. Available: https://www.thepoultrysite.com/articles/market-trends-slow-growing-breeds-and-the-impact-on-efficiency [Accessed 19 May 2025].

Havenstein, G. B., et al. 1994. Growth, livability, and feed conversion of 1957 vs 1991 broilers when fed "typical" 1957 and 1991 broiler diets. *Poultry Science*, 73, 1785–1794.

Havenstein, G. B., et al. 2003. Growth, livability, and feed conversion of 1957 versus 2001 broilers when fed representative 1957 and 2001 broiler diets. *Poultry Science*, 82, 1500–1508.

Hill, J. E., et al. 1988. Skeletal and tendon lesions of male broiler breeders reared on elevated slats. *Avian Diseases*, 32, 540–543.

Hopcroft, R. L., et al. 2019. Changes to mineral levels in the yolk of meat chicken embryos during incubation. *Poultry Science*, 98, 1511–1516.

Hunter, J. M., et al. 2017. Practical assessment and management of foot pad dermatitis in commercial broiler chickens: A field study. *Journal of Applied Poultry Research*, 26, 593–604.

Ibelli, A. M. G., et al. 2022. Down regulation of growth plate genes involved with the onset of femoral head separation in young broilers *Frontiers in Physiology*, 13, 941134. https://doi.org/10.3389/fphys.202.941134.

Iqbal, M., et al. 2023. Baicalin inhibits apoptosis and enhances chondrocyte proliferation in thiram-induced tibial dyschondroplasia in chickens regulating Bcl-2/Caspaase-9 and Sox-9/collagen-II expression. *Ecotoxicology and Environmental Safety*, 268, 11568, 15pp.

Jacob, G., et al. 2016. Risk of incidence of hock burn and pododermatitis in broilers reared under commercial conditions *Brazilian Journal of Poultry Science*, 18, 357–361.

Kang, A., et al. 2020. The fungicide thiram perturbs gut microbiota community and causes lipid metabolism disorder. *Ecotoxicology Environmental Safety*, 206. https://doi.org/10.1016/j.ecoenv.2020.111400.

Kaup, B. 1918. Osteology/appendicular skeleton/hind limb. *In: The Anatomy of the Domestic Fowl* Philadelphia, PA: W.B. Saunders Company.

Kember, N. F., et al. 1990. Comparative cell kinetics of avian growth plates. *Research in Veterinary Science*, 49, 283–288.

Kestin, S. C., et al. 1992. Prevalence of leg weakness in broiler chickens and its relationship with genotype. *Veterinary Record*, 131, 190–194.

Kestin, S. C., et al. 1999. Different commercial broiler crosses have different susceptibilities to leg weakness. *Poultry Science*, 78, 1085–1090.

Khan, K., et al. 2023. Impact of litter on femur and tibial morphology, bone biomechanics, and leg health parameters in broiler chickens. *Animal Bioscience*, 36, 1393–1402.

Kidd, M. T. 2003. A treatise on chicken dam nutrition that impacts on progeny. *World's Poultry Science Journal*, 59, 475–494.

Kidd, M. T., et al. 1992. Progeny performance when dams and chicks are fed supplemental zinc. *Poultry Science*, 71, 1201–1206.

Kierończyk, B., et al. 2017. Infectious and non-infectious factors associated with leg disorders in poultry – a review. *Annals of Animal Science*, 17, 645–669.

Kittelsen, K. E., et al. 2017. Associations among gait score, production data, abattoir registrations, and postmortem tibia measurements in broiler chickens. *Poultry Science*, 96, 1033–1040.

Knowles, T. G., et al. 2008. Leg disorders in broiler chickens: Prevalence, risk factors and prevention. *PLoS One*, 3. https://doi.org/10.1371/journal.pone.0001545.

Ko, M. L., et al. 2010. Circadian profiles in the embryonic chick heart: L-type voltage-gated calcium channels and signalling pathways. *Chronobiology International*, 27, 1673–1696.

Kpodo, K. R. & Proszkowiec-Weglarz, M. 2023. Physiological effects of *in ovo* delivery of bioactive substances in broiler chickens *Frontiers in Veterinary Science*, 10. https://doi.org/10.3389/fvets.2023.1124007.

Lacey, D. L. & Huffer, W. E. 1982. Studies on the pathogenesis of avian rickets. I. Changes in epiphyseal and metaphyseal vessels in hypocalcaemic and hypophosphatemic rickets. *American Journal of Pathology*, 109, 288–301.

Leach, R. M. & Monsonego-Ornan, E. 2007. Tibial dyschondroplasia 40 years later. *Poultry Science*, 86, 2053–2058.

Lister, S. 2011. Leg health in chickens: assessing the assessment. *Veterinary Record*, 168, 75–76.

Liu, K.-L., et al. 2023. Leg disorders in broiler chickens: a review of current knowledge. *Animal Biotechnology*, 34, 5124–5138.

Lopez, J. C. & Hergott, T. 2015. Incorrect incubation can generate leg problems. *International Hatchery Practice*, 30, 29.

Lourens, A., et al. 2005. Effect of egg shell temperature during incubation on embryo development, hatchability and posthatch development. *Poultry Science*, 84, 914–920.

Lourens, A., et al. 2006. Effect of egg size on heat production and the transition of energy from egg to hatchling. *Poultry Science*, 85, 770–776.

Maatjens, C., et al. 2009. Relation between incubation, chick quality and later performance. *Proceedings Worlds' Poultry Science Association*, Egypt. Available at https://hatchtech.com/wp-content/uploads/2019/10/HatchTech-research-abstract-relation-between-incubation-chick-quality-and-later-performance-EN-Web.pdf [Accessed 16 April 2025].

Maharjan, P., et al. 2021. Review: Physiological growth trend of current meat broilers and dietary protein and energy management approaches for sustainable broiler production. *Animal*, 15. https://doi.org/10.1016/j.animal.2021.100284.

Mandal, R. K., et al. 2016. An investigation into blood microbiota and its potential association with bacterial chondronecrosis with osteomyelitis (BCO) in broilers. *Nature*. https://doi.org/10.1038/srep25882.

Manolagas, S. C. 1998. The role of IL-6 type cytokines and their receptors in bone. *Annals of the New York Academy of Sciences*, 840, 194–204.

Massé, P. G., et al. 1996. Pyridoxine deficiency affects biomechanical properties of chick tibial bone. *Bone*, 18, 567–574.

McGeown, D., et al. 1999. Effect of carprofen on lameness in broiler chickens. *Veterinary Record*, 144, 668–671.

McNamee, P. T. & Smyth, J. A. 2000. Bacterial chondronecrosis with osteomyelitis ('femoral head necrosis'). A review. *Avian Pathology*, 29(4), 253–270.

Mireles, A. J., et al. 2005. An acute inflammatory response alters bone homeostasis, body composition, and the humoral immune response of broiler chickens. *Poultry Science*, 84, 553–560.

Mottet, A. & Tempio, G. 2017. Global poultry production: current state and future outlook and challenges. *World's Poultry Science Journal*, 73, 245–256.

Muir, W. I. & Groves, P. J. 2018. Incubation and hatch management: consequences for bone mineralization in Cobb 500 meat chickens. *Animal*, 12, 794–801.

Muir, W. I. & Groves, P. J. 2019. The leg strength of two commercial strains of meat chicken subjected to different incubation profiles. *Animal*, 13, 1489–1497.

Nääs, I. A., et al. 2009. Impact of lameness on broiler well-being. *Journal of Applied Poultry Research*, 18, 432–439.

Nairn, M. E. & Watson, A. R. A. 1972. Leg weakness in poultry: a clinical and pathological characterisation. *Australian Veterinary Journal*, 48, 645–656.

Nangsuay, A., et al. 2016. Effects of breeder age, broiler strain, and eggshell temperature on development and physiological status of embryos and hatchlings. *Poultry Science*, 95, 1666–1679.

Neeteson, A.-M., et al. 2023. Evolutions in commercial meat poultry breeding. *Animals* [Online], 13. https://doi.org/10.3390/ani1319315.

Nelson, T. S., et al. 1992. Effect of age of maternal flock and strain on the incidence of tibial dyschondroplasia in growing chicks. *Avian Diseases*, 36, 1015–1018.

Newberry, R. C. 1995. Environmental enrichment: Increasing the biological relevance of captive environments. *Applied Animal Behaviour Science*, 44, 229–243.

Nicol, C. J., et al. 2024. An analysis of the welfare of fast-growing and slower-growing strains of broiler chicken. *Frontiers in Animal Science*, 5. https://doi.org/10.3389/fanim.2024.1374609.

Norris, L.C. 1958. The significant advances of the past fifty years in poultry nutrition. *Proceedings of the American Feed Manufacturers Association*, May 21–23, Chicago, Illinois, USA, 256–274.

Nuamah, E., et al. 2024. Unlocking phytate with phytase: A meta-analytic view of meat-type chicken muscle growth and bone mineralization potential. *Animals*, 14. https://doi.org/10.3390/ani14142090

OECD/FAO. 2023. *OECD-FAO Agricutural Outlook 2023–2032, Chapter 6. Meat* [Online]. Paris: OECD Publishing. Available: https://www.oecd.org/en/publications/oecd-fao-agricultural-outlook-2023-2032_08801ab7-en/full-report/meat_7b036d52.html [Accessed 26 November 2024].

Oliveira, T. F. B., et al. 2015a. Effects of *in-ovo* injection of organic trace minerals and post-hatch holding time on broiler performance and bone characteristics. *Poultry Science*, 94, 2677–2685.

Oliveira, T. F. B., et al. 2015b. Effects of *in-ovo* injection of organic zinc, manganese, and copper on the hatchability and bone parameters of broiler hatchlings. *Poultry Science*, 94, 2488–2494.

Ono, T. & Tuan, R. S. 1991. Vitamin D and chick embryonic yolk calcium mobilization: identification and regulation of expression of vitamin D-dependent Ca2+-binding protein, calbindin-D28K, in the yolk sac. *Developmental Biology*, 144, 167–176.

Orban, J. I., et al. 1993. Factors influencing bone mineral content, density, breaking strength, and ash as response criteria for assessing bone quality in chickens. *Poultry Science*, 72, 437–446.

Orth, M. W. & Cook, M. E. 1994. Avian tibial dyschondroplasia: a morphological and biochemical review of the growth plate lesions and causes. *Veterinary Pathology*, 31, 403–414.

Oviedo-Rondón, E. O. 2007. Predisposing factors that affect walking ability in turkeys and broilers. *Proceedings 34th Annual Carolina Poultry Nutrition Conference*. North Carolina State University, Raleigh NC. https://www.thepoultrysite.com/articles/predisposing-factors-that-affect-walking-ability-in-turkeys-and-broilers.

Oviedo-Rondón, E. O. 2008. Leg health in large broilers. *North Carolina Broiler Supervisor's Short Course*, North Carolina State University, Sanford. Available: https://www.researchgate.net/profile/Edgar-Oviedo-3/publication/242376063_Leg_Health_in_Large_Broilers/links/567808a908ae125516ee55c5/Leg-Health-in-Large-Broilers.pdf?__cf_chl_tk=w495KFFVapVZJRBXiB7vNnbHUSpJk8clhcQ

Qk1FlnBk-1745560245-1.0.1.1-1FFDknrLv2a25_EGbyyYpW396flKcpuksSSXg8hU2Cl [Accessed 25 April 2025].

Oviedo-Rondón, E. O., et al. 2008. Effects of incubator temperature and oxygen concentration during the plateau stage of oxygen consumption on turkey embryo long bone development. *Poultry Science*, 87, 1484–1492. https://doi.org/10.3382/ps.2007-00470.

Oviedo-Rondón, E. O., et al. 2009a. Effect of incubation temperatures and chick transportation conditions on bone development and leg health. *Journal of Applied Poultry Research*, 18, 671–678. https://doi.org/10.3382/japr.2008-00135.

Oviedo-Rondón, E. O., et al. 2009b. Incubation conditions affect leg health in large, high-yield broilers. *Journal of Applied Poultry Research*, 18, 640–646. https://doi.org/10.3382/japr.2008-00127.

Oviedo-Rondón, E. O. & Wineland, M. J. 2011. Incubation distress easily leads to splayed legs. *Poultry World*. https://www.poultryworld.net/poultry/incubation-distress-easily-leads-to-splayed-legs/?register=true [Accessed 16 April 2025].

Packialakshmi, B., et al. 2015. Poultry femoral head separation and necrosis: a review. *Avian Diseases*, 59, 349–354.

Paxton, H., et al. 2010. The effects of selective breeding on the architectural properties of the pelvic limb in broiler chickens: a comparative study across modern and ancestral populations. *Journal of Anatomy*, 217, 153–166.

Paxton, H., et al. 2013. The gait dynamics of the modern broiler chicken: a cautionary tale of selective breeding. *Journal of Experimental Biology*, 216, 3237–3248.

Petek, M., et al. 2010. Physical chick parameters and effects on growth performance in broiler. *Archiv fur Tierzucht*, 53,108–115.

Phibbs, D. V., et al. 2021. Leg health of meat chickens: impact on welfare, consumer behaviour, and the role of environmental enrichment. *Animal Production Science*, 61, 1203–1212.

Pratt, I. V. & Cooper, D. M. L. 2018. The effect of growth rate on the three-dimensional orientation of vascular canals in the cortical bone of broiler chickens. *Journal of Anatomy*, 233, 531–541.

Praul, C. A., et al. 2000. Gene expression and tibial dyschondroplasia. *Poultry Science*, 79(7), 1009–1013.

Randall, C.J. 1991. Rickets. *In: A Colour Atlas of Diseases and Disorders of the Domestic Fowl and Turkey. 2nd Edition*, London: Wolfe Pub Ltd, pp. 114–116.

Rath, N.C., et al. 1998. Cell death in tibial dyschondroplasia. *Avian Diseases*, 42, 72–79.

Rath, N. C., et al. 2000. Factors regulating bone maturity and strength in poultry. *Poultry Science*, 79, 1024–1032.

Rawlinson, S. C. F., et al. 2009. Genetic selection for fast growth generates bone architecture characterised by enhanced periosteal expansion and limited consolidation of the cortices but a diminution in the early responses to mechanical loading. *Bone*, 45(7), 357–366.

Rayner, A. C., et al. 2020. Slow-growing broilers are healthier and express more behavioural indicators of positive welfare. *Scientific Reports*, 10. https://doi.org/10.1038/s41598-020-72198-x.

Reece, R. L., et al. 1985. Diseases diagnosed in broiler chicken flocks in Victoria, Australia, 1977 to 1984. *Veterinary Record*, 116, 315–320.

Reiland, S., et al. 1978. Normal and pathologic skeletal development in broiler and leghorn chickens: a comparative investigation. *Acta Radiologica*, 358(Supplement S), 277–298.

Reiter, K. & Bessei, W. 2009. Effect of locomoter activity in leg disorder in fattening chicken *Berliner und Munchener Tierarztliche Wochenschrft*, 122, 264–270.

Riber, A. B., et al. 2018. Review of environmental enrichment for broiler chickens. *Poultry Science*, 97, 378–396.

Riber, A. B. & Wurtz, K. E. 2024. Impact of growth rate on the welfare of broilers. *Animals* [Online], 14. https://doi.org/10.3390/ani14223330.

Riddell, C. 1975. Studies on the pathogenesis of tibial dyschondroplasia in chickens. II. Growth rate of long bones. *Avian Diseases*, 19(3), 490–496.

Roux, S. & Orcel, P. 2000. Bone loss. Factors that regulate osteoclast differentiation: an update. *Arthritis Research*, 2, 451–456.

RSPCA Australia 2020. RSPCA Approved Farming Scheme Standard: Meat Chickens, V1.1 RSPCA Australia: Australian Capital Territory.

Ruiz-Feria, C.A., et al. 2014. Effects of distance and barriers between resources on bone and tendon strength and production performance of broiler chickens. *Poultry Science*, 93, (7), 1608–1617.

Sanchez-Rodriguez, E., et al. 2019. Changes with age (from 0 to 37 D) in tibial bone mineralization, chemical composition and structural organisation in broiler chickens. *Poultry Science*, 98, 5215–5225.

Sanotra, G. S., et al. 2003. A comparison between leg problems in Danish and Swedish broiler production. *Animal Welfare*, 12, 677–683.

Santos, M. N., et al. 2022. In pursuit of a better broiler: tibial morphology, breaking strength, and ash content in conventional and slower-growing strains of broiler chickens. *Poultry Science*, 101. https://doi.org/10.1016/j.psj.2022.10175.

Saunders-Blades, J. L. & Korver, D. R. 2014. The effect of maternal vitamin D source on broiler hatching egg quality, hatchability, and progeny bone mineral density and performance. *Journal of Applied Poultry Research*, 23, 773–783.

Sauveur, B. & Mongin, P. 1978. Tibial dyschondroplasia: a cartilage abnormality in poultry. *Annals of Biology and Animal Biochemistry and Biophysiology*, 18(1), 87–98.

Sharma, M. K., et al. 2023. Osteoimmunology: a link between gastrointestinal diseases and skeletal health in chickens. *Animals* [Online], 13. https://doi.org/10.3390/ani13111816.

Shaw, A. L., et al. 2010. Effects of flesh attachment on bone breaking and of phosphorus concentration on performance of broilers hatched from young and old flocks. *Poultry Science*, 89, 295–302.

Sherlock, L., et al. 2010. The relationship between physical activity and leg health in the broiler chicken. *British Poultry Science*, 51, 22–30.

Shi, K., et al. 2024. Transcriptome analysis reveals the pathogenesis of spontaneous tibial dyschondroplasia in broilers. *Frontiers of Genetics*, 15. https://doi.org/10.3389/fgene.2024.1434532.

Shim, M. Y. & Pesti, G. M. 2011. Effects of incubation temperature on bone development of broilers. *Poultry Science*, 90, 1867–1877. https://doi.org/10.3382/ps.2010-01242.

Shim, M. Y., et al. 2012. The effects of broiler chicken growth rate on valgus, varus and tibial dyschondroplasia. *Poultry Science*, 91,62–65. https://doi.org/10.3382/ps.2011-01599.

Shynkaruk, T., et al. 2023. Impact of stocking density on the welfare and productivity of broiler chickens reared to 34 d of age. *Journal of Applied Poultry Research*, 32. https://doi.org/10.1016/j.japr.2023.100344.

Singh, M., et al 2021. Comparison of performance and carcass composition of a novel slow-growing crossbred broiler with fast-growing broiler for chicken meat in Australia. *Poultry Science*, 100. https://doi.org/10.1016/j.psj.2020.12.063.

Su, G., et al. 2000. A note on the effects of perches and litter substrate on leg weakness in broiler chickens. *Poultry Science*, 79, 1259–1263.

Suarez, J. C., et al. 2023. Dietary supplementation of 25-hydroxycholecalciferol and its impact on performance, intestinal morphology, and vitamin D status in broilers subjected to coccidiosis vaccine. *Journal of Applied Poultry Research*, 32. https://doi.org/10.1016/j.japr.2023.100360.

Tahamtani, F. M., et al. 2020. Effects of environmental complexity on welfare indicators of fast-growing broiler chickens. *Poultry Science*, 99, 21–29.

Tahamtani, F. M., et al. 2021. Assessment of mobility and pain in broiler chickens with identifiable gait defects. *Applied Animal Behaviour Science*, 234. https://doi.org/10.1016/j.applanim.2020.105183.

Thanabalan, A. & Kiarie, E. 2021. Influence of feeding omega-3 polyunsaturated fatty acids to broiler breeders on indices of immunocompetence, gastrointestinal and skeletal development in broiler chickens *Frontiers in Veterinary Science*, 8. https://doi.org/10.3389/fvets.2021.653152.

Thorp, B. H., et al. 1993a. Osteochondrosis/dyschondroplasia: a failure of chondrocyte differentiation. *Equine Veterinary Journal*, 16, 13–18.

Thorp, B. H., et al. 1993b. Proximal femoral degeneration in growing broiler fowl. *Avian Pathology*, 22, 325–342.

Tompkins, Y. H., et al. 2022. Long bone mineral loss, bone microstructural changes and oxidative stress after *Eimeria* challenge in broilers. *Frontiers in Physiology*, 13. https://doi.org/10.3389/fphys.2022.945740.

Tompkins, Y. H., et al. 2023. Reduced bone formation and increased bone resorption drive bone loss in *Eimeria* infected broilers. *Scientific Reports*, 13. https://doi.org/10.1038/s41598-023-27585-5.

Tona, K., et al. 2004. Comparison of embryo physiology parameters during incubation, chick quality, and growth performance of three lines of broiler breeders differing in genetic composition and growth rate. *Poultry Science*, 83, 507–513.

Tona, K., et al. 2010. Comparison of Cobb and Ross strains in embryo physiology and chick juvenile growth. *Poultry Science*, 89, 1677–1683.

Tong, X., et al. 2020. LPS-induced inflammation disorders bone modeling and remodeling by inhibiting angiogenesis and disordering osteogenesis in chickens. *Inflammation Research*, 69, 765–777.

Torella, K. 2024. How a shipping error more than a century ago launched the $30 billion chicken industry. *VOX USA Future Perfect*. News Journalism site.

Torres, C. A. & Korver, D. R. 2018. Influences of trace mineral nutrition and maternal flock age on broiler embryo bone development. *Poultry Science*, 97, 2996–3003.

Triyuwanta, L., C. & Nys, Y. 1992. Dietary phosphorus and food allowance of Dwarf breeders affect reproductive performance of hens and bone development of their progeny. *British Poultry Science*, 33, 363–379.

Turk, D. E. 1973. Calcium absorption during coccidial infections in chicks. *Poultry Science*, 52, 854–857.

Turk, D. E. & Stephens, J. F. 1969. Coccidial infections of the ileum, colon, and ceca of the chick and nutrient absorption. *Poultry Science*, 48, 586–589.

Uni, Z., et al. 2012. Nutritional limitations during poultry embryonic development. *Journal of Applied Poultry Research*, 21, 175–184.

Vaishya, R., et al. 2016. Apophysitis of the tibial tuberosity (Osgood-Schlatter Disease): a review. *Cereus*, 13(8), e780. https://doi.org/10.7795/cereus.780.

van den Brand, H., et al. 2022. Research Note: Comparing methods to assess Valgus-Varus deformity in broiler chickens. *Poultry Science*, 101. https://doi.org/10.1016/j.psj.2022.101907.

van den Brand, H., et al. 2023. Effects of trace minerals in the broiler breeder diet and eggshell translucency on embryonic development of the offspring. *Poultry Science*, 102. https://doi.org/10.1016/j.psj.2022.102455.

van der Berge, J. C. 1979. Mycologia. *In:* Bamuel, J.J., et al. (eds) *Nomina Anatomica Avium*. London: Academic Press, p. 185.

van der Eijk, J. A. J., et al. 2023. Fast- and slower-growing broilers respond similarly to a reduction in stocking density with regard to gait, hock burn, skin lesions, cleanliness, and performance. *Poultry Science*, 102. https://doi.org/10.1016/j.psj.2023.102603.

van der Pol, C. W., et al. 2019. Light-dark rhythms during incubation of broiler chicken embryos and their effects on embryonic and post hatch leg bone development. *PLoS One*, 14. https://doi.org/10.1371/journal.pone.0210886.

Van Wettere, A. J. 2025. Non infectious skeletal diseases of poultry. *MSD Veterinary Manual*. Available at: https://www.msdvetmanual.com/poultry/disorders-of-the-skeletal-system-in-poultry/noninfectious-skeletal-disorders-in-poultry?query=tibial%20dyschondroplasia [Accessed 16 April 2025].

Varol Avcilar, Ö., et al. 2024. Egg mineral levels, hatching results and bone properties in embryos and chicks obtained from broiler breeders at different ages. *Journal of Animal Physiology and Animal Nutrition*, 108, 139–147. https://doi.org/10.1111/jpn.13874.

Velleman, S.G. 2007. Muscle development in the embryo and hatchling. *Poultry Science*, 86, 1050–1054.

Ventura, B. A., et al. 2010. Effects of barrier perches and density on broiler leg health, fear, and performance. *Poultry Science*, 89, 1574–1583.

Vestergaard, S. & Sanotra, G. S. 1999. Relationships between leg disorders and changes in the behaviour of broiler chickens. *Veterinary Record*, 144, 205–209.

Weber, P. 1999. The role of vitamins in the prevention of osteoporosis-a brief status report. *International Journal of Vitamin Nutrition Research*, 69, 194–197.

Weeks, C. A., et al. 2000. The behaviour of broiler chickens and its modification by lameness. *Applied Animal Behaviour Science*, 67, 111–125.

Weeks, C. A., et al. 2002. New method for objectively assessing lameness in broiler chickens. *Veterinary Record*, 151, 762–764.

Welfare Quality® 2009. *Welfare Quality® Assessment Protocol for Poultry (Broilers, Laying Hens)*. Lelystad, Netherlands: Welfare Quality® Consortium.

Whitehead, C. C. 1997. Dyschondroplasia in poultry. *Proceedings of the Nutrition Society*, 56, 957–966.

Whitehead, C. C. 2002. Nutrition and poultry welfare. *World's Poultry Science Journal*, 58, 349–356.

Whitehead, C. C., et al. 2004. High vitamin D3 requirements in broilers for bone quality and prevention of tibial dyschondroplasia and interactions with dietary calcium, available phosphorus and vitamin A. *British Poultry Science*, 45, 425–436.

Wideman, R. F. & Prisby, R. D. 2013. Bone circulatory disturbances in the development of spontaneous bacterial chondronecrosis with osteomyelitis: a translational model for the pathogenesis of femoral head necrosis. *Frontiers in Endocrinology*, 3. https://doi.org/10.3389/fendo.2012.00183.

Willemsen, H. 2022. *Slower-Growing Broilers and Sustainable Animal Welfare: An Introduction* [Online]. Available: https://www.petersime.com/expertise/slower-growing-broilers-and-sustainable-animal-welfare-an-introduction. [Accessed 19 May 2025].

Williams, B. 2000. Skeletal development in the meat-type chicken. *British Poultry Science*, 41(2), 141–149.

Williams, B., et al. 2004. Bone strength during growth: Influence of growth rate on cortical porosity and mineralization. *Calcified Tissue International*, 74(3), 236–245.

Wong-Valle, J., et al. 1993. Correlated responses to selection for high and low incidence of tibial dyschondroplasia in broilers. *Poultry Science*, 72, 1621–1629.

Wu, K. & Hocking, P. M. 2011. Turkeys are equally susceptible to foot pad dermatitis 1 to 10 weeks of age and foot pad scores were minimised when the litter moisture was less than 30%. *Poultry Science*, 90, 1170–1178.

Wurtz, K., et al. 2025. Research Note: Testing the validity of latency-to-lie tests without water for objective on-farm assessment of walking ability of broiler chickens. *Poultry Science*, 104. https://doi.org/10.1016/j.psj.2024.104577.

Xu, J., et al. 2010. Blood supply to the chicken femoral head. *Comparative Medicine*, 60(4), 295–299.

Yair, R. & Uni, Z. 2011. Content and uptake of minerals in the yolk of broiler embryos during incubation and effect of nutrient enrichment. *Poultry Science*, 90, 1523–1531.

Yair, R., et al. 2012. Bone characteristics of late term embryonic and hatching broilers: bone development under extreme growth rate. *Poultry Science*, 91(10), 2614–2620.

Yair, R., et al. 2015. *In-ovo* feeding with minerals and vitamin D3 improves bone properties in hatchlings and mature broilers. *Poultry Science*, 94, 2695–2707.

Yair, R., et al. 2017. Maternal and genetic effects on broiler bone properties during incubation period. *Poultry Science*, 96, 2301–2311.

Yalçin, S., et al. 2001. Effects of strain, maternal age and sex on morphological characteristics and composition of tibial bone in broilers. *British Poultry Science*, 42, 184–190.

Yalçin, S., et al. 2007. Effect of temperature during the incubation period on tibial growth plate chondrocyte differentiation and the incidence of tibial dyschondroplasia. *Poultry Science*, 86, 1772–1783.

Yi, G. F., et al. 2007. Determining the methionine activity of Mintrex organic trace minerals in broiler chicks by using radiolabel tracing or growth assay. *Poultry Science*, 86, 877–887.

Zuidhof, M. J., et al. 2014. Growth, efficiency, and yield of commercial broilers from 1957, 1978, and 2005. *Poultry Science*, 93, 2970–2982.

# Chapter 22

# Managing health issues in broilers: respiratory health

*Elena Catelli, Giulia Graziosi and Caterina Lupini, University of Bologna, Italy*

1 Introduction
2 Main respiratory conditions of broilers
3 Diagnosis and lab testing
4 Key tools for respiratory disease prevention and control
5 Case study: Molecular approach to respiratory pathogens epidemiology and control
6 Future trends in research
7 Conclusion
8 Where we look for further information
9 References

## 1 Introduction

The sustainability of broiler production is increasingly under scrutiny due to the growing global demand for poultry meat, which has made broilers one of the most widely produced and consumed animal protein sources (Kleyn and Ciacciariello, 2021). A critical pillar of sustainable broiler farming is the effective management of health, particularly respiratory health, which has profound implications for animal welfare, productivity, antibiotic usage, and environmental footprint (Bist et al., 2024). Due to their multifactorial nature, involving pathogens, environmental stressors, and management practices, respiratory problems are among the most common issues faced in intensive farming settings (Pantin-Jackwood and Spackman, 2020). Consequently, understanding and controlling respiratory diseases is essential for ensuring sustainable productivity and enhancing broiler health and welfare.

The aim of this chapter is to provide a comprehensive overview of the current state of the art related to understanding and preventing the principal respiratory diseases affecting broilers.

http://dx.doi.org/10.19103/AS.2025.0165.28

Viruses, bacteria, or fungi can each be primary causes of respiratory diseases in broilers. However, uncomplicated infections with a single agent are rare in the field. In commercial settings, complex infections involving multiple factors other than infectious – such as immunosuppressive agents, and adverse environmental or management conditions – are more commonly observed than simple infections.

Accurate diagnosis is crucial for effective disease control, as clinical signs in avian respiratory diseases are rarely specific. Therefore, laboratory methods – both direct (e.g. molecular tests, pathogen isolation) and indirect (e.g. serology) – play a key role in identifying the causative agent. The main diagnostic tests in this chapter will be listed and thoroughly examined, highlighting their strengths and limitations.

Given the multifactorial nature of respiratory disorders, a comprehensive approach to their prevention and control is essential. A specific section will discuss the key tools currently available, with a focus on the importance of biosecurity, environmental management, vaccination, and both conventional and alternative therapeutic strategies.

The chapter includes a case study on the molecular approach to respiratory pathogens epidemiology and control.

## 2 Main respiratory conditions of broilers

Transmissible agents involved in the so-called avian respiratory disease complex in broilers will be described in this section, with a focus on disease etiology, epidemiology, and peculiar clinical and pathological features.

### *2.1 Viruses*

Among endemic viral respiratory diseases, infectious bronchitis (IB) is considered to be the most important. Avian metapneumovirus (aMPV) infections follow, although they generally receive less attention despite their capability of causing economic losses and their global distribution and impact on poultry production. Infectious laryngotracheitis (ILT) is also highly relevant and can cause production losses during severe epizootics, with outbreaks occurring in countries where large broiler populations are present (Jones, 2010). Additionally, low pathogenic avian influenza (LPAI) should also be included in the list (Umar et al., 2017). All these pathogens are globally distributed, with challenges unique to each virus.

Common features of these viral infections include high contagiousness, which can result in up to 100% morbidity under intensive farming conditions, and the involvement of the upper respiratory tract of broilers, although clinical signs can vary depending on the specific virus. Mortality rates can differ based on the virus and/or viral strain involved, the age and immune status of the broilers, and the presence of secondary or concurrent bacterial or viral infections (Villegas, 1998).

### 2.1.1 *Infectious bronchitis*

IB is not only a respiratory disease affecting broilers of all ages, but it is also a major cause of reduced egg production (up to 35%), poor eggshell quality in laying hens (Broadfoot et al., 1954), as well as of nephropathy in young chickens. This is due to the virus's ability to replicate and produce lesions in various epithelial cells, including those of the respiratory tract, kidney, and reproductive organs.

The infectious bronchitis virus (IBV) is an enveloped virus characterized by a surface fringe of club-shaped spike proteins, which determine the typical crown-like appearance of coronaviruses under electron microscopy. It is currently classified within the species *Gammacoronavirus galli*, genus *Gammacoronavirus*, family *Coronaviridae* (Simmonds et al., 2024). IBV has an RNA genome of approximately 27 kb that encodes for both non-structural and structural proteins. The Spike (S) is the most well-characterized structural protein. This transmembrane protein is post-translationally cleaved into two sub-units, S1 and S2. The S1 sub-unit is considered the main determinant of host and tissue tropism, and the most relevant epitopes are located on this protein (Promkuntod et al., 2014; Wickramasinghe et al., 2014). The selective pressure driven by the host immune response acting on S1 is probably responsible for its remarkable genetic heterogeneity, resulting in a wide range of antigenically and genetically distinct viral types (Valastro et al., 2016). Mutation is the primary source of genetic polymorphism in the S gene and represents the fundamental basis for virus evolution. Another crucial source of genetic variability is recombination, which refers to the process where different IBV strains exchange genetic material, leading to novel viruses (Wells et al., 2023). Several studies have reported recombination events occurring under field conditions, involving both field and vaccine strains (Lee and Jackwood, 2000; Mardani et al., 2010; Xu et al., 2019).

Different methods have been used to differentiate and classify IBV isolates; however, serotyping and genetic typing are the most used. Traditionally, serotypes were defined by cross-virus neutralization testing in laboratory host systems. Currently, most laboratories apply molecular approaches to classify isolates by genetic type using the Spike-coding gene. Valastro et al. (2016) proposed a simple and reproducible S1 phylogeny-based classification system, combined with an unambiguous lineage nomenclature, for the consistent assignment of IBV strains. Based on this system, eight genotypes have been described worldwide (GI to GVIII) (Valastro et al., 2016; Ma et al., 2019; Domanska-Blicharz et al., 2020), comprising at least 35 lineages. Some of the different IBV antigenic and genetic types recognized so far have had a substantial economic impact on the global poultry industry, while others have remained restricted to specific geographical areas (de Wit et al., 2011; Jackwood, 2012).

Regardless of the chicken's age, the disease has a short incubation period, with clinical signs usually developing in 24–48 hours (Jackwood and de Wit, 2020). Respiratory signs of IB in non-immune young broilers, aged up to 2-week-old (Animas et al., 1994), where the disease is most severe, include nasal discharge, watery eyes, sneezing, gasping, coughing, and tracheal rales and, occasionally, swollen sinuses. Broilers may appear depressed and be seen huddling under a heat source. Decreased feed consumption and significantly reduced weight gain can be recorded. In uncomplicated cases, these signs last up to 5–7 days and disappear within 10 to 14 days (Raj and Jones, 1997). In chickens older than 6 weeks of age, clinical signs are typically less pronounced and even unnoticed (Jackwood and de Wit, 2020).

During the acute stage of IBV infection, high virus concentrations are shed from the respiratory tract (3–5 days post infection), making it readily detectable in tracheal samples or swabs (de Wit, 2000). During the chronic stage, the virus can be more readily detectable in the intestinal tract (cecal tonsils or cloacal swabs) and kidneys and can be excreted in the feces for a longer time, up to 2 months of age, than in the respiratory tract (Naqi et al., 2003).

### 2.1.2 Avian metapneumovirus

Among respiratory viruses, aMPV has generally been considered a minor problem in broilers compared to other pathogens. However, aside from its primary pathogenetic role, which has been long debated (Cook, 2000a; Aung et al., 2008), its presence can facilitate the entry of bacterial pathogens, whose involvement might lead to more severe losses. When aMPV affects turkeys, which are considered the most sensitive species, the disease is called Turkey Rhinotracheitis (TRT) (Suarez et al., 2020).

Avian metapneumovirus, renamed in 2022 as *Metapneumovirus avis* (Simmonds et al., 2024), is a member of the family *Pneumoviridae*, genus *Metapneumovirus*. It is an enveloped RNA virus with a genome of approximately 14 kb, classified into four subtypes (A, B, C, and D) based on nucleotide and deduced amino acid sequences (Cook and Cavanagh, 2002). Two additional subtypes have been more recently identified in gulls and parakeets in North America (Canuti et al., 2019; Retallack et al., 2019); however, their ability to infect broilers has not yet been clarified.

With respect to broilers, subtypes A and B have been circulating worldwide; however, subtype A has only been sporadically detected in recent years, with subtype B becoming increasingly predominant (Chacón et al., 2011; Franzo et al., 2017, 2020; Tucciarone et al., 2017; Franzo et al., 2020; Mescolini et al., 2021; Nguyen et al., 2021; Al-Hasan et al., 2022; Kariithi et al., 2022; Lupini et al., 2022; Escobar-Alfonso et al., 2024; Mernizi et al., 2025). Regarding North America, first detections of subtype A and B in broilers were reported in 2024, with renewed challenges for the poultry industry (Franzo et al., 2024; Goraichuk et al., 2024; Luqman et al., 2024). Subtype C has only been detected once in a local breed of meat-type chickens in southeastern China (Wei et al., 2013). Lastly, subtype D was identified only once in turkeys in France during the 1980s (Bäyon-Auboyer et al., 2000) and has never been detected in broilers.

Broilers can become infected by aMPV at any age, but infections occur most frequently between 2 and 6 weeks of age (Tucciarone et al., 2024). Affected broilers typically exhibit respiratory signs such as sneezing, coughing, nasal discharge, head shaking, and frothy eyes (Aung et al., 2008). However, infection may not always be associated with obvious clinical signs (Cook, 2000b; Suarez et al., 2020). A small proportion of broilers within the affected flock may develop Swollen Head Syndrome (SHS), characterized by swelling around the head and eyes (Fig. 1). Field cases of SHS have been associated with combined infection with aMPV and *E. coli* (Pattison et al., 1989), which appear to play a role in the pathogenesis of the SHS (Cook, 2000b).

Avian metapneumovirus infection is generally short-lived and remains localized to the upper respiratory tract, where the virus replicates quickly and abundantly following entry (Catelli et al., 1998), being rarely detected beyond

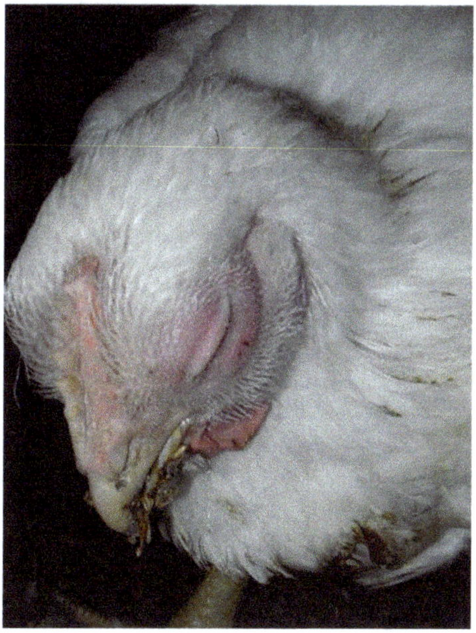

**Figure 1** Avian metapneumovirus infection: clinical signs of Swollen Head Syndrome.

2 weeks post infection, and only when highly sensitive molecular methods are used (Cook and Cavanagh, 2002). Field data showed that broilers are particularly exposed to aMPV, especially in regions where IB is well-controlled by vaccination (Tucciarone et al., 2018a; Lupini et al., 2022).

### 2.1.3 Infectious laryngotracheitis

ILT is caused by *Iltovirus gallidalpha1*, previously named *Gallid alphaherpesvirus 1* (GaHV-1), a member of the genus *Iltovirus*, subfamily *Alphaherpesvirinae* of the *Orthoherpesviridae* family (Simmonds et al., 2024). The GaHV-1 genome is a DNA molecule with a length ranging from 150 to 155 kb (Lee et al., 2011a). The virion has a nucleocapsid of icosahedral symmetry surrounded by a protein tegument layer, encapsulated by an outer envelope with incorporated virus-encoded glycoproteins. GaHV-1 strains have shown a conserved antigenicity over time, therefore discrimination among viruses has been done mostly at the genome level (García and Spatz, 2020). However, virulence may vary between strains, resulting in clinical signs from mild to severe (Nazir et al., 2020). Recombination events have been reported to occur: since late 2007, a large number of ILT outbreaks have been documented in Australia (Agnew-Crumpton et al., 2016), and the origin of the outbreak-related viruses was traced to multiple genome recombination events between different vaccine strains (Lee et al., 2012).

Chickens are considered the primary host of GaHV-1, but natural infections have been reported in other *Galliformes* species (Crawshaw and Boycott, 1982; Portz et al., 2008). GaHV-1 infection is followed by intense viral replication in tracheal tissues and persistence of virus in tissues and secretions up to 1 week post infection (Bagust et al., 1986). As with other herpesviruses, GaHV-1 establishes latent infections (Hughes et al., 1991; Williams et al., 1992), and recovered chickens do become long-term carriers of the virus. During latency, the viruses (field strains or live vaccines) are dormant in the trigeminal ganglia (Bagust et al., 1986; Williams et al., 1992). Stressful conditions, such as the onset of reproduction, transport, re-housing, and vaccination, are known to reactivate latent GaHV-1 infections, leading to intermittent virus shedding.

Clinically recovered asymptomatic chickens should therefore be recognized as potential sources of infection for their lifetime (Bagust et al., 2000). This aspect of GaHV-1 pathobiology is relevant for long-life chickens; however, it remains essential for understanding the viral epidemiology and the main sources of infections for broilers.

Chickens of almost any age can become infected and develop disease. In the field, ILT outbreaks have been reported as early as three weeks of age (Dufour-Zavala, 2008a). Severe forms are best characterized by increased nasal discharge, conjunctivitis, moist rales, and depression, followed by labored breathing and marked dyspnea (Fig. 2), often accompanied by the expectoration of blood-stained mucus (Jordan, 1966). Mortality, ranging from 5% to 20%, can occur mainly due to asphyxia caused by obstruction of the airways. At post-mortem examination, gross lesions are most consistently observed in the larynx and trachea, with hemorrhage and/or diphtheritic changes observed in the later stages of severe cases. Hemorrhages into the tracheal lumen can result in the formation of blood clots. In contrast, low-virulence strains cause mild or inapparent infections that are clinically indistinguishable from the other respiratory conditions hereby described. These may present with mild tracheitis and conjunctivitis, swelling of the infraorbital sinuses, persistent nasal discharge, and mild respiratory rales (Linares et al., 1994; Timurkaan et al., 2003; Sellers et al., 2004).

### *2.1.4 Low pathogenic avian influenza and lentogenic strains of Avian orthoavulavirus 1*

Among respiratory viruses affecting broilers, LPAIV should also be included. Avian influenza viruses (AIVs) belong to the species *Alphainfluenzavirus influenzae*, genus *Alphainfluenzavirus*, family *Orthomyxoviridae* (Schoch et al., 2020; Walker et al., 2020). These are enveloped viruses with a segmented RNA genome (approximately 13 kb); the molecular properties of hemagglutinin (HA) and neuraminidase (NA) surface glycoproteins determine AIV classification into subtypes (H1-H16; N1-N9) (Lamb and Krug, 2001). AIVs are highly contagious viruses of chickens, responsible for high morbidity and mortality depending on the characteristics of the strain (Swayne, D.E., et al. 2020). Based on disease severity in chickens, AIVs are further classified into highly pathogenic (HPAIV) or LPAIV. Wild waterfowl are the natural reservoir of AIVs, and they carry the infection mostly subclinically (Alexander, 2000). Spillover events at

**Figure 2** Infectious laryngotracheitis: broiler exhibiting marked dyspnea.

the wild-domestic bird interface can therefore result in virus transmission to poultry (Spackman, 2009). Over the past two decades, LPAIVs of the H9N2 subtype have become enzootic in Asia, the Middle East, and parts of North and Central Africa, where they cause significant economic losses to the poultry industry (Carnaccini and Perez, 2020). In experimental settings, H9N2 infections typically result in subclinical to mild respiratory disease in broilers (Kye et al., 2021; Begum et al., 2023; Bóna et al., 2023). However, field outbreaks of H9N2 infection are frequently associated with high mortality rates, ranging from 10% to 60%, primarily due to mixed infections with other pathogens (Nili and Asasi, 2002; Banet-Noach et al., 2007; Zhang et al., 2008; Pan et al., 2012; Umar et al., 2017; Samy et al., 2018; Pusch et al., 2018).

Among the viral causes that could challenge the respiratory health of broilers, the impact of live vaccines using lentogenic strains of *Avian orthoavulavirus 1* (such as Hitchner B1, LaSota, and their derivatives) should also be mentioned. These strains form the basis of worldwide control measures against Newcastle Disease (ND) (Senne et al., 2004), a devastating viral disease affecting poultry, which is caused by highly virulent strains of the same viral species (Suarez et al., 2020). The ND attenuated viruses replicate in the bird's respiratory tract and, despite their weakened state, can still induce a certain degree of cellular damage. The clinical respiratory signs resulting from this viral replication and the associated tissue pathology are collectively referred to as the 'vaccination reaction'. When this reaction is unusually severe or persists longer than expected, it is commonly termed a 'rolling vaccination reaction' (Pantin-Jackwood and Spackman, 2020).

### 2.1.5 Epidemiology of respiratory viruses

All respiratory viral diseases are highly contagious, with viruses spreading rapidly among chickens within a flock. During the acute stage of infections, high virus concentrations are shed from the respiratory tract of chickens, and aerosol transmission becomes a significant mode of spread. Infection can occur through inhalation or ocular/conjunctival contact with infectious virus particles. Transmission by aerosol is most likely effective over short distances, such as within a flock or a premise, because the enveloped RNA virions (IBV, aMPV, and LPAIV) are inactivated relatively quickly in the environment. With respect to IBV, long-term excretion of infectious viral particles can occur through feces and in the urine, and this phenomenon may enhance the possibility of flock-to-flock transmission by contact with contaminated fomites and litter (Jackwood and de Wit, 2020). To date, there is no experimental evidence of long-term persistence of aMPV in chickens (Catelli et al., 1998). Anyway, indirect contact by exposure to virus-contaminated fomites and litter is generally recognized as a possible mode of transmission for all viruses, particularly true for ILT virus that has been shown to be more resistant in the environment and able to persist for long periods in deep litter or buried carcasses (Gowthaman et al., 2020). Particularly, the spread of non-composted litter from vaccinated and/or infected flocks as fertilizer for fields is considered a practice that can cause ILT epizootics in broilers (Dufour-Zavala, 2008a; Moreno et al., 2010). Similarly, live transportation of infected broilers has been associated with viral spread, as infected chickens likely release the virus on fomites during hauling (García and Spatz, 2020).

The risk of respiratory viral infection is generally heightened in densely poultry-populated areas, where long-life chickens or turkeys (especially for aMPV) are reared near broilers, setting the premises for an intense virus exchange and increased environmental infectious pressure (Sims, 2007). Because of artificial ventilation systems of poultry houses, a dust-mediated transmission of respiratory pathogens between farms may also be taken into account (Filaire et al., 2022). Backyard flocks are also universally considered potential carriers of viruses due to their unknown vaccination status and clinical history (Johnson et al., 2004).

Direct or indirect contact with wild birds can contribute to the introduction of respiratory viruses, such as LPAIV and aMPV, into poultry holdings. LPAIV has recognized reservoirs in waterfowl and other migratory birds, such as shorebirds and gulls, and the proximity to wetlands or coastline can serve as a proxy for a heightened risk of virus introduction into farms (Liu et al., 2024). For aMPV, wild ducks, geese, pheasants, and gulls have been shown to be susceptible to the infection and may therefore play a role in the viral circulation at the domestic-wild bird interface (Graziosi et al., 2022a).

## 2.2 Bacteria

Among respiratory bacterial diseases, mycoplasmosis and respiratory-origin colisepticemia can have a major impact on broiler flocks.

### 2.2.1 Mycoplasmosis

Mycoplasmas are very small bacteria belonging to the class *Mollicutes*, order *Mycoplasmatales*. Although the main manifestation of mycoplasma disease includes respiratory signs, other systems, such as the reproductive or locomotory, can be affected. Encouraged by poor husbandry, as well as co-infections and synergism with other pathogens, *Mycoplasma gallisepticum* (MG) and, to a lesser extent, *Mycoplasma synoviae* (MS) are recognized as the most economically important species for causing respiratory disorders in broilers (Ferguson-Noel et al., 2020). Losses resulting from carcass condemnations and reduced feed efficiency, along with the costs associated with prevention and control measures, such as surveillance, medication, vaccination, and elimination of infected breeding flocks, make mycoplasmosis one of the most economically impactful infectious diseases in commercial poultry production worldwide (Ferguson-Noel et al., 2020). These prokaryotes are characterized by the complete absence of cell walls and are efficiently vertically transmitted from naturally infected hens to their progeny (Lin and Kleven, 1982). Infection of broiler breeders has been shown to be the main source of MG and MS for siblings and broiler progeny. Mycoplasmas seldom survive for more than a few days outside of a host; therefore, clinically or subclinically infected carrier chickens are essential to their epizootiology. The establishment of chronic infection, despite the presence of an active immune response, is a feature of MG and MS diseases. This may be achieved through several recognized mechanisms, including the ability to modulate the host's immune response (Lam, 2002; Kleven and Ferguson-Noel, 2008; Ley and Yoder Jr, 2008; Cizelj et al., 2011; Szczepanek and Silbart, 2014). Although mycoplasmas have poor survival outside the host, they may spread by mechanical transmission between flocks within a farm and further afield, protected by mucous exudate or other organic matter. Their ability to survive for up to several days on contaminated fomites has been demonstrated (Chandiramani et al., 1966; Christensen et al., 1994; Marois et al., 2000; Marois et al., 2005) and provides an important mechanism for indirect horizontal transmission. Strains of MG vary in virulence, tropism, and antigenicity, with young chickens being more susceptible and more likely to develop disease. Chickens usually show few clinical signs in uncomplicated infection, but may develop conjunctivitis (Fig. 3) with frothy ocular exudate, nasal exudate, and tracheal rales. MS infection most frequently occurs subclinically, although it is possible for only a few chickens

**Figure 3** Conjunctivitis in a broiler affected by mycoplasmosis.

to develop respiratory signs (Feberwee et al., 2022). Gross lesions primarily consist of mucosal congestion and catarrhal or caseous exudate in the upper respiratory tract and air sacs (Ferguson-Noel et al., 2020). Retarded growth, carcass condemnations, and downgrading at processing constitute additional losses. Overt respiratory disease and airsacculitis are exacerbated in mixed and complicated infections, as will be addressed later.

Mycoplasmosis in broilers is uncommon in countries with well-developed poultry industries, where eradication programs are implemented. However, occasional outbreaks may still occur, leading to significant economic losses. Endemically infected commercial multi-age egg layer farms (Kleven, 2008), as well as backyard, hobby, and free-range or village poultry flocks (McBride et al., 1991; Xavier et al., 2011; Chaka et al., 2012; Haesendonck et al., 2014; Messa Júnior et al., 2017; Felice et al., 2020) may serve as potential permanent sources of infection, especially under conditions of poor biosecurity in intensive farming systems.

### 2.2.2 *Escherichia coli*

Respiratory-origin colisepticemia is one of the major causes of economic losses in the broiler industry worldwide, due to chicken mortality and increased condemnation at the processing. It is caused by bacteria belonging to the Enterobacteriaceae family, genus *Escherichia* (Strockbine et al., 2015). *E. coli* involved in respiratory-origin colisepticemia are classified as Avian pathogenic *E. coli* (APEC), strains able to cause disease because equipped with specific virulence factors (Dho-Moulin and Fairbrother, 1999; Dziva and Stevens, 2008; Guabiraba and Schouler, 2015; Nawaz et al., 2024). The condition usually occurs in broilers between 4 and 6 weeks of age, when APEC gains access to the blood stream following damage to the respiratory mucosa caused by infectious or non-infectious agents (Gross, 1961; Ginns et al., 1998), such as the respiratory viruses previously listed (including live vaccine strains), mycoplasmas, or ammonia (see later in the chapter). Inhaled dust particles contaminated with APEC, highly concentrated in poultry environments, where the chicken intestine serves as a major reservoir (Ewers et al., 2009), have been identified as a key source of respiratory tract infection in broilers. This exposure can lead to the development of subacute fibrinopurulent airsacculitis, pericarditis, and perihepatitis (Peighambari et al., 2000; Kathayat et al., 2021; Paudel et al., 2021).

### 2.2.3 Other bacterial infections

Other respiratory bacterial infections reported occasionally in broilers include infections by *Avibacterium paragallinarum,* the causative agent of infectious coryza (Droual et al., 1990; Sandoval et al., 1994; El-Gazzar et al. 2025),

*Ornithobacterium rhinotracheale* (El-Ghany, 2021; Mohammad Karim, 2022; Stępień-Pyśniak et al., 2024), and *Riemerella anatipestifer* (Tzora et al., 2021; Bashashati et al., 2023). *A. paragallinarum* causes an acute respiratory disease of chickens associated with nasal discharge, facial edema, and conjunctivitis (Blackall and Soriano-Vargas, 2020). Broilers infected with *O. rhinotracheale* can manifest non-specific and upper respiratory clinical signs, including facial edema, with a mortality rate up to 10% (Hafez and Chin, 2020). Common gross lesions include tracheitis, airsacculitis, and pneumonia, often leading to high condemnation rates (50%) at slaughterhouse (van Veen et al., 2000, 2005). These pathogens are often found in association with other respiratory viruses and bacteria, and the severity of clinical signs, duration of the disease, and mortality depend on multiple factors, including poor management, concurrent infections, inadequate environmental conditions, and high stocking density (El-Sukhon et al., 2002; De Boeck et al., 2015; Morales-Erasto et al., 2016).

## 2.3 Fungi

Aspergillosis by *Aspergillus fumigatus* is the major fungal respiratory infection of birds (Beernaert et al., 2010; Eassa et al., 2017). Initial contamination of broiler farms may occur through the use of moldy litter or the introduction of one-day-old chickens harboring conidia from hatchery facilities (Hamet et al., 1991). Humidity and temperature conditions in the farm environment can then promote the copious production of airborne conidia, which can easily be dispersed in the air and inhaled by the chickens (Arné et al., 2011; Nururrozi et al., 2020). In young chickens, aspergillosis is also referred to as 'brooder pneumonia', because it is most commonly diagnosed in neonatal poultry. The disease is associated with high morbidity and mortality rates up to from 89% in spontaneous outbreaks (Eassa et al., 2017). Clinical signs are characterized by severe respiratory distress, dyspnea, and gasping. The main lesions are granulomatous airsacculitis and pneumonia (Sultana et al., 2014; Nururrozi et al., 2020).

## 2.4 Multicausal respiratory conditions

Under field conditions, complicated infections involving various etiologies, including viruses, bacteria, immunosuppressive agents, and poor environmental conditions due to poor husbandry (Figure 4), are the norm in commercial poultry settings (Kleven, 1998; Pantin-Jackwood and Spackman, 2020).

Co-infection with multiple respiratory pathogens often leads to more severe clinical signs compared to infections with a single agent (Sid et al., 2015; Al-Natour et al., 2024). Even infections with relatively mild agents may result in a substantial impact on production when combined with other agents under the right circumstances. Numerous, often synergistic, interactions have

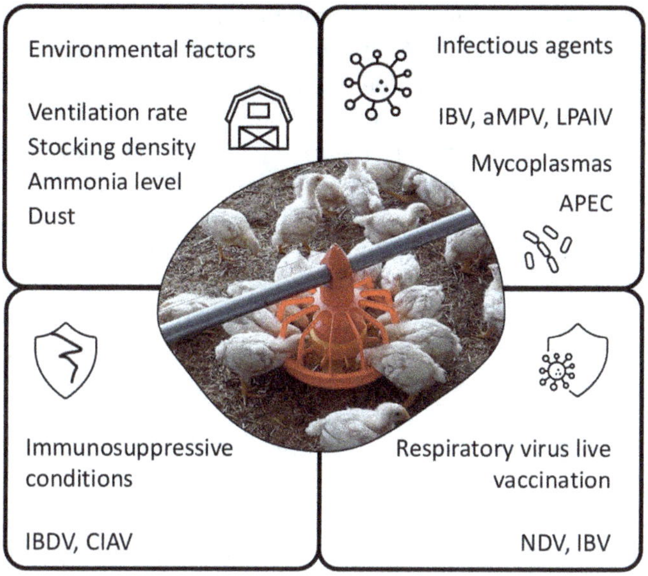


**Figure 4** Key components of the avian respiratory disease complex. The figure illustrates the multifactorial nature of the condition, which involves a combination of infectious agents and non-infectious contributing factors.

been observed in field conditions and demonstrated in experimental studies between almost all the common viral and bacterial respiratory pathogens of chicken. Some of the mechanisms of these interactions are known, including damage to the respiratory epithelial integrity and reduction of the mucociliary transport. However, other factors involved are not well understood.

Common examples of multiple respiratory infections in broilers are those involving viruses, mycoplasmas, and *E. coli*. Co-infections with IBV or NDV field or vaccine strains or LPAIV are known to increase the severity of MG (Omuro et al., 1971; Timms, 1972; Stipkovits et al., 2012a, 2012b) or MS infections (Kleven et al., 1972, 1978; King et al., 1973; Kleven et al., 1978; Hopkins and Yoder, 1982, 1984). Increased susceptibility to *E. coli* and severity of clinical signs and lesions was observed after previous IBV or aMPV infections (Yoder et al., 1977; Smith et al., 1985; Picault et al., 1987; Nakamura et al., 1996; Majó et al., 1997; Al-Ankari et al., 2001; Dwars et al., 2009; Matthijs et al., 2009). Mycoplasma infection can also increase susceptibility to *E. coli* (Gross, 1990; Nakamura et al., 1994). Three-way interaction between vaccine viruses (NDV and/or IBV), Mycoplasma (MG or MS), and *E. coli* resulted in more severe respiratory disease than either agent alone (Springer et al., 1974; Nakamura et al., 1992).

Immunosuppression is another important player in the avian respiratory disease complex. In intensive production environments, chickens can be exposed to stressors and infectious agents that impair both innate and acquired immunity (reviewed in Hoerr 2010). Increased incidence, duration, and severity of respiratory diseases have been shown to result from co-infection with viruses that specifically target immune cells. Among these, Infectious Bursal Disease virus (IBDV) is a major agent that can increase the susceptibility of chickens to *E. coli* infection irrespective of prior viral infections (Dhillon, 1986; Nakamura et al., 1990; Mitra et al., 2004). Furthermore, all immunosuppressive viral agents, including Marek's Disease virus (MDV) and Chicken anemia virus (CIAV), have been shown to affect the efficacy of vaccination against respiratory pathogens (Valentine, 1964 ; Kleven et al., 1972; De Boer et al., 1994). Moreover, thermal stress can also affect the host immune response and cause a transitory immunosuppression, which could increase susceptibility to infections (Matsumoto and Huang, 2000). It has been demonstrated that chickens exposed to MS and IBV at temperatures of 7–10°C had more extensive air sac lesions than chickens housed at 29–32°C (Yoder et al., 1977).

Environmental factors can also impair the immune defense of the respiratory system (Gabal and Azzam, 1998). Ammonia is an irritant gas produced by microflora from uric acid and amino acids present in the lumen of the cecum of chicken (Karasawa et al., 1988) and in poultry houses, resulting from the decomposition of fecal matter by the action of litter microflora (Carlile, 1984). High levels of atmospheric ammonia have been shown to affect mucociliary clearance of animals, primary innate mechanism of the upper respiratory system defense (Anderson et al., 1966; Nagaraja et al., 1983, 1984), and to influence the ultrastructure of chicken lungs (Al-Mashhadani and Beck, 1985). Exposure to 35 ppm of ammonia can significantly affect the respiratory system of poultry, causing damage to tracheal cilia and histopathological changes in the tracheal epithelium (Liu et al., 2020). Moreover, high ammonia levels are able to exacerbate and prolong the clinical course of MG infection (Kempf et al., 1988) and determine the occurrence of air sac lesions after IBV live vaccination (Kling and Quarles, 1974). Several management factors, including litter type, chickens' activity, stocking density, manure handling, ventilation rate, and the pH and humidity of the litter, can significantly influence ammonia concentration in poultry houses (Tasistro et al., 2007; David et al., 2015; Barbosa et al., 2024). Respiratory disease and airsacculitis condemnations are well known to increase during the winter months, when ammonia concentrations are highest in poultry houses (Davis and Morishita, 2005).

Additionally, dust has been shown to have a detrimental effect on the response to respiratory infections. Exposure of chickens to either ammonia or dust resulted in mild to moderate macroscopic and microscopic lesions in the trachea, lungs, and particularly in the air sacs. Chickens exposed to a combination of *E. coli* and either ammonia or dust manifested the most marked pathological changes (Oyetunde et al., 1978).

Finally, the use of formaldehyde fumigation in the hatchery to control environmental microbial load has been demonstrated to impair the integrity of the epithelial lining of the respiratory tract in neonatal chicks (Sander et al., 1995), potentially predisposing to respiratory infection and contributing to compromised performance.

Broiler respiratory diseases are frequently complex, multifactorial conditions involving a combination of infectious and non-infectious factors. Pathogens can act alone but often cause more severe outcomes when present in co-infections, particularly under poor environmental or management conditions. Viral pathogens are characterized by high contagiousness and the potential for long-term persistence or shedding, particularly IBV and ILTV. Bacterial infections often arise secondary to viral-induced damage of the respiratory epithelium, and environmental stressors further impair respiratory defenses. From a clinical perspective, these conditions are often indistinguishable, necessitating a thorough and accurate laboratory diagnosis.

## 3 Diagnosis and lab testing

Accurate diagnosis is essential for effective disease control and prevention. While clinical signs and lesions observed in respiratory conditions in affected chickens can provide valuable initial clues, they are rarely pathognomonic (Fig. 5). Therefore, diagnostic methods, such as laboratory-based analyses, are crucial to detect the presence of the pathogen or evidence of exposure to it. Diagnostic techniques include direct, such as molecular tests and viral or

**Figure 5** Group of broilers showing non-specific respiratory signs and depression.

bacterial isolation, and indirect methods, such as serology. Histopathology may also serve in some instances.

Proper sampling is crucial for achieving an accurate diagnosis. The choice of the appropriate sample depends not only on the suspected pathogen and the stage of infection but also on the type of diagnostic test to be performed. Common sample types used for the direct diagnosis of respiratory diseases include: oropharyngeal/choanal (Fig. 6) and tracheal swabs (from live chickens), sinuses, tracheal and lung tissues, or samples of other affected organs (at necropsy). Samples should be collected promptly at the onset of respiratory signs, as viral titers peak in respiratory samples usually occurs before the appearance of overt clinical signs. Cloacal swabs may also serve as an appropriate sample, for IBV and LPAIV detection (de Wit, 2000; Spackman et al., 2013). Virus detection can be unsuccessful when samples are collected from chickens when secondary bacterial infections have occurred. This may explain the difficulty in detecting aMPV or IBV in chickens affected by SHS or respiratory-origin colisepticemia (Suarez et al., 2020).

Currently, biomolecular assays are the most widely used tools for pathogen detection due to their high sensitivity and rapid response time. In addition to detecting the genome, these assays enable the genetic characterization of the identified strains, which is essential for properly planning and evaluating vaccination protocols. For molecular analysis, samples can be collected and transported as well using FTA (Flinders Technology Associates) cards, preserving nucleic acids and inactivating pathogens, while reducing storage constraints, though they may slightly reduce sensitivity (Legnardi et al., 2020). Molecular techniques include PCR (Polymerase Chain Reaction), which amplifies DNA, and qPCR (quantitative PCR), which enables real-time monitoring of

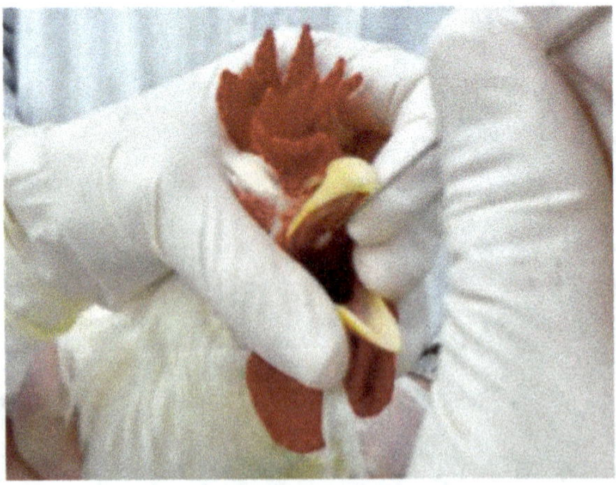

**Figure 6** Collection of an oropharyngeal/choanal swab for diagnostic purposes.

DNA amplification; when detecting RNA viruses, such as IBV, LPAIV, and aMPV, reverse transcription is required to first convert RNA into complementary DNA. A variety of molecular methods have been validated, ranging from assays targeting conserved regions for broad detection of pathogens, subtypes, or genotypes (García et al., 2005; Zhao et al., 2013; Lemaitre et al., 2018; Mo et al., 2020) to highly specific assays designed for the precise identification of individual strains, including live vaccines (Cavanagh et al., 1999; García et al., 2005; Cecchinato et al., 2013; Domanska-Blicharz et al., 2017; Tucciarone et al., 2018b). These methods also include multiplex assays, which allow for the simultaneous detection of multiple co-endemic respiratory pathogens, providing a more efficient and comprehensive diagnostic approach (Rashid et al., 2009; Mo et al., 2022). Sequence analysis is crucial for the characterization and classification of detected strains, gaining a deeper understanding of the pathogen's epidemiology. In this context, phylogenetic and phylodynamic analyses, techniques that investigate the evolutionary relationships and temporal dynamics of pathogen spread, are essential for shedding light on the origins, transmission, and evolutionary patterns of pathogens (Lee et al., 2013; Franzo et al., 2020; Mescolini et al., 2021; Chen et al., 2023). Pathogen characterization can be further refined using advanced sequencing techniques, such as next-generation sequencing (NGS), which provides a more comprehensive genetic profile of the pathogen. This approach not only facilitates species and strain identification but also offers valuable insights into genetic diversity, virulence factors, and potential antimicrobial resistance mechanisms (Afonso and Afonso, 2023).

Nowadays, viral isolation is not routinely performed for diagnostic purposes because of its lengthiness and stringent requirements, but it is still useful for several other scopes, such as performing *in vivo* trials and vaccine production. Specimens for virus isolation should be chilled immediately after collection and maintained at fridge temperature or frozen, at −20°C or below −70°C, until processing. Various substrates can be used for virus isolation, including cell cultures, embryonated eggs from specific pathogen-free (SPF) chickens, inoculated by the corion-allantoic sac route (for IBV and LPAIV) or the chorioallantoic membrane (for ILTV) (Williams et al., 2016). Chicken embryo tracheal organ cultures (TOC) are preferred for primary isolation of aMPV subtypes A and B and IBV (Cook et al., 1976; Catelli et al., 1998). Immunological or molecular techniques should be employed to confirm virus isolation.

For bacterial pathogens, culture is still considered the 'gold standard'. In samples for bacterial isolation, an appropriate transport medium is used to preserve the viability of the bacteria until reaching the laboratory. Samples should ideally be transported at refrigerated temperatures (4–8°C) to preserve bacterial integrity and slow down growth. Samples are cultured on selective media. For example, *E. coli* can be grown on MacConkey agar, which allows for the differentiation of lactose-fermenting colonies (Nolan et al., 2020), while Mycoplasmas require specialized media, such as PPLO agar, which is selective for their slow-growing, fastidious nature (World Organisation for Animal Health, 2021). An advanced diagnostic tool for identifying bacterial species is Matrix-Assisted Laser Desorption/Ionization Time-of-Flight Mass Spectrometry (MALDI-TOF) (Baudler et al., 2019). This method is faster and more cost-effective compared to traditional culture-based techniques, and it eliminates the need for complex biochemical tests. Once bacterial isolation is achieved, performing an antibiotic susceptibility test (antibiogram) is essential, providing crucial data for choosing effective antibiotics and guiding therapeutic strategies. In further characterization of APEC, molecular methods, detecting specific virulence genes are particularly valuable (Subedi et al., 2018).

Demonstrating a rise in antibody titers between preclinical and convalescent sera is a useful diagnostic method. For accurate interpretation, it is essential to establish baseline values through continuous monitoring of the local epidemiological context, as antibody titers can be affected by multiple factors, including breed, age at sampling, and vaccination programs (de Wit et al., 2010). Observed results should always be compared to these baseline values. When titers are significantly elevated beyond expected levels, a field infection should be suspected. Currently, the ELISA (Enzyme-Linked Immunosorbent Assay) is the most widely employed serologic test for detecting antibodies, due to its cost-effectiveness and suitability for high-throughput testing. However, the accuracy of ELISA–specifically its sensitivity and specificity–depends heavily on the reagents used, particularly the quality and nature of the antigens or antibodies included in the assay. Serological screening, typically performed at regular intervals, along with molecular tests, is commonly used to test broiler breeder flocks for MG and MS infection, in order to maintain mycoplasma-free status and prevent vertical transmission to offspring (World Organisation for Animal Health, 2021). Serological testing is also widely used to assess vaccine coverage. High, consistent, and sustained antibody titers are indicative of successful immunization, whereas low, variable, or short-lived titers may signal vaccination failure, potentially due to incorrect administration or substandard vaccine quality (Toro et al., 1996).

Despite its reduced routine use in the diagnosis of respiratory diseases, histopathology remains an important ancillary tool in specific contexts. It is particularly valuable for the diagnosis of ILT, where respiratory and conjunctival epithelial tissues can demonstrate the presence of pathognomonic intranuclear inclusion bodies (García and Spatz, 2020).

Laboratory-based diagnostics are essential for identifying pathogens and informing control measures. The reliability of these tests depends on proper sample collection, timing, and preservation. Molecular techniques enable sensitive, rapid pathogen detection, and strain characterization, while traditional methods like viral and bacterial isolation remain valuable in certain contexts. Serology helps monitor infection dynamics and assess vaccine efficacy, although interpretation must consider local epidemiology. Histopathology remains a useful adjunct tool in select cases, such as ILT diagnosis. Ultimately, combining multiple diagnostic approaches enhances accuracy and guides appropriate intervention strategies.

# 4 Key tools for respiratory disease prevention and control

Given the multifactorial nature of respiratory disorders in broilers, effective control in intensive production systems necessitates a comprehensive approach that addresses various interrelated factors.

## *4.1 Biosecurity*

Biosecurity is a crucial preventive tool in poultry farming to reduce the risk of introduction of infectious diseases, as well as their establishment and spread from and within the production site (World Organisation for Animal Health, 2024). Proper implementation and evaluation of external and internal biosecurity measures should therefore be prioritized to ensure the health of flocks (Gelaude et al., 2014; Van Limbergen et al., 2018; Meirhaeghe et al., 2019; Tilli et al., 2022). General biosecurity principles include geographical location and structural layout of the poultry farm and buildings and well-planned operational protocols focusing on the control of potential sources of infection. These include the regulation of the flow of vehicles, people, materials and waste, use of hygiene locks, application of proper cleaning and disinfection procedures, water quality control, and pest management (Meirhaeghe et al., 2019).

Considering that poultry farm proximity significantly impacts the airborne spread of viral diseases, adequate distance between farms is recommended (Van Steenwinkel et al., 2011). Closed feed and bedding storage are pivotal to prevent pathogens' introduction through wild birds (Ssematimba et al., 2013; Sawicka et al., 2020; Graziosi et al., 2022a). Specific sites on poultry farms, such as feed silo areas and chicken manure collection points, can serve as attractants for foraging wild birds and rodents. It is therefore crucial to regularly clean the farm area in order to avoid direct or indirect contact with wild species that can act as reservoirs or carriers of respiratory viruses or bacteria (Graziosi et al., 2024a; Graziosi et al., 2024b). Pest control programs are also considered important measures to be implemented, given the role of insects and rodents as vectors (Ou et al., 2012; Himsworth et al., 2016; Tamburro et al., 2022).

To avoid pathogens' persistence and drag from one cycle to the next, management practices, including the use of an adequate empty period, should be implemented (Tablante et al., 1999; García and Spatz, 2020). Longer downtime between consecutive cycles has been required to clear ILT virus (Chin et al., 2009). In countries where built-up litter is commonly used (e.g. the USA), appropriate litter treatments, such as heating and in-house composting, are recommended to avoid carryover of pathogens between consecutive cycles (Giambrone et al., 2008; Jones and Rautenschlein, 2013). The use of non-composted litter as fertilizer for pastures should be avoided (Kyakuwaire et al., 2019; Hubbard et al., 2020), as it can potentially trigger disease epizootics (e.g. observed during ILT outbreaks) (Dufour-Zavala, 2008a; Moreno et al., 2010).

For the control of mycoplasmosis, as these bacteria are vertically transmitted, obtaining broiler chickens from mycoplasma-free breeders is the first fundamental step. Subsequently, adequate biosecurity should follow in order to prevent mycoplasma introduction by other means (Kleven, 2008). Serological and molecular monitoring plans are implemented in broiler breeders' flocks to facilitate early detection of mycoplasmas and prevent the deleterious consequences of egg transmission. As broiler congenitally infected with mycoplasmas are more prone to complicated respiratory syndromes, especially after live NDV or IBV vaccinations, in countries with well-developed poultry industries, infected broiler-breeder flocks are not tolerated and typically eliminated (generally by early slaughter) and their hatching eggs are destroyed (Kleven, 2008; Landman, 2014).

### 4.2 Vaccination and treatments

Another critical component in preventing broiler respiratory diseases is vaccination, as discussed elsewhere in this book. It is widely acknowledged that broilers globally are routinely vaccinated against IBV. In contrast, vaccinations for aMPV and ILT virus are less commonly applied. This disparity arises from the need to assess the cost-effectiveness of such practices, considering factors like the local epidemiological landscape, the short production cycle of broilers, and the associated risks (García and Zavala, 2019; Tucciarone et al., 2024). Even though vaccination cannot fully prevent virus infection, control of clinical signs, improved health of chickens, and decrease in infectious pressures can be achieved (de Wit et al., 1998).

Protection from respiratory viruses is mostly dependent on local and cell-mediated immunity (Rubbenstroth et al., 2010; Cook et al., 2012); attenuated live vaccines are therefore almost exclusively the preferred choice. Moreover, attenuated vaccines can be administered by mass vaccination methods, spray/aerosol or drinking water, making the practice economically sustainable (Guérin et al., 2024). Spray cabinet vaccination in the hatchery is currently preferred to that in the field, since the administration of the vaccine is more standardized and guarantees a more accurate coverage of the chickens (Abdul-Cader et al., 2018). Recently, gel administration has garnered interest as an alternative method for live attenuated vaccine delivery, either at the hatchery (Tucciarone et al., 2018c) or in the field (Legnardi et al., 2021).

The relevance of IBV control for broiler respiratory health justifies the implementation of massive vaccination strategies. The selection of vaccines of appropriate IBV variants is essential to decide the success of the vaccination, given the existence and the continual emergence of new genetic and antigenic variants, showing limited cross-protection between them (Jones, 2010; de Wit et al., 2021). However, only a limited number of different IBV live vaccine types are currently available globally; consequently, there are variant viruses for which no vaccines exist. Research has shown that broader protection of the respiratory tract against challenges with viruses of heterologous serotypes can be achieved by using a combination of live-attenuated vaccines of different serotypes, either simultaneously in the hatchery or alternately first in the hatchery and then in the field. This so-called protectotype vaccination strategy is widely adopted (Jordan, 2017). IBV vaccine variants used in vaccination programs may vary from country to country, or even from region to region within the same country, depending on local IBV epidemiology (Cook et al., 2012).

AMPV vaccination is not commonly practiced in broilers for issues mainly related to perceived unfavorable cost-benefit ratios (Suarez et al., 2020). However, recent observations indicate an increase in aMPV circulation and impact in broilers, particularly in densely populated poultry areas where different species and/or productive types of poultry are reared closely (Franzo et al., 2017; Tucciarone et al., 2018a, 2024; Lupini et al., 2022; Luqman et al., 2024). Therefore, it may be necessary that broiler vaccination strategies be reevaluated, including aMPV, at least in areas where the infectious pressure is high. When used, aMPV vaccination in broilers is applied in the first days of life. Heterologous good cross-protection between vaccines of subtypes A and B has been demonstrated, although homologous protection seems to be superior (Cook et al., 1995; Van de Zande et al., 2000); this knowledge should be considered in the selection of the implemented vaccination strategy. Interferences in aMPV vaccine replication can occur after simultaneous live vaccination with IBV and NDV, but no adverse effect of immune protection achieved seems to occur (Cook et al., 2001).

ILT vaccination of broilers is limited to certain countries (e.g. USA), and is primarily used in densely populated production sites highly exposed to ILT virus challenge, while routinely used in long-lived chickens' categories (García, 2017; García and Zavala, 2019). Live attenuated vaccines, for preferably individual vaccination, and recombinant DNA vaccines, using *turkey herpesvirus* or *Fowlpox* virus (rHVT-LT and rFPV-LT, respectively) in-ovo or day-old applied, are available commercially against ILTV (García and Spatz, 2020).

Live vaccinations for respiratory viruses offer significant advantages, such as lower administrative costs and the feasibility of mass administration procedures. However, drawbacks arise particularly when partial coverage is achieved due to poor vaccine administration. All currently used live respiratory vaccines are attenuated through time-consuming process of serial passage of a virulent isolate through SPF eggs or cell cultures. A careful balance is obtained between attenuation and immunogenicity, and the risk of reversion to virulence of the vaccines after several back passages in unprotected, poorly vaccinated chickens is always present. Respiratory disease outbreaks due to IBV, aMPV, or ILT vaccine-origin viral strains have been reported (Catelli et al., 2006; Moreno et al., 2010; Cecchinato et al., 2014). Indeed, the empirically obtained vaccines are composed of numerous viral subpopulations (Guy et al., 1991; McKinley et al., 2008; Ndegwa et al., 2012; Franzo et al., 2015), and virus circulation in the natural host can lead to the possibility of selection and establishment of more virulent viral subpopulations.

ILTV vaccines can undergo long-term latent infection and potentially reactivate and spread through biosecurity breaches in naïve bird populations, therefore regaining virulence and causing disease (García and Spatz, 2020). Alongside the occurrence of recombination events between vaccine and field IBV strains (McKinley et al., 2011; Jackwood et al., 2012), or even between two different ILTV vaccine strains (Lee et al., 2012), has been reported, and IBV or ILTV variants with high fitness can emerge and effectively disseminate in poultry populations (Agnew-Crumpton et al., 2016; Moreno et al., 2017). Moreover, evidence of vaccine-driven immune selection generating immune escape viral variants has been provided for aMPV (Cecchinato et al., 2010; Catelli et al., 2010; Mescolini et al., 2021) and IBV (Franzo et al., 2019), as assessed by molecular studies or animal trials.

The safeguard of an efficient immune status of the broilers is also crucial for disease prevention. Therefore, the control of respiratory diseases is indirectly entrusted to adequate immune control for immunosuppressive conditions, including viral diseases of chickens, such as IBD, Marek's disease, and Chicken infectious anemia (Hoerr, 2010). For the first two, broilers are mainly vaccinated, while for the latter, reliance is placed on maternal immunity provided by vaccination of breeders (Miller and Schat, 2004).

Controlling colibacillosis in the broiler industry has always been a significant challenge due to the substantial economic impact of disease-related losses (Kathayat et al., 2021). For its control in the past, antibiotics have been largely used worldwide either as an antimicrobial growth promoter (AGP) or for metaphylaxis. However, growing concern over antibiotic resistance, especially multidrug resistance, and the potential of bacterial strains affecting people acquiring transmissible resistance factors from bacteria in animals, has led to the disuse of AGP and the imperative that antimicrobial use in food-producing animals be limited to judicious use to reduce the risks (World Health Organization, 2016).

Prevention of respiratory-origin colibacillosis in broilers cannot rely on vaccination, even though live and inactivated or autogenous vaccines are available (Landman et al., 2017). Apart from the cost-benefit ratio concern of the practice, APEC strains, however, are highly diverse antigenically, and vaccines made with specific APEC serotypes protect against homologous strains, but are less or not effective against heterologous strains (Dho-Moulin and Fairbrother, 1999). More importantly, as previously discussed, colibacillosis occurs when there are deficiencies in the zootechnical, hygiene, and health management of intensive broiler production units. Therefore, establishing optimal poultry housing conditions in terms of temperature, humidity, air ammonia levels, maintaining proper stocking density, and controlling environmental stressors is fundamental, alongside sourcing MG- and MS-free day-old animals (Nolan et al., 2020). Moreover, colibacillosis occurrence can be indirectly controlled by controlling other respiratory agents, ensuring adequate immunization, while minimizing the negative effects of live vaccines. Alongside monitoring and controlling, the vertical transmission of APEC from breeders, proper hatchery management and sanitation, and implementation of biosecurity measures at both breeder and broiler levels are crucial to prevent APEC entry into the flock (Nolan et al., 2020). All these aspects are essential and largely under the control of the farm manager.

When bacterial infections occur and the flock begins to exhibit clinical signs, the administration of antimicrobial drugs becomes necessary to safeguard the well-being and welfare of the broiler population. However, with the increasing occurrence of multidrug-resistant *E. coli* isolates (Nhung et al., 2017), treatment may be unsuccessful if the susceptibility of the isolate involved in the disease outbreak has not been previously assessed through diagnostic processes.

In high-income countries, broiler production is increasingly shifting towards antibiotic-free programs. Consequently, many poultry producers are adopting organic and antibiotic-free practices, eliminating antimicrobial use throughout the production cycle. This transition introduces new health challenges (Fancher et al., 2020). In this context, there is a growing interest in alternative methods for preventing and treating colibacillosis. These alternatives include prebiotics, probiotics, enzymes, organic acids, vitamins, immune enhancers, anti-inflammatory drugs, bacteriophages (Karami et al., 2024), essential oils (Kovačević et al., 2024), and other antimicrobial products (Suresh et al., 2018; Nolan et al., 2020). Additionally, research is underway to develop new vaccine strategies for APEC control that are applicable on a large scale (Van Goor et al., 2017; Hu et al., 2019).

Preventing and controlling respiratory diseases in broilers requires an integrated approach. Biosecurity remains the cornerstone of prevention, reducing the risk of disease introduction and spread. Vaccination plays a central role in disease control, with live attenuated vaccines offering practical mass administration but requiring careful handling to avoid safety issues like reversion to virulence, as demonstrated for aMPV. Vaccine selection must align with local epidemiology, and protectotype strategies are commonly used for broader IBV protection. Although less frequent, vaccination against aMPV and ILTV may be justified in high-risk settings. The control of colibacillosis, a major secondary complication, cannot rely solely on antibiotics or vaccines; it demands optimal environmental management and robust biosecurity. As antibiotic use declines due to resistance concerns, alternative strategies are gaining attention. Ultimately, disease prevention depends on a well-coordinated, evidence-based health management strategy tailored to local conditions and challenges.

## 5 Case study: Molecular approach to respiratory pathogens epidemiology and control

Molecular techniques have provided powerful tools for the diagnosis, epidemiology, and control of human and animal transmissible diseases in recent decades. Molecular diagnosis performed via PCR or qPCR allows for the identification of the pathogen not only at the species level but also provides, depending on the test used, information about the strain involved. Moreover, the possibility to perform genomic sequencing of parts or the entire genome at affordable prices has made further and deeper molecular characterization of respiratory pathogens accessible to most research and diagnostic laboratories. These findings can also be shared on public web databases, enhancing collaborative research and data availability. Genome phylogenetic analysis and phylodynamic reconstruction have enabled the identification of molecular relationships among IBV variants (Laconi et al., 2019) and allowed for the tracking of aMPV subtype B spread across Europe, as well as its evolution over time (Franzo et al., 2020; Mescolini et al., 2021).

Molecular surveys in wild birds (Graziosi et al., 2022b, 2024c; Tucciarone et al., 2022; Lupini et al., 2025) or in backyard poultry (Felice et al., 2020) have provided insights into pathogens' circulation patterns within local ecosystems at the wildfowl-poultry interface, offering essential information for pathogen introduction risk assessment and for identifying potential hotspots for the emergence of novel respiratory viruses or mycoplasma strains.

Characterization at the molecular level has been applied to live viral attenuated vaccines, enabling differentiation from field strains. Indeed, during the attenuation process, generally achieved through serial passages of field viruses in laboratory substrates, such as SPF embryonated eggs or cell cultures, the selection/modification of viral populations occurs, which, while maintaining immunogenicity, are attenuated in terms of virulence. These changes in phenotype are rooted in molecular alterations, which can be detected in specific genomic traits unique to each viral species and strain. The ability to differentiate between field and vaccine viruses at the molecular level (Listorti et al., 2014, 2017) has demonstrated useful practical applications in evaluating the efficacy of control measures for IBV and aMPV. First and foremost, this includes the ability to assess the vaccine take, soon after vaccination and, over time, vaccine kinetics and coverage (Cecchinato et al., 2012; Tucciarone et al., 2018b; Lupini et al., 2022). Moreover, for aMPV subtype A or B, the detection of the vaccine strain for 2–4 weeks after vaccination has been associated with vaccine reversion to virulence after back passages of the vaccine virus in poorly mass vaccinated flocks (Catelli et al., 2006; Cecchinato et al., 2014). Viral subpopulations in a live aMPV subtype B vaccine empirically attenuated have also been investigated as a possible cause of vaccine instability (Franzo et al., 2015).

Molecular studies have highlighted the occurrence of field virus evolution following application of extensive field vaccination over the years. Field strains of aMPV subtype B have been reported to undergo both molecular and antigenic evolution over time, enabling them to avoid vaccine-induced immunity (Cecchinato et al., 2010), and thereby reducing the protective efficacy of vaccines against these evading field strains under experimental conditions (Catelli et al., 2010).

Molecular-level knowledge of local IBV epidemiology has become increasingly valuable, offering important insights into how vaccination strategies may influence the circulation of variants. Such knowledge has proven to be essential for the correct interpretation of diagnostic results and supports the selection of the most appropriate vaccine (Franzo et al., 2014; Legnardi et al., 2019, 2022).

Molecular techniques have revolutionized the understanding and control of avian respiratory diseases. Tools such as PCR, qPCR, genomic sequencing, phylogenetic, and phylodynamic analyses provide critical insights into pathogen genetic diversity, virus evolution, transmission dynamics, and vaccine efficacy. Molecular differentiation between vaccine and field strains has become pivotal for evaluating vaccination outcomes and detecting potential vaccine-related issues, such as reversion to virulence. Ultimately, these approaches enhance diagnostic accuracy, improve risk assessment, and contribute to the refinement of vaccination programs based on local epidemiological data.

## 6 Future trends in research

Research on respiratory diseases is increasingly focused on integrating early warning systems for disease outbreaks, advancements in diagnostic tools, and the development of novel vaccines.

The integration of precision livestock farming systems in the poultry sector involves mechanical and digital tools, such as biosensors (Bhuiyan et al., 2025), that can be used to monitor animal health (He et al., 2022; Brassó et al., 2025). Among these systems, precision tools for sound analysis of chickens' vocalizations may have a great potential (Banakar et al., 2016; Carpentier et al., 2019; Mahdavian et al., 2021) in early detection of respiratory diseases.

Recent advancements in molecular diagnostic technology have introduced isothermal amplification assays, such as LAMP and Penn-RAMP, which are well-suited for on-farm use (El-Tholoth and Bau, 2024). These methods, enabling real-time pathogen detection directly at the point of care, will significantly reduce the time required for diagnosis. Beyond clinical specimens, preliminary investigations have demonstrated that pathogens can also be detected directly from environmental matrices, such as air and dust collected within poultry housing facilities, including via the deployment of automated sampling technologies (Hietala et al., 2005; Chen et al., 2021; Andersen et al., 2022).

Next-generation sequencing is likely to become the molecular diagnostic method of choice in the future (Afonso and Afonso, 2023). While targeted NGS, as previously mentioned, allows for precise characterization of a suspected agent, non-targeted NGS permits the identification of most of the infectious agents present in the collected samples (Diao et al., 2022; Kariithi et al., 2022), broadening the diagnostic capability of the test. This technique is also useful for the characterization of large microbial communities, such as respiratory and gut microbiome, that act as gatekeeper for broiler health and appear to minimize pathological changes induced by respiratory pathogens and could lead to the discovery of new treatments for respiratory infections (Wang et al., 2021; Feng et al., 2024; Oladokun and Sharif, 2024; Herath Mudiyanselage et al., 2025).

Vaccination is undoubtedly the most rapidly evolving tool for controlling respiratory diseases in broilers. A particularly promising direction in vaccine research is the development of recombinant viral vector vaccines, engineered via reverse genetics, which can also confer protection against multiple pathogens (Tang et al., 2020; Hein et al., 2021; Wang et al., 2024). NDV and IBV have already been successfully employed as viral vectors for the expression of heterologous proteins (Ge et al., 2010; Kanabagatte Basavarajappa et al., 2014; Zhao et al., 2014; Yang et al., 2016; Shao et al., 2025). Additionally, reverse genetics systems have been developed for aMPV (Naylor et al., 2004; Govindarajan et al., 2006; Falchieri et al., 2013; Laconi et al., 2016; Zhang et al., 2016; Meng et al., 2025), opening new avenues for disease control. Another promising strategy is the design of virus-like particles that display heterologous viral antigens on their surface, offering a versatile platform for vaccine development (Lee et al., 2011b; Schädler et al., 2019; Sepotokele et al., 2023; Zhang et al., 2023). Lastly, innovative molecular approaches, such as One-pot Golden Gate Assembly (Bilotti et al., 2024) and mRNA-based technologies (Rcheulishvili et al., 2022), could enable a rapid response to emerging viral variants. Overall, next-generation vaccines developed through genome engineering are designed not only to overcome the limitations of traditional live vaccines – such as the risk of reversion to virulence and rolling reactions – but also to unlock new possibilities for safer, more targeted, and more adaptable immunization strategies.

In the future, alternatives to antibiotics are expected to gain increasing traction, driven by the growing shift toward antibiotic-free production systems. Correspondingly, research in this area has expanded significantly (Balta et al., 2020; Rossi et al., 2020; Gholami-Ahangaran et al., 2022; Adjei-Mensah et al., 2023; Phillips et al., 2023), aiming to identify and validate effective natural or innovative substitutes for conventional therapies, with antimicrobial and antiviral or immune-boosting properties.

## 7 Conclusion

Respiratory diseases in broilers represent a major challenge for animal health, welfare, and production worldwide, due to their multifactorial nature, economic impact, and complex control dynamics. These conditions often result from interactions between viral agents, bacterial pathogens, and environmental or management-related stressors. Reliable diagnosis is fundamental for appropriate control but is rarely achievable based on clinical signs alone; therefore, laboratory methods must be integrated to identify pathogens.

Effective prevention relies on a combination of strict biosecurity, tailored vaccination strategies, and improved environmental management. While live attenuated vaccines are widely used for IBV, and to a lesser extent for aMPV and ILTV, their efficacy depends on correct administration, strain selection, and flock-specific risk factors. The rise of antimicrobial resistance has further highlighted the need for alternative approaches to the control of respiratory bacterial infection, including new-generation vaccines and non-antibiotic interventions. Looking to the future, research and innovation are reshaping how respiratory diseases are diagnosed and managed. Precision livestock farming technologies, biosensors, and rapid molecular diagnostics promise earlier, more accurate detection. Advances in genomics and vaccine technology – including recombinant vectors and mRNA platforms – offer new preventive options aligned with evolving field challenges.

The case study on the molecular approach to respiratory pathogens exemplifies the transformative role of genomics and phylogenetics in understanding viral infection dynamics. Molecular tools not only improve diagnostic resolution but also reveal patterns of viral evolution, transmission pathways, and vaccine-virus interactions. The ability to distinguish vaccine from field strains and to monitor genetic evolution under vaccine pressure provides essential support for evidence-based vaccination strategies. Moreover, molecular surveys in wild birds and backyard flocks help identify ecological interfaces and potential hotspots for disease emergence.

In summary, the control of broiler respiratory diseases must be approached through a comprehensive, integrated strategy that bridges traditional and molecular diagnostics, targeted vaccination, improved management, and ongoing surveillance. Continued investment in research and the adoption of innovative tools are key to improving broiler health and welfare, reducing antimicrobial use, and ensuring sustainable production in an increasingly demanding global market.

## 8 Where we look for further information

Key Conferences on the topic:

- International Symposium on Avian Viral Respiratory Diseases – It is the top symposium on the topic, highly attended by the most renowned scientists and key stakeholders. It is held alternately every 2 years, rotating between Europe (Utrecht, the Netherlands) and the USA (Athens, Georgia).
- The World Veterinary Poultry Association (WVPA, *https://wvpa.info*) organizes every 2 years a World Congress and Asia and Africa-Middle East Meetings. WVPA serves as a global platform for professionals, researchers, and industry leaders to converge, share knowledge, and discuss the latest advancements in the field of poultry health, including respiratory diseases.
- International Avian Mycoplasma Conference (https://www.poultrymycoplasma.com/) – The symposium has proven to be a vital platform for the exchange of knowledge and ideas within the field of mycoplasma research. The symposium coincides with the conclusion of the International Organization of Mycoplasma (IOM) conference and is organized every 3 years.
- Symposium on Marek's Diseases and Avian Herpesviruses (https://www.aaap.info/2024-international-symposium-on-marek-s-disease-and-avian-herpesviruses) – The symposium covers both Marek's Disease and Infectious Laryngotracheitis, and it is organized every 3 years.
- International Symposium on Avian Influenza (https://www.avianinfluenzasymposium.com/ISAI2025) – A pivotal moment in the changing epidemiology of avian influenza is usually organized every 3 to 5 years.

## 9 References

Abdul-Cader, M.S., et al. (2018), 'Hatchery vaccination against poultry viral diseases: potential mechanisms and limitations', *Viral Immunol.*, 31(1), 23–33.

Adjei-Mensah, B., et al. (2023), 'Antiviral potentials of garlic (*Allium sativum*) in poultry production: a mini review', *Vet. Med. Sci.*, 9(6), 2711–2718.

Afonso, C.L. and Afonso, A.M. (2023), 'Next-generation sequencing for the detection of microbial agents in avian clinical samples', *Vet. Sci.*, 10(12), 690.

Agnew-Crumpton, R., et al. (2016), 'Spread of the newly emerging infectious laryngotracheitis viruses in Australia', *Infect. Genet. Evol.*, 43, 67–73.

Al-Ankari, A.R., et al. (2001), 'Avian pneumovirus infection in broiler chicks inoculated with *Escherichia coli* at different time intervals', *Avian Pathol.*, 30(3), 257–267.

Al-Hasan, B.A., et al. (2022), 'First report of avian metapneumovirus type B in Iraqi broiler flocks with swollen head syndrome', *Vet. World*, 15(1), 16–21.

Al-Mashhadani, E.H. and Beck, M.M. (1985), 'Effect of atmospheric ammonia on the surface ultrastructure of the lung and trachea of broiler chicks', *Poult. Sci.*, 64(11), 2056–2061.

Al-Natour, M.Q., et al. (2024), 'Respiratory disease complex due to mixed viral infections in chicken in Jordan', *Poult. Sci.*, 103(4), 103565.

Alexander, D.J. (2000), 'A review of avian influenza in different bird species', *Vet. Microbiol.*, 74(1–2), 3–13.

Andersen, K.B., et al. (2022), 'A new tool for air sample-based surveillance of Campylobacter and Salmonella in poultry flocks', *J. Appl. Poultry Res.*, 31(2), 100236.

Anderson, D.P., et al. (1966), 'Influence of poultry house dust, ammonia, and carbon dioxide on the resistance of chickens to Newcastle disease virus', *Avian Dis.*, 10(2), 177–188.

Animas, S.B., et al. (1994), 'Experimental infection with avian infectious bronchitis virus (Kagoshima-34 strain) in chicks at different ages', *J. Vet. Med. Sci*, 56(3), 443–447.

Arné, P., et al. (2011), '*Aspergillus fumigatus* in poultry', *Int. J. Microbiol.*, 2011, 746356.

Aung, Y.H., et al. (2008), 'Reproducibility of swollen sinuses in broilers by experimental infection with avian metapneumovirus subtypes A and B of turkey origin and their comparative pathogenesis', *Avian Pathol.*, 37(1), 65–74.

Bagust, T.J., et al. (1986), 'Gallid-1 herpesvirus infection in the chicken. 3. Reinvestigation of the pathogenesis of infectious laryngotracheitis in acute and early post-acute respiratory disease', *Avian Dis.*, 30(1), 179–190.

Bagust, T.J., et al. (2000), 'Avian infectious laryngotracheitis', *Rev. Sci. Tech.*, 19(2), 483–492.

Balta, I., et al. (2020), 'Antiviral activity of a novel mixture of natural antimicrobials, in vitro, and in a chicken infection model in vivo', *Sci. Rep.*, 10(1), 16631.

Banakar, A., et al. (2016), 'An intelligent device for diagnosing avian diseases: newcastle, infectious bronchitis, avian influenza', *Comput. Electr. Agric.*, 127, 744–753.

Banet-Noach, C., et al. (2007), 'H9N2 influenza viruses from Israeli poultry: a five-year outbreak', *Avian Dis.*, 51(1 Suppl), 290–296.

Barbosa, L.V.S., et al. (2024), 'Predicting risk of ammonia exposure in broiler housing: correlation with incidence of health issues', *Animals*, 14(4), 615..

Bashashati, M., et al. (2023), 'Pathogenic bacteria associated with outbreaks of respiratory disease in Iranian broiler farms', *Vet. Med. Sci.*, 9(4), 1675–1684.

Baudler, L., et al. (2019), 'Identification and differentiation of avian Mycoplasma species using MALDI-TOF MS', *J. Vet. Diagn. Invest.*, 31(4), 620–624.

Bäyon-Auboyer, M.H., et al. (2000), 'Nucleotide sequences of the F, L and G protein genes of two non-A/non-B avian pneumoviruses (APV) reveal a novel APV subgroup', *J. Gen. Virol.*, 81(Pt 11), 2723–2733.

Beernaert, L.A., et al. (2010), 'Aspergillus infections in birds: a review', *Avian Pathol.*, 39(5), 325–331.

Begum, J.A., et al. (2023), 'Experimental pathogenicity of H9N2 avian influenza viruses harboring a tri-basic hemagglutinin cleavage site in Sonali and broiler chickens', *Viruses*, 15(2), 461.

Bhuiyan, M.S.A., et al. (2025), 'Current trends and future potential in the detection of avian coronaviruses: an emphasis on sensors-based technologies', *Virology*, 604, 110399.

Bilotti, K., et al. (2024), 'One-pot golden gate assembly of an avian infectious bronchitis virus reverse genetics system', *PLoS One*, 19(7), e0307655.

Bist, R.B., et al. (2024), 'Sustainable poultry farming practices: a critical review of current strategies and future prospects', *Poult. Sci.*, 103(12), 104295.

Blackall, P.J. and Soriano-Vargas, E. (2020), 'Infectious coryza and related bacterial infections', In Swayne, D.E., et al. (Eds), *Diseases of poultry*, 14th ed., Hoboken, NJ, USA: Wiley-Blackwell, pp. 890–906.

Bóna, M., et al. (2023), 'Evaluation of the virulence of low pathogenic H9N2 avian influenza virus strains in broiler chickens', *Vet. Sci.*, 10(12), 671.

Brassó, L.D., et al. (2025), 'Modern technologies for improving broiler production and welfare: a review', *Animals*, 15(4), 493.

Broadfoot, D., et al. (1954), 'Effect of infectious bronchitis on egg production', *J. Am. Vet. Med. Assoc.*, 124(923), 128–130.

Canuti, M., et al. (2019), 'Discovery and characterization of novel RNA viruses in aquatic North American wild birds', *Viruses*, 11(9), 768.

Carlile, F.S. (1984), 'Ammonia in poultry houses: a literature review', *World's Poult. Sci. J.*, 40, 99–113.

Carnaccini, S. and Perez, D.R. (2020), 'H9 Influenza Viruses: an Emerging Challenge', *Cold Spring Harb. Perspect. Med.*, 10(6), a038588.

Carpentier, L., et al. (2019), 'Development of sound-based poultry health monitoring tool for automated sneeze detection', *Comput. Electron. Agric.*, 162, 573–581.

Catelli, E., et al. (1998), 'The use of virus isolation, histopathology and immunoperoxidase techniques to study the dissemination of a chicken isolate of avian pneumovirus in chickens', *Avian Pathol.*, 27(6), 632–640.

Catelli, E., et al. (2006), 'Demonstration of loss of attenuation and extended field persistence of a live avian metapneumovirus vaccine', *Vaccine*, 24(42–43), 6476–6482.

Catelli, E., et al. (2010), 'Field avian metapneumovirus evolution avoiding vaccine induced immunity', *Vaccine*, 28(4), 916–921.

Cavanagh, D., et al. (1999), 'Longitudinal field studies of infectious bronchitis virus and avian pneumovirus in broilers using type-specific polymerase chain reactions', *Avian Pathol.*, 28(6), 593–605.

Cecchinato, M., et al. (2010), 'Avian metapneumovirus (AMPV) attachment protein involvement in probable virus evolution concurrent with mass live vaccine introduction', *Vet. Microbiol.*, 146(1–2), 24–34.

Cecchinato, M., et al. (2012), 'Italian field survey reveals a high diffusion of avian metapneumovirus subtype B in layers and weaknesses in the vaccination strategy applied', *Avian Dis.*, 56(4), 720–724.

Cecchinato, M., et al. (2013), 'Development of a real-time RT-PCR assay for the simultaneous identification, quantitation and differentiation of avian metapneumovirus subtypes A and B', *Avian Pathol.*, 42(3), 283–289.

Cecchinato, M., et al. (2014), 'Reversion to virulence of a subtype B avian metapneumovirus vaccine: is it time for regulators to require availability of vaccine progenitors?', *Vaccine*, 32(36), 4660–4664.

Chacón, J.L., et al. (2011), 'Avian metapneumovirus subtypes circulating in Brazilian vaccinated and nonvaccinated chicken and turkey farms', *Avian Dis.*, 55(1), 82–89.

Chaka, H., et al. (2012), 'Seroprevalence of Newcastle disease and other infectious diseases in backyard chickens at markets in Eastern Shewa zone, Ethiopia', *Poult. Sci.*, 91(4), 862–869.

Chandiramani, N.K., et al. (1966), 'Viability studies with *Mycoplasma gallisepticum* under different environmental conditions', *Poult. Sci.*, 45(5), 1029–1044.

Chen, G., et al. (2021), 'Aerosol concentrations and fungal communities within broiler houses in different broiler growth stages in summer', *Front. Vet. Sci.*, 8, 775502.

Chen, L., et al. (2023), 'Insights into the epidemiology, phylodynamics, and evolutionary changes of lineage GI-7 infectious bronchitis virus', *Transbound. Emerg. Dis.*, 2023(1), 9520616.

Chin, R.P., et al. (2009), 'Intervention strategies for laryngotracheitis: impact of extended downtime and enhanced biosecurity auditing', *Avian Dis.*, 53(4), 574–577.

Christensen, N.H., et al. (1994), 'Investigations into the survival of *Mycoplasma gallisepticum*, *Mycoplasma synoviae* and *Mycoplasma iowae* on materials found in the poultry house environment', *Avian Pathol.*, 23(1), 127–143.

Cizelj, I., et al. (2011), '*Mycoplasma gallisepticum* and Mycoplasma synoviae express a cysteine protease CysP, which can cleave chicken IgG into Fab and Fc', *Microbiology (Reading)*, 157(Pt 2), 362–372.

Cook, J.K.A. (2000a), 'Avian pneumovirus infections of turkeys and chickens', *Vet. J.*, 160(2), 118–125.

Cook, J.K.A. (2000b), 'Avian rhinotracheitis', *Rev. Sci. Tech. Off. Int. Épizoot.*, 19(2), 602–613.

Cook, J.K.A. and Cavanagh, D. (2002), 'Detection and differentiation of avian pneumoviruses (metapneumoviruses)', *Avian Pathol.*, 31(2), 117–132.

Cook, J.K.A., et al. (1976), 'The use of chicken tracheal organ cultures for the isolation and assay of avian infectious bronchitis virus', *Arch. Virol.*, 50(1–2), 109–118.

Cook, J.K.A., et al. (1995), 'Protection provided by a commercially available vaccine against different strains of turkey rhinotracheitis virus', *Vet. Rec.*, 136(15), 392–393.

Cook, J.K.A., et al. (2001), 'Infectious bronchitis virus vaccine interferes with the replication of avian pneumovirus vaccine in domestic fowl', *Avian Pathol.*, 30(3), 233–242.

Cook, J.K.A., et al. (2012), 'The long view: 40 years of infectious bronchitis research', *Avian Pathol.*, 41(3), 239–250.

Crawshaw, G.J. and Boycott, B.R. (1982), 'Infectious laryngotracheitis in peafowl and pheasants', *Avian Dis.*, 26(2), 397–401.

David, B., et al. (2015), 'Air quality in alternative housing systems may have an impact on laying hen welfare. Part II-Ammonia', *Animals*, 5(3), 886–896.

Davis, M. and Morishita, T.Y. (2005), 'Relative ammonia concentrations, dust concentrations, and presence of Salmonella species and *Escherichia coli* inside and outside commercial layer facilities', *Avian Dis.*, 49(1), 30–35.

De Boeck, C., et al. (2015), 'Longitudinal monitoring for respiratory pathogens in broiler chickens reveals co-infection of *Chlamydia psittaci* and *Ornithobacterium rhinotracheale*', *J. Med. Microbiol.*, 64(Pt 5), 565–574.

De Boer, G.F., et al. (1994), 'Interaction between chicken anaemia virus and live Newcastle disease vaccine', *Avian Pathol.*, 23(2), 263–275.

de Wit, J.J. (2000), 'Detection of infectious bronchitis virus', *Avian Pathol.*, 29(2), 71–93.
de Wit, J.J., et al. (1998), 'Transmission of infectious bronchitis virus within vaccinated and unvaccinated groups of chickens', *Avian Pathol.*, 27(5), 464–471.
de Wit, J.J., et al. (2010), 'Efficacy of infectious bronchitis virus vaccinations in the field: association between the alpha-IBV IgM response, protection and vaccine application parameters', *Avian Pathol.*, 39(2), 123–131.
de Wit, J.J., et al. (2011), 'Infectious bronchitis virus variants: a review of the history, current situation and control measures', *Avian Pathol.*, 40(3), 223–235.
de Wit, J.J., et al. (2021), 'Infectious bronchitis virus types affecting European countries: a review', *Avian Dis.*, 65(4), 643–648.
Dhillon, A.S. (1986), 'Pathology of avian adenovirus serotypes in the presence of Escherichia coli in infectious-bursal-disease-virus-infected specific-pathogen-free chickens', *Avian Dis.*, 30(1), 81–86.
Dho-Moulin, M. and Fairbrother, J.M. (1999), 'Avian pathogenic *Escherichia coli* (APEC)', *Vet. Res.*, 30(2–3), 299–316.
Diao, Z., et al. (2022), 'Metagenomics next-generation sequencing tests take the stage in the diagnosis of lower respiratory tract infections', *J. Adv. Res.*, 38, 201–212.
Domanska-Blicharz, K., et al. (2017), 'Specific detection of GII-1 lineage of infectious bronchitis virus', *Lett. Appl. Microbiol.*, 65(2), 141–146.
Domanska-Blicharz, K., et al. (2020), 'New PA/1220/98-like variant of infectious bronchitis virus in Poland', *Avian Pathol.*, 49(4), 380–388.
Droual, R., et al. (1990), 'Infectious coryza in meat chickens in the San Joaquin Valley of California', *Avian Dis.*, 34(4), 1009–1016.
Dufour-Zavala, L. (2008a), 'Epizootiology of infectious laryngotracheitis and presentation of an industry control program', *Avian Dis.*, 52(1), 1–7.
Dwars, R.M., et al. (2009), 'Progression of lesions in the respiratory tract of broilers after single infection with *Escherichia coli* compared to superinfection with *E. coli* after infection with infectious bronchitis virus', *Vet. Immunol. Immunopathol.*, 127(1–2), 65–76.
Dziva, F. and Stevens, M.P. (2008), 'Colibacillosis in poultry: unravelling the molecular basis of virulence of avian pathogenic *Escherichia coli* in their natural hosts', *Avian Pathol.*, 37(4), 355–366.
Eassa, S., et al. (2017), 'Case Report: Prevalence and significance of aspergellosis in commercial broiler chicken: Pathological study', *Iraqi J. Vet. Sci.*, 31, 113–116.
El-Gazzar, M., et al. (2025), 'Avibacterium paragallinarum, the Causative Agent of Infectious Coryza: A Comprehensive Review', *Avian Dis.*, 68(S1), 362–379.
El-Ghany, W.A.A. (2021), 'An updated comprehensive review on ornithobacteriosis: a worldwide emerging avian respiratory disease', *Open Vet. J.*, 11(4), 555–568.
El-Sukhon, S.N., et al. (2002), 'Studies on the bacterial etiology of airsacculitis of broilers in northern and middle Jordan with special reference to *Escherichia coli*, *Ornithobacterium rhinotracheale*, and *Bordetella avium*', *Avian Dis.*, 46(3), 605–612.
El-Tholoth, M. and Bau, H.H. (2024), 'Molecular detection of respiratory tract viruses in chickens at the point of need by loop-mediated isothermal amplification (LAMP)', *Viruses*, 16(8), 1248.
Escobar-Alfonso, S., et al. (2024), 'Avian metapneumovirus subtype b circulation in poultry and wild birds of Colombia', *Pathogens*, 13(10), 882.

Ewers, C., et al. (2009), 'Intestine and environment of the chicken as reservoirs for extraintestinal pathogenic *Escherichia coli* strains with zoonotic potential', *Appl. Environ. Microbiol.*, 75(1), 184–192.

Falchieri, M., et al. (2013), 'Avian metapneumoviruses expressing Infectious Bronchitis virus genes are stable and induce protection', *Vaccine*, 31(22), 2565–2571.

Fancher, C.A., et al. (2020), 'Avian pathogenic *Escherichia coli* and *Clostridium perfringens*: challenges in no antibiotics ever broiler production and potential solutions', *Microorganisms*, 8(10), 1533.

Feberwee, A., et al. (2022), 'Clinical expression, epidemiology, and monitoring of *Mycoplasma gallisepticum* and *Mycoplasma synoviae*: an update', *Avian Pathol.*, 51(1), 2–18.

Felice, V., et al. (2020), 'Molecular detection and characterization of *Mycoplasma gallisepticum* and *Mycoplasma synoviae* strains in backyard poultry in Italy', *Poult. Sci.*, 99(2), 719–724.

Feng, H., et al. (2024), 'The alleviating effect of Phillygenin on the regulation of respiratory microbiota and its metabolites in IBV-infected broilers by inhibiting the TLR7/MyD88/NF-κB axis', *FASEB J.*, 38(16), e23882.

Ferguson-Noel, N., et al. (2020), 'Mycoplasmosis', In: Swayne, D.E., et al. (Eds), *Diseases of poultry,* 14th ed., Hoboken, NJ, USA: Wiley-Blackwell, 907–965.

Filaire, F., et al. (2022), 'Highly pathogenic avian influenza A(H5N8) clade 2.3.4.4b virus in dust samples from poultry farms, France, 2021', *Emerg. Infect. Dis.*, 28(7), 1446–1450.

Franzo, G., et al. (2014), 'Continued use of IBV 793B vaccine needs reassessment after its withdrawal led to the genotype's disappearance', *Vaccine*, 32(50), 6765–6767.

Franzo, G., et al. (2015), 'Subpopulations in aMPV vaccines are unlikely to be the only cause of reversion to virulence', *Vaccine*, 33(21), 2438–2441.

Franzo, G., et al. (2017), 'First report of avian metapneumovirus subtype B field strain in a romanian broiler flock during an outbreak of respiratory disease', *Avian Dis.*, 61(2), 250–254.

Franzo, G., et al. (2019), 'Evolution of infectious bronchitis virus in the field after homologous vaccination introduction', *Vet. Res.*, 50(1), 92.

Franzo, G., et al. (2020), 'Avian Metapneumovirus subtype B around Europe: a phylodynamic reconstruction', *Vet. Res.*, 51(1), 88.

Franzo, G., et al. (2024), 'Tracing the flight: investigating the introduction of avian metapneumovirus (aMPV) A and B', *Animals*, 14(12), 1786.

Gabal, M.A. and Azzam, A.H. (1998), 'Interaction of aflatoxin in the feed and immunization against selected infectious diseases in poultry. II. Effect on one-day-old layer chicks simultaneously vaccinated against Newcastle disease, infectious bronchitis and infectious bursal disease', *Avian Pathol.*, 27(3), 290–295.

García, M. (2017), 'Current and future vaccines and vaccination strategies against infectious laryngotracheitis (ILT) respiratory disease of poultry', *Vet. Microbiol.*, 206, 157–162.

García, M. and Spatz, S. (2020), 'Infectious laryngotracheitis', In Swayne, D.E., et al. (Eds), *Diseases of poultry*, 14th ed., Hoboken, NJ, USA: Wiley-Blackwell, pp. 189–209.

García, M. and Zavala, G. (2019), 'Commercial vaccines and vaccination strategies against infectious laryngotracheitis: what we have learned and knowledge gaps that remain', *Avian Dis.*, 63(2), 325–334.

García, M., et al. (2005), 'Evaluation and comparison of various PCR methods for detection of Mycoplasma gallisepticum infection in chickens', *Avian Dis.*, 49(1), 125–132.

Ge, J., et al. (2010), 'Generation and evaluation of a newcastle disease virus-based H9 avian influenza live vaccine', *Avian Dis.*, 54(s1), 294–296.

Gelaude, P., et al. (2014), 'Biocheck.UGent: a quantitative tool to measure biosecurity at broiler farms and the relationship with technical performances and antimicrobial use', *Poult. Sci.*, 93(11), 2740–2751.

Gholami-Ahangaran, M., et al. (2022), 'Thymol and carvacrol supplementation in poultry health and performance', *Vet. Med. Sci.*, 8(1), 267–288.

Giambrone, J.J., et al. (2008), 'Management practices to reduce infectious laryngotracheitis virus in poultry litter', *J. Appl. Poul. Res.*, 17(1), 64–68.

Ginns, C.A., et al. (1998), 'Development and application of an aerosol challenge method for reproduction of avian colibacillosis', *Avian Pathol.*, 27(5), 505–511.

Goraichuk, I.V., et al. (2024), 'Introduction of Avian metapneumovirus subtype A to the United States: molecular insights and implications', *Front. Microbiol.*, 15, 1428248.

Govindarajan, D., et al. (2006), 'Recovery of avian metapneumovirus subgroup C from cDNA: cross-recognition of avian and human metapneumovirus support proteins', *J. Virol.*, 80(12), 5790–5797.

Gowthaman, V., et al. (2020), 'Infectious laryngotracheitis: etiology, epidemiology, pathobiology, and advances in diagnosis and control – a comprehensive review', *Vet Q*, 40(1), 140–161.

Graziosi, G., et al. (2022a), 'Disentangling the role of wild birds in avian metapneumovirus (aMPV) epidemiology: a systematic review and meta-analysis', *Transbound. Emerg. Dis.*, 69(6), 3285–3299.

Graziosi, G., et al. (2022b), 'First detection of avian metapneumovirus subtype C Eurasian lineage in a Eurasian wigeon (*Mareca penelope*) wintering in Northeastern Italy: an additional hint on the role of migrating birds in the viral epidemiology', *Avian Pathol.*, 51(3), 283–290.

Graziosi, G., et al. (2024a), 'Potential biosecurity breaches in poultry farms: presence of free-ranging mammals near laying-hen houses assessed through a camera-trap study', *Vet. Anim. Sci.*, 26, 100393.

Graziosi, G., et al. (2024b), 'Characterizing the domestic-wild bird interface through camera traps in an area at risk for avian influenza introduction in Northern Italy', *Poult Sci.*, 103(8), 103892.

Graziosi, G., et al. (2024c), 'Genetic diversity of avian influenza viruses detected in waterbirds in Northeast Italy Using Two Different Sampling Strategies', *Animals*, 14(7), 1018.

Gross, W.B. (1961), 'The development of "Air Sac Disease"', *Avian Dis.*, 5(4), 431–439.

Gross, W.B. (1990), 'Factors affecting the development of respiratory disease complex in chickens', *Avian Dis.*, 34(3), 607–610.

Guabiraba, R. and Schouler, C. (2015), 'Avian colibacillosis: still many black holes', *FEMS Microbiol. Lett.*, 362(15), fnv118.

Guérin, J.L., et al. (2024), 'Vaccination technology in poultry: principles of vaccine administration', *Avian Dis.*, 67(4), 489–494.

Guy, J.S., et al. (1991), 'Increased virulence of modified-live infectious laryngotracheitis vaccine virus following bird-to-bird passage', *Avian Dis.*, 35(2), 348–355.

Haesendonck, R., et al. (2014), 'High seroprevalence of respiratory pathogens in hobby poultry', *Avian Dis.*, 58(4), 623–627.

Hafez, H.M. and Chin, R.P. (2020), '*Ornithobacterium rhinotracheale* infection', in Swayne, D.E., et al. (Eds), *Disease of poultry*, Hoboken, NJ, USA: Wiley-Blackwell, pp. 853–860.

Hamet, N., et al. (1991), 'Contribution to the prophylaxis of chicks aspergillosis: study of the contamination of a hatchery by *Aspergillus fumigatus*', *Zentralbl Veterinarmed B.*, 38(7), 529–537.

He, P., et al. (2022), 'Research progress in the early warning of chicken diseases by monitoring clinical symptoms', *Appl. Sci.*, 12(11), 5601.

Hein, R., et al. (2021), 'Review of poultry recombinant vector vaccines', *Avian Dis.*, 65(3), 438–452.

Herath Mudiyanselage, H., et al. (2025), 'Delmarva (DMV1639) infectious bronchitis virus infection alters the microbiome of gastrointestinal and respiratory tracts of broiler chickens', *Virology*, 604, 110428.

Hietala, S.K., et al. (2005), 'Environmental air sampling to detect exotic Newcastle disease virus in two California commercial poultry flocks', *J. Vet. Diagn. Invest.*, 17(2), 198–200.

Himsworth, C.G., et al. (2016), 'Avian pathogenicity genes and antibiotic resistance in escherichia coli isolates from Wild Norway Rats (*Rattus norvegicus*) in British Columbia, Canada', *J. Wildl. Dis.*, 52(2), 418–421.

Hoerr, F.J. (2010), 'Clinical aspects of immunosuppression in poultry', *Avian Dis.*, 54(1), 2–15.

Hopkins, S.R. and Yoder, H.W., Jr. (1982), 'Influence of infectious bronchitis strains and vaccines on the incidence of *Mycoplasma synoviae* airsacculitis', *Avian Dis.*, 26(4), 741–752.

Hopkins, S.R. and Yoder, H.W., Jr. (1984), 'Increased incidence of airsacculitis in broilers infected with mycoplasma synoviae and chicken-passaged infectious bronchitis vaccine virus', *Avian Dis.*, 28(2), 386–396.

Hu, J., et al. (2019), 'Use of a modified bacterial ghost lysis system for the construction of an inactivated avian pathogenic *Escherichia coli* vaccine candidate', *Vet. Microbiol.*, 229, 48–58.

Hubbard, L.E., et al. (2020), 'Poultry litter as potential source of pathogens and other contaminants in groundwater and surface water proximal to large-scale confined poultry feeding operations', *Sci. Total Environ.*, 735, 139459.

Hughes, C.S., et al. (1991), 'Latency and reactivation of infectious laryngotracheitis vaccine virus', *Arch Virol*, 121(1–4), 213–218.

Jackwood, M.W. (2012), 'Review of infectious bronchitis virus around the world', *Avian Dis.*, 56(4), 634–641.

Jackwood, M.W. and de Wit, S. (2020), 'Infectious bronchitis', In Swayne, D.E., et al. (Eds), *Diseases of poultry*, 14th ed., Hoboken, NJ, USA: Wiley-Blackwell, pp. 167–188.

Jackwood, M.W., et al. (2012), 'Molecular evolution and emergence of avian gammacoronaviruses', *Infect. Genet. Evol.*, 12(6), 1305–1311.

Johnson, Y.J., et al. (2004), 'Application of commercial and backyard poultry geographic information system databases for the identification of risk factors for clinical infectious laryngotracheitis in a cluster of cases on the Delmarva Peninsula', *Int. J. Poul. Sci.*, 3, 201–205.

Jones, R.C. (2010), 'Viral respiratory diseases (ILT, aMPV infections, IB): are they ever under control?', *Br. Poult. Sci.*, 51(1), 1–11.

Jones, R.C. and Rautenschlein, S. (2013), 'Avian metapneumovirus', In Swayne, D.E., et al. (Eds), *Diseases of poultry*, 13th ed., Ames, IA, USA: Wiley-Blackwell, pp. 112–138.

Jordan, B. (2017), 'Vaccination against infectious bronchitis virus: a continuous challenge', *Vet. Microbiol.*, 206, 137–143.

Jordan, F. (1966), 'A review of the literature on infectious laryngotracheitis (ILT)', *Avian Dis.*, 10(1), 1–26.

Kanabagatte Basavarajappa, M., et al. (2014), 'A recombinant Newcastle disease virus (NDV) expressing infectious laryngotracheitis virus (ILTV) surface glycoprotein D protects against highly virulent ILTV and NDV challenges in chickens', *Vaccine*, 32(28), 3555–3563.

Karami, M., et al. (2024), 'In vitro evaluation of two novel Escherichia bacteriophages against multiple drug resistant avian pathogenic *Escherichia coli*', *BMC Infect. Dis.*, 24(1), 497.

Karasawa, Y., et al. (1988), 'Ammonia production from amino acids and urea in the caecal contents of the chicken', *Comp. Biochem. Physiol. B*, 90(1), 205–207.

Kariithi, H.M., et al. (2022), 'Detection and genome sequence analysis of avian metapneumovirus subtype A viruses circulating in commercial chicken flocks in Mexico', *Vet. Sci.*, 9(10), 579.

Kathayat, D., et al. (2021), 'Avian pathogenic *Escherichia coli* (APEC): an overview of virulence and pathogenesis factors, zoonotic potential, and control strategies', *Pathogens*, 10(4), 467.

Kempf, I., et al. (1988), 'Experimental infection with *Mycoplasma gallisepticum*: influence of ammonia as an exacerbating factor', *Avian Pathol.*, 17(3), 601–615.

King, D.D., et al. (1973), 'Field studies with *Mycoplasma synoviae*', *Avian Dis.*, 17(4), 722–726.

Kleven, S.H. (1998), 'Mycoplasmas in the etiology of multifactorial respiratory disease', *Poult. Sci.*, 77(8), 1146–1149.

Kleven, S.H. (2008), 'Control of avian mycoplasma infections in commercial poultry', *Avian Dis.*, 52(3), 367–374.

Kleven, S.H. and Ferguson-Noel, N. (2008), '*Mycoplasma synoviae* infection', In Saif, Y.M., et al. (Eds), *Diseases of poultry*, 12th ed. Ames, IA, USA: Wiley-Blackwell, pp. 845–856.

Kleven, S.H., et al. (1972), 'Decrease of antibody response to Mycoplasma synoviae in chickens infected with Marek's disease herpesvirus', *Am. J. Vet. Res.*, 33(10), 2037–2042.

Kleven, S.H., et al. (1978), 'Airsacculitis induced in broilers with a combination of *Mycoplasma gallinarum* and respiratory viruses', *Avian Dis.*, 22(4), 707–716.

Kleyn, F.J. and Ciacciariello, M. (2021), 'Future demands of the poultry industry: will we meet our commitments sustainably in developed and developing economies?', *World's Poult. Sci. J.*, 77(2), 267–278.

Kling, H.F. and Quarles, C.L. (1974), 'Effect of atmospheric ammonia and the stress of infectious bronchitis vaccination on leghorn males', *Poult. Sci.*, 53(3), 1161–1167.

Kovačević, Z., et al. (2024), 'Natural solutions to antimicrobial resistance: the role of essential oils in poultry meat preservation with focus on gram-negative bacteria', *Foods*, 13(23).

Kyakuwaire, M., et al. (2019), 'How safe is chicken litter for land application as an organic fertilizer? A review', *Int. J. Environ. Res. Public Health*, 16(19), 3521.

Kye, S.-J., et al. (2021), 'Pathogenicity of H9N2 low pathogenic avian influenza viruses of different lineages isolated from live bird markets tested in three animal models: SPF chickens, Korean native chickens, and ducks', *Poult. Sci.*, 100(9), 101318.

Laconi, A., et al. (2016), 'A comparison of AMPV subtypes A and B full genomes, gene transcripts and proteins led to reverse-genetics systems rescuing both subtypes', *J. Gen. Virol.*, 97(6), 1324–1332.

Laconi, A., et al. (2019), 'Molecular characterization of whole genome sequence of infectious bronchitis virus 624I genotype confirms the close relationship with Q1 genotype', *Transbound. Emerg. Dis.*, 66(1), 207–216.

Lam, K.M. (2002), 'The macrophage inflammatory protein-1beta in the supernatants of Mycoplasma gallisepticum-infected chicken leukocytes attracts the migration of chicken heterophils and lymphocytes', *Dev. Comp. Immunol.*, 26(1), 85–93.

Lamb, R. and Krug, R. (2001), 'Orthomyxoviridae: the viruses and their replication', In Knipe, D.M. and Howley, P.M. (Eds), *Fields virology*. Philadelphia, PA, USA: Lippincott Williams & Wilkins, pp. 1487–1531.

Landman, W.J. (2014), 'Is Mycoplasma synoviae outrunning *Mycoplasma gallisepticum*? A viewpoint from the Netherlands', *Avian Pathol.*, 43(1), 2–8.

Landman, W.J.M. and van Eck, J.H.H. (2017), 'The efficacy of inactivated *Escherichia coli* autogenous vaccines against the *E. coli* peritonitis syndrome in layers', *Avian Pathol.*, 46(6), 658–665.

Lee, C.W. and Jackwood, M.W. (2000), 'Evidence of genetic diversity generated by recombination among avian coronavirus IBV', *Arch. Virol.*, 145(10), 2135–2148.

Lee, S.W., et al. (2011a), 'First complete genome sequence of infectious laryngotracheitis virus', *BMC Genomics*, 12, 197.

Lee, D.H., et al. (2011b), 'H9N2 avian influenza virus-like particle vaccine provides protective immunity and a strategy for the differentiation of infected from vaccinated animals', *Vaccine*, 29(23), 4003–4007.

Lee, S.W., et al. (2012), 'Attenuated vaccines can recombine to form virulent field viruses', *Science*, 337(6091), 188.

Lee, S.W., et al. (2013), 'Phylogenetic and molecular epidemiological studies reveal evidence of multiple past recombination events between infectious laryngotracheitis viruses', *PLoS One*, 8(2), e55121.

Legnardi, M., et al. (2019), 'Vaccine or field strains: the jigsaw pattern of infectious bronchitis virus molecular epidemiology in Poland', *Poult. Sci.*, 98(12), 6388–6392.

Legnardi, M., et al. (2020), 'Infectious bronchitis virus evolution, diagnosis and control', *Vet. Sci.*, 7(2), 79.

Legnardi, M., et al. (2021), 'Infectious bronchitis hatchery vaccination: comparison between traditional spray administration and a newly developed gel delivery system in field conditions', *Vet. Sci.*, 8(8), 145.

Legnardi, M., et al. (2022), 'Effect of changes of vaccination strategies on IBV epidemiology, diagnosis and control: an Italian retrospective study', *Vet. Ital.*, 58(1), 41–45.

Lemaitre, E., et al. (2018), 'Single reaction, real time RT-PCR detection of all known avian and human metapneumoviruses', *J. Virol. Methods*, 251, 61–68.

Ley, D.H. and Yoder Jr, H. (2008), '*Mycoplasma gallisepticum* infection', In Saif, Y.M., et al. (Eds), *Diseases of poultry*. Ames, IA, USA: Wiley-Blackwell, pp. 807–834.

Lin, M.Y. and Kleven, S.H. (1982), 'Egg transmission of two strains of *Mycoplasma gallisepticum* in chickens', *Avian Dis.*, 26(3), 487–495.

Linares, J.A., et al. (1994), 'An outbreak of infectious laryngotracheitis in California broilers', *Avian Dis.*, 38(1), 188–192.

Listorti, V., et al. (2014), 'Rapid detection of subtype B avian metapneumoviruses using RT-PCR restriction endonuclease digestion indicates field circulation of vaccine-derived viruses in older turkeys', *Avian Pathol.*, 43(1), 51–56.

Listorti, V., et al. (2017), 'Identification of IBV QX vaccine markers : Should vaccine acceptance by authorities require similar identifications for all live IBV vaccines?', *Vaccine*, 35(42), 5531–5534.

Liu, Y., et al. (2024), 'Risk factors for avian influenza in Danish poultry and wild birds during the epidemic from June 2020 to May 2021', *Front. Vet. Sci.*, 11, 1358995.

Liu, Q.X., et al. (2020), 'Ammonia induce lung tissue injury in broilers by activating NLRP3 inflammasome via Escherichia/Shigella', *Poult. Sci.*, 99(7), 3402–3410.

Lupini, C., et al. (2022), 'Longitudinal Survey on aMPV Circulation in French Broiler Flocks following Different Vaccination Strategies', *Animals*, 13(1), 57.

Lupini, C., et al. (2025), 'Avian metapneumovirus subtype B in a Northern shoveler (Spatula clypeata) wintering in Italy: implications for the domestic-wild bird interface?', *Avian Pathol.*, 54(3), 351–358.

Luqman, M., et al. (2024), 'Geographical expansion of avian metapneumovirus subtype B: first detection and molecular characterization of avian metapneumovirus subtype B in US poultry', *Viruses*, 16(4), 508.

Ma, T., et al. (2019), 'Novel genotype of infectious bronchitis virus isolated in China', *Vet. Microbiol.*, 230, 178–186.

Mahdavian, A., et al. (2021), 'Acoustic features of vocalization signal in poultry health monitoring', *Appl. Acoust.*, 175, 107756.

Majó, N., et al. (1997), 'Turkey rhinotracheitis virus and Escherichia coli experimental infection in chickens: histopathological, immunocytochemical and microbiological study', *Vet. Microbiol.*, 57(1), 29–40.

Mardani, K., et al. (2010), 'Naturally occurring recombination between distant strains of infectious bronchitis virus', *Arch. Virol.*, 155(10), 1581–1586.

Marois, C., et al. (2000), 'Detection of *Mycoplasma synoviae* in poultry environment samples by culture and polymerase chain reaction', *Vet. Microbiol.*, 73(4), 311–318.

Marois, C., et al. (2005), 'Experimental evidence of indirect transmission of *Mycoplasma synoviae*', *Vet. Res.*, 36(5–6), 759–769.

Matsumoto, M. and Huang, H.J. (2000), 'Induction of short-term, nonspecific immunity against *Escherichia coli* infection in chickens is suppressed by cold stress or corticosterone treatment', *Avian Pathol.*, 29(3), 227–232.

Matthijs, M.G., et al. (2009), 'Course of infection and immune responses in the respiratory tract of IBV infected broilers after superinfection with E. coli', *Vet. Immunol. Immunopathol.*, 127(1–2), 77–84.

McBride, M.D., et al. (1991), 'Health survey of backyard poultry and other avian species located within one mile of commercial California meat-turkey flocks', *Avian Dis.*, 35(2), 403–407.

McKinley, E.T., et al. (2008), 'Avian coronavirus infectious bronchitis attenuated live vaccines undergo selection of subpopulations and mutations following vaccination', *Vaccine*, 26(10), 1274–1284.

McKinley, E.T., et al. (2011), 'Attenuated live vaccine usage affects accurate measures of virus diversity and mutation rates in avian coronavirus infectious bronchitis virus', *Virus Res.*, 158(1–2), 225–234.

Meirhaeghe, H.V., et al. (2019), 'Transmission of poultry diseases and biosecurity in poultry production', In Dewulf, J. and van Immerseel, F. (Eds), *Biosecurity in animal production and veterinary medicine: from principles to practice*. Oxfordshire, UK: CABI, pp. 329–356.

Meng, L., et al. (2025), 'Development of an improved reverse genetics system for avian metapneumovirus (aMPV): a novel vaccine vector protects against aMPV and infectious bursal disease virus', *J. Integr. Agric.*, 24(5), 1972–1986.

Mernizi, A., et al. (2025), 'Revisiting avian metapneumovirus subtype B in broiler chickens and Turkeys in Morocco: first molecular characterization', *Poultry*, 4(1), 3.

Mescolini, G., et al. (2021), 'What is new on molecular characteristics of Avian metapneumovirus strains circulating in Europe?', *Transbound. Emerg. Dis.*, 68(3), 1314–1322.

Messa Júnior, A., et al. (2017), 'Serological screening suggests extensive presence of mycoplasma gallisepticum and mycoplasma synoviae in backyard chickens in Southern Mozambique', *J. Vet. Med.*, 2017, 2743187.

Miller, M.M. and Schat, K. (2004), 'Chicken infectious anemia virus: an example of the ultimate host–parasite relationship', *Avian Dis.*, 48(4), 734–745.

Mitra, M., et al. (2004), 'Spontaneous colibacillosis in infectious bursal disease-affected broiler flocks', *Trop. Anim. Health Prod.*, 36, 627–632.

Mo, J., et al. (2020), 'Validation of specific quantitative real-time RT-PCR assay panel for Infectious Bronchitis using synthetic DNA standards and clinical specimens', *J. Virol. Methods.*, 276, 113773.

Mo, J., et al. (2022), 'Quantitative real-time PCR assays for the concurrent diagnosis of infectious laryngotracheitis virus, Newcastle disease virus and avian metapneumovirus in poultry', *J. Vet. Sci.*, 23(2), e21.

Mohammad Karim, M. (2022), 'Molecular characterization of *Ornithobacterium rhinotracheale* strains isolated from chickens in the Northwestern Iran', *Vet. Ital.*, 58(2), 189–197.

Morales-Erasto, V., et al. (2016), 'Coinfection of *Avibacterium paragallinarum* and *Ornithobacterium rhinotracheale* in Chickens from Peru', *Avian Dis.*, 60(1), 75–78.

Moreno, A., et al. (2010), 'Epidemic of infectious laryngotracheitis in Italy: characterization of virus isolates by PCR-restriction fragment length polymorphism and sequence analysis', *Avian Dis.*, 54(4), 1172–1177.

Moreno, A., et al. (2017), 'A novel variant of the infectious bronchitis virus resulting from recombination events in Italy and Spain', *Avian Pathol.*, 46(1), 28–35.

Nagaraja, K.V., et al. (1983), 'Scanning electron microscopic studies of adverse effects of ammonia on tracheal tissues of turkeys', *Am. J. Vet. Res.*, 44(8), 1530–1536.

Nagaraja, K.V., et al. (1984), 'Effect of ammonia on the quantitative clearance of *Escherichia coli* from lungs, air sacs, and livers of turkeys aerosol vaccinated against *Escherichia coli*', *Am. J. Vet. Res*, 45(2), 392–395.

Nakamura, K., et al. (1990), 'Effect of infectious bursal disease virus on infections produced by *Escherichia coli* of high and low virulence in chickens', *Avian Pathol.*, 19(4), 713–721.

Nakamura, K., et al. (1992), 'The effect of mixed live vaccines of Newcastle disease and infectious bronchitis on the chicken respiratory tract', *J. Comp. Pathol.*, 106(4), 341–350.

Nakamura, K., et al. (1994), 'Effect of mixed live vaccine (Newcastle disease and infectious bronchitis) and *Mycoplasma gallisepticum* on the chicken respiratory tract and on Escherichia coli infection', *J. Comp. Pathol.*, 111(1), 33–42.

Nakamura, K., et al. (1996), 'Comparison of the effects of infectious bronchitis and infectious laryngotracheitis on the chicken respiratory tract', *J. Comp. Pathol.*, 114(1), 11–21.

Naqi, S., et al (2003), 'Establishment of persistent avian infectious bronchitis virus infection in antibody-free and antibody-positive chickens', *Avian Dis.*,47(3), 594–601.

Nawaz, S., et al. (2024), 'Avian pathogenic *Escherichia coli* (APEC): current insights and future challenges', *Poult. Sci.*, 103(12), 104359.

Naylor, C.J., et al. (2004), 'Development of a reverse-genetics system for Avian pneumovirus demonstrates that the small hydrophobic (SH) and attachment (G) genes are not essential for virus viability', *J. Gen. Virol.*, 85(Pt 11), 3219–3227.

Nazir, S., et al. (2020), 'Marked differences in virulence of three Australian field isolates of infectious laryngotracheitis virus in meat and layer chickens', *Avian Pathol.*, 49(6), 600–610.

Ndegwa, E.N., et al. (2012), 'The proportion of specific viral subpopulations in attenuated Arkansas Delmarva poultry industry infectious bronchitis vaccines influences vaccination outcome', *Avian Dis.*, 56(4), 642–653.

Nguyen, V.-G., et al. (2021), 'Serological and molecular characterization of avian metapneumovirus in chickens in Northern Vietnam', *Vet. Sci.*, 8(10), 206.

Nhung, N.T., et al. (2017), 'Antimicrobial resistance in bacterial poultry pathogens: a review', *Front. Vet. Sci.*, 4, 126.

Nili, H. and Asasi, K. (2002), 'Natural cases and an experimental study of H9N2 avian influenza in commercial broiler chickens of Iran', *Avian Pathol.*, 31(3), 247–252.

Nolan, L.K., et al. (2020), 'Colibacillosis', In Swayne, D.E., et al. (Eds), *Diseases of poultry*, 14th ed., Hoboken, NJ, USA, pp. 770–830.

Nururrozi, A., et al. (2020), 'Clinical and pathological features of aspergillosis due to Aspergillus fumigatus in broilers', *Vet. World*, 13(12), 2787–2792.

Oladokun, S. and Sharif, S. (2024), 'Exploring the complexities of poultry respiratory microbiota: colonization, composition, and impact on health', *Anim. Microbiome*, 6(1), 25.

Omuro, M., et al. (1971), 'Interaction of *Mycoplasma gallisepticum*, mild strains of Newcastle disease virus and infectious bronchitis virus in chickens', *Natl. Inst. Anim. Health. Q. (Tokyo)*, 11(2), 83–93.

Ou, S.C., et al. (2012), 'Detection of infectious laryngotracheitis virus from darkling beetles and their immature stage (lesser mealworms) by quantitative polymerase chain reaction and virus isolation', *J. App. Poult. Res.*, 21(1), 33–38.

Oyetunde, O.O., et al. (1978), 'Aerosol exposure of ammonia, dust and Escherichia coli in broiler chickens', *Can. Vet. J.*, 19(7), 187–193.

Pan, Q., et al. (2012), 'Co-infection of broilers with *Ornithobacterium rhinotracheale* and H9N2 avian influenza virus', *BMC Vet. Res.*, 8(1), 104.

Pantin-Jackwood, M. and Spackman, E. (2020), 'Multicausal respiratory diseases', In Swayne, D.E., et al. (Eds), *Diseases of poultry*, 14th ed., Ames, IA, USA: Wiley-Blackwell, pp. 1386–1389.

Pattison, M., et al. (1989), 'Observations on swollen head syndrome in broiler and broiler breeder chickens', *Vet. Rec.*, 125(9), 229–231.

Paudel, S., et al. (2021), 'Aerosol is the optimal route of respiratory tract infection to induce pathological lesions of colibacillosis by a lux-tagged avian pathogenic *Escherichia coli* in chickens', *Avian Pathol.*, 50(5), 417–426.

Peighambari, S.M., et al. (2000), 'Experimental *Escherichia coli* respiratory infection in broilers', *Avian Dis.*, 44(4), 759–769.

Phillips, C.J., et al. (2023), 'Immunomodulatory effects of natural feed additives for meat chickens', *Life*, 13(6), 1287.

Picault, J.P., et al. (1987), 'Isolation of a TRTV-like virus from chickens with swollen-head syndrome', *Vet. Rec.*, 121(6), 135.

Portz, C., et al. (2008), 'Natural infection of turkeys by infectious laryngotracheitis virus', *Vet. Microbiol.*, 131(1–2), 57–64.

Promkuntod, N., et al. (2014), 'Mapping of the receptor-binding domain and amino acids critical for attachment in the spike protein of avian coronavirus infectious bronchitis virus', *Virology*, 448, 26–32.

Pusch, E.A. and Suarez, D.L. (2018), 'The multifaceted zoonotic risk of H9N2 avian influenza', *Vet. Sci.*, 5(4), 82.

Raj, G.D. and Jones, R.C. (1997), 'Infectious bronchitis virus: immunopathogenesis of infection in the chicken', *Avian Pathol.*, 26(4), 677–706.

Rashid, S., et al. (2009), 'Multiplex polymerase chain reaction for the detection and differentiation of avian influenza viruses and other poultry respiratory pathogens', *Poult. Sci.*, 88(12), 2526–2531.

Rcheulishvili, N., et al. (2022), 'Promising strategy for developing mRNA-based universal influenza virus vaccine for human population, poultry, and pigs- focus on the bigger picture', *Front. Immunol.*, 13, 1025884.

Retallack, H., et al. (2019), 'Genome sequence of a divergent avian metapneumovirus from a monk parakeet (*Myiopsitta monachus*)', *Microbiol. Resour. Announc.*, 8(16), e00284–19.

Rossi, B., et al. (2020), 'Single components of botanicals and nature-identical compounds as a non-antibiotic strategy to ameliorate health status and improve performance in poultry and pigs', *Nutr. Res. Rev.*, 33(2), 218–234.

Rubbenstroth, D., et al. (2010), 'Effects of cyclosporin A induced T-lymphocyte depletion on the course of avian Metapneumovirus (aMPV) infection in turkeys', *Dev. Comp. Immunol.*, 34(5), 518–529.

Samy, A. and Naguib, M.M. (2018), 'Avian respiratory coinfection and impact on avian influenza pathogenicity in domestic poultry: field and experimental findings', *Vet. Sci.*, 5(1), 23.

Sander, J.E., et al. (1995), 'Formaldehyde vaporization in the hatcher and the effect on tracheal epithelium of the chick', *Avian Dis.*, 39(1), 152–157.

Sandoval, V.E., et al. (1994), 'Complicated infectious coryza outbreaks in Argentina', *Avian Dis.*, 38(3), 672–678.

Sawicka, A., et al. (2020), 'Occurrence of *Mycoplasma gallisepticum* in wild birds: a systematic review and meta-analysis', *PLoS One*, 15(4), e0231545.

Schädler, J., et al. (2019), 'Virus-like particles in a new vaccination approach against infectious laryngotracheitis', *J. Gen. Virol.*, 100(6), 1013–1026.

Schoch, C.L., et al. (2020), 'NCBI taxonomy: a comprehensive update on curation, resources and tools', *Database (Oxford)*, 2020, baaa062.

Sellers, H.S., et al. (2004), 'Mild infectious laryngotracheitis in broilers in the southeast', *Avian Dis.*, 48(2), 430–436.

Senne, D.A., et al. (2004), 'Control of Newcastle disease by vaccination', *Dev. Biol.*, 119, 165–170.

Sepotokele, K.M., et al. (2023), 'Optimization of infectious bronchitis virus-like particle expression in Nicotiana benthamiana as potential poultry vaccines', *PLoS One*, 18(7), e0288970.

Shao, G., et al. (2025), 'Development of a recombinant infectious bronchitis virus vaccine expressing infectious laryngotracheitis virus multiple epitopes', *Poult Sci.*, 104(1), 104578.

Sid, H., et al. (2015), 'Co-infection with multiple respiratory pathogens contributes to increased mortality rates in Algerian poultry flocks', *Avian Dis.*, 59(3), 440–446.

Simmonds, P., et al. (2024), 'Changes to virus taxonomy and the ICTV statutes ratified by the international committee on taxonomy of viruses', *Arch. Virol.*, 169(11), 236.

Sims, L. (2007), 'Risks associated with poultry production systems', *International Poultry Conference: Poultry in the 21st century*, 5–7 November 2007, Bangkok, Thailand, pp. 355–356.

Smith, H.W., et al. (1985), 'The experimental infection of chickens with mixtures of infectious bronchitis virus and *Escherichia coli*', *J. Gen. Virol.*, 66(Pt 4), 777–786.

Spackman, E. (2009), 'The ecology of avian influenza virus in wild birds: what does this mean for poultry?', *Poult Sci.*, 88(4), 847–850.

Spackman, E., et al. (2013), 'Optimal specimen collection and transport methods for the detection of avian influenza virus and Newcastle disease virus', *BMC Vet. Res.*, 9, 35.

Springer, W.T., et al. (1974), 'Infectious bronchitis and mixed infections of *Mycoplasma synoviae* and *Escherichia coli* in gnotobiotic chickens. I. Synergistic role in the airsacculitis syndrome', *Infect. Immun.*, 10(3), 578–589.

Ssematimba, A., et al. (2013), 'Avian influenza transmission risks: analysis of biosecurity measures and contact structure in Dutch poultry farming', *Prev. Vet. Med.*, 109(1), 106–115.

Stępień-Pyśniak, D., et al. (2024), 'Case reports involving coinfection with *Avibacterium paragallinarum* and *Ornithobacterium rhinotracheale* in broiler chickens and *Avibacterium endocarditis* in broiler breeding hens in Poland', *Avian Pathol.*, 53(4), 291–302.

Stipkovits, L., et al. (2012a), 'Effect of low-pathogenicity influenza virus H3N8 infection on *Mycoplasma gallisepticum* infection of chickens', *Avian Pathol.*, 41(1), 51–57.

Stipkovits, L., et al. (2012b), 'Pathologic lesions caused by coinfection of *Mycoplasma gallisepticum* and H3N8 low pathogenic avian influenza virus in chickens', *Vet. Pathol.*, 49(2), 273–283.

Strockbine, N.A., et al. (2015), 'Escherichia, Shigella, and Salmonella', In Jorgensen, J.H., et al. (Eds), *Manual of clinical microbiology*, Washington, DC, USA: ASM Press, pp. 685–713.

Suarez, D.L., et al. (2020), 'Newcastle Disease, Other Avian Paramyxoviruses, and Avian Metapneumovirus Infections', In Swayne, D.E., et al. (Eds), *Diseases of poultry*, 14th ed., Hoboken, NJ, USA: Wiley-Blackwell, pp. 109–166.

Subedi, M., et al. (2018), 'Comparative virulence and genomic diversity of avian pathogenic *Escherichia coli* (APEC) associated with different clinical manifestations in poultry', *PLoS One*, 13(2).

Sultana, S., et al. (2014), 'Pathological investigation of avian Aspergillosis in commercial broiler chicken at Chittagong district', *IJARBS*, 1, 74–85.

Suresh, G., et al. (2018), 'Alternatives to antibiotics in poultry feed: molecular perspectives', *Crit. Rev. Microbiol.*, 44(3), 318–335.

Swayne, D.E., et al. (2020), 'Influenza', In Swayne, D.E., et al. (Eds), *Diseases of poultry*, 14th ed., Hoboken, NJ, USA: Wiley-Blackwell, pp. 109–166.

Szczepanek, S.M. and Silbart, L.K. (2014), 'Host immune responses to mycoplasmas', In Browning, G.F. and Citti, C. (Eds), *Mollicutes: molecular biology and pathogenesis*, Norfolk, UK: Caister Academic Press, 273–288.

Tablante, N.L., et al. (1999), 'Risk factors associated with early respiratory disease complex in broiler chickens', *Avian Dis.*, 43(3), 424–428.

Tamburro, M., et al. (2022), '*Alphitobius diaperinus* Panzer (Insecta, Coleoptera) in a single house of a broiler production facility as a potential source of pathogenic bacteria for broilers and humans', *Lett. Appl. Microbiol.*, 74(6), 883–892.

Tang, N., et al. (2020), 'Generation of A triple insert live avian herpesvirus vectored vaccine using CRISPR/Cas9-based gene editing', *Vaccines*, 8(1), 97.

Tasistro, A.S., et al. (2007), 'Ammonia emissions from broiler litter: response to bedding materials and acidifiers', *Br. Poult. Sci.*, 48(4), 399–405.

Tilli, G., et al. (2022), 'Assessing biosecurity compliance in poultry farms: a survey in a densely populated poultry area in North East Italy', *Animals*, 12(11), 1409.

Timms, L. (1972), 'The effects of infectious bronchitis superimposed on latent *M. gallisepticum* infection in adult chickens', *Vet. Rec.*, 91(8), 185–190.

Timurkaan, N., et al. (2003), 'Pathological and immunohistochemical findings in broilers inoculated with a low virulent strain of infectious laryngotracheitis virus', *J. Vet. Sci.*, 4(2), 175–180.

Toro, H., et al. (1996), 'Local and systemic specific antibody response of different chicken lines after ocular vaccination against infectious bronchitis', *Zentralbl Veterinarmed B.*, 43(8), 449–454.

Tucciarone, C.M., et al. (2017), 'First identification and molecular characterization of avian metapneumovirus subtype B from chickens in Greece', *Avian Dis.*, 61(3), 409–413.

Tucciarone, C.M., et al. (2018a), 'Avian Metapneumovirus circulation in Italian broiler farms', *Poult. Sci.*, 97(2), 503–509.

Tucciarone, C.M., et al. (2018b), 'Evaluation of 793/B-like and Mass-like vaccine strain kinetics in experimental and field conditions by real-time RT-PCR quantification', *Poult. Sci.*, 97(1), 303–312.

Tucciarone, C.M., et al. (2018c), 'Infectious bronchitis virus gel vaccination: evaluation of Mass-like (B-48) and 793/B-like (1/96) vaccine kinetics after combined administration at 1 day of age', *Poult. Sci.*, 97(10), 3501–3509.

Tucciarone, C.M., et al. (2022), 'Molecular survey on A, B, C and new avian metapneumovirus (aMPV) subtypes in wild birds of Northern-Central Italy', *Vet. Sci.*, 9(7), 373.

Tucciarone, C.M., et al. (2024), 'Research note: indirect evidence of avian Metapneumovirus circulation in broilers in Italy', *Poult. Sci.*, 103(11), 104182.

Tzora, A., et al. (2021), 'Identification by MALDI-TOF MS and antibiotic resistance of *Riemerella anatipestifer*, isolated from a clinical case in commercial broiler chickens', *Vet. Sci.*, 8(2), 29.

Umar, S., et al. (2017), 'Low Pathogenic Avian Influenza and Coinfecting Pathogens: a Review of Experimental Infections in Avian Models', *Avian Dis.*, 61(1), 3–15.

Valastro, V., et al. (2016), 'S1 gene-based phylogeny of infectious bronchitis virus: an attempt to harmonize virus classification', *Infect. Genet. Evol.*, 39, 349–364.

Valentine, H. (1964), 'A study of the effect of different ventilation rates on the ammonia concentrations in the atmosphere of broiler houses', *Br. Poult. Sci.*, 5(2), 149–159.

Van de Zande, S., et al. (2000), 'Duration of cross-protection between subtypes A and B avian pneumovirus in turkeys', *Vet. Rec.*, 147(5), 132–134.

Van Goor, A., et al. (2017), 'A recombinant multi-antigen vaccine with broad protection potential against avian pathogenic *Escherichia coli*', *PLoS One*, 12(8), e0183929.

Van Limbergen, T., et al. (2018), 'Scoring biosecurity in European conventional broiler production', *Poult. Sci.*, 97(1), 74–83.

Van Steenwinkel, S., et al. (2011), 'Assessing biosecurity practices, movements and densities of poultry sites across Belgium, resulting in different farm risk-groups for infectious disease introduction and spread', *Prev. Vet. Med.*, 98(4), 259–270.

van Veen, L., et al. (2000), 'Increased condemnation of broilers associated with *Ornithobacterium rhinotracheale*', *Vet. Rec.*, 147(15), 422–423.

van Veen, L., et al. (2005), 'Diagnosis and incidence of *Ornithobacterium rhinotracheale* infections in commercial broiler chickens at slaughter', *Vet. Rec.*, 156(10), 315–317.

Villegas, P. (1998), 'Viral diseases of the respiratory system', *Poult. Sci.*, 77(8), 1143–1145.

Walker, P.J., et al. (2020), 'Changes to virus taxonomy and the Statutes ratified by the International Committee on Taxonomy of Viruses', *Arch. Virol.*, 165(11), 2737–2748.

Wang, H., et al. (2024), 'Current status of poultry recombinant virus vector vaccine development', *Vaccines*, 12(6), 630.

Wang, J., et al. (2021), 'Gut microbiota dysbiosis aggravates *Mycoplasma gallisepticum* colonization in the chicken lung', *Front. Vet. Sci.*, 8, 788811.

Wei, L., et al. (2013), 'Avian metapneumovirus subgroup C infection in chickens, China', *Emerg. Infect. Dis.*, 19(7), 1092–1094.

Wells, H.L., et al. (2023), 'The coronavirus recombination pathway', *Cell. Host. Microbe*, 31(6), 874–889.

Wickramasinghe, I.N., et al. (2014), 'The avian coronavirus spike protein', *Virus Res.*, 194, 37–48.

Williams, R.A., et al. (1992), 'Demonstration of sites of latency of infectious laryngotracheitis virus using the polymerase chain reaction', *J. Gen. Virol.*, 73(Pt 9), 2415–2420.

Williams, S.M. et al. (2016), 'Isolation, identification, and characterization of avian pathogens', 6th ed., Jacksonville, FL, USA: American Association of Avian Pathologists.

World Health Organization (2016), 'Global action plan on antimicrobial resistance', Available at: https://www.who.int/publications/i/item/9789241509763 (accessed 18 May 2025).

World Organisation for Animal Health (2021), 'Chapter 3.3.5. Avian mycoplasmosis (Mycoplasma gallisepticum, M. synoviae)', Available at: https://www.woah.org/fileadmin/Home/fr/Health_standards/tahm/3.03.05_AVIAN_MYCO.pdf (accessed 18 May 2025).

World Organisation for Animal Health (2024), 'Terrestrial Animal Health Code', Available at: https://www.woah.org/en/what-we-do/standards/codes-and-manuals/terrestrial-code-online-access/?id=169&L=1&htmfile=glossaire.htm (accessed 15 May 2025).

Xavier, J., et al. (2011), 'Seroprevalence of Salmonella and Mycoplasma infection in backyard chickens in the state of Entre Rios in Argentina', *Poult Sci.*, 90(4), 746–751.

Xu, L., et al. (2019), 'Genetic and biological characteristics of four novel recombinant avian infectious bronchitis viruses isolated in China', *Virus Res.*, 263, 87–97.

Yang, X., et al. (2016), 'Recombinant infectious bronchitis virus (IBV) H120 vaccine strain expressing the hemagglutinin-neuraminidase (HN) protein of Newcastle disease virus (NDV) protects chickens against IBV and NDV challenge', *Arch. Virol.*, 161(5), 1209–1216.

Yoder, H.W. Jr., et al. (1977), 'Influence of environment on airsacculitis: effects of relative humidity and air temperature on broilers infected with *Mycoplasma synoviae* and infectious bronchitis', *Avian Dis.*, 21(2), 195–208.

Zhang, P., et al. (2008), 'Characterization of H9N2 influenza viruses isolated from vaccinated flocks in an integrated broiler chicken operation in eastern China during a 5 year period (1998–2002)', *J. Gen. Virol.*, 89(12), 3102–3112.

Zhang, P., et al. (2023), 'Development and Immunoprotection of Bacterium-like Particle Vaccine against Infectious Bronchitis in Chickens', *Vaccines*, 11(8), 1292.

Zhang, Y., et al. (2016), 'A reverse genetics approach for the design of methyltransferase-defective live attenuated avian metapneumovirus vaccines', *Methods Mol. Biol.*, 1404, 103–121.

Zhao, W., et al. (2014), 'Newcastle disease virus (NDV) recombinants expressing infectious laryngotracheitis virus (ILTV) glycoproteins gB and gD protect chickens against ILTV and NDV challenges', *J. Virol.*, 88(15), 8397–8406.

Zhao, Y., et al. (2013), 'Detection of infectious laryngotracheitis virus by real-time PCR in naturally and experimentally infected chickens', *PLoS One*, 8(6), e67598.

# Part 7

## Managing the end of life stage in broiler production

# Chapter 23

# Humane end-of-life management of broilers: catching, crating and loading

*Helen Louton, Jenny Wessel and Julia Mönch, Ludwig Maximilian University of Munich, Germany*

1 Introduction
2 Legislation and recommendations for catching and loading
3 Possible methods and perceptions of farmers
4 Manual catching: different methods
5 Mechanical loading systems
6 Loading-related stress reactions
7 Loading-related injuries
8 Factors impacting loading quality or loading-related injuries
9 Loading staff: issues to consider
10 Crating
11 Conclusion
12 Summary
13 Where to look for further information
14 References

## 1 Introduction

To be transported to the slaughterhouse, broilers need to be caught and loaded into crates or containers, which are then placed on lorries. The process of crating, loading and subsequent transport to the slaughterhouse has been associated with stress (Duncan et al., 1986; Wolff et al., 2019) and can cause injuries or death of broilers (Kittelsen et al., 2015; Jacobs et al., 2017; Cockram and Dulal, 2018). Manual catching also represents a considerable physical effort for the catching crew itself and can lead to pain of the crew loading the broilers (Delanglez et al., 2025).

Chicken meat production is usually integrated, meaning that the stages of hatchery, rearing or fattening of the broilers, and slaughtering are performed

http://dx.doi.org/10.19103/AS.2025.0165.29

within one company with several members, whereby each member of the supply chain produces a different product or part of the chain. The National Chicken Council (2023) has stated that this integration can improve welfare of the chicken and at the same time lower prices for production. Consequently, structural change in the chicken meat sector towards a production in an integrated and a rising demand has led to a massive increase in the number of ready-to-slaughter broilers that have to be caught during a single shipment. At the same time, there is time pressure due to a tightly organised processing in the slaughterhouse and the previous steps of production. It is therefore necessary to find a suitable method to manage the loading of large numbers of broilers as economically as possible without impairing broiler welfare.

It is therefore important to develop a loading method that is as animal friendly as possible and, at the same time, economically viable. There are various ways of loading the broilers into containers so that they can then be transported to the slaughterhouse. On the one hand, there is manual catching, in which several chickens are typically lifted from the ground by one or both legs and then carried to the container (Langkabel et al., 2015; Wessel et al., 2022). On the other hand, there is mechanised loading, using special-purposed machines (Knierim and Gocke, 2003; Mönch et al., 2020).

Already in the 1960s, machine loading methods were developed, followed by further development, construction and testing (de Koning et al., 1987; Bayliss and Hinton, 1990; Kettlewell and Mitchell, 1994; Lacy and Czarick, 1998; Knierim and Gocke, 2003; Nijdam et al., 2005; Mönch et al., 2020; Werner et al., 2023). Nevertheless, in some parts of the world, for example Germany, the most common form of broiler loading currently practised is still manual catching by commercial catching crews (Knierim and Gocke, 2003; Louton et al., 2023a). These aspects are addressed in more detail in Section 5 of this chapter.

The aim of mechanisation is to improve animal welfare, animal health, economic efficiency, and occupational health and safety (Knierim and Gocke, 2003). During mechanical loading, the contact between chicken and humans is reduced to a minimum, which improves welfare aspects because the broilers may perceive humans as predators. However, human staff is still necessary during loading, for tasks such as moving the broilers away from the barn walls or loading and distributing the animals in the containers. Staff performing machine loading need to be well trained (Mönch et al., 2020).

This chapter aims to provide an overview of the methods and challenges involved in the humane end-of-life management of broilers, with a focus on catching, crating, and loading practices. It addresses the welfare implications of manual and mechanical loading methods, reviews current legislation and recommendations, explores farmers' perspectives on loading practices, and discusses loading-related stress and injuries in broilers. By presenting research findings and practical insights, the chapter highlights strategies to optimise animal welfare.

## 2 Legislation and recommendations for catching and loading

The minimum rules for the protection of chickens kept for meat production in Europe are defined in Council Directive 2007/43/CE (2007). All member states of the European Union are bound by these rules. In the Directive, a maximum stocking density and details on specific requirements concerning lighting, litter, feeding, and ventilation are noted to ensure that the welfare of broilers is assured (Council Directive 2007/43/CE, 2007). In Article 4 of the Directive, training and guidance for people dealing with chickens are stated as mandatory. Training needs to include welfare aspects, in particular the matters of practical aspects of careful handling of chickens as well as catching, loading, and transporting broilers.

Carrying animals in an upside-down position is not recommended (Council of Europe, 1995). The current legislation on the protection of animals during transport prohibits hitting or kicking animals during loading, as well as dragging and pulling animals by the legs (Council Regulation EC No 1/2005, 2004). The European Commission is currently revising this regulation with a proposal brought forward in 2023 (European Commission, 2023).

In that proposal, the ban on pulling up and dragging animals by the legs is to be lifted for poultry. Instead, an aspect to be added is that chickens are to be caught, lifted, and carried by two legs, with the broiler's breast being supported. No more than three chickens per hand may be caught. Additionally, the European Food Safety Authority (EFSA) stated that chickens should not be carried upside-down during loading but rather in an upright position with support on the chest and with the wings resting against the body. To ensure careful handling of broilers during loading, education and training of the involved staff is recommended (EFSA AHAW Panel, 2023).

There is a court ruling from 2022 from the Netherlands which judged the catching of poultry by one leg. The court identified an infringement of the European legislation on the protection of animals during transport following the report on an inspection of three catching companies by the Dutch inspection authorities the NVWA (District Court of Rotterdam, 2022), where laying hens and broilers were caught and carried by one leg and in an inverted position. The court ruled that the ban on lifting animals by one leg in the European Transport Regulation from 2005 is clear and does not constitute an exception for poultry. Due to the specific regulations of the different continents and legislation often being written in the national language, the authors have no information of legislation considering loading of broilers outside of Europe.

### 2.1 Take-home message

Proper handling of broilers and training of staff are not only ethical and operational imperatives but also legal requirements under EU regulations. Training must cover animal welfare principles, including careful catching and loading techniques. Handling broilers upside down by one leg is discouraged and may soon be legally restricted. Legislation continues to evolve towards better protection of broiler welfare during transport.

## 3 Possible methods and perceptions of farmers

The process of loading broilers is currently performed at a low level of automation in many countries (Louton et al., 2023a). In Germany, for example, only 5% of the broilers are loaded by loading machines, with the rest loaded manually. In Scandinavian countries, such as Finland, almost 100% of the broilers are caught and loaded by machine (Personal communication, 2022). It is likely that these numbers differ between countries because of the different structures of broiler production operations around the world.

Farmers in Germany who decided not to load their broilers manually stated various reasons for their decision. Poultry farmers were asked in an online survey (SoSciSurvey, 21.02.2022 to 09.08.2022) to name the loading methods they use and the advantages and disadvantages of that method. Ten people completed the questionnaire consisting of 15 questions and, despite a low participation rate, provided interesting insights (Louton et al., 2023b).

Of the ten respondents, two used a loading machine to load their broilers and described their experiences with this method in the free comments section of the survey. In particular, animal welfare standards, profitability, hygiene, and the biosecurity of the farm were important points in favour, but also against, the decision to use a machine. According to the farmers who responded, the loading speed of animals per hour was higher with mechanical loading than with manual catching. More workers were used for manual catching than for mechanical loading, which corresponds with other studies (Mönch et al., 2020).

### 3.1 Farmers' reasons for choosing mechanical loading

Reasons for using loading machines included staff health and safety (100%, 2 of 2 respondents), animal welfare (100%), cost (50%), time savings (50%), and personal reasons (50%). Marketing reasons (e.g. less rejection at the slaughterhouse) were not mentioned. It became clear that hygiene and biosecurity on the farm were key factors for their decision. The two farmers who opted for the loading machine said that they did so for reasons of hygiene and less reliance on manual catching crews or because they were dissatisfied with the hygiene standards demonstrated by manual catchers.

These two farmers also answered the question of whether they would recommend loading machines with a clear 'yes'. The advantages mentioned were that the containers would not be placed on the barn floor, the broilers could be loaded quietly, and that thinning could also be done by the company's own staff. The latter aspect eliminates the need for external personnel to be present in the barn, while placing containers on a platform ensures hygiene. The fact that the chickens are loaded in an upright position during machine loading and do not have to be grasped by a leg was mentioned as a further advantage.

### 3.2 Farmers' reasons for choosing manual catching

Farmers who relied on manual catching gave various reasons for not using a loading machine. Lack of contractors (62.5%, 5 of 8 respondents), structural constraints within the integrated production system (37.5%), economic viability (25.0%), better animal control with manual catching (25.0%), lack of information (25.0%), and/or no interest in loading machines (12.5%) played a role. One respondent reared laying hen brother cockerels and emphasised that these chickens are very mobile, and it was therefore questionable whether a loading machine would be suitable for loading these at all. There are currently no scientific findings on the mechanical loading of laying hen brother cockerels. Following the ban of culling day-old chicks, this topic should be emphasised in research.

One aspect that farmers were very concerned about is that it would not be possible to load chickens if the loading machine were to break down, resulting in further logistical problems (possibly no feed, no time for cleaning and disinfection etc.) and additional costs if the broilers needed to remain in the barn for longer than planned. Reports from Finland (Personal communication, 2022), and our experiences from projects in Germany, showed that it clearly makes sense to have several catching machines available, so that a spare machine can be used. The following quote from a farmer is interesting in this regard: 'It is very time consuming to clean the machines after each final loading. A certain understanding of technology is required to be able to rectify faults of the loading machine. The loading staff (note: when loading by machine) has a much better motivation to load the animals carefully.' It is interesting to note that hygiene was not only viewed positively by farmers who use loading machines but also as a potential obstacle to people who load the flocks manually. One person commented that they do not use a machine for reasons of operational safety and hygiene.

### 3.3 Further thoughts on the choice of method

The survey also investigated under what circumstances farmers who had not previously used a machine would decide to do so. They answered that they would do so if manual catching crews were no longer available, if the financial outlay was proportionate to the costs, if there was enough experience that the use of a loading machine would guarantee animal welfare–friendly loading, if the technology proved reliable, and if the costs were justified by the expected benefits. Furthermore, the respondents mentioned the need for loading machines that can be used in a smaller or double-deck barn, for example.

It can be concluded that mechanical loading will become more important as labour costs rise and the availability of personnel decreases. Mechanical loading requires fewer personnel than manual catching. Currently, costs of mechanised loading do not exceed those of manual catching. Past projects have shown that machine loading can be performed in an animal welfare–friendly manner, but that the staff operating the machine must be properly trained (Wolff et al., 2019; Mönch et al., 2020; Werner et al., 2023; Unterholzner et al., under review). Training of the staff is also essential in manual catching and has a positive effect on the reduction of injuries to chicken or staff (Pilecco et al., 2013).

### 3.4 Take-home message

Farmers choose between manual and mechanical loading based on factors like animal welfare, hygiene, operational independence, and financial viability. Mechanical loading is often associated with better hygiene and staff safety but is limited by cost and availability. Manual catching remains widespread due to structural barriers and lack of information. Overall, farmer perspectives highlight the need for practical and economical solutions that improve broiler welfare.

## 4 Manual catching: different methods

Care must be taken to ensure that no injuries occur to the animals during catching, crating, and loading. This requirement implies that the animals are carefully grasped by both legs to be carried, that the transport containers are placed as close as possible to the animals, and that neither head nor wings hit hard objects (Wessel et al., 2022).

### 4.1 Carrying the broilers: inverted, upright or in bags

In Germany, manual catching is still the most common method of loading broilers (Louton et al., 2023a). The broilers can be caught either in an inverted position by being picked up by one leg (1LCM, LCM = leg catching method, Fig. 1) or two legs (2LCM, Fig 2) or in an upright position. If caught by the 1LCM, a maximum of three to four broilers per hand can be carried upside down, in contrast to the 2LCM, where a maximum of two broilers per hand can

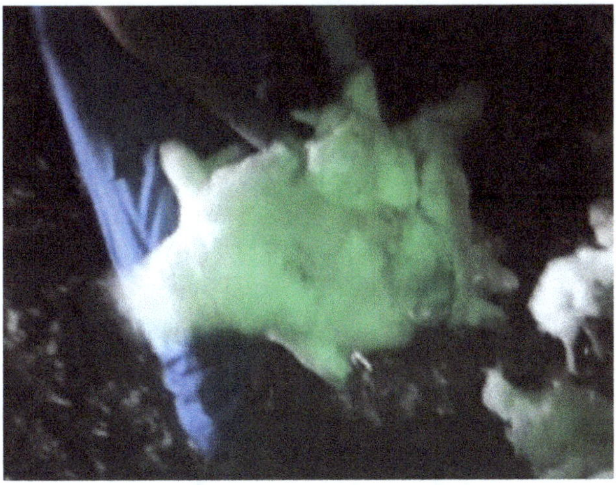

**Figure 1** Leg catching method: Broilers are grasped by one leg, and several broilers (max. three per hand) are placed in an inverted position in the container. Photo: LMU Munich.

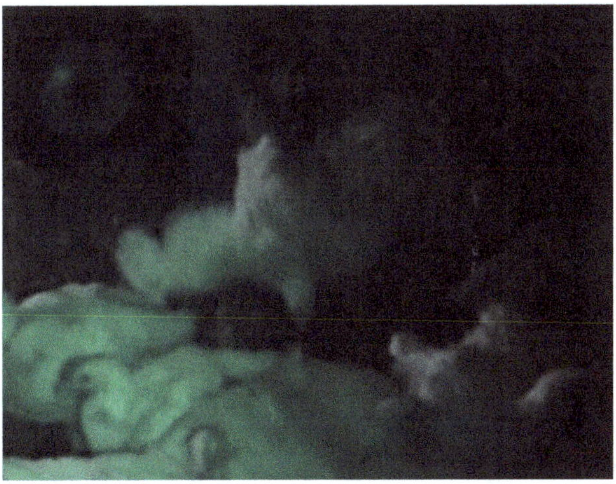

**Figure 2** Leg catching method: Broilers are grasped by both legs, and several broilers (max. two per hand) are placed in an inverted position in the container. Photo: LMU Munich.

be caught (Langkabel et al., 2015; Wessel et al., 2022). If caught in an upright position, one or two broilers can be carried at once (de Lima et al., 2019; Wolff et al., 2019).

Another manual method of catching broilers is described by Queiroz et al. (2015) as the 'bag' method. In this method, seven broilers per bag are caught by the back ('Japanese method') in groups of two and placed in bags (50 cm long,

30 cm wide, maximum top opening 60 cm) (Queiroz et al., 2015). The bags with the broilers are then carried outside, weighed, and placed in transport crates. The authors observed wing flapping and hitting the crates after the broilers had been placed in the crates, which could lead to injuries (Queiroz et al., 2015). Nevertheless, they concluded that the 'bag' method reduces excessive handling and carrying by the catchers, and the occurrence of injuries (immobility due to a physical limitation and low light intensity in the bags) and thus contributes to an improvement in animal welfare (Queiroz et al., 2015). To avoid injuries such as scratches, attention should be paid to the number of broilers in the bags (Dutra et al., 2021). Additionally, the length of stay in the bags should not be too long to prevent possible heat stress and suffocation of the broilers (Queiroz et al., 2015; Dutra et al., 2021). However, this 'bag' method seems to be rarely used (Dutra et al., 2021).

### 4.2 Studies comparing inverted and upright carrying

Two methods of manual catching of broilers were compared by Kittelsen et al. (2018): a method in which the broilers were caught by two legs with a total of four to six broilers carried at once (the 2LCM) and a method in which the broilers were caught and carried in an upright position with one to two broilers held simultaneously ('upright' method). The authors found that the 'upright' method resulted in better compliance with the stocking density in the drawers, a shorter loading time (2LCM 252 s per transport module, 'upright' method: 231 s per transport module), and fewer fractures in the wing area (2LCM: seven animals of 2010 with a fracture on the wing, 0.35%, 'upright' method: one animal with a fracture on the wing of 1941, 0.05%).

Wolff et al. (2019) and de Lima et al. (2019) described a similar catching method in which the broilers were also caught in an upright position while containing the wings and carefully placing the broilers in crates. Less wing flapping (24.0%) occurred with this method than with the two-legged inverted method (74.0%) in the study by Wolff et al. (2019). Similarly, de Lima et al. (2019) observed less agitation and less striking against the container if the broilers were carried upright. The authors attributed this difference to the fact that the catcher holding the broiler has more control of the movement of the broilers.

Kannan and Mench (1997) found a lower stress response in broilers after upright handling than in an inverted position, expressed in higher plasma corticosterone of broilers handled in an inverted position. Langkabel et al. (2015) found no significant difference in the occurrence of injuries if broilers were caught inverted and grasped by one leg (1LCM) or inverted by two legs (2LCM). However, the loading time was significantly longer for the 2LCM (2.05–4.17 minutes per container) than for the 1LCM (1.04–3.10 minutes per container).

Wessel et al. (2022) also found that the 1LCM and 2LCM differed significantly in loading duration (1LCM = 8,048 broilers per hour, 2LCM = 3,408 broilers per hour). The catchers in the study by Wessel et al. (2022) needed twice as long for the 2LCM than for loading broilers with the 1LCM, whereby the stocking density rather complied with the planned stocking density in the 1LCM. However, the broilers loaded by the 2LCM showed less wing flapping than those loaded by the 1LCM (1LCM = 80.6% of the broilers, 2 LCM = 88.0% of the broilers).

Kittelsen et al. (2018) found the 'upright' method (231 s [3.9 min] per container) to be a faster catching method than the 2LCM (252 s [4.2 min] per container). However, the loading of the broilers by Kittelsen et al. (2018) took longer than that reported by Langkabel et al. (2015).

The 'upright' method is physically more demanding for the catching crew than the 2LCM, because fewer animals can be caught at the same time and the catchers therefore have to bend down more often (Kittelsen et al., 2018). Delanglez et al. (2025) made similar observations, reporting that the upright loading was physically strenuous and associated with more pain for the catching crew. Nevertheless, they concluded that, owing to a lower wing flapping frequency, the 'upright' method reveals benefits of animal welfare aspects. Yet, more person-hours per 1,000 broilers were necessary for the upright loading. Accordingly, the loading time of an entire barn could be significantly extended by the 'upright' method (Kittelsen et al., 2018).

### *4.3 Considerations about broiler welfare, worker safety and economic efficiency*

All three methods ('bag' method, 'upright' method [231 s or 3.9 min/container], and 2LCM [252 s or 4.2 min/container] Kittelsen et al., 2018 [2.05–4.17 min/container] Langkabel et al., 2015) appear to be significantly slower than the 1LCM [1.04–3.10 min/container] (Langkabel et al., 2015). A correspondingly longer loading time also increases the stress load on the animals and the duration of water and feed withdrawal, influencing several metabolic processes, and leading to a loss of body weight in the broilers (Delezie et al., 2007). It is also a challenge for the catchers to ensure concentrated and careful handling of the animals during longer loading times (Mönch et al., 2020). To ensure careful loading, the number of team members in the catching crew would have to be increased accordingly. This in turn would lead to increased labour costs and have an economic impact (Kittelsen et al., 2018; Delanglez et al., 2025).

The catching of broilers should be performed in darkened barns to ensure calmer broiler behaviour (Nicol and Scott, 1990). If catchers stay longer in the barn and there is more movement and excessive handling of animals, more stress-related behaviour could be the result (de Lima et al., 2019). Owing to the thousands of broilers, the catcher team, and a forklift truck, there is constant movement in the barn (Quandt et al., 2013). The forklift truck moves through the dark barn, often with its lights switched off and often at high speeds, which means that the catchers must be constantly vigilant to avoid being injured (Quandt et al., 2013). If the number of catchers or the size of the catching team is increased, the number of forklift trucks has to be increased as well, and the aforementioned aspects need to be taken into consideration.

### 4.4 Take-home message

Different manual catching techniques have distinct impacts on broiler welfare. Carrying broilers upright reduces stress and injuries but is more physically demanding for the staff and slower. Inverted catching methods are faster but linked to more wing flapping and injuries in the broilers. Balancing animal welfare and operational efficiency is a central challenge in manual catching.

## 5 Mechanical loading systems

Various systems have been developed for mechanical loading of broilers and are commonly used worldwide (de Koning et al., 1987; Bayliss and Hinton, 1990; Kettlewell and Mitchell, 1994; Lacy and Czarick, 1998; Knierim and Gocke, 2003; Delezie et al., 2006; Nijdam et al., 2005). The way in which the broilers are picked up from the ground can vary greatly between the different technologies, whereas the passage to the containers follows similar principles. For the passage of the broilers within the loading machines, several machines use conveyor belts or in early years mats placed or pushed underneath the chickens (de Koning et al., 1987; Bayliss and Hinton, 1990; Kettlewell and Mitchell, 1994; Lacy and Czarick, 1998; Knierim and Gocke, 2003; Nijdam et al., 2005).

### 5.1 Currently used loading machines

Currently, two types of machines are used which differ in how they pick up broilers from the floor of the barn. The 'Apollo Generation 2' (CMC Calabria, Italy), which has been examined in several studies (Wolff et al., 2019; Mönch et al., 2020; Werner et al., 2023; Unterholzner et al., under review), has six parallel collection belts to pick broilers up from the ground at the front of the machine. Figures 3 and 4 show photos of the Apollo Generation 2.

The other commonly used loading machine is the 'Chicken Cat' harvester (JTT Conveying A/S, Bredsten, Denmark). The Chicken Cat is equipped with a front part with three units consisting of rotating rubber-fingers to pick up broilers from the ground (Knierim and Gocke, 2003; Nijdam et al., 2005). Both

**Figure 3** Front part of the Apollo Generation 2 loading machine (CMC Calabria, Italy). The broilers are picked up by six conveyor belts. Photo: LMU Munich.

**Figure 4** Back of the Apollo Generation 2 loading machine (CMC Calabria, Italy). The containers are placed on a platform in which a scale is included. The broilers are loaded by weight and not by number. This complies with the Council Regulation (EC) No 1/2005 (2004). Photo: LMU Munich.

machines have the advantage that the broilers have no direct contact with humans and are loaded into the containers in an upright body position.

## 5.2 Advantages and disadvantages of mechanical loading

Loading broilers by machines can improve animal welfare and animal health, as measured by reduction of injuries, stress and animal losses. At the same time, it reduces labour costs, making mechanical loading more economic than manual catching methods (Knierim and Gocke, 2003; Wolff et al., 2019; Delanglez et al., 2025). The risk for wing flapping, for example, was twice as high during manual catching as during mechanical loading with the Apollo Generation 2 (Wolff et al., 2019).

Compared with manual catching, mechanical loading is less influenced by environmental and flock-specific factors such as average weight or catching duration (Wolff et al., 2019). Bruising of wings and legs can be reduced by using loading machines (Farsaie et al., 1983; Lacy and Czarick, 1998; Knierim and Gocke, 2003; Delezie et al., 2006). However, Mönch et al. (2020) showed that the number of broilers with a haematoma on the wing was smaller after manual catching than after mechanical loading, and the number of broilers with severe wing injuries did not differ between loading methods (Table 1).

Mönch et al. (2020) also found the number of broilers dead on arrival at the slaughterhouse to be larger in mechanically than in manually loaded flocks, although the overall percentage was lower than in comparable studies (Table 1). In that same study, the average stocking density in the crates of mechanically loaded containers was lower than targeted. The most likely explanation was that the loading machine can adapt to short-term changes in weight and adjust the stocking rate during the loading process.

Interestingly, Mönch et al. (2020) found a longer total loading duration in manually loaded flocks associated with an increase in the occurrence of wing haematomas. At the same time, the involvement of more working staff per 10,000 broilers during manual loading was associated with a lower occurrence of haematomas. In contrast, total loading duration in mechanically loaded flocks had no notable influence on the occurrence of injuries. The authors assumed that physical conditions and and fatigue of the involved staff might play a larger role in manual catching than in mechanical loading (Mönch et al., 2020).

## 5.3 Options for optimising mechanical loading

Werner et al. (2023) have identified possibilities to optimise the use of loading machines. They found that an increase in rotation and conveyor belt speeds was associated with an elevated risk of behaviours or impacts, such as wing flapping, bumping into an animal, and bumping against the machine or container. The use of a SmartStack container type can reduce the risk of

**Table 1** Summary of results of animal-based measures of injuries in broilers. Highlighted in grey are studies where data was collected at the slaughterhouse, if not highlighted in grey, data was collected on farm after catching and crating (n/a = not assessed)

| Authors | Type of loading | Haematoma wing % | Fracture wing % | Haematoma leg % | Fracture leg % | Dead on arrival % |
|---|---|---|---|---|---|---|
| **Mechanical loading** | | | | | | |
| Knierim and Gocke (2003) | Chicken Cat, JTT Conveying A/S, Denmark | 0.93 | 0.66 (broken)<br>0.47 (dislocation) | 0.62 | 0.03 (broken)<br>0.01 (dislocation) | 0.54 |
| Nijdam et al. (2005) | Chicken Cat, JTT Conveying A/S, Denmark | 7.8 | n/a | 0.54–0.69 | n/a | 0.32–0.38 |
| Musilová et al. (2013) | Chicken Cat, JTT Conveying A/S, Denmark | 3.80 | 2.15 (luxation)<br>0.15 (broken) | 0.02 | 0 (luxation)<br>0.01 (broken) | n/a |
| Mönch et al. (2020) | Apollo Generation 2, CMC Industries, Italy | 7.19 | 1.36 | 0 | 0 | 0.16 |
| Delanglez et al. (2025) | Apollo Generation 2, CMC Industries, Italy | 7.38 | 0.88 | 4.53 | 0.84 | 0.20 |
| Unterholzner et al. (under review) | Apollo Generation 2, CMC Industries, Italy | 4.61–6.56 | 0.36–1.79 | n/a | n/a | n/a |
| **Manual catching** | | | | | | |
| Knierim and Gocke (2003) | Manual | 1.27 | 0.77 (broken)<br>0.59 (dislocation) | 1.20 | 0.10 (broken)<br>0.02 (dislocation) | 0.39 |
| Nijdam et al. (2005) | Manual | 6.7–8.4 | n/a | 0.51–0.74 | n/a | 0.13–0.17 |
| Musilová et al. (2013) | Manual | 3.77 | 1.38 (luxation)<br>0.03 (broken) | 0.04 | 0 (luxation)<br>0.02 (broken) | n/a |

(*Continued*)

**Table 1** *Continued*

| Authors | Type of loading | Haematoma wing % | Fracture wing % | Haematoma leg % | Fracture leg % | Dead on arrival % |
|---|---|---|---|---|---|---|
| Junghans et al. (2022) | Manual | n/a | n/a | n/a | n/a | 0.17 |
| Langkabel et al. (2015) (only data for heavy broilers presented) | 1LCM<br>2LCM | 0.32 | 11.62–11.70<br>13.15–13.97 | n/a | n/a | n/a |
| Mönch et al. (2020) | 2LCM | 1.49 | 1.02 | 0 | 0 | 0.06 |
| Wessel et al. 2022 | 1LCM<br>2LCM | 0.99<br>0.60 | 1.10<br>0.43 | 0<br>0 | 0<br>0 | n/a |
| Delanglez et al. (2025) | Inverted<br>Upright | 6.53<br>5.70 | 0.70<br>0.73 | 4.45<br>4.39 | 0.570.57 | 0.25<br>0.14 |

bumping against the machine or container (Werner et al., 2023). A picture of the SmartStack container can be viewed in Werner et al. (2023). This is a container with no middle partition. The risk for haematomas is increased if broilers bump against the container. Thus, this is a risk of loading by machine. Care should therefore be taken that broilers do not bump into the container or the machine (Wolff et al., 2019; Mönch et al., 2020). A reduction in injuries was achieved by a slower belt speed in the study by Unterholzner et al. (2025). Unterholzner et al. (2025) observed mean belt speeds of 1.02 m/s in the slow and 1.20 m/s in the fast setting.

The results from studies indicate that the choice of loading method alone is not the main determinant of injury risk. Multiple factors are associated with broiler welfare during loading, and it is important that the chosen method is carried out under appropriate conditions (Mönch et al., 2020; Wessel et al., 2022).

### 5.4 Take-home message

Mechanical loading can improve broiler welfare by reducing stress and direct human-animal contact. Machines like the Apollo Generation 2 and Chicken Cat have different methods to collect broilers but share similar principles of gentler handling. However, the speed of the conveyor belts and operation of the machine must be carefully adjusted to minimise injury risks. Staff training remains essential for optimal welfare of broilers with mechanical loading systems.

## 6 Loading-related stress reactions

The loading of broilers for transport is a critical point in poultry production, as it exposes chickens to significant stressors, including handling, confinement, and transport-related injuries. Stress is described as a situation that can have a negative impact on the 'health and performance' of poultry (Rosales, 1994). During manual catching, handling by humans is one of the major stressors that can affect animal welfare and lead to stress (Wolff et al., 2019). The elicited stress-related behaviour can then cause injuries such as haematomas, fractures, and dislocations, especially in rough handling (Jones, 1992; Kannan and Mench, 1997). Fraser (2008) states that for the assessment of animal welfare, one needs to consider the health, behaviour, and affective states of an animal. This approach would also be desirable for the assessment of loading, particularly taking into account the broiler's perspective. To date, however, there are no studies that consider the affective states of chickens during loading.

Jones (1992) assessed the stress response of broilers by measuring tonic immobility (TI - where fear induces a 'playing dead' state) and latency to head movement (the time it takes to recover from TI) after manual catching, and found that holding broilers in an inverted position resulted in more stress compared to broilers held upright. Similarly, numerous studies have demonstrated that holding broilers in an upright position is associated with less stress for the animals than holding them in inverted positions (whether single or with multiple broilers) and with calmer behaviour during loading (Broom and Knowles, 1989; Kannan and Mench, 1997; Kittelsen et al., 2018; Langkabel et al., 2015; Wolff et al., 2019). Stress responses in poultry during/after transportation can be measured, for example by plasma corticosterone or plasma glucose (Broom and Knowles, 1989), heterophil/lymphocyte ratio (Wolff et al., 2019), heart and respiration rate, tonic immobility, or injuries (Broom and Knowles, 1989). Studies have indicated that stress (e.g. plasma corticosterone or behavioural aspects such as wing flapping) can be reduced by mechanical loading compared to manual catching (Duncan et al., 1986; Weeks, 2007; Wolff et al., 2019). Results have also shown that regular positive handling or applying a gentling programme to young animals can lead to higher stress resistance and less anxiety (Zulkifli and Azah, 2004; Stocker et al., 2025). The gentling programme (Stocker et al., 2025) is a procedure in which positive contact is intended to reduce chickens' fear of humans. To this end, increased (positive) contact with the animal in the first 72 h after housing leads to reduced fear later in the rearing period.

A key behavioural response to stress during loading is wing flapping, which occurs because of fear responses, imbalance, or external pressures during handling (Wolff et al., 2019; Wessel et al., 2022; Werner et al., 2023). The risk of wing flapping was found to be twice as high during manual catching (68.6% of the broilers showing wing flapping) as during mechanical loading (50.9% of the broilers showing wing flapping) with the Apollo Generation 2 (Wolff et al., 2019). In manual catching, wing flapping is more frequent when chickens are carried by one leg rather than two, and improper grasping positions exacerbate this response (Wessel et al., 2022). Broilers should be grasped as close to the body as possible, taking care not to squeeze the thighs. Recommended grasping positions are shown in Wessel et al. (2022). Additionally, striking of wings against crates during loading further increases stress and injury risk (Wessel et al., 2022).

Other studies have shown that increased conveyor belt speed in mechanical loading systems correlates with a higher incidence of wing flapping, which can lead to abrasions and haematomas (Werner et al., 2023).

Research results highlight the importance of reducing the risk of agitation in chickens. Mitigating stress-related behaviours, such as wing flapping, requires optimising handling techniques, reducing rotation and conveyor belt speeds, and ensuring proper worker training. Adjustments in container design (e.g. no middle partition) and transport procedures can help improve broiler welfare and reduce loading-related stress responses and the associated injuries.

### 6.1 Take-home message

Loading is a major stressor for broilers, leading to defensive behaviours like wing flapping and increased stress responses. Upright handling and mechanical loading are associated with lower stress responses than inverted manual catching. Environmental factors such as lighting and temperature strongly influence stress levels. Reducing handling time and improving handling techniques are key strategies to mitigate stress.

## 7 Loading-related injuries

The frequency of injuries that can occur during catching and crating is one measurable parameter for determining animal welfare. Numerous studies have shown that loading and transport of broilers can cause injuries (Taylor and Helbacka, 1968; de Koning et al., 1987; Lacy and Czarick, 1998; Knierim and Gocke, 2003; Nijdam et al., 2005; Langkabel et al., 2015; Mönch et al., 2020; Wessel et al., 2022; Unterholzner et al., 2025). In the European Union injured animals that cannot move without pain are no longer considered fit for transport, especially where there are fractures and dislocations, as loading-related injuries are considered an important welfare indicator (Council Regulation [EC] No 1/2005, 2004).

Catching-related injuries include haematomas on wings, legs, and/or chest, and frequencies of occurrence have been used to assess loading quality (Knierim and Gocke, 2003; Nijdam et al., 2004; Gouveia et al., 2009; Musilová et al., 2013; Langkabel et al., 2015; Mönch et al., 2020; Wessel et al., 2022; van Emous et al., 2024; Unterholzner et al., under review). Additionally, several authors have recorded the frequency of wing (Knierim and Gocke, 2003; Musilová et al., 2013; Langkabel et al., 2015; Mönch et al., 2020; Wessel et al., 2022) and leg fractures (Knierim and Gocke, 2003; Musilová et al., 2013; Mönch et al., 2020; Wessel et al., 2022) or abrasions (Unterholzer et al., under review). Injuries of internal organs may also be a possible consequence of loading (Kittelsen et al., 2015). Louton et al. (2018) found that the majority of slaughterhouses for broilers in Germany participating in a survey record injuries due to loading and transport and report this to veterinary officials and the farm.

### 7.1 Haematomas

In general, a haematoma is an extravascular accumulation of blood, e.g. because of trauma or associated injury to subcutaneous blood vessels, regardless of location (Northcutt et al., 2000; Cockram and Dulal, 2018). Nijdam et al. (2005) defined a haematoma as a discolouration of the skin that indicates the presence of blood in the tissue. The colour of the haematoma is considered important for age assessment and thus possible time of formation (Northcutt et al., 2000). Greenish discolouration indicates the breakdown of the red blood pigment haemoglobin and thus older formation (Hamdy et al., 1961). For this reason, greenish haematomas cannot be regarded as being related to previous loading.

We performed a study to demonstrate the colour and size progression of haematomas resulting from blood sampling in five broilers (not published). A clear age allocation of a haematoma in the first 12 h after formation did not appear to be possible with certainty using our approach. The colour of the haematoma was dark purple in all five broilers immediately after blood collection (RGB: red: 37, green: 14, blue: 98; the smaller the value, the stronger the colour component). There was only little change in colour during the first 12 h after emergence. In all five broilers, the haematoma showed a slight 'shimmer' from 24 h after blood collection, becoming lighter over time. From 48 hours after blood collection, the haematoma became progressively fainter, reflected in an increase in the proportion of white (higher RGB values) in all five broilers. Please note that only haematomas caused by venous blood sampling were examined. Traumatic haematomas, e.g. in connection with fractures or arterial vascular damage, could possibly result in a different colour and size development. Nevertheless, venous vessel rupture as loading-related injury seems possible and could lead to respective change of colour (Togami et al., 2024).

Nijdam et al. (2005) based their definition of haematoma size on the legal requirements applicable in the Netherlands. Accordingly, a discolouration of the skin on the wing was only considered as a haematoma when it was at least 2.0 $cm^2$ in size. This size definition also applies to studies by Knierim and Gocke (2003), Gouveia et al. (2009), and van Emous et al. (2024). In the study by Langkabel et al. (2015), the relative size of the haematoma to the body part was scored. In the studies by Mönch et al. (2020), Wessel et al. (2022), and Unterholzner et al. (2025), haematomas were scored as such if they were at least 0.5 cm in diameter. We recommend assessing haematoma after catching and crating of at least 0.5 cm and differing in assessment between proximal wing and wing tip (sum = wing) because the causes for these may differ. This also ensures that also small lesions are recorded.

Some other studies do not provide precise information on the definition of the colour or size of the evaluated haematomas (Musilová et al., 2013). This means that a wide range of loading-related haematomas (and other injuries) have been recorded in literature. The decisive factor for the comparability and interpretation of the results is the size at which a haematoma is considered as such, the time of assessment (on-farm, at slaughter), and thus a clear description of research methods.

Prevalences of haematomas on the wing are summarised in Table 1. Haematomas on the wing (6.55%) were the most common type of injuries found by Unterholzner et al. (2025), followed by abrasions on the body (4.92%), abrasions on the wing tip (4.25%), and severe wing injuries (1.13%). Haematomas of the wing were divided into haematomas on the wing tip (6.17%) (Fig. 5) and haematomas on the wing proximal to the wing tip (0.38%) (Fig. 6). Haematomas have also been reported on the breast and legs, albeit at a lower level than on the wings in most studies (Knierim and Gocke, 2003; Nijdam et al., 2005;

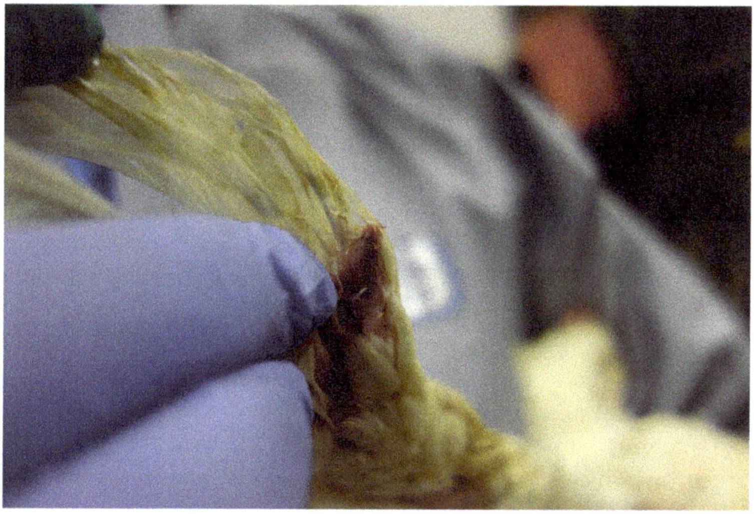

**Figure 5** Haematoma on the wing tip of a broiler after loading. Photo: LMU Munich.

**Figure 6** Haematoma of the wing and (closed) fracture of the epiphysis during an assessment on the farm. Photo: LMU Munich.

Musilová et al., 2013). Wing haematomas are related to wing flapping or bumping into containers or conspecifics (Werner et al., 2023).

### 7.2 Fractures and dislocations

As a result of loading and transportation, wing fractures (Knierim and Gocke, 2003; Musilová et al., 2013; Langkabel et al., 2015; Mönch et al., 2020; Wessel et al., 2022; Unterholzner et al., 2025) and leg injuries (Knierim and Gocke, 2003; Musilová et al., 2013) have been observed as loading-related injuries. Comparing 1LCM and 2LCM showed an average of up to 4.74% of broilers at first loading (so-called 'thinning') and 13.97% of the broilers at final unloading with a wing fracture. These percentages varied depending on the catching crew and were recorded at the slaughterhouse (Langkabel et al., 2015). Fractures in these high amounts have not been recorded in on-farm assessments (Mönch et al., 2020; Wessel et al., 2022; Unterholzner et al., 2025). This could be because processes between crating and slaughter incluse several critical points where fractures could occur. Another important point is that fractures can occur during the slaughter process. These can be clearly differentiated from those that occurred before the animal's death due to the absence of hematomas (Figs. 7 and 8). However, in the case of fractures with hematomas at the slaughterhouse, it is not possible to say for certain which stage of the loading and transportation process caused them.

The figures found by Langkabel et al. (2015) differ significantly from the results of other studies, which give average values of 0.77% (Knierim and Gocke, 2003) or lower (Musilová et al., 2013) for the prevalence of wing fractures in manually loaded animals. Knierim and Gocke (2003) assessed leg fractures in both mechanical loading and manual catching of broilers: 0.03% of the broilers after mechanical loading and 0.10% of the broilers after manual catching had a leg fracture. In contrast, Musilová et al. (2013) were unable to diagnose leg fractures after either manual or mechanical loading. Similarly, in our studies on manual catching and mechanical loading, we did not find diagnosable leg fractures (Mönch et al., 2020; Wessel et al., 2022; Unterholzner et al., 2025).

In addition to the frequency of fractures, some authors also investigated the frequency of joint dislocations on wings and legs. Average values of 2.15% after mechanical loading and 1.38% after manual catching were found for wings (Musilová et al., 2013). The prevalence of leg dislocations reported in the literature is close to zero (Knierim and Gocke, 2003; Musilová et al., 2013). In our studies, we found that most wing fractures (93% of all fractures) were epiphysiolysis of the distal humerus (see Mönch et al., 2020, for visualisation). An association with a potential underlying disease, e.g. of the bone, is assumed

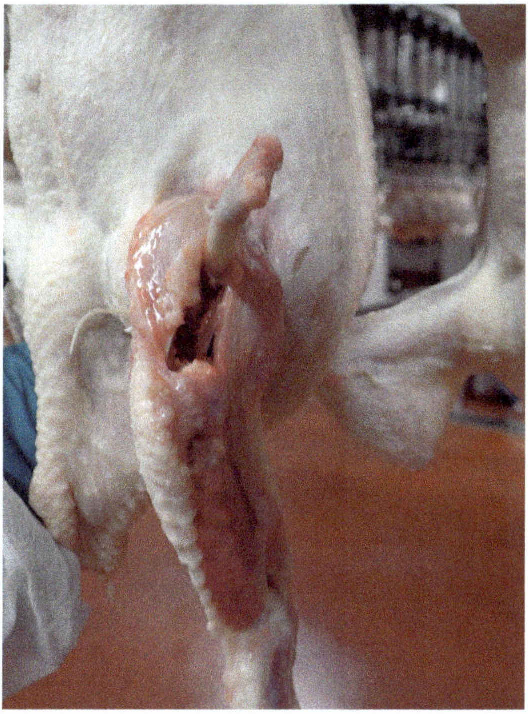

**Figure 7** Fracture with haematoma documented at the slaughterhouse. It is not possible to clearly state the cause and point in time of the fracture from this picture. Photo: LMU Munich.

because a correlation to a higher risk for condemnation is observed at slaughter in flocks with higher risks for fractured bones (Unterholzner et al., 2025).

### *7.3 Dead on arrival*

Another indicator from an animal welfare perspective is the number of animals that arrive dead at the slaughterhouse, known as 'dead on arrival' (DOA). These animals can no longer be used for human consumption and must be rejected for further processing (Bayliss and Hinton, 1990). As early as 1990 Nicol and Scott identified the number of DOA broilers as an important indicator of animal welfare during the loading phase up to slaughter. This has been confirmed by several authors who have shown that the number of DOA broilers can be influenced by various loading- and transportation-related factors (Nijdam et al., 2004; Vecerek et al., 2006; Lund et al., 2013; Kittelsen et al., 2015). However, animal health during the rearing period also has an influence on the DOA rate and should therefore be considered in assessment (Junghans et al., 2022). DOA levels are not only an indicator of loading but of the whole process.

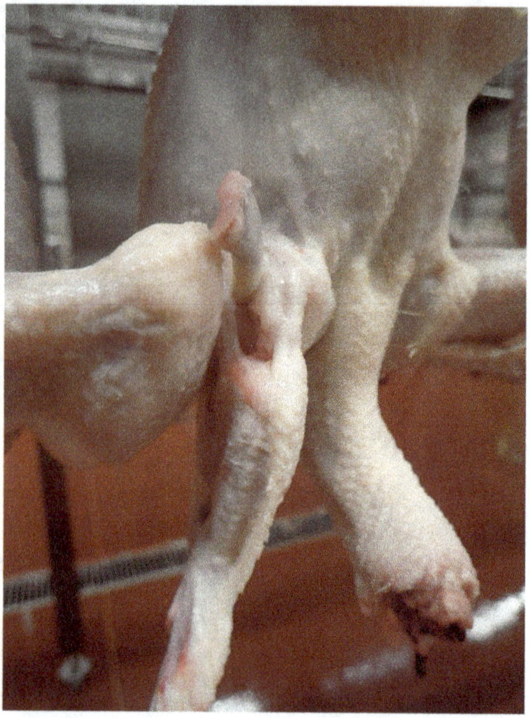

**Figure 8** Fracture of the wing of a broiler without haematoma. Most likely this fracture occurred after death of the broiler during processing at the slaughterhouse. Photo: LMU Munich.

Various rates are provided in the literature for losses due to DOA. Junghans et al. (2022) found 0.17% DOA in a retrospective study of slaughter data from conventional, manually-loaded broiler flocks. Haslam et al. (2008) reported losses of up to 0.64% DOA, with an average of 0.12%. Mönch et al. (2020) found an average 0.12% DOA, which is comparable to the results of Warriss et al. (1992) (0.16%), but slightly higher than Kittelsen et al. (2015) who found an average of 0.09% DOA. The values given by other authors for the average losses due to DOA are between 0.30% and 0.68% (Knierim and Gocke, 2003; Nijdam et al., 2004; Ritz et al., 2005; Lund et al., 2013; Jacobs et al., 2017).

However, not all studies used the same method to determine the number of DOA. Some authors calculated the number of DOA broilers at the flock level (Knierim and Gocke, 2003; Nijdam et al., 2004; Lund et al., 2013; Kittelsen et al., 2015; Jacobs et al., 2017), whereas Ritz et al. (2005) calculated the number of DOA broilers per truck. In Germany, the Animal Welfare Ordinance (2006) requires that if there are any anomalies during the slaughter process that indicate a breach of animal welfare, a report must be made to the authority to determine the cause. To this end, thresholds have been set in the relevant implementation notes. For example, if the DOA rate exceeds 0.5%, the cause of this must be investigated (Implementation instructions TierSchNutztV AGT, 2021).

There is limited literature on the frequency of loading-related injuries of internal organs of broilers. The frequency of traumatic injuries to internal organs has so far been investigated mainly in studies that have dealt with the pathological changes in broilers that arrived dead at the slaughterhouse (Nijdam et al., 2006; Lund et al., 2013; Kittelsen et al., 2015).

Liver ruptures, in association with intra-abdominal haemorrhages, are one of the most frequently cited injuries to the internal organs of DOA broilers (Nijdam et al., 2006; Kittelsen et al., 2015; Cockram and Dulal, 2018). In a study from Norway, liver rupture was the major cause of trauma, found in 51.40% of DOA broilers (Kittelsen et al., 2015). Lund et al. (2013) found liver ruptures in 14.60% of all DOA broilers examined and severe pulmonary congestion in 51.5%. The authors highlight the importance of handling to reduce the DOA rate.

### *7.4 Take-home message*

Injuries like haematomas, fractures, and dislocations are important welfare indicators during loading. Both manual and mechanical methods carry risks, but correct technique and conditions can reduce injury rates. Slow and careful handling, appropriate container design, and machine settings in mechanical loading all play roles in injury prevention. Monitoring and recording injury rates help to identify and improve critical points and thus animal welfare.

## 8 Factors impacting loading quality or loading-related injuries

The environment in broiler barns remains relatively constant throughout the rearing period. Broilers have minimal contact with humans, aside from routine checks by the farmer. However, on the day of loading, these conditions change significantly. The increased presence of workers, the noise generated by forklifts and loading machines, fluctuating temperature, light intensity, and increased dust levels can all contribute to stress in broilers. Light intensity for example plays a critical role in influencing broilers' defensive behaviour or stress reaction (Jones et al., 1998; Wolff et al., 2019).

The correlation between stress-induced behaviour such as wing flapping or escape behaviour and the occurrence of injuries has been well documented (Wolff et al., 2019). Heightened defensive reactions increase the risk of injuries and, consequently, the likelihood of mortality during transport (assessed as broilers dead on arrival, DOA) (Kittelsen et al., 2015; Werner et al., 2023).

### 8.1 Intrinsic and extrinsic risk factors for injuries

Key risk factors affecting loading-related injuries can be categorised into intrinsic, flock-specific factors and extrinsic loading conditions. Among flock-specific factors, body weight at slaughter has been identified as a significant determinant of injury risk (Langkabel et al., 2015; Cockram and Dulal, 2018). Additionally, sex and age influence injury prevalence, with heavier and older broilers showing a higher susceptibility to fractures and haematomas (Cockram and Dulal, 2018). Mönch et al. (2020) and Wessel et al. (2022) observed a significantly higher risk for fractures in female broilers and discussed the potential influence of bone reformation in the growing female chicken.

Extrinsic conditions, including the loading method, play a crucial role in injury rates. Whereas manual and mechanical loading methods have been compared widely (Knierim and Gocke, 2003; Nijdam et al., 2005), manual catching methods vary in terms of animal welfare impact. For manual catching, we can differentiate between inverted one-legged, inverted two-legged, and upright loading (Langkabel et al., 2015; Wessel et al., 2022; Delanglez et al., 2025). In addition to the choice of catching technique, the loading speed (Cockram and Dulal, 2018) as well as the total duration of loading and the personnel themselves can have an influence on the risk of injury (Mönch et al., 2020).

Studies comparing one-legged (1LCM) and two-legged (2LCM) catching methods suggested that 1LCM increases wing flapping and injury risk (Wessel et al., 2022). However, multiple studies emphasise that the loading method alone does not solely determine animal welfare outcomes.

Werner et al. (2023) demonstrated that higher conveyor belt speeds in mechanical loading can lead to an increased frequency of wing flapping and bumping against containers which, in turn, elevates the risk of haematomas and fractures. The authors also found that the risk of haematomas was increased if broilers bumped against each other during the machine loading process. Their analysis revealed that the behaviour and impacts on the animals during loading are influenced by each factor analysed (loading belt speed, husbandry system/rearing method, season, and container type) and that these factors in turn can affect loading-related injuries (Werner et al., 2023).

Timing and environmental conditions during loading further influence injury rates. Seasonal variations, outdoor temperature, and time of day have been linked to the prevalence of injuries (Nijdam et al., 2004). Low and high ambient temperatures both contribute to thermal stress, which elevates corticosterone levels (Wolff et al., 2019). Studies also indicate that night-time loading reduces stress, injury prevalence, and death, owing to lower temperatures and reduced light exposure (Haslam et al., 2008; Junghans et al., 2022). Mönch et al. (2020) also reported that lower light intensities during loading lowered the incidence of the injuries of broilers, most likely because of reduced panic reactions. Corticosterone concentration as an indicator of stress was strongly influenced by light intensity (high light intensities) and outdoor temperature (high temperatures) in the study by Wolff et al. (2019), highlighting the importance of managing environmental factors.

The husbandry system and rearing method also influence loading-related injuries, as recently demonstrated by Unterholzner et al. (2025). They grouped husbandry systems according to the specifications of a retail trade's label into 'Indoor Plus', an indoor housing system for a fast-growing broiler strain that includes environmental enrichment such as straw bales or pecking blocks, and 'Outdoor Climate', an outdoor housing system that includes environmental enrichment and uses a slow-growing broiler strain. The rearing methods were classified according to specifications of the 'Standard', 'Standard Premium', and 'Premium' labels, which differ, among other aspects, in stocking density (maximum of 35 kg/m$^2$ for Standard, 30 kg/m$^2$ for Standard Premium, 25 kg/m$^2$ for Premium), dark period (minimum of uninterrupted 6 h for Standard and 8 h for Standard Premium and Premium), and access to a wintergarden (only for Standard Premium and Premium).

The broilers raised in Outdoor Climate showed significantly lower rates of severe wing injuries and abrasions than those raised in Indoor Plus. Furthermore, the broilers from the Standard fattening method were at higher risk for severe injuries (e.g. fractures) than those from Standard Premium and Premium (Unterholzner et al., 2025). These differences suggest that not only genetic factors but also stocking density, dark periods, and access to a winter garden play essential roles in reducing loading-related injuries. The stocking density and the winter garden could be related to bone health in that activity can be promoted with low stocking densities and access to an outdoor run (Bergmann et al., 2017). Increased walking can have a training effect and a positive impact on the health of the animals (Reiter and Bessei, 2009).

### *8.2 Risk factors for 'dead on arrival' rates*

DOA rates are primarily influenced by three factors: thermal stress, pre-existing health status, and injuries sustained during loading (Chauvin et al., 2011). Flock-specific variables, including body weight at slaughter, flock size, cumulative mortality during rearing, and antibiotic treatments, have been correlated with DOA rates (Haslam et al., 2008; Junghans et al., 2022). Smaller flocks tend to have shorter loading times, which in turn reduce stress levels. Additionally, loading at night mitigates heat stress and prevents light-induced agitation (Junghans et al., 2022). Transport time and stocking density also contribute to mortality rates, with longer journeys and higher densities leading to increased DOA levels (Nijdam et al., 2004). Antibiotic treatments indicate an incidence of disease during the rearing phase, which could make the animals more susceptible to mortality during transportation.

## 8.3 Options to minimise loading-related stress and injuries

Prolonged loading times negatively affect animal welfare, because of increasing time of feed withdrawal (among other factors). A longer loading time during manual and mechanical loading can have a negative effect on animal health (Jacobs et al., 2017; Mönch et al., 2020) and on stress-related defensive behaviour, such as wing flapping (Wolff et al., 2019). The duration of the loading process depends on multiple factors, including conveyor belt speed, container type, and stocking density.

For example, Mönch et al. (2020) found that slower belt speeds correlated with longer individual loading times, increasing overall stress. However, Knierim and Gocke (2003) reported no direct correlation between conveyor belt speed and overall loading efficiency, suggesting that additional logistical elements, such as coordination of forklift operators, also play a role. Unterholzner et al. (2025) found that, using SmartStack containers instead of GP containers, reduced the frequency of haematomas and fractures, presumably because of improved loading dynamics and reduced handling stress.

To minimise stress and injuries during loading, several preventive measures should be taken. A fasting period exceeding 12 h should be avoided (Animal Welfare Ordinance, 2006), and access to water should be maintained until transport begins. In warm weather, additional ventilation near transport trucks can help prevent heat stress. Darkening the barn as much as possible and utilising blue or green light lamps can also reduce stress reactions in broilers. These lamps allow sufficient visibility for workers, while maintaining a low light environment for animals, thereby improving handling conditions and overall welfare (Mönch et al., 2020). Werner et al. (2023) and Wessel et al. (2022) have also suggested that training workers in stress-minimising techniques, such as gentle handling and optimised catching methods, can further improve welfare outcomes during loading.

## 8.4 Take-home message

Intrinsic factors like broiler weight and age, and extrinsic factors like loading speed, season, and flock management, influence injury risks. Higher body weight and poor environmental conditions increase vulnerability to injuries. Appropriate lighting, reduced handling stress, and optimised catching methods improve animal welfare outcomes. Each loading situation requires specific adjustments to ensure humane treatment.

## 9 Loading staff: issues to consider

Manual catching of broilers is not only a critical issue in terms of the health and welfare of the animals, but also for catching staff. Loading broilers can cause injuries and even disease in catching staff (Quandt et al., 2013) and may be related to pain (Delangez et al., 2025). It is stated to be an 'unhealthy, dirty and dangerous job' (Dutra et al., 2021) and has also been referred to as a '3-D' (dirty, dangerous and demanding) (Quandt et al., 2013). On top of this, it is often inadequately paid (Dutra et al., 2021).

With an average weight of 2.5 kg per broiler and standard barn sizes of 40,000 animals in Germany, one team member of a catching crew of eight persons (a standard size of catching crew) can load 12.5 tons within a few hours. This has already been the case as early as 1985 (Kettlewell and Turner, 1985). Owing to technical reasons relating to slaughterhouse and supply chain logistics, loading usually takes place at night. The barn should be darkened as far as possible to keep chickens calm. However, limited visibility can easily lead to injuries among the catchers.

Dust and ammonia exposure, as well as high humidity, can also have a negative impact on catchers' health (Quandt et al., 2013; Millman et al., 2017; Dutra et al., 2021). Protective clothing is often perceived to be impractical in certain cases (Millman et al., 2017). Previous studies have revealed that fatigue sets in after a few hours of manual catching (Lacy and Czarick, 1998; Delezie et al., 2006). Fatigue and exertion lead to a poor body posture during loading and can lead to chronic discomfort in the back, legs, arms, and hands (Dutra et al., 2021). Specific manual catching methods, such as the upright catching method, might cause more pain than others (Delangez et al., 2025). Careful handling of animals might be neglected as loss of concentration and physical discomfort set in (Kettlewell and Mitchell, 1994; Lacy and Czarick, 1998; Nijdam et al., 2004; Mönch et al., 2020).

If the catching crew is paid by the number of loaded broilers, rather than by the hour, this could also lead to chickens being loaded as quickly as possible (Quandt et al., 2013). The way employees are trained, and how animals are handled during loading, has an influence on manual catching. Rough handling and carrying as many broilers as possible can increase injury risk (Wessel et al., 2022). It is therefore important that members of catching crews receive detailed training, especially younger and less experienced team members, to improve their catching technique and thus avoid injuries (Pilecco et al., 2013; Wessel et al., 2022). This necessity is also given by European law as mentioned in Section 4 (Council Directive 2007/43/ CE, 2007). Mechanical loading can improve working conditions for staff members and does not raise the price of the loaded chicken (Lacy and Czarick, 1998).

### 9.1 Take-home message

Catching and loading broilers is physically demanding, hazardous, and often poorly rewarded work. Staff face risks of injury, respiratory issues, and fatigue, which can compromise careful animal handling. Improving ergonomic conditions, providing proper training, and promoting mechanical loading can protect both worker health and broiler welfare. Supporting staff well-being is essential for achieving high animal welfare standards.

## 10 Crating

The crating of the animals is a moment of risk for broilers to bump into a container or become agitated. In a study performed by Wessel et al. (2022), precise observations of the process of crating and of behavioural effects were made in commercial setting with fast-growing broilers. Crating was scored with a 3-point scoring system: settling, dropping, or throwing. Of the observed broilers, 80.2% were settled, 17.3.% dropped, and 2.5% thrown (Wessel et al., 2022). If the broilers were caught by the two-legged method, they had a higher chance of being settled. The chance of being settled also increased in case of longer catching duration, a higher weight of the broilers, and if they were placed in high or middle crates (Wessel et al., 2022).

Wessel et al. (2022) also observed that broilers showed a tendency for more wing flapping after being crated with the one-legged method as compared with the two-legged method (albeit the difference was not statistically significant). The catching method, grasping position (on the leg), and the number of broilers carried simultaneously did not affect the risk of wing flapping after crating. However, the crating method and 'striking against the container' did affect the risk of wing flapping after crating. A higher risk was found if broilers were not settled but dropped or thrown in the crate, or if they hit parts of the container. Similarly, de Lima et al. (2019) found that if broilers were settled in the crate, they showed less agitation.

### 10.1 Take-home message

Careful crating is crucial to avoid stress and injuries in broilers during loading. Gently settling the broilers into crates rather than dropping or throwing them significantly reduces wing flapping and agitation. The catching method and the way broilers are placed into containers both influence animal welfare outcomes. Training staff to handle crating properly is a key step to improve broiler handling and minimise injuries.

## 11 Conclusion

For broilers, the final hours before slaughter are a critical period that can either exacerbate or alleviate stress and suffering. The way they are caught, crated and loaded determines whether they experience fear, pain, and injury, or whether their welfare is protected until slaughter. From the broilers' point of view, gentle, careful handling – ideally without rough grasping or inversion – significantly reduces fear responses such as wing flapping and minimises the risk of fractures, haematomas, and mortality.

Mechanical loading offers broilers the advantage of reduced human contact, calmer handling, and less physical trauma when properly managed. However, even manual catching can be less stressful when broilers are crated carefully and handled by trained, attentive staff. Factors such as noise, light, temperature, and barn conditions are immediately perceived by broilers and can either intensify or alleviate their stress.

Ensuring humane loading is thus not just a matter of technical efficiency but a matter of seeing the experience through the eyes of the animals themselves. Improvements in machine design, catching techniques, and staff welfare ultimately lead to a better outcome for broilers – allowing them to complete their lives with minimal additional distress. Ongoing commitment to research, training, and respectful handling is essential for truly protecting broiler welfare at the end of their lives.

## 12 Summary

The chapter discusses the humane handling of broilers at the end of their production cycle, focusing on catching, crating, and loading processes before transport to slaughterhouses. These procedures are critical for animal welfare, as they can cause significant stress and injuries if not conducted properly.

There are two primary methods for catching broilers: manual catching and mechanical loading. On the one hand, manual catching is still widely practised, particularly in Germany, where workers catch multiple chickens at once by their legs. This method is currently intensively discussed, and non-governmental organisations (NGOs) request that broilers are caught in an upright position. On the other hand, mechanical loading, which is common practice in Scandinavian countries, reduces human contact, thereby minimising stress and potentially also injuries. Despite its benefits, the use of loading machines is limited because of financial constraints, lack of contractors, and farm-specific conditions.

The European Union regulates broiler handling through Council Directive 2007/43/CE (2007), which sets guidelines for animal welfare, including training for workers involved in catching and loading. Additionally, the European Commission has proposed changes to regulations that could impact manual handling practices. Experts recommend avoiding handling broilers in an inverted position and promoting methods that support the breast and wings during transport.

A survey of German poultry farmers revealed mixed opinions about mechanical loading. Those using machines stated better hygiene, higher efficiency, and reduced labour dependency as advantages. However, some farmers preferred manual methods owing to concerns about machine reliability and lack of contractors.

Broilers experience stress during handling, leading to behaviours such as wing flapping, which increases the risk of haemorrhages and fractures. Mechanical loading generally results in fewer injuries than does manual catching. However, factors such as catching duration, flock size, and environmental conditions (e.g. temperature, lighting) influence injury rates. Proper training of workers and optimised handling techniques can improve outcomes.

Manual catching is a physically demanding job, exposing workers to health risks such as musculoskeletal injuries and respiratory issues due to dust and ammonia exposure. Fatigue and financial pressure can lead to rough handling, increasing the likelihood of injuries to both workers and chickens. Mechanical loading can improve animal welfare and working conditions.

To ensure humane end-of-life management for broilers, a balance between economic efficiency and animal welfare must be achieved. Increased use of mechanical loading, worker training, and adherence to best practices can improve both animal welfare and operational efficiency. Further research and development of catching and loading technologies are needed to address animal welfare.

## 13 Where to look for further information

- Louton, H., et al. (2023a), Which methods of loading broilers are used in Germany – and why? XI European Symposium on Poultry Welfare, 26.-29.06.2023, Prague, Czech Republic. Hrsg. GUARANT International spol. s r.o. for Czech branch of WPSA. ISBN: 978-80-908114-1-6: 151.
- Mönch, J., et al. (2020), The welfare impacts of mechanical and manual broiler catching and of circumstances at loading under field conditions. *Poult. Sci.*, 99(11), 5233–5251. https://doi.org/10.1016/j.psj.2020.08.030.
- Unterholzner, J., et al. (under review), Loading-related injuries of mechanically loaded broilers under field conditions. *Poult. Sci.*
- Werner, A., et al. (2023), Behavior of broilers and impacts occurring to them during mechanical loading under field conditions. *Poult. Sci.*, 102(7), 102688. https://doi.org/10.1016/j.psj.2023.102688.

- Wessel, J., et al. (2022), A comparison of two manual catching methods of broiler considering injuries and behavior. *Poult. Sci.*, 101(11), 102127. https://doi.org/10.1016/j.psj.2022.102127.
- Wolff, I., et al. (2019), Harvesting-induced stress in broilers: Comparison of a manual and a mechanical harvesting method under field conditions. *Appl. Anim. Behav. Sci.*, *221*, 104877. https://doi.org/10.1016/j.applanim.2019.104877.

## 14 References

Animal Welfare Ordinance (2006), Animal Welfare Livestock Farming Ordinance in the version published on August 22, 2006 (Federal Law Gazette I p. 2043), last amended by Article 1a of the Ordinance of January 29, 2021 (Federal Law Gazette I 146).

Bayliss, P.A. and Hinton, M.H. (1990), Transportation of broilers with special reference to mortality rates, *Appl. Anim. Behav. Sci.*, 28, 93–118.

Bergmann, S., et al. (2017), Behavior as welfare indicator for the rearing of broilers in an enriched husbandry environment–a field study. *J. Vet. Behav.* 19:90–101.

Broom D.M. and Knowles T.G. (1989), The assessment of welfare during the handling and transport of spent hens, In Faure, J.M. and Mills, A.D. (eds) *Proceedings of the 3rd European Symposium on Poultry Welfare*, 79–91. Tour: World Poultry Science Association.

Chauvin, C., et al. (2011), Factors associated with mortality of broilers during transport to slaughterhouse, *Animal*, 5(2), 287–293. https://doi.org/10.1017/S1751731110001916.

Cockram, M.S. and Dulal, K.J. (2018), Injury and mortality in broilers during handling and transport to slaughter, *Can. J. Anim. Sci.*, 98(3), 416–432. https://doi.org/10.1139/cjas-2017-0076.

Council Directive 2007/43/CE (2007), Council Directive 2007/43/EC of 28 June 2007 laying down minimum rules for the protection of chickens kept for meat production, *OJ L* 182, 19–28.

Council Regulation (EC) No 1/2005 (2004), Council Regulation (EC) No 1/2005 of 22 December 2004 on the protection of animals during transport and related operations and amending Directives 64/432/EEC and 93/119/EC and Regulation (EC) No 1255/97.

Council of Europe (1995), European Convention for the protection of animals kept for farming purposes. Recommendation concerning domestic fowls of the species Gallus gallus. Adopted by the Standing Committee at its 30th meeting on November 28, 1995.

De Koning, K., et al. (1987), Mechanized harvesting and transport of broilers, *J. Agric. Eng. Res.*, 38(2), 105–111.

Delanglez, F., et al. (2025), Comparing methods for catching and crating broiler chicken flocks: A trade-off between animal welfare, ergonomics and economics, *Poult. Sci.*, 104(2), 104704. https://doi.org/10.1016/j.psj.2024.104704.

Delezie, E., et al. (2006), Is the mechanisation of catching broilers a welfare improvement?, *Anim. Welf.*, 15(2), 141–147. https://doi.org/10.1017/S0962728600030220.

Delezie, E., et al. (2007), The effect of feed withdrawal and crating density in transit on metabolism and meat quality of broilers at slaughter weight, *Poult. Sci.*, 86, 1414–1423.

de Lima, V.A., et al. (2019), Effect of different catching practices during manual upright handling on broiler welfare and behavior. *Poult. Sci.*, 98(10), 4282–4289. https://doi.org/10.3382/ps/pez284.

District Court of Rotterdam (2022), Judgment of the Chamber in extended composition of November 24, 2022 in the cases between Stichting Wakker Dier, Amsterdam, (the Foundation) (Agent: Mr. J.L. Baar), and the Minister of Agriculture, Nature and Food Quality, (the Minister) (Agent: Mr. M.M. de Vries).

Duncan, I.J., et al. (1986), Comparison of the stressfulness of harvesting broiler chickens by machine and by hand, *Br. Poult. Sci.*, 27(1), 109–14. https://doi.org/10.1080/00071668608416861.

Dutra, F.M., et al. (2021), What do we know about the impacts of poultry catching?, *Worlds Poult. Sci. J.*, *77*(4), 983–999. https://doi.org/10.1080/00439339.2021.1976056.

EFSA AHAW Panel (EFSA Panel on Animal Health and Animal Welfare), et al. (2023), Scientific Opinion on the welfare of broilers on farm. *EFSA J.*, 21(2), 7788. https://doi.org/10.2903/j.efsa.2023.7788.

European Commission, (2023), Proposal for a Regulation of the European Parliament and of the Council on the protection of animals during transport and related operations, amending Council Regulation (EC) No 1255/97 and repealing Council Regulation (EC) No 1/2005.

Farsaie, A., et al. (1983), Mechanical harvest of broilers. *Transactions of the ASAE*, 26(6), 1650–1653.

Fraser, D. (2008), Understanding animal welfare. *Acta Vet. Scand.*, 50, 1–7. https://doi.org/10.1186/1751-0147-50-S1-S1.

Gouveia, K., et al. (2009), Welfare assessment of broilers through examination of haematomas, foot-pad dermatitis, scratches and breast blisters at processing. *Anim. Welf.*, 18(1), 43–48. https://doi.org/10.1017/S0962728600000051.

Hamdy, M.K., et al. (1961), Determination of the age of bruises in chicken broilers. *Poult. Sci.*, 40(3), 787–789.

Haslam, S.M., et al. (2008), Prevalence and factors associated with it, of birds dead on arrival at the slaughterhouse and other rejection conditions in broiler chickens. *Br. Poult. Sci.*, 49(6), 685–696. https://doi.org/10.1080/00071660802433719.

Implementation instructions TierSchNutztV AGT (2021), Implementation instructions for broiler chickens Animal Welfare Livestock Farming Ordinance - TierSchNutztV - in the version published on August 22, 2006 (BGBl. I p. 2043), last amended by Art. 1a V of 29.1.2021 (BGBl. I p. 146), Section 4, Requirements for the keeping of chickens for fattening. https://www.openagrar.de/servlets/MCRFileNodeServlet/openagrar_derivate_00042568/H-III-Ausfuehrungshinweise-Masthuehner-2021-10.pdf. (Accessed 18.02.2025).

Jacobs, L., et al. (2017), Broiler chickens dead on arrival: associated risk factors and welfare indicators. *Poult. Sci.*, 96(2), 259–265. https://doi.org/10.3382/ps/pew353.

Jones, R. (1992), The nature of handling immediately prior to test affects tonic immobility fear reactions in laying hens and broilers, *Appl. Anim. Behav. Sci.*, 34, 247–254.

Jones, R.B., et al. (1998), Adoption of immobility by shackled broiler chickens: effects of light intensity and diverse hooding devices, *Appl. Anim. Behav. Sci.*, 55, 327–335.

Junghans, A., et al. (2022), Data evaluation of broiler chicken rearing and slaughter–an exploratory study, *Front. Vet. Sci.*, 9, 957786.

Kannan, G. and Mench, J.A. (1997), Prior handling does not significantly reduce the stress response to pre-slaughter handling in broiler chickens, *Appl. Anim. Behav. Sci.*, 51, 87–99.

Kettlewell, P.J. and Mitchell, M.A. (1994), Catching, handling and loading of poultry for road transportation, *Worlds Poult. Sci. J.*, 50(1), 54–56. https://doi.org/10.1079/WPS19940005.

Kettlewell, P.J. and Turner, M.J. (1985), A review of broiler chicken catching and transport systems, *J. Agric. Eng. Res.*, 31, 93–114.

Kittelsen, K.E., et al. (2015), A comparison of post-mortem findings in broilers dead-on-farm and broilers dead-on-arrival at the abattoir, *Poult. Sci.*, 94(11), 2622–9. https://doi.org/10.3382/ps/pev294.

Kittelsen, K.E., et al. (2018), An evaluation of two different broiler catching methods, *Animals* (*Basel*), 15(8), 141. https://doi.org/10.3390/ani8080141.

Knierim, U. and Gocke, A. (2003), Effect of catching broilers by hand or machine on rates of injuries and dead-on-arrivals. *Anim. Welf.*, 12(1), 63–73. https://doi.org/10.1017/S0962728600025380.

Lacy, M.P. and Czarick, M. (1998), Mechanical harvesting of broilers, *Poult. Sci.*, 77(12), 1794–7. https://doi.org/10.1093/ps/77.12.1794.

Langkabel, N., et al. (2015), Influence of two catching methods on the occurrence of lesions in broilers, *Poult. Sci.*, 94(8), 1735–41. https://doi.org/10.3382/ps/pev164.

Louton, H., et al. (2018), Acquisition of animal-based welfare measures at slaughter of poultry, *Fleischwirtschaft*, 11, 94-98.

Louton, H., et al. (2023a), Which methods of loading broilers are used in Germany – and why? XI European Symposium on Poultry Welfare, 26.-29.06.2023, Prague, Czech Republic. Hrsg. GUARANT International spol. s r.o. for Czech branch of WPSA.

Louton, H., et al. (2023b), Welche Methoden zum Verladen von Masthühnern werden in Deutschland eingesetzt – und warum? 28. Internationale DVG-Fachtagung zum Thema Tierschutz in München, Verlag der DVG Service GmbH, Gießen, 295-300.

Lund, V.P., et al. (2013), Pathological manifestations observed in dead-on-arrival broilers at a Danish abattoir, *Br. Poult. Sci.*, 54(4), 430–440. https://doi.org/10.1080/00071668.2013.804173.

Millman, C., et al. (2017), "Catch 22": biosecurity awareness, interpretation and practice amongst poultry catchers, *J. Prev. Vet. Med.*, 141, 22–32.

Mönch, J., et al. (2020), The welfare impacts of mechanical and manual broiler catching and of circumstances at loading under field conditions. *Poult. Sci.*, 99(11), 5233–5251. https://doi.org/10.1016/j.psj.2020.08.030.

Musilová, A., et al. (2013), The effect of broiler catching method on quality of carcasses, *Mendelnet*, 251–255.

National Chicken Council (2023), National Chicken Council | Vertical Integration, https://www.nationalchickencouncil.org/industry-issues/vertical-integration/. (Accessed 16.02.2025).

Nicol, C.J. and Scott, G.B. (1990), Pre-slaughter handling and transport of broiler chickens, *Appl. Anim. Behav. Sci.*, 28, 57–73.

Nijdam, E., et al. (2004), Factors influencing bruises and mortality of broilers during catching, transport, and lairage, *Poult. Sci.* 83(9), 1610–5. https://doi.org/10.1093/ps/83.9.1610.

Nijdam, E., et al. (2005), Comparison of bruises and mortality, stress parameters, and meat quality in manually and mechanically caught broilers, *Poult. Sci.*, 84(3), 467–74. https://doi.org/10.1093/ps/84.3.467.

Nijdam, E., et al. (2006), Pathological features in dead on arrival broilers with special reference to heart disorders, *Poult. Sci.*, 85(7), 1303–1308.

Northcutt, J.K., et al. (2000), Relationship of broiler bruise age to appearance and tissue histological characteristics, *J. Appl. Poult. Res.*, 9(1), 13–20.

Personal communication (2022), Personal communication by Hanna Hamina 2022, Executive Director of the Finish Poultry Association.

Pilecco, M., et al. (2013), Training of catching teams and reduction of back scratches in broilers, *Rev. Bras. Cienc. Avic.*, 15, 283–286. https://doi.org/10.1590/S1516-635X2013000300017.

Quandt, S.A., et al. (2013), 3-D jobs and health disparities: the health implications of latino chicken catchers' working conditions. *Am. J. Ind. Med.*, 56(2), 206–215. https://doi.org/10.1002/ajim.22072.

Queiroz, M.L., et al. (2015), Environmental and physiological variables during the catching of broilers, *Rev. Bras. Cienc. Avic.*, 17(1), 37–44. https://doi.org/10.1590/1516-635x170137-44.

Reiter, K. and Bessei, W. (2009), Einfluss der Laufaktivität auf die Beinschäden beim Mastgeflügel, *Berl. Muench. Tieraerztl. Wochenschr.*, 122, 2754–2767.

Ritz, C.W., et al. (2005), Evaluation of hot weather thermal environment and incidence of mortality associated with broiler live haul, *J. Appl. Poult. Res.*, 14(3), 594–602.

Rosales, A.G. (1994), Managing stress in broiler breeders: a review, *Appl. Poult. Res.*, 3, 199–207.

Stocker, T., et al. (2025), A gentling program decreases fear-related behaviour towards humans but not towards novel objects in broiler chickens, *Appl. Anim. Behav. Sci.*, 290, 106710. https://doi.org/10.1016/j.applanim.2025.106710.

Taylor, M.H. and Helbacka, N.V. (1968), A study of various factors affecting the bruising of broilers. *Poult. Sci.*, 47, 1616–1623.

Togami, Y., et al. (2024), Hämatome bei Masthühnern: Eine Frage der Farbe. *DGS Magazin für Geflügelwirtschaft* 03/2024. Hrsg.: Eugen Ulmer KG, Stuttgart: 32–33.

Unterholzner, J., et al. (2025), Loading-related injuries of mechanically loaded broilers under field conditions. *Poult. Sci.*

van Emous, R., et al. (2024), Effects of season, catching method, and thinning on carcass quality and production parameters in 4 different broiler production systems in the Netherlands, *Poult. Sci.*, 103, 103688. https://doi.org/10.1016/j.psj.2024.103688.

Vecerek, V., et al. (2006), Effects of travel distance and the season of the year on death rates of broilers transported to poultry processing plants. *Poult. Sci.*, 85(11), 1881–1884. https://doi.org/10.1093/ps/85.11.1881.

Warriss, P.D., et al. (1992), Longer journeys to processing plants are associated with higher mortality in broiler chickens, *Br. Poult. Sci.*, 33(1), 201–206. https://doi.org/10.1080/00071669208417458.

Weeks, C.A. (2007), Poultry handling and transport. In Grandin, T. (eds.) *Livestock handling and transport*, 295–311. Wallingford: CABI Publishing. https://doi.org/10.1079/9781845932190.0295.

Werner, A., et al. (2023), Behavior of broilers and impacts occurring to them during mechanical loading under field conditions. *Poult. Sci.*, 102(7), 102688. https://doi.org/10.1016/j.psj.2023.102688.

Wessel, J., et al. (2022), A comparison of two manual catching methods of broiler considering injuries and behavior. *Poult. Sci.*, 101(11), 102127. https://doi.org/10.1016/j.psj.2022.102127.

Wolff, I., et al. (2019), Harvesting-induced stress in broilers: comparison of a manual and a mechanical harvesting method under field conditions. *Appl. Anim. Behav. Sci.*, *221*, 104877. https://doi.org/10.1016/j.applanim.2019.104877.

Zulkifli, I. and Azah, A.S. (2004), Fear and stress reactions, and the performance of commercial broiler chickens subjected to regular pleasant and unpleasant contacts with human being. *Appl. Anim. Behav. Sci.*, 88, 77–87.

# Chapter 24

## Managing breast myopathies in broiler production

*Martina Bordini, Massimiliano Petracci and Francesca Soglia, University of Bologna, Italy*

1 Introduction
2 Suggested mechanisms and predisposing factors underlying the onset of breast myopathies
3 Potential strategies to mitigate breast myopathies
4 Implication on broiler welfare
5 Conclusion and future trends in research
6 Where to look for further information
7 References

## 1 Introduction

Since the 1960s, poultry meat production and consumption have been rising steadily (OECD-FAO, 2024). This has driven poultry companies to implement intensive breeding programs aimed at achieving a faster growth rate and higher breast yield in commercial meat-type chickens. Overall, the remarkable advancements in poultry genetic selection, nutrition, housing, and meat processing (from slaughtering to packaging) have allowed poultry meat to become the leading contributor to global meat production. However, these improvements have also introduced new challenges. Indeed, the outstanding results gained in terms of productive performance have resulted in the emergence of several breast meat defects. Early examples include deep pectoral myopathy (also known as 'green muscle disease') (Siller, 1985) and the pale soft and exudative-like defects (Solomon et al., 1998). More recently (around fifteen years ago), a group of breast meat abnormalities known as 'growth-related myopathies' started to appear worldwide. These quality defects mainly affect the *Pectoralis major* (PM) muscle of broiler chickens selected for fast growth rate, increased muscle mass accretion, and high feed efficiency. Incidence levels can reach as high as 90% with varying degrees of severity (Petracci et al., 2019), both depending on several factors described later in this chapter.

http://dx.doi.org/10.19103/AS.2025.0165.32

These breast muscle myopathies (BMM), specifically white striping (WS), wooden breast (WB), and spaghetti meat (SM) defects, have raised concerns within both the scientific and industry communities of the poultry sector. Indeed, these defects adversely impact the overall chicken meat quality with detrimental effects on consumer acceptability, thus leading to massive economic losses associated with their occurrence (Petracci et al., 2019). It has been estimated that these myopathies directly and indirectly damage the global poultry industry by over $1 billion per year (Barbut et al., 2024). Rising consumer concern for animal welfare, including meat-type chickens, draws the attention to the potential implications of these meat quality defects on chickens' health and welfare. However, there is a profound lack of knowledge concerning the relationship between BMM and broiler welfare, probably due to the substantial challenges in evaluating welfare parameters in broiler production (Castilho Heiss et al., 2024). This highlights the urgent need for studies addressing this topic, other than practical strategies aimed at controlling the onset of BMM and their negative effects on broiler production.

Within this context, this chapter explores the main features characterizing these meat quality defects, their causes and associated factors, as well as the proposed strategies aimed at mitigating their incidence level and severity. Finally, the noteworthy lack of investigations about the effects of BMM on broiler welfare together with the limited information concerning this subject are addressed at the end of this chapter.

## 1.1 White striping

As shown in Fig. 1, the WS defect is characterized by the macroscopic appearance of white striations on the breast surface at different degrees of thickness. These run parallel to the direction of muscle fibers within the breast (Kuttappan et al., 2012a). Soon after its first description (Kuttappan et al., 2009), researchers put great effort into characterizing WS and establishing suitable criteria to classify this defect. This muscular abnormality was first described as a lesion that occurs on the cranial part of the ventral PM surface. It has also been shown that the thickness and distribution of the white striations can vary depending on the severity of the defect (Kuttappan et al., 2013b). When WS is manifested in mild to moderate degrees of severity, the white stripes are thinner than 2 mm and mainly cover the cranial part of the PM surface. In more severe cases, their thickness exceeds 2 mm, and their distribution affects the entire surface of the breast muscle. Based on this knowledge, the following criteria are widely used to classify WS-affected breasts by using a four-point scale (Aviagen, 2023):

- Score 0: normal – no white stripes on any section of the fillet;
- Score 1: mild – noticeable white stripes less than 1 mm thick, but visually detectable and covering approximately 30% of the fillet surface;

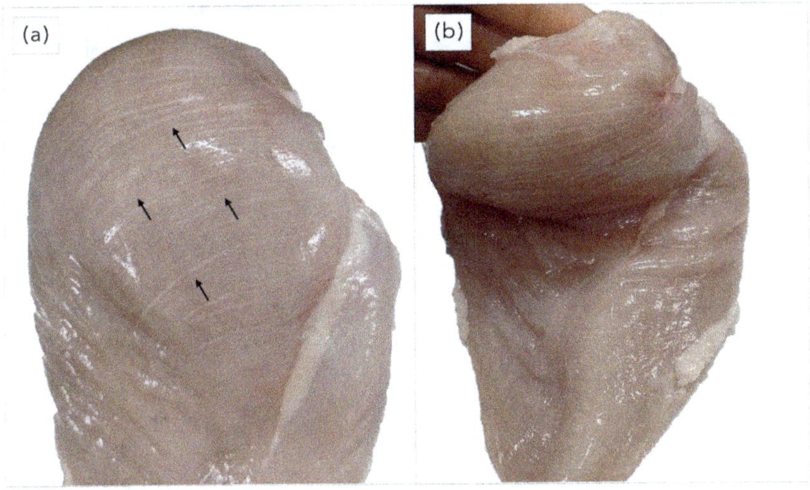


**Figure 1** Representative images of breast fillets affected by white striping. White stripes are characterized by different degrees of thickness (Panel a; black arrows) and mainly occur on the cranial part of the breast (Panel b).

- Score 2: moderate – white stripes <2 mm thick covering the entire surface of the fillet or thicker stripes covering approximately 40% of the fillet; and
- Score 3: severe – white stripes thicker than 2 mm and covering almost the entire surface of the fillet.

Other than characterizing the WS defect in terms of macroscopic features, histological investigations have also been carried out to microscopically characterize the affected breasts (Kuttappan et al., 2013b; Russo et al., 2015; Soglia et al., 2016). At the microscopic level, WS-affected fillets exhibit extensive degeneration of muscle fibers and inflammatory cells infiltration (Russo et al., 2015), mainly composed of macrophages and lymphocytes (Kuttappan et al., 2013b). Muscle fibers going through myodegeneration also undergo regeneration processes, resulting in abnormal lipid (lipidosis) and collagen (fibrosis) deposition (Soglia et al., 2016). These histological lesions (e.g. myodegeneration, lipidosis, and fibrosis) tend to increase as the animals grow (Griffin et al., 2018). Not surprisingly, the histological changes characterizing WS breasts determine profound alterations in the proximate composition of abnormal fillets and, in turn, have detrimental effects on overall chicken meat quality (Soglia et al., 2016, 2019).

As for the main effects of WS on poultry meat quality, this defect significantly alters the composition of breast fillets. Compared to normal breast fillets, affected meat contains a higher level of fat (2.53 vs 0.78%) and total collagen (1.30 vs 1.4%), along with a lower level of protein (20.9 vs 22.9%) (Petracci et al.,

2014). This results in a reduced nutritional value in WS-affected fillets (Petracci et al., 2014; Mudalal et al., 2015; Soglia et al., 2016; Kuttappan et al., 2021). It has been shown that the manifestation of WS, with particular reference to the severe cases (score 3), determines the impairment of technological properties of abnormal chicken meat. Specifically, WS-affected fillets have a reduced water holding capacity (WHC) compared to unaffected ones (26.7% vs 21.3%) (Mudalal et al., 2015). This is likely due to the degeneration of myofibrillar proteins, which are key components of muscle fibers and play an essential role in protein functionality. This deterioration results in functionality impairment and leads to poorer technological properties (e.g. WHC) in poultry meat (Soglia et al., 2016).

In terms of prevalence, the occurrence of WS mainly depends on genetic growth profile, live weight, slaughter age, sex, and breast yield of chickens. Studies from the USA, Canada, Europe, and Thailand estimated the prevalence of moderate WS (score 2) to range from 20% to 98% and from 5% to 30% for severe WS (score 3) cases (Barbut, 2019; Che et al., 2022; Kuttappan et al., 2013a; Lorenzi et al., 2014; Malila et al., 2018; Russo et al., 2015; Trocino et al., 2015). As suggested by the high prevalence of moderate WS (up to 98%), this defect is considered the 'current normal' breast meat coming from fast-growing broilers as most of the chicken breast fillets sold on the market exhibit this defect (Barbut et al., 2024).

### 1.2 Wooden breast

The WB defect is mainly characterized by the hardening of the breast muscle and the presence of out-bulging and pale areas, especially in its thicker part. Depending on the severity of this defect, pale and hard areas, viscous materials, and hemorrhagic petechiae may occur at the surface level of the cranial and caudal part of the pectoral muscle (mild to moderate cases) (Sihvo et al., 2014) or diffuse throughout the whole fillet (the most severe cases) (Fig. 2). These distinctive WB features result in a remarkable inflexibility of the severely affected muscle (Fig. 3).

In the matter of macroscopic appearance, visual examination and/or palpation of muscles still represent the gold standard for the detection and classification of WB myopathy (Barbut et al., 2024). In this regard, the abovementioned macroscopic features have been widely used to classify broiler breast muscles for the presence/absence and degree of severity of WB (Petracci et al., 2019). A four-point score scale (from 0 to 3) is currently used as the most common method to grade the manifestation and severity of WB. This scoring method is based on the degree of hardness/stiffness of the muscle, as well as the presence and distribution of pale areas, viscous materials, and petechial hemorrhages (Aviagen, 2023):

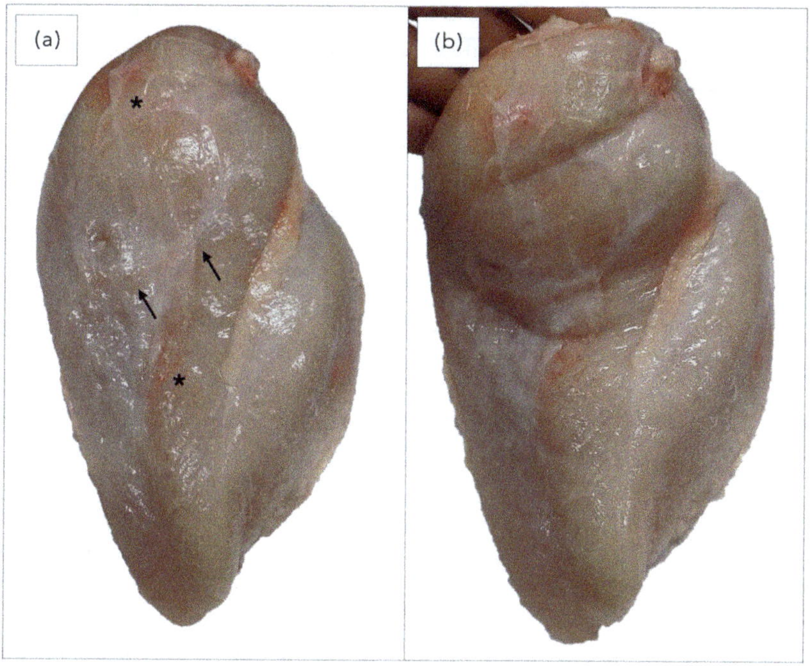


**Figure 2** Representative images of breast fillets affected by wooden breast. Severe woody breast defect is characterized by pale areas with viscous materials (Panel a; black arrows), hemorrhagic petechiae (Panel a; asterisks), and hardening of the breasts with diffuse stiffness (Panel b).

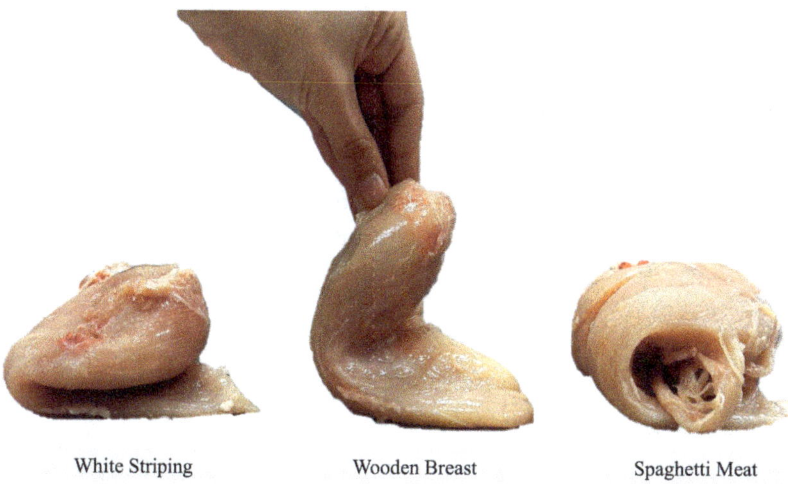


**Figure 3** Representative image of the different degrees of flexibility characterizing pectoral muscles affected by white striping, wooden breast, and spaghetti meat defects, particularly showing the remarkable inflexibility characterizing woody breast fillets.

- Score 0: normal – no hardness/stiffness in any part of the fillet;
- Score 1: mild – firm areas affecting no more than 30% of the breast;
- Score 2: moderate – hardening accounting for an area of approximately 50–60% of the breast, or hardness affecting the entire fillet without compromising its complete flexibility; and
- Score 3: severe – extremely hard and pale areas diffused throughout the muscle (not able to be flexed across its central part – Fig. 3) with the presence of hemorrhages and viscous materials.

At the microscopic level, histological studies showed that, despite their distinctive macroscopic phenotypes, the WB defect essentially shares common histological features with WS (Sihvo et al., 2014; Russo et al., 2015). However, extensive proliferation of connective tissue (fibrosis) appears to be a distinctive feature of WB (Soglia et al., 2016).

Similarly to WS, the occurrence of WB has detrimental effects on both the nutritional value and technological properties of affected fillets (Tijare et al., 2016; Tasoniero et al., 2016; Bošković Cabrol et al., 2024). WB-affected fillets have a significantly higher ($P < 0.001$) level of collagen (1.18 vs 1.09%), moisture (75.3 vs 74.1%), and fat (1.25 vs 0.87%), coupled with a lower protein content (21.4 vs 22.8%) when compared to normal breasts (Soglia et al., 2016). As for some of the most important technological properties, WB meat is characterized by a significantly reduced WHC compared to unaffected ones, as shown by the significantly higher cooking losses in affected fillets (30.1 vs 26.0%; WB vs normal) (Bošković Cabrol et al., 2024).

A few years ago, the prevalence of WB was estimated to range from 60% to 70% for moderate cases and from 11% to 48% for severe cases (Sihvo et al., 2017; Malila et al., 2018). More recently, Che et al. (2022) reported prevalence levels of 70.5% for mild and 11.8% for severe WB. Similarly to WS, estimations of WB occurrence vary largely depending on the genetic profile, sex, age, live weight, and breast yield at slaughter of chickens. For example, in Italy, it has been estimated that levels of occurrence of severe WB were 8% for female broilers and 16% for male broilers (Trocino et al., 2015). In line with these results, its higher prevalence in male broilers than in females can be considered well-established knowledge (Barbut et al., 2024).

### *1.3 Spaghetti meat*

Broiler pectoral muscles affected by SM exhibit reduced muscle integrity due to the detachment and separation of the muscle fiber bundles (Fig. 4). This defect was previously described as 'poor cohesion' of the muscle fiber bundles in broiler pectoral muscle (Sirri et al., 2016) but was subsequently renamed

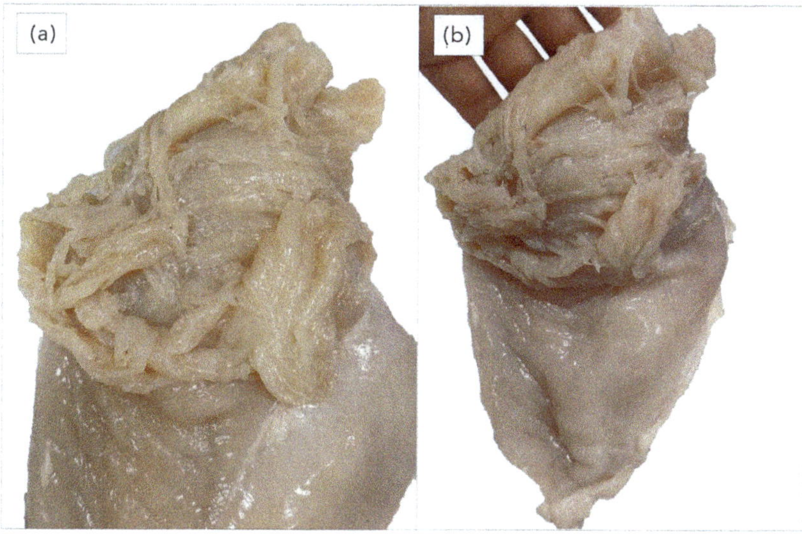


**Figure 4** Representative images of the spaghetti meat defect.

due to its resemblance to the 'spaghetti pasta' (Baldi et al., 2018; Bailey et al., 2020).

Similar to WS and WB, this condition mainly occurs in the cranial and superficial portion of PM muscles and can be classified depending on different degrees of severity evaluated by the visual examination and tactile manipulation of the muscles. In this regard, the presence of SM can be detected and graded based on a three-score scale (from 0 to 2), depending on the visual perception of the detachment of the fiber bundles, as well as the presence/absence of soft and stringy structure perceived by pinching the muscle on its surface (Aviagen, 2023):

- Score 0: normal – no signs of stringy-sponginess/fiber bundles detachment;
- Score 1: mild – no visible detached fiber bundles, but a stringy and soft texture occurs;
- Score 2: severe – detached fiber bundles affecting at least two areas of the breast and/or more than 50% of the surface.

Fillets affected by SM exhibit myodegeneration, necrosis progressing to lysis and concurrent regenerative processes similar to those observed in WS and WB defects. However, SM fillets are distinctively characterized by progressive rarefaction of the endomysial and perimysial connective tissues, which may contribute to the poor cohesion of muscle fiber bundles (Baldi et al., 2018; Wu et al., 2024).

When considering the main effects of SM on meat quality, Baldi et al. (2018) reported a higher moisture level (76.3 vs 75.0 %; $P < 0.001$) along with a decline of total protein content (21.9 vs 23.6 %; $P < 0.001$) in SM-affected breasts compared to non-SM ones. This evidence is supported by the results published by Tasoniero et al. (2020) reporting significant differences in the moisture level (76.9 vs 75.6%; $P < 0.05$) and total crude protein content (20.3 vs 22.5%; $P < 0.001$) between SM-affected and non-affected fillets. The SM defect has an impact on both the nutritional value and visual appearance of fillets (Baldi et al., 2021).

To date, the SM defect can be considered less investigated than the other two myopathies (Baldi et al., 2021). Compared to WS and WB, not many research studies have investigated the incidence level of SM. However, Che et al. (2022) recently showed a prevalence of mild SM (score 1) of 36.3%, while Barbut et al. (2024) reported an incidence level of 21% for mildly affected SM (score 1) and 7% for severe SM (score 2), which is in line with what was previously reported by Baldi et al. (2021), showing an incidence level that can reach up to 20% for mild SM breasts.

## 2 Suggested mechanisms and predisposing factors underlying the onset of breast myopathies

WS, WB, and SM are common myopathies with a worldwide distribution, their frequent occurrence has a negative economic impact on the poultry industry (Kuttappan et al., 2016; Barbut, 2019; Barbut et al., 2024). Significant efforts have been made to understand the underlying mechanisms and the main factors involved in their occurrence. It is now understood that numerous genetic and non-genetic factors influence their occurrence and progression (e.g. genotype, average body weight, nutrition, sex, growth rate, breast yield, etc.) (Bailey, 2023). The following sections will explore the suggested underlying mechanisms responsible for the occurrence and progression of BMM, as well as the predisposing factors associated with their onset.

### *2.1 Suggested underlying mechanisms*

A good understanding of the physiological and biological mechanisms responsible for the manifestation and progression of these meat quality defects is essential in identifying strategies useful to mitigate their incidence and severity.

Despite a rich body of literature on the initial causes of BMM, their precise etiology is still not fully understood. Nonetheless, remarkable improvements have been achieved by the scientific community in terms of understanding and describing the primary mechanisms involved in their onset and progression. Numerous studies dealing with omics technologies, such as genomics

(Pampouille et al., 2018; Lake et al., 2021), transcriptomics (Mutryn et al., 2015; Malila et al., 2021; Wang et al., 2023a; Bordini et al., 2024a; Che et al., 2024), proteomics (Kuttappan et al., 2017; Zhang et al., 2020; Bottje et al., 2021; Kong et al., 2024), and metabolomics (Boerboom et al., 2018, 2023; Kang et al., 2022; Wang et al., 2023b; Wu et al., 2024; Zhao et al., 2024) have been carried out to meticulously characterize the mechanisms underpinning the occurrence of these breast quality defects. These studies have led to a deeper understanding of the molecular and cellular processes associated with the onset of these conditions. Furthermore, they demonstrated both the simultaneous and subsequent occurrence of several biological and physiological phenomena, thus highlighting the complexity of their etiology and the intricacy of the subject. Hypoxia, oxidative stress, impaired muscle regeneration processes, dysregulated lipid and glucose metabolisms, mitochondrial dysfunction, and altered extracellular matrix composition have been identified as conditions relevant to the cascade of events leading to the occurrence of BMM (Soglia et al., 2021a; Wang et al., 2025). Soglia et al. (2021a) reported a thorough schematic representation of the potential time series of events resulting in WS, WB, and SM in fast-growing broilers. The schematic took into consideration the evidence available at the time, concerning both the proposed theories about their etiology and their intricacy in terms of timing and causative consequential sequences of these events. Figure 5 shows a summarized version of the schematic proposed by Soglia et al. (2021a). Some of the evidence supporting theories about the primary causes underlying these myopathies are detailed below.

Some of the most discussed and widely supported hypotheses about the causes of these myopathies rely on the extensive hypertrophic growth (i.e. increased muscle fiber diameter after chickens' hatch) characterizing breast fillets of fast-growing broilers. This type of post-hatch muscle growth boosted by genetic selection determines, at the microscopic level of broiler breast muscles, a reduction of the perimysium (i.e. collagenous connective tissue surrounding fiber bundles) and endomysium (i.e. collagenous connective tissue surrounding individual muscle fibers) layers, which are crucial for the proper stability and functionality of skeletal muscles (Velleman, 2019). The reduced space between fiber bundles and muscle cells can also cause reduced muscle capillarization. This reduction can lead to impaired metabolic waste clearance and oxygen delivery, resulting in hypoxia and oxidative stress in the pectoral muscle (Petracci et al., 2019). These cellular conditions have been extensively explored in muscles affected by WS and WB, numerous studies have identified molecular and biological pathways associated with hypoxia and oxidative stress as potential drivers for the onset and progression of these myopathies (Abasht et al., 2016; Papah et al., 2017; Hasegawa et al., 2022; Alnahhas et al., 2023). Many hypotheses concerning

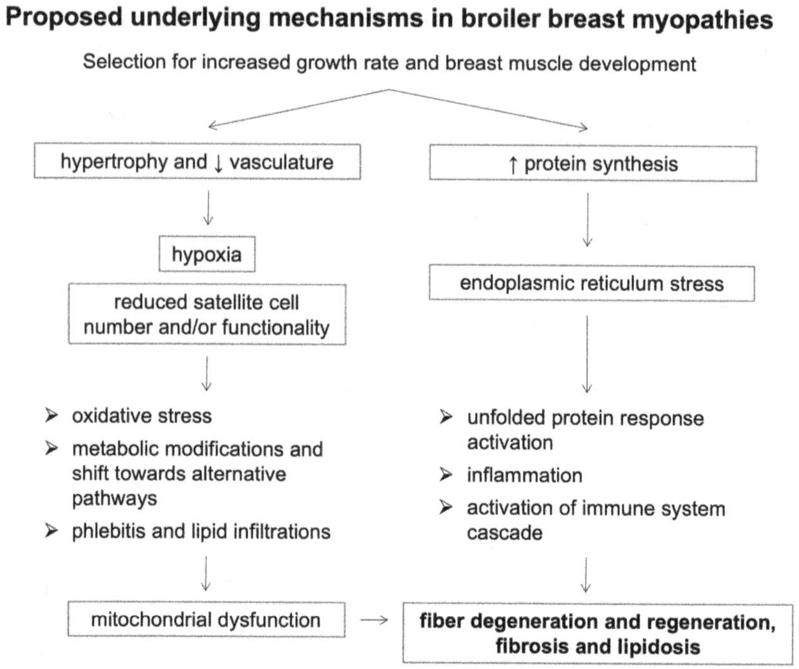


**Figure 5** Schematic representation of proposed underlying mechanisms of the white striping, wooden breast, and spaghetti meat in fast-growing broilers. Source: Adapted from Soglia et al. (2021a).

the cellular and biological alterations determining the manifestation of BMM have been raised in the literature (Soglia et al., 2021a).

For example, the activation of the hypoxia-inducible factor 1 (HIF-1) pathway, a major regulator of cellular responses to hypoxic conditions and oxidative stress, has been suggested to play a key role in the development of these defects (Abasht et al., 2016; Alnahhas et al., 2023). It has also been suggested that its activation could promote some of the major metabolic dysfunctions characterizing BMM (Lake and Abasht, 2020; Papah et al., 2017), such as suppressing canonical pathways of energy metabolisms (e.g. glycolysis and tricarboxylic acid cycle), as well as promoting fatty acid synthesis and lipid accumulation. Additionally, the HIF-1 pathway significantly influences mitochondrial functionality (Alnahhas et al., 2023), which has been described as being severely disrupted in WB-affected muscles (Wang et al., 2023a; Bordini et al., 2024a).

Other than promoting hypoxic conditions, the diminished perimysial and endomysial layers negatively affect the activity of satellite cells (SCs) (Clark and Velleman, 2017; Velleman, 2023). Adult myoblast stem cells that support fiber muscle hypertrophy (i.e. post-hatch growth) and, in case of damage, are responsible for muscle repair and regeneration (Velleman, 2015). These stem cells require an appropriate environment – consisting of a niche between the basal lamina and the sarcolemma (Mauro, 1961) – and sufficient muscular vascularization to exploit their role in sustaining skeletal muscle structure and functionality (Velleman, 2019). Velleman (2015, 2019, 2023) pointed out the impaired condition of SCs – in terms of number and activity – characterizing pectoral muscles belonging to modern fast-growing broilers. The decreased proliferation and differentiation of SCs has been suggested to be associated with the failure of muscle regeneration, leading to an abnormal accumulation of degenerated and necrotic muscle fibers characterizing muscles affected by BMM (Emami et al., 2021; Velleman, 2023).

Furthermore, alterations in the extracellular matrix composition (a functional network surrounding cells of several tissues) are potentially involved in the onset of BMM (Sihvo, 2019; Velleman, 2020; Bordini et al., 2024b, 2021; Pejšková et al., 2023). It has been recently suggested that alterations of extracellular matrix composition can cause endoplasmic reticulum (ER) stress at the endothelial and/or muscle fiber level in abnormal muscles (Bordini et al., 2021, 2024a; Greene et al., 2023; İpek et al., 2024). The authors also suggest that an abnormal accumulation of misfolded extracellular matrix proteins within the ER acts as a trigger for ER stress. As a result, cells activate the unfolded protein response pathway, leading to inflammation and apoptosis within the affected muscles (Soglia et al., 2021b; Bordini et al., 2021, 2024b).

It has also been discovered that there are striking similarities between WB and type 2 diabetes in terms of their underlying mechanisms. Papah et al. (2017), Abasht et al. (2019), and Papah and Abasht (2019) suggested that the WB defect may be considered a phenotypic consequence of lipotoxicity and glucotoxicity resulting from the disruption of canonical glucose and lipid metabolism (Lake and Abasht, 2020). Furthermore, Lake and Abasht (2020) suggested that abnormal lipid accumulation at the perivascular level may induce phlebitis (i.e. inflammation of veins) which, in turn, may lead to the myodegeneration characterizing abnormal fillets.

The scientific community has achieved outstanding results in terms of understanding and describing the metabolic pathways and molecular processes involved in the onset and progression of BMM. The initial mechanisms and subsequent chain of events leading to their onset still requires further research. Considering all the above, it can be stated that despite the increase in knowledge of the complex etiology related to the occurrence of WS, WB, and SM, there is still a need to further explore the intricated biophysical mechanisms underlying their occurrence and progression.

## 2.2 Predisposing factors

Literature showed the higher prevalence of BMM in chicken hybrids selected for fast growth rate compared to the breeds characterized by a slower growth rate (Petracci et al., 2019; Santos et al., 2021). Santos et al. (2021) indicated that moderate and severe (score 2 and 3) cases of WS and WB were significantly higher in conventional fast-growing broilers ($ADG_{0\text{-}48}$ > 60 g/d; 80 and 70% for WB and WS, respectively, at 48 d) compared to those with moderate ($ADG_{0\text{-}62}$ = 50–51 g/d; 30 and 25% for WB and WS, respectively, at 62 d) and slow ($ADG_{0\text{-}62}$ < 50 g/d; 10 and 5% for WB and WS, respectively, at 62 d) growth rates. It is also well-established that a higher incidence and severity of BMM are associated with higher chicken weights at slaughter (Cruz et al., 2017; Papah et al., 2017; Barbut et al., 2024). Bordignon et al. (2022) highlighted that increased slaughter weight can increase the likelihood of SM development (OR: 2.10; 95% CI: 1.18 -3.80), while male chickens with higher daily weight gain (>78 g/d) were more likely to develop WB (OR: 1.75; 95% CI: 1.12 -2.73). As for WS, carcass traits such as breast weight and breast yield significantly affect the occurrence of this defect, with chickens characterized by a high-weight breast (>960 g) showing higher odds of WS occurrence (OR: 1.49; 95% CI: 1.03–2.15). Similar results were found by Kuttappan et al. (2012b) and Bowker et al. (2019), who reported increased prevalence and severity of WS as the fillet weight increased. Bordignon et al. (2022) also evaluated the effect of sex and genotype (i.e. Ross 308, Ross 708, and Cobb 500 as considered genotypes) on myopathies occurrence by using logistic regression analysis. As for chicken sex, the results showed females being less prone to manifest WB (OR: 0.55; 95% CI: 0.40–0.75), but were more susceptible to SM compared to males (OR: 15.4; 95% CI: 8.55–27.8). In males, higher odds of being affected by WB were found (Bordignon et al., 2022). Overall, results suggested a different probability between males and females of manifesting WB or SM. Considering the genotype effect, only WS and WB were found significantly influenced by it, with higher odds of WS occurrence in Cobb 500 than Ross 308 broilers, while Ross 708 genotype showed higher odds of having WB compared to Ross 308 (Bordignon et al., 2022).

Since WS and WB manifest with higher prevalence and degree of severity in heavy broilers selected for faster growth rate, some of the major predisposing factors are linked to the selection of modern chicken hybrids (Trocino et al., 2015; Soglia et al., 2019; Petracci et al., 2019). However, there are several non-genetic factors that can also influence broilers' growth performances and the onset of BMM, such as incubation conditions and management at hatcheries, nutrition and farming practices (e.g. environmental temperature and ventilation), and procedures during *peri-mortem* phases (Bailey, 2023; Barbut et al., 2024). Overall, it can be speculated that any interventions in broiler management determining accelerated growth of chickens and/or larger chickens could place the BMM development at risk, with special reference to WS and WB. For example, evaluating the effect of different nutritional strategies on the prevalence and severity of BMM – especially WS and WB – suggested that, in most cases, a significant reduction in their prevalence was indirectly related to slowing down chickens' growth curve and/or decreasing broilers performance and carcass yields (Petracci et al., 2019; Castilho Heiss et al., 2024). However, even though using nutritional strategies aimed at reducing the incidence level of BMM requires thorough examination in view of their influence on animals' productive traits, a wealth of literature has provided valuable insights into dietary strategies suitable for broiler production, offering potential approaches to mitigate their manifestation. Some of them are discussed later in this chapter (Section 3.2.3).

With particular reference to SM predisposing factors, different studies pointed out its higher prevalence in female chickens than in males (Bordignon et al., 2022; Che et al., 2022) compared to WS and WB. This piece of evidence contributed to supporting the idea that the onset of this myopathy may be associated with factors differing from the other two conditions. In this regard, some researchers hypothesized that, compared to WS and WB, this meat defect could have a stronger relationship with slaughter operations (e.g. chilling method) than chickens' performances and slaughter age (Baldi et al. 2021; Che et al, 2022). In general, data concerning its predisposing factors are still considered fragmentary and under debate, even though Che et al. (2022) reported an association between SM and heavier chickens.

Che et al. (2022) applied multivariate logistic regression models to evaluate risk factors associated with the onset of BMM. They analyzed the risk factors associated with the occurrence of SM and WB in different flocks of Canadian broilers by considering variables related to flock demographics (e.g. age, breed, sex, and average live weight at slaughter), flock health (e.g. mortality rate during entire growing period, usage/not usage of antibiotics, and coccidiosis vaccine administration), transportation from farm to slaughter plant (e.g. chickens per crate, truck travel time, percentage of dead on arrival, and loading duration), processing procedures (e.g. hold time on lairage at plant, exsanguination to deboning time, and type of chilling method), and environmental variables (e.g. season and average temperature during the entire growing period and transport of chickens). They concluded that both SM and WB have a positive association with live weight of chickens at slaughter (SM: OR = 1.30, 95% CI 1.01–1.69; and WB: OR = 1.23, 95% CI 1.03–1.47), which supports research by Bordignon et al. (2022). Furthermore, a higher external temperature during the growing period (between 15.8°C and 22.7°C vs equal or lower 0°C) showed higher incidence rates of SM (OR = 1.75, 95% CI 1.31–2.34), whereas slaughtering chickens during summer was positively associated with higher WB incidence compared to winter (OR = 1.33, 95% CI 1.05–1.70). As for the flock health variables, it is interesting to note that administering a *coccidiosis vaccine* had opposite effects on SM and WB development. Chickens who received the vaccine seemed to be more susceptible to SM (OR = 1.46, 95% CI 1.07 -2.00), while those that didn't had a higher risk of developing WB compared to vaccinated chickens (OR = 1.86, 95% CI 1.51 -2.29). The authors speculated that the effect of administering a *coccidiosis vaccine* may have had an impact on the initial growth performance of broilers, even though the final weight at slaughter was the same. Therefore, this may have influenced the growth trajectory of chickens and the prevalence of SM and WB. The authors also showed a positive correlation between truck travel time and lairage hold times at the plant with the occurrence of SM (OR = 1.64, 95% CI 1.21 -2.22 and OR = 1.89, 95% CI 1.43 -2.49, respectively). The number of chickens per crate and chickens dead on arrival, loading and lairage hold times at the plant were also found to influence the occurrence of WB. The effects of feed withdrawal prior to slaughter can lead to glycogen depletion and influence proteolytic activity within the muscle. Because proteolysis includes collagenases activity, this may play a role in the degradation of connective tissue components, which might be speculated to be associated with different odds of manifesting SM or WB defect (Che et al., 2022). The authors also discovered a higher prevalence of SM in the processing plant adopting the water chilling method (39.8%, 95% CI 38.4 -41.3) compared to the air-chilling method (32.1%, 95% CI 30.8 -33.4). Therefore, it can be suggested that the mechanical effect of water-chilling can increase the likelihood of SM development (Che et al., 2022).

Overall, taking into consideration some of the major factors influencing the incidence and severity of BMM, different strategies built on adjusting nutritional plans and management factors that may contribute to their onset were proposed as potential practical interventions to mitigate broiler breast abnormalities. These strategies are discussed later in this chapter.

## 3 Potential strategies to mitigate breast myopathies

Over the past decade, the poultry community has made significant efforts to better understand the breast myopathies affecting commercial fast-growing chickens. These efforts have led to the development of strategies to reduce their prevalence and minimize their negative effects on broiler production.

Therefore, based on the factors associated with their occurrence (Section 2.2), extensive research has been conducted to identify strategies for reducing the incidence level of BMM and improving the utilization of affected meat. As recently reviewed by Barbut et al. (2024), these strategies can be grouped into long-term and short-term strategies, depending on whether or not they have direct or long-lasting effects on the current poultry industry framework. As for long-lasting solutions, the core strategy relies on the genetic selection, which is considered a key area of opportunity to successfully manage broiler breast myopathies (Barbut et al., 2024). Short-term solutions mainly involve measures that can be implemented at the hatchery (i.e. incubation conditions), farm (e.g. management and dietary interventions), and processing plant, where severely affected meat can be sorted and used for different purposes (e.g. incorporated into further processed products). Some of the main results concerning these strategies are discussed in the following subsections and are also schematically reported in Fig. 6.

### 3.1 Long-term solutions

Since the genetic selection of broiler productive traits has largely been associated with the occurrence of BMM, applying breeding programs targeted at reducing their incidence level could be considered a potential long-lasting solution. However, due to the moderate-to-low heritability ($h^2$) of these meat defects (i.e. the genetic variance of these conditions only explains a low proportion of their phenotypic variance), this strategy is considered challenging and requires a lot of time to see appreciable effects.

More specifically, studies conducted on pure lines have reported low to moderate $h^2$ for each BMM (i.e. 0.04 for SM, 0.024–0.097 for WB, and 0.185–0.338 for WS) (Bailey, 2023). This suggests that, even though genetic factors can influence their occurrence, non-genetic factors (e.g. management and environment) may play a greater role in their development. In contrast, other studies have reportedhigher heritability estimates (from 0.34 to 0.65) of WS and WB in commercial high-breast-yielding chickens (Alnahhas et al., 2016; Lake et al., 2021), thus giving more importance to the genetic determinism involved in broiler myopathies.

Despite the low to moderate $h_2$ values reported by Bailey (2023), a two year genetic selection program in 'higher generation' broilers resulted in an 18.4% reduction in WB incidence while simultaneously improving breast yield by 2.4% compared with commercial broilers (Bailey et al., 2020). These broilers were selectively bred to improve breast yield whilst reducing the genetic predisposition to WB. These results demonstrate that breeding programs targeting a balance between production efficiency and meat quality can effectively reduce the prevalence of breast myopathies and may represent a promising long-term strategy for their management.

WB can be phenotypically detected in live chickens through methods such as breast palpation and wing lift examination (Kawasaki et al., 2016). It is therefore easier to incorporate into a breeding program compared to WS and SM. For WS and SM, the development of biomarkers based on their distinctive metabolic, biological and physiological characteristics, could strengthen

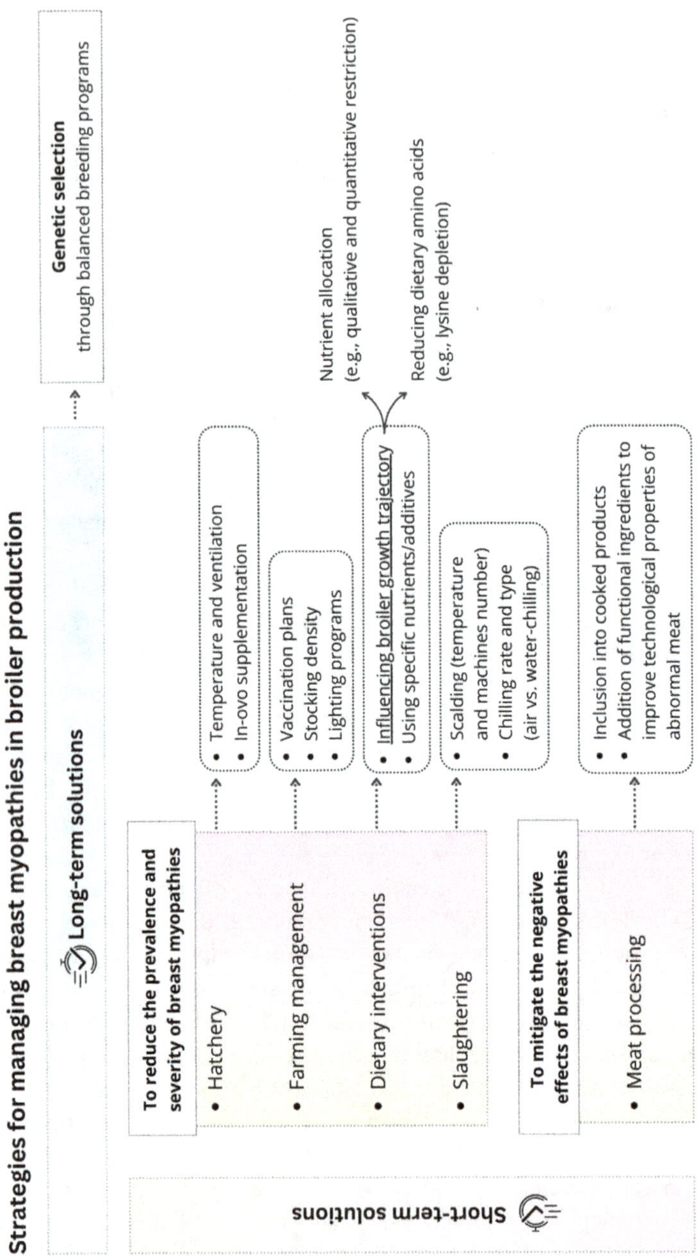


**Figure 6** Schematic representation of strategies proposed and reported in the literature to mitigate the incidence level and severity as well as the negative effects of the occurrence of breast myopathies.

genetic selection efforts. Omics-based approaches have shown promise in identifying biomarkers, including differentially expressed metabolites, proteins, and genes, that may enable the in vivo detection of BMM-affected birds and facilitate their inclusion in breeding strategies (Pampouille et al., 2018; Praud et al., 2020; Kong et al., 2021; Qin et al., 2023; Zhang et al., 2024). Understanding the causes underlying these breast meat defects may also lead to the development of new strategies to control their onset and reduce their negative impact in poultry production. For instance, Velleman (2023) recently pointed out how selection for muscle growth has affected SCs' function in fast-growing, high breast-yield broilers compared to the turkey, which has negligible manifest BMM. Therefore, it was suggested that redirecting the selection for breast muscle accretion, which is mainly based on muscle hypertrophic growth (Berri et al., 2007), to specifically select for hyperplasia and SCs activity may offer a potential solution to mitigating the development of BMM.

### *3.2 Short-term solutions*

Practical interventions in terms of incubation conditions, farming management, dietary plans, and meat processing can be managed to alleviate the detrimental consequences of BMM in broiler production (Bailey, 2023). These approaches, framed as short-term solutions and designated either for reducing the occurrence and severity of BMM or improving the sensory and technological properties of abnormal meat by redirecting it for further processing, are discussed in the following sections.

#### *3.2.1 Hatchery*

Embryo development is a critical phase in broiler production lasting almost one-third of the overall life cycle of fast-growing meat-type chickens (i.e. 21 days of incubation followed by 5–6 weeks of rearing period). Since this phase is crucial for embryonic myogenesis as well as post-hatching muscle development, modulating incubation conditions such as temperature, humidity, ventilation, and lighting can have a strong influence on final meat quality (Oviedo-Rondón et al., 2020; Pierzchala-Koziec et al., 2020; Wang et al., 2022a). Halevy et al. (2006) pointed out how exposing embryos to green light during the incubation period positively affected skeletal muscle development by specifically enhancing differentiation of adult myoblasts. Furthermore, it has been suggested that thermal manipulation (temperature increased up to 39°C, 24 h per day) during mid-term embryogenesis (from d 7 to 16) significantly promoted embryonic muscle development with lasting effects on post-hatch muscle growth by enhancing myoblast proliferation (Piestun et al., 2015). Similarly, Janisch et al. (2015) showed that increasing incubation temperature (38.8°C from d 7 to 10) positively influenced broiler muscle growth. Based on this evidence, experimental studies have been carried out to investigate how modulating incubation conditions can influence muscle development, meat quality, and ultimately the occurrence of pectoral abnormalities. Particularly, Brannan et al. (2021) assessed the effect of thermal manipulation during embryo development on BMM prevalence, following similar experimental conditions tested by Piestun et al. (2015) (i.e. temperature increased to 39.5°C from d 7 to 16, 12 h per day). The authors reported that thermally manipulated embryos showed a lower incidence of WS (1.71 vs 1.96 in control group; $P < 0.01$) and WB (1.70 vs 1.95 in control group; $P < 0.01$), accompanied by a reduced body weight compared to the control group. Consequently, it remains unclear whether the lower prevalence of BMM was directly influenced by the reduced body weight or by physiological changes induced by the thermal manipulation that influenced WS and WB development. These findings support a strong association between the occurrence of these myopathies and breast muscle accretion. However, potential benefits of thermal manipulation on myoblast proliferation and SC activity cannot be excluded and may also have contributed to the observed effects (Brannan et al., 2021).

A different approach has been suggested by Bogucka et al. (2022), who investigated the effect of administering synbiotics – specifically composed of prebiotic galactooligosaccharide (GOS) and *Lactobacillus salivarius* – during embryo incubation on the histological characteristics of broiler breast muscles at slaughter age (42 d). Particularly, the research study tested the in-ovo injection of synbiotics on d 12 of egg incubation and evaluated its effect on the histopathological features characterizing the breast fillets of fast-growing chickens. The results demonstrated a positive effect of the treatment on the breast muscle architecture, as shown by the higher number of normal muscle fibers composing the fillets of in-ovo injected chickens, together with a lower percentage of necrotic fibers in the same group.

In line with the above, Bailey (2023) recently emphasized that optimizing incubation conditions is of utmost importance to managing BMM. In fact, it has been pointed out that temperature variations affect myoblast proliferation and overall muscle growth (Oviedo-Rondón et al., 2020). Particularly, optimal incubation conditions (e.g. temperature and ventilation) are crucial to avoid embryo hypoxia and overheating, which can adversely affect muscle development (Clark et al., 2017; Oviedo-Rondón et al., 2020). For example, prolonged high temperatures (more than 6 h; >38°C) or low oxygen concentrations (17% vs 21%) during incubation have a negative impact on myogenesis and reduces the number of SCs, with increased signs of muscle degeneration (Oviedo-Rondón et al., 2020). Low temperature (36.9°C) during the first 3 days of incubation coupled with high temperature (38.9°C) for the last 3 days of incubation has been shown to be associated with a 2.18 times higher prevalence of moderate WB (score 2) than in broilers subjected to standard incubation temperature (Oviedo-Rondón et al., 2020). Additionally, a key aspect of this phase is represented by its crucial impact on SCs development (Velleman et al., 2014), which, in turn, plays an essential role in supporting muscle development and potentially mitigating the occurrence and progression of BMM. Optimizing early nutrition (i.e. during the first week post hatch) is crucial in maximizing the development of SCs, leading to optimized muscle development and as a consequence, reducing the risk of BMM (Bailey, 2023).

### *3.2.2 Farming management*

In addition to the hatchery, farm management plays a crucial role in growth performance and overall meat quality in poultry production. Not surprisingly, many factors related to management practices during the broiler's growing period can significantly influence the prevalence and severity of myopathies.

Since lighting programs have an influence on broiler performance and welfare (Olanrewaju et al., 2006; Sanotra et al., 2002), primarily due to their effects on feed intake and growth rate (Olanrewaju et al., 2006), these lighting programs have recently been explored as a means of mitigating BMM. Gratta et al. (2023) implemented an experiment aimed at investigating the effects of light restriction (14 h instead of 18 h of light during 24 h) on broiler performance and behavior, as well as on the rate of occurrence of WS and WB. Overall, the study highlighted a reduction in broiler growth rate and final weight at slaughter (45 d) of chickens subjected to light restriction. As expected, the depressed growth rate resulted in a decrease of WS-affected breasts in the experimental group. Despite the negligible effect on mitigating the occurrence of WS and WB without compromising broilers' performance, the authors pointed out that light-restricted chickens (14 h of lights of 24 h) exhibited lower inactive behaviors compared to the control group (18 h of lights of 24 h), thus suggesting a positive effect of light restriction on broiler welfare.

Stocking density also has an impact on broiler growth (Dozier et al., 2005; Estevez, 2007), specifically by influencing broiler feed intake and body weight gain (Tong et al., 2012), as well as animal welfare. Therefore, some authors investigated the effect of increasing stocking density, in combination with dietary modifications (e.g. reduced dietary amino-acid density), on the development of myopathies. Aside from the dietary effect, increasing stocking density determined lower breast weights but had no positive effects on reducing WS and WB (Cônsolo et al., 2022; Pekel et al., 2020).

Another critical aspect of broiler production is represented by gut health (Kogut et al., 2017). A few studies have explored the impact of dietary antibiotics and probiotics, such as *Bacillus subtilis*, on the incidence of WB in broiler chickens. Notably, Jia et al. (2022) reported that such dietary interventions could potentially increase WB occurrence, likely due to their effects on enhancing growth performance. Other evidence suggests that managing gut health may affect the development of WS and WB without compromising productive traits (e.g. body weight and breast weight) (Bean-Hodgins et al., 2022). More specifically, Bean-Hodgins et al. (2022) showed that chickens raised without the use of antibiotics established as 'medically important' by the World Health Organization (WHO) (e.g. Bacitracin), but supplemented with feed additives promoting a stable microbial community (e.g. *Bacillus subtilis*; 0.8 billion CFU/Kg of feed) exhibited a remarkable decrease in both the prevalence and severity of WS (0.22 vs 0.75 in control group, as average mean myopathy score) and WB (0.94 vs 0.44 in control group, as average mean myopathy score) without showing a lower performance (Bean-Hodgins et al., 2022). Taking into consideration chickens' health, it has been reported earlier in this chapter (Section 2.2) that administering/not administering the coccidiosis vaccine might be considered a risk factor for SM and WB occurrence (Che et al., 2022). Additional research is therefore needed to investigate potential associations between BMM and both gut and general health management practices in broiler production. Lastly, considering the importance of gut health on influencing muscle growth (Lahiri et al., 2019), some research studies investigated the role of gut microbiota in WB defect development (Maharjan et al., 2020; Kang et al., 2022; Zhang et al., 2021). These studies highlighted a reduced diversity in terms of microbial species of the intestinal microflora belonging to chickens severely affected by WB. Differences in gut microbial metabolites (e.g. anserine and choline) were found between broilers affected and not affected by the defect. Specifically, WB-affected broilers showed higher levels of these metabolites, which can influence skeletal muscle development (Maharjan et al., 2020; Kang et al., 2022). Therefore, driving gut microbiota of fast-growing broilers through feeding and nutritional strategies may have direct and indirect effects on the development of BMM.

### *3.2.3 Dietary interventions*

Given the remarkably high number of studies investigating the effect of nutritional and feeding interventions, excellent effort has been made by several researchers to summarize as extensively as possible the dietary strategies reported in the literature, in order to efficiently identify those potentially exploitable to reduce the prevalence and severity of BMM (Zampiga and Sirri, 2022; Choi et al., 2023; Trocino et al., 2023; Castilho Heiss et al., 2024). For practical reasons, in this chapter, just some of the most interesting results achieved within this research area have been described.

Most of these experimental studies can be divided into two main groups: the first focused on influencing the broiler growth curve, and the second focused on the use of specific nutrients and/or additives potentially beneficial to the optimal development of chicken breast muscles (e.g. reducing oxidative stress) (Kuttappan et al., 2021).

Regarding the first group of feeding and nutritional interventions, approaches influencing broiler growth curve, such as nutrient allocation (qualitative and quantitative feed restriction), reduced nutrient and dietary amino acids (AA) content, have been widely tested as potential strategies to control the development of BMM, with variable degrees of success on mitigating their severity and occurrence (Cruz et al., 2017; Livingston et al., 2019; Meloche et al., 2018a, 2018b, 2018c; Soglia et al., 2021b; Bertechini et al., 2024). For instance, Meloche et al. (2018c) reported that quantitative feed restriction strategies can lead to a lower incidence and/or a less severe degree of WB and WS when compared to broilers fed *ad libitum*. However, the positive effects on BMM prevalence were accompanied by poor final productive performance of chickens (Meloche et al., 2018c).

To explore the effect of feed restriction on BMM, Radaelli et al. (2017) investigated the effects of quantitative feed restriction (80% of *ad libitum*) on muscle fiber degeneration at day 14, 21, 28, 35, and 46 for broilers. The study showed a decrease in histological alteration in feed restricted chickens compared to those only fed *ad libitum* during the restriction period. However, no positive effects on muscle fiber condition were found after the refeeding phase. Following the same experimental feeding strategy, Trocino et al. (2015) also showed that altering the growth curve through quantitative restriction (80% of *ad libitum*; from 13 to 21 days) not only affected broilers' live performance, but also slightly increased the occurrence of WS (+10%) due to the faster compensatory growth occurring during the refeeding phase. Overall, these studies predominantly suggested that the mitigation of BMM development is possible, but is generally accompanied by decreased final slaughter weight and/or breast yield (Petracci et al., 2019; Kuttappan et al., 2021). In fact, most of the dietary programs based on feed restriction have generally been found to be potentially valuable for controlling BMM but, at the same time, negatively impact the overall efficiency of broilers' productive traits. Furthermore, it is worth noting that an early feed restriction, particularly during the first week post hatch (Velleman et al., 2014), may negatively influence the proliferation and differentiation of SCs, which are essential for proper muscle growth, thus posing a risk for BMM development (Velleman, 2023). In fact, avoiding delayed feeding and/or inadequate nutrition at the very early stage of broilers' lifecycle (from 0 to 7 days of age) is of utmost importance to guarantee proper muscle growth development and thus lowering the risk of muscular abnormalities (Iwasaki et al., 2021; Bailey, 2023).

In terms of strategies that rely on the manipulation of nutrient and/or amino acid content in the diet, modifications of dietary digestible lysine have been investigated as a mitigation strategy for broiler growth trajectory considering its importance for breast muscle accretion (Meloche et al., 2018b; Johnson et al., 2020; Soglia et al., 2021b; Bertechini et al., 2024; Flees et al., 2024). On the one hand, Meloche et al. (2018b) suggested that temporary reductions of digestible lysine density (85% and 75% of the requirement) at different stages of broiler production (12–18 and 19–26 days of age) can have positive effects on mitigating WB and WS severity. However, as expected, even short-term reductions of dietary digestible lysine adversely affect broilers' productive traits. On the other hand, Soglia et al. (2021b) reported that lysine depletion (77% of the requirement) during growing I (10–20 days of age) and growing II (21–34 days of age) feeding phases did not exert significant effects on controlling WS, WB and SM, and, at the same time, had only marginal effects on broiler productive performance. Following a different approach, Cruz et al. (2017) investigated the effect of increased levels of digestible lysine (63% and 70% of the requirement) during two different stages of broiler production (12–28 and 28–42 days of age) on the prevalence and severity of WS and WB. The researchers highlighted a positive correlation between increasing lysine in diets and the occurrence of the defects, probably due to the higher body weight registered for broilers fed a diet formulated with the higher lysine content.

The use of specific nutrients and/or additives as functional ingredients for optimized muscle growth and meat quality has been largely explored in the past ten years, as reported recently by Trocino et al. (2023) and Castilho Heiss et al. (2024). Several researchers investigated the supplementation of diets with antioxidants (e.g. vitamin E) as a strategy to alleviate oxidative stress characterizing breasts affected by WS, WB, and SM (Guetchom et al., 2012; Sirri et al., 2016; Bodle et al., 2018; Estevez and Petracci, 2019; Oviedo-Rondón and Córdova-Noboa, 2020; Kuttappan et al., 2021; Vieira et al., 2021, 2022; Lee et al., 2024; Dayan et al., 2024). Specifically, it has been suggested that supplementing diets with antioxidants may help decrease inflammation and oxidative stress in affected fillets, which could potentially lead to a reduction of the myodegeneration processes occurring in abnormal muscles (Kuttappan et al., 2012c; Wang et al., 2020). However, the use of specific antioxidants and other additives (e.g. vitamins E and C, selenium, and trace minerals) only partially exerted a beneficial influence on mitigating BMM, and there is a lack of consistency among the results obtained from different studies (Sirri et al., 2016; Bodle et al., 2018; Wang et al., 2020; Kuttappan et al., 2021). Kuttappan et al. (2021) observed a decrease in severe WB prevalence (score 3; 15%) when combining dietary antioxidants (such as ethoxyquin), chelated trace minerals (Zn, Cu, and Mn), and organic Se, which are acknowledged as additives preventing oxidative stress compared to the control group (30% of score 3 WB). Wang et al. (2020) showed that supplementing starter diets with vitamin E and lipoic acid could effectively reduce the severity of WB in view of the anti-inflammatory activity. On the other hand, Vieira et al. (2022) showed that supplementing diets with hydroxy-selenomethionine, an organic form of Se, does not influence the occurrence or severity of WB, while Cemin et al. (2018) highlighted that the use of diets supplemented with increasing dietary levels of another organic source of Se (Zinc-L-selenomethionine; 0, 0.2, 0.4, 0.6, 0.8, and 1.0 ppm) led to higher degrees of severity of WS and WB. Lee et al. (2024) reported that the use of dietary chitosan (at least 0.4%) in female broilers' diet during growing and finishing phases showed a reduction in WS occurrence (0.71 vs 1.00 in control group, as means of WS scores; $P < 0.05$), without compromising the overall broilers' growth.

With regard to other feed additives, supplementing diets with guanidine acetic acid (GAA) has been found as a fruitful strategy to reduce the incidence level and severity of SM (Dayan et al., 2024) and WB (Khalil et al., 2021; Maynard et al., 2023; Oviedo-Rondón and Córdova-Noboa, 2020), as well as the degree of severity of WS and WB (de Souza et al., 2021) with no adverse effects on broiler productivity. It has been hypothesized that GAA, by being a precursor of creatine, may have positive effects on improving breast muscle vascularization, vasodilation, and oxygenation (Oviedo-Róndon and Córdova-Noboa, 2020; Maynard et al., 2023; Castilho Heiss et al., 2024; Barbut et al., 2024). Similarly, considering the link between BMM and reduced vasodilation and blood flow (Boerboom et al., 2023), the potential benefits of supplementing additional L-arginine to affect WS and WB prevalence and severity were explored. Bodle et al. (2018) found that dietary supplementation with arginine, combined with vitamin C, during the grower phase could alleviate WB. However, this improvement in WB was once again accompanied by decreased chicken body weight at slaughter and carcass yield. Conversely, Zampiga et al. (2018) reported that supplementing diets with L-arginine did not appear to effectively reduce the incidence of WB and WS, while increasing the arginine-to-lysine ratio by up to 30% lowered the prevalence of BMM without significantly impacting broiler productivity traits (Zampiga et al., 2019).

A different nutritional strategy aimed at specifically mitigating WB incidence was investigated by Greene et al. (2019). They pointed out that a high level of dietary phytase (2000 FTU/kg of feed) significantly reduced the severity of WB. These results seemed to be achieved by a boosted breakdown of phytate in the feed, thus resulting in a greater availability of myo-inositol for the chickens. At the metabolic level, the authors showed that the improvement likely exploited by the increased dietary phytase levels could be related to a reduction in oxidative stress achieved by enhancing oxygen levels and counteracting the hypoxic environment in the muscles (Greene et al., 2019).

Overall, it could be stated that, to date, there is a lack of practical nutritional solutions allowing to achieve a relevant reduction in BMM prevalence without negatively affecting the live and slaughter performance of commercial fast-growing chickens.

### 3.2.4 Slaughtering

In view of their critical role in influencing final meat quality, procedures taking place as part of the slaughtering process, such as stunning, scalding, defeathering, and carcass chilling, could have an impact on the degree of severity with which the BMM manifests. As an example, these processes can determine mechanical stress applied to the muscle surface (e.g. defeathering), as well as influence the *post-mortem* metabolism of breast meat (e.g. scalding temperature and rate of chilling). Cahaner (2024) strengthened the idea already reported in the literature (Baldi et al., 2021) that an aggressive defeathering process can have a strong impact on the SM condition due to the physical/mechanical effects of the machines. In particular, it was reported that reducing the vehemence by which the defeathering machines impact the carcass (e.g. by reducing their number) can decrease the incidence and severity of SM (Cahaner, 2024). Similarly, Che et al. (2022) suggested that the mechanical effect exploited by the water agitation during water-chilling may determine a higher prevalence of SM when compared to the air-chilling method (Section 2.2). Carcass chilling can also determine alterations of the *post-mortem* metabolic activity of breast muscles, thus having direct implications on meat quality (Aviagen, 2023). Specifically, a trial by Aviagen (2023) that examined two different cooling rates showed a relative increase in incidence of SM in slow-chilled broilers (64.85% increase of SM rate; confidence level not shown). It has also been proposed that the rate of chilling (slow vs fast chilling) can influence the incidence of SM by modulating the *post-mortem* activity of proteolytic enzymes and glycogen breakdown and in turn might influence the integrity of muscle fibers. Additional studies should be performed to better identify the association between slaughter operations, the occurrence/severity of BMM, and the potential physiological mechanisms involved to establish the best practices useful to mitigate their manifestation.

### 3.2.5 Meat processing

Among the practical interventions aimed at mitigating the impressive adverse impact of breast myopathies in broiler production, exploiting meat affected by growth-related myopathies for further processing has been proposed and put into practice by the poultry meat industry. This can be possible, for example, through the addition of functional ingredients during meat-processing preparations (i.e. phosphates, starches, hydrocolloids, vegetable proteins), which improves the technological properties of abnormal meat (Petracci et al., 2013; Barbut, 2019). This is particularly important in view of the detrimental effects that the BMM have on the main technological and sensory properties of both raw and processed meat, such as water holding capacity, texture, and general appearance (Tasoniero et al., 2020, 2016).

First of all, proper sorting of carcasses and fillets at the processing plant is critical to detect abnormal meat that can be eventually redirected for further processing. For this purpose, the identification of online methods able to get objective, rapid, and reliable evaluations of fillets is a key point for the industry. In this regard, non-destructive detection methods applying image analysis, air pressure, radiofrequency spectra, and near-infrared hyperspectral imaging, alongside magnetic resonance imaging combined with computer vision techniques (Traffano-Schiffo et al., 2017; Caldas-Cueva et al., 2021a, 2021b; Geronimo et al., 2019; Sun et al., 2021, 2022a), have been proposed to detect fillets severely affected by BMM. This then allows the affected fillets to be diverted to further processing such as chopping or mechanical tenderization of abnormal meat (Barbut et al., 2024). Furthermore, integration of WS and WB meat into cooked products, such as patties, nuggets, sausages, and meatballs (Bordignon et al., 2020; Caldas-Cueva and Owens, 2020; Sun et al., 2022b) can be considered to mitigate the economic losses and food waste potentially resulting from BMM occurrence (Barbut et al., 2024). In this regard, Bordignon et al. (2020) suggested that the occurrence of WS has slight effects on cooking loss attributes of burgers and no negative effects on the quality properties of nuggets obtained by including meat affected by WS. Conversely, Chen et al. (2018) highlighted the difficulties in using WB meat in emulsion-type meat products, due to the severely impaired functional properties of WB meat, which lead to lower water holding/binding capacity and textural characteristics compared to processed products made with normal breasts. Similarly, Caldas-Cueva et al. (2021a) reported that totally cooked products (100%) as well as partially (33% and 67%) made with severe WB showed higher cooking loss in respect to products made with 100% of normal breast meat. On the other hand, Sun et al. (2022b) recently demonstrated that using up to 25% WB meat could produce chicken meatballs still suitable for the industry and acceptable for consumers. Likewise, Barbut et al. (2024) reported industry data showing that partial inclusion (up to 50%) of severe WB-affected meat in breaded chicken nuggets preparation did not affect their technological properties and overall consumer liking.

Proposed strategies for alleviating the negative effects of BMM in further processing include vacuum-tumbling marination, which is widely recognized for significantly improving meat tenderness and is a common approach for adding value to poultry products (Bowker et al., 2018; Lima et al., 2022). However, it is well established that the manifestation of BMM negatively impacts the meat's ability to absorb marinade solutions (Tijare et al., 2016; U-Chupaj et al., 2017; Bowker et al., 2018; Maxwell et al., 2018). Consequently, the reduced marinade uptake and diminished water-holding capacity of WB and WS fillets limit the effectiveness of processing strategies that, under normal conditions, would enhance the sensory and technological qualities of poultry meat, such as texture. Nonetheless, U-Chupaj et al. (2021) showed that a non-phosphate and low-sodium marination, combined with vacuum-tumbling, can enhance the water-holding capacity of WS fillets, as evidenced by the higher moisture content and total cooked yield, coupled with the lower cook losses characterizing marinated WS fillets compared to the non-marinated WS counterparts. In terms of improving the textural properties of WB fillets, Baldi et al. (2023) proposed the use of High-Intensity Ultrasonication to attenuate the toughness of meat affected by this defect, by showing ameliorated textural properties (reduced compression force) of the treated WB meat compared to the untreated one.

As for the use of functional ingredients to ameliorate the technological properties of affected meat, Lima et al. (2022) demonstrated that including papain enzyme (from 0.10% to 0.30%), a heat-stable cysteine protease, in marinated WB meat can significantly improve its texture by breaking the chemical bonds between myofibrillar proteins and collagen, thereby reducing the hardness of the affected tissue.

Even in terms of further meat processing, SM has not been investigated as much as the other two meat quality defects. Because of its opposite features with respect to WB and WS, processing strategies generally used for their mitigation are not suitable for fillets characterized by the SM condition. Despite that, in the past few years, some studies investigating alternative formulations for SM products have been carried out. In this context, the use of dairy proteins (e.g. whole milk powder) in the formulation of cooked SM minced meat has been proposed to obtain products having texture profiles comparable with normal samples (Wang et al., 2022b). Furthermore, Wang et al. (2023a) recently suggested the use of alginate gels (i.e. linear polymers formed by guluronic and mannuronic acid) to positively influence the technological properties of SM-affected meat. Overall, the study showed positive effects related to their inclusion in cooked SM products, especially in relation to their effects on the textural properties and water-holding capacity of affected meat.

The significant negative effects of BMM on the sensory and technological qualities of chicken meat highlight the importance of developing of processing strategies to improve affected fillets, especially with regard to texture and WHC. Despite the progress that has been made to help mitigate some of these meat quality issues, these approaches still represent a cost for the industry and emphasize the importance of reducing BMM occurrence.

## 4 Implication on broiler welfare

To date, very few studies have taken into consideration the implications of breast meat defects on broilers' health and welfare (Che et al., 2023; Kawasaki et al., 2016; Norring et al., 2019). While Kawasaki et al. (2016) and Che et al. (2023) explored clinical symptoms associated with WB, the first study investigating how it may affect animal welfare was published in 2019 (Norring et al., 2019). Particularly, the researchers investigated potential connections between the WB defect and animal behavior and activity, specifically investigating broilers' mobility by using the gait score. The study pointed out an association between a general impairment in broilers' walking ability and WB, as evidenced by the poorer gait in chickens with WB compared to healthy ones. The authors also suggested that the heavy weight could be a common underlying factor for both poor gait and the meat defect, thus making it difficult to differentiate the effects of chickens' weight and the myopathy. Overall, the study raised concerns on the potential effects of WB on broiler welfare and demanded additional studies to further investigate this aspect. Despite that, no other studies directly exploring the effect of BMM on broilers' mobility have been published since then.

Because BMM are characterized by severe myodegenerative processes, which in human medicine have been reported to cause substantial pain, physical distress associated with WB has been hypothesized (Norring et al., 2019). Since scoring pain caused by WB can be considered very difficult to implement, the investigation of biological markers associated with inflammation and other biological/cellular processes directly associated with the myodegenerative process could be considered likely exploitable in the future to investigate the implications of BMM on broiler welfare.

## 5 Conclusion and future trends in research

Scientists and industry personnel belonging to the poultry community consider the BMM a major challenge for broiler production. Besides, there is a broad consensus regarding the urgency of identifying strategies and approaches useful to reduce their occurrence. As a long-lasting solution, balanced breeding programs are considered a core strategy to reduce the prevalence of these myopathies without completely renouncing to improve broiler productive performance. However, it may require some time to obtain noticeable effects. In the meantime, there is a need for practical interventions on the non-genetic factors influencing the occurrence and severity of BMM to be used as short-term strategies. Although numerous approaches implemented throughout the whole production process (from the hatchery to the processing plant) to reduce the incidence of BMM and mitigate their negative effects on meat quality have been tested, all can be considered only partially effective when it comes to productive performance and economic loss. In fact, in terms of farming and dietary strategies, it could be generally stated that there is a lack of solutions consistently allowing poultry producers to achieve a relevant reduction in BMM incidence and severity without negatively affecting the live and slaughter performance of fast-growing chickens. As for meat processing, sorting out abnormal meat and redirecting it for further processing can be considered a fruitful approach to mitigate the adverse effects of BMM, but having an economic impact on the poultry industry. Finally, a concern for possible effects of wooden breast myopathy on animal welfare has been raised, but it still needs to be addressed. Indeed, the limited availability of studies investigating the effects of pectoral myopathies on animal welfare highlights a critical gap that warrants immediate attention from the scientific community.

## 6 Where to look for further information

The following articles provide a good overview of the subject:

- Wang, Y., et al. 2025. Oxidative stress-induced changes in wooden breast and mitigation strategies: a review. *Compr Rev Food Sci Food Saf* (in press).
- Barbut, S., et al. 2024. Review: Myopathies in broilers: supply chain approach to provide solutions to challenges related to raising fast growing birds. *Poult Sci* 103(8), 103801. https://doi.org/10.1016/j.psj.2024.103801.
- Rimmer, L. A. & Zumbaugh, M. D. 2024. Skeletal muscle metabolic characteristics and fresh meat quality defects associated with wooden breast. *Front Physiol* 15, 1501362. https://doi.org/10.3389/fphys.2024.1501362.
- Soglia, F., et al. 2021. A critical review of the mechanisms involved in the occurrence of growth-related abnormalities affecting broiler chicken breast muscles. *Poult Sci* 100(6), 101180. https://doi.org/10.1016/J.PSJ.2021.101180.

The following book provides a good overview of the subject:

- Petracci, M. and Estévez, M. 2023. *Improving poultry meat quality*. Burleigh Dodds Science Publishing Limited, Cambridge, UK.

Key research in this area can be found at the following associations:

- World's Poultry Science Association (WPSA) (https://www.wpsa.com/).
- Poultry Science Association (PSA) (https://poultryscience.org/home).
- National Chicken Council (NCC) (https://www.nationalchickencouncil.org/).

- International Poultry Congress (IPC) (https://internationalpoultrycouncil.org/).
- Association of Poultry Processors and Poultry Trade in the EU Countries (AVEC) (https://avec-poultry.eu/).

## 7 References


Abasht, B., et al., 2016. Oxidative stress and metabolic perturbations in wooden breast disorder in chickens. *PLoS One* 11(4), e0153750. https://doi.org/10.1371/journal.pone.0153750.

Abasht, B., et al., 2019. The metabolic characteristics of susceptibility to wooden breast disease in chickens with high feed efficiency. *Poult Sci* 98, 3246–3256. https://doi.org/10.3382/ps/pez183.

Alnahhas, N., et al., 2016. Genetic parameters of white striping in relation to body weight, carcass composition, and meat quality traits in two broiler lines divergently selected for the ultimate pH of the pectoralis major muscle. *BMC Genet* 17, 61. https://doi.org/10.1186/s12863-016-0369-2.

Alnahhas, N., et al., 2023. The hypoxia-inducible factor 1 pathway plays a critical role in the development of breast muscle myopathies in broiler chickens: a comprehensive review. *Front Physiol* 14, 1260987. https://doi.org/10.3389/fphys.2023.1260987.

Aviagen, 2023. Broiler myopathies handbook. https://aviagen.com/assets/Tech_Center/Broiler_Breeder_Tech_Articles/English/BroilerMyopathieHandbook-2023_EN.pdf (Accessed December 1, 2024).

Bailey, R.A., et al., 2020. Characterising the influence of genetics on breast muscle myopathies in broiler chickens. *Front Physiol* 11, 01041. https://doi.org/10.3389/FPHYS.2020.01041.

Bailey, R.A., 2023. Strategies and opportunities to control breast myopathies: an opinion paper. *Front Physiol* 14, 1173564. https://doi.org/10.3389/FPHYS.2023.1173564.

Baldi, G., et al., 2018. Implications of white striping and spaghetti meat abnormalities on meat quality and histological features in broilers. *Animal* 12, 164–173. https://doi.org/10.1017/S1751731117001069.

Baldi, G., et al., 2021. Spaghetti meat abnormality in broilers: current understanding and future research directions. *Front Physiol* 12, 684497. https://doi.org/10.3389/FPHYS.2021.684497.

Baldi, G., et al., 2023. High-intensity ultrasonication as an innovative approach for the softening of wooden breast meat in broilers. *MMB* 7, 16092–16093. https://doi.org/10.22175/MMB.16092.

Barbut, S., 2019. Recent myopathies in broiler's breast meat fillets. *Worlds Poult Sci J* 75, 559–582. https://doi.org/10.1017/S0043933919000436.

Barbut, S., et al., 2024. Review: myopathies in broilers: supply chain approach to provide solutions to challenges related to raising fast growing birds. *Poult Sci* 103(8), 103801. https://doi.org/10.1016/j.psj.2024.103801.

Bean-Hodgins, L., et al., 2022. Comparative impact of conventional and alternative gut health management programs on growth performance and breast meat quality in broiler chickens raised in commercial and research settings. *J Appl Poult Res* 31, 100228. https://doi.org/10.1016/J.JAPR.2021.100228.

Berri, C., et al., 2007. Consequence of muscle hypertrophy on characteristics of Pectoralis major muscle and breast meat quality of broiler chickens. *J Anim Sci* 85, 2005–2011. https://doi.org/10.2527/JAS.2006-398.

Bertechini, A.G., et al., 2024. Dietary nutritional density, calcium pidolate, and lysine source on performance, occurence of myopathies and other carcass quality properties in broilers. *Livest Sci* 280, 105403. https://doi.org/10.1016/J.LIVSCI.2024.105403.

Bodle, B.C., et al., 2018. Evaluation of different dietary alterations in their ability to mitigate the incidence and severity of woody breast and white striping in commercial male broilers. *Poult Sci* 97, 3298–3310. https://doi.org/10.3382/PS/PEY166.

Boerboom, G., et al., 2018. Unraveling the cause of white striping in broilers using metabolomics. *Poult Sci* 97, 3977–3986. https://doi.org/10.3382/PS/PEY266.

Boerboom, G.M., et al., 2023. Metabolomic analysis of wooden breast myopathy shows a disturbed lipid metabolism. *Metabolites* 13, 20. https://doi.org/10.3390/metabo13010020.

Bogucka, J., et al., (Eds.). 2022. Effect of synbiotics administered in ovo on microvascularization and histopathological changes in pectoral muscle and the biochemical profile of broiler chicken blood. *Poult Sci* 101, 101628. https://doi.org/10.1016/J.PSJ.2021.101628.

Bordignon, S., et al., 2020. The use of white striped chicken breasts on the quality of nuggets and hamburgers. *Food Sci Technol* 41, 570–575. https://doi.org/10.1590/FST.16320.

Bordignon, F., et al., 2022. Factors affecting breast myopathies in broiler chickens and quality of defective meat: a meta-analysis. *Front Physiol* 13, 933235. https://doi.org/10.3389/FPHYS.2022.933235.

Bordini, M., et al., 2021. Weighted gene co-expression network analysis identifies molecular pathways and hub genes involved in broiler White Striping and Wooden Breast myopathies. *Sci Rep* 11. https://doi.org/10.1038/s41598-021-81303-7.

Bordini, M., et al., 2024a. RNA-sequencing revisited data shed new light on wooden breast myopathy. *Poult Sci* 103, 103902. https://doi.org/10.1016/J.PSJ.2024.103902.

Bordini, M., et al., 2024b. Time course evaluation of collagen type IV in Pectoralis major muscles of broiler chickens selected for different growth-rates. *Poult Sci* 103, 103179. https://doi.org/10.1016/J.PSJ.2023.103179.

Bošković Cabrol, M., et al., 2024. Nutritional composition, technological quality, and sensory attributes of chicken breast meat affected by white striping, wooden breast, and spaghetti meat: a comprehensive evaluation. *Foods* 13, 4007. https://doi.org/10.3390/foods13244007.

Bottje, W.G., et al, 2021. Upstream regulator analysis of wooden breast myopathy proteomics in commercial broilers and comparison to feed efficiency proteomics in pedigree male broilers. *Foods* 10, 104. https://doi.org/10.3390/FOODS10010104.

Bowker, B.C., et al., 2018. Marination and cooking performance of portioned broiler breast fillets with the wooden breast condition. *Poult Sci* 97, 2966–2970. https://doi.org/10.3382/PS/PEY144.

Bowker, B., et al., 2019. relationships between attributes of woody breast and white striping myopathies in commercially processed broiler breast meat. *J Appl Poult Res* 28, 490–496. https://doi.org/10.3382/japr/pfz007.

Brannan, K.E., et al., 2021. Embryonic thermal manipulation and dietary fat source during acute heat stress: 2. Effect on broiler carcass characteristics and breast muscle myopathies. *J Appl Poult Res* 30, 100174. https://doi.org/10.1016/J.JAPR.2021.100174.

Cahaner, A., 2024. *Defeathering and its implications on the spaghetti meat syndrome*. 113th Poultry Science Association Annual Meeting. Champaign, IL: Louisville, KY, 283–284.

Caldas-Cueva, J.P. and Owens, C.M., 2020. A review on the woody breast condition, detection methods, and product utilization in the contemporary poultry industry. *J Anim Sci* 98(8), skaa207. https://doi.org/10.1093/jas/skaa207.

Caldas-Cueva, J.P., et al., 2021a. Detection of woody breast condition in commercial broiler carcasses using image analysis. *Poult Sci* 100, 100977. https://doi.org/10.1016/J.PSJ.2020.12.074.

Caldas-Cueva, J.P., et al., 2021b. Use of image analysis to identify woody breast characteristics in 8-week-old broiler carcasses. *Poult Sci* 100, 100890. https://doi.org/10.1016/J.PSJ.2020.12.003.

Castilho Heiss, V.A.R., et al. 2024. Chickens. *Worlds Poult Sci J* 80, 897–914. https://doi.org/10.1080/00439339.2024.2338350.

Cemin, H.S., et al., 2018. Broiler responses to increasing selenium supplementation using Zn-L-selenomethionine with special attention to breast myopathies. *Poult Sci* 97, 1832–1840. https://doi.org/10.3382/PS/PEY001.

Che, S., et al., 2022. Prevalence of breast muscle myopathies (spaghetti meat, woody breast, white striping) and associated risk factors in broiler chickens from Ontario Canada. *PLoS One* 17, 0267019. https://doi.org/10.1371/journal.pone.0267019.

Che, S., et al., 2023. Characterization of dorsal recumbency syndrome associated with woody breast in broiler flocks from Ontario, Canada. *Poult Sci* 102, 102307. https://doi.org/10.1016/J.PSJ.2022.102307.

Che, S., et al., 2024. Transcriptomic profiles of pectoralis major muscles affected by spaghetti meat and woody breast in broiler chickens. *Animals* 14, 176. https://doi.org/10.3390/ANI14020176.

Chen, H., et al., 2017. Chicken breast quality – normal, pale, soft and exudative (PSE) and woody – influences the functional properties of meat batters. *Int J Food Sci Technol* 53, 654–664. https://doi.org/10.1111/IJFS.13640.

Choi, J., et al., 2023. Nutritional strategies to improve meat quality and composition in the challenging conditions of broiler production: a review. *Animals* 13, 1386. https://doi.org/10.3390/ANI13081386.

Clark, D.L. and Velleman, S.G., 2017. Physiology and reproduction: Spatial influence on breast muscle morphological structure, myofiber size, and gene expression associated with the wooden breast myopathy in broilers. *Poult Sci* 95, 2930–2945. https://doi.org/10.3382/ps/pew243.

Clark, D. L., et al., 2017. Incubation temperature and time of hatch impact broiler muscle growth and morphology. *Poult Sci*, 96, 4085–4095. https://doi.org/10.3382/ps/pex202.

Cônsolo, N.R.B., et al., 2022. Using TD-NMR relaxometry to assess the effects of diet type and stocking rate on the incidence and degree of severity of myopathies in broilers. *Microchem J* 181, 107745. https://doi.org/10.1016/J.MICROC.2022.107745.

Cruz, R.F.A., et al., 2017. Occurrence of white striping and wooden breast in broilers fed grower and finisher diets with increasing lysine levels. *Poult Sci* 96, 501–510. https://doi.org/10.3382/PS/PEW310.

Dayan, J., et al., 2024. Dietary guanidinoacetate reduces spaghetti meat myopathy risk in the breast muscle of broiler chickens. *Animal* 18, 101144. https://doi.org/10.1016/J.ANIMAL.2024.101144.

de Souza, C., et al., 2021. Effect of dietary guanidinoacetic acid or nucleotides supplementation on growth performances, carcass traits, meat quality and occurrence of myopathies in broilers. *Livest Sci* 251, 104659. https://doi.org/10.1016/J.LIVSCI.2021.104659.

Dozier, W.A., et al., 2005. Stocking density effects on growth performance and processing yields of heavy broilers. *Poult Sci* 84, 1332–1338. https://doi.org/10.1093/PS/84.8.1332.

Emami, N.K., et al., 2021. Hypoxia further exacerbates woody breast myopathy in broilers via alteration of satellite cell fate. *Poult Sci* 100, 101167. https://doi.org/10.1016/J.PSJ.2021.101167.

Estevez, I., 2007. Density Allowances for broilers: Where to set the limits? *Poult Sci* 86, 1265–1272. https://doi.org/10.1093/PS/86.6.1265.

Estevez, M. and Petracci, M., 2019. Benefits of magnesium supplementation to broiler subjected to dietary and heat stress: improved redox status, breast quality and decreased myopathy incidence. *Antioxidants* 8, 456. https://doi.org/10.3390/ANTIOX8100456.

Flees, J.J., et al., 2024. Effects of light intensity and reduction of starter diet digestible lysine and metabolizable energy on broiler chicken growth performance, breast meat yield, and meat quality defects. *Poult Sci* 103, 103222. https://doi.org/10.1016/J.PSJ.2023.103222.

Gao, J., et al., 2008. Corticosterone alters meat quality by changing pre- and post-slaughter muscle metabolism. *Poult Sci* 87, 1609–1617. https://doi.org/10.3382/ps.2007-00007.

Geronimo, B.C., et al., 2019. Computer vision system and near-infrared spectroscopy for identification and classification of chicken with wooden breast, and physicochemical and technological characterization. *Infrared Phys Technol* 96, 303–310. https://doi.org/10.1016/J.INFRARED.2018.11.036.

Gratta, F., et al., 2023. Effect of light restriction on productive results and behavior of broiler chickens. *Poult Sci* 102, 103084. https://doi.org/10.1016/J.PSJ.2023.103084.

Greene, E., et al., 2019. Quantum blue reduces the severity of woody breast myopathy via modulation of oxygen homeostasis-related genes in broiler chickens. *Front Physiol* 10, 1251. https://doi.org/10.3389/FPHYS.2019.01251.

Greene, E.S., et al, S., 2023. Potential role of endoplasmic reticulum stress in broiler woody breast myopathy. *Am J Physiol Cell Physiol* 324(3), C679–C693. https://doi.org/10.1152/AJPCELL.00275.2022.

Griffin, J.R., et al., 2018. Onset of white striping and progression into wooden breast as defined by myopathic changes underlying Pectoralis major growth. Estimation of growth parameters as predictors for stage of myopathy progression. *Avian Pathol* 47(1), 2–13. https://doi.org/10.1080/03079457.2017.1356908.

Guetchom, B., et al., 2012. Effect of extra dietary vitamin E on preventing nutritional myopathy in broiler chickens. *J Appl Poult Res* 21, 548–555. https://doi.org/10.3382/JAPR.2011-00440.

Halevy, O., et al., 2006. In ovo exposure to monochromatic green light promotes skeletal muscle cell proliferation and affects myofiber growth in posthatch chicks. *Am J Physiol Regul Integr Comp Physiol* 290(4), R1062-R1070. https://doi.org/10.1152/AJPREGU.00378.2005.

Hasegawa, Y., et al., 2022. Mitochondrial characteristics of chicken breast muscle affected by wooden breast. *Comp Biochem Physiol A Mol Integr Physiol* 273, 111296. https://doi.org/10.1016/j.cbpa.2022.111296.

İpek, E., et al., 2024. Endoplasmic reticulum stress and associated apoptosis are linked with the pathogenesis of white striping in broiler breast muscles. *Poult Sci* 103, 104103. https://doi.org/10.1016/J.PSJ.2024.104103.

Iwasaki, T., et al., 2021. Nutrition during the early rearing period affects the incidence of wooden breasts in broilers. *J Poult Sci* 58, 177–185. https://doi.org/10.2141/jpsa.0200034.

Janisch, S., et al., 2015. Changing the incubation temperature during embryonic myogenesis influences the weight performance and meat quality of male and female broilers. *Poult Sci* 94, 2581–2588. https://doi.org/10.3382/PS/PEV239.

Jia, L., et al., 2022. Bacitracin, Bacillus subtilis, and Eimeria spp. challenge exacerbates woody breast incidence and severity in broilers. *Poult Sci* 101, 101512. https://doi.org/10.1016/j.psj.2021.101512.

Johnson, C.A., et al., 2020. Effects of amino acid and energy density on growth performance and processing yield of mixed-sex Cobb 700 × MV broiler chickens. *J Appl Poult Res* 29, 269–283. https://doi.org/10.1016/J.JAPR.2019.10.014.

Kang, K., et al., 2022. Multi-omics analysis of the microbiome and metabolome reveals the relationship between the gut microbiota and wooden breast myopathy in broilers. *Front Vet Sci* 9, 922516. https://doi.org/10.3389/fvets.2022.922516.

Kawasaki, T., et al., 2016. Simple method for screening the affected birds with remarkably hardened pectoralis major muscles among broiler chickens. *J Poult Sci* 53, 291–297. https://doi.org/10.2141/jpsa.0160036.

Khalil, S., et al., 2021. Effects of guanidinoacetic acid supplementation on productive performance, pectoral myopathies, and meat quality of broiler chickens. *Animals* 11, 3180. https://doi.org/10.3390/ANI11113180.

Kogut, M.H., et al., 2017. Gut health in poultry. *CAB Rev* 12, 031. https://doi.org/10.1079/PAVSNNR201712031.

Kong, F., et al., 2021. Serum creatine kinase as a biomarker to predict wooden breast in vivo for chicken breeding. *Front Physiol* 12, 711711. https://doi.org/10.3389/fphys.2021.711711.

Kong, B., et al., 2024. Proteomic analyses on chicken breast meat with white striping myopathy. *Poult Sci* 103, 103682. https://doi.org/10.1016/J.PSJ.2024.103682.

Kuttappan, V., et al., 2009. Effect of white striping on the histological and meat quality characteristics of broiler fillets. *Poult Sci*, 88, 136–137.

Kuttappan, V.A., et al., 2012a. Consumer acceptance of visual appearance of broiler breast meat with varying degrees of white striping. *Poult Sci* 91, 1240–1247. https://doi.org/10.3382/ps.2011-01947.

Kuttappan, V.A., et al., 2012b. Influence of growth rate on the occurrence of white striping in broiler breast fillets. *Poult Sci* 91, 2677–2685. https://doi.org/10.3382/ps.2012-02259.

Kuttappan, V.A., et al., 2012c. Effect of different levels of dietary vitamin E (dl-α-tocopherol acetate) on the occurrence of various degrees of white striping on broiler breast fillets. *Poult Sci* 91, 3230–3235. https://doi.org/10.3382/PS.2012-02397.

Kuttappan, V.A., et al., 2013a. Estimation of factors associated with the occurrence of white striping in broiler breast fillets. *Poult Sci* 92, 811–819. https://doi.org/10.3382/ps.2012-02506.

Kuttappan, V.A., et al., 2013b. Pathological changes associated with white striping in broiler breast muscles. *Poult Sci* 92, 331–338. https://doi.org/10.3382/ps.2012-02646.

Kuttappan, V.A., et al., 2016. White striping and woody breast myopathies in the modern poultry industry: A review. *Poult Sci* 95, 2724–2733. https://doi.org/10.3382/ps/pew216.

Kuttappan, V.A., et al., 2017. Proteomic analysis reveals changes in carbohydrate and protein metabolism associated with broiler breast myopathy. *Poult Sci* 96, 2992–2999. https://doi.org/10.3382/ps/pex069.

Kuttappan, V.A., et al., 2021. Nutritional intervention strategies using dietary antioxidants and organic trace minerals to reduce the incidence of wooden breast and other carcass quality defects in broiler birds. *Front Physiol* 12, 428. https://doi.org/10.3389/FPHYS.2021.663409.

Lahiri, S., et al., 2019. The gut microbiota influences skeletal muscle mass and function in mice. Sci Transl Med 11. https://doi.org/10.1126/SCITRANSLMED.AAN5662.

Lake, J.A. and Abasht, B., 2020. Glucolipotoxicity: A proposed etiology for wooden breast and related myopathies in commercial broiler chickens. *Front Physiol* 11, 00169. https://doi.org/10.3389/fphys.2020.00169.

Lake, J.A., et al., 2021. Genetic basis and identification of candidate genes for wooden breast and white striping in commercial broiler chickens. *Sci Rep* 11(1), 6785. https://doi.org/10.1038/s41598-021-86176-4.

Lee, J., et al., 2024. The effects of supplemental dietary chitosan on broiler performance and myopathic features of white striping. *Poult Sci* 103, 103396. https://doi.org/10.1016/J.PSJ.2023.103396.

Lima, J.L., et al., 2022. Improving the poor texture and technological properties of chicken wooden breast by enzymatic hydrolysis and low-frequency ultrasound. *J Food Sci* 87, 2364–2376. https://doi.org/10.1111/1750-3841.16149.

Livingston, M.L., et al., 2019. White striping and wooden breast myopathies of broiler breast muscle is affected by time-limited feeding, genetic background, and egg storage. *Poult Sci* 98, 217–226. https://doi.org/10.3382/ps/pey333.

Lorenzi, M., et al., 2014. Incidence of white striping under commercial conditions in medium and heavy broiler chickens in Italy. *J App Poul Res* 23, 754–758. https://doi.org/10.3382/japr.2014-00968.

Maharjan, P., et al., 2020. Characterizing woody breast myopathy in a meat broiler line by heat production, microbiota, and plasma metabolites. *Front Vet Sci* 6, 00497. https://doi.org/10.3389/fvets.2019.00497.

Malila, Y., et al., 2018. Monitoring of white striping and wooden breast cases and impacts on quality of breast meat collected from commercial broilers (Gallus gallus). *Asian-Australas J Anim Sci* 31, 1807–1817. https://doi.org/10.5713/AJAS.18.0355.

Malila, Y., et al., 2021. Insights into transcriptome profiles associated with wooden breast myopathy in broilers slaughtered at the age of 6 or 7 weeks. *Front Physiol* 12, 691194. https://doi.org/10.3389/FPHYS.2021.691194.

Mauro, A., 1961. Satellite cell of skeletal muscle fibers. *J Biophys Biochem Cytol* 9, 493–495. https://doi.org/10.1083/JCB.9.2.493.

Maxwell, A.D., et al., 2018. Descriptive sensory analysis of marinated and non-marinated wooden breast fillet portions. *Poult Sci* 97, 2971–2978. https://doi.org/10.3382/PS/PEY145.

Maynard, C.J., et al., 2023. Reducing broiler breast myopathies through supplementation of guanidinoacetic acid in broiler diets. *J Appl Poul Res* 32, 100324. https://doi.org/10.1016/J.JAPR.2022.100324.

Meloche, K.J., et al., 2018a. Effects of reduced dietary energy and amino acid density on Pectoralis major myopathies in broiler chickens at 36 and 49 days of age. *Poult Sci* 97, 1794–1807. https://doi.org/10.3382/ps/pex454.

Meloche, K.J., et al., 2018b. Effects of reduced digestible lysine density on myopathies of the Pectoralis major muscles in broiler chickens at 48 and 62 days of age. *Poult Sci* 97, 3311–3324. https://doi.org/10.3382/PS/PEY171.

Meloche, K.J., et al., 2018c. Effects of quantitative nutrient allocation on myopathies of the Pectoralis major muscles in broiler chickens at 32, 43, and 50 days of age. *Poult Sci* 97, 1786–1793. https://doi.org/10.3382/PS/PEX453.

Mudalal, S., et al., 2015. Implications of white striping and wooden breast abnormalities on quality traits of raw and marinated chicken meat. *Animal* 9, 728–734. https://doi.org/10.1017/S175173111400295X.

Mutryn, M.F., et al., 2015. Characterization of a novel chicken muscle disorder through differential gene expression and pathway analysis using RNA-sequencing. *BMC Genom* 16, 1–19. https://doi.org/10.1186/s12864-015-1623-0.

Norring, M., et al., 2019. Wooden breast myopathy links with poorer gait in broiler chickens. *Animal* 13, 1690–1695. https://doi.org/10.1017/S1751731118003270.

OECD/FAO, 2024. OECD-FAO Agricultural Outlook 2024–2033, Paris and Rome. https://doi.org/10.1787/4c5d2cfb-en.

Olanrewaju, H.A., et al., 2006. A review of lighting programs for broiler production. *Int J Poult Sci* 5, 301–308. https://doi.org/10.3923/IJPS.2006.301.308.

Oviedo-Rondón, E.O. and Córdova-Noboa, H.A., 2020. The potential of guanidino acetic acid to reduce the occurrence and severity of broiler muscle myopathies. *Front Physiol* 11, 539481. https://doi.org/10.3389/FPHYS.2020.00909.

Oviedo-Rondón, E.O., et al., 2020. The role of incubation conditions in the onset of avian myopathies. *Front Physiol* 11, 545045. https://doi.org/10.3389/FPHYS.2020.545045.

Pampouille, E., et al., 2018. Mapping QTL for white striping in relation to breast muscle yield and meat quality traits in broiler chickens. BMC Genomics 19, 202. https://doi.org/10.1186/s12864-018-4598-9.

Papah, M.B. and Abasht, B., 2019. Dysregulation of lipid metabolism and appearance of slow myofiber-specific isoforms accompany the development of Wooden Breast myopathy in modern broiler chickens. *Sci Rep* 9(1). https://doi.org/10.1038/s41598-019-53728-8.

Papah, M.B., et al., 2017. evidence and role of phlebitis and lipid infiltration in the onset and pathogenesis of wooden breast disease in modern broiler chickens. *Avian Pathol* 46, 623–643. https://doi.org/10.1080/03079457.2017.1339346.

Pejšková, L., et al., 2023. Characterization of wooden breast myopathy: a focus on syndecans and ECM remodeling. *Front Physiol* 14, 1301804. https://doi.org/10.3389/FPHYS.2023.1301804.

Pekel, A.Y., et al., 2020. Effects of reducing dietary amino acid density and stocking density on growth performance, carcass characteristics, meat quality, and occurrence of white striping in broiler chickens. *Poult Sci* 99, 7178–7191. https://doi.org/10.1016/J.PSJ.2020.08.077.

Petracci, M., et al., 2013. Functional ingredients for poultry meat products. *Trends Food Sci Technol* 33, 27–39. https://doi.org/10.1016/J.TIFS.2013.06.004.

Petracci, M., et al., 2014. Effect of white striping on chemical composition and nutritional value of chicken breast meat. *J Anim Sci* 13, 3138. https://doi.org/10.4081/ijas.2014.3138.

Petracci, M., et al., 2019. Wooden breast, white striping, and spaghetti meat: causes, consequences and consumer perception of emerging broiler meat abnormalities. *Compr Rev Food Sci Food Saf* 18, 565–583. https://doi.org/10.1111/1541-4337.12431.

Piestun, Y., et al., 2015. Thermal manipulation during embryogenesis affects myoblast proliferation and skeletal muscle growth in meat-type chickens. *Poult Sci* 94, 2528–2536. https://doi.org/10.3382/PS/PEV245.

Praud, C., et al., 2020. Molecular Phenotyping of White Striping and Wooden Breast Myopathies in Chicken. *Front Phys*, 11, 633. https://doi.org/10.3389/fphys.2020.00633.

Qin, J., et al., 2023. Gas chromatography-mass spectrometry-based untargeted metabolomics analysis reveals circulating biomarkers related to wooden breast myopathy in broilers: a preliminary study. *Poult Sci*, 102, 102718. https://doi.org/10.1016/j.psj.2023.102718.

Radaelli, G., et al., 2017. Effect of age on the occurrence of muscle fiber degeneration associated with myopathies in broiler chickens submitted to feed restriction. *Poult Sci* 96, 309–319. https://doi.org/10.3382/PS/PEW270.

Russo, E., et al, 2015. Evaluation of White Striping prevalence and predisposing factors in broilers at slaughter. *Poult Sci* 94, 1843–1848. https://doi.org/10.3382/ps/pev172.

Sanotra, G.S., et al., 2002. Influence of light-dark schedules and stocking density on behaviour, risk of leg problems and occurrence of chronic fear in broilers. *Br Poult Sci* 43, 344–354. https://doi.org/10.1080/000716601201036023611.

Santos, M.N., et al., 2021. In pursuit of a better broiler: carcass traits and muscle myopathies in conventional and slower-growing strains of broiler chickens. *Poult Sci* 100, 101309. https://doi.org/10.1016/j.psj.2021.101309.

Sihvo, H.K., et al., 2014. Myodegeneration with fibrosis and regeneration in the pectoralis major muscle of broilers. *Vet Pathol* 51, 619–623. https://doi.org/10.1177/0300985813497488.

Sihvo, H.K., et al., 2017. Wooden breast myodegeneration of pectoralis major muscle over the growth period in broilers. *Vet Pathol* 54, 119–128. https://doi.org/10.1177/0300985816658099.

Sihvo, H., 2019. Pathology of wooden breast myopathy in broiler chickens. *Dissertation*.

Siller, W.G., 1985. Deep pectoral myopathy: a penalty of successful selection for muscle growth. *Poult Sci* 64, 1591–1595. https://doi.org/10.3382/PS.0641591.

Sirri, F., et al., 2016. Effect of different levels of dietary zinc, manganese, and copper from organic or inorganic sources on performance, bacterial chondronecrosis, intramuscular collagen characteristics, and occurrence of meat quality defects of broiler chickens. *Poult Sci* 95, 1813–1824. https://doi.org/10.3382/ps/pew064.

Soglia, F., et al., 2016. Histology, composition, and quality traits of chicken *Pectoralis major* muscle affected by wooden breast abnormality. *Poult Sci* 95, 651–659. https://doi.org/10.3382/ps/pev353.

Soglia, F., et al., 2019. Spotlight on *Avian Pathol*ogy: current growth-related breast meat abnormalities in broilers. *Avian Pathol* 48, 1508821. https://doi.org/10.1080/03079457.2018.1508821.

Soglia, F., et al., 2021a. A critical review of the mechanisms involved in the occurrence of growth-related abnormalities affecting broiler chicken breast muscles. *Poult Sci* 100, 101180. https://doi.org/10.1016/J.PSJ.2021.101180.

Soglia, F., et al., 2021b. Lysine depletion during different feeding phases: effects on growth performances and meat quality of broiler chickens. *Animals* 11, 1499. https://doi.org/10.3390/ANI11061499.

Solomon, M.B., et al., 1998. biophysical basis of pale, soft, exudative (pse) pork and poultry muscle: a review. *J Muscle Foods* 9, 1–11. https://doi.org/10.1111/J.1745-4573.1998.TB00639.X.

Sun, X., et al., 2021. Using air deformation of raw fillet surfaces to identify severity of woody breast myopathy in broiler fillets. *Lwt* 141, 110904. https://doi.org/10.1016/j.lwt.2021.110904.

Sun, X., et al., 2022a. Use of image analysis to evaluate morphometric measurements of broiler breast fillets affected by the woody breast condition. *J Food Sci Technol* 59, 4457–4465. https://doi.org/10.1007/s13197-022-05525-x.

Sun, X., et al., 2022b. Quality properties of chicken meatballs prepared with varying proportions of woody breast meat. *J Food Sci Technol* 59, 666–676. https://doi.org/10.1007/s13197-021-05060-1.

Tasoniero, G., et al., 2016. Technological quality, mineral profile, and sensory attributes of broiler chicken breasts affected by White Striping and Wooden Breast myopathies. *Poult Sci* 95, 2707–2714. https://doi.org/10.3382/PS/PEW215.

Tasoniero, G., et al., 2020. Effect of spaghetti meat abnormality on broiler chicken breast meat composition and technological quality. *Poult Sci* 99, 1724–1733. https://doi.org/10.1016/J.PSJ.2019.10.069.

Tijare, V.V., et al., 2016. Meat quality of broiler breast fillets with white striping and woody breast muscle myopathies. *Poult Sci* 95, 2167–2173. https://doi.org/10.3382/ps/pew129.

Tong, H.B., et al., 2012. Effects of stocking density on growth performance, carcass yield, and immune status of a local chicken breed. *Poult Sci* 91, 667–673. https://doi.org/10.3382/PS.2011-01597.

Traffano-Schiffo, M.V., et al., 2017. Development of a spectrophotometric system to detect white striping physiopathy in whole chicken carcasses. *Sensors* 17, 1024. https://doi.org/10.3390/S17051024.

Trocino, A., et al., 2015. Effect of genotype, gender and feed restriction on growth, meat quality and the occurrence of white striping and wooden breast in broiler chickens. *Poult Sci* 94, 2996–3004. https://doi.org/10.3382/ps/pev296.

Trocino, A., et al., 2023. Nutritional and feeding strategies for controlling breast muscle myopathy occurrence in broiler chickens: a survey of the published literature. *Meat Technol* 64, 30–35. https://doi.org/10.18485/meattech.2023.64.2.5.

U-Chupaj, J., et al., 2017. Effect of tumbling marination on marinade uptake of chicken carcass and parts quality. *Rev Bras Cienc Avic* 19, 61–68. https://doi.org/10.1590/1806-9061-2016-0380.

U-Chupaj, J., et al., 2021. Influence of non-phosphate and low-sodium salt marination in combination with tumbling process on properties of chicken breast meat affected by white striping abnormality. *J Food Sci* 86, 319–326. https://doi.org/10.1111/1750-3841.15565.

Velleman, S.G., et al., 2014. Effect of the timing of posthatch feed restrictions on the deposition of fat during broiler breast muscle development. *Poult Sci* 93, 2622–2627. https://doi.org/10.3382/ps.2014-04206.

Velleman, S.G., 2015. Relationship of skeletal muscle development and growth to breast muscle myopathies: a review. *Avian Dis* 59, 525–531. https://doi.org/10.1637/11223-063015-review.1.

Velleman, S.G., 2019. Recent developments in breast muscle myopathies associated with growth in poultry. *Annu Rev Anim Biosci* 7, 289–308. https://doi.org/10.1146/annurev-animal-020518-115311.

Velleman, S.G., 2020. Pectoralis major (breast) muscle extracellular matrix fibrillar collagen modifications associated with the wooden breast fibrotic myopathy in broilers. *Front Physiol* 11, 1–7. https://doi.org/10.3389/fphys.2020.00461.

Velleman, S.G., 2023. Broiler breast muscle myopathies: association with satellite cells. *Poult Sci* 102, 102917. https://doi.org/10.1016/J.PSJ.2023.102917.

Vieira, V., et al., 2021. Effect of age and different doses of dietary vitamin E on breast meat qualitative characteristics of finishing broilers. *Anim Nut* 7, 163–167. https://doi.org/10.1016/J.ANINU.2020.08.004.

Vieira, S.L., et al., 2022. Broiler meat production as affected by dietary supplemental hydroxy-selenomethionine. *Livest Sci* 259. https://doi.org/10.1016/j.livsci.2022.104912.

Wang, C., et al., 2022a. Restoring functionalities in chicken breast fillets with spaghetti meat myopathy by using dairy proteins gels. *Gels* 8(9), 558. https://doi.org/10.3390/gels8090558.

Wang, C., et al., 2023a. Textural restoration of broiler breast fillets with spaghetti meat myopathy, using two alginate gels systems. *Gels* 10(1), 7. https://doi.org/10.3390/GELS10010007.

Wang, J., et al., 2020. Effect of early posthatch supplementation of vitamin E and omega-3 fatty acids on the severity of wooden breast, breast muscle morphological structure, and gene expression in the broiler breast muscle. *Poult Sci* 99, 5925–5935. https://doi.org/10.1016/J.PSJ.2020.08.043.

Wang, Y., et al. 2025. Oxidative stress-induced changes in wooden breast and mitigation strategies: a review. *Compr Rev Food Sci Food Saf* (in press).

Wang, Y.H., et al., 2022b. the role of incubation conditions on the regulation of muscle development and meat quality in poultry. *Front Physiol* 13, 883134. https://doi.org/10.3389/FPHYS.2022.883134.

Wang, Z., et al., 2023b. Integrative transcriptomic and metabolomic analysis reveals alterations in energy metabolism and mitochondrial functionality in broiler chickens with wooden breast. *Sci Rep* 13, 4747. https://doi.org/10.1038/s41598-023-31429-7.

Wu, T., et al., 2024. Broiler spaghetti meat abnormalities: muscle characteristics and metabolomic profiles. *Animals* 14, 1236. https://doi.org/10.3390/ANI14081236/S1.

Zampiga, M., et al., 2018. Effect of dietary arginine to lysine ratios on productive performance, meat quality, plasma and muscle metabolomics profile in fast-growing broiler chickens. *J Anim Sci Biotechnol* 9, Article 63. https://doi.org/10.1186/s40104-018-0294-5.

Zampiga, M., et al., 2019. Effect of different arginine-to-lysine ratios in broiler chicken diets on the occurrence of breast myopathies and meat quality attributes. *Poult Sci* 98, 2691–2697. https://doi.org/10.3382/PS/PEY608.

Zampiga, M. and Sirri, F., 2022. Nutritional strategies and management practices to improve poultry meat quality, in: Petracci, M., et al. (eds) *Improving poultry meat quality*, pp. 59–106. Academic Press. https://doi.org/10.19103/AS.2022.0112.03.

Zhang, X., et al., 2020. Proteomic characterization of normal and woody breast meat from broilers of five genetic strains. *Meat Muscle Biol* 4, 1–17. https://doi.org/10.22175/MMB.8759.

Zhang, X., et al., 2021. Cecal microbiota contribute to the development of woody breast myopathy. *Poult Sci* 100, 101124. https://doi.org/10.1016/J.PSJ.2021.101124.

Zhang, X., et al., 2024. Transcriptomic meta-analysis and exploration of differentially expressed gene functions in wooden breast myopathy of broilers. *Poult Sci* 103, 104047. https://doi.org/10.1016/j.psj.2024.104047.

Zhao, D., et al., 2024. Single-cell transcriptomics and tissue metabolomics uncover mechanisms underlying wooden breast disease in broilers. *Poult Sci* 103, 104433. https://doi.org/10.1016/J.PSJ.2024.104433.

# Index

Adaptive immunity 418, 558, 561, 562, 564, 565, 566, 567
Agricultural Research Service (ARS) 32
Alanine aminotransferase (ALT) 416
Albumen 113, 116, 151, 207, 563
Amnion 207
Animal welfare concepts 5
  ability to adapt 7
  adaptations 6
  DAWCon 7
  environmental challenges 7
  *vs.* ethics 11–13
  Five Domains Model 5
  quality of life
    biological functioning 6
    natural living 6
    subjective experiences 6
  value positions 7
Anti-nutritional factors (ANFs) 266, 268, 372
Antibiotic use 334–335, 556
Antibiotic-free broiler production
  challenges 337–339
  circularity indicators 363–364
    efficiency indicators 366
    nutrient and material 364–366
    resource use efficiency 366–369
  genesis of 335–336
  improved production efficiency 369–370
  improved recycling 370–377
  LCOB (*see* Low opportunity-cost biomass (LOCB))
  management
    breeder farm 339–340
    chick quality 341–342
    hatching eggs 340–341
    houses 342–343
  nutrition 343–344
    digestive capacity 344–345
    immunocompetence 347–348
    intestinal integrity 346–347
    mitigating enteric pathogens 345–346
    vaccines 348
  restructure livestock production
    integrated crop-broiler 377–379
    recoupling 379
  state of 336–337
Antimicrobial resistance (AMR) 333, 350, 557
Antimicrobial use (AMU) 334, 737
Apollo Generation 2, 772, 778
Arginine 198, 204, 284, 821
Artificial intelligence (AI) 53, 319, 478
Aspartate aminotransferase (AST) 416
Automated deboning machines 58
Automated poultry farming process 319
Avian pathogenic *E. coli* (APEC) 725, 732, 737

Back-view pose estimation 468–469
Bacterial chondronecrosis with osteomyelitis (BCO) 252
Bacterial vaccines 603
Bag method 769, 770
Balanced diet 265
Behavioral expression
  abnormal behaviors 250–252
  behavioral needs 238–239
  fast-growing broiler chickens
    dustbathing 242–243
    eating and drinking 239–240
    foraging 240–242
    perching 244–246
    play and exploration 247–249
    preening 244
    reproductive behaviors 250
    social behaviors 249–250
  impacted by genetics
    body weight 253–254
    clinically healthy broilers 252
    health modification 253
    lameness 253
  slower-growing strains 237
Behavioural interactions 6
Biosecurity compliance 556, 602
Black soldier fly (BSF) 271
Blockchain-based traceability systems 53–54
Bone mineralisation 662, 688, 689, 691, 693, 695, 696
Bootstrap aggregating (Bagging) 484

Breast fillet deboning machines 58
Breast muscle myopathies (BMM)
effects of pectoral myopathies 825–826
growth-related myopathies 799–800
spaghetti meat 804–806
white striping 800–802
wooden breast 802–804
implications of 825
potential strategies 812–813
long-term solutions 813–815
short-term solutions 815–824
suggested mechanisms 806
physiological and biological 806–809
predisposing factors 810–812
Broiler breeder paradox 16–18, 94
Broiler breeding
developments and consequences
climate change 89
demand for animal protein 88–89
health and welfare 89–90
diversification 91–93
genomic prediction 87
goals 87–88
heat tolerance 97–98
improving genomic selection 100–101
optimizing the breeding goal 93–94
phenotyping 98–99
pyramid 86
resilience 95–97
resistance against infectious diseases 101–102
robustness 95–97
trait antagonisms 94–95
Broiler chicken supply chain
advanced processing techniques 57
blockchain technology 68
chemical contamination risks 69–70
cloud-based traceability systems 66
cold chain logistics 59
commercial 56
Data Matrix barcodes 67
data sharing and integration 70–73
educational programs 75–76
extension services 74–75
extruders 57
food traceability 65–66
GP farms 55
hatchery management 56
horizontal integration 62, 64
ingredients suppliers 56–57
meat processing plants 57–59
parent stock (PS) generations 55
precision formulation software 57
products for consumers 73–74
real-time monitoring tools 68
vertical integration 62, 64
veterinary services 56–57
Broiler production
air quality challenges 41–43
co-products 26, 27
cold storage facilities 59–61
cycle 27–28
data integration 59–61
digital technologies 53
environmental awareness 27
global 61–62
improvement opportunities 44–45
land application 28
nutrient management planning (*see* Nutrient management planning (NMP))
on-farm broiler operations 28
sustainability/circularity 43–44
Brooder pneumonia 154, 528, 726
By-products 371

CEALVET S.L.U., 492
Cell-mediated immunity 558, 559, 565, 584, 593, 594
Chick quality 111
egg component 116–117
egg contamination/sweating 126–127
egg size and breeder age 119–122
egg storage 123–124
eggs set upside down 127–128
eggshell conductance 117–119
eggshell quality 124–126
heat production 112
incubation conditions 112
methods of measuring 112–113
nutrients to develop embryo 113–116
Chicken Cat 772
Circular food system 361–362
and bioeconomy 363
definition 362
FOODSOM 379
potential of broilers 379–382
role of animals 362
Classification and regression trees (CART) 484
*Clostridium perfringens* 283
Commercial broilers 52, 86, 412
Computer vision 319
Contact dermatitis 252

Convolutional neural networks (CNNS) 484
Council Directive 2007/43/CE (2007) 765, 791
Cross-virus neutralization 717
Current grading systems 58
Cytochrome P450-mediated catabolic pathways 506

Day-of-hatch vaccination techniques 567, 578
Day-old chick quality
  immunoglobulins
    endogenous IgY 148–150
    immune system 146–148
  microbiological 150–151
    bacterial penetration 152–153
    early-life microbiota establishment 151–152
    egg collection and sanitization 154–155, 156
    eggshell contamination 152–153
    floor eggs 155
    fungal contamination 153–154
  physical quality
    pre-hatching factors 142–146
    qualitative chick quality 140–142
    quantitative chick quality 138–140
  quantitative criteria 137
Dead on arrival (DOA) 641, 783–785
Dead-reckoning/path integration 454
Deep neural networks (DNN) 484
Dempster-Shafer evidence theory 486
Dempster-Shafer infusion of sound data 486
Dietary interventions 818–822
Digital tools 53
Direct embryo injections 208
Discrete Wavelet Transform 486
Drinking water vaccination 587–588
Dust accumulation 319
Dynamic animal welfare concept (DAWCon) 7, 17

Early feeding 163
  body composition 174
  case studies 180–183
  delayed feeding 165–168
  future research 183–184
  growth 174–177
  growth rate 172–174
  management procedures 164
  residual yolk 168–170
    nutrients needed for development 171–172
    nutrients needed for survival 170–171
  welfare
    affective state 177–178
    natural behaviour 179
    physical well-being 178–179
Eggshell mottling 124
Egyptian local breeds 419
Embryogenesis 557, 559, 561, 566
Embryonic development 193
Endoplasmic reticulum (ER) stress 809
Environmental needs
  air quality 307
    ammonia 308
    carbon dioxide 308
    particle matter 308–309
  dimensions 293
  flooring system 313–314
  geographic locations 295
  geostatistical techniques 310
  housing 293
  lighting 314–316
  modern technologies 319
  space management
    enrichment 317–318
    stocking densities 316–317
  temperature 295–299
    hematological parameters 303
    thermal needs 299–301
    thermal stress 301–303
  variability 310–313
  ventilation 294, 303–304
    brooding phase 304–306
    tunnel fan ventilation 306–307
Enzyme-Linked Immunosorbent Assay (ELISA) 732
European Food Safety Authority (EFSA) 765
Extreme Learning Machine (ELM) 485

Farm management 817–818
Farm Management Software (FMS) tools 54
Fast Fourier Transform 486
Fayoumi chickens 410
Feed additives 284
Feed conversion ratio (FCR) 265, 409
First-week mortality
  external factors
    chick transportation conditions 639–641
    environmental factors 645

nipple drinkers 646
pre-placement holding time 641–644
seasonal variations 644–645
stocking density 646
vaccination 646–647
mortality patterns 628–629
risk factors
breeder flock age 631–633
chick quality 637–639
egg shell contamination 635
egg storage duration 634–635
egg weight 633–634
genetic selection 630–631
hatching time 637–639
incubation conditions 636–637
sex 631
Fisher Discriminant Analysis (FDA) 483
Footpad dermatitis 11, 12
Fractures 782–783
Free-roaming chickens 240–242
Fructo-oligo-saccharides (FOS) 274

Galacto-oligo-saccharides (GOS) 274
Gas chromatography-mass spectrometer (GC-MS) 535, 539
Gene editing 91
Genetic selection 85, 86
Genome-editing technologies 423
Genome-Wide Association Study (GWAS) 421, 422
Gentling programme 778
Global Food Safety Initiative (GFSI)-recognized standards 68
Grandparent (GP) stock 51
Great-Grandparent (GGP) stock 51
Green leg 680
Gut-associated lymphoid tissue (GALT) 564, 565, 577, 600

Haematomas 779–782
Hardware component 458
Hatchery management 56, 815–817
Hatchery vaccination 568, 578, 584
Health and performance 777
Heat shock proteins (HSPs) 419
Heat stress 302, 407–409
advances 434–435
economic implications 413
factors
age 411
breed 412
physical activity 412
sex 411
size and weight 412
stocking density 412–413
genomics and biotechnology
genome editing 423–426
genome-wide association study (GWAS) 422
next-generation sequencing 422–423
quantitative trait loci (QTL) 421–422
single nucleotide polymorphisms (SNPs) 421
impacts of
activity patterns 433–434
altered feeding behavior 429–433
carcass yield 428–429
drinking patterns 433
growth rate and performance 423, 427
meat quality 427–428
metabolic and physiological changes
blood components and metabolites 416
body temperature 414–416
immune system 417–418
metabolic processes 418–419
neuroendocrine changes 416–417
oxidative stress 417
transcriptional changes 419
selective breeding 420
susceptibility of broilers 409–410
Hemorrhages 721
Heterophil ratio 777
Heterophil/lymphocyte (H/L) 416
Hidden Markov Models (HMM) 479, 487
Hierarchical clustering analysis (HCA) 540
Humoral immunity 558, 559, 571
Hyperthermia 302
Hypothalamic-pituitary-thyroid (HPT) 417
Hypothalamic–pituitary–adrenal (HPA) 416, 418
Hypoxia stress 807
Hypoxia-inducible factor 1 (HIF-1) 808

Identification switches 99
Immune complex vaccines 572, 575, 578, 580–581, 589, 593, 603, 604
Independent sample t-test 493
Infectious bronchitis virus (IBV) 283
Infectious Bursal Disease virus (IBDV) 583, 728
Innate immunity 558, 561, 562, 564, 565, 566, 700

Intelligent systems 319
Intestinal pathogens 283
Ion mobility spectrometer (IMS) 539

K-nearest neighbour (KNN) 484

Lectins 268
Leg catching method 769
Leg problems/disorders
  assessment method 681–684
  embryology 662–668
  fertile egg incubation *vs.* broiler leg weakness
    broiler diet 698–699
    broiler health 700
    crooked toes 695–696
    housing environment enrichment 699–700
    incubation temperature effects 690–695
    muscle/tendon weakness 697–698
    splayed legs/spraddle legs 695
    tibial dyschondroplasia 696–697
    varus-valgus deformity (VVD)/slipped tendon 696
  future trends 700–701
  gastrocnemius tendon rupture 679–681
  infectious problems
    bacterial osteomyelitis 670–671
    femoral head necrosis 672–674
    pododermatitis 675
    skin damage 675
  leg health improvement
    breeder age 688
    breeder diet 686–688
    in-ovo administration 689
    skeletal characteristics 685
    stocking density 685–686
  osteodystrophies 668–670
  overview 661–662
  rickets 668–670
  tibial dyschondroplasia 675–679
Leghorn chickens 410
Licuicel® Complex 492
Low input farming systems (LIFS)
  environmental impact 392
  off-farm resources 392
  on-farm resources 392
  outdoor access
    broiler welfare 399
    consequences for farmers and consumers 398
    environmental conditions 399–400
    free-range conditions 394–395
    risk of disease 395–396
    thermoregulation capacity 393–394
    weather impacts 396–398
Low opportunity-cost biomass (LOCB) 362
  feeding 370–371
  increasing the potential 375–376
  potential of feeding 373–375
  risks of feeding 372–373
  slaughter wastes 376–377
  valorizing manure 376–377
Low pathogenic avian influenza (LPAI) 716
Lymphocyte ratio 777

Machine learning algorithms 99
Machine loading methods
  advantages and disadvantages of 774
  catching staff 789–790
  crating 790
  definition of 772
  different methods of
    loading broilers 768–770
    take-home message 772
    upright carrying 770–771
    worker safety 771–772
  effectiveness of 785
    DOA rates (*see* Dead on arrival (DOA) rates)
    intrinsic and extrinsic factors 786–787
    stress-minimising 788
    take-home message 788
  frequency of injuries 779–785
  humane handling of 791–792
  legality of catching 765–766
  management of broilers 763–765
  optimise use of 774–777
  poultry farmers 766
    choice of method 767–768
    hygiene and biosecurity 766–767
    manual catching 767
    take-home message 768
  slaughter 791
  stress response of 777–779
  types of machines 772–774
Macroscopic appearance 800, 802
Mannan-oligo-saccharides (MOS) 274
Maternal antibodies 147
Matrix-Assisted Laser Desorption/Ionization Time-of-Flight Mass Spectrometry (MALDI-TOF) 732

Maximal overlap discrete wavelet transform (MODWT) 479
Meat processes 822–824
Medium-chain fatty acids (MCAs) 273–274
Mel Frequency Cepstral Coefficients (MFCC) 481
Monounsaturated fatty acids (MUFAs) 508
Mucosal immunity 558, 565, 577, 581, 584, 589, 600
Muscle fiber 808

N use efficiency (NUE) 366
National Oceanic and Atmospheric Administration–National Centers for Environmental Information (NOAA-NCEI) 413
Newcastle Disease (ND) 583, 722
Next-generation sequencing (NGS) 731, 740
Non-governmental organisations (NGOs) 791
Non-starch polysaccharides (NSP) 269, 272
Nutrient management planning (NMP)
  appropriate storage 33–34
  calibration of spreading equipment 34–35
  challenges 39–41
  dead broiler disposal 37–39
  erosion control 37
  land application 35–37
  litter analysis 31–33
  nutrient generation assessment 30–31
  principles of 30
  record keeping 39
  site-specific conservation 37
  soil testing 31–33
Nutritional needs
  feed additives
    animal nutrition 274
    emulsifiers 273
    health- and performance-promoting effects 273
    plants phenolic compounds 273
    prebiotics are carbohydrates 274
    zootechnical 272
  raw materials
    alternative protein sources 269–272
    ANFs 268
    common feed ingredients 267–269
    cost of 270
    macronutrients 267
    micronutrients 267
    nutritional content 270
    seaweeds 270
    technological aspects 270
Nutritional requirements
  amino acid 275–277
  energy 277
  factors influencing
    genetic selection 277
    glycogen 279
    lipids 279
    production system 278
    residual yolk sac 279
    sex 278
  feed ingredients 280–282
  feed texture 280–282
  impact of circularity 282–283
  particle size 280–282
  phase feeding 279–280
  welfare and disease control 283–284

Object detection 456–457
On-farm hatching 56
Optical flow 456
Oregano EO (OEO) 225
Outdoor Climate 787
Ovo delivery
  acute heat stress 222
  commercial settings 194
  diluent type 223
  early-stage injections 194–195
  Eimeria challenge 222
  embryonic aflatoxin 222
  embryonic mortality 222
  environmental conditions 224
  environmental stressors 218
  of essential oils 224–227
  hatchability 223
  injection temperature 224
  manual injection techniques 223
  methodological factors
    combining bioactive compound 212
    compound properties 208–211
    control groups 212–218
    dose 197–204
    embryonic target compartment 206–208
    experimental conditions 218–223
    injection volume 204–205
    time of injection 205–206
    time of measurement 218–223
  necrotic enteritis 222
  nutrients/biochemical compounds 195

pre-hatch stage 194
Salmonella enteritidis challenge 222
sterility of injection 223
vaccination 194
Oxidative stress 807
Oxidative stress mechanisms 506

Partial least squares regression (PLS-R) 540
Passive immunity 562, 563, 564, 567, 581, 603
Pathogen-associated molecular patterns (PAMPs) 558
*Pectoralis major* (PM) 799
Personal Digital Assistants (PDAs) 67
Peyer's patch 565
Phenolic compounds 273
Physical/functional domains 5
Polyunsaturated fatty acids (PUFAs) 508
Positive animal welfare 18
Poultry meat 50
Precision Livestock Farming (PLF) technologies 53, 478, 479, 495, 497
Precision tools 319
Precision vaccination 603
Principal component analysis (PCA) 483, 540
Processed animal proteins (PAPs) 282
Protectotype vaccination strategy 571, 735
Pyruvate 513

Quantitative trait loci (QTL) 421–422
Quick response (QR) codes 54

Radio frequency identification (RFID) 451, 454–455, 465–467
Random Forest (RF) 484
Rapid Alert System for Food and Feed (RASFF) 53
Reactive oxygen species (ROS) 417
Recombinant vaccines 567, 568, 570, 575, 578
Red jungle fowl 412
Reference population 100
Remote data transfer technology 451
Respiratory diseases
accurate diagnosis 716, 729–733
ammonia exposure 728
bacteria infections 725–726
Escherichia coli (E.coli) 725
mycoplasmosis 723–725
fungi 726
future trends 739–741
molecular techniques 732–733, 738–739
multicausal respiratory conditions 726–729
prevention
biosecurity 733–734
vaccination 734–738
viruses
avian influenza viruses (AIVs) 721–722
avian metapneumovirus *(Metapneumovirus avis)* 718–720
epidemiology 722–723
infectious bronchitis virus (IBV) 717–718
infectious laryngotracheitis 720–721
Respiratory pathogens 283
Rolling vaccination reaction 722

Saponins 268
Satellite cells (SCs) 808
Selected ion flow tube mass spectrometers (SIFT-MS) 521, 539
Selection index 88
Sensor technologies 449–450
advances 470
body-worn sensor
accelerometers 451–454
locomotor activity tracking 465–468
radio frequency identification (RFID) 454–455
*vs.* remote sensors 459–460
ultrawideband (UWB) 454–455
combining sensors 460
remote sensor
cameras 456–458
force plate sensors 458
remote sensors
*vs.* body-worn sensor 459–460
walking ability monitoring 468–470
welfare indicators
context decision 461–463
recording data 463–464
Sensors 319
Single cell protein production (SCP) 271
Single-nucleotide polymorphism (SNP) 86–87, 421
Single-stage incubation systems 56
Slaughterhouse 763, 783, 822
Small- and medium-sized enterprises (SMEs) 52

Software component 458
Sound Pressure Level (SPL) 480, 481
Soybean meal (SBM) 268
Spaghetti meat (SM) 800, 804–806
Spanish commercial hatchery 492
Spatial agricultural resources 362
Spectral bandwidth 480
Spectral centroid 480
Spectral entropy 480
Spectral flatness 481
Spectral rolloff 481
Spectrogram 481
Spray cabinet vaccination 586, 734
Spray vaccination 568, 586–587
Support Vector Machine (SVM) 484, 485–486, 491
Swollen Head Syndrome (SHS) 647, 719
Sympathetic adrenomedullary (SAM) pathway 416, 418

Temperature-humidity index (THI) 97, 491
Tempograms 481
Thermoneutral zone 302
Thermoregulatory stress 8
Threat Assessment and Critical Control Points (TACCP) 68
Thyroxine (T4) 417
Top-view tracking 468–469
Transgenerational immunity 347
Triiodothyronine (T3) 417
Trypsin 268
Turkey Rhinotracheitis (TRT) 718

Ultra-wideband (UWB) 451, 465–467, 545–545
Unbalanced diet 266
Upright method 770, 771

Vaccination strategies 568, 569
- ammonia exposure 601
- avian immune system
  - maternal immunity transfer 562–564
  - microbiota role 564–565
  - ontogeny development 559–562
  - trained immunity 566–567
- biosecurity 556, 557, 568, 572, 578, 582, 602, 604
- delivery methods
  - individual vaccination methods 588–590
  - mass application methods 586–588
- enteric diseases 572
- factors influencing
  - environmental and managerial conditions 599–603
  - host 591–594
  - pathogens 594–596
  - vaccines 596–599
- fast- *vs.* slow-growing broilers 564, 573–576
- future perspectives 603–604
- genetic selection 556, 604
- highly pathogenic avian influenza (HPAI) 573
- immunosuppression 556, 557, 571, 573, 580, 584, 591, 593, 594, 596, 599, 601, 602, 604
- inactivated vaccines 559, 568, 580, 581
- infectious bronchitis virus (IBV) 570–571, 580
- infectious bursal disease virus (IBDV) 571, 572, 580, 593
- live attenuated vaccines 558, 559, 566, 567, 568, 570, 571, 579, 580, 585, 597, 598
- major broiler diseases 556, 569–574
- Marek disease virus (MDV) 571, 572, 580, 591, 593
- newcastle disease virus (NDV) infection 570
- on-farm hatching systems 576–578
- *ovo* vaccination 556, 557, 561, 562, 567, 568, 577, 585, 588, 589, 603
- recombinant herpesvirus of turkey (rHVT) 582, 584, 589
- vaccine types
  - conventional poultry vaccines 579–581
  - new generation vaccines 582–585

Vaccine interference 557, 568–569, 598
Village fowl 412
Vitamin D3 (VD3) 668–670, 687, 689
Vocalisation technology 477
- acoustic data collection 492
- acoustic features of broilers
  - audio samples 479
  - automatic detection 483–484
  - bioacoustic monitoring 479
  - dimensionality reduction 483
  - feature selection 483
  - fine-tuning process 482
  - frequency features 480–481
  - pattern recognition algorithms 479
  - signal processing techniques 479

spectrotemporal features 481–482
statistical analysis 483
syrinx 478
temporal features 480
acoustic recording data 492
advances 494–495, 497
animal behaviour track
environmental conditions 490
hen vocalisation detection method 491
Mel-frequency cepstral coefficients (MFCCs) 490–491
pecking sounds and feed 488–489
use of bioacoustics 489
animal disease detector
animal welfare 484
artificial intelligence method 485
deep learning techniques 487
extreme learning machine (ELM) 485
Gaussian Naive Bayes (GNB) model 487–488
health status of animals 485
laryngotracheitis (ILT) 487
non-invasive acoustic detection 487
non-invasive acoustic sensors 484–485
post-reaction phase 487
pre-reaction phase 487
prediction model 486
reaction phase 487
support vector machine (SVM) 485
wavelet entropy (WET) 486
chicken vocalisation 492–493
European Food Safety Authority (EFSA) 491
hatching 492
Levene's test 493
nutrient-enriched hydrated gels 492
pitch frequency values per minute 493–496
Shapiro-Wilk test 493
Volatile organic compounds (VOCs) 503–505
advances 540–541
in chickens 515
blood 519–520
breath 518
feathers and skin 519
feces and gut content 516–518
external factors
bedding material 534
diet 533
humidity 534–535
rearing systems 534
temperature 534–535
internal factors 531
age 530
breeds 530, 532
circadian rhythms 533
emotional state 532
sex and reproductive status 532
measurement and analysis 535–540
pathogen biomarkers 529
pathogen influence 520, 522–526
bacteria 521, 527
fungi 528–529
parasites 528
viruses 527
primary metabolic pathways
carbohydrate metabolism 508, 512–513
fat metabolism 508–511
protein metabolism 506–508
secondary metabolic pathways 513–515
Vulnerability Assessment and Critical Control Points (VACC) 68

Water holding capacity (WPC) 802
Water-based vaccines 568
Welfare challenges
behavioural restrictions 10
environmental factors 10
gastrointestinal disorders 9
integument damage 9–10
locomotory disorders 9
thermoregulatory stress 8–9
Welfare-centred approaches 13–16
White striping (WS) 800
Wooden breast (WB) 800, 802–804

Xylooligo-saccharides (XOS) 274

Yolk sac injections 207
Yolk-free body mass (YFBM) 170

Zero Crossing Rate (ZCR) 480
Zone of provisional mineralisation (ZPM) 664, 668

www.ingramcontent.com/pod-product-compliance
Lightning Source LLC
LaVergne TN
LVHW010619300926
849162LV00001B/7